Major Geological, Climatic, and Biological Events

ate time since beginning terval in millions of years present	
0.01 1.7	There were extensive and repeated periods of glaciation in the Northern Hemisphere. The Neogene midlatitude savanna faunas became extinct, and hominids expanded throughout the Old World. Near the end of the interval, hominids reached the New World. There was an extinction of many large mammals, especially in the New World and in Australia. The remaining carnivorous flightless birds also became extinct.
5.2 23 33.4	Cooler and more arid climates persisted, resulting from mountain uplift and the formation of the Isthmus of Panama near the end of the interval. The Arctic ice cap formed by the end of the interval. The first grasslands spread in the middle latitudes. Modern families of mammals and birds radiated, and marine mammals and birds diversified in the oceans. The first hominids were seen near the end of the interval.
55 **65**	Global climate was warm in the early part of the interval, with forests above the Arctic Circle, but later in the interval temperatures fell in the higher latitudes, with the formation of the Antarctic ice cap. Mammals diversified into larger body sizes and a greater variety of adaptive types, including predators and herbivores. Mammal radiations included archaic forms, now extinct, and the earliest members of living orders. Giant carnivorous flightless birds were common as predators.
144	Further separation of the continents occurred, including the breakup of the southern continent, Godwanaland. Teleost fishes radiated, but marine reptiles flourished. Angiosperms first appeared, and rapidly diversified to become the dominant land plants by the end of the period. Dinosaurs remained the dominant tetrapods, but small mammals diversified. Air space and shorelines were shared by birds and pterosaurs, and the first snakes appeared. A major mass extinction at the end of the period, defining the end of the Mesozoic, claimed dinosaurs, pterosaurs, and marine reptiles, as well as many marine invertebrates.
206	The world continent began to break up, with the formation of the Atlantic Ocean. Marine invertebrates began to take on a modern aspect with the diversification of predators, modern sharks and rays appeared, and marine reptiles diversified. Conifers and other gymnosperms were the dominant terrestrial vegetation, and insects diversified. Dinosaurs diversified while mammals remained small and relatively inconspicuous. The first birds, lizards, and salamanders were seen at the end of the period.
248	The world continent was relatively high, with few shallow seas. No evidence of glaciation existed, and the interior of the continent was arid. Seed fern terrestrial vegetation was replaced by conifers in the later part of the period. Mammal-like reptiles declined, while archosaurian reptiles (including dinosaur ancestors) diversified. Remaining large nonamniote tetrapods now all specialized aquatic forms. First appearances by the end of the period included true mammals, dinosaurs, pterosaurs, marine reptiles, crocodiles, lepidosaurs, frog-like amphibians, and teleost fishes.
290	A single world continent, Pangea, was formed at the end of the period. Glaciation ceased early in the period. The large terrestrial nonamniote tetrapods declined and the amniotes radiated. Amniote diversification included the ancestors of modern reptiles and the ancestors of mammals, the mammal-like reptiles, which were the dominant large terrestrial tetrapods. The first herbivorous tetrapods were known. The largest known mass extinction event occurred on both land and sea at the end of the period, defining the end of the Paleozoic.
354	There was a major glaciation in the second half of the period, with low atmospheric levels of CO2. Coal swamps were prevalent in the then-tropical areas of North America and Europe. Major radiation of insects, including flying forms. Diversification of jawed fishes, including shark-like forms and primitive bony fishes, and first appearance of modern types of jawless fishes. Extensive radiation of nonamniote tetrapods, with the appearance of the first amniotes (including the earliest mammal-like reptiles) by the late part of the period.
417	There was major mountain building in North America and Europe. Major freshwater basins preserved, containing the first tetrapods at the end of the period in equatorial regions. About the same time there were the first forests with tall trees on land, and terrestrial arthopods diversified. Both jawed and jawless fishes diversified, but both experienced major extinctions toward the end the end of the period, with the disappearance of the ostracoderms, the armored jawless fishes.
443	The extensive shallow seas continued, but on dry land there was the first evidence of vascular plants and arthropods. Jawless fishes radiated, and jawed fishes (shark-like forms) were now definitely known.
490	There were widespread shallow seas over the continents, and the global climate was equable until a sharp glaciation at the end of the period. First evidence of complex plants on land. Major radiation of marine animals, including the first well-known jawless fishes and fragmentary evidence of jawed fishes.
540	Continental masses of the late Proterozoic now broken up in to smaller blocks, covered by shallow seas. Explosive radiation of animals at the beginning of the period, with first appearance of forms with shells or other hard coverings. First appearance of chordates and great diversification of arthopods, including trilobites. First vertebrates appeared early in the period.
2,500	Formation of large continental masses. Oxygen first appears in the atmosphere. First eukaryotic organisms appeared around 2 billion years ago. Major diversification of life at 1 billion years ago, with multicellular organisms, including algae. First animals appeared around 600 million years ago, just after a major glaciation.
4,600	Formation of the Earth. Major bombardment of the Earth by extraterrestrial bodies, precluding formation of life until 4 billion years ago (first fossils known at 3.5 billion years ago). Small continents. Hydrosphere definite at 3.8 billion years, atmosphere without free oxygen.

Vertebrate Life

Vertebrate Life

Sixth Edition

F. Harvey Pough

Arizona State University West

Christine M. Janis

Brown University

John B. Heiser

Cornell University

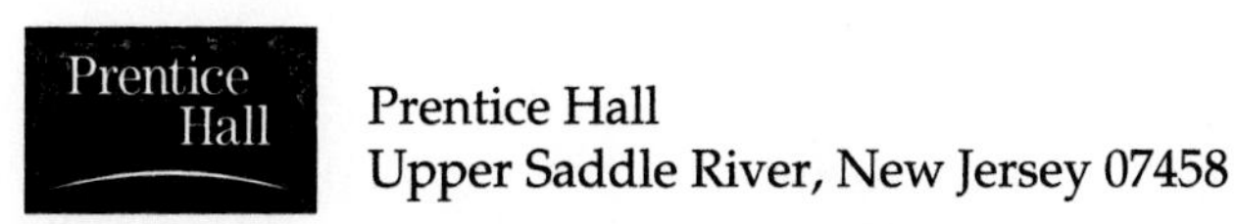

Prentice Hall

Upper Saddle River, New Jersey 07458

Library of Congress Cataloging-in-Publication Data

Pough, F. Harvey.
Vertebrate life / F. Harvey Pough, Christine M. Janis, John B. Heiser.—6th ed.
p. cm.
Includes bibliographical references.
ISBN 0–13–041248–1
1. Vertebrates. 2. Vertebrates, Fossil. I. Janis, Christine M. (Christine Marie), II. Heiser, John B. III. Title.

QL605.P68 2002
596—dc21

2001018535

Executive Editor: Teresa Ryu
Editor in Chief: Sheri L. Snavely
Project Manager: Travis Moses-Westphal
Executive Managing Editor: Kathleen Schiaparelli
Assistant Managing Editor: Beth Sturla
Executive Marketing Manager: Jennifer Welchans
Manufacturing Manager: Trudy Pisciotti
Assistant Manufacturing Manager: Michael Bell
Vice President of Production and Manufacturing: David W. Riccardi
Production Editor/Composition: Pine Tree Composition
Managing Editor, Audio/Video Assets: Grace Hazeldine
Director of Creative Services: Paul Belfanti
Director of Design: Carole Anson
Art Director: Jonathan Boylan
Interior Designer: John Christiana
Cover Designer: John Christiana
Photo Researcher: Reynold Reiger
Editorial Assistant: Colleen Lee
Art Studios: Kandis Elliot; Artworks:
Senior Manager: Patty Burns
Production Manager: Ronda Whitson
Manager, Production Technologies: Matt Haas
Project Coordinator: Jessica Einsig
Illustrator: Kathryn Anderson; Jay McElroy

Printed in the United States of America

10 9 8 7 6 5 4 3 2 1

ISBN 0-13-041248-1

Pearson Education Ltd., *London*
Pearson Education Australia Pty., Limited, *Sydney*
Pearson Education Singapore, Pte. Ltd.
Pearson Education North Asia Ltd., *Hong Kong*
Pearson Education Canada, Ltd., *Toronto*
Pearson Educación de Mexico, S.A. de C.V.
Pearson Education—Japan, *Tokyo*
Pearson Education Malaysia, Pte. Ltd.

About the Authors

F. Harvey Pough began his biological career at the age of fourteen when he conducted his first research project on the ecology of turtles in Rhode Island. His research now focuses on organismal biology, blending physiology, morphology, behavior, and ecology in an evolutionary context. He especially enjoys teaching undergraduates and has taught courses in vertebrate zoology, functional ecology, herpetology, and the ecology, environmental physiology, and the biology of humans. After 23 years at Cornell University, he moved to Arizona State University West as Chair of the Department of Life Sciences to focus on the challenges of teaching undergraduates at a university that emphasizes community involvement. When not slaving over a hot computer revising *Vertebrate Life,* he enjoys walking in the desert with his labrador retriever, Martha.

Christine M. Janis is a Professor of Biology at Brown University where she teaches comparative anatomy and vertebrate evolution. She obtained her bachelor's degree at Cambridge University and then crossed the pond to get her Ph.D. at Harvard University. She is a vertebrate paleontologist with a particular interest in mammalian evolution (especially hoofed mammals) and faunal responses to climatic change. She first became interested in vertebrate evolution after seeing the movie *Fantasia* at the impressionable age of seven. That critical year was also the year that she began riding lessons, and she has owned at least one horse since the age of 12. She is still an active rider, although no longer as aggressive a competitor (she used to do combined training events). She attributes her lifestyle to the fact that she has failed to outgrow either the dinosaur phase or the horse phase.

John B. Heiser was born and raised in Indiana and completed his undergraduate degree in biology at Purdue University. He earned his Ph.D. in ichthyology from Cornell University for studies of the behavior, evolution and ecology of coral reef fishes, research which he continues today. For fifteen years he was Director of the Shoals Marine Laboratory operated by Cornell University and the University of New Hampshire on the Isles of Shoals in the Gulf of Maine. While at the Isles of Shoals his research interests focused on opposite ends of the vertebrate spectrum—hagfish and baleen whales. J.B. enjoys teaching vertebrate morphology, evolution, and ecology both in the campus classroom and in the field and is recipient of the Clark Distinguished Teaching Award from Cornell Univerisity. His hobbies are natural history, travel and nature photography and videography, especially underwater scuba. He has pursued his natural history interests on every continent and all the world's major ocean regions. Because of his experience he is a popular ecotourism leader having led Cornell Adult University groups to the Caribbean, Sea of Cortez, French Polynesia, Central America, the Amazon, Bornea, Antarctica, and Spitsbergen in the High Arctic.

Brief Contents

Contents

PART III Terrestrial Ectotherms: Amphibians, Turtles, Lepidosaurs, and Archosaurs 177

Preface

The sixth edition of *Vertebrate Life* incorporates a large number of changes, many of them the result of suggestions by colleagues who are using the book. The most conspicuous change is elimination of the overview of vertebrate anatomy and physiology that was provided in Chapters 3 and 4. With the concurrence of reviewers, we decided that a phylogenetic perspective is better emphasized by moving material from those introductory chapters into later chapters that cover the particular taxa whose characters are being discussed. Most of the material and figures in Chapters 3 and 4 have been retained, but they appear now in different locations.

Also in response to suggestions from colleagues, we have increased the emphasis on conservation, which has become a central theme of many vertebrate biology courses. We have included information about conservation issues in the chapters on extant taxa, and we have added a new final chapter that draws on information from earlier chapters to present an overview of conservation in the context of humans and other vertebrates. The historical and organismal perspectives of the book are retained in this chapter, which considers both the record of extinctions that coincides with the spread of modern humans in the past 100,000 years and current problems and conservation efforts.

We have changed the literature citations from a bibliography of papers actually cited in each chapter to a list of additional readings. This change allows us to include useful references that are not actually cited in the text. We have made greater use of websites as sources of information, and Prentice Hall maintains a web page for this book *http://prenhall.com/pough/* that includes links to Internet sites for each chapter.

As in previous editions, we have included cladograms illustrating the postulated relationships of vertebrates. In doing so, we have tried to point out areas of controversy. The cladograms include synopses of the character states on which they are based. New in this edition are simplified cladograms that emphasize the relationships of extant taxa.

Acknowledgments

As always, our editor Teresa Ryu combined support and decisiveness to keep the project on schedule, and she was ably assisted by Colleen Lee. It has been a pleasure to work with both of them. The production manager, Patty Donovan of Pine Tree Composition, kept all the parts organized. Without her skill and experience, this book would still be in pieces at sites around the Northern Hemisphere. We are fortunate to have Kandis Elliot as the artist for this edition. Her work blends so well with earlier illustrations by Laura Schuett and Carol Abrazincas that the extent of her contribution is not readily apparent.

Writing a book with a scope as broad as this one requires the assistance of many people. We are grateful to the following colleagues for their generous response to requests for information and their comments and suggestions:

David Armstrong, *University of Colorado, Boulder*
Joseph Beatty, *Oregon State University*
Willy Bemis, *University of Massachusetts at Amherst*
Alfred Beulig, *New College of University of South Florida*
Karen Bjorndahl, *University of Florida*
Beth Broadman, *Arizona State University West*
Brooks M. Burr, *Southern Illinois University at Carbondale*
Mark Chappell, *University of California, Riverside*
Jenny Clack, *University of Cambridge (UK)*
Mike Coates, *University College, London*

Davide Csermely, *Universita de Parma*
John Fitzpatrick, *Laboratory of Ornithology, Cornell University*
Margaret Fusari, *University of California, Santa Cruz*
Graeme Hays, *University of Wales, Swansea*
Virginia Hayssen, *Smith College*
Graham C. Hickman, *Texas A & M University, Corpus Christi*
James Hopson, *University of Chicago*
David G. Huckaby, *California State University, Long Beach*
Jon Mallatt, *Washington State University*
Moya Meredith-Smith, *Kings College, London*
Ken Mowbray, *American Museum of Natural History*
Caroline Pond, *The Open University (UK)*
Donald Prothero, *Occidental College*
Marilyn Renfree, *University of Melbourne*
Rob Slotow, *University of Natal*
Ellen M. Smith, *Phoenix, AZ*
Thomas F. Turner, *University of New Mexico*
Gus van Dyk, *Pilanesberg National Park, South Africa*
Blaire Van Valkenburgh, *University of California, Los Angeles*
George H. Waring, *Southern Illinois University at Carbondale*
Kelly Zamudio, *Cornell University*

PART 1

Vertebrate Diversity, Function, and Evolution

The more than 50,000 living species of vertebrates inhabit nearly every part of the Earth, and other kinds of vertebrates that are now extinct lived in habitats that no longer exist. Increasing knowledge of the diversity of vertebrates was a product of the European exploration and expansion that began in the fifteenth and sixteenth centuries. In the middle of the eighteenth century, Swedish naturalist Carolus Linnaeus developed a binominal classification to catalog the varieties of animals and plants. The Linnean system remains the basis for naming living organisms today.

A century later, Charles Darwin explained the diversity of plants and animals as the product of natural selection and evolution. In the early twentieth century, Darwin's work was coupled with the burgeoning information about mechanisms of genetic inheritance. This combination of genetics and evolutionary biology, known as the New Synthesis or neo-Darwinism, continues to be the basis for understanding the mechanics of evolution. Methods of classifying animals have also changed during the twentieth century; and classification, which began as a way of trying to organize the diversity of organisms, has become a powerful tool for generating testable hypotheses about evolution.

Vertebrate biology and the fossil record of vertebrates have been at the center of these changes in our view of life. Comparative studies of the anatomy, embryology, and physiology of living vertebrates have often supplemented the fossil record. These studies reveal that evolution acts by changing existing structures. All vertebrates share basic characteristics that are the products of their common ancestry, and the progress of evolution can be analyzed by tracing the modifications of these characters. Thus, an understanding of vertebrate form and function is basic to understanding the evolution of vertebrates and the ecology and behavior of living species.

CHAPTER

1 The Diversity, Classification, and Evolution of Vertebrates

Evolution is central to vertebrate biology because it provides a principle that organizes the diversity we see among living vertebrates and helps to fit extinct forms into the context of living species. Classification, initially a process of attaching names to organisms, has become a method of understanding evolution. Current views of evolution stress natural selection operating at the level of individuals as a predominant mechanism that produces change over time. The processes and events of evolution are intimately linked to the changes that have occurred on Earth during the history of vertebrates. These changes have resulted from the movements of continents and the effects of those movements on climates and geography. In this chapter we present an overview of the scene, the participants, and the rules governing the events that have shaped the biology of vertebrates.

1.1 The Vertebrate Story

Mention "animal" and most people will think of a vertebrate. Vertebrates are often abundant and conspicuous parts of people's experience of the natural world. Vertebrates are also very diverse: The more than 50,000 extant (currently living) species of vertebrates range in size from fishes weighing as little as 0.1 gram when fully mature to whales weighing over 100,000 kilograms. Vertebrates live in virtually all the habitats on Earth. Bizarre fishes, some with mouths so large they can swallow prey larger than their own bodies, cruise through the depths of the sea, sometimes luring prey to them with glowing lights. Fifteen kilometers above the fishes, migrating birds fly over the crest of the Himalayas, the highest mountains on Earth.

The behaviors of vertebrates are as diverse and complex as their body forms. Vertebrate life is energetically expensive, and vertebrates get the energy they need from food they eat. Carnivores eat the flesh of other animals and show a wide range of methods of capturing prey. Some predators search the environment to find prey, whereas others wait in one place for prey to come to them. Some carnivores pursue their prey at high speeds, others pull prey into their mouths by suction. In some cases the foraging behaviors vertebrates use appear to be exactly the ones that maximize the amount of energy they obtain for the time they spend hunting; in other cases vertebrates appear to be remarkably inept predators. Many vertebrates swallow their prey intact, sometimes while it is alive and struggling, but other vertebrates have very specific methods of dispatching prey. Venomous snakes inject complex mixtures of toxins, and cats (of all sizes from house cats to tigers) kill their prey with a distinctive bite on the neck. Herbivores eat plants. Plants cannot run away when an animal approaches, but they are hard to digest and they frequently contain toxic compounds. Herbivorous vertebrates show an array of specializations to deal with the difficulties of eating plants. These specializations include elaborately sculptured teeth and digestive tracts that provide sites where symbiotic microorganisms digest compounds that are impervious to the digestive systems of vertebrates.

Reproduction is a critical factor in the evolutionary success of an organism, and vertebrates show an astonishing range of behaviors associated with mating and reproduction. In general, males court females and females care for the young; but these roles are reversed in many species of vertebrates. The forms of reproduction employed by vertebrates range from laying eggs to giving birth to babies that are largely or entirely independent of their parents (precocial young). These variations range across almost all kinds of vertebrates. Many fishes and amphibians produce precocial young, and a few mammals lay eggs. At the time of birth or hatching some vertebrates are entirely self-sufficient and never see their parents, whereas other vertebrates (including humans) have extended periods of obligatory parental care. Extensive parental care is found in seemingly unlikely groups of vertebrates—fishes that incubate eggs in their mouths, frogs that incubate eggs in their stomachs, and birds that feed their nestlings a fluid called crop milk that is very similar in composition to mammalian milk.

The diversity of living vertebrates is fascinating, but the species now living are only a small proportion of the species of vertebrates that have existed. For each living species there may be more than one hundred extinct species, and some of these have no counterparts among living forms. The dinosaurs, for example, that dominated the Earth for 180 million years are so entirely different from any living animals that it is hard to reconstruct the lives they led. Even mammals were once more diverse than they are now. The Pleistocene epoch saw giants of many kinds—ground sloths as big as modern rhinoceroses, and raccoons as large as bears. Humans are great apes, close relatives of chimpanzees, and much of the biology of humans is best understood in the context of our vertebrate heritage. The number of species of vertebrates probably reached its maximum in the middle Miocene, 12 to 14 million years ago, and has been declining since then.

The story of vertebrates is fascinating. Where they originated, how they evolved, what they do, and how they work provide endless intriguing details. In preparing to tell this story we must introduce some basic information, including what the different kinds of vertebrates are called, how they are classified, and what the world was like as the story of vertebrates unfolded.

The modern approach to biological classification is popularly known as cladistics, from the Greek word *cladus,* meaning "branch."

Cladistics recognizes only groups of organisms that are related by common descent, or phylogeny (Greek *phyla* = tribe and *genesis* = origin). The application of cladistic methods has made the study of evolution rigorous. The natural groups recognized by cladistics are easier to understand than the artificial groups we are used to, except that we are familiar with the names of the artificial groups and the names of the new groups are sometimes strange. At this point we need to establish a basis for talking about particular animals by naming them, and to relate the old, familiar names to the new, less familiar ones. Figure 1–1 shows the major kinds of vertebrates and the relative numbers of living species. In the following sections we describe briefly the different kinds of living vertebrates.

Hagfishes and Lampreys—Myxinoidea and Petromyzontoidea Lampreys and hagfishes are elongate, scaleless, and slimy and have no internal hard tissues. They are scavengers and parasites and are specialized for those roles. Hagfishes (about 40 species) are marine, living on the continental shelf and open ocean at depths around 100 meters. In contrast, many of the 41 species of lampreys are migratory forms that live in oceans and spawn in rivers.

Hagfishes and lampreys are unique among living vertebrates because they lack jaws; this feature makes them important in the study of vertebrate evolution. They have traditionally been grouped as agnathans (Greek *a* = without and *gnath* = jaw) or cyclostomes (Greek *cyclo* = round and *stoma* = mouth), but they probably represent two independent evolutionary lineages. Hagfishes lack many of the features characterizing most vertebrates; for example, they have no trace of vertebrae and sometimes are classified in the Craniata, but not in the Vertebrata. In contrast, lampreys have rudimentary vertebrae as well as many other characters they share with jawed vertebrates. The jawless condition of both lampreys and hagfishes, however, is ancestral.

Sharks, Rays, and Ratfishes—Elasmobranchii and Holocephali Sharks have a reputation for ferocity that most of the 350 to 400 species would have difficulty living up to. Many sharks are small (15 centimeters or less); and the largest species, the whale shark—which grows to 10 meters—is a filter feeder that subsists on plankton it strains from the water. The 450 species of rays are dorsoventrally flattened, frequently bottom dwellers that swim with undulations of their extremely broad pectoral fins. The approximately 30 species of ratfish are bizarre marine fishes with long, slender tails and buck-

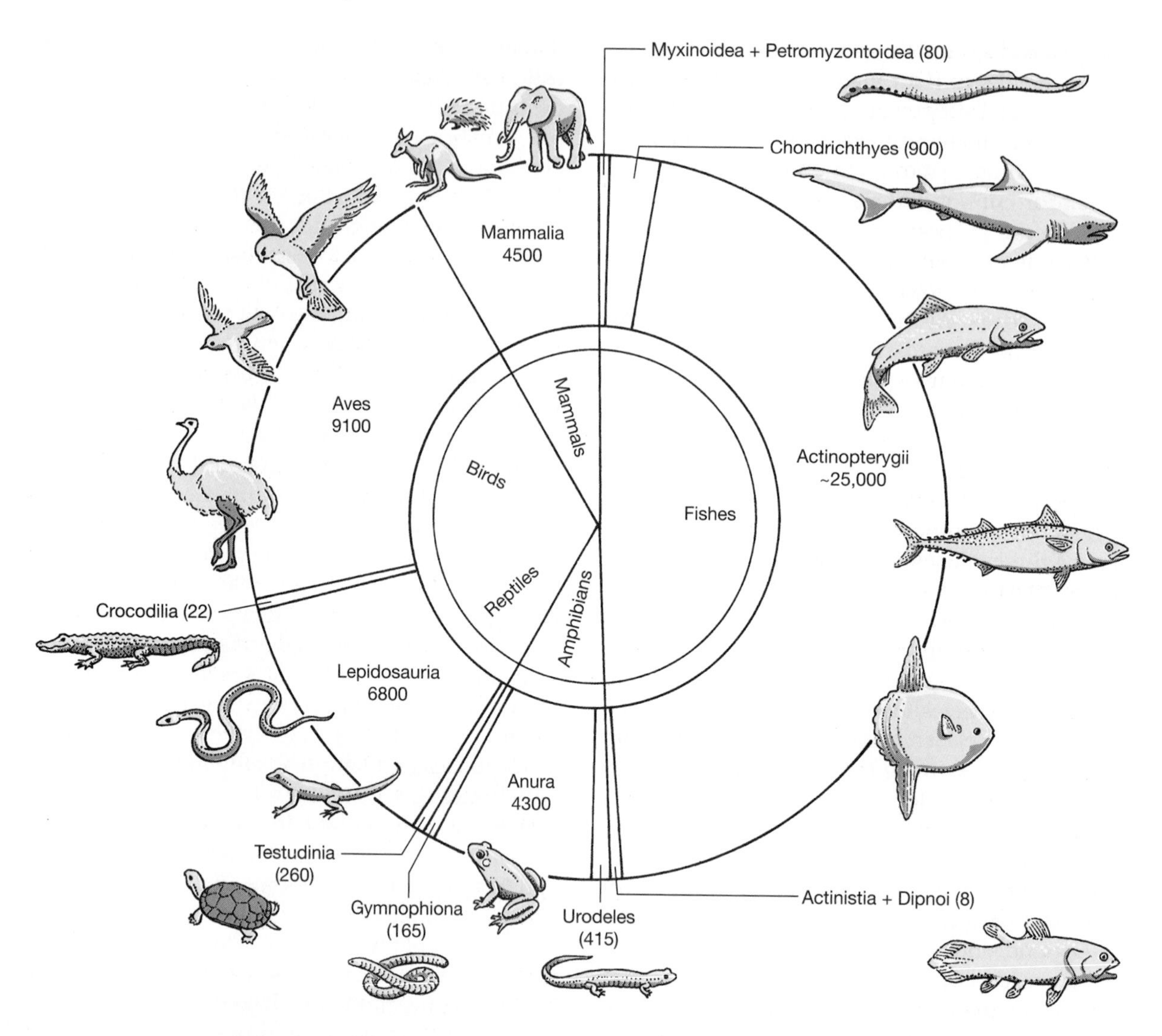

▲ Figure 1–1 Diversity of vertebrates. Areas in the diagram correspond to approximate numbers of living species in each group. Common names are in the center circle, and formal names for the groups are on the outer circle.

toothed faces that look rather like rabbits. The name Chondrichthyes (Greek *chondro* = cartilage and *ichthyes* = fish) refers to the cartilaginous skeletons of these fishes.

Bony Fishes—Osteichthyes Bony fishes are so diverse that any attempt to characterize them briefly is doomed to failure. Two broad categories can be recognized: the lobe-finned or fleshy-finned fishes (sarcopterygians; Greek *sarco* = flesh) and the ray-finned fishes (actinopterygians; Greek *actino* = ray and *ptero* = wing or fin).

Only eight species of lobe-finned fishes survive, the six species of lungfishes (Dipnoi) found in South America, Africa, and Australia and the two species of coelacanth (Coelacanthiformes), one from deep waters off the east coast of Africa and a second species recently discovered near Indonesia. These are the living fishes most closely related to terrestrial vertebrates.

In contrast to the lobe-fins, the ray-finned fishes have radiated extensively in fresh and salt water. More than 22,000 species of ray-finned fishes have been named, and several thousand additional species may await discovery (Eschmeyer 1998). Two major groups can be distinguished among actinopterygians. The Chondrostei (bichirs, sturgeons, and paddlefishes) includes about 35 species that are survivors of an early radiation of bony fishes. Bichirs (10 species) are swamp- and river-dwellers from Africa; they are known as African reed fish in the aquarium trade. Sturgeons (25 species) are large fishes with protrusible, toothless mouths that are used to suck food items from the bottom. Sturgeons are the source of caviar—eggs taken from the female before being laid—and many species have been driven close to extinction by overfishing. Paddlefish (two species, one in the Mississippi drainage of North America and the other in the Yangtze River of China) have a paddle-like snout with organs that locate prey by sensing electrical fields.

The Neopterygii, the modern radiation of bony fishes, can be divided into three lineages. Two of these—the gars (Lepisosteiformes, seven species) and the bowfin (one species, *Amia calva*)—are relicts of earlier radiations. These fishes have cylindrical bodies, thick scales, and jaws armed with sharp teeth. They seize prey in their mouths with a sudden rush or gulp, and they lack the specializations of the jaw apparatus that allow later bony fishes to use more complex feeding modes.

The third lineage of neopterygians, the Teleostei, includes more than 20,000 species of fishes covering every imaginable combination of body size, habitat, and habits. Most familiar fishes are teleosts—the trout, bass, and panfish that dedicated anglers seek; the sole (a kind of flounder) and swordfish featured by seafood restaurants; and the salmon and tuna whose by-products find their way into canned catfood. Modifications of the body form and jaw apparatus have allowed many teleosts to be highly specialized in their swimming and feeding habits.

Salamanders, Frogs, and Caecilians—Urodela, Anura, and Gymnophiona These three groups of vertebrates are popularly known as amphibians (Greek *amphi* = double and *bios* = life) in recognition of their complex life histories, which often include an aquatic larval form (the larva of a salamander or caecilian and the tadpole of a frog) and a terrestrial adult. All amphibians have bare skins (that is, lacking scales, hair, or feathers) that are important in the exchange of water, ions, and gases with their environment. Salamanders (about 425 species) are elongate animals, mostly terrestrial and usually with four legs; anurans (frogs, toads, treefrogs—about 4300 species in all) are short bodied with large heads and large hind legs used for walking, jumping, and climbing; and caecilians (about 165 species) are legless aquatic or burrowing animals.

Turtles—Testudinia The 260 species of turtles are probably the most immediately recognizable of all vertebrates. The shell that encloses a turtle has no exact duplicate among other vertebrates, and the morphological modifications associated with the shell make turtles extremely peculiar animals. They are, for example, the only vertebrates with the shoulders (pectoral girdle) and hips (pelvic girdle) inside the ribs.

Tuatara, Lizards, and Snakes—Lepidosauria These three kinds of vertebrates can be recognized by their scale-covered skin as well as by characteristics of the skull. The two species of tuatara, stocky-bodied animals found only on some islands near New Zealand, are the sole living remnant of a lineage of animals called Sphenodontida that were more diverse in the Mesozoic. In contrast, lizards (more than 4000 species) and snakes (more than 2700 species) are now at the peak of their diversity.

Alligators and Crocodiles—Crocodilia These impressive animals (the saltwater crocodile has the potential to grow to a length of 7 meters) are in the same lineage (the Archosauria) as dinosaurs and birds. The 22 species of crocodilians, as they are known collectively, are semiaquatic predators with long snouts armed with numerous teeth. Their skin contains many bones (osteoderms; Greek *osteo* = bone and *derm* = skin) that lie beneath the scales and provide a kind of armor plating. Crocodilians are noted for the parental care they provide for their eggs and young.

Birds—Aves The birds are a lineage of dinosaurs that evolved flight in the Mesozoic. They have diversified into more than 9100 species. Feathers are the distinguishing characteristic of birds, and feathered wings are the structures that power a bird's flight. Recent discoveries of dinosaur fossils with what might be traces of feathers may indicate that feathers evolved before flight. Birds are conspicuous animals in our world because they are active during the day (diurnal) and often attract attention with loud vocalizations, bright colors, and flamboyant behaviors. As a result, birds have been studied extensively, and much of our information about the behavior and ecology of terrestrial vertebrates is based on those studies.

Mammals—Mammalia The living mammals can be traced to an origin in the late Paleozoic, from some of the earliest fully terrestrial vertebrates. Extant mammals include about 4500 species, most of which are eutherian (placental) mammals. Both eutherians and marsupials possess a placenta, a structure that transfers nutrients from the mother to the embryo and removes the waste products of the embryo's metabolism. Eutherians have a more extensive system of placentation and a long gestation period, while marsupials have a short gestation period and give birth to very immature young that continue their development in an external pouch on the mother's abdomen. Marsupials dominate the mammalian fauna only in Australia. Kangaroos, koalas, and wombats are familiar Australian marsupials. The strange monotremes, the platypus and the echidnas—also from Australia—are mammals whose young are hatched from eggs.

1.2 Classification of Vertebrates

The diversity of vertebrates (more than 50,000 living species and at least 10 times that number of species now extinct) makes the classification of vertebrates an extraordinarily difficult task. Yet classification has long been at the heart of evolutionary biology. Initially, classification of species was seen as a way of managing the diversity of organisms, much as an office filing system manages the paperwork of the office. Each species could be placed in a pigeonhole marked with its name, and when all species were in their pigeonholes the diversity of vertebrates would have been encompassed. This approach to classification was satisfactory as long as species were regarded as static and immutable: Once a species was placed in the filing system it was there to stay.

Acceptance of the fact of evolution has made that kind of classification inadequate. Now biologists must express evolutionary relationships among species by incorporating evolutionary information in the system of classification. Ideally, a classification system should not only attach a label to each species, it should also encode the evolutionary relationship between that species and other species. Modern techniques of systematics (the evolutionary classification of organisms) are moving beyond filing systems and have become methods for generating testable hypotheses about evolution.

Classification and Names

Our system of naming species is pre-Darwinian. It traces back to methods established by the naturalists of the seventeenth and eighteenth centuries, especially those of Carl von Linné, a Swedish naturalist, better known by his Latin pen name, Carolus Linnaeus. The Linnaean system employs binominal nomenclature to designate species, and arranges species into hierarchical categories (**taxa,** singular *taxon*) for classification. This system is incompatible in some respects with evolutionary biology (de Queiroz and Gauthier 1992), but it is still widely used. Thus it is important to understand the traditional classification, because it forms the basis for much of the literature of vertebrate biology.

Binominal Nomenclature

The scientific naming of species became standardized when Linnaeus's monumental work, *Systema Naturae* (*The System of Nature*), was published in sections between 1735 and 1758. Linnaeus attempted to give an identifying name to every known species of plant and animal. His method assigns a binominal (two-word) name to each species. Familiar examples include *Homo sapiens* for human beings (Latin *hom* = human and *sapien* = wise), *Passer domesticus* for the house sparrow (Latin *passer* = sparrow and *domesticus* = belonging to the house), and *Canis familiaris* for the domestic dog (Latin *canis* = dog and *familiaris* = of the family).

Why use Latin words? Latin was the early universal language of European scholars and scientists. It has provided a uniform usage that scientists, regardless of their native language, continue to recognize worldwide. The same species may have different colloquial names, even in the same language. For example, *Felis concolor* (Latin for "the uniformly colored cat") is known in various parts of North America as cougar, puma, mountain lion, American panther, painter, and catamount. In Central and South America it is called león colorado, onça-vermelha, poema, guasura, or yaguá-pitá. But biologists of all nationalities recognize the name *Felis concolor* as referring to a specific kind of cat.

Hierarchical Groups: The Higher Taxa

Linnaeus and other naturalists of his time developed what they called a natural system of classification. All similar species are grouped together in one

genus (plural *genera*), based on characters that define the genus. The most commonly used characters were anatomical, because they can be most easily preserved in museum specimens. Thus all dog-like species—various wolves, coyotes, and jackals—were grouped together in the genus *Canis* because they all share certain anatomical features, such as an erectile mane on the neck and a skull with a long, prominent sagittal crest on which massive temporal (jaw-closing) muscles originate. Linnaeus's method of grouping species was functional because it was based on anatomical (and to some extent on physiological and behavioral) similarities and differences. Linnaeus lived before there was any knowledge of genetics and the mechanisms of inheritance. But he used taxonomic characters that we understand today are genetically determined biological traits that generally express the degree of genetic similarity or difference among groups of organisms.

Subsequent development of biological classification has employed seven basic taxonomic categories, listed here in decreasing order of inclusiveness:

Kingdom
  Phylum (called a Division in plant classification)
    Class
      Order
        Family
          Genus
            Species

These are the taxonomic levels (categories) used in most of the primary and secondary literature of vertebrate biology.

1.3 Traditional and Cladistic Classifications

All methods of classifying organisms are based on similarities among the included species, but some differences are more significant than others. For example, nearly all vertebrates have paired limbs, but only a few kinds of vertebrates have mammary glands. Consequently, knowing that the species in question have mammary glands tells you more about the closeness of their relationship than knowing that they have paired limbs. You would thus give more weight to the presence of mammary glands than to paired limbs.

Phylogenetic Systematics or Cladistics

In 1966, Willi Hennig introduced **phylogenetic systematics** by forcefully emphasizing that the history of an evolutionary lineage can be reconstructed only on the basis of **derived characters.** An evolutionary lineage is a **clade** (from the Greek word for a branch), and phylogenetic systematics is also called **cladistics.**

Derived means "different from the ancestral condition." A derived character is called an **apomorphy** (Greek *apo* = away from—i.e., derived from—and *morph* = form). For example, the feet of terrestrial vertebrates have distinctive bones—the carpals, tarsals, and digits. This arrangement of foot bones is different from the ancestral pattern seen in lobe-finned fishes, and all lineages of terrestrial vertebrates had that derived pattern of foot bones at some stage in their evolution. (Some groups of terrestrial vertebrates—snakes, for example—have subsequently modified that pattern of foot bones; but the significant point is that the evolutionary lineage of snakes once had the pattern.) Thus, the terrestrial pattern of foot bones is a **shared derived character** of terrestrial vertebrates. In cladistic terminology, shared derived characters are called **synapomorphies** (Greek *syn* = together).

Of course, organisms also have shared ancestral characters, that is, characters that they have inherited unchanged from their ancestors. These are called **plesiomorphies** (Greek *plesi* = near in the sense of "similar to the ancestor"). Terrestrial vertebrates have a vertebral column, for example, that was inherited essentially unchanged from lobe-finned fishes. Hennig called shared ancestral characters **symplesiomorphies.** (*sym*, like *syn*, is a Greek root that means "together"). Symplesiomorphies tell us nothing about degrees of relatedness. The principle that only *shared derived* characters can be used to determine genealogies is the core of cladistics.

The conceptual basis of cladistics is straightforward, although applying cladistic criteria to real organisms can become very complicated. To illustrate the conceptual basis of cladistic classification, consider the examples presented in Figure 1–2. Each of the three **cladograms** (diagrams showing hypothetical sequences of branching during evolution) is a possible evolutionary relationship (i.e., a **phylogeny**) for the three **taxa** (plural of **taxon**, which means a species or group of species) identified as 1, 2, and 3. To make the example a bit more concrete, we can

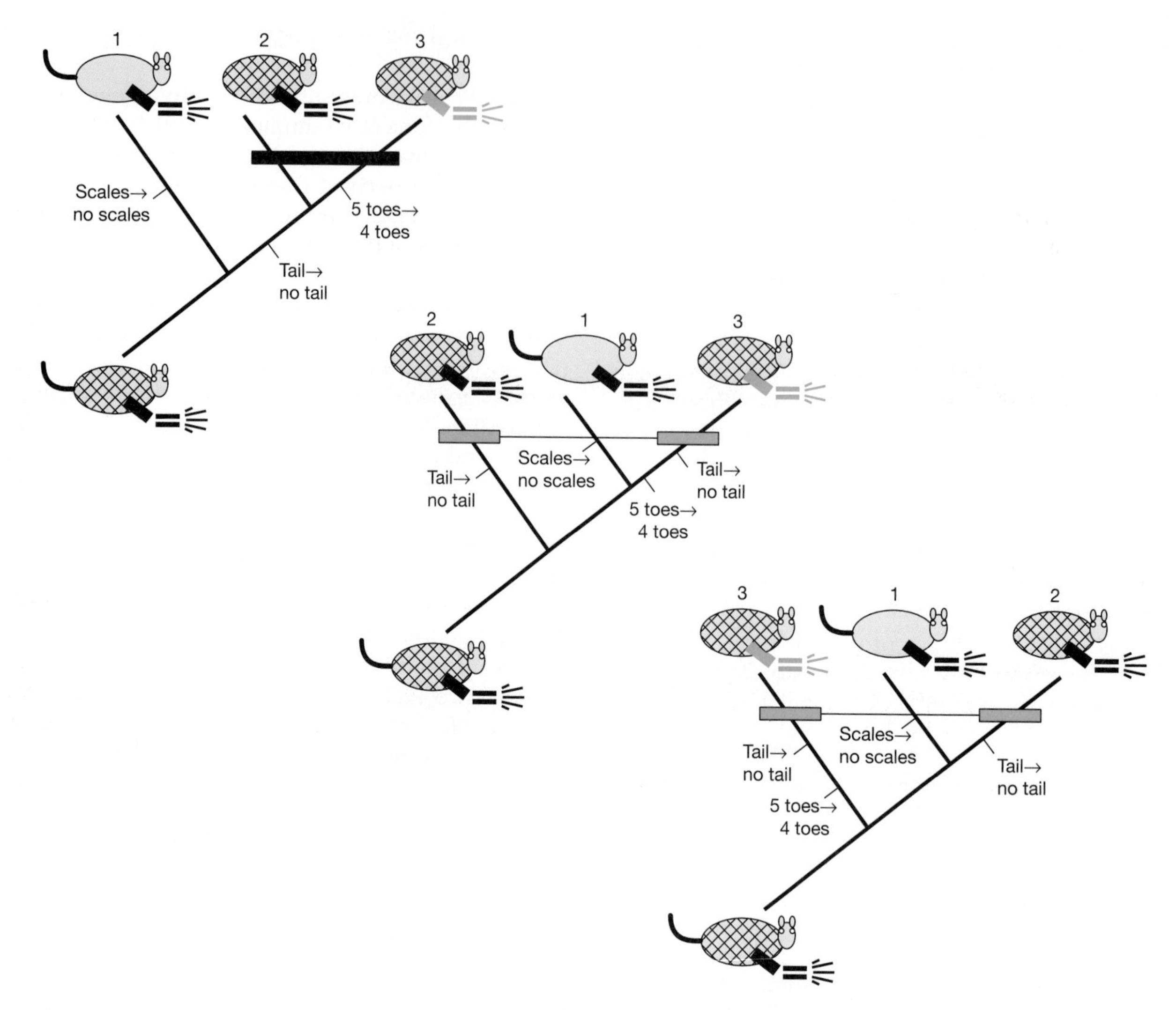

▲ Figure 1–2 Hypothetical distribution of three character states in three taxa (1, 2, and 3). Bars connect derived characters (apomorphies). The black bar shows a shared derived character (a synapomorphy) of the lineage that includes taxa 2 and 3. Blue bars represent two independent origins of the same derived character state that must be assumed to have occurred if there was no apomorphy in the most recent common ancestor of taxa 2 and 3. The labels identify changes from the ancestral character state to the derived condition.

consider three characters: the number of toes on the front foot, the skin covering, and the tail. For this example, let's say that in the ancestral character state there are five toes on the front foot and in the derived state there are four toes. We'll say that the ancestral state is a scaly skin and the derived state lacks scales. As for the tail, the ancestral state is present and the derived state is absent.

Figure 1–2 shows the distribution of those three character states in the three taxa. The animals in taxon 1 have five toes on the front feet, lack scales, and have a tail. Animals in taxon 2 have five toes, scaly skins, and no tails. Animals in taxon 3 have four toes, scaly skins, and no tails.

How can we use this information to decipher the evolutionary relationships of the three groups of animals? Notice that the ancestral number of toes occurs only in taxon 1, and the derived tail condition (absent) is found in taxa 2 and 3. The most **parsimonious** phylogeny (i.e., the phylogeny requiring the fewest number of evolutionary changes) is represented by the leftmost diagram in Figure 1–2. Only three changes are needed to produce the derived character states:

1. In the evolution of taxon 1, scales are lost.
2. In the evolution of taxon 2 + taxon 3, the tail is lost.
3. In the evolution of taxon 3, a toe is lost from the front foot.

The other two phylogenies shown in Figure 1–2 are possible, but they would require tail loss to occur independently in taxon 2 and in taxon 3. Any change in a structure is an unlikely event, so the most plausible phylogeny is the one requiring the fewest changes. The second and third phylogenies require four evolutionary changes, so they are less parsimonious than the first phylogeny we considered.

A phylogeny is an hypothesis about the evolutionary relationships of the groups included. Like any scientific hypothesis, it can be tested when new data become available. If it fails that test it is falsified; that is, it is rejected and a different hypothesis (a different cladogram) takes its place. The process of testing hypotheses about vertebrates and evolution and replacing those that are falsified is a continuous one, and changes in the cladograms in successive editions of this book show where new information has generated new hypotheses. The most important contribution of phylogenetic systematics is that it enables us to frame testable hypotheses about the sequence of events during evolution.

So far we have avoided a central issue of phylogenetic systematics: How do scientists know which character state is ancestral (plesiomorphic) and which is derived (apomorphic)? That is, how can we determine the direction (**polarity**) of evolutionary transformation of the characters? For that, we need additional information. Increasing the number of characters we are considering can help, but comparing the characters we are using with an **outgroup** that consists of the closest relatives of the **ingroup** (i.e., the organisms we are studying) is the preferred method. A well-chosen outgroup will possess ancestral character states compared to the ingroup. For example, lobe-finned fishes are an appropriate outgroup for terrestrial vertebrates.

Cladists establish and name evolutionary lineages solely on the basis of **monophyly** (a single evolutionary origin, from the Greek *mono* = one, single). For example, every vertebrate depicted in Figure 1–3 that arises from a branch point (i.e., each lineage) along the cladogram is related by the shared derived characters used to diagnose that lineage. The name assigned to the lineage identifies all members in the clade. The cladogram depicted in Figure 1–3 is a hypothesis of the evolutionary relationships of the major living groups of vertebrates. There are 14 dichotomous branches leading from the origin of vertebrates from other chordates to birds and mammals. Cladistic terminology assigns names to the lineages originating at each branch point. This process produces a nested series of groups, starting with the most inclusive. Thus, the name Gnathostomata includes all vertebrate animals that have jaws; that is, every taxon to the right of the number 2 in Figure 1–3 is included in the Gnathostomata, every taxon to the right of number 4 is included in the Osteichthyes, and so on. Because the lineages are nested, it is correct to say that humans are bony fishes (Osteichthyes, number 4). Humans are also included in the lineages below branch 4 and in most of those above it. After number 7 the cladogram divides into Lissamphibia and Amniota, and we are in the Amniota lineage. The cladogram divides again above number 9 into the Sauropsida and Synapsida lineages, and humans are in the Synapsida lineage.

At the top of Figure 1–3 are traditional names that have been applied to different groups of living vertebrates. Unlike cladistic terminology, traditional names were not restricted to monophyletic groups. The Agnatha, for example, included two evolutionary lineages—the Myxinoidea (hagfishes) and Petromyzontidoidea (lampreys)—that are not each other's closest relatives, because lampreys are more closely related to gnathosomes than hagfishes are.

Two names are used in both systems—Osteichthyes (bony fish) and Aves (birds). This dual usage can be confusing, and it is important to distinguish which use is meant. In traditional classification amd common use Osteichthyes means only the bony fish, but in cladistic terminology it includes all more derived forms. Thus, in cladistic classification, Aves is included in Osteichthyes.

Evolutionary Hypotheses

Phylogenetic systematics is based on the assumption that organisms that are grouped together share a common heritage, which accounts for their similarities. Because of that common heritage we can use cladograms to ask questions about evolution. By examining the origin and significance of characters of living animals, we can make inferences about the biology of extinct species. For example, the phylogenetic relationship of crocodilians, dinosaurs, and birds is shown in Figure 1–4. We know that

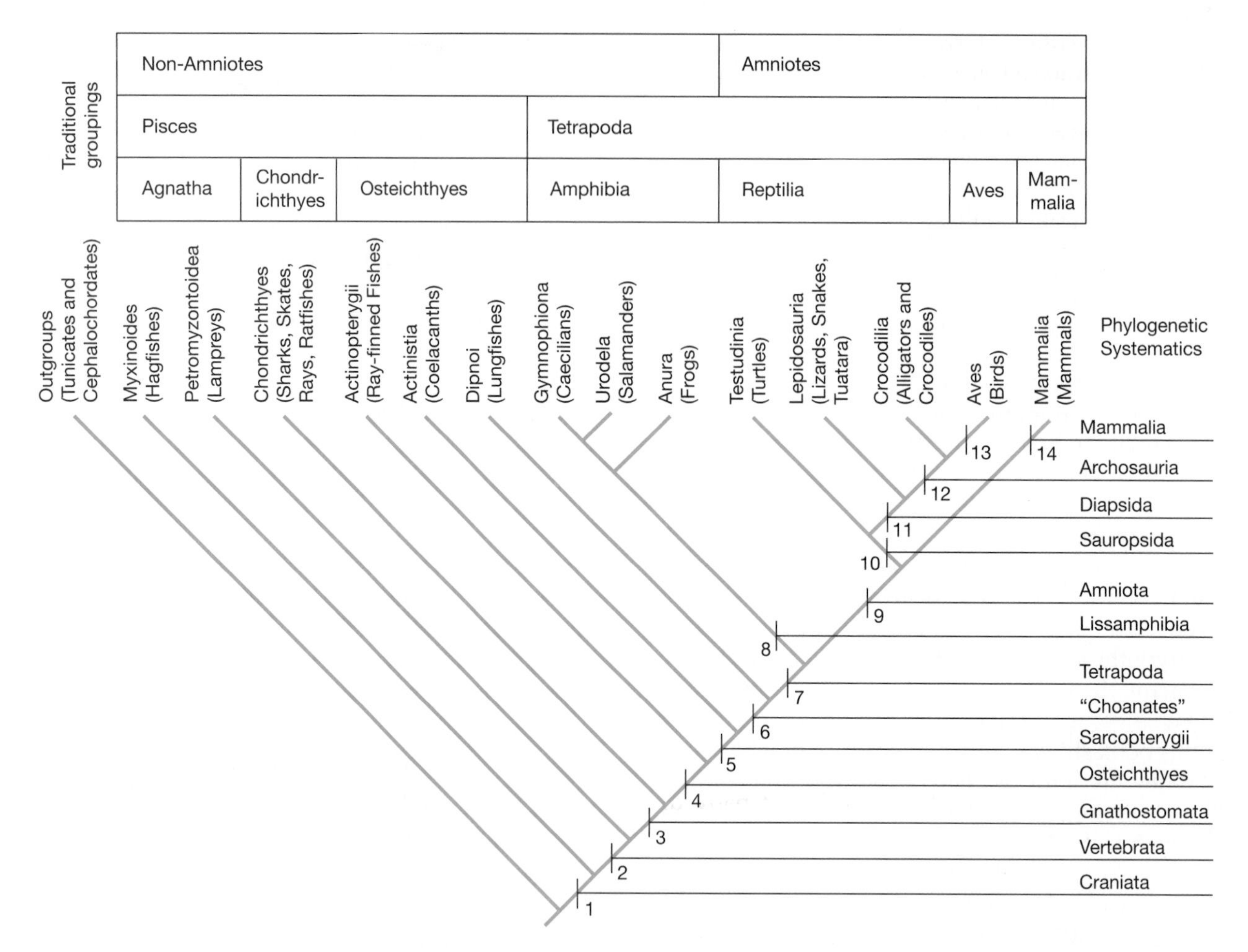

Legend: **1. Craniata**—Distinct head region skeleton incorporating anterior end of notochord, one or more semicircular canals, brain consisting of three regions, paired kidneys, gill bars, neural crest tissue. **2. Vertebrata**—Arcualia or their derivatives form vertebrae, two or three semicircular canals. **3. Gnathostomata**—Jaws formed from mandibular arch, teeth containing dentine, three semicircular canals, branchial arches containing four elements on each side plus one unpaired ventral median element, paired fins with internal skeleton and muscles supported by girdles in the body wall. **4. Osteichthyes**—Presence of lung or swimbladder derived from the gut, unique pattern of dermal bones of the head and shoulder region, unique characters of jaw and branchial muscles. **5. Sarcopterygii**—Unique supporting skeleton in fins. **6.—"Choanates"**—Choanae* present, derived paired appendage structure, conus arteriosus of heart partly divided, unique dermal bone pattern of head, loss of interhyal bone. **7. Tetrapoda**—Limbs with carpals, tarsals, and digits. **8. Lissamphibia**—Structure of the skin, papilla amphibiorum and opercular-plectrum complex in inner ear, green rods in retina, pedicellate teeth, presence of levator bulbi muscle. **9. Amniota**—A distinctive arrangement of extraembryonic membranes (the amnion, chorion, and allantois). **10. Sauropsida**—Tabular and supratemporal bones small or absent, simple coronoid, centrum of atlas and intercentrum of axis fused, medial centrale of ankle absent. **11. Diapsida**—Skull with a dorsal temporal fenestra, upper temporal arch formed by triradiate postorbital and triradiate squamosal bones. **12. Archosauria**—Presence of a fenestra anterior to the orbit of the eye, orbit shaped like an inverted triangle. **13. Aves**—Feathers, in derived forms loss of teeth and metabolic heat production used to regulate body temperature (endothermy). **14. Synapsida**—Lower temporal fenestra present. **15. Mammalia**—Hair, lower jaw formed only by dentary bone, mammary glands, development of endothermy independent of Aves.

*Many authorities question the homology of the choanae (internal nostrils) of dipnoans with those of tetrapods. Thus the name "Choanates" may not be appropriate, but the grouping appears to be valid.

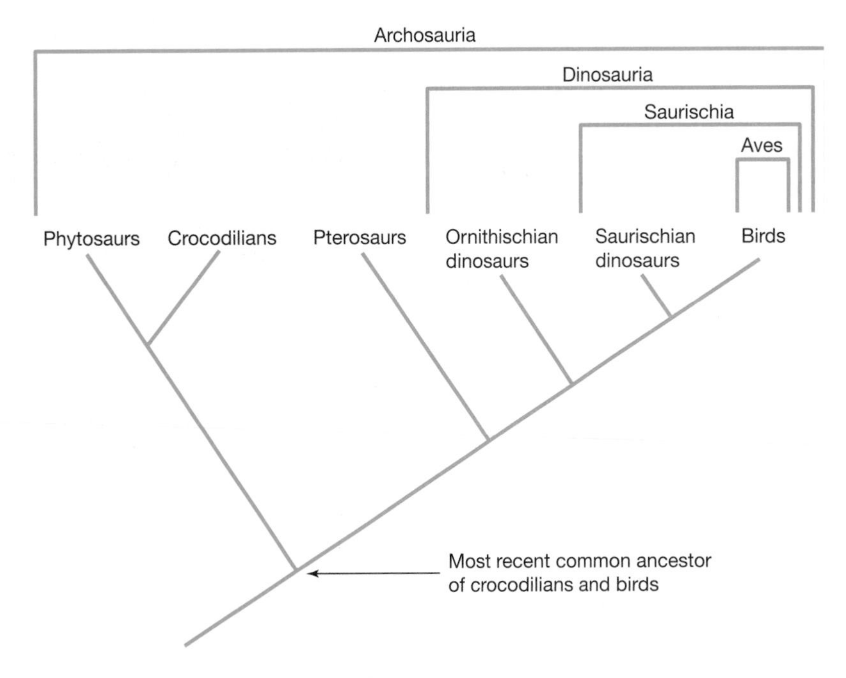

▶ Figure 1–4 Using a cladogram to make inferences about behavior. The cladogram shows the relationships of the Archosauria, the evolutionary lineage that includes living crocodilians and birds. (Phytosaurs were crocodile-like animals that disappeared at the end of the Triassic, and pterosaurs were the flying reptiles of the Jurassic and Cretaceous.) Both extant groups, crocodilians and birds, display extensive parental care of eggs and young. The most parsimonious explanation of this situation assumes that parental care is an ancestral character of the archosaur lineage.

both crocodilians and birds display extensive parental care of their eggs and young. Some fossilized dinosaur nests contain remains of baby dinosaurs, suggesting that at least some dinosaurs may also have cared for their young. Is that a plausible inference?

Obviously there is no direct way to determine what sort of parental care dinosaurs had. The intermediate lineages in the cladogram (pterosaurs and dinosaurs) are extinct, so we cannot observe their reproductive behavior. But the phylogenetic diagram in Figure 1–4 provides an indirect way to approach the question. Both of the closest living relatives of the dinosaurs, crocodilians and birds, do have parental care. Looking at living representatives of more distantly related lineages (outgroups), we see that parental care is not universal among fishes, amphibians, or reptiles other than crocodilians. With that information, the most parsimonious explanation of the occurrence of parental care in both crocodilians and birds is that it had evolved in that lineage *before* the crocodilians separated from dinosaurs + birds. (We cannot prove that parental care did not evolve separately in crocodilians and in birds, but one change to parental care is more parsimonius than two changes instead of one.) Thus, the most parsimonious hypothesis is that parental care is a derived character of the evolutionary lineage containing crocodilians + dinosaurs + birds (the Archosauria). That means we are probably correct when we interpret the fossil evidence as showing that dinosaurs did have parental care.

Figure 1–4 also shows how cladistics has made talking about restricted groups of animals more complicated than it used to be. Suppose you wanted to refer to just the two lineages of animals that are popularly known as dinosaurs—ornithischians and saurischians. What could you call them? Well, if you

◀ Figure 1–3 Phylogenetic relationships of extant vertebrates. This diagram shows the probable relationships among the major groups of vertebrates. The boxes across the top of the diagram show the traditional groupings and the names along the right side show how the lineages are grouped by phylogenetic systematics (cladistics). Note that the cladistic groupings are nested progressively; that is, all mammals are amniotes, all amniotes are tetrapods, all tetrapods are sarcopterygians, all sarcopterygians are osteichthyans, and so on. The numbers indicate derived characters that distinguish the lineages. For clarity this cladogram uses names of both node-based crown clades (e.g., Mammalia) and stem-based clades (e.g., Synapsida). This is not strictly correct, but an explanation of the significance of the difference in the names is beyond the scope of this text. See Figure 10–2 for alternative hypotheses about the affinities of turtles.

call them dinosaurs, you're not being PC (phylogenetically correct), because the taxon Dinosauria includes birds. So if you say dinosaurs, you are including ornithischians + saurischians + birds, even though any seven-year-old would understand that you are trying to restrict the conversation to extinct Mesozoic animals.

In fact, in cladistic terminology there is no correct taxonomic name for just the animals popularly known as dinosaurs. That's because cladistics recognizes only monophyletic lineages, and a monophyletic lineage includes an ancestral form and all its descendants. The most recent common ancestor of ornithischians, saurischians, and birds in Figure 1–4 lies at the intersection of the lineage of ornithischians with saurischians + birds, so Dinosauria is a monophyletic lineage. But if birds are omitted, all the descendants of the common ancestor are no longer included; and ornithischians + saurischians minus birds does not fit the definition of a monophyletic lineage. It would be called a **paraphyletic** group (Greek *para* = beside, beyond).

Biologists who are interested in how organisms live often want to talk about paraphyletic groups. After all, the dinosaurs (in the popular sense of the word) differed from birds in many ways. The only correct way of referring to the animals popularly known as dinosaurs is to call them nonavian dinosaurs, and you will find that and other examples of paraphyletic groups later in the book. Sometimes even this construction does not work, because there is no appropriate name for the part of the lineage you want to distinguish. In this situation we will use quotation marks (e.g., "ostracoderms") to indicate that the group is paraphyletic.

Another important bit of terminology is **sister group.** The sister group is the monophyletic lineage most closely related to the monophyletic lineage being discussed. In Figure 1–4, for example, the lineage that includes crocodilians and phytosaurs is the sister group of the lineage that includes pterosaurs + ornithischians + saurischians + birds. Similarly, pterosaurs are the sister group of ornithischians + saurischians + birds, ornithischians are the sister group of saurischians + birds, and saurischians are the sister group of birds.

Determining Phylogenetic Relationships

We've established that the derived characters systematists used to group species into higher taxa must be inherited through common ancestry. That is, they are **homologous** similarities. In principle that notion is straightforward; but in practice, the determination of common ancestry can be complex. For example, birds and bats have wings that are modified forelimbs, but the wings were not inherited from a common ancestor with wings. The evolutionary lineages of birds (Sauropsida) and bats (Synapsida) diverged long ago, and wings evolved independently in the two groups. This process is called **convergent evolution.** (**Parallel evolution** describes the situation in which species that have diverged relatively recently develop similar specializations. The long hind legs that allow the North American kangaroo rats and the African jerboa to jump are an example of parallel evolution in these two lineages of rodents.) A third mechanism, **reversal,** can produce similar structures in distantly related organisms. Sharks and cetaceans (porpoises and whales) have very similar body forms, but sharks have retained an ancestral aquatic body form. Cetaceans, however, arise from a lineage of terrestrial mammals with well-developed limbs that returned to an aquatic environment and reverted to the aquatic body form. Convergence, parallelism, and reversal are forms of **homoplasy** (Greek *homo* = same and *plas* = form, shape). Homoplastic similarities do not indicate common ancestry. Indeed, they complicate the process of deciphering evolutionary relationships. Convergence and parallelism give an appearance of similarity (as in the wings of birds and bats) that is not the result of common evolutionary origin. Reversal, in contrast, conceals similarity (e.g., between cetaceans and their four-legged terrestrial ancestors) that is the result of common evolutionary origin.

1.4 Earth History and Vertebrate Evolution

Since their origin in the early Paleozoic, vertebrates have been evolving in a world that has changed enormously and repeatedly. These changes have affected vertebrate evolution directly and indirectly. Understanding the sequence of changes in the positions of continents, and the significance of those positions regarding climates and interchange of faunas, is central to understanding the vertebrate story. These events are summarized in a table at the front of the book, and Chapters 7, 13, and 18 give details.

The history of the Earth has occupied three geological **eons:** the, Archean, Proterozoic, and Phanerozoic. Only the Phanerozoic, which began 545 million years ago, contains vertebrate life, and it is divided into three geological eras: the Paleozoic (Greek *paleo* = ancient and *zoo* = animal), Mesozoic (Greek *meso* = middle), and Cenozoic (Greek *cen* = recent). These eras are divided into periods, which can be further subdivided in a variety of ways. Here we will be considering only the subdivisions called epochs within the Cenozoic period.

Movements of landmasses have been a feature of Earth's history at least since the Proterozoic. The course of vertebrate evolution has been molded largely by continental drift. By the early Paleozoic, roughly 545 million years ago, a recognizable scene had appeared. The seas had formed, continents floated atop the Earth's mantle, life had become complex, and an atmosphere of oxygen had formed, signifying that the photosynthetic production of food resources had become a central phenomenon of life.

The continents still drift today—North America is moving westward and Australia northward at approximately 4 centimeters per year. Because the movements are so complex, their sequence, their varied directions, and the precise timing of the changes are difficult to summarize. When the movements are viewed broadly, however, a simple pattern unfolds during vertebrate history: fragmentation—coalescence—fragmentation (Figure 1–5).

Continents existed as separate entities over 2 billion years ago. Some 300 million years ago, all of these separate continents combined to form a single landmass known as Pangaea, which was the birthplace of terrestrial vertebrates. Persisting and drifting northward as an entity, this huge continent began to break apart about 150 million years ago. Its separation occurred in two stages: first into Laurasia in the north and Gondwana in the south, and then into a series of units that have drifted and become the continents we know today.

The complex movements of the continents through time have had major effects on evolution of vertebrates. Most obvious is the relationship between the location of landmasses and their climates. At the start of the Paleozoic much of Pangaea was located on the equator; this situation persisted through the middle of the Mesozoic. Solar radiation is most intense at the equator, and climates at the equator are correspondingly warm. During the Paleozoic and much of the Mesozoic, large areas of land enjoyed tropical conditions. Terrestrial vertebrates evolved and spread in these tropical regions. By the end of the Mesozoic, much of Earth's landmass had moved out of equatorial regions; by the late Eocene, most climates in the Northern and Southern Hemispheres were temperate instead of tropical.

A less obvious effect of the position of continents on terrestrial climates comes from changes in patterns of ocean circulation. For example, the Arctic Ocean is now largely isolated from the other oceans, and it does not receive warm water via currents flowing from more equatorial regions. High latitudes are cold because they receive less solar radiation than do areas closer to the equator, and the Arctic Basin does not receive enough warm water to offset the lack of solar radiation. As a result, the Arctic Ocean is permanently frozen, and cold climates extend well southward across the continents. The cooling of climates in the Northern Hemisphere at the end of the Eocene may have been a factor leading to the extinction of archaic mammals, and it is partly the result of changes in oceanic circulation at that time.

Another factor that influences climates is the relative level of the continents and the seas. At some periods in Earth's history, most recently in the late Mesozoic and again in the first part of the Cenozoic, shallow seas have flooded large parts of the continents. These epicontinental seas extended across the middle of North America and the middle of Eurasia in the Cretaceous and early Cenozoic. Water absorbs heat as air temperature rises and releases that heat as air temperature falls. Thus, areas of land near bodies of water have maritime climates—they do not get very hot in summer or very cold in winter, and they are usually moist because water that evaporates from the sea falls as rain on the land. Continental climates, which characterize areas of land far from the sea, are usually dry with cold winters and hot summers. The draining of the epicontinental seas at the end of the Cretaceous probably contributed to the demise of the dinosaurs by making climates in the Northern Hemisphere more continental.

In addition to changing climates, continental drift has formed and broken land connections between the continents. Isolation of different lineages of vertebrates on different landmasses has produced dramatic examples of the independent evolution of similar types of organisms. These are well shown by mammals in the early Cenozoic, a period when the Earth's continents were more widely separated than they ever have been during the history of vertebrates.

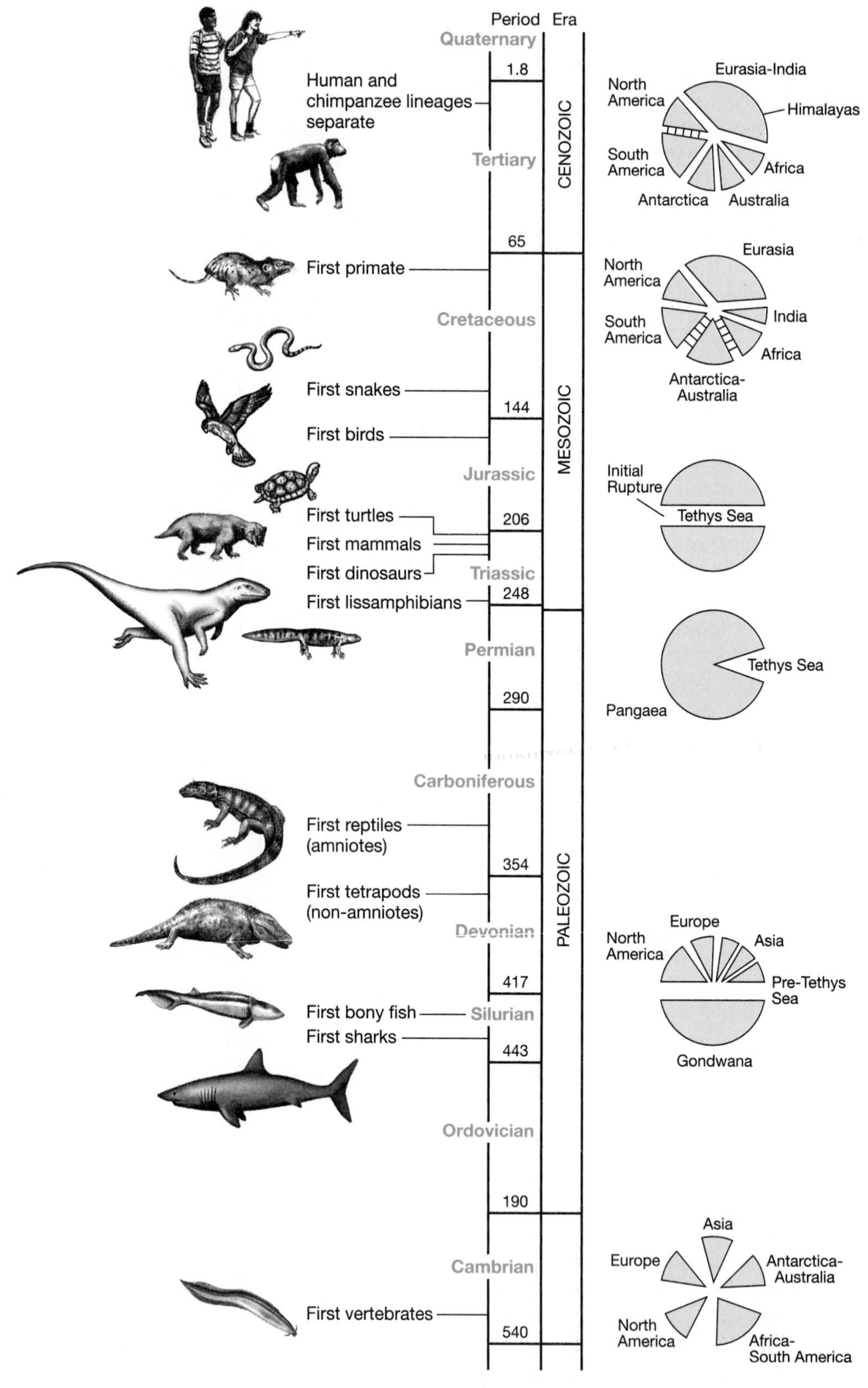

▲ Figure 1–5 A summary of continental drift from the Cambrian to the present. The first appearances in the fossil record of major groups in evolution are shown.

Saber-toothed carnivores evolved among placental mammals in the Northern Hemisphere as well as among marsupial mammals in South America. Specialized herbivorous animals were represented by entirely different lineages of placental mammals in North America and South America, and by kangaroos and wombats (both marsupials) in Australia.

Much of evolutionary history appears to depend on whether a particular lineage was in the right place at the right time. This random element of evolution is assuming increasing prominence as more detailed information about the times of extinction of old groups and radiation of new groups suggests that competitive replacement of one group by another is not the usual mechanism of large-scale evolutionary change. The movements of continents and their effects on climates and the isolation or dispersal of animals are taking an increasingly central role in our understanding of vertebrate evolution.

Summary

The 50,000 species of living vertebrates span a size range from less than a gram to more than 100,000 kilograms. They live in habitats from the bottom of the sea to the tops of mountains. This extraordinary diversity is the product of more than 500 million years of evolution.

Phylogenetic systematics, usually called cladistics, classifies animals on the basis of shared derived character states. Natural evolutionary groups can be defined only by these derived characters; retention of ancestral characters does not provide information about evolutionary lineages. Application of this principle produces groupings of animals that reflect evolutionary history as accurately as we can discern it, and forms a basis for making hypotheses about evolution.

The Earth has changed dramatically during the half-billion years of vertebrate history. Continents were fragmented when vertebrates first appeared; coalesced into one enormous continent, Pangaea, about 300 million years ago; and began to fragment again about 100 million years ago. This pattern of fragmentation—coalescence—fragmentation has resulted in isolation and renewed contact of major groups of vertebrates on a worldwide scale. On a continental scale, the advance and retreat of glaciers in the Pleistocene caused homogeneous habitats to split and merge repeatedly, isolating populations of widespread species and leading to the evolution of new species.

Additional Readings

de Queiroz, K., and J. Gauthier. 1992. Phylogenetic taxonomy. *Annual Review of Ecology and Systematics* 23:449–480.

Eschmeyer, W. N. (editor). 1998. *Catalog of Fishes* (3 vol.). San Francisco: California Academy of Sciences.

Gould, S. J. 1993. *The Book of Life: An Illustrated History of the Evolution of Life on Earth*. New York: W.W. Norton.

Hennig, W. 1966. *Phylogenetic Systematics*. Urbana, Ill: University of Illinois Press.

Web Explorations

On-line resources for this chapter are on the World Wide Web at **http://www.prenhall.com/pough** (click on the Table of Contents link and then select Chapter 1).

CHAPTER

2 Vertebrate Relationships and Basic Structure

In this chapter we explain the structures that are characteristic of vertebrates, discuss the relationship of vertebrates to other members of the animal kingdom, and describe the systems that make vertebrate functional animals. We need an understanding of the fundamentals of vertebrate design to appreciate the changes that have occurred during the evolution of vertebrates, and to trace homologies between primitive vertebrates and derived ones.

2.1 Vertebrates in Relation to Other Animals

Vertebrates are a diverse and fascinating group of animals. Because we are vertebrates ourselves that statement may seem chauvinistic, but vertebrates are remarkable in comparison with most other animal groups. Vertebrates are in the subphylum **Vertebrata** of the phylum **Chordata.** About 30 other phyla have been named, but only the phylum Arthropoda (insects, crustaceans, spiders, etc.) rivals the vertebrates in diversity of forms and habitat. And it is only in the phylum Mollusca (snails, clams, squid) that we find animals (such as octopus and squid) that approach the very large size of some vertebrates and have a capacity for complex learning.

The tunicates (subphylum Urochordata) and cephalochordates (subphylum Cephalochordata) are placed with vertebrates in the phylum Chordata. The shared derived features of chordates, which are seen in all members of the phylum at some point in their lives, include a **notochord** (a dorsal stiffening rod that gives the phylum Chordata its name), a dorsal hollow nerve cord, a muscular postanal tail, and an endostyle. The endostyle is a ciliated, glandular groove on the floor of the **pharynx** (the throat region) that secretes mucus for trapping food particles during filter feeding. The endostyle is present in adult tunicates and cephalochordates and in larval lampreys. It is homologous with the thyroid gland of vertebrates, an endocrine gland in the throat involved with regulating metabolism. Within the chordates, tunicates appear more primitive than cephalochordates and vertebrates, which are linked by several derived features.

The relationship of chordates to other kinds of animals is revealed by anatomical, biochemical, and embryological characters as well as by the fossil record. Figure 2–1 shows the relationships of animal phyla. Vertebrates superficially resemble other active animals such as insects in having a distinct head end, jointed legs, and bilateral symmetry (i.e., one side is the mirror image of the other). Perhaps surprisingly, the phylum Chordata is closely related to the phylum Echinodermata (starfishes, sea urchins, and the like), which are marine forms without distinct heads and with pentaradial (fivefold and circular) symmetry as adults.

The exact relationships among chordates, echinoderms, and hemichordates (a small, rather obscure

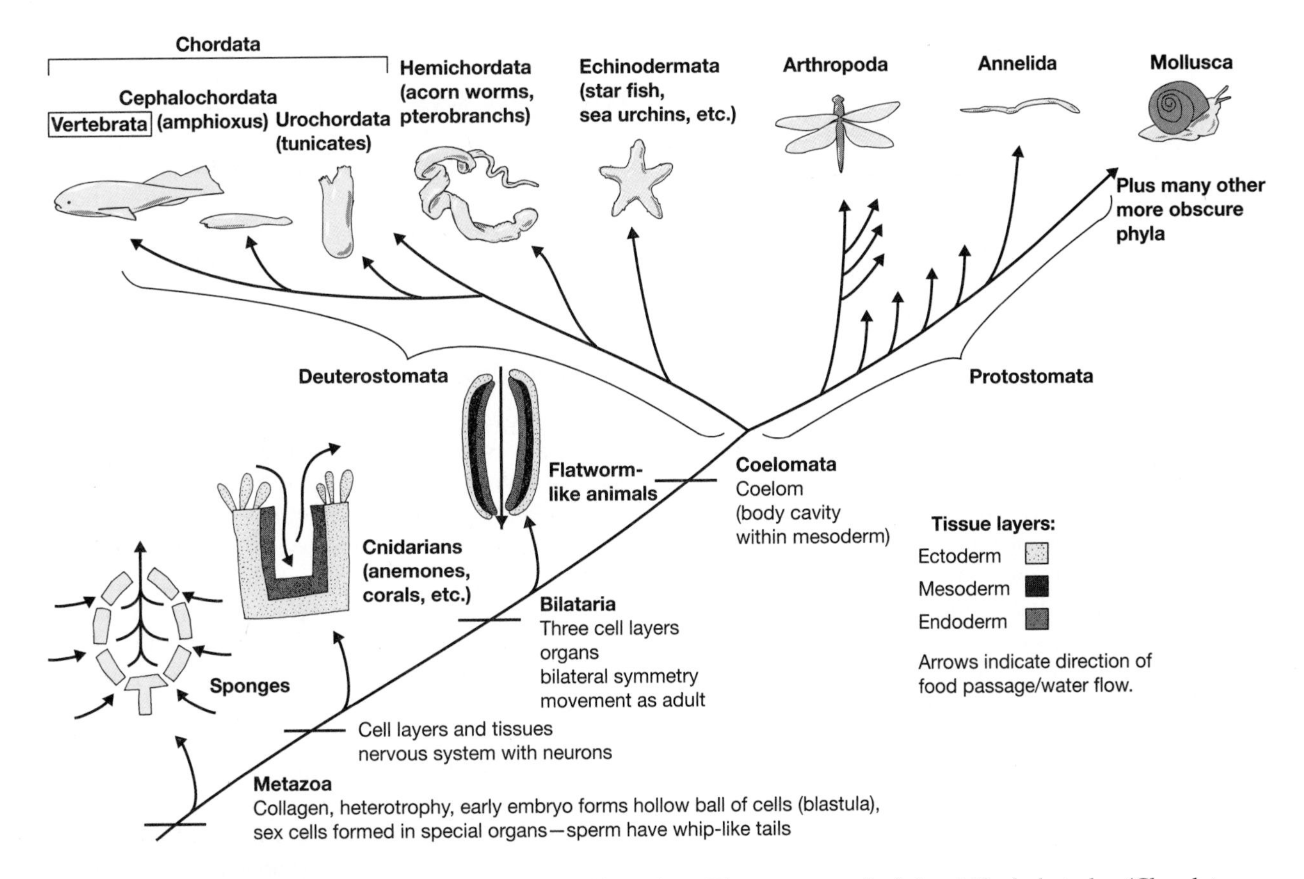

▲ Figure 2–1 A simplified phylogeny of the animal kingdom. There are a total of about 30 phyla today (Chordata, Echinodermata, Annelida, etc. all represent phyla). Approximately 15 additional phyla are known from the early Paleozoic, and became extinct at the end of the Cambrian period.

phylum of marine animals containing the earthworm-like acorn worms and the fern-like pterobranchs) are not clearly understood. The three groups are linked as deuterostomes by several unique embryonic features, such as the way in which their eggs divide after fertilization (egg cleavage), their larval form, and some other features discussed later. Hemichordates are often considered the sister group of chordates because both groups have **pharyngeal slits,** which are openings in the pharynx (throat region) that were originally used for filtering food particles out of the water. Modern echinoderms lack pharyngeal slits, but some extinct echinoderms may have had them. Furthermore, primitive echinoderms may have had bilateral symmetry, meaning that the fivefold symmetry of modern echinoderms may be a derived character of that lineage (see Gee 1996). Thus, the sister group of vertebrates may lie among extinct, free-swimming echinoderms.

To consider how deuterostomes are related to other animals, we will take a different approach, starting at the bottom of the tree and working upward. All animals (metazoans) are multicellular and share common embryonic and reproductive features: The embryo initially forms a hollow ball of cells (the blastula), they have sex cells formed in special organs, and they have motile sperm with whip-like tails.

Above the level of sponges, all animals have a nervous system and their bodies are made of distinct layers of cells, or germ layers, that are laid down early in development at a stage called gastrulation. Gastrulation occurs when the hollow ball of cells forming the blastula folds in upon itself, producing two distinct layers of cells and an inner gut with an opening to the outside at one end. The outer layer of cells is the **ectoderm** (Greek *ecto* = outside and *derm* = skin), and the inner layer forms the **endoderm** (Greek *endo* = within).

Jellyfishes and related animals have these two layers of body tissue, making them diploblastic (Greek *diplo* = two and *blast* = a bud or sprout). Jellyfishes are probably also the first animals to have the ***Hox*** **genes** that determine anterior/posterior identity within the developing embryo. Animals above the level of jellyfishes and their kin add an additional, middle cell layer of **mesoderm** (Greek *mesos* = middle), making them triploblastic (Greek *triplo* = three). Triploblasts also have a gut that opens at both ends (i.e., with a mouth and an anus), and are bilaterally symmetrical with a distinct head end at some point in their life. The mesoderm forms the body's muscles, and only animals with a mesoderm are able to be motile as adults.

The **coelom,** an inner body cavity that forms as a split within the mesoderm, is another derived character of most, but not all, triploblastic animals. Coelomate animals (i.e., animals with a coelom) are split into two groups on the basis of how the mouth and anus form. When the blastula folds in on itself to form a gastrula, it leaves an opening to the outside called the blastopore (Latin *porus* = a small opening). In jellyfish the blastopore is the only opening into the interior of the body, and it serves as mouth and anus. But during the embryonic development of coelomates, a second opening develops. In the lineage called protostomes (Greek *proto* = first and *stoma* = mouth) the blastopore (which was the first opening in the embryo) becomes the mouth, whereas in deuterostomes (Greek *deutero* = second) the second opening becomes the mouth and the blastopore becomes the anus. The way that the coelom forms during development also differs between protostomes and deuterostomes. Molluscs (snails, clams, squid), arthropods (insects, crabs, spiders), and annelids (earthworms) and some other phyla are protostomes; chordates, hemichordates, and echinoderms are deuterostomes.

Nonvertebrate Chordates

The two groups of extant nonvertebrate chordates are small, marine animals. There might have been more nonvertebrate chordates in the past, but these soft-bodied animals are rarely preserved as fossils. There are about 60 specimens of the animal known as *Pikaia,* which may be a cephalochordate, from the Middle Cambrian Burgess Shale in British Columbia (Figure 2–2). This animal has a small bilobed head with a pair of short, slender tentacles, and bilaterally paired rows of about 12 tufts behind the head may be evidence of gill slits. The most obvious chordate features of *Pikaia* are **myomeres** (blocks of striated muscle fibers arranged along both sides of the body and separated by sheets of connective tissue) and a notochord running along the posterior two-thirds of the body. Another chordate-like animal, *Cathaymyrus,* has been reported from Early Cambrian sediments in China that are about 10 million years earlier

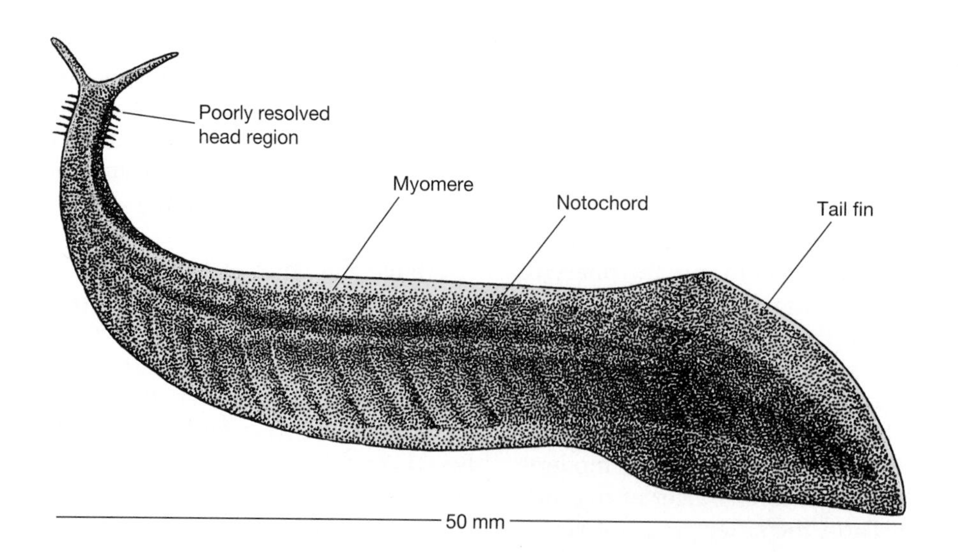

▲ **Figure 2–2** *Pikaia,* an early chordate from the Middle Cambrian Burgess Shale of British Columbia.

than the Burgess Shale (Shu et al. 1996). The only other fossilized evidence of cephalochordates is *Palaeobranchiostoma* from the Early Permian of South Africa (Blieck 1992). This animal is related to the living form known as amphioxus, and the large dorsal and ventral fins of *Palaeobranchiostoma* suggest that it was a swimmer. If that is so, the burrowing habit of modern amphioxus is probably a derived feature.

Urochordates

Present-day **tunicates** (subphylum Urochordata) are marine animals that filter particles of food from the water with a basketlike perforated pharynx. There are about 2000 extant species, and all but 100 or so are sedentary as adults, attaching themselves to the substrate either singly or in colonies.

Most adult tunicates (also known as sea squirts) bear little similarity to cephalochordates and vertebrates, except for having an endostyle and a pharynx with slits that is used for filter feeding (Figure 2–3b). However, their tadpole-like, free-swimming larvae (Figure 2–3a) look more like forms that belong within the phylum Chordata. Tunicate larvae have a notochord, a dorsal hollow nerve cord, and a muscular postanal tail that moves in a fish-like swimming pattern. Most species have a brief free-swimming larval period (a few minutes to a few days) after which the larvae metamorphose into sedentary adults attached to the substrate.

A popular and long-held theory was that cephalochordates and vertebrates evolved from an ancestor that resembled a tunicate larva. In the early twentieth century Walter Garstang proposed a complicated evolutionary scenario by which larval forms could become sexually mature, giving rise to mobile adults. However, more recent workers have suggested that the tunicates with the sessile adult stage are the derived forms, and that the few species of extant tunicates that remain in a permanent larval form are the ones that resemble the ancestral chordate.

Cephalochordates

The subphylum Cephalochordata contains some 22 species, all of which are small, superficially fish-like, marine animals usually less than 5 centimeters long. The best-known cephalochordate is the lancelet (*Branchiostoma lanceolatum*), more commonly known as **amphioxus.** (Greek *amphi* = both and *oxy* = sharp; *amphioxus* means "sharp at both ends," an appropriate term for an animal lacking a distinct head, so that both head and tail end appear pointed.) Lancelets are widely distributed in marine waters of the continental shelves and are usually burrowing, sedentary animals as adults. In a few species the adults retain the active, free-swimming behavior of the larvae.

A notable characteristic of amphioxus is its fish-like locomotion. This results from contraction of the segmental myomeres, which contract in sequence and bend the body in a way that causes forward propulsion. The notochord acts as an incompressible elastic rod, extending the full length of the body and preventing the body from shortening when the myomeres contract. The notochord of amphioxus extends from the tip of the snout to the end of the tail, projecting well beyond the region of the myomeres at both ends. The anterior elongation of the notochord apparently is a specialization that aids in burrowing.

Figure 2–3c shows some details of the internal structure of amphioxus. Amphioxus and vertebrates differ in the use of the pharyngeal slits. Amphioxus has no gill tissue associated with these slits; it is small enough that oxygen uptake and carbon dioxide loss occur by diffusion over the body surface. Instead, the gill slits are used for filter feeding. Water is moved over the gill slits by cilia on the gill bars between the slits, aided by the features of the buccal cirri and the wheel organ, while the velum is a flap helping to control the one-way flow of water.

An external body cavity called the atrium distinguishes amphioxus from vertebrates. (This atrium is not the same as the atrium of the vertebrate heart.) The atrium is formed by outgrowths of the body wall (metapleural folds), which enclose the body rather like a cloak. The atrium opens to the outside world via the atriopore. The atrium appears to control passage of substances through the pharynx, in combination with the beating of the cilia on the gill bars and the wheel organ in the head. An atrium is also present in tunicates, and it may be a primitive feature of all chordates that is lost in vertebrates.

Cephalochordates have several derived characters that are absent from tunicates. In addition to the myomeres, amphioxus has a circulatory system similar to that of vertebrates, with a dorsal aorta and a ventral pumping heart (or heart-like structure) that forces blood through the gills. Additionally, although amphioxus lacks a distinct kidney, it shares with vertebrates specialized excretory cells called podocytes, and it has a vertebrate-like tail fin.

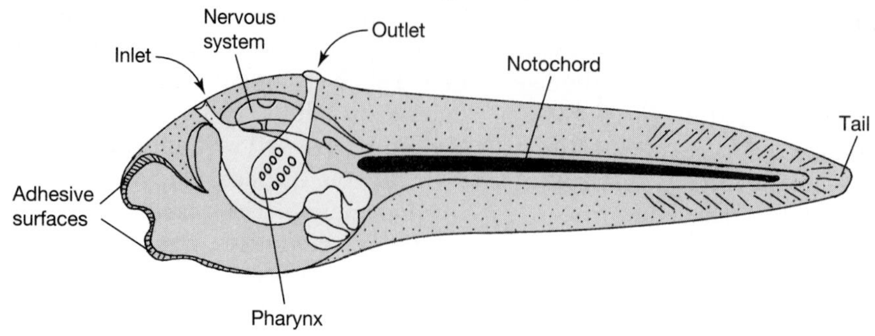

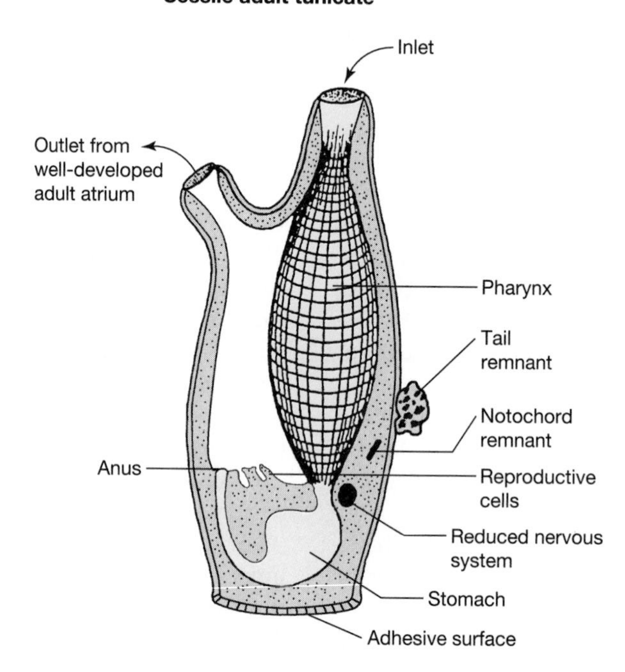

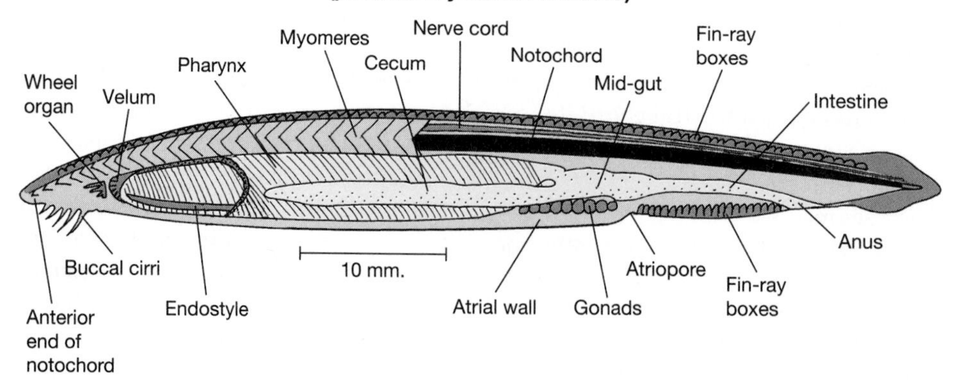

▲ Figure 2–3 Nonvertebrate chordates.

Amphioxus and vertebrates also share some embryonic features, including lateral plate mesoderm (body tissue derived from the ventral, unsegmented portion of the mesoderm) and embryonic induction of various neural (brain and nerve) tissues by the notochord during development (see Gans 1989). These characters indicate that cephalocordates are the sister group of vertebrates.

2.2 Definition of a Vertebrate

The term *vertebrate* is obviously derived from the vertebrae that are serially arranged to make up the spinal column, or backbone. In ourselves, as in other land vertebrates, the vertebrae form around the notochord during development and also encircle the nerve cord. The bony vertebral column replaces the original notochord after the embryonic period. In many fishes the vertebrae are made of cartilage rather than bone.

Not all animals included within the traditional subphylum Vertebrata have vertebrae, however. Among living **agnathans** (jawless vertebrates), hagfishes lack vertebral elements entirely and lampreys have only cartilaginous rudiments flanking the nerve cord. Fully formed vertebrae, with a **centrum** (plural = *centra*) surrounding the notochord, are found only in **gnathostomes** (jawed vertebrates) among the array of living forms. Many jawed fishes retain a functional notochord as adults, with centra composed of several separate elements rather than a single block of bone. We will probably never know whether extinct agnathans had vertebrae, because if they were present they would probably have been cartilaginous; but some extinct forms may have had vertebrae of some sort.

All vertebrates do have the uniquely derived feature of a **cranium,** or skull, which is a bony, cartilaginous, or fibrous structure surrounding the brain. They also have a prominent head, containing complex sense organs. Classical embryology holds that the front part of the vertebrate head is a new feature, consisting of three segments anterior to the original tip of the chordate head. However, recent genetic studies identified a segmental series of genes (*Hox* genes) in both amphioxus and vertebrates, suggesting that the head may have been enlarged in vertebrates, but that it is not an entirely new addition.

Some people prefer the older term, **Craniata,** to Vertebrata. The distinction is based on the observation that all animals in the subphylum currently called Vertebrata have a cranium, but hagfishes alone among living vertebrates lack any trace of vertebrae. The question is, did hagfishes once have vertebrae that have now been lost, or did they never have vertebrae? If hagfish ever had vertebrae, it is correct to include them with the vertebrates. But if they never had vertebrae, the name Vertebrata should be reserved for those vertebrates that actually have vertebrae of some sort (i.e., all living forms with the exception of hagfishes). Until that question is resolved, we will continue to use the familiar term *vertebrate* in this book, including hagfishes with vertebrates.

Two embryonic features may account for many of the differences between vertebrates and other chordates. The first is the duplication of the *Hox* gene complex (= *homeobox* genes) that characterizes most animals, vertebrate and invertebrate (Marx 1992, Garcia-Fernandez and Holland 1994, Monastersky 1996). More complex animals usually have a greater amount of genetic material, and it is thought that a doubling of this gene sequence enabled the evolution of a more complex type of animal. The second feature is the development of a type of embryonic tissue called **neural crest** that forms many new structures in vertebrates, especially in the head region (Northcutt and Gans 1983). It was originally thought that neural crest was derived from the ectoderm germ layer. But recently it has been proposed to be a truly independent germ layer, on a par with ectoderm, endoderm, and mesoderm, meaning that vertebrates are the only animals that have four germ layers, making them quadroblastic (Hall 2000).

Embryonic tissue that may be related to neural crest forms the epidermal placodes (= thickenings), which give rise to the complex sensory organs of vertebrates, including the nose, eyes, inner ear, and taste buds. Some placode cells migrate caudally to contribute, along with the neural crest cells, to the lateral line system, and to the cranial nerves that innervate it.

The brain of vertebrates is larger than the brains of primitive chordates and has three parts—the forebrain, midbrain, and hindbrain. The brain of amphioxus is not obviously divided; but genetic studies show that it may be homologous to the vertebrate brain, with the exception of the front part of the forebrain (the **telencephalon**). The telencephalon is the portion of the brain that contains the cerebral cortex, the area of higher processing in vertebrates.

2.3 Basic Vertebrate Structure

At the whole-animal level an increase in body size and increased activity distinguish vertebrates from more primitive chordates. Early vertebrates generally had body lengths of 10 cm or more, which is about an order of magnitude larger than nonvertebrate chordates. Because of their relatively large size, vertebrates need specialized systems to carry out processes that are accomplished by diffusion or ciliary action in nonvertebrate chordates. Vertebrates are also more active animals than chordates, so they are also in need of more specialized organ systems that can carry out physiological processes at a greater rate. Table 2.1 and Figure 2–4 summarize some basic structural and physiological differences between vertebrates and nonvertebrate chordates.

Table 2.1 Comparison of Features in Nonvertebrate Chordates and in Primitive Vertebrates

Generalized Nonvertebrate Chordate	*Primitive Vertebrate*
(Based on features of the living cephalochordate amphioxus)	(Based on features of the living jawless vertebrates—hagfishes and lampreys)
A. Brain and Head End	
Notochord extends to tip of head (may be derived condition).	Head extends beyond tip of notochord.
No cranium (skull).	Cranium—skeletal supports around brain, consisting of capsules surrounding the main parts of the brain and their sensory components.
Simple brain (= cerebral vesicle) or specialized sense organs (except photoreceptive frontal organ, probably homologous with the vertebrate eye).	Tripartite brain and multicellular sense organs (eye, nose, inner ear).
Poor distance sensation (although the skin is sensitive).	Improved distance sensation: in addition to the eyes and nose, also have a lateral line system along the head and body that can detect water movements (poorly developed lateral line system on the head only in hagfishes).
No electroreception.	Electroreception may be a primitive vertebrate feature (but absent in hagfishes).
B. Pharynx and Respiration	
Gill arches for filter feeding (respiration is by diffusion over the body surface).	Gill arches (= pharyngeal arches) support gills used for respiration.
Numerous gill slits and arches (up to 100 on each side).	Fewer gill slits (6–10 on each side), individual gills with highly complex internal structure (gill filaments).
Pharynx not muscularized (except in wall of atrium, or external body cavity).	Pharynx with specialized (branchiomeric) musculature.
Water moved through pharynx and over gill arches by ciliary action.	Water moved through pharynx and over gills by active muscular pumping.
Gill arches made of collagen.	Gill arches made of cartilage (allows for elastic recoil—aids in pumping).

Table 2.1 *Continued*

Generalized Nonvertebrate Chordate	*Primitive Vertebrate*
C. Feeding and Digestion	
Gut not muscularized: Food passage by means of ciliary action.	Gut muscularized: Food passage by means of muscular peristalsis.
Digestion of food is intracellular: Individual food particles taken into cells lining gut.	Digestion of food is extracellular: Enzymes poured onto food in gut lumen, then breakdown products absorbed by cells lining gut.
No discrete liver and pancreas: Structure called the midgut cecum or diverticulum is probably homologous to both.	Discrete liver and pancreatic tissue.
D. Heart and Circulation	
Ventral pumping structure (no true heart, just contracting regions of vessels, = sinus venosus of vertebrates). Also accessory pumping regions elsewhere in the system.	Ventral pumping heart only (but accessory pumping regions retained in hagfishes). Three-chambered heart: (in order of blood flow) sinus venosus, atrium, and ventricle.
No neural control of the heart to regulate pumping.	Neural control of the heart (except in hagfishes).
Circulatory system open: Large blood sinuses, capillary system not extensive.	Circulatory system closed: Without blood sinuses (some remain in hagfishes and lampreys) and with an extensive capillary system.
Blood not specifically involved in the transport of respiratory gases (O_2 and CO_2 mainly transported via diffusion). No red blood cells or respiratory pigment.	Blood specifically involved in the transport of respiratory gases. Have red blood cells containing the respiratory pigment hemoglobin (binds with O_2 and CO_2 and aids in their transport).
E. Excretion and Osmoregulation	
No specialized kidney. Coelom filtered by solenocytes (flame cells) that work by creating negative pressure within cell. Cells empty into the atrium (false body cavity) and then to the outside world via the atriopore.	Specialized glomerular kidney, segmental structure along dorsal body wall, works by ultrafiltration of blood. Empties to the outside via the archinephric ducts leading to the cloaca.
Body fluids same concentration and ionic composition as seawater. No need for volume control or ionic regulation.	Body fluids more dilute than seawater (except for hagfishes). Kidney important in volume regulation, especially in freshwater environment. Monovalent ions regulated by the gills (also the site of nitrogen excretion), divalent ions regulated by the kidney.
F. Support and Locomotion	
Notochord provides main support for body muscles.	Notochord provides main support for body muscles, vertebral elements around nerve cord at least in all vertebrates except hagfishes.
Myomeres with simple V-shape.	Myomeres with more complex W-shape.
No lateral fins, or median fins besides tail fin.	Primitively, no lateral fins. Caudal (tail) fin has dermal fin rays. Dorsal fins present in all except hagfishes.

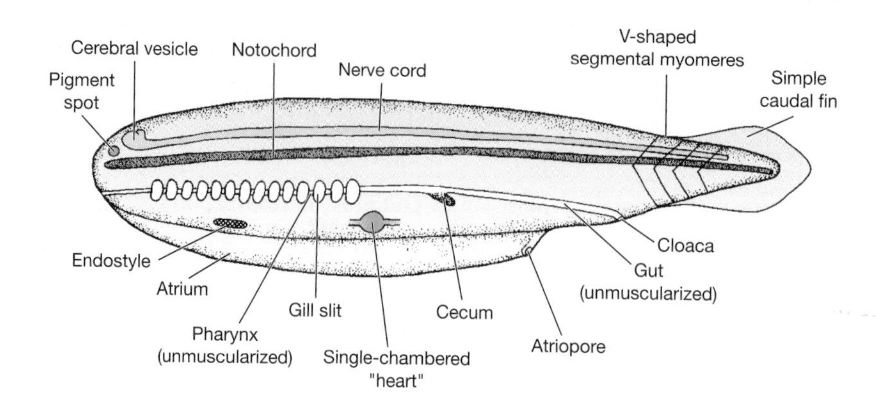

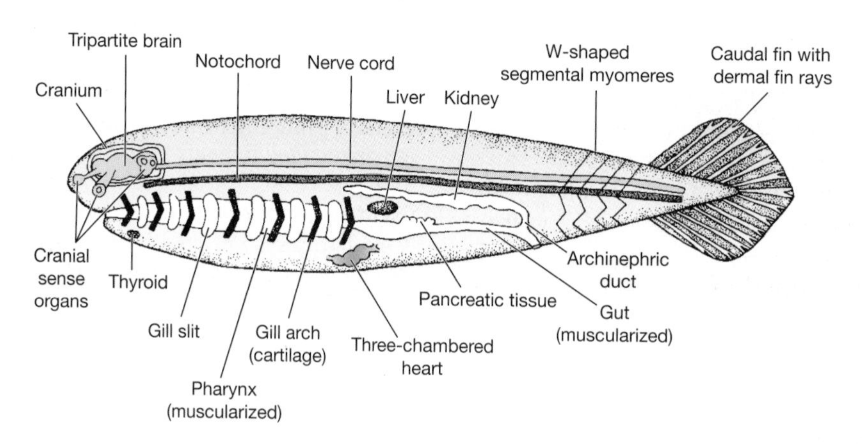

▲ Figure 2–4 Design of a generalized nonvertebrate chordate compared with that of a primitive vertebrate.

Vertebrates are characterized by mobility, and the ability to move requires muscles and an endoskeleton. Mobility brings vertebrates into contact with a wide range of environments and objects in those environments, and a vertebrate's external protective covering must be tough but flexible. Bone and other mineralized tissues that we consider characteristic of vertebrates had their origins in this protective integument.

Embryology

Studying embryos can show how systems develop and how the form of the adult is related to functional and historical constraints during development. Modern scientists no longer adhere rigidly to the biogenetic law that "ontogeny recapitulates phylogeny" proposed by nineteenth-century embryologists such as von Baer and Haeckel (i.e., the idea that the embryo faithfully passes through its ancestral evolutionary stages in the course of its development). Nevertheless, embryology can provide clues about the ancestral condition and about homologies between structures in different animals (Northcutt 1990).

The development of vertebrates from a single fertilized cell (the zygote) to the adult condition will be summarized only briefly. This is important background information for many studies, but a detailed treatment beyond the scope of this book. Summaries of vertebrate embryology can be found in textbooks dealing specifically with vertebrate anatomy, such as Romer and Parsons (1986), Walker and Liem (1994), Hildebrand (1995), and Kardong (1998). Hall (1992) provides an excellent review of recent work integrating genetic evolutionary studies with traditional morphological ones.

We saw earlier that all animals, with the exception of sponges, are formed of distinct tissue layers, or germ layers. The fates of germ layers have been very conservative throughout vertebrate evolution.

The outermost germ layer, the ectoderm, forms the adult superficial layers of skin (the epidermis); the linings of the most anterior and most posterior parts of the digestive tract; and the nervous system, including most of the sense organs, such as the eye and the ear. The innermost layer, the endoderm, forms the rest of the digestive tract's lining, as well as the lining of glands associated with the gut—including the liver and the pancreas—and most respiratory surfaces of vertebrate gills and lungs.

The middle layer, the mesoderm, is usually the last of the three layers to appear in development. It forms everything else: muscles, skeleton (including the notochord), connective tissues, and circulatory and urogenital systems. A little later in development there is a split within the originally solid mesoderm layer, forming a coelom or body cavity. The coelom is the cavity containing the internal organs, and it is divided into the **pleuroperitoneal cavity** (around the viscera) and the **pericardial cavity** (around the heart). These cavities are lined by thin sheets of lateral-plate mesoderm—the **peritoneum** (= **pericardium** around the heart). The gut is suspended in the peritoneal cavity by sheets of peritoneum called **mesenteries.**

As previously mentioned, vertebrates also have a unique type of developmental tissue, called neural crest. Neural crest forms many of the structures in the anterior head region, including bones and muscles, which were previously thought to be formed by mesoderm. It also forms almost all of the peripheral nervous system (i.e., that part of the nervous system outside of the brain and the spinal cord), and additionally contributes to portions of the brain. Some structures in the body that are new features of vertebrates are also formed from neural crest. These include the adrenal glands, pigment cells in the skin, secretory cells of the gut, and smooth muscle tissue lining the aorta.

At the pharyngula stage, early in embryonic development, the ancestral chordate feature of pharyngeal pouches in the head region makes at least a fleeting appearance in the embryos of all vertebrates (Figure 2–5). In fish the grooves between the pouches (the pharyngeal clefts) perforate to become

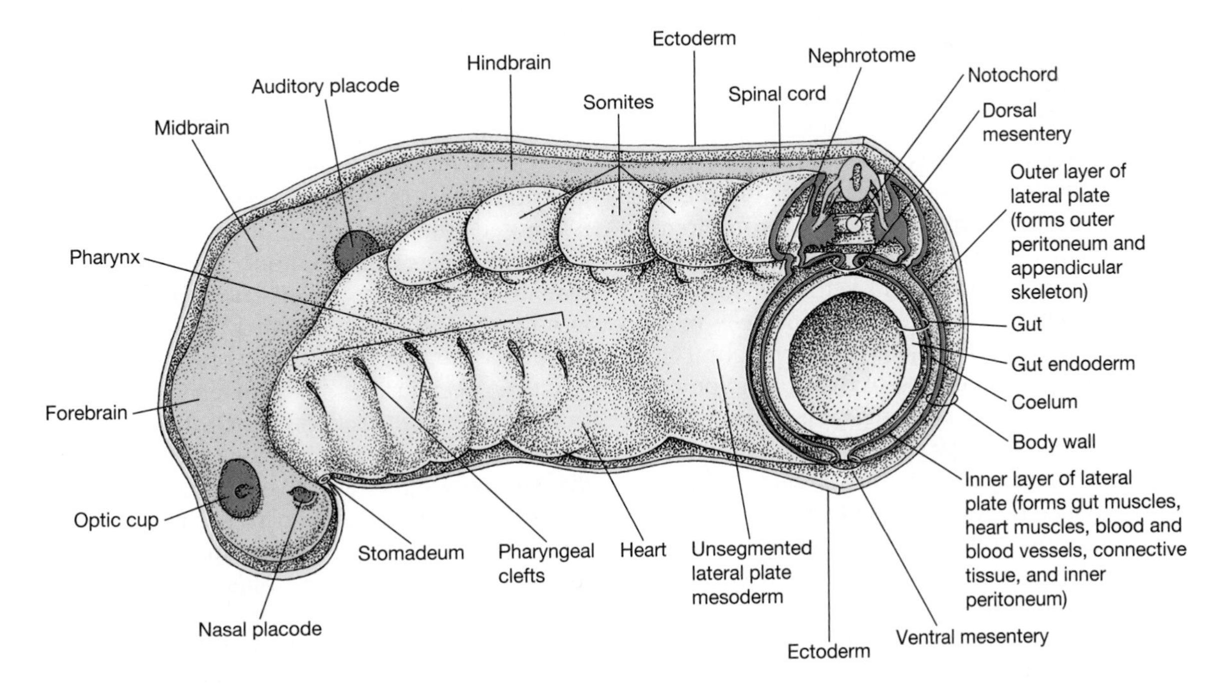

▲ Figure 2–5 Three-dimensional view of a portion of a generalized vertebrate embryo. The ectoderm is stripped off, showing segmentation of the mesoderm in the trunk region and pharyngeal development. The stomadeum is the developing mouth.

the gill slits, whereas in land vertebrates these clefts disappear in the adult. The linings of the pharyngeal pouches give rise to half a dozen or more glandular structures often associated with the lymphatic system, including the thymus gland, parathyroid glands, carotid bodies, and tonsils.

The dorsal hollow nerve cord typical of vertebrates and other chordates is formed by the infolding, and subsequent pinching off and isolation of, a long ridge of ectoderm running dorsal to the developing notochord. The notochord itself appears to contain the developmental instructions for this critical embryonic event, which is probably why the notochord is retained in the embryos of vertebrates (such as ourselves) that no longer have the structure in the adult. The cells that will form the neural crest arise next to the developing nerve cord (the neural tube) at this stage. Slightly later in development, these neural crest cells disperse laterally and ventrally, ultimately settling and differentiating throughout the embryo.

Embryonic mesoderm becomes divided into three distinct portions, as shown in Figure 2–5, with the result that adult vertebrates are a strange mixture of segmented and unsegmented components. The dorsal (upper) part of the mesoderm, lying above the gut and next to the nerve cord, forms a series of thick-walled segmental buds, or **somites** (= the epimere), running from the head end to the tail end. The ventral (lower) part of the mesoderm, surrounding the gut and containing the coelom, is thin-walled and unsegmented, and is called the **lateral plate** (= the hypomere). Small segmental buds linking the somites and the lateral plate are called nephrotomes (= the mesomere or the intermediate mesoderm).

The segmental somites will eventually form the dermis of the skin, the trunk and limb muscles of the body, and portions of the skeleton (the vertebral column and portions of the back of the skull). Some of these segmental muscles later migrate ventrally from their originally dorsal (**epaxial**) position to form the layer of striated muscles on the underside of the body (the **hypaxial** muscles), and from there they form the muscles of the limbs in **tetrapods** (four-footed land vertebrates). The lateral plate forms all the internal, nonsegmented portions of the body, such as the connective tissue, the blood vascular system, the mesenteries (tissue connecting the gut and other structures to the body wall), the peritoneal and pericardial linings of the coelomic cavities, and the reproductive system. It also forms the smooth muscle of the gut and the cardiac muscle. The nephrotomes form the kidney, an elongated segmental structure in the primitive vertebrate condition, and the kidney drainage duct (the **archinephric duct**).

The developing nervous system also follows this segmental (somite) versus unsegmented (lateral plate) division of the body. Vertebrates are unique in possessing a dual type of nervous system, the **somatic nervous system** (the so-called voluntary nervous system) and the **visceral nervous system** (the so-called involuntary nervous system). The somatic nervous system innervates muscles that we can move consciously (e.g., the muscles of the limbs) and relays information from sensation that we are consciously aware of (e.g., temperature and pain receptors in the dermis of the skin). The visceral nervous system innervates muscles that we normally cannot move consciously (e.g., the smooth muscles of the gut). This involves the **sympathetic** and **parasympathetic** nerves, and this portion of the visceral nervous system is known as the **autonomic nervous system.** The visceral nervous system also relays information from sensations that we are not consciously aware of (e.g., the receptors monitoring the levels of carbon dioxide in the blood).

Some exceptions exist to this segmented versus nonsegmented division of the vertebrate body. The locomotory muscles, both **axial** (i.e., within the trunk region) and **appendicular** (i.e., within the limbs), and the axial skeleton are derived from the somites. But curiously, the appendicular skeleton (with the exception of the scapula bone in the shoulder girdle) is derived from the lateral plate, as are the tendons and ligaments of the appendicular muscles. The explanation for this apparent anomaly may lie in the fact that limbs are add-ons to the basic vertebrate design.

Other peculiarities are found in the expanded front end of the head of vertebrates, which has a complex pattern of development and does not follow the simple segmentation of the body (Northcutt 1990). The head mesoderm is not divided into somites and lateral plate, but instead contains undifferentiated paraxial mesoderm, which gives rise to the striated eye muscles and **branchiomeric** muscles powering the visceral arches (gills and jaws). Within the brain, the forebrain and midbrain do not appear to be segmented; but the hindbrain is divided into segmental structures called rhombomeres.

Adult Tissue Types

There are five kinds of tissue in vertebrates: epithelial, connective, blood (= **vascular**), muscular, and nervous. These tissues are combined to form larger units called organs. Organs often contain most or all of the five basic tissues. The functions of vertebrate life are supported by groups of organs united into one of the 10 vertebrate organ systems.

A fundamental component of most animal tissues is the fibrous protein **collagen.** Collagen also forms the organic matrix of bone, and the tough tissue of tendons and ligaments as well as the softer tissues of organs. Collagen is stiff and does not stretch easily. In some tissues collagen is combined with the protein **elastin,** which can stretch and recoil. Another important fibrous protein, seen only in vertebrates, is **keratin.** Keratin is mainly found in tetrapods, but it also forms the horny tooth-like structures of the living agnathans. Keratin is found in epidermal structures (hair, scales, feathers, claws, horns, beaks, etc.), whereas collagen is primarily in the mesoderm. Keratin, like collagen, can become calcified—as in the baleen (whalebone) that forms the filtering apparatus in filter-feeding whales.

The Integument The external covering of vertebrates, the integument, is a single organ, making up 15 to 20 percent of the body weight of many vertebrates and much more in armored forms. It includes the skin and its derivatives, such as glands, scales, dermal armor, and hair. The skin protects the body and receives information from the outside world. The major divisions of the vertebrate skin are the **epidermis** (the superficial cell layer derived from embryonic ectoderm) and the unique vertebrate **dermis** (the deeper cell layer of mesodermal and neural crest origin). The dermis extends deeper into a subcutaneous tissue (hypodermis) that is derived from mesoderm and overlies the muscles and bones.

The epidermis forms the boundary between the vertebrate and its environment, and is of paramount importance in protection, exchange, and sensation. It often contains secretory glands, and may play a significant role in osmotic and volume regulation. The dermis, the main structural layer of the skin, includes many collagen fibers that provide elasticity and help to maintain its strength and shape. The dermis contains blood vessels, and blood flow within these vessels is under neural and hormonal control.

Hormonal control of the blood flow can be illustrated by human emotional responses, such as blushing, in which the vessels are dilated and blood rushes to the skin. Smooth muscle fibers may also occur in the dermis, such as the ones in mammals that produce skin wrinkling around the nipples. In tetrapods the dermis houses most of the sensory structures and nerves associated with sensations of temperature, pressure, and pain. The dermis also houses **melanocytes,** pigment cells containing melanin, derived from the neural crest.

The hypodermis, or subcutaneous tissue layer, is not functionally a part of the skin; it lies between the dermis and the fascia overlying the muscles. This region contains collagenous and elastic fibers, and is the area where subcutaneous fat is stored by birds and mammals. The subcutaneous striated muscles of mammals, such as those that enable them to make facial expressions and to flick the skin to get rid of a fly, are found in this area.

Mineralized Tissues Vertebrates have a unique type of mineral called hydroxyapatite, a complex compound of calcium and phosphorous. Hydroxyapatite is more resistant to acid than is calcite (calcium carbonate), which forms the shells of mollusks. This resistance to acid may be important when vertebrates engage in muscular activity strenuous enough to release lactic acid into the blood.

Four major types of tissues can become mineralized in vertebrates. Three of them, **enamel, dentine,** and bone, are found only in the mineralized condition in the adult. The fourth, cartilage, is usually unmineralized in tetrapods, but is the main mineralized internal skeletal tissue in sharks. (Sharks and other cartilaginous fishes appear to have lost true bone.) Enamel and dentine are the most mineralized of the tissues—(about 99 percent and 90 percent mineral, respectively)—and are found in teeth of extant vertebrates and in the dermal skeleton of some primitive fishes. This high degree of mineralization explains why teeth are more likely to be found as fossils than are bones, which are only about 50 percent mineralized.

The different types of hard tissues are formed from different cell lineages in development. Thus, although bone may replace cartilage in development, bone is not simply cartilage to which minerals have been added. Rather, it is composed of different types of cells, osteocytes (called osteoblasts while they are actually making the bone), as opposed to the chondrocytes that form cartilage. The cells that form bone and cartilage cells are derived from the mesoderm, those that form dentine are derived from neural crest tissue, and those that form enamel are derived from the ectoderm.

A fifth type of vertebrate hard tissue, enameloid, is found in fishes (with the exception of the lobe-finned fish ancestral to tetrapods). This tissue appears similar to enamel, being hard and glassy, and is found in the outer layers of dermal bones like enamel. But enameloid is actually more closely related to dentine than to enamel in terms of the cell lineage that it is derived from. The final type of hard tissue is **cementum,** a bone-like substance that fastens the teeth in their sockets, and that may grow to become part of the tooth structure itself.

Vertebrate mineralized tissues are composed of a complex matrix of collagenous fibers, cells that secrete a protinaceous tissue matrix, and crystals of hydroxyapatite. The hydroxyapatite crystals are aligned on the matrix of collagenous fibers in layers with alternating directions, much like the structure of plywood (Figure 2–6a). This combination of cells, fibers, and minerals gives bone its complex latticework appearance that combines strength with relative lightness and helps to prevent cracks from spreading.

Bone remains highly vascularized even when it is mineralized (ossified). Bone in **amniotes** (reptiles, birds, and mammals) is arranged in concentric layers around blood vessels within the bone, forming cylindrical units called **Haversian systems** (Figure 2–6a). The calcium and phosphorus in bone are in dynamic turnover with calcium and phosphate ions in the blood.

The vascular nature of bone also enables bone to remodel itself. Old bone is eaten away by specialized blood cells (osteoclasts), which are derived from the same cell lines as the macrophage white blood cells that engulf foreign bacteria in the body. New osteoblasts enter behind the osteoclasts and deposit new bone. In this way, a broken bone can mend itself and bones can change their shape to suit the mechanical stress imposed on the animal. This is why exercise builds up bone and why astronauts lose bone in the zero gravity of space. Calcified cartilage, as seen in the skeleton of sharks, is unable to remodel itself because it does not contain blood vessels.

A bone is not uniform in structure. If it were, animals would be very heavy. The external layers of a bone are formed of dense, compact or lamellar bone, but the internal layers are spongy or cancellous bone. The joints at the ends of bones are covered by a smooth layer of articular cartilage that reduces friction as the joint moves. (Arthritis occurs when this cartilage is damaged or worn.) The bone within the joint is comprised of cancellous (= spongy) bone rather than dense bone. The entire joint is enclosed in a joint capsule, containing synovial fluid for lubrication (Figure 2–6b).

There are two main types of bone in vertebrates: **dermal bone,** which as its name suggests is formed in the skin, and **endochondral bone,** which is formed inside cartilage. Dermal bone is the primitive type of vertebrate bone, first seen in the fossil jawless vertebrates called the **ostracoderms** (Figure 2–6c). Dermal bone originally was formed around the outside of the body like a suit of armor (*ostracoderm* means "shell-skinned"), basically forming a type of exoskeleton. We think of vertebrates as possessing only an internal skeleton (endoskeleton), but most of our skull bones are dermal bones and they lie externally, forming a shell around our brains. The endoskeletal structure of vertebrates, initially consisting only of the cranium and later adding the vertebrae and the limb bones, was originally formed from cartilage. Thus, the condition in many early vertebrates was a bony exoskeleton and a cartilaginous endoskeleton (Figure 2–7).

Only in the bony fishes and their descendants is the endoskeleton comprised primarily of bone. In these vertebrates the endoskeleton is initially laid down in cartilage that is replaced by bone later in development. This process allows the long bones to grow, because after bone has become ossified it can no longer grow from within. A zone of cartilage in a long bone remains as an area where bones can expand in length. In mammals this zone is found between the shaft of the bone (the **diaphysis**) and the joints at the end (the **epiphyses**). Clear lines can be seen in the long bones of juvenile mammals between the bone shaft and the ends, whereas these lines have vanished in adults due to epiphyseal fusion.

The basic structure of the teeth of gnathostomes is like that of the **odontodes,** the original tooth-like components of primitive vertebrate dermal armor. Teeth are composed of an inner layer of dentine and an outer layer of enamel or enameloid, around a central pulp cavity (Figure 2–6d). Shark scales (dermal denticles) have a similar structure.

Teeth form on dermal bones or in the skin, places where there is a juxtaposition of epidermal tissues (which form the enamel) and dermal neural crest tissues (which form the dentine). When the tooth is

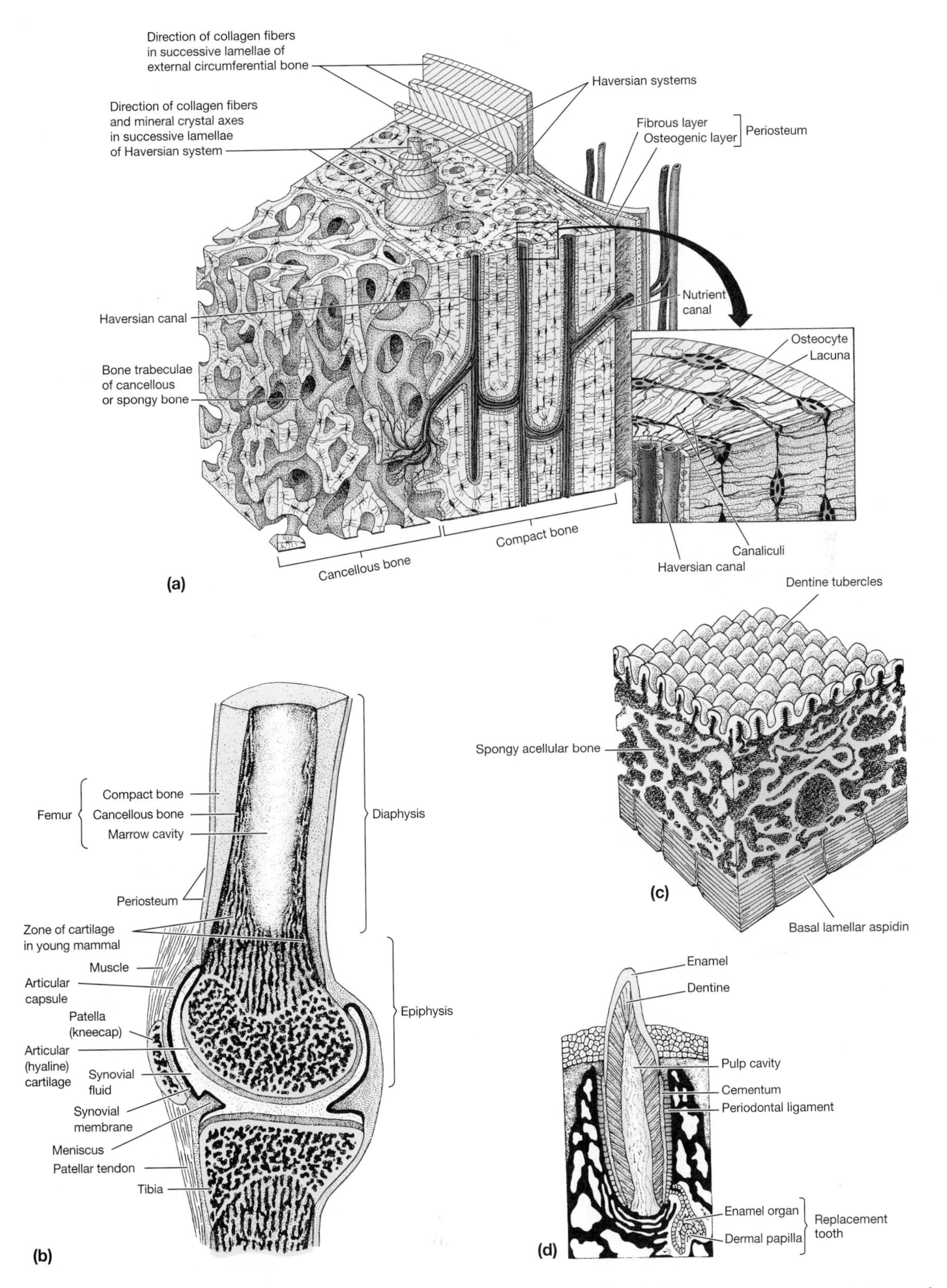

▲ Figure 2–6 Organization of vertebrate mineralized tissues. (a) Bone from a section of the shaft of a long bone of a mammal, showing Haversian canal systems. (b) Section through a human knee joint, showing the gross internal structure of a long bone and the structure of a joint capsule. Note that the patella (kneecap) is an example of a sesamoid bone (a bone formed within a tendon). (c) Three-dimensional block diagram of ostracoderm (heterostracan) dermal bone. (d) Section through developing tooth.

The dermal skeleton or exoskeleton

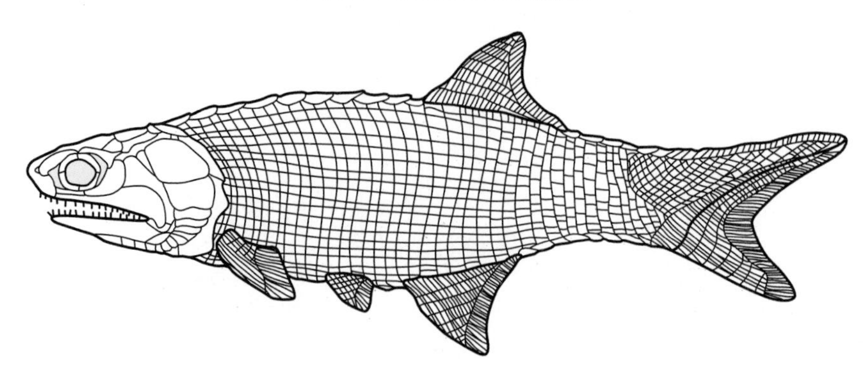

The endodermal skeleton or endoskeleton

▲ Figure 2–7 Vertebrate skeletons.

fully formed it erupts through the gum line. Replacement teeth may start to develop to one side of the main tooth even before its eruption. A tooth can continue to get longer after it erupts if its roots have not yet been formed, but no more enamel can be added to top of the tooth.

Only mammals and some reptiles (archosaurs) have truly rooted teeth (**thecodont** teeth) that are set in sockets and held in place by periodontal ligaments. Other types of teeth can be called acrodont (fused to the jaw bone) or pleurodont (set in a shelf on the inner side of the jawbone; Figure 2–8).

The Skeletomuscular System

The basic endoskeletal structural feature of chordates is the notochord, acting as a dorsal stiffening rod running along the length of the body. Vertebrates added the cranium surrounding the brain, the axial skeleton, and the visceral (pharyngeal) skeleton of the gill arches and their derivatives.

The Notochord The notochord is made up of a core of large, closely packed cells distended with incompressible fluid-filled vacuoles that make the notochord rigid. The notochord is wrapped in a complex fibrous sheath that is the site of attachment for segmental muscle and connective tissues. In all vertebrates, the notochord ends anteriorly just posterior to the pituitary gland and continues posteriorly to the tip of the fleshy portion of the tail.

The more familiar vertebrate feature of a vertebral column made of cartilage or bone that originally surrounds and later replaces the notochord is not found in the most primitive vertebrates. Only gnathostomes have true vertebrae, ribs, and appendages with internal skeletons. The original form of the notochord is lost in adult tetrapods, but portions remain as intervertebral discs between the vertebrae.

The cranium, visceral skeleton, notochord, vertebrae, and ribs, as well as the median fin supports of fishes, are together called the axial skeleton. The paired fin or limb skeletons and girdles make up the appendicular skeleton.

The Cranial Skeleton and Musculature The skull or cranium is formed by three basic compo-

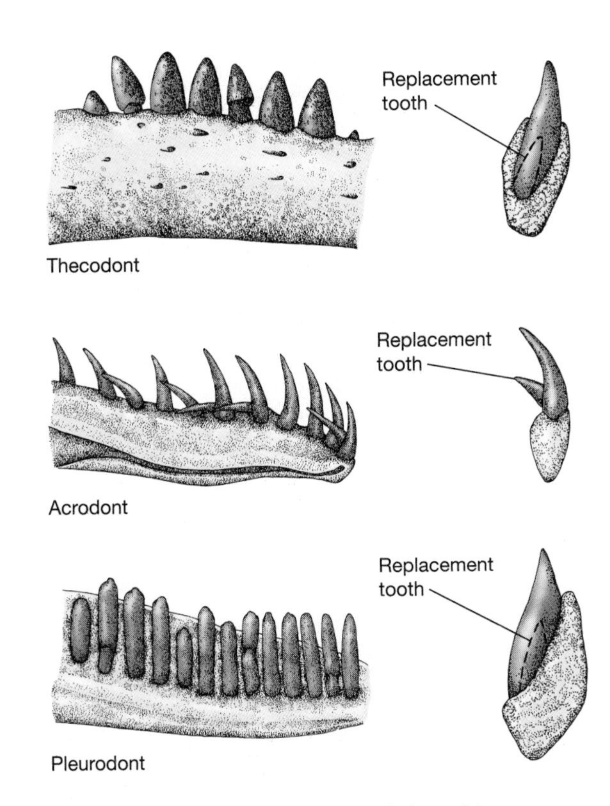

▲ Figure 2–8 Types of rooted teeth found in vertebrates.

nents: the chondrocranium, the splanchnocranium, and the dermatocranium (Figure 2–9).

The **chondrocranium,** also known as the neurocranium, surrounds the brain. It is initially formed as cartilage, derived mainly from neural crest tissue, and it may be partly or entirely replaced by endochondral ossification in more derived vertebrates such as bony fishes and tetrapods. The chondrocranium is made up of underlying supporting bars—the parachordals at the back and the anterior trabeculae (in gnathostomes) in the front—and overlying paired sensory capsules that form around the sense organs. These consist of the olfactory (nasal) capsules around the nose in the region of the forebrain, the optic capsules around the eyes, and the **otic capsules** around the inner ear regions of the hindbrain.

The **splanchnocranium,** also known as the visceral skeleton or the pharyngeal skeleton, includes the **pharyngeal arches** (gill supports between the pharyngeal gill slits or pouches, also known as visceral arches or gill arches). Visceral arches form initially in cartilage and may be replaced by bone in bony fishes and tetrapods. Each visceral arch is made up of several elements, hinged so that the attached muscles can change their shape and thus pump water through the mouth and over the gills. In gnathostomes the anterior two pairs of arches become the jaws and their supporting structures, including the support of the tongue in tetrapods. The posterior arches no longer form gill supports in air-breathing vertebrates, but their remnants form part of the larynx and other structures in the throat.

The **dermatocranium** is derived from dermal bones formed in the integument, covering the other portions of the skull (Figure 2–9c). In bony vertebrates the dermatocranium surrounds, unites with, or replaces portions of the other components of the skull so closely as to have usurped many of the functions of chondrocranial and visceral elements. The dermatocranium includes the roof of the skull, the superficial area around the orbits, the gill covers (opercula) when they are present, and the roof of the mouth (the palate and portions of the floor of the skull). It also contributes to the jaws, which are formed primarily (in mammals entirely) by dermal bone.

Although it is seen only in bony fishes and tetrapods among living vertebrates, a dermatocranium was initially seen in the ostracoderms and in the extinct agnathans, as well as in some extinct gnathostome groups. However, in living bony vertebrates the pattern and distribution of dermatocranium bones are distinctive for this group, and different from any extinct vertebrate. Because most fossil vertebrates had a dermatocranium, the chondrichthyian fishes, which do not have this structure, likely were descended from an ancestral form that did have it. That is, the lack of a dermatocranium in modern sharks and other cartilaginous fishes probably represents the secondary loss of this structure.

There are two main types of striated muscles in the head of vertebrates, both derived from the paraxial mesoderm, but with different patterns of innervation. The first type is the extrinsic eye muscles that rotate the eyeball. There are six of these to each eye in all vertebrates except hagfishes, where their absence may represent secondary loss. These muscles are innervated by somatic motor nerves from the ventral cranial nerves, following a similar pattern to the innervation of the striated muscles of the body.

The second type of striated muscle is the branchiomeric muscles, the muscles associated with the splanchnocranium and involved with feeding and respiration, which are innervated by fibers from the dorsal cranial nerves (Figure 2–10a). In gnathostomes,

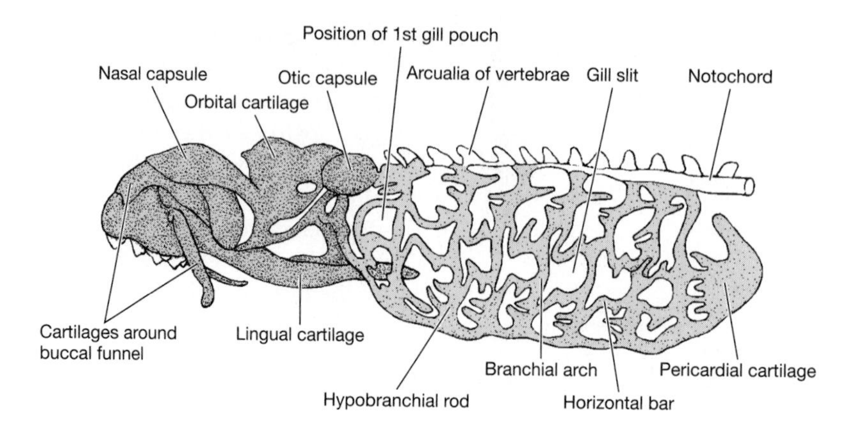

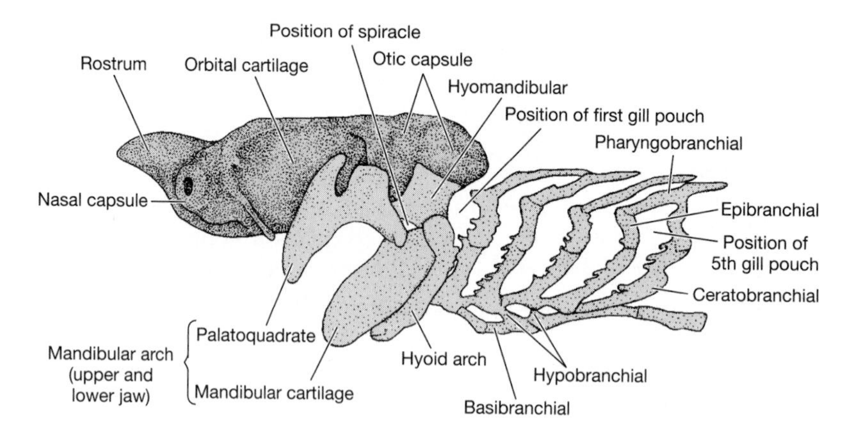

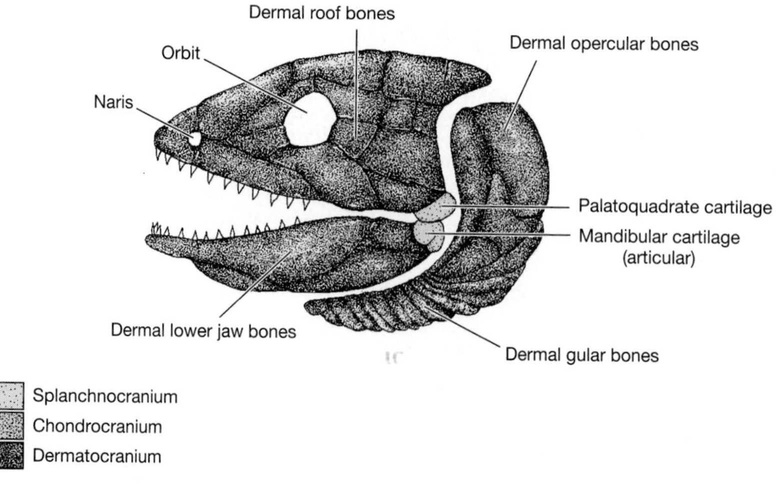

▲ Figure 2–9 Basic form of the vertebrate skull.

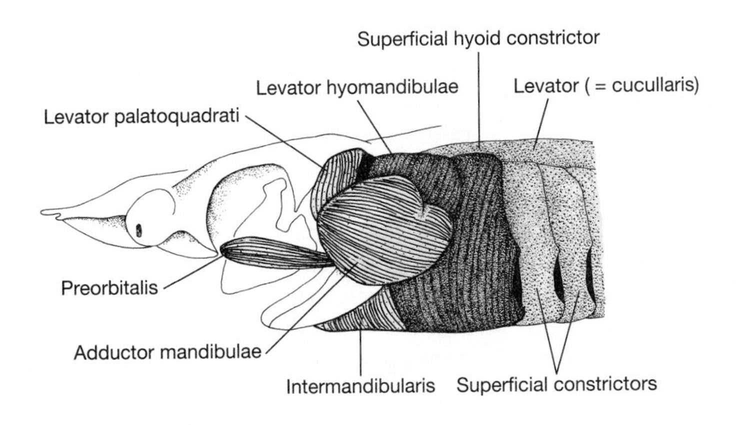

▲ Figure 2–10 Form of head and neck musculature in a shark.

the branchiomeric muscles are associated with the two anterior gill arches (the **mandibular** and **hyoid** arches, respectively) that form the jaws and jaw supports. These muscles are primarily involved in feeding (the major muscle being the **adductor mandibularis** that closes the jaws), but they also aid in ventilation by sucking water into the mouth. The muscles associated with the posterior arches that support the gills form a characteristic series of levators, adductors, and constrictors in association with each arch. Their action aids in forcing water through the gills. Living agnathans have a similar system of branchiomeric muscles powering the gill arches, although their organization is not precisely like that of gnathostomes.

The mandibular and hyoid arches and their branchiomeric musculature may have had a role in feeding even before the evolution of jaws (Mallatt 1996). A new feature in gnathostomes is the addition of hypobranchial muscles that run from the mandibular and hyoid arches to the pectoral girdle and aid in opening the mouth (Figure 2–10b). These muscles are derived from trunk somites that migrate forward in development and are innervated by spinal nerves, in contrast to the cranial nerve innovation of the branchiomeric muscles.

The Axial Skeleton and Musculature Vertebrae are composed of a centrum, which originally surrounds and later obliterates the notochord, and a neural arch that surrounds the spinal nerve cord (Figure 2–11). The caudal vertebrae may also have a hemal arch, enclosing the caudal artery and vein. Vertebrae have a variable number of bilaterally symmetrical projections (apophyses) that are used for the attachment of muscles and ligaments, or to link the vertebrae to other skeletal elements. Ribs are functionally and anatomically related to the vertebral apophyses. They provide sites for muscle attachment and strengthen the body wall.

Vertebrae are segmentally arranged structures. Each vertebra lies at the junction between two segments, with the rib lying in the myoseptum—the fibrous sheet that divides the segmental muscle blocks. With this arrangement, the segmental muscles (myomeres) lying within each segment can run from one vertebra at the front of the segment to another vertebra at the rear of the segment. When the muscles contract, the vertebrae are moved relative to each other and the spinal column bends.

The vertebrae and ribs of fishes (Figure 2–11a) usually have a purely locomotory function, although occasionally some elements may be uniquely modified for specialized functions, such as hearing in catfish and their relatives (Chapter 6). The tetrapod function of the axial skeleton for support on land is the derived vertebrate condition. The vertebrae of tetrapods interlock by means of processes called **zygapophyses** (Figure 2–11b) that allow the vertebral column to act as an analogue of a suspension bridge to support the weight of the viscera on land. (Tetrapods that have permanently returned to the water, such as whales and many of the extinct Mesozoic marine reptiles, have secondarily lost the zygapophyses.)

The myomeres are complexly folded in three dimensions, so that each one extends anteriorly and posteriorly over several body segments (Figure 2–12). Sequential muscle blocks overlap and produce undulation of the body. In amphioxus, myomeres have a simple V shape; in vertebrates they have a more complex W shape as in the lamprey. The myomeres of gnathostomes (e.g., the shark and teleost in Figure 2–12) are more complex than those of agnathans and show the distinctive division of the axial blocks into epaxial and hypaxial portions, divided by a sheet of fibrous tissue called the horizontal septum. This horizontal septum also marks the placement of the lateral line along the side of jawed fishes.

This segmental patterning of the axial muscles is clearly visible in fishes. It is easily seen in a piece of raw or cooked fish, where the flesh flakes apart in zigzag blocks, each block representing a myomere. This pattern is similar to the fabric pattern of inter-

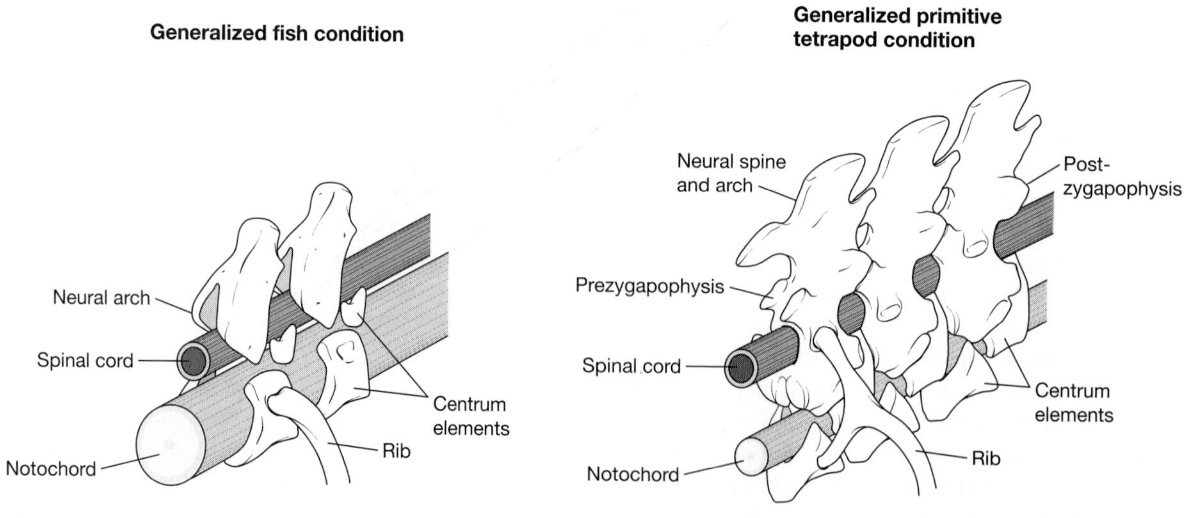

▲ Figure 2–11 Gnathostome vertebrae and ribs. Note that in more derived tetrapods the central elements form a single solid structure, obliterating the notochord in the adult form.

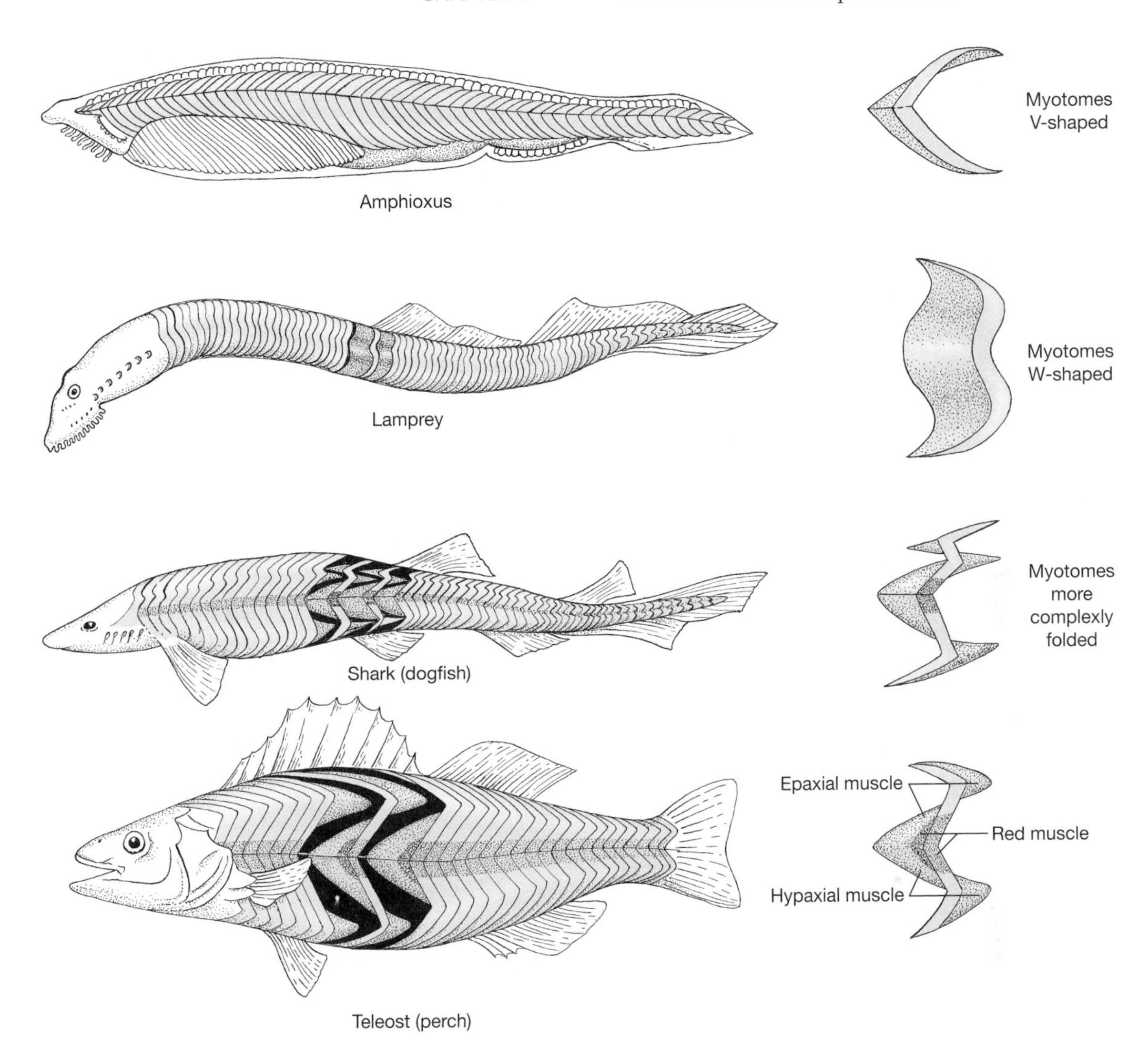

▲ Figure 2–12 Vertebrate myomeres.

locking V shapes known as herringbone (perhaps it would be better termed herring-muscle). In tetrapods the pattern is less obvious; but the segmental pattern can be observed on the washboard stomach of body builders, where each ridge of the washboard represents a segment of the rectus abdominus muscle.

The Appendicular Skeleton and Musculature The appendicular skeleton includes the limbs and limb girdles. In the primitive gnathostome condition, illustrated by sharks, the pectoral girdle (supporting the fore fins) is a simple cartilaginous bar, called the **coracoid** bar. In bony fishes the **pectoral** girdle is attached to the opercular bones that form the posterior portion of the dermal skull roof. The **pelvic** girdle is represented by the puboishiatic plate, which has no connection with the vertebral column but merely anchors the hind fins in the body wall.

Within the fin itself, fanlike **basal** elements support one or more ranks of cylindrical **radials,** which usually articulate with ray-like structures that support most of the surface of the fin web (Figure 2–13a). The primitive gnathostome condition is represented by a shark, where the basals and radials are robust but are restricted to the **proximal** (close to the body) part of the fin. Bony fishes diverge from this pattern in two different ways: Primitive ray-finned fishes such as *Polypterus* have an internal fin structure that is rather shark-like, but more derived ones reduce the basals and radials so that the fin is almost

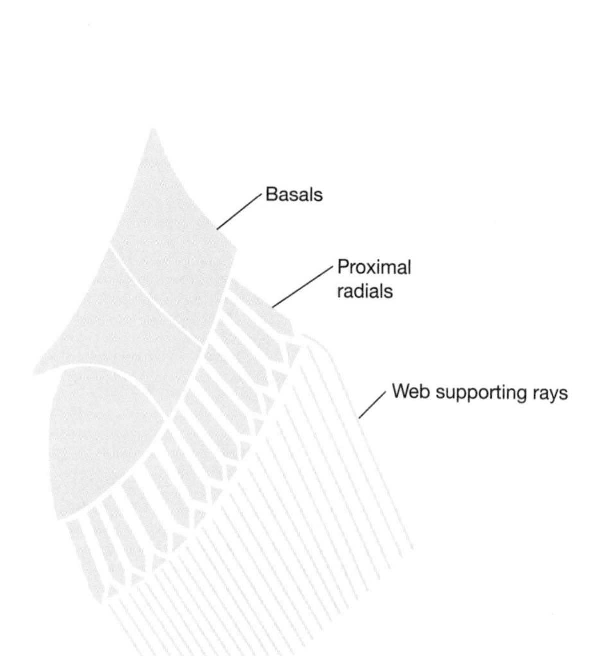

Tetrapods (primitive mammalian condition)

Appendicular segments		Pectoral	Pelvic
Girdle		Scapula (shoulder blade)	Pelvis (hip bone)
Propodium		Humerus (upper arm)	Femur (thigh)
Epipodium		Ulna/radius (forearm)	Tibia/fibula (lower leg)
Mesopodium		Carpus (wrist)	Tarsus (ankle)
Metapodium		Metacarpus (hand)	Metatarsus (foot)
Phalanges		Phalanges (fingers)	Phalanges (toes)

▲ Figure 2–13 Basic divisions of the endoskeletal supports of vertebrate paired appendages, with common and anatomical terminology compared.

entirely composed by the fin rays. Lobe-finned fishes (sarcopterygians) extend the basals and radials **distally** into the body of the fin itself, and the fin rays may be reduced. The basic tetrapod limb is made up of the limb girdle and five segments, articulating end to end (Figure 2–13b, and see Chapter 8).

Locomotion Many small aquatic animals, especially larval forms, move by using cilia (small hair-like structures) on their surfaces to beat against the water. However, ciliary propulsion works only at very small body sizes, and adult chordates are in general too big to use ciliary locomotion. The basis of their locomotion is the chordate feature of the postanal tail (possibly first evolved for a startle response in larval forms). This mechanism works by the serial contraction of segmental muscle bands in conjunction with the notochord, which acts as a dorsal stiffening rod for the muscles to work against. Without the notochord the contraction of these muscles would result only in shortening of the body, not in forward propulsion.

The basic vertebrate mode of locomotion is the use of the tail, and to a lesser extent the trunk in a series of axial undulations for swimming in the water. This type of locomotion is still seen in most fishes today. The paired fins of fishes are generally used for steering, braking, and providing lift, but not for propulsion—except in some specialized fishes such as skates and rays, which have wing-like pectoral fins.

Energy Acquisition and Support of Metabolism

Food energy gleaned from the environment must be processed by the digestive system to release energy and nutrients that must then be carried to the tissues. Oxygen is required for the process of energy release, and the functions of gas exchange surfaces and the circulatory system are closely intertwined with those of the digestive system.

Feeding and Digestion Feeding includes getting food into the oral chamber (i.e., the mouth), oral or pharyngeal processing (i.e., chewing in the broad sense—although today only mammals truly chew, or masticate, their food), and swallowing. Digestion includes the breakdown of complex compounds into small molecules that are absorbed across the wall of the gut. Both feeding and digestion are two-part processes; each has a physical component and a chemical component, although physical components dominate in feeding, and chemical components dominate in digestion.

Protovertebrates probably filtered small particles of food from the water, as amphioxus and larval lampreys still do. Most vertebrates are particulate feeders; they take in their food as bite-sized pieces rather than as tiny particles. Vertebrates have a larger volume of gut than amphioxus, and vertebrates digest their food by secreting digestive enzymes onto the digesta in the gut lumen. (Amphioxus digests its food within the gut cells themselves.)

Vertebrates have a muscularized gut, which moves food along the digestive tract by means of peristaltic contractions (rhythmical contractions of the smooth muscles of the gut). A stomach is present only in gnathostomes among extant vertebrates; it stores food, secretes acid that aids in killing bacteria, and is the site of protein digestion. The liver and pancreas are digestive glands that secrete their products into the gut just posterior to the stomach. The liver produces digestive secretions—bile, produced in the gall bladder, and fat-emulsifying enzymes—which process absorbed nutrients and metabolites, and detoxify harmful substances. The pancreas, present only as diffuse tissue in living agnathans, produces enzymes that break down carbohydrates and fats. The pancreas also secretes the hormone insulin, which is involved in the regulation of glucose metabolism and blood sugar levels.

In the primitive vertebrate condition there is no division of the intestine into small and large portions, as we see in ourselves, nor is there a distinct rectum. The intestine opens to the **cloaca,** a common opening in most vertebrates for the urinary, reproductive, and digestive systems.

Respiration and Ventilation Ancestral chordates probably relied on oxygen absorption and carbon dioxide loss by diffusion across a thin skin (cutaneous respiration). This is the mode of respiration of amphioxus, an animal that is still small and sluggish enough that it does not require specialized respiratory organs. Although amphioxus has pharnygeal slits, these are employed in filter feeding rather than in respiration.

Cutaneous loss of carbon dioxide is still a major component of respiration for many vertebrates, and modern amphibians rely on the skin for much of their oxygen uptake as well. For other vertebrates the combination of large body size and high levels of activity make specialized gas exchange structures essential. Gills are effective in water, whereas lungs work better in air (see Chapter 4). Both gills and lungs have large surface areas that allow oxygen to diffuse from the surrounding medium (water or air) into the blood.

Cardiovascular System Blood carries oxygen and nutrients through the vessels to the cells of the body, removes carbon dioxide and other metabolic waste products, and stabilizes the internal environment. Blood also carries hormones from their sites of release to their target tissues.

Blood is a fluid tissue composed of liquid **plasma** and cellular constituents known as red blood cells (**erythrocytes**) that contain the iron-rich protein **hemoglobin,** and several different types of white blood cells (**leucocytes**), which are involved in the body's immune defense system. Cells specialized to promote clotting of blood (thrombocytes) are present in all vertebrates except mammals, where they are replaced by noncellular **platelets.** The interstitial fluid (= tissue fluid) is essentially blood plasma minus the large proteins, which do not pass through the linings of the blood vessels.

The blood of vertebrates is contained within specialized vessels and organs in a closed circulatory system; that is, one where the arteries and veins are linked up via capilliaries. Arteries carry blood away from the heart and veins return blood to the heart (Figure 2–14a). Blood pressure is higher in the arterial system than in the venous system, and arteries have thicker walls than veins, with a layer of smooth muscle and an outer layer of fibrous connective tissue.

Interposed between the smallest arteries (arterioles) and the smallest veins (venules) are the capillaries, which are the sites of exchange between blood and tissues. Capillaries pass close to every living cell and provide an enormous surface area for the exchange of gases, nutrients, and waste products. Their walls are only one cell layer thick. Capillaries form

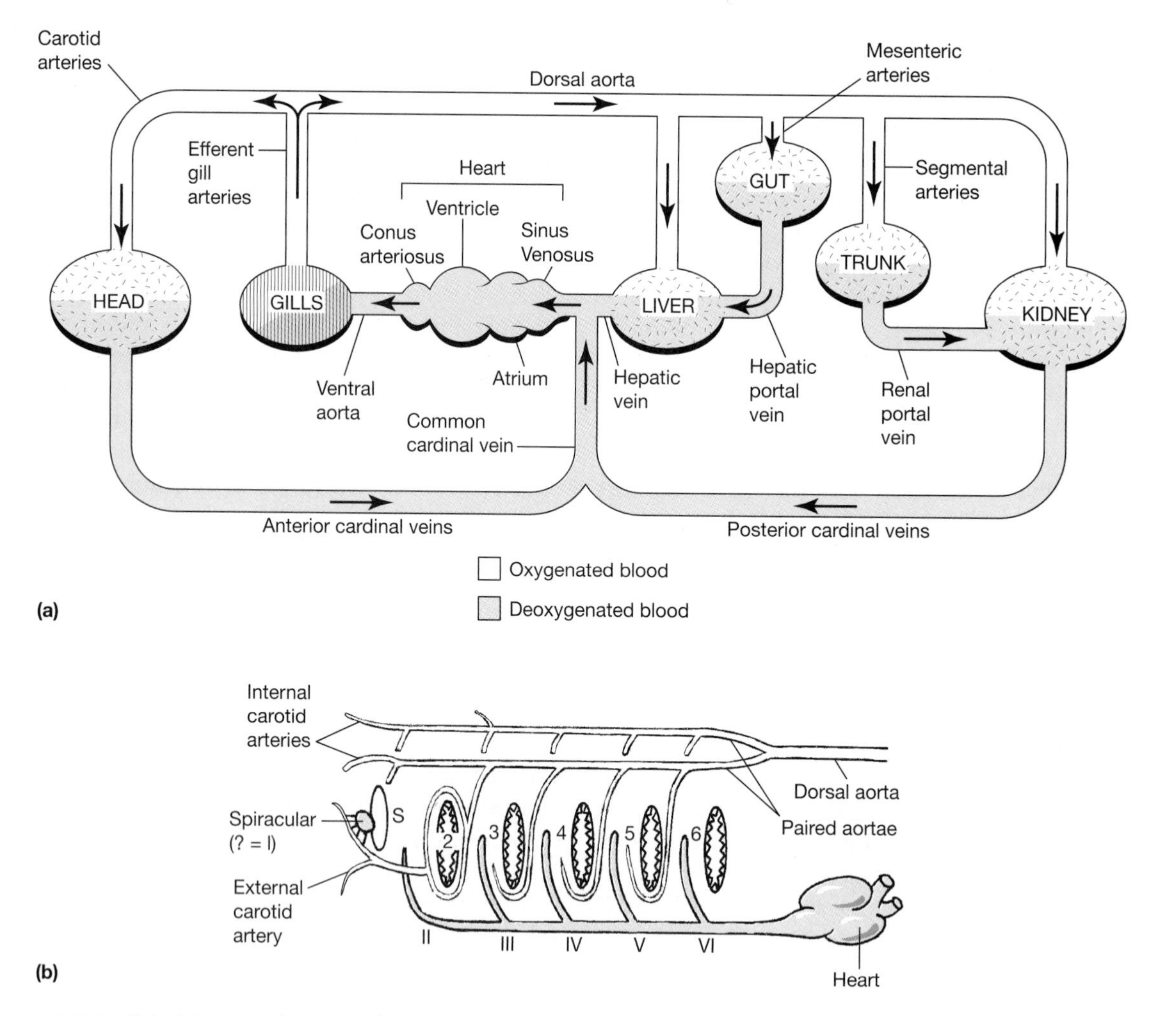

▲ Figure 2–14 Primitive vertebrate cardiovascular system. (a) Basic plan of vertebrate (gnathostome) cardiovascular circuit. (b) Generalized scheme of aortic arches in a gill-breathing primitive gnathostome, retained by most chondrichthyans. Bony fishes have numerous independently derived specializations that do not depart significantly from this basic plan.

dense beds in metabolically active tissues and are sparsely distributed in tissues with low metabolic activity.

Blood flow through capillary beds is regulated by precapillary sphincter muscles. Arteriovenous **anastomoses** connect some arterioles directly to venules, allowing blood to bypass a capillary bed. Normally only a fraction of the capillaries in a tissue have blood flowing through them. When the metabolic activity of a tissue increases—when a muscle becomes active, for example—waste products of metabolism stimulate precapillary sphincters to dilate, increasing blood flow to that tissue.

Portal vessels lie between two capillary beds. The **hepatic portal vein,** seen in all vertebrates, lies between the capillary beds of the gut and the liver (Figure 2–14a). Substances absorbed from the gut are transported directly to the liver, where toxins are rendered harmless and some nutrients are processed or removed for storage. Most gnathostomes also have a **renal portal vein** between the kidneys and the veins returning from the posterior trunk and the tail (Figure 2–14a). This setup presumably helps in processing the waste metabolites returning from the axial muscles that are used in locomotion. This system is lost in mammals and greatly reduced in birds.

The vertebrate heart is a muscular tube folded on itself, and primitively constricted into three sequential chambers: the **sinus venosus,** the **atrium,** and the **ventricle.** Gnathostomes add a fourth chamber, the

conus arteriosus, situated between the ventricle and the ventral aorta (Figure 2–14a). Our so-called four-chambered heart represents the combination of an atrium and a ventricle, both divided into two halves (left and right). We no longer have distinct structures indentifiable as the sinus venosus and the conus arteriousus.

The sinus venosus is a thin-walled sac with few **cardiac** muscle fibers. It is filled by pressure in the veins, along with pulsatile drops in pressure in the pericardial cavity as the heart beats. Suction produced by muscular contraction draws blood anteriorly into the atrium, which has valves at each end that prevent backflow. The ventricle is thick walled, and the muscular walls have an intrinsic pulsatile rhythm, which can be speeded up or slowed down by the autonomic nervous system. Contraction of the ventricle forces the blood into the conus arteriosus, and from there into the ventral aorta.

The basic vertebrate circulatory plan consists of a heart that pumps blood into the single midline ventral **aorta.** Paired sets of aortic arches (one of each pair supplying each side of the head) branch from the aorta (Figure 2–14b). In aquatic vertebrates the aortic arches lead to the gills, where the blood is oxygenated and returns to the dorsal aorta. The dorsal aorta is paired above the gills and the anterior portion runs forward to the head as the **carotid arteries.** Behind the gill region, the two vessels unite into a single dorsal aorta that carries blood posteriorly.

The aorta is flanked by paired **cardinal veins** that return blood to the heart (Figure 2–14a). Anterior cardinal veins (the jugular veins) draining the head, and posterior cardinal veins draining the body, unite into a common cardinal vein that enters the atrium of the heart. Blood is also returned separately to the heart from the gut and liver via the hepatic portal system.

The primitive vertebrate plan appears to be six pairs of aortic arches, presumably supplying an original system of six gill slits. Six aortic arches can be seen in the embryonic development of all vertebrates, but no living vertebrate has a first aortic arch as an adult. The original first gill slit has been lost, or in some jawed fishes converted into a small opening called the **spiracle.** In sharks the spiracular artery is probably the homologue of the original first aortic arch (Figure 2–14b).

Excretory and Reproductive Systems The excretory and reproductive systems share ducts through which metabolic wastes and gametes are released to the outside world, but the two systems have different developmental origins. The kidneys are segmental, derived from the nephrotome or intermediate mesoderm, which forms the embryonic nephric ridge (Figure 2–15). The gonads are formed from the genital ridge, which lies beneath the nephric ridge and is nonsegmented.

The kidneys dispose of waste products, primarily nitrogenous waste from protein metabolism, and

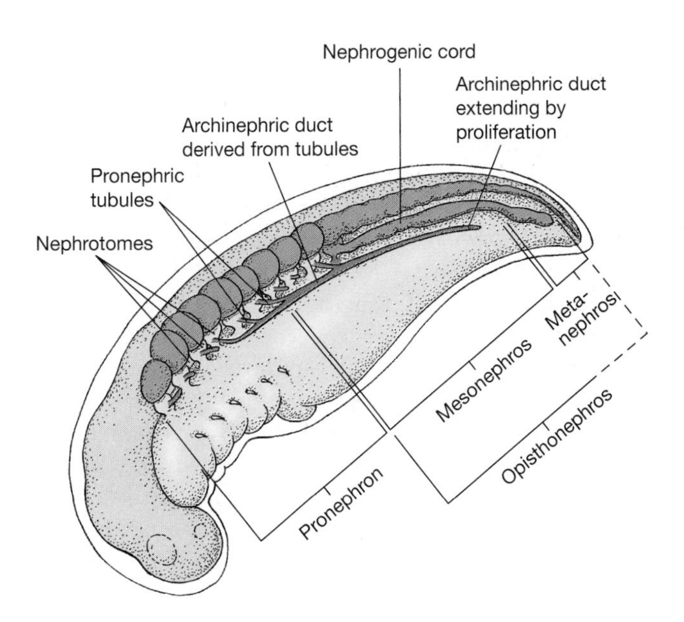

▶ Figure 2–15 Kidney development in a generalized vertebrate embryo, showing the nephrotome regions.

regulate the body's water and minerals—especially sodium, chloride, calcium, magnesium, potassium, bicarbonate, and phosphate. In tetrapods the kidneys are responsible for almost all these functions, but in primitive vertebrates the skin and the gills play important roles. The original role of the kidney in vertebrates may have been primarily regulation of divalent ions such as calcium and phosphate.

The kidney of fishes is a long, segmental structure extending along the entire length of the dorsal body wall, drained by the archinephric duct, and composed of three portions: pronephros, mesonephros, and metanephros (Figure 2–15). The same structure is seen in all vertebrate embryos. The pronephros is functional only in the embryos of living vertebrates, with the possible exception of hagfishes. The primitive adult vertebrate kidney includes the mesonephric and metanephric portions and is known as an **opisthonephric kidney.** The compact kidney seen in adult amniotes (the **metanephric kidney**) includes only the metanephros, drained by a new tube, the **ureter.**

Vertebrate kidneys work by ultrafiltration: Blood is brought directly to the kidney by renal vessels and the high blood pressure forces water, ions, and small molecules through tiny gaps in the capillary walls. The basic units of the kidney are microscopic structures called **nephrons** (see Chapter 4). Other chordates lack true kidneys. Amphioxus has excretory cells called solenocytes (also called flame cells or nephridia). These cells are associated with the pharyngeal blood vessels that empty individually into the false body cavity (the atrium). The effluent is finally discharged to the outside via the atriopore. These excretory solenocyte cells may be homologous with cells in the nephrons of vertebrate kidneys.

Early in embryonic development the kidney tubules (nephrons) of vertebrates are segmentally arranged and open through funnel-like ciliated mouths to the coelom, from which they drain fluid. This fluid is conveyed along the archinephric ducts (one for each kidney), which empty into the cloaca. Later in development the individual tubules lose their direct connection to the coelom, and derive their filtrate directly from the capillary system.

Reproduction is the means by which gametes are produced, released, and combined with gametes from a member of the opposite sex to produce a fertilized zygote. Vertebrates usually have two sexes, and sexual reproduction is the norm—although unisexual species occur among fishes, amphibians, and lizards.

The organs that produce **gametes** (sex cells) are the gonads—**ovaries** in females and **testes** in males. The gonads are paired and usually lie on the posterior body wall, behind the peritoneum (only in some mammals are the testes found outside the body in a scrotum). Ovaries contain large primary sex cells called **follicles.** As they mature the follicular cell layer becomes much larger, nurturing the developing egg (**ovum;** plural, *ova*), stimulating the development of yolk in the egg, and producing the hormone **estrogen.** When the eggs mature the follicle ruptures, releasing the completed egg (**ovulation**). The production of eggs by the ovary may be almost continuous (as in humans), seasonal (in the vast majority of vertebrates), or occur only once in a lifetime (some fishes, such as the eel, and some mammals, such as marsupial mice).

The testes are composed of interconnecting **seminiferous tubules** where sperm develop. In these tubules, sperm are supported, nourished, and conditioned by cells—the supporting or Sertoli cells—that remain permanently attached to the tubule walls. The testes also produce the hormone **testosterone.**

In the primitive vertebrate condition, which is retained in living agnathans, sperm or eggs erupt from the gonad and move through the coelom to pores that open to the outside near the cloaca. In gnathostomes, however, the gametes are always transported to the cloaca via specialized ducts (paired, one for each gonad). In the males, sperm are released directly into a duct that is usually formed from the archinephric duct that originally drained the kidney. In females, the egg is still released into the coelom, but is then transported via a new structure, the **oviduct.** The oviducts produce the jelly surrounding amphibian eggs and the shells of shark, reptile, and bird eggs, as well as the albumen (= egg white) surrounding the embryo and yolk in amniotes. The oviducts can become enlarged and fused in various ways to form a uterus or paired uteri in which eggs are stored or young develop.

Vertebrates may deposit eggs that develop outside the body (**oviparity**) or retain the eggs within the mother's body until embyronic development is complete (**viviparity**). Oviparous vertebrates are called lecithotrophic when the yolk provides the energy for embryonic growth, and matrotrophic when the nutrients come from maternal secretions. Shelled eggs must be fertilized in the oviduct before the shell and albumen are deposited. Vertebrates that lay shelled eggs, and vertebrates that are viviparous, must have some sort of intromittent organ—such as the pelvic claspers of sharks and the amniote penis—by which internal fertilization is accomplished. However, unlike most amniotes, most birds have no intromittent organ; the male transfers sperm to the female by pressing his cloaca against hers.

Coordination and Integration

The nervous system provides an organism with information about the outside world and controls the actions and functions of the organs and muscles. Organs must be coordinated to work in concert if the action of one is not to cancel the action of another. Signals transmitted by the nervous system, and endocrine secretions distributed by the blood, are significant coordinators in this regard.

General Features of the Nervous System The basic unit of the nervous system is the **neuron.** In gnathostomes the axons of neurons are encased in a fatty insulating coat, the **myelin sheath,** which increases the conduction velocity of the nerve impulse. Generally, **axons** (extensions of the neurons) are collected together like wires in a cable (Figure 2–16). Such collections of axons in the peripheral nervous system (= PNS; i.e., in the body) are called nerves; within the central nervous system (= CNS; i.e., within the brain and spinal cord) they are called tracts.

The spinal cord (= nerve cord) is composed of a hollow tube with an inner core of gray matter (cell bodies) and an outer layer of white matter (axons covered by myelin sheaths). The spinal cord receives sensory inputs, integrates them with other portions of the CNS, and sends impulses that cause muscles to contract. The spinal cord has considerable autonomy in many vertebrates—even such complex movements as swimming are controlled by the spinal cord rather than the brain, and fishes continue coordinated swimming movements when the brain is severed from the spinal cord. The familiar knee-jerk reaction is produced by the spinal cord as a reflex arc. Similarly, a spinal reflex causes you to jerk your hand away from a hot surface before the pain signals reach your brain.

The trend in vertebrate evolution has been to develop more complex connections within the spinal cord and between the spinal cord and the brain. The nerves of the PNS are segmentally arranged, exiting from the spinal cord between the vertebrae. Each spinal nerve complex is made up of four types of fibers: somatic sensory fibers from the body wall; somatic motor fibers to the body; visceral sensory fibers from the gut wall and blood vessels; and visceral motor fibers to the muscles and glands of the gut and to the blood vessels.

In mammals there is a clear division of the autonomic portion of the visceral nervous system into the

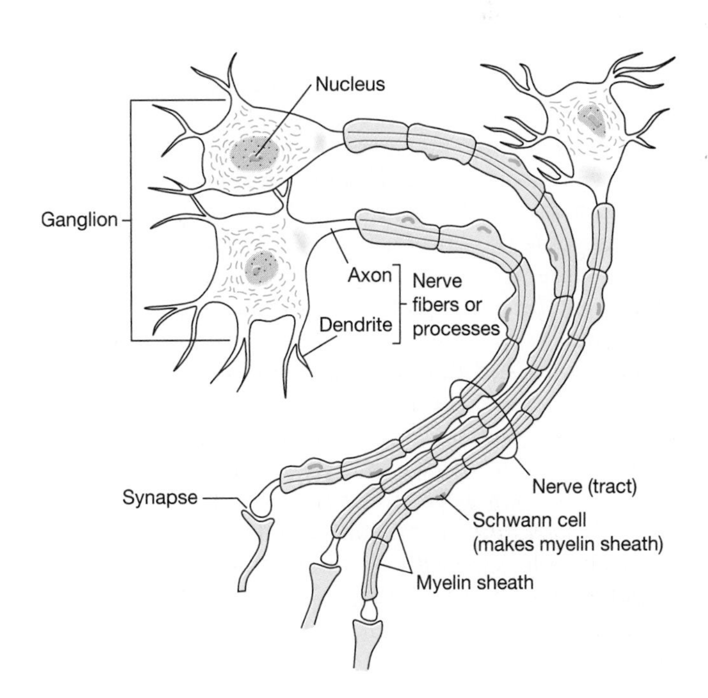

▶ Figure 2–16 Generalized vertebrate (gnathostome) neurons.

sympathetic nerves (emerging from the trunk region, generally speeding things up) and parasympathetic nerves (emerging from the cranial or sacral region, generally slowing things down). But in other vertebrates this division is less clear.

Vertebrates also have nerves that emerge directly from the brain, called **cranial nerves** (10 pairs in the primitive vertebrate condition, 12 in amniotes), which scientists designate by Roman numerals. Some of these nerves, such as the ones supplying the nose (the olfactory nerve, I) or the eyes (the optic nerve, II) are not true nerves at all, but outgrowths of the brain. Motor fibers in cranial nerves V, VII, IX, and X innervate the branchiomeric muscles, and those in nerves III, IV, and VI innervate the muscles that move the eyeball.

There is also a system of special visceral sensory fibers in nerves VII, IX, and X, conveying information to the brain from the taste buds. (We are unaware of most visceral sensory sensations, but taste is obviously a conscious phenomenon.) Some evidence suggests that these fibers may in fact be separate cranial nerves in their own right. The special sensory nerves that supply the lateral line in fishes may also be distinct cranial nerves.

The **vagus nerve** (cranial nerve X) ramifies through all but the most posterior part of the trunk, carrying the parasympathetic nerve supply to various organs. People who break their necks may be paralyzed from the neck down (i.e., lose the function of their skeletal muscles), but still may retain their visceral functions (workings of the gut, heart, etc.) because the vagus nerve is independent of the spinal cord and exits above the break.

Brain Anatomy and Evolution The brain of all vertebrates is a tripartite structure (Figure 2–17). In the most simple condition the forebrain (also known as the prosencephalon) is associated with olfaction (the sense of smell), the midbrain (also known as the mesencephalon) with vision, and the hindbrain (also known as the rhombencephalon) with balance and detection of vibrations (hearing). These portions of the brain are associated with the nasal, optic, and otic capsules of the chondrocranium, respectively (see Figure 2–9). The brain of amphioxus appears to be little more than a swelling at the end of the nerve chord, but recent studies show it to be homologous with all of the vertebrate brain with the exception of the anterior forebrain—the telencephalon (Zimmer 2000).

Posteriorly, two regions differentiate from the embryonic hindbrain region associated with the developing ear. The most posterior, the myelencephalon or medulla oblongata, controls functions such as respiration and acts as a relay station for receptor cells from the inner ear. The anterior portion of the hindbrain, the metencephalon, develops an important dorsal outgrowth, the **cerebellum**—present as a distinct structure only in gnathostomes among living vertebrates. The cerebellum coordinates and regulates motor activities whether they are reflexive, such as maintenance of posture, or directed, such as escape movements. The midbrain, or mesencephalon, develops in conjunction with the eyes and receives input from the optic nerve, although in mammals the forebrain has taken over much of the task of vision.

The forebrain has two parts. The posterior region is the **diencephalon,** which acts as a major relay station between sensory areas and the higher brain centers. The **pituitary gland** or hypophysis, an important endocrine organ, is a ventral outgrowth of the diencephalon. The floor of the diencephalon (the **hypothalamus**) and the pituitary gland form the primary center for neural-hormonal coordination and integration. Another endocrine gland, the **pineal organ**—also known as the epithalamus, is a dorsal outgrowth of the diencephalon. Its original function was as a median photoreceptor (light-sensitive organ). Many early tetrapods had a hole in the skull over the pineal to admit light, and this condition is still seen today in the primitive reptile *Sphenodon* (the tuatara).

The most anterior region of the adult forebrain, the telencephalon, develops in association with the olfactory capsules and coordinates inputs from other sensory modalities. In various ways in different vertebrate groups, the telencephalon becomes enlarged—in which condition it is also known as the **cerebrum** or cerebral hemispheres. Tetrapods develop an area called the **neocortex** or neopallium: In mammals this region becomes the primary seat of sensory integration and nervous control. On a different branch of the vertebrate evolutionary lineage, bony fishes also evolved a larger, more complex telencephalon—but by a completely different mechanism. In bony fishes the forebrain appears to be inside out in comparison with other vertebrates. Sharks and, perhaps surprisingly, hagfishes have also independently evolved relatively large fore-

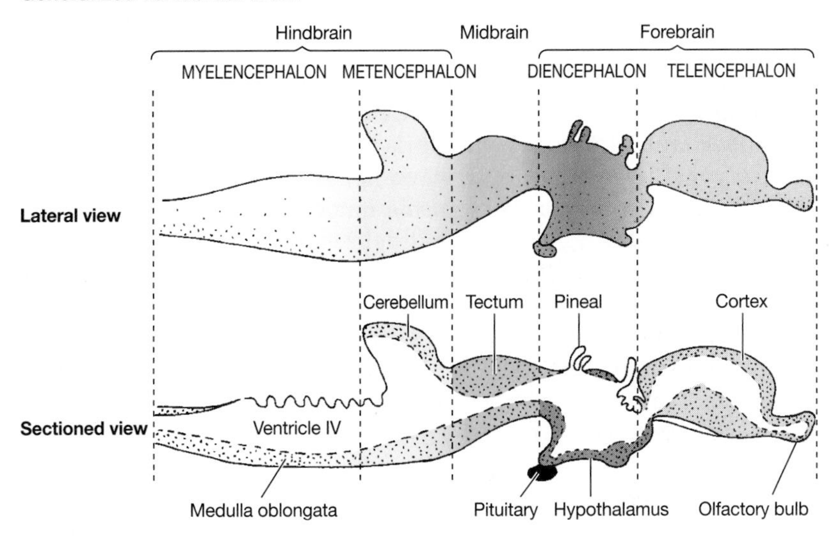

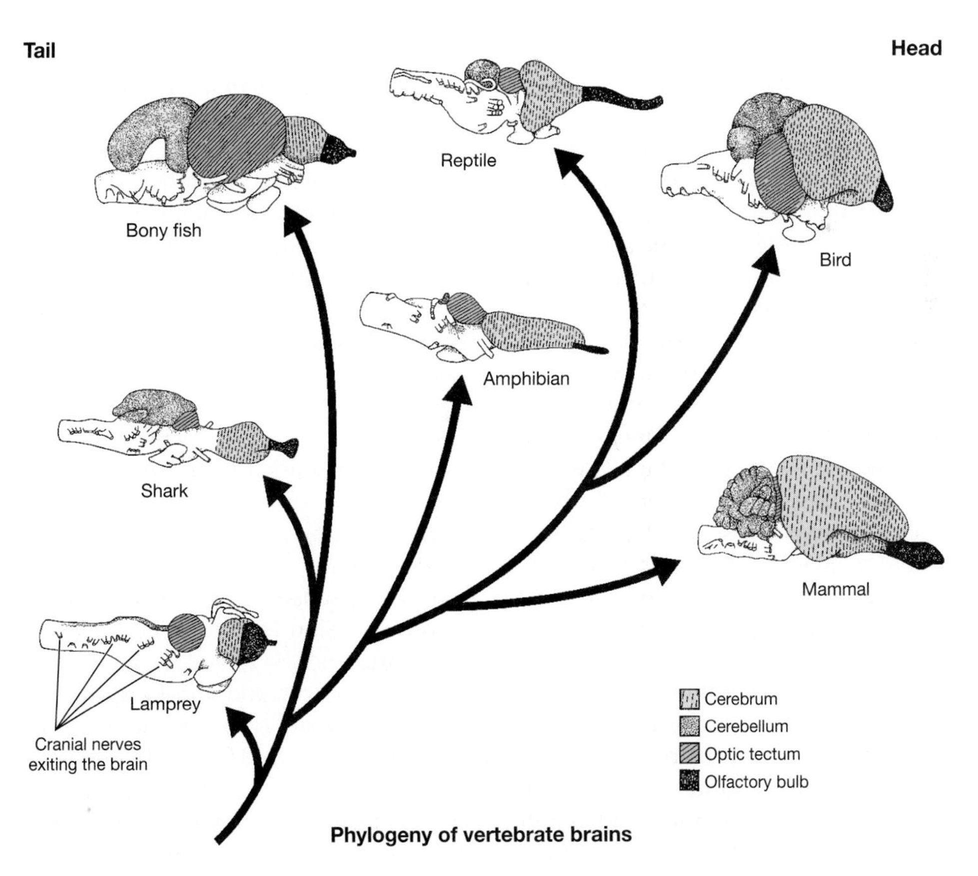

▲ Figure 2–17 The vertebrate brain.

brains, although a large cerebrum is primarily a feature of tetrapods (Figure 2–17b).

The Sense Organs We think of vertebrates as having five senses—taste, touch, sight, smell, and hearing—but this list does not reflect the primitive condition, nor all the senses of extant vertebrates. Complex, multicellular sense organs, formed from epidermal placodes and tuned to the sensory worlds of the species that possess them, are a derived feature of vertebrates. For example, detecting airborne sound waves is not something aquatic vertebrates need to do, and the **lateral line** system, which is a major sensory mode of fishes, is not functional on land.

The senses of smell and taste both involve the detection of dissolved molecules by specialized receptors. We think of these two senses as being closely interlinked; for example, our sense of taste is poorer if our sense of smell is blocked by having a cold. But the two senses are actually very different in their embryonic origin. Smell is a somatic sensory system—sensing items at a distance, with the sensations projecting to the forebrain. Taste is a visceral sensory system—sensing items on direct contact, with the sensations projecting initially to the hindbrain.

Vision is considered the distance sense par excellence, sensitive to those wavelengths of electromagnetic radiation (light) that reach the surface of the Earth with the least interference from air and water. The receptor field of the vertebrate eye is arrayed in a hemispherical sheet, the **retina,** which originates as an outgrowth of the diencephalon of the brain. Nerves from the retina project to the midbrain and, especially in mammals, to the visual cortex of the telencephalon. The retina contains two types of light-sensitive cells, **cones** and **rods,** which are distinguished from each other by morphology, photochemistry, and neural connections. Each point on the retina corresponds to specific neural connections and a different visual axis in space. Thus, a vertebrate can determine where an object is and whether it is stationary or moving.

The capacity to perceive electrical impulses generated by the muscles of other organisms is also a form of distance reception, but one that is foreign to most terrestrial animals. **Electroreception** was probably an important feature of early vertebrates and is seen today primarily in fishes. Related to electroreception is the ability seen in many fishes today to produce electric discharge for communication with other individuals or for protection from predators (see Chapter 4).

Originally the inner ear detected an animal's position in space, and it retains that function today in both aquatic and terrestrial vertebrates. The inner ear is also used for hearing (reception in sound waves) primarily by tetrapods. The basic sensory cell in the inner ear is the **hair cell**, a mechanoreceptor that detects the movement of fluid resulting from a change of position or the impact of sound waves. In the lateral line system of fishes and aquatic amphibians, hair cells are aggregated into **neuromast organs** that detect the movement of water around the body (see Chapter 4).

The inner ear contains the **vestibular apparatus** (also known as the **membranous labyrinth**), which includes the organs of balance and, in tetrapods only, the **cochlea** (organ of hearing). The vestibular apparatus is comprised of a series of sacs and tubules containing the fluid endolymph. It is enclosed within the otic capsule of the skull, surrounded by the fluid perilymph. The vestibular sensory cells are variants of the basic hair cell and detect the movement of endolymph.

The lower parts of the vestibular apparatus, the sacculus and utriculus, house sensory organs called macculae which contain tiny crystals of calcium carbonate resting on hair cells. Sensations from the maculae tell the animal which way is up and detect linear acceleration. The upper part of the vestibular apparatus contains the **semicircular canals.** Sensory areas in the ampullae of these canals detect angular acceleration through cristae, hair cells embedded in a jellylike substance, by monitoring the displacement of endolymph during motion. Gnathostomes have three semicircular canals on each side of the head, hagfishes have one, and lampreys and fossil agnathans have two (Figure 2–18).

We often fail to realize the importance of the vestibular senses in ourselves, since we usually depend on vision to determine position. However, we can sometimes be fooled, as when sitting in a stationary train or car and thinking that we are moving, only to realize from the lack of input from our vestibular system that it is the vehicle *next* to us that is moving.

The Endocrine System The endocrine system transfers information from one part of the body to another via the release of a chemical messenger (**hormone**) that produces a response in the target cells.

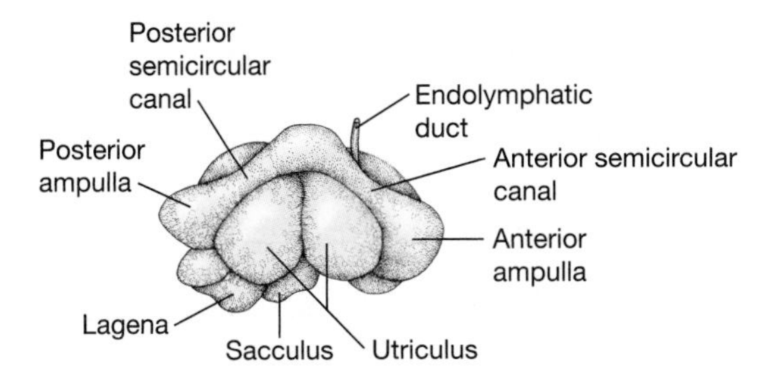

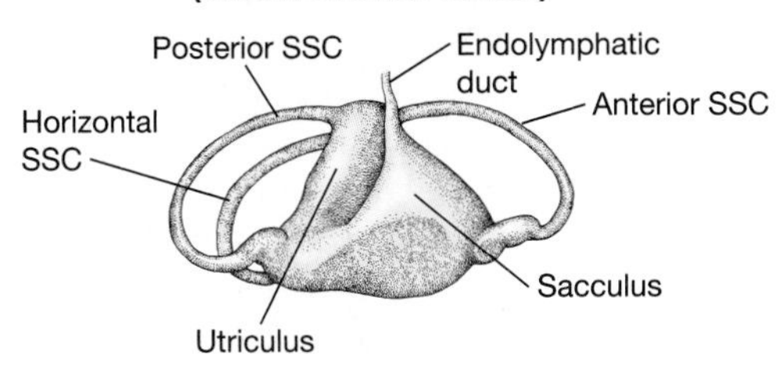

▲ Figure 2–18 Design of the vestibular apparatus in fishes.

The time required for an endocrine response ranges from seconds to hours. Hormones are produced in discrete endocrine glands, whose primary function is hormone production and excretion (e.g., the pituitary, thyroid, thymus, and adrenals), and by organs with other major bodily functions—such as the gonads, kidney, and gastrointestinal tract.

The trend in the evolution of vertebrate endocrine glands has been consolidation from scattered clusters of cells or small organs in fishes to larger, better-defined organs in amniotes. Endocrine secretions are predominantly involved in controlling and regulating energy use, storage, and release, as well as in allocating energy to special functions at critical times.

Summary

Vertebrates are members of the phylum Chordata, a group of animals whose other members (tunicates and lancelets) are quite different from most modern vertebrates, being small, marine, and sluggish or entirely sessile as adults. Chordates share with many derived animal phyla the features of being bilaterally symmetrical, with a distinct head and tail end and a tube-within-a-tube internal design. Within this range of more complex animals, embryological as well as molecular evidence shows that the affinities of chordates lie with echinoderms rather than with superficially more similar animals like arthropods.

Chordates are generally distinguished from other animals by the presence of a notochord, a dorsal hollow nerve chord, a muscular postanal tail, and an endostyle (= thyroid gland). Vertebrates have the unique features of an expanded head with multicellular sense organs and a cranium (skull) housing an enlarged, tripartite brain. The features that distinguish vertebrates from other chordates appear to be related to two critical embryonic innovations: a doubling of the *Hox* gene complex and the development of neural crest tissue.

Vertebrates appear to be most closely related to lancelets (cephalochordates) among nonvertebrate chordates, and most of the differences in structure and physiology between lancelets and primitive vertebrates appear to reflect an evolutionary change to larger body size, a greater level of activity, and a switch from filter feeding to predation.

The complex activities of vertebrates are supported by a complex morphology. Patterns of embryonic development are generally phylogenetically conservative, and many of the shared derived characters of vertebrates can be traced to the embryo. The neural crest cells, unique to vertebrates, form many of the derived characters of vertebrates, especially those of the new anterior portion of the head.

An adult vertebrate can be viewed as interacting systems involved in protection, support and movement, acquisition of energy, excretion, reproduction, coordination, and integration. These systems underwent profound functional and structural changes at several key points in vertebrate evolution. The most important transition is from the prevertebrate condition, as represented today by the lancelet—also known as amphioxus—to the vertebrate condition shown by the living jawless vertebrates. Other important transitions, to be considered in later chapters, include the shift from jawless to jawed vertebrates and from fish to tetrapod, reflecting the change in habitat from water to land.

Additional Readings

Blieck, A. 1992. At the origin of chordates. *Geobios* 25(1): 101–103.

Gans, C. 1989. Stages in the origin of vertebrates: Analysis by means of scenarios. *Biological Reviews* 64:221–268.

Garcia-Fernandez, J., and P. W. H. Holland. 1994. Archetypal organization of the amphioxus *Hox* gene cluster. *Nature* 370:504–505.

Garstang, W. 1928. The morphology of the Tunicata and its bearing on the phylogeny of the Chordata. *Quarterly Journal of the Microscopical Society* 72:51–87.

Gee, H. 1996. *Before the Backbone*. London: Chapman & Hall.

Hall, B. K. 1992. *Evolutionary Developmental Biology*. London: Chapman & Hall.

Hall, B. K. 2000. The neural crest as a fourth germ layer and vertebrates as quadroblastic and triploblastic. *Evolution and Development* 2:3–5.

Hildebrand, M. 1995. *Analysis of Vertebrate Structure,* 4th ed., New York, NY: Wiley.

Kardong, K. V. 1998. *Vertebrates—Comparative Anatomy, Function, Evolution,* 2nd ed. Dubuque, Ia: Wm. C. Brown.

Mallatt, J. 1996. Ventilation and the origin of jawed vertebrates: A new mouth. *Zoological Journal of the Linnean Society* 117:329–404.

Marx, J. 1992. Homeobox genes go evolutionary. *Science* 255:399–401.

Monastersky, R. 1996. Jump-start for the vertebrates: New clues to how our ancestors got a head. *Science News* 149: 74–75.

Northcutt, R. G. 1990. Ontogeny and phylogeny: A re-evaluation of conceptual relationships and some applications. *Brain, Behavior and Evolution* 36:116–140.

Northcutt, R. G., and C. Gans. 1983. The genesis of neural crest and epidermal placodes: A reinterpretation of vertebrate origins. *Quarterly Review of Biology* 58:1–28.

Romer, A. S., and T. S. Parsons 1986. *The Vertebrate Body*, 6th ed. Philadelphia, Penn: Saunders College Publishing.

Shu, D.-G., S. Conway Morris, and X.-L. Xang. 1996. A *Pikaia*-like chordate from the Lower Cambrian of China. *Nature* 384:157–158.

Walker, W. F., Jr., and K. F. Liem. 1994. *Functional Anatomy of the Vertebrates—an Evolutionary Perspective,* 2nd ed. Philadelphia, Penn: Saunders College Publishing.

Zimmer, C. 2000. In search of vertebrate origins: Beyond brain and bone. *Science* 287:1576–1579.

Web Explorations

On-line resources for this chapter are on the World Wide Web at: http://www.prenhall.com/pough (click on the Table of Contents link and then select Chapter 2).

CHAPTER

3 Early Vertebrates: Jawless Vertebrates and the Origin of Jawed Vertebrates

The earliest vertebrates represented an important advance over the nonvertebrate chordate filter feeders. Their most conspicuous new feature was a distinct head end, containing a tripartite brain enclosed by a cartilaginous cranium (skull) and complex sense organs (eyes, nose). Instead of using cilia to move water over the gill bars, they used the newly acquired pharyngeal musculature to draw water into the mouth. This water current was used for respiration rather than for filter feeding, and tissues specialized for gas exchange were present on the gill bars. Early vertebrates were active predators rather than sessile filter feeders. Many of them also had external armor made of bone, which is a distinctive form of mineralized tissue.

We know a remarkable amount about the anatomy of some of these early vertebrates, because the internal structure of their bony armor reveals the positions and shapes of many parts of their soft anatomy. Gnathostomes (jawed vertebrates), which evolved a little later than the agnathans (jawless vertebrates), represented a further advance in the vertebrate design for high levels of activity and predation. Jaws themselves are homologous with the structures that form the gill arches, and probably first evolved as devices to improve the strength and effectiveness of gill ventilation. Later they were modified for seizing and holding prey. In this chapter we trace the earliest steps in the radiation of vertebrates, beginning some 500 million years ago, discuss the biology of both the Paleozoic agnathans (the ostracoderms) and the extant forms (hagfishes and lampreys). We also consider the transition from the jawless condition to the jawed one and the biology of some of the early types of jawed fishes that did not survive the Paleozoic (placoderms and acanthodians).

3.1 Reconstructing the Biology of the Earliest Vertebrates

The Earliest Evidence of Vertebrates

Until very recently our oldest evidence of vertebrates consisted of fragments of the dermal armor of the armored jawless vertebrates known as **ostracoderms** (Greek *ostrac* = shell and *derm* = skin). These animals were very different from any vertebrate alive today. They were essentially fishes encased in bony armor, quite unlike the living jawless vertebrates that lack bone completely. Bone fragments are known from the Ordovician, some 480 million years ago. This was about 80 million years before whole-body vertebrate fossils became abundant, in the Late Silurian. Other recent evidence includes some mineralized tissue, possibly from vertebrates in the Late Cambrian of North America and Australia, around 500 million years ago.

More recent finds of early vertebrates from the Early Cambrian of China (Shu et al. 1999) extend the fossil record back by another 50 million years or so. Two different types of Early Cambrian vertebrates, *Myllokumingia* and *Haikouichthys*, are found in the same fossil deposit. Both were small, about 3 centimeters long, and shaped like fishes (Figure 3–1a). Evidence of a cranium and W-shaped myomeres mark these animals as true vertebrates. However, unlike the later ostracoderms, they lack any evidence

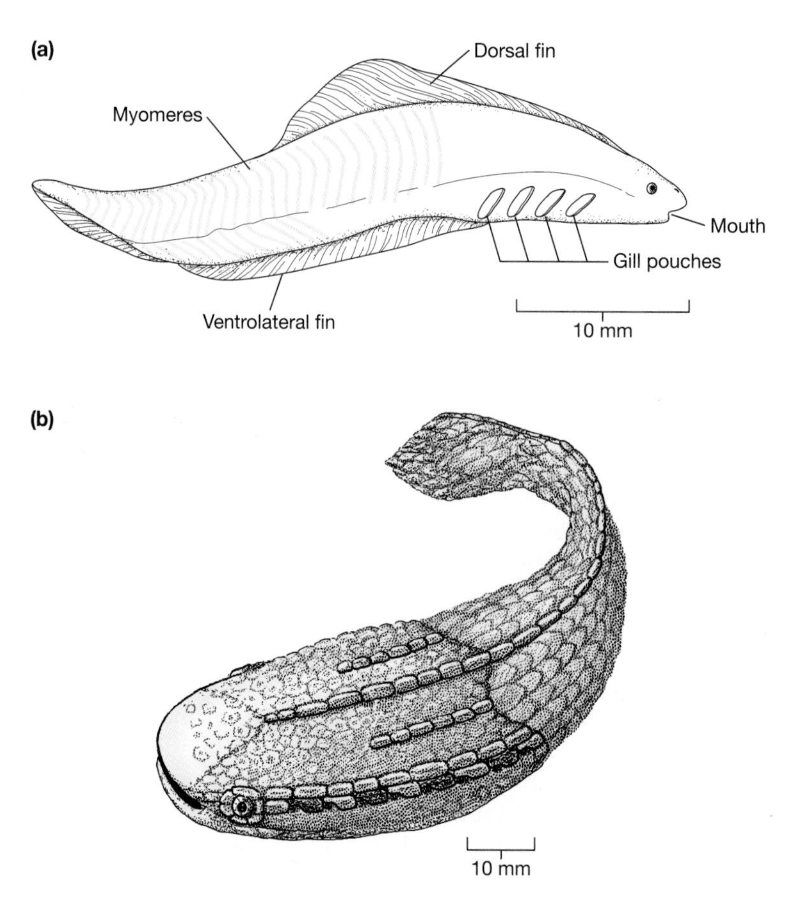

▲ Figure 3–1 Some of the earliest vertebrates. (a) The Early Cambrian *Myllokumingia* from China. (b) The Ordovician pteraspidomorph *Astraspis* from North America.

of bone or mineralized scales. Both animals had a dorsal fin and ribbon-like pairs of ventrolateral fins. Despite the great age of these early vertebrates, they were evidently more derived than present-day hagfishes, as shown by the presence of a dorsal fin. This suggests that the vertebrate lineage had considerably diversified before this time. *Haikouichthys*, which may be the more derived form, appears to have lamprey-like cartilaginous gill supports (a branchial basket) and may be the sister taxon to the lampreys.

The next good evidence of early vertebrates is from the Early Ordovician; several sites have yielded vertebrate fossils in the form of bone fragments. These sites are scattered around the world, suggesting that by this time vertebrates had a worldwide distribution. The earliest vertebrates represented by complete articulated fossils are from the Late Ordovician of Bolivia, Australia, and North America. These are armored, torpedo-shaped jawless fishes, ranging from 12 to 35 cm in length. These early fishes were externally armored with many small, close-fitting, polygonal bony plates 3 to 5 millimeters long. These plates abut each another in the head and gill region, forming a head shield. Posteriorly they overlap, as do scales in extant fishes. These bony plates show the presence of sensory canals, special protection around the eye, and—in the reconstruction of the North American *Astraspis* (Figure 3–1b)—as many as eight gill openings on each side of the head.

Similar types of early vertebrates are known from the Late Silurian to Middle Devonian formations known as Old Red Sandstone in southwestern England and Wales, and in rocks of equivalent age in Scotland, Norway, Spitzbergen, and North America, dating from around 400 million years ago. The more abundant Silurian fossils are similar to those of the Ordovician, but they are known for a greater variety of whole-body fossils. The diversity of vertebrates in the Late Ordovician suggests that a significant radiation of vertebrates must have occurred earlier in the Ordovician. The Ordovician was also a time for great radiation and diversification among marine invertebrates, following the extinctions at the end of the Cambrian. The early radiation of vertebrates involved both jawed and jawless groups. Based on the evidence provided by scales or denticles, ostracoderms first occurred in the Early Ordovician, and jawed fish first occurred a little later—in the Middle Ordovician.

The Problem of Conodonts

Curious microfossils known as **conodont** elements are widespread and abundant in marine deposits from the Late Cambrian to the Late Triassic. Conodonts are small (generally less than 1 mm long) spine-like or comb-like structures composed of apatite (the particular mineralized structure of calcium carbonate and calcium phosphate that is characteristic of vertebrate hard tissues). They were variously described as skeletal parts of marine algae, phosphatic annelid jaws, fish teeth, gill rakers, gastropod radulae, nematode copulatory spicules, arthropod spines, and denticles of free-swimming lophophorate animals. However, recent studies of conodont mineralized tissues have shown that they are histologically similar to dentine and enamel, which are uniquely vertebrate tissues. Thus conodont elements are considered to be the tooth-like elements of true vertebrates. This interpretation has been confirmed by the discovery of impressions of complete conodont animals with conodont elements arranged within the pharynx in a complex apparatus (Figure 3–2).

Conodonts appear to have had a notochord, a cranium, myomeres, fin rays and large eyes protected by sclerotic rings of cartilage with evidence of extrinsic eye muscles to move the eyes. Although the pharynx may have been muscular (to operate the conodont apparatus), there are no signs of pharyngeal slits. However, at their very small size (40 mm long) respiration could have been accomplished by diffusion of gas across the body surface. Thus, assuming that their ancestors (as true vertebrates) had gills, these structures may have been unnecessary for the tiny conodonts and thus lost.

Accepting conodonts as vertebrates has changed our ideas about early vertebrate interrelationships, particularly the importance—in the evolutionary hierarchy—of having mineralized tissues. Mineralized

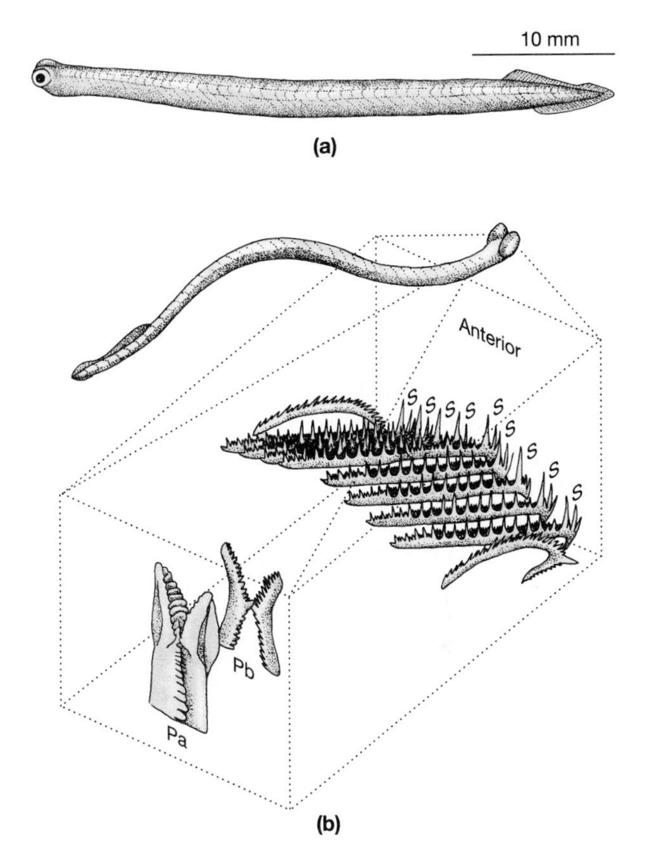

▲ Figure 3–2 Conodonts. (a) *Clydagnathus*. (b) Close-up of the feeding apparatus (conodont elements) inside the head of *Idiognathus*. The anterior S elements appear to be designed for grasping, and the posterior P elements appear to be designed for crushing.

tissues, as seen in tooth-like structures of conodonts and the bony dermal skeletons of ostracoderms, are now thought to make these animals more derived types of vertebrates than the soft-bodied, jawless fishes living today (see Figure 3–3, later in this section).

The Origin of Bone and Other Mineralized Tissues

Mineralized tissues composed of calcium phosphate—such as bone, dentine, and enamel—are a major new feature of vertebrates. Although only vertebrates have bone, not all vertebrates have bone, so bone does not characterize vertebrates. As we have just seen, the earliest known vertebrates lacked bone, as do the living jawless fishes. Bone is only one type of mineralized tissue in vertebrates: Enamel (or enameloid) and dentine are more highly mineralized than bone. These tissues, which occur primarily in the teeth among living vertebrates, are at least as old as bone. In the ostracoderms enamel and dentine were originally found in intimate association with bone in the dermal armor.

The origin of vertebrate mineralized tissues remains a puzzle, because they cannot be homologized with any tissues in invertebrate phyla. Bone results from the deposition of minerals in an organic matrix produced by special cells, laid down in a supporting framework of the fibrous protein collagen. In this respect it is like the shells of mollusks and hard parts of other animals, but vertebrate bone is unique in its combination of characteristic cells, matrices, and minerals.

The earliest known mineralized vertebrate tissues include several different tissue types and were no less complex in structure than the mineralized tissues of living vertebrates. The basic units of mineralized tissue in most early vertebrates appear to be **odontodes**, little tooth-like elements formed in the skin. They consist of projections of dentine, covered in some cases with an outer layer of enameloid (an enamel-like tissue), with a base of bone. Our own teeth are very similar to these structures and are probably homologous with them. Odontodes occur in almost unmodified form as the sharp denticles in the skin of sharks—and the larger scales, plates, and shields on the heads of many ostracoderms and early bony fishes are interpreted as aggregations of these units. Note that these bony elements would not have been external to the skin like a snail's shell. Rather, they were formed within the dermis of the skin and overlain by a layer of epidermis, as with our own skull bones. The primitive condition for vertebrate bone is to lack cells in the adult form. This type of acellular bone is also known as aspidin. Cellular bone is found only in gnathostomes and in the osteostracan ostracoderms, and may even have evolved independently in these forms (Donoghue et al. 2000).

We do not as yet understand quite how conodont teeth fit into this evolutionary picture. They may represent an early experiment in tissue mineralization prior to the evolution of the dermal skeleton in other vertebrates. However, pharyngeal tooth-like structures have also been reported in thelodonts, poorly known ostracoderms that lack a well-mineralized skeleton.

What could have been the original selective advantage of mineralized tissues in vertebrates? The head shields of ostracoderms were originally thought to be defensive structures against attack from predators—particularly the large, scorpion-like creatures known as eurypterids. However, the detailed structure of the bony tissues suggests that they had a more complex function than mere protection. The dentine of the head shield has a characteristic system of pores and tubercules, suggesting that the tissue had a sensory function. The bony head shield may have originated in the heads of early vertebrates as a protective and electrically insulating coating around the electroreceptors that enhanced detection of their prey. Subsequently, regulation of phosphorus, which is a relatively rare element in natural environments, may have been one of the early selective forces involved in the evolution of bone. Although we think of bone as primarily supportive or protective, it is also a store of calcium and phosphorus. Mineral regulation involves deposition and mobilization of calcium and phosphorus ions. Vertebrates rely on anaerobic metabolism during activity, producing lactic acid that lowers blood pH. A skeleton made of hydroxyapatite may be more resistant to acidity of the blood during anaerobic metabolism than is the calcium carbonate (calcite) that forms the shells of mollusks.

The explanations proposed for the origin of bone—protection, electroreception, and mineral storage—are not mutually exclusive. They all may have been involved in the evolutionary origin of this complex tissue.

The Environment of Early Vertebrate Evolution

By the Late Silurian, armored ostracoderms and early jawed fishes were abundant in both freshwater and marine environments. Under what conditions did the first vertebrates evolve? Researchers originally supposed that vertebrates evolved in freshwater. This was because the vertebrate kidney is clearly advantageous in that habitat, acting to rapidly eliminate excess water entering the body by osmosis. However, a marine origin of vertebrates is now widely accepted; it is considered that the kidney was fortuitously preadapted for this role in freshwater, rather than specifically evolved for that purpose.

The first line of evidence for a marine origin of vertebrates is paleontological—the earliest vertebrate fossils are found in marine sediments. The second line of evidence comes from comparative physiology—all nonvertebrate chordates and deuterostome invertebrate phyla are exclusively marine forms, with body fluids in the same concentration as their surroundings. Thus the concentrated body fluids of the very primitive hagfish most likely represent the original vertebrate condition.

3.2 Extant Jawless Fishes

The extant jawless vertebrates—hagfishes and lampreys—have long been placed with the ostracoderms in the class "Agnatha" because they lack the derived (gnathostome) features of jaws and two sets of paired fins. However, phylogenetic systematics no longer groups organisms on the basis of shared primitive characters, and it is now clear that the "Agnatha" is a paraphyletic assemblage. Living agnathans have often been linked as **cyclostomes** because they have round, jawless mouths (Greek *cyclo* = circle and *stoma* = mouth). But this grouping is also paraphyletic because lampreys appear to be more closely related to gnathostomes than are hagfishes (Figures 3–3 and 3–4). Both hagfishes and lampreys appear to be more primitive than the armored ostracoderms of the Paleozoic, so we will look at them before considering the extinct agnathans.

The fossil record of the modern types of jawless vertebrates is sparse. Lampreys are known from the Carboniferous—*Hardistiella* from Montana and *Mayomyzon* from Illinois. *Myxinikela*, an undisputed hagfish, and a second possible hagfish relative, *Gilpichthys*, have been found in the same deposits as *Mayomyzon*.

Hagfishes—Myxinoidea

There are around 60 recognized species of hagfishes (Figure 3–5) in two major genera (*Eptatretus* and *Myxine*). Adult hagfishes are generally under 1 meter in length, and they are elongated, scaleless, and pinkish to purple in color. Hagfishes are entirely marine, with a nearly worldwide distribution—except for the polar regions. They are primarily deep-sea, cold-water inhabitants. They are the major scavengers of the deep-sea floor, drawn to carcasses in large numbers by their sense of smell.

A unique feature of hagfishes is the large mucus glands that open through the body wall to the outside. These so-called slime glands secrete enormous quantities of mucus and tightly coiled proteinaceous threads. The threads straighten on contact with seawater to entrap the slimy mucus close to the hagfish's body. An adult hagfish can produce enough slime within a few minutes to turn a bucket of water into a gelatinous mess (Martini 1998). This obnoxious behavior is apparently a deterrent to predators. When danger is past, the hagfish makes a knot in its body and scrapes off the mass of mucus, then sneezes sharply to blow its nasal passage clear.

Hagfishes lack any trace of vertebrae, which is one reason they are placed as the sister group of all other Vertebrata. Their internal anatomy shows many additional primitive features. The kidneys are simple, and there is only one semicircular canal on each side of the head. This characteristic has been the subject of much debate: Is this single semicircular canal a genuinely primitive feature, or does it represent a simplification of the vestibular apparatus with two canals seen in other agnathans? Northcutt (1985) pointed out that despite the single hagfish canal system, neurologically the canal represents double sensory patches just as in lampreys, which have two semicircular canals. This observation suggests that the single canal of hagfishes is a simplification of a system that originally had two canals, not a primitive character. Hagfishes have long been thought to lack a lateral line system. Recent work suggests that at least one genus has traces of the system, but whether this is an ancestral condition or a secondary reduction of ancestrally well developed structures is not known.

Hagfishes have a single terminal nasal opening that connects with the pharynx via a broad tube. The eyes are degenerate or rudimentary and covered with a thick skin. The mouth is surrounded by six tentacles that can be spread and swept to and fro by movements of the head when the hagfish is searching for food. In the mouth are two horny plates bearing sharp, horny (keratinous) tooth-like structures. These tooth plates lie to each side of a protrusible tongue and spread apart when the tongue is protruded. When the tongue is retracted the plates fold together, and the teeth interdigitate in a pincer-like action. The feeding apparatus of hagfishes has been described as "extremely efficient at reeling in long worms, because the keratin plates alternately flick in and out of the oral cavity" (Mallatt 1985).

When feeding on something the size and shape of another fish, hagfishes concentrate their pinching

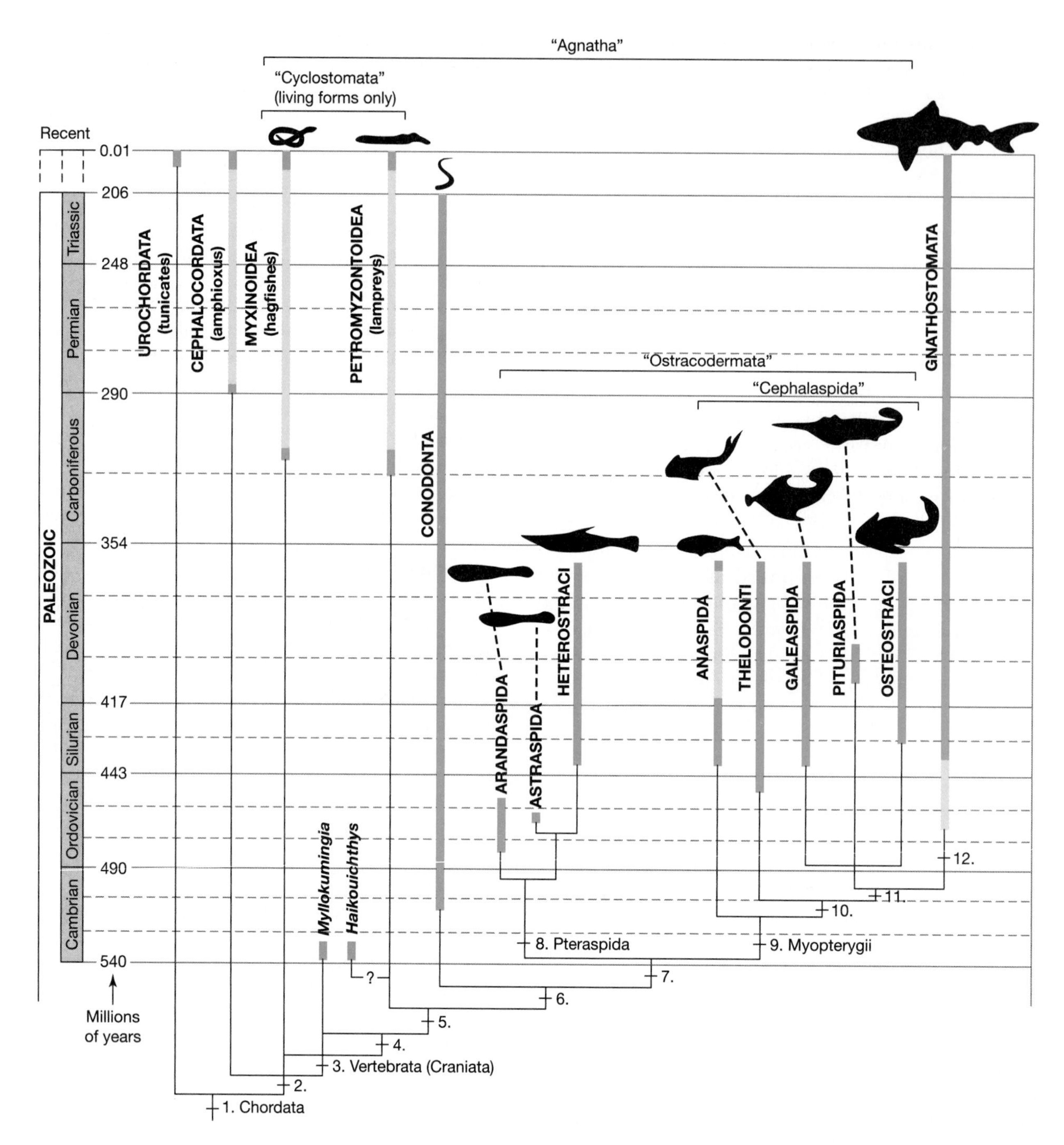

Legend: **1.** Chordata—Notochord; hollow dorsal nerve cord; postanal tail; segmental muscle bands in at least the tail region; endostyle. **2.** Myomeres in trunk and tail region; excretory system with podocyte cells; lateral plate mesoderm; embryological induction of neural tissues by the notochord; caudal fin fold, ventral to dorsal pattern of blood circulation through the gills; brain of some sort at end of nerve cord. **3.** Craniata (= Vertebrata in the broad sense)—Cranium incorporating anterior end of the notochord and enclosing brain and paired sensory organs; distinct head region with these characters—tripartite division of brain with a telencephalon and cranial nerves differentiated from neural tube, paired optic, auditory, and probably olfactory organs, one or more semicircular canals; neural crest cells and structures derived from these tissues (including cartilaginous gill skeleton); W-shaped myotomes; distinctive endocrine glands; lateral line system; well-developed heart; paired kidneys. **4.** Dorsal fin. **5.** Vertebrata (in the more restricted sense, possibly should be placed at node no. 5)—Presence of arcualia (vertebral rudiments surrounding the nerve cord); dorsal nasohypophysial opening; eyes well developed with extrinsic eye musculature; pineal eye; hypoglossal nerve; Mauthner neurons in the brain stem; sensory-line neuromasts; sensory lines on head and body; capacity for electroreception; two semicircular canals; autonomic innervation of the heart; renal collecting ducts; spleen or splenic tissue; three (versus one) types of granular white blood cells; dilute body fluids; blood comprises more than 10% of body volume; ion transport in gills; higher metabolic rate; pituitary control of melanophores and gameteogenesis. **6.** Physiological capacity to form mineralized tissues in the dermis; horny teeth lost. **7.** Dermal-bone head shield; olfactory tract; cerebellum in brain. **8.** Pteraspidomorpha—Paired nasal openings (= loss of nasohypophysial opening); dermal skeleton with characteristic three layers, including spongy bone. **9.** Myopterygii—Paired lateral fin folds, dorsal and anal fin; endolymphatic duct in inner ear opens to surface (lost in gnathostomes above level of Chondrichthyes). **10.** Sclerotic ossicles. **11.** Perichondral bone (at least in head); calcified cartilage; cellular dermal bone; three-layered exoskeleton; pectoral fins with a narrow, concentrated base (lost in galeaspids); large orbits; large head vein (dorsal jugular). **12.** Gnathostomata—Jaws and many other derived characters (see text), including paired nasal openings (= loss of nasohypophysial opening)

◀ Figure 3–3 Phylogenetic relationships of vertebrates within the Chordata. This diagram depicts probable relationships among primitive vertebrates, including living and extinct jawless vertebrates and the earliest jawed vertebrates. Dotted lines show interrelationships only; they do not indicate times of divergence nor the unrecorded presence of taxa in the fossil record. Lightly shaded bars indicate ranges of time when the taxon is known to be present, but is unrecorded (or poorly recorded) in the fossil record. The numbers indicate derived characters that distinguish the lineages.

efforts on surface irregularities, such as the gills or the anus, where they can more easily grasp the flesh. Once attached, they can tie a knot in their tail and pass it forward along their body until they are braced against their prey and can tear off the flesh in their pinching grasp. Hagfishes attack dead or dying vertebrate prey. They often begin by eating only enough outer flesh to enter the prey's coelomic cavity, where they dine on soft parts. Once a food parcel reaches the hagfish's gut, it is enfolded in a mucoid

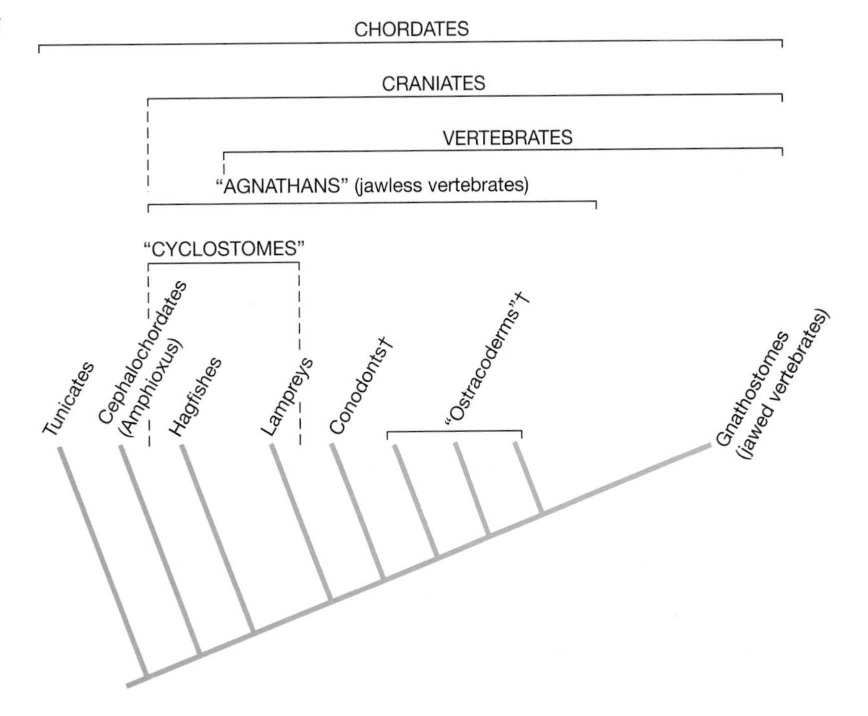

▶ Figure 3–4 Simplified cladogram of vertebrates within the Chordata, showing living taxa and major extinct groups only. Quotation marks indicate paraphyletic groups. A dagger (†) indicates an extinct taxon.

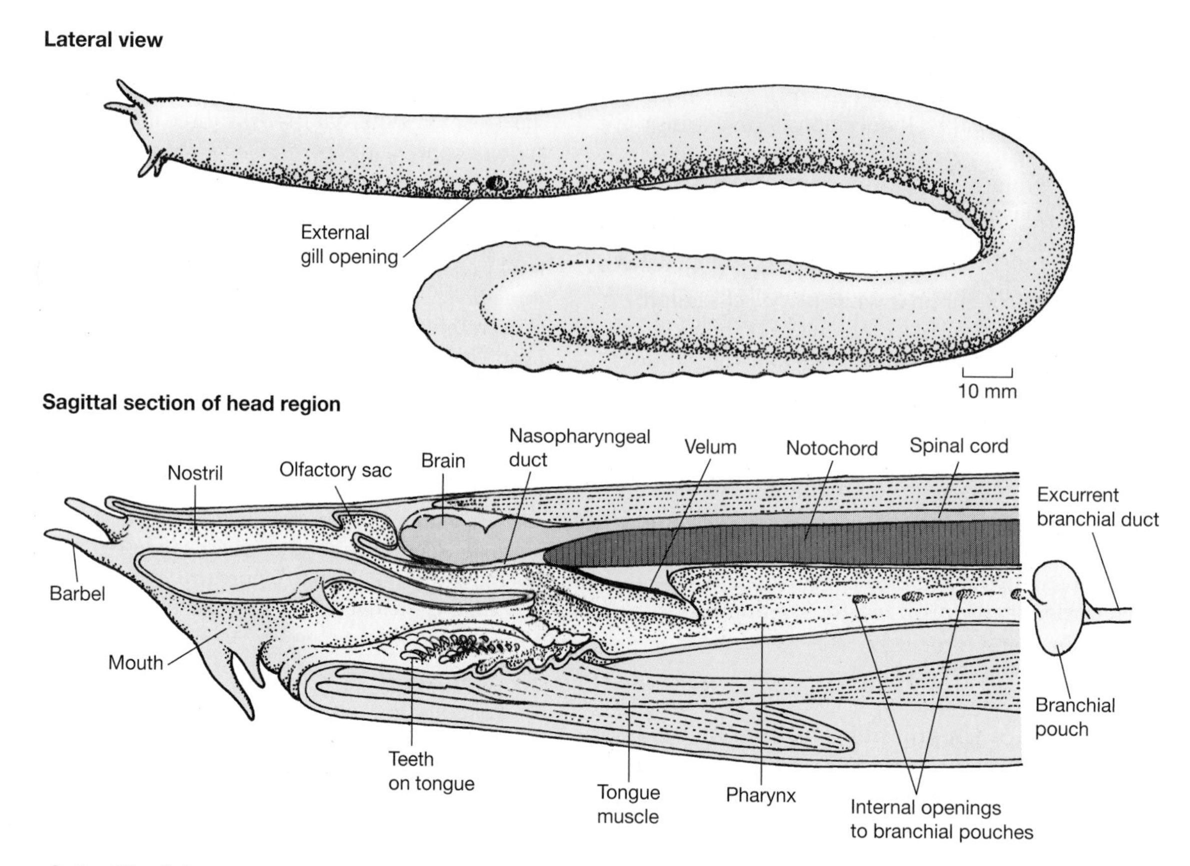

▲ Figure 3–5 Hagfishes.

bag secreted by the gut wall. This membrane is permeable to digestive enzymes and the products of digestion diffuse out to be absorbed by the gut. The indigestible parts of the prey are excreted still enclosed in the mucoid bag. The functional significance of this curious feature is unknown.

Different genera and species of hagfishes have variable numbers of external gill openings. From 1 to 15 openings occur on each side, but they do not correspond to the number of internal gills. The external openings occur as far back as the midbody, although the pouch-like gill chambers are just posterior to the head. In some genera, the long tubes leading from the gills fuse to reduce the number of external openings to which they lead. The posterior position of the gill openings may be related to burrowing.

Hagfishes have large blood sinuses and very low blood pressure. In contrast to all other vertebrates, hagfishes have accessory hearts in the liver and caudal regions in addition to the true heart near the gills. The several hearts of hagfishes are aneural, meaning that their pumping rhythm is intrinsic to the hearts themselves rather than coordinated via the central nervous system. In all these features hagfishes resemble the primitive condition of amphioxus—although, like other vertebrates, their blood does have red blood cells containing hemoglobin. The blood vascular system demonstrates few of the immune reactions characteristic of other vertebrates.

In most species female hagfishes outnumber males by a hundred to one; the reason for this strange sex ratio is unknown. Examination of the gonads suggests that at least some species are hermaphroditic, but nothing is known of mating. Neither hagfishes nor lampreys have specialized reproductive ducts; eggs and sperm are released into the coelom, and leave the body through pores opening into the cloaca. The yolky eggs, which are oval and over a centimeter long, are encased in a tough, clear covering that is secured to the sea bottom by hooks. The eggs are believed to hatch into small, completely formed hagfishes, bypassing a larval stage. Unfortunately, almost nothing is known of the embryology or early life history of any hagfish. Fertile eggs from which the development of hagfish anatomy can be studied have not been found since the nineteenth century!

Hagfishes are never caught much above the bottom of the ocean, and they are often found in deep regions of the continental shelf. Some live in colonies, each individual in a mud burrow marked in some species by a volcano-like mound at the entrance. Polychaete worms and shrimps are found in the guts of many species, and hagfishes probably live a mole-like existence, finding their prey beneath the ooze and at its surface. They must be active when out of their burrows, for they are quickly attracted to bait and fishes caught in gill nets. The small amount of morphological differences between populations indicates that hagfishes are not wide ranging, but rather tend to live and breed locally.

An increased economic interaction between hagfishes and humans over the past two decades is but one example of how vertebrate species are threatened by a burgeoning and highly consumptive human society. Fishermen using stationary gear such as gill nets have long been pulling in fishes damaged beyond sale by scavenging hagfishes. It is not surprising that the fishermen responded quickly when the use of hagfish skins for leather made hagfishes as valuable as the fish they had previously been seeking. Almost all so-called eel-skin leather products are made from hagfish skin. Worldwide demand for this leather has eradicated economically harvestable hagfish populations in Asian waters and in some sites along the West Coast of North America. Current fishing efforts are focusing on South America and the North Atlantic, where hagfishes may still be highly abundant. By one estimate, the Gulf of Maine contains a population density of hagfishes of 500,000 per square kilometer (Martini 1998).

Human exploitation of natural resources, such as fisheries, typically depletes stocks—because no attention is given to the biology of the resource and its renewable, sustainable characteristics. For example, we do not know how long hagfishes live; how old they are when they first begin to reproduce; exactly how, when, or where they breed; where the youngest juveniles live; what are the diets and energy requirements of free-living hagfishes; or virtually any of the other information needed for good management. As a result, eel-skin wallets will probably become as rare as items made of whalebone (baleen), tortoiseshell, and ivory.

Lampreys—Petromyzontoidea

There are 40 to 50 species of lampreys (Figure 3–6) in two major genera (*Petromyzon* and *Lampetra*). Although lampreys are similar to hagfishes in size and shape, they are radically different in certain other respects. They have many features lacking in hagfishes, but shared with gnathostomes. Perhaps most important, they have vertebral structures (arcualia)—although these cartilaginous skeletal elements are minute and homologous only with the neural arches of gnathostome vertebrae.

Most lampreys are parasitic on other fishes. They attach to the body of another vertebrate (usually a larger bony fish) by suction, and rasp a shallow, seeping wound through the integument of the host. The round mouth and tiny esophagus are located at the bottom of a large fleshy funnel (the oral hood), the inner surface of which is studded with horny conical spines. The protrusible tongue-like structure is covered with similar spines, and together these structures allow tight attachment and rapid abrasion of the host's integument. This tongue is not homologous with the tongue of gnathostomes, because it is innervated by cranial nerve V rather than nerve XII as in gnathostomes. An oral gland secretes an anticoagulant that prevents the victim's blood from clotting. Feeding is probably continuous when a lamprey is attached to its host. The bulk of an adult lamprey's diet consists of body fluids of fishes. The digestive tract is straight and simple, as one would expect for an animal that feeds on such a rich and easily digested diet as blood and tissue fluids. Lampreys generally do not kill their hosts, but leave a weakened animal with an open wound. At sea, lampreys feed on several species of whales and porpoises in addition to fishes. Swimmers in the Great Lakes, after having been in the water long enough for their skin temperature to drop, have reported initial attempts by lampreys to attach to their bodies.

Lampreys are unique among living vertebrates in having a single nasal opening situated on the top of the head, combined with a duct leading to the hypophysis (pituitary) and known as a nasohypophysial opening (Figure 3–6). Development of this structure involves distortion of the front of the head, and its function is not known. Several groups of ostracoderms had an apparently similar structure, which may have evolved convergently in those groups. The eyes of lampreys are large and well developed, as is the pineal body, which lies under a

Lateral view of an adult

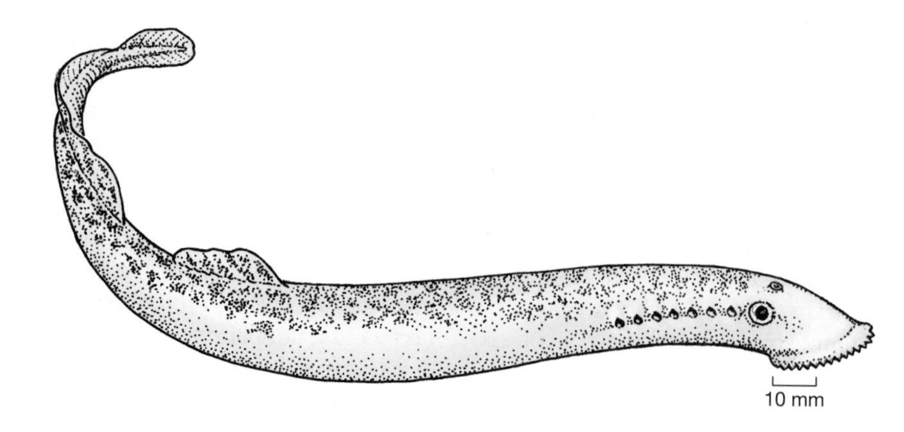

Sagittal section of head region

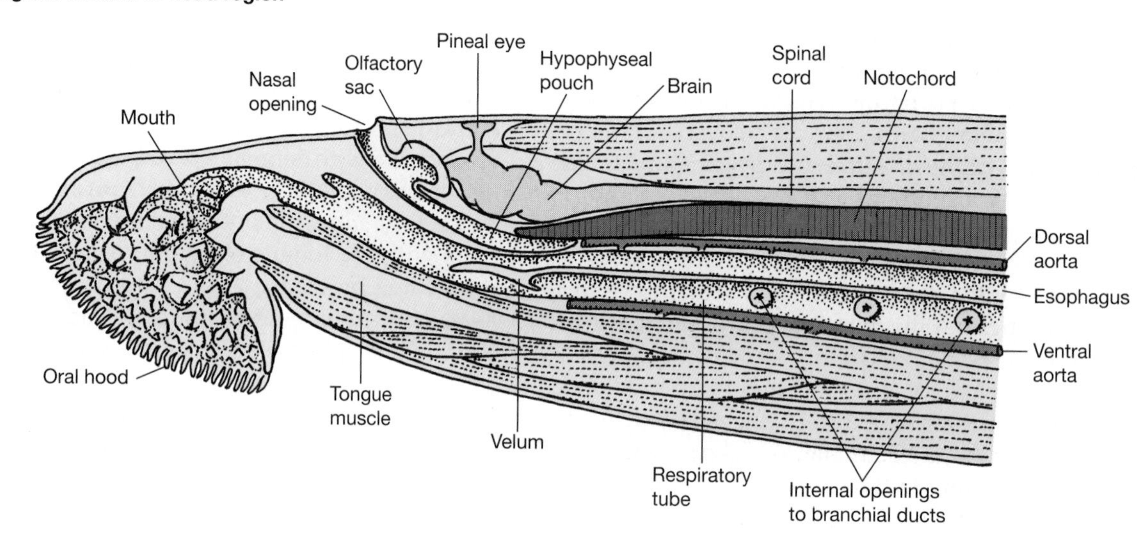

Larval lamprey (ammocete)

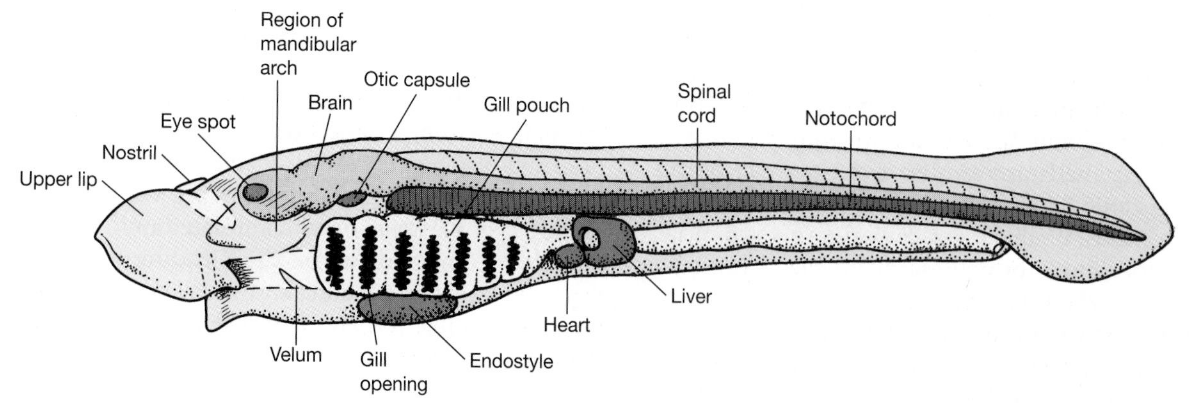

▲ Figure 3–6 Lampreys.

pale spot just posterior to the nasal opening. In contrast to hagfishes, lampreys have two semicircular canals on each side of the head—a condition shared with the extinct ostracoderms. In addition, the heart is not aneural as in hagfishes, but is innervated by the parasympathetic nervous system. In lampreys these nerves cause cardiac acceleration, not deceleration—as do the neural regulators of heart rate in all other vertebrates. Chloride cells in the gills and well-developed kidneys regulate ions, water, and nitrogenous wastes and maintain the osmolality of the body fluids, allowing the lamprey to exist in a variety of salinities.

Lampreys have seven pairs of gill pouches that open to the outside just behind the head. Ventilation of these gills differs from the condition in all other vertebrates, where water is drawn in the mouth and then pumped out over the gills in continuous or **flow-through ventilation**. Adult lampreys spend much of their time with their sucker-like mouths affixed to the bodies of other fishes, and during this time they would be unable to ventilate the gills in this flow-through fashion. Instead they employ **tidal ventilation**; water is both drawn in and expelled through the gill slits, and there is a respiratory tube leading off from the pharynx, separated from the esophagus. A flap called the velum prevents water from flowing out of the respiratory tube into the mouth (Figure 3–6). This design is analogous to the separation of the trachea (leading to the lungs) and esophagus (leading to the stomach) in ourselves and in other tetrapods, but our mode of tidal ventilation is in and out of the mouth. In comparison with jawed fishes, the lamprey's mode of ventilation is not very efficient at oxygen extraction. However, it is a necessary compromise given lampreys' specialized parasitic mode of feeding.

Lampreys have a worldwide distribution except for the tropics and high polar regions. Nearly all lampreys are **anadromous**; that is, they live as adults in oceans or big lakes, and ascend rivers and streams to breed. Some of the most specialized species are known only from freshwater, where the adults neither feed nor migrate, and act solely as a reproductive stage in the life history of the species. Anadromous species that spend some of their life in the sea attain the greatest size, although 1 meter is about the upper limit. The smallest species are less than one-fourth that size. Little is known of the habits of adult lampreys because they are generally observed only during reproductive activities or when captured with their host. Despite their anatomically well developed senses, no clear picture has emerged of how a lamprey locates or initially attaches to its prey. In captivity, lampreys swim sporadically with exaggerated, rather awkward lateral undulations.

Female lampreys produce hundreds to thousands of eggs, about a millimeter in diameter and devoid of any specialized covering such as that found in hagfishes. Like hagfishes, however, lampreys have no duct to transport the specialized products of the gonads to the outside of the body. Instead, the eggs or sperm fill the coelom, and contractions of the body wall expel them from pores located near the openings of the urinary ducts. Fertilization is external.

Lampreys spawn after a temperature-triggered migration to the upper reaches of streams where the current flow is moderate and the stream bed is composed of cobbles and gravel. Male and female lampreys construct a nest by attaching themselves by their mouths to large rocks and thrashing about violently. Smaller rocks are dislodged and carried away by the current. The nest is complete when a pit is rimmed upstream by large stones, downstream by a mound of smaller stones, and by sand that produces eddies. Water in the nest is oxygenated by this turbulence, but does not flow strongly in a single direction. The weary nest builders spend the last of their energy depositing eggs and sperm, a process that may take two days. The female attaches to one of the upstream rocks, laying eggs, and the male wraps around her, fertilizing them as they are extruded. Adult lampreys die after breeding once.

The larvae hatch in about 2 weeks. The larvae are radically different from their parents, and were originally described as a distinct genus, *Ammocoetes* (Figure 3–6c). This name has been retained as a vernacular name for the larval form. A week to 10 days after hatching, the tiny 6- to 10-millimeter-long ammocoetes leave the nest. They are wormlike organisms with a large, fleshy oral hood and nonfunctional eyes hidden deep beneath the skin. Currents carry the ammocoetes downstream to backwaters and quiet banks, where they burrow into the soft mud or sand and spend three to seven years as sedentary filter feeders. The protruding oral hood funnels water through the muscular pharynx, where food particles are trapped in mucus and carried to the esophagus. An ammocoete may spend its entire larval life in the same bed of sediment, with no major morphological or behavioral change until it is 10 cm or more in length and several years old. Metamorphosis begins in midsummer, and produces a silver-gray juvenile ready to begin its life as a parasite. Downstream migration to a lake or the sea may not occur until the spring following metamorphosis. Adult life is usually no more than two years, and many lampreys return to spawn after one year. Some lamprey species lack parasitic adults. The larvae of these species metamorphose, and the new adults leave their burrows to spawn immediately and die.

During the past 100 years humans and lampreys have increasingly been at odds. Although the sea lamprey, *Petromyzon marinus*, seems to have been indigenous to Lake Ontario, it was unknown from the other Great Lakes of North America before 1921. The St. Lawrence River—flowing from Lake Ontario to the Atlantic Ocean—was no barrier to colonization by sea lampreys, and the rivers and streams that fed into Lake Ontario held landlocked populations. During their spawning migrations, lampreys negotiate waterfalls by slowly creeping upward using their sucking mouth, but the 50-meter height of Niagara Falls (between Lake Ontario and Lake Erie) was too much for even the most amorous lampreys. Even after the Welland Canal connected Lakes Erie and Ontario in 1829, lampreys did not immediately invade Lake Erie; it took a century for lampreys to establish themselves in Lake Erie's drainage basin.

From the 1920s to the 1950s lampreys expanded rapidly across the entire Great Lakes basin. The surprising fact is not that they were able to invade the upper Great Lakes, but that it took them so long to initiate the invasion. Environmental conditions that vary among the lakes may provide the answer to this curious delay. Lake Erie is the most eutrophic and warmest of all the lakes, and it has the least appropriate feeder streams for spawning. Many of these streams run through flat agricultural land that has been under intensive cultivation since the early nineteenth century. The streams are silty and frequently have had their courses changed by human activities. Because of the terrain, flow is slow and few rocky or gravel bottoms occur. Perhaps lampreys simply could not find appropriate spawning sites in Lake Erie to develop a strong population.

Once they reached the upper end of Lake Erie, however, lampreys quickly gained access to the other lakes. There they found suitable conditions, and by 1946 they inhabited all the Great Lakes. Lampreys were able to expand unchecked until sporting and commercial interests became alarmed at the reduction of economically important fish species such as lake trout, turbot, and lake whitefish. Chemical lampricides as well as electrical barriers and mechanical weirs at the mouths of spawning streams have been employed to bring the Great Lakes lamprey populations down to their present level. Although the populations of large-jawed fish species, including those of commercial value, are recovering, it may never be possible to discontinue these antilamprey measures, costly though they are. Human mismanagement (or initial lack of management) of lampreys has been to our own disadvantage. The story of the demise of the Great Lakes fishery is but one of hundreds in the recent history of vertebrate life where human failure to understand and appreciate the interlocking nature of the biology of our nearest relatives has led to gross changes in our environment. Introduction of exotic (not indigenous) species is a primary cause for the decline of many vertebrate species worldwide, especially in aquatic habitats.

The Importance of Extant Primitive Vertebrates in Understanding Ancient Ones

The fossil record of the first vertebrates reveals little about their pre-Silurian evolution, and it yields no undisputed clues about the evolution of vertebrate organization from the condition in nonvertebrate chordates. However, hagfishes and lampreys provide examples of surviving primitive vertebrates, representatives of the early agnathous vertebrate radiation.

Although the biology and development of hagfishes is less well known than that of lampreys, hagfishes are especially important to our understanding of the earliest vertebrate condition. Hagfishes retain more primitive features than any known vertebrate, living or fossil. In contrast, some researchers would still prefer to link hagfishes and lampreys together as sister taxa, either on molecular or morphological grounds. Jon Mallatt (1996, Mallatt and Sullivan 1998) especially notes in both hagfishes and lampreys the similarity of the elongated pharyngeal cavity, housing a powerful muscular tongue supported by a cartilaginous lingual apparatus, and a respiratory flap called the velum. An alternative interpretation is that these oral features evolved convergently, or perhaps represent a retained primitive feature inherited from the earliest vertebrates.

Table 3.1 describes some features of hagfishes that are more primitive than those of other living vertebrates, including lampreys. Many apparently primitive features of hagfishes, such as their virtual lack of eyes and associated nerves and muscles, are probably secondary features associated with mud-burrowing habits. Variation in character states among hagfishes is more common than many researchers have realized. *Myxine* and its relatives appear to be considerably more derived than other hagfishes, and comparative studies offer the most promising path to understanding what is primitive about hagfishes and what is not. However, two additional features of hagfishes suggest an extremely primitive phylogenetic position.

First, hagfish body fluids are isosmolal with seawater, while other living vertebrates have dilute body fluids. Concentrated body fluids would preclude survival in freshwater habitats, but all major groups of ostracoderms have freshwater representatives—implying that they, too, had dilute body fluids. Thus, in this feature, hagfishes are more primitive than all ostracoderms.

A second important feature is the anatomy of the hagfish taste sensory system. In other vertebrates the taste buds are in the oropharyngeal epithelium and innervated by cranial nerves VII, IX, and X. In hagfishes, however, specialized sensory buds are found in the epidermis and are innervated by mainly by cranial nerve V and by spinal nerves. In addition, the brain area that receives nervous input from the taste sensors is different in hagfishes and other vertebrates. The hagfishes' system must have evolved independently from the sense of taste in lampreys and gnathostomes, suggesting an early divergence for hagfishes from the lineage leading to other vertebrates.

3.3 The Radiation of Paleozoic Jawless Vertebrates—"Ostracoderms"

Ostracoderms encompass several distinct lineages, each with recognizable subdivisions, but our understanding of exactly how the different ostracoderm taxa are related to one another and to living

Table 3.1 Features of Hagfishes that are More Primitive than those of Lampreys and Gnathostomes

1. Have very small, paired eyes; extrinsic eye muscles and associated cranial nerves to move the eyes are lacking (but see discussion in text).
2. Cranium made of a sheath of fibrous tissue rather than of cartilage.
3. Lack a pineal eye.
4. Lack the facility of electroreception.
5. Apparent lack of lateral line sensory system (but canals, lacking neuromasts, are present in the head of some species).
6. A single semicircular canal only (organs of balance, one on each side of the head) in the inner ear (versus two in lampreys and ostracoderms, and three in gnathostomes)—see discussion in text.
7. No hypoglossal nerve.
8. No Mauthner neurons in brain stem.
9. No true taste buds.
10. Circulatory system retains accessory "hearts."
11. Only one type of granular white blood cell (three types in lampreys and gnathostomes).
12. Lack autonomic innervation of the heart (but vagal innervation speeds, rather than slows, the heart in lampreys; may have evolved convergently with gnathostome condition).
13. Lack any type of spleen or splenic precursor.
14. Lack muscles in the midline fins.
15. Lack cartilaginous vertebral elements.
16. Lack renal collecting ducts in the kidney.
17. Have an open connection between the pericardial cavity and the coelom (closed in lampreys and gnathostomes).
18. Have body fluids isosmolal with seawater (see text).

vertebrates is in a considerable state of flux (Janvier 1996, Maisey 1996, Donoghue et al. 2000). All ostracoderms are characterized by the presence of a covering of dermal bone, usually in the form of an extensive armored shell or **carapace**, but sometimes in the form of smaller plates or scales.

The ostracoderms represent a paraphyletic assemblage, because some more derived types are clearly more closely related to the gnathostomes (jawed vertebrates) than others (see Figure 3–3). Thus the term *ostracoderm* is often in quotation marks. Ostracoderms are more derived than extant agnathans: They had dermal bone; and impressions on the dorsal head shield suggest that they had an olfactory tract connecting the olfactory bulb with the forebrain, and a cerebellum in the hindbrain. Living agnathans lack a cerebellum; and their olfactory bulbs are incorporated within the rest of the forebrain, rather than placed more anteriorly and linked to the head via the olfactory tract (= cranial nerve I, but actually an outgrowth of the brain rather than a true nerve).

Except for a few small early forms, ostracoderms ranged in length from about 10 cm to more than 50 cm. Although they lacked jaws, some apparently had various types of movable mouth plates that lack analogues in any living vertebrates. These plates were arranged around a small, circular mouth that appears to have been located farther forward in the head than the larger, more gaping mouth of jawed vertebrates. Although most ostracoderms had some sort of midline dorsal fin, only the more derived forms had any form of lateral paired appendages. Where lateral fins are preserved, only anterior (pectoral) fins are known. No trace of an endoskeleton has been found within the fins, but they may have had a cartilaginous internal fin skeleton that was not fossilized, and some ostracoderms have evidence of a pectoral limb girdle. Their respiratory apparatus consisted of a variable number of separate pharyngeal gill pouches that opened along the side of the head, usually independently—but opening through a common passage in one group (heterostracans). As in living jawless vertebrates, the notochord must have been the main axial support throughout adult life. Figure 3–7 depicts some typical ostracoderms.

The fossil record offers us some indisputable information about ostracoderms; they had bone and other mineralized tissues, and we can make inferences about their soft anatomy by analogy with the living jawless vertebrates. The living forms appear to be more primitive types of vertebrates than ostracoderms, so any characters they share with gnathostomes would have been present in the ostracoderms too. For example, ostracoderms must have had a muscular pump rather than a ciliary one for ventilating the gills, and they must have used their gills for respiration rather than for filter feeding.

It was once assumed that, lacking jaws, early vertebrates also engaged in some form of filter feeding. However, more recent interpretations of the anatomy of ostracoderms indicate that the adults, at least, were predators. The muscular pharynx would have produced a flow of water powerful enough to suck in large objects. Ostracoderms probably ate small, slow-moving, bottom-living prey.

During the Late Silurian and the Devonian most major known groups of extinct agnathans coexisted with early gnathostomes. Approximately 50 million years of coexistence makes it highly unlikely that ostracoderms were pushed into extinction by the radiation of gnathostomes. Agnathous and gnathostomatous vertebrates appear to represent two different basic types of animals, possibly exploiting different types of resources. The extinction of the ostracoderms in the Late Devonian occurred at the same time as mass extinctions among many marine invertebrates and was probably a chance event. Gnathostomes also suffered in the Late Devonian mass extinctions, and an entire lineage (the placoderms) became extinct at the end of the Devonian.

Heterostracans and Other Primitive Ostracoderms

The **Heterostraci** (Greek *hetero* = different and *ostrac* = shell), are grouped with the Ordovician ostracoderms previously discussed in the Pteraspidomorpha (Figure 3–3). The pteraspidomorphs (Greek *ptera* = wing and *aspid* = shield) are united by a distinctive three-layered exoskeleton containing spongy bone (= spongy aspidin), and they also had oral plates. Heterostracans are known with certainty from the Early Silurian. They had their major radiation in the Late Silurian and Early Devonian in North America, Europe, and Siberia, and survived into the Late Devonian. Impressions on the inside of the dorsal plate suggest that the brain had two separate olfactory bulbs. Because it is assumed that these bulbs were connected with two separate nasal openings, as seen in gnathostomes, these primitive vertebrates are also called the Diplorhina (Greek *diplo* = two and *rhin* = nostril) in some early classifications.

The heterostracan *Pteraspis* (Late Silurian)

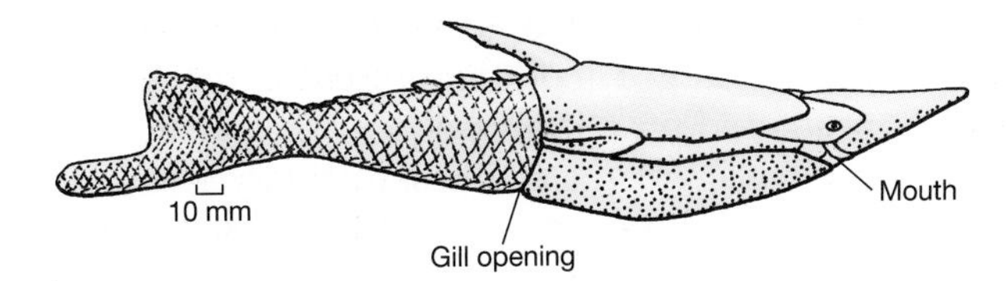

The psammosteid heterostracan *Drepanaspis* (dorsal view) (Early Devonian)

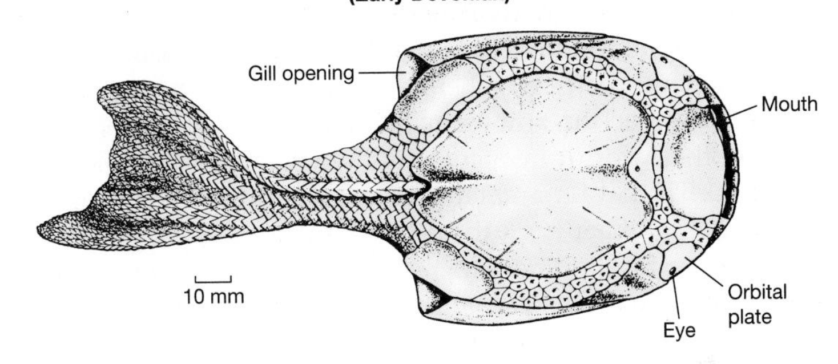

The anaspid *Pharyngolepis* (Late Silurian)

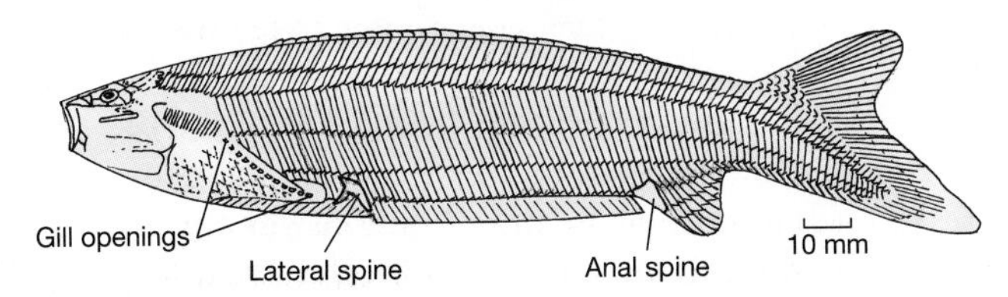

The osteostracan *Hemicyclaspis* (Late Devonian)

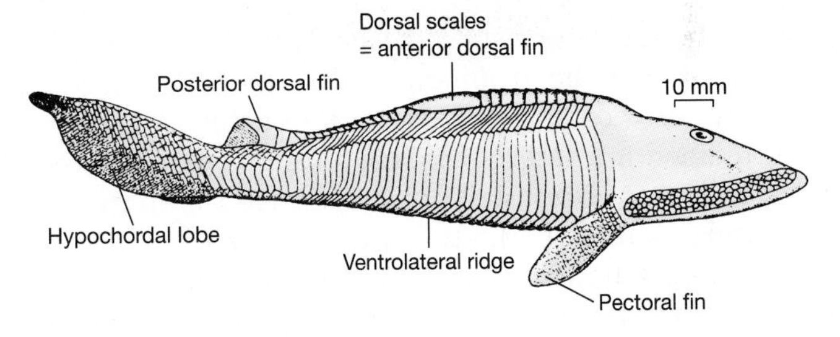

Reconstruction of the brain and cranial nerves of the Early Devonian osteostracan *Kiaeraspis*

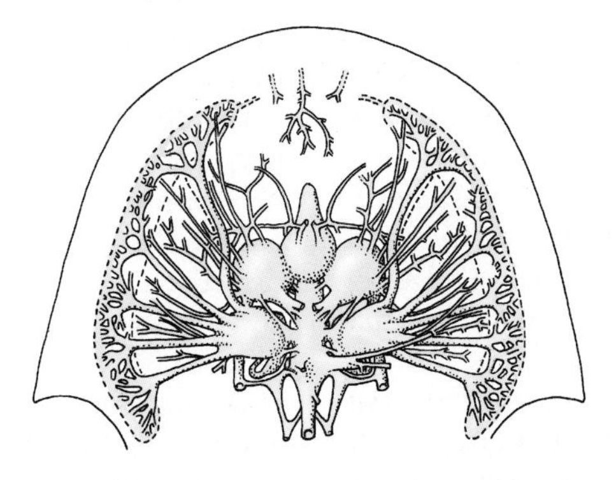

▲ Figure 3–7 Ostracoderm diversity 1.

Current opinion holds that the double nasal openings in heterostracans and gnathostomes were derived separately from the more primitive condition of a single opening, as seen in other ostracoderms and in living agnathans.

Heterostracans, which ranged in size from 10 cm to 2 m, were encased by articulating bony plates that extended to the anus in some groups (Figure 3–7). Over the anterior one-third of their body, all heterostracans had an essentially solid carapace, pierced by the mouth and a single pair each of eyes and external gill openings. They lacked any form of paired appendages, and their dorsal and anal fins were not well developed. There was a small opening in the middle of the dorsal plate for a third (median) eye, also known as the pineal organ. The terminal mouth was rimmed on its lower border by as many as two rows of small plates and is thought to have opened like a scoop with a V-shaped cross section.

The head carapace had an ornamented dorsal plate, one or more lateral plates, and several large ventral elements. Thus heterostracans provide the earliest recorded solution to the growth-in-a-suit-of-armor problem of these early shelled vertebrates. Either the large articulating plates did not form until maximum size was attained, or numerous centers of bone formation enlarged circumferentially and fused into plates or a solid shield only as the animal reached maximum size.

Posterior to the anus was a short, probably mobile tail covered by smaller, protruding barb-like plates. The lower lobe of the tail was disproportionately large and contained the notochord. This tail-fin construction is called **hypocercal.** In early heterostracans—the pteraspids—the body was generally round in cross section, like that of a tadpole, with little sign of stabilizing projections. Possibly these fishes were erratic swimmers and lacked precisely controlled locomotion. While feeding they may have oriented head down and plowed their jawless mouths along the bottom sediments.

There were a number of evolutionary trends within the heterostracans. During their later history, the bottom-dwelling (**benthic**) species had a body that was flattened ventrally and arched or rounded dorsally. The head shield developed solid, lateral, wing-like stabilizing projections called **cornua** (= horns). The head shield became shorter, and the bony covering became restricted to the anterior end except for a dorsal ridge of plates. Although specialized edges around the mouth for biting and grasping did not evolve, some of the oral plates developed enlarged tooth-like projections that may have been used for scraping.

The heterostracan lineage also gave rise to species of bizarre appearance. Some developed enormous cornua that may have acted as water-planing surfaces (hydrofoils) to produce lift for the heavy head when swimming. Some forms, the amphiaspids, fused the bones of their head shield to form a single solid carapace. The eyes were reduced or lost, and some had a snorkel-like tube carrying the mouth out in front of the carapace. These features suggest that amphiaspids were burrowing mud-dwellers. Other forms, the psammosteids, were flattened with widespread eyes and a dorsally directed mouth. These were probably bottom-dwelling forms, as are modern flattened fishes such as skates and rays.

More Derived Ostracoderms—Cephalaspida

Other groups of vertebrates appeared at the same time as the heterostracans, and they also had their major diversification in the Late Silurian and Early Devonian. Several distinct groups have sometimes been united as the **Cephalaspida** (Greek *cephal* = head and *aspid* = shield) or Cephalaspidomorpha. Currently, however, it seems that this is a paraphyletic assemblage, and that these taxa represent the sister group to jawed vertebrates (see Figure 3–3). These fishes share with gnathostomes some sort of paired appendages, either lateral fin folds or pectoral fins. For this reason some paleontologists place the Cephalaspida with the gnathostomes, in a group called the Myopterygii.

The Cephalaspida had a single, large, dorsal nasal opening—lying in the center of the head anterior to the eyes, and usually associated with the pineal gland. This single nasal opening is the basis for another name, Monorhina, sometimes applied to this assemblage of ostracoderms (Greek *mono* = one and *rhin* = nostril). However, because lampreys have the nasal opening in a similar position, this condition may represent a relatively primitive vertebrate feature. A reversal of this feature, and the acquisition of paired nasal openings, of course, had to have occurred in the immediate ancestry of gnathostomes, and also convergently in the ancestry of the heterostracans.

A major group of cephalaspids is the **Osteostraci** (Greek *osteo* = bone) from the mid-Silurian to Late Devonian of North America, Europe, and Siberia. A second is the **Galeaspida** (Greek *gale* = helmet and *aspid* = shield) from the Early Silurian to Late Devonian of China and northern Vietnam. A third group is the recently described **Pituriaspida** from the Middle Devonian of Australia. (*Pituri* is an Aboriginal word for a hallucinogenic plant, and refers to the bizarre shape of these animals.) The fourth group is the **Anaspida** (Greek *an* = without), primarily from the Early Silurian of North America, Europe, and China, although a few Late Devonian forms are known from Canada.

The anaspids appear to be the least derived of the Cephalaspida; more derived ostracoderms share with gnathostomes the feature of sclerotic ossicles (little bones around the eyes) and a lateral line system set in definite canals. Anaspids are found predominantly in freshwater sediments. They had minnow-like body proportions (Figure 3–7) and were about 15 cm long. Their bodies were covered in narrow scale rows, and the heads of most species were either naked or covered by a complex of small plates. Their tails were hypoceral, like the tails of heterostracans but unlike those of the osteostracans. Anaspids are thought to have been benthic detritus feeders or scrapers of algal film from rocks, and they fed in a head-down position. Their stabilizing dorsal, anal, and lateral projections or folds, the spines and scutes associated with these projections, and the compressed shape of their fusiform bodies probably allowed an agility and locomotor capacity not known in the heterostracans or osteostracans.

Like heterostracans, osteostracans were heavily armored (Figure 3–7). However, the bone contained lacunae or spaces for bone cells; and their head shield was a single, solid element devoid of sutures on its dorsal surface. While most heterostracans showed evidence of periodic growth around the margins of their shield plates, the solid construction of the shield in osteostracans—and the absence of growth marks—indicate that their head shield did not grow throughout life. Furthermore, all individuals in a species of osteostracan were about the same size. Perhaps osteostracans had naked larvae, like a lamprey ammocoete, and then metamorphosed into a stage where a head shield and other bony armor were deposited without further growth.

Osteostracans had a **heterocercal** tail, in which the lobe above the midline of the body was larger and stiffer than the lower lobe (a condition more precisely called **epicercal**). Their heterocercal tail may have resulted in a locomotor system that provided considerable lift, increasing their overall mobility. Early osteostracans had extensive shields and no paired lateral stabilizers. Later types had short shields, movable paddle-like extensions of the body in the position of pectoral appendages, and horn-like extensions of the head shield just anterior to these paddles (Figure 3–7).

The osteostracans are used as models of early jawless vertebrates even though they are fairly derived forms. Two features have made them better known than other extinct agnathans. The first is their single-piece head shield, which resists disintegration better than a series of articulated plates. Second, within this shield the inner surface of the braincase and the channels are lined by thin layers of periosteal bone, a type of bone related to endochondral bone. (The presence of this derived character is one feature grouping osteostracans more closely with gnathostomes than other ostracoderms.) The internal features of the braincase, with channels and **foramina** (small openings for the passage of nerves and blood vessels; singular = **foramen**) are preserved in sufficient detail to allow reconstruction of the soft anatomy of the head. Eric Stensio and his collaborators in Sweden and England patiently polished and ground away layer after layer of the these fossils, taking photographs of each successive layer until they had serial photographs through the complete head shield. From these photos, the researchers could trace in three dimensions the canals and cavities that in life had been lined with bone (Figure 3–7). The internal anatomy of the brain and nervous system of osteostracans 400 million years old is very similar to that found in modern lampreys, suggesting that both these animals preserve the primitive vertebrate condition.

There were peculiar fields of small, thin, irregular plates along the dorsolateral edges of the head shield, and sometimes in the center behind the pineal opening. These fields formed depressions connected to the inner ear cavity inside the braincase by huge canals that ran through the shield. This combination of features has been interpreted as an electroreceptive system or as a sensory expansion of the membranous labyrinth of the inner ear to detect vibrations in the water outside of the animal. The osteostracans' success may relate, in part, to these unique adjuncts to their nervous system.

Two apparently geographically isolated groups of Devonian cephalaspids have come to light during the past few decades. Galeaspids are known from southern China and northern Vietnam. They differed from the osteostracans in lacking paired fins—possibly representing secondary loss—but they had lateral spines that may have provided body stability in the manner of a pectoral fin. They also had a large slit-, bean-, or even heart-shaped inhalant opening on the dorsal surface of their head shield, connecting to the pharynx.

The pituriaspids are the most recently described group of ostracoderms, found in the mid-Devonian of Australia (Figure 3–8). Superficially they resemble galeaspids; they are also thought to have had paired pectoral appendages containing muscles like those of osteostracans.

Problematic Ostracoderms—Thelodonti

Thelodonts differ from other ostracoderms in their body covering. Rather than having large plates, these small fishes (Figure 3–8) were covered by numerous tiny **denticles**—small tooth-like structures not unlike those of living sharks. Isolated denticles from the Ordovician also belong to these fishes, and whole-body fossils are common in the Late Silurian and Early Devonian. On the assumption that all poorly armored ostracoderms of this type are related, several names have been given to this group of fishes, all referring to characteristics of the scales: Thelodonti (Greek *thelo* = nipple and *dont* = tooth) and Coelolepida (Greek *coel* = hollow and *lepida* = scale) are the names most commonly used.

The phylogenetic relationships of thelodonts are uncertain. Their distinctive small, separate scales are of no value in determining who they are related to. In fact, a body covered by small bony elements is exactly what has been hypothesized as the most primitive condition for all ostracoderms. Perhaps "thelodonts" represent an unrelated assemblage of the remains of ostracoderms lacking heavy dermal armor. Other schemes place them as the sister taxon to the osteostracans as well as the gnathostomes, since some forms appear to have sclerotic ossicles (Figure 3–3).

Thelodonts were between 10 and 20 cm long. They were most numerous in coastal estuaries, but eventually also radiated into freshwater. Some forms were fusiform (torpedo-shaped) with ridges corresponding to the dorsal and anal fins in more derived fishes, lateral broad-based flanges, and a hypocercal tail (Figure 3–8). Other forms recently discovered in northwestern Canada have deep, laterally compressed bodies with symmetrical and deeply forked tails like those of midwater fishes (Figure 3–8). Most interesting in these forms is evidence of a well-developed stomach, a feature previously thought to be a derived character of gnathostomes. These fork-tailed forms have been placed within their own order, the Furcacaudiformes, related to other thelodonts but separate from them (Wilson and Caldwell 1998).

The small, lightly armored thelodonts were probably very different behaviorally from the other larger, armored, heavy-bodied ostracoderms; and they were probably better swimmers. Living fishes shaped like thelodonts feed by swimming along the bottom of the sea. Some of the thelodonts had pharyngeal denticles and are thought to have crushed prey and trapped the fragments as some extant fishes do with their pharyngeal teeth and gill rakers.

3.4 The Transition from Jawless to Jawed Vertebrates

Gnathostomes were once viewed as an entirely separate radiation from the agnathans, but we now consider that they originated from within the agnathan radiation. Gnathostomes are considerably more derived than agnathans, not only in their possession of jaws but also in many other ways.

Albert Sherwood Romer (1967) proposed that "perhaps the greatest of all advances in vertebrate history was the development of jaws and the consequent revolution in the mode of life of early fishes." Jaws allow a variety of new feeding behaviors, including the ability to grasp objects firmly, and along with teeth enable the animal to cut food to pieces small enough to swallow or to grind hard foods. New food resources became available when vertebrates evolved jaws, and many gnathostomes became larger than contemporary jawless vertebrates. A grasping, movable jaw also permits manipulation of objects: Jaws are used to dig holes, to carry pebbles and vegetation to build nests, and to grasp mates during courtship and juveniles during parental care. Later in this chapter we will discuss the proposal that jaws were originally evolved for more forceful ventilation of the gills.

The homologies of jaws with the anterior gill arches (= branchial or visceral arches) of jawless vertebrates were described in Chapter 2. Figure 3–9

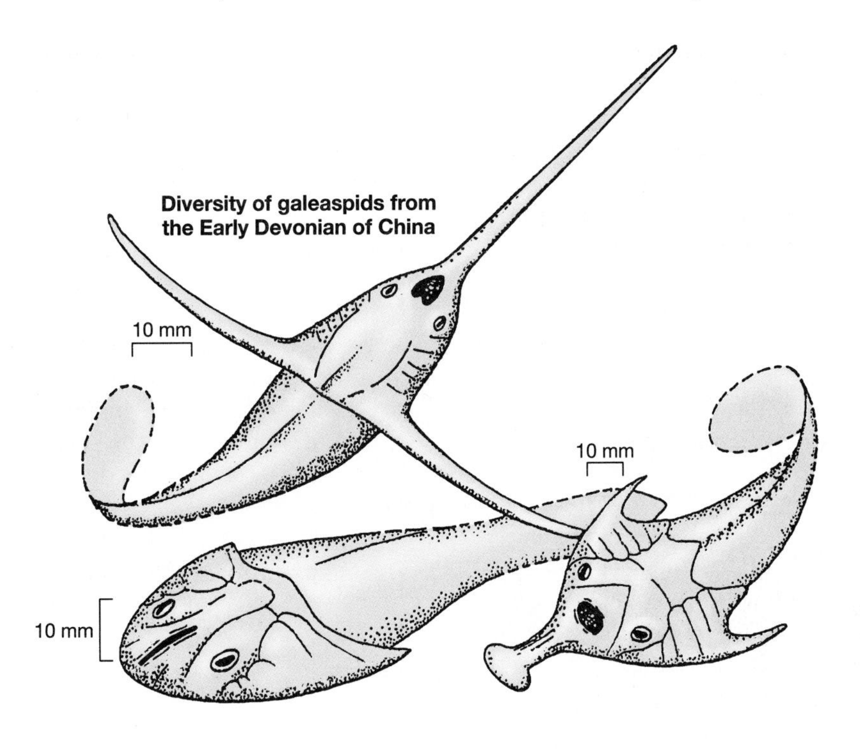

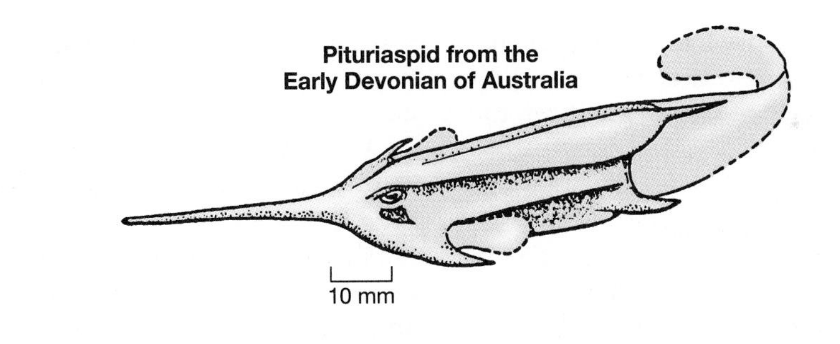

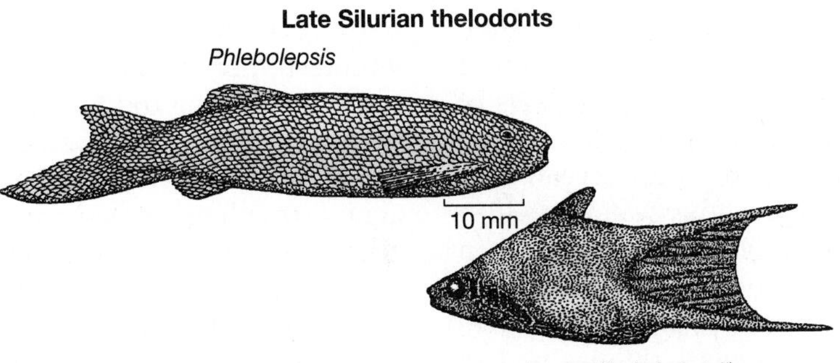

▲ Figure 3–8 Ostracoderm diversity 2.

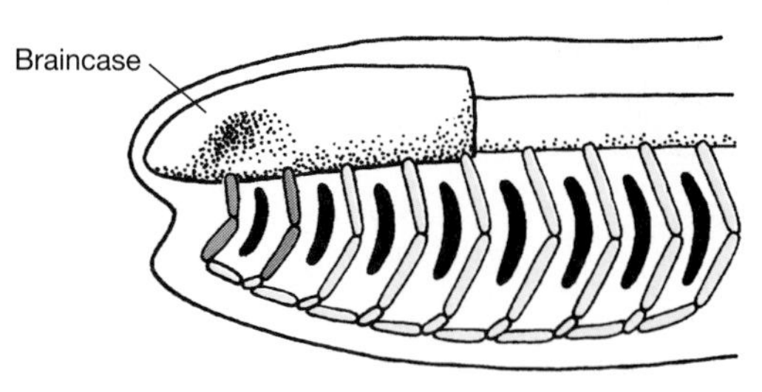

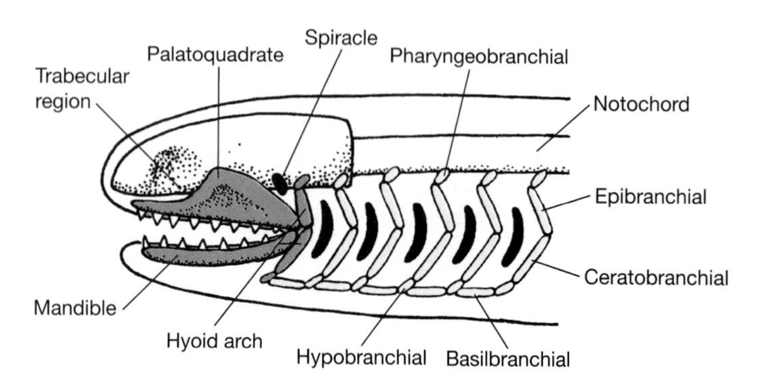

◀ Figure 3–9 Evolution of the vertebrate jaw from anterior branchial arches.

summarizes the differences in the gill arches between jawed and jawless vertebrates, and illustrates the major components of the hinged gnathostome gill arches.

The Basic Gnathostome Design

Gnathostomes are first known with certainty from the Early Silurian, but isolated shark-like scales suggest that they date back to the Middle Ordovician. The difference between gnathostomes and agnathans is traditionally described as the possession of jaws that bear teeth in most forms and two sets of paired fins or limbs (pectoral and pelvic). The gnathostome body plan (Figure 3–10) reveals that they are characterized by many other features, which imply that gnathostomes represent a basic step-up in level of activity and complexity from the jawless vertebrates (Table 3.2). These features include improvements in locomotor and predatory abilities and in the sensory and circulatory systems. Just as the transition from nonvertebrate chordate to vertebrate was characterized by a duplication of the *Hox* gene complex, the transition from jawless to jawed vertebrates may have involved a second duplication event (Monastersky 1996). Gene duplication would have resulted in a greater amount of genetic material, perhaps necessary for building a more complex type of animal.

An obvious change in the locomotor system is the acquisition of more complete vertebrae. The vertebrae initially consisted of neural arches flanking the nerve cord and hemal arches below the notochord, as seen in the posterior portion of the trunk in the animal depicted in Figure 3–10. Later, more derived gnathostomes had a vertebral centrum or central elements, with attached ribs. More complete vertebrae would support the notochord and eventually replace it as a supporting rod for strengthened axial muscles used for locomotion. Ribs lie in the connective tissue between successive segmental muscles (myomeres), again providing increased anchorage for axial muscles. There is also now a clear distinction between the dorsal (epaxial) and ventral (hypaxial) blocks of the axial (trunk) muscles, divided by a horizontal septum running the length of the animal. The lateral line canal—containing the neuromast organs that sense vibrations in the surrounding water—lies in the plane of this septum, perhaps reflecting improved integration between locomotion and sensory feedback. The neuromasts of the lateral line now lie in canals (in lampreys they lie in unconnected pits,

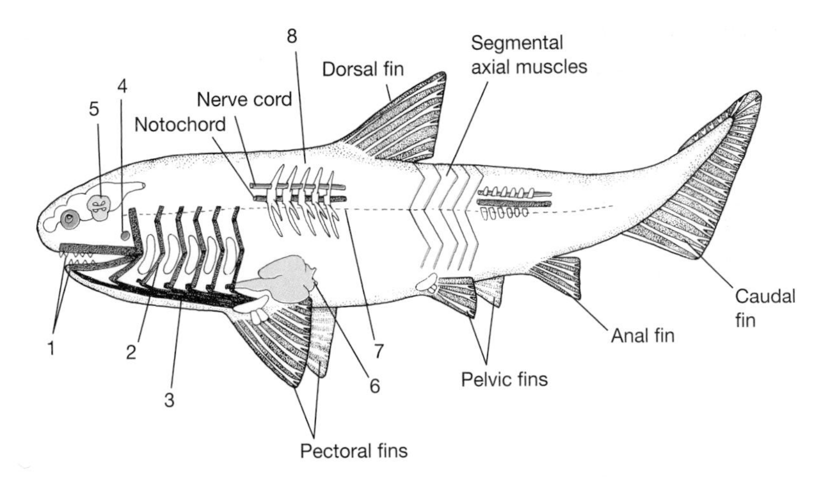

Legend: **1.** Jaws (containing teeth) formed from the mandibular gill arch. **2.** Gill skeleton consists of jointed branchial arches, and contains internal gill rakers that stop particulate food from entering the gills. Gill musculature is also more robust. **3.** Hypobranchial musculature allows strong suction in inhalation and suction feeding. **4.** Original first gill slit squeezed to form the spiracle, situated between mandibular and hyoid arches. **5.** Three semicircular canals in the inner ear (addition of horizontal canal). **6.** Addition of a conus arteriosus to the heart, between the ventricle and the ventral aorta. (Note that the position of the heart is actually more anterior than shown here, right behind the most posterior gill arch.) **7.** Horizontal septum divides trunk muscle into epaxial (dorsal) and hypaxial (ventral) portions. It also marks the position of the lateral line canal, containing the neuromast sensory organs. **8.** Vertebrae now have centra (surrounding the notochord) and ribs, but note that the earliest gnathostomes have only neural and hemal arches, as shown in the posterior trunk.

▲ Figure 3–10 Generalized jawed vertebrate (gnathostome) showing derived features compared to the jawless vertebrate (agnathan) condition.

although canals may have been present in some ostracoderms). In the inner ear there is a third (horizontal) semicircular canal, which may reflect an improved ability to navigate and orient in three dimensions.

The features just described can be observed in fossils, so we know that they are unique to gnathostomes. There are many other new features of gnathostomes within the soft anatomy. We cannot know for sure whether these features characterize gnathostomes alone, or whether they were adopted somewhere within the ostracoderm lineage. We know that some features of the nervous system, which are seen only in gnathostomes among living vertebrates, were acquired by the earliest ostracoderms. For example, impressions on the inner surface of the dermal head shield reveal the presence of a cerebellum in the brain and an olfactory tract. Other new features of the nervous system in extant gnathostomes include insulating sheaths of myelin on the nerve fibers that increase the speed of nerve impulses. The eyes have intrinsic musculature for the accommodation of the lens, allowing better focusing. Within the spinal column the dorsal and ventral spinal nerve roots are linked to form compound spinal nerves.

There are also new features within the circulatory, digestive, and urogenital systems. The heart of gnathostomes has an additional small chamber in front of the pumping ventricle, the **conus arteriosus,** which acts as an elastic reservoir. Its presence is probably due to the strong ventricular pumping and high blood pressures of gnathostomes and the need to smooth out the pulsatile nature of the flow of blood. Gnathostomes also have a renal portal vein. A true stomach is a new feature in gnathostomes among living vertebrates; but as we have seen, some thelodont ostracoderms also appear to have had a stomach. Gnathostomes also have a distinct spleen, a pancreas with both exocrine and endocrine portions, and a spiral valve within the intestine to increase the surface area available for the absorption of food. The gonads now have distinct ducts linking them to the cloaca. In the males the testes are linked by

Table 3.2 Derived Features of Gnathostomes

Cranial Characters

1. Cranium enlarged anteriorly to the end in a precerebral fontinelle.
2. Cranium elongated posteriorly, so that one or more occipital neural arches are incorporated in the rear of the skull.
3. Development of a postorbital process on the cranium, separating the functions of supporting the jaws and enclosing the eyes.
4. Intrinsic musculature in the eye for lens accommodation.

Intenal Anatomical Characters

5. Atrium lies posterodorsally (versus laterally) to ventricle.
6. Renal portal vein present.
7. Spiral valve primitively formed within intestine.
8. Pancreas with both endocrine and exocrine functions.
9. Distinct spleen.
10. Kidneys formed only by more posterior sections in adult (mesonephros and metanephros).
11. Male gonads linked by ducts to excretory (archinephric) duct.
12. Female gonads with distinct oviducts.
13. Two (versus one) contractile actin proteins (one specific to striated muscle and one specific to smooth muscle).

Sensory Characters

14. Nerves enclosed in myelinated sheaths.
15. Large, distinct cerebellum in the hind brain.
16. Two distinct olfactory tracts leading to widely separated olfactory bulbs.
17. Thicker spinal cord with "horns" of gray matter in section.
18. Dorsal and ventral spinal nerve roots linked to form compound spinal nerves.
19. Unique and evolutionarily conservative pattern of head lateral line canals (lost in adult amphibians and amniotes).
20. Lateral line on trunk region flanked or enclosed by specialized scales (lateral line lost in adult amphibians and amniotes).

tubules to the archinephric duct that originally drained the kidney. In the female the ovaries are associated with oviducts (= paramesonephric ducts).

The Origin of Fins

Guidance of a body in three-dimensional space is complicated. Fins act as hydrofoils, applying pressure to the surrounding water. Because water is practically incompressible, force applied by a fin in one direction against the water is opposed by an equal force in the opposite direction (Figure 3–11). The fins must correct the tendency of a fish in water to roll (rotate around the body axis), pitch (tilt up or down), or yaw (swing to the right or left). Roll can be controlled by pressing fins against the water, pitch is counteracted by fins projecting horizontally near the anterior end of the body, and yaw is controlled by vertical fins along the middorsal and midventral lines. Rapid adjustments of the body position in the water may be especially important for active, predatory fishes like the early gnathostomes. The paired fins (pectoral and pelvic fins) can act as brakes, can be used to produce lift, and are occasionally specialized to provide thrust during swimming (as with the enlarged pectoral fins of skates and rays). A tail fin increases the area of the tail,

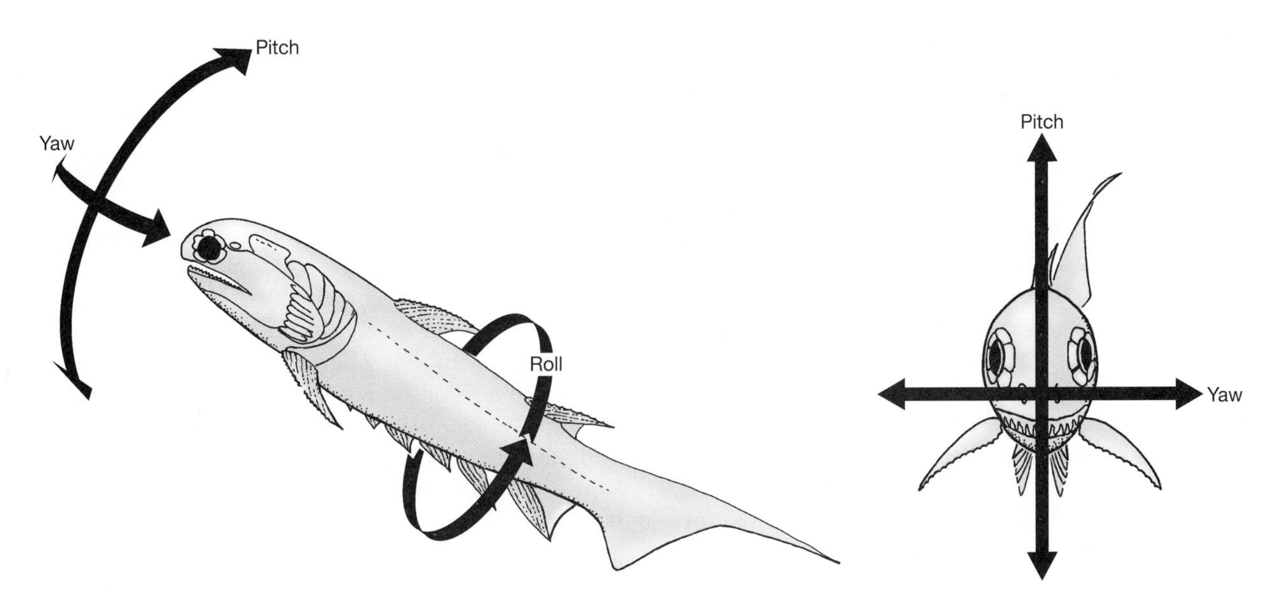

▲ Figure 3–11 A primitive jawed fish (the acanthodian, *Climatius*), shown from the side and front to illustrate pitch, yaw, and roll and the fins that counteract these movements.

giving more thrust during propulsion. Fins have functions in addition to controlling locomotion. Spiny fins are used in defense, and they may become systems to inject poison when combined with glandular secretions. Colorful fins are used to send visual signals to potential mates, rivals, and predators.

Even before the gnathostomes appeared, fish had structures that served the same purpose as fins. Many ostracoderms had spines or enlarged scales derived from dermal armor that acted like immobile fins. Some anaspids had long fin-like sheets of tissue running along the flanks. Osteostracans had pectoral fins, but they apparently lacked internal supports. Although a complete cartilaginous fin endoskeleton with a characteristic pattern of basal and radial elements is known only from gnathostomes, osteostracans probably had an endoskeletal pectoral girdle.

Most gnathostome fishes have a well-developed heterocercal caudal fin. Heterocercal tails are found in some ostracoderms, but others had hypocercal tails or tails of other shapes. A notochord that turns abruptly up or down increases the depth of the caudal fin and allows it to exert the force needed for rapid acceleration. All fishes with a fin-strengthening upturned or downturned axial skeleton tip have a noncollapsible caudal fin that is effective for burst swimming. Burst swimming is important in predator avoidance and can save energy when bursts of acceleration are alternated with glides.

A Problem Posed by the Gills of Early Vertebrates

For many years, researchers considered jawless and jawed vertebrates as two separate evolutionary radiations. Both the extant agnathans and the ostracoderms were seen as very different types of animals from jawed vertebrates, representing a quite separate evolutionary lineage. Features such as the pectoral fins of osteostracans, which we now consider to be a derived characteristic shared with gnathostomes, were interpreted as examples of convergent evolution. However, as shown in Figure 3–3, osteostracans are now perceived as having numerous other derived characteristics that place them closer to gnathostomes than to other ostracoderms. Lampreys are also interpreted as being closer to gnathostomes than are hagfishes. Thus the Agnatha is a paraphyletic assemblage, and any understanding of the origin of gnathostomes must encompass the view that at some point a jawless vertebrate was transformed into a jawed one.

This narrative sounds like a simple, straightforward story of progress in our understanding of evolutionary events. Yet until quite recently, many researchers perceived a major barrier to the notion that any jawless vertebrate could be ancestral to the jawed ones: the structure of the gills and cartilaginous gill supports (branchial arches or visceral arches). Living jawless vertebrates have what are termed pouched gills, as distinguished from the flatter, more lens-shaped spaces between the gills of gnathostomes. In the gills of lampreys and hagfishes, the gill filaments themselves (the sites of gas exchange) are *internal* to the branchial arches and their surface is formed from endodermal tissue. In contrast, in gnathostomes the filaments are *external* to the branchial arches and their surface is formed from ectodermal tissue.

This difference in structure was seen as strong evidence that jawless and jawed vertebrates represent two radiations from an original situation in which there were branchial arches but as yet no actual gill tissue. Note that the situation in amphioxus cannot resolve the issue of what might be primitive and what derived regarding the position of the gill supports. Amphioxus lacks gill filaments and has gill supports made from collagen; these features may not be homologous with the cartilaginous branchial arches of vertebrates, which are formed from the vertebrate innovation of neural crest tissue.

This issue was addressed by Jon Mallatt (e.g., Mallatt 1985, 1996, 1997). Much of the previous argument about the issue of agnathan/gnathostome interrelationships had taken place among paleontologists concerned about the phylogenetic position of the ostracoderms. But Mallatt focused on the anatomy and feeding behavior of living primitive vertebrates, using sharks as an example of primitive gnathostomes. He pointed out that the type of tissue forming the gills (endodermal versus ectodermal) is not a fixed aspect of inherited morphology. Instead, the tissue layer that forms the gills is related to exactly where in the pharynx the gills were developed, and could easily have changed during evolution. Mallat also noted that the detailed structure of lamprey gills was actually more similar to the condition in sharks than the condition in hagfishes. These similarities extended to such features as the position of gill filaments, the entrance and exit of blood vessels supplying the gills, how water passes through the gill itself, and numerous features of the gill microstructure.

Thus it seemed that the gills of lampreys are homologous with those of gnathostomes, and the supporting branchial arches (external versus internal) are not. But how could the difference in the position of the gill supports be explained? Had they somehow moved from an original external position in jawless vertebrates to a new internal position in jawed ones?

Mallatt (1996) resolved this issue by examining the detailed anatomical structures of these primitive vertebrates (Figure 3–12). It appears that the original vertebrate condition (inferred from the anterior part of the pharynx in hagfishes) was to have both sets of gill supports, one internal and one external, with the afferent and efferent arteries are both internal. In hagfishes (Figure 3–12b) the first four branchial arches are modified to support a velum (a pumping flap), but parts of both external and internal arches remain. The first three gill pouches are lost, but other gills migrate posteriorly and only external arches form there. The gills are more highly pouched, and the arteries are rearranged in a complex way, probably to maximize oxygen extraction in oxygen-poor water of deep sea sediment. In the lamprey (Figure 3–12c) the first branchial arch is modified to support the velum, and the first gill pouch is lost. The internal arches on the other gills are lost perhaps in association with the new, tidal, ventilation. In a hypothetical gnathostome ancestor (Figure 3–12d) the internal branchial arches would have enlarged to support strengthened breathing muscles that attach to these arches. In an early true gnathostome (Figure 3–12e) the first branchial arch is modified into a jaw, and the mouth opening enlarges. Small external arches are present in extant sharks, and the external arches are lost entirely in bony fishes. Mallatt proposed that the progressive enlargement of the internal support in gnathostomes was related to strengthening muscles that move these arches back and forth during the strong inhalation and exhalation stages of gnathostome ventilation.

Stages in the Origin of Jaws

It has long been known that vertebrate jaws are made of the same material (cartilage derived from the neural crest) as the branchial arches, and they clearly develop as the first arch of this series in vertebrates. The arch that gives rise to the jaws (the mandibular arch) does not form a functional gill arch in any living vertebrate, although it is presumed to have done so in some extinct forms. The notion that

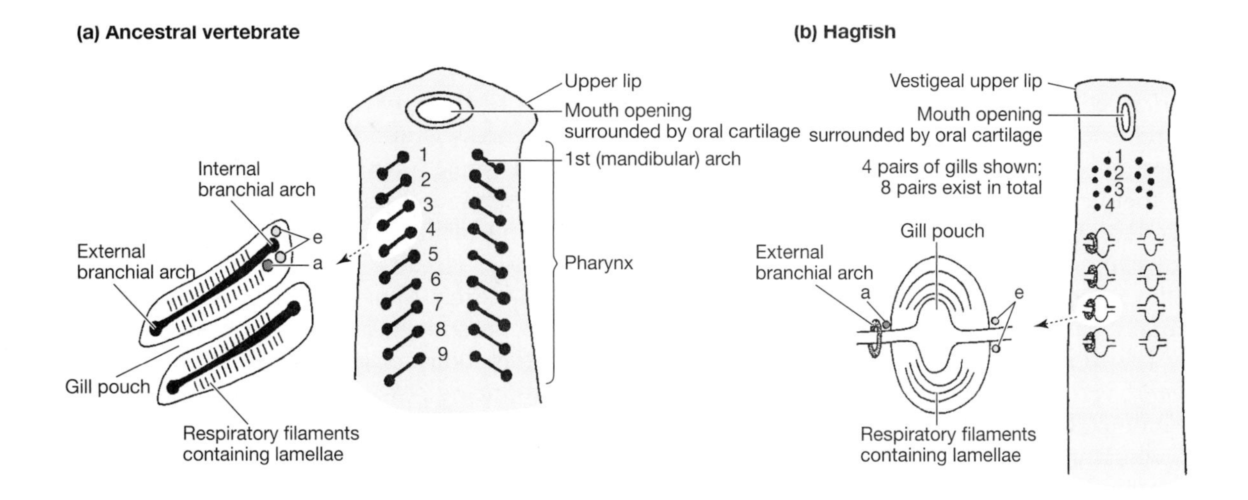

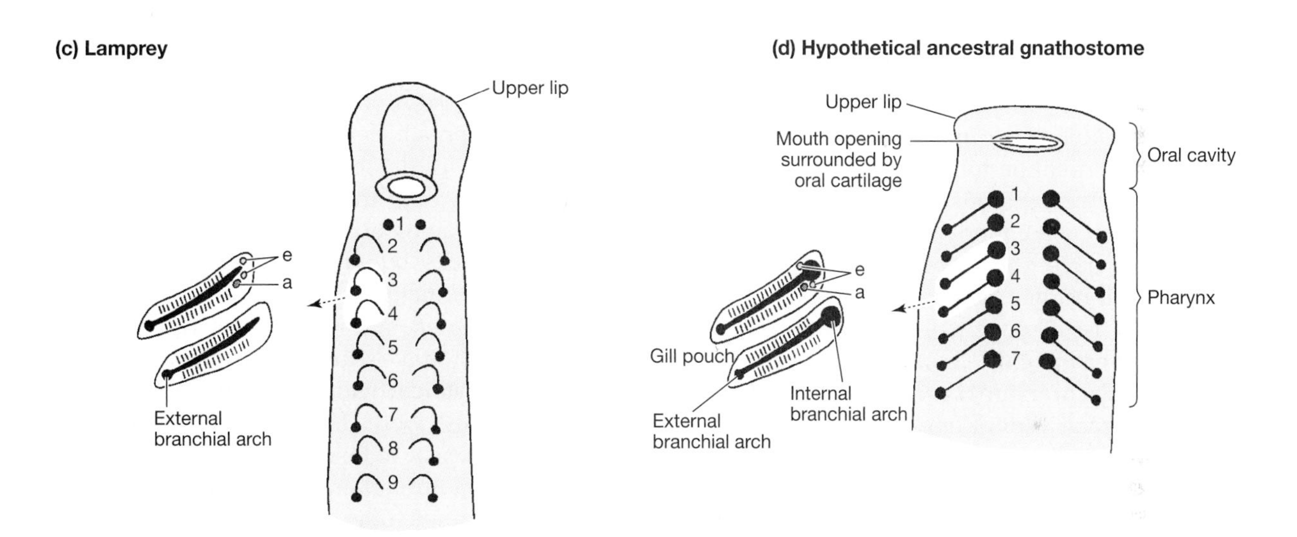

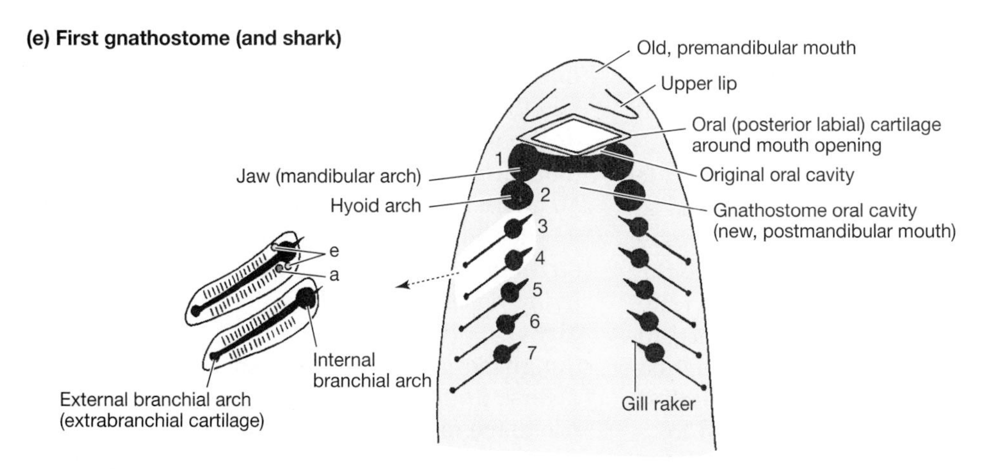

▲ Figure 3–12 Jon Mallatt's scenario of the evolution of vertebrate gills and jaws. Animals are portrayed in ventral view. "A" = afferent artery (carrying deoxygenated blood to the gills); "E" = efferent artery (carrying oxygenated blood away from the gills). (a) Ancestral vertebrate condition (consistent with ostracoderms). (b) Hagfishes. (c) Lamprey. (d) Hypothetical gnathostome ancestor. (e) Early true gnathostome.

the more derived, predatory vertebrates should convert gill arches into toothed jaws has been more or less unquestioned. It is a common assumption that jaws are superior devices for feeding, and thus more derived vertebrates were somehow bound to obtain them. However, this simplistic approach does not address the issue of how the evolutionary event might actually have taken place: What use would a proto-jaw be prior to its full transformation? And even if early vertebrates had needed some sort of superior mouth design, why modify a pharyngeal gill arch, which initially was located some distance behind the mouth opening? Why not just modify the existing cartilages and plates surrounding the mouth? Living agnathans have specialized oral cartilages, and various ostracoderms apparently had oral plates.

Mallatt has suggested that the initial enlargement of the mandibular arch into proto-jaws was for improved gill ventilation rather than for feeding (Mallatt 1996, 1997). Numerous features of gnathostomes suggest that they are more active than ancestral vertebrates and have greater metabolic demands. One derived gnathostome feature associated with such high activity is the powerful mechanism for pumping water over the gills. Gnathostomes have a characteristic series of internal branchial muscles as well as new, external, ventral ones (the hypobranchials). These muscles not only push water through the pharynx in exhalation but also draw (suck) water into the pharynx during inhalation. Gnathostome fishes can generate much stronger suction than known agnathans, and powerful suction is also a way to draw food into the mouth. Living agnathans derive a certain amount of suction from their pumping velum, but this pump mostly just pushes water and its action is weak.

Mallatt suggested that the mandibular branchial arch enlarged into proto-jaws because it played an essential role in forceful ventilation—rapidly closing and opening the entrance to the pharynx. During strong exhalation, as the pharynx squeezed water back across the gills, water was kept from regurgitating out of the mouth by bending the mandibular arch sharply shut. Next, during forceful inspiration, the mandibular arch was rapidly straightened to reopen the pharynx and allow the inhaled water to enter. To accommodate the forces of the powerful muscles that bent the arch (the adductor mandibulae) and straightened it (the hypobranchial muscles), the mandibular arch enlarged and became more robust. The advantage of using the mandibular arch for this process would be that the muscles involved were of the same functional series as the other ventilatory muscles, and their common origin would ensure that all of the muscles were controlled by the same nerve circuits. Perhaps that is why the mandibular arch, rather than the more anterior oral cartilages, became the jaws of gnathostomes.

With the ability of jaws to grasp prey as well as aid in producing powerful suction, the stage was set for gnathostomes to enter a new realm of feeding ecology—attacking large, actively swimming mobile prey.

3.5 Extinct Paleozoic Jawed Fishes

The earliest possible evidence of gnathostomes dates back to the Middle Ordovician, and they are known with certainty from the Early Silurian. However, it is not until the Devonian that they are well known as entire body fossils, by which time they can be divided into four distinctive clades: two extinct groups—**placoderms** and **acanthodians**—and two groups that survive today—**chondrichthyans** (cartilaginous fishes) and **osteichthyans** (bony vertebrates).

Placoderms were highly specialized, armored fishes that appear to be more primitive than all other gnathostomes in certain respects. The cartilaginous fishes, which include sharks, rays, and chimeras, evolved distinctive specializations of dermal armor, internal calcification, jaw and fin mobility, and reproduction. The acanthodians and the bony fishes may be closely related and are sometimes grouped together as the teleostomes. Bony fishes evolved endochondral bone, a distinctive dermal head skeleton that included an operculum covering the gills, and an internal air sac forming a lung or a swim bladder.

The bony fishes include the ray-finned fishes (**actinopterygians**), which comprise the majority of living fishes, and the lobe-finned fishes (**sarcopterygians**). Only a few lobe-finned fishes survive today (lungfishes and the coelacanths), but they were more diverse in the Paleozoic and are the group that gave rise to the tetrapods. Note that in the proper cladistic sense Osteichthyes should also include tetrapods, since bony fishes by themselves constitute a paraphyletic group (because their common ancestor is also the ancestor of tetrapods). The same is true of the Sarcopterygii.

Before studying the extant groups of jawed fishes, we turn to the placoderms and acanthodians to examine the variety of early gnathostomes. Figures 3–13 and 3–14 show the interrelationships of gnathostome fishes. The various extinct groups of chondrichthyans are discussed in Chapter 5, and the extinct osteichthyans are discussed in Chapter 6.

Placoderms—The Armored Fishes

As the name placoderm (Greek *placo* = plate and *derm* = skin) implies, placoderms were covered with a thick, often ornamented bony shield over the anterior one-half to one-third of their bodies (Figure 3–15). This dermal shield differed from the condition in ostracoderms; in placoderms the shield was divided into a separate head portion and a trunk portion, linked by a mobile joint. The earliest forms were the most heavily armored, and the body armor in particular tended to be reduced in later, more derived forms. Placoderm vertebrae had neural and hemal arches, but none are known that had central elements or ribs.

Placoderms are known from the Early Silurian to the end of the Devonian. Like the ostracoderms, they suffered massive losses in the Late Devonian extinctions. But unlike any ostracoderm group, a few placoderm lineages continued for a further five million years until the very end of the period. Placoderms were very diverse in their morphological specializations and were the most numerous fishes (in number of species) of the Devonian.

Several elements of placoderm morphology appear to isolate them from all other jaw-bearing vertebrates, suggesting that they were an early offshoot off the main stem, not closely related to other gnathostomes. Placoderms lacked teeth that correspond to those of any other gnathostome; their tooth-like structures were actually projections of the dermal jaw bones that were subject to wear and breakage without replacement. In addition, there is a difference in the inferred position of the jaw musculature. In all other gnathostomes the jaw muscles lie external to the upper jaws. In those placoderms where it can be determined, it appears that the main mass of the jaw musculature was medial to the upper jaw elements.

Placoderms were mostly creatures of the Devonian. During that period they radiated into a large number of lineages and types, and were the most diverse vertebrates of their time. Ancestral placoderms were primarily marine, but a great many lineages became adapted to freshwater and estuarine habitats. Placoderms have no modern analogues, and their massive external armor makes interpreting their lifestyle particularly difficult. They must have been primarily benthic fishes; in many species the body was dorsoventrally depressed with flattened ventral surfaces.

In some species of placoderms, only some individuals had pelvic appendages. A similar situation occurs among living chondricthyans—only males have pelvic appendages, which they use for internal fertilization. We can infer that the placoderms, like living chondrichthyans, had internal fertilization and probably complex courtship behaviors.

Placoderms had several peculiarities related to their mode of feeding. A mobile connection between the anterior vertebrae and the skull allowed the head to be lifted. This craniovertebral joint permitted the mouth to open wider than it could by lowering only the mandible. Placoderms' upper jaws were often immovably bound to the cranium or tightly articulated to the rest of the head shield. This limited their participation in any sucking action—a very important feeding process, as shown by its elaboration in the vast majority of extant jawed fishes.

More than half of the known placoderms, nearly 200 genera, belonged to the predatory arthrodires (Greek *arthros* = a joint and *dira* = the neck). As their name suggests, they had specializations of the joint between the head shield and the trunk shield. The space between the head and the thoracic shields was widened and a pair of joints evolved—one above each pectoral fin on a line passing through the older craniovertebral joint of the axial skeleton. This arrangement provided great flexibility between the shields and allowed an enormous head-up gape, proably also increasing respiratory efficiency. This extreme ability to move the upper head on the body may be related to a bottom-dwelling feeding mode in which the lower jaw rested in the substrate. Arthrodires tended to reduce the size of the trunk shield, and their jaws were projected into sharp-bladed, tooth-like plates. *Dunkleosteus* was a voracious, 10-meter-long, predatory arthrodire, undoubtably the fiercest predator of the Devonian.

There were several other types of placoderms, most of them flattened, bottom-dwelling forms. The antiarchs, such as *Bothriolepis*, looked rather like armored catfishes. Their pectoral fins were also

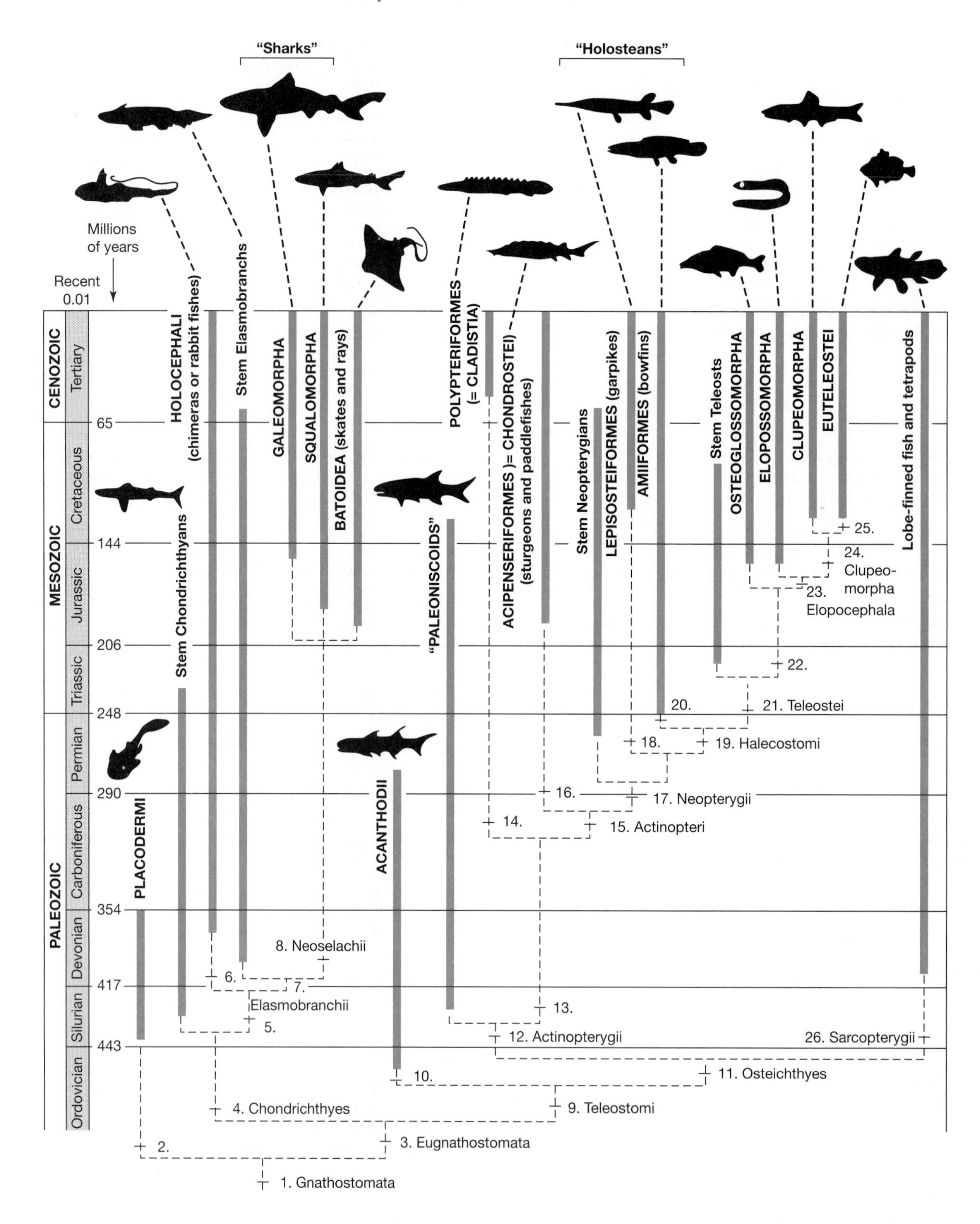

"Sharks"
"Holosteans"
Millions of years
Recent 0.01
CENOZOIC
Tertiary
65
MESOZOIC
Cretaceous
144
Jurassic
206
Triassic
248
PALEOZOIC
Permian
290
Carboniferous
354
Devonian
417
Silurian
443
Ordovician
HOLOCEPHALI (chimeras or rabbit fishes)
Stem Elasmobranchs
GALEOMORPHA
SQUALOMORPHA
BATOIDEA (skates and rays)
POLYPTERIFORMES (= CLADISTIA)
ACIPENSERIFORMES)= CHONDROSTEI) (sturgeons and paddlefishes)
Stem Neopterygians
LEPISOSTEIFORMES (garpikes)
AMIIFORMES (bowfins)
Stem Teleosts
OSTEOGLOSSOMORPHA
ELOPOSSOMORPHA
CLUPEOMORPHA
EUTELEOSTEI
Lobe-finned fish and tetrapods
Stem Chondrichthyans
"PALEONISCOIDS"
PLACODERMI
ACANTHODII
25.
24. Clupeo-morpha
23. Elopocephala
22.
21. Teleostei
20.
18.
19. Halecostomi
16.
17. Neopterygii
14.
15. Actinopteri
8. Neoselachii
6.
7.
Elasmobranchii
5.
13.
12. Actinopterygii
26. Sarcopterygii
10.
11. Osteichthyes
4. Chondrichthyes
9. Teleostomi
2.
3. Eugnathostomata
1. Gnathostomata

Legend: **1.** Gnathostomata—Jaws formed of bilateral palatoquadrate (upper) and mandibular (lower) cartilages of the mandibular visceral arch at some stage of development, modified hyoid gill arch, major branchial elements internal to gill membranes; branchial arches contain four elements on each side plus one unpaired ventral median element; three semicircular canals; internal supporting girdles associated with pectoral and pelvic fins; and many features of the soft anatomy (see Table 6.2). **2.** Placodermi—Tentatively placed as the sister group of all other gnathostomes, but see Gardiner (1984) for a different view. Provisionally united by the following derived characters: a specialized joint in the neck vertebrae, a unique arrangement of dermal skeletal plates of the head and shoulder girdle, a distinctive articulation of the upper jaw, and a unique pattern of lateral line canals on the head. **3.** Chondrichthyes plus Teleostomi (Eugnathostoma)—Epihyal element of second visceral arch modified as the hyomandibula, which is a supporting element for the jaw; true teeth rooted to the jaw (Osteichthyes) or in a tooth whorl. **4.** Chondrichthyes—Unique perichondral and endochondral mineralization as prismatic hydroxyapatite tesserae, placoid scales, unique teeth and tooth replacement mechanisms, distinctive characters of the basal and radial elements of the fins, inner ear labyrinth opens externally via the endolymphatic duct, distinctive features of the endocrine system. **5.** Elasmobranchii plus Holocephali—Claspers on male pelvic fins and at least four additional fin-support characters. **6.** Holocephali—Hyostylic jaw suspension, gill arches beneath the braincase, dibasal pectoral fin, dorsal fin articulates with anterior elements of the axial skeleton. **7.** Elasmobranchii—Tribasal pectoral, shoulder joint narrowed, basibranchial separated by gap from basihyal. **8.** Neoselachii—Pectoral fin with three basal elements, the anteriormost of which is supported by the shoulder girdle; plus characteristics of the nervous system, cranium, and gill arches. **9.** Teleostomi—Hemibranchial elements of gills not attached to interbranchial septum, bony opercular covers, branchiostegal rays. **10.** Acanthodii—Fin spines on anal and paired fins as well as on dorsal fins; paired intermediate fin spines between pectoral and pelvic fins. **11.** Osteichthyes—A unique pattern of dermal head bones, including dermal marginal mouth bones with rooted teeth, a unique pattern of ossification of the dermal bones of the shoulder girdle, presence of lepidotrichia (fin rays), differentiation of the muscles of the branchial region, presence of a lung or swim bladder derived from the gut, medial insertion of the mandibular muscle on the lower jaw. **12.** Actinopterygii—Basal elements of pectoral fin enlarged, median fin rays attached to skeletal elements that do not extend into fin, single dorsal fin, scales with unique arrangement, shape, interlocking mechanism, and histology (outer layer of ganoine). **13.** Cladistia plus Actinopteri (plus fossils such as *Moythomasia* and *Mimia*)—A specialized dentine (acrodin) forms a cap on the teeth, details of posterior braincase structure, specific basal elements of the pelvic fin are fused, and numerous features of the soft anatomy of extant forms that cannot be verified for fossils. **14.** Polypteriformes (Cladistia)—Unique dorsal-fin spines, facial bone fusion, and pectoral-fin skeleton and musculature. **15.** Actinopteri—Derived characters of the dermal elements of the skull and pectoral girdle and fins, a spiracular canal formed by a diverticulum of the spiracle penetrating the postorbital process of the skull, other details of skull structure, three cartilages or ossifications in the hyoid below the interhyal. Swim bladder connects dorsally to the foregut, fins edged by specialized scales (fulcra). **16.** Acipensiformes (Chondrostei)—Fusion of premaxillae, maxillae, and dermopalatines; unique anterior palatoquadrate symphysis. **17.** Neopterygii—Rays of dorsal and anal fins reduced to equal the number of endoskeletal supports, upper lobe of caudal fin containing axial skeleton reduced in size to produce a nearly symmetrical caudal fin, upper pharyngeal teeth consolidated into tooth-bearing plates, characters of pectoral girdle and skull bones. **18.** Lepisosteiformes (Ginglymodi)—Vertebrae with convex anterior faces and concave posterior faces (opisthocoelus), toothed infraorbital bones contribute to elongate jaws. See character state 19. **19.** Halecostomi—Modifications of the cheek, jaw articulation, and opercular bones including a mobile maxilla. Relationships of the Lepisosteiformes, Amiiformes, their fossil relatives, and the Teleostei are currently subject to many differing opinions with no clear resolution based on unique shared derived characters. More conservative phylogenies than those here would represent them as an unresolved tricotomy. Others would unite the lepisosteiformes and the amiiformes as the "Holostei." **20.** Amiidae (recent Amiiformes)—Jaw articulation formed by both the quadrate and the symplectic bones. **21.** Teleostei—Elongate posterior neural arches (uroneurals) contributing to the stiffening of the upper lobe of the internally asymmetrical caudal fin (the caudal is externally symmetrical, = homocercal, at least primitively in teleosts), unpaired ventral pharyngeal toothplates on basibranchial elements, premaxillae mobile, urohyal formed as an unpaired ossification of the tendon of the sternohyoideus muscle, details of skull foramina, jaw muscles, and axial and pectoral skeleton. **22.** Recent Teleosts—Presence of an endoskeletal basihyal, four pharyngobranchials and three hypobranchials, median toothplates overlying basibranchials and basihyals. **23.** Elopocephala—Two uroneural bones extend anteriorly to the second ural (tail) vertebral centrum; abdominal and anterior caudal epipleural intermuscular bones present. **24.** Clupeocephala—Pharyngeal toothplates fused with endoskeletal gill-arch elements, neural arch of first caudal centrum reduced or absent, distinctive patterns of ossification and articulation of the jaw joint. **25.** Euteleostei—This numerically dominant group of vertebrates is poorly characterized with no known unique shared derived character present in all or perhaps even in most forms. But, the following have been used in establishing monophyly: presence of an adipose fin posteriorly on the mid-dorsal line, presence of nuptial tubercles on the head and body, paired anterior membranous outgrowths of the first uroneural bones of the caudal fin. (These characters are usually lost in the most derived euteleosts.) The nature of these characters leads to a lack of consensus on the interrelationships of the basal clupeocephalids, although the group's monophyly is still generally accepted. **26.** Sarcopterygii—Fleshy pectoral and pelvic fins have a single basal skeletal element, muscular lobes at the bases of those fins, enamel (versus enameloid) on surfaces of teeth, cosmine (unique type of dentine) in body scales, unique characters of jaws, articulation of jaw supports, gill arches, and shoulder girdles..

◄ Figure 3–13 Phylogenetic relationships of jawed fishes. This diagram depicts the probable interrelationships among the major groups of basal gnathostomes. Extinct lineages are marked by a dagger (†). Dotted lines show interrelationships only; they do not indicate times of divergence nor the unrecorded presence of taxa in the fossil record. The bold numbers indicate derived characters that distinguish the lineages. Only the best-corroborated relationships are shown.

encased in the bony shield, so that their front fins looked more like those of a crab. Ptyctodontid placoderms looked rather like present-day chimeras, and like chimeras they had solid tooth plates for crushing shellfish. Other groups of placoderms, such as the rhenanids and the petalichthyids, were very flattened dorsoventrally with the eyes on top of the head and subterminal mouths—all features indicating benthic specialization. The rhenanids had a body covering of small plates that might have allowed for greater mobility; *Gemuendina* bears a striking resemblance to a modern skate, with a whip-like tail and large pectoral fins that it may have used to swim.

Acanthodians

Acanthodians are so named because of the stout spines *(acanthi)* anterior to their well-developed dorsal, anal, and often numerous paired fins. They are distinguished from other vertebrates by the possession of up to six pairs of ventrolateral fins in addition to the pectoral and pelvic fins of gnathostomes (Figure 3–16). Had land vertebrates evolved from these fishes, they might have been hexapods (six-legged) or octopods (eight-legged) rather than tetrapods (four-legged)! Acanthodians ranged from the Late Ordovician through the Early Permian, with their major diversity in the Early Devonian. The earliest forms were marine, but by the Devonian they were predominantly a freshwater group.

Acanthodians had slender bodies with a heterocercal tail fin, suggesting a preference for mid-water conditions in contrast to the bottom-dwelling placoderms. They were usually not more than 20 cm long although some species are known that were 2m long. Acanthodians were covered by small, square-crowned scales, each of which grew in size as the animal grew. The head was large and blunt, housing large eyes and a wide-gaping mouth. As in the placoderms, acanthodian vertebrae are known to consist only of neural and hemal arches. The teeth and scales lacked enamel, unlike the condition in other gnathostomes. Earlier acanthodians had a basic fusiform fish shape, and some—the ischnacanthids—were apparently specialized predators. The acanthodids, the only group to survive into the Permian, became more eel-like in body form and lost their teeth. These fishes had long gill rakers and the capacity for a very wide gape, suggesting a life-style of skimming the water for plankton and straining them out of the water through their gill rakers.

Most workers now accept acanthodians as the sister group of the Osteichthyes. The acanthodians + osteichthyans (sometimes grouped as the Teleostomei) are diagnosed by the following features: a unique mechanism of opening the mouth by lowering the mandible through movements of a hyoid apparatus transmitted to the lower jaw by ligaments; dermal branchiostegal rays associated with the gill covering; a similarly-shaped braincase; calcifications

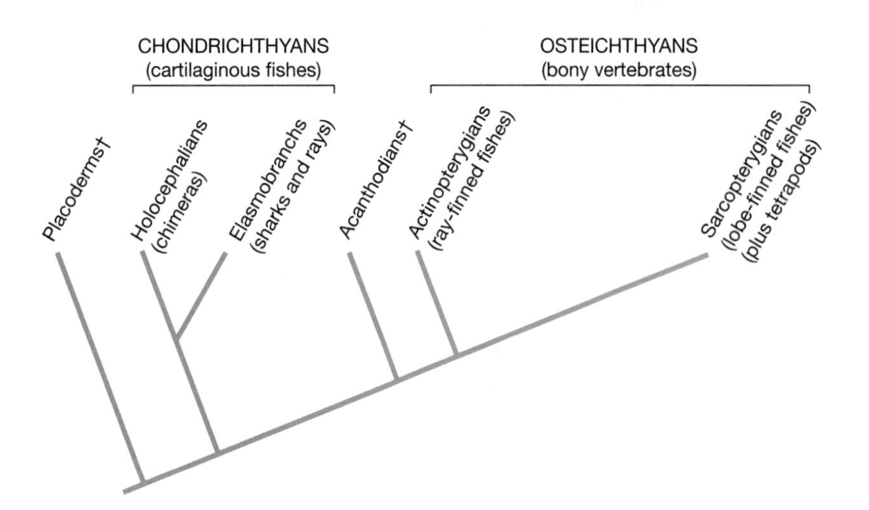

◄ Figure 3–14 Simplified cladogram of gnathostomes, showing living taxa and major extinct groups only. A dagger (†) indicates an extinct taxon.

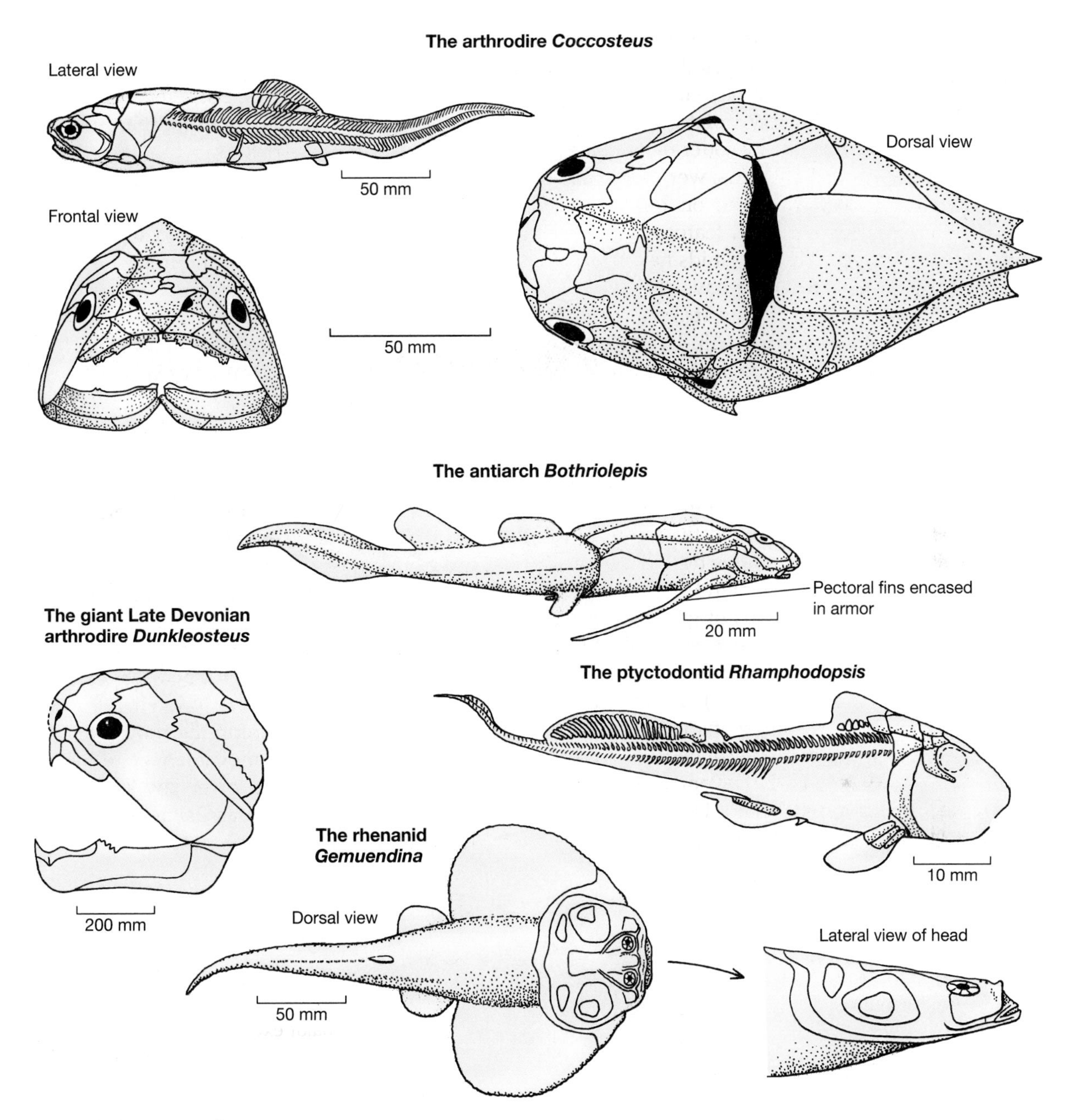

▲ Figure 3–15 Placoderms.

called otoliths in the inner ear; and a similar type of scales that lack the tooth-like pulp cavity seen in the scales of chondricthyans. However, acanthodians also bore a resemblance to chondrichthyans in having spines associated with their fins, and many species had a shark-like tooth whorl rather than teeth embedded in the jaws. These were probably primitive gnathostome features, rather than features indicating a close relationship of acanthodians to chondrichthyans.

The Early Devonian *Climatius*, a generalized ancestral form

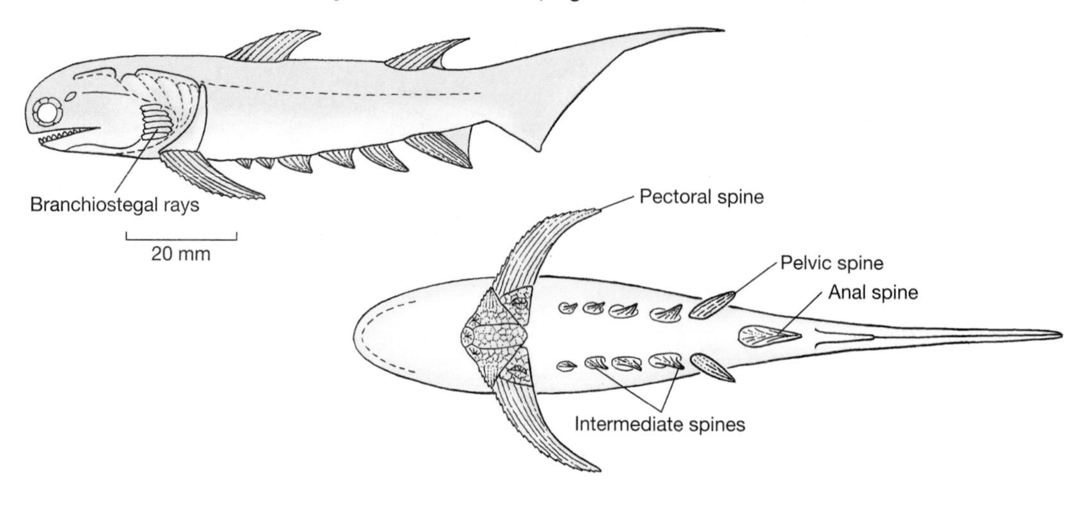

***Ischnacanthus*, an Early Devonian predacious form**

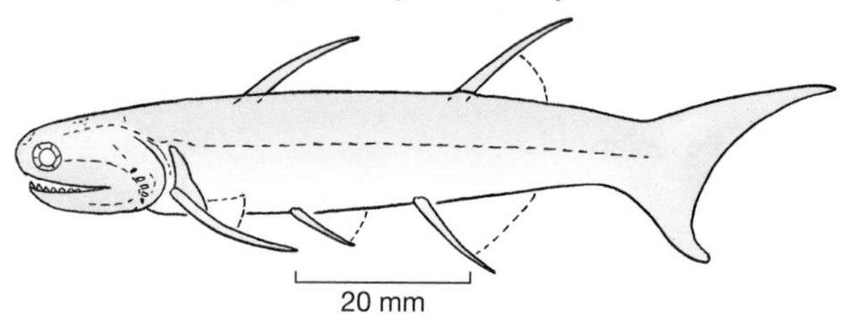

***Acanthodes*, a Permian filter-feeding form**

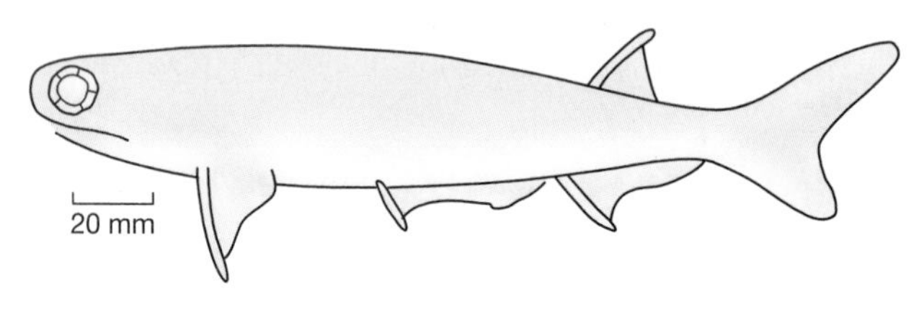

▲ Figure 3–16 Acanthodians.

Summary

Fossil evidence indicates that vertebrates evolved in a marine environment and had appeared by the Early Cambrian. The first vertebrates were jawless forms (agnathans). They would have been more active than their ancestors, with a switch from filter feeding to more active predation, and with a muscular pharyngeal pump for gill ventilation. Bone is a feature of many early vertebrates, although it was absent from the Early Cambrian forms and is also absent in the living jawless vertebrates, the hagfishes and lampreys. The first mineralized tissues were seen in the teeth of conodonts, enigmatic animals that have only recently been considered true vertebrates. Bone was first found with accompanying external layers of dentine and enamel-like tissue in the dermal armor of early jawless fishes called the ostracoderms, present either as complete sheets or as separate small elements. Current explanations for the original evolutionary use of bone include protection, a store for calcium and phosphorus, and housing for electroreceptive sense organs.

Ostracoderms are widely known from the Silurian and Early Devonian, and none survived past the end of the Devonian. All ostracoderms are more closely related to gnathostomes than are the living jawless vertebrates. Among the living jawless forms, hagfishes are more primitive than the lampreys and all other known vertebrates, living or fossil. Ostracoderms were not a unified evolutionary group: Some forms were more closely related to gnathostomes than were other forms. There were two main groups: the Pteraspidomorpha, and the more derived Cephalaspidomorpha. Some members of the latter group shared with gnathostomes the feature of a pectoral fin.

While jawless vertebrates are first known from the Early Cambrian, there is evidence that the first jawed vertebrates (gnathostomes) evolved as long ago as Early Ordovician. Both ostracoderms and gnathostomes flourished together for 50 million years in the Late Silurian and Devonian periods. Thus, there is little evidence in the fossil record to support the idea that jawed vertebrates outcompeted and replaced jawless ones. Ostracoderms were victims of the Late Devonian extinctions, along with many marine invertebrates and some gnathostome groups.

Just as the evolution of vertebrates from nonvertebrate chordates represented a step up in anatomical and physiological design, so did the evolution of jawed vertebrates from jawless ones. Jaws may have evolved initially to improve gill ventilation rather than to bite prey. In addition to jaws, gnathostomes have a number of derived anatomical features (such as true vertebrae, ribs, and a complete lateral line sensory system) suggesting a sophisticated and powerful mode of locomotion and sensory feedback.

The early radiation of jawed fishes, first known in detail from the fossil record in the Late Silurian, included four major groups. Two groups, the chondrichthyans (cartilaginous fishes) and osteichthyans (bony fishes) survive today. Osteichthyans were the forms that gave rise to tetrapods in the Late Devonian. The other two groups, placoderms and acanthodians, are now extinct; placoderms did not survive the Devonian period, while acanthodians survived almost until the end of the Paleozoic. Placoderms were armored fishes, superficially like the ostracoderms in their appearance, and were the most diverse fishes of the Devonian period. Placoderms are considered to be the sister taxon to other gnathostomes, largely because they lacked true teeth. Acanthodians were more derived fishes, probably forming the sister taxon to the osteichthyans. They had the unique feature of additional pairs of ventral paired fins to the usual gnathostome complement of pectoral and pelvic fins.

Additional Readings

Bardack, D., and R. Zangerl. 1971. Lampreys in the fossil record. In M. W. Hardisty and I. C. Potter (Eds.), *The Biology of Lampreys*, vol. 1. London: Academic Press, 67–84.

Braun, C. B. 1996. The sensory biology of the living jawless fishes: A phylogenetic assessment. *Brain, Behavior and Evolution* 48:262–276.

Briggs, D. E. G. 1992. Conodonts: A major extinct group added to the vertebrates. *Science* 256:1285–1286.

Briggs, D. E. G., E. N. K. Clarkson, and R. J. Aldridge. 1983. The conodont animal. *Lethaia* 26:275–287.

Donoghue, P. C. J., P. L. Forey, and R. J. Aldridge. 2000. Conodont affinity and chordate phylogeny. *Biological Reviews* 75:191–251.

Forey, P., and P. Janvier. 1994. Evolution of the early vertebrates. *American Scientist* 82:554–565.

Gans, C. 1989. Stages in the origin of vertebrates: Analysis by means of scenarios. *Biological Reviews* 64:221–268.

Janvier, P. 1996. *Early Vertebrates*. Oxford Monographs on Geology and Geophysics—33. Oxford: Clarendon Press.

Long, J. A. 1995. *The Rise of Fishes*. Baltimore: Johns Hopkins University Press.

Maisey, J. G. 1986. Heads and tails: A chordate phylogeny. *Cladistics* 2:201–256.

Maisey, J. G. 1994. Gnathostomes (jawed vertebrates). In D. R. Prothero and R. M. Schoch (Eds.), *Major Features of Vertebrate Evolution*, Short Courses in Palentology, no. 7. Knoxville, TN: University of Tennessee and the Paleontological Society, 38–56.

Maisey, J. G. 1996. *Discovering Fossil Fishes*. New York: Henry Holt and Company.

Mallatt, J. 1985. Reconstructing the life cycle and the feeding of ancestral vertebrates. In R. E. Foreman et al. (Eds.), *Evolutionary Biology of Primitive Fishes*. New York: Plenum, 59–68.

Mallatt, J. 1996. Ventilation and the origin of jawed vertebrates: A new mouth. *Zoological Journal of the Linnean Society* 117:329–404.

Mallatt, J. 1997/98. Crossing a major morphological boundary: The origin of jaws in vertebrates. *Zoology—Analysis of Complex Systems* 100:128–140.

Mallatt, J., and J. Sullivan. 1998. 28S and 18S rDNA sequences support the monophyly of lampreys and hagfishes. *Molecular Biology and Evolution* 15:1706–1718.

Martini, F., J. B. Heiser, and M. P. Lesser. 1997. A population profile for Atlantic hagfish, *Myxine glutinosa* (L.), in the Gulf of Maine. Part I: Morphometrics and reproductive state. *Fishery Bulletin* 95:311–320.

Martini, F. H. 1998. Secrets of the slime hag. *Scientific American* 279(4):70–75.

Monastersky, R. 1996. Jump-start for the vertebrates: New clues to how our ancestors got a head. *Science News* 149:74–75.

Northcutt, R.G. 1985. The brain and sense organs of the earliest vertebrates: Reconstruction of a morphotype. In R. E. Foreman et al. (Eds.), In *Evolutionary Biology*.

Romer, A. S. 1967. Major steps in vertebrate evolution. *Science* 158:1629–1637.

Ruben, J. A., and A. F. Bennett. 1987. The evolution of bone. *Evolution* 41:1187–1197.

Shu, D-G., H-L. Luo, S. Conway Morris, X-L. Zhang, S-X. Hu, L. Chen, J. Han, M. Zhu, Y. Li, and L-Z. Chan. 1999. Lower Cambrian vertebrates from south China. *Nature* 402:42–46.

Smith, M. M., and B. K. Hall. 1990. Development and evolutionary origins of vertebrate skeletogenic and odontogenic tissues. *Biological Reviews* 65:277–373.

Smith, M. P., I. J. Sansom, and M. P. Smith. 1995. Diversity of the dermal skeleton in Ordovician to Silurian vertebrate taxa from North America: Histology, skeletogenesis and relationships. *Geobios*, Special Memoir 19:65–70.

Smith, M. M., I. J. Sansom, and M. P. Smith. 1996. "Teeth" before armour: The earliest vertebrate mineralized tissues. *Modern Geology* 20:303–319.

Wilson, M. V. H., and M. W. Caldwell. 1998. The Furcacaudiformes: A new order of jawless vertebrates with thelodont scales, based on articulated Silurian and Devonian fossils from Northern Canada. *Journal of Vertebrate Paleontology* 18:10–29.

Web Explorations

On-line resources for this chapter are on the World Wide Web at: http://www.prenhall.com/pough (click on the Table of Contents link and then select Chapter 3).

PART II

Aquatic Vertebrates: Cartilaginous and Bony Fishes

Vertebrates originated in the sea, and more than half of the species of living vertebrates are the products of evolutionary lineages that have never left an aquatic environment. Water now covers 73 percent of the Earth's surface (the percentage has been higher in the past) and provides habitats extending from deep oceans, lakes, and mighty rivers to fast-flowing streams and tiny pools in deserts. Fishes have adapted to all these habitats, and the nearly 26,000 species of cartilaginous and bony fishes currently living are the subject of this portion of the book.

Life in water poses challenges for vertebrates but offers many opportunities. Aquatic habitats are some of the most productive on Earth, and energy is plentifully available in many of them. Some aquatic habitats (coral reefs are an example) have enormous structural complexity, whereas others (like the open ocean) have virtually none. The diversity of fishes reflects specializations for this variety of habitats.

The diversity of fishes and the habitats in which they live have offered unparalleled scope for variations in life history. Some fishes produce millions of eggs that are released into the water to drift and develop on their own, other species of fishes produce a few eggs and guard both the eggs and the young, and numerous fishes give birth to precocial young. Males of some species of fishes are larger than females, in others the reverse is true; some species have no males at all, and a few species of fishes change sex partway through life. Feeding mechanisms have been a central element in the evolution of fishes, and the specializations of modern fishes range from species that swallow prey longer than their own bodies to species that rapidly extend their jaws like a tube to stick up minute invertebrates from tiny crevices. In this part of the book we consider the evolution of this extraordinary array of vertebrates and the ecological conditions in the Devonian that contributed to the next major step of evolution, the origin of terrestrial vertebrates.

CHAPTER

4

Living in Water

Although life evolved in water and the earliest vertebrates were aquatic, the physical properties of water create some difficulties for aquatic animals. To live successfully in open water, a vertebrate must adjust its buoyancy to remain at a selected depth and force its way through a dense medium to pursue prey or to escape its own predators. Heat flows rapidly between an animal and the water around it, and it is difficult for an aquatic vertebrate to maintain a body temperature that is different from water temperature. (That phenomenon was dramatically illustrated when the Titanic sank—in the cold water of the North Atlantic, most of the victims died from hypothermia rather than by drowning.) Ions and water molecules move readily between the external environment and an animal's internal body fluids, so maintaining a stable internal environment can be difficult. On the plus side, ammonia is extremely soluble in water so disposal of nitrogenous waste products is easier in aquatic environments than on land. The concentration of oxygen in water is lower than it is in air, however, and the density of water imposes limits on the kinds of gas-exchange structures that can be effective. Despite these challenges, many vertebrates are entirely aquatic. Fishes in particular, especially the bony fishes, have diversified into an enormous array of sizes and ways of life. In this chapter we will examine some of the challenges of living in water and the ways aquatic vertebrates (especially fishes) have responded to them.

4.1 The Aquatic Environment

Seventy-three percent of the surface of the Earth is covered by fresh or salt water. Most of this water is held in the ocean basins, which are populated everywhere by vertebrates. Freshwater lakes and rivers hold a negligible amount of the water on Earth—about 0.01 percent. This is much less than the water tied up in the atmosphere, ice, and groundwater, but freshwater habitats are exceedingly rich biologically, and nearly 40 percent of all bony fishes live in freshwater.

Water and air are both liquids, but they have different physical properties that make them drastically different environments for vertebrates to live in (Box 4-1). In air, for example, gravity is an important force acting on an animal; but fluid resistance to movement (air resistance) is trivial for all but the fastest-flying birds. In water the opposite relationship holds—gravity is negligible, but fluid resistance to movement is a major factor that vertebrates must contend with. Although each major clade of aquatic vertebrates solved environmental challenges in somewhat different ways, the basic specializations needed by all aquatic vertebrates are the same.

Obtaining Oxygen in Water—Gills

Most aquatic vertebrates have gills, which are specialized structures where oxygen and carbon dioxide are exchanged. Fish gills are enclosed in pharyngeal pockets (Figure 4–1). The flow of water is usually unidirectional—in through the mouth and out through the gills. Buccal flaps just inside the mouth and flaps at the margins of the **operculae** (gill covers) of bony fish act as valves to prevent backflow. The respiratory

BOX 4-1 WATER—A NICE PLACE TO VISIT, BUT WOULD YOU WANT TO LIVE THERE?

Water and air are both fluids in which animals live, but the different physical properties of the two fluids make aquatic and terrestrial environments quite different places. Compared to air, water has high density, high viscosity, low oxygen content, high heat capacity and heat conductivity, and high electrical conductivity. These physical characteristics are reflected in the sizes and shapes of aquatic and terrestrial animals and in their physiology and behavior.

Density Water is more than 800 times as dense as air. A liter of water weighs 1 kilogram, whereas a liter of air weighs about 1.25 grams. Because of its density, water supports an animal's body. Aquatic animals do not need weight-bearing skeletons, because they are close to neutral buoyancy in water. Aquatic vertebrates can grow larger than terrestrial forms because gravity has little effect on their body structure.

Viscosity Water is approximately 18 times as viscous as air. Viscosity is a measure of how readily a fluid flows across a surface—the higher the viscosity, the slower the fluid flows. (To visualize the effect of viscosity, think of how rapidly water flows compared to a more viscous fluid such as syrup.)

The effect of the high density and high viscosity of water can be seen in the streamlined shape of most aquatic animals compared to the lack of streamlining of terrestrial animals. Only the fastest birds need to worry about air resistance, but even slow-moving fish must be streamlined.

The path followed by fluid during breathing also reflects the differences in density and viscosity between water and air. Because air is light and flows easily, it can be pumped in and out of closed sacks (the lungs) during respiration. (This is called tidal ventilation because the fluid moves in and out.) The density and viscosity of water are too large for tidal ventilation to be feasible, and the gills of bony fishes and chondrichthyans employ a one-way flow of water.

Oxygen Content Oxygen makes up approximately 20.9 percent of the volume of air. In other words, there are 209 milliters of oxygen in a liter of air. The oxygen content of water varies, but is never more than 50 ml of oxygen per liter of water and is often 10 ml or less. The low oxygen content of water compared to air is an additional reason why fishes do not use tidal ventilation.

Heat Capacity and Heat Conductivity The specific heat of water (defined as the energy needed to produce a 1-degree change in temperature in 1 gm of the fluid) is nearly 3400 times that of air, and water conducts heat almost 24 times as fast as air. Those differences are reflected in the thermal (temperature) characteristics of aquatic compared to terrestrial environments. The high specific heat of water means that the water temperature in a pond changes far less from day to night than does the air termperature on the shore of the pond. Thus, an animal living in the pond has a more stable thermal regime than an animal living on the shore. However, because the heat conductivity of water is so high, the temperature of the water in a shallow pond scarcely varies from place to place. If the water gets too hot, the only route of escape for an aquatic animal is to go to deeper water. That could require a dangerous long-distance movement for an animal living near the shore. In contrast, temperature varies substantially over short distances on land because the low heat conductivity of air allows temperature differences to develop between sunny and shady spots. Thus terrestrial animals have a mosaic of temperatures to choose from, whereas aquatic animals have much less temperature variation in their habitats.

Electrical Conductivity Water is an electrical conductor, and aquatic animals can use electricty to detect the presence of other animals and as an offensive or defensive weapon to stun their prey or predators. Because air does not conduct electricity (within the range of voltages that animals can generate), terrestrial animals cannot use electricity in these ways.

---

surfaces of the gills are delicate projections from the lateral side of each gill arch. Two columns of gill filaments extend from each gill arch. The tips of the filaments from adjacent arches meet when the filaments are extended. As water leaves the buccal cavity, it passes over the filaments. Gas exchange takes place at the numerous microscopic projections from the filaments called **secondary lamellae.**

The pumping action of the mouth and opercular cavities (**buccal pumping**) creates a positive

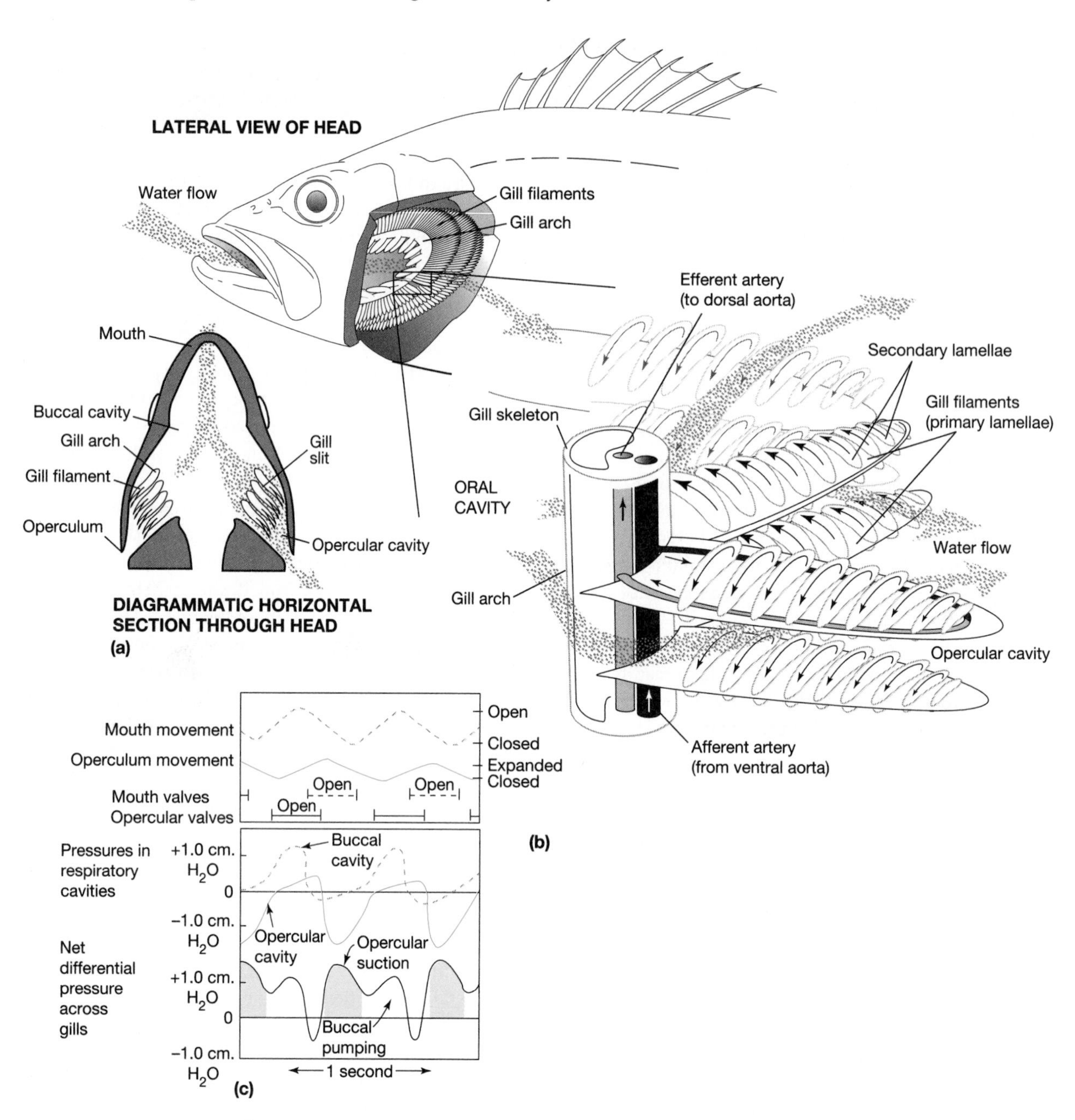

▲ Figure 4–1 Anatomy of teleost gills. (a) Position of gills in head and general flow of water; (b) countercurrent flow of water (blue arrows) and blood (black arrows) through the gills; (c) water pressure changes maintain a unidirectional flow of water across the gills during most phases of ventilation. The operculum forms a valve that closes to prevent backflow during the brief time that pressure in the buccal cavity is below external pressure.

pressure across the gills, so that the respiratory current is only slightly interrupted during each pumping cycle. The filter-feeding paddlefishes and many open-ocean fishes—such as mackerel, certain sharks, tunas, and swordfishes—have reduced or even lost the ability to pump water across the gills. A respiratory current is created by swimming with the mouth slightly open, a method known as **ram ventilation,** and these fishes must swim perpetually. Many other fishes rely on buccal pumping when they are at rest and switch to ram ventilation when they are swimming.

The vascular arrangement in the gills maximizes oxygen exchange. Each gill filament has two arteries, an afferent vessel running from the gill arch to the filament tip and an efferent vessel returning blood to the arch. Each secondary lamella is a blood space connecting the afferent and efferent vessels (Figure 4–2). The direction of blood flow through the lamellae is opposite to the direction of water flow across the gill. This structural arrangement, known as a **countercurrent exchanger,** assures that as much oxygen as possible diffuses into the blood. Pelagic fishes such as tunas, which sustain high levels of activity for long periods, have skeletal tissue reinforcing the gill filaments, large gill-exchange areas, and a high oxygen-carrying capacity per milliliter of blood compared with sluggish benthic fishes, such as toadfishes and flatfishes (Table 4.1).

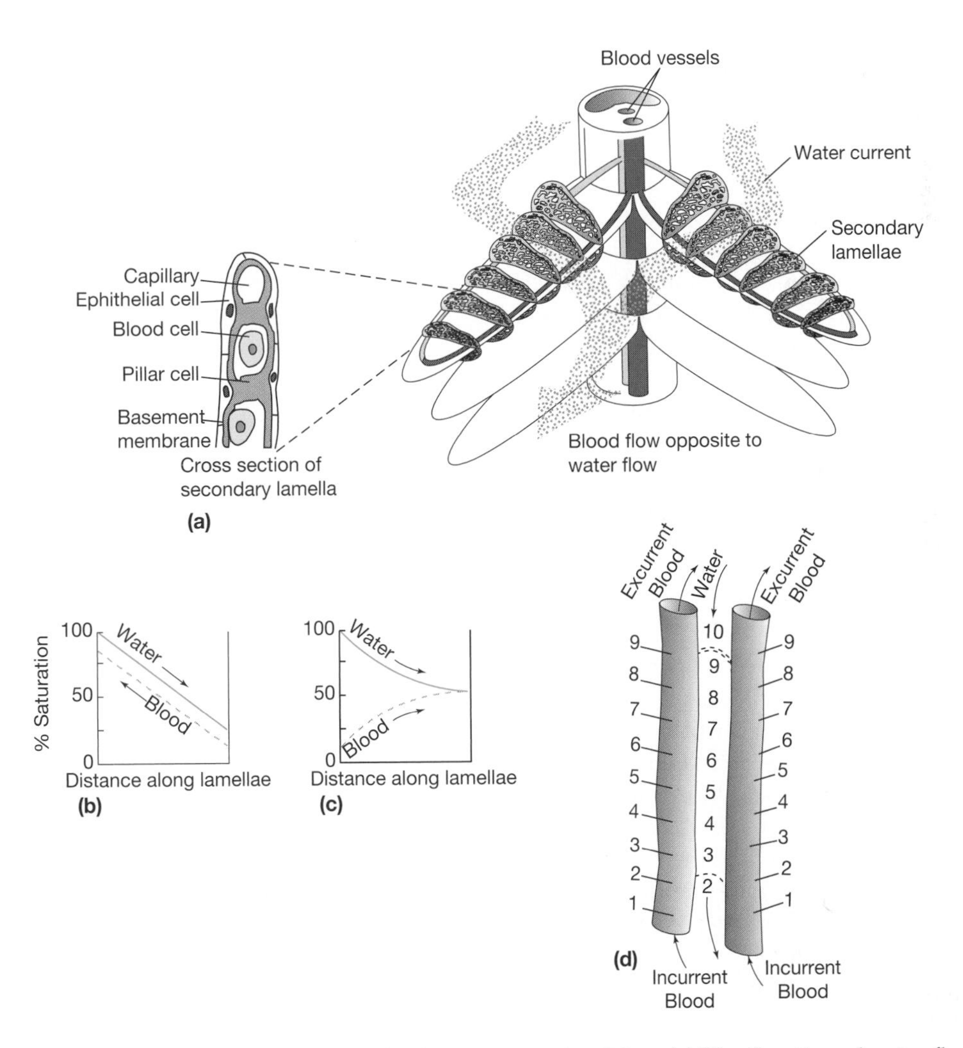

▲ Figure 4–2 Countercurrent exchange in the gills of actinopterygian fishes. (a) The direction of water flow across the gill opposes the flow of blood through the secondary lamellae. Blood cells are separated from oxygen-rich water only by the thin epithelial cells and capillary wall, as shown in the cross section of a secondary lamella. (b) Countercurrent flow results in a high oxygen concentration in the blood leaving the gills. (c) If water and blood flowed in the same direction, the blood leaving the gills would have a lower oxygen concentration. (d) Relative oxygen content of blood in secondary lamellae and the water passing over them.

Table 4.1 The relation among general level of activity, rate of oxygen consumption, respiratory structures, and blood characteristics in fishes of three activity levels

Activity Level	*Species of Fish*	*Oxygen Consumption ($mL\ O_2/g\ h^{-1}$)*	*No. Secondary Gill Lamellae (mm^{-1} of primary gill lamella)*	*Gill Area (mm^2/gm body mass)*	*Oxygen Capacity ($ml\ O_2/100\ mL$ blood)*
High	Mackerel* *(Scomber)*	0.73	31	1160	14.8
Intermediate	Porgy *(Stenotomus)*	0.17	26	506	7.3
Sluggish	Toadfish+ *(Ospanus)*	0.11	11	197	6.2

*Modified carangiform swimmers; swim continuously.
+Benthic fish.

Although the vast majority of fishes depend on gills to extract dissolved oxygen from water and to release carbon dioxide, fishes that live in low oxygen conditions cannot obtain enough oxygen via gills. These fishes have accessory respiratory structures that enable them to breathe the air. In the Amazon basin, where seasonal or even day-to-night variations in oxygen levels can be large, at least nine families of fishes have air-breathing representatives (Graham 1997). The surfaces used to take up oxygen from air in these Amazon fishes include enlarged lips, which are extended just above the water surface, and a variety of internal structures into which air is gulped. Among the air-breathing fishes are species that have modified the lining of the mouth, lung-like vascularized swim bladders, and even portions of the stomach and intestine into gas exchange sites. The anabantid fishes of tropical Asia (including the bettas and gouramies seen in pet stores) have vascularized chambers in the rear of the head, called labyrinths. Air is sucked into the mouth and transferred to the labyrinth, where gas exchange takes place. Many fishes are facultative air breathers; that is, oxygen uptake switches from gills to accessory respiratory structures when oxygen in the surrounding water becomes low. Others, like the electric eel and the anabantids, are obligatory air breathers. The gills alone cannot meet the respiratory needs of the fish, even if the surrounding water is saturated with oxygen. These fishes drown if they cannot reach the surface to breathe air.

South American and African lungfish are obligate air breathers. Lungs are evaginations of the gut, and similar structures can be found in the most primitive living ray-finned fishes, the polypterids. Air breathing is an ancestral characteristic of all osteichthyans, perhaps extending back to the Early Devonian acanthodians. The Early Devonian freshwater fishes probably encountered stresses like those faced by tropical freshwater fishes today. Early lungs may also have served as floats to regulate a fish's position in the water, as does the homologous swim bladder of teleosts and chondrosteans.

Adjusting Buoyancy

Because body tissues are heavier than water, animals sink unless vertical lift is created by swimming or their total density is reduced by a flotation device (Pelster 1998). Sharks and some specialized pelagic bony fishes, such as the scombroids (tunas, mackerels, and swordfishes), swim constantly to overcome gravity. Some of these fishes have large, wing-like pectoral fins that extend out from the body. The leading edges of the fins are angled upward, giving the fin a positive angle of attack and generating lift.

Many bony fishes are neutrally buoyant (i.e., have the same density as water). Neutral buoyancy results from an internal float—a gas-filled swim bladder lying below the spinal column (Figure 4–3). These fish do not have to swim to maintain their vertical

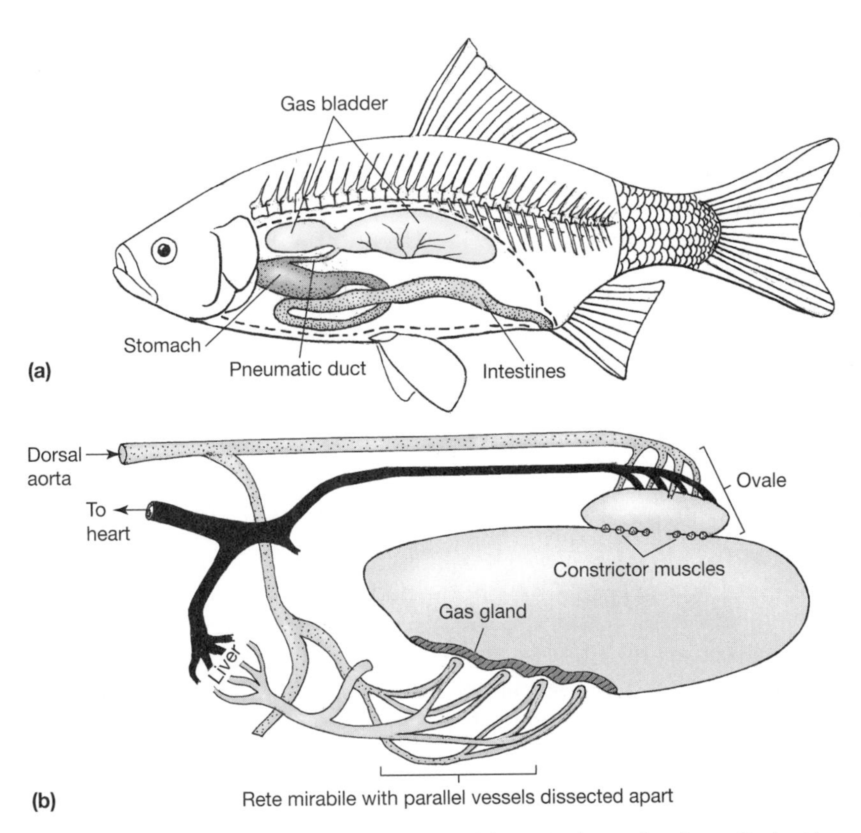

▲ Figure 4–3 Gas bladder of actinopterygians. (a) The gas bladder is in the coelomic cavity just beneath the vertebral column. In this physostomous fish, the gas bladder retains its ancestral connection to the gut via the pneumatic duct. (b) The vascular connections of a physoclistous swim bladder, which has no connection to the gut.

position in the water column. The only movement they make when at rest is backpedaling of the pectoral fins to counteract the forward thrust produced by water as it is ejected from the gills. Fishes capable of hovering in the water like this usually have well-developed swim bladders.

The swim bladder is located between the peritoneal cavity and the vertebral column. It is a gas-filled sac that arises as an evagination from the embryonic gut. The bladder occupies about 5 percent of the body volume of marine teleosts and 7 percent of the volume of freshwater teleosts. The difference in volume corresponds to the difference in density of salt water and freshwater—salt water is more dense, so a smaller swim bladder is needed. The swim bladder wall, which is composed of interwoven collagen fibers, is virtually impermeable to gas.

Neutral buoyancy produced by a swim bladder works as long as a fish remains at one depth; but if a fish swims vertically up or down, the water pressure changes and the volume of the gas bladder changes. For example, if a fish swims deeper the additional weight of the water column above it compresses the gas in its swim bladder, reducing its buoyancy. If the fish swims toward the surface, water pressure decreases, the swim bladder expands, and the fish becomes more buoyant. To maintain neutral buoyancy a fish must adjust the volume of gas in the swim bladder as it changes depth.

The fish regulates the volume of its swim bladder by secreting gas into the bladder when it swims down and removing gas when it swims up. Primitive teleosts, such as bony tongues, eels, herrings, anchovies, salmons, and the minnows, and their kin, retain a connection, the pneumatic duct, between the gut and swim bladder (Figure 4–3a). These fishes are called **physostomous** (Greek *phys* = bladder and *stom* = mouth), and goldfish are a familiar example of this group. Because they have a connection between the gut and the swim bladder, they can gulp air at the surface to fill the bladder and can burp gas out to reduce its volume.

The pneumatic duct is absent in adult teleosts from more derived clades, a condition termed **physoclistous** (Greek *clist* = closed). Physoclists regulate the volume of the swim bladder by secreting gas from the blood into the bladder. Both physostomes and physoclists have a gas gland, which is located in the anterior ventral floor of the swim bladder (Figure 4–3b). Underlying the gas gland is an area with many capillaries arranged to give countercurrent flow of blood entering and leaving the area. This structure, which is known as a **rete mirabile** (plural *retia mirabilia*), moves gas (especially oxygen) from the blood to the gas bladder. It is remarkably effective at extracting oxygen from the blood and releasing it into the swim bladder, even when the pressure of oxygen in the bladder is many times higher than its pressure in blood. Gas secretion occurs in many deep-sea fishes despite the hundreds of atmospheres of gas pressure within the bladder.

The gas gland secretes oxygen by releasing lactic acid, which acidifies the blood in the rete mirabile. Acidification causes hemoglobin to release oxygen into solution (the Bohr effect). Because of the anatomical relations of the rete mirabile, which folds back upon itself in a countercurrent multiplier arrangement, oxygen released from the hemoglobin accumulates and is retained within the rete until its pressure exceeds the oxygen pressure in the swim bladder. At this point oxygen diffuses into the bladder, increasing its volume. The maximum multiplication of gas pressure that can be achieved is proportional to the length of the capillaries of the rete mirabile, and deep-sea fishes have very long rete.

Physoclists have no connection between the swim bladder and the gut, so they cannot burp to release excess gas from the bladder. Instead, physoclists open a muscular valve, called the **ovale,** located in the posterior dorsal region of the bladder adjacent to a capillary bed. The high internal pressure of oxygen in the bladder causes it to diffuse into the blood of this capillary bed when the ovale sphincter is opened.

Sharks, rays, and ratfish do not have swim bladders. Instead, these fish use their liver to create neutral buoyancy. The average tissue densities of sharks with their livers removed are heavier than water—1.06 to 1.09 grams per milliliter. The liver of a shark, however, is well known for its high oil content (shark-liver oil). Shark liver tissue has a density of only 0.95 gm/mL, which is lighter than water, and the liver may contribute as much as 25 percent of the body mass. Thus, sharks use an oil-filled liver instead of a gas-filled swim bladder to regulate their buoyancy. A 4-meter tiger shark (*Galeocerdo cuvieri*) weighing 460 kg on land may weigh as little as 3.5 kg in the sea. Not surprisingly, bottom-dwelling sharks, such as nurse sharks, have livers with fewer and smaller oil vacuoles in their cells and these sharks are negatively buoyant.

Nitrogen-containing compounds in the blood of chondricthyans also contribute to their buoyancy (Withers et al. 1994). Urea and trimethylamine oxide in the blood and muscle tissue provide positive buoyancy because they are less dense than an equal volume of water. Chloride ions, too, are lighter than water and provide positive buoyancy, whereas sodium ions and protein molecules are denser than water and are negatively buoyant. The net effect of these solutes is a significant positive buoyancy.

Many deep-sea fishes, including coeloacanths, have deposits of light oil or fat in the gas bladder, and others have reduced or lost the gas bladder entirely and have lipids distributed throughout the body. These lipids provide static lift, just like the oils in shark livers. Because a smaller volume of the bladder contains gas, the amount of secretion required for a given vertical descent is less. Nevertheless, a long rete mirabile is needed to secrete oxygen at high pressures, and the gas gland in deep-sea fishes is very large. Fishes that migrate large vertical distances depend more on lipids such as wax esters than on gas for buoyancy, whereas their close relatives that do not undertake such extensive vertical movements depend more on gas for buoyancy.

Air in the lungs of air-breathing aquatic vertebrates reduces their density. Unlike fishes, air-breathing vertebrates must return to the surface at intervals, so they do not hover at one depth in the water column. Deep-diving animals, such as elephant seals and some whales and porpoises, face a different problem, however (Kooyman 1989). These animals dive to depths of 1000 m or more and are subjected to pressures more than 100 times higher than at the surface. Under those conditions nitrogen would be forced into solution in the blood and carried to the tissues at high pressure. When the animal rose toward the surface, the nitrogen would be released from solution. If the animal moved upward too fast, the nitrogen would not have time to diffuse into the blood and would form bubbles in the tissues—this is what happens when human deep-sea divers get the bends (decompression sickness). Specialized diving mammals avoid the problem by allowing the thoracic cavity to collapse as external pressure rises. Air is forced out of the lungs as they collapse, reducing the amount of nitrogen that diffuses into the blood. Even these specialized divers would have problems if they made repeated deep dives, however; a deep dive is normally followed by a period during which the animal remains near the surface and makes only shallow dives until the nitrogen level in its blood has equilibrated with the atmosphere.

4.2 Water and the Sensory World of Fishes

Water has properties that influence the behaviors of fishes and other aquatic vertebrates. Light is absorbed by water molecules and scattered by suspended particles. Objects become invisible at a distance of a few hundred meters even in the very clearest water, whereas distance vision is virtually unlimited in clean air. Fishes supplement vision with other senses, some of which can operate only in water. The most important of these aquatic senses is detecting water movement via the lateral line system. Small currents of water can stimulate the sensory organs of the lateral line because water is dense and viscous. Electric sensitivity depends on the properties of water and does not operate in air. In this case it is the electrical conductivity of water that is the key. And even vision is different in water and air because of the different refractive properties of the two media.

Vision

Vertebrates generally have well-developed eyes, but the way an image is focused on the retina is different in terrestrial and aquatic animals. Air, by definition, has an index of refraction of 1.00. The cornea in the eye of both terrestrial and aquatic vertebrates has an index of refraction of about 1.37, which is only slightly greater than the index of refraction of water (1.33). Light rays are bent as they pass through a boundary between media with different refractive indices, and the amount of bending is proportional to the difference in indices of refraction. Because the index of refraction of the cornea is substantially different from that of air, light rays are bent as they pass from air into the cornea, making the cornea an important part of the focusing system of the eyes of terrestrial vertebrates. This relationship does not hold in water, because the refractive index of the cornea is too close to that of water for the cornea to have much effect in bending light. Terrestrial vertebrates rely on the lens, which often is flattened and pliable, for detailed focus. Fishes have a less pliable spherical lens with high refractive power. The entire lens is moved toward or away from the retina to focus images. A similar spherical lens has evolved in aquatic mammals, such as cetaceans. Otherwise, a fish's eye is similar to the eye of a terrestrial vertebrate.

Fishes have taste-bud organs in the mouth and around the head and anterior fins. In addition, receptors of general chemical sense detect substances that are only slightly soluble in water, and olfactory organs on the snout detect soluble substances. Sharks and salmon can detect odors at concentrations of less than 1 part per billion. Homeward-migrating salmon are directed to their stream of origin from astonishing distances by a chemical signature from the home stream that was permanently imprinted when they were juveniles. Plugging the nasal olfactory organs of salmon destroys their ability to home.

Mechanical receptors provide the basis for detection of displacement—touch, sound, pressure, and motion. Like all vertebrates, fishes have an internal ear (the labyrinth organ, not to be confused with the organ of the same name that assists in respiration in anabantid fishes) that detects changes in speed and direction of motion. Fishes also have gravity detectors at the base of the semicircular canals that allow them to distinguish up from down. Most terrestrial vertebrates also have an auditory region of the inner ear that is sensitive to sound-pressure waves. These diverse functions of the labyrinth depend on basically similar types of sense cells, the hair cells (Figure 4–4). In fishes and aquatic amphibians, clusters of hair cells and associated support cells form **neuromast** organs that are dispersed over the surface of the head and body. In jawed fishes neuromast organs are often located in a series of canals on the head, and one or more canals pass along the sides of the body onto the tail. This surface receptor system of fishes and aquatic amphibians is referred to as the

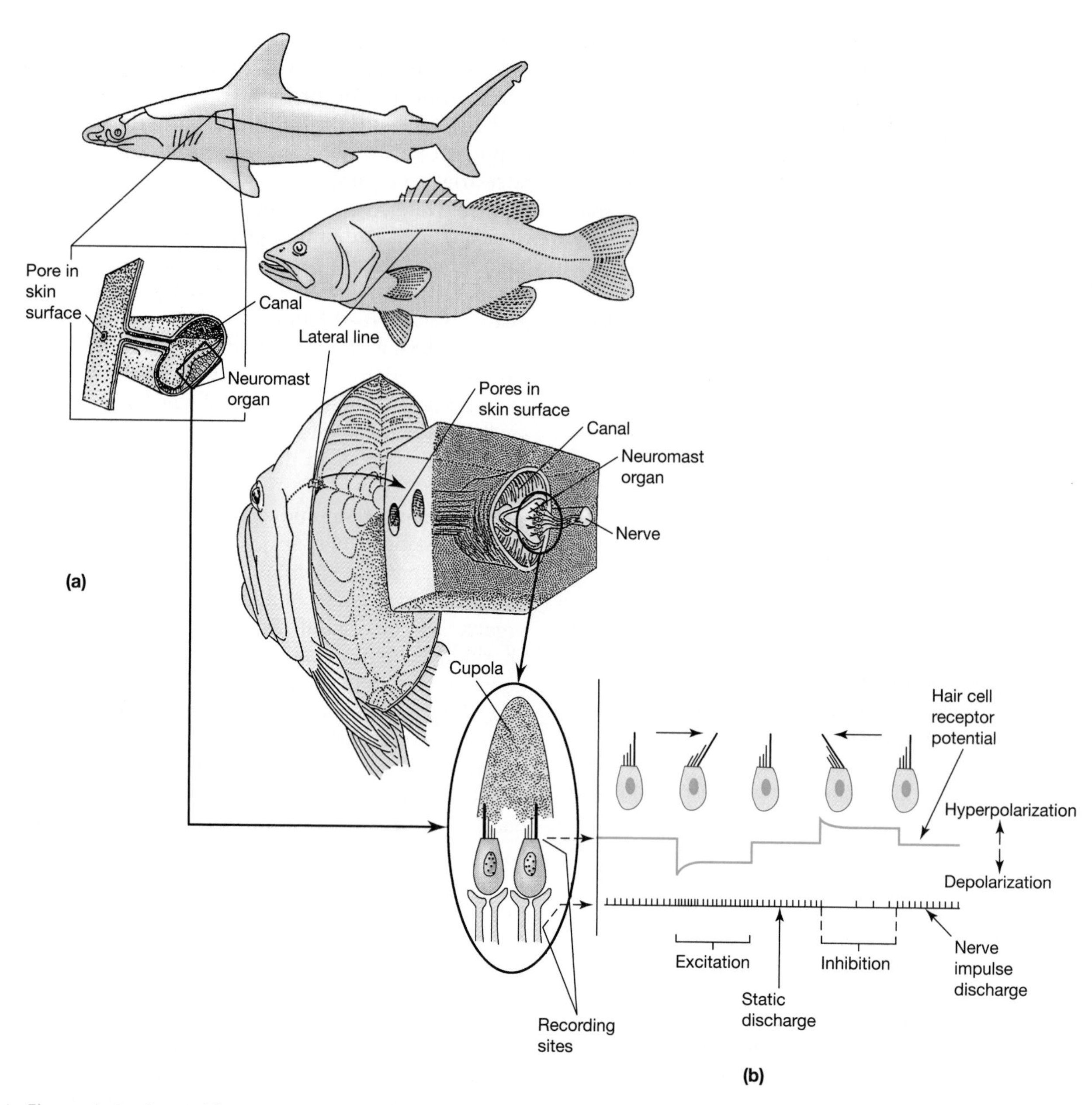

▲ Figure 4–4 Lateral line systems. (a) Semidiagrammatic representations of the two configurations of lateral line organs in fishes. (b) Hair cell deformations and their effect on hair cell transmembrane potential (receptor potential) and afferent nerve-cell discharge rates. Deflection of the kinocilium (dark line) in one direction reduces discharge, and deflection in the opposite direction increases discharge.

lateral line system. Lateral line systems are found only in aquatic vertebrates, because air is not dense enough to stimulate the neuromast organs. Amphibian larvae have lateral line systems, and lateral lines are retained in adults of aquatic amphibians such as African clawed frogs and mudpuppies, but are lost in the adults of terrestrial species.

Detecting Water Displacement—The Lateral Line

Neuromasts of the lateral line system are distributed in two configurations—within tubular canals or exposed in epidermal depressions. Many kinds of fishes have both arrangements. Hair cells have a

kinocilium placed asymmetrically in a cluster of **microvilli.** Hair cells are arranged in pairs with the kinocilia positioned on opposite sides of adjacent cells. A neuromast contains many such hair cell pairs. Each neuromast has two afferent nerves: One transmits impulses from hair cells with kinocilia in one orientation, and the other carries impulses from cells with kinocilia positions reversed by 180 degrees. This arrangement allows a fish to determine the direction of displacement of the kinocilia.

All kinocilia and microvilli are embedded in a gelatinous structure, the **cupula.** Displacement of the cupula causes the kinocilia to bend. The resultant deformation either excites or inhibits the neuromast's nerve discharge. Each hair cell pair, therefore, signals the direction of cupula displacement. The excitatory output of each pair has a maximum sensitivity to displacement along the line joining the kinocilia and falling off in other directions. The net effect of cupula displacement is to increase the firing rate in one afferent nerve and to decrease it in the other nerve. These changes in lateral line nerve firing rates thus inform a fish of the direction of water currents on different surfaces of its body.

Water currents of only 0.025 millimeter per second are detected by the exposed neuromasts of the African clawed frog, *Xenopus laevis,* with maximum response to currents of 2 or 3 mm per second. Similar responses occur in fishes. The lateral line organs also respond to low-frequency sound, but controversy exists as to whether sound is a natural lateral line stimulus. Sound induces traveling pressure waves in the water and also causes local water displacement as the pressure wave passes. It has been difficult to be sure whether neuromast output results from the water motions on the body surface or the sound's compression wave.

Several surface-feeding fishes and African clawed frogs provide vivid examples of how the lateral line organs act under natural conditions. These animals find insects on the water surface by detecting surface waves created by the prey's movements. Each neuromast group on the head of the killifish, *Aplocheilus lineatus,* provides information about surface waves coming from a different direction (Figure 4–5). All groups, however, have overlapping stimulus fields. Removing a neuromast group from one side of the head disturbs the directional response to stimuli, showing that a fish combines information from groups on both sides of the head to interpret water movements.

The large numbers of neuromasts on the heads of some fishes might be important for sensing vortex

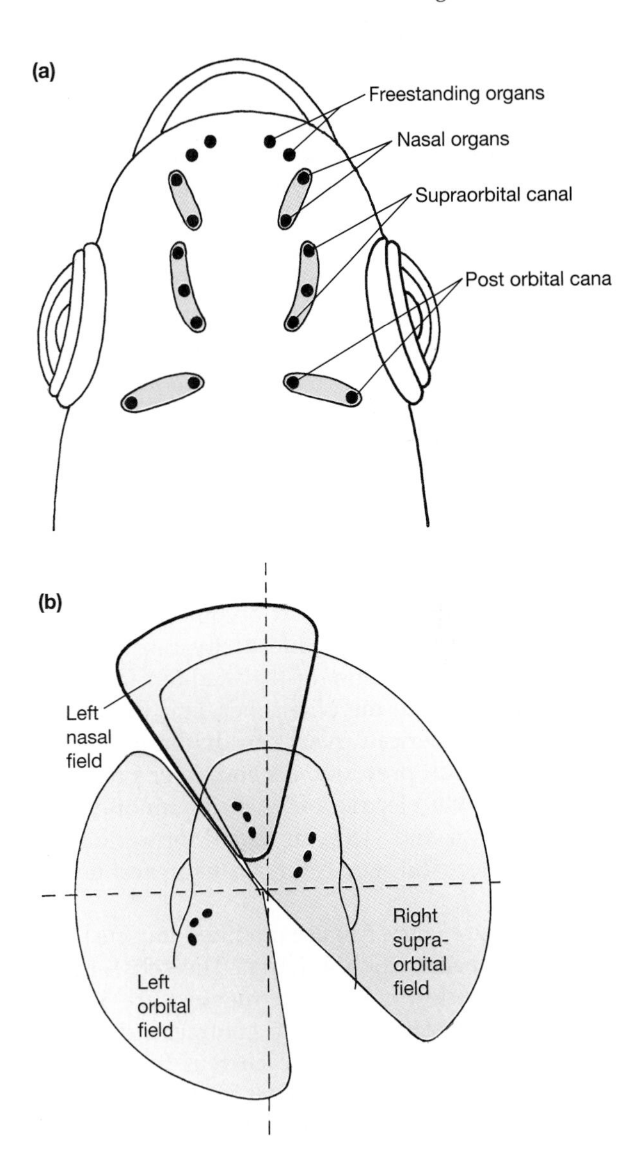

▲ Figure 4–5 Distribution of the lateral line canal organs. (a) On the dorsal surface of the head of the killifish *Fundulus notatus,* and (b) the perceptual fields of the head canal organs in another killifish, *Aplocheilus lineatus.* The wedge-shaped areas indicate the fields of view for each group of canal organs. Note that fields overlap on opposite sides as well as on the same side of the body, allowing the lateral line system to localize the source of a water movement.

trails in the wakes of adjacent fishes in a school. Many of the fishes that form extremely dense schools (herrings, atherinids, mullets) lack lateral line organs along the flanks and retain canal organs only on the head. These well-developed cephalic canal organs concentrate sensitivity to water motion in the head region, where it is needed to sense the turbulence into which the fish is swimming, and the reduction of flank lateral line elements would reduce noise from turbulence beside the fish.

Electric Discharge

Unlike air, water conducts electricity, and seawater is a better conductor of electricity than freshwater because it contains dissolved salts. The high conductivity of seawater makes it possible for sharks to detect the electrical activity that accompanies muscle contractions of their prey. Electricity can also be a weapon—the torpedo ray of the Mediterranean, the electric catfish from the Nile River, and the electric eel of South American rivers can discharge enough electricity to stun prey animals and deter predators. And the weakly electric knifefish (Gymnotidae) of South America and elephant fish (Mormyridae) of Africa use electrical signals for courtship and territorial defense.

All of these electric fish use modified muscle tissue to produce the electric discharge. The cells of such modified muscles, called **electrocytes,** are muscle cells that have lost the capacity to contract and are specialized for generating an ion current flow (Figure 4–6). When at rest, the membranes of muscle cells and nerve cells are electrically charged, with the intracellular fluids about 84 mV, negative relative to the extracellular fluids. The imbalance is primarily due to sodium ion exclusion. When the cell is stimulated, sodium ions flow rapidly across the smooth surface, sending its potential to a positive 67 mV. Only the smooth surface depolarizes; the rough surface remains at –84 mV, so the potential difference across the cell is 151 mV (from –84 to +67 mV). Because electrocytes are arranged in stacks like the batteries in a flashlight, the potentials of many layers of cells combine to produce high voltages. The South American electric eel has up to 10,000 layers of cells and can generate potentials in excess of 600 volts.

Most electric fish are found in tropical freshwaters of Africa and South America. Few marine forms can generate specialized electric discharges—among marine elasmobranchs only the torpedo ray *(Torpedo),* the ray genus *Narcine,* and some skates are electric; and among marine teleosts, only the stargazers (family Uranoscopidae) produce specialized discharges.

Electroreception by Elasmobranchs

Many fishes, especially elasmobranchs, are able to detect electric fields. Sharks have structures known as the **ampullae of Lorenzini** on their heads, and rays have them on the pectoral fins as well. The ampullae are sensitive electroreceptors (Figure 4–7). The canal connecting the receptor to the surface pore is filled with an electrically conductive gel, and the wall of the canal is nonconductive. Because the canal runs for some distance beneath the epidermis, the sensory cell can detect a difference in electrical potential between the tissue in which it lies (which reflects the adjacent epidermis and environment) and the distant pore opening. Thus, it can detect electrical fields, which are changes in electrical potential in space. The ancestral electroreceptor cell was a modification of the hair cells of the lateral line. Electroreceptors of elasmobranchs respond to minute changes in the electrical field surrounding an animal. They act like voltmeters, measuring a difference in electric potentials at discrete locations across the body surface. Voltage sensitivities are remarkable: Ampullary organs have thresholds lower than 0.01 microvolt per centimeter, a level of detection achieved only by the best voltmeters.

Elasmobranchs use their electric sensitivity to detect prey. A shark can locate and attack a hidden fish by relying only on the electrical discharges produced by the prey (Figure 4–8). All muscle activity generates electric potential: Motor nerve cells produce extremely brief changes in electrical potential, and muscular contraction generates changes of longer duration. In addition, a steady potential issues from an aquatic organism as a result of the chemical imbalance between the organism and its surroundings.

Elasmobranchs may use electroreception for navigation as well. The electromagnetic field at the Earth's surface produces tiny voltage gradients, and a swimming shark could encounter gradients as large as 0.4 mV per centimeter—well above the level that can be detected by ampullary organs. In addition, ocean currents generate electric gradients as large as 0.5 mV/cm as they carry ions through the Earth's magnetic field.

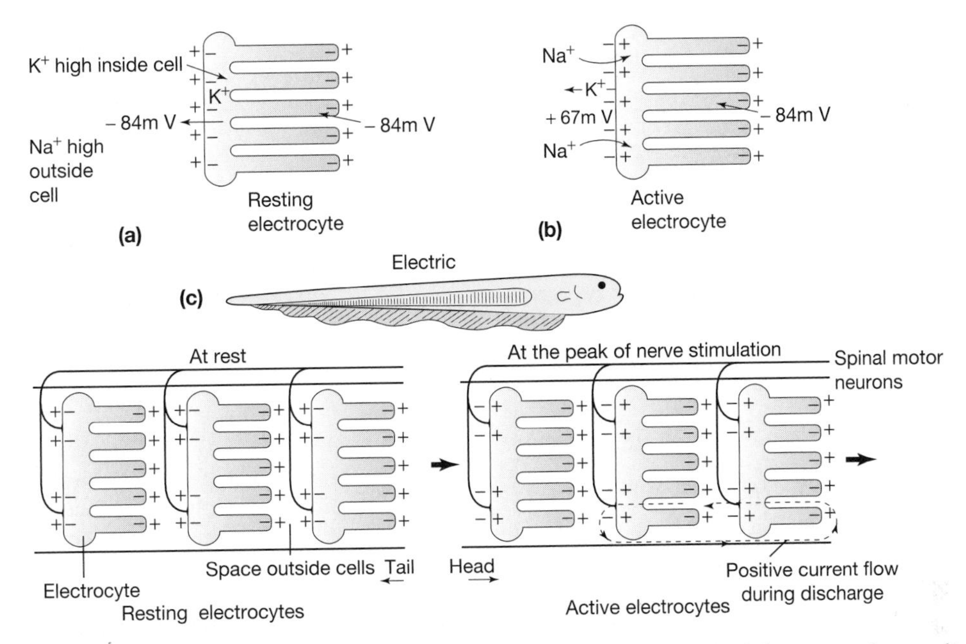

▲ Figure 4–6 Use of transmembrane potentials of modified muscle cells by electric fishes to produce a discharge. In this diagram the smooth surface is on the left and the rough surface on the right. Only the smooth surface is innervated. (a) At rest, K^+ (potassium ion) is maintained at a high internal concentration and Na^+ (sodium ion) at a low internal concentration by the action of a Na^+/K^+ cell-membrane pump. Permeability of the membrane to K^+ exceeds the permeability to Na^+. As a result, K^+ diffuses outward faster than Na^+ diffuses inward (arrow) and sets up the –84-mV resting potential. (b) When the smooth surface of the cell is stimulated by the discharge of the nerve, sodium diffuses into the cell and potassium diffuses out of the smooth surface, changing the potential to +67 mV. The rough surface does not depolarize and retains a –84-mV potential, creating a potential difference of 151 mV across the cell. (c) By arranging electrocytes in series to sum the potentials of individual cells, some electric fishes can generate very high voltages. Electric eels, for example, have 10,000 electrocytes in series and produce potentials in excess of 600 volts.

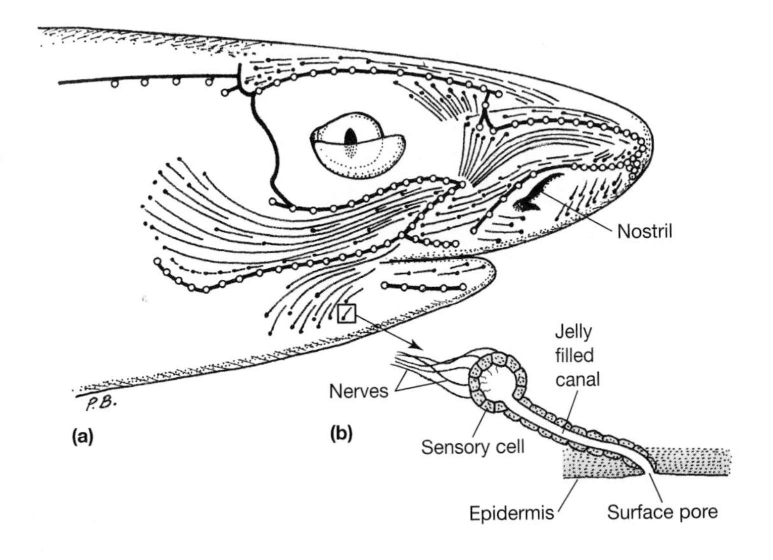

▶ Figure 4–7 Ampullae of Lorenzini. (a) Distribution of the ampullae on the head of a spiny dogfish, *Squalus acanthius*. Open circles represent the surface pores; the black dots are positions of the sensory cells. (b) A single ampullary organ consists of a sensory cell connected to the surface by a pore filled with a substance that conducts electricity.

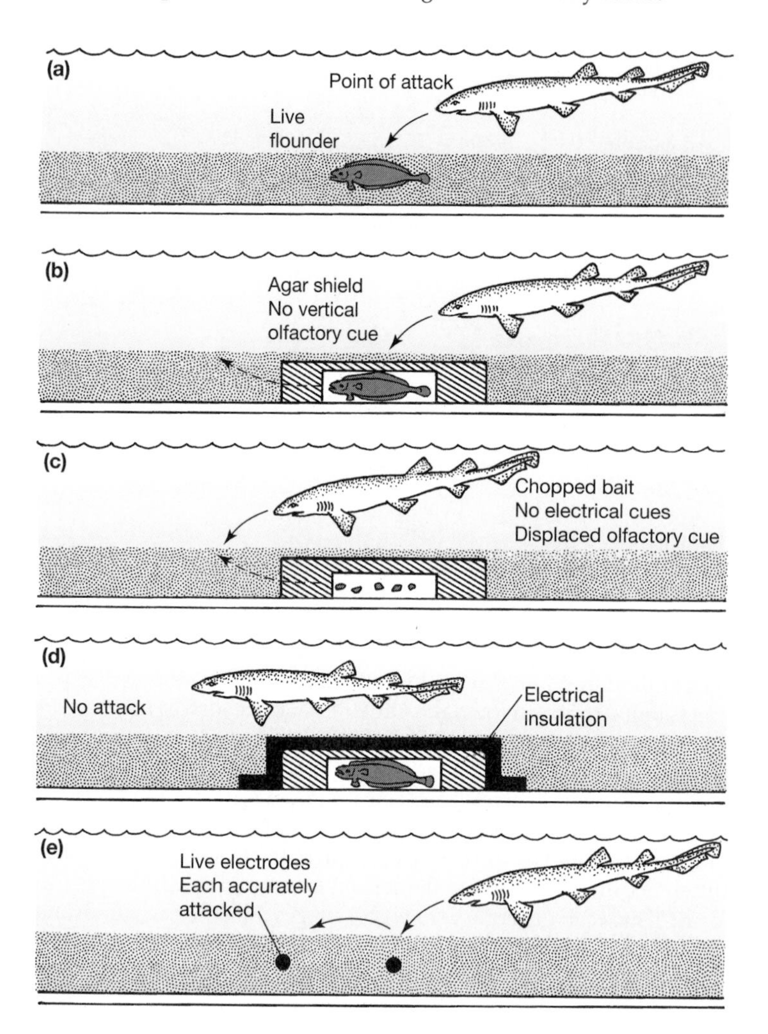

Figure 4–8 Electrolocation capacity of elasmobranchs. (a) A shark can locate a live fish concealed from sight beneath the sand. (b) The shark can still detect the fish when it is covered by an agar shield that blocks olfactory cues but allows the electrical signal to pass. (c) The shark follows the olfactory cues (displaced by the agar shield) when the live fish is replaced by chopped bait that produces no electrical signal. (d) The shark is unable to detect a live fish when it is covered by a shield that blocks both olfactory cues and the electrical signal. (e) The shark attacks electrodes that give off an electrical signal duplicating a live fish without producing olfactory cues. These experiments indicate that when the shark was able to detect no electrical signal, it used that to locate the fish—and it was also capable of homing on a chemical signal when no electrical signal was present. This dual system allows sharks to find both living and dead food items.

Electrolocation by Teleosts

Unusual arrangements of electrocytes are present in several species of fishes that do not produce electric shocks. In these fishes—which include the knifefishes (Gymnotidae) of South America and the elephant fish (Mormyridae) of Africa—the discharge voltages are too small to be of direct defensive or offensive value. Instead, weakly electric teleost fishes use their discharges for electrolocation and social communications (von der Emde 1998). When a fish discharges its electric organ, it creates an electric field in its immediate vicinity (Figure 4–9). Because of the high energy costs of maintaining a continuous discharge, electric fishes produce a pulsating discharge. Most weakly electric teleost fishes pulse at rates between 50 and 300 cycles per second, but the knifefishes of South America reach 1700 cycles per second, which is the most rapid, continuous firing rate known for any vertebrate muscle or nerve. African and South American electric fishes are mostly nocturnal and usually live in turbid waters where vision is limited to short distances even in daylight.

The electric field from even weak discharges may extend outward for a considerable distance in freshwater, because electric conductivity is relatively low. The electric field the fish creates will be distorted by the presence of conductive and resistant objects. Rocks are highly resistive, whereas other fishes, invertebrates, and plants are conductive. Distortions of the field cause a change in the distribution of electric potential across the fish's body surface. An electric fish detects the presence, position, and movement of objects by sensing where on its body maximum distortion of its electric field occurs.

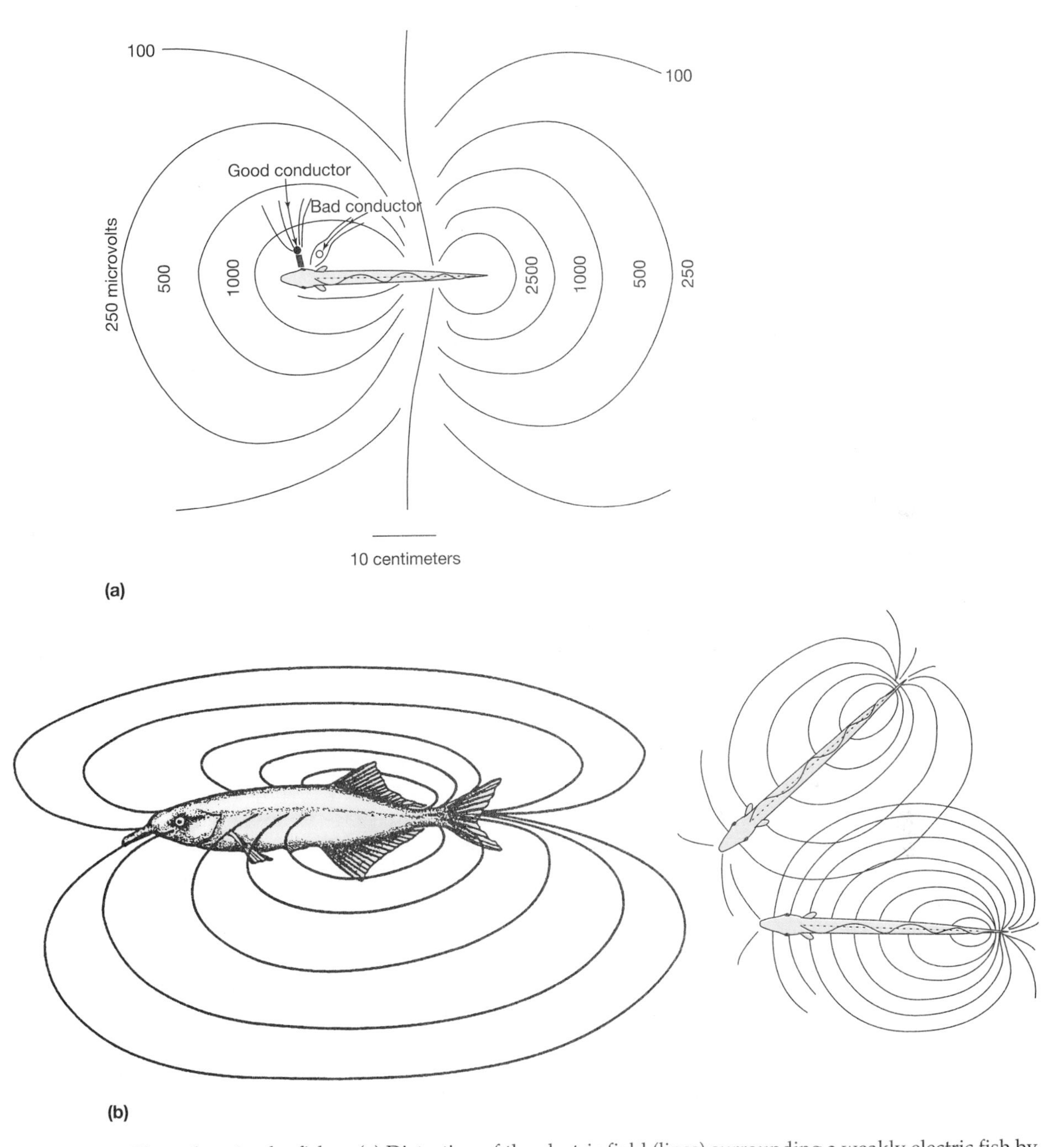

▲ Figure 4–9 Electrolocation by fishes. (a) Distortion of the electric field (lines) surrounding a weakly electric fish by conductive and nonconductive objects. (Reference electrode 150 cm lateral to fish.) Conductive objects concentrate the field on the skin of the fish, where the increase in electrical potential is detected by the electroreceptors. Nonconductive objects spread the field and diffuse potential differences along the body surface. (b) The electric field surrounding a weakly electric fish can be modulated by variations in the electric organ discharge for communication (left). When two electric fish swim close enough to one another, significant interference can occur, requiring changes in the electric organ discharge (right).

The skin of weakly electric teleosts contains special sensory receptors: ampullary organs and tuberous organs. These organs detect tonic (steady) and phase (rapidly changing) discharges, respectively. Electroreceptors of teleosts are modified lateral line neuromast receptors. Like lateral line receptors, they have double innervation—an afferent channel that sends impulses to the brain and an efferent channel that causes inhibition of the receptors. During each electric organ discharge, an inhibitory command is sent to the electroreceptors and the fish is rendered insensitive to its own discharge. Between pulses, electroreceptors report distortion in the electric field or the presence of a foreign electric field to the brain.

Electric organ discharges vary with habits and habitat. Species that form groups or live in shallow, narrow streams generally have discharges with high frequency and short duration. These characteristics reduce the chances of interference from the discharges of neighbors. Territorial species, in contrast, have long electric organ discharges. Electric organ discharges vary from species to species. In fact, some species of electric fishes were first identified by their electric organ discharges. During the breeding season, electric organ discharges distinguish immature individuals, ripe females, and sexually active males of some species.

An electrical discharge advertises the presence of a male fish to potential mates, but predators can listen in as well. Catfish and electric eels are important predators of South American knifefish in some parts of their geographic range. These predators use the electrosensitivity of their ampullary organs to locate a knifefish by homing on its courtship signals. Ampullary organs are most sensitive to low discharge frequencies, however, and knifefish have used this limitation to avoid detection. The knifefish have shifted from a monophasic discharge to a disphasic discharge at a higher frequency, and they have evolved a second set of electroreceptors—tuberous organs—that are sensitive to the higher frequency. As a result, the electrical signals of the knifefish are emitted at a frequency that other individuals of their species can detect but that is above the range of greatest sensitivity for their predators (Stoddard 1999).

Electrogenesis and electroreception are not restricted to a single group of aquatic vertebrates, and monotremes (the platypus and the echidna, early offshoots off the main mammalian lineages that still lay eggs) use electroreception to detect prey (Figure 4–10). Electrosensitivity was probably an early feature of vertebrate evolution. The brain of the lamprey responds to electric fields, and it seems likely that the earliest vertebrates had electroreceptive capacity. All fish-like vertebrates of lineages that evolved before the earliest neopterygians (represented today by gars and *Amia*) have electroreceptor cells. These cells, which have a prominent kinocilium, fire when the environment around the kinocilium is negative relative to the cell. Their impulses pass to the medial region of the posterior third of the brain. Electrosensitivity was apparently lost in neopterygians, and teleosts have at least two separate new evolutions of electroreceptors. Electrosensitivity in teleosts is distinct from that of other vertebrates: Teleost electroreceptors *lack* a kinocilium and fire when the environment is *positive* relative to the cell, and nerve impulses are sent to the *lateral* aspect of the rhombencephalon.

4.3 The Internal Environment of Vertebrates

Seventy to eighty percent of the body mass of most vertebrates is water, and the chemical reactions that release energy or synthesize new chemical compounds take place in an aqueous environment. The body fluids of vertebrates contain a complex mixture of ions and other solutes. Some ions are cofactors that control the rates of metabolic processes; others are involved in the regulation of pH, the stability of cell membranes, or the electrical activity of nerves. Metabolic substrates and products must diffuse from sites of synthesis to the sites of utilization. Almost everything that happens in the body tissues of vertebrates involves water, and maintaining the concentrations of water and solutes within narrow limits is a vital activity. Water sounds like an ideal place to live for an animal that is itself mostly made of water, but in some ways an aquatic environment can be too much of a good thing. Freshwater vertebrates—especially fishes and amphibians—face the threat of being flooded with water that flows into them from their environment, and saltwater vertebrates must prevent the water in their bodies from being sucked out into the sea.

Temperature, too, is a critical factor for living organisms, because chemical reactions are temperature sensitive. In general, the rates of chemical reactions

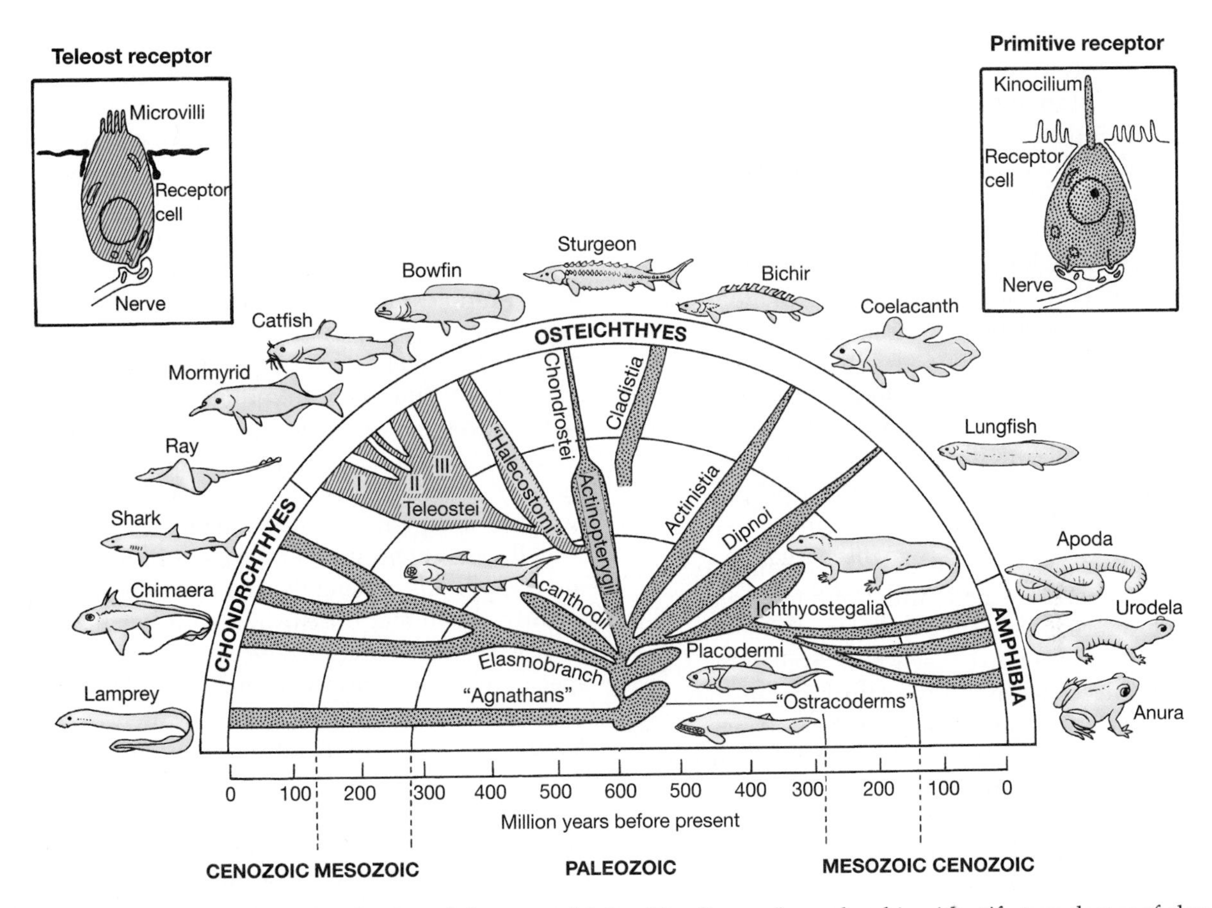

▲ Figure 4–10 Phylogenetic distribution of electrosensitivity. Stippling and crosshatching identify two classes of electroreceptors as indicated in the upper right and left panels.

increase as temperature increases, but not all reactions have the same sensitivity to temperature. Furthermore, the permeability of cell membranes and other features of the cellular environment are sensitive to temperature. A metabolic pathway is a series of chemical reactions in which the product of one reaction is the substrate for the next, yet each of these reactions may have a different sensitivity to temperature, so a change in temperature can mean that too much or too little substrate is produced to sustain the next reaction in the series. To complicate the process of regulation of substrates and products even more, the chemical reactions take place in a cellular milieu that itself is changed by temperature. Clearly, the smooth functioning of metabolic pathways is greatly simplified if an organism can limit the range of temperatures its tissues experience.

Water temperature is more stable than air temperature, because water has a much higher specific heat than air. The stability of water temperature simplifies the task of maintaining a constant body temperature, as long as the body temperature the animal needs to maintain is the same as the temperature of the water around it. An aquatic animal has a hard time maintaining a body temperature different from water temperature, however, because water conducts heat so well. Heat flows out of the body if an animal is warmer than the surrounding water and into the body if the animal is cooler than the water.

In the following sections we discuss in more detail how and why vertebrates regulate their internal environments and the special problems faced by aquatic animals.

4.4 Exchange of Water and Ions

An organism can be described as a leaky bag of dirty water. That is not an elegant description, but it accurately identifies the two important characteristics of a living animal—it contains organic and inorganic substances dissolved in water and this fluid is enclosed by a permeable body surface. Exchange of matter and energy with the environment is essential to the survival of the organism, and much of that exchange is regulated by the body surface. Water molecules and ions pass through the skin quite freely, whereas larger molecules move less readily. The significance of this differential permeability to various compounds is particularly conspicuous in the case of aquatic vertebrates, but it applies to terrestrial vertebrates as well. Vertebrates use both active and passive exchange to regulate their internal concentrations in the face of varying external conditions.

The Vertebrate Kidney

An organism can tolerate only a narrow range of concentrations of the body fluids and must eliminate waste products before they reach harmful levels. The molecules of ammonia that result from protein catabolism are especially important because they are toxic. Vertebrates have evolved superb capacities for controlling water balance and excreting wastes, and the kidney plays a crucial role in these processes.

The adult vertebrate kidney consists of hundreds to millions of tubular **nephrons,** each of which produces urine. The primary function of a nephron is removing excess water, salts, waste metabolites, and foreign substances from the blood. In this process, the blood is first filtered through the **glomerulus,** a structure unique to vertebrates (Figure 4–11). Each glomerulus is composed of a leaky arterial capillary tuft encapsulated within a sieve-like filter. Arterial blood pressure forces fluid into the nephron to form an **ultrafiltrate,** composed of blood minus blood cells and larger molecules. The ultrafiltrate is then processed to return essential metabolites (glucose, amino acids, and so on) and water to the general circulation.

Regulation of Ions and Body Fluids

The salt concentrations in the body fluids of many marine invertebrates are similar to those in seawater, as are those of hagfishes (Table 4.2). It is likely that the first vertebrates also had ion levels similar to those in seawater. In contrast, solute levels are greatly reduced in the blood of all other vertebrates, a characteristic shared only with invertebrates that have penetrated estuaries, fresh waters, or the terrestrial environment.

The presence of solutes in seawater or blood plasma lowers the kinetic activity of water. Therefore, water flows from a dilute solution (high kinetic activity of water) to a more concentrated solution (low kinetic activity)—a phenomenon called **osmosis.** The osmotic concentrations of various animals and of seawater are shown in Table 4.2. Seawater has a concentration of approximately 1000 millimoles per kg of water [mmoles $\cdot$ kg^{-1}]. Most marine invertebrates and hagfishes have body fluids that are in osmotic equilibrium with seawater; that is, they are **isosmolal** to seawater. Body fluid concentrations in marine teleosts and lampreys are between 350 and 450 mmoles $\cdot$ kg^{-1}. Therefore, water flows outward from their blood to the sea (i.e., from a region of high kinetic activity of water to a region of lower kinetic activity). Chondrichthyans retain urea and other nitrogen-containing compounds, raising the osmolality of their blood slightly above that of seawater so water flows from the sea into their bodies. These osmolal differences are specified by the terms **hyposmolal** (lower solute concentrations than the surrounding water, as seen in marine teleosts and lampreys) and **hyperosmolal** (higher solute concentrations than the surrounding water, as seen in coelacanths and chondrichthyans). Freshwater fishes are hyperosmolal to the medium, but through reduction in NaCl their blood osmolality is lower than that of their marine counterparts (Table 4.2). Salt ions can also diffuse through the surface membranes of an animal, so the water and salt balance of an aquatic vertebrate is constantly threatened by outflow of water and inflow of salt in seawater and by inflow of water and outflow of salts in freshwater.

Most fishes are **stenohaline** (Greek *steno* = narrow and *haline* = salt); they inhabit either freshwater or seawater and tolerate only modest changes in salinity. Because they remain in one environment, the magnitude and direction of the osmotic gradient to which they are exposed is stable. Some fishes, however, are **euryhaline** (Greek *eury* = wide); they inhabit both freshwater and seawater and tolerate large changes in salinity. The water and salt gradients are reversed in euryhaline species as they move from one medium to the other.

Freshwater Organisms—Teleosts and Amphibians Several mechanisms are involved in the salt and water regulation of vertebrates that live in freshwater. The body surface of fishes has low permeability to water and to ions. However, fishes cannot entirely prevent osmotic exchange. Gills, which must be permeable to oxygen and carbon dioxide, are also

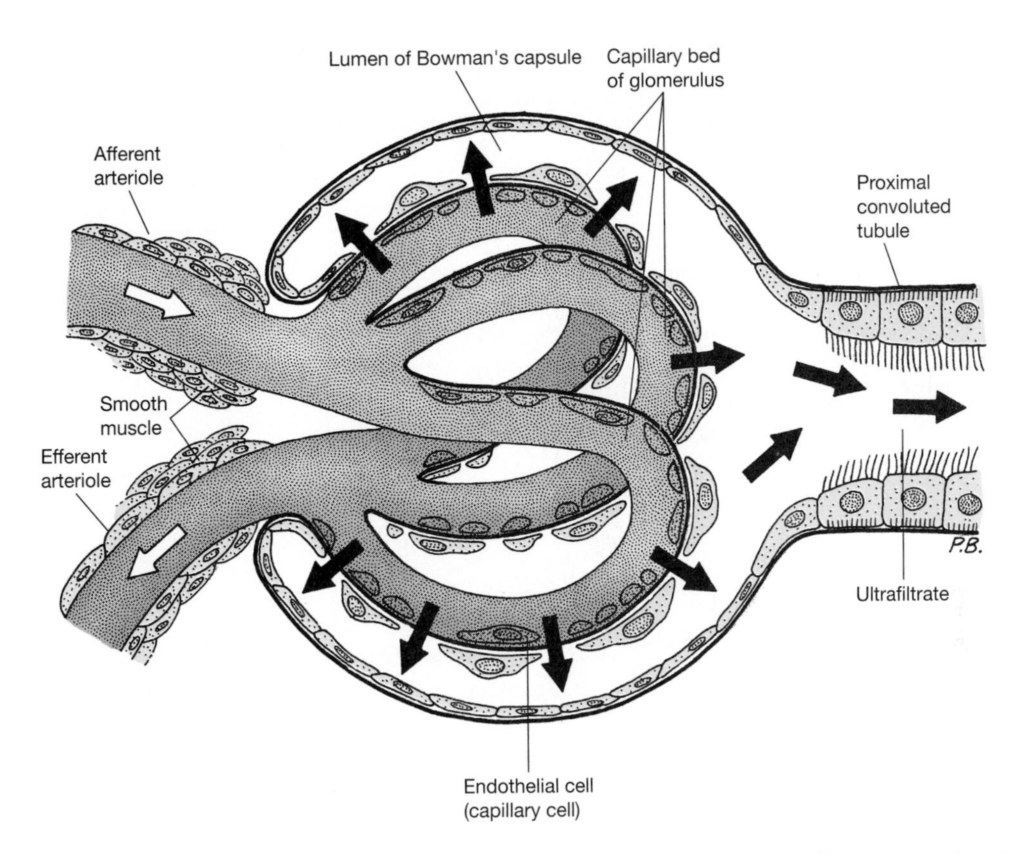

▲ Figure 4–11 Detail of a typical mammalian glomerulus, illustrating the morphological basis for its function. Blood pressure forces an ultrafiltrate of the blood through the walls of the capillary into the lumen of Bowman's capsule. The blood flow to each glomerulus is regulated by smooth muscles that can close off the afferent and efferent arterioles to adjust the glomerular filtration rate of the kidney as a whole. The ultrafiltrate, which consists of water, ions, and small molecules, passes from Bowman's capsule into the proximal convoluted tubule, where the process of adding and removing specific substances begins.

permeable to water. As a result, most water and ion movements take place across the gill surfaces. Water is gained by osmosis, and ions are lost by diffusion. A freshwater teleost does not drink water, because osmotic water movement is already providing more water than it needs—drinking would only increase the amount of water it had to excrete via the kidneys. To compensate for this influx of water, the kidney produces a large volume of urine. Salts are actively reabsorbed to reduce salt loss. Indeed, urine processing in a freshwater teleost provides a simple model of vertebrate kidney function.

The large glomeruli of freshwater teleosts produce a copious flow of urine, but the glomerular ultrafiltrate is isosmolal to the blood and contains essential blood salts (Figure 4–12). To conserve salt, ions are reabsorbed across the **proximal and distal convoluted tubules.** (Reabsorption is an active process that consumes metabolic energy.) Because the distal convoluted tubule is impermeable to water, the urine becomes less concentrated as ions are removed from it. Ultimately, the urine becomes hyposmolal to the blood. In this way the water that was absorbed across the gills is removed and ions are conserved. Nonetheless, some ions are lost in the urine in addition to those lost by diffusion across the gills. Salts from food compensate for some of this loss, and teleosts have cells in the gills (**chloride cells**) that actively transport chloride ions. In freshwater these cells take up chloride ions from the water. The chloride ions are moved by active transport against a concentration gradient, and this process requires energy. Sodium ions also enter the gills, passively following the chloride ions.

Freshwater amphibians face the same osmotic problems as freshwater fishes. The entire body sur-

Table 4.2 Representative concentrations of sodium and chloride and osmolality of the blood in vertebrates and marine invertebrates. Concentrations are expressed in millimoles per kilogram of water; all values are reported to the nearest 5 units

Type of Animal	*mmole · kg^{-1}*	*Na^+*	*Cl^{-1}*	*Other Major Osmotic Factor*
Seawater	~1000	475	550	
Freshwater	<10	~5	~5	
Marine invertebrates				
Coelenterates, mollusks, etc.	~1000	470	545	
Crustacea	~1000	460	500	
Marine vertebrates				
Hagfishes	~1000	535	540	
Lamprey	~300	120	95	
Teleosts	<350	180	150	
Coelacanth	<1000 to 1180	180	200	Urea 375
Elasmobranch (bull shark)	1050	290	290	Urea 360
Holocephalian	~1000	340	345	Urea 280
Freshwater vertebrates				
Polypterids	200	100	90	
Acipenserids	250	130	105	
Primitive neopterygians	280	150	130	
Dipnoans	240	110	90	
Teleosts	<300	140	120	
Elasmobranch (bull shark)	680	245	220	Urea 170
Elasmobranch (freshwater rays)	310	150	150	
Amphibians*	~250	~100	~80	
Terrestrial vertebrates				
Reptiles	350	160	130	
Birds	320	150	120	
Mammals	300	145	105	

*Ion levels and osmolality highly variable, but tend toward 200 mmole · kg^{-1} in freshwater.

face of amphibians is involved in the active uptake of ions from the water. Like freshwater fishes, aquatic amphibians do not drink. Acidity inhibits this active transport of ions in both amphibians and fishes, and inability to maintain internal ion concentrations is one of the causes of death of these animals in habitats acidified by acid rain and snow.

Marine Organisms—Teleosts and Other Fishes The osmotic and ionic gradients of vertebrates in seawater are basically the reverse of those experienced by freshwater vertebrates. Seawater is more concentrated than the body fluids of vertebrates, so there is a net outflow of water by osmosis and a net inward diffusion of ions.

Teleosts The integument of marine fishes, like that of freshwater teleosts, is highly impermeable, so that most osmotic and ion movements occur across the gills (Figure 4–13). The kidney glomeruli are small and the glomerular filtration rate is low. Little urine is formed and the water lost in urine is reduced. Marine teleosts lack a water-impermeable distal convoluted tubule. As a result, urine leaving the nephron is less copious but more concentrated than that of freshwater teleosts, although it is always hyposmolal to blood. To compensate for osmotic dehydration, marine teleosts do something unusual—they drink seawater. Sodium and chloride ions are actively absorbed across the lining of the gut, and

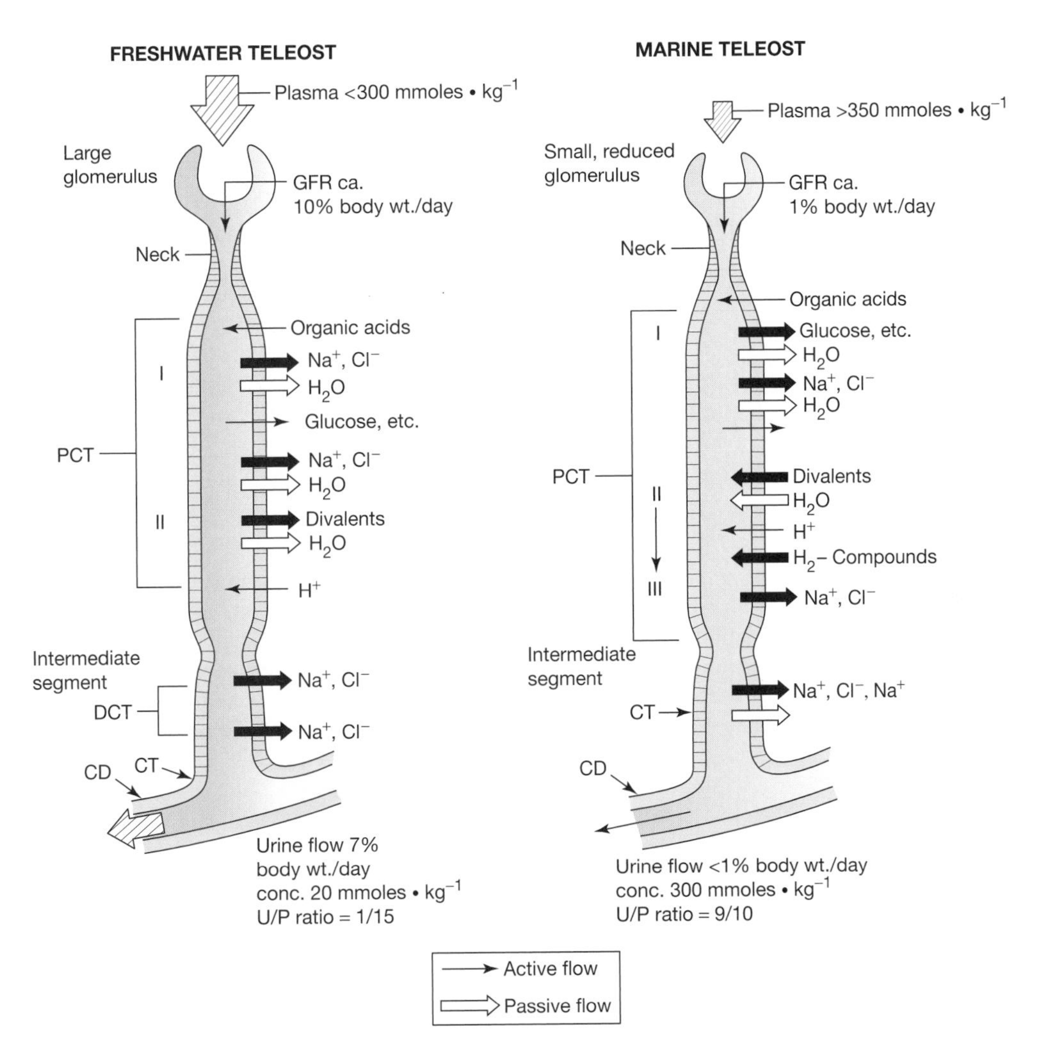

▲ Figure 4–12 Kidney structure and function of marine and freshwater teleosts. GFR is the glomerular filtration rate, that is, the rate at which the ultrafiltrate is formed. It is expressed in percentage of body weight per day. Freshwater teleosts are flooded by water that they must excrete; consequently they have high GFRs. Marine teleosts have the opposite problem; they lose water by osmosis to their surroundings and must conserve water in the kidney, consequently they have low GFRs. PCT is the proximal convoluted tubule (in fishes sometimes referred to as the proximal tubule). The ultrafiltrate is processed in the PCT to remove substances the fish needs to conserve (sodium, chloride, and divalent ions, glucose, and water) and to add nitrogen-containing waste products the fish is excreting. Two segments (I and II) of the PCT are recognized in both freshwater and marine teleosts. Segment III of the PCT of marine teleosts is sometimes equated with the DCT (distal convoluted tubule) of freshwater teleosts. Darkened arrows represent active movements of substances, open arrows passive movement, and hatched arrows indicate by their size the relative magnitude of fluid flow. Note that Na$^+$ and Cl$^-$ are reabsorbed in the PCT segment I and in the CT (collecting tubule) in both freshwater and marine teleosts; water flows passively across the PCT in both freshwater and marine teleosts, but only across the CT of marine teleosts. Water permeability of the CT (and the DCT) is therefore low in freshwater teleosts. Freshwater fish, which must rid themselves of water and conserve Na$^+$ and Cl$^-$, excrete a large volume of dilute urine. Marine fish, which must conserve water and rid themselves of Na$^+$ and Cl$^-$, excrete a small volume of concentrated urine. U/P, ultrafiltrate-to-blood-plasma concentration ratio, is a measure of the concentrating power of a nephron.

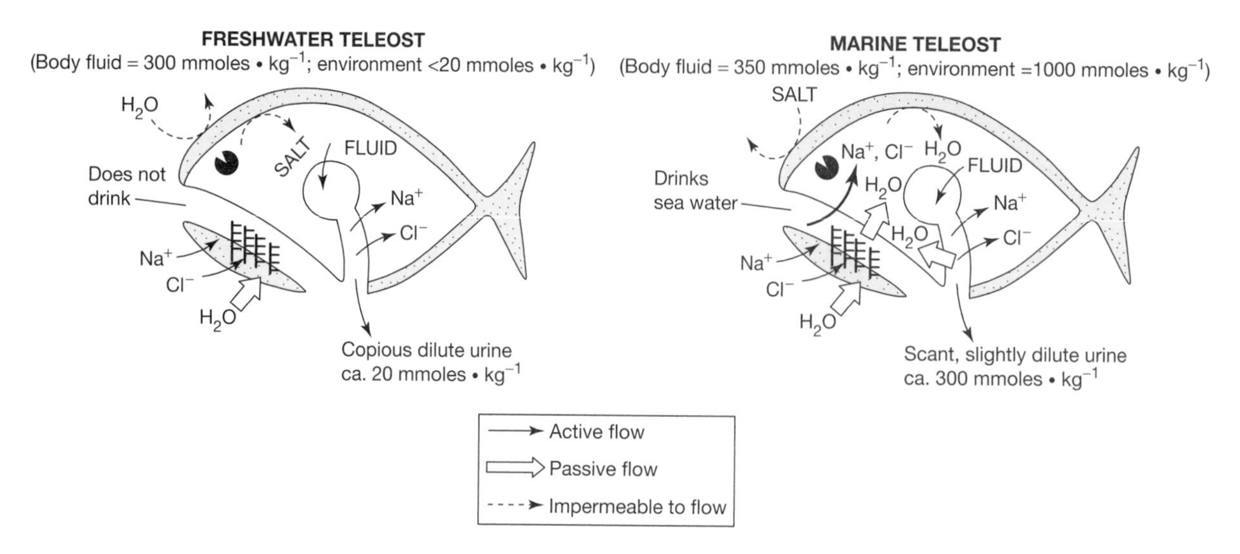

▲ Figure 4–13 General scheme of the osmolal and ionic gradients encountered by freshwater and marine teleosts. Freshwater fish are more concentrated than the water surrounding them; consequently they gain water by osmosis and lose Na^+ and Cl^- by diffusion. They do not drink water, they take up Na^+ and Cl^- via the gills, and they have kidneys with large glomeruli that produce a large volume of dilute urine. Marine fish are less concentrated than the water they live in; consequently they lose water by osmosis and gain Na^+ and Cl^- by diffusion. Marine fish drink water and actively excrete the Na^+ and Cl^- via the gills. They have kidneys with small glomeruli that produce small volumes of concentrated urine.

water flows by osmosis into the blood. Estimates of seawater consumption vary, but many species drink in excess of 25 percent of their body weight per day and absorb 80 percent of this ingested water. Of course, drinking seawater to compensate for osmotic water loss increases the influx of sodium and chloride ions. To compensate for this salt load, chloride cells in the gills actively pump chloride ions outward against a large concentration gradient.

Hagfishes and Chondricthyans Hagfishes have few problems with ion balance because they regulate only divalent ions and reduce osmotic water movement by being nearly isosmolal to seawater. Chondrichthyans and coelacanths also minimize osmotic flow by maintaining the internal concentration of the body fluid close to that of seawater. These animals retain nitrogen-containing compounds (primarily urea and trimethylamine oxide) to produce osmolalities that are usually slightly hyperosmolal to seawater (Table 4.2). As a result, chondrichthyans gain water by osmotic diffusion across the gills and do not need to drink seawater. This net influx of water permits large kidney glomeruli to produce high filtration rates and therefore rapid cleansing of the blood. Urea is very soluble and diffuses through most biological membranes, but the gills of chondrichthyans are nearly impermeable to urea and the kidney tubules actively reabsorb it. With internal ion concentrations that are low relative to seawater, chondrichthyans experience ion influxes across the gills as do marine teleosts. Unlike the gills of marine teleosts, those of chondrichthyans have low ion permeabilities (less than 1 percent those of teleosts). Chondrichthyans generally do not have highly developed salt-excreting cells in the gills. Rather, they achieve ion balance by secreting from the rectal gland a fluid that is approximately isosmolal to body fluids and seawater, but contains higher concentrations of sodium and chloride ions than the body fluids.

Freshwater Elasmobranchs and Marine Amphibians Some elasmobranchs are euryhaline—sawfishes, some sting rays, and bull sharks are examples. In seawater, bull sharks retain high levels of urea; but in freshwater their blood urea levels decline. Sting rays in the family Potamotrygonidae spend their entire lives in freshwater and have very low blood urea concentrations. Their blood sodium and chloride ion concentrations are 35 to 40 percent below those in sharks that enter freshwater, and only slightly above levels typical of freshwater teleosts (Table 4.2). The potamotrygonids may have lived in the Amazon basin for tens of millions of years, and their reduced salt and water gradients may reflect long adaptation to freshwater. When exposed to increased salinity, potamotrygonids do not increase the concentration of urea in the blood as euryhaline elasmobranchs do, even though the enzymes required to produce urea are present. Apparently their long evolution in freshwater has led to an increase in the permeability of their gills to urea and reduced the ability of their kidney tubules to reabsorb it.

Most amphibians are found in freshwater or terrestrial habitats. One of the few species that occurs in salt- water is the crab-eating frog, *Rana cancrivora*. This frog inhabits intertidal mudflats in southeast Asia and is exposed to 80 percent seawater at each high tide. During seawater exposure, the frog allows its blood ion concentrations to rise and thus reduces the ionic gradient. In addition, proteins are deaminated and the ammonia is rapidly converted to urea, which is released into the blood. Blood urea rises from 20 to 30 mmoles $\cdot$ kg^{-1}, and the frogs become hyperosmolal to the surrounding water. In this sense *Rana cancrivora* functions like an elasmobranch and absorbs water osmotically. Frog skin, unlike that of elasmobranchs, is permeable to urea; and urea is thus rapidly lost. To compensate for this loss, the activity of the urea-synthesizing enzymes is very high. The tadpoles of *Rana cancrivora*, like most tadpoles, lack urea-synthesizing enzymes until late in their development. Thus, tadpoles of crab-eating frogs must use a method of osmoregulation different from that of adults. The tadpoles have extrarenal salt-excreting cells in the gills, and by pumping ions outward as they diffuse inward, the tadpoles maintain their blood hyposmolal to seawater in the same manner as do marine teleosts.

Nitrogen Excretion by Aquatic Vertebrates

Carbohydrates and fats are composed of carbon, hydrogen, and oxygen and their waste products are carbon dioxide and water molecules that are easily voided. Proteins and nucleic acids are another matter, for they contain nitrogen. When protein is metabolized, the nitrogen is enzymatically reduced to ammonia through a process called deamination. Ammonia is very soluble in water and diffuses readily, but is also extremely toxic. Rapid excretion of ammonia is therefore crucial. Differences in how ammonia is excreted are partly a matter of the availability of water and partly the result of differences among phylogenetic lineages. Nitrogen is eliminated by most vertebrates as ammonia, as urea, or as uric acid. Many vertebrates excrete all three of these substances, but the proportions of the three compounds differ among the groups of vertebrates (Figure 4–14).

Many aquatic invertebrates excrete ammonia directly, as do vertebrates with gills, permeable skins, or other permeable membranes that contact water. Excretion of nitrogenous wastes as ammonia is called **ammonotelism,** excretion as urea is **ureotelism,** and excretion as uric acid is **uricotelism.** Urea is synthesized from ammonia in a cellular enzymatic process called the **urea cycle.** Urea synthesis requires more energy than does ammonia production, but urea is less toxic than ammonia.

Urea has two advantages. First, it is retained by some marine vertebrates to counter osmotic dehydration. A second function of urea synthesis is the detoxification of ammonia when there is not enough water available to allow it to be excreted as fast as it is produced. Because urea is not very toxic, it can be concentrated in urine, thus conserving water.

Ureotelism probably evolved independently several times. Perhaps ureotelism developed in early freshwater fishes to avoid osmotic dehydration as they reinvaded the sea. Chondrichthyans and coelacanths illustrate the use of urea and other nitrogen-containing compounds to regulate internal osmotic concentrations at high levels. Ureotelism was an important factor for the evolution of terrestrial vertebrates, because it allowed nitrogen to be retained in a detoxified state until water was available to excrete it.

4.5 Responses to Temperature

Vertebrates occupy habitats from cold polar latitudes to hot deserts. To appreciate this adaptability, we must consider how temperature affects a vertebrate such as a fish or amphibian that has little capacity to maintain a difference between its body temperature and the temperature of the water around it (a poikilotherm). Organisms have been called bags of chemicals catalyzed by enzymes. This description emphasizes that living systems are subject to the laws of physics and chemistry just as nonliving systems are. Because temperature influences the rates at which chemical reactions proceed, temperature vitally affects the life

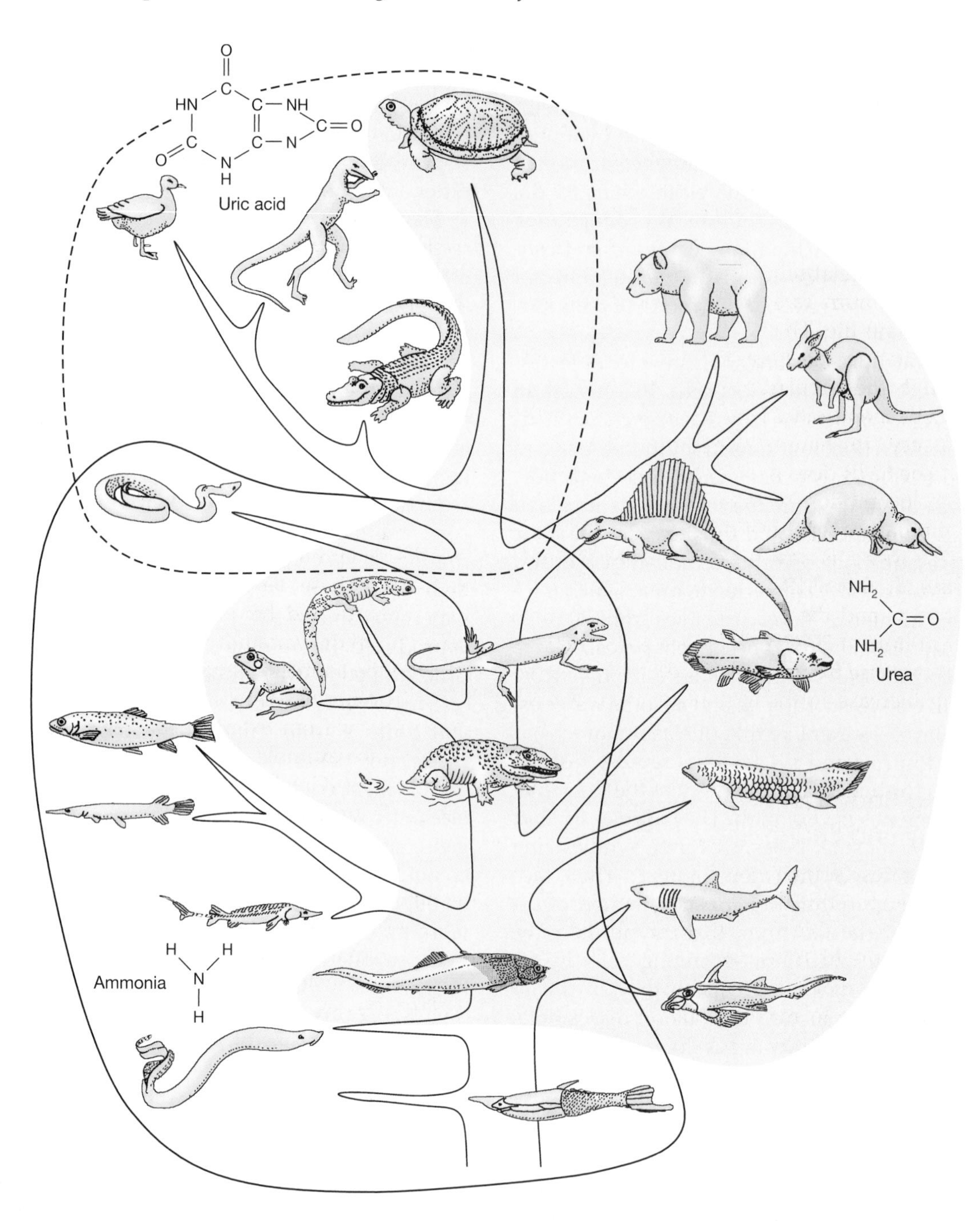

▲ Figure 4–14 Phylogenetic distribution of the three major nitrogenous wastes in vertebrates. The types of wastes excreted by extinct vertebrates are unknown; examples merely provide visual continuity to the diagram.

processes of organisms. Most chemical reactions double or triple in rate for every rise of 10°C. We describe this change in rate by saying that the reaction has a Q_{10} of 2 or 3, respectively. Q_{10} is the ratio of the rate at one temperature compared to the rate at a higher temperature, so a Q_{10} of 1.0 means that the rate stays the same and a Q_{10} less than 1.0 indicates a decrease in rate (Figure 4–15).

The **standard metabolic rate** (SMR) of an organism is the minimum rate of oxygen consumption needed to sustain life. That is, the SMR includes the costs of ventilating the lungs or gills, of pumping blood through the circulatory system, of transporting ions across membranes, and of all the other activities necessary to maintain the integrity of an organism. The SMR does not include the costs of activities like locomotion or the cost of growth. The SMR is temperature sensitive, and that means the energy cost of living is affected by changes in body temperature. If the SMR of a fish is 2 joules per minute at 10°C and the Q_{10} is 2, the fish will use 4 joules per minute at 20°C and 8 joules per minute at 30°C. That increase in energy use translates to a corresponding increase in the amount of food the fish must eat.

Controlling Body Temperature—Ectothermy and Endothermy

Because the rates of many biological processes are affected by temperature, it would be advantageous for any animal to be able to control its body temperature. However, the high heat capacity and heat conductivity of water make it difficult for most fishes or aquatic amphibians to maintain a temperature difference between their bodies and their surroundings. Air has both a lower heat capacity and a lower conductivity than water, and the body temperatures of most terrestrial vertebrates are at least partly independent of the air temperature. Some aquatic vertebrates also have body temperatures substantially above the temperature of the water around them. Maintaining these temperature differences requires thermoregulatory mechanisms, and these are well developed among vertebrates.

The classification of vertebrates as poikilotherms (Greek *poikilo* = variable and *therm* = heat) and homeotherms (Greek *homeo* = the same) was widely used through the middle of the twentieth century, but this terminology has become less appropriate as our knowledge of the temperature-regulating capacities of a wide variety of animals has become more sophisticated. Poikilothermy and homeothermy describe the variability of body temperature, and these terms cannot readily be applied to groups of animals. For example, mammals have been called homeotherms and fishes poikilotherms, but some mammals become torpid at night or in the winter and allow their body temperatures to drop 20°C or more from their normal levels, whereas many fishes live in water that changes temperature less than 2°C in an entire year. That example presents the contradictory situation of a homeotherm that experiences 10 times as much variation in body temperature as a poikilotherm.

Complications like these make it very hard to use the words homeotherm and poikilotherm rigorously. Most biologists concerned with temperature regulation prefer the terms **ectotherm** and **endotherm.** These terms are *not* synonymous with poikilotherm and homeotherm because, instead of referring to the variability of body temperature, they refer to the sources of energy used in thermoregulation. Ectotherms (Greek *ecto* = outside) gain their heat largely from external sources—by basking in the sun, for example, or by resting on a warm rock. Endotherms (Greek *endo* = inside) largely depend on metabolic production of heat to raise their body temperatures. The source of heat used to maintain body temperature is the major difference between ectotherms and endotherms. Terrestrial ectotherms—like lizards and turtles—and endotherms—like birds and mammals—all have activity temperatures ranging from 30 to 40°C.

Endothermy and ectothermy are not mutually exclusive mechanisms of temperature regulation, and many animals use them in combination. In general, birds and mammals are endothermal, but some species make extensive use of external sources of heat. For example, roadrunners are predatory birds living in the deserts of the southwestern United States and adjacent Mexico. On cold nights roadrunners become hypothermic, allowing their body temperatures to fall from the normal level of 38 or 39°C down to 35°C or less. In the morning they bask in the sun, raising the feathers on the back to expose an area of black skin. Calculations indicate that a roadrunner can save 132 joules per hour by using solar energy instead of metabolism to raise its body temperature.

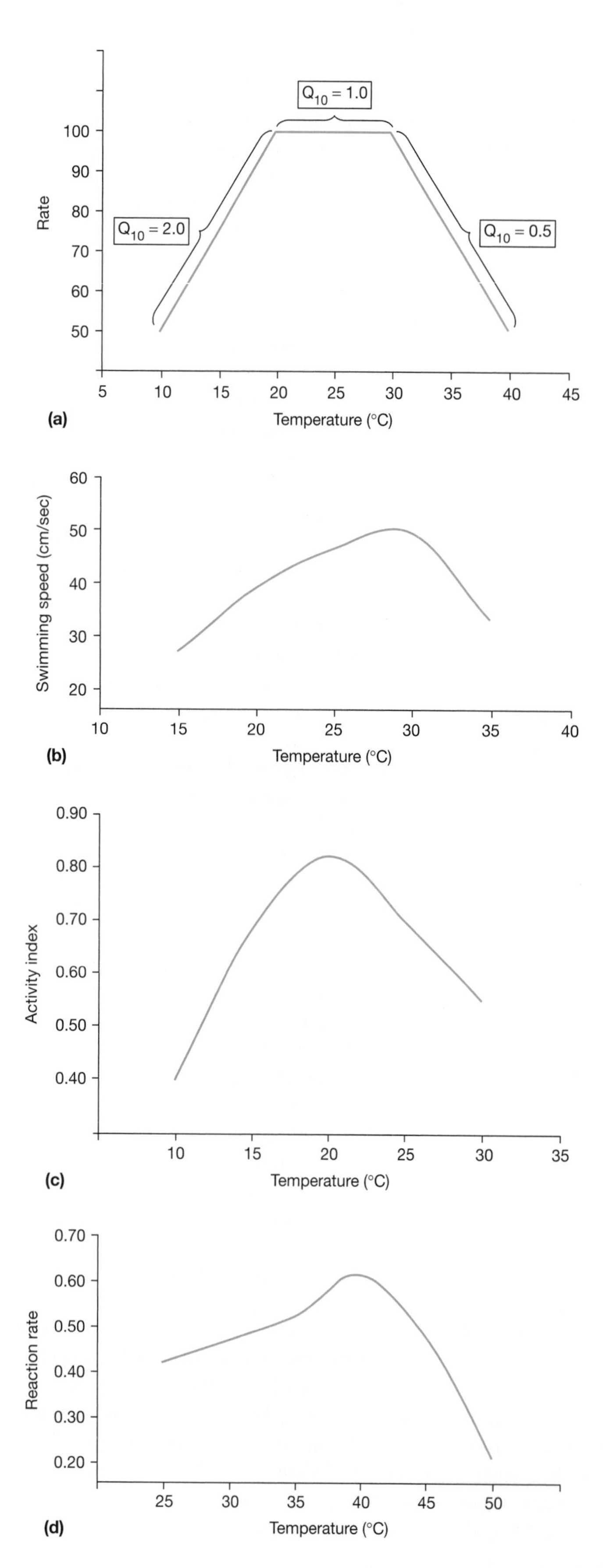

Figure 4–15 The effect of temperature on organisms. (a) A hypothetical reaction for which the reate initially increases and then falls as temperature rises. The Q_{10} of a reaction can be calculated between any two temperatures that are far enough apart to be biologically meaningful by using the formula $Q_{10} = (R_2/R_1)^{10/(T_2-T_1)}$ where R_1 and R_2 are the rates at temperatuers T_1 and T_2, respectively. (b) The maximum swimming speed of a goldfish. (c) Spontaneous activity by goldfish. (d) The activity of the enzyme lactic dehydrogenase from a lungfish.

Deviations from general patterns of temperature regulation go the other way as well. Snakes are normally ectothermal, but the females of several species of python coil around their eggs and produce heat by rhythmic contraction of their trunk muscles. The rate of contraction increases as air temperature falls, and a female Indian python is able to maintain her eggs close to 30°C at air temperatures as low as 23°C. This heat production entails a substantial increase in the python's metabolic rate—at 23°C, a female python uses about 20 times as much energy when she is brooding as she does normally. Thus, generalizations about the body temperatures and thermoregulatory capacities of vertebrates must be made cautiously, and the actual mechanisms used to regulate body temperature must be studied carefully.

Regional Heterothermy—Warm Fishes

Regulation of body temperature is not an all-or-nothing phenomenon for vertebrates. **Regional heterothermy** is a general term used to refer to different temperatures in different parts of an animal's body. Dramatic examples of regional heterothermy are found in several fishes that maintain some parts of their bodies at temperatures 15°C warmer than the water they are swimming in. That's a remarkable accomplishment for a fish, because each time the blood passes through the gills it comes into temperature equilibrium with the water. Thus, to raise its body temperature by using endothermal heat production, a fish must prevent the loss of heat to the water via the gills.

The mechanism used to retain heat is a countercurrent system of blood flow in retia mirabilia. As cold arterial blood from the gills enters the warm part of the body, it flows through a rete and is warmed by heat from the warm venous blood that is leaving the tissue. This arrangement is found in some sharks, especially species in the family Lamnidae (including the mako, great white shark, and porbeagle), which have retia mirabilia in the trunk. These retia retain the heat produced by activity of the swimming muscles, with the result that those muscles are kept 5 to 10°C warmer than water temperature.

Scombroid fishes, a group of teleosts that includes the mackerels, tunas, and billfishes (swordfish, sailfish, spearfish, and marlin), have also evolved endothermal heat production. Tuna have an arrangement of retia that retains the heat produced by myoglobin-rich swimming muscles located close to the vertebral column (Figure 4–16). The temperature of these muscles is held near 30°C at water temperatures from 7 to 23°C. Additional heat exchangers are found in the brains and eyes of tunas and sharks, and these organs are warmer than water temperature but somewhat cooler than the swimming muscles.

The billfishes have a somewhat different arrangement, in which only the brain and eyes are warmed, and the source of heat is a muscle that has changed its function from contraction to heat production. The superior rectus eye muscle of these billfishes has been extensively modified. Mitochondria occupy more than 60 percent of the cell volume, and changes in cell structure and biochemistry result in the release of heat by the calcium-cycling mechanism that is usually associated with contraction of muscles. A related scombroid, the butterfly mackerel, has a thermogenic organ with the same structural and biochemical characteristics found in billfishes, but in the mackerel it is the lateral rectus eye muscle that has been modified.

An analysis of the phylogenetic relationships of scombroid fishes by Barbara Block and her colleagues (Block et al. 1993) suggests that endothermal heat production has arisen independently three times in the lineage—once in the common ancestor of the living billfishes (by modification of the superior rectus eye muscle), once in the butterfly mackerel lineage (modification of the lateral rectus eye muscle), and a third time in the common ancestor of tunas and bonitos (involving the development of countercurrent heat exchangers in muscle, viscera, and brain, and development of red muscle along the horizontal septum of the body).

The ability of these fishes to keep parts of the body warm may allow them to venture into cold water that would otherwise interfere with body functions. Block has pointed out that modification of the eye muscles and the capacity for heat production among scombroids is related to the temperature of the water in which they swim and capture prey. The oxidative capacity of the heater cells of the butterfly mackerel, which is the species that occurs in the coldest water, is the highest of all vertebrates. Swordfishes, which dive to great depths and spend several hours in water temperatures of 10°C or less, have better-developed heater organs than do marlins, sailfishes, and spearfishes, which spend less time in cold water.

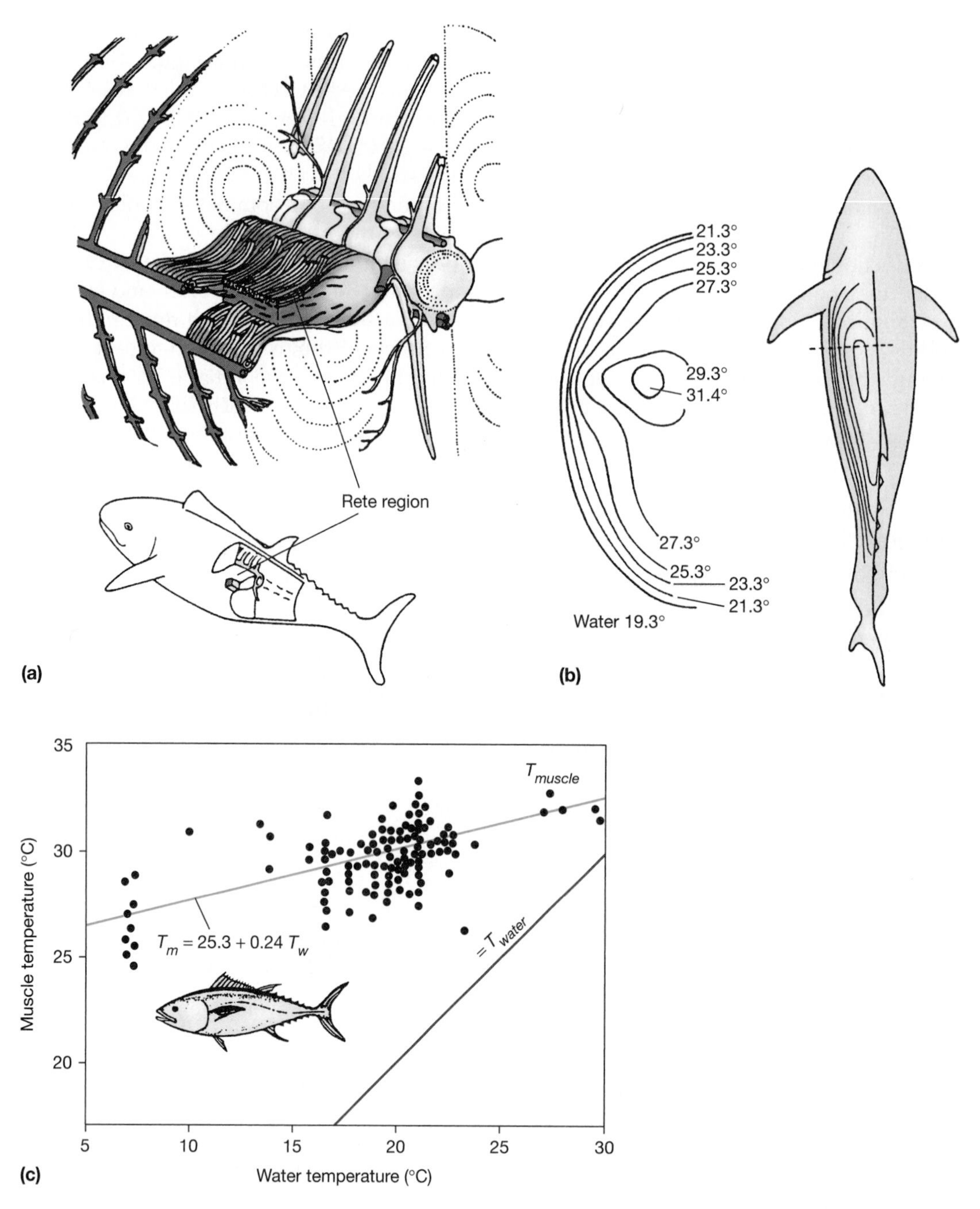

▲ Figure 4–16 Details of body temperature regulation by the bluefin tuna. (a) The red muscle and retia are located adjacent to the vertebral column. (b) Cross-sectional views showing the temperature gradient between the core (at 31.4°C) and water temperature (19.3°C). (c) Core muscle temperatures of bluefins compared to water temperature.

Warm Bodies, Cold Seas—Marine Mammals and Sea Turtles

The temperature equilibration of blood and water that occurs in the gills is the primary obstacle to whole-body endothermy for fish. Countercurrent systems allow them to keep critical parts of their bodies warm, but other parts are at water temperature. Air-breathing aquatic tetrapods avoid that problem, because they have lungs instead of gills. In addition, fully aquatic mammals have a layer of insulation that helps to retain metabolic heat in the body. As a result, marine mammals can maintain their entire bodies at normal mammalian temperatures even though some of them spend their lives in water that is 30 degrees or more below body temperature.

Mammals are endotherms, and their high metabolic rates combined with the muscular activity that accompanies swimming release heat that warms the body. **Blubber** (a layer of fat beneath the skin) is a better insulator than fur in water. The insulation provided by a covering of fur depends on the presence of air trapped between the hairs. This dead air space slows the movement of heat out of the body, just as the air trapped between strands of fiberglass insulation slows the movement of heat through the wall of a building. A fur-covered aquatic mammal must groom its coat on land frequently to renew the layer of trapped air, and in water a coat of fur works as an insulator only as long as water does not displace air in the coat. Many semi-aquatic mammals, such as otters and beavers, have water-repellent pelts that retain air very well. Air clinging to the outer surface of their fur gives these animals a silvery appearance when they are underwater.

The most specialized marine mammals, cetaceans (whales and proposes) and pinnipeds (seals, sea lions, and walruses) use blubber instead of fur as insulation. Blubber is an extremely effective insulator, so good that some seals risk death by overheating if they undertake prolonged strenuous activity on land. Even in water, strenuous activity can lead to overheating. Aquatic mammals have countercurrent exchange systems in their flippers that allow them to retain heat in the body or release it to the ocean. The venous blood returning from the flipper is cold because it has passed through a flat structure with a large surface area that is in contact with the frigid ocean water. Those are ideal conditions for heat exchange, and the blood that leaves the flippers is very close to water temperature. When a marine mammal needs to retain heat in the body, blood returning from the flippers flows through veins that are closely associated with the arteries that carry blood from the body to the flippers. Cold venous blood is heated by warm arterial blood flowing out from the core of the body. By the time the venous blood reaches the body, it is nearly back to body temperature.

When a bout of rapid swimming produces enough heat to increase body temperature above normal, the animal changes the route that venous blood takes as it returns from the flipper. Instead of flowing through veins that are pressed closely against the walls of the arteries, the returning blood is shunted through vessels distant from the arteries. The arterial blood is still hot when it reaches the flipper, and that heat is dissipated into the water, cooling the animal.

Body Size and Surface-to-Volume Ratio

Body size is an extremely important element in the exchange occurring between an organism and its environment. For objects of the same shape, volume increases as the cube of linear dimensions, whereas surface area increases only as the square of linear dimensions. Consider a cube that is 1 cm on each side. Each face of the cube is 1 cm^2 (1 cm · 1 cm) and a cube has six faces (numbered 1 through 6 if the cube comes from a set of dice), so the total surface area is 6 cm^2. The volume of that cube is one cubic centimeter (1 cm · 1 cm · 1 cm), and the ratio of surface to volume is 6 cm^2 per 1 cm^3.

If we double the linear dimensions of the cube, each side becomes 2 cm long. Each face has a surface area of 4 cm^2 (2 cm · 2 cm) and the total surface area is 24 cm^2 (6 · 4 cm^2). The volume of this larger cube is 8 cm^3 (2 cm · 2 cm · 2cm) and the ratio of surface to volume is 24 cm^2 per 8 cm^3, which reduces to 3 cm^2 per 1 cm^3. Thus the larger cube has only half as much surface area per unit volume as the smaller cube. A cube 3 cm on a side has a surface-to -volume ratio of 54 cm^2 per 27 cm^3 (2 cm^2 per 1 cm^3); a cube 4 cm on a side has a ratio of 96 cm^2 per 64 cm^3 (1.5 cm^2 per 1 cm^3).

The pattern emerging from these calculations is shown in Figure 4–17: As an object gets larger, it has progressively less surface area in relation to its volume. Exchange between an animal and its environment occurs through its body surface, and larger species have proportionally less area for exchange in relation to the volume (or mass) of their bodies. The biological significance of that relationship lies in the

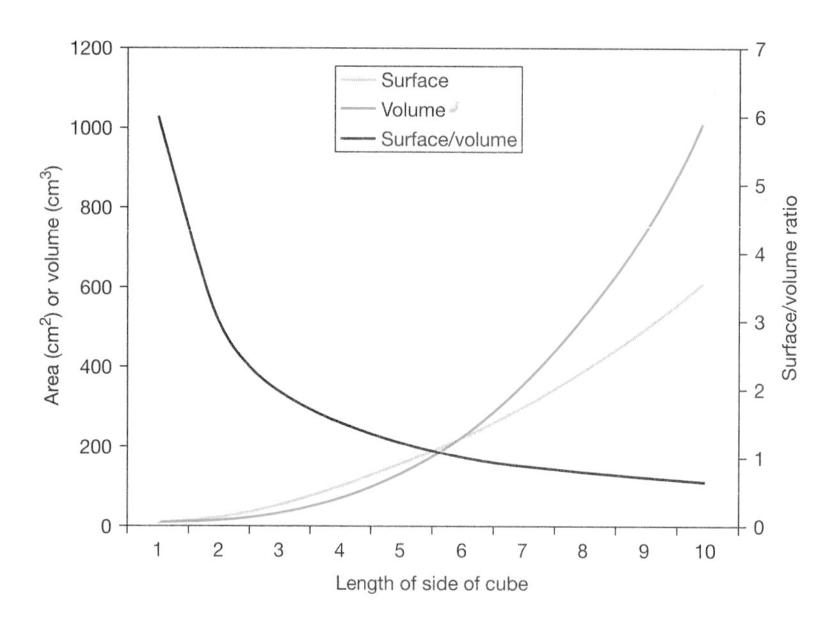

◀ Figure 4–17 Relationships between linear dimensions, surface area, and volume of an object. These are data for a cube. As the length of the side of the cube is increased from 1 cm to 10 cm, total surface area of the cube increases as the square of that length, whereas the volume of the cube increases as the cube of the length of a side. Because volume increases more rapidly than surface area, the surface-to-volume ratio of the cube decreases as the size of the cube increases. Functionally this means that as an object becomes larger, the less surface area it has relative to its volume. Thus the rate of exchange with the environment decreases. For example, if you take two cubes—one that is 1 cm on a side and the other 10 cm on a side—then heat them to the same temperature and put them side by side on a table, the small cube will cool to room temperature faster than the large cube.

conclusion that bigger species exchange energy with the environment less rapidly than smaller species, merely because of the difference in surface-to-volume ratio.

Simply being big gives an animal some independence of external temperature, because heat cannot flow rapidly in or out of a large body through its relatively small surface. The enormous dinosaurs that lived in the Jurassic and Cretaceous would have had very stable body temperatures just by virtue of their size, and it would take many days for a brontosaurus-like dinosaur to warm or cool as its environment changed temperature. Even elephants (which are only one-twentieth the size of the largest dinosaurs) are big enough to feel the consequences of surface-to-volume ratio in body temperature regulation. Elephants can easily overheat when they are active. When that happens, they dump heat by sending large volumes of blood flowing through the ears and waving their ears to promote cooling. (We know from cave paintings made by Pleistocene humans that mammoths, the far northern species of elephant that is now extinct, had smaller ears than the elephants living in warm climates in Asia and Africa today.)

Being big makes temperature regulation in water easier, as leatherback sea turtles dramatically illustrate. Leatherbacks (*Dermochelys coriacea)* are the largest living turtles, reaching adult body masses of 850 kg or more. They are also the most specialized sea turtles, having lost the bony shell that covers most turtles and replaced it with a thick, leather-like external body covering. Leatherback turtles are **pelagic** (live in the open ocean); their geographic range extends north to Alaska and halfway up the coast of Norway, and south past the southern tip of Africa almost to the tip of South America. Water temperatures in these areas are frigid, and the body temperatures of the turtles are as much as 18°C higher than the water. Turtles are reptiles, with metabolic rates much lower than those of mammals. Nonetheless, the combination of large body size and a correspondingly small surface-to-volume ratio with countercurrent heat exchangers in the flippers allows leatherback sea turtles to retain the heat produced by muscular activity. Large body size is an essential part of the turtle's temperature-regulating mechanism. Other sea turtles are half the size of leatherbacks or less, and their geographic ranges are limited to warm water because they are not big enough to maintain a large difference between their body temperatures and water temperature.

Summary

The properties of water offer both advantages and disadvantages for aquatic vertebrates. Water is some 800 times more dense than air, and vertebrates in water are close to neutral buoyancy. That means the skeletons of aquatic vertebrates do not have to resist the force of gravity, and the largest aquatic vertebrates are substantially larger than the largest terrestrial forms. Animal body tissues (especially muscle and bone) are denser than water, and aquatic animals offset that weight with lighter tissues (air-filled swim bladders and lungs or oily livers) to achieve neutral buoyancy. A teleost fish can adjust its buoyancy so precisely that it can hang stationary in the water with only a backpedaling of its pectoral fins to counteract the forward propulsion generated by water leaving the gills.

The density and viscosity of water create problems for animals trying to move through water or to move water across gas-exchange surfaces. Even slow-moving aquatic vertebrates must be streamlined, and a tidal respiratory system (moving the respiratory fluid in and out of a lung) takes too much energy to be practical for vertebrates that breathe water. These animals either have flow-through ventilation (the gills of most fishes) or use the entire body surface for aquatic gas exchange (amphibians).

Water is less transparent than air, and vision is often limited to short distances. Distributed over the bodies of fishes and aquatic amphibians are sensors that are exquisitely sensitive to the movement of water. In addition, elasmobranchs can find hidden prey by detecting the electrical discharges from contracting muscles as the prey breathes. Other fishes create electric fields in the water to detect the presence of prey or predators, and some use powerful electric discharges to stun prey or deter predators.

Aquatic animals are continuously gaining or losing water and ions to their surroundings. Water flows from areas of high kinetic activity (dilute solutions) to low kinetic activity, and ions move down their own activity gradients. Freshwater fishes are more concentrated than the surrounding water, so their body fluids are flooded by an inward flow of water and further diluted by an outward diffusion of ions. Marine fishes are less concentrated than seawater, so they must contend with an outward flow of water and an inward diffusion of ions. Some marine fishes (especially chondrichthyans and coelacanths) accumulate urea in the body to raise their internal osmotic concentration close to that of seawater.

Ammonia is a waste product from deamination of proteins. It is toxic, but very soluble in water, so it is easy for aquatic vertebrates to excrete. Urea and uric acid are less-toxic compounds vertebrates use to dispose of waste nitrogen. Some vertebrates produce mixtures of all three compounds, changing the proportions as the availability of water changes.

Water has a high heat capacity and conducts heat readily. Because of these properties, water temperature is more stable than air temperature, and it is hard for an aquatic animal to maintain a difference between its own body temperature and the temperature of the water surrounding it. Fish have a particularly difficult time maintaining a body temperature different from water, because the temperature of their blood comes into equilibrium with water temperature as the blood passes through the gills. Nonetheless, some fish use countercurrent heat-exchange systems to keep parts of their bodies at temperatures well above water temperature. The largest extant sea turtle, the leatherback, is able to maintain a body temperature 18°C above water temperature. Part of the temperature difference can be traced to countercurrent heat exchangers, which minimize loss of heat from the flippers; the enormous body size of these turtles is also a critical factor. Body surface area increases as the square of linear dimensions, whereas body volume increases as the cube of linear dimensions. Consequently, large animals exchange heat energy with the environment more slowly than do smaller animals. Merely being big confers a degree of stability to the internal environment of an animal.

Additional Readings

Block, B. A. 1991. Evolutionary novelties: How fish have built a heater out of muscle. *American Zoologist* 31:726–742.

Block B. A., J. R. Finnerty, F. R. Stewart, and J. Kidd. 1993. Evolution of endothermy in fish: Mapping physiological traits on a molecular phylogeny. *Science* 260:210–214.

Bodznick, D., and R. G. Northcutt. 1981. Electroreception in lampreys: Evidence that the earliest vertebrates were electroreceptive. *Science* 212:465–467.

Carey, F. G. 1982. Warm fish. In C. R. Taylor, K. Johansen, and L. Bolis (Eds.), *A Companion to Animal Physiology*. Cambridge, UK: Cambridge University Press, 216–233.

Gilmour, K. M. 1998. Gas exchange. In D. H. Evans (Ed.), *The Physiology of Fishes*, 2nd ed. Boca Raton, Fla.: CRC Press, 101–127.

Graham, J. B. 1997. *Air-Breathing Fishes.* San Diego, Calif.: Academic Press.

Johnston, A. A., and A. F. Bennett (eds.). 1996. *Animals and Temperature.* Society for Experimental Biology, Seminar Series 59. Cambridge, UK: Cambridge University Press.

Karnaky, K. J. Jr. 1998. Osmotic and ionic regulation. In D. H. Evans (Ed.) *Physiology of Fishes*, 159–178.

Kirschner, L. B. 1995. Energetics of osmoregulation in freshwater vertebrates. *Journal of Experimental Zoology* 271: 243–252.

Kooyman, G. L. 1989. *Diverse Divers*. Berlin: Springer-Verlag.

Pelster, B. 1998. Buoyancy. In D. H. Evans (Ed.), *Physiology of Fishes*, 25–42.

Scheich, H. et al. 1986. Electroreception and electrolocation in the platypus. *Nature* 319:401–402.

Schellart, N. A. M., and R. J. Wubbels. 1998. The auditory and mechanosensory lateral line system. In D. H. Evans (Ed.), *Physiology of Fishes*, 285–314.

Stoddard, P. K. 1999. Predation enchances complexity in the evolution of electric fish signals. *Nature* 400:254–256.

von der Emde, G. 1998. Electroreception. In D. H. Evans (Ed.) *The Physiology of Fishes* 313–314.

Westby, G. W. M. 1988. The ecology, discharge diversity, and predatory behavior of gymnotiform electric fish in the coastal streams of French Guiana. *Behavioral Ecology and Sociobiology* 22:341–354.

Withers, P. C., G. Morrison, and M. Guppy. 1994. Buoyancy role of urea and TMAO in an elasmobranch fish, the Port Jackson shark, *Heterodontus portjacksoni. Physiological Zoology* 67:693–705.

Web Explorations

On-line resources for this chapter are on the World Wide Web at: **http://www.prenhall.com/pough** (click on the Table of Contents link and then select Chapter 4).

CHAPTER

5

Radiation of the Chondrichthyes

Relatively soon after the first evidence of vertebrates in the fossil record, the next major step in vertebrate evolution appeared: jaws and internally supported, paired appendages. The diversity of predatory specializations available to a vertebrate with jaws and precise steering is great, and the appearance of these characters signaled a new radiation of vertebrates. The cartilaginous fishes (sharks, rays, and ratfishes) are the descendants of one clade of this radiation, and they combine derived characters such as a cartilaginous skeleton with a generally primitive anatomy. Sharks have undergone three major radiations, which can be broadly associated with increasingly specialized feeding mechanisms, and extant sharks are a diverse and successful group of fishes. In this chapter we consider the origins and success of extant cartilaginous fishes.

5.1 Chondrichthyes:—The Cartilaginous Fishes

The sharks and their relatives make a definite first appearance in the fossil record in the Late Silurian. Some of their systems have evolved to levels surpassed by few other extant vertebrates. At the same time, these fishes retain many primitive anatomical characters; sharks have long been used to exemplify an ancestral vertebrate body form. Identified by a cartilaginous skeleton, extant forms can be divided into two groups: those with a single gill opening on each side of the head and those with multiple gill openings on each side. The clade with one gill opening is the Holocephali (Greek *holo* = whole and *cephalo* = head), so named for the undivided appearance of the head that results from having a single gill opening. The common names of this group, ratfish, rabbitfish, and chimera, come from their bizarre form—a long flexible tail, a fish-like body, and a head with big eyes and buckteeth that resembles a caricature of a rabbit. The Elasmobranchii (Greek *elasmo* = plate and *branch* = gill) have multiple gill openings on each side of the head and include the sharks (most often cylindrical forms with five to seven gill openings on each side of the head) as well as skates and rays (flattened forms with gill openings on the ventral surface of the head).

5.2 Evolutionary Specializations of Chondrichthyes

Despite a rather good fossil record, the phylogeny of cartilaginous fishes remains unclear. Early Chondrichthyes, like extant species, were diverse in form and habitats. (In the Late Devonian, sharks had invaded freshwater habitats to a greater extent than they do today.) Their initial radiation from a common ancestor emphasized changes in teeth, jaws, and fins. Changes in feeding structures (teeth and jaws) and locomotor apparatus (fins and tail) evolved at different rates within different lineages. In some lineages a derived dentition is found with an ancestral fin structure; in other lineages the opposite combination is seen. As a result, fossil chondrichthyans display mosaics of ancestral and derived characters. The relationships of ancient and modern chondrichthyans are shown in Figure 3–13.

Through time, different lineages of Chondrichthyes developed similar but not identical modifications in feeding and locomotor structures, presumably because of similar selective pressures. This pattern of similar adaptations in related lineages is an example of parallel evolution. When similar selective forces act on similar body forms and developmental mechanisms, certain modifications appear independently and often repeatedly in the course of time.

In the following sections, we will trace three radiations of chondrichthyans, focusing on the fusiform (torpedo-shaped) predators popularly known as sharks, and describing progressive changes in tooth and jaw structure, and in the form of the fins and tail. This analysis focuses on the parallel evolution of characters in multiple lineages during the Paleozoic and Mesozoic. Next we will examine skates and rays, and finally we will describe the bizarre and little-known forms called holocephalians.

Table 5.1 Classification of Chondrichthyes, the cartilaginous fishes

Major groups are shown in boldface. The "Squamolmorpha" lineage is not monophyletic, but the relationships of the four to six lineages it contains are not yet understood.

Chondrichthyes (cartilaginous fishes)—About 840 extant species

Neoselachii (sharks, skates and rays)—About 810 living species, mostly marine

Galeomorpha (Galea)—About 250 species of extant shark-like fishes with an anal fin. Includes 4 orders encompassing horn sharks, wobbegons, nurse sharks, whale sharks, sand tigers, mackerel sharks, megamouth shark, thresher sharks, basking shark, and requiem sharks.

"Squalomorpha"—About 110 extant species of shark-like fishes without an anal fin and all skates and rays, which also lack an anal fin. The shark-like forms are placed in 4 to 6 orders encompassing most deep-sea sharks, dogfish sharks, angel sharks, and saw sharks.

Batoidea (skates and rays)—About 450 extant species, a few entirely freshwater. Includes electric rays, stingrays, manta rays, and a large number of skates.

Holocephali (chimeras)—About 30 extant species, all marine

Most live near the bottom at depths where they are rarely observed by humans.

5.3 The Paleozoic Chondrichthyan Radiation

The stem chondrichthyans are identified by the form of the teeth common to most members of the species—basically three-cusped with little root development (Figure 5–1). Although there is evidence of bone around their bases, the teeth are primarily dentine capped with an enameloid coat. The central cusp is the largest in *Cladoselache*, the best-known genus, and smallest in *Xenacanthus*, a more specialized form.

Cladoselache was shark-like in appearance, about 2 meters long when fully grown, with large fins and mouth and five separate external gill openings on each side of the head. The mouth opened terminally, and the chondrocranium had several large areas for tight ligamentous attachment of the palatoquadrate. The jaw also obtained some support from the second visceral arch, the hyoid arch. The name **amphistylic** (Greek *amphi* = both and *styl* = pillar or support) is applied to this mode of multiple sites of upper jaw suspension. The gape was large, the jaws extending well behind the rest of the skull. The three-pronged teeth were probably especially efficient for feeding on fishes or cephalopods that could be swallowed whole or severed by the knife-edged cusps.

As teeth are used, they become worn; cusps break off and cutting edges grow dull. In sharks each tooth on the functional edge of the jaw is but one member of a tooth whorl, attached to a ligamentous band that coursed down the inside of the jaw cartilage deep

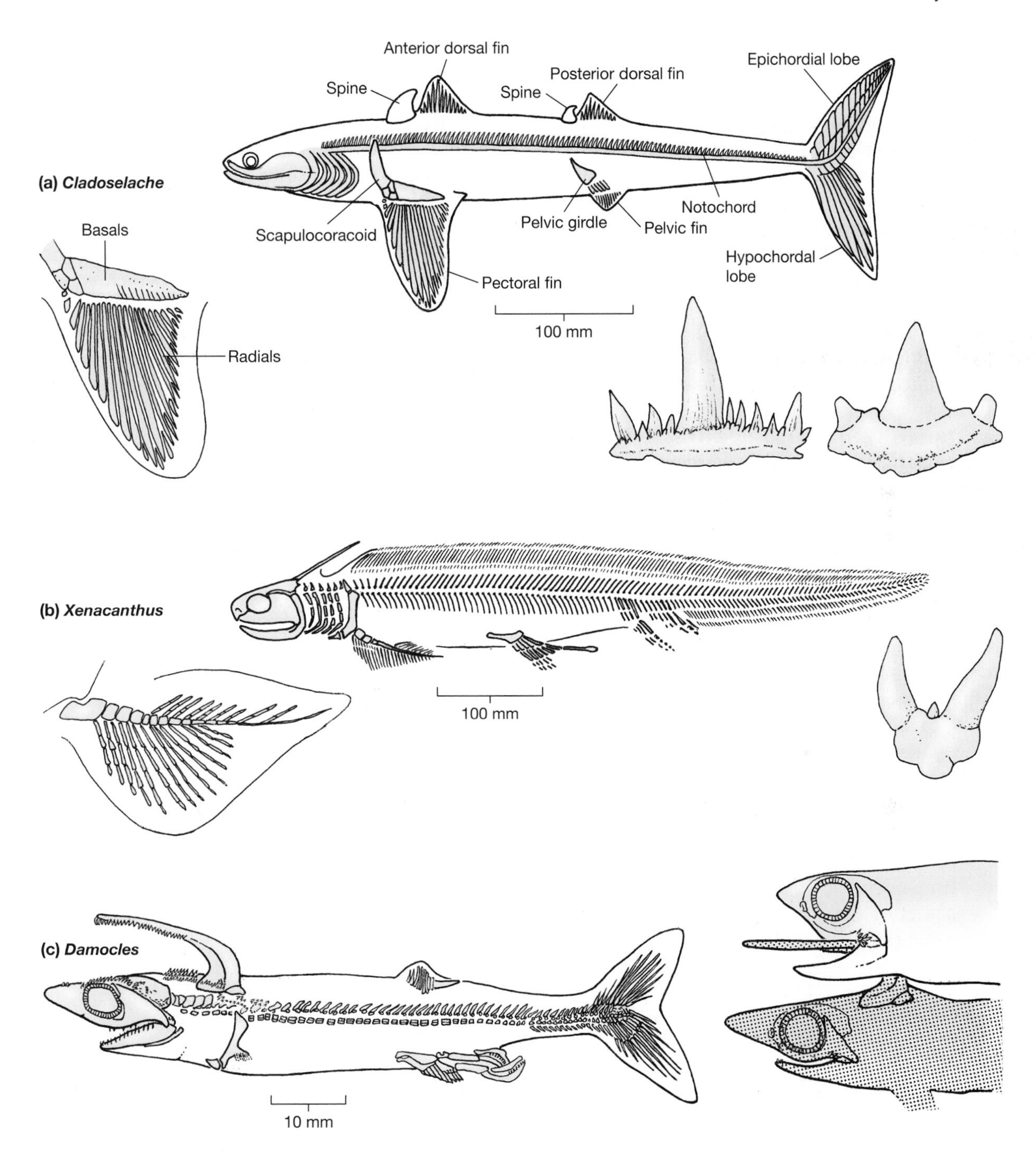

▲ Figure 5–1 Early chondrichthyans. (a) *Cladoselache.* (b) *Xenacanthus,* a freshwater elasmobranch with details of its archipterygial pectoral fin structure and peculiar teeth. (c) Left—male *Damocles serratus,* a 15-cm shark from the Late Carboniferous showing sexually dimorphic nuchal spine and pelvic claspers; right—male (below) and female (above) as fossilized, possibly in courtship position.

below the fleshy lining of the mouth (Figure 5–2). Aligned in each whorl in a file directly behind the functional tooth are a series of developing teeth. Essentially the same dental apparatus is present in all sharks, living and extinct. Tooth replacement is rapid: Young and growing modern sharks under ideal conditions replace each lower-jaw tooth every 8.2 days and each upper-jaw tooth every 7.8 days.

The body of *Cladoselache* was supported only by a notochord, but cartilaginous neural arches gave added protection to the spinal cord. The fins of *Cladoselache* consisted of two dorsal fins, paired pectoral and pelvic fins, and a well-developed forked tail. The first and sometimes second dorsal fins were preceded by stout spines, triangular in cross section and thought to have been covered by soft tissue during the life of the shark. The dorsal fins were broad triangles with an internal structure consisting of a triangular basal cartilage and a parallel series of long radial cartilages extending to the margin of the fin. The pectoral fins were larger but similar in construction.

Among the early radiations of sharks, almost every type seems to have had a different sort of internal pectoral fin arrangement (see Figure 5–1), but all possessed basal elements that anchored the pectoral fins in place. The pectoral fins appear to have had little capacity for altering their angle of contact with the water. The pelvic fins were smaller, but otherwise shaped like the pectorals. Some species in other genera had pelvic fins with claspers, which are male copulatory organs. No anal fin is known, and it is also lacking in many extant sharks.

The caudal fin of *Cladoselache* is distinctive (Figure 5–1a). Externally symmetrical, its internal structure was asymmetrical and contained elements resembling the hemal arches that protect the caudal blood vessels in extant sharks. Long, unsegmented radial cartilages extended into the hypochordal (lower) lobe of the fin. At the base of the caudal fin were paired lateral keels that are identifying characteristics of extant rapid pelagic (open-water) swimmers.

The skin had only a few scales, which were limited to the fins, the circumference of the eye, and within the mouth behind the teeth. These scales resembled the teeth—cusps of dentine were covered with an enamel-like substance and contained a cellular core or pulp cavity. Scales of this sort are called **placoid** scales. Unlike a tooth, each scale of these most ancient sharks had several pulp cavities corresponding to the several cusps.

We can piece together the lives of many of the early chondrichthyans from their morphology and fossil localities. Most early sharks, and *Cladoselache* in particular, were probably pelagic predators that swam after their prey in a sinuous manner, engulfing it whole or slashing it with dagger-like teeth. The lack of body denticles and calcification suggests a tendency to reduce weight and thereby increase buoyancy.

The presence of pelvic claspers shows that at least some Paleozoic sharks had internal fertilization. Successful fertilization would have been accompanied by reproductive behavior that ensured successful mating. Two species of small (15 cm) shark-like forms from the Early Carboniferous of Montana may show

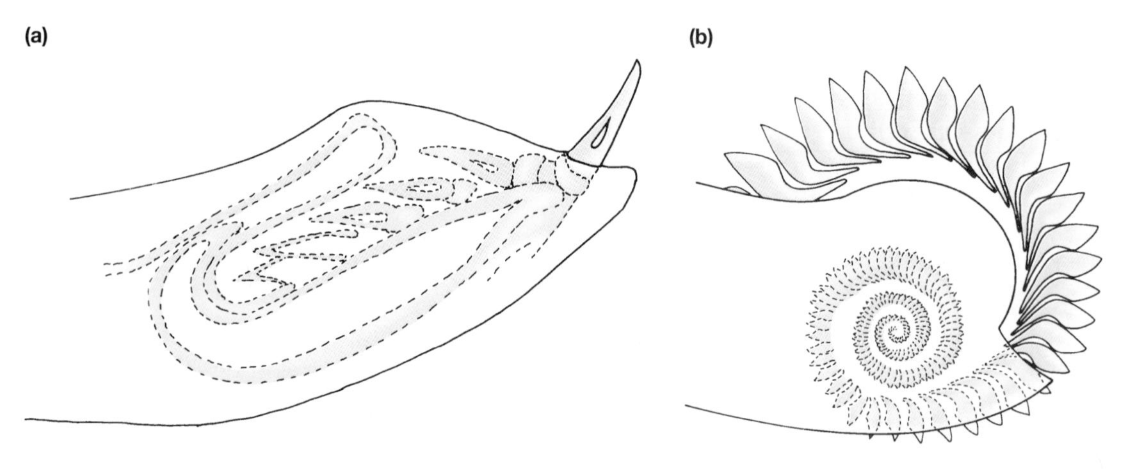

▲ Figure 5–2 Tooth replacement by chondrichthyans. (a) Cross section of the jaw of an extant shark, showing a single functional tooth backed by a band of replacement teeth in various stages of development. (b) Lateral view of the symphysial (middle of the lower jaw) tooth whorl of the edestoid cladodont *Helicoprion*, showing the chamber into which the lifelong production of teeth spiraled.

how complex reproductive behavior was in early chondrichthyes. Males of these species can be identified by pelvic claspers, a sharp rostrum, and an enormous forwardly curved middorsal spine just behind the head (Figure 5–1c), whereas females had neither claspers nor spines. One of the fossils might be a pair in a precopulatory courtship position, with the female grasping the male's dorsal spine in her jaws.

Another early group of chondrichthyan fishes, the edestoids, had large pectoral fins and stiff, symmetrical, deeply forked tails. Among extant sharks, these features are characteristic of fast-swimming oceanic species. The edestoids had a peculiar dentition (Figure 5–2b). Most of the tooth whorls were greatly reduced, but the central (symphysial) tooth row of the mandible was tremendously enlarged, and each tooth interlocked with adjacent teeth at its base. Several members of this tooth whorl were functional at the same time. Some forms had blunt teeth for crushing shelled prey, and others had teeth with a series of knife-edged blades. The mandibular tooth row bit against small, flat teeth in the palatoquadrate. Most edestoids replaced their teeth rapidly, the oldest worn teeth being shed from the tip of the mandible. In contrast, *Helicoprion* retained all its teeth in a specialized chamber into which the lifelong production of teeth spiraled (Figure 5–2b). Perhaps the teeth in this chamber provided a solid foundation for the functional teeth.

One of the score or so of genera produced in the early radiation of chondrichthyans was *Xenacanthus,* which had a braincase, jaws, and jaw suspension very similar to those of *Cladoselache.* But there the resemblance ends. The xenacanths were freshwater bottom-dwellers with very robust fins and heavily calcified, cartilaginous skeletons that would have decreased their buoyancy. The xenacanths appeared in the Devonian and survived until the Triassic, when they died out without leaving direct descendants. Details of their gills and fin skeletons indicate that xenacanths are at the base of the elasmobranch lineage.

5.4 The Early Mesozoic Elasmobranch Radiation

Further chondrichthyan evolution involved changes in feeding and locomotor systems. Species exhibiting these modifications appear in the Carboniferous, and this radiation of stem elasmobranchs flourished until the Late Cretaceous. *Hybodus* was a well-known genus of the Late Triassic and Cretaceous. We have complete skeletons 2 meters in length that look very much like modern sharks except that the mouth is terminal, not underslung beneath a sensory rostrum (Figure 5–3a).

The heterodont dentition (that is, different-shaped teeth along the jaw) of hybodont sharks (*Hybodus* and its relatives) seems pivotal to their success. The anterior teeth had sharp cusps and appear to have been used for piercing, holding, and slashing softer foods. The posterior teeth were stout, blunt versions of the anterior teeth in batteries consisting of several teeth from each individual tooth whorl. The living horn sharks of the genus *Heterodontus* have similar dentition (Figure 5–3b). Horn sharks feed on small fishes, crabs, shrimp, sea urchins, clams, mussels, and oysters. The sharp anterior teeth seize and kill soft-bodied food, and the pavement-like posterior teeth crush the shells of crustaceans and mollusks.

The hybodont sharks also showed advances in the structure of the pectoral and pelvic fins that made them more mobile than the broad-based fins of Paleozoic sharks. Both pairs of fins were supported on narrow stalks formed by three narrow, plate-like basal cartilages that replaced the long series of basals seen in earlier sharks. The narrow base allowed the fin to be rotated to different angles as the shark swam up or down. The blade of the fin also changed: The cartilaginous radials were segmented and did not extend to the fin margin. Proteinaceous, flexible fin rays called ceratotrichia extended from the outer radials to the margin of the fin. Intrinsic fin muscles could curve the fin from front to back and from base to tip. The mobility and flexibility of these fins allowed them to be used for steering in ways that seem impossible for the more rigid fin construction characteristic of *Cladoselache.* By assuming different shapes, the pectoral fins could produce lift anteriorly, aid in turning, or stabilize straight-line movement.

Along with changes in the paired fins, the caudal fin assumed new functions and an anal fin appeared. Caudal fin shape was altered by reduction of the hypochordal lobe, division of its radials, and addition of flexible ceratotrichia. This tail-fin arrangement is generally known as **heterocercal** (Greek *hetero* = different and *kerkos* = tail), although **epicercal** (Greek *epi* = above) is a more precise name for this tail shape—known from as far back as some Paleozoic jawless fishes (Chapter 3). The value of the elasmobranch heterocercal tail lies in its flexibility

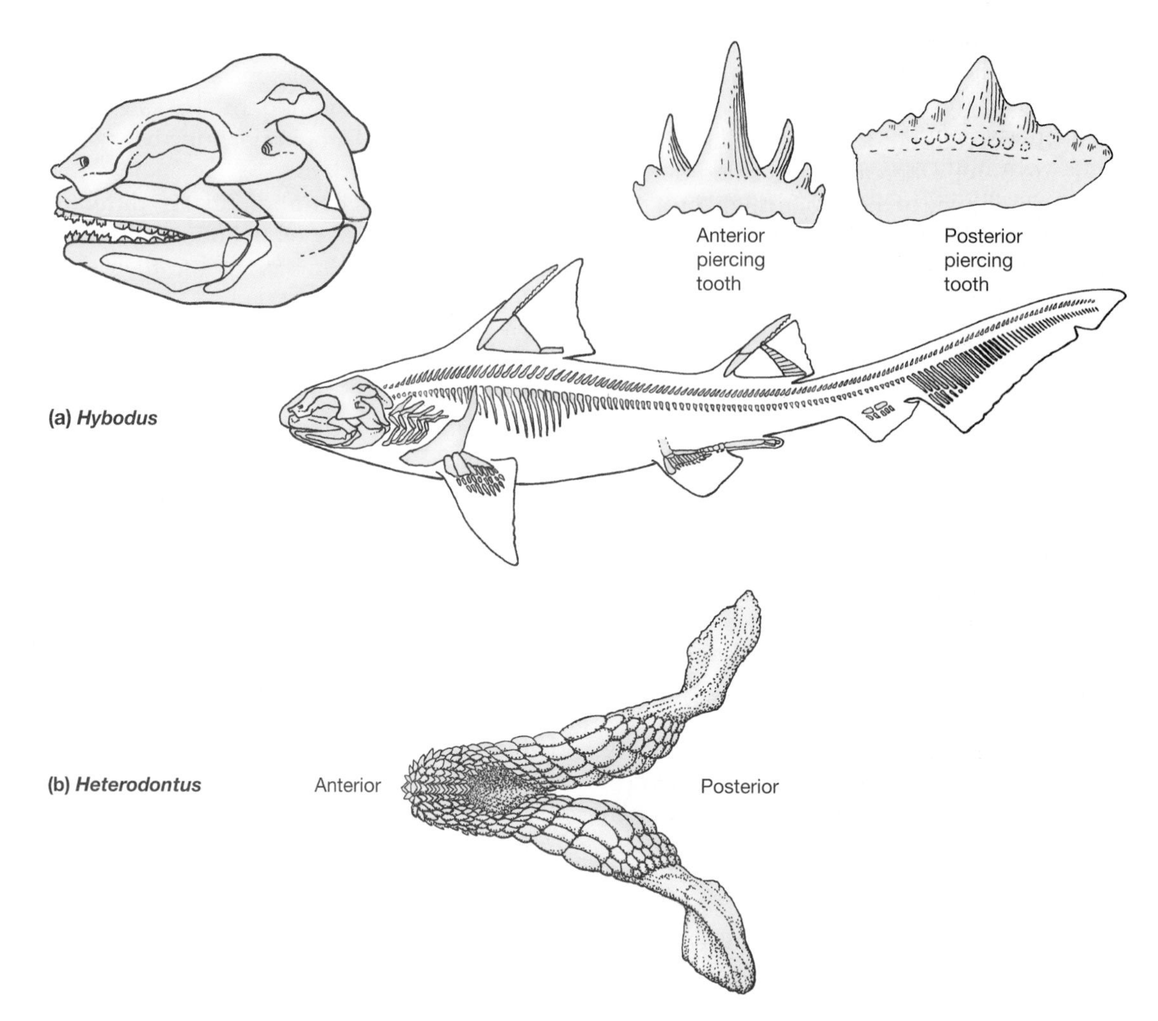

▲ Figure 5–3 *Hybodus* and *Heterodontus.* (a) The fossil elasmobranch *Hybodus.* (b) Upper jaw of the extant hornshark, *Heterodontus,* with a dentition similar to that of many elasmobranchs during the late Paleozoic and early Mesozoic.

(because of the more numerous radial skeletal elements) and the control of shape made possible by the intrinsic musculature. When it was undulated from side to side, the fin twisted so that the flexible lower lobe trailed behind the stiff upper one. This distribution of force produced forward thrust that could counter the shark's tendency to sink or lift it from a resting position.

Other morphological changes in sharks of the second major radiation include the appearance of a complete set of hemal arches that protected the arteries and veins running below the notochord; well-developed ribs; and narrow, more pointed dorsal-fin spines closely associated with the leading edges of the dorsal fins. These spines were ornamented with ridges and grooves and studded with barbs on the posterior surface, suggesting that they were used in defense. Claspers are found in all species, leaving little doubt that they had elaborate courtship and internal fertilization.

Hybodus and its relatives resembled their presumed *Cladoselache*-like ancestors in having terminal mouths, an amphistylic jaw suspension, unconstricted notochords, and multicusped teeth; but a direct line cannot be drawn between the two in time or in morphology. Some forms considered related to *Hybodus* had *Cladoselache*-like dentition combined with tribasal pectoral fins; others developed a very tetrapod-like support for highly mobile pectoral fins. Their caudal fin was reduced, and they probably moved around on the seafloor using their limb-like pectoral fins. Another form, known only from a 5-centimeter juvenile, had a paddle-shaped rostrum one-third its body length. Other types were 2.5-meter giants with blunt snouts and enormous jaws. Despite their variety and success during the Mesozoic, this second radiation of elasmobranchs became increasingly rare and disappeared at the end of the Mesozoic or in the early Cenozoic.

5.5 The Extant Radiation—Sharks, Skates, and Rays

The first representatives of the extant radiation of elasmobranchs appeared at least as early as the Triassic. By the Jurassic, sharks of modern appearance had evolved, and a surprising number of Jurassic and Cretaceous genera are still extant. The most conspicuous difference between most members of the earlier radiations and extant sharks is the rostrum or snout that overhangs the ventrally positioned mouth in most extant forms. Less obvious, but of major importance, was the development of solid, calcified vertebrae which constricted—and in some species even replaced—much of the notochord. Another innovation of the extant elasmobranchs is a thicker and more structurally complex enamel-like material on the teeth than seen in earlier groups.

Sharks

The technical characters distinguishing the clades of extant sharks are subtle. There are about 360 species of pleurotremate (Greek *pleur* = the side and *trem* = a hole) elasmobranchs—the sharks with gill openings on the sides of the head (Figure 5–4). Although similar in overall appearance, the extant pleurotremate sharks actually come from two lineages (see Figure 3–13). More ancestral in their general anatomy (especially the smaller size of their brain) are the 80 species of squaloid sharks and their numerous relatives, all grouped together as the Squalea. Squaloids include the spiny and green dogfish, the cookie-cutter shark, and the basking and megamouth sharks. These species usually live in cold, deep water. The other 280 or so species of sharks are almost all members of the galeoid (Galea) lineage that includes the hornshark, the nurse and carpet sharks, the whale shark, the mackerel sharks (including the great white shark), the carcharhinid or requiem sharks, and the hammerhead sharks. Galeoid sharks are the dominant carnivores of shallow, warm, species-rich regions of the oceans.

Throughout their evolutionary history, sharks have been consummate carnivores. In the third adaptive radiation of the elasmobranchs, locomotor, trophic, sensory, and behavioral characteristics evolved together in the mid-Mesozoic to produce forms that still dominate the top levels of marine food webs. This position has gone hand in hand with the evolution of gigantism, one advantage of which is avoiding predation. A typical shark is about two meters long, but a few interesting miniature forms only 25 cm long have evolved and inhabit mostly deeper seas off the continental shelves.

Despite their enormous range in size, all extant elasmobranchs have common skeletal characteristics that earlier shark radiations lacked. The cartilaginous vertebral centra of extant sharks are distinctive. Between centra, spherical remnants of the notochord fit into depressions on the opposing faces of adjacent vertebrae. Thus, the axial skeleton can flex from side to side with rigid central elements swiveling on ball-bearing joints of calcified cartilage and notochordal remnants. In addition to the neural and hemal arches, extra elements that are not found in the axial skeleton of other vertebrates (the intercalary plates) protect the spinal cord above and the major arteries and veins below the centra.

Shark placoid scales also changed in the third radiation. Although scales of the same general type are known from earlier chondrichthyes, they are often in clusters or fused into larger plates. The shagreen (sharkskin) body covering of modern elasmobranchs is a unique armor that is flexible yet very protective. The placoid scales of extant sharks have a single cusp and a single pulp cavity. The size, shape, and arrangement of these placoid scales reduce turbulence in the flow of water next to the body surface and increase the efficiency of swimming. (Swimsuits duplicating the surface properties of sharkskin appear to cut a few hundredths of a second from a swimmer's time—enough to be the margin of victory in world-level competition.) Individual scales in more ancestral sharks often fused to form larger scales as the shark grew larger. Extant sharks add more scales to their skin as they grow, and these new scales may be larger than those developed earlier in proportion to the increase in size of the shark.

Sensory Systems and Prey Detection The sensory systems of extant sharks, skates, and rays

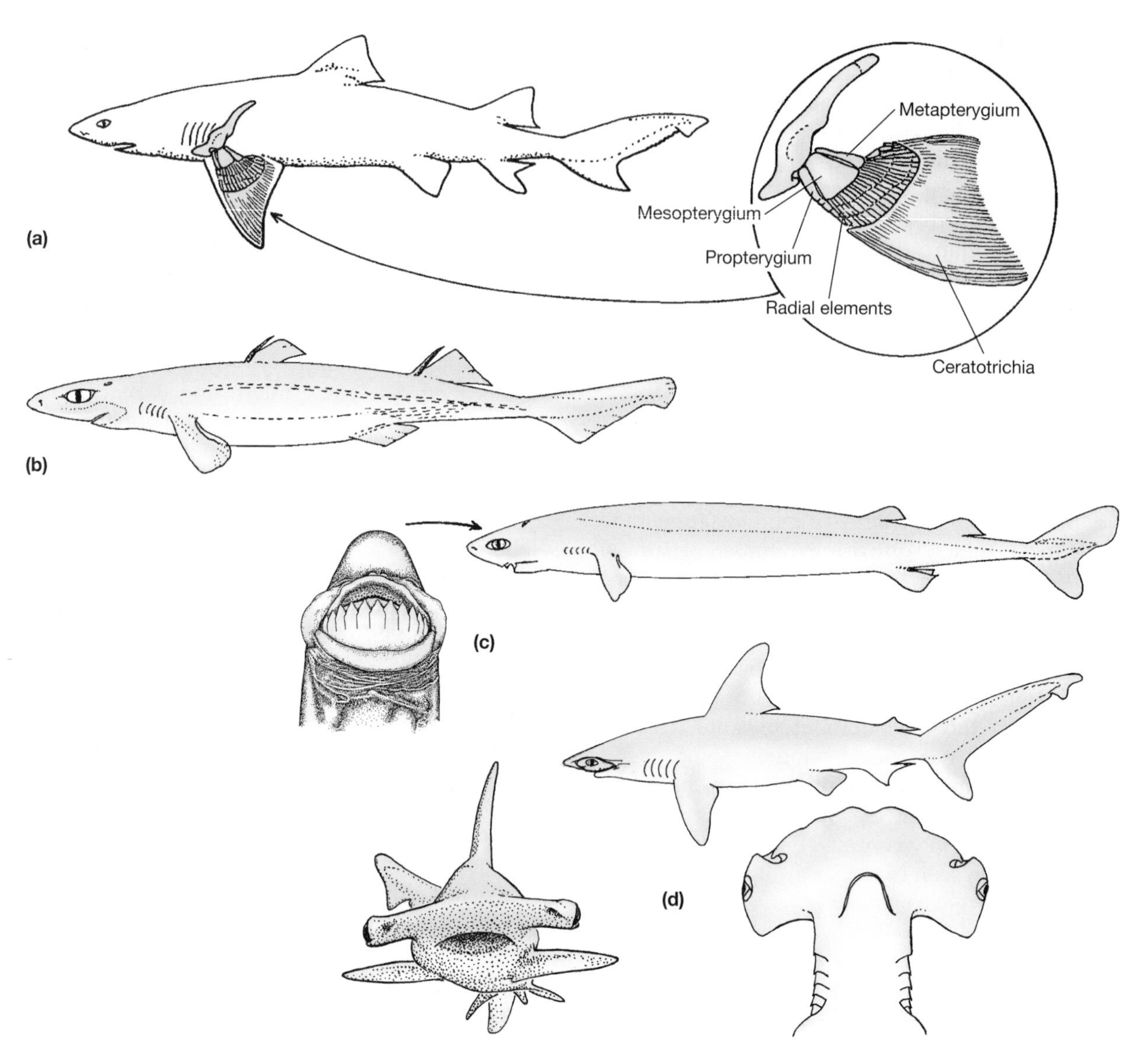

▲ Figure 5–4 Some extant sharks. (a) *Megaprion brevirostris,* the lemon shark, grows to a length of 2m. The internal anatomy of the pectoral girdle and fin are shown superimposed in their correct relative positions. (b) *Etmopterus vierens,* the green dogfish, is a miniature shark only 25 cm long, yet it feeds on much larger prey items. (c) *Isistius brasiliensis,* the cookie-cutter shark, is another miniature species whose curious mouth (left) is able to take chunks from fish and cetaceans much larger than itself. (d) Hammerhead shark (*Sphyrna*), which grows to lengths of 3m or more, in lateral, ventral, and frontal views.

are refined and diverse. Sharks may detect prey via mechanoreceptors of their lateralis system, an interconnected series of superficial tubes, pores, and patches of sensory cells distributed over the head and along the sides that respond to vibrations transmitted through the water. The basic units of mechanoreceptors are the **neuromast organs,** a cluster of sensory and supporting cells that are found on the surface and within the lateral line canals. The anatomically related **ampullae of Lorenzini,** mucus-filled tubes with sensory cells and afferent neurons at their base, are exquisitely sensitive to electrical potentials and can even detect prey by detecting the weak electrical fields they create (Chapter 4).

Chemoreception is another important sense. In fact, sharks have been described as swimming noses, so acute is their sense of smell. Experiments have shown that some sharks respond to chemicals in concentrations as low as 1 part in 10 billion! Hammerhead sharks of the genus *Sphyrna* (Figure 5–4) may have enhanced the directionality of their olfactory apparatus by placing the nostrils far apart on the odd lateral expansions of their heads.

Finally, vision is important to the feeding behavior of sharks. Vision at low light intensities is especially well developed. This sensitivity is due to a rod-rich retina and cells with numerous plate-like crystals of guanine that are located just behind the retina in the choroid layer. Collectively called the tapetum lucidum, the cells containing the crystals act like mirrors to reflect light back through the retina and increase the chance that light will be absorbed. This mechanism, although of great benefit at night or in the depths, has obvious disadvantages in the bright sea surface of midday. In this situation cells containing the dark pigment melanin expand over the reflective surface to absorb light passing through the retina. With so many sophisticated sensory systems, it is not surprising that the brains of many species of sharks are proportionately heavier than the brains of other fishes and approach the brain-to-body-mass ratios of some tetrapods.

Anecdotal and circumstantial evidence suggests that sharks regularly use their various sensory modalities in an ordered sequence to locate, identify, and attack prey. Olfaction is often the first of the senses to alert a shark to potential prey, especially when the prey is wounded or otherwise releasing body fluids. A shark employs its sensitive sense of smell to swim up-current through an increasing odor gradient. Because of its exquisite sensitivity, a shark can use smell as a long-distance sense.

Not as useful over great distances, but much more directional over a wide range of environmental conditions, is another distance sense—a vibration sensitivity called mechanoreception. The lateralis system and the sensory areas of the inner ear are related forms of mechanoreceptors, both highly efficient in detecting vibrations such as those produced by a struggling fish. The effectiveness of mechanoreception in drawing sharks from considerable distances to a sound source has been demonstrated by using hydrophones to broadcast vibrations like those produced by a struggling fish. The same phenomenon has been demonstrated unintentionally in macabre sea-rescue operations in which sharks were apparently attracted to downed pilots by vibrations from the rotors of rescue helicopters.

Whether by olfaction or vibration detection, once a shark is close to the stimulus source, vision takes over as the primary mode of prey detection. If the prey is easily recognized, a shark may proceed directly to an attack. Unfamiliar prey is treated differently, as studies aimed at developing shark deterrents have discovered. A circling shark may suddenly turn and rush toward unknown prey. Instead of opening its jaws to attack, however, the shark bumps or slashes the surface of the object with its rostrum. Opinions differ on whether this is an attempt to determine texture through mechanoreception, make a quick electrosensory appraisal, or use the rough placoid scales to abrade the surface, releasing fresh olfactory cues.

Following further circling and apparent evaluation of all sensory cues from the potential prey, the shark may either wander off or attack. In the latter case the rostrum is raised, the jaws protruded, and in the last moments before contact many sharks draw an opaque eyelid (the nictitating membrane) across each eye to protect it. At this point it appears that sharks shift entirely to electroreception to track prey. This hypothesis was developed while studying the attacks by large sharks on bait suspended from boats. Divers in submerged protective cages watched the attacks. After occluding their eyes with the nictitating membranes, attacking sharks frequently veered away from the bait and bit some inanimate object near the bait (including the observer's cage, much to the dismay of the divers). Apparently, olfaction and vibration senses are of little use at very close range, leaving only electroreception to guide the attacking shark. The unnatural environment of the bait stations included strong influences on local electric fields from the cages, and the sharks mistakenly attacked these artificial sources of electrical activity.

Jaws and Feeding Mobility within the head skeleton, known as **cranial kinesis,** allows consumption of large food items. Cranial kinesis permits inclusion of large items in the diet without excluding smaller foodstuff. Extant elasmobranchs have an enlarged hyomandibular cartilage (= hyomandibula), which braces the posterior portion of the palatoquadrate and attaches firmly but movably to the otic region of the cranium (Figure 5–5). A second connection to the chondrocranium is via paired palatoquadrate projections to either side of the braincase just behind the eyes and attached to the braincase by elastic ligaments. Jaw suspension of this type is

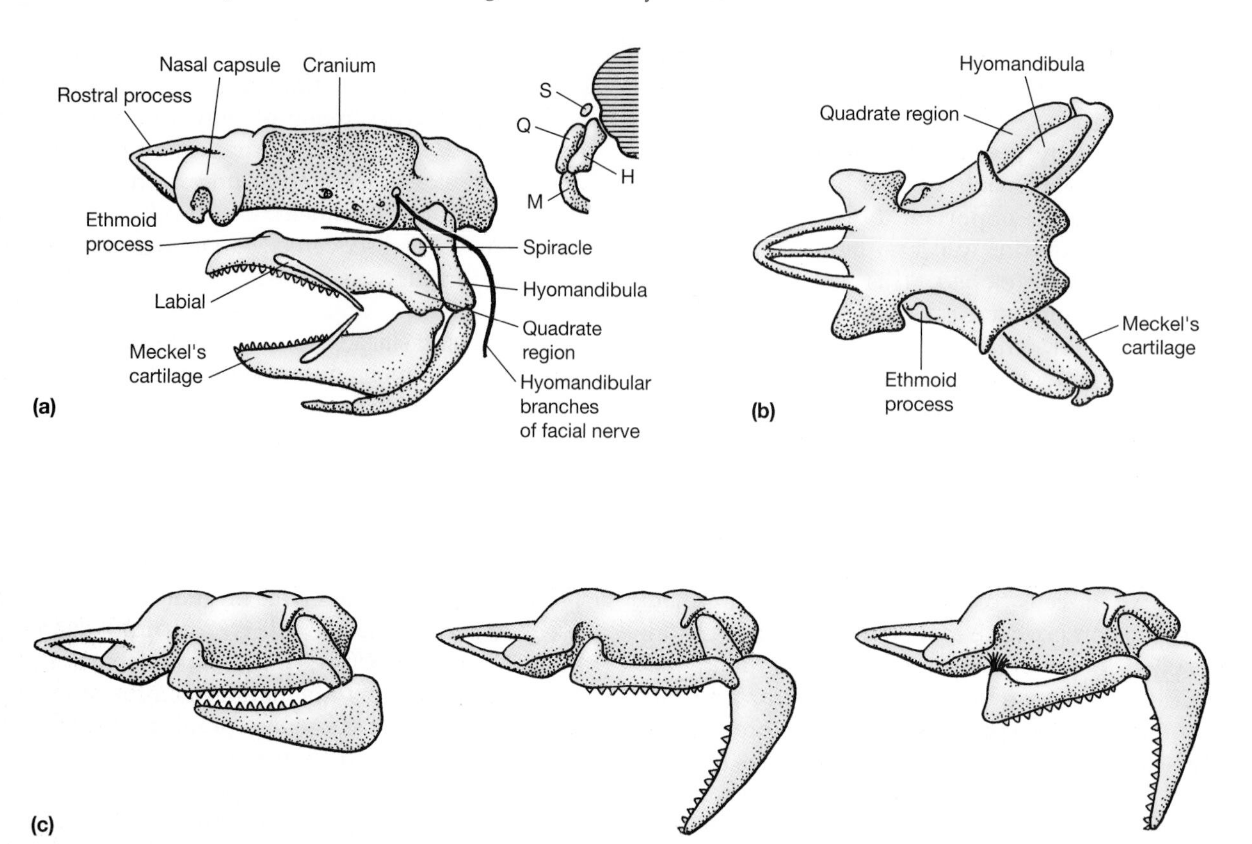

▲ Figure 5–5 Anatomical relationships of the jaws and chondrocranium of extant hyostylic sharks based on *Scyllium* and *Carcharhinus*. (a) Lateral and cross-sectional views of the head skeleton of *Scyllium* with the jaws closed. S = spiracle; Q = quadrate region of the palatoquadrate; H = hyomandibula; M = mandible. (b) Dorsal view of *Carcharhinus*. (c) During jaw opening and upper jaw protrusion the hyomandibula rotates from a position parallel to the long axis of the cranium to a 45° angle to that axis.

known as hyostylic. Hyostyly permits multiple jaw positions, each appropriate to different feeding opportunities.

The right and left halves of the pectoral girdle are fused together ventrally into a single U-shaped scapulocoracoid cartilage. Muscles running from the ventral coracoid portion to the symphysis of the lower jaw open the mouth. The advantages of the jaws of extant elasmobranchs are displayed when the upper jaw is protruded. Muscles swing the hyomandibula laterally and anteriorly to increase the distance between the right and left jaw articulations and thereby increase the volume of the orobranchial chamber. This expansion, which sucks water and food forcefully into the mouth, is not possible with an amphistylic jaw suspension because the palatoquadrate is tightly attached to the chondrocranium.

With hyomandibular extension, the palatoquadrate is protruded to the limits of the elastic ligaments on its orbital processes. Protrusion drops the mouth away from the head to allow a shark to bite an organism much larger than itself. The dentition of the palatoquadrate is specialized for attacking prey too large to be swallowed whole—the teeth on the palatoquadrate are stouter than those in the mandible and often recurved and strongly serrated. When feeding on large prey a shark opens its mouth, sinks its lower and upper teeth deeply into the prey, and then protrudes its upper jaw ever more deeply into the slash initiated by the teeth. As the jaws reach their maximum initial penetration, the shark throws its body into exaggerated lateral undulations, which results in a violent side-to-side shaking of the head. The head movements bring the serrated upper teeth into action as saws to sever a large piece of flesh from the victim.

Rare observations of sharks feeding under natural conditions indicate that these fishes are versatile and effective predators. The great white shark (*Carcharodon carcharias*) kills mammalian prey, such as seals, by exsanguination—bleeding them to death. A shark holds a seal tightly in its jaws until it is no longer bleeding and then bites down, removing an enormous chunk of flesh. The carcass floats to the surface and the shark returns to it for another bite. A white shark may feed rather leisurely on an acceptable carcass and defend it from other white sharks with typical fish-like side-by-side tail slaps, as well as tail lobbing and breaching into the air—behaviors more familiar among dolphins and whales.

Attacks by great white sharks on sea lions are quite different from those on seals, because sea lions—unlike seals—have powerful front flippers that can be used effectively in defense. Sharks often seize and release sea lions repeatedly until they die from blood loss. White sharks quickly release prey they find unacceptable after initial mouthing. Klimley (1994) suggested that lack of blubber is a primary rejection criterion. This behavior may explain why great white sharks seize and then release sea otters and humans.

Reproduction Much of the success of the extant grade of elasmobranchs may be attributed to their sophisticated breeding mechanisms. Internal fertilization is universal. The pelvic claspers of males have a solid skeletal structure that may increase their effectiveness. During copulation (Figure 5–6a) a single clasper is bent at 90 degrees to the long axis of the body, so the dorsal groove on the clasper lies directly under the cloacal papilla from which sperm are emitted. The flexed clasper is inserted into the female's cloaca and locked there by an assortment of barbs, hooks, and spines near the clasper's tip. Sperm from the genital tract are ejaculated into the clasper groove. Simultaneously, a muscular subcutaneous sac extending anteriorly beneath the skin of the male's pelvic fins contracts. This siphon sac has a secretory lining and is filled with seawater by the pumping of the male's pelvic fins before copulation. Seminal fluid from the siphon sac washes sperm down the groove into the female's cloaca, from which point the sperm swim up the female's reproductive tract.

Male sharks of small species secure themselves *in copulo* by wrapping around the female's body. Large sharks swim side by side, their bodies touching, or enter copulation in a sedentary position with their heads on the substrate and their bodies angled upward. Many male sharks and skates bite the female's flanks or hold onto one of her pectoral fins with their jaws. In these species females may have skin on the back and flanks that is twice as thick as the skin of a male the same size.

Reproductive strategies among vertebrates illustrate a trade-off between quantity and quality of offspring. At one extreme are species like the cod, which produce millions of small eggs. A mother cod invests little time or energy in an individual egg, but she produces so many offsping that some survive to become adults. At the opposite end of the scale are

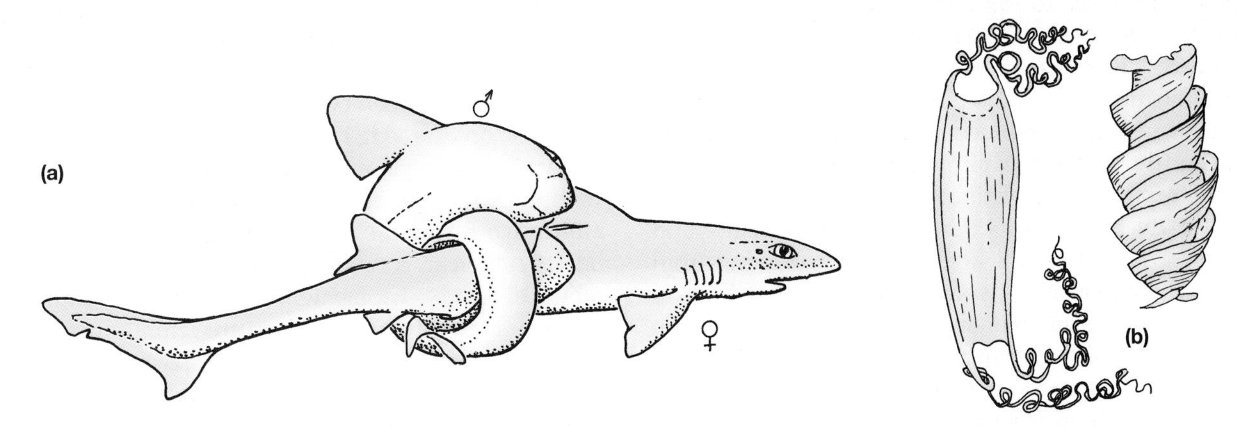

▲ Figure 5–6 Reproduction by sharks. (a) Copulation in the European spotted catshark *Scyliorhinus*. Only a few other species of sharks and rays have been observed *in copulo*, but all assume postures so that one of the male's claspers can be inserted into the female's cloaca. (b) (Not to same scale.) The egg cases of two oviparous sharks, *Scyliorhinus* (left) and *Heterodontus* (right).

animals (humans, for example) that produce very few young and invest a great deal of time and energy to ensure the survival of each one.

With the evolution of internal fertilization elasmobranchs adopted a reproductive strategy favoring the production of a small number of offspring that are retained, protected, and nourished for varying periods within the female's body. This mode of reproduction requires a significant investment of energy by the female, and it succeeds when adults have long life expectancies.

Most elasmobranch eggs are large (the size of a chicken yolk or larger) and contain a very substantial store of nutritious yolk. Oviparous (Latin *ovum* = egg and *pari* = to bring forth) elasmobranchs have large eggs. A specialized structure at the anterior end of the oviduct, the nidimental gland, secretes a proteinaceous case around the fertilized egg. Protuberances on the cases become tangled with vegetation or wedged into protected sites on the substrate (Figure 5–6b). The zygote obtains nutrition exclusively from the yolk during the 6- to 10-month developmental period. Inorganic molecules and dissolved oxygen are taken from a flow of water through the shell openings induced by movements of the embryo. On hatching, the young are generally miniature replicas of the adults and seem to live much as they do when mature.

A significant step in the evolution of elasmobranch reproduction was prolonged retention of the fertilized eggs in the reproductive tract, a reproductive mode called **ovoviviparity.** The only notable differences between oviparous and ovoviviparous forms are reduction in the nidimental gland's shell production and increased vascularization of the oviducts of the female and yolk sacs of the embryo. All nutrition comes from the yolk, and only inorganic ions and dissolved gases are exchanged between the maternal circulation and that of the developing young. This ovoviviparity is properly called **lecithotrophy** (Greek *lecith* = egg and *troph* = nourishment). The eggs often hatch within the oviducts, and the young may spend as long in their mother after hatching as they did within the shell. Most ovoviviparous elasmobranchs have about a dozen young at a time.

A natural step from the lecithotropic condition is full viviparity or **matrotrophy** (Latin *matro* = mother), the situation in which the nutritional supply is not limited to the yolk. Elasmobranchs have independently evolved matrotrophy several times. Some elasmobranchs develop long spaghetti-like extensions of the oviduct walls that penetrate the mouth and gill openings of the internally hatched young and secrete a milky nutritive substance. Other species simply continue to ovulate, and the young that hatch in the oviducts feed on the new eggs. The most common and most complex form of viviparity is found among sharks that develop a yolk sac placenta that allows an embryo to obtain nourishment from the maternal uterine bloodstream via its highly vascular yolk sac. This mode of reproduction is called **placentotrophic viviparity** (Latin *placenta* = a round flat cake). No matter which form of nourishment has brought young elasmobranchs to their free-living size, there is no evidence of further parental investment once the eggs are laid or the young born.

Social Behavior Elasmobranchs have long been considered solitary and asocial, but this view is changing. Accumulating field observations, often from aerial surveys or by scuba divers in remote areas, indicate that elasmobranchs of many species aggregate in great numbers periodically, perhaps annually. More than 60 giant basking sharks have been observed milling together and occasionally circling in head-to-tail formations in an area off Cape Cod in summer, and an additional 40 individuals were nearby. Over 200 hammerhead sharks have been seen near the surface off the eastern shore of Virginia in successive summers. Divers on seamounts that reach to within 30 meters of the surface in the Gulf of California have observed enormous aggregations of hammerheads schooling in an organized manner around the seamount tip. Some of these hammerhead observations include behavior thought to be related to courtship. More than 1000 individuals of the blue shark have been observed near the surface over canyons on the edge of the continental shelf off Ocean City, Maryland. Fishermen are all too familiar with the large schools of spiny dogfishes that seasonally move through shelf regions, ruining fishing by destroying gear, consuming bottom fishes and invertebrates, and displacing commercially valuable species. These dogfish schools are usually made up of individuals that are all the same size and the same sex. The distribution of schools is also peculiar: Female schools may be inshore and males offshore, or male schools may all be north of some point and the females south. Our understanding of these phenomena is slim, but it is clear that not all elasmobranchs are solitary all the time.

Life History and Conservation of Sharks Whatever their social behavior and reproductive mode, sharks produce relatively few young during an individual female's lifetime. This is a life-history pattern that depends on high rates of survival for young animals and long life expectancies for adults. Internal fertilization and the life-history characteristics that accompany it evolved in sharks fully 350 million years ago and have been a successful strategy throughout the world's oceans ever since. Now alterations of the habitat and heavy predation by humans threaten the survival of many species. Although young sharks are relatively large compared with other fishes, they are subject to predation, especially by other sharks. Many species depend on protected nursery grounds—usually shallow inshore waters, which are the areas most subject to human disturbance and alteration. In addition, adult sharks are increasingly falling prey to humans. A rapid expansion in recreational and commercial shark fishing worldwide threatens numerous species of these long-lived, slowly reproducing top predators.

People like to eat sharks, whether they know what they are eating not. Spiny dogfish (*Squalus acanthias*) has long been served up as "fish and chips" in Europe and began to make its unheralded way into the American prepared food market in the 1980s. "Mako" shark steaks (sliced from sharks of two genera, *Isurus* and *Lamna*) have become an alternative to swordfish in the fresh seafood cases of nearly every American supermarket. Europe has already seen drastic drops in the populations of these sharks from commercial fisheries. For example, Norway targeted *Lamna* for intensive fishing and initially harvested as much as 8060 tons in a single year from the northeastern Atlantic. But within seven years, the catch had fallen to 207 tons, and since the 1970s the Norwegians have been unable to catch 100 tons per year. Despite such histories, the United States did not begin regulation of any shark fishery until 1993 and did nothing to control the spiny dogfish catch until years later. By the year 2000 emergency seasonal bans on all dogfish catches were being implemented as annual harvest quotas were filled months before the quota year ended.

At about the same time as domestic consumption of shark increased, an even more powerful economic force exploded on the scene—export of shark fins to Asian markets. Shark-fin soup, which is reputed to have medicinal properties, may fetch $90 or more a bowl in restaurants in Asia. Dried shark fins wholesale in Hong Kong for $256 a pound; in the United States, fresh wet fins sell for $100 a pound. The success of Asian economies during the 1980s and 1990s created an almost unlimited market for fins alone, because the rest of the carcass was worthless by comparison. Shark finning—the practice of catching sharks, cutting off the fins, and throwing the rest of the animal, dead or alive, back into the sea—became a worldwide phenomenon. This wasteful and cruel business moved into American waters in a very big way. Long liners set 8.9 million hooks along the Eastern and Gulf coasts of the United States in 1995. In 1997, long liners in Hawaii caught over 100,000 sharks and discarded 98.6 percent of the mass of the catch. Shark, skate, and ray catches worldwide more than quadrupled to over 800,000 metric tons each year (70 million individual sharks contributing most of the tonnage). It is little wonder that in the United States alone, various species of coastal sharks have experienced population reductions of 50 to 85 percent over the past 20 years.

Because of their biological characteristics, all elasmobranchs are particularly susceptible to near extirpation by fishing. They grow slowly, mature late in their lives, have few young at a time, and because of the great amount of energy invested in those young, females do not reproduce every year (Figure 5–7). Relatively few individuals occur in any area, except perhaps at times of breeding or other social aggregations. For example, only nine to 14 individual great white sharks were observed over a period of five years in the South Farallon Islands near San Francisco. The same individuals returned each fall when prey numbers were high. When only four great white sharks were killed near the sea lion colony that attracted them, attacks on seals and sea lions in the area dropped by half for two years. Fisheries management specialists say that there is little chance that populations of many species of overfished sharks will show significant recovery in less than half a century, even with strict fishing limitations.

To make things worse, elasmobranchs receive very little international fisheries attention, and most nations have no effective fisheries management at all. Many commercially important sharks are pelagic, migrating across multiple political boundaries and spending significant portions of their lives in international waters where there is no political jurisdiction. Effective preservation measures are hard to implement. As of the year 2000, the U.S. House of Representatives had approved a nationwide ban on

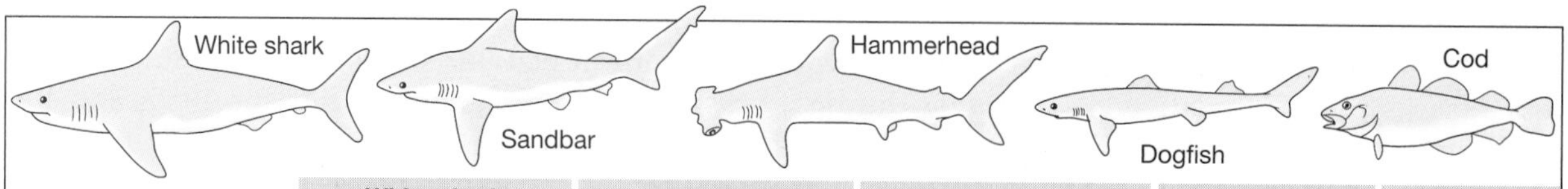

	White shark *Carcharodon carcharias*	**Sandbar** *Carcharhinus plumbeus*	**Scalloped hammerhead** *Sphyrna lewini*	**Spiny dogfish** *Squalus acanthius*	**Atlantic cod** *Gadus morhua*
Age to maturity (years)	m 9–10, f 12–14	m 13–16	m 4–10, f 4–15	m 6–14, f 10–12	m 2–4
Size at maturity (centimeters)	m 350–410, f 400–430	m 170, f >180	m 140–280, f 150–300	m 60, f 70	m 32–41
Life span (years)	m 15(?)	m 25–35	m 35	m 35, f 40–50	m 20+
Litter size	2–10 pups	8–13 pups	12–40 pups	2–14 pups	2 million–11 million eggs
Reproductive frequency	Biennial(?)	Biennial	(?)	Biennial	Annual
Gestation period (months)	>12	9–12	9–12	18–24	n/a

▲ Figure 5–7 Some shark life-history parameters compared to that of another important object of intensive fisheries, the Atlantic cod. Their life-history characteristics make sharks vulnerable to over exploitation.

shark finning, but even before the bill was introduced to the Senate (which it must pass it before it can become law), long liners were simply moving their base of operations from Hawaii to islands out of American jurisdiction.

Skates and Rays

The hypotremate (Greek *hypo* = below and *trema* = plate) elasmobranchs—the skates and rays—are more diverse than sharks. Approximately 456 extant species of skates and rays are currently recognized (Figure 5–8). These fishes have a long history of phylogenetic isolation from the lineages of extant sharks but appear to be derived from the squaloids and form the bulk of the species in the Squalea. The suite of specializations characteristic of skates and rays relates to their early assumption of a benthic (bottom-dwelling), durophagous (Latin *duro* = hard and Greek *phagus* = to eat) habit. The teeth are almost all flat, crowned plates that form a pavement-like dentition. The mouth is often highly and rapidly protrusible to provide powerful suction used to dislodge shelled invertebrates from the substrate.

Skates and rays are derivatives of the extant shark radiation adapted for benthic habitats. A handful of specialized shark species belonging to the squatinoid lineage of the Squalea are bottom-living specialists and illustrate what an intermediate stage in the transition to skates and rays may have been like. Skates and rays are distinguished by their tails and modes of reproduction, Skates have an elongate but thick tail stalk supporting two dorsal fins and a terminal caudal fin (Figure 5–8a). A typical ray (Figure 5–8b) has a whip-like tail stalk with fins replaced by one or more enlarged, serrated, and venomous dorsal barbs. Derived forms, such as stingrays (family Dasyatidae), have a few greatly elongated and venomous spines derived from modified placoid scales at the base of the tail. The electric skates (family Rajidae) have specialized tissues in their long tails that are capable of emitting a weak electric discharge. Each species appears to have a unique pattern of discharge, and the discharges may identify conspecifics in the gloom of the seafloor. The electric rays and torpedo rays (family Torpedinidae) have modified gill muscles, producing electrical discharges of up to 200 volts that are used to stun prey. Skates are oviparous, laying eggs enclosed in horny shells popularly called "mermaid's purses," whereas rays are viviparous.

Skates and rays are similar in being dorsoventrally flattened, with radial cartilages extending to the tips of the greatly enlarged pectoral fins. The anteriormost basal elements fuse with the chondrocranium in front of the eye and with one another in front of the rest of the head. Skates and rays swim by undulating these massively enlarged pectoral fins. The placoid scales so characteristic of the integument of a shark are absent from large areas of the bodies

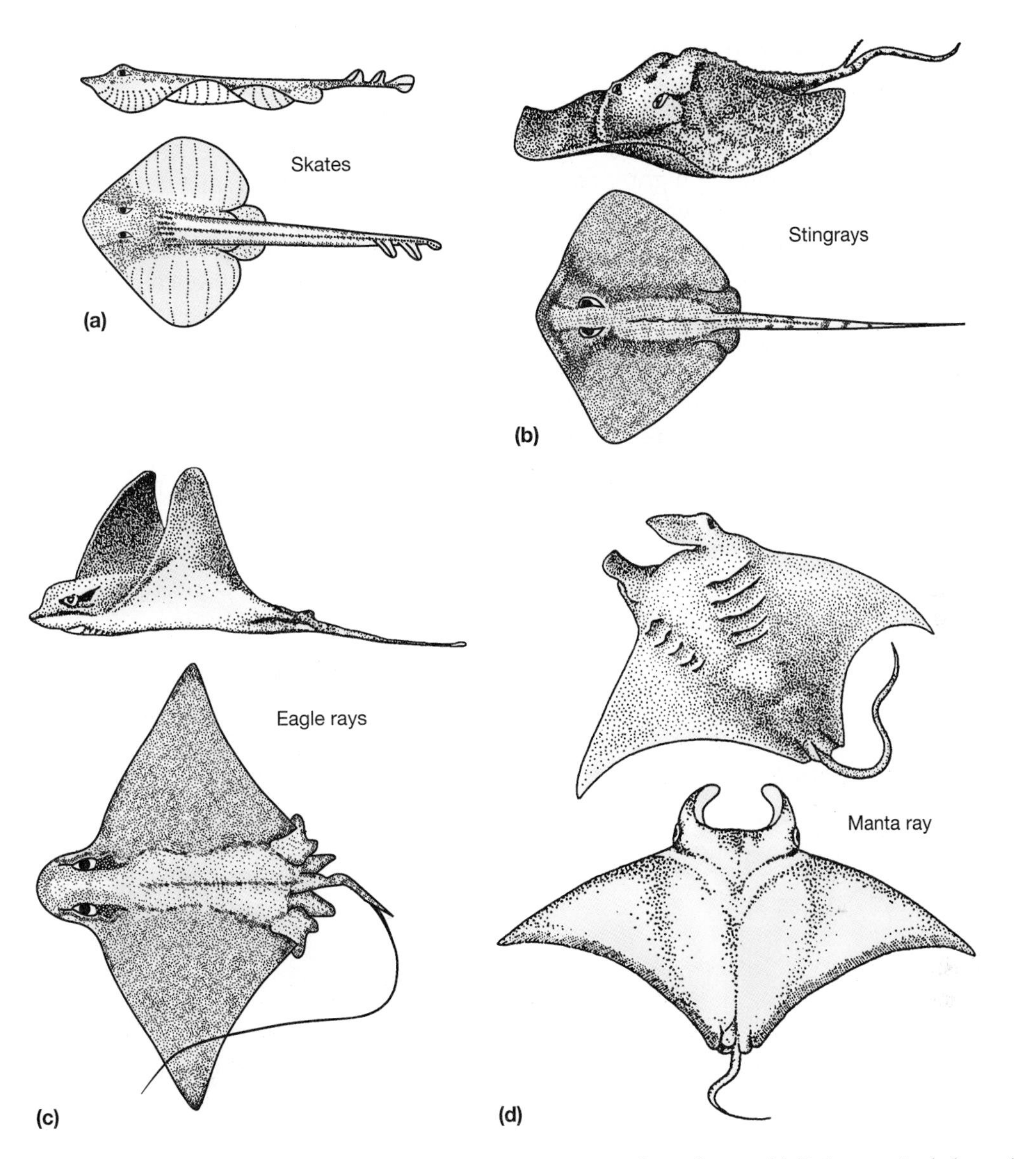

▲ Figure 5–8 Some extant skates and rays: (a, b) benthic forms; (c, d) pelagic forms. (a) *Raja,* a typical skate. (b) *Dasyatis,* a typical ray. The following pelagic batoids are closely related to the rays: (c) *Aetobatus* is representative of the eagle rays, wide-ranging shelled invertebrate predators. (d) *Manta* is representative of its family of gigantic fishes that feed exclusively on zooplankton. Extensions of the pectoral fins anterior to the eyes (the horns of these devil rays) help funnel water into the mouth during filter feeding.

and pectoral fins of skates and rays. The few remaining denticles are often greatly enlarged to form sharp, stout bucklers along the dorsal midline.

Skates and rays are primarily benthic invertebrate feeders (occasionally managing to capture small fishes). Many skates and rays rest on the seafloor and cover themselves with a thin layer of sand. They spend hours partially buried and nearly invisible except for their prominent eyes, surveying their surroundings. The largest rays, like the largest sharks, are plankton strainers. Devilfishes or manta rays of the family Mobulidae are up to 6 meters in width (Figure 5–8d). These highly specialized rays swim through the open sea with flapping motions of the pectoral fins, filtering plankton from the water as they go.

The dentition of many benthic rays is sexually dimorphic. Different dentitions coupled with the generally larger size of females might reduce competition for food resources between the sexes, but no difference in stomach contents has been found. Since the male uses its teeth to hold or stimulate a female before and during copulation, sexual selection may be at work. Males of the stingray *Dasyatis sabina* have blunt teeth like those of females for most of the year, but during the breeding season males grow sharp-cusped teeth that are used for courtship.

5.6 Holocephali—The Bizarre Chondrichthyans

The approximately 34 extant forms of chimeras (none much over a meter in length) have a soft anatomy more similar to sharks and rays than to any other extant fishes (Figure 5–9). They have long been grouped with elasmobranchs as Chondrichthyes because of these shared specializations, but they have a bizarre suite of unique features. Generally found deeper than 80 meters and thus not well known in the wild, the Holocephali move into shallow water to deposit their 10-cm horny-shelled eggs, from which hatch miniature chimeras. Several holo-

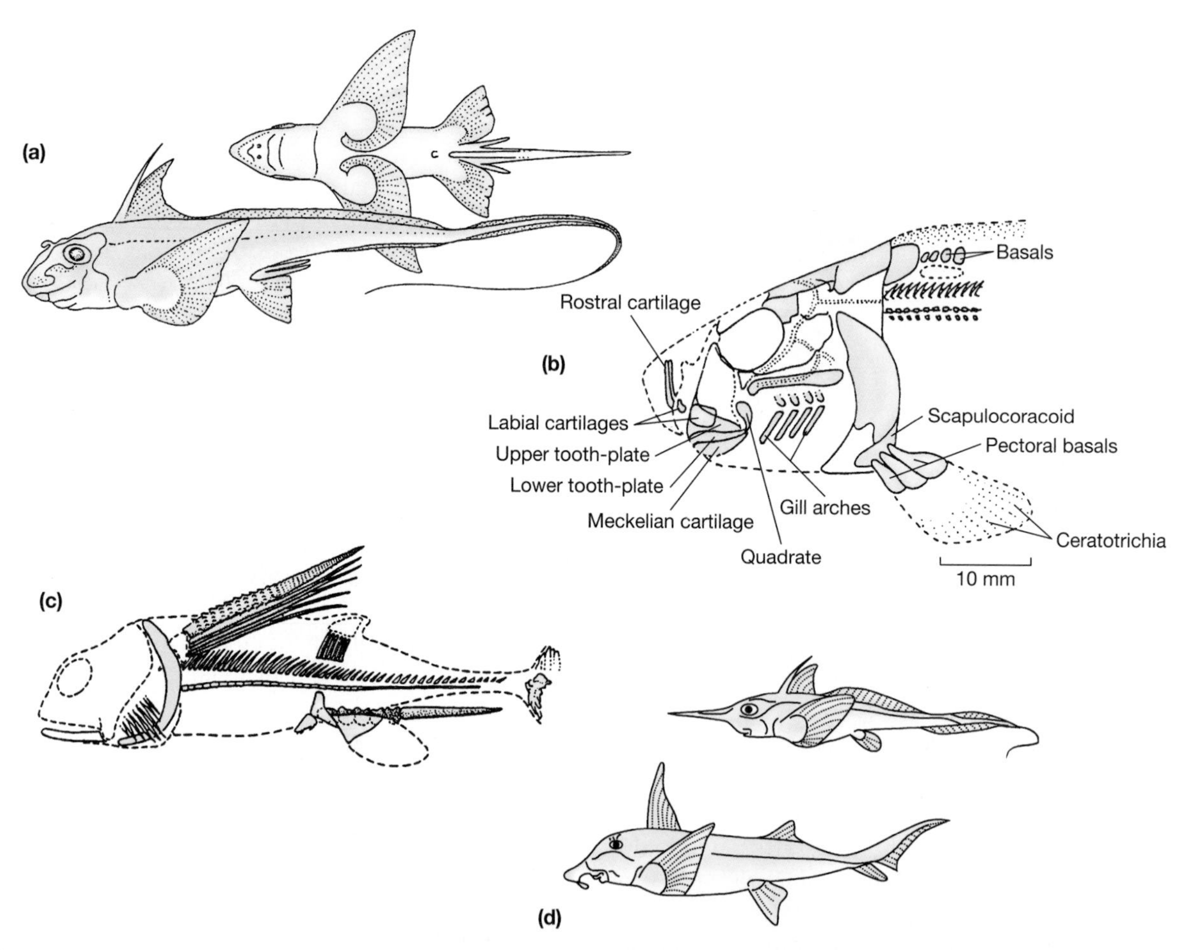

▲ Figure 5–9 Chimeras. (a) Common chimera *Hydrolagus colliei,* an extant holocephalan. (b) Representatives of two other extant groups: plownose chimera (lower) and longnose chimera (upper). (c) *Ctenurella,* a Late Devonian ptyctodont placoderm once thought to be related to chimeras. (d) *Phomeryele,* a Carboniferous iniopterygian shark favored by most paleontologists as a relative of the extant chimeras.

cephalians have elaborate rostral extensions of unknown function. From what is known, most species appear to feed on shrimp, gastropod mollusks, and sea urchins. Their locomotion is produced by lateral undulations of the body that throw the long tail into sinusoidal waves and by fluttering movements of the large, mobile pectoral fins. The solidly fused nipping and crushing tooth plates grow throughout life, adjusting their height to the wear they suffer. Of special interest are the armaments—a mace-like cephalic clasper of males and a poison gland associated with the stout dorsal spine in some species.

There is little agreement about holocephalian ancestry. Two distinct fossil groups have been proposed as ancestors—ptyctodonts and iniopterygians. A group of peculiar placoderms, the ptyctodontids are known from the mid-Devonian. Rarely exceeding 20 cm long, they showed a reduction in the extent and number of head and thoracic shield plates. A short palatoquadrate bound to the cranium carried a single pair of large upper tooth plates that were opposed by a smaller mandibular pair. The gills were covered by a single operculum. Postcranial characters—such as fin spines, large paired appendages, claspers, and caudal development—are strongly reminiscent of extant holocephalians. The first undoubted modern holocephalians are of Jurassic age, and fossils proposed as earlier members of the chimaera lineage do not seem to link the ptyctodont placoderms to the extant forms. Ptyctodonts are more like extant holocephalians than they are like the truly ancestral holocephalians, which are in most cases rather shark-like (Janvier 1996).

Since the 1950s Rainer Zangerl and co-workers have described a group of bizarre and obviously specialized forms (Figure 5–9c). These Iniopterygia (Greek *inio* = back of the neck and *ptero* = wing) have characteristics that Zangerl considered to be evidence for a link between the earliest sharks and holocephalians. As in extant holocephalians, the palatoquadrate is fused to the cranium (autostylic suspension), but the teeth, unlike those of extant chimeras, are in replacement families like those of elasmobranchs. The earliest close relative of the modern holocephalians is now thought to be of Late Carboniferous age, too old to be descended from the contemporaneous iniopterygians. Current thinking is that both iniopterygians and holocephalians arose from ancestral sharks following a somewhat parallel line of evolution. Now that a phylogenetic connection between holocephalians and ancestral-grade elasmobranchs seems clearly established, the ptyctodont placoderms and chimeras represent one of the most outstanding examples of convergent evolution in groups from different geologic eras.

Summary

The elasmobranchs first appear in the fossil record in the Late Silurian and are distinguished by a cartilaginous endoskeleton. Three radiations of shark-like elasmobranchs can be traced through increasingly sophisticated characters of the jaws and fins. Paleozoic forms had broad-based fins that were probably immovable and an upper jaw that was firmly fixed to the skull, limiting the gape of the mouth. In the early Mesozoic a second radiation produced species in which the bases of the pelvic and pectoral fins were narrow, allowing them to swivel, and the fins had intrinsic muscles that could change their curvature. The third radiation of sharks appeared in the middle of the Mesozoic, and descendants of these forms remain the dominant predators in shallow seas. Further modifications of skull allow the jaws to be protruded, and a shark can bite chunks of flesh from prey too large to be swallowed whole.

Skates and rays are derived from the extant shark radiation and are adapted for life on the sea bottom. They are dorsoventrally flattened, with eyes and spiracles on tops of their heads. Many skates and rays lie buried in sand and ambush their prey, but the largest species—the manta rays—are plankton feeders.

The holocephalians (chimeras or ratfish) are a small group of bizarre fishes that probably branched off from ancestral elasmobranchs in the Paleozoic. They generally occur in deep water, and little is known of their natural history and behavior.

Elasmobranchs probably evolved internal fertilization in the Paleozoic. The life histories of most species are based on producing a few relatively large young at a time. This reproductive strategy depends on high survival of young and long life expectancies for adults. It worked well for 350 million years; but within the past 50 years loss of coastal habitat and outrageous overfishing has brought many species of elasmobranchs to the edge of extinction.

Additional Readings

Bond, C. E. 1996. *Biology of Fishes,* 2d ed. Philadelphia, Penn.: Saunders College Publishing.

Carroll, R. L. 1987. *Vertebrate Paleontology and Evolution.* New York: Freeman.

Didier, D. A. 1995. Phylogenetic systematics of extant chimaeroid fishes (Holocephali, Chimaeroidei). *American Museum Novitates,* no. 3119.

Ferry, L. A., and G. V. Lauder. 1996. Heterocercal tail function in leopard sharks: A three-dimensional kinematic analysis of two models. *Journal of Experimental Biology* 199:2253–2268.

Helfman, G. S., B. B. Collette, and D. E. Facey. 1997. *The Diversity of Fishes.* Malden, Mass.: Blackwell.

Janvier, P. 1996. *Early Vertebrates.* Oxford, UK: Clarendon Press.

Kajiura S. M., and T. C. Tricas. 1996. Seasonal dynamics of dental sexual dimorphism in the Atlantic stingray *Dasyatis sabina. Journal of Experimental Biology* 199:2297–2306.

Kenney, R. D., R. E. Owen, and H. E. Winn. 1985. Shark distributions off the Northeast United States from marine mammal surveys. *Copeia* 1985:220–223.

Klimley, A. P. 1987. The determinants of sexual segregation in the scalloped hammerhead shark, *Sphyrna lewini. Environmental Biology of Fishes* 18:27–40.

Klimley, A. P. 1993. Highly directional swimming by scalloped hammerhead sharks, *Sphyrna lewini,* and subsurface irradiance, temperature, bathymetry, and geomagnetic field. *Marine Biology* 117:1–22.

Klimley, A. P. 1994. The predatory behavior of the white shark. *American Scientist* 82:122–133.

Klimley, A. P. 1999. Sharks beware. *American Scientist* 87: 488–491.

Klimley, A. P., and D. G. Ainley. 1996. *Great White Sharks: The biology of* Caraharodon carcharias. San Diego: Academic Press.

Litman, G. W. 1996. The origins of vertebrate immunity. *Scientific American* 275(5):67–71.

Maisey, J. G. 1996. *Discovering Fossil Fishes.* New York: Henry Holt.

Martin, A. P., G. J. P. Naylor, and S. R. Palumbi. 1992. Rates of mitochondrial DNA evolution in sharks are slow compared with mammals. *Nature* 357:153–155.

Moyle, P. B., and J. J. Cech, Jr. 2000. *Fishes: An Introduction to Ichthyology,* 4th ed. Upper Saddle River, N.J.: Prentice Hall.

Nelson, J. S. 1994. *Fishes of the World,* 3d edition. New York: Wiley.

Schmidt, E. M., and G. V. Lauder. 1995. Kinematics of locomotion in sturgeon: Do heterocercal tails function similarly? *American Zoologist* 35(5):62A, Abstract no. 250.

Stiassny, M. L., L. R. Parenti, and G. D. Johnson (Eds). 1996. *Interrelationships of Fishes.* San Diego, Calif.: Academic Press.

Web Explorations

On-line resource for this chapter are on the World Wide Web at: http://www.prenhall.com/pough (click on the Table of Contents link and then select Chapter 5).

CHAPTER

6

Dominating Life in Water: The Major Radiation of Fishes

By the end of the Silurian the agnathous fishes (Chapter 3) had diversified and the cartilaginous gnathostomes (Chapter 5) were in the midst of their first radiation. The stage was set for the appearance of the largest extant group of vertebrates, the bony fishes. The first fossils of bony fishes (Osteichthyes) occur in the Late Silurian. Osteichthyans are well represented from the Early Devonian. Their radiation was in full bloom by the middle of the Devonian, with two major groups diverging: the ray-finned fishes (Actinopterygii) and lobe-finned fishes (Sarcopterygii). Specialization of feeding mechanisms is a key feature of the evolution of these major groups of vertebrates. An increasing flexibility among the bones of the skull and jaws allowed ray-finned fishes in particular to exploit a wide range of prey types and predatory modes. Specializations of locomotion, habitat, behavior, and life histories have accompanied the specializations of feeding mechanisms. The body forms, behaviors, mechanical functions, and physiology of bony fishes are intimately related to the characteristics of the aquatic habitat and to the properties of water as an environment for life.

6.1 The Appearance of Bony Fishes

Because all major lineages of fishes, extant and extinct, coexisted in the fresh and marine waters of the planet for its 48-million-year duration, the Devonian is known as the Age of Fishes. Most groups of gnathostome fishes either made their first appearance or significantly diversified during this period. Among them was the most species-rich and morphologically diverse lineage of vertebrates, the Osteichthyes or bony fishes (Figure 3–13).

The Earliest Osteichthyes and the Major Groups of Bony Fishes

Fragmentary remains of bony fishes are known from the Late Silurian. It is not until the Devonian that more complete remains are found. These animals resemble acanthodians in details of cranial structure (see Chapter 3). The similarities suggest a common ancestor for acanthodians and osteichthyans in the Early Silurian.

Remains of the bony fishes representing a radiation of forms already in full bloom appear in the Early to Middle Devonian. Two major and distinctive types of osteichthyans (Greek *osteo* = bone and *ichthys* = fish) possessed unique locomotor and feeding characters and were the dominant fishes during the Devonian. The sarcopterygian lineage (Greek *sarco* = fleshy and *pteryg* = fin) had fins containing bones and muscles from which the tetrapod limb was derived, whereas the actinopterygian lineage (Greek *actinos* = ray) had fins supported primarily by rays. The derived modern bony fishes are the teleosts, the largest group of living vertebrates that arose among the actinopterygians. (Table 6.1).

Fossils of the two basic types of osteichthyans are abundant from the Middle Devonian onward. The Sarcopterygii (Figure 6–1, a–d) and the Actinopterygii (Figure 6–1e) are sister groups. Shared derived characters of these Osteichthyes include patterns of lateral line canals, similar opercular and pectoral girdle dermal bone elements, and fin webs supported by bony dermal rays. A fissure allowed movement between

Table 6.1 Classification and geographic distribution of osteichthyes, the bony fishes. Major groups are shown in boldface. Only the evolutionarily or numerically most important groups are listed. The subdivision of the Neopterygii varies greatly from author to author as does the number of living species claimed for the larger groups. Groups in quotation marks are not monophyletic, but relationships are not yet understood.

Sarcopterygii (fleshy-finned fishes and tetrapods)—7 extant species

Actinistia [Coelacanthiformes] (coelacanths)—2 living species; Western Indian Ocean and central Indonesia, deep-water marine.

Dipnoi (lungfishes)—6 living species; Southern Hemisphere, freshwater.

Tetrapods

Actinopterygii (ray-finned fishes)—About 24,000 extant species

Polypteriformes [Cladistia] (bichirs)—11 living species; Africa, freshwater.

Acipenseriformes [Chondrostea] (sturgeons and paddlefishes)—26 living species; Northern Hemisphere, coastal and freshwater.

Neopterygii—About 23,800 extant species

Lepisosteiformes [Ginglymodi] (gars)—7 living species; North and Central America, fresh and brackish water.

Amiiformes (bowfins)—1 living species; North America, freshwater.

Teleostei—About 23,800 extant species

Osteoglossomorpha (bony tongues)—About 220 living species; worldwide, tropical freshwater.

Elopomorpha (tarpons and eels)—About 800 living species; worldwide, mostly marine.

Clupeomorpha (herrings and anchovies)—About 360 living species; worldwide, especially marine.

Euteleostei—About 22,400 extant species

Ostariophysi (catfish and minnows)—About 6500 living species; worldwide, freshwater.

"Protacanthopterygii" (trouts and relatives)—320 living species; temperate Northern and Southern Hemisphere, freshwater.

"Basal Neoteleosts" (lanternfishes and relatives)—About 820 living species; worldwide, majority mesopelagic (middle depth) or bathypelagic (deep), marine.

Paracanthopterygii (cods and anglerfishes)—About 1220 living species; Northern Hemisphere, primarily marine.

Acanthopterygii (spiny-rayed fishes)—About 13,500 living species, including the Atherinomorpha (silversides, killifishes, and relatives), about 1220 living species; worldwide, surface-dwelling, freshwater and marine; Perciformes (perches and relatives), about 9300 living species; worldwide, primarily marine; and other smaller groups.

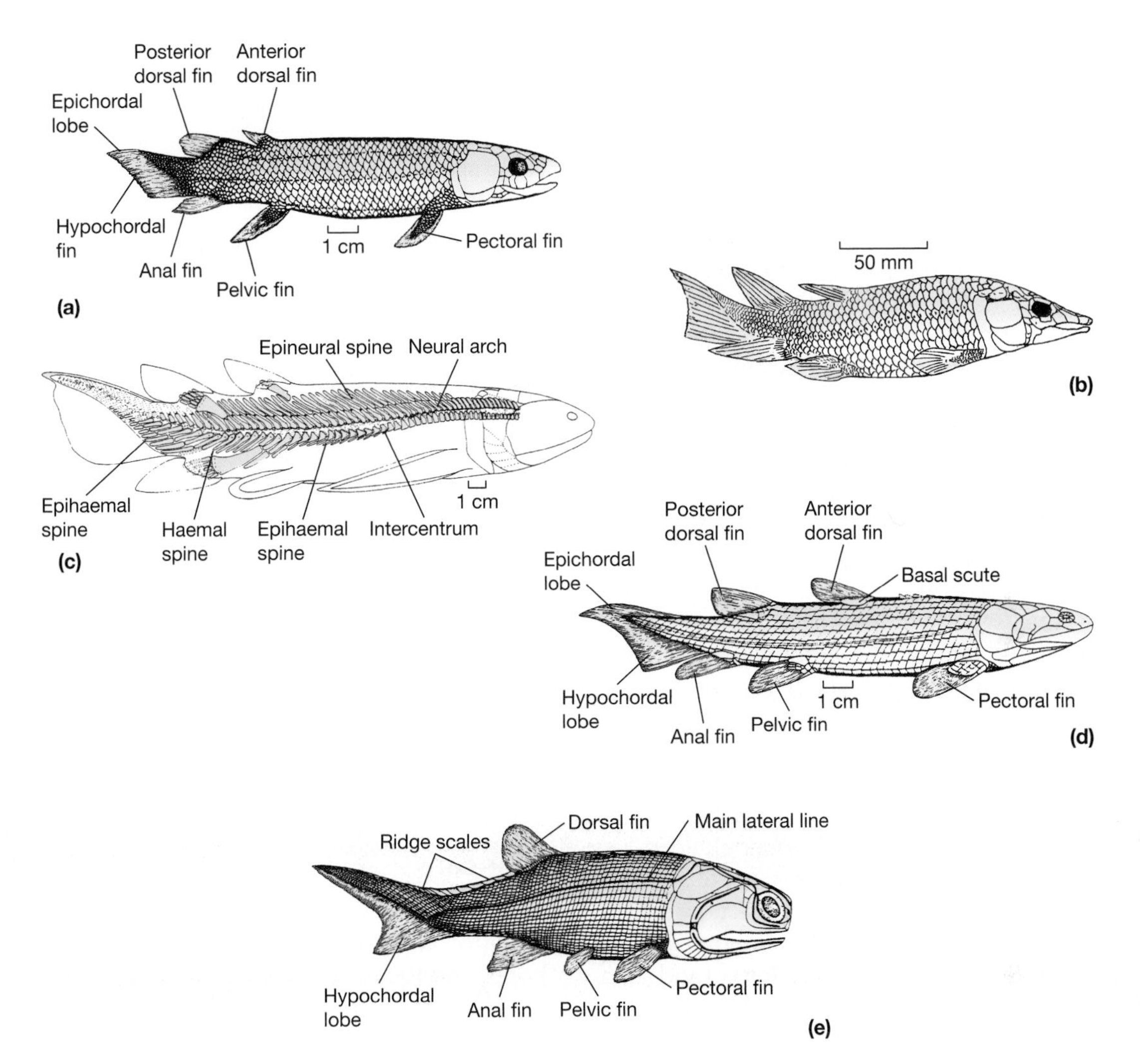

▲ Figure 6–1 Primitive osteichthyans. Dipnoans: (a) relatively unspecialized dipnoan *Dipterus,* Middle Devonian; (b) long-snouted dipnoan *Griphognathus,* Late Devonian. Other Sarcopterygian fishes: (c) porolepiform *Holoptychius,* Late Devonian to Early Carboniferous which was laterally compressed in life; (d) cylindrical osteolepiform *Osteolepis,* Middle Devonian; (e) typical early actinopterygian *Moythomasia,* Late Devonian.

the anterior and posterior halves of the chondrocranium in many forms. The presence of bone is not a unifying osteichthyan characteristic, because agnathans, placoderms, and acanthodians also possess bone, and the absence of bone in chondrichthyans is derived. What is unique to the bony fishes is endochondral bone (bone that replaces cartilage ontogenetically found in the internal skeleton). In addition to endochondral ossification, osteichthyes retained the more primitive dermal and perichondral bone-forming mechanisms as well. The name Osteichthyes was coined before the occurrence of bone in other primitive vertebrates was recognized. Likewise, the various names long in use for the extant actinopterygian subgroups imply an increase in the ossification of the skeleton as an evolutionary trend (for example, chondrosteans ("cartilaginous bony fishes"—sturgeons and paddlefishes) are the sister group of a radiation often called the holosteans ("entirely bony fishes"—gar and *Amia*), which culminated in teleosteans ("final bony fishes"). The fossil record does not reveal a regular sequence of increasing ossification. On the contrary, a tendency to reduce ossification, especially in the skull and scales, is apparent when the full array of early Osteichthyes is compared with their derived descendants.

Although the monophyletic relationship of the ray-finned fishes as a whole is supported by several shared derived characters, the relationships among early osteichthyans are disputed and no phylogenetic hypothesis is yet widely accepted. The phylogenetic relationships of the Sarcopterygii are also controversial.

Evolution of the Sarcopterygii

Primitive Sarcopterygii have similar body shapes and sizes (20 to 70 centimeters), two dorsal fins, an epichordal lobe (a fin area supported by the dorsal side of the vertebral column) on the heterocercal caudal fin, and paired fins that were fleshy, scaled, and had a bony central axis. The rays of the paired fins extend from a central shaft in a feather-like or leaf-like manner in contrast to the fan-like form of the rays in the paired fins of actinopterygians.

The jaw muscles of sarcopterygian fishes were massive by comparison with those of actinopterygians, and the size of these muscles produced skull characters that set sarcopterygians apart from the actinopterygians. Finally, the early sarcopterygians were coated with a peculiar layer of dentine-like material, **cosmine,** that spread across the sutures between dermal bones and shows indications of being periodically reabsorbed and then rebuilt.

The Dipnoi (lungfishes) have unique derived features clearly indicating that the lineage is monophyletic. The other sarcopterygian fishes have been variously combined as a single taxon, the Crossopterygii, now considered by most workers to be paraphyletic, or two separate lineages, rhipidistians (entirely extinct forms) and actinistians (for the sole surviving form *Latimeria*, the living coelacanth and its undisputed fossil relatives). Even this arrangement has been brought into question by the splitting of the rhipidistians into several different lineages, including one (the porolepiformes) that is related to lungfishes and another (the osteolepiformes) that is more closely related to tetrapods.

6.2 Extant Sarcopterygii—Lobe-Finned Fishes

Although they were abundant in the Devonian, the number of primarily aquatic sarcopterygians dwindled in the late Paleozoic and Mesozoic. (All terrestrial vertebrates are technically sarcopterygians, so the total number of extant sarcopterygians is enormous.) The early evolution of sarcopterygians resulted in a significant radiation in fresh and marine waters. Today only four nontetrapod genera remain: the dipnoans or lungfishes (*Protopterus* in Africa, *Lepidosiren* in South America, and *Neoceratodus* in Australia; see Figure 6–2), and the actinistian *Latimeria* (the coelacanths) in waters 100 to 300 meters deep off East Africa and central Indonesia (Figure 6–3). We will discuss fossil sarcopterygian fishes in more detail with their sister group, the tetrapods, in Chapter 8.

Dipnoans

Living Dipnoi are distinguishable by the lack of articulated tooth-bearing premaxillary and maxillary bones and the fusion of the palatoquadrate to the undivided cranium (**autostyly**). Teeth are scattered over the palate and fused into tooth ridges along the

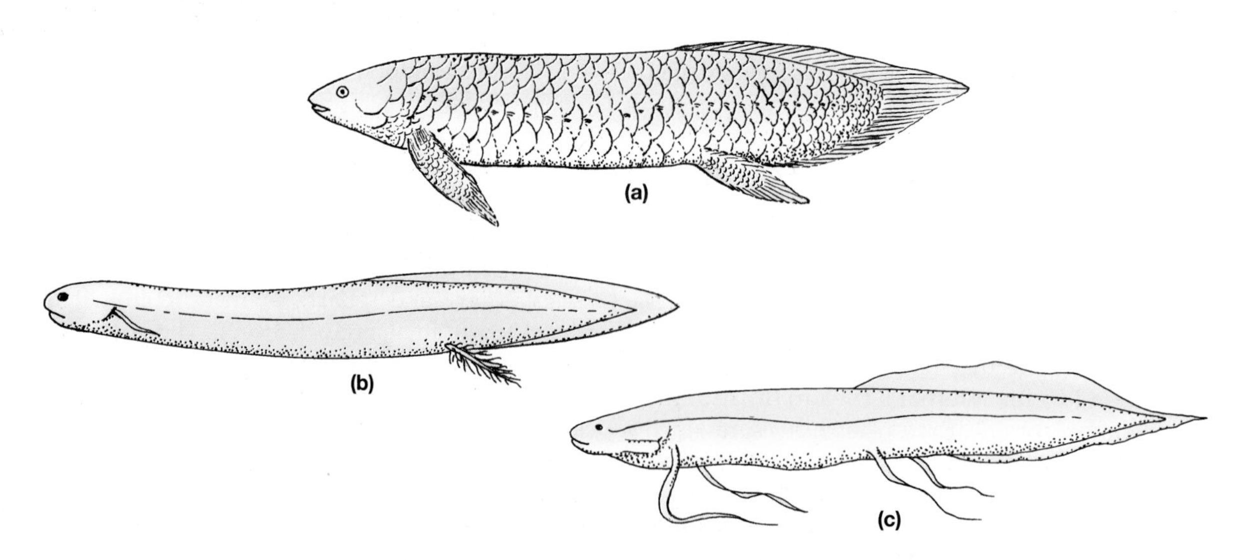

▲ Figure 6–2 Extant dipnoans. (a) Australian lungfish, *Neoceratodus forsteri.* (b) South American lungfish, *Lepidosiren paradoxa*, male. (c) African lungfish, *Protopterus.*

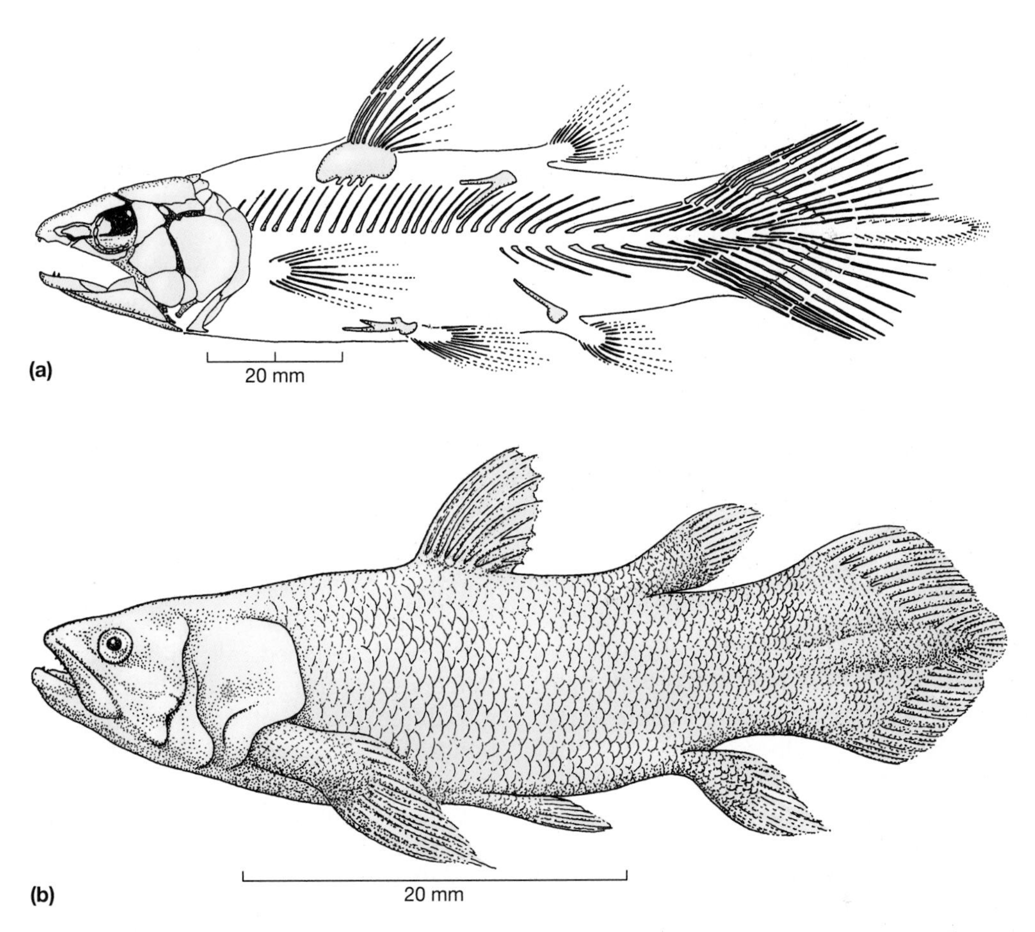

▲ Figure 6–3 Representative Actinistia (coelacanths). (a) *Rhabdoderma*, a Carboniferous actinistian. (b) *Latimeria chalumnae*, the best-known of the extant coelacanths.

lateral palatal margins. Powerful adductor muscles of the lower jaw spread upward over the chondrocranium. Throughout their evolution, this durophagous (feeding on hard foods) crushing apparatus has persisted. The earliest dipnoans were marine. During the Devonian, lungfishes evolved a body form quite distinct from the other Osteichthyes. The median fins fused around the posterior third of the body; the caudal fin, originally heterocercal, became symmetrical; and the mosaic of small dermal bones of the earliest dipnoan skulls (often covered by a continuous sheet of cosmine) evolved a pattern of fewer large elements without the cosmine cover. Most of this transformation can be explained as a result of **paedomorphosis** (the appearance of juvenile characters in an adult).

The monotypic Australian lungfish, *Neoceratodus forsteri,* is morphologically most similar to Paleozoic and Mesozoic Dipnoi. Like all other extant dipnoans, *Neoceratodus* is restricted to fresh waters; naturally occurring populations are limited to southeastern Queensland. The Australian lungfish may attain a length of 1.5 meters and a reported weight of 45 kilograms. It swims by body undulations or slowly walks across the bottom of a pond on its pectoral and pelvic appendages. Chemical senses seem important to lungfishes, and their mouths contain numerous taste buds. The nasal passages are located near the upper lip, with the incurrent openings on the rostrum just outside the mouth and the excurrent openings within the oral cavity. Thus, gill ventilation draws water across the nasal epithelium. *Neoceratodus* respires almost exclusively via its gills and uses its single lung only when stressed. Little is known of its behavior. Although lungfishes go through a complex courtship that may include male territoriality, and they are selective about the vegetation upon which they lay their adhesive eggs, no parental care has been observed after spawning. The jelly-coated eggs, 3 millimeters in diameter, hatch in 3 to 4 weeks, but the young have proved elusive and nothing is known of their juvenile life.

Surprisingly little is known about the single South American lungfish, *Lepidosiren paradoxa;* but the closely related African lungfishes, *Protopterus,* with four recognized species, are better known. These two genera are distinguished by different numbers of weakly developed gills. Because their gills are very small, these lungfishes drown if they are prevented from using their paired lungs. Nevertheless, the gills are important in eliminating carbon dioxide. Males of these thin-scaled, heavy-bodied, elongate fishes, 1 or 2 meters long, have unique filamentous and highly mobile paired appendages. These vascularized extensions develop during the breeding season and are probably used to supply oxygen from the male's blood to the young in the nest cavity. For a time after their discovery 150 years ago, lungfishes were considered to be specialized urodele amphibians (salamanders)—which they do superficially resemble, especially newly hatched individuals that have external gills like some aquatic salamanders. Although the skeletons of these lungfishes are mostly cartilaginous, their tooth plates are heavily mineralized and fossilize readily.

One habit of some species of African dipnoans, estivation, considerably increases the chance of fossilization. Similar in some ways to hibernation, estivation is induced by drying of the habitat rather than by cold. African lungfishes frequent areas that flood during the wet season and bake during the dry season—habitats not available to actinopterygians except by immigration during floods. The lungfishes enjoy the flood periods, feeding heavily and growing rapidly. When the flood waters recede, the lungfish digs a vertical burrow in the mud. The burrow ends in an enlarged chamber that varies in depth in proportion to the size of the animal—the deepest chambers are less than a meter. As drying proceeds, the lungfish becomes more lethargic and breathes air from the burrow opening. Eventually the water in the burrow dries up, and the lungfish enters the final stages of estivation—folded into a U-shape with its tail over its eyes. Heavy mucus secretions the fish has produced since entering the burrow condense and dry to form a protective envelope around its body. Only an opening at its mouth remains to permit breathing.

Although the rate of energy consumption during estivation is very low, metabolism continues, using muscle proteins as an energy source. Lungfishes normally spend less than 6 months in estivation, but they have been revived after 4 years of enforced estivation. When the rains return, the withered and shrunken lungfish becomes active and feeds voraciously. In less than a month it regains its previous size.

Estivation is an ancient trait of dipnoans. Fossil burrows containing lungfish tooth plates have been found in Carboniferous and Permian deposits of North America and Europe. Without the unwitting assistance of the lungfishes, which initiated fossilization by burying themselves, such fossils might not exist.

Actinistians

Actinistians are unknown before the Middle Devonian. Their hallmarks are fleshy, lobed fins—except for an unlobed first dorsal fin—and a unique symmetrical three-lobed tail with a central fleshy lobe that ends in a fringe of rays (Figure 6–3). Actinistians also differ from all other sarcopterygians in the head bones (they lack, among other elements, a maxilla), in details of the fin structure, and in the presence of a curious rostral organ. Following rapid evolution during the Devonian, the actinistians show a history of stability. Late Devonian actinistians differ from the more recent Cretaceous fossils, mostly in the degree of skull ossification. Some early actinistians lived in shallow fresh waters, but the fossil remains of actinistians during the Mesozoic are largely marine. While other osteichthyans radiated into a variety of niches, the actinistians retained their peculiar form. Fossil actinistians are not known after the Cretaceous, and until a little over 60 years ago they were thought to be extinct.

In 1938, an African fisherman bent over an unfamiliar catch from the Indian Ocean and nearly lost his hand to its ferocious snap. Imagine the astonishment of the scientific community when J. L. B. Smith of Rhodes University announced that the catch was an actinistian. This large fish was so similar to Mesozoic fossil coelacanths that its systematic position was unquestionable. Smith named this living fossil *Latimeria chalumnae* in honor of his former student Marjorie Courtenay Latimer, who saw the strange catch, recognized it as unusual, and brought the specimen to his attention.

Despite public appeals, no further specimens of *Latimeria* were captured until 1952. Since then more than 150 specimens, ranging in size from 75 cm to slightly over 2m and weighing from 13 to 80 kilograms, have been caught in the Comoro Archipelago or in nearby Madagascar or Mozambique. Coelacanths are hooked near the bottom, usually in 260 to 300m of water about 1.5 kilometers offshore. Strong and aggressive, *Latimeria chalumnae* is steely blue-gray with irregular white spots and reflective golden eyes. The reflection comes from a tapetum lucidum that enhances visual ability in dim light. The swim bladder is filled with fat and has ossified walls. A large cavity in the midline of the snout communicates with the exterior by three pairs of rostral tubes, enclosed by canals in the wall of the chondrocranium. These tubes are filled with gelatinous material and open to the surface through a series of six pores. The rostral organ is almost certainly an electroreceptor. *Latimeria chalumnae* is a predator—stomachs have contained fishes and cephalopods.

A fascinating glimpse of the life of the coelacanth was reported by Hans Fricke and his colleagues, who used a small submarine to observe the fishes. They saw six coelacanths at depths between 117 and 198m off a short stretch of the shoreline of one of the Comoro Islands. Coelacanths were seen only in the middle of the night and only on or near the bottom. Unlike extant lungfishes, the coelacanths did not use their paired fins as props or to walk across the bottom. However, when they swam the pectoral and pelvic appendages were moved in the same sequence as tetrapods move their limbs.

In 1927, D. M. S. Watson described two small skeletons from inside the body cavity of *Undina*, a Jurassic coelacanth, and suggested that coelacanths are viviparous. In 1975 dissection of a 1.6-meter *Latimeria* confirmed this prediction, revealing five young, each 30 cm long and at an advanced stage of embryonic development. Several other females with young inside in advanced stages of development have now been collected. Internal fertilization must occur, but how copulation is achieved is unknown, since males show no specialized copulatory organs.

In 1998 an announcement was made that astounded ichthyologists: Another coelacanth was discovered 10,000 km (6200 miles) to the east of the Comoro Islands. Two specimens were caught in shark-fishing nets set 100 to 150m deep off the northeast tip of Sulawesi, a large central Indonesian island positioned between Borneo, the Philippines, and New Guinea. Subsequently named *Latimeria menadoensis*, after the nearest major city, the new coelacanth appears from DNA data to have separated from the East African species 1.8 to 11.0 million years ago. The Comoro Islands are volcanic and much younger than the date of divergence of the two species of *Latimeria*, and it is clear that there was an ancestral population of coelacanths elsewhere. That population may still exist, awaiting discovery. If that discovery occurs, it will probably begin once again with an ichthyologist's visit to a fish market.

Despite some excellent fossils and numerous specimens of the extant coelacanths, there has never been stable agreement about the evolutionary relationships of the Actinistia with other gnathostomes. Workers have disagreed about coelacanths more than about most other vertebrate taxa, because *Latimeria* has a puzzling combination of derived morphological and physiological characters. Some of its derived characters are similar to Chondrichthyes, others to Dipnoi, and others to Actinopterygii, In addition, *Latimeria* have a curious collection of unique features. Currently most workers agree that coelacanths are the sister group of the lineage that produced the lungfishes and the tetrapods.

Evolution of the Actinopterygii

Basal actinopterygians include a variety of diverse taxa, formerly placed in a group of extinct fishes, the "paleoniscoids," which is no longer considered to be monophyletic. Although fragments of Late Silurian Actinopterygii exist, complete fossil skeletons are not found earlier than the Middle to Late Devonian. Early actinopterygians were small fishes (usually 5 to 25 cm long, although some were over a meter) with a single dorsal fin and a strongly heterocercal, forked caudal fin with little fin web. Paired fins with long bases were common, but several taxa had lobate pectoral fins (Figure 6–1e). The interlocking scales, although thick like those of sarcopterygians, were otherwise distinct in structure and in growth pattern. In sarcopterygians the hard outer coating of the scales is cosmine (which is derived from dentine), and in actinopterygians it is ganoine (derived from enamel). Parallel arrays of closely packed radial bones supported the bases of the fins. The number of bony rays supporting the fin membrane was greater than the number of supporting radials, and these rays were clearly derived from elongated scales aligned end to end. Two morphological aspects of the early ray-finned fishes deserve special attention: specializations for locomotion and for feeding. Unfortunately, the otherwise primitive extant cladistians (polypterids) and chondrosteans are so specialized in these respects that they shed little light on the biology of early actinopterygians.

These primitive actinopterygians were diverse and successful in freshwater and marine habitats from 380 to 280 million years ago. There is no evidence that the primitive actinopterygians (often grouped together as "paleoniscoids") were a monophyletic group, with characters distinguishing them from later, more derived fishes. The different types share numerous characters, but these are all primitive relative to later members of the Actinopterygii and thus of no use in defining an evolutionary lineage. Near the end of the Paleozoic, the ray-finned fishes showed signs of change. The upper and lower lobes of the caudal fin were often nearly symmetrical, and all fin membranes were supported by fewer bony rays. For example, the dorsal and anal fins had about one bony ray for each internal supporting radial. This morphological reorganization probably increased the flexibility of the fins.

The dermal armor of late Paleozoic ray-finned fishes was also reduced compared to that of their ancestors. The changes in fins and armor may have been complementary—more mobile fins mean more versatile locomotion, and increased ability to avoid predators may have permitted a reduction in heavy armor. This reduction of weight could have further stimulated the evolution of increased locomotor ability, which was probably enhanced by perfection of the swim bladder as a delicately controlled hydrostatic device.

The lower jaw of early actinopterygians was supported by the hyomandibula, and in most forms the jaw was snapped closed in a scissors action by the adductor mandibulae muscles. The adductor mandibulae muscle originated in a narrow enclosed cavity between the maxilla and the palatoquadrate and inserted toward the rear of the lower jaw, near the jaw's articulation with the quadrate. This arrangement produced a lever system that moved the jaw rapidly (a quick snap), but the force created was low. The close-knit dermal bones of the cheeks permitted little expansion of the **orobranchial chamber** (the mouth and gills) beyond that required for respiration, and there was no space for a larger adductor mandibulae muscle.

The food-gathering apparatus of several lineages of bony fish underwent radical changes in the Permian. In the Late Permian, one lineage produced a new clade of actinopterygians—the neopterygians—distinguished by a new jaw mechanism. Neopterygians have jaws with a short maxilla, with its posterior end freed from the other bones of the cheek. Because the cheek was no longer solid, the nearly vertically oriented hyomandibula could swing out laterally when the mouth opened, rapidly increasing the volume of the orobranchial chamber and producing a powerful suction that drew prey into the mouth. The crushing power of the sharply toothed jaw could be increased because the adductor muscle, no longer limited in size by a solid bony cheek, expanded dorsally through the space opened by the freeing of the maxilla. In addition, an extra lever arm developed at the site of insertion of the adductor muscle on the lower jaw. This lever, called a **coronoid process,** appears on different jaw bones in different lineages of bony fishes. In most modern fishes the coronoid process is on either the dentary or the angular bone. A coronoid process adds torque, and thus power, to the jaws of many forceful biters.

The neopterygians first appeared in the Late Permian (Figure 6–4) and became the dominant fishes of the Mesozoic. During the Late Triassic, basal neopterygians gave rise to fishes with further feeding and locomotor specializations. In the more derived neopterygians, the bones of the gill cover (operculum) were connected to the mandible so that expansion of the orobranchial chamber aided in opening the mouth. The anterior, articulated end of the maxilla developed a ball-and-socket joint with the neurocranium. Because of its ligamentous connection to the mandible, the free posterior end of the maxilla was rotated forward as the mouth opened (Figure 6–4b). This pointed the maxilla's marginal teeth forward and helped to grasp prey. The folds of skin covering the maxilla changed the shape of the gape from a semicircle to a circular opening. These changes increased the suction produced during opening of the mouth. The result was greater directionality of suction and elimination of a possible side-door escape route for small prey.

Specializations of the Teleosts

Fishes with these adaptations constitute the largest vertebrate radiation, the **Teleostei**. Although teleosts probably evolved in the sea, they soon radiated into freshwater. By the Late Cretaceous, teleosts seem to

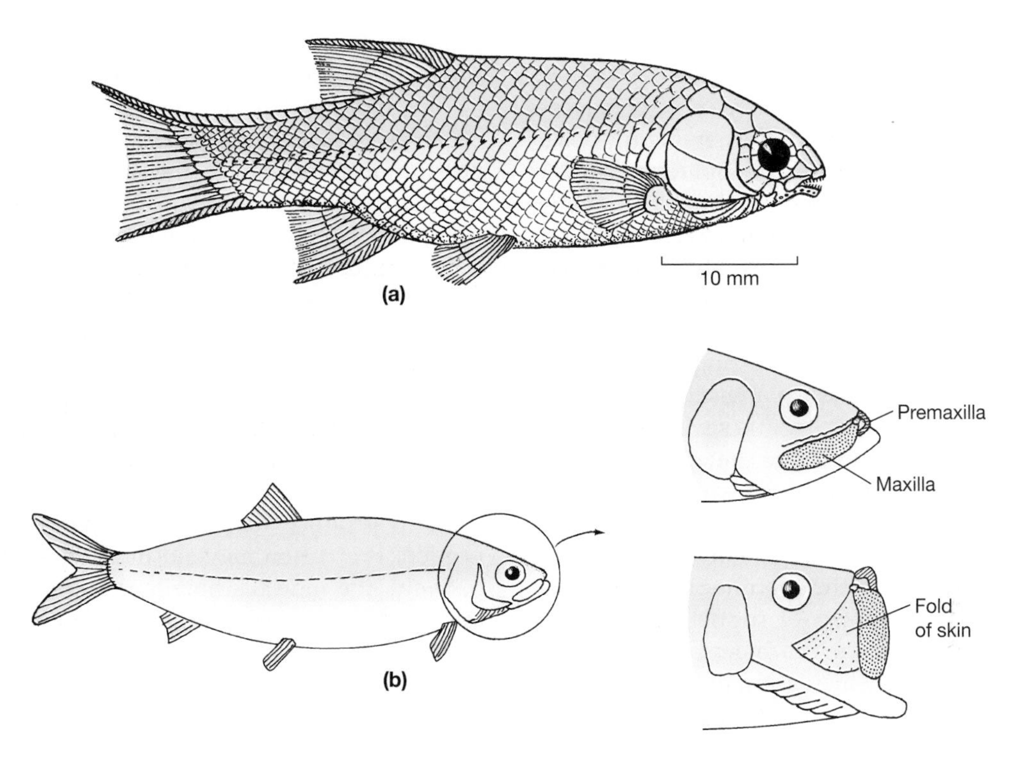

▲ Figure 6–4 Teleosts. (a) *Acentrophorus* of the Permian illustrates an early member of the late Paleozoic neopterygian radiation. (b) *Leptolepis,* an Early Jurassic teleost with enlarged mobile maxillae that form a nearly circular mouth when the jaws are fully opened. Membranes of skin close the gaps behind the protruded bony elements. Modern herrings have a similar jaw structure.

have replaced most of the more primitive neopterygians, and most of the more than 400 families of modern teleosts had evolved. The first specializations of teleosts involved changes in the fins.

Specializations of the Fins The caudal fin of adult actinopterygians is supported by a few enlarged and modified hemal spines, called hypural bones, that articulate with the tip of the abruptly upturned vertebral column. In general, the number of hypural bones decreases during the transition from the earliest actinopterygians to the more derived teleosts. Modified posterior neural arches—the uroneurals—add further support to the dorsal side of the tail. These uroneurals are a derived character of teleosts. Thus supported, the caudal fin of teleosts is symmetrical and flexible. This type of caudal structure is known as homocercal. Along with a swim bladder that adjusts buoyancy, a homocercal tail allows a teleost to swim horizontally without using its paired fins for control, as sharks must. Recent studies suggest that the action of the symmetrical tail may be more complex than had been realized previously (Lauder 1994). In burst and sprint swimming the tail produces a symmetrical force; but during steady-speed swimming, intrinsic muscles in the tail may produce an asymmetric action that increases maneuverability without requiring use of the lateral fins. Relieved of responsibility for controlling lift, the paired fins of teleosts could be more flexible, mobile, and diverse in shape, size, and position. Because of the changes in the tail, the paired fins of teleosts have become specialized for activities from food gathering to courtship and from sound production to walking and flying.

As we saw among earlier groups, improvements in locomotion were accompanied by reduction of armor. Modern teleosts are thin-scaled by Paleozoic and Mesozoic standards, and many lack scales entirely. The few heavily armored exceptions generally show a secondary reduction in locomotor abilities.

Specializations of the Jaws Teleosts, like most fishes that preceded them, also evolved improvements in their feeding apparatus. Trophic (feeding) specializations in the earliest teleosts involved only a slight loosening of the premaxillae, so that they moved during jaw opening to accentuate the round mouth shape. One early clade of teleosts showed an enlargement of the free-swinging posterior end of the maxilla to form a nearly circular mouth when the jaws were fully opened. Later in the radiation of teleosts, distinctive changes in the jaw apparatus permitted a wide variety of feeding modes based on the speed of opening the jaws and the powerful suction produced by the highly integrated jaw, gill arch, and cranium.

Most of the main themes in actinopterygian evolution involve changes in the jaws, from simple prey-grabbers to highly sophisticated suction devices. Suction is important in prey capture in water, because a rapid approach by a predator pushes a wave of water in toward the prey. This wave flows around and away from the mouth and could push prey away from the grasp of the predator's jaws. Neopterygians (and some marine mammals) solve this problem by rapidly increasing the volume of the orobranchial chamber, creating a flow of water that carries prey into the mouth.

In addition to rapid and forceful suction, many teleosts have evolved a great deal of mobility in the skeletal elements that rim the mouth opening. This mobility allows the grasping margins of the jaws to be extended forward from the head, often at remarkable speed. The functional result, called the **protrusible jaw,** has evolved three or four times in different derived teleost clades.

Although jaw protrusion is generally associated with perches and their relatives (perciform fishes, Figure 6–5), it occurs in the related atherinid and paracanthopterygian fishes and in the cypriniform ostariophysans as well. Jaw morphology differs significantly among fishes with protrusible jaws, which clearly shows that these groups have evolved protrusion independently.

All jaw protrusion mechanisms involve complex ligamentous attachments that allow the ascending processes of the premaxilla to slide forward on top of the cranium without dislocation. In addition, since no muscles are in position to pull the premaxillae forward, they must be pushed by leverage from behind. Two sources provide the necessary leverage. First, opening the lower jaw may protrude the premaxillae through ligamentous ties between the mandible and the posterior tip of the premaxillae (Figure 6–5c). Second, leverage can be provided by complex movements of the maxillae, which become isolated from the rim of the mouth by long, posterior projections of the premaxillae that often bear teeth.

The independent movement of the protrusible upper jaw also permits closure of the mouth through maximum extension of the premaxillae while the orobranchial cavity is still expanded. Thus, engulfed prey are trapped before the orobranchial cavity volume is reduced as water flows out the gill openings.

With so many groups converging on the same complex function, the adaptive significance of protrusion must be great. Surprisingly, no single hypothesis of advantage has much experimental support. Protrusion may have enhanced the hydrodynamic efficiency of the circular mouth opening of primitive teleosts, but this hypothesis seems insufficient to produce the complex anatomical changes needed. Protrusion may aid in gripping prey. Alternatively, the mouth's mobile jaws may be fitted to the substrate during feeding while the body remains in the horizontal position required for rapid escape from the fish's own predators.

Further advantages of protrusible jaws may lie in the functional independence of the upper jaws relative to other parts of the feeding apparatus. Some fishes, such as the silversides and killifishes, can greatly protrude, moderately protrude, or not protrude the upper jaw while opening the mouth and creating suction. These modulations direct the mouth opening and direction of suction ventrally, straight ahead, or dorsally, allowing the fish to feed from substrate, water column, or surface with equal ease. Perhaps the most broadly applicable hypothesis for jaw protrusion is that shooting out the jaws in front of the head increases the predator's approach velocity by 39 to 89 percent in the crucial last instant of its approach.

Pharyngeal Teeth Powerful mobile pharyngeal jaws evolved several times among actinopterygians. Ancestrally, ray-finned fishes had numerous dermal tooth plates in the pharynx. These plates were aligned with (but not fused to) both dorsal and ventral skeletal elements of the gill arches. A general trend of fusion of these tooth plates to one another and to a few gill arch elements above and below the esophagus can be traced in the Neopterygii. These consolidated pharyngeal jaws were not very mobile ancestrally; they were used primarily to hold and manipulate prey in preparation for swallowing it whole. In the ostariophysan minnows and their relatives the suckers, the primary jaws are toothless but protrusible. The pharyngeal jaws are greatly en-

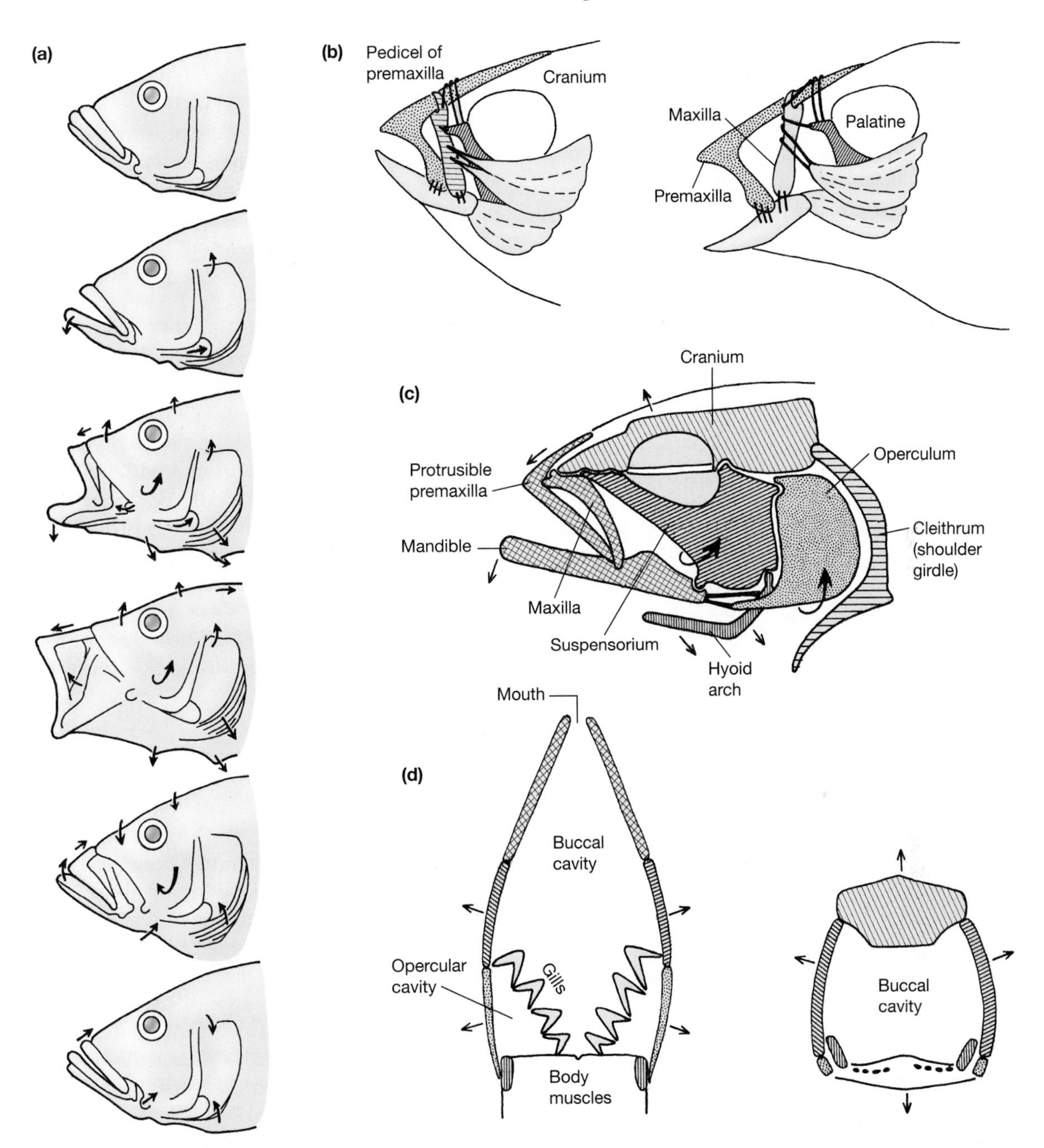

▲ Figure 6–5 Jaw protrusion in suction feeding. (a) Top to bottom: Sequence of jaw movements in an African cichlid fish, *Serranochromi*. (b) Muscles, ligaments, and bones involved in movements during premaxillary protrusion. (c) Skeletal movements and ligament actions during jaw protrusion. (d) Frontal section (left) and cross section (right) of buccal expansion during suction feeding.

larged and close against a horny pad on the base of the skull. These feeding and digestive specializations allow extraction of nutrients from thick-walled plant cells; the minnows and suckers represent one of the largest radiations of herbivores among vertebrates.

In the Neoteleostei the muscles associated with the branchial skeletal elements supporting the pharyngeal jaws have undergone radical evolution, resulting in a variety of powerful movements of the pharyngeal jaw tooth plates. Not only are the movements of these second jaws completely unrelated to the movements and functions of the primary jaws, but in a variety of derived teleosts the upper and lower tooth plates of the pharyngeal jaws move quite independently of each other. With so many separate systems to work with, it is little wonder that some of the most extensive adaptive radiations among teleosts have been in fishes endowed with protrusible primary jaws and specialized mobile pharyngeal jaws.

6.3 Extant Actinopterygii—Ray-Finned Fishes

With an estimated 24,000 extant species so far described—and new species being discovered regularly—the extant actinopterygians present a fascinating, even bewildering, diversity of forms of vertebrate life. Because of their numbers we are forced to survey them briefly, focusing on the primary characteristics of selected groups and their evolution.

The study of the phylogenetic relationships of actinopterygians entered its current active state in 1966 with a major revised scheme of teleostean phylogeny proposing several new relationships. In the following decades our understanding of the interrelationships of ray-finned fishes has grown phenomenally. Nevertheless, some of the relationships are uncertain and should be considered hypotheses (Lauder and Liem 1983, Nelson 1994, Stiassny et al. 1996).

Why are there so many actinopterygians? For one thing, their habitats cover 73 percent of the planet. But why then are the chondrichthyians and the sarcopterygian fishes relatively species poor? Recent studies by A. R. McCune and her colleagues (1996–1998) indicate that both fossil and living actinopterygians may have rates of speciation higher than those estimated for birds or even some insects. Fish species flocks evolving in the isolation of lakes appear to take the least time for speciation of all the highly derived organisms surveyed (vertebrates and insects).

Polypteriformes and Acipenseriformes

Although primitive actinopterygians were replaced during the early Mesozoic by neopterygians, a few specialized forms of primitive ray-finned fishes have survived. The most primitive surviving lineage of actinopterygian fishes are the **Polypteriformes,** the bichirs and African reed fish (Figure 6–6). Sometimes called the Cladistia, the extant Polypteriformes are 11 species of elongate, heavily armored fishes. They are modestly sized (less than a meter), slow-moving fishes with modified heterocercal tails. Polypteriforms differ from other extant basal actinopterygians in having well-ossified skeletons. In addition to a full complement of dermal and endochondral bones, polypteriforms are covered by thick, interlocking, multilayered scales. These ganoid scales are covered with a coat of ganoine, an enamel-like tissue characteristic of primitive actinopterygians. Larval bichirs (*Polypterus*) have external gills, possibly an ancestral condition for Osteichthyes. *Erpetoichthys,* the reedfish, is eel-like although armored with a full complement of ganoid scales. All polypteriforms are predatory, and their jaw mechanics provide our best model of the original actinopterygian condition. Their peculiar flag-like dorsal finlets and the fleshy bases of their pectoral fins are among their unique specializations, but so little is known about their natural history that the significance of these features cannot yet be appreciated.

Acipenseriformes (sturgeons and paddlefishes, also known as chondrosteans) includes two living and two fossil families of specialized actinopterygians. The 24 species of sturgeons, family Acipenseridae, are large (1 to 6m), active, benthic fishes. They lack endochondral bone and have lost much of the dermal skeleton of more primitive actinopterygians. Sturgeons have a strongly heterocercal tail armored with a specialized series of scales extending from the dorsal margin of the caudal peduncle along the upper edge of the caudal fin (Figure 6–6b). These caudal fulcral scales are an armored caudal cutwater, an ancestral character of the earliest Paleozoic actinopterygians. Most sturgeons have five rows of enlarged armor-like scales along the body. Sturgeons have a swim bladder derived from the lung, but it has a separate evolutionary origin from the swim bladder of teleosts. The protrusible jaws of sturgeons make them effective suction feeders. The mode of jaw protrusion is unique and derived independently from that of teleosts. Protrusible jaws plus adaptations for benthic life are the basis for considering sturgeons specialized compared to early fossil actinopterygians.

Sturgeons are found only in the Northern Hemisphere; they are either **anadromous** (ascending into freshwater to breed) or entirely freshwater in habit.

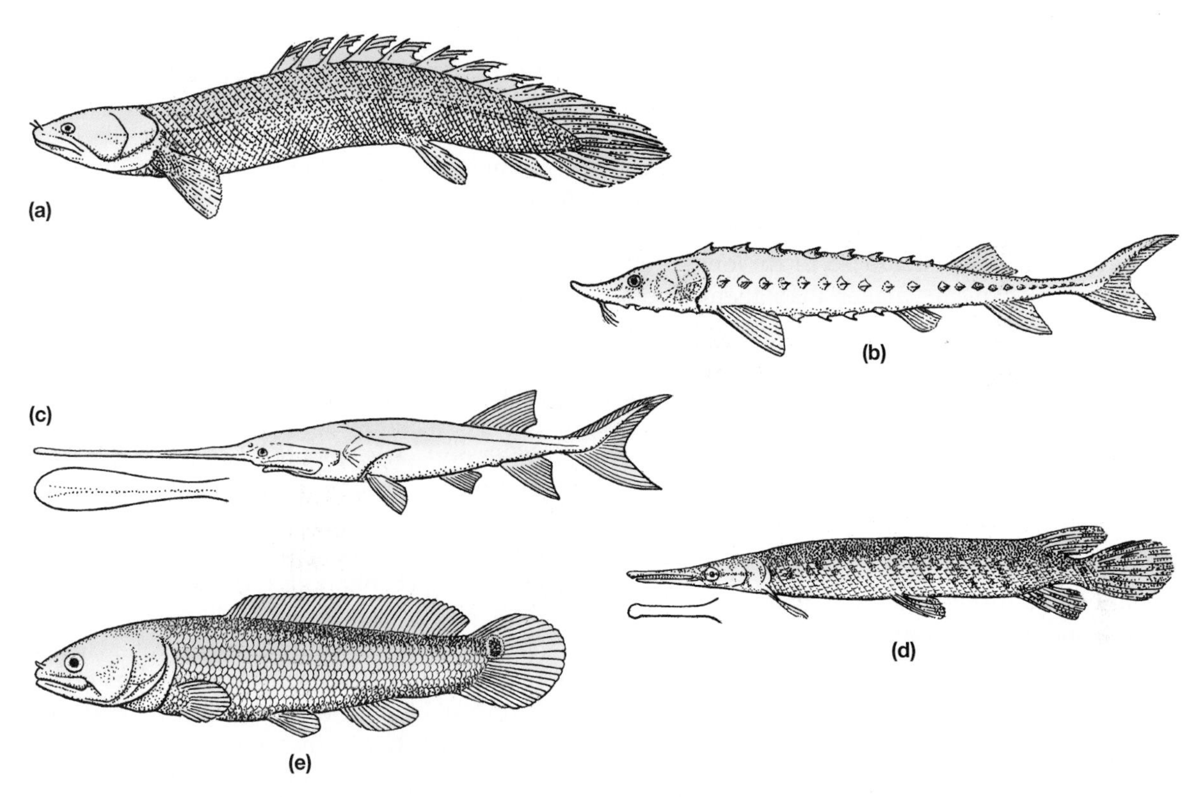

▲ Figure 6–6 Extant nonteleostean actinopterygian fishes and primitive neopterygians (not to scale). Actinopterygians: (a) *Polypterus,* a bichir; (b) *Acipenser,* a sturgeon; and (c) *Polyodon spathula,* one of two extant species of paddlefishes. Primitive neopterygians: (d) *Lepisosteus*, a gar; and (e) *Amia calva,* the bowfin.

Commercially important for their rich flesh and as a source of the best caviar, they have been severely depleted by intensive fisheries in much of their range. Dams and river pollution have also taken their toll on anadromous sturgeon. With intensive watershed-wide conservation measures, they appear to be making a successful comeback in some areas such as New York's Hudson River.

The two surviving species of paddlefishes, Polyodontidae, are closely related to the sturgeons but have a still greater reduction of dermal ossification. Their most outstanding feature is a greatly elongate and flattened rostrum, which extends nearly one-third of their 2-meter length (Figure 6–6c). The rostrum is richly innervated with ampullary organs that are believed to detect minute electric fields. Contrary to the common notion that the paddle is used to stir food from muddy river bottoms, the American paddlefish is a planktivore that feeds by swimming with its prodigious mouth agape, straining crustaceans and small fishes from the water using modified gill rakers as strainers. The two species of paddlefishes have a disjunct zoogeographic distribution similar to that of alligators: One is found in the Chang River valley of China, where it feeds on fishes; the planktivorous species is in the Mississippi River valley of the United States. Fossil paddlefishes are known from western North America.

Primitive Neopterygians

The two extant genera of primitive neopterygians, formerly grouped together as "holosteans," are currently limited to North America and represent widely divergent types. The **Lepisosteiformes** is composed of seven species of gars (*Lepisosteus)*. They are medium- to large-size (1 to 4m) predators of warm temperate fresh and brackish (estuarine) waters. The elongate body, jaws, and teeth are specialized features; but their interlocking multilayered scales are similar to those of many Paleozoic and Mesozoic actinopterygians (Figure 6–6d). Gars feed on other fishes taken unaware when the seemingly lethargic and excellently camouflaged gar dashes alongside them and, with a sideways flip of the body, grasps prey with its needle-like teeth. Alligators are the only natural predators able to cope with the thick armor of an adult gar.

Sympatric with gars is the single species of **Amiiformes**, the bowfin, *Amia calva* (Figure 6–6e). The head skeleton shows modifications of the jaws as a suction device. *Amia*, which are 0.5 to 1m long, prey on almost any organism smaller than themselves. Scales of the bowfin are comparatively thin and made up of a single layer of bone as in teleost fishes; however, the asymmetric caudal fin is very similar to the heterocercal caudal fin of more primitive fishes. The interrelationships of gars, *Amia*, and the teleosts have long been controversial.

Teleosteans

Most extant fishes are teleosts. They share many characters of caudal and cranial structure and are grouped into four clades of varying size and diversity.

The **Osteoglossomorpha**, which appeared in Late Jurassic seas, are now restricted to about 220 species in tropical fresh waters. *Osteoglossum* (Figure 6–7a) is a 1m-long predator from the Amazon, familiar to tropical fish enthusiasts as the arawana. *Arapaima* is an even larger Amazonian predator, perhaps the largest strictly freshwater fish. Before intense fishing reduced the populations they were known to reach a length of at least 3m and perhaps as much as 4.5m. *Mormyrus* (Figure 6–7b), one of the African elephant-nosed fishes, is representative of the smaller African bottom feeders that use weak electric discharges to communicate with other members of their species. As dissimilar as they may seem, the osteoglossomorph fishes are united by unique bony characters of the mouth and by the mechanics of their jaws.

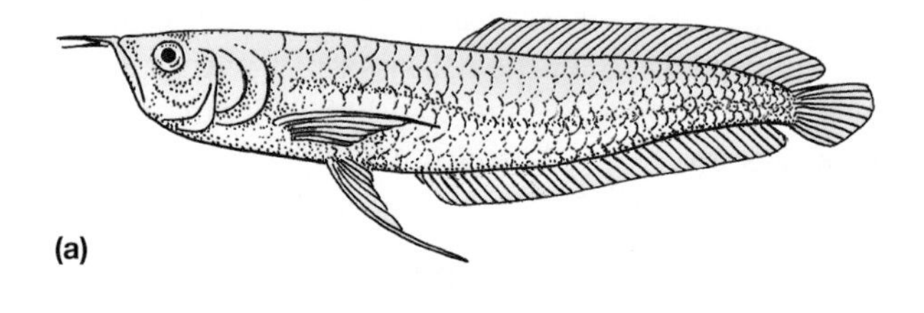

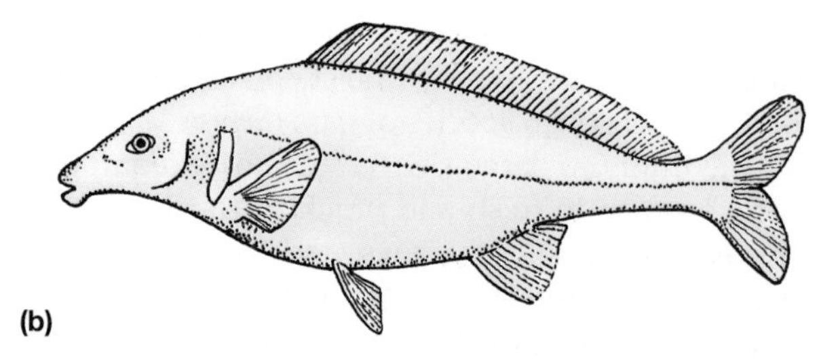

▲ Figure 6–7 Extant osteoglossomorphs (not to scale). (a) *Osteoglossum*, the arawana, from South America. (b) *Mormyrus*, an elephant-nose from Africa.

The **Elopomorpha** (Figure 6–8a and b) had appeared by the Late Jurassic. A specialized leptocephalous (Greek *lepto* = small and *cephal* = head) larva is a unique character of elopomorphs. These larvae spend a long time adrift, usually at the ocean surface, and are widely dispersed by currents. Elopomorphs include about 35 species of tarpons (Megalopidae), ladyfish (Elopidae), and bonefish (Albulidae) and 764 species of true eels (Anguilliformes and Saccopharyngiformes).

Most elopomorphs are eel-like and marine, but some species are tolerant of fresh waters. The common American eel, *Anguilla rostrata*, has one of the most unusual life histories of any fish. After growing to sexual maturity (which takes as long as 10 to 12 years) in rivers, lakes, and even ponds, the **catadromous** (downstream migrating) eels enter the sea. The North Atlantic eels migrate to the Sargasso Sea. Here they are thought to spawn and die, presumably at great depth. The eggs and newly hatched leptocephalous larvae float to the surface and drift in the currents. Larval life continues until the larvae reach continental margins, where they transform into miniature eels and ascend rivers to feed and mature.

Most of the more than 360 species of **Clupeomorpha** are specialized for feeding on minute plankton gathered by a specialized mouth and gill-straining apparatus. They are silvery, mostly marine schooling fishes of great commercial importance. Common examples are herrings, shad, sardines, and anchovies (Figure 6–8c). Several clupeomorphs are anadromous, and the springtime migrations of American shad (*Alosa sapidissima*) from the North Atlantic into rivers in eastern North America once involved millions of individuals. The great shad runs of the recent past have been greatly depleted by dams and pollution of aquatic environments.

The vast majority of extant teleosts belong to the fourth clade, the **Euteleostei**, which evolved before the Late Cretaceous. With so many thousands of species, it is impossible to give more than scant information about them here. For more about these fishes, refer to Bond (1996), Helfman et al.

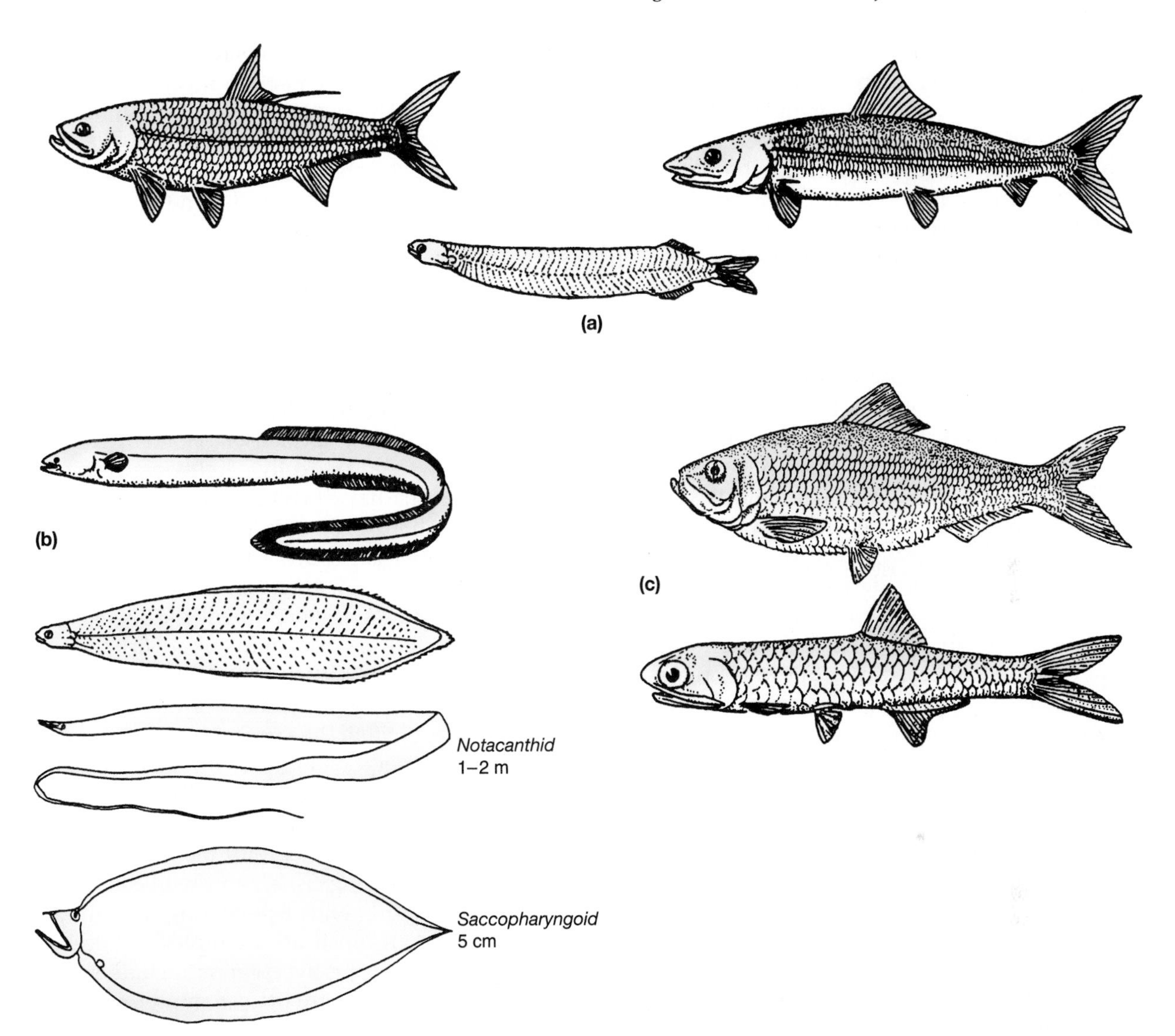

▲ Figure 6–8 Extant teleosts of isolated phylogenetic position (not to scale). (a) Elopomorpha, represented by a tarpon (left), a bonefish (right), and a typical fork-tailed leptocephalous larva (below). (b) Anguilliform elopormorphs, represented by the common eel, *Anguilla rostrata* (above), its leptocephalous larva (immediately below), and two other very different eel leptocephali. (c) Clupeomorpha, represented by a herring (above) and an anchovy (below).

(1997), Moyle and Cech (2000), Nelson (1994), and the beautifully illustrated Paxton and Eschmeyer (1998). Here we will consider the basal euteleostian stock as represented today by the specialized ostariophysans and the generalized salmoniforms, but how these two monophyletic groups relate to each other is a matter of much dispute (Stiassny et al. 1996).

The **Ostariophysi**, the predominant fishes of the world's fresh waters, represent perhaps 25 to 30 percent of all living fishes and about 80 percent of the fish species in freshwater. As a group, ostariophysans display diverse traits. For example, many ostariophysans have protrusible jaws and are adept at obtaining food in a variety of ways. In addition, pharyngeal teeth act as second jaws. Many forms have fin spines or special armor for protection, and the skin typically contains glands that produce substances used in olfactory communication. Although they have diverse reproductive habits, most lay sticky eggs or otherwise guard the eggs, preventing their loss downstream.

Ostariophysans have two distinctive derived characters. Their name refers to small bones (Greek *ost* = a bone) that connect the swim bladder (Greek *physa* = a bladder) with the inner ear (Figure 6–9). Using the swim bladder as an amplifier and the chain of bones as conductors, this **Weberian apparatus** greatly enhances hearing sensitivity of these fishes. Sound (pressure) waves impinging on the fish cause the swim bladder to vibrate. The tripus is in contact with the swim bladder; as the bladder vibrates, the tripus pivots on its articulation with the vertebra. This motion is transmitted by ligaments to the intercalarium and scaphum. Movement of the scaphum compresses an extension of the membranous labyrinth (inner ear) against the claustrum, stimulating the auditory region of the brain. The ostariophysans are more sensitive to sounds and have a broader frequency range of detection than other fishes. The second derived character uniting the ostariophysans is the presence of a fright or alarm substance in the skin. Chemical signals (pheromones) are released into the water when the skin is damaged, and they produce a fright reaction in nearby members of their own species and other ostariophysan species. The fright reaction may cause fish to rush for cover or form a tighter school.

Although all Ostariophysi have an alarm substance in the skin and a Weberian apparatus (or a rudimentary precursor of it), in other respects they are a diverse taxon of 6500 species. Ostariophysans include the characins (piranhas, neon tetras, and other familiar aquarium fishes) of South America and Africa, the carps and minnows (all continents except South America, Antarctica, and Australia), the catfishes (all continents except Antarctica and in many shallow marine areas as well), and the highly derived electric knifefishes of South America.

The other basal group of euteleostians, the esocid and salmonid fishes (Figure 6–10a), include important commercial and game fishes. These fishes have often been lumped into a taxon, the "**Protacanthopterygii**," but the basis for this classification was often no more than shared ancestral euteleostean characters that are not valid for determining phylogenetic relationships. A new usage of the term includes only salmonids and some deep-water relatives and may be monophyletic. The salmonids include the anadromous salmon, which usually spend their adult lives at sea, as well as the closely related trout which live in freshwater. Among the most primitive extant euteleosteans, at the base of the radiation of all other more derived teleosts, may be the esocids. These temperate freshwater fishes of the Northern Hemisphere include game species such as pickerel, pikes, and muskellunges and their relatives. The Southern Hemisphere galaxiids are also primitive euteleosts that live in habitats similar to those occupied by salmonids.

Another, more diverse group of euteleosts is not recognized as monophyletic. The eight hundred or more species in this category include the marine mesopelagic lanternfishes and marine hatchetfishes, the lizard fishes, and related fishes—mostly from mesopelagic or bathypelagic waters and thus not well known except to specialists (see Figure 6–20).

Mobile jaws and protective, lightweight spines in the median fins have evolved in several groups of euteleosts. About 1200 species of fishes, including cods and anglerfishes, are grouped as the **Paracanthopterygii**, although their similarities may represent convergence (Figure 6–10b) However, the **Acanthopterygii**, or true spiny-rayed fishes, whose 13,500 species dominate the open ocean surface and shallow marine waters of the world, do appear to form a monophyletic lineage within the euteleosts. Among the acanthopterygians, the atherinomorphs have protrusible jaws and specializations of form and behavior that suit them to shallow marine and freshwater habitats. This group includes the silversides, grunions, flying fishes, and halfbeaks, as well as the egg-laying and live-bearing cyprinodonts (Figure 6–10c). Killifish are examples of egg-laying cyprinodonts, and the live bearers include the guppies, mollies, and swordtails commonly maintained in home aquaria.

Most species of acanthopterygians—and the largest order of fishes—are Perciformes, with over 9300 extant species. A few of the well-known members are snooks, sea basses, sunfishes, perches, darters, dolphins (mahi mahi), snappers, grunts, porgies, drums, cichlids, barracudas, tunas, billfish, and most of the fishes found on coral reefs.

Locomotion in Water

Perhaps the single most recognizable characteristic of the enormous diversity of fishes is their mode of locomotion. Fish swimming is immediately recognizable, aesthetically pleasing, and—when first considered—rather mysterious, at least when compared to the locomotion of most land animals. Fish swimming results from anterior to posterior sequential

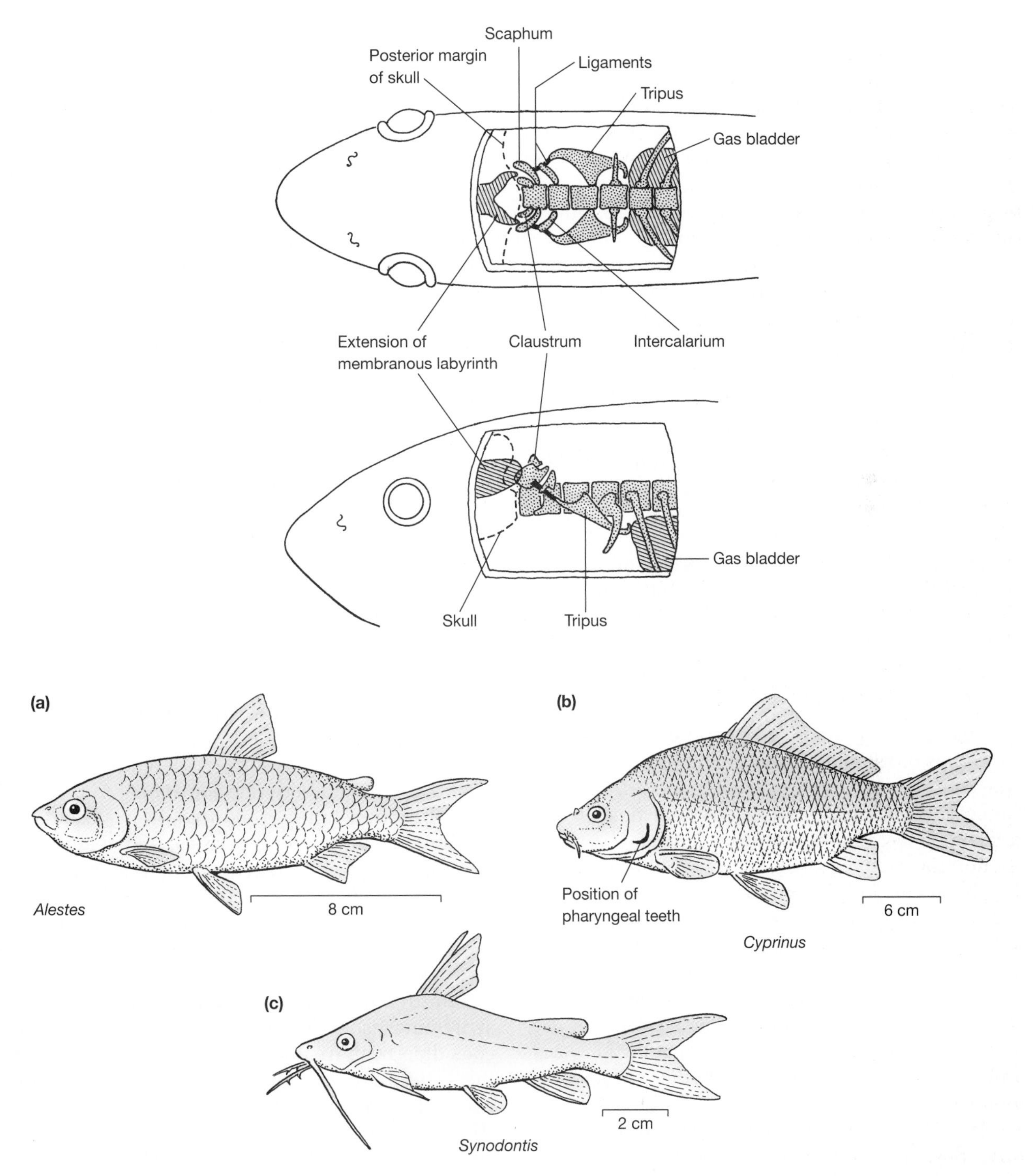

▲ Figure 6–9 Ostariophysan fishes have a sound-detection system, the Weberian apparatus, which is a modification of the swim bladder and the first few vertebrae and their processes. Typical ostariophysans include (a) characins, (b) minnows, and (c) catfishes.

contractions of the muscle segments along one side of the body and simultaneous relaxation of those of the opposite side. Thus a portion of the body momentarily bends, the bend is propagated posteriorly, and a fish oscillates from side to side as it swims. These lateral undulations are most visible in elongate fishes, such as lampreys and eels (Figure 6–11). Most of the power for swimming comes from muscles in the posterior region of the fish.

In 1926, Charles Breder classified the undulatory motions of fishes into three types:

- Anguilliform—Typical of highly flexible fishes capable of bending into more than half a sinusoidal wavelength. Named for the locomotion seen in the true eels, the anguilliformes.
- Carangiform—Undulations limited mostly to the caudal region, the body bending into less than half a wavelength. Named for *Caranx,* the genus of most jacks or trevallies that swim in this mode.
- Ostraciiform—The body is inflexible, undulation is limited to the caudal fin. Named for the boxfishes, trunkfishes, and cowfishes (family Ostraciidae) whose fused scales form a rigid box around the body, preventing undulations (Figures 6–11 and 6–13).

Although these forms of locomotion were named for groups of fishes that exemplify them, many other types of fishes use these swimming modes. Hagfish, lampreys, most sharks, sturgeons, arawanas, many catfishes, and countless elongate spiny-rayed fishes are not eels; yet they swim in an anguilliform mode. The basic categories have been extensively subdivided and redefined since 1926. For example, subcarangiform and modified carangiform variants have been added for those swimming styles slightly more and somewhat less eel-like, respectively, than carangiform swimming. But Breder's categories are still useful for understanding locomotion in water and are central to most modern studies of fish locomotion.

Many specializations of body form, surface structure, fins, and muscle arrangement increase the efficiency of the different modes of swimming. A swimming fish must overcome the effect of gravity by producing lift, and the drag of water by producing thrust (Figure 6–12). To overcome gravity, fishes generate vertical lift. Teleosts most often do this by generating bouyancy with a gas-filled swim bladder.

Overcoming Drag—The Generation of Thrust In general, fishes swim forward by pushing backward on the water. For every *active* force there is an opposite *reactive* force (Newton's third law of motion). Undulations produce an active force directed backward, and also a lateral force. The overall reactive force is directed forward and at an angle to the side.

Anguilliform and carangiform swimmers increase speed by increasing the frequency of their body undulations. Increasing the frequency of body undulations applies more power (force per unit time) to the water. Different fishes achieve very different maximum speeds—some (eels) are slow and others (tunas) are very fast.

An eel's long body limits speed, because it induces drag from the friction of water on the elongate surface of the fish. Fishes that swim rapidly are shorter and less flexible. Force from the contraction of anterior muscle segments is transferred through ligaments to the caudal peduncle and the tail. Morphological specializations of this swimming mode reach their zenith in fishes like tunas, where the caudal peduncle is slender and the tail greatly expanded vertically.

Other fishes seldom flex the body to swim, but undulate the median fins (referred to as amiiform, gymnotiform, or balistiform swimming). Usually, several complete waves are observed along the fin (Figure 6–13), and very fine adjustment in the direction of motion can be produced. Many fishes—for example, chimeras, surf perches, and many coral-reef fishes, such as surgeonfishes, wrasses, and parrot fishes—generally do not oscillate the body or median fins to generate most of their thrust, but row with the pectoral fins to produce movement (labriform swimming).

Improving Thrust: Minimizing Drag A swimming fish experiences drag of two forms: viscous drag from friction between the fish's body and the water, and inertial drag from pressure differences created by the fish's displacement of water. Viscous drag is relatively constant over a range of speeds, but inertial drag is low at slow speeds and increases rapidly with increasing speed. Viscous drag is affected by surface smoothness, whereas inertial drag is influenced by body shape. A thin body has high viscous drag, because it has a large surface area relative to its muscle mass. A thick body induces high inertial drag because it displaces a large volume of water as it moves forward. Streamlined

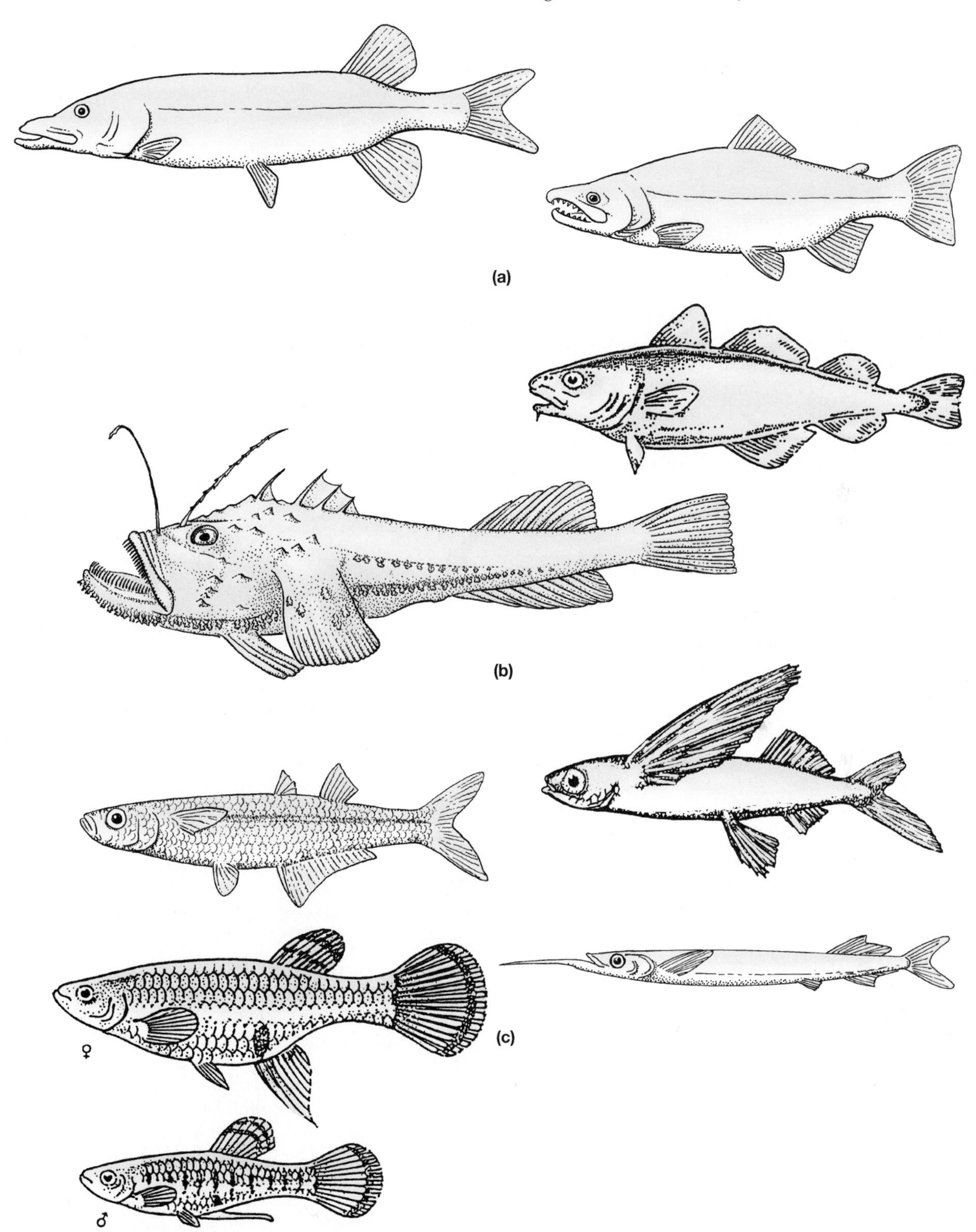

▲ Figure 6–10 Euteleosts (not to scale). (a) Primitive euteleosts represented by the pike (left) and the salmon (right). (b) Paracanthopterygians represented by the cod (right) and the goosefish angler (left). (c) Atherinomorph fishes represented by (clockwise from upper left) an Atlantic silverside (*Menidia*), a flying fish, a halfbeak, and live-bearing killifish, the male of which has a modified anal fin used in internal fertilization.

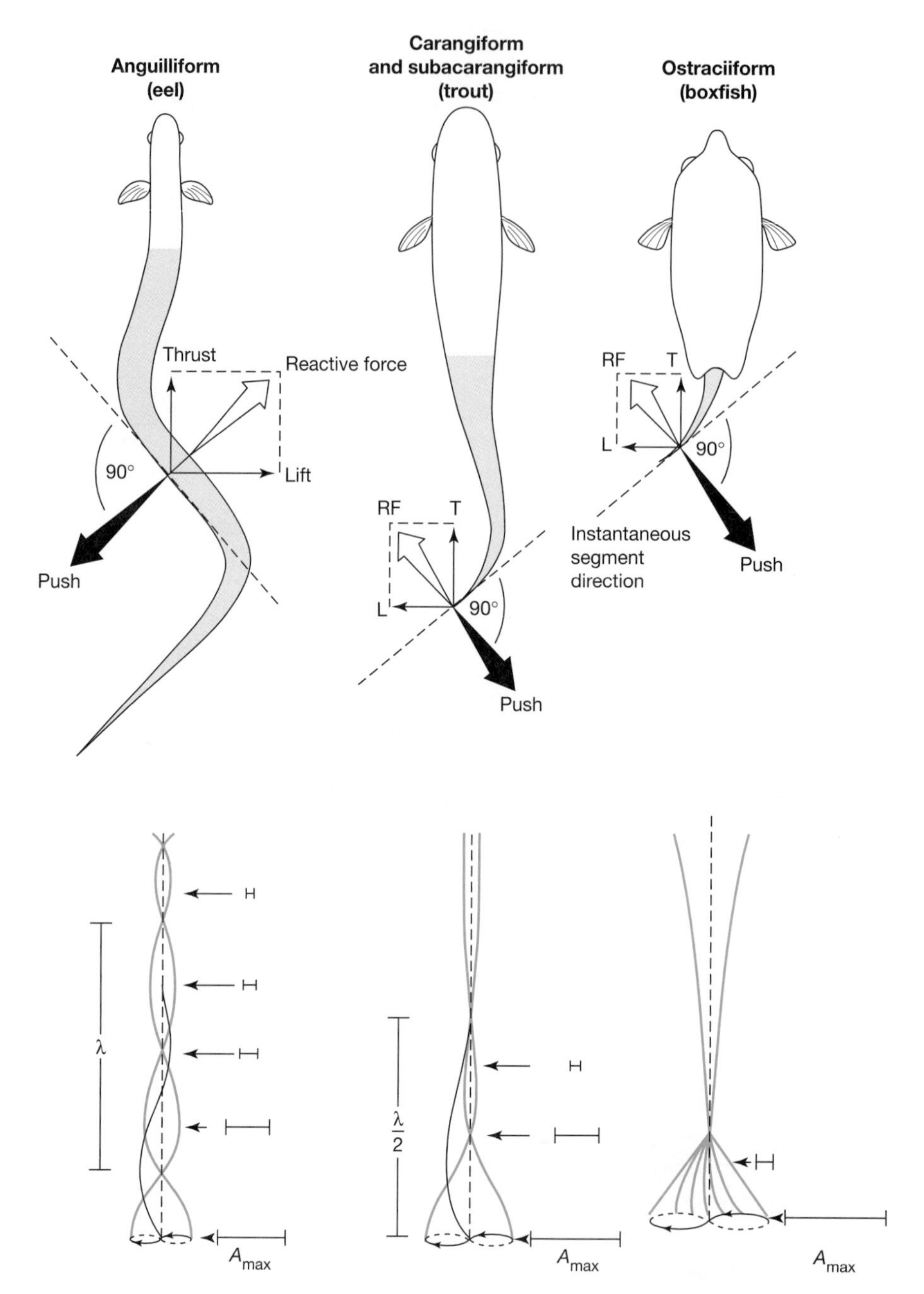

▲ Figure 6–11 Basic movements of swimming fishes. (upper) Outlines of some major swimming types; body regions that undulate are shaded blue. The lift component of the reactive force produced by one undulation's push on the water is canceled by that of the next, oppositely directed undulation. The thrust from each undulation is in the same direction and thus is additive. (lower) Waveforms created by undulations of points along the body and tail. A_{max} represents the maximum lateral displacement of any point. Note that A_{max} increases posteriorly; λ is the wavelength of the undulatory wave.

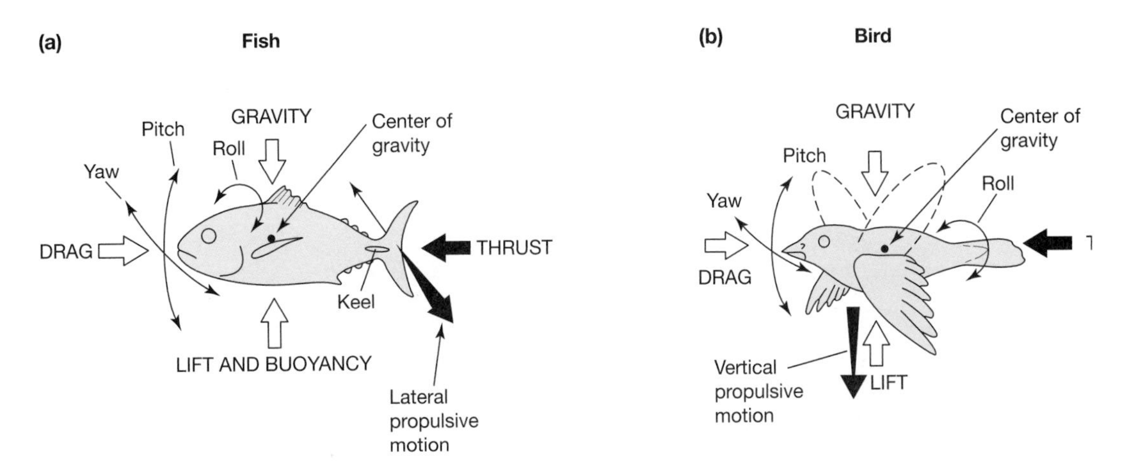

▲ Figure 6–12 Comparison of forces associated with locomotion for a swimming and a flying vertebrate. (a) By moving its caudal fin, a fish produces lateral movement far from the center of gravity. The reactive force of the water to this movement causes the head to yaw in the same direction as the tail. (b) In a bird, the major propulsive stroke of the wings is downward. Since both the propulsive stroke and the reactive force (lift) act near the center of gravity, the bird does not pitch.

(teardrop) shapes produce minimum inertial drag when their maximum width is about one-fourth of their length and is situated about one-third of the length from the leading tip (Figure 6–14). The shape of many rapidly swimming vertebrates closely approximates these dimensions. Usually, fast-swimming fishes have small scales or are scaleless, with smooth body contours lowering viscous drag. (Of course, many slow-swimming fishes also are scaleless, which shows that universal generalizations are difficult to make.) Mucus also contributes to the reduction of viscous drag.

Swimming movements in which only the caudal peduncle and caudal fin undulate are usually called modified carangiform motion. Scombroids (mackerels, tunas) and many pelagic sharks have a caudal peduncle that is narrow dorsoventrally but is relatively wide from side to side. The peduncle of carangids is often studded laterally with bony plates called scutes. These structures present a knife-edge profile to the water as the peduncle undulates from side to side; they also contribute to the reduction of drag on the laterally sweeping peduncle. The importance of these seemingly minor morphological changes is underscored by the tail stalk of whales and porpoises, which is also narrow and has a double knife-edge profile. But the tail stalk of cetaceans is narrow laterally, whereas the caudal peduncle of scombroid fishes is narrow vertically. The difference reflects the plane of undulation—up and down for cetaceans and side to side for fishes. These strikingly similar specializations of modified carangiform swimmers—whether shark, scombroid, or cetacean—produce efficient conversion of muscle contractions into forward motion.

The tail creates turbulent vortices of swirling water in a fish's wake, which may be a source of inertial drag, or the vortices may be modified to produce beneficial thrust (Stix 1994, Triantafyllou and Triantafyllou 1995). The total drag created by the caudal fin depends on its shape. When the aspect ratio of the fin (dorsal-to-ventral length divided by the anterior-to-posterior width) is large, the amount of thrust produced relative to drag is high. The stiff sickle-shaped fin of scombroids and of certain sharks (mako, great white) results in a high aspect ratio and efficient forward motion. Even the cross section of the forks of these caudal fins assumes a streamlined teardrop shape, further reducing drag. Many species with these specializations swim continuously.

The caudal fins of trout, minnows, and perches are not stiff and seldom have high aspect ratios. These subcarangiform swimmers change caudal fin area and regional stiffness to modify thrust and to produce vertical movements of the posterior part of the body. The latter action is achieved when the fish is at rest by propagating an undulatory wave up or down the flexible caudal fin (Figure 6–13). In these

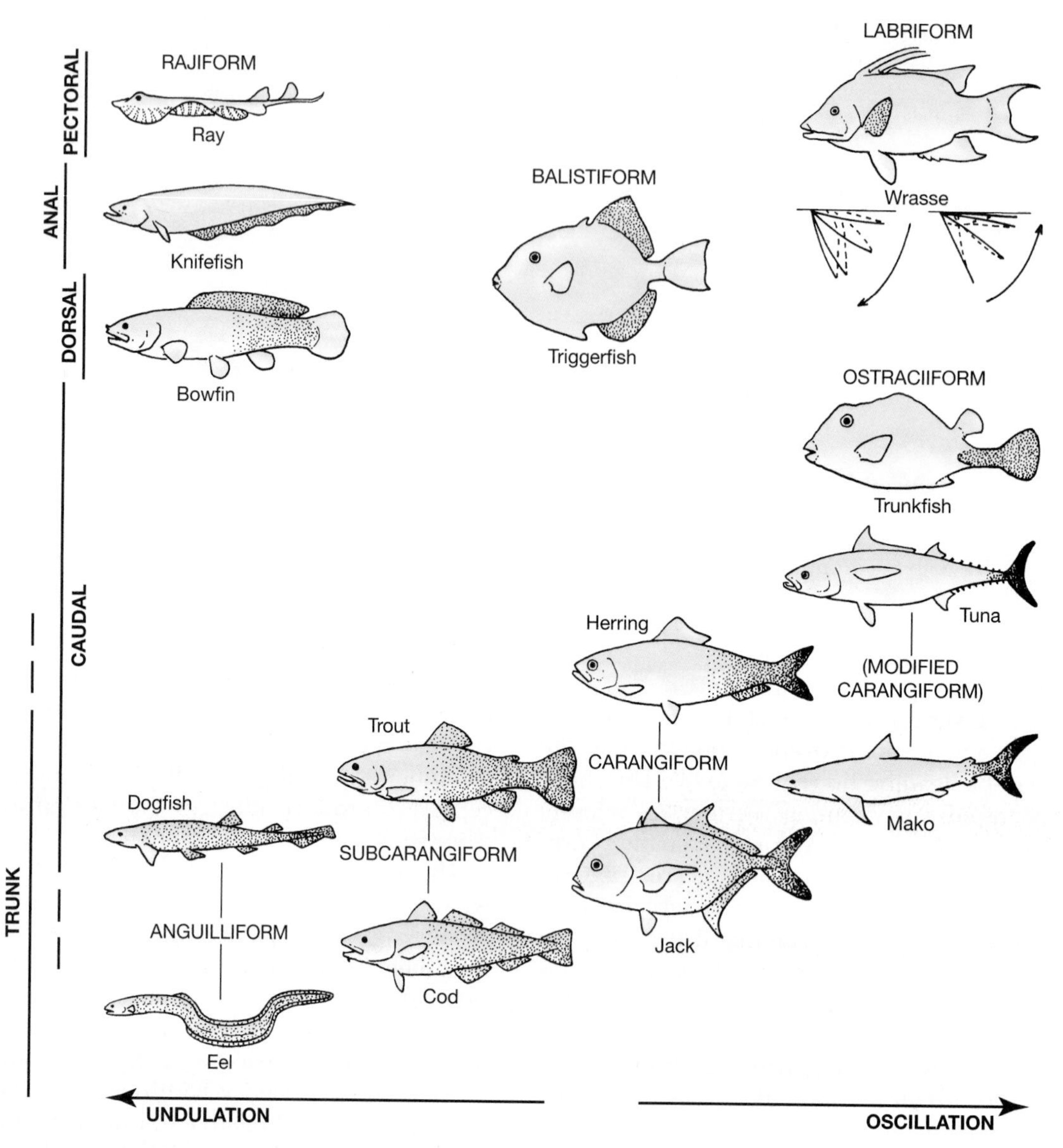

▲ Figure 6–13 Location of swimming movements in various fishes. Stippled areas of body undulate or move in swimming. Names such as carangiform describe the major types of locomotion found in fishes; they are not a phylogenetically based identification of all fishes using a given mode.

subcarangiforms propulsion often proceeds in bursts, usually from a standstill with rapid acceleration initiated by special neural systems, the Mauthner cells. The caudal peduncle of these fishes, unlike that of modified carangiform swimmers, is laterally compressed and deep. Under these circumstances the peduncle contributes a substantial part of the total force of propulsion.

Current understanding of these varied adaptations is the result of extensive study of vertebrate swimming (especially that of bony fishes) over the past several decades. Nevertheless, we still do not fully understand how the extraordinary propulsion efficiencies, accelerations, and maneuverability of fishes are achieved. Naval architects have not been able to design boats that perform nearly as well as fishes.

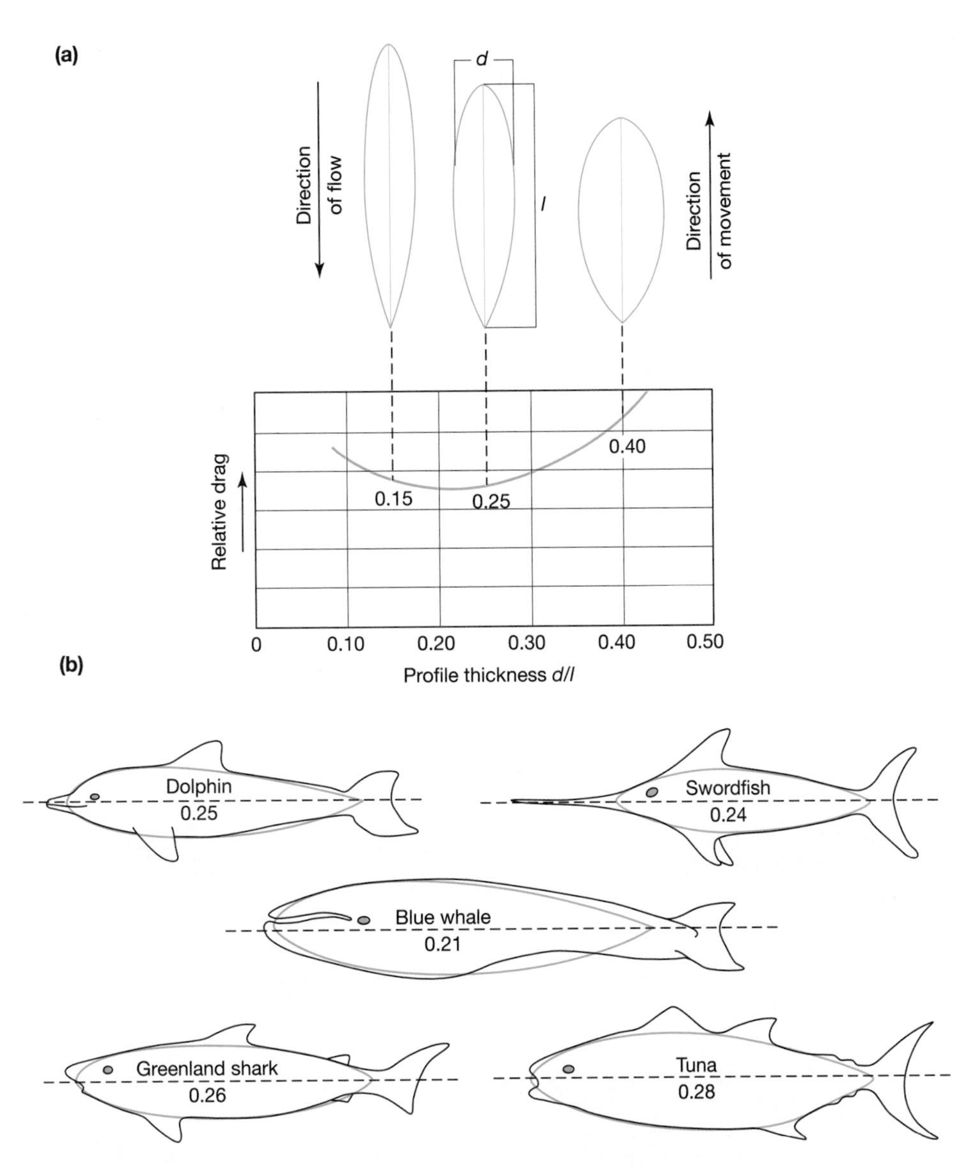

▲ Figure 6–14 Effect of body shape on drag. (a) Streamlined profiles with width (*d*) equal to approximately one-fourth of length (*l*) minimize drag. The examples are for solid, smooth test objects with thickest section about two-fifths of the distance from the tip. (b) Width-to-length ratios (d/l) for several swimming vertebrates. Like the test objects, these vertebrates tend to be circular in cross section. Note that the ratio is near 0.25, and the general body shape approximates a fusiform shape.

6.4 Actinopterygian Reproduction and Conservation

Reproductive modes of actinopterygians show a greater diversity than is known in any other vertebrate taxon. Despite this diversity, the vast majority of ray-finned fishes are oviparous (producing eggs that develop outside the body of the mother). Within oviparous teleosts, freshwater and marine species show contrasting specializations. Some marine species of fishes and most freshwater species lay adhesive eggs on rocks or plants, or in gravel or sand, and one or both parents guard the eggs and young. Nests vary from depressions in sand or gravel to

BOX 6-1 WHAT A FISH'S EARS TELL ABOUT ITS LIFE

Following minute fish eggs or translucent larvae in the open sea or turbid rivers presents insurmountable problems. However, there is an indirect method of tracing the details of the life history of an individual fish. A characteristic of bony fishes is the presence of up to three compact, mineralized structures suspended within each inner ear. These structures are especially well developed in the majority of teleosts, where these otoliths are often curiously shaped, fitting into the spaces of the membranous labyrinth very exactly and growing in proportion to the growth of the fish. They are important in orientation and locomotion, and they are formed in most teleosts during late embryonic stages

Otoliths grow in concentric layers, much like the layers of an onion (Figure 6–15). Each of these layers reflects a day's growth. The relative width, density, and interruptions of the layers show the environmental conditions the individual encountered daily, from hatching, including variations in temperature and food capture. Minute quantities of substances characteristic of the environment where the fish has spent each day are incorporated in the bands, further enhancing the information provided by the otolith. A day-by-day record of the individual is written in its otoliths. It is even possible to imprint a permanent code in the otoliths by subjecting young fish to a series of temperature increases and decreases. Fishery managers are using this method to mark juvenile fishes before they are released. Years later, when the fishes are recaptured as adults, the code embedded in the otoliths shows which brood they belong to. While still fraught with problems of of interpretation, the study of otolith microstructure is significantly advancing our understanding of reproduction and early life history of fishes.

▲ Figure 6–15 Scanning electron micrograph of an otolith from a juvenile French grunt, *Haemulon flavolineatum*. The central area represents the focus of the otolith from which growth proceeds. The alternating dark and light rings are daily growth increments. Note that the width of the growth increments varies, signifying day-to-day variation in the rate of growth.

elaborate constructions of woven plant material held together by parental secretions. Species of marine fishes that construct nests are smaller than species that are pelagic spawners, perhaps because small species are better able to find secure nest sites than are large species. Parental guarding of eggs may be unwittingly assisted by other organisms near, on, or in which the eggs are laid. Among these are stinging anemones, mussels, crabs, sponges, and tunicates. Perhaps the ultimate in protection of eggs is for one of the parents to carry them. Species are known that carry eggs on fins, under lips, in the mouth or gill cavities, on specialized protuberances, skin patches, and even in pouches.

Reproductive Characteristics of Freshwater Teleosts

Freshwater teleosts generally produce and care for a relatively small number of large, yolk-rich **demersal** eggs (i.e., eggs that are buried in gravel, placed in a nest, or attached to the surface of a rock of plant). Attachment is important, because flowing water could easily carry a pelagic (i.e., free-floating) egg away from habitats suitable for development. The eggs of freshwater teleosts hatch into babies that have body forms and behaviors similar to those of adults.

Conservation Concerns for Freshwater Fishes

Nearly 40 percent of fish species live in the world's fresh waters, and all of them are threatened by alteration and pollution of lakes, rivers, and streams. Draining, damming, canalization, and diversion of rivers create habitats that no longer sustain indigenous fishes. Perhaps the most imperiled fish species in the world is the one with the smallest range of any vertebrate. The tiny Devil's Hole pupfish (*Cyprinodon diabolis*, an atherinomorph killifish) lives in a single spring-fed pool in a specially designated portion of Death Valley National Monument in southern Nevada. While the long, slit-like pool is approximately 17 by 3m at the surface, the pupfish live their entire lives over a single shallow algae covered shelf only 18 square meters in area. Although no other species is so extremely limited in range, many other species of fishes in the region are limited to a single spring or pool. Human activity in the region is threatening the Devil's Hole pupfish and the other isolated fishes. Local ranchers and farmers, and cities as far away as Las Vegas, are pumping water from the underground aquifers that feed the pools. The water in the aquifers is fossil water—it has been there since the end of the Pleistocene glaciations—and it is not replaced by rain and snow melt. Pumping fossil water is like pumping oil; the level drops and eventually the entire reserve is gone. Pool levels and spring flows are dropping; and unless the aquifers are protected, the pupfish's habitat is threatened—it will dry up and the species will become extinct.

In addition to the loss and physical degradation of fish habitat, fresh waters in much of the world are polluted by silt and toxic chemicals of human origin. This is especially true of the Western nations and urbanized regions elsewhere. The United States has had, in recent years, over 2400 instances *annually* of beaches and flowing waterways closed to human use because of pollution. Sites that are too dangerous for people to play in them are often lethal to organisms trying to live in them! Of the nearly 800 species of native freshwater fishes in the United States, almost 20 percent are considered imperiled. As much as 85 percent of the fish fauna of some states is endangered, threatened, or of special concern (Figure 6–16).

Reproductive Characteristics of Marine Teleosts

Demersal eggs may be ancestral for actinopterygians, and producing plegic eggs and larvae is probably a derived characteristic of euteleosts. Most marine teleosts release large numbers of small, buoyant, transparent eggs into the water. These eggs are fertilized externally and left to develop and hatch while drifting in the open sea. The larvae are also small, and usually have little yolk reserve. They begin feeding on microplankton soon after hatching. Marine larvae are generally very different in appearance from their parents, and many larvae have been described for which the adult forms are unknown. Such larvae are often specialized for life in the oceanic plankton, feeding and growing while adrift at sea for weeks or months, depending on the species. The larvae eventually settle into the juvenile or adult habitats appropriate for their species. It is not yet generally understood whether arrival at the appropriate adult habitat (deep-sea floor, coral reef, or river mouth) is an active or passive process on the part of larvae. However, the arrival coincides with metamorphosis from larval to juvenile morphology

▲ Figure 6–16 Number of known native freshwater fish species in each of the contiguous states. In parenthesis, the number of species from that state considered by fisheries professionals as endangered, threatened, or of special concern.

in a matter of hours to days. Although juveniles are usually identifiable to species, relatively few premetamorphic larvae have successfully been reared in captivity to resolve unknown larval relationships. Studies of bones called otoliths are contributing to our understanding of the early life history of fishes in the wild (Box 6–1).

The strategy of producing planktonic eggs and larvae exposed to a prolonged and risky pelagic existence appears to be wasteful of gametes. Nevertheless, complex life cycles of this sort are the principal mode of reproduction of marine fishes. One advantage that fishes may achieve by spawning pelagically is reduction of some types of predation on fertilized eggs. Predators that would capture the eggs may be abundant in the parental habitat but relatively absent from the pelagic realm. Pelagic spawning fishes often migrate to areas of strong currents to spawn, or spawn in synchrony with maximum monthly or annual tidal currents, thus assuring rapid offshore dispersal of their zygotes. These behaviors would help to ensure that the eggs are quickly carried away from potential predators.

A second advantage of pelagic spawning may involve the high biological productivity of the sunlit surface of the pelagic environment. Microplankton (bacteria, algae, protozoans, rotifers, and minute crustaceans) are abundant where sufficient nutrients reach sunlit waters. If energy is limiting to the parent fishes, it could be advantageous to invest the minimum possible amount of energy by producing eggs that hatch into specialized larvae, which feed on pelagic food items that are too small for the adults to eat.

A final hypothesis involves species-level selection. Producing floating, current-borne eggs and larvae increases the chances of colonizing all patches of appropriate adult habitat in a large area. A species that is widely dispersed is not vulnerable to local environmental changes, which could extinguish a species with a restricted geographic distribution. Perhaps the predominance of pelagic spawning species in the marine environment reflects the results of millions of years of extinctions of species with reproductive behaviors that did not disperse their young as effectively.

Conservation Concerns for Marine Fishes

The vast majority of marine teleosts produce eggs that are shed into the environment with little further parental investment. These eggs tend to be small relative to the size of the adult fish, and they are produced in prodigious numbers. Once the pelagic eggs are launched, many things can determine how many survive. Some will always be lost to predation by filter-feeding fishes, but weather-related events are less predictable. Excessively hot or cold water can kill eggs and larvae outright or disrupt their development. Changes in ocean currents produced by the action of wind and global events such as the El Niño/La Niña phenomenon affect the abundance of nutrients and the amount of microplankton. Because so many variables are involved, the number of individuals breeding in any given year (breeding stock size) bears no clear relationship to the number of individuals in the next generation. Thus, a breeding season rich with spawning adults may produce few or no offspring that survive to breed in subsequent seasons if environmental factors prevent the survival of eggs, larvae, or juveniles. Conversely, under exceptionally favorable circumstances a few breeding adults could produce a very large number of offspring that survive to maturity.

The low predictability of future stock size based on current stock size has been a major stumbling block to effective fisheries management. Because so much of a population's future size depends on the environment experienced by eggs and larvae—conditions not usually obvious to fishermen or scientists—it is difficult to demonstrate the effects of overfishing in its early stages or the direct results of conservation efforts.

The problems inherent in fisheries management have resulted in destruction of commercial fish populations and entire marine food webs by overfishing. Many of the world's richest fisheries are on the verge of collapse. The Georges Bank, which lies east of Cape Cod, is an example of what overfishing can do. For years, conservation organizations called for a reduction in catches of cod, yellowtail flounder, and haddock. Many of these bottom-dwelling fishes are also caught unintentionally by vessels fishing for other species. The conservationists' cries were not heeded, and populations of these fishes crashed dramatically in the 1990s. By October 1994, the situation was so bad that a government and industry group, the New England Fishery Management Council, directed its staff to devise measures that would reduce the catch of those species essentially to zero. Much of Georges Bank remains closed to all fishing by methods targeting the one abundant species. Thus thousands of fishermen have been affected by draconian measures such as complete moratoria on fishing. Many of them have moved their boats to other heavily fished areas, such as the mid-Atlantic coast and the Gulf of Mexico, and the depletion process will repeat itself. Many smaller fishing operations, some of which have been family businesses for several generations, have gone out of business. After a decade of complete protection, some of the species endangered by overfishing are again to be found on Georges Bank, but others show no signs of recovery.

6.5 The Adaptable Fish—Teleost Communities in Contrasting Environments

Given the great numbers of both species and individuals of living fishes, especially the teleosts, it is little wonder that they inhabit an enormous diversity of watery environments. Examining two of the most distinctive, the deep sea and coral reefs, allows us to better understand the amazing adaptability of the teleost fishes.

Deep-Sea Fishes

Of all regions on Earth, the deep sea is the least studied. Two major life zones exist in the sea: the **pelagic**, where organisms live a free-floating or swimming existence, and the **benthic**, where organisms associate with the bottom (Figure 6–17). Solar light is totally extinguished at a depth of 1000 meters in the clearest oceans and at much shallower depths in the murkier coastal seas. As a result, about 75 percent of the ocean is perpetually dark, illuminated only by the flashes and glow of bioluminescent organisms. A distinctive and bizarre array of deep-sea fishes, with representatives from no fewer than five orders, has evolved in those zones.

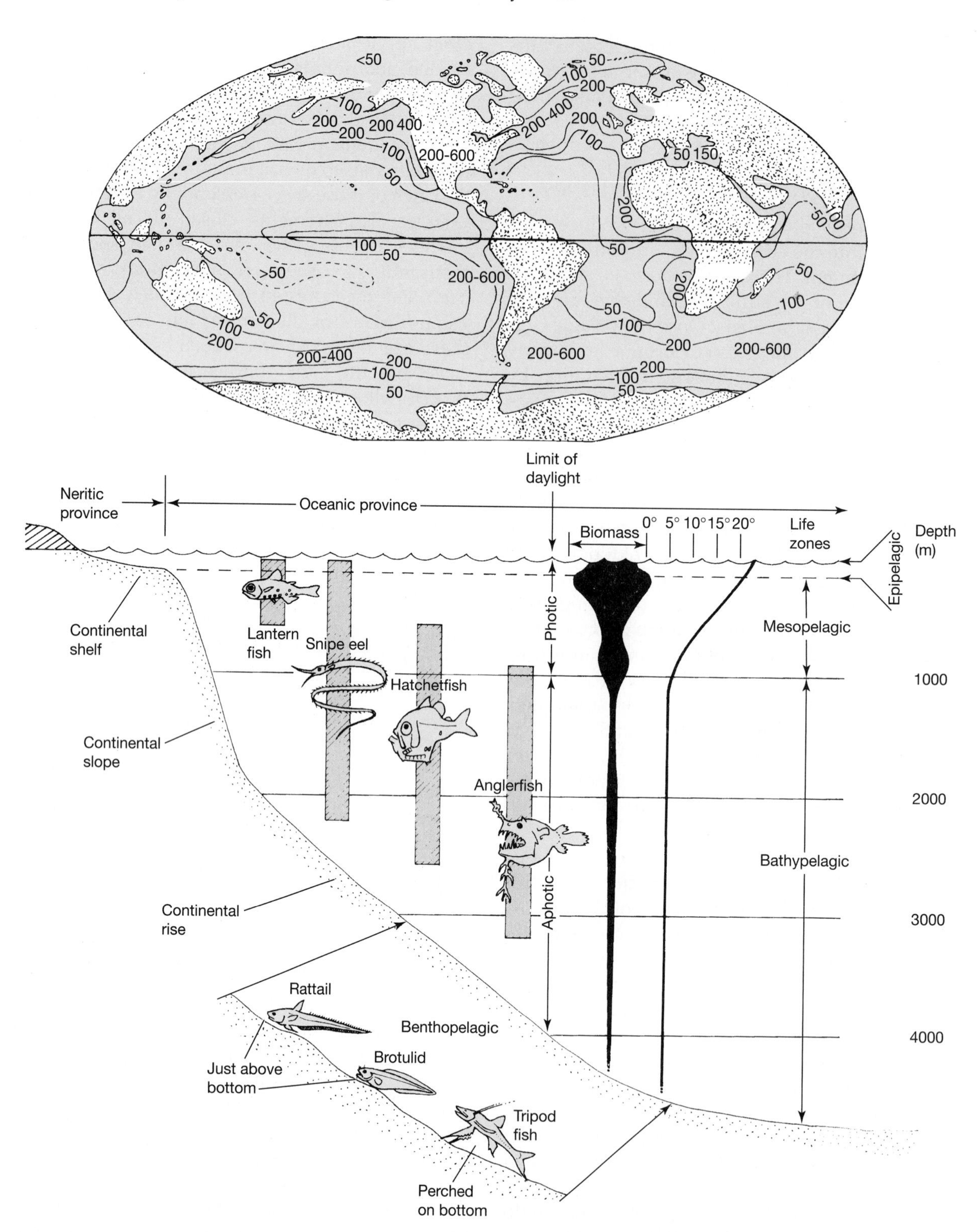

▲ Figure 6–17 Life zones of the ocean depths. (a) Annual productivity at the ocean surface. Numbers are grams of carbon produced per square meter per year. Rich assemblages of deep-sea fishes occur where highly productive waters overlie deep waters. (b) Schematic cross section of the life zones within the deep sea. Various pertinent physical and biotic parameters are superimposed on the arbitrary life zones, as are the vertical ranges of several characteristic fish species, some of which migrate daily.

Fishes decrease in abundance, size, and species diversity at greater depths. These trends are not surprising, for all animals ultimately depend on plant photosynthesis, which is limited to the epipelagic regions. Below the epipelagic, animals must depend on a rain of detritus from the surface into the deep sea. The amount of food available diminishes because it is consumed during descent. Sampling confirms this decrease in food: Surface plankton can reach biomass levels of 500 milligrams wet mass per cubic meter, but at 1000m plankton biomass is only 25mg/m^3. At 3000 to 4000m, the value falls to 5mg/m^3, and at 10,000m there is only 0.5 mg of plankton per cubic meter of water.

Fish diversity parallels this decrease: About 800 species of deep-sea fishes are estimated to occupy the mesopelagic zone (from 100 to 1000m below the surface), and only 150 species inhabit the bathypelagic regions (below 1000m). Deep seas lying under areas of high surface productivity contain more and larger fish species than do regions underlying less productive surface waters. High productivity occurs in areas of upwelling, where currents recycle nutrients previously removed by the sinking of detritus. In these places deep-sea fishes tend to be most diverse and abundant.

In tropical waters, photosynthesis continues throughout the year. Away from the tropics it is more cyclic, following seasonal changes in light, temperature, and sometimes currents. Diversity and abundance of deep-sea fishes decrease away from the tropics. Over 300 species of mesopelagic and bathypelagic fishes occur in the vicinity of tropical Bermuda, whereas only 50 species have been described in the entire Antarctic region. The high productivity of Antarctic waters is restricted to a few months of each year, and the detritus that sinks through the rest of the year is not sufficient to nourish a diverse assemblage of deep-sea fishes.

We emphasize that availability of food (energy) is the most formidable environmental problem that deep-sea fishes encounter; indeed, many of their specific characteristics may have been selected by food scarcity. Another variable factor is hydrostatic pressure, whereas low temperature and the absence of light are constants.

Mesopelagic Fishes With increased depth, each species is further removed from a primary source of food. In general, mesopelagic fishes and invertebrates migrate vertically. At dusk they ascend toward the surface, only to descend again near dawn, apparently following light-intensity levels (Figure 6–18). Sonar signals readily reflect off the swim bladders of mesopelagic fishes, and aggregations of them produce an acoustic scattering layer. At sunset the intensity of light at the surface decreases (right axis in Figure 6–18), as does the light penetrating the sea. Plotting the depth of a single light intensity (called an isolume) against time illustrates the progressively shallower depth at which a given light intensity occurs during dusk. The light intensity chosen for this example falls between that of starlight and full moonlight. It corresponds closely with the upper boundary of the deep scattering layer, which rises rapidly with the upward migration of mesopelagic fishes.

Several benefits and costs probably result from this behavior. By rising at dusk, mesopelagic fishes enter a region of higher productivity, where food is more concentrated. But they also increase their exposure to predators. Furthermore, ascending mesopelagic fishes are exposed to temperature increases that can exceed 10°C. The energy costs of maintenance at these higher temperatures can double or even triple. In contrast, daytime descent into cooler waters lowers metabolism, which conserves

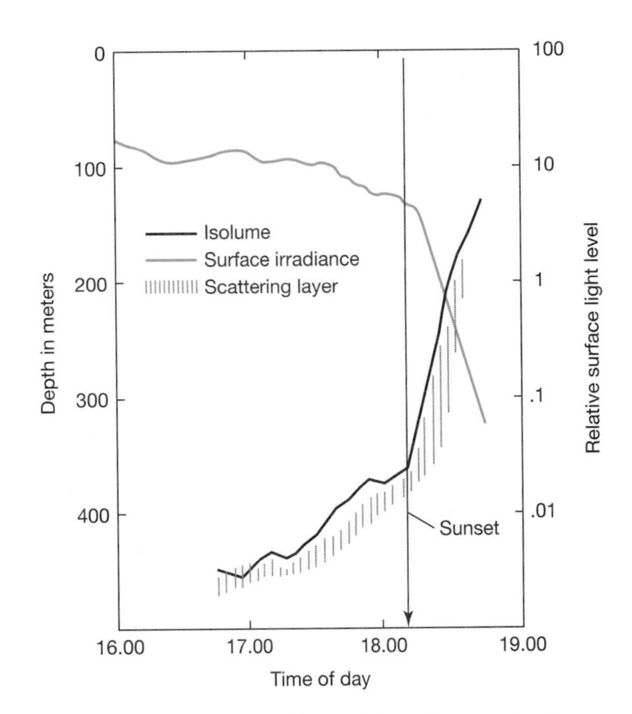

▲ Figure 6–18 Upward migration of mesopelagic fishes as indicated by changes in depth of deep scattering layer.

energy and reduces the chance of predation because fewer predators exist at increased depth.

Bathypelagic Fishes It is less certain that bathypelagic fishes undertake daily vertical migrations. There is little metabolic economy from vertical migration within this lightless zone, because temperatures are uniform (about 5°C). Further, the cost and time of migration over the several thousand meters from the bathypelagic region to the surface would probably outweigh the energy gained from invading the rich surface waters. Instead of migrating, bathypelagic fishes are specialized to live less active lives than those of their mesopelagic counterparts.

Pelagic deep-sea fishes have a series of structures related to their particular life-styles. For example, eye size and light sensitivity correlate with depth. Mesopelagic fishes have larger eyes than pelagic fishes. The retina of mesopelagic fishes contains a high concentration of visual pigment, the photosensitive chemical that absorbs light in the process of vision. The visual pigments of deep-sea fishes are most efficient in absorbing blue light, which is the wavelength of light that is most readily transmitted through clear water. Many deep-sea fishes and invertebrates are emblazoned with startling bioluminescent designs, formed by photophore organs that emit blue light.

Deep-sea fishes have less dense bone and less skeletal muscle than do fishes from shallower depths (Figure 6–19). Surface fishes have strong, ossified skeletons and large red muscles especially adapted for continuous cruising. Mesopelagic fishes, which swim mostly during vertical migration, have a more delicate skeleton and less axial red muscle. In bathypelagic fishes (species that live on the bottom at great depths), the axial skeleton and the mass of muscles are greatly reduced and locomotion is limited.

The jaws and teeth of deep-sea fishes are often enormous in proportion to the rest of the body. Many bathypelagic fishes can be described as a large mouth accompanied by a stomach. If a fish rarely encounters potential prey, it is important to have a mouth large enough to engulf nearly anything it does meet—and a gut that can extend to accommodate a meal (Figure 6–20). Increasing the chances of encountering prey is also important to survival in the deep sea. Rather than searching for prey through the blackness of the depths, the ceratioid anglerfishes dangle a bioluminescent bait in front of them.

◀ Figure 6–19 Reduction of bone in deep-sea fishes. X rays show the reduced bone density of mesopelagic and bathypelagic fishes compared to a surface dweller. Top: The surface-dwelling herring (Clupeidae). Middle: A mesopelagic, vertically migrating lanternfish (Myctophidae). Bottom: A mesopelagic, deep-living bristlemouth (Gonostomatidae).

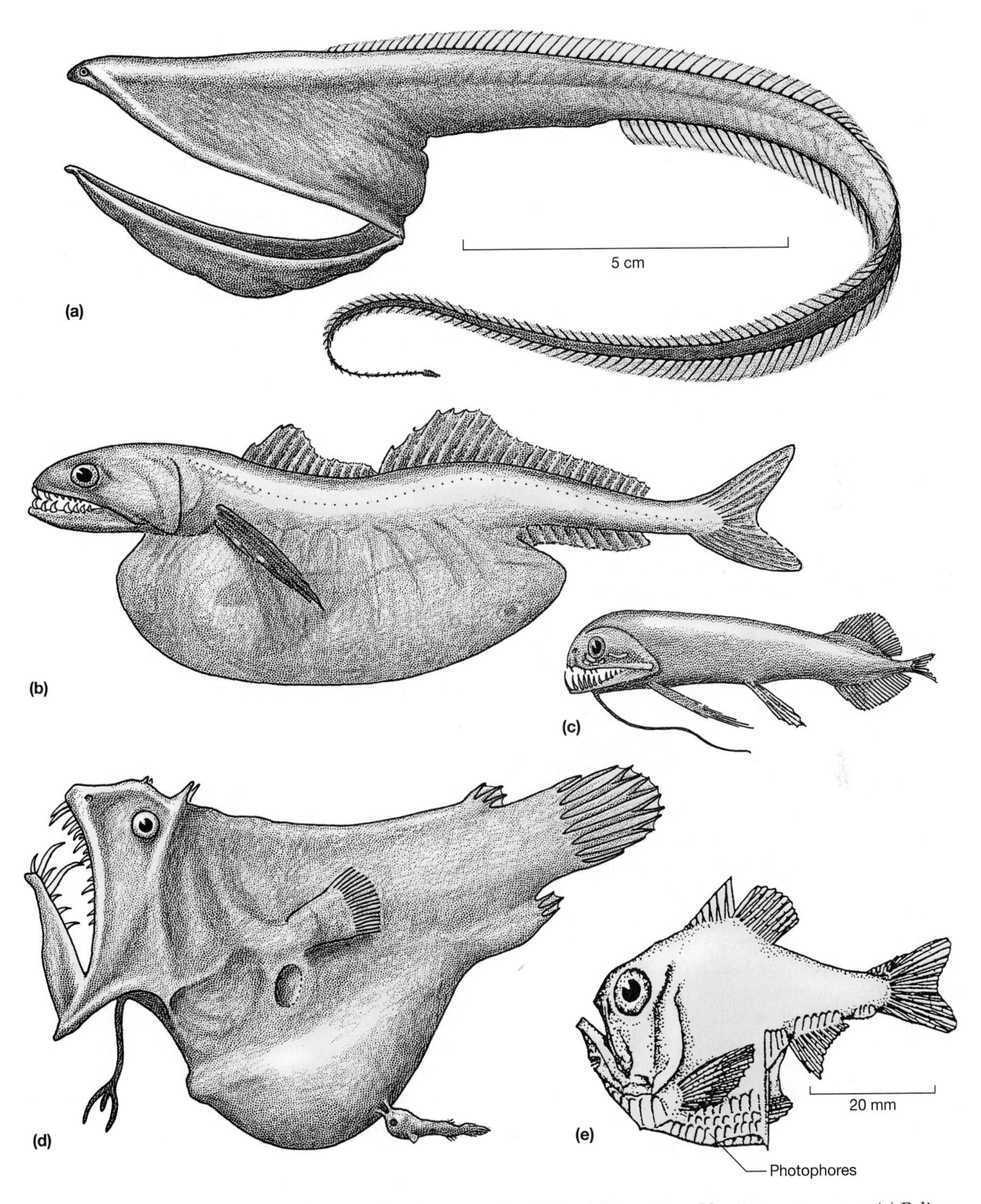

▲ Figure 6–20 Deep-sea fishes showing their large mouths, distendable guts, and luminescent organs. (a) Pelican eel, *Eurypharynx pelecanoides*. (b) Deep-sea perch, *Chiasmodus niger*, its belly distended by a fish bigger than itself. (c) Stomiatoid, *Aristostomias grimaldii*. (d) Female anglerfish, *Liophryne argyresca*, with a parasitic male attached to her belly. (e) Hatchet fish, *Sternoptyx diaphana* with photophores on the ventrolateral region.

The bioluminescent lure is believed to mimic the movements of zooplankton and to lure fishes and larger crustaceans to the mouth. Prey is sucked in with a sudden opening of the mouth, snared in the teeth, and then swallowed. The jaws of many anglerfishes expand and the stomach stretches to accommodate prey larger than the predator. Thus deep-sea fishes, like most teleosts, show major specializations in locomotor and feeding structures. Unlike surface teleosts, however, in deep-sea teleosts the scarcity of food has selected for structures that minimize the costs of foraging and maximize the capture of prey.

Most features of the biology of deep-sea fishes are directly related to the scarcity of food, but in some cases the relationship is indirect. For example, most species are small (the average length is less than 5 centimeters), and individuals of a species are not abundant. In so vast a habitat the number of anglerfishes is very small. The density of females in the most common species is typically less than one female per cubic mile. Imagine trying to find another human under similar circumstances! Yet to reproduce, each fish must locate a mate and recognize it as its own species.

Distinctive bioluminescent patterns characterize the males and females of many bathypelagic fishes. Tiny, light-producing organs called photophores are arranged on their bodies in species- and even sex-specific patterns (Figure 6–20e). The light is produced by symbiotic species of *Photobacterium* and previously unknown groups of bacteria related to *Vibrio*. Some photophores probably act as signals to conspecifics in the darkness of the deep sea, where mates may be difficult to find. Others, such as those in the modified fin-ray lures of anglerfishes, probably attract prey. Female anglerfishes have a bioluminescent lure that looks species-specific. Although it is used to lure prey, the bait probably also attracts males. Because light does not travel far in water, visual detection of other individuals much beyond 40 or 50m is not possible and other senses, such as scent trails, must be used. The females secrete a pheromone, and males usually have enlarged olfactory organs. Sensing the pheromone during searching movements, males swim upstream to an intimate encounter.

The life history of ceratioid anglerfishes dramatizes how selection adapts a vertebrate to its habitat. The adults typically spend their lives in lightless regions below 1000m. Fertilized eggs, however, rise to the surface, where they hatch into larvae. The larvae remain mostly in the upper 30m where they grow, and later descend to the lightless region. Descent is accompanied by metamorphic changes that differentiate females and males. During metamorphosis, young females descend to great depths, where they feed and grow slowly, reaching maturity after several years.

Female anglerfishes feed throughout their lives, whereas males feed only during the larval stage. Metamorphic changes in males prepare them for a different future; their function is reproduction, literally by lifelong matrimony. The body may elongate and axial red muscles develop. The males cease eating, and the enlarged liver apparently provides energy for an extended period of swimming. The olfactory organs of males enlarge at metamorphosis and the eyes continue to grow. These changes suggest that adolescent males spend a brief, but active, free-swimming period concentrated on finding a female. The journey is precarious, for males must search vast, dark regions for a single female while running a gauntlet of other deep-sea predators. In the young adults there is an unbalanced sex ratio—often more than 30 males for every female. Apparently, at least 29 of those males will not locate a virgin female.

Having found a female, a male does not want to lose her. He ensures a permanent union by attaching himself as a parasite to the female, biting into her flesh and attaching himself firmly. Preparation for this encounter begins during metamorphosis when the male's teeth degenerate and strong tooth-like bones develop at the tips of the jaws. A male remains attached to the female for life, and in this parasitic state he grows and his testes mature. Monogamy prevails in this pairing, for females invariably have but one attached male. This life-style is unknown among other vertebrates, but it has been successful for anglerfish—some 200 ceratioid species exist. The lesson these unique fishes provide is that vertebrate life is a plastic venture capable of adapting to extreme conditions. In the ceratioids, two features stand out: efficient energy use and reproduction. Adaptations in all vertebrates are related to these two goals, but they are not often painted in such bold relief.

Fishes in Coral-Reef Communities—Specialization, Coexistence, and Threats

A coral reef is one of the most spectacular displays of animal life on Earth. The diversity of invertebrate and vertebrate animals in such concentrations occurs nowhere else, and the vertebrate component of a coral-reef community is drawn almost exclusively from a single taxon, the acanthopterygian teleosts.

The unparalleled diversity in feeding modes that characterizes actinopterygians requires precise control of jaw actions and body positions. This precision was achieved through strong evolutionary pressures on the interactions between feeding and locomotion. Nowhere are these interactions more obvious than in coral-reef fishes. Over 600 species may be found on a single Indo-Pacific reef. The most primitive spiny-rayed fishes in coral reefs are predators of invertebrates (for example, squirrelfishes, cardinalfishes; Figure 6–21). As a response to predation, many reef invertebrates became **nocturnal**, limiting their activity to night and remaining concealed during the day. In response to the nocturnal activity of their prey, early acanthopterygians evolved the capacity to locate prey at night. To this day their descendants disperse over the reef at night to feed, but during the daylight hours they congregate in holes and crevices in the reef. The large, sensitive eyes of these nocturnal predators are effective at low light intensities. They use the irregular contours of the reef to conceal their approach and rely on a large protrusible mouth and suction to capture prey.

The evolution of fishes specialized to take food items hidden in the complex reef surface was a major advance among reef acanthopterygians Some species rely on suction, whereas others use a forceps action of their protrusible jaws. This mode of predation demands sensory specializations, the most important being high visual acuity that can be achieved only in the bright light of day. In addition, delicate positioning is required to direct the jaws. These selection pressures produced fishes capable of maneuvering through the complex three-dimensional reef in search of food. So accurate are their locomotion, visual surveillance, and memory for hiding places and escape routes that these fishes can expose themselves and feed in broad daylight (**diurnal**). The refined feeding specializations of diurnal reef fishes allow them to extract small, nocturnally active invertebrates from their daytime hiding places or snip off coral polyps, bits of sponges, and other exposed reef organisms.

Released from heavy predation during daylight because of their ability to find safety in the labyrinthine structure of the reef, many reef fishes have evolved gaudy patterns and colors. These visual signals communicate a wide range of information.

A coral reef has day and night shifts—at dusk the colorful diurnal fishes seek nighttime refuge in the reef, and the nocturnal fishes leave these hiding places and replace the diurnal fishes in the water column (Figure 6–21). The temporal precision with which each species leaves or enters the protective cover of the reef day after day indicates that this is a strongly selected behavioral and ecological characteristic. Space, time, and the food resources available on a reef are partitioned through this activity pattern.

Many reef fishes belong to a small number of families and genera with many coexisting species. The cardinalfishes, damselfishes, angelfishes, butterflyfishes, wrasses, and parrotfishes are examples. The great number of closely related species is thought to be the result of the repeated partial isolation of sections of the great pantropical ocean. (For example, the tropical Atlantic Ocean was isolated from the tropical Pacific a mere three to five million years ago by the uplift of the Isthmus of Panama) The fossil record shows that many genera of reef fish have existed for over 50 million years. During this long time, sea levels have varied, volcanic islands have grown and sunk, and continents and archipelagos have drifted with the spreading of the seafloor. Populations of reef fishes, almost all of which have pelagic eggs and/or larvae, have sometimes found themselves isolated in a corner of the vast tropical sea—and at other times, close enough for their larvae to drift and mingle with those of other populations. Periods of isolation probably led to the simultaneous formation of species in multiple isolated reef systems. The result has been a reef fish fauna of great variety, with new species entering the fauna after each period of isolation. Currently, the fauna is in a period of rather low isolation. The geographic ranges of many species extend from the east coast of Africa to Hawaii, and a few of these same species reach the Pacific coast of the Americas. Since many groups of reef fish species are closely related, they share most of their adaptations through inheritance of shared derived characters. One would expect these species groups to experience intense competition from other species within their group. Such competition should ultimately lead either to elimination of most of the closely related species or to driving related species to exploit different resources despite their original similarity. Surprisingly, neither of these results of competition seems to have occurred extensively. Instead, predation and events

(a)

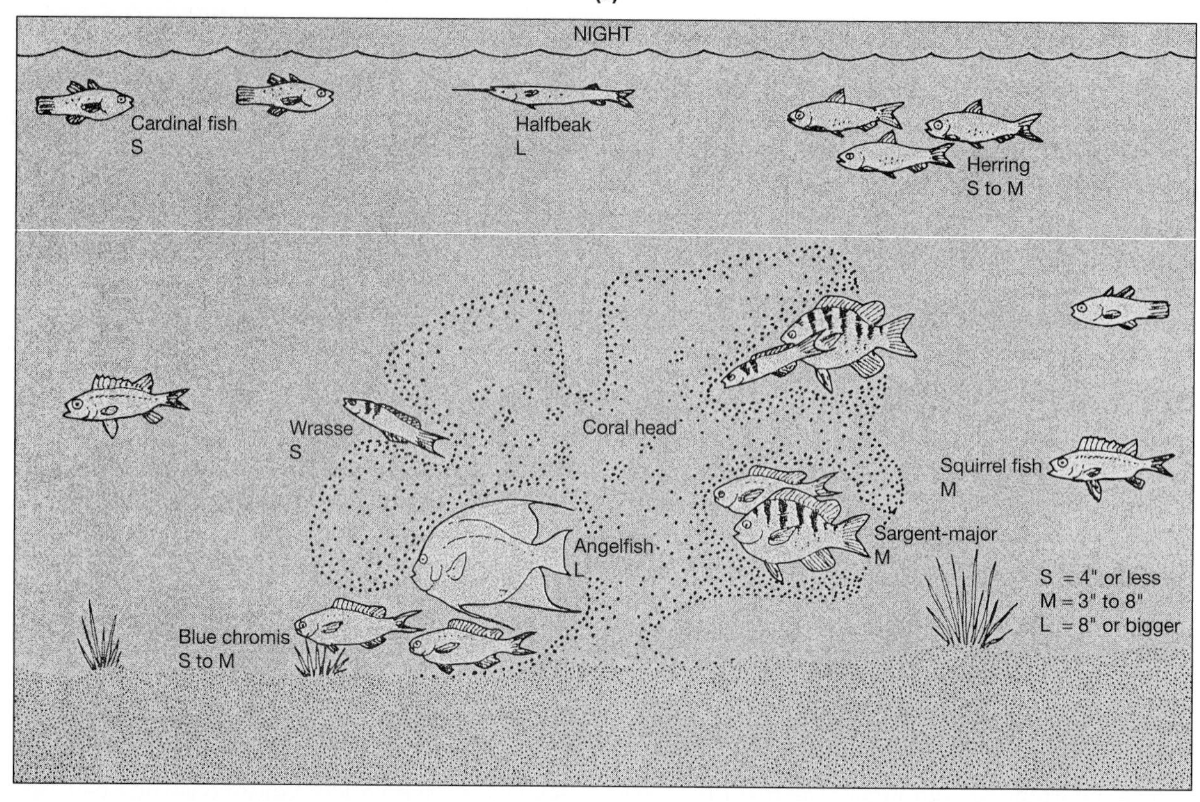

(b)

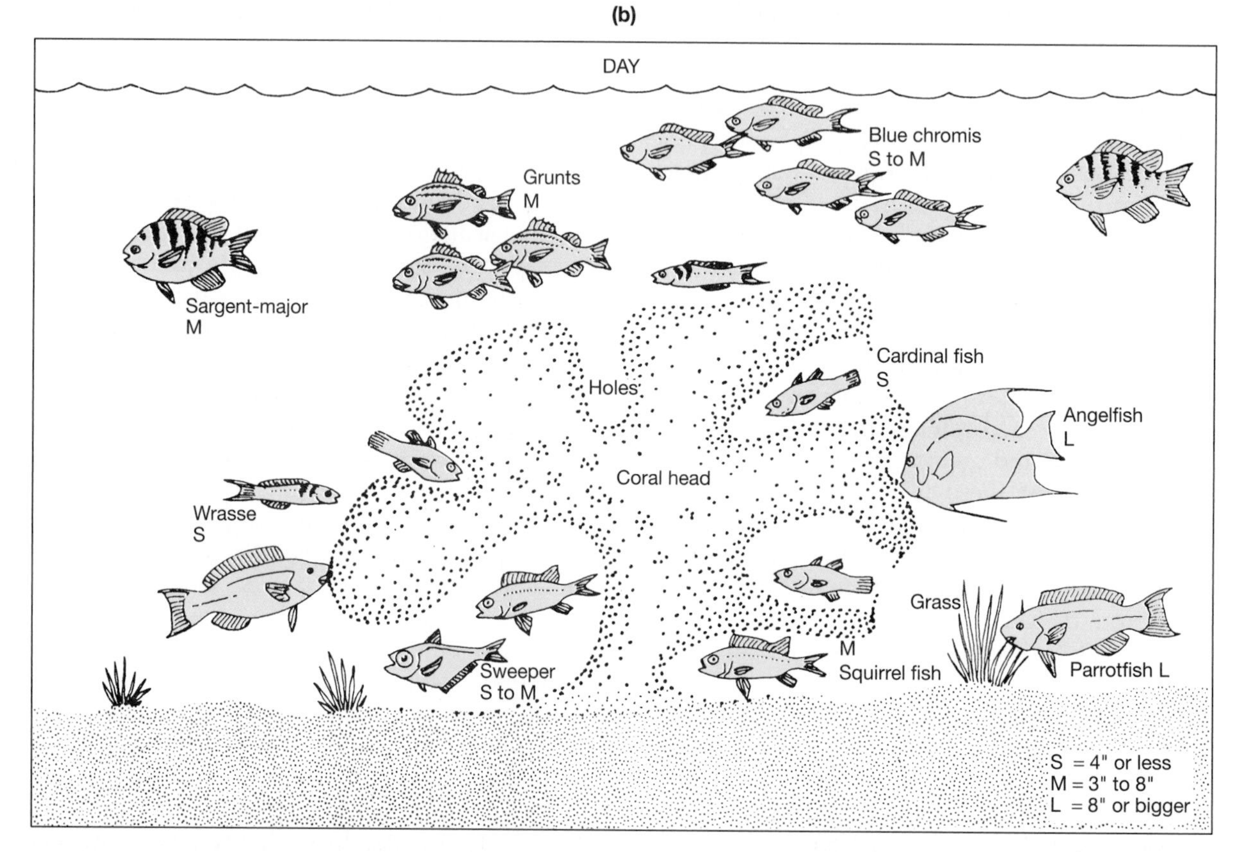

▲ Figure 6–21 Fish activity, diversity, and relative population differences on a Caribbean coral reef (a) at midnight and (b) at midday.

(such as storms) that reduce populations appear to have prevented competitive interactions from running their full course.

Conservation Concerns for Coral-Reef Fishes

The evolution and maintenance of a coral reef's rich vertebrate fauna depends on the complex three-dimensional structure of the reef. When this structure collapses, as in severe storm damage or when the reef is mined for limestone to make concrete, fish populations and diversity plummet. In the last three to five years an ominous phenomenon has been observed worldwide. Coral reefs are showing signs of physiological stress and dying en masse. The cause appears to be unusually high sea surface temperatures worldwide. The coral animals that build the reef succumb to prolonged exposure to higher-than-normal temperatures, leaving their dead skeletons exposed to organisms that erode limestone and storms that shatter the three-dimensional structure of the reef.

A two-month period of exceptionally high temperatures in late summer of 1998 left less than 5 percent of inshore reefs of Belize on Central America's Caribbean coast covered by live coral. In the Southern Hemisphere summer of 1998–1999, heat-induced coral stress was documented on the inner reefs along the entire length of Australia's Great Barrier Reef. Although it is still too recent a phenomenon to predict the consequences, if reefs around the globe physically collapse, the greatest vertebrate diversity on Earth will also dwindle. Reefs grow slowly. Even if conditions permit regeneration, reefs will be greatly changed over wide geographic areas for decades or longer. If warmer waters prevail into the future, the new reefs that might form in newly warm waters to the north and south of the current occurrence of corals will take centuries to become mature and three-dimensionally complex. Without new habitats, reef organisms face the possibility of massive extinction over the next few years. Will the disappearance of coral reefs and their marvelous vertebrate fauna be the planet's "canary in the mine"—the earliest warning signal of global warming?

Summary

At their first appearance in the fossil record, osteichthyans, the largest vertebrate taxon, are separable into distinct lineages: the Sarcopterygii (fleshy-finned fishes including actinistians, lungfishes and, strictly speaking, tetrapods and the Actinopterygii (ray-finned fishes). Extant sarcopterygian fishes offer exciting glimpses of adaptations evolved in Paleozoic environments. Actinopterygian fishes were distinct as early as the Silurian. Actinopterygians inhabit the 73 percent of the Earth's surface that is covered by water, and they are the most numerous and species-rich lineages of vertebrates. Several levels of development in food-gathering and locomotory structures characterize actinopterygian evolution. The radiations of these levels are represented today by relict groups: cladistians (bichirs and reedfishes), chondrosteans (sturgeons and paddlefishes), and the primitive neopterygians (gars and *Amia*). The most derived level of bony fishes, teleosteans, may number close to 24,000 extant species—with two groups, ostariophysans in freshwater and acanthopterygians in seawater, constituting a large proportion of these species. Examining how teleost fish communities are adapted to very specialized habitats like the deep sea and coral reefs helps us to understand how evolution has acted upon the basic teleost body plan. But looking at how humans have changed fish habitats explains why so many fishes are now in danger of extinction.

Additional Readings

Allan, J. D., and A. S. Flecker. 1993. Biodiversity conservation in running water. *Bioscience* 43:32–43.

Bemis, W. E., W. W. Burggren, and N. E. Kemp (Eds.). 1987. *The Biology and Evolution of Lungfishes*. New York: Liss.

Birstein, V. J., J. R. Waldman, and W. E. Bemis (Eds.). 1997. Sturgeon biodiversity and conservation. *Environmental Biology of Fishes* 48 (1–4):9–435.

Bond, C. E. 1996. *Biology of Fishes*, 2d ed. Fort Worth, Tex.: Saunders College Publishing.

Brothers, E. B. 1990. Otolith marking. *American Fisheries Society Symposium* 7:183–202.

Fricke, H. 1988. Coelacanths, the fish that time forgot. *National Geographic* 173:824–838.

Fricke, H., O. Reinicke, H. Hofer, and W. Nachtigall. 1987. Locomotion of the coelacanth *Latimeria chalumnae* in its natural environment. *Nature* 324:331–333.

Grande L., and W. E. Bemis. 1997. A comprehensive phylogenetic study of amiid fishes (Amiidae) based on comparative skeletal anatomy. *Journal of Vertebrate Paleontology* (Memoir no. 4): 17.

Haygood, M. C., and D. L Distel. 1993. Bioluminescent symbionts of flashlight fishes and deep-sea anglerfishes form unique lineages related to the genus *Vibrio*. *Nature* 363:154–156.

Helfman, G. S., B. B. Collette, aand D. E. Facey. 1997. *The Diversity of Fishes*. Malden Mass.: Blackwell Science.

Kingsmill, S. 1993. Ear stones speak volumes to fish researchers. *Science* 260:1233–1234.

Lauder, G. V. 1994. Caudal fin locomotion by teleost fishes: Function of the homocercal tail. *American Zoologist* 34(5): 13A, abst. no. 66.

Lauder, G. V., and K. F. Liem. 1983. The evolution and interrelationships of the actinopterygian fishes. *Bulletin of the Museum of Comparative Zoology* 150:95–197.

Lauder, G. V., and J. H. Long (Eds.). 1996. Aquatic locomotion: New approaches to invertebrate and vertebrate biomechanics. *American Zoologist* 36:535–709.

Long, J. H. Jr., M. E. Hale, M. J. McHenry, and M. W. Westneat. 1996. Functions of fish skin: Flexural stiffness and steady swimming of longnose gar *Lepisosteus osseous*. *Journal of Experimental Biology* 199:2139–2151.

McCune, A. R. 1996. Biogeographic and stratigraphic evidence for rapid speciation in semionotid fishes. *Paleobiology* 22(1):34–48.

McCune, A. R. 1997. How fast is speciation: Molecular, geologic and phylogenetic evidence from adaptive radiations of fishes. In T. Givnish and K. Sytsma (Eds.), *Molecular Evolution and Adaptive Radiation*. Cambridge, Mass.: Cambridge University Press.

McCune, A. R., and N. R. Lovejoy. 1998. The relative rate of sympatric and allopatric speciation in fishes: Tests using DNA sequence divergence between sister species and among clades. In D. Howard and S. Berlocher (Eds.), *Endless Forms: Species and Speciation*. Oxford, UK: Oxford University Press.

Moyle, P. B., and J. J. Cech Jr. 2000. *Fishes: An Introduction to Ichthyology*, 4th ed. Upper Saddle River, N.J.: Prentice Hall.

Musick, J. A., M. N. Bruton, and E. K. Balon (Eds.). 1991. The biology of *Latimeria chalumnae* and evolution of coelacanths. *Environmental Biology of Fishes* 32(1–4):9–435.

Nelson, J. S. 1994. *Fishes of the World*, 3d ed. New York: Wiley.

Pauly, D., V. Christensen, J. Dalsgaard, R. Froese, and F. Torres, Jr. 1998. Fishing down marine food webs. *Science* 279:860–863.

Paxton, J. R., and W. N. Eschmeyer (Eds.). 1994. *Encyclopedia of Fishes*. San Diego: Academic.

Popper, A. N., and S. Coombs. 1980. Auditory mechanisms in teleost fishes. *American Scientist* 68:429–440.

Safina, C. 1998. *Song for the Blue Ocean: Encounters along the World's Coasts and beneath the Seas*. New York: Henry Holt.

Stiassny, M. L., L. R. Parenti, and G. D. Johnson. 1996. *Interrelationships of Fishes*. San Diego: Academic.

Stix, G. 1994. Robotuna. *Scientific American* 270(1):142.

Thompson, K. S. 1992. *Living Fossil: The Story of the Coelacanth*. New York: Norton.

Thresher, R. E. 1984. *Reproduction in Reef Fishes*. Neptune City, N.J.: T. F. H. Publications.

Triantafyllou, M. S., and G. S. Triantafyllou. 1995. An efficient swimming machine. *Scientific American* 274(3):64–70.

Warren, M. L. Jr., and B. M. Burr. 1994. Status of fresh water fishes of the United States: Overview of an imperiled fauna. *Fisheries* 19:6–18.

Weinberg, S. 1999. *A Fish Caught in Time: The Search for the Coelacanth*. London: Fourth Estate.

Web Explorations

On-line resources for this chapter are on the World Wide Web at: http://www.prenhall.com/pough (click on the Table of Contents link and then select Chapter 6).

CHAPTER

7

Geography and Ecology of the Paleozoic

The Paleozoic world was very different from the one we know—the continents were in different places, climates were different, and initially there was little structurally complex life on land. By the Early Devonian terrestrial environments supported a substantial diversity of plants and invertebrates, setting the stage for the first terrestrial vertebrates (tetrapods) in the Late Devonian. Early plants were represented by primitive groups such as horsetails, mosses, and ferns. The evolution of plants had a profound effect on the Earth's atmosphere as well its surface, resulting in the production of soils that trapped carbon dioxide. Removal of carbon dioxide from the atmosphere caused a reverse greenhouse effect—global cooling. Extensive glaciation occurred during the Late Carboniferous and Early Permian periods, and the earliest tetrapods lived in equatorial regions. Terrestrial ecosystems became more complex in the Carboniferous and Permian; some modern types of plants such as conifers appeared, both nomamniote and amniote tetrapods diversified, and the first flying insects took to the air. Major extinctions occurred in terrestrial ecosystems at the end of the Early Permian and again at the end of the Paleozoic in the transition between Permian and Triassic periods.

7.1 Earth History, Changing Environments, and Vertebrate Evolution

To understand the patterns of vertebrate evolution, it is important to realize that the world of today is very different from the world of times past. Our particular pattern of global climates, including such features as ice at the poles and the directions of major winds and water currents, results from the present-day positions of the continents. The world today is in general rather cold and dry in comparison with many past times. It is also unusual because the continents are widely separated from one another, and the main continental land mass is in the Northern Hemisphere. Neither of these conditions existed for most of vertebrate evolution.

The Earth's Time Scale and the Early History of the Continents

Vertebrates are known from the portion of Earth's history called the **Phanerozoic eon** (Greek *phanero* = visible and *zoo* = animal) in which we still live. The Phanerozoic began 540 million years ago, and contains the **Paleozoic** (Greek *paleo* = old), **Mesozoic** (Greek *meso* = middle), and **Cenozoic** (Greek *ceno* = recent) eras. Our own portion of time, the **Recent,** lies within the Cenozoic era. Each era contains a number of periods, and each period contains a number of epochs (see the figure on the front end paper). At least 99 percent of described fossil species occur in the Phanerozoic, although the oldest known fossils are from around 3.5 billion years ago and the origin of life is estimated to be around 4 billion years ago.

The time before the Phanerozoic is often loosely referred to as the **Precambrian,** because the Cambrian is the first period in the Paleozoic era. Thus the beginning of the Cambrian is the same as the beginning of the Phanerozoic, and the earlier part of Earth's history is indeed "before the Cambrian." This terminology minimizes the fact that the Precambrian actually represents seven-eighths of the entire history of the Earth! Precambrian time is better perceived as two eons, comparable to the Phanerozoic eon. The earliest eon is the Archaen (Greek *archeo* = first or beginning), commencing with the Earth's formation around 4.5 billion years ago. The later one is the Proterozoic (Greek *protero* = fomer), which began around 2.5 billion years ago. The start of the Proterozoic is marked by the appearance of the large continental blocks seen in today's world. (The pre-Proterozoic world would have looked rather like the South Pacific does now—many little volcanic islands separated by large tracts of ocean.)

Although life dates from the Archean, it is not until the Proterozoic that organisms more complex than bacteria are known, and multicellular organisms are not known until near the end of the eon about 1 billion years ago. The evolution of eukaryotic organisms, which depend on oxygen for respiration, followed shortly after the first appearance of atmospheric oxygen in the middle Proterozoic (around 2.2 billion years ago). By the start of the Phanerozoic a major biotic shift occurred—the evolution of forms capable of secreting calcareous (calcium-containing) skeletal parts.

Continental Drift—History of Ideas and Effects on Global Climate

The Earth's climate results from the interaction of sunlight, temperature, rainfall, evaporation, and wind during the annual passage of the Earth in its orbit around the Sun. Knowledge of paleoclimates helps us to understand the conditions under which plants and animals evolved, because climate profoundly affects the kinds of plants and animals that occupy an area. We discussed in Chapter 1 how the positions of the continents could influence global climates and oceanic circulation, and how mountain ranges could influence climate by exerting a rain shadow effect. The primary factors determining the climate of a particular portion of a continent include its latitude (i.e., how far north or south of the equator it is, because its position affects the amount of solar energy it receives); its proximity to an ocean (because water buffers temperature change and provides moisture); and the presence of barriers like mountains, which influence the movement of air masses.

Our understanding of the dynamic nature of the Earth and the variable nature of the Earth's climate over time are fairly recent occurrences. The notion of mobile continents, or **continental drift,** dates back to the middle of the nineteenth century, and was formally proposed by Alfred Wegener in 1924. The theory was not generally accepted at that time because no mechanism was known that could cause entire continents to move. It was not until the late 1960s, following new oceanographic research demonstrating the spreading of the seafloor as a plausible mechanism for continental movement, that the theory of **plate tectonics** became established. (Plate tectonics is essentially the same as continental drift, but focuses on the tectonic plates that continents sit on.) Even then, the theory was not universally accepted, and plate tectonics was not embraced by mainstream science until the 1970s.

Continents move because they float. Surface rocks are less dense than the underlying mantle rock, so continental blocks float in the mantle much as an ice cube floats in water. Heat in the Earth's core produces slow convective currents in the mantle. Upwelling plumes of molten basalt rise toward the Earth's surface, forming midoceanic ridges where they reach the top of the lithosphere (the rocky shell of the Earth) and spread horizontally (Figure 7–1). The seafloor is covered by a chain of midoceanic ridges that extend around the globe. The youngest seafloor crust is found in the centers of the ridges; moving away from the axis of the ridge, the seafloor becomes older. Subduction zones form where the lithosphere sinks back down into the mantle. The ocean floor is continuously renewed by this cycle of upwelling at midoceanic ridges and sinking back into the mantle at subduction zones, and rocks older than 200 million years do not occur anywhere on the ocean floor.

Movements of the tectonic plates are responsible for the fragmentation, coalescence, and refragmenta-

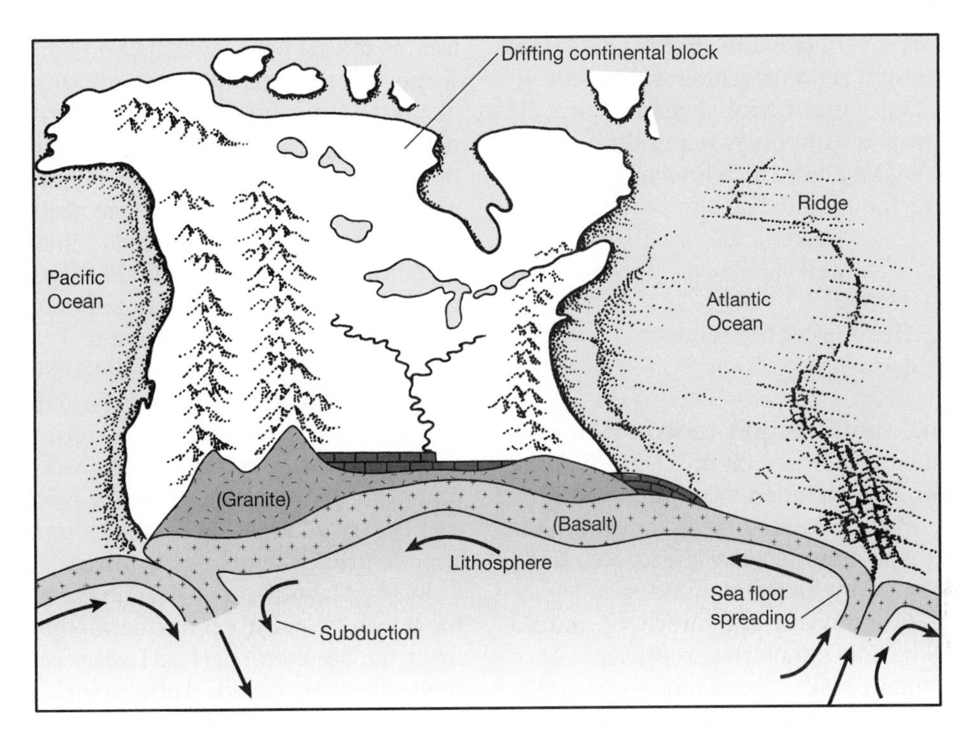

▲ Figure 7–1 Generalized geological structure of the North American continent. The continental blocks float on a basaltic crust (the bottom part of the picture shows a cross-sectional view of the continent). Arrows show the movements of crustal elements and the interactions with the mantle that produce continental drift.

tion of the continents that have occurred during the Earth's history. Plants and animals were carried along as continents slowly drifted, collided, and separated. When continents moved toward the poles, they carried organisms into cooler climates. As once-separate continents collided, terrestrial floras and faunas that had evolved in isolation mixed, and populations of marine organisms were separated. A recent example of this phenomenon is the joining of North and South America around 2.5 million years ago. (That is recent in geological terms!) The faunas and floras of the two continents mingled, which is why we now have armadillos (of South American origin) in Texas and deer (of North American origin) in Argentina. In contrast, the marine organisms originally found in the sea between North and South America, which were identical 3 million years ago, have been evolving in isolation since the land bridge formed and today are rather different from each other.

The position of continents affects the flow of ocean currents, and because ocean currents transport enormous quantities of heat, changes in their flow affect climates worldwide. For example, the breakup and northern movements of the continents in the late Mesozoic and Cenozoic eventually led to the isolation of the Arctic Ocean, with the formation of Arctic ice by the start of the Pliocene epoch, around 5 million years ago. The Arctic ice cap is not just a cold habitat in a more equable world. The appearance of the ice cap influenced global climatic conditions in a variety of ways, and the world today is colder and drier than it was before the Pliocene.

The Arctic ice cap also plays critical roles in the dynamic systems that influence today's global climates. Some researchers have proposed that should the world become warmer and the ice cap become smaller, parts of western Europe (such as England) might actually become *colder.* Arctic waters are a key part of the dynamic conveyer belt of global oceanic currents. This system of currents includes the Gulf Stream, which transports warm water from the equatorial Atlantic and the Gulf of Mexico across the north Atlantic Ocean to Europe. With a smaller ice cap this conveyer belt might be disrupted; and without the Gulf Stream, England would probably have the same cold climate as Newfoundland, which is at the same latitude (reviewed in Kunzig 1996). This example shows how very labile the Earth's climate is and how dependent it is on the configuration of continental masses that influence ice cover and oceanic currents. The British can only hope that global warming won't make England's climate a test case in demonstrating the validity of this particular hypothesis.

7.2 Continental Geography of the Paleozoic

The world of the early Paleozoic contained at least six major continental blocks (Figure 7–2). A large block called **Laurentia** included most of present-day North America, plus Greenland, Scotland, and part of northwestern Asia. Four smaller blocks contained other parts of what are now the Northern Hemisphere: Baltica—Scandinavia and much of central Europe; Kazakhstania—central southern Asia; Siberia—northeastern Asia; and China—Mongolia, North China, and Indochina. **Gondwana** (also known as Gondwanaland) included most of what is now the Southern Hemisphere (South America, Africa, Antarctica, and Australia) as well as India, Tibet, South China, Iran, Saudi Arabia, Turkey, southern Europe, and part of the southeastern United States.

In the Late Cambrian, the time when vertebrates probably first appeared, Gondwana and Laurentia straddled the equator; Siberia, Kazakhstania, and China were slightly to the south of the equator; and Baltica was positioned far to the south (Figure 7–2a). Note that the position of the modern continents within Gondwana was different from today; for example, Africa and South America appear to be upside down. Over the next hundred million years, Gondwana drifted south and rotated clockwise. By the Late Silurian the eastern portion of Gondwana was over the South Pole, and Africa and South America were in positions similar to those they assume today. Laurentia was still in approximately the same position, although it had rotated slightly counterclockwise. Baltica had by now moved north and collided with Laurentia, to form a united block called Laurussia. Kazakhstania, Siberia, and China had also moved north and were now situated in the Northern Hemisphere.

The most dramatic radiation of metazoan life had occurred by the start of the Cambrian. However, many of the groups known from the Cambrian were unique to that period and left no living descendants. A more profound, although less dramatic, radiation of marine animals occurred in the Ordovician. No new phyla appeared, but there was a near tripling in the number of families, and many of the groups that were to dominate the ecosystem of the rest of the Paleozoic appeared and radiated at this time. Although vertebrates date from the Early Cambrian, their major early diversification apparently took place during this Ordovician radiation.

From the Devonian through the Permian, the continents were drifting together. The continental blocks that correspond to parts of modern North America, Greenland, and western Europe had come into proximity along the equator (Figure 7–3). With the later addition of Siberia, these blocks formed a northern supercontinent known as **Laurasia**.

Most of Gondwana was in the far south, overlying the South Pole, but its northern edge was separated from the southern part of Laurentia only by a narrow extension of the **Tethys Sea.** The arm of the Tethys Sea between Laurentia and Gondwana did not close completely until the Late Carboniferous, when Africa moved northward to contact the east coast of North America (Figure 7–4). During the Carboniferous, the process of coalescence continued, and by the Permian most of the continental surface was united in a single continent, **Pangaea** (sometimes spelled Pangea). At its maximum extent, the land area of Pangaea covered 36 percent of the Earth's surface, compared with 31 percent for the present arrangement of continents. This supercontinent persisted for 160 million years, from the mid-Carboniferous to the mid-Jurassic, and profoundly influenced the evolution of terrestrial plants and animals.

7.3 Paleozoic Climates

During the early Paleozoic, sea levels were at or near an all-time high for the Phanerozoic, and atmospheric carbon dioxide levels were also apparently very high. There was a major glaciation in the Late Ordovician, which would have created cool overall global conditions. In the Silurian the ice sheets retreated and by the Late Silurian the extent of the shallow continental seas was reduced, restricting oceanic circulation. These climatic ameliorations may have set the scene for the development of the Late Silurian terrestrial ecosystems.

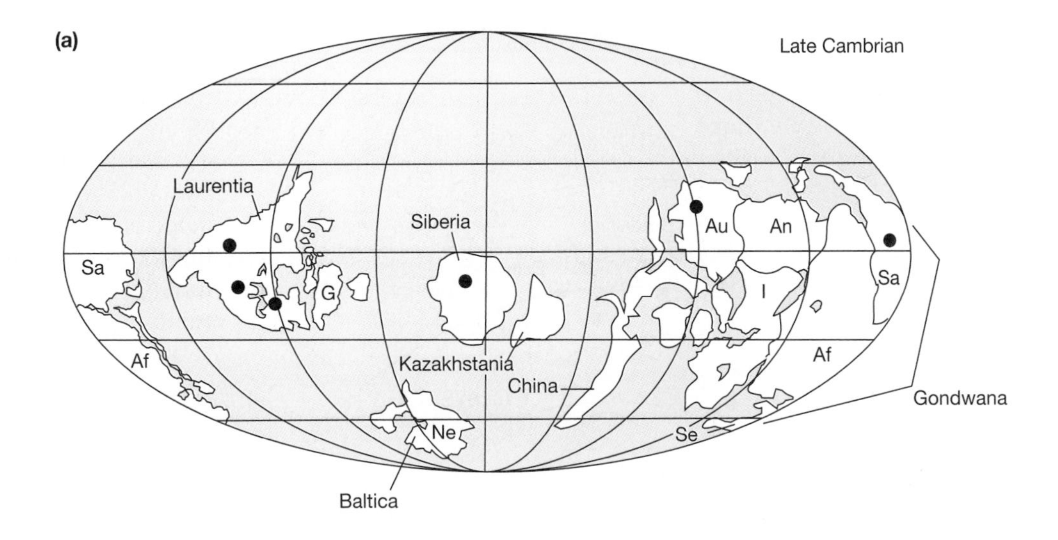

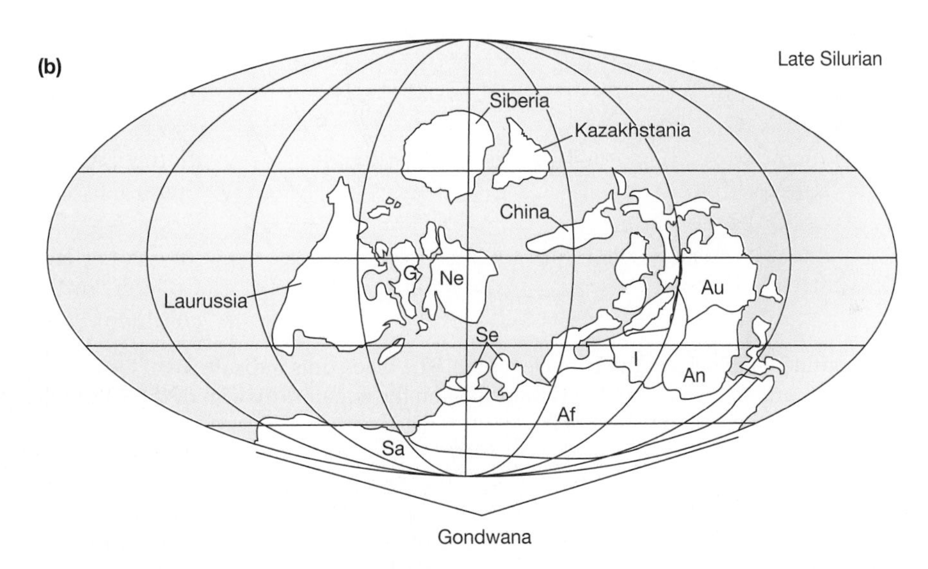

▲ Figure 7–2 Location of continental blocks in (a) the Late Cambrian and (b) the Late Silurian. The black dots in (a) indicate fossil localities where Ordovician vertebrates have been found. Positions of modern continents are indicated as follows: Af = Africa, An = Antarctica, Au = Australia, G = Greenland, I = India, Ne = northern Europe, Sa = South America, Se = southern Europe.

A relatively equable climate appears to have continued through the mid-Devonian. However, further glacial episodes characterized the late Paleozoic. Some glaciation was evident in the Late Devonian, and ice sheets covered much of Gondwana from the mid-Carboniferous until the mid-Permian. The waxing and waning of the glaciers created oscillations in sea level, which resulted in the cyclic formation of coal deposits of eastern North America, western Europe, and other parts of the Earth. The climate over Pangaea was fairly uniform in the Early Carboniferous, but in the Late Carboniferous and Early Permian it was highly differentiated as the result of glaciation, with significant regional differences in the flora. Most vertebrates were found in equatorial regions during this time.

The later Paleozoic glaciations may have been indirectly related to the rise of land plants. The spread

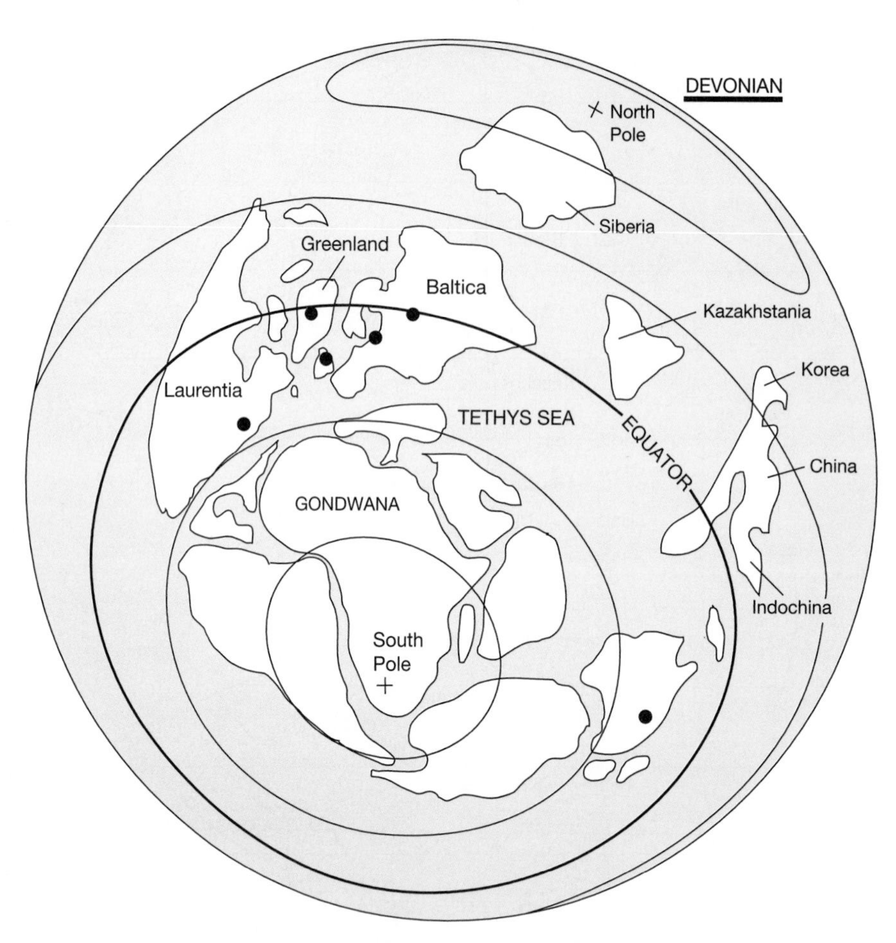

▲ Figure 7–3 Location of continental blocks in the Late Devonian. The black dots indicate fossil localities where Devonian tetrapods have been found. Laurentia, Greenland, and Baltica lie on the equator. An arm of the Tethys Sea extends westward between Gondwana and the northern continents.

of land plants in the Devonian would have profoundly affected the Earth's atmosphere and climate. Fossil soils with deep-rooted plants first appear in the Devonian, and the formation of soils would speed the breakdown of the underlying rocks as plant roots penetrated them and organic secretions and the decomposition of dead plant material dissolved minerals in the rock. The chemical processes of weathering trap atmospheric carbon dioxide, and evidence points to a sharp decrease in atmospheric carbon dioxide during the Devonian that may have been caused by the spread of land plants (Berner 1997). Atmospheric levels of carbon dioxide reached an extreme low during the Late Carboniferous and Early Permian, resembling the levels of today's world which, even now, are much lower than in most of the Phanerozoic. In contrast, oxygen levels were high during this time. The reverse greenhouse effect of this low atmospheric carbon dioxide probably caused the extensive Permo-Carboniferous glaciations.

7.4 Paleozoic Terrestrial Ecosystems

The evolution of terrestrial ecosystems has been traced throughout the Phanerozoic (reviewed by Behrensmeyer et al. 1992). Photosynthesizing bacteria (cyanobacteria) probably existed in wet terrestrial habitats from their origin in the Archean, and algae, lichens, and fungi probably occurred on land since the late Proterozoic. Fossilized soils from the Ordovician

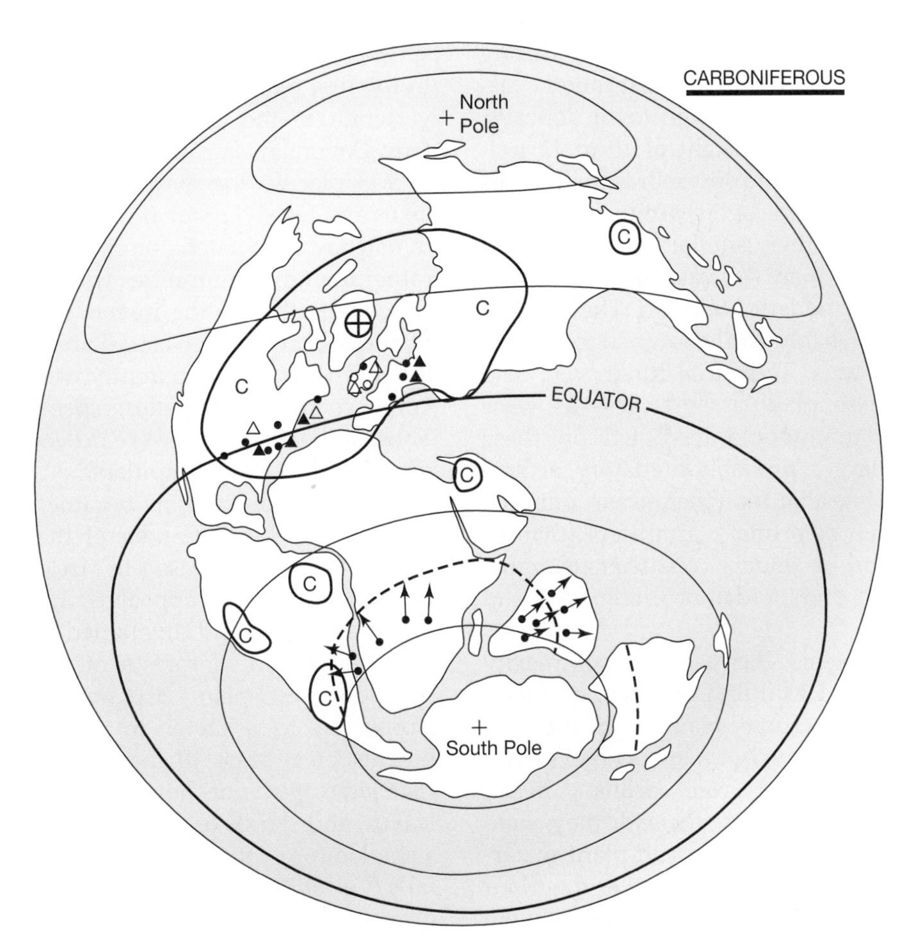

▲ Figure 7–4 Location of continental blocks in the Carboniferous. This map illustrates an early stage of Pangaea. The location and extent of continental glaciation in the Late Carboniferous is shown by the dashed lines and arrows radiating out from the South Pole. The extent of coal forests is marked by the heavy lines enclosing the "C"s. The small circles and triangles mark the locations of Carboniferous tetrapods. The triangles are the lepospondyls, and the circles are the other nonamniote tetrapods. Filled symbols represent Early Carboniferous localities; open symbols represent Late Carboniferous ones. The circle enclosing a cross marks the major Late Devonian tetrapod locality.

have mottled patterns that seem to indicate the presence of bacterial mats, and traces of erosion suggest that some of the soil surface was covered by algae, but there is no evidence of rooted plants. Land plants appear to represent a single terrestrial invasion from a particular group of green algae. The first major radiation of plants onto land probably took place in the Middle to Late Ordovician, as shown by fossilized plant spores, although we have no direct evidence (i.e., macrofossils) of land plants until the Late Silurian. These pioneers included bryophytes, represented now by mosses, liverworts, and hornworts. The landscape would have looked bleak by our standards—mostly barren, with a few kinds of low-growing vegetation limited to moist areas.

As was the case with the evolution of land vertebrates, land plants had to cope with the transition from water to land. The earliest land plants were small and simple, bearing a resemblance to the spore-producing phase of living primitive plants such as mosses. Adaptive responses to life on land included the evolution of an impenetrable outer surface (to prevent water loss), water-conducting internal tubes, and spore-bearing organs for reproduction.

The diversity of terrestrial life increased during the Silurian, and a rootless, leafless plant called *Cooksonia* was abundant in Late Silurian fossil deposits. *Cooksonia,* which grew to a height of 10 to 15 millimeters, consisted of a group of unbranched stems topped by pinhead-sized spore-producing structures. *Cooksonia* and other, similar Silurian plants were more primitive than the vascular plants that started to appear in the latest Silurian. The Silurian plant cover was probably fairly low, concentrated along river floodplains. Terrestrial fungi were also known among these plant assemblages—as were small arthropods that would have fed on these fungi—and some larger, probably predatory, arthropods. Thus by the latest Silurian there was a minimal terrestrial food web of primary producers (plants), decomposers (fungi), secondary consumers (fungus-eating arthropods), and predators (millipedes and scorpions).

Terrestrial ecosystems increased in complexity through the Early and Middle Devonian, but food webs remained simple. Today plants form the base of the terrestrial food chain, but there is no evidence that Devonian invertebrates were primary herbivores, feeding on living plants. Instead, they were probably detritivores, consuming dead plant material and fungi. This in turn would recycle the plant nutrients to the soil. Millipedes and scorpions were abundant, and springtails and mites were also present.

In the Early Devonian the land would still have appeared barren, although the changes that had occurred since the Silurian would be apparent. The diversity of plant species was greater than it had been in the Silurian, and increased heights were possible for the vascular plants (which could transport water from the site of uptake to other locations). By the Middle Devonian, these plants probably attained heights of 2 meters, and the canopy they created would have modified microclimatic conditions on the ground. Tree-like forms evolved independently among several ancient plant lineages, and by the Middle Devonian there were stratified forest communities. However, these plants were not related to modern trees and were not really like modern trees in their structure. It would not be possible to make furniture out of Devonian trees; they had narrow trunks and would not have provided enough woody tissue.

The terrestrial environment of the mid-Devonian had substantially more species of plants and animals than any previous period, but it differed in many respects from the modern ecosystems we are used to. In the first place, of course, there were no terrestrial vertebrates—the earliest of those appeared in the Late Devonian. Furthermore, plant life was limited to wet places—the margins of streams, rivers, and lakes and low-lying areas. In these areas plants grew in patches, each patch dominated by a single species, rather than in communities with a mixture of species as we see now. Flying insects were absent, as were plant-eating ones. Terrestrial animal communities in the mid-Devonian apparently were based on detritivores such as millipedes, springtails, and mites. Those animals in turn were preyed on by scorpions, pseudoscorpions, and spiders.

Terrestrial ecosystems became increasingly complex during the remainder of the Paleozoic. Plants diversified and increased in size, invertebrates with new specializations appeared, and terrestrial vertebrates appeared and diversified. The Late Devonian saw the spread of forests of the progymnosperm (primitive seed plant) *Archaeopteris,* large trees with trunks up to a meter in diameter and reaching heights of at least 10 meters. Most species of *Archaeopteris* put out horizontal branches that bore leaves, and stands of these trees would have created a shaded forest floor in low-lying areas. Giant horsetails (*Calamites*), relatives of the living horsetails that grow in moist areas today, reached heights of several meters. There were also many species of giant clubmosses (lycophytes), a few of which survive today as small ground plants. Both male and female lycophytes produced large cones, and spent most of their lives as unbranched trunks looking rather like telephone poles. Other areas were apparently covered by bush-like plants, vines, and low-growing ground cover.

The diversity and habitat specificity of Late Devonian floras continued to expand in the Carboniferous. Most of the preserved habitats represent swamp environments and vegetation buried in these swamps formed today's coal beds. (The word *Carboniferous* means "coal-bearing.") Most of the major taxonomic groups of plants evolved during this time, although the flowering seed plants (**angiosperms**) that dominate the world today were as yet unknown. Seed ferns (the extinct pteridosperms) and ferns (which survive today) lived in well-drained areas, and swamps were dominated by lycophytes, with horsetails, ferns, and seed ferns also present. Vegetation became structurally modern in form during the Early Carboniferous. Plants covered a variety of lowland habitats. There were forests consisting of trees of varying

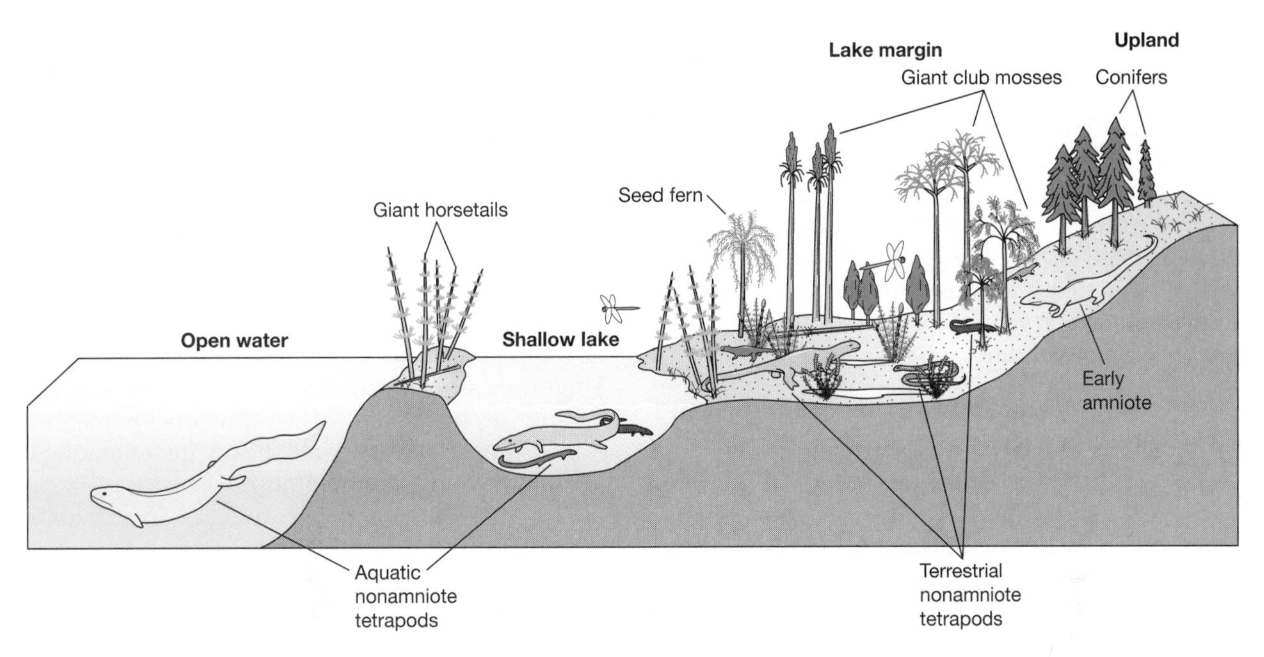

▲ Figure 7–5 A reconstuction of scene from a Late Carboniferous lake in Europe and its surroundings.

heights, giving stratification to the canopy, and vine-like plants hung from their branches. The terrestrial vegetation would have looked superficially like it does today, although the actual types of plants present were completely different.

The taxonomic composition of the plant communities changed somewhat in the Late Carboniferous and Permian. Seed plants such as conifers became an important component of the flora, and by the end of the Paleozoic they were the major group of plants. In the Late Carboniferous the more primitive spore-bearing plants that had been present earlier declined in numbers, and ferns became the dominant plants in upland habitats as well as lowland ones. In the Permian the newer, seed-plant-dominated flora spread worldwide, possibly responding to a greater diversity of drier, more upland types of habitats.

Terrestrial invertebrates diversified during the Carboniferous. Millipedes, arachnids, and insects were common. Flying insects are known from late in the Early Carboniferous. Detritivores continued to be an important part of the food web, but insect herbivory appears to have been well established by the end of the Carboniferous. Fossil leaves from the Late Carboniferous have ragged holes, seeds and wood are penetrated by tunnels, and pollen is found in the guts and feces of fossilized insects. High levels of atmospheric oxygen at this time enabled insects to attain much larger body sizes than seen today. Dragonflies (one species with a wingspan of 63 cm) flew through the air, and predatory arachnids such as scorpions and spiders prowled the forest floor. The extinct arthropleurids were terrestrial predators, with body lengths that reached nearly 2 meters.

Late Paleozoic arthropod communities were rather different from those of today. Predatory forms were larger, but species diversity was lower. Although spiders were abundant, most appear to have been burrow-building forms rather than web builders. The earliest evidence of silk production by a spider is from the mid-Devonian. New arthropod types appearing in the Permian include hemipterans (bugs), beetles, and forms resembling mosquitoes but not closely related to true mosquitoes.

Terrestrial vertebrates appeared in the Late Devonian and diversified during the Carboniferous. The first **amniotes** (tetrapods laying shelled eggs) were mid-Carboniferous in age, and by the Late Carboniferous amniotes had split into two major lineages—one leading to the mammals, and the other to modern reptiles and birds. The Carboniferous was dominated by a diversity of semiaquatic primitive tetrapods, but by the Permian the more terrestrially adapted amniotes were common, and many fossil vertebrate communities appear to represent upland habitats.

By the Early Permian several vertebrate lineages had given rise to small insectivorous predators, rather like modern salamanders and lizards. Larger vertebrates (up to 1.5m long) were probably predators of these small species, and still larger predators topped the food web. An important development in the Permian was the appearance of herbivorous vertebrates. For the first time vertebrates were able to exploit the primary production of terrestrial plants directly, and the food web was no longer based solely on invertebrates. By the end of the Permian the structure and function of terrestrial ecosystems were essentially modern, although the kinds of plants and animals in those ecosystems were almost entirely different from the ones we know today.

7.5 Early Paleozoic Extinctions

There was a major extinction event among marine invertebrates in the Ordovician, but the record of vertebrates from that time is too poor to know if this event affected them as well. The next major extinction, in the Late Devonian, had severe effects on marine vertebrates. Thirty-five families of fish (76 percent of the existing families) became extinct, including all of the remaining ostracoderms, most of the placoderms (with complete extinction of this group by the end of the Devonian), and many of the lobe-finned fishes.

Major extinctions occurred at the end of the Paleozoic in both terrestrial and marine environments. These were the most severe extinctions of the Phanerozoic, affecting approximately 57 percent of marine invertebrate families and 95 percent of all marine species, including 12 families of fishes (Erwin 1993). Theories for the cause of this extinction abound, including volcanic eruptions and global warming following the return of high levels of atmospheric carbon dioxide. High carbon dioxide could have led to stagnant oceanic circulation and/or toxic levels of dissolved carbon dioxide in water. These factors could have acted in concert. There were also significant casualties on land: Twenty-seven families of tetrapods (49 percent) became extinct, with especially heavy losses among the mammal-like reptiles (Benton 1989).

A significant extinction among vertebrates occurred at the end of the Early Permian. This extinction was not paralleled in the marine invertebrate record. On land, 15 families of tetrapods became extinct, including many nonamniote tetrapods ("amphibians" in the broad sense) and pelycosaurs (mammal-like reptiles). These extinctions may have been related to climatic changes associated with the end of the Permo-Carboniferous period of glaciation, and perhaps also to the accompanying changes in the atmosphere, with a decrease in the levels of oxygen and an increase in the levels of carbon dioxide.

Additional Readings

Behrensmeyer, A. K. et al. (Eds.) 1992. *Terrestrial Ecosystems through Time.* Chicago: University of Chicago Press.

Benton, M. J. 1989. Patterns of evolution and extinction in vertebrates. In K. C. Allen and D. E. G. Briggs (Eds.), *Evolution and the Fossil Record.* London, U.K.: Bellhaven Press.

Berner, R. A. 1997. The rise of plants and their effect on weathering and atmospheric CO_2. *Science* 276:544–545.

Cox, C. B., and P. D. Moore. 1993. *Biogeography: An Ecological and Evolutionary Approach,* 5th ed. Oxford, U.K.: Blackwell Scientific.

Crowley, T. J., and G. R. North. 1991. *Paleoclimatology.* Oxford, U.K.: Oxford University Press.

Droser, M. L., R. A. Fortey, and X. Li. 1996. The Ordovician radiation. *American Scientist* 84:122–131.

Erwin, D. H. 1993. *The Great Paleozoic Crisis.* New York: Columbia University Press.

Gould, S. J. 1989. Wonderful Life: *The Burgess Shale and the Nature of History.* New York: Norton.

Hallam, A. 1994. *An Outline of Phanerozoic Biogeography.* Oxford, U.K.: Oxford University Press.

Holland, H. D. 1984. *The Chemical Evolution of the Atmosphere and Oceans.* Princeton, N.J.: Princeton University Press.

Kenrick, P., and P. R. Crane. 1991. The origin and early evolution of plants on land. *Nature* 389:33–39.

Knoll, A. et al. 1996. Comparative Earth history and Late Permian mass extinction. *Science* 273:452–451.

Kunzig, R. 1996. In deep water. *Discover* 11(2):86–96.

Retallack, G. J. 1997. Early forest soils and their role in Devonian climate change. *Science* 276:583–585.

Shear, W. A., and J. Kukalová–Peck. 1990. The ecology of Paleozoic terrestrial arthropods: The fossil evidence. *Canadian Journal of Zoology* 68:1801–1834.

Ward, P. D. 1998. The Greenhouse extinction. *Discover* 19(6): 54–58.

Web Explorations

On-line resources for this chapter are on the World Wide Web at: http://www.prenhall.com/pough (click on the Table of Contents link and then select Chapter 7).

Terrestrial Ectotherms Amphibians, Turtles, Lepidosaurs, and Archosaurs

PART III

The spread of plants and then invertebrates across the land provided a new habitat for vertebrates. The evolutionary transition from water to land is complex because water and air have such different properties: Aquatic animals are supported by water, whereas terrestrial animals need skeletons and limbs for support. Aquatic animals extract oxygen from a unidirectional flow of water across the gills, whereas terrestrial animals breathe air that they must pump in and out of sac-like lungs. Aquatic animals face problems of water and ion balance as the result of osmotic flow, whereas terrestrial animals lose water by evaporation. Even sensory systems like eyes and ears work differently in water and air. The transition from aquatic to terrestrial habitats must have been facilitated by characteristics of fishes that were functional both in water and in air, although the functions may not have been exactly the same in the two fluids.

Once they were in terrestrial habitats, vertebrates radiated into some of the most remarkable animals that have ever lived. The dinosaurs are the best known of these, but many smaller groups contained forms that were just as bizarre, although not as large as many of the dinosaurs. One contribution of phylogenetic systematics has been the emphasis it has placed on the close evolutionary relationship of birds, crocodilians, and dinosaurs. This perspective indicates that the complex behaviors we consider normal for birds are ancestral characters of their lineage. Living crocodilians display parental care that is quite like that of birds (allowing for the differences in the size and body form of birds and crocodilians), and evidence is accumulating that at least some dinosaurs also showed extensive parental care and probably other behaviors we now associate with birds.

The distinction between ectotherms (animals that obtain the heat needed to raise their body temperature from outside their bodies) and endotherms (animals that use metabolic heat production for thermoregulation) has important functional considerations that cut across phylogenetic lines. The relationship between an ectothermal organism and its physical environment (solar radiation, air temperature, wind speed, and humidity) is often an important factor in its ecology and behavior. One consequence of relying on outside sources of energy to raise body temperature is efficient use of metabolic energy, and ectotherms transform a high proportion of the food they eat into their own body tissue. This characteristic gives them a unique position in the flow of energy through terrestrial ecosystems.

In this part of the book we describe the radiation of vertebrates into terrestrial habitats in the Paleozoic and Mesozoic and various theories to account for the extinction of dinosaurs and for mass extinctions generally, and we consider the advantages and disadvantages of being an ectotherm in the modern world.

CHAPTER

8 Origin and Radiation of Tetrapods and Modifications for Life on Land

By the Middle Devonian the stage was set for the appearance of terrestrial vertebrates, and the first vertebrate to venture onto the land was a sarcopterygian. The sarcopterygian fishes, introduced with the other bony fishes in Chapter 6, did not undergo the extensive evolutionary radiations seen among the ray-finned fishes. Indeed, the only surviving sarcopterygian fishes are the lungfishes and the coelacanth. However, all the terrestrial vertebrates are sarcopterygians.

The demands of life on land are quite different from those in the aquatic environment, and many changes were needed in the anatomical structures and the physiological functions of the early terrestrial vertebrates. These changes required for life on land were barely complete when some lineages of tetrapods became secondarily aquatic, returning to freshwater or marine habitats. However, other lineages became increasingly specialized for terrestrial life, showing progressive changes in the jaws that allowed new ways of feeding and changes in the limbs that appear to have increased the agility of terrestrial locomotion. These structural changes were widespread; but only one of the terrestrial lineages of Paleozoic tetrapods made the next major transition in vertebrate history, developing the embryonic membranes that define the amniote vertebrates. Amniote diversification shows an initial early split between the synapsids, the lineage that includes mammals, and the sauropsids, the lineage that includes reptiles and birds.

8.1 Modifications for Life on Land

Life on the land presents different challenges and demands than life in the water. Perhaps most important, terrestrial vertebrates must deal with the demands of gravity. Gravity has little significance for a fish living in water. The bodies of vertebrates are approximately the same density as water, and hence they are buoyed up by the water. However, vertebrates are much denser than air, and gravity is a very important factor. Thus the skeletal system of tetrapods must be designed differently from that of fishes, and the modes of locomotion are different.

Some physiological processes change as well. The gills of fishes are the sites of exchange with the environment: for respiratory gases (oxygen and carbon dioxide loss) and monovalent ions such as sodium and chloride, and for the excretion of nitrogen as ammonia. Gills do not work in air, because they clump together and the effective surface area for exchange is greatly reduced. Thus in tetrapods these functions must be taken over by other organs—the lungs and the kidneys.

Another difference between fish and tetrapods is found in the storage of fat. Adipose tissue, where fat is stored in specialized cells called adipocytes, is mainly a tetrapod feature. In most fishes fats are stored as lipid droplets in the liver and the muscles. Note that the type of fat deposition seen in humans—subcutaneous layers and in association with the muscles, gut, heart, and other organs—is seen only in mammals and birds. Reptiles and amphibians store fat in discrete abdominal fat bodies, and around the tail base.

Figure 8–1 illustrates some differences between the fishes ancestral to tetrapods, early primitive tetrapods, and more derived tetrapods such as primitive amniotes.

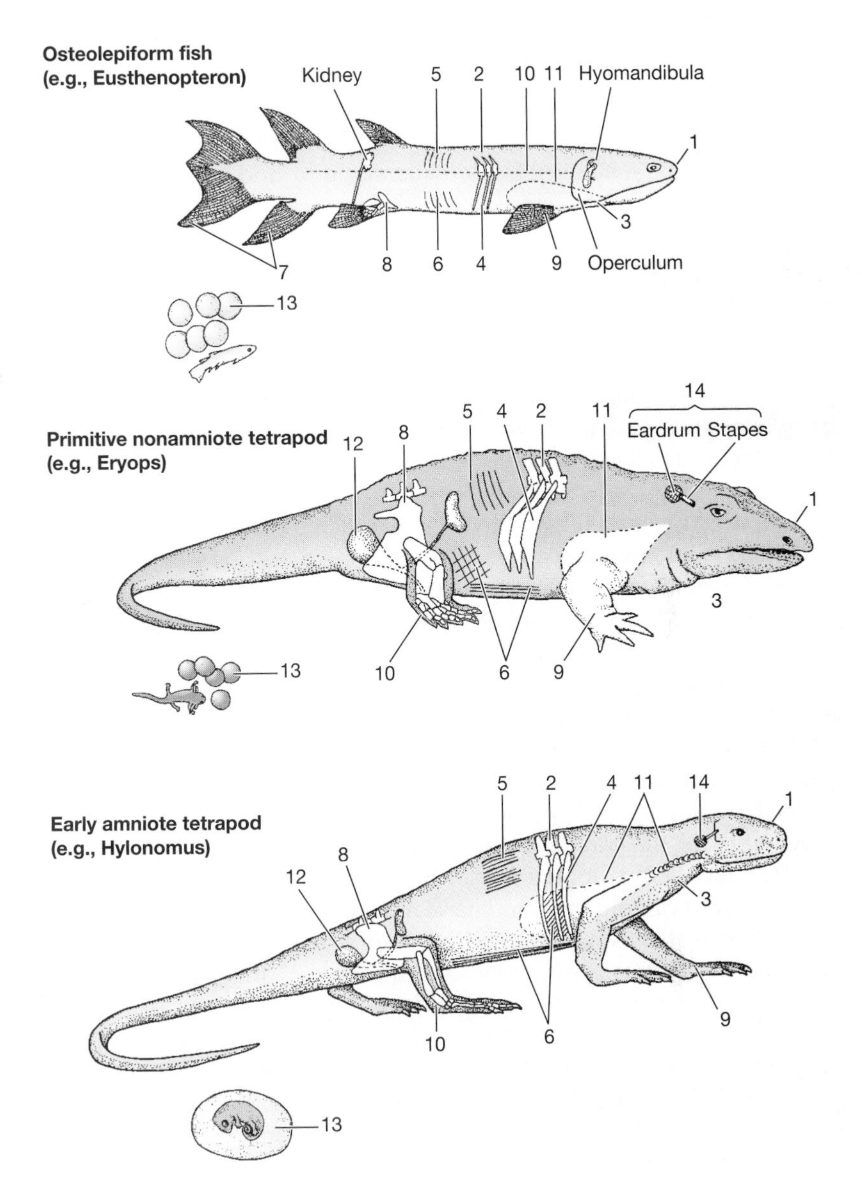

▶ Figure 8–1 Morphological and physiological differences among fishes, primitive tetrapods, and amniotes: Numbers indicate the various systems that are referred to in text. 1. Length of snout. 2. Interlocking of vertebral column. 3. Length of neck. 4. Form of ribs. 5. Differentiation of epaxial muscles. 6. Differentiation of hypaxial muscles. 7. Presence of midline fins. 8. Attachment of pelvic girdle to vertebral column. 9. Form of the limbs. 10. Form of the ankle joint. 11. Form of lungs and trachea. 12. Presence of urinary bladder. Note that the kidneys of fishes and nonamniote tetrapods are in fact elongated structures lying along the dorsal body wall. The kidneys have been portrayed in all the animals as a mammalian bean-shaped form, for familiarity and convenience. 13. Mode of reproduction. 14. Form of the acousticolateralis system and middle ear.

Skeletomuscular System

The Craniodental System—Skull and Teeth The skull of early tetrapods is much like that of primitive bony fishes, with an extensive dermatocranium that is retained in most extant tetrapods. However, the gill skeleton, the operculum, and most of the bones connecting the operculum to the pectoral girdle are lost in all but the very earliest known tetrapods. The skull of bony fishes has a short snout, and movements of the jaws cause water to be sucked into the mouth for gill ventilation and feeding. Suction feeding is not an option on land, because air is much less dense than the food particles. (You can suck up the noodles in soup along with the liquid, but you're unable to suck in the same noodles placed on the side of the plate.) On land the head must be moved over the prey, and in tetrapods the skull is greatly lengthened (Figure 8–1[1]). Early tetrapods had wide, flat skulls that were probably associated with buccal pumping for lung ventilation.

The tongue of fishes is small and bony, whereas the tongue of tetrapods is large and muscular. The tetrapod tongue works in concert with the hyoid apparatus, the lower part of the hyoid arch that was integrated with the gill arches in fishes, and is probably a key innovation for feeding on land. Most tetrapods use the tongue to manipulate food in the mouth and transport it to the pharynx. In addition, some tetrapods—such as frogs, salamanders, and true chameleon lizards—can project their tongue to capture prey.

Salivary glands are known only in terrestrial vertebrates, probably because lubrication is required to swallow food on land. Saliva contains enzymes that begin the chemical digestion of food while it is still in the mouth. Some insectivorous mammals, two species of lizards, and several lineages of snakes have elaborated these secretions into venoms that kill prey.

With the loss of the gills in tetrapods, much of the associated branchiomeric musculature is also lost, but the gill levators are a prominent exception. In fishes these muscles are combined into a single unit, the cucullaris, and this muscle in tetrapods becomes the **trapezius**, which runs from the top of the neck and shoulders to the shoulder girdle. In mammals this muscle helps to rotate and stabilize the scapula (shoulder blades) in locomotion, and we use it when we shrug our shoulders. Understanding the original homologies of the trapezius muscle explains an interesting fact about human spinal injuries. Because the trapezius is an old branchiomeric muscle, it is innervated directly from the brain by cranial nerves, not from the nerves exiting from the spinal cord in the neck. Thus people who are paralyzed from the neck down by a spinal injury can still shrug their shoulders. Small muscles in the throat, for example those powering the larynx and the vocal cords, are other remnants of the branchiomeric muscles associated with the gill arches. Ingenious biomedical engineering allows paraplegic individuals to use this remaining muscle function to control prosthetic devices.

The major branchiomeric muscles of tetrapods are associated with the mandibular and hyoid arches, and are now solely involved in feeding (Figure 8–2). The adductor mandibularis remains the major jaw-closing muscle, and it becomes increasingly complex in more derived tetrapods. The hyoid musculature forms two new important muscles in tetrapods. One is the depressor mandibulae, running from the back of the jaw to the skull and helping the hypobranchials to open the mouth. The other is the constrictor colli (also known as the sphincter colli) that surrounds the neck (as its name suggests) and acts primarily in swallowing food.

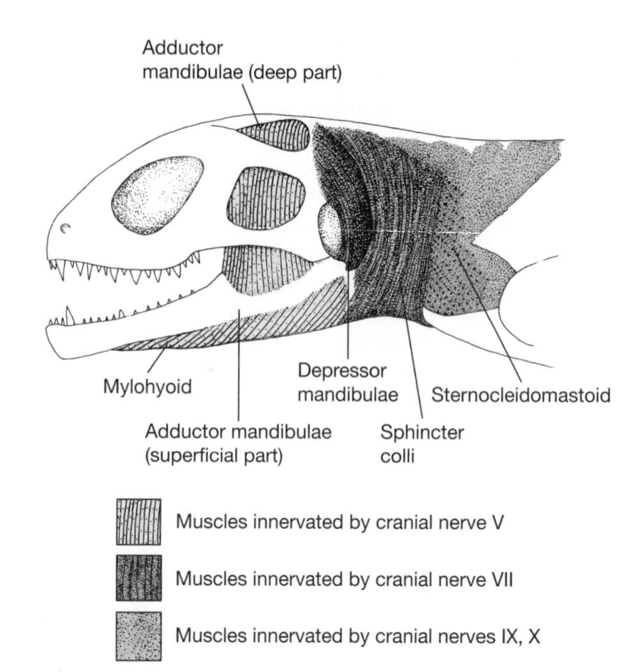

▲ Figure 8–2 Form of head and neck musculature in a generalized tetrapod condition (the reptile *Sphenodon*).

The Axial System While the axial skeletomuscular system (vertebrae, ribs, and axial muscles) of fishes has a purely locomotor function, the axial skeletomuscular system of tetrapods must also support the body on land. The vertebral column suspends the weight of the body underneath it, the ribs help to protect the viscera and prevent them from sagging, and the axial musculature is involved in postural control as well as in locomotion.

The vertebrae of tetrapods interlock by means of processes called **zygapopheses** (Figure 8–1[2], Figure 8–3). Zygapophyses transform the vertebral column into a stiff rod that withstands twisting (torsion) as well as bending (compression), and allows the spine to act like a suspension bridge to support the weight of the viscera on land (Figure 8–4). Tetrapods that have permanently returned to the water—whales and Mesozoic reptiles such as ichthyosaurs and plesiosaurs, for example—have secondarily lost the zygapophyses because their vertebral columns no longer need to support their body weight.

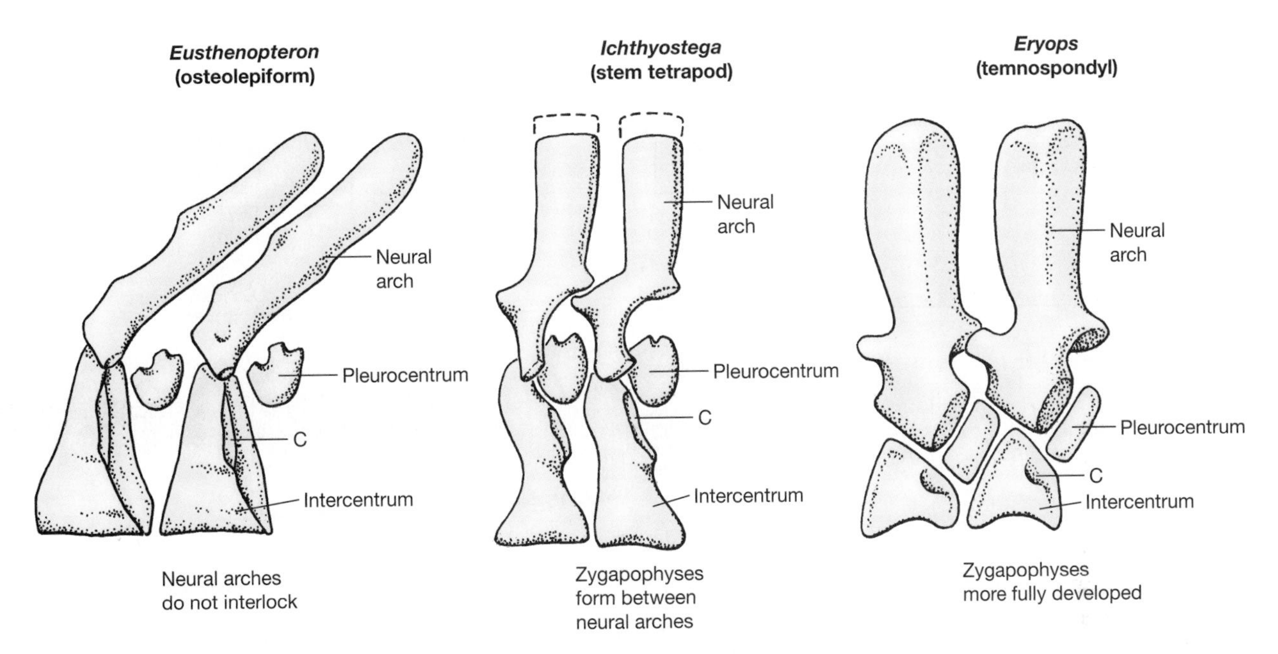

▲ Figure 8–3 Vertebral structure of sarcopterygian fishes and primitive tetrapods (anterior is to the left): c = capitulum (articulation for head of rib).

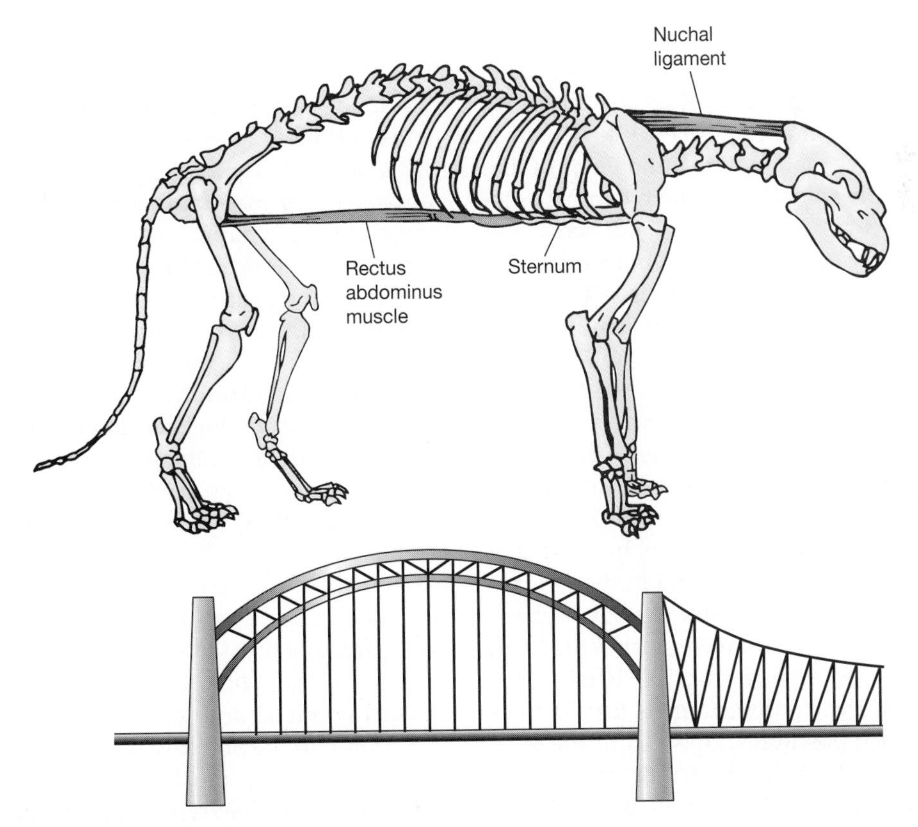

▲ Figure 8–4 Skeleton of a cat, showing muscles and ligaments of postural support, in comparison with a suspension bridge.

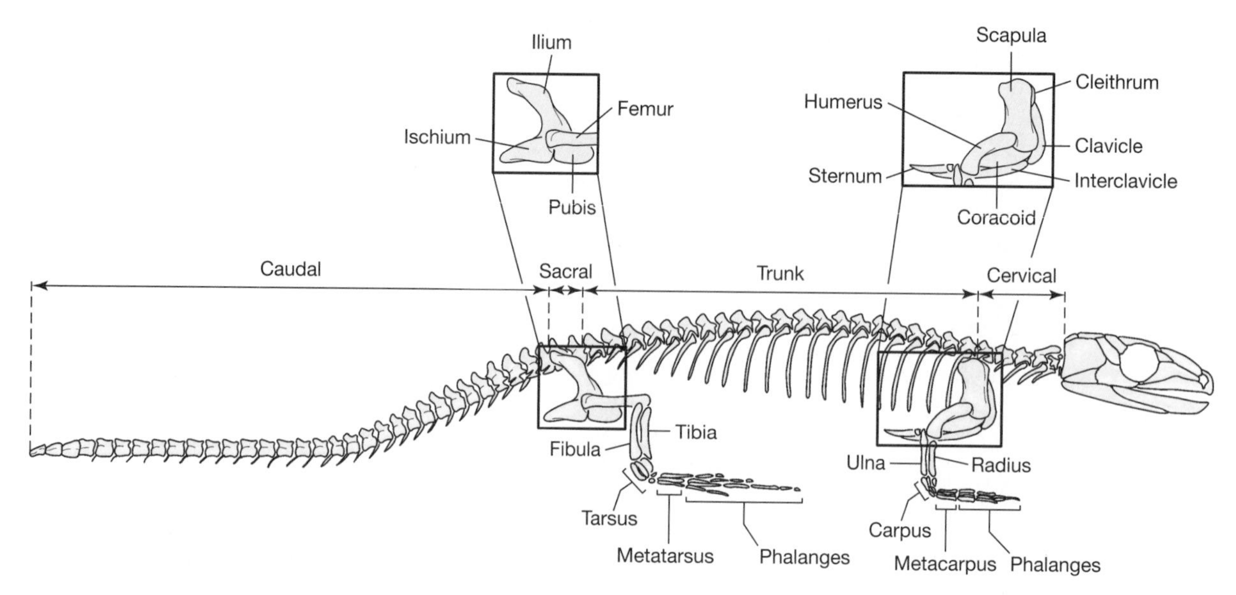

▲ Figure 8–5 Generalized tetrapod skeleton, as illustrated by the primitive amniote *Hylonomus*.

The trend toward differentiation of the axial skeleton into four or five regions from anterior to posterior can be seen in tetrapods (Figure 8–5). With the loss of the operculum of the bony fishes, which connected the head to the pectoral girdle, tetrapods have a distinct neck region (Figure 8–1[3]). The neck or **cervical** vertebrae enhance the mobility of the head and protect the spinal cord. The two most anterior cervical vertebrae are the **atlas** and **axis,** and they are most highly differentiated in mammals. The **trunk** vertebrae are in the middle region of the body: In mammals these are further differentiated into **thoracic** vertebrae (those that bear ribs) and **lumbar** vertebrae (those that have lost ribs). The **sacral** vertebrae, which are derived from the trunk vertebrae, fuse with the pelvic girdle and transfer force to the appendicular skeleton. Early tetrapods and living amphibians have a single sacral vertebra, mammals have three to five, and some dinosaurs had a dozen or more. The **caudal** vertebrae, found in the tail, are usually simpler in structure than the trunk vertebrae

The ribs of early tetrapods were fairly stout and more prominently developed than in fishes. They may have stiffened the trunk in animals that had not yet developed much postural support from the axial musculature (Figure 8–1[4]). The trunk ribs are the most prominent ones in tetrapods in general, and many primitive tetrapods have small ribs extending throughout the entire vertebral column. Modern amphibians have almost entirely lost their ribs, and ribs are confined to the anterior trunk (thoracic) vertebrae in the mammalian skeleton.

The axial muscles assume new roles in tetrapods—postural support of the body and ventilation of the lungs—and become increasingly differentiated in structure and function. Muscles are important for maintaining posture on land, because the body is not supported by water; without muscular action, the skeleton would buckle and collapse. Likewise, the process of ventilating the lungs is very different if the chest is surrounded by air rather than by water. The axial muscles still participate in locomotion in primitive tetrapods, producing the lateral flexion of the backbone seen during movement in many amphibians and reptiles. However, in birds and mammals the locomotory role of the axial muscles has largely been replaced by limb movements.

In fishes and modern amphibians, the epaxial muscles form an undifferentiated single mass. This was probably their condition in the earliest tetrapods (Figure 8–1[5]). In unspecialized amphibians, such as salamanders, both epaxial and hypaxial muscles contribute to the bending motions of the trunk while walking on land, much as they do in fishes swimming in water. The hypaxial muscles of all tetrapods show a more complex pattern of differentiation into layers than is seen in fishes. The hypaxials form two layers in bony fishes (the external and internal obliques), but in tetrapods a third inner layer is added, the transversus abdominus (Figure 8–1[6]). This muscle is responsible for exhalation of air from the lungs of modern amphibians (which do not use their ribs to breathe), and it may have been an essential acquisition for respiration on land by early tetrapods. Air-breathing fish use the pressure of the water column on the body to force air from the lungs, but land-dwelling tetrapods need muscular action.

The rectus abdominus, which runs along the ventral surface from the pectoral girdle to the pelvic girdle, is another new hypaxial muscle in tetrapods. This is the muscle responsible for the washboard stomach of human body builders, and its role appears to be primarily postural. The costal muscles in the rib cage of amniotes are formed by all three layers of the hypaxial muscles, and are responsible for inhalation as well as for exhalation (Figure 8–1[6]). Amniote epaxial muscles are also distinctly differentiated into three major components (Figure 8–1[5]), and their primary role is now postural rather than locomotory. The use of the ribs and their associated musculature as devices to ventilate the lungs was probably an amniote innovation.

Thus in tetrapods the axial skeleton and its muscles assume very different roles from their original functions in aquatic vertebrates. The skeleton now participates in postural support and ventilation of the lungs, as well as in locomotion, and some of these functions are incompatible. For example, the bending of the trunk when a lizard runs means that it has difficulty in using its ribs for lung ventilation, creating a conflict between locomotory and respiratory functions of the axial skeleton and muscles. More derived tetrapods such as mammals and birds have addressed this conflict by a change in posture from sprawling limbs to limbs that are held more directly underneath the body. These tetrapods are propelled by limb movements rather than by trunk bending.

The Appendicular System Tetrapods have lost the median fins of fishes, although remnants of a tail fin are seen in the earliest tetrapods (Figure 8–1[7]). While sarcopterygian fishes have bones in their fins, they have no structures that are equivalent to wrists, ankles, hands, or feet. The basic form of the tetrapod skeleton is illustrated in Figure 8–5.

In tetrapods the pelvic girdle is fused directly to modified sacral vertebrae, and the hindlimbs are the primary propulsive mechanism. The pelvic girdle is tripartite, consisting of the **ilium** (plural *ilia*), **pubis**, and **ischium**. The ilia on each side connect the pelvic limbs to the vertebral column, forming an attachment at the sacrum via one or more modified ribs (Figure 8–1[8]).

The pectoral girdle is freed from the dermal skull roof with the loss of the operculum and the opercular series of bones of bony fishes. The main endochondral bones are the **scapula** and the **coracoid**; the **humerus** (upper arm bone) articulates with the pectoral girdle where these two bones meet. However, some dermal post-opercular bones (the cleithrum, the clavicle, and the interclavicle) become incorporated into the pectoral girdle. The cleithrum runs along the anterior border of the scapula, and is seen only in extinct primitive tetrapods. The **clavicle** (the collar bone of humans) connects the scapula to the **sternum** or to the interclavicle. The interclavicle is a single, midline element lying ventral to the sternum. It has been lost in birds and in most mammals, but is still present in the monotremes (egg-laying mammals).

The pectoral girdle does not articulate directly with the vertebral column; only in pterosaurs (extinct flying reptiles) is there an equivalent of a sacrum in the anterior vertebral column (the notarium). In all other vertebrates the connection between the pectoral girdle and the vertebral column is indirect via the sternum and the ribs. The sternum is a midventral structure, usually segmented, that links the lower ends of right and left thoracic ribs in amniotes. Its evolutionary origins are obscure. The sternum is extensively ossified only in birds and mammals.

The basic tetrapod limb is made up of the limb girdle and five segments articulating end to end. In fishes mobility is limited within the joints of the fin, and fin rays are present. All tetrapods have jointed limbs (with a forwardly pointing knee and a backwardly pointing elbow), wrist/ankle joints, and hands and feet with digits (Figure 8–1[9]). The fin rays are lost. The feet of primitive tetrapods are used mainly as holdfasts, providing frictional contact with the ground. Propulsion is mainly generated by the body axial musculature. The feet of amniotes are used as levers to propel the animal, and the ankle forms a distinct hinge joint (mesotarsal joint; Figure 8–1[10]). (Some non-amniotes, e.g., frogs, parallel amniotes in this condition.)

In fishes the limbs are not usually used for propulsion (except in rays and some coral-reef teleosts), but they are used for lift, steering, and braking. The original form of vertebrate appendicular muscles, as seen today in sharks, is fairly simple: a set of muscles to lift up the fins and draw them outward and backward (pectoral and pelvic levators) and another set of muscles to move the fin down, in, and forward (pectoral and pelvic depressors).

In tetrapods the limbs are increasingly used for propulsion in more derived forms. The appendicular muscles have become increasingly complicated and differentiated, but the old pattern of a major levator and a major depressor is still retained. In our arms (= pectoral fins) the levator corresponds to the deltoids (running from the scapula over the shoulder down to the upper arm), and the depressor corresponds to the pectoralis (running from the upper arm to the chest). But we have many additional muscles in the shoulder region alone—not to mention the ones that move the elbow, the wrist, and the multitude of complex muscles that move the fingers in the hand.

Locomotion A fish-like mode of swimming works only in a dense medium, and energy must be expended to overcome the drag produced by moving though a dense fluid. Overcoming drag from the surrounding medium is not a problem in air, but friction must be generated between the feet and the ground for propulsion (think of the difficulty of walking on ice where there is very little friction between your foot and the substrate). Locomotion on land is energetically more expensive than in water. Providing an animal is suitably modified for the particular mode of locomotion (e.g., streamlined for swimming), travel for a given distance is most expensive for a walker, less expensive for a flier, and least expensive for a swimmer. The expense of terrestrial locomotion poses another challenge to tetrapods.

Figure 8–6 illustrates various modes of locomotion by tetrapods from a phylogenetic perspective. The basic form of tetrapod limb movement consists of moving diagonal pairs of legs together. The right front and left hind move as one unit, and the left front and right hind move as another, in a type of gait known as the walking-trot. Even though humans are bipedal, relying entirely on the hind legs for locomotion, we retain this primitive coupling of the limbs in walking, swinging the right arm forward when striding with the left leg and vice versa. This type of coupled, diagonally paired limb movement is probably a primitive feature for gnathostomes, since sharks also move their fins in this fashion when bottom-walking over the submerged substrate.

The initial mode of tetrapod locomotion, as seen today in salamanders, probably combined axial flexion of the body with limbs moving in this walking-trot fashion, and with the feet acting primarily as holdfasts on the substrate rather than propelling the body. (That is, the force used to move the animal came from the trunk muscles rather than from the limb muscles). Lizards retain a modified version of this mode of locomotion, although their limbs are more important for propulsion. The swimming, walking, and jumping modes of locomotion by frogs rely on limb muscles only, and are highly specialized.

A new specialization of amniotes is the walk, which is different from the more primitive walking-trot gait. In walking, each leg moves independently in succession, usually with three feet being on the ground at any one time. However, it is possible for mammals to employ a speeded-up walk in which only one foot or two feet may be on the ground at any time, as seen in the **amble** of elephants and some horses (Figure 8–6o). All amniotes can also use a form of the **trot** for faster movement—by moving diagonal pairs of limbs together, as in the primitive tetrapod condition.

More derived amniotes such as mammals and archosaurs (birds, dinosaurs, and crocodiles) are specialized compared to the primitive, sprawling tetrapod condition. These amniotes have an upright posture, and hold their limbs more directly underneath the body. While archosaurs tended toward bipedality, mammals devised some new modes of locomotion with the evolution of the dorsoventral flexion of the vertebral column. The characteristic new fast gait of mammals is the **bound** (Figure 8–6k), which involves jumping off the hind legs and landing on the fore legs, with the flexion of the back contributing to the length of the stride.

If you watch domestic pets, you can see that cats usually bound and dogs usually gallop, although many animals larger than dogs (such as cheetahs and antelope) use the bound. Larger mammals must move more cautiously to avoid injury (Box 8.1), and the bound is modified into the **gallop**, as seen for example in horses. (The canter gait of horses is essentially a slow version of the gallop.) Here the period of suspension in the air with all four feet off the ground is not with legs stretched out in midleap, as in the bound, but in the bunched-up recovery phase where the hind legs move forward for the next stride. Galloping also involves less bending of the back than does bounding (see Figure 8–6l and Fig-ure 8–8).

Another gait that is typical of large mammals is the trot, used at medium speeds between the walk and the bound. Basically the trot is diagonal pairs of limbs acting in sequence. But unlike the walking-trot, the trot is a distinct jump from one pair of legs to the other with a period where all four legs are off the ground (see Figure 8–6m). A more specialized type of mammalian locomotion is the **ricochet**, or bipedal hopping, probably derived from the bound at some evolutionary point (Figure 8–6n). Although kangaroos are famous for this type of gait, several rodents (e.g., kangaroo rats, jerboas, and spring hares) have also, independently of one another, evolved this locomotory mode. Note that the mode of human locomotion, bipedal striding with an upright trunk (Figure 8–6p), is unique among vertebrates (although penguins waddle with an upright trunk).

Respiratory System

The lung is a primitive feature of bony fishes, and so lungs were not evolved for the express purpose of breathing on land. For many years it was assumed lungs evolved in fishes living in stagnant, oxygen-depleted water where gulping oxygen-rich air would supplement oxygen uptake by the gills. However, although some lungfishes are found in stagnant, anoxic environments, other air-breathing fishes (e.g., the bowfin) are active animals found in oxygen-rich habitats. Recent work by Colleen Farmer (Farmer 1997) provided an alternative explanation for the evolution of lungs. She suggested that air breathing could have evolved in well-aerated waters, in active fishes where the additional oxygen is needed primarily to supply the heart muscle itself, rather than the body tissues.

Fishes and amphibians use the hyoid apparatus to ventilate the lungs. The oral cavity is expanded, sucking air into the mouth, and then the floor of the mouth is raised, squeezing the air into the lungs. This method of lung ventilation is called a pulse pump or positive pressure mechanism, also known as buccal pumping. Amniotes use an aspiration pump or negative pressure mechanism of lung ventilation via the ribs (costal aspiration). Expansion of the rib cage by the intercostal hypaxial muscles creates a negative pressure (that is, below atmospheric pressure) in the pleuroperitoneal (abdominal) cavity and sucks air into the lungs. Air is expelled (exhaled) by compressing the abdominal cavity, primarily through elastic return of the rib cage to a resting position and contraction of the elastic lungs, as well as by contraction of the transversus abdominus muscle.

The lungs of many modern amphibians are simple sacs with few internal divisions. They have only a simple short orobranchial chamber leading directly into the lungs, with little development of a distinct trachea. In contrast, amniotes have lungs that are subdivided, sometimes in very complex ways. They also have a long **trachea** (windpipe), strengthened by cartilage rings, that branches into a series of bronchii in each lung (Figure 8–1[11]). The form of lung complication is somewhat different in mammals and other amniotes, suggesting independent evolutionary innovation. The combination of a trachea and costal aspiration allows many amniotes to develop longer necks than those seen in modern amphibians or in extinct nonamniote tetrapods. Amniotes also possess a **larynx** (derived from pharyngeal arch elements), at the junction of the pharynx and the trachea, that is used for sound production.

Cardiovascular System

An important feature of tetrapods is the presence of a **lymphatic system**. A lymphatic system is also well developed in teleost fishes, but it is of critical importance on land, where the cardiovascular system is subject to the forces of gravity. The lymphatic system is a one-way system of blind-ending, vein-like vessels that parallel the pathways of the veins and allow fluid in the tissues to drain into the venous system at the base of the neck. In tetrapods lymph is kept moving through the contraction of muscles and tissues, and valves in the tubes prevent backflow. **Lymph nodes**, concentrations of lymphatic tissues, are found in mammals and some birds at intervals along the lymph channels. Lymphatic tissue is also involved in the immune system; white blood cells (macrophages) travel along this route, and the lymph tissue can intercept foreign or unwanted material, such as migrating cancer cells.

The sinus venosus and conus arteriosus are reduced or absent in the hearts of tetrapods. With the advent of lungs, vertebrates evolved a double circulation in which the **pulmonary** circuit supplies the lungs with deoxygenated blood and the **systemic** circuit supplies oxygenated blood to the body. Lungfish and tetrapods have a heart that is at least

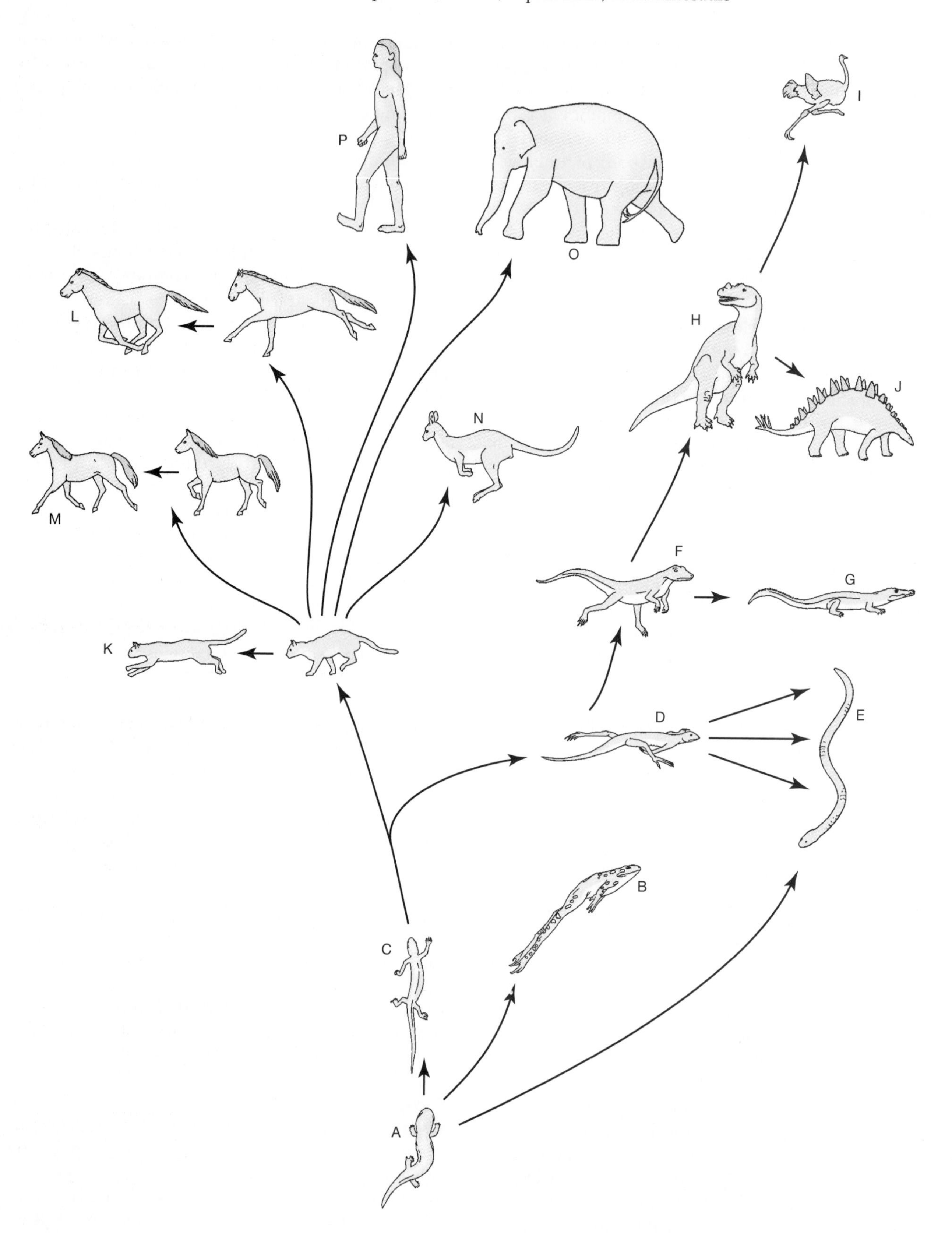
P
O
I
L
H
J
N
M
F
G
K
D
E
B
C
A

partially divided. The atrium is always completely subdivided in lungfish and tetrapods, and the ventricle is at least partially subdivided in lungfish and amniotes. The right side of the heart receives deoxygenated blood returning from the body via the systemic veins, and the left side of the heart receives oxygenated blood returning from the lungs via the pulmonary veins. The double circulation of tetrapods can be pictured as a figure eight with the heart at the intersection of the loops (see Figure 8–9).

The aortic arches have undergone considerable change in association with the loss of the gills in tetrapods (see Figure 8–10). Arches two and five are lost in adult tetrapods (although arch number five is retained in salamanders). Three major arches are retained: the third (**carotid** arch) going to the head, the fourth (**systemic** arch) going to the body, and the sixth (**pulmonary** arch), going to the lungs. Modern amphibians retain the fish-like condition, where the aortic arches do not arise directly from the heart. This condition, with a retention of a prominent conus arteriosus, was probably the condition among the earliest tetrapods. In amniotes the pulmonary artery connects directly to the right ventricle, and the aorta (systemic and carotid arches) connects directly to the left ventricle, although details of the heart anatomy suggest that this condition evolved independently in mammals and in other amniotes.

In modern amphibians the skin is of prime importance in the exchange of oxygen and carbon dioxide. In these animals the pulmonary arch is actually a pulmocutaneous arch, with a major cutaneous artery branching off the pulmonary artery to supply the skin (Figure 8–10). The cutaneous vein, now carrying oxygenated blood, feeds back into general systemic system and hence into the right atrium. Thus oxygenated blood feeds into the amphibian ventricle from both the left atrium (supplied by the pulmonary vein), and the right atrium. This type of heart in modern amphibians, with the absence of any ventricular division, is probably a derived condition adapted for using the skin as well as the lungs for respiration. The skins of the earliest tetrapods contained scale-like dermal bones that would have limited cutaneous respiration.

A ventricular septum of some sort is present in all amniotes, but the form is different in the various amniote lineages. A horizontal ventricular septum is found in turtles and lizards, whereas a vertical septum is present in mammals as well as crocodiles and birds (where it must represent an independent evolutionary event). These differences suggest that a complex ventricular septum evolved independently in different amniote groups.

The completely divided ventricle of birds and mammals may reflect a constraint imposed by their high metabolic rates. The endothermal metabolism of birds and mammals necessitates a high systemic blood pressure, and the lungs must be protected from this blood pressure. However, if the ventricle were completely divided by a septum, the right ventricle would be exposed only to deoxygenated blood returning from the body and would have no access to the oxygenated blood returning from the lungs in the left ventricle. Coronary arteries must be developed to supply oxygenated blood to the ventricles, especially

◀ Figure 8–6 Phylogenetic view of tetrapod terrestrial stance and locomotion. (a) Primitive tetrapod condition, retained today in salamanders: movement mainly via axial movements of the body, limbs moved in diagonal pairs (basic walk-trot gait). (b) Derived jumping form of locomotion in the frog. (c) Primitive amniote condition: limbs used more for propulsion, with development of the walk gait (limbs moved one at a time independently). (d) Diapsid amniote condition with hindlimbs longer than forelimbs, tendency for bipedal running. (e) Derived limbless condition with anguilliform (eel-like) locomotion. Evolved convergently several times among primitive tetrapods (e.g., several types of lepospondyls), amphibians (caecilians and limb-reduced salamanders), and squamates (snakes, amphisbaenids, and limb-reduced or limbless lizards). (f) Primitive archosaur condition, with upright posture and tendency to bipedalism. (g) Secondary return to sprawling posture and quadrupedalism in crocodiles. (h) Obligate bipedality in early dinosaurs and (i) birds. (j) Return to quadrupedality several times within dinosaurs. (k) Primitive mammalian condition: upright posture and the use of the bound as a fast gait with dorsoventral flexion of the vertebral column (all mammals use the walk as a slow gait). (l) Condition in larger mammals where the bound is turned into the gallop. (m) The true trot, as seen in larger mammals. (n) The ricochet, a derived hopping gait of kangaroos and some rodents. (o) The amble, a speeded-up walk gait seen as the fast gait of elephants, and in some horses. (p) The human condition of upright bipedality.

BOX 8-1 Size and Scaling in Terrestrial Vertebrates

Body size is one of the most important things to be known about an organism. Because all structures are subject to the laws of physics, the absolute size of an animal profoundly affects its anatomy and physiology. The design of the skeletomuscular system is especially sensitive to absolute body size in tetrapods because of the effects of gravity on land.

The study of scaling, or of how shape changes with size, is also known as **allometry** (Greek *allo* = other and *metric* = measure). If the features of an animal show no relative changes with increasing body size (i.e., if a larger animal appeared just like a photo enlargement of a smaller one), all of its component parts would be scaled with **isometry** (Greek *iso* = the same). However, animals are not built this way, and very few body components scale isometrically (i.e., with a one-to-one relationship with body size). One notable example of isometric scaling is in the dimensions of the dentition (tooth length, width, and height) in mammals.

Most body components scale in some disproportionate way, or in an allometric fashion. For example, although eye size is absolutely bigger in large animals, large animals have eyes that are relatively smaller in relation to the size of their head than do smaller animals. Thus eye size scales with less than a one-to-one relationship to body size; that is, with negative allometry. On the other hand, the limb bones of larger animals are not only absolutely larger than those of smaller animals, but the bones are also proportionally thicker. Thus limb diameter scales with positive allometry.

Underlying all scaling relationships is the issue of how the surface area of an object relates to its volume: When linear dimensions double (i.e., a twofold change), the surface area increases as the square of the change in linear dimensions (a fourfold change) and the volume increases as the cube of the change in linear dimensions (an eightfold change; Figure 4–7). **Linear dimensions** refer to the height or length of the animal, its **surface** is the area it exposes to the environment, and its **volume** is proportional to its body mass (or weight). An animal that is twice as tall as another is eight times as heavy. In considering the range of size of living organisms, a bacterium (length 10^{-6} m) has 100,000,000 as much external surface relative to its volume as a blue whale (length 10^{2} m).

Differences in surface area have a profound effect on animal function. For example, bigger animals take much longer to heat up or cool down than smaller ones, because there is less surface area for exchange relative to the volume of the animal. This relationship affects considerations of body temperature and metabolic rate (see Chapters 12 and 21). Absolute size also affects how animals survive falls from a height. A small animal has a proportionally larger amount of surface area to cushion its mass on impact. A mouse falling out of a second-storey window would be stunned, but otherwise little hurt, whereas a human would sustain broken bones, and a horse would be killed outright (don't try this experiment at home).

The cross-sectional area of the bones actually supports an animal's weight. If an animal increased in size isometrically, its weight would rise as the cube of its linear dimensions, but the cross section of its bones would increase only as the square of its linear dimensions. Thus the bones must increase their surface area by a disproportional amount to keep up with increases in weight. Bone diameter scales with positive allometry: Bigger animals have proportionally thicker limb bones than smaller ones. The skeleton of a bigger animal can easily be distinguished from a smaller one even when they are drawn to the same size, by virtue of the proportionally thicker bones (Figure 8–7).

However, animals do not simply change the proportions of their limb bones to the extent that might be predicted using simple allometric laws. If they did, their skeletons would become disproportionately heavy. The skeleton of a cat makes up about 5 percent of its total body mass, and if an elephant were built to the same structural proportions as a cat its skeleton would be 78 percent of its total mass. In fact, the skeleton of an elephant is about 13 percent of its mass, almost three times more than that of a cat—but still within the realm of biological feasibility!

The skeletal mass of terrestrial vertebrates does scale with positive allometry, but not as much as you might think. Interestingly enough, the skeletal mass of both fishes and whales (animals that do not have to worry about the skeleton supporting their body weight) scales approximately isometrically. Consequently, the skeleton of larger animals is proportionally more fragile than that of smaller ones. So how do larger terrestrial animals cope with living in the real world? Figures 8–7 and 8–8 depict some other critical differences between

BOX 8-1 SIZE AND SCALING IN TERRESTRIAL VERTEBRATES (CONTINUED)

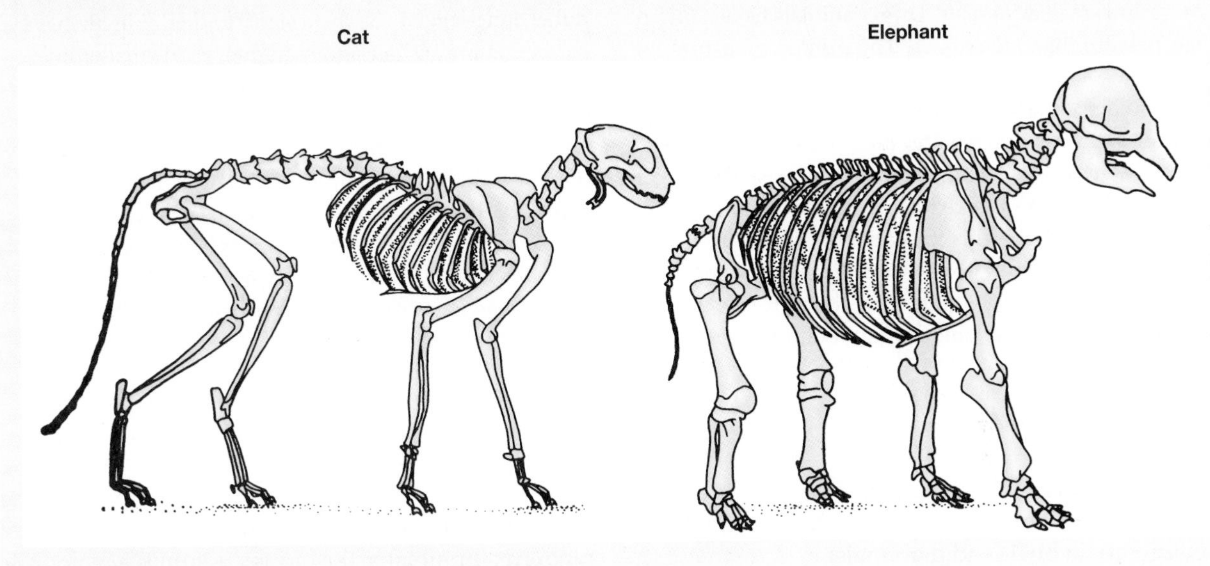

▲ Figure 8–7 Comparison of skeletons of animals of different sizes.

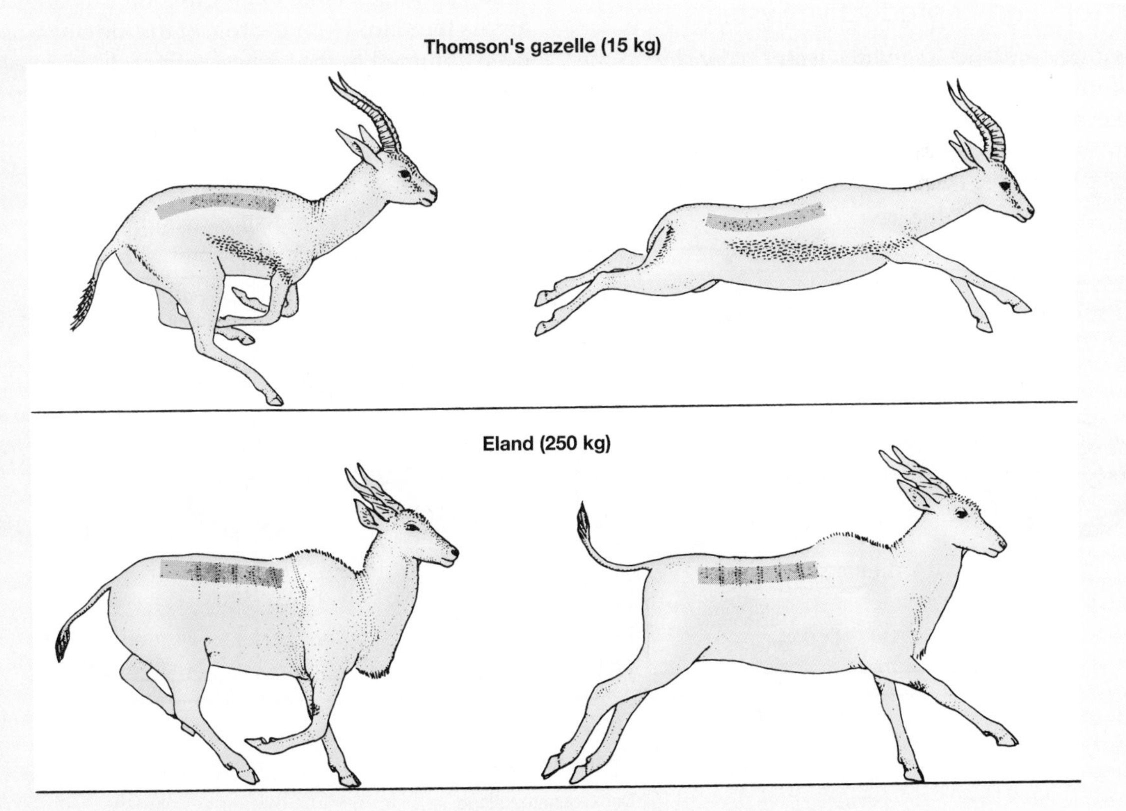

▲ Figure 8–8 Galloping locomotion compared between small and large antelope.

BOX 8-1 SIZE AND SCALING IN TERRESTRIAL VERTEBRATES (CONTINUED)

large and small mammals. Larger animals have a different posture than small ones. The primitive posture for mammals is to stand with flexed joints. A bone withstands compressive forces (forces exerted parallel to the axis of the bone) much better than shearing or torsional forces exerted at an angle to the long axis. The larger the animal, the less flexed are its limb joints, resulting in a more pillar-like, directly weight-bearing stance that reduces the effect of torsion on the limb bones. Torsional stresses are also reduced by the relatively shorter limb bones of larger animals (although elephants are rather long-legged for their size). Larger animals have backs that are shorter and straighter than smaller ones. Finally, larger animals move in a different fashion than smaller ones: They engage in less leaping behavior, moving their limbs through smaller angles of excursion (Figure 8–8). The straight-legged gait of elephants ensures that most of the forces on their legs will be compressive rather than shearing. An animal as large as an elephant is actually too big to trot or gallop, and its fastest gait is an amble, or speeded-up walk (Figure 8-6o).

to the right ventricle. Coronaries are absent in modern amphibians and reptiles and must have been evolved convergently in birds and mammals.

Excretory and Reproductive Systems

The kidney of fishes regulates water balance and the concentration of some divalent ions. The gills regulate uptake and excretion of monovalent ions, and nitrogenous wastes (in the form of ammonia) diffuse across the gills and skin. In the terrestrial adults of modern amphibians, the kidney assumes a major role in the excretion of monovalent ions and nitrogen (as urea), although some ion regulation exchange occurs across the skin. This new role of the kidney must have been assumed in the earliest tetrapods, in association with the loss of the gills. Amniotes excrete all their nitrogenous wastes via the kidney as urea or uric acid.

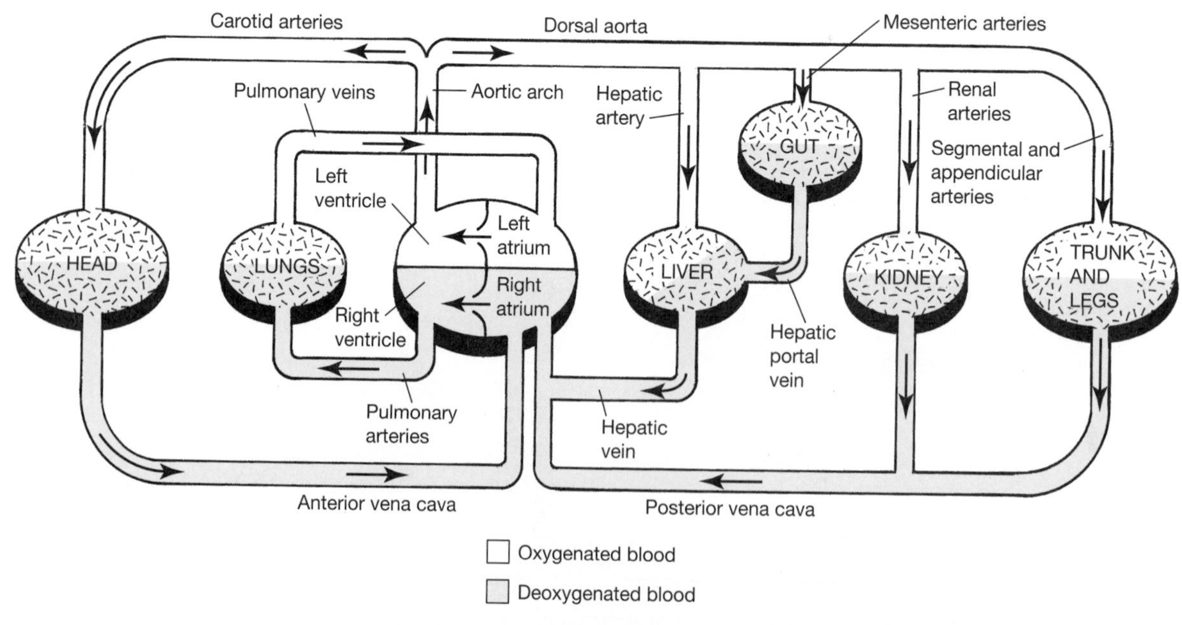

▲ Figure 8–9 Double-circuit cardiovascular system in a tetrapod.

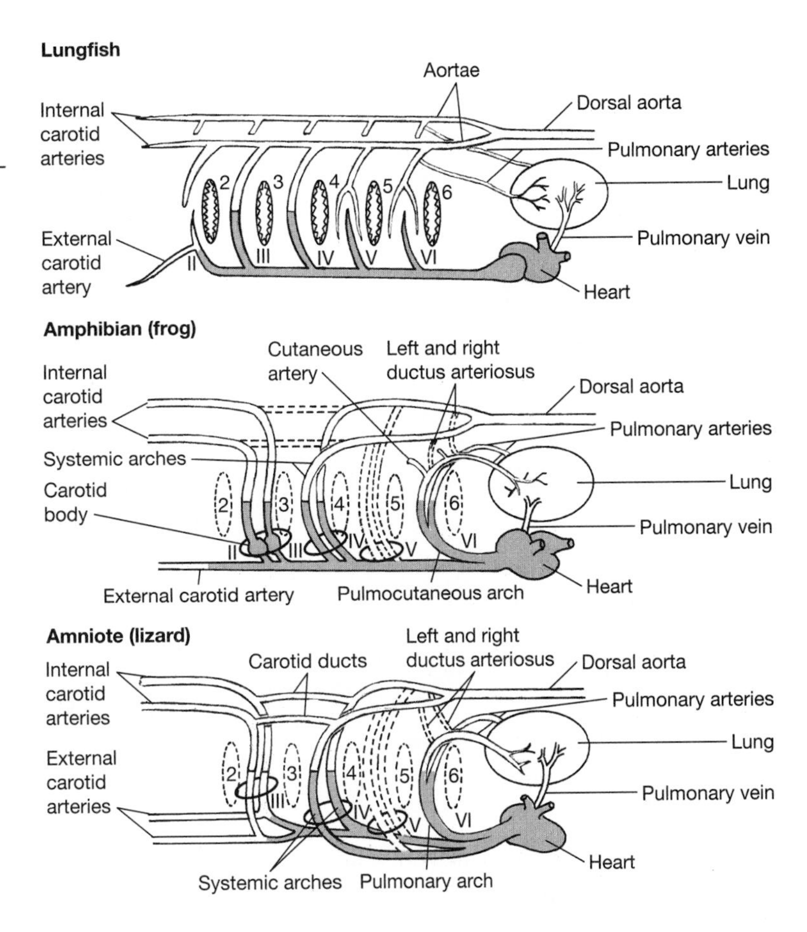

▶ Figure 8–10 Form of the aortic arches in jawed fishes and tetrapods. Arabic numbers: gill slits/visceral pouches. Roman numerals: aortic arches. Numberings represent presumptive primitive condition for vertebrates.

Adult amphibians resemble bony fishes in having an opisthonephric kidney, comprised of the embryonic mesonephric and metanephric portions. This kidney is an elongated structure lying along the dorsal body wall and was presumably the type of kidney possessed by the earliest tetrapods. The smaller and more compact kidney seen in adult amniotes is the **metanephric kidney**, composed only of the embryonic metanephros. The tubules of the metanephric kidney become highly compacted and numerous, achieving the capacity to concentrate urine. Modern amphibians also resemble fishes in having a single duct, the archinephric duct, draining both the kidney and the gonads. Amniotes have a new duct draining the kidney—the **ureter**—derived from the base of the archinephric duct. In male adult amniotes the archinephric duct drains the testis. In female adult amphibians the archinephric duct is lost, and the ovary is drained by the oviduct.

A urinary bladder for the storage of urine is a new feature of tetrapods, although some bony fishes have a bladder-like extension of the kidney duct (Figure 8–1[12]). The bladder of modern amphibians has the capacity for water resorption, and this may have been the case for the earliest tetrapods. But in amniotes most or all water resorption is done via the kidney, and in these animals the bladder has been described as an organ of social convenience. The bladder is actually lost in many reptiles and birds that excrete a semisolid paste of uric acid.

In most vertebrates the urinary, reproductive, and alimentary systems reach the outside via a single common opening, the **cloaca** (see Figure 8–11). There may be some division within the cloaca to separate the different functions, but there is still only one common opening. Only in therian mammals (marsupials and eutherians) is the cloaca replaced by separate openings for the urogenital and alimentary systems.

The obvious reproductive difference between amphibians and amniotes is the presence of the **amniotic egg** in amniotes (Figure 8–1[13]). This egg has a shell and a characteristic series of extraembryonic membranes. The amniote bladder, and the duct that links it to the cloaca (the **urethra**), are homologous

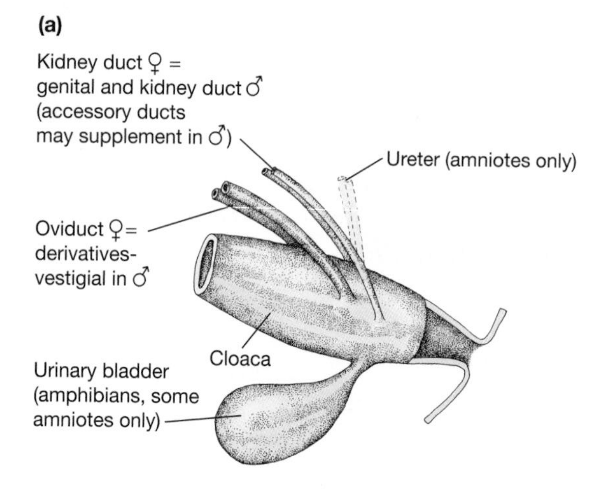

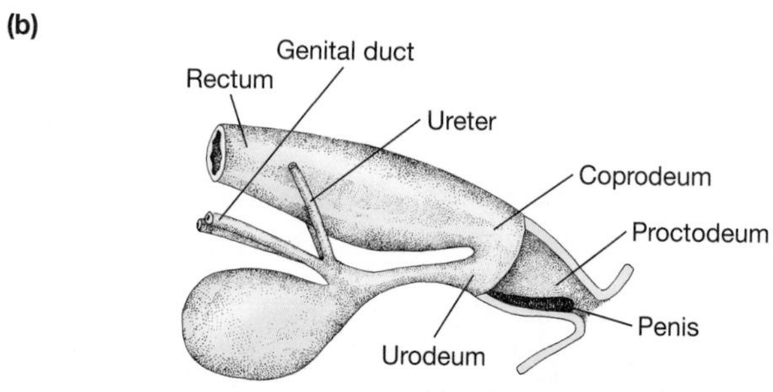

▲ Figure 8–11 Anatomy of urogenital ducts in tetrapods. (a) Generalized jawed vertebrate condition, including nonamniote tetrapods and some primitive amniotes. (b) More derived amniote condition (except for therian mammals), as illustrated by a male monotreme.

with the allantoic membrane of the amniote egg. The gilled larval form is lost in amniotes, and amniotic eggs are not usually laid in the water. When an amniote lineage returns to the water the females either come onto land to lay eggs, or else give birth to live young.

The detailed structure and evolution of the amniotic egg are discussed later in the chapter. However, note that a shelled egg requires internal fertilization—because the sperm must reach the egg in the mother's oviduct before the shell is applied—and amniotes have evolved a penis for this purpose. (The penis has been secondarily lost in many birds.) In most amniotes the penis is used solely as a conduit for sperm. The use of the penis for the transmission of urine as well as sperm is seen only in therian mammals.

Nervous System

In the anterior roof of the mouth, tetrapods have a unique organ of olfaction—the **vomeronasal organ** or **Jacobson's organ**. When snakes flick their tongues in and out of their mouth, they are transferring molecules directly from the air to this organ. Many male ungulates (hoofed mammals) sniff or taste the urine of females, a behavior that permits them to determine the stage of her reproductive cycle. This sniffing is usually followed by *flehmen*, a behavior in which the male curls the upper lip and often holds his head high, probably inhaling molecules of pheromones into the vomeronasal organ. Primates, with their relatively flat faces, were thought to have lost their vomeronasal organs; but some recent work suggests the presence of a remnant of this structure in humans that is used for pheromone detection.

The sense of vision is actually easier to use on land than in the water. Light is transferred through the air with less disturbance than in water—air is rarely murky in the way that water can frequently be. As a result, vision is more useful as a distance sense in air than in water. The formation of an image is easier in air due to the refraction of light at the surface between the air and the cornea (the transparent covering of the front of the eye). This is why it is easier for us see underwater if we use goggles. Because of this role of the cornea in air, the lens of the eye needs to do less focusing in tetrapods than it does in fishes, and the refractive index of the lens is different. Additionally, fishes focus light by moving the position of the lens within the eye, while tetrapods focus light by changing the shape of the lens (snakes are an odd exception here).

However, a problem with vision on land is that the eye's surface must be protected, and kept moist and free of particles. New features in tetrapods include eyelids, various glands that produce tears to lubricate the eye and keep it moist (including lacrimal and Harderian glands), and a nasolacrimal duct to drain the tears from the eyes into the nose. Interestingly enough, permanently aquatic living amphibians lack eyelids and lacrimal glands.

The lateral line system, which is a significant sensory mode in fishes, does not work on land. It is lost in all tetrapods except for larval or permanently aquatic amphibians. The inner ear assumes the function of hearing airborne sounds in tetrapods, with the transmission of sound waves through a bone (or a chain of bones) in a middle ear (Figure 8–1[14]). Considerably more energy is needed to set the fluids of the inner ear in motion than most airborne sounds impart. The middle ear is a sound amplifier. It receives the relatively low energy of sound waves on its outer membranous end, the **tympanic membrane** or eardrum, and these vibrations are transmitted by the bones of the middle ear to the oval window of the otic capsule in the skull. Movement of the oval window produces waves of compression in the fluids of the inner ear, and in turn these waves stimulate the hair cells in the **lagena** (= the **cochlea** in mammals), a structure unique to tetrapods (Figure 8–12). This organ discriminates the frequency and intensity of vibrations it receives by means of a sensory system based on the hair cell (as in the lateral lines of fishes), the organ of Corti, and transmits this information to the central nervous system in the form of neurally encoded firing patterns.

The middle ear of tetrapods has clearly evolved convergently several times, although in each case the **stapes** (= the old fish hyomandibula, often called the **columella** in nonmammalian tetrapods) is used as an auditory ossicle. A middle ear enables the reception of high-frequency sounds. Modern amphibians have an organization of the inner ear that is different from that of amniotes, indicating a completely independent evolution of hearing. Among amniotes, a functional middle ear with an eardrum has clearly been evolved independently in mammals and other amniotes. Within nonmammalian amniotes, differences in the detailed anatomy suggest that evolution of the middle ear occurred independently in the ancestors of turtles and lizards, and is derived in archosaurs (crocodiles, dinosaurs, and birds).

The middle ear is not an airtight cavity; if it were, the movement of the eardrum would be resisted by pressure changes in the middle ear. The **eustachian tube**, derived from the first embryonic visceral pouch (= the spiracle of fishes), connects the mouth with the middle ear. Air flows in or out of the middle ear as air pressure changes. (These tubes sometimes become blocked. When that happens, changes in external air pressure can produce a painful sensation in addition to reduced auditory sensitivity. Anyone who flies in an airplane while they have a bad cold knows about this.)

Another new feature of tetrapods is **proprioception** of the appendicular muscles. Proprioception provides information about where your limbs are in space relative to the body: It's the proprioceptors in your arm that enable you to touch your finger to your nose when your eyes are closed. Proprioceptors include muscle spindles, which detect the amount of stretch in the muscle, and tendon organs, which convey information about the position of the joints. Proprioceptors are found only in the limbs of tetrapods, and they are important for determining posture and balance on land. (A fish, buoyed up by the water, has little need to worry about posture.)

8.2 Tetrapod Origins

Our understanding of the origin of tetrapods is advancing rapidly. The earliest known tetrapods are from the Late Devonian, some 360 million years ago. Until fairly recently, the Late Devonian genus *Ichthyostega* (originally found in East Greenland in 1932) was the only well-known representative of early tetrapods. In the past couple of decades, however, we have discovered new material from this fossil site, including both skulls and skeletons of a different genus, *Acanthostega*, an animal that was more fish-like than *Ichthyostega*. More fragmentary fossil material of other Late Devonian tetrapods has also been found in Latvia, Scotland, Australia, and North America. Analysis of the new specimens has focused on derived characters, and this perspective has emphasized the sequence in which the characteristics of tetrapods were acquired. The gap between fishes and tetrapods has narrowed, and the earliest tetrapods now appear to have been much more fish-like than we had previously realized. That information provides a basis for hypotheses about the ecology of animals at the transition between aquatic and terrestrial life.

The next stage in the history of tetrapods was their radiation into different lineages and different ecological types during the late Paleozoic and Mesozoic. By the Early Carboniferous, tetrapods had split into two lineages that are distinguished in part by the way the roof of the skull is fastened to the posterior portion of the braincase. One of these lineages is the **batrachomorphs**, which includes the temnospondyls—the largest and longest-lasting group of primitive, extinct nonamniote tetrapods. Some lineages of temnospondyls extended into the Cretaceous, and at least some of the living amphibians may be derived from temnospondyls.

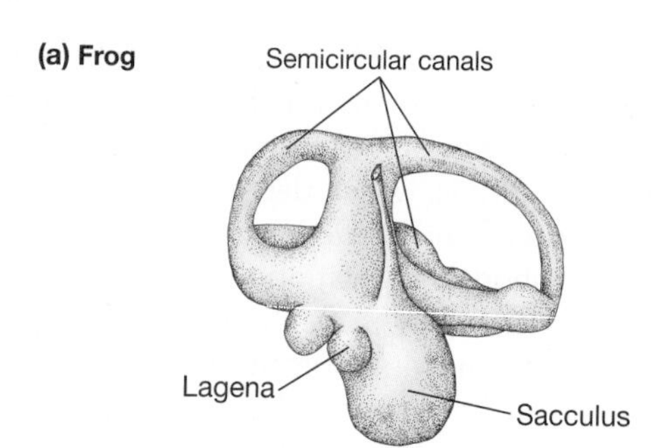

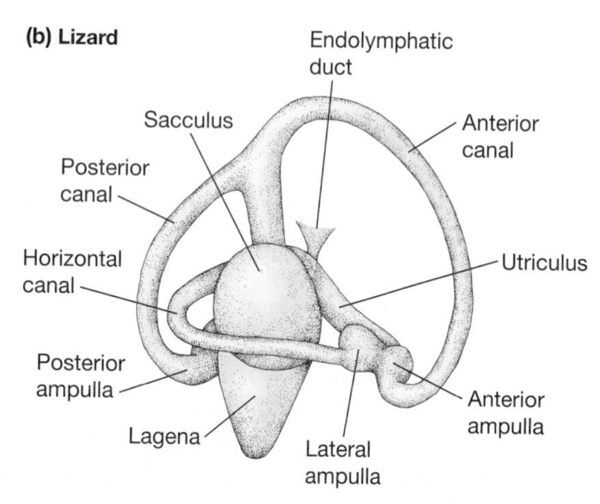

▲ Figure 8–12 Design of the vestibular apparatus in tetrapods. (a) Generalized amphibian condition. Note the acquisition of a small lagena for hearing airborne sounds. (b) Generalized non-mammalian amniote condition. Note the increase in size of the lagena.

The second early lineage of tetrapods, the **reptilomorphs**, contains a diverse array of animals, both nonamniotes and amniotes. The immediate ancestors of the major amniote groups appeared in the Late Carboniferous. They were small, agile animals, showing modifications of the skeleton and jaws that suggest that they fed on terrestrial invertebrates. The diversity of nonamniote tetrapods waned during the Late Permian and Triassic. By the start of the Cenozoic the only remaining nonamniotes were the lineages of amphibians that we see today: frogs, salamanders, and caecilians. Amniotes have been the dominant tetrapods since the late Paleozoic. They have radiated into many of the terrestrial life zones that were previously occupied by nonamniotes, and have developed feeding and locomotor specializations that had not previously been seen among tetrapods. Figure 8–13 shows a detailed phylogeny of primitive tetrapods, and Figure 18–14 shows a simplified cladogram.

Fish-Tetrapod Relationships

Tetrapods are clearly related to the **sarcopterygian** fishes. There are two major types of extant sarcopterygian fishes: the dipnoans (six species of lungfishes) and the actinistians (two species of coelacanths). The discovery of lungfishes seemed to provide an ideal model of a prototetrapod—what more could one ask for than an air-breathing fish? However, lungfishes are very specialized animals, and Devonian lungfishes were scarcely less specialized than the living species. The coelacanth lacks the specializations of lungfishes and for a while after its discovery in 1938 was hailed as a surviving member of the group ancestral to tetrapods. However, most scientists now consider that the lungfishes are more closely related to tetrapods than is the coelacanth (see Figure 8–13).

Both dipnoans and actinistians have an extensive Paleozoic fossil record, along with a third group called the **rhipidistians**, which were considered ancestral to tetrapods. Traditionally the actinistians and the rhipidistians were grouped together as the crossopterygians (= tassel-fins), and crossopterygians were considered ancestral to tetrapods. We now consider the term *crossopterygian* invalid, because it groups together coelacanths and other sarcopterygians lacking the specializations of lungfishes on the basis of primitive characters. The term *rhipidistian* also has fallen into disuse because it too is a paraphyletic group, including some groups that are more closely related to lungfishes (porolepiforms and rhizodontiforms) and others that are more closely related to tetrapods (osteolepiforms).

► Figure 8–13 Phylogenetic relationships of sarcopterygian fishes and early tetrapods. Dotted lines show interrelationships only, and are not indicative of the times of divergence of or the unrecorded presence of taxa in the fossil record. Lightly shaded bars indicate ranges of time when the taxon is known to be present, but is unrecorded in the fossil record. Numbers indicate derived characters that distinguish the lineages.

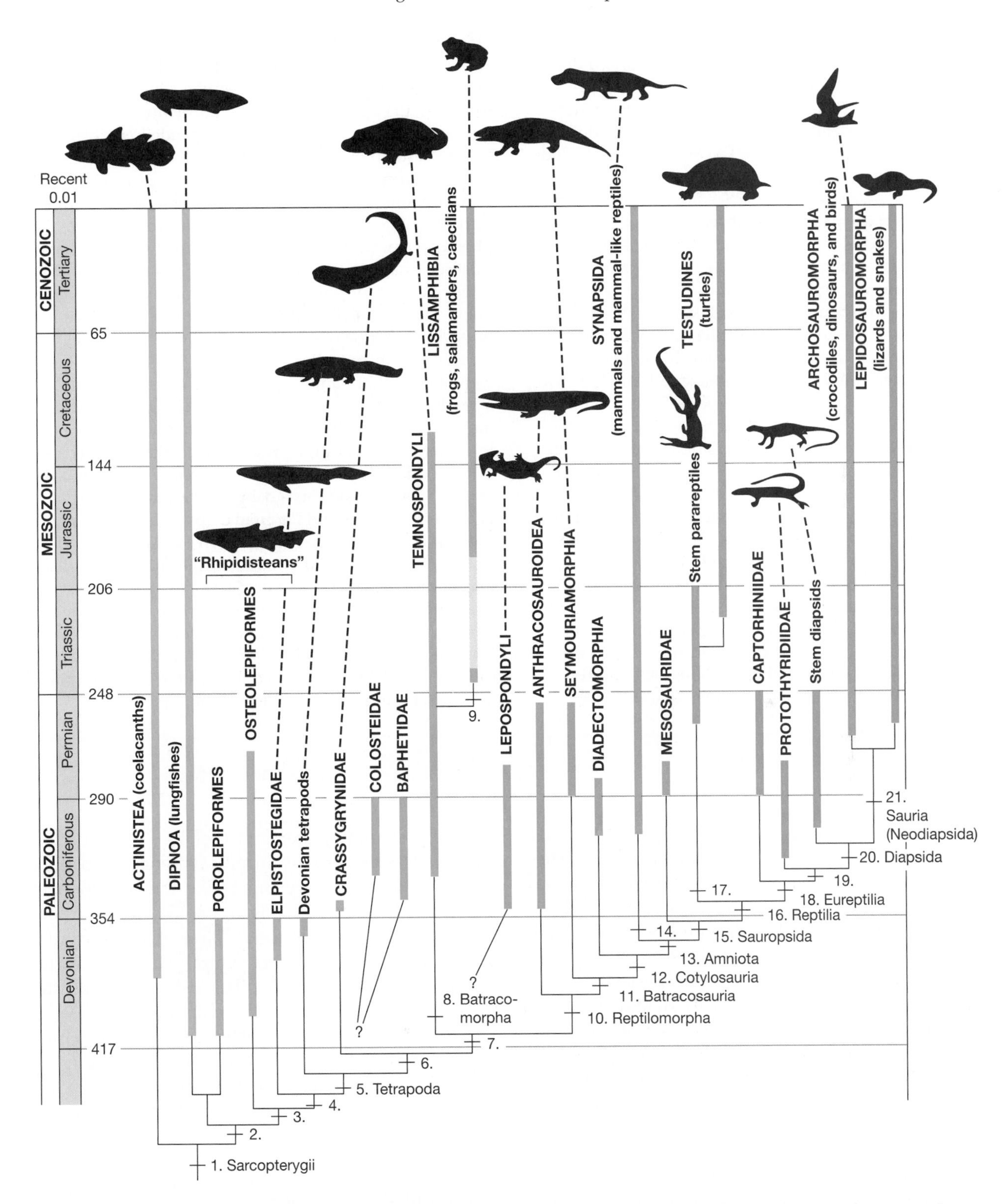

Legend: 1. Sarcopterygii—Fleshy pectoral and pelvic fins with a single basal element, muscular lobes at the base of those fins, true enamel on teeth, plus features of jaws and limb girdles. **2.** Heart with partial ventricular septum and separated pulmonary and systemic circulations. **3.** True choana (internal nostril), labyrinthine folding of tooth enamel, and details of limb skeleton. **4.** Flattened head with elongate snout, orbits on top of skull, body flattened, absence of dorsal and anal fins, enlarged ribs. **5.** Tetrapoda (= Stegocephalia)—Limbs with carpals, tarsals, and digits, vertebrae with zygapophyses, iliac blade of pelvis attached to vertebral column, loss of contact between dermal skull and pectoral girdle. ("Devonian Tetrapoda" is a paraphyletic assemblage of Late Devonian genera, including [in order of more primitive to more derived] *Acanthostega, Ichthyostega*, and *Tulerpeton*, plus others known from fragmentary material.) **6.** Absence of anocleithrum, five or fewer digits. **7.** Crown group Tetrapoda—Occipital condyles present, notochord excluded from braincase in adult. **8.** Batrachomorpha—Skull roof attached to braincase via the exoccipitals, loss of skull kinesis, only four fingers in hand. **9.** Lissamphibia—Pedicellate teeth, teeth bicuspid or multicuspid. (Lissamphibia includes the Salientia [frogs], the Caudata [salamanders], and the Apoda [caecilians].) **10.** Reptilomorpha—Several skull characters, plus vertebrae with the pleurocentrum as the predominant element. **11.** Batrachosauria—Intercentrum reduced in size, enlarged maxillary canine-like tooth. **12.** Cotylosauria—Sacrum with more than one vertebra, robust claws on feet, more derived atlas-axis complex, plus other skull characters. **13.** Amniota—Loss of labyrinthodont teeth, hemispherical and well-ossified occipital condyle, frontal bone contacts orbit in skull, transverse pterygoid flange present (reflects differentiation of pterygoideus muscle), three ossifications in scapulocoracoid (shoulder girdle), distinct astragalus in ankle. **14.** Synapsida—Presence of lower temporal fenestra. **15.** Sauropsida—Single centrale in ankle, maxilla separated from quadratojugal, single coronoid bone in jaw. **16.** Reptilia—Suborbital foramen in palate, tabular small or absent, large post-temporal fenestra. **17.** Parareptilia—Loss of caniniform maxillary teeth, posterior emargination of skull, quadratojugal expanded dorsally, expanded iliac blade. ("Stem Parareptiles" includes the Late Permian Millerettidae and Pareiasauridae, and the Late Permian and Triassic Procolophonidae. Opinions vary as to whether Testudines [turtles] are derived from pareiasaurs or from procolophonids, or even if they might be included with the diapsids [see Chapter 10].) **18.** Eureptilia—Supratemporal small, parietal and squamosal broadly in contact, tabular not in contact with opisthotic, horizontal ventral margin of postorbital portion of skull, ontogenetic fusion of atlas pleurocentrum and axis intercentrum. **19.** Postorbital region of skull short, anterior pleurocentra keeled ventrally, limbs long and slender, hands and feet long and slender, metapodials overlap proximally. **20.** Diapsida—Upper and lower temporal fenestrae present, exoccipitals not in contact on occipital condyle, ridge-and-groove tibia–astralagal joint. ("Stem Diapsids" is a paraphyletic assemblage of Late Carboniferous and Permian diapsids including [in order of more primitive to more derived] Araeoscelidia, Coelurosauravidae, and Younginiformes.) **21.** Sauria (Neodiapsida)—Dorsal origin of temporal musculature, quadrate exposed laterally, tabular bone lost, unossified dorsal process of stapes, loss of caniniform region in maxillary tooth row, sacral ribs oriented laterally, ontogenetic fusion of caudal ribs, modified ilium, short and stout fifth metatarsal, small proximal carpals and tarsals.

The **osteolepiforms** were cylindrical-bodied, large-headed fishes with thick scales. Many osteolepiforms had paired, crescentic vertebral centra similar to the vertebrae found in the earliest tetrapods, and some had teeth with the labyrinthine infolding of enamel seen in many early tetrapods. Although all osteolepiforms were basically free-swimming predators that lived in shallow waters, some may have been specialized for life at the water's edge. The most likely sister group of tetrapods is a newly defined lineage of Late Devonian osteolepiforms called the **Elpistostegidae** (formerly known as the Panderichthyidae).

The elpistostegids resemble other osteolepiforms in most respects, but have lost their dorsal and anal fins and have a greatly reduced tail fin. Only two genera are known, *Panderichthys* and *Elpistostege.* These fishes had eyes on the top of their heads, suggesting a shallow-water mode of life. Their bodies and heads were dorsoventrally flattened, and their snouts were long, very much like the condition in the earliest tetrapods (Figure 8–15). In addition, elpistostegids and tetrapods have distinct frontal bones, and their ribs project ventrally from the vertebral column; whereas in other osteolepiforms the frontal bone is medial, and the ribs project dorsally. These shared derived characters suggest that elpistostegids should be placed as the sister group of tetrapods, with other osteolepiforms as the paraphyletic sister group of elpistostegids plus tetrapods (see Figure 8–13).

The Earliest Tetrapods

The new fossils, especially specimens of *Acanthostega* from East Greenland, have shed light on the charac-

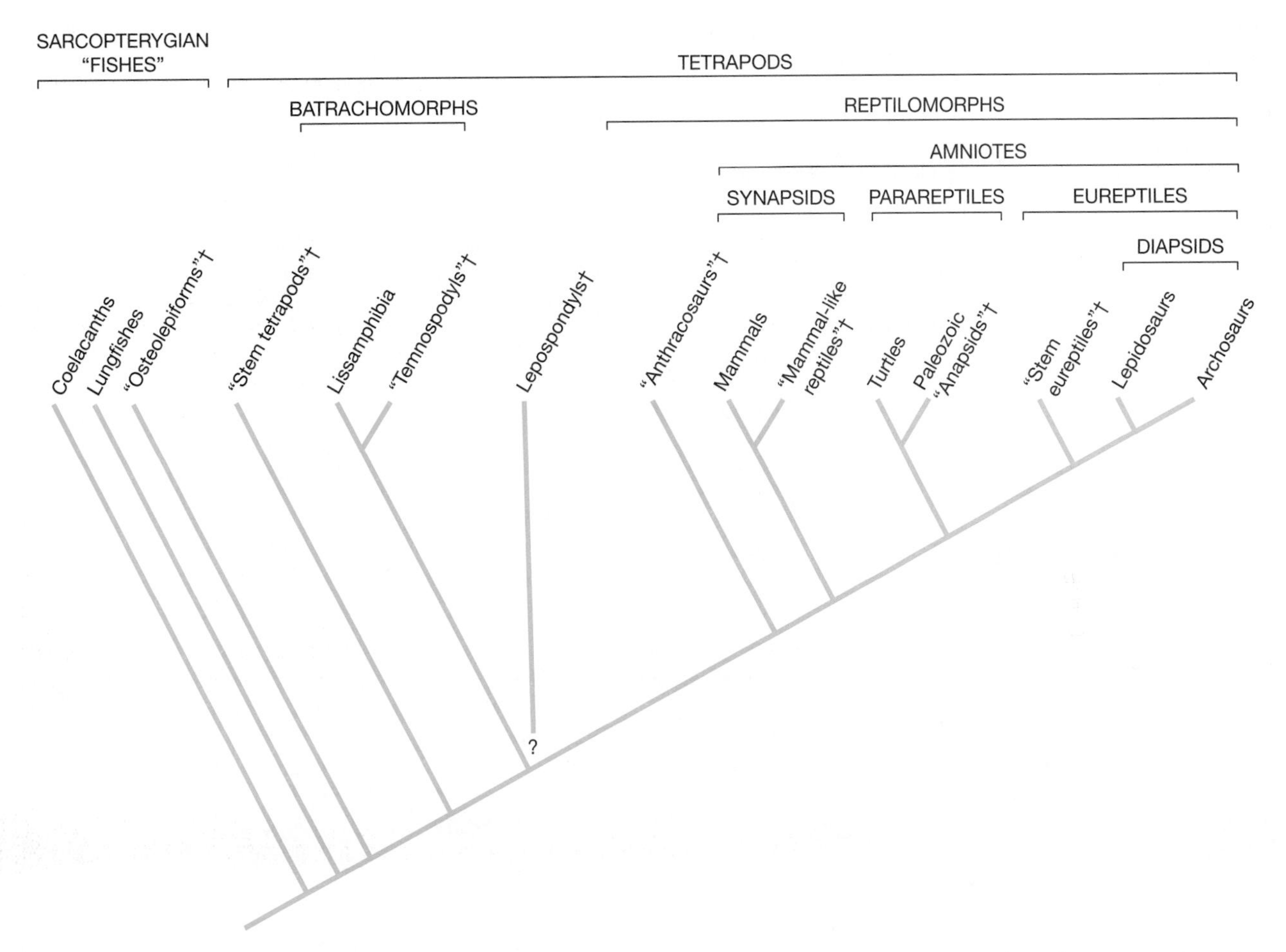

▲ Figure 8–14 Simplified cladogram of sarcopterygian fishes and tetrapods. Quotation marks indicate paraphyletic groups. A dagger (†) indicates an extinct taxon.

teristics of early tetrapods, suggesting that they were more aquatic than once thought (Coates and Clack 1995). In addition, one of the most widespread features of tetrapods, the pentadactyl (five-fingered) limb, turns out not to be an ancestral character.

The evidence for an aquatic way of life for early tetrapods comes partly from the presence of a groove on the ventral surface of the ceratobranchials. The ceratobranchials are part of the branchial apparatus, which supports the gills of fishes. Elements of the branchial apparatus are retained in all tetrapods, including birds and mammals, so it is not merely the presence of ceratobranchials in *Acanthostega* that is important—it is the groove on their ventral surface. In derived fishes that groove accommodates the afferent branchial aortic arches, which carry blood to the gills. The presence of a similar groove on the ceratobranchials of *Acanthostega* strongly suggests that these tetrapods also had internal fish-like gills, which are different from the external gills found in the larvae of modern amphibians and in some adult salamanders. In addition, the cleithrum (a bone in the shoulder girdle) of *Acanthostega* has a flange, the postbranchial lamina, on its anterior margin. In fishes, this ridge supports the posterior wall of the opercular chamber. The picture of *Acanthostega* that emerges from these features is an animal with fish-like internal gills and an open opercular chamber; that is, an animal with aquatic respiration, like a fish.

A second unexpected feature of *Acanthostega* is found in its feet—it had eight toes on its front feet.

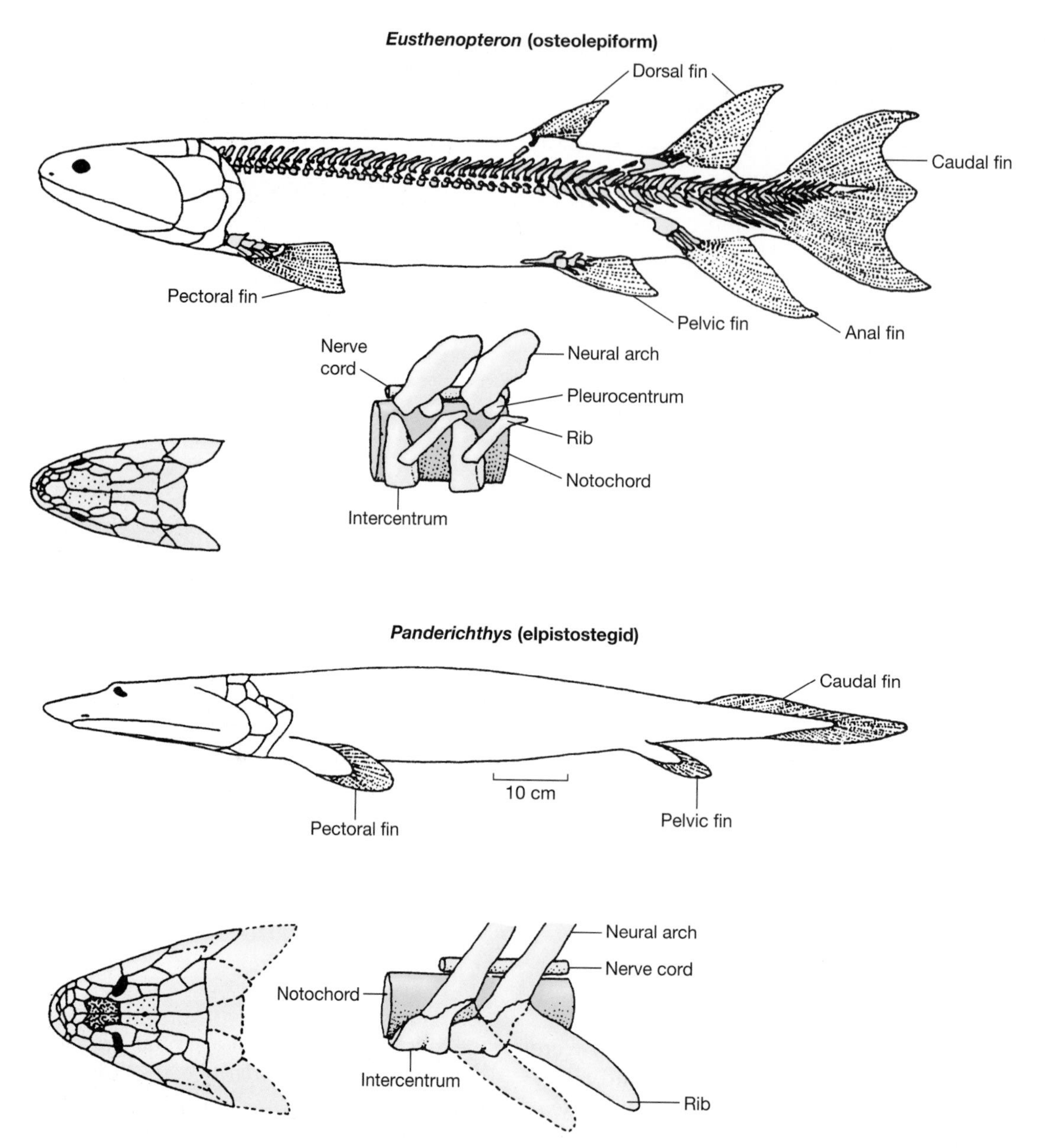

▲ Figure 8–15 A Devonian osteolepiform and elpistostegid: The osteolepiform *Eusthenopteron* has a cylindrical body, a short snout, and four unpaired fins in addition to the paired pectoral and pelvic appendages. The elpistostegid *Panderichthys* has a dorsoventrally flattened body with a long, broad snout and eyes on top of the head. The dorsal and anal fins have been lost, and the caudal fin has been reduced in size. In the vertebral column of *Eusthenopteron* the ribs are short and probably extended dorsally. The ribs are larger in *Panderichthys* and project laterally and ventrally. In the skull roof of *Eusthenopteron* the area anterior to the parietals (sparsely stippled) is occupied by a single, median element. In the skull roof of *Panderichthys* there is a single pair of large frontal bones (densely stippled) immediately anterior to the parietals, as in tetrapods.

Furthermore, *Acanthostega* was not alone in its polydactyly (having more than five toes). *Ichthyostega* had seven toes on its hind foot; its fore foot is unknown (Figure 8–16). *Tulerpeton,* a Devonian tetrapod from Russia, had six toes. In fact, not one of the Devonian tetrapods yet known had five toes. These discoveries confound long-standing explanations of the supposed homologies of bones in the fins of sarcopterygian fins with those in tetrapod hands and feet, but they correspond beautifully with predictions based on a study of the embryology of limb development (Box 8-2).

These new discoveries leave us with a paradoxical situation: Animals with well-developed limbs and other structural features that suggest they were capable of locomotion on land appear to have retained gills that would function only in water. How does a land animal evolve in water?

Evolution of Tetrapod Characters in an Aquatic Habitat

Tantalizingly incomplete as the skeletal evidence is, it is massive compared to the information we have about the ecology of elpistostegids and early tetrapods. It is not possible even to be certain if the evolution of tetrapods occurred in purely freshwater habitats. Greenland and Australia, the two landmasses from which Devonian tetrapods are known, are believed to have been separated by marine environments in the Devonian. Furthermore, *Tulerpeton* was found in a deposit that formed in a large, shallow marine basin, a considerable distance from the nearest known landmass. Tetrapods may have evolved in brackish or saline lagoons, or even in marine habitats.

What sorts of lives did the elpistostegids lead? What was their habitat like? What were their major competitors and predators? What were the earliest tetrapods able to do that elpistostegids could not? This is the sort of information we need to assess the selective forces that shaped the evolution of tetrapods.

How Does a Land Animal Evolve in Water? Certain inferences can be drawn from the fossil material available. We start from the basis that any animal must function in its habitat. If it does not, it does not leave descendants and its lineage becomes extinct. Evolutionary change occurs because the naturally occurring variation in organisms is subject to selection. Looking backward, we may see an evolutionary trend if progressive changes conferred an advantage on the individuals possessing them; but it is merely a fortunate coincidence if some features turn out to be advantageous for a different way of life. Tetrapod characteristics did not evolve because they would someday be useful to animals that would live on land; they evolved because they were advantageous for animals that were still living in water.

Elpistostegids were large fish—as much as a meter long—with heavy bodies, long snouts, and large teeth. They presumably could breathe atmospheric oxygen by swimming to the surface and gulping air, or by propping themselves on their pectoral fins in shallow water to lift their heads to the surface. Elpistostegids probably either stalked their prey or lay in ambush and made a sudden rush. Imagine a fish prowling through the dense growth of plants on the bottom of a Devonian pond or estuary. The flexible lobed fins would support it as it waited motionless for prey. Recently a rhizodontid fish has been discovered with the development of rather finger-like structures at the end of its fins (Daeschler and Shubin 1988). Rhizodontids are a radiation of lobe-finned fishes separate from the osteolepiforms (although fairly closely related to them), thus representing an independent evolution of a limb-like fin in a fish. This similarity bolsters the hypothesis that extinct sarcopterygian fishes were experimenting with the evolution of fin morphologies for specialized behavior in the water.

A group of living fishes, the frogfishes, provides a model for the usefulness of a tetrapod-like limb in water. The pectoral fins of frogfishes are modified into structures that look remarkably like the limbs of tetrapods (Figure 8–18), and are used to walk over the substrate. An analysis of frogfish locomotion revealed that they employ two gaits that are used by tetrapods—a walk and a (slow) gallop.

What Were the Advantages of Terrestrial Activity? This question has fascinated biologists for a century or more, and there is no shortage of theories. The classic theory is that the Devonian was a time of seasonal droughts. Shallow ponds that formed during the monsoon period often evaporated during the dry season, stranding their inhabitants in rapidly shrinking bodies of stagnant water. Fishes trapped in such situations are doomed unless the next rainy season begins before the pond is completely dry. We know that the living African and South American lungfishes cope with this situation by estivating in the mud of their dry pond until the

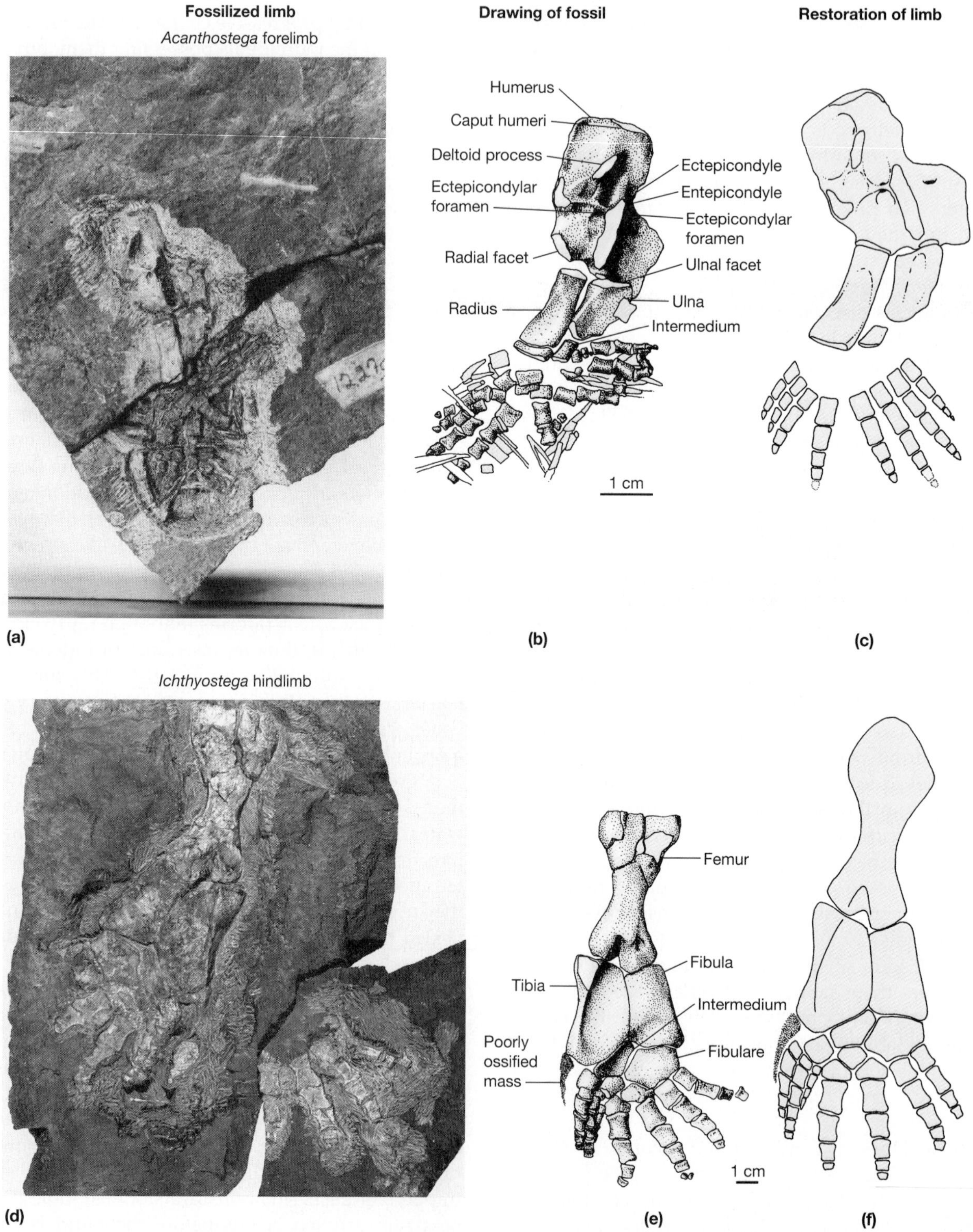

Fossilized limb

Acanthostega forelimb

Drawing of fossil

Restoration of limb

Humerus
Caput humeri
Deltoid process
Ectepicondylar foramen
Radial facet
Radius
Ectepicondyle
Entepicondyle
Ectepicondylar foramen
Ulnal facet
Ulna
Intermedium
1 cm

(a) (b) (c)

Ichthyostega hindlimb

Femur
Tibia
Fibula
Intermedium
Poorly ossified mass
Fibulare
1 cm

(d) (e) (f)

▲ Figure 8–16 The polydactyl feet of *Acanthostega* and *Ichthyostega*.

BOX 8-2 EARLY FEET

How a tetrapod limb could evolve from the fin of a sarcopterygian fish has been hotly debated for more than a century. Until recently, the various theories were based on equating specific bones in the limbs of tetrapods with their presumed counterparts in the limbs of sarcopterygian fishes. Thus, Gegenbaur in the nineteenth century suggested that extension and additional segmentation of the radials seen in the limb of *Eusthenopteron* could produce a limb with digits like those seen in tetrapods (Figure 8–17a, b). The same speculation about equivalent structures has been applied to the fin skeletons of lungfishes (Figure 8–17c, d).

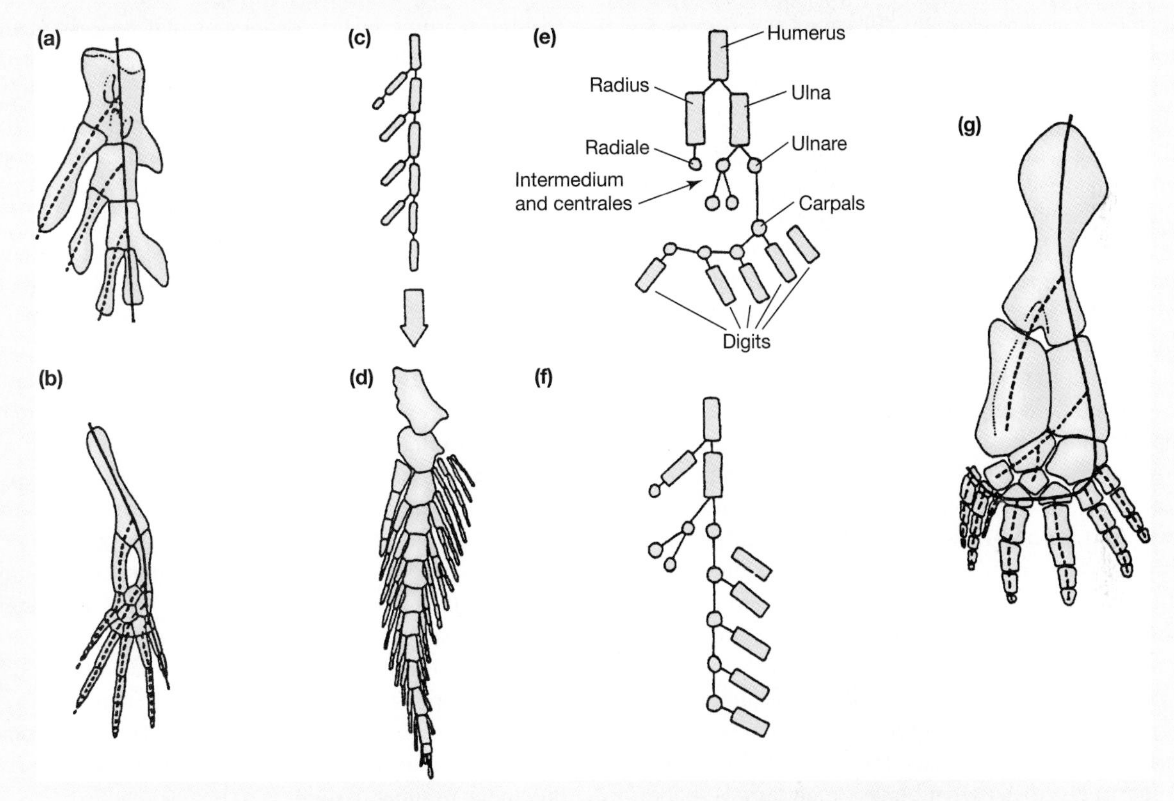

▲ Figure 8–17 Three hypotheses of the origin of tetrapod limbs. In every case the head of the animal is to the left. Thus, the preaxial direction (i.e., anterior to the axis of the limb) is to the left. (a) The pectoral fin skeleton of *Eusthenopteron.* The limb has a longitudinal axis (solid line) and preaxial radials. (b) Gegenbaur's nineteenth-century hypothesis of the origin of the vertebrate fore foot and fingers from preaxial radials. (c, d) The pectoral fin skeleton of the Australian lungfish *Neoceratodus.* (c) The fin axis and the preaxial radials that appear early in embryonic development by branching from the axis. The radials of *Eusthenopteron* may have developed in the same way. (d) The postaxial radials seen in the adult fin of *Neoceratodus* develop by condensation of tissue, not by branching. (e, f) A diagram of the forelimb skeleton of a mouse during early development. (e) The proximal (= nearest to the shoulder) parts of the limb skeleton consist of an axis with preaxial branches (radials) as seen in *Eusthenopteron* and *Neoceratodus;* compare the upper part of (e) with (a) and (c). The digits, however, are formed as postaxial branches. This pattern is unknown in fishes. (f) The embryonic limb skeleton of a mouse, straightened for comparison with *Eusthenopteron* (a) and *Neoceratodus* (c). (g) The hindlimb of *Ichthyostega* showing the inferred position of the axis (solid line) and radials. Note that the tibia and fibula and the first two rows of tarsal bones in the foot are formed by preaxial radials, whereas the third row of tarsals, the metatarsals, and the digits originate as postaxial branches.

BOX 8-2 EARLY FEET (CONTINUED)

A new perspective on the evolution of tetrapod limbs has been supplied by studies of the embryonic development of the limbs of derived vertebrates (Shubin and Alberch 1986). Recent work detailing the involvement of *Hox* genes in tetrapod limb formation is reviewed in Shubin et al. (1997) and Coates and Cohn (1998). *Hox* genes determine anteroposterior patterning in the vertebrate embryo. All jawed vertebrates have four different *Hox* gene clusters, labeled HOXA-HOXD. Fin development in fishes involves the 5′ *Hox* genes that are expressed in the posterior part of the body of these four *Hox* gene clusters. These four gene clusters are also involved in the development of tetrapod limbs, but similarities in gene expression with fishes extends only to the proximal (nearest to the body) part of the limb. The genes expressed in the distal limb of tetrapods (i.e., the hands and feet) appear to have been initially derived from the HOXA cluster, with later recruitment of the HOXD cluster.

All tetrapod limbs have a common pattern of development, and the same sequence of events applies to the forelimbs and the hindlimbs. In the developing forelimb, the humerus branches to form the radius (anteriorly) and the ulna (posteriorly). The developmental axis runs through the ulna. Subsequent development on the anterior side of the limb occurs (1) by segmentation of the radius, which produces the radiale (shown in Figure 8–17e) and sometimes two additional segments, and (2) by both branching and segmentation of the ulna. A preaxial branch of the ulna produces the proximal bones in the wrist—the intermedium and centrales. The carpals (distal wrist bones), metacarpals (bones of the palm), and digits (fingers) result from postaxial branching. The postaxial digits are a new feature of tetrapods—nothing like them is know in any fish.

For pentadactyl tetrapods, the formation of digits starts with digit 4 and concludes with digit 1 (which is the thumb of humans). Digit 5 (our little finger) forms at different times in different lineages, and one of the attractive features of this developmental process is the ease with which more or fewer than five digits can develop. If the process of segmentation and branching continues, a polydactylous foot is produced with the extra digits forming beyond the thumb. If the developmental process is shortened, fewer than five digits are produced, and the thumb is the first digit to be lost.

Changes in the timing of development do not have to produce an all-or-none addition or loss of a digit; a reduction in size is common. Dogs, for example, have four well-developed digits (numbers 2 through 5), plus a vestigial digit (number 1) called a dewclaw. Many dogs are born without external dewclaws (a carpal may be present internally), and some breeds of dogs are required by their breed standards to have double dewclaws. This pattern of increase or reduction in the number of digits results from a change in the timing of development during evolution. Reduction in the number of digits in evolutionary lineages of birds and mammals is frequently associated with specialization for high-speed running—ostriches and some artiodactyls (e.g., antelope) have two digits, and a some perissodactyls (horses) have a single digit. The embryonic process by which toes are formed explains why variation in the number of digits is so widespread.

---

rains return. Perhaps certain Devonian sarcopterygians had limb-like fins that allowed them to crawl from a drying pond and move overland to larger ponds that still held water. Could millions of years of selection of the fishes best able to escape death by finding their way to permanent water produce a lineage showing increasing ability on land?

This theory has been criticized on several grounds. After all, a fish that succeeds in moving from a drying pond to one that still holds water has enabled itself to go on leading the life of a fish. That seems a backward way to evolve a terrestrial animal. Various alternative theories have been proposed that stress positive selective values associated with increasing terrestrial activity. One theory emphasizes the contrast between terrestrial and aquatic habitats in the Devonian. The water was swarming with a variety of fishes that had radiated to fill a multiplicity of ecological niches. Active, powerful predators and competitors abounded. In contrast, the land was free of vertebrates. Any sarcopterygian that could occupy terrestrial situations had a predator- and competitor-free environment at its disposal. The exploitation of this habitat can be seen as proceeding by gradual steps.

The reinterpretation of early tetrapods as basically aquatic animals suggests that we should base speculations about selection for terrestrial activity on *Acanthostega* and *Ichthyostega,* rather than on elpistostegids. Juvenile *Ichthyostega* and *Acanthostega* might have congregated in shallow water, as do juveniles of living fishes and amphibians, to escape the attention of larger predatory fishes that are restricted to deeper water. At the edge of a lake or estuary many of the morphological and physiological characteristics of terrestrial vertebrates would have been useful to a still-aquatic tetrapod. Warm water holds little oxygen, and shallow pond edges are likely to be especially warm during the day. Thus lungs are important to a vertebrate in that habitat, whether it be fish or tetrapod. Similarly, legs would have borne the weight of the animal in the absence of water deep enough to float. In shallow water an air-breathing, upstanding tetrapod could have lifted its head above water, and due to the differences in the refractive indices of water and air, changes in the shape of the lens of the eye could have started to occur. These behavioral and morphological features can be found among several species of living fishes, such as the mudskippers, climbing perches, and walking catfishes, which make extensive excursions out of the water—even climbing trees and capturing food on land.

Starting from aquatic tetrapods snapping up terrestrial invertebrates that fell into the water, we can envision a gradual progression of increasingly agile forms capable of exploiting the terrestrial habitat for food as well as shelter from aquatic predators. Terrestrial agility might have developed to the stage where juvenile tetrapods moved overland from the pond of their origin to other ponds. Many vertebrates include a dispersal stage in their life history, usually in the juvenile period. In this stage individuals spread from the place of their origin to colonize suitable habitats, sometimes long distances from their starting point. This type of behavior is so widespread among living vertebrates that we may assume it occurred in early tetrapods as well. Because the earliest tetrapods were relatively large animals, the smaller body size of a juvenile would have greatly simplified the difficulties of support, locomotion, and respiration in the transition from an aquatic to a terrestrial habitat.

8.3 Radiation and Diversity of Nonamniote Paleozoic Tetrapods

For over 200 million years, from the Late Devonian to the Early Cretaceous, nonamniote tetrapods radiated into a great variety of terrestrial and aquatic forms. Parallel and convergent evolution were widespread, and it is hard to separate phylogenetic relationship from convergent evolution. Large gaps in the fossil record, especially the lack of early representatives of many groups, still obscure relationships.

Nonamniote tetrapods are often called amphibians, but that term is now reserved for the extant nonamniote tetrapods, the lissamphibians (frogs, salamanders, and caecilians). It is misleading to think of the primitive Paleozoic tetrapods as being amphibians, for several reasons. First, many of them were much larger than any living amphibians, and would have been more crocodile-like in appearance and habits. Second, they would have lacked the specializations of modern amphibians: For example, many forms had dermal scales, making it unlikely that they relied on their skin for respiration, as do the modern forms. Finally, and most important, many of them were actually more closely related to amniotes than to modern amphibians.

Table 8.1 lists the different types of Paleozoic nonamniote tetrapods, and Figure 8–13 illustrates a current consensus of their interrelationships. Note that some authors prefer to restrict the term *tetrapod* to the crown group, which encompasses only extant taxa and extinct taxa that fall within the phylogenetic space bracketed by the extant taxa. Thus, in the cladogram portrayed in Figure 8–13 the taxonomic term Tetrapoda would be shifted to node 7. The taxonomic term at node 5 would be Stegocephalia (i.e., crown-group tetrapods and stem tetrapods). Note, however, that under this mode of nomenclature different cladograms will result in different extinct taxa being excluded from the group defined as Tetrapoda.

Devonian Tetrapods

Specimens collected mostly in the past decade have greatly increased our knowledge of tetrapods in the

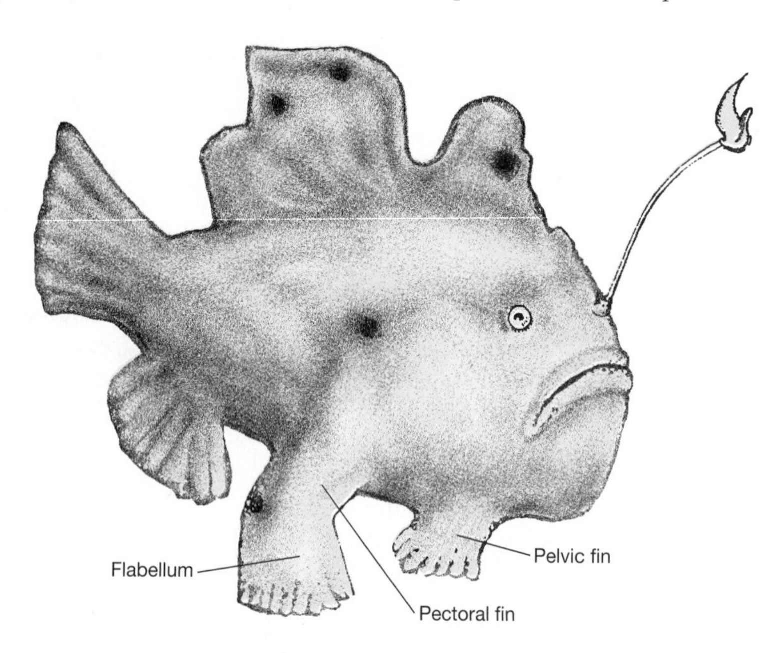

◀ Figure 8–18 The frogfish *Antennarius pictus* in its typical posture, with its pectoral and pelvic fins planted firmly on the substrate. Only the right pectoral and pelvic fins can be seen in this view. The small pelvic fins are in an anterior position, but are not fused. When the animal walks, the left and right pelvic fins make contact with the substrate independently, allowing the gait of the fish to be compared with the gaits of tetrapods.

Late Devonian. It is clear that substantial diversity had evolved among tetrapods by this time. *Ichthyostega* and *Acanthostega* from Greenland have been known since the 1930s, but only in the past couple of decades has more complete material of *Acanthostega* been described. *Acanthostega* has now supplanted *Ichthyostega* as the best known Devonian tetrapod (Figure 8–19).

Other Devonian tetrapods are known from less complete material. The Russian *Tulerpeton*, first described in 1984, is known from a skeleton, but lacks a skull. Other tetrapods are known only from fragmentary material. *Metaxygnathus,* found in Australia in 1977, and *Obruchevichthys,* found in Latvia in 1995, are represented only by lower jaws, and originally were thought to be fishes. *Hynerpeton,* found in Pennsylvania in 1994, is known only from a shoulder girdle. *Ventustega,* found in Latvia in 1994, is known from fragments of a skull and skeleton. *Elginerpeton* is known from skeletal fragments found in Scotland in 1995 and is apparently even more primitive than *Acanthostega*.

These Devonian tetrapods ranged from about 0.5 to 1.2 meters in length, and they differ enough from each other to show that by the end of the Devonian, some 7 million years after their first appearance, tetrapods had diversified into several niches. *Tulerpeton* was more lightly built than *Ichthyostega,* and had proportionally longer limbs, suggesting that it might have led a more active life. The robust shoulder girdle of *Hynerpeton* suggests that it could walk quite well on land. *Acanthostega* was more aquatic than *Ichthyostega.* Its forelimb is less robust than that of *Ichthyostega* and more finlike than its hindlimb, and the elbow was unable to bend in a way that would have supported the body out of the water. The articulating surfaces of the vertebrae are small, the neural arches are weakly ossified, the ribs are short and straight (not broad and overlapping like the ribs of *Ichthyostega*), and the fin rays on the tail are longer than those of *Ichthyostega.* Most important, *Acanthostega* appears to have retained internal gills, but *Hynerpeton* and *Ichthyostega* had both lost the postbranchial lamina on the cleithrum, suggesting that they did not have an internal gill chamber like that of *Acanthostega.*

The combination of fish-like and terrestrial characters seen in Devonian tetrapods is puzzling. Coates and Clack (1995) called attention to similarities between *Ichthyostega* and extant pinnipeds such as the elephant seal (*Mirounga leonina*). The forelimbs of pinnipeds are props for the body on land and the pelvic limbs are used as paddles and rudders in the water. The proportions of humerus and femur of the elephant seal are very similar to those of *Ichthyostega,* and the permanently bent forelimb of *Ichthyostega* forms a prop like the forelimb of a seal. Figure 8–19 shows a restoration of *Ichthyostega* that reflects these seal-like features.

The mosaic of ancestral and derived features of Devonian tetrapods suggests that the origin of

Table 8.1 Major Groups of Paleozoic Nonamniotic Tetrapods

Stem Tetrapods

Devonian taxa: e.g., *Acanthostega, Ichthyostega* (Figure 8–20a, b) and *Tulerpeton.*

Enigmatic Carboniferous taxa: e.g., *Crassygyrinus* from Europe (Figure 8–20c) and *Whatcheeria* from North America.

Colosteidae: Aquatic mid to Late Carboniferous forms, possibly secondarily so, known from North America and Europe, with elongate, flattened bodies, small limbs, and lateral line grooves (e.g., *Greererpeton, Pholidogaster, Colosteus*).

Baphetidae (formerly Loxomattidae): Mid and Late Carboniferous forms from North America and Europe, with crocodile-like skulls and distinctive keyhole-shaped orbits (e.g., *Eucritta, Megalocephalus*).

Batrachomorphs

Temnospondyli: The most diverse, longest-lived group, ranging worldwide from the mid Carboniferous to the Early Cretaceous. Possessed large heads with akinetic skulls. Paleozoic forms (e.g., *Eryops, Cacops*) were terrestrial or semiaquatic (Figure 8–20d); Mesozoic forms (e.g., *Cyclotosaurus, Trematosaurus, Gerrothorax*) were all secondarily fully aquatic (Figure 8–20e, f, g).

Reptilomorphs

Anthracosauroidea: The other diverse, long-lived group, although to a lesser extent than the temnospondyls. Known from the Early Carboniferous to the Late Permian of North America and Europe. Anthracosauroids had deeper skulls than temnospondyls with prominent tabular horns, and retained cranial kinesis. Some forms (e.g., *Gephyrostegus*) were terrestrial (Figure 8–21b). Others, grouped together as embolomeres, were secondarily aquatic (e.g., *Pholiderpeton, Archeria;* Figure 8–21a).

Seymouriamorpha: Known from the Permian only. Early Permian forms known from North America (e.g., *Seymouria*, Figure 8–21c) were large and fully terrestrial. Later Permian forms known from Europe and China (discosauriscids and kotlassiids) were secondarily fully aquatic.

Diadectomorpha: Known from Late Carboniferous and the Early Permian of North America and Europe. Large, fully terrestrial forms, now considered to be the sister group of amniotes. Diadectidae (e.g., *Diadectes*, Figure 8–21d) had laterally expanded cheek teeth suggestive of a herbivorous diet. Limnoscelidae and Tseajaiidae had sharper, pointed teeth and were probably carnivorous.

"Lepospondyls"

Microsauria: Distinguished by a single bone in the temporal series termed the tabular. Many (e.g., the tuditanomorphs) were terrestrial and rather lizard-like with deep skulls and elongate bodies (*microsaur* = small reptile; e.g., *Pantylus*, Figure 8–22a). Some forms (microbranchomorphs) had evidence of external gill supports and lateral-line grooves and were probably aquatic. Known from Late Carboniferous and the Early Permian of North America and Europe.

Aïstopoda: Limbless forms, lacking limb girdles, with elongate bodies (up to 200 trunk vertebrae) and rather snake-like skulls that may have allowed them to swallow large prey items (e.g., *Ophiderpeton*, Figure 8–22c). They may have been aquatic or have lived in leaf litter. Known from the mid to Late Carboniferous of North America and Europe.

Adelogyrinidae: Limbless, long-trunked forms, but retaining the dermal shoulder girdle. Known from the Early Carboniferous of Europe.

Lysorophia: Elongate forms with greatly reduced limbs. Known from the Late Carboniferous and the Early Permian of North America.

Nectridia: Also rather elongate, but with a long tail rather than a long trunk. Distinguished by having fan-shaped neural and haemal arches in the vertebral column. Limbs small and poorly ossified, indicative of an aquatic mode of life. Some nectridians (keraterpetontids) had broad flattened skulls with enlarged tabular bones (e.g., *Diplocaulus*, Figure 8–22b). Known from Late Carboniferous and Early Permian of North America, Europe, and North Africa.

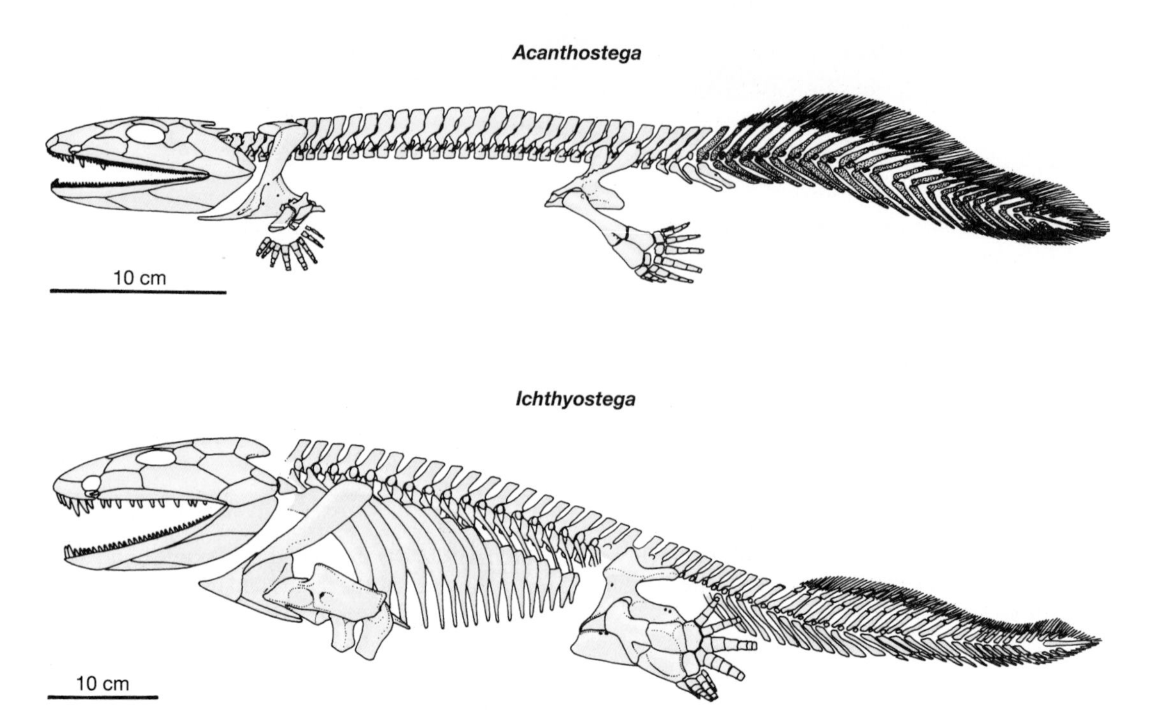

▲ Figure 8–19 Skeletal reconstructions of Devonian tetrapods from East Greenland (ribs of *Acanthostega* omitted, fore foot of *Ichthyostega* is unknown).

tetrapods and the origin of terrestrial life were two separate events. But the loss of the internal gills in all tetrapods more derived than *Acanthostega* suggests that the immediate ancestry of tetrapods must lie in a more recent—and more terrestrial—form than *Acanthostega*.

Carboniferous-Mesozoic Nonamniote Tetrapods

General Patterns of Radiation The groups listed in Table 8.1 are all well established; the problem comes in trying to understand how these different groups are related to one another and to the modern groups of tetrapods, the lissamphibians and the amniotes. A major problem is that we are missing a critical piece of the geological record of tetrapod history. Although fossils are known from the Late Devonian, the subsequent record is a complete blank for 20 to 30 million years, with no further fossils known until the later part of the Early Carboniferous. Thus, for the period when the major groups of tetrapods were diversifying, the fossil record provides us with little or no information. This gap may bias our understanding of how the groups known from the later Carboniferous are related to one another. The interrelationship of early tetrapods shown in Figure 8–13 broadly agrees with the 1994 review by Ahlberg and Milner; but since that time a plethora of new phylogenies has been suggested, most offering dramatically different opinions. The discovery of more Early Carboniferous tetrapod fossils is crucially needed to help confirm or refute the various competing hypotheses. For the purposes of this volume, we will treat the phylogeny in Figure 8–13 as a working hypothesis.

The Paleozoic tetrapods were originally divided into groups called "**labyrinthodonts**" and "**lepospondyls**" (see Table 8.1). "Labyrinthodonts" were mainly larger forms (large lizard- to crocodile-size; Figures 8–20 and 8–21) with a multipartite vertebral centrum, and teeth with complexly infolded enamel (labyrinthodont teeth). "Lepospondyls" were small forms (small lizard- or salamander-size; Figure 8–22) with a single, spool-shaped vertebral centrum and without the labyrinthine form of enamel. The various groups of lepospondyls may be related to each other, and this term remains in current usage. But

▲ Figure 8–20 Stem tetrapods and temnospondyls. (a) *Acanthostega*, an aquatic Late Devonian stem tetrapod. (b) *Ichthyostega*, an aquatic Late Devonian stem tetrapod. (c) *Crassigyrinus*, an aquatic Early Carboniferous stem tetrapod. (d) *Eryops*, a semiterrestrial Early Permian eryopoid temnospondyl. (e) *Trematosaurus*, an aquatic (marine) Early Triassic trematosaurid temnospondyl. (f) *Cyclotosaurus*, an aquatic Middle Triassic capitosaurid temnospondyl. (g) *Gerrothorax*, an aquatic Late Triassic plagiosaurid temnospondyl. *Eryops* was around 2m long (the size of a medium-size crocodile). The other animals are drawn approximately to scale.

▲ Figure 8–21 "Anthracosaur" (= nonamniote reptilomorph) tetrapods. (a) *Pholiderpeton*, an aquatic Late Carboniferous embolomere. (b) *Gephyrostegus*, a terrestrial Late Carboniferous anthracosauroid. (c) *Seymouria*, a terrestrial Early Permian seymouriamorph. (d) *Diadectes*, a terrestrial Early Permian diadectomorph. (e) *Westlothiana*, an Early Carboniferous form, probably closely related to amniotes. *Seymouria* is around 1m long (size of a golden retriever). The other animals are drawn approximately to scale (*Diadectes* should be a little larger and *Westlothiana* a little smaller).

larger-size tetrapods are a diverse taxonomic grouping; some were stem tetrapods, some were closely related to modern amphibians, and some were more closely related to amniotes. For this reason the term "labyrinthodont" is no longer employed.

Differences in the form of vertebral centra and the enamel of teeth most likely relate to body size. The larger tetrapods probably required the bipartite centrum for twisting movements of the spinal column on land, because the heavy head would tend to rotate around the trunk axis during locomotion. The labyrinthodont teeth probably reflect the reliance on inertial feeding. (That is, feeding with only a simple snap of the jaws, without fine control of jaw movements. The teeth might need to be strong to resist the forces generated during this kind of feeding.) The smaller absolute size of lepospondyls would reduce the forces experienced in locomotion and feeding, and thus reduce the need for specializations of the teeth and a more mobile vertebral column.

Current consensus perceives two main groups of large nonamniote tetrapods: the temnospondyls (batrachosaurs; Figure 8–20) and the "anthracosaurs" (i.e., nonamniote reptilomorphs: anthracosauroids, seymouriamorphs and diadectids; Figure 8–21). Temnospondyls were in general more aquatic, characterized by flat, immobile skulls and a reduction of the hand to four fingers. The modern amphibians (**Lissamphibia**), or at least the frogs, may have their origin within this group. "Anthracosaurs" were in general more terrestrial, characterized by domed skulls retaining some kinetic ability, and had a five-fingered hand. Amniotes may originate within this group.

The affinities of the lepospondyls (Figure 8–22) are open to question. Figure 8–13 shows them as a monophyletic group possibly allied with the Batrachomorpha, mainly for convenience of illustration. But the different lepospondyl groups may not be closely related, and some or all of them may be more closely related to the living amphibians or even to the amniotes. Also somewhat questionable are the origins of the Lissamphibia. Although one current hypothesis has them derived from within the temnospondyls (from a group of terrestrial, mainly

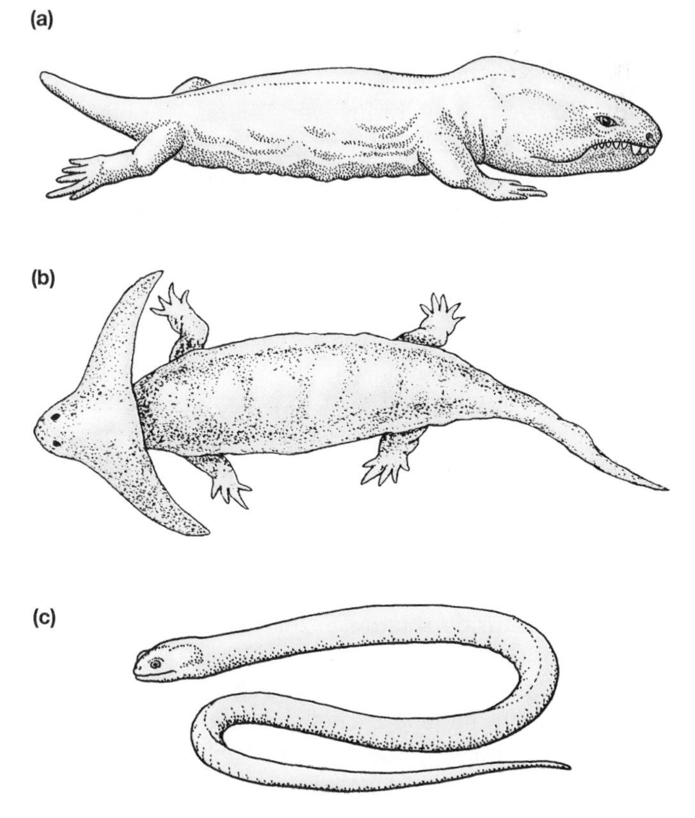

▶ Figure 8–22 Lepospondyl tetrapods. (a) *Pantylus*, a terrestrial Early Permian microsaur. (b) *Diplocaulus*, an aquatic Early Permian nectridian. (c) *Ophiderpeton*, an aquatic (or possibly terrestrial burrowing form) Late Carboniferous aïstopod. *Pantylus* is around 20 cm long (the size of a hamster). The other animals are drawn approximately to scale.

small forms—the dissorophids), other researchers would prefer to derive only frogs from this group, claiming that the origin of other modern amphibians lies within various lepospondyl groups. The study of the interrelationships of nonamniote tetrapods is ongoing, amid much debate and dissension.

Nonamniote tetrapods reached their peak of generic diversity in the Late Carboniferous and Early Permian, when they consisted of fully aquatic, semiaquatic, and terrestrial forms. Most lineages were extinct by the mid-Permian. The only groups to survive into the Mesozoic were the ancestors of the modern groups of amphibians and the fully aquatic temnospondyls. Temnospondyls were extinct in most of the world by the end of the Triassic but persisted into the Early Cretaceous in Australia. Although all living lissamphibian groups had their origins in the Mesozoic, the generic diversity of nonamniote tetrapods did not return to Permian levels until the mid Tertiary.

Ecological and Adaptive Trends One of the most striking aspects of early tetrapods is the number and diversity of forms that returned to a fully aquatic mode of life. Forms that were apparently fully aquatic as adults include colosteids, embolomerous anthracosauroids, and numerous groups within the temnospondyls and lepospondyls. Among living amphibians, many salamanders and some frogs are fully aquatic as adults (see Chapter 9).

Some the most bizarre aquatic forms were found among the lepospondyls. The keraterpetonid nectridians had broad, flattened skulls with enlarged tabular bones (Figure 8–22b). These tabular horns were up to five times the width of the anterior part of the skull, and skin imprints show that they were covered by a flap of skin extending back to the shoulder. These horns may have acted as a hydrofoil to help in underwater locomotion, and they may have supported highly vascularized skin to help in underwater respiration.

The temnospondyls were the only group of nonamniote tetrapods (aside from the lissamphibians) to survive the Paleozoic, and all of the Mesozoic forms were large, flattened, fully aquatic predators with apparently independent origins from less specialized forms. The capitosaurids (Figure 8–20f) were large, crocodile-like forms, but with very small legs. The plagiosaurids (Figure 8–20g) had broad, short heads and were extremely flattened. They retained external gills as adults, and may have evolved via the process of neotony (the retention of juvenile characters into adult life). The trematosaurids were among the most remarkable temnospondyls. They evolved the elongated snout characteristic of specialized fish eaters (Figure 8–20e) and are found in marine beds, making them the only known fully marine nonamniote tetrapods. How did these animals osmoregulate in the marine environment? Even if the adults had evolved a reptile-like impermeable skin, the larvae still would have gills as a site of water loss. Perhaps they retained high levels of urea to raise their internal osmotic pressure, like some modern estuarine frogs.

In contrast to the temnospondyls, the "anthracosaurs" appear to have been predominantly terrestrial as adults, and many have been mistaken for early reptiles (especially animals such as *Seymouria*, Figure 8–21c, and *Diadectes*, Figure 8–21d). Terrestriality also evolved convergently among other early tetrapods, predominately in the microsaurs (Figure 8–22a) and the dissorophid temnospondyls that may have been ancestral to frogs.

A final trend in nonamniote tetrapods is the evolution of forms with elongated bodies and limbs that are either greatly reduced or lost entirely (Figure 8–22c). Several groups of lepospondyls independently acquired this body form (see Table 8.1). It is probably associated with an aquatic or burrowing mode of life, as in the living limbless caecilian amphibians.

8.4 Amniotes

Amniotes are more derived tetrapods than those groups already discussed, and include most of the tetrapods alive today (mammals, birds, and reptiles). Their name refers to the **amniotic egg**, which is one of the most obvious features distinguishing amniotes from current amphibians. Amniotes appeared somewhat later in the fossil record than the earliest tetrapods of the Devonian, but they seem to have been established by the later radiation of nonamniote tetrapods. Their initial major radiation occurred in the Permian. The earliest known candidates for the status of early amniote—or near-amniote—are from the Early Carboniferous of Scotland, and they are only 20 million years younger than the earliest known tetrapods. These include *Casineria* and the slightly younger *Westlothiana* (Figure 8–21e), both discovered in the 1990s. Both were small animals; *Casineria* was only about 85 mm long the size of a salamander, and *Westlothiana* was only a bit larger.

Amniotes began to radiate in the Carboniferous and Early Permian into many of the life zones originally occupied by nonamniote tetrapods. A key event in their radiation may have been the great diversification of insects in the Late Carboniferous, probably in response to the increasing quantity and diversity of terrestrial vegetation (see Chapter 7). Most, if not all, terrestrial vertebrates at this time were carnivorous (including fish- and invertebrate-eaters). Carnivorous vertebrates could not respond directly to the energy supply offered by terrestrial plants, but they could and apparently did respond to the opportunities presented by the radiation of insects. Probably for the first time in evolutionary history, the food supply was adequate to support a diverse fauna of fully terrestrial vertebrate predators. No adult amphibian among living forms is herbivorous, and there is little evidence in the fossil record to suggest that Paleozoic nonamniote tetrapods were herbivores. *Diadectes* is the only likely exception to this generalization.

The morphological and ecological diversity of the amniotes that had developed by the late Paleozoic (see Figure 8–23) testifies to the success of this clade. As the terrestrial amniotes diversified, the terrestrial nonamniotes became less varied. Terrestrial nonam-

▲ Figure 8–23 Diversity of Paleozoic amniotes: Early amniotes varied in size from a few centimeters long to a couple of meters, and their ecological roles were equally diverse. (a) *Hylonomus,* a protorothyridid (lizard-size), represents the typical lizard-like body form of many early amniotes. (b) *Haptodus,* a synapsid (dog-size). (c) *Mesosaurus,* a mesosaur (cat-size). (d) *Captorhinus,* a captorhinid (lizard-size). (e) *Petrolacosaurus kansensis,* a stem diapsid (araeoscelidan; lizard-size). (f) *Procolophon,* a procolophonid (dog-size). (g) *Pareiasaurus,* a pareiasaur (cow-size).

niote tetrapods were at their peak in the Early Permian, and the groups that survived through the Triassic were mostly flattened, fully aquatic forms. From the start of the Mesozoic onward, terrestrial habitats were dominated by a series of radiations of amniote tetrapods.

Derived Features of Amniotes

Traditionally, amniotes have been distinguished by the amniotic egg and a waterproof skin. The egg is a good character, the skin not so good. The amniotic egg (Figure 8–24) is characteristic of turtles, lepidosaurs (lizards and their relatives), crocodilians, birds, monotremes (egg-laying mammals). Further, embryonic membranes that contribute to the placenta of therian mammals are homologous to certain membranes in the egg (see Chapter 20). The amniotic egg is assumed to have been the reproductive mode of Mesozoic diapsids, and fossil dinosaur eggs are relatively common. But in many other ways, amniotes represent a more derived kind of tetrapod than both the living amphibians and the Paleozoic nonamniote tetrapods (see Figure 8–1).

Skin permeability varies widely among living amphibians and amniotes. Although amniotes have a thicker skin than amphibians, and they have a keratinized epidermis, the presence of lipids in the skin is probably the most important determining factor in making the skin relatively impermeable to water. Compared to amphibians, amniotes have a greater variety of skin elaborations—scales, hair, and feathers—all formed from keratin. The lack of similar structures in living amphibians may be related to their use of the skin in respiration. Another important derived amniote feature is costal (rib) ventilation of the lungs. This also helps with conservation of water; amniotes rely on the lungs for gas exchange, and the skin need not be moist.

The amniotic egg is sometimes referred to as the land egg, but this is a misnomer. Many species of living amphibians and even some fishes have nonamniotic eggs that develop quite successfully on land, and terrestrial invertebrates also lay nonamniotic eggs. Both amniotic and nonamniotic eggs must have relatively moist conditions to avoid desiccation. The term *cleidoic egg* (Greek *cleido* = closed or locked) is often used as a synonym for the amniotic egg; but this, too, is not entirely correct. Technically, this term should be reserved for the rigid-shelled calcareous egg of birds that, unlike the eggs of most other amniotes, does not allow water uptake through the shell.

An amniotic egg is a remarkable example of biological complexity (Figure 8–24). The shell, which may be leathery and flexible (as in many lizards and turtles), or calcified and rigid (as in other lizards, turtles, and birds), provides mechanical protection while being porous enough to allow movement of respiratory gases and water vapor. The albumin (egg white) gives further protection against mechanical damage and provides a reservoir of water and protein. The large yolk is the energy supply for the developing embryo. At the beginning of embryonic development, the embryo is represented by a few cells resting on top of the yolk. As development proceeds these multiply and endodermal and mesodermal tissue surrounds the yolk, enclosing it in a **yolk sac** that is part of the developing gut. Blood vessels differentiate rapidly in the mesodermal tissue surrounding the yolk sac and transport food and gases to the embryo. By the end of development, only a small amount of yolk remains, and this is absorbed before or shortly after hatching.

In these respects the amniotic egg does not differ greatly from the nonamniotic eggs of amphibians and fishes: All vertebrates have a yolk sac extraembryonic membrane. The significant differences lie in three additional extraembryonic membranes—the **chorion**, **amnion**, and **allantois**. The chorion and amnion develop from outgrowths of the body wall at the edges of the developing plate-like embryo. These two pouches spread outward and around the embryo until they meet. At their junction, the membranes merge and leave an outer membrane—the chorion—which surrounds the embryo and yolk sac, and an inner membrane—the amnion—which surrounds the embryo itself. The allantoic membrane develops as an outgrowth of the hindgut posterior to the yolk sac and lies within the chorion. The allantois appears to have evolved as a storage place for nitrogenous wastes produced by the metabolism of the embryo, and the urinary bladder of the adult grows out from its base. The allantois also serves a respiratory organ because it is vascularized and can transport oxygen from the surface of the egg back to the embryo (and transport carbon dioxide in the opposite direction). The allantois is left behind in the egg when the embryo emerges, and the nitrogenous wastes stored in it do not have to be reprocessed.

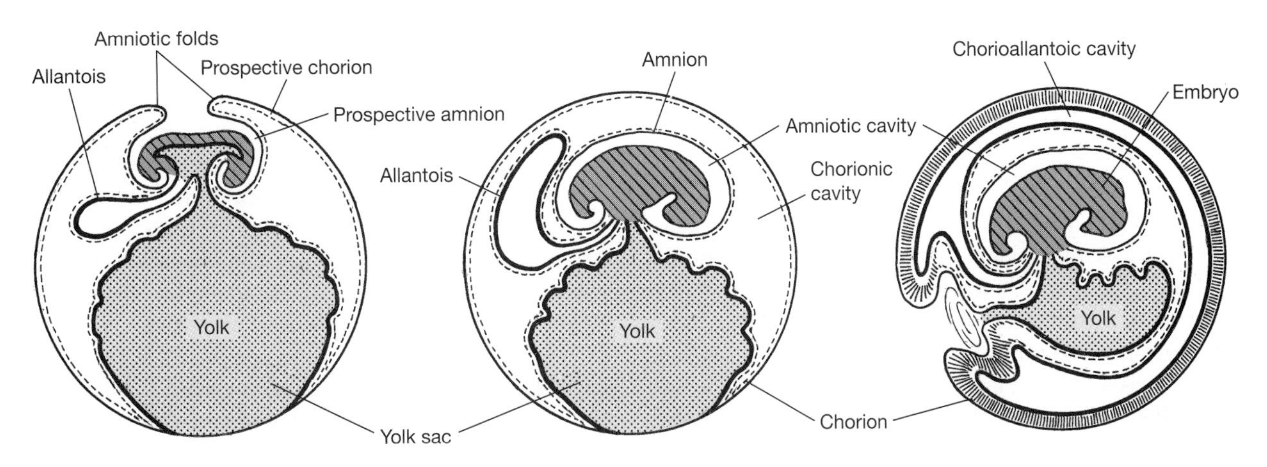

▲ Figure 8–24 Distinctive features of the amniotic egg. Progressive stages in development are illustrated from left to right.

How and why might the amniotic egg have evolved? It is possible for nonamniotic eggs to be laid on land, but probably only very small ones. Larger eggs (more than 10 millimeters in diameter) may be too large for diffusion of oxygen across their surface to satisfy the needs of the developing embryo, and they may also collapse under their own weight. The presence of additional extraembryonic membranes would enable larger eggs to be supported (as would a shell) and to provide exchange of respiratory gases. Robert Carroll (1970) developed an evolutionary scenario in which the ancestors of amniotes would have to have been small animals. The precursor to the amniote egg would have to be a nonamniote egg laid on land, otherwise there would be no evolutionary pressure to improve the gas exchange. But as egg size and adult size are correlated, only animals of small adult size (6 to 7 cm long) would be able to lay eggs small enough to survive. Note that Early Carboniferous candidates for amniote ancestry, *Casineria* and *Westlothiana*, were indeed small animals. Evolution of the amniotic egg might have been most important in allowing an increase in the size of adults that could lay eggs on the land.

How could we tell if a given fossil animal laid an amniotic egg? We call modern amniotes by that name because they lay eggs that contain extraembryonic membranes. We infer that some extinct tetrapods may have been amniotes because of their bony morphology, but clearly we cannot determine for certain whether they possessed extraembryonic membranes. How might we infer the reproductive mode of fossil animals called amniotes?

The capacity to lay an amniotic egg is not preserved in features of the skeleton, but we can make a fair estimate of its point of origin from the tetrapod phylogeny. As shown in Figure 8–13, the synapsids (mammals and extinct relatives) branched off from other reptiles very early on, and all other fossil animals that we consider to be amniotes are actually more closely related to living reptiles and birds. Because mammals have (or had) the same type of egg as other living amniotes, all animals higher than node number thirteen in the cladogram must have inherited this type of egg from the common ancestor of mammals and other amniotes.

A more difficult question is whether any fossil tetrapods lower in the phylogeny might have laid an amniotic egg. We know that this was not true of seymouriamorphs, because larval forms with external gills and lateral lines are known, showing that seymouriamorphs still had aquatic larvae. That leaves the diadectomorphs as the only possible candidates. Diadectomorphs such as *Diadectes* have morphological features indicative of herbivory, and terrestrial reproduction might be essential for a herbivorous tetrapod. Herbivorous young must ingest the symbiotic microorganisms needed to aid in digesting herbage immediately after hatching, and these organisms must be obtained from terrestrial microbial decomposers. Thus, it might be inferred (although we cannot be certain) that diadectomorphs had a fully terrestrial form of reproduction and hence possessed the amniotic egg.

Patterns of Amniote Temporal Fenestration

Amniotes have traditionally been subdivided by features of the skull, specifically by the number of holes in their head, or **temporal fenestration** (Latin *fenestra* = window). The major configurations (giving the name to different amniote lineages) are as follows: **anapsid** (Greek *an* = without and *apsid* = junction), seen in primitive amniotes and in turtles; **synapsid** (single arch, Greek *syn* = joined), seen in mammals and their ancestors; and **diapsid** (double arch, Greek *di* = two), seen in other reptiles and in birds. The arch in the name refers to the temporal bars (also known as temporal arcades) lying below and between the holes. Figure 8–25 illustrates the different patterns in the different groups. Note that the phylogenetic pattern of acquisition of these holes suggests that the condition arose independently in the synapsid and diapsid lineages, because more primitive sauropsids than diapsids lack holes entirely.

Even though the skull of living mammals is highly modified from the primitive synapsid condition, you can still feel these skull features in yourselves. If you put your hands on either side of your eyes, you can feel your cheek bone (the zygomatic arch)—that is the temporal bar that lies below your synapsid skull opening. Then, if you clench your jaw, you can feel the muscles bulging above the arch. They are passing through the temporal opening, running from their origin from the top of the skull to the insertion on your lower jaw (see Figure 8–25j).

What is the function of these holes? As you just discovered, they provide room for muscles to bulge. In general, amniotes have larger and more differentiated jaw muscles than nonamniotes, and the notion of room for bulging was originally the preferred evolutionary explanation. However, Frazzetta (1968) pointed out that the *initial* evolutionary reason for developing these holes must have been something different. Only a large hole will allow enough room for a bulging muscle, so what could be the evolutionary advantage of an initial, small hole? And why does no nonamniote ever develop temporal fenestration?

Frazzetta suggested that the key to the evolution of the temporal fenestrae lies in changes in the complexity and orientation of the jaw muscles. In fishes and nonamniote tetrapods the jaw muscles are a simple, little-differentiated mass, and the feeding movements consist of a simple inertial snap. The more domed skull of amniotes allows muscles to originate directly from the underside of the skull roof and to run vertically down to the jaw. This new orientation of one portion of the jaw adductor complex allows the animal to apply static pressure between the teeth when the jaws are closed. The other portion of the original adductor complex is now the pterygoideus muscle, originating from the pterygoid flange on the palate, a new feature in amniotes (Figure 8–26). The skull of nonamniotes is flat and shallow (possibly because they use their head for the buccal-pumping mode of lung ventilation), and a flat skull does not allow this orientation of the muscles. A muscle in this position in a nonamniote's skull would be too short to allow the jaw to open very far because muscles can be stretched only one third of their resting length.

Increased differentiation of the adductor muscle mass in amniotes would allow for more complex jaw movements, and thus permit a greater diversity of feeding types. But muscles with this new 90-degree orientation to the bone of attachment would result in considerable stress on the periosteal covering of the bone. A way of compensating for this would be to leave portions of the skull partially unossified (the temporal fenestrae always form at the junction of three or four skull bones), and to have the muscle originate from the connective tissue covering the fenestra (Figure 8–27). In this way a small hole would still reflect an important structural change to reduce stresses on the skull. Later evolutionary developments would then include the enlargement of this hole, with the muscles running through the hole to attach to the outside of the skull roof, as they do in ourselves and other extant amniotes (Figure 8–25j).

This evolutionary scenario may explain how a small, initial temporal fenestra might still have an important function. But it does not explain why the pattern is different in synapsids and diapsids, nor why some primitive, early amniotes do not have any fenestration. Although turtles have skulls that are technically anapsid, they have a posterior emargination of the skull (Figure 8–25b) that also allows expansion of the jaw muscles, and is analogous to a fenestra. Frogs and salamanders also have emarginated skulls. Early synapsids and diapsids were rather different kinds of animals; synapsids were

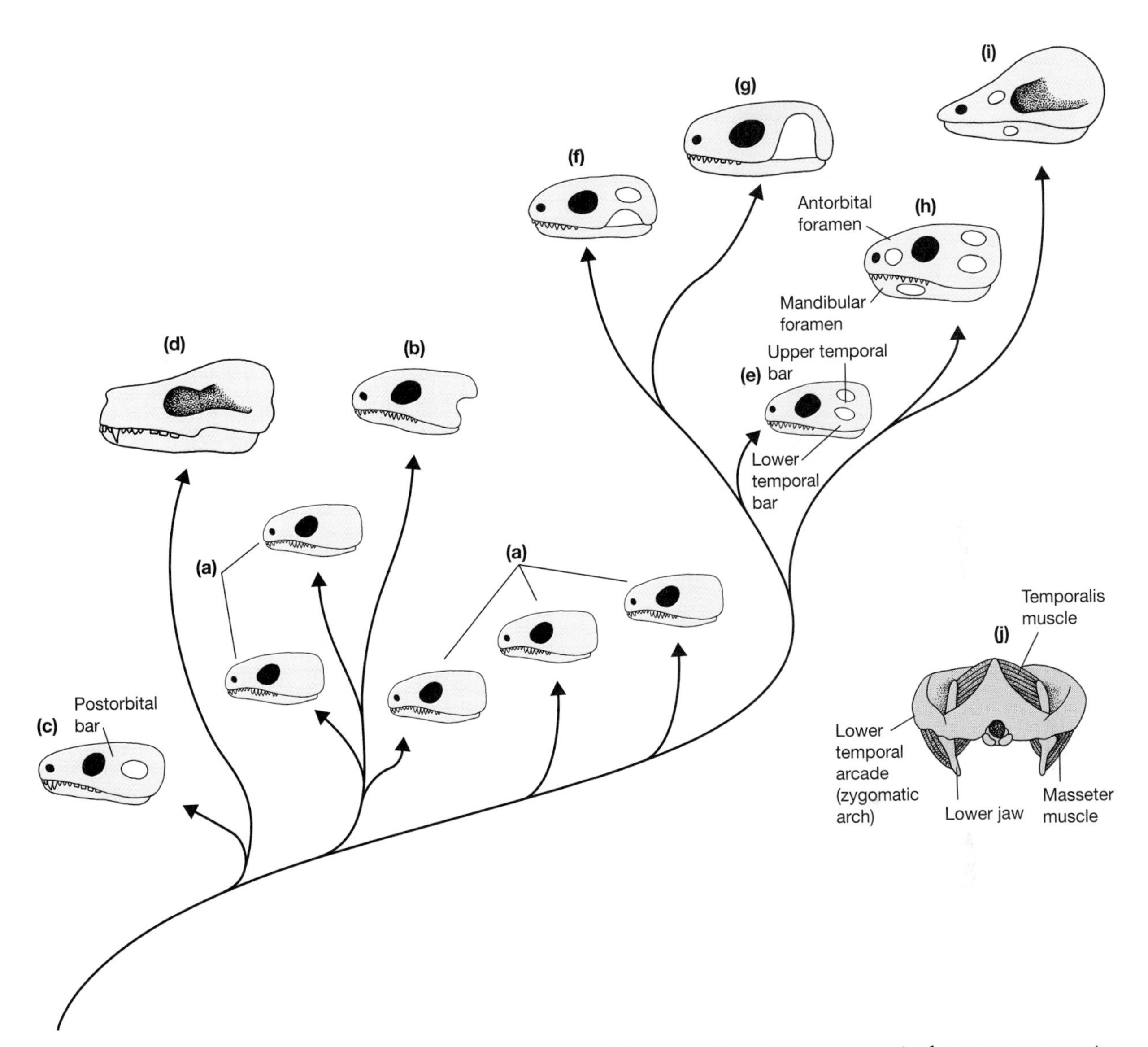

▲ Figure 8–25 Patterns in amniote skull fenestration. (a) Primitive anapsid condition, as seen in the common ancestor of all amniotes, and in basal members of the parareptiles and eureptiles. (b) Modified anapsid condition with emargination of the posterior portion of the skull, as seen in turtles. (c) Primitive synapsid condition, with lower temporal fenestra only. (d) Derived mammalian synapsid condition, where the orbit has become merged with the temporal opening, and dermal bone has grown down from the skull roof to surround the braincase. (e) Primitive diapsid condition, as seen today in the reptile *Sphenodon*; both upper and lower temporal fenestrae are present. (f) Lizard-like condition, typical of most squamates, where lower temporal bar has been lost. (g) Snake condition; upper temporal bar has been lost in addition to the lower bar. (h) Primitive archosaur diapsid condition, as seen in thecodonts and most dinosaurs; an antorbital foramen and a mandibular foramen have been added to the basic diapsid pattern (note that the antorbital foramen is secondarily reduced or lost in crocodiles). (i) Derived avian archosaur condition; convergently with the condition in mammals, the orbit has become merged with the temporal openings and the braincase is enclosed in dermal bone. (j) Posterior view through the skull of a synapsid (a cynodont mammal-like reptile) showing how the temporal fenestra allows muscles to insert on the outside of the skull roof. The temporalis and masseter muscles are divisions of the original amniote adductor muscle complex. Note that no separate "euryapsid" or "parapsid" condition is shown here. This condition, with the supposed existence of an upper temporal fenestra only, was thought to characterize marine reptiles. It is now realized that the so-called euryapsid condition is merely a modified diapsid one.

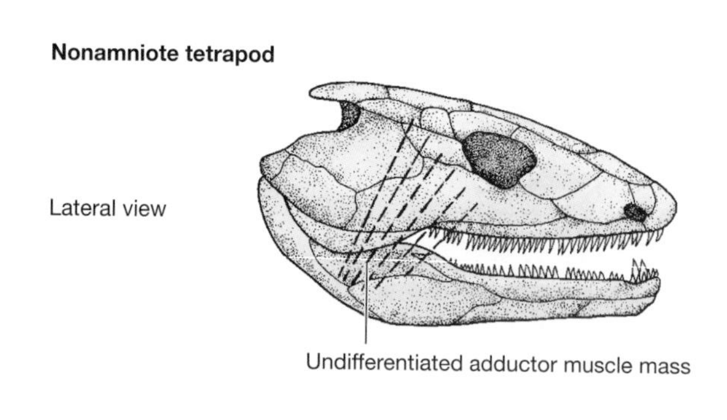

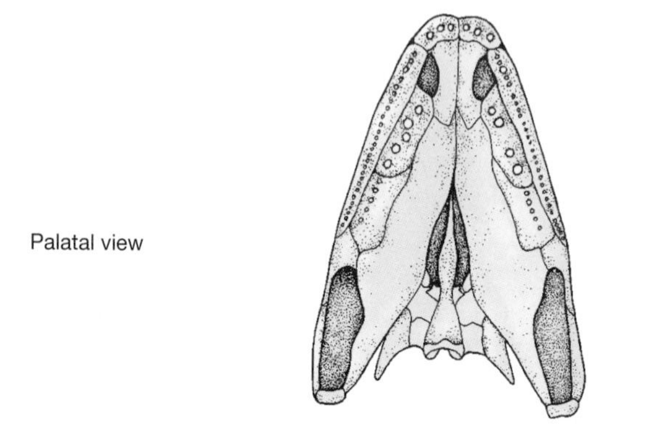

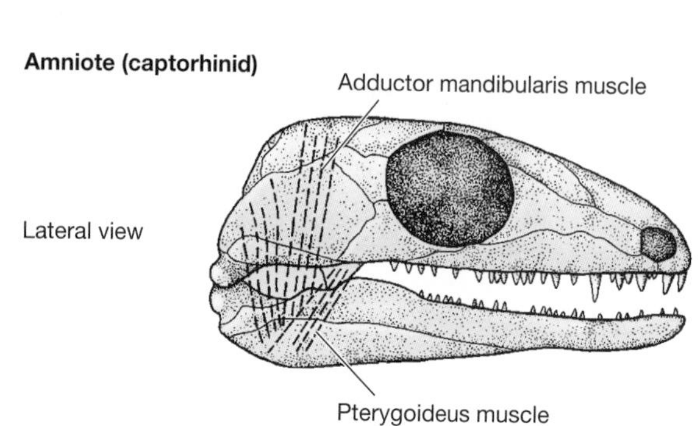

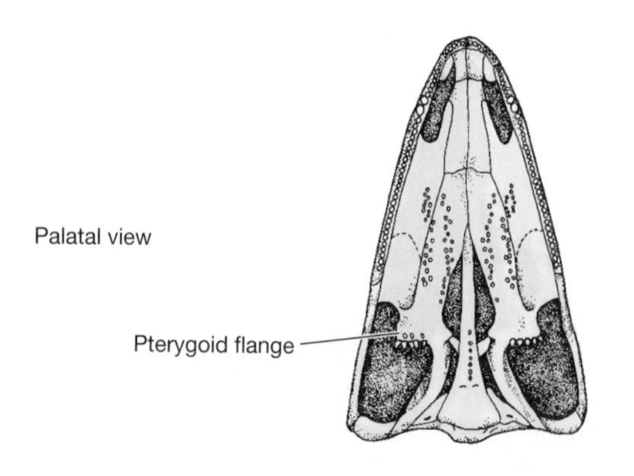

▲ Figure 8–26 Skulls showing differences in jaw muscles between nonamniote and amniote tetrapods.

more carnivorous and diapsids more insectivorous (Figure 8–23). Perhaps differences in muscle actions, relating to different feeding styles, encouraged temporal fenestration to take a different form in the two groups. An emphasis on static pressure in synapsids (as seen in modern mammals during feeding) might have allowed them to seize and crush large, slow prey, while an emphasis on inertial snapping in early diapsids (as seen in modern diapsids during feeding) might have allowed them to capture small, agile prey.

Differences among Major Amniote Groups

By looking at the phylogeny in Figure 8–13, you can see that the major splits in diversification of amniote lineages happened very early, soon after the origin of the amniotes themselves. The initial split was into synapsids (mammals and their extinct relatives) and sauropsids (reptiles and birds). The sauropsid lineage split into the parareptiles (probably including turtles) and the eureptiles (including other reptiles and birds; see Table 8.2). Although there is some debate about the monophyletic nature of the parareptiles, this grouping emphasizes that the late Paleozoic radiation included many lineages with little or no relationship to the living groups of reptiles (except for turtles). All amniotes share the basic derived features shown in Figure 8–1, but nevertheless different amniote lineages have long and independent evolutionary histories, and they have some profound differences from one another.

First, it is important to remember how different the mammal lineage is from the other amniotes. Although we loosely talk about mammals coming from reptiles, they branched off from the other amniotes very early. Mammals lack some of the derived features that we usually attribute to reptiles in general. Sauropsids (= reptiles plus birds) have several features that distinguish them from mammals, in addition to those of their skull as described in Figure 8–25. In particular, sauropsids all have good color vision, and the anatomy of their eyes suggests that they all came from a diurnal ancestry. All sauropsids also have a new type of harder keratin (called beta keratin) forming the scales or feathers.

The split between parareptiles and eureptiles also reveals some fundamental differences. All living eureptiles, at least, have hindlimbs longer than the forelimbs, owing them to run swiftly and even bipedally. Anatomical evidence also indicates that

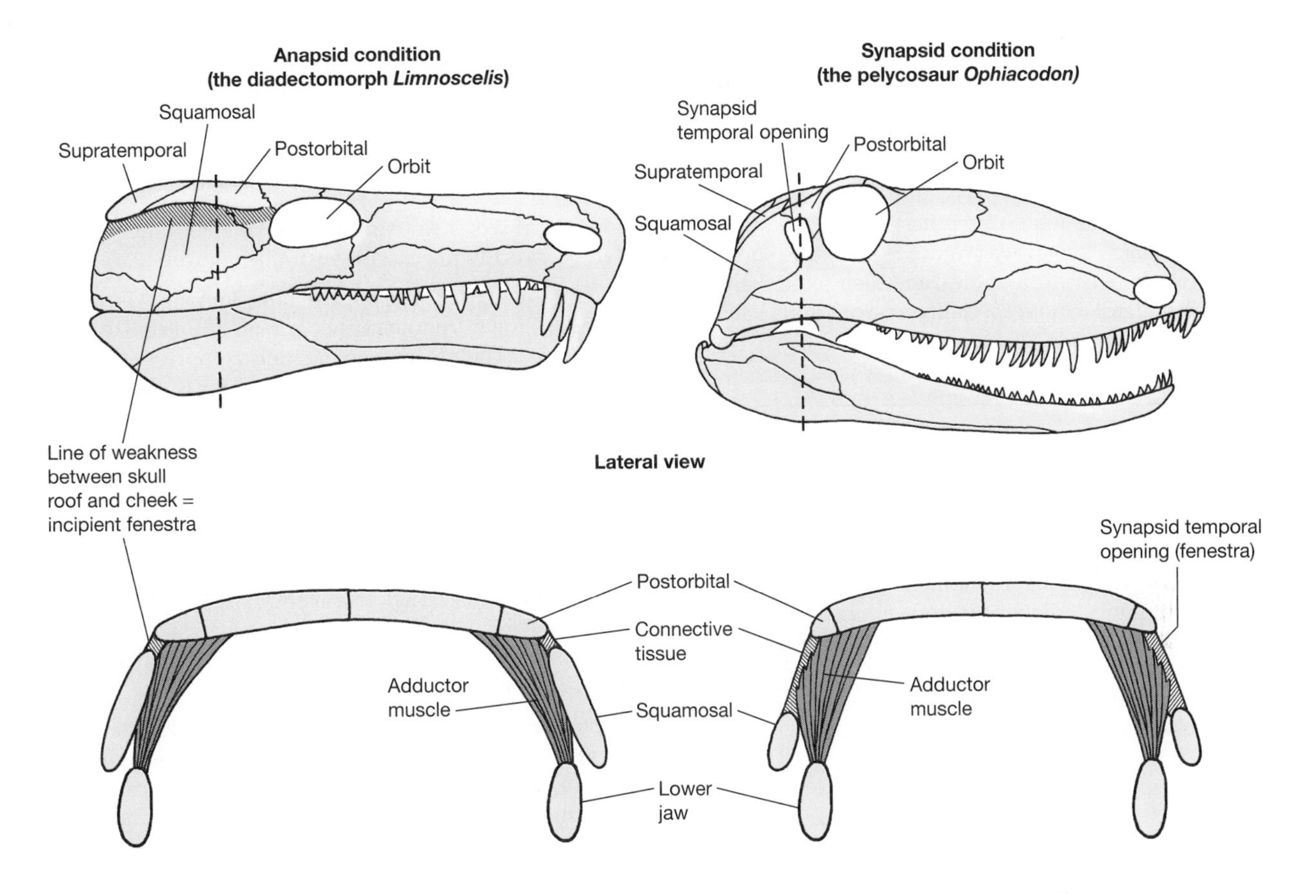

▲ Figure 8–27 Hypothetical origin of the temporal fenestra in amniotes.

the middle ear and eardrum evolved separately in the different sauropsid lineages; turtles, lepidosaurs, and archosaurs (crocodiles plus birds). It might seem strange that a middle-ear system, using the stapes as an auditory ossicle, could evolve convergently so many times. But it must be remembered that the stapes abuts the otic capsule at one end and the cheek at the other. Its original function as the hyomandibula in fishes was to support the jaw and operculum. Once freed from that role, it will transmit airborne sound vibrations to the inner ear just because of its location. The stapes of early tetrapods was a short, stout bone bracing the dermal skull roof against the braincase. It did not have a specific auditory function. In more derived tetrapods it became lightened to form an auditory ossicle, and this process occurred convergently in different groups.

Eureptiles split fairly early, certainly by the mid-Permian, into lepidosaur and archosaur lineages. We often think of lizards as typical amniotes because their general body form is probably rather like those of the most primitive known forms (see Figure 8–23a). But lizards, like other living lepidosaurs, are highly specialized in various ways. The forked tongue (seen in certain lizards and snakes) is a derived feature. (Old movies that portray dinosaurs with forked tongues are completely incorrect.) Lizards and snakes (squamates) have paired bifurcate penes (= plural of penis), the hemipenes, that lie in the tail base. Lepidosaurs shed their skin in its entirety, rather than in small fragments. (We all know that snakes shed their skins, but we rarely consider that we do as well. In fact, most house dust is made up of shed pieces of human skin!) Archosaurs too have some unique specializations, including the presence of a muscular gizzard, a chamber situated in the gut behind the stomach that helps in food processing. Archosaurs too have a more upright stance and a more stiffened torso than

Table 8.2 Major Groups Of Paleozoic Amniotes

Synapsida

Synapsids, or "mammal-like reptiles" are covered in Chapter 17. Early synapsids (Figure 8–23b) were somewhat larger than early eureptiles, and their larger heads and teeth suggest a more specialized carnivorous habit.

Sauropsida

Mesosaurs: The first secondarily aquatic amniotes (Figure 8–23c). Known from freshwater deposits in the Early Permian of South Africa and South America, they provide one of the classic pieces of evidence for continental drift, because these continents were united in Gondwana at that time. Swimming adaptations include large, probably webbed, hind feet, a laterally flattened tail, and heavily ossified ribs that may have acted as ballast in diving. The long jaws and slender teeth may have been used to strain small crustaceans from the surrounding water.

Parareptilia

Millerettids: Rather like the eureptile protorothyridids shown in Figure 8–23a. Known from the Late Permian of South Africa.

Procolophonids: Medium size, with peg-like teeth that were laterally expanded in later members of the group, apparently specialized for crushing or grinding, suggestive of herbivory. Known from the Late Permian to Late Triassic worldwide except Australia (Figure 8–23f).

Pareiasaurs. Large size, approaching 3m long (Figure 8–23g). Known from the Late Permian of Europe, Asia, and Africa. Their teeth were laterally compressed and leaf-shaped, like the teeth of herbivorous lizards. Pareiasaurs were evidently the dominant terrestrial herbivores of the later Permian.

Eureptilia

Protorothyridids: Small, relatively short-legged, rather lizard-like forms, probably insectivorous in habits (Figure 8–23a). Known from Mid Carboniferous to Early Permian, North America and Europe.

Captorhinids: Tetrapods with more robust skulls and flatter teeth than protorothyridids and early diapsids, and may have had more of an omnivorous diet that required crushing (Figure 8–23d). Known from the Early and Mid Permian of North America and Europe, and the Late Permian of East Africa.

Araeoscelidans (stem diapsids): Early diapsids with shorter bodied and longer legs than protorothyridids, but probably also insectivorous (Figure 8–23e). Araeoscelidans were known from the Late Carboniferous and Early Permian of North America and Europe.

---

lepidosaurs, and larger and more complexly subdivided lungs.

Posture and Locomotion in Derived Amniotes

A new role of the axial muscles in tetrapods is that of ventilation on land. However, as originally pointed out by David Carrier (1987), tetrapods such as lizards that still move via axial movements experience a conflict between the use of the axial muscles for locomotion and for ventilation. In primitive tetrapods the trunk muscles are used for both respiration and locomotion, and these animals are unable to run and breathe at the same time. This is an acceptable situation for animals with fairly low metabolic rates, such as present-day ectothermal amniotes, which rely primarily on anaerobic metabolism for short bursts of activity. But for a more active animal, especially an endotherm, this arrangement of the muscles limits the capacity for sustained aerobic activity. This locomotory conflict is discussed in more detail in Chapter 17 in connection with mammalian evolution, but basically the same problem was faced by all more-active amniotes, especially more-derived archosaurs such as dinosaurs and birds.

The conflict arising from the dual use of the trunk muscles is solved in part by the adoption of an upright posture, with the limbs held more directly underneath the body, and with the trunk held more rigidly in locomotion. With this posture, the limbs (rather than the trunk) provide the predominant propulsive force. This new posture was associated solely with a quadrupedal stance in the mammalian (synapsid) lineage of amniotes, but in the archosaurs upright posture was combined with a tendency toward bipedality (with obligate bipedality being a feature of early dinosaurs). Running on two legs rather than on four also isolates the trunk from the effects of locomotor bending. Crocodiles are archosaurs with a sprawling stance today, but this posture is probably associated with their aquatic mode of life. The fossil record suggests that crocodilians were derived from ancestors with an upright stance, and some early crocodiles appear to have been bipedal.

Summary

The origin of tetrapods from elpistostegid lobe-finned fishes in the Devonian is inferred from similarities in the bones of the skull and braincase, vertebral structure, and limb skeleton. Paleozoic tetrapods comprise about a dozen distinct lineages of uncertain relationships. One current view divides them into three major groups; batrachomorphs, reptilomorphs, and lepospondyls. The batrachomorphs (mainly the temnospondyls) were predominantly aquatic, and a few were as large as crocodiles. Temnospondyls radiated extensively in the Late Carboniferous and Permian, and several lineages extended through the Triassic into the Early Cretaceous. Modern amphibians—the salamanders, frogs, and caecilians—may be derived from the temnospondyl lineage. The nonamniote reptilomorphs (= "anthracosaurs") were never as diverse as the temnospondyls. They included medium- to large-size terrestrial and aquatic forms that radiated during the Carboniferous, and became extinct in the Permian. Amniotes may be derived from this group. The lepospondyls were small forms of uncertain phylogenetic affiliation.

The amniotic egg, with its distinctive extraembryonic membranes, is a shared derived character that distinguishes the amniotes (turtles, squamates, crocodilians, birds, and mammals) from the nonamniotes (fishes and amphibians). The earliest amniotes were small animals, and their appearance coincided with a major radiation of terrestrial insects in the Carboniferous. Progressive modifications of the postcranial skeleton of early amniotes appear to show increased agility, and simultaneous changes in the jaws may be related to predation on insects. By the end of the Carboniferous, amniotes had begun to radiate into most of the terrestrial life zones that had been occupied by nonamniotes, and only the relatively aquatic groups of nonamniote tetrapods maintained much diversity through the Triassic.

The major groups of amniotes can be distinguished by different patterns of temporal fenestration—holes in the dermal skull roof that reflect increasing complexity of jaw musculature. The major division of amniotes is into synapsids (mammals and their relatives) and sauropsids (reptiles and birds). The sauropsids can be subdivided into the parareptiles (probably including the turtles) and the eureptiles. The basic division within eureptiles is lepidosaurs (including lizards and snakes) and archosaurs (including crocodiles, birds, and dinosaurs).

Some derived amniote features probably evolved convergently in the different groups. An enclosed middle ear was evolved independently within synapsids, turtles, lepidosaurs, and archosaurs (and also in temnospondyls and frogs). The conflict between breathing and running was solved independently in synapsids and archosaurs with the adoption of an upright posture, but only archosaurs combined this with a predominately bipedal stance.

Additional Readings

Ahlberg, P. E., and A. R. Milner. 1994. The origin and early diversification of tetrapods. *Nature* 368:507–514.

Benton, M. J. 1997. *Vertebrate Palaeontology*, 2d ed. London: Chapman & Hall.

Brainerd, E. L., J. S. Ditelberg, and D. M. Bramble. 1993. Lung ventilation in salamanders and the evolution of vertebrate air-breathing mechanisms. *Biological Journal of the Linnean Society* 49:163–183.

Bray, A. A. 1985. The evolution of terrestrial vertebrates: Environmental and physiological considerations. *Philosophical Transactions of the Royal Society of London* B309:289–322.

Carrier, D. R. 1987. The evolution of locomotor stamina in tetrapods: Circumventing a mechanical constraint. *Paleobiology* 12:326–341.

Carroll, R. L. 1970. Quantitative aspects of the amphibian-reptilian transition. *Forma et Functio* 3:165–178.

Clack, J. A. 1997. The evolution of tetrapod ears and the fossil record. *Brain, Behavior and Evolution* 50:198–212.

Coates, M. I., and J. A. Clack. 1990. Polydactyly in the earliest known tetrapod limbs. *Nature* 347:66–69.

Coates, M. I., and J. A. Clack. 1991. Fish-like gills and breathing in the earliest known tetrapod. *Nature* 352:234–235.

Coates, M. I., and J. A. Clack. 1995. Romer's gap: Tetrapod origins and terrestriality. *Bulletin de Musée national Histoire naturelle, Paris, 4ᵉ series* 17:373–388.

Coates, M. I. and M. J. Cohn. 1998. Fins, limbs, and tails: Outgrowths and axial patterning in vertebrate evolution. *BioEssays* 20:371–381.

Daeschler, E. B., and N. H. Shubin. 1998. Fish with fingers? *Nature* 391:133.

Daeschler, E. B. et al. 1994. A Devonian tetrapod from North America. *Science* 265:639–642.

Erdman, M. V., R. L. Caldwell, and M. Moosa, 1998. Indonesian "king of the sea" discovered. *Nature* 395:335.

Farmer, C. 1997. Did the lungs and intracardiac shunt evolve to oxygenate the heart in vertebrates? *Paleobiology* 23:358–372.

Frazzetta, T. H. 1968. Adaptive problems and possibilities in the temporal fenestration of tetrapod skulls. *Journal of Morphology* 125:145–158.

Gauthier, J. A. 1994. The diversification of the amniotes. In D. R. Prothero and R. M. Schoch (Eds.), *Major Features of Vertebrate Evolution*, pp. 129–159. Short Courses in Paleontology no. 7. Knoxville: Paleontological Society and University of Tennessee Press.

Hildebrand, M. 1980. The adaptive significance of tetrapod gait selection. *American Zoologist* 20:255–267.

Janis, C. M., and C. Farmer. 1999. Proposed habits of early tetrapods. Gills, kidneys, and the water-land transition. *Zoological Journal of the Linnean Society* 126:117–126.

Janis, C. M. and J. C. Keller. 2001. Modes of ventilation in early tetrapods: Costal aspiration as a key feature of tetrapods. *Acta Palaeontologica Polonica* 46:137–170.

Laurin, M., M. Girondot, and A. de Ricqlès. 2000. Early tetrapod evolution. *Trends in Ecology and Evolutionary Biology* 15:118–123.

Monastersky, R. 1999. Out of the swamps. *Science News* 155:328–330.

Paton, R. L., T. R. Smithson, and J. A. Clack. 1999. An amniote-like skeleton from the Early Carboniferous of Scotland. *Nature* 398:508–513.

Reisz, R. R. 1997. The origin and early evolutionary history of amniotes. *Trends in Ecology and Evolution* 12:218–222.

Shubin, N. H., and P. Alberch. 1986. A morphogenetic approach to the origin and basic organization of the tetrapod limb. *Evolutionary Biology* 20:319–387.

Shubin, N. H., C. Tabin, and S. Carroll. 1997. Fossils, genes and the evolution of animal limbs. *Nature* 388:639–648.

Sumida, S. S. and K. L. M. Martin (Eds.). 1997. *Amniote Origins: Completing the Transition to the Land*. San Diego: Academic.

Thomson, K. S. 1993. The origin of the tetrapods. *American Journal of Science* 293A:33–62.

Vogel, S. 1988. *Life's Devices: The Physical World of Plants and Animals*. Princeton, N.J.: Princeton University Press.

Zimmer, C. 1998. *At the Water's Edge*. New York: Free Press.

Web Explorations

On-line resources for this chapter are on the World Wide Web at: http://www.prenhall.com/pough (click on the Table of Contents link and then select Chapter 8).

CHAPTER

9

Salamanders, Anurans, and Caecilians

The three lineages of extant amphibians (salamanders, frogs, and caecilians) have very different body forms, but they are identified as a monophyletic evolutionary lineage by several shared derived characters. Some of these characters—especially the moist, permeable skin—have channeled the evolution of the three lineages in similar directions. Frogs are the most successful amphibians, and it is tempting to think that the variety of locomotor modes permitted by their specialized morphology may be related to their success: Frogs can jump, walk, climb, and swim. In contrast to frogs, salamanders retain the ancestral tetrapod locomotor pattern of lateral undulations combined with limb movements. The greatest diversity among salamanders is found in the Plethodontidae, many species of which project the tongue to capture prey on its sticky tip.

The range of reproductive specializations of amphibians is nearly as great as that of fishes, a remarkable fact considering that there are more than five times as many species of fishes as amphibians. The ancestral reproductive mode of amphibians probably consisted of laying large numbers of eggs that hatched into aquatic larvae, and many amphibians still reproduce this way. An aquatic larva gives a terrestrial species access to resources that would not otherwise be available to it. Modifications of the ancestral reproductive mode include bypassing the larval stage, viviparity, and parental care of eggs and young, including females that feed their tadpoles.

The permeable skin of amphibians is central to many aspects of their lives. The skin is a major site of respiratory gas exchange, and must be kept moist. Evaporation of water from the skin limits the activity of most amphibians to relatively moist microenviroments. The skin contains glands that produce substances used in courtship and other glands that produce toxic substances that deter predators. Many amphibians advertise their toxicity with bright warning colors, and some nontoxic species deceive predators by mimicking the warning colors of toxic forms.

9.1 Amphibians

The extant amphibians, or Lissamphibia, are tetrapods with moist, scaleless skins. The group includes three distinct lineages: anurans (frogs), urodeles (salamanders), and gymnophionans (caecilians or apodans). Most amphibians have four well-developed limbs, although a few salamanders and all caecilians are limbless. Frogs lack tails (hence the name anura, which means "without a tail"), whereas most salamanders have long tails. The tails of caecilians are short, as are those of other groups of elongate, burrowing animals.

At first glance, the three lineages of amphibians appear to be very different kinds of animals: Frogs have long hindlimbs and short, stiff bodies that don't bend when they walk, salamanders have forelimbs and hindlimbs of equal size and move with lateral undulations, and caecilians are limbless and employ serpentine locomotion. These obvious differences are all related to locomotor specializations, however, and closer examination shows that amphibians have many derived characters in common, indicating that they form a monophyletic evolutionary lineage (Table 9.1). We will see that many of these shared characters play important roles in the

Table 9.1 Shared derived characters of amphibians

1. *Structure of the skin and the importance of cutaneous gas exchange.* All amphibians have mucus glands that keep the skin moist. A substantial part of an amphibian's exchange of oxygen and carbon dioxide with the environment takes place through the skin. All amphibians also have poison (granular) glands in the skin.
2. *Papilla amphibiorum.* All amphibians have a special sensory area, the papilla amphibiorum, in the wall of the sacculus of the inner ear. The papilla amphibiorum is sensitive to frequencies below 1000 hertz (cycles per second), and a second sensory area, the papilla basilaris, detects sound frequencies above 1000 Hz.
3. *Operculum-plectrum complex.* Most amphibians have two bones that are involved in transmitting sounds to the inner ear. The columella (plectrum), which is derived from the hyoid arch, is present in salamanders, caecilians, and most frogs. The operculum develops in association with the fenestra ovalis of the inner ear. The columella and operculum are fused in anurans and caecilians and in some salamanders.
4. *Green rods.* Salamanders and frogs have a distinct type of retinal cell, the green rod. Caecilians apparently lack green rods, but the eyes of caecilians are extremely reduced and these cells may have been lost.
5. *Pedicellate teeth.* Nearly all modern amphibians have teeth in which the crown and base (pedicel) are composed of dentine and are separated by a narrow zone of uncalcified dentine or fibrous connective tissue. A few amphibians lack pedicellate teeth, including salamanders of the genus *Siren* and frogs of the genera *Phyllobates* and *Ceratophrys;* the boundary between the crown and base is obscured in some other genera. Pedicellate teeth also occur in some actinopterygian fishes, which are not thought to be related to amphibians.
6. *Structure of the levator bulbi muscle.* This muscle is a thin sheet in the floor of the orbit that is innervated by the fifth cranial nerve. It causes the eyes to bulge outward, thereby enlarging the buccal cavity. This muscle is present in salamanders and anurans and in modified form in caecilians.

functional biology of amphibians. Perhaps the most important derived character of extant amphibians is a moist, permeable skin. The name applied to the lineage, Lissamphibia, refers to the texture of the skin (Greek *liss* = smooth). Many of the Paleozoic nonamniote tatrapods had dermal armor in the form of bony scutes in the skin; a permeable, unadorned skin is a derived character shared by lissamphibians.

All living adult amphibians are carnivorous, and relatively little morphological specialization is associated with different dietary habits within each group. Amphibians eat almost anything they are able to catch and swallow. The tongue of aquatic forms is broad, flat, and relatively immobile, but some terrestrial amphibians can protrude the tongue from the mouth to capture prey. The size of the head is an important determinant of the maximum size of prey that can be taken, and sympatric species of salamanders frequently have markedly different head sizes, suggesting that this is a feature that reduces competition. Frogs in the tropical American genera *Lepidobatrachus* and *Ceratophrys,* which feed largely on other frogs, have such large heads that they are practically walking mouths.

The anuran body form probably evolved from a more salamander-like starting point. Both jumping and swimming have been suggested as the mode of locomotion that made the change advantageous. Salamanders and caecilians swim as fishes do—by passing a sine wave down the body. Anurans have inflexible bodies and swim with simultaneous thrusts of the hind legs. Some paleontologists have proposed that the anuran body form evolved because of the advantages of that mode of swimming. An alternative hypothesis traces the anuran body form to the advantage gained by an animal that could rest near the edge of a body of water and escape aquatic or terrestrial predators with a rapid leap followed by locomotion on either land or water. The stem anuran *Triadobatrachus* may be an example of that body form.

The oldest fossils that may represent modern amphibians are isolated vertebrae of Permian age that appear to include both salamander and anuran types. The oldest true frog, *Prosalirus,* is from the Early Jurassic of North America. Salamanders and caecilians also are known from the Jurassic. Clearly, the modern orders of amphibians have had separate evolutionary histories for a long time. The continued presence of such common characteristics as a permeable skin, after at least 250 million years of independent evolution, suggests that the shared characteristics are critical in shaping the evolutionary success of modern amphibians. In other characters, such as reproduction, locomotion, and defense, amphibians show tremendous diversity.

Salamanders—Urodela

The salamanders have the most generalized body form and locomotion of the living amphibians. Salamanders are elongate, and all but a very few species of completely aquatic salamanders have four functional limbs (Figure 9–1). Their walking-trot gait is probably similar to that employed by the earliest tetrapods. It combines the lateral bending characteristic of fish locomotion with leg movements. The 10 families, containing approximately 415 species, are almost entirely limited to the Northern Hemisphere; their southernmost occurrence is in northern South America (Table 9.2). North and Central America have the greatest diversity of salamanders—more species of salamanders are found in Tennessee than in all of Europe and Asia combined. Paedomorphosis is widespread among salamanders, and several families of aquatic salamanders are constituted solely of such paedomorphic forms. These forms can be recognized by the retention of larval characteristics, including larval tooth and bone patterns, the absence of eyelids, retention of a functional lateral line system, and (in some cases) retention of external gills.

The largest living salamanders are the Japanese and Chinese giant salamanders (*Andrias*), which reach lengths of 1 meter or more. The related North American hellbenders (*Cryptobranchus*) grow to 60 centimeters. All are members of the Cryptobranchidae and are paedomorphic and permanently aquatic. As their name indicates (Greek *crypto* = hidden and *branchus* = gill), they do not retain external gills, although they do have other larval characteristics. Another group of large aquatic salamanders are the mudpuppies (*Necturus*, Proteidae and *Siren*, Sirenidae), which consist of paedomorphic species that retain external gills. Mudpuppies occur in lakes and streams in eastern North America. The congo eels (three species of aquatic salamanders in the genus *Amphiuma*) live in the lower Mississippi Valley and coastal plain of the United States. They have well developed lungs, and can estivate in the mud of dried ponds for up to two years.

Several lineages of salamanders have adapted to life in caves. The constant temperature and moisture of caves makes them good salamander habitats, and food is supplied by cave-dwelling invertebrates. The brook salamanders (*Eurycea*, Plethodontidae) include species that form a continuum from those with fully metamorphosed adults inhabiting the twilight zone near cave mouths to fully paedomorphic forms in the depths of caves or sinkholes. The Texas blind salamander, *Typhlomolge*, is a highly specialized cave dweller—blind, white, with external gills, extremely long legs, and a flattened snout used to probe underneath pebbles for food. The unrelated European olm (*Proteus*, Proteidae) is another cave salamander that has converged on the same body form.

Terrestrial salamanders like the North American mole salamanders (*Ambystoma*) and the European salamanders (*Salamandra*) have aquatic larvae that lose their gills at metamorphosis. The most fully terrestrial salamanders, the lungless plethodontids (such as the slimy salamander, *Plethodon glutinosus*, include species in which the young hatch from eggs as miniatures of the adult and there is no aquatic larval stage.

Feeding Specializations of Plethodontid Salamanders Lungs seem an unlikely organ for a terrestrial vertebrate to abandon, but among salamanders the evolutionary loss of lungs has been a successful tactic. The Plethodontidae is characterized by the absence of lungs and contains more species and has a wider geographic distribution than any other lineage of salamanders. Furthermore, many plethodontids have evolved specializations of the hyobranchial apparatus that allow them to protrude the tongue a considerable distance from the mouth to capture prey. This ability has not evolved in salamanders with lungs, probably because the hyobranchial apparatus in these forms is an essential part of the respiratory system.

Salamanders lack ribs, so they cannot expand and contract the rib cage to move air in and out of the lungs. Instead, they employ a buccal pump that forces air from the mouth into the lungs. A sturdy hyobranchial apparatus in the floor of the mouth and throat is an essential part of this pumping system, whereas tongue protrusion requires that parts of the hyobranchial apparatus be elongated and lightened. The modification of the hyobranchial apparatus that allows tongue protrusion is not compatible with buccal pump respiration. Reliance on the skin instead of the lungs for gas exchange may have been a necessary first step in the evolution of tongue protrusion by plethodontids.

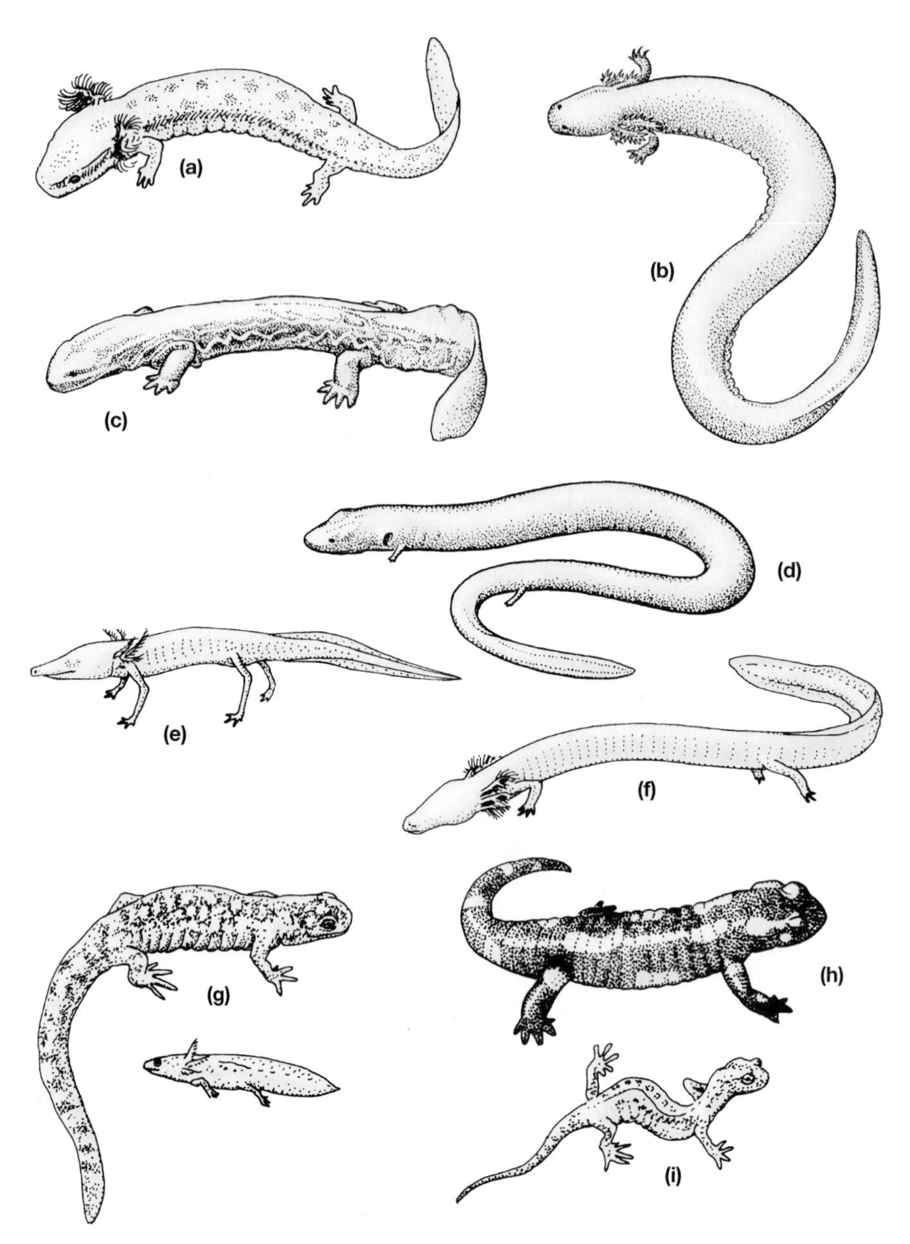

▲ Figure 9–1 Diversity of salamanders. The body forms of salamanders reflect differences in their life histories and habitats. Aquatic salamanders may retain gills as adults: (a) North American mudpuppy and (b) North American siren. They may have folds of skin that are used for gas exchange, or they may rely on lungs and the body surface. (c) North American hellbender. (d) North American congo eel. Specialized cave-dwelling salamanders are white and lack eyes: (e) Texas blind salamander and (f) European olm. Terrestrial salamanders usually have sturdy legs: (g) North American tiger salamander and its aquatic larva, (h) European fire salamander and (i) North American slimy salamander—*Plethodon*, Plethodontidae.

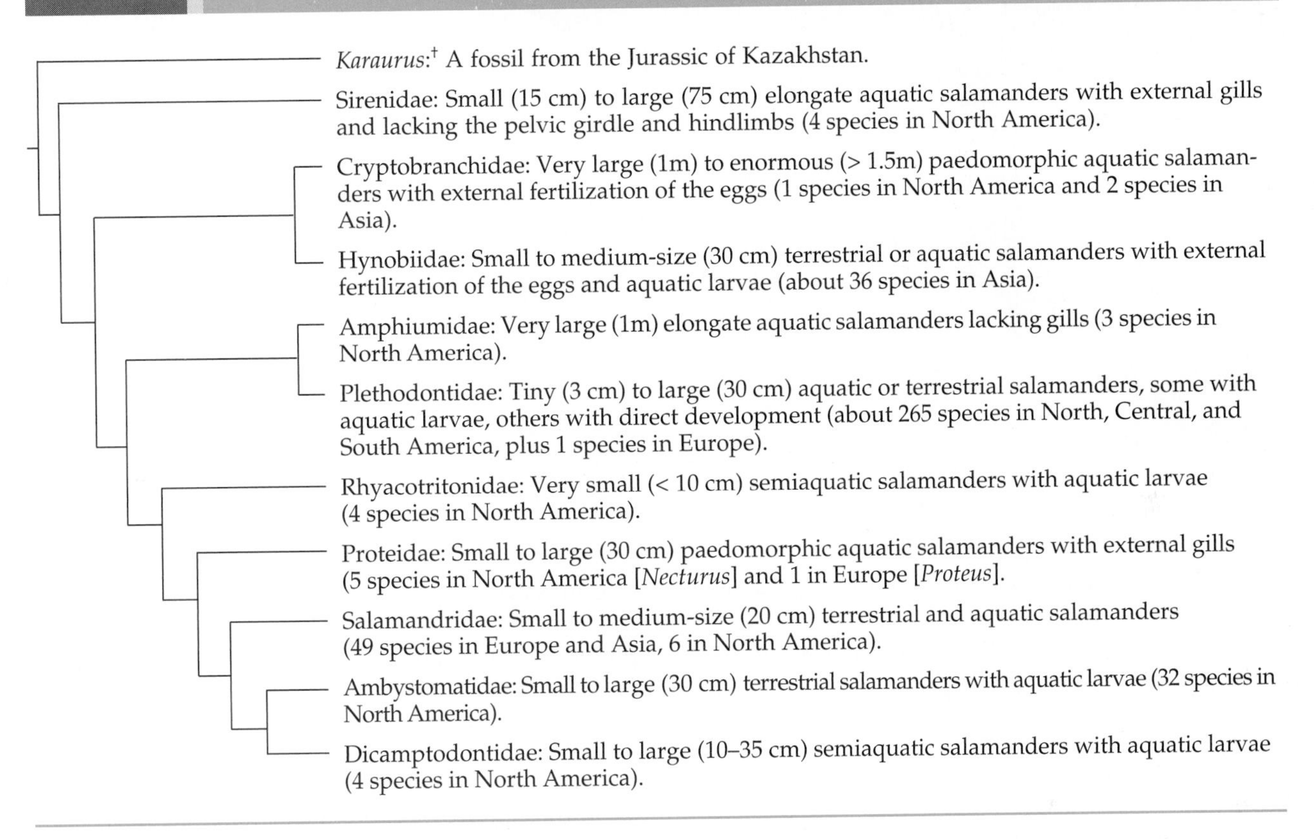

Table 9.2 Families of salamanders

Karaurus:† A fossil from the Jurassic of Kazakhstan.

Sirenidae: Small (15 cm) to large (75 cm) elongate aquatic salamanders with external gills and lacking the pelvic girdle and hindlimbs (4 species in North America).

Cryptobranchidae: Very large (1m) to enormous (> 1.5m) paedomorphic aquatic salamanders with external fertilization of the eggs (1 species in North America and 2 species in Asia).

Hynobiidae: Small to medium-size (30 cm) terrestrial or aquatic salamanders with external fertilization of the eggs and aquatic larvae (about 36 species in Asia).

Amphiumidae: Very large (1m) elongate aquatic salamanders lacking gills (3 species in North America).

Plethodontidae: Tiny (3 cm) to large (30 cm) aquatic or terrestrial salamanders, some with aquatic larvae, others with direct development (about 265 species in North, Central, and South America, plus 1 species in Europe).

Rhyacotritonidae: Very small (< 10 cm) semiaquatic salamanders with aquatic larvae (4 species in North America).

Proteidae: Small to large (30 cm) paedomorphic aquatic salamanders with external gills (5 species in North America [*Necturus*] and 1 in Europe [*Proteus*].

Salamandridae: Small to medium-size (20 cm) terrestrial and aquatic salamanders (49 species in Europe and Asia, 6 in North America).

Ambystomatidae: Small to large (30 cm) terrestrial salamanders with aquatic larvae (32 species in North America).

Dicamptodontidae: Small to large (10–35 cm) semiaquatic salamanders with aquatic larvae (4 species in North America).

The symbol † indicates an extinct taxon.

The modifications of the respiratory system and hyobranchial apparatus that allow tongue protrusion appear to be linked with several other characteristics of the biology of plethodontids (Roth and Wake 1985, Wake and Marks 1993). These associations can be seen most clearly in the bolitoglossine plethodontids, which have the most specialized tongue-projection mechanisms (Figure 9–2). Bolitoglossine plethodontids (Greek *bola* = dart and *glossa* = tongue) can project the tongue a distance equivalent to their head plus trunk length and can pick off moving prey. This ability requires fine visual discrimination of distance and direction, and the eyes of bolitoglossines are placed more frontally on the head than the eyes of less specialized plethodontids. Furthermore, the eyes of bolitoglossines have a large number of nerves that travel to the ipsilateral (= same side) visual centers of the brain as well as the strong contralateral (= opposite side) visual projection that is typical of salamanders. Because of this neuroanatomy, bolitoglossines have a complete dual projection of the binocular visual fields to both hemispheres of the brain. They can very exactly and rapidly estimate their distance from a prey object.

Tongue projection is reflected in many different aspects of the life-history characteristics of plethodontid salamanders, including their reproductive modes. Aquatic larval salamanders employ suction feeding, opening the mouth and expanding the throat to create a current of water that carries the prey item with it. The hyobranchial apparatus is an essential part of this feeding mechanism, and the first ceratobranchial becomes well developed during the larval period. In contrast, enlargement of the second ceratobranchial is associated with the tongue-projection mechanism of adult plethodontids. Furthermore, larval

◀ Figure 9–2 Prey capture by a bolitoglossine salamander, *Hydromantes*.

salamanders have laterally placed eyes, and the optic nerves project mostly to the contralateral side of the brain. Thus the morphological specializations that make aquatic plethodontid larvae successful are different from the specializations of adults that allow tongue projection, and this situation creates a conflict between the selective forces that act on juveniles and adults.

The bolitoglossines do not have aquatic larvae, and the morphological specializations of adult bolitoglossines appear during embryonic development. In contrast, hemidactyline plethodontids do have aquatic larvae that use suction feeding. As adults, hemidactylines have considerable ability to project the tongues, but they retain the large first ceratobranchial that appears in the larvae. This is a mechanically less efficient arrangement than the large second ceratobranchial of bolitoglossines, and the ability of hemidactylines to project their tongues is correspondingly less than that of bolitoglossines. Thus, the development of a specialized feeding mechanism by plethodontid salamanders has gone hand in hand with such diverse aspects of their biology as respiratory physiology and life history, and demonstrates that organisms evolve as whole functioning units, not as collections of independent characters.

Social Behavior of Plethodontid Salamanders Plethodontid salamanders can be recognized externally by the nasolabial groove that extends ventrally from each external naris to the lip of the upper jaw (Figure 9–3). These grooves are an important part of the chemosensory system of plethodontids. As a plethodontid salamander moves about, it repeatedly presses its snout against the substrate. Fluid is drawn into the grooves and moves upward to the external nares, into the nasal chambers, and over the chemoreceptors of the vomeronasal organ.

Studies of plethodontid salamanders have contributed greatly to our understanding of the roles of competition and predation in shaping the structure of ecological communities. Much of the recent work in this area has focused on experimental manipulations of animals in the field or laboratory. Because plethodontid salamanders have small home ranges and often remain in a restricted area for their entire lives, they are excellent species to use for these studies.

Males of many plethodontid salamanders defend all-purpose territories that are used for feeding and

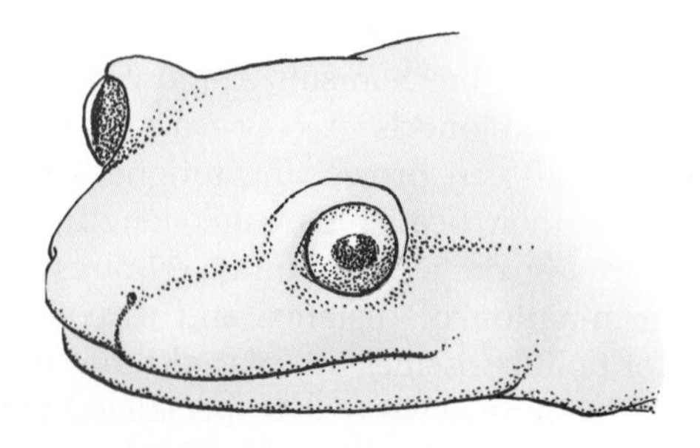

▲ Figure 9–3 Nasolabial grooves of a plethodontid salamander.

reproduction. Studies of these salamanders have revealed patterns of social behavior and foraging that seem remarkably complex for animals with skulls the size of a match head and brains little larger than the head of a pin. Robert Jaeger and his colleagues have studied the territorial behavior of the red-backed salamander, *Plethodon cinereus,* a common species in woodlands of eastern North America. Male red-backed salamanders readily establish territories in cages in the laboratory. A resident male salamander marks the substrate of its cage with pheromones (chemical substances that are released by an individual and stimulate responses by other individuals of the species). A salamander can distinguish between substrates it has marked and those marked by another male salamander or by a female salamander. Male salamanders can also distinguish between the familiar scent of a neighboring male salamander and the scent of a male they have not previously encountered, and they react differently to those scents.

In laboratory experiments, red-backed salamanders select their prey in a way that maximizes their energy intake: When equal numbers of large and small fruit flies are released in the cages, the salamanders first capture the large flies. This is the most profitable foraging behavior for the salamanders because it provides the maximum energy intake per capture. In a series of experiments, Jaeger and his colleagues showed that territorial behavior and fighting can interfere with the ability of salamanders to select the most profitable prey (Jaeger et al. 1983). These experiments used surrogate salamanders that were made of a roll of moist filter paper the same length and diameter as a salamander. The surrogates were placed in the cages of resident salamanders to produce three experimental conditions: a control surrogate, a familiar surrogate, and an unfamiliar surrogate. In the control experiment, male red-backed salamanders were exposed to a surrogate that was only moistened filter paper; it did not carry any salamander pheromone. For both of the other groups, the surrogate was rolled across the substrate of the cage of a different male salamander to absorb the scent of that salamander before being placed in the cage of a resident male.

The experiments lasted 7 days; the first 6 days were conditioning periods, and the test itself occurred on the seventh day. For the first 6 days, the resident salamanders in both of the experimental groups were given surrogates bearing the scent of another male salamander. The residents thus had the opportunity to become familiar with the scent of that male. On the seventh day, however, the familiar and unfamiliar surrogate groups were treated differently. The familiar surrogate group once again received a surrogate salamander bearing the scent of the same individual it had been exposed to for the previous 6 days, whereas the resident salamanders in the unfamiliar surrogate group received a surrogate bearing the scent of a different salamander, one they had never been exposed to before. After a 5-minute pause, a mixture of large and small fruit flies was placed in each cage, and the behavior of the resident salamander was recorded.

The salamanders in the familiar surrogate group showed little response to the now-familiar scent of the other male salamander. They fed as usual, capturing large fruit flies. In contrast, the salamanders that were exposed to the scent of an unfamiliar surrogate began to give threatening and submissive displays, and their rate of prey capture decreased as a result of the time they spent displaying. In addition, salamanders exposed to unfamiliar surrogates did not concentrate on catching large fruit flies, so the average energy intake per capture also declined. The combined effects of the reduced time spent feeding and the failure to concentrate on the most profitable prey items caused an overall 50-percent decrease in the rate of energy intake for the salamanders exposed to the scent of an unfamiliar male.

The ability of male salamanders to recognize the scent of another male after a week of habituation in the laboratory cages suggests that they would show the same behavior in the woods. That is, a male salamander could learn to recognize and ignore the scent of a male in the adjacent territory, while still being able to recognize and attack a strange intruder. Learning not to respond to the presence of a neighbor may allow a salamander to forage more effectively, and it may also help to avoid injuries that can occur during territorial encounters. Resident male red-backed salamanders challenge strange intruders, and the encounters involve aggressive and submissive displays and biting. Bites on the body (Figure 9–4a) can drive another male away, but are not likely to do permanent damage. A bite to the tail (Figure 9–4b) may cause the bitten salamander to autotomize (break off) its tail. Salamanders store fat in their tails, and this injury may delay reproduction for a year while the tail is regenerated. Most bites are directed at the snout of an opponent (Figure 9–4c),

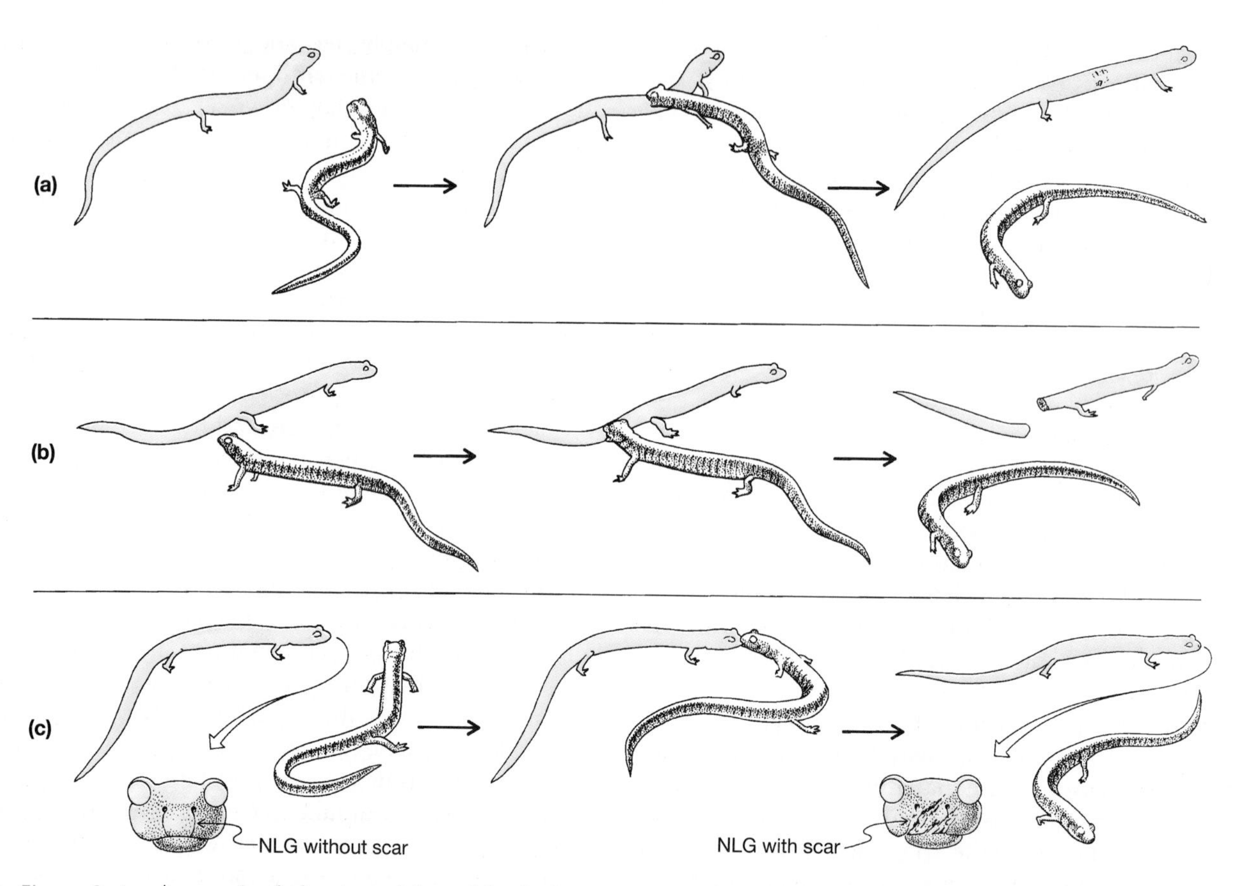

▲ Figure 9–4 Aggressive behaviors of the red-backed salamander, *Plethodon cinereus.* The resident salamander is dark and the intruder is blue in these drawings. (a) Resident bites the intruder on the body. (b) Bitten on the tail by the resident, the intruder autotomizes its tail to escape. (c) Resident bites intruder on the snout, injuring the nasolabial grooves.

and may damage the nasolabial grooves. The nasolabial grooves are used for olfaction, and these injuries can reduce a salamander's success in finding prey. Twelve salamanders that had been bitten on the snout were able to capture an average of only 5.8 fruit flies in a 2-hour period compared with an average of 18.6 flies for 12 salamanders that had not been bitten. In a sample of 144 red-backed salamanders from the Shenandoah National Forest, 11.8 percent had been bitten on the nasolabial grooves, and these animals weighed less than the unbitten animals, presumably because their foraging success had been reduced (Jaeger 1981).

The possibility of serious damage to an important sensory system during territorial defense provides an additional advantage for a red-backed salamander in being able to distinguish neighbors (which are always there and are not worth attacking) from intruders (which represent a threat and should be attacked). The phenomenon of being able to recognize territorial neighbors has been called dear enemy recognition, and may be generally advantageous because it minimizes the time and energy that territorial individuals expend on territorial defense and also minimizes the risk of injury during territorial encounters. Similar dear enemy recognition has been described among territorial birds that show more aggressive behavior on hearing the songs of strangers than they do when hearing the songs of neighbors.

Anurans

In contrast to the limited number of species of salamanders and their restricted geographic distribution, the anurans (Greek *an* = without and *uro* = tail) include 27 families with 4300 species and occur on all the continents except Antarctica (Table 9.3). Specialization of the body for jumping is the most conspicuous skeletal feature of anurans. The hindlimbs and muscles form a lever system that can catapult an anuran into the air (Figure 9–5), and numerous morphological specializations are associated with this type of locomotion; the hind legs are elongate and the tibia and fibula are fused. A powerful pelvis strongly fastened to the vertebral column is clearly necessary, as is stiffening of the vertebral column. The ilium is elongate and reaches far anteriorly, and the posterior vertebrae are fused into a solid rod, the **urostyle.** The pelvis and urostyle render the posterior half of the trunk rigid. The vertebral column is short, with only five to nine presacral vertebrae, and these are strongly braced by zygapophyses that restrict lateral bending. The strong forelimbs and flexible pectoral girdle absorb the impact of landing. The eyes are large and are placed well forward on the head, giving binocular vision.

The hindlimbs generate the power to propel the frog into the air, and this high level of power production results from structural and biochemical features of the limb muscles. The internal architecture of the semimembranosus muscle and its origin on the ischium and insertion below the knee allow it to operate at the length that produces maximum force during the entire period of contraction. In addition, the muscle shortens faster and generates more power than muscles from most other animals. Furthermore, the intracellular physiological processes of muscle contraction continue at the maximum level throughout contraction, rather than declining as is the case in muscles of most vertebrates.

Specializations of the locomotor system can be used to distinguish many different kinds of anurans (Figure 9–6). The difficulty is finding names for them, since the diversity of anurans exceeds the number of common names that can be used to distinguish various ecological specialties (Figure 9–7). Animals called frogs usually have long legs and move by jumping. Many species of ranids have this body form, and very similar jumping frogs are found in other lineages as well. Semiaquatic forms are moderately streamlined and have webbed feet. Stout-bodied terrestrial anurans that make short hops instead of long leaps are often called toads. They usually have blunt heads, heavy bodies, relatively short legs, and little webbing between the toes. This body form is represented by members of the Bufonidae, and very similar body forms are found in other families, including the spadefoot toads of western North America and the horned frogs of South America. Spadefoot toads take their name from a keratinized structure on the hind foot that they use for digging backward into the soil with rapid movements of their hind legs. The horned frogs have extremely large heads and mouths. They feed on small vertebrates, including birds and mammals, but particularly on other frogs. The tadpoles of horned frogs also are carnivorous and feed on other tadpoles. Many frogs that burrow headfirst have pointed heads, stout bodies, and short legs.

Arboreal frogs usually have large heads and eyes, and often slim waists and long legs. Arboreal frogs in many different families move by quadrupedal walking and climbing as much as by leaping. Many arboreal species of hylids and rhacophorids have enlarged toe disks and are called tree frogs. The surfaces of the toe pads consist of an epidermal layer with peg-like projections separated by spaces or canals (Figure 9–8). Mucus glands distributed over the disks secrete a viscous solution of long-chain, high-molecular-weight polymers in water. Arboreal species of frogs use a mechanism known as wet adhesion to stick to smooth surfaces. (This is the same mechanism by which a wet scrap of paper sticks to glass.) The watery mucus secreted by the glands on the toe disks forms a layer of fluid between the disk and the surface and establishes a meniscus at the interface between air and fluid at the edges of the toes. As long as no air bubble enters the fluid layer, a combination of surface tension (capillarity) and viscosity holds the toe pad and surface together.

Frogs can adhere to vertical surfaces and even to the undersides of leaves. Cuban tree frogs (*Osteopilus septentrionalis,* Hylidae) can cling to a sheet of smooth plastic as it is rotated past the vertical; the frogs do not begin to slip until the rotation reaches an average of 151 degrees—that is, 61 degrees past vertical. Adhesion and detachment of the pads alternate as a frog walks across a leaf. When a frog moves forward its pads are peeled loose, starting at the rear, as the toes are lifted. The two most specialized families of tree frogs, the Hylidae and Rhacophoridae, have a cartilage (the intercalary cartilage) that lies between the last two bones in the toes. The intercalary cartilage may promote adhesion by increasing the angle through which the toe can move before peeling begins.

Tree frogs are not able to rest facing downward, because in that orientation the frog's weight causes

Table 9.3 Families of anurans

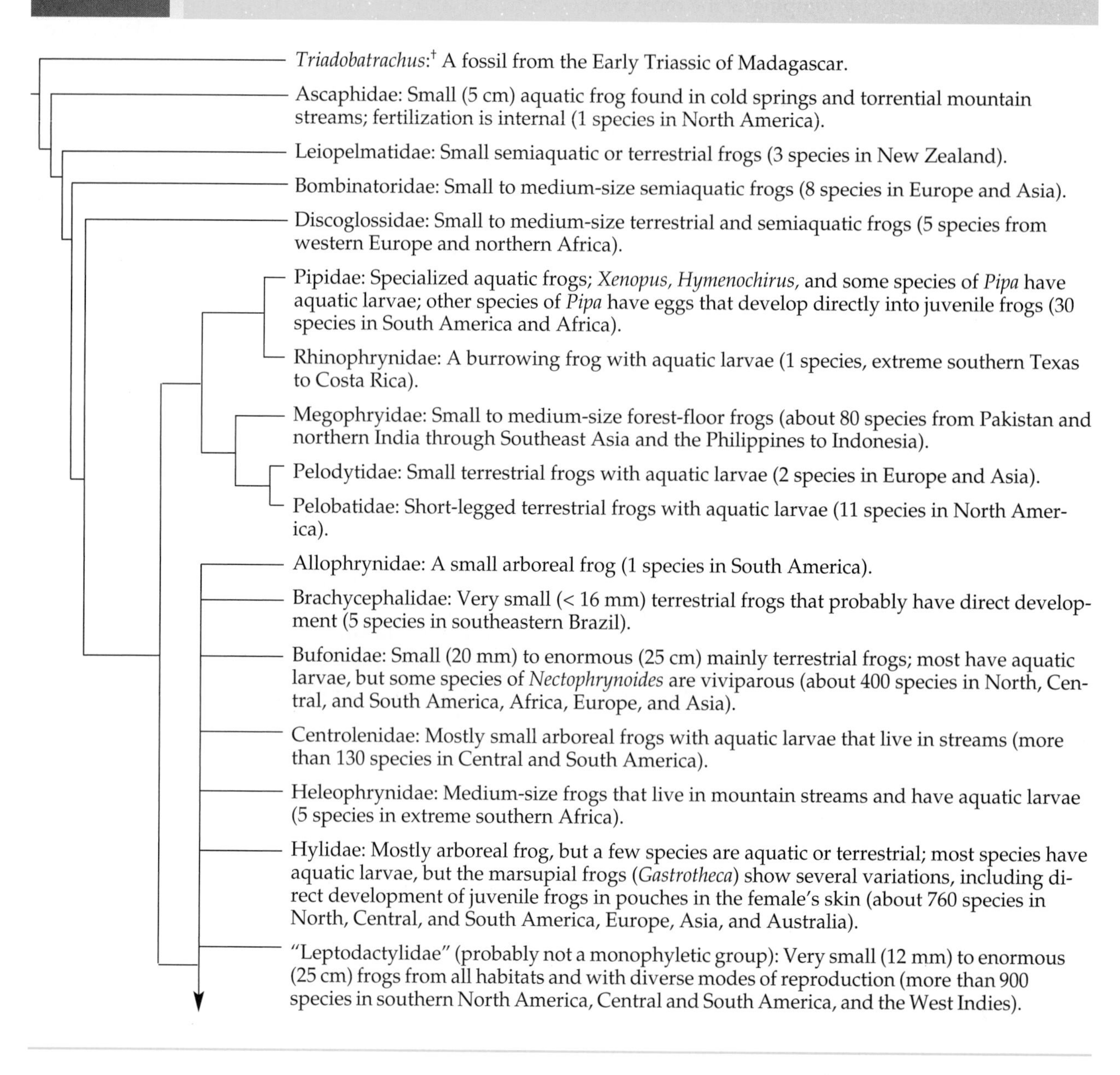

Triadobatrachus:† A fossil from the Early Triassic of Madagascar.

Ascaphidae: Small (5 cm) aquatic frog found in cold springs and torrential mountain streams; fertilization is internal (1 species in North America).

Leiopelmatidae: Small semiaquatic or terrestrial frogs (3 species in New Zealand).

Bombinatoridae: Small to medium-size semiaquatic frogs (8 species in Europe and Asia).

Discoglossidae: Small to medium-size terrestrial and semiaquatic frogs (5 species from western Europe and northern Africa).

Pipidae: Specialized aquatic frogs; *Xenopus, Hymenochirus,* and some species of *Pipa* have aquatic larvae; other species of *Pipa* have eggs that develop directly into juvenile frogs (30 species in South America and Africa).

Rhinophrynidae: A burrowing frog with aquatic larvae (1 species, extreme southern Texas to Costa Rica).

Megophryidae: Small to medium-size forest-floor frogs (about 80 species from Pakistan and northern India through Southeast Asia and the Philippines to Indonesia).

Pelodytidae: Small terrestrial frogs with aquatic larvae (2 species in Europe and Asia).

Pelobatidae: Short-legged terrestrial frogs with aquatic larvae (11 species in North America).

Allophrynidae: A small arboreal frog (1 species in South America).

Brachycephalidae: Very small (< 16 mm) terrestrial frogs that probably have direct development (5 species in southeastern Brazil).

Bufonidae: Small (20 mm) to enormous (25 cm) mainly terrestrial frogs; most have aquatic larvae, but some species of *Nectophrynoides* are viviparous (about 400 species in North, Central, and South America, Africa, Europe, and Asia).

Centrolenidae: Mostly small arboreal frogs with aquatic larvae that live in streams (more than 130 species in Central and South America).

Heleophrynidae: Medium-size frogs that live in mountain streams and have aquatic larvae (5 species in extreme southern Africa).

Hylidae: Mostly arboreal frog, but a few species are aquatic or terrestrial; most species have aquatic larvae, but the marsupial frogs (*Gastrotheca*) show several variations, including direct development of juvenile frogs in pouches in the female's skin (about 760 species in North, Central, and South America, Europe, Asia, and Australia).

"Leptodactylidae" (probably not a monophyletic group): Very small (12 mm) to enormous (25 cm) frogs from all habitats and with diverse modes of reproduction (more than 900 species in southern North America, Central and South America, and the West Indies).

the toe pads to peel off the surface. Frogs invariably orient their bodies facing upward or across a slope, and they rotate their feet if necessary to keep the toes pointed upward. When a frog must descend a vertical surface, it moves backward. This orientation keeps the toes facing upward. During backward locomotion, toes are peeled loose from the tip backward by a pair of tendons that insert on the dorsal surface of the terminal bone of the toe.

Toe disks have evolved independently in several lineages of frogs, and show substantial convergence in structure. Expanded toe disks are not limited exclusively to arboreal frogs; many terrestrial species that move across fallen leaves on the forest floor also have toe disks.

Several aspects of the natural history of anurans appear to be related to their different modes of locomotion. In particular, short-legged species that move

Table 9.3 Continued

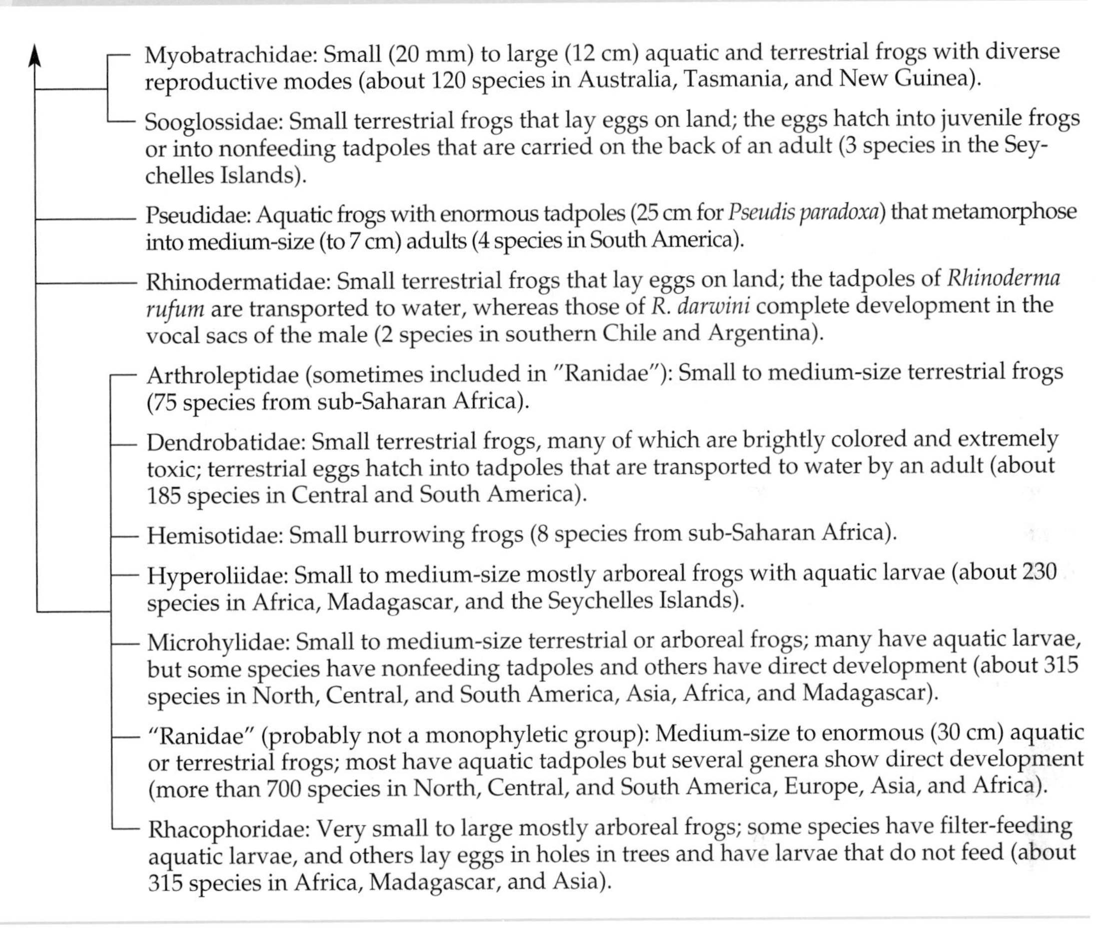

Myobatrachidae: Small (20 mm) to large (12 cm) aquatic and terrestrial frogs with diverse reproductive modes (about 120 species in Australia, Tasmania, and New Guinea).

Sooglossidae: Small terrestrial frogs that lay eggs on land; the eggs hatch into juvenile frogs or into nonfeeding tadpoles that are carried on the back of an adult (3 species in the Seychelles Islands).

Pseudidae: Aquatic frogs with enormous tadpoles (25 cm for *Pseudis paradoxa*) that metamorphose into medium-size (to 7 cm) adults (4 species in South America).

Rhinodermatidae: Small terrestrial frogs that lay eggs on land; the tadpoles of *Rhinoderma rufum* are transported to water, whereas those of *R. darwini* complete development in the vocal sacs of the male (2 species in southern Chile and Argentina).

Arthroleptidae (sometimes included in "Ranidae"): Small to medium-size terrestrial frogs (75 species from sub-Saharan Africa).

Dendrobatidae: Small terrestrial frogs, many of which are brightly colored and extremely toxic; terrestrial eggs hatch into tadpoles that are transported to water by an adult (about 185 species in Central and South America).

Hemisotidae: Small burrowing frogs (8 species from sub-Saharan Africa).

Hyperoliidae: Small to medium-size mostly arboreal frogs with aquatic larvae (about 230 species in Africa, Madagascar, and the Seychelles Islands).

Microhylidae: Small to medium-size terrestrial or arboreal frogs; many have aquatic larvae, but some species have nonfeeding tadpoles and others have direct development (about 315 species in North, Central, and South America, Asia, Africa, and Madagascar).

"Ranidae" (probably not a monophyletic group): Medium-size to enormous (30 cm) aquatic or terrestrial frogs; most have aquatic tadpoles but several genera show direct development (more than 700 species in North, Central, and South America, Europe, Asia, and Africa).

Rhacophoridae: Very small to large mostly arboreal frogs; some species have filter-feeding aquatic larvae, and others lay eggs in holes in trees and have larvae that do not feed (about 315 species in Africa, Madagascar, and Asia).

The symbol † indicates an extinct taxon.

by hopping are frequently wide-ranging predators that cover large areas as they search for food. This behavior makes them conspicuous to predators, and their short legs prevent them from fleeing rapidly enough to escape. Many of these anurans have potent defensive chemicals that are released from glands in the skin when they are attacked. Species of frogs that move by jumping, in contrast to those that hop, are usually sedentary predators that wait in ambush for prey to pass their hiding places. These species are usually cryptically colored, and they often lack chemical defenses. If they are discovered by a predator, they rely on a series of rapid leaps to get away. Anurans that forage widely encounter different kinds of prey from those that wait in one spot, and differences in dietary habits may be associated with differences in locomotor mode.

Aquatic anurans use suction to engulf food in water, but most species of semiaquatic and terrestrial anurans have sticky tongues that can be flipped out to trap prey and carry it back to the mouth (Figure 9–9). Most anurans use a catapult-like mechanism to project the tongue. As the mouth is opened, contraction of the genioglossus muscles causes the front of the tongue to stiffen (Figure 9–9a). Simultaneously, contraction of a short muscle at the front of the jaws (the submentalis) provides a fulcrum, and the stiffened tongue rotates forward over the submentalis and flips out of the mouth (Figure 9–9b). Inertia causes the rear portion of the tongue to

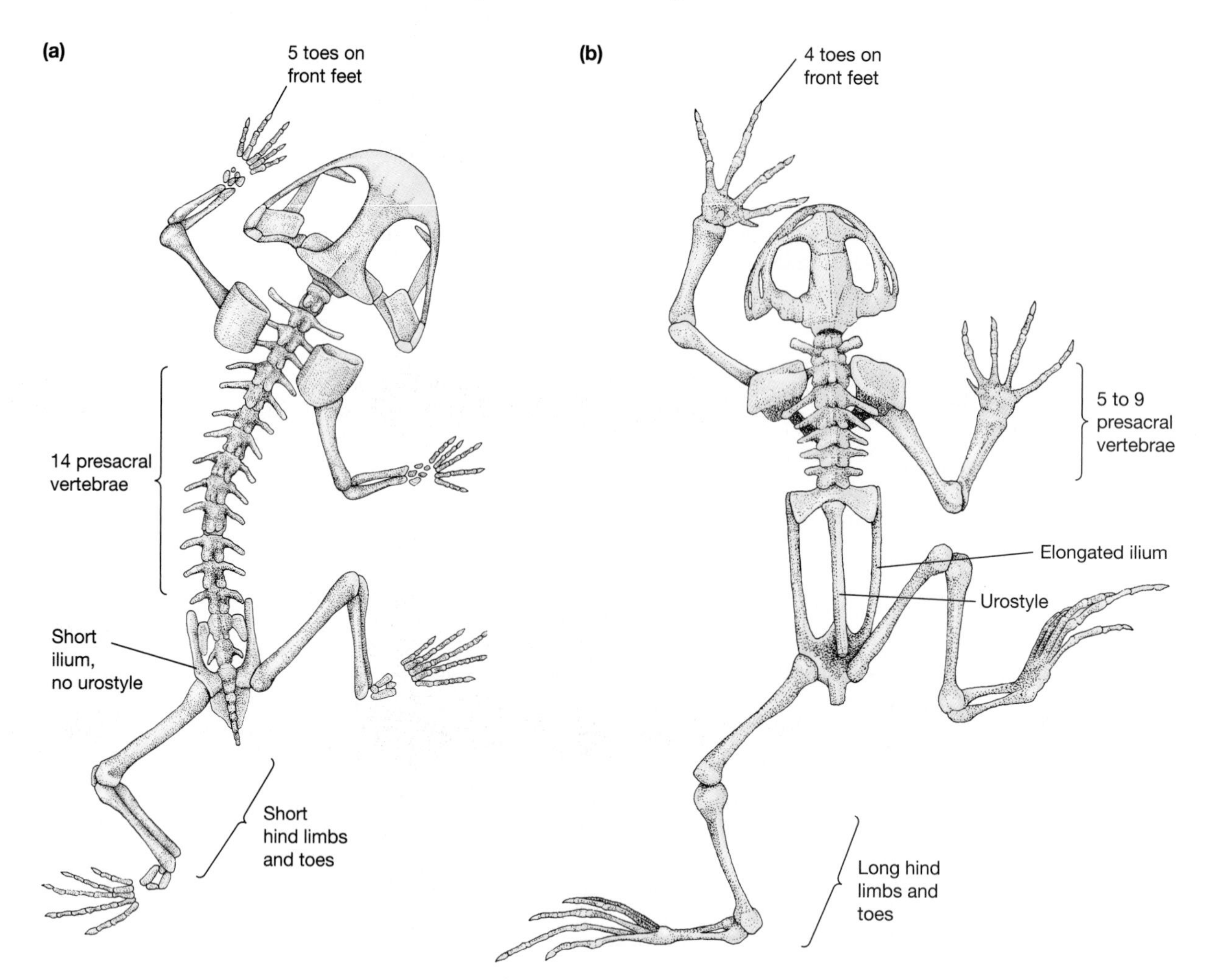

▲ Figure 9–5 *Triadobatrachus* and a modern anuran. The Triassic fossil *Triadobatrachus* is considered the sister group of anurans. Derived characters of anurans visible in this comparison include shortening of the body, elongation of the ilia, and fusion of the posterior vertebrae to form a urostyle.

elongate as it emerges; and because the tongue has rotated, its dorsal surface slams down on the prey. The tongue is drawn back into the mouth by the hyoglossus muscle, which originates on the hyoid apparatus and inserts within the tongue (Figure 9–9c, d).

Caecilians

The third group of living amphibians is the least known and does not even have an English common name (Figure 9–10). These are the caecilians (Gymnophiona), legless burrowing or aquatic amphibians that occur in tropical habitats around the world (Table 9.4). The eyes of caecilians are covered by skin or even by bone. Some species lack eyes entirely, but the retinae of other species have the layered organization that is typical of vertebrates and appear to be able to detect light. Conspicuous dermal folds (annuli) encircle the bodies of caecilians. The primary annuli overlie vertebrae and myotomal septa and reflect body segmentation. Many species of caecilians have dermal scales in pockets in the annuli; scales are not known in the other groups of living amphibians. A second unique feature of caecilians is a pair of protrusible tentacles, one on each side of the snout between the eye and nostril. Some structures that are associated with the eyes of other vertebrates have become associated with the tentacles of caecilians. One of the eye muscles, the retractor bulbi, has become the retractor muscle for the

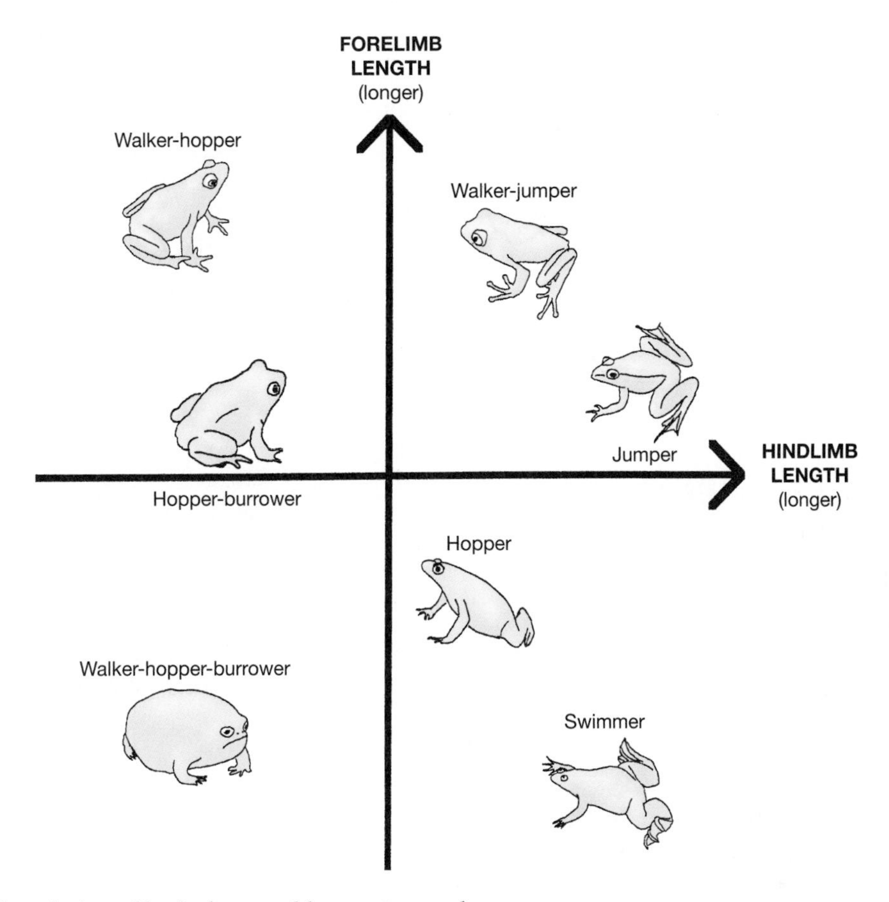

▲ Figure 9–6 The relation of body form and locomotor mode among anurans.

tentacle; the levator bulbi moves the tentacle sheath; and the Harderian gland lubricates the channel of the tentacle. The tentacle is probably a sensory organ that allows chemical substances to be transported from the animal's surroundings to the vomeronasal organ on the roof of the mouth. The eye of caecilians in the African family Scolecomorphidae is attached to the side of the tentacle near its base. When the tentacle is protruded, the eye is carried along with it, moving out the tentacular aperature beyond the roofing bones of the skull.

The earliest caecilian known is *Eocaecilia,* an Early Jurassic fossil from the Kayenta formation of western North America. It has a combination of ancestral and derived characters, and is the sister taxon of extant caecilians (Table 9.4). *Eocaecilia* has a fossa for a chemosensory tentacle, which is a unique derived character of apodans, but it also has four legs, whereas all living apodans are legless.

Caecilians feed on small or elongate prey—termites, earthworms, and larval and adult insects—and the tentacle may allow them to detect the presence of prey when they are underground. Females of some species of caecilians brood their eggs, whereas other species give birth to living young. The embryos of terrestrial species have long, filamentous gills, and the embryos of aquatic species have sac-like gills.

9.2 Diversity of Life Histories of Amphibians

Of all the characteristics of amphibians, none is more remarkable than the variety they display in modes of reproduction and parental care. It is astonishing that the range of reproductive modes among the 4800 species of amphibians far exceeds that of any other

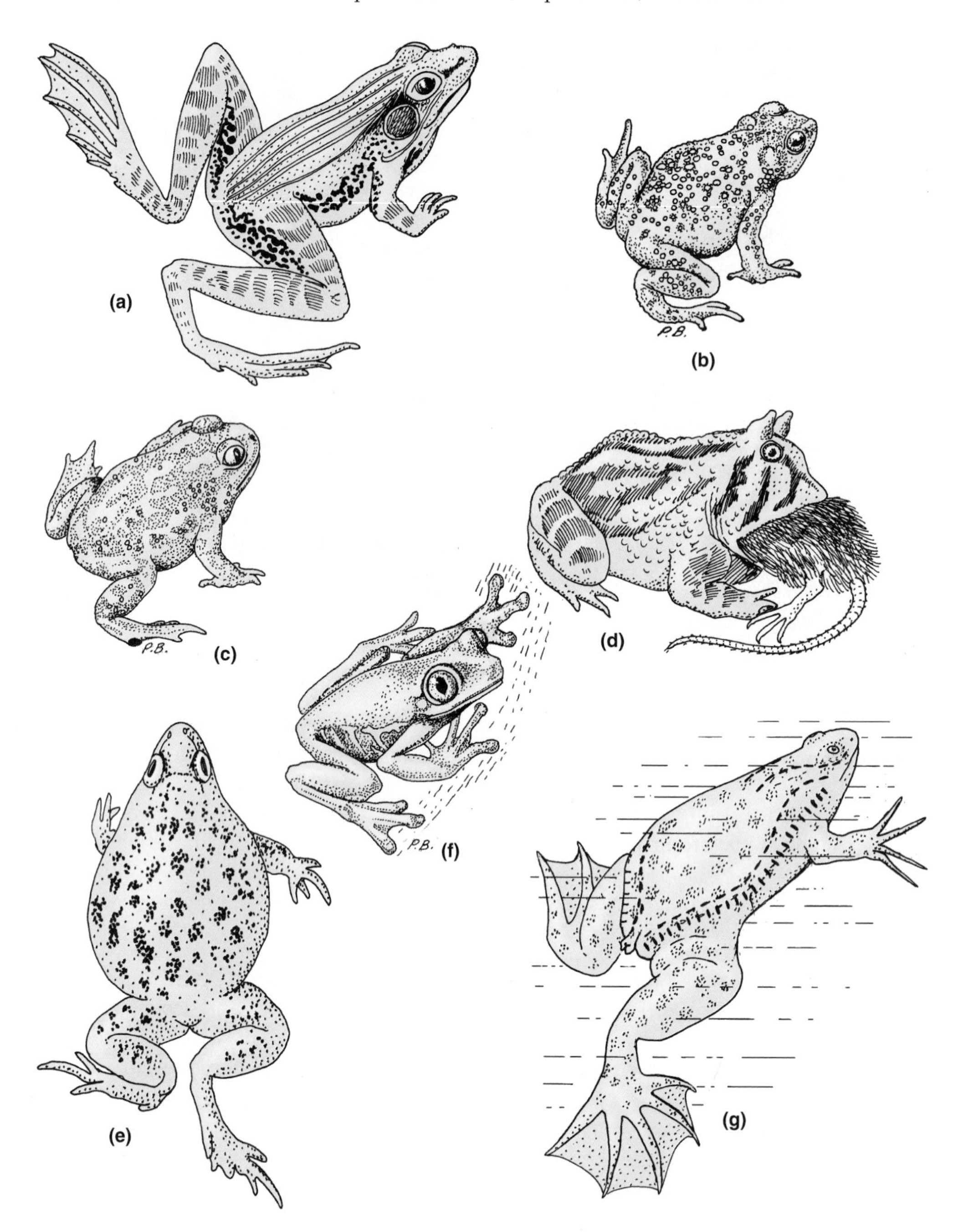

▲ Figure 9–7 Anuran body forms reflect specializations for different habitats and different methods of locomotion. Semiaquatic form: (a) African ridged frog (*Ptychadena,* Ranidae). Terrestrial anurans: (b) toad (*Bufo,* Bufonidae); (c) spadefoot toad (*Scaphiopus,* Pelobatidae); (d) horned frog (*Ceratophrys,* Leptodactylidae). Burrowing species: (e), African shovel-nosed frog (*Hemisus,* Hemisotidae). Arboreal frog: (f) Central American leaf frog (*Agalychnis,* Hylidae). Specialized aquatic frog: (g) African clawed frog (*Xenopus,* Pipidae).

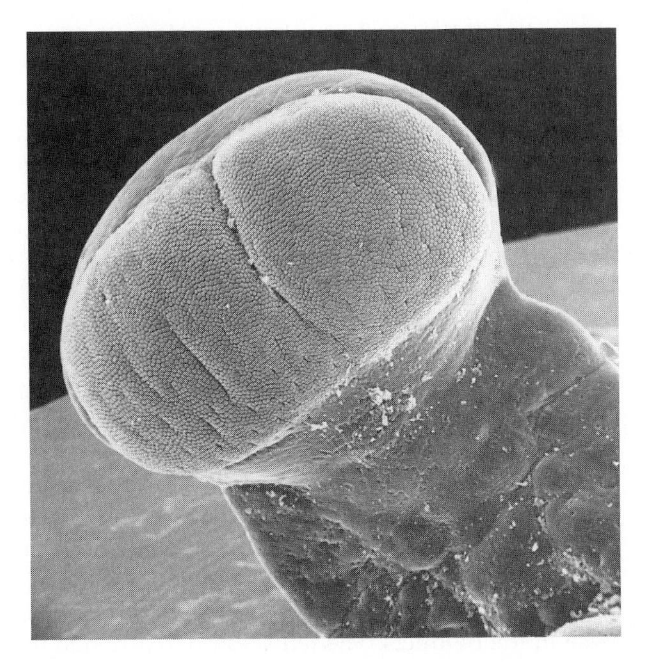

▲ Figure 9–8 Toe disks of a hylid frog. (a) A single toe pad. (b) Detail of the polygonal plates.

group of vertebrates except for fishes, which outnumber amphibian species by five to one. Most species of amphibians lay eggs. The eggs may be deposited in water or on land, and they may hatch into aquatic larvae or into miniatures of the terrestrial adults. The adults of some species of frogs carry eggs attached to the surface of their bodies. Others carry their eggs in pockets in the skin of the back or flanks, in the vocal sacs, or even in the stomach. In still other species the females retain the eggs in the oviducts and give birth to metamorphosed young. Many amphibians have no parental care of their eggs or young, but in many other species a parent remains with the eggs and sometimes with the hatchlings, transports tadpoles from the nest to water, and in a few species an adult even feeds the tadpoles.

Amphibians have two characteristics that make their population ecology hard to study. First, fluctuation in size appears to be a normal feature of amphibian populations. Many species of amphibians lay hundreds of eggs, and the vast majority of these eggs never reach maturity. In a good year, however, survival may be unusually high and a large number of individuals may be added to the population. Conversely, in a year of drought the entire reproductive output of a population may die and recruitment will be zero for that year. Thus, year-to-year variation in recruitment creates natural fluctuations in populations that can obscure long-term trends in population size. In addition, many species of amphibians live in metapopulations in which individual animals move among local populations that are often centered around breeding sites. In the shifting existence of a metapopulation, breeding populations may disappear from some sites while a healthy metapopulation continues to exist and breed at other sites. A limited study might conclude that a species was vanishing, whereas a broader analysis would show that the total population of the species had not changed. *Measuring and Monitoring Biological Diversity: Standard Methods for Amphibians* (Heyer et al. 1993) was compiled to provide standard methods for studies of amphibian populations so that data from different studies could be compared and combined.

Caecilians

The reproductive adaptations of caecilians are as specialized as their body form and ecology. Internal fertilization is accomplished by a male intromittent organ that is protruded from the cloaca. Some species of caecilians lay eggs, and the female may coil around the eggs, remaining with them until they hatch (see Figure 9–10). Viviparity is widespread,

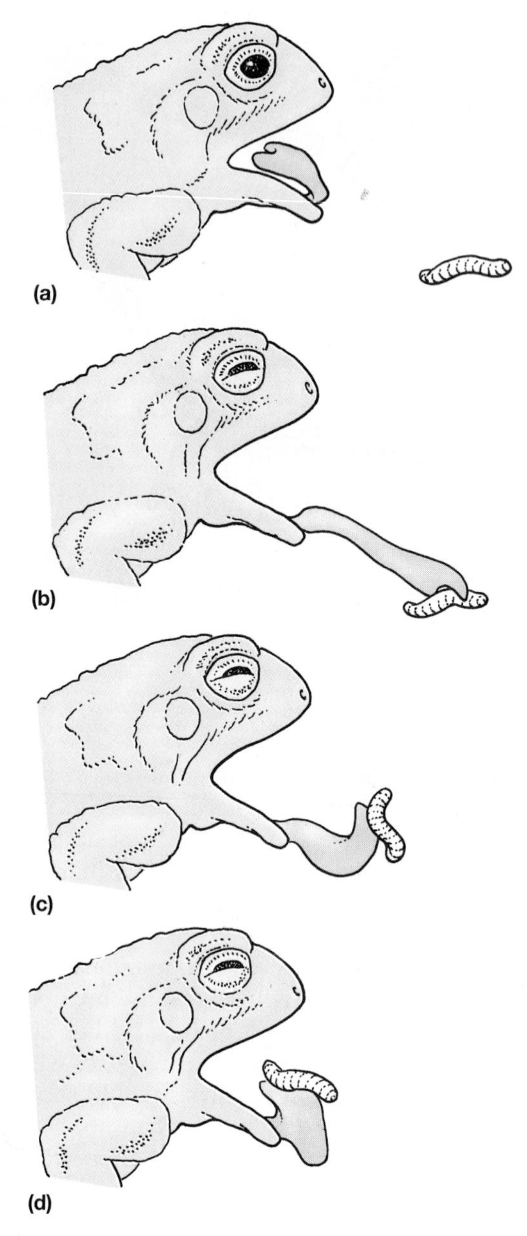

▲ Figure 9–9 Prey capture by a toad.

however, and about 75 percent of the species are viviparous (Wake 1993). At birth young caecilians are 30 to 60 percent of their mother's body length. A female *Typhlonectes* 500 millimeters long may give birth to nine babies, each 200 millimeters long. The initial growth of the fetuses is supported by yolk contained in the egg at the time of fertilization, but this yolk is exhausted long before embryonic development is complete. *Typhlonectes* fetuses have absorbed all of the yolk in the eggs by the time they are 30 mm long. Thus, the energy they need to grow to 200 mm (a 6.6-fold increase in length) must be supplied by the mother. The energetic demands of producing nine babies, each one increasing its length 6.6 times and reaching 40 percent of the mother's length at birth, must be considerable.

The fetuses obtain this energy by scraping material from the walls of the oviducts with specialized embryonic teeth. The epithelium of the oviduct proliferates and forms thick beds surrounded by ramifications of connective tissue and capillaries. As the fetuses exhaust their yolk supply, these beds begin to secrete a thick, white, creamy substance that has been called uterine milk. When their yolk supply has been exhausted, the fetuses emerge from their egg membranes, uncurl, and align themselves lengthwise in the oviducts. The fetuses apparently bite the walls of the oviduct, stimulating secretion and stripping some epithelial cells and muscle fibers that they swallow with the uterine milk. Small fetuses are regularly spaced along the oviducts. Large fetuses have their heads spaced at intervals, although the body of one fetus may overlap the head of the next. This spacing probably gives all the fetuses access to the secretory areas on the walls of the oviducts.

Gas exchange appears to be achieved by close contact between the fetal gills and the walls of the oviducts. All the terrestrial species of caecilians have fetuses with a pair of triple-branched filamentous gills. In preserved specimens the fetuses frequently have one gill extending forward beyond the head and the other stretched along the body. In the aquatic genus *Typhlonectes,* the gills are sac-like but are usually positioned in the same way. Both the gills and the walls of the oviducts are highly vascularized, and it seems likely that exchange of gases, and possibly of small molecules such as metabolic substrates and waste products, takes place across the adjacent gill and oviduct. The gills are absorbed before birth, and cutaneous gas exchange may be important for fetuses late in development.

Details of fetal dentition differ among species of caecilians, suggesting that this specialized form of fetal nourishment may have evolved independently in different phylogenetic lines. Analogous methods of supplying energy to fetuses are known in some elasmobranch fishes.

▲ Figure 9–10 Caecilians. (a) Adult, showing body form. (b) A female coiled around her eggs. Embryos of terrestrial (c) and aquatic (d) species.

Salamanders

Most groups of salamanders use internal fertilization, but the Cryptobranchoidea (Cryptobranchidae and Hynobiidae) and probably the Sirenidae retain external fertilization. Internal fertilization in salamanders is accomplished not by an intromittent organ but by the transfer of a packet of sperm (the **spermatophore**) from the male to the female (Figure 9–11). The form of the spermatophore differs in various species of salamanders, but all consist of a sperm cap on a gelatinous base. The base is a cast of the interior of the male's cloaca, and in some species it reproduces the ridges and furrows in accurate detail. Males of the Asian salamandrid *Euproctus* deposit a spermatophore on the body of a female and then, holding her with their tail or jaws, use their feet to

Table 9.4 Families of caecilians

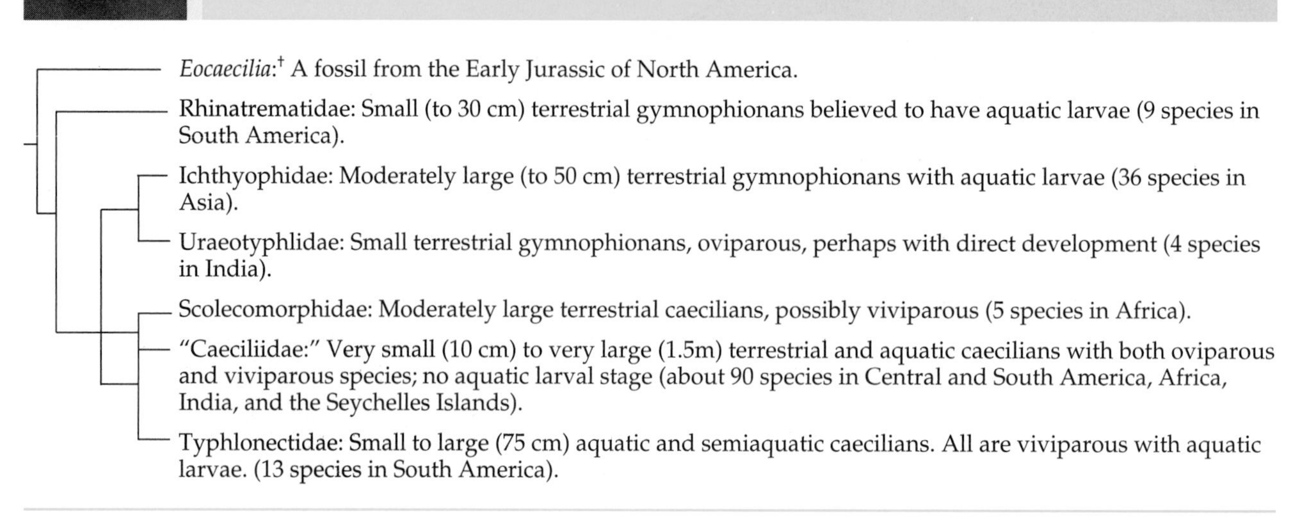

Eocaecilia:† A fossil from the Early Jurassic of North America.

Rhinatrematidae: Small (to 30 cm) terrestrial gymnophionans believed to have aquatic larvae (9 species in South America).

Ichthyophidae: Moderately large (to 50 cm) terrestrial gymnophionans with aquatic larvae (36 species in Asia).

Uraeotyphlidae: Small terrestrial gymnophionans, oviparous, perhaps with direct development (4 species in India).

Scolecomorphidae: Moderately large terrestrial caecilians, possibly viviparous (5 species in Africa).

"Caeciliidae:" Very small (10 cm) to very large (1.5m) terrestrial and aquatic caecilians with both oviparous and viviparous species; no aquatic larval stage (about 90 species in Central and South America, Africa, India, and the Seychelles Islands).

Typhlonectidae: Small to large (75 cm) aquatic and semiaquatic caecilians. All are viviparous with aquatic larvae. (13 species in South America).

The symbol † indicates an extinct taxon.

insert the spermatophore into her cloaca. Females of the hynobiid salamander *Ranodon sibiricus* deposit egg sacs on top of a spermatophore. In derived species of salamanders the male deposits a spermatophore on the substrate, and the female picks off the cap with her cloaca. The sperm are released as the cap dissolves, and fertilization occurs in the oviducts.

Courtship Courtship patterns are important for species recognition, and they show great interspecific variation. Males of some species have elaborate secondary sexual characters that are used during courtship. Pheromones (chemicals used for communication) are released primarily by males and play a large role in the courtship of salamanders; they probably contribute to species recognition, and may stimulate endocrine activity that increases the receptivity of females.

Pheromone delivery by most salamanders that breed on land involves physical contact between a male and female, during which the male applies secretions of specialized courtship glands (hedonic

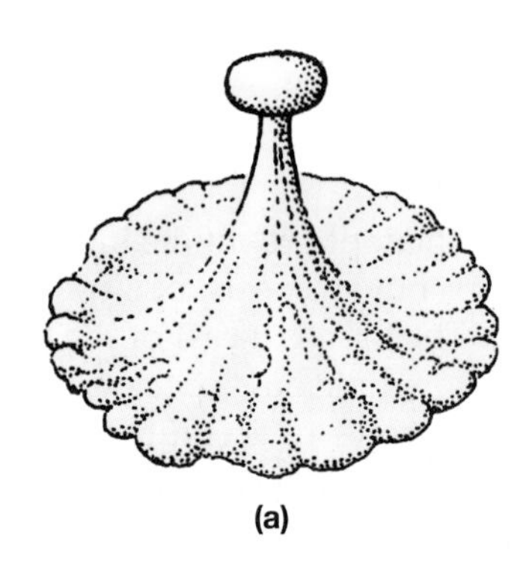

(a)

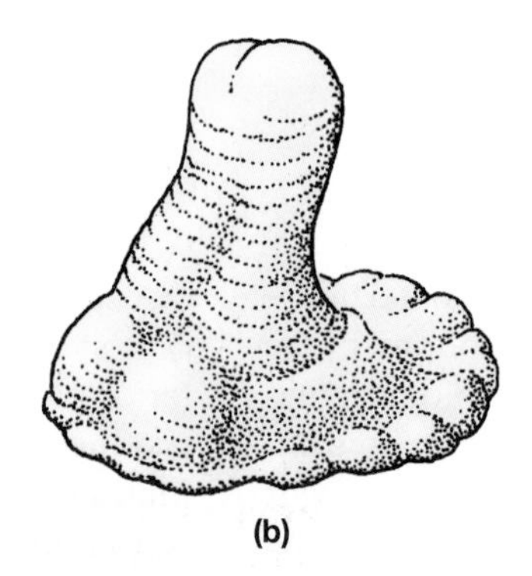

(b)

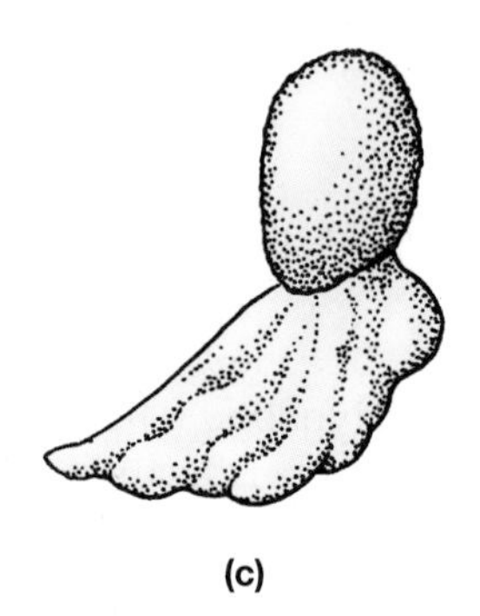

(c)

▲ Figure 9–11 Spermatophores. (a) Red-spotted newt, *Notophthalmus viridescens*, Salamandridae. (b) Dusky salamander, *Desmognathus fuscus*, Plethodontidae. (c) Two-lined salamander, *Eurycea bislineata*, Plethodontidae.

glands) to the nostrils or body of the female. Several modes of pheromone delivery have been described. Males of many plethodontids (e.g., *Plethodon jordani*) have a large gland beneath the chin (the mental gland), and secretions of the gland are applied to the nostrils of the female with a slapping motion (Figure 9–12b). The anterior teeth of males of many species of *Desmognathus* and *Eurycea* (both members of the Plethodontidae) hypertrophy during the breeding season. A male of these species spreads secretion from his mental gland on the female's skin, and then abrades the skin with his teeth, inoculating the female with the pheromone (Figure 9–12c). Males of two small species of *Desmognathus* use specialized mandibular teeth to bite and stimulate the female. Male salamandrids (Salamandridae) rub the female's snout with hedonic glands on their cheeks (the red-spotted newt, *Notophthalmus viridescens*), chin (the rough-skinned newt, *Taricha granulosa* [Figure 9–12a]), or cloaca (the Spanish newt, *Pleurodeles waltl*). Newts in the genera *Triturus* and *Cynops* transfer pheromones without physical contact

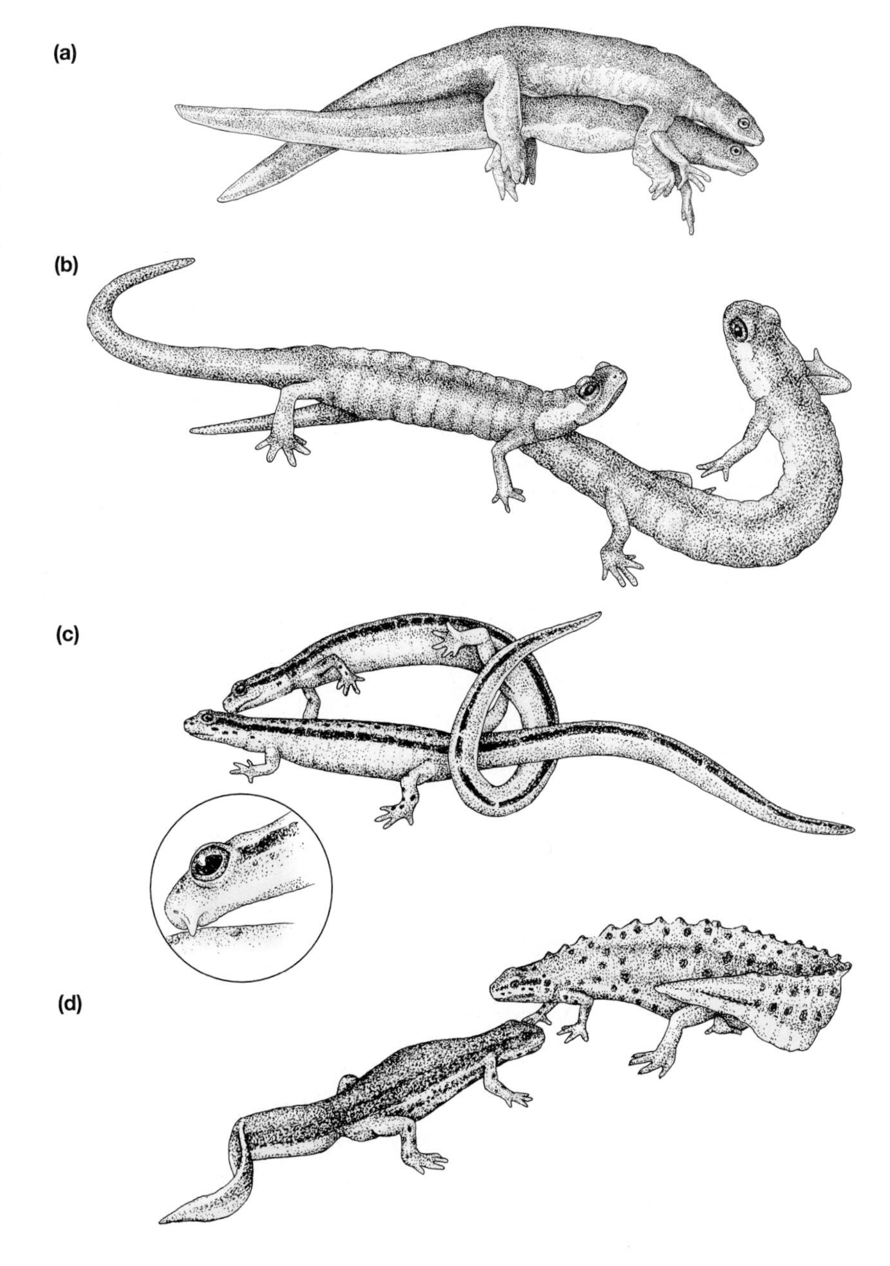

▶ Figure 9–12 Transfer of pheromones by male salamanders. (a) The rough-skinned newt, *Taricha granulosa*, Salamandridae. (b) Jordan's salamander, *Plethodon jordani*, Plethodontidae. (c) The two-lined salamander, *Eurycea bislineata*, Plethodontidae. (d) The smooth newt, *Triturus vulgaris*, Salamandridae.

between the male and female. The males of these species perform elaborate courtship displays in which the male vibrates its tail to create a stream of water that wafts pheromones, secreted by a gland in his cloaca, toward the female (Figure 9–12d).

Three groups of *Triturus,* probably representing evolutionary lineages, are tentatively recognized within the genus, and the evolution of courtship probably reflects this phylogenetic relationship. Two trends are apparent: an increase in diversity of the sexual displays performed by the male, and an increase in the importance of positive feedback from the female. The behaviors seen in *Triturus alpestris* may represent the ancestral condition. This species shows little sexual dimorphism (Figure 9–13c), and the male's display consists only of fanning (a display in which the tail is folded back against the flank nearest the female and the tail tip is vibrated rapidly). The male's behavior is nearly independent of response by the female—a male *T. alpestris* may perform his entire courtship sequence and deposit a spermatophore without active response by the female he is courting.

A group of large newts, including *Triturus cristatus* and *T. vittatus* (Figure 9–13a, b), show derived morphological and behavioral characters. They are highly sexually dimorphic, and males defend display sites. Their displays are relatively static, and lack the rapid fanning movements of the tail that characterize other groups of *Triturus.* A male of these species does not deposit a spermatophore unless the female he is courting touches his tail with her snout.

A group of small-bodied species includes *Triturus vulgaris* and *T. boscai* (Figure 9–13d, e). These newts show less sexual dimorphism than the large species, and have a more diverse array of behaviors—including a nearly static lateral display, whipping the tail violently against the female's body, fanning with the tail tip, and other displays with names like wiggle and flamenco that occur in some species in the group. Response by the female is an essential component of courtship for these species—a male will not move on from the static display that begins courtship to the next phase unless the female approaches him repeatedly, and he will not deposit a spermatophore until the female touches his tail.

These trends to greater sexual dimorphism, more diverse displays, and more active involvement of the female in courtship may reflect sexual selection by females within the derived groups. Halliday (1990) has suggested that in the ancestral condition there was a single male display, and females mated with the males that performed it most vigorously. That kind of selection by females would produce a population of males, all of which displayed vigorously, and males that added new components to their courtship might be more attractive to females than their rivals.

Eggs And Larvae In most cases, salamanders that breed in water lay their eggs in water. The eggs may be laid singly, or in a mass of transparent gelatinous material. The eggs hatch into gilled aquatic larvae that, except in paedomorphic forms, transform into terrestrial adults. Some families, including the lungless salamanders (Plethodontidae), have a number of species that have dispensed in part or entirely with an aquatic larval stage. The dusky salamander, *Desmognathus fuscus,* lays its eggs beneath a rock or log near water, and the female remains with them until after they have hatched. The larvae have small gills at hatching, and may either take up an aquatic existence or move directly to terrestrial life. The red-backed salamander, *Plethodon cinereus,* lays its eggs in a hollow space in a rotten log or beneath a rock. The embryos have gills, but these are reabsorbed before hatching and the hatchlings are miniatures of the adults.

Viviparity Only four species of salamanders give birth to young. The European salamander (*Salamandra salamandra*) produces 20 or more small larvae, each about one-twentieth the length of an adult. The embryos probably get all the energy needed for growth and development from egg yolk. The larvae are released in water and have an aquatic stage that lasts about 3 months. The closely related alpine salamander (*S. atra*) gives birth to one or two fully developed young, each about one-third the adult body length. The embryos of this species are nourished by oviductal secretions, as is the case for caecilians.

Paedomorphosis Paedomorphosis is the rule in families like the Cryptobranchidae and Proteidae, and it characterizes most cave dwellers. It also appears as a variant in the life history of species of salamanders that usually metamorphose, and can be a short-term response to conditions in aquatic or terrestrial habitats. The life histories of two species of salamanders from eastern North America provide examples of the flexibility of paedomorphosis.

The small-mouthed salamander, *Ambystoma talpoideum,* is the only species of mole salamander in eastern North America that displays paedomorpho-

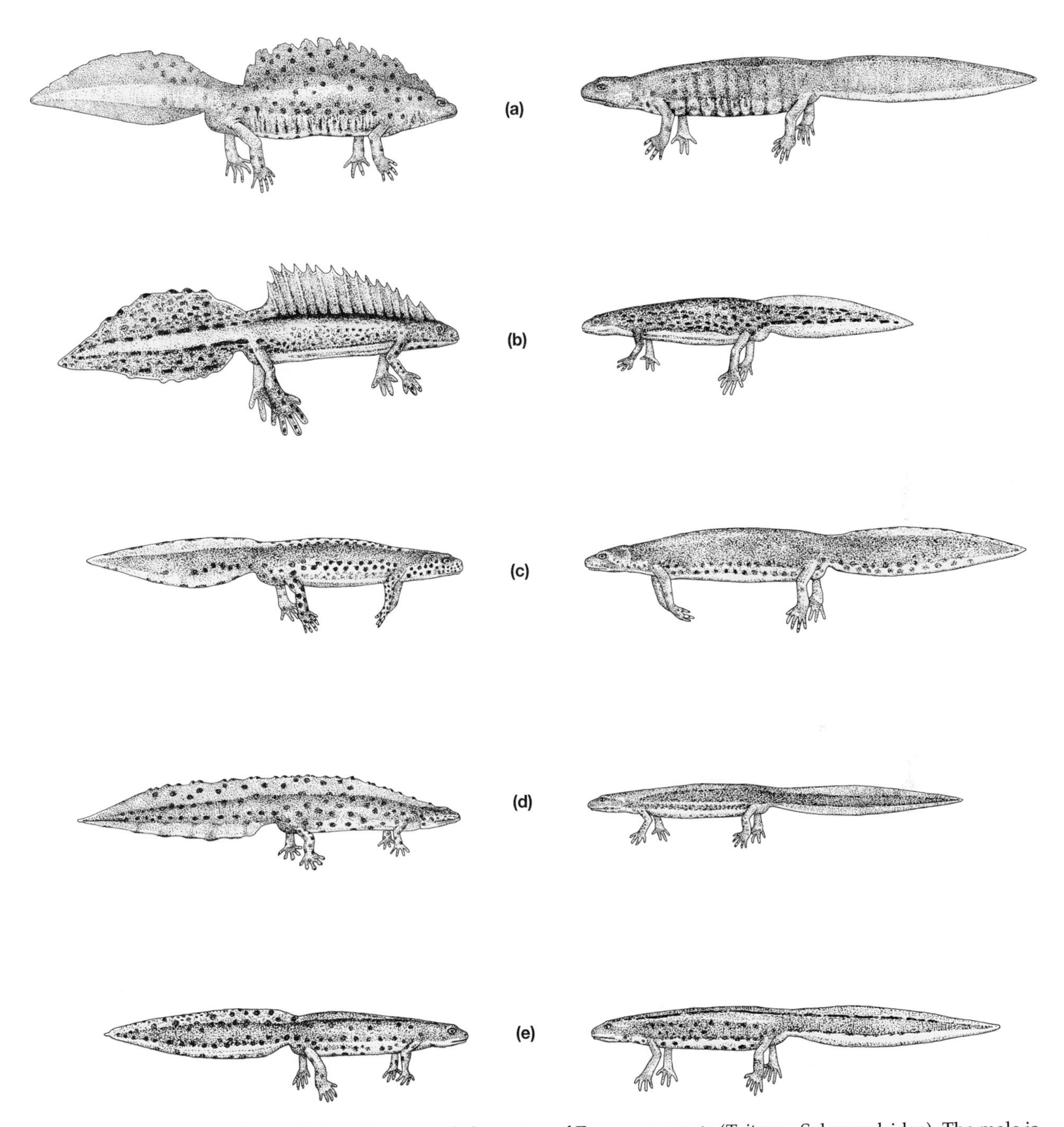

▲ Figure 9–13 Dimorphism of secondary sexual characters of European newts (*Triturus*, Salamandridae). The male is on the left, the female is on the right. (a) The great crested newt, *T. cristatus.* (b) The banded newt, *T. vittatus.* (c) The alpine newt, *T. alpestris.* (d) The smooth newt, *T. vulgaris.* (e) Bosca's newt, *T. boscai.*

sis, although a number of species of *Ambystoma* in the western United States and in Mexico are paedomorphic. Small-mouthed salamanders breed in the autumn and winter, and during the following summer some larvae metamorphose to become terrestrial juveniles. These animals become sexually mature by autumn and return to the ponds to breed when they are about a year old. Ponds in South Carolina also contain paedomorphic larvae that remain in the ponds through the summer and mature and breed in the winter. Some of these paedomorphs metamorphose after breeding, whereas others do not metamorphose and remain in the ponds as permanently paedomorphic adults.

Anurans

Anurans are the most familiar amphibians, largely because of the vocalizations associated with their reproductive behavior. It is not even necessary to get outside a city to hear them. In springtime a weed-choked drainage ditch beside a highway or a trash-filled marsh at the edge of a shopping center parking lot is likely to attract a few toads or tree frogs that have not yet succumbed to human usurpation of their habitat.

The mating systems of anurans can be divided roughly into *explosive breeding,* in which the breeding season is very short, sometimes only a few days, and *prolonged breeding,* with breeding seasons that may extend for several months. Explosive breeders include many species of toads and other anurans that breed in temporary aquatic habitats such as vernal ponds or pools created in the desert. Because these bodies of water do not last very long, breeding congregations of anurans usually form as soon as the site is available. Males and females arrive at the breeding sites nearly simultaneously and often in very large numbers. Because the entire population breeds in a short time, the number of males and females present is approximately equal. Time is the main constraint on how many females a male is able to court, and mating success is usually approximately the same for all the males in a chorus.

In species with prolonged breeding seasons, the males usually arrive at the breeding sites first. Males of some species, such as green frogs (*Rana clamitans*), establish territories in which they spend several months, defending the spot against the approach of other males. The males of other species move between daytime retreats and nocturnal calling sites on a daily basis. Females come to the breeding site just to breed, and leave when they have finished. Just a few females arrive every day, and the number of males at the breeding site is greater than the number of females every night. Mating success may be very skewed, with many of the males not mating at all, and a few males mating several times. Males of anuran species with prolonged breeding seasons compete to attract females, usually by vocalizing. The characteristics of a male frog's vocalization (pitch, length, or repetition rate) might provide information that a female frog could use to evaluate his quality as a potential mate. This is an active area of study in anuran behavior.

Vocalizations Anuran calls are diverse; they vary from species to species, and most species have two or three different sorts of calls used in different situations. The most familiar calls are the ones usually referred to as mating calls, although a less specific term such as **advertisement calls** is preferable. These calls range from the high-pitched *peep* of a spring peeper to the nasal *waaah* of a spadefoot toad or the bass *jug-o-rum* of a bullfrog. The characteristics of a call identify the species and sex of the calling individual. Many species of anurans are territorial, and males of at least one species, the North American bullfrog (*Rana catesbeiana*) recognize one another individually by voice.

An advertisement call is a conservative evolutionary character, and among related taxa there is often considerable similarity in advertisement calls. Superimposed on the basic similarity are the effects of morphological factors, such as body size, as well as ecological factors that stem from characteristics of the habitat. Most toads (*Bufo*) have a trilled advertisement call that consists of a train of repeated pulses (a trill), but the pitch of the call varies with the body size. The oak toad (*B. quercicus*), which has a body length of only 2 or 3 centimeters, has a dominant frequency of 5200 Hz. The larger southwestern toad (*B. microscaphus*), which is 8 cm long, has a lower dominant frequency, 1500 Hz, and the giant toad (*B. marinus*), with a body length of nearly 20 cm, has the lowest pitch of all, 600 Hz.

Female frogs are responsive to the advertisement calls of males of their species for a brief period when their eggs are ready to be laid. The hormones associated with ovulation are thought to sensitize specific cells in the auditory pathway that respond to the species-specific characteristics of the male's call. Mixed choruses of anurans are common in the mating season; a dozen species may breed simultaneously in one pond. A female's response to her own species' mating call is a mechanism for species recognition in that situation.

Costs and Benefits of Vocalization The vocalizations of male frogs are costly in two senses. The actual energy that goes into call production can be very large, and the variations in calling pattern that accompany social interactions among male frogs in a breeding chorus can increase the cost per call (see Box 9-1). Another cost of vocalization for a male frog is the risk of predation. A critical function of vocalization is to permit a female frog to locate a male, but female frogs are not the only animals that can use vocalizations as a cue to find male frogs; predators of frogs also find that calling males are easy to locate. The túngara frog (*Physalaemus pustulosus*) is a small terrestrial leptodactylid that occurs in Central America (Figure 9–14).

Túngara frogs breed in small pools, and breeding assemblies range from a single male to choruses of several hundred males. The advertisement call of a male túngara frog is a strange noise, a whine that sounds as if it would be more at home in an arcade of video games than in the tropical night. The whine starts at a frequency of 900 Hz and sweeps downward to 400 Hz in about 400 milliseconds (Figure 9–15). The whine may be produced by itself, or it may be followed by one or several *chucks* with a dominant frequency of 250 Hz. When a male túngara frog is calling alone in a pond, it usually gives only the whine portion of the call; but as additional males join a chorus, more and more of the frogs produce calls that include chucks. Male túngara frogs calling in a breeding pond added chucks to their calls when they heard playbacks of calls of other males. That observation suggested that it was the presence of other calling males that stimulated frogs to make their calls more complex by adding chucks to the end of the whine.

What advantage would a male frog in a chorus gain from using a whine-chuck call instead of a whine? Perhaps the complex call is more attractive to female frogs than the simple call. Michael Ryan and Stanley Rand tested that hypothesis by placing female túngara frogs in a test arena with a speaker at each side. One speaker broadcast a whine call, and the second speaker broadcast a whine-chuck. When female frogs were released individually in the center of the arena 14 of the 15 frogs tested moved toward the speaker broadcasting the whine-chuck call.

If female frogs are attracted to whine-chuck calls in preference to whine calls, why do male frogs give whine-chuck calls only when other males are present? Why not always give the most attractive call possible? One possibility is that whine-chuck calls require more energy than whines, and males save energy by using whine-chucks only when competition with other males makes the energy expenditure necessary. However, measurements of the energy expenditure of calling male túngara frogs showed that energy cost was not related to the number of chucks. Another possibility is that male frogs giving whine-chuck calls are more vulnerable to predators than frogs giving only whine calls. Túngara frogs in breeding choruses are preyed upon by frog-eating

▲ Figure 9–14 Male túngara frog, *Physalaemus pustulosus* (Leptodactylidae), vocalizing. Air is forced from the lungs (a) into the vocal sacs (b).

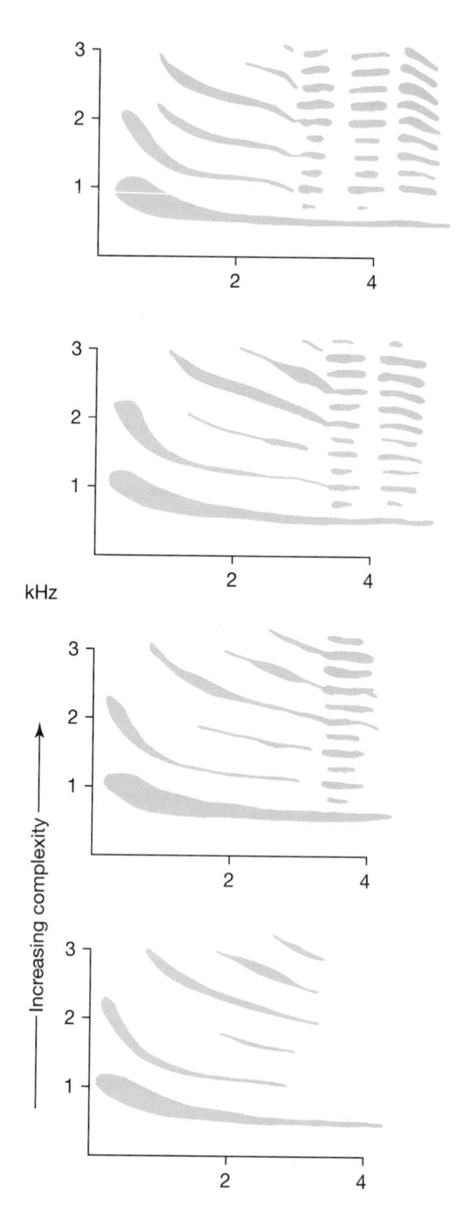

▲ Figure 9–15 Sonograms of the advertisement call of *Physalaemus pustulosus* (Leptodactylidae). The calls increase in complexity from bottom (a whine only) to top (a whine followed by three chucks). A sonogram is a graphic representation of a sound: Time is shown on the horizontal axis and frequency on the vertical axis.

bats, *Trachops cirrhosus,* and the bats locate the frogs by homing on their vocalizations.

In a series of playback experiments, Ryan and Merlin Tuttle placed pairs of speakers in the forest and broadcast vocalizations of túngara frogs. One speaker played a recording of a whine and the other a recording of a whine-chuck. The bats responded as if the speakers were frogs: They flew toward the speakers and even landed on them. In five experiments at different sites, the bats approached speakers broadcasting whine-chuck calls twice as frequently as those playing simple whines (168 approaches versus 81). Thus, female frogs are not alone in finding whine-chuck calls more attractive than simple whines—an important predator of frogs also responds more strongly to the complex calls. Predation can be a serious risk for male túngara frogs. Ryan and his colleagues measured the rates of predation in choruses of different sizes. The major predators were frog-eating bats, a species of opossum (*Philander opossum*), and a larger species of frog (*Leptodactylus pentadactylus*); the bats were the most important predators of the túngara frogs. Large choruses of frogs did not attract more bats than small choruses, and consequently the risk of predation for an individual frog was less in a large chorus than in a small one. Predation was an astonishing 19 percent of the frogs per night in the smallest chorus and a substantial 1.5 percent per night even in the largest chorus. When a male frog shifts from a simple whine to a whine-chuck call, it increases its chances of attracting a female, but it simultaneously increases its risk of attracting a predator. In small choruses the competition from other males for females is relatively small, and the risk of predation is relatively large. Under those conditions it is apparently advantageous for a male túngara frog to give simple whines. However, as chorus size increases, competition with other males also increases—while the risk of predation falls. In that situation the advantage of giving a complex call apparently outweighs the risks.

Modes of Reproduction Fertilization is external in most anurans; the male uses his fore legs to clasp the female in the pectoral region (**axillary amplexus**) or pelvic region (**inguinal amplexus**). Amplexus may be maintained for several hours or even days before the female lays eggs. Males of the tailed frog of the Pacific Northwest (*Ascaphus truei*) have an extension of the cloaca (the "tail" that gives them their name) that is used to introduce sperm into the cloaca of the female. Internal fertilization has been demonstrated for the Puerto Rican coquí (*Eleutherodactylus coquí*) and may be widespread among frogs that lay eggs on land. Fertilization must also be internal for the few species of anurans that give birth to living young.

Anurans show even greater diversity in their modes of reproduction than salamanders. Similar reproductive habits have clearly evolved independently in different groups. Large eggs produce large offspring that probably have a better chance of surviving than smaller ones, but large eggs also require more time to hatch and are exposed to predators for a longer period. Thus, the evolution of large eggs and hatchlings has often been accompanied by the simultaneous evolution of behaviors that protect the eggs, and sometimes the tadpoles as well, from predation. A study of Amazon rain forest frogs revealed a positive relationship between the intensity of predation on frogs eggs in a pond and the proportion of frog species in the area that laid eggs in terrestrial situations (Magnusson and Hero 1991). Many arboreal frogs (represented in Figure 9–18a by *Centrolenella*) lay their eggs in the leaves of trees overhanging water. The eggs undergo their embryonic development out of the reach of aquatic egg predators, and when the tadpoles hatch they drop into the water and take up an aquatic existence. Other frogs, such as *Physalaemus pustulosus* (Figure 9–18b), achieve the same result by constructing foam nests that float on the water surface. The female emits a copious mucus secretion during amplexus that the pair of frogs beat into a foam with their hind legs. The eggs are laid in the foam mass, and when the tadpoles hatch they drop through the foam into the water.

Although these methods reduce egg mortality, the tadpoles are subjected to predation and competition. Some anurans avoid both problems by finding or constructing breeding sites free from competition and predation. Some frogs, for example, lay their eggs in the water that accumulates in bromeliads—epiphytic tropical plants that grow in trees and are morphologically specialized to collect rainwater. A large bromeliad may hold several liters of water, and the frogs pass through egg and larval stages in that protected microhabitat. Many tropical frogs lay eggs on land near water. The eggs or tadpoles may be released from the nest sites when pond levels rise after a rainstorm. Other frogs construct pools in the mud banks beside streams. These volcano-shaped structures are filled with water by rain or seepage and provide a favorable environment for the eggs and tadpoles. Some frogs have eliminated the tadpole stage entirely. These frogs lay large eggs on land that develop directly into little frogs. This pattern is characteristic of about 20 percent of all anuran species.

Parental Care Adults of many species of frogs guard the eggs specifically. In some cases it is the male that protects the eggs, in others it is the female. In most cases it is not clearly known which sex is involved, because external sex identification is difficult with many anurans. Some of the frogs that lay their eggs over water remain with them. Some species sit beside the eggs; others rest on top of them. Many of the terrestrial frogs that lay direct-developing eggs remain with the eggs and will attack an animal that approaches the nest. Removing the guarding frog frequently results in the eggs desiccating and dying before hatching or being eaten by predators. Male African bullfrogs (*Pyxicephalus adspersus*) guard their eggs, and then continue to guard the tadpoles after they hatch. The male frog moves with the school of tadpoles, and will even dig a channel to allow the tadpoles to swim from one pool in a marsh to an adjacent one. Tadpoles of several species of the tropical American frog genus *Leptodactylus* follow their mother around the pond. *Leptodactylus* are large and aggressive frogs that are able to deter many potential predators.

Some of the dart-poison frogs of the American tropics deposit their eggs on the ground, and one of the parents remains with the eggs until they hatch into tadpoles. The tadpoles adhere to the adult and are transported to water (Figure 9–18d). Females of the Panamanian frog *Colostethus inguinalis* carry their tadpoles for more than a week, and the tadpoles increase in size during this period. The largest tadpoles being carried by females had small amounts of plant material in their stomachs, suggesting that they had begun to feed while they were still being transported by their mother. Females of another Central American dart-poison frog, *Dendrobates pumilio*, release their tadpoles in small pools of water, and then return at intervals to the pools to deposit unfertilized eggs that the tadpoles eat.

Other anurans, instead of remaining with the eggs, carry the eggs with them. The male of the European midwife toad (*Alytes obstetricians*) gathers the egg strings about his hind legs as the female lays them. He carries them with him until they are ready to hatch, at which time he releases the tadpoles into water. The male of the terrestrial Darwin's frog

BOX 9-1 The Energy Cost of Vocalization by Frogs

The vocalizations of frogs, like most acoustic signals of tetrapods, are produced when air from the lungs is forced over the vocal cords, causing them to vibrate. Contraction of trunk muscles provides the pressure in the lungs that propels the air across the vocal cords, and these contractions require metabolic energy. Measurement of the actual energy expenditure by frogs during calling is technically difficult because a frog must be placed in an airtight metabolism chamber to measure the amount of oxygen it consumes, and that procedure can frighten the frog and prevent it from calling. Ted Taigen and Kent Wells (1985) at the University of Connecticut overcame that difficulty in studies of the gray tree frog, *Hyla versicolor,* by taking the metabolism chambers to the breeding ponds. Calling male frogs were placed in the chambers early in the evening and then left undisturbed. With the stimulus of the chorus around them, frogs would call in the chambers. Their vocalizations were recorded with microphones attached to each chamber, and the amount of oxygen they used during calling was determined from the decline in the concentration of oxygen in the chamber over time (Figure 9–16).

The rates at which individual frogs consumed oxygen were directly proportional to their rates of vocalization (Figure 9–17). At the lowest calling rate, 150 calls per hour, oxygen consumption was barely above resting levels. However, at the highest calling rates, 1500 calls per hour, the frogs were consuming oxygen even more rapidly than they did during high levels of locomotor activity. Examination of the trunk muscles of the male frogs, which hypertrophy enormously during the breeding season, revealed biochemical specializations that appear to permit this high level of oxygen consumption during vocalization (Marsh and Taigen 1987).

The advertisement call of the gray tree frog is a trill that lasts from 0.3 to 0.6 second. During their studies, Wells and Taigen found that gray tree frogs gave short calls when they were in small choruses, and lengthened their calls when many other males were calling near them (Wells and Taigen 1986). It has subsequently been shown that long calls are more attractive to female frogs than short calls (Klump and Gerhardt 1987), but the long calls require more energy. A 0.6-second call requires about twice as much energy as a 0.3-second call,

▲ Figure 9–16 Measuring oxygen consumption of a calling frog. Gray tree frog (*Hyla versicolor,* Hylidae) in a metabolism chamber beside a breeding pond. A microphone in the chamber records the frog's calls, and a thermocouple measures the temperature inside the chamber. Gas samples are drawn from the tube for measurements of oxygen consumption.

BOX 9-1 **THE ENERGY COST OF VOCALIZATION BY FROGS** *(CONTINUED)*

and the rate of oxygen consumption during calling increases as the length of the calls increases. That relationship suggests that, all else being equal, a male gray tree frog that increased its call duration to be more attractive to female frogs would pay a price for its attractiveness with a higher rate of energy expenditure.

Indirect evidence suggests that the energy cost of calling might limit the time a male gray tree frog could spend in a breeding chorus. The tree frogs call for only 2 to 4 hours each night, and stores of glycogen (the metabolic substrate used by calling frogs) decreased by 50 percent in that time. Wells and Taigen were able to simulate the effects of different chorus sizes by playing tape recordings of vocalizations to frogs. The frogs matched their own calls to the recorded calls they heard—short responses to short calls, medium to medium calls, and long responses to long calls. As the length of their calls increased, the frogs reduced the rate at which they called. The reduction in rate of calling approximately balanced the increased length of each call, so the overall calling effort (the number of seconds of vocalization per hour) was nearly independent of call duration. Thus, it appears that male gray tree frogs compensate for the higher energy cost of long calls by giving fewer of them. However, that compromise may not entirely eliminate the problem of high energy costs for frogs giving long calls. Even though the calling effort was approximately the same, males giving long calls at slow rates spent fewer hours per night calling than did frogs that produced short calls at higher rates.

The high energy cost of calling offers an explanation for the pattern of short and long calls produced by male gray tree frogs. The length of time that an isolated male can call may be the most important determinant of his success in attracting a female, and the trade-off between rate of calling and call duration suggests that male frogs are performing at or near their physiological limits. Giving short calls and calling for several hours every night may be the best strategy if a male has that option available. In a large chorus, however, competition with other males is intense and giving a longer and more attractive call may be important, even if the male can call for only a short time.

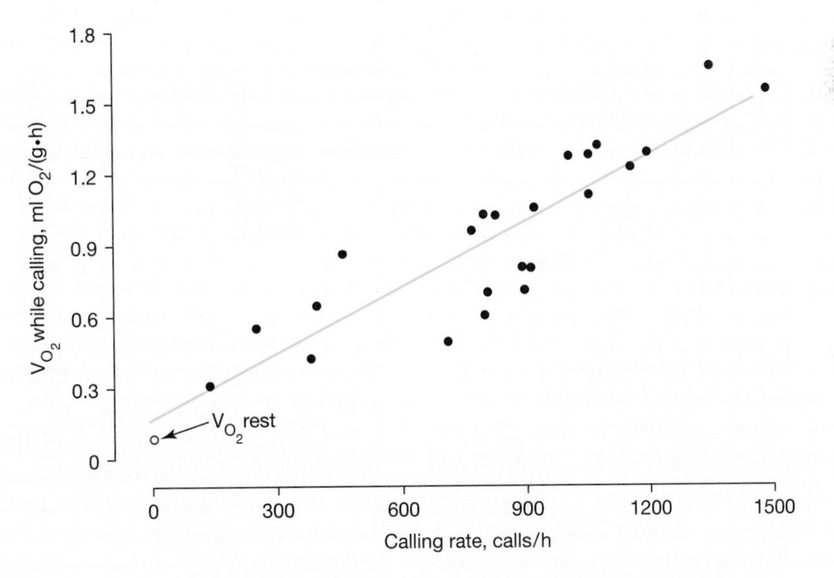

▲ Figure 9–17 Energetic cost of calling. Rates of oxygen consumption of frogs calling inside metabolism chambers (on the vertical axis) as a function of the rate of calling (on the horizontal axis). The energy expended by a calling frog increases linearly with the number of times it calls per hour.

▲ Figure 9–18 Egg development and parental care among anurans. (a) Eggs laid over water, *Centrolenella*, Centrolenidae. (b) Eggs in a nest of foam, *Physalaemus*, Leptodacylidae. (c) Eggs carried by the adult, *Rhinoderma*, Rhinodermatidae. (d) Tadpoles carried by the adult, *Colostethus*, Dendrobatidae. Eggs carried on the back of an adult. (e) *Hemiphractus*, Hylidae. (f) *Pipa*, Pipidae.

(*Rhinoderma darwinii*) of Chile snaps up the eggs the female lays and carries them in his vocal pouches, which extend back to the pelvic region (Figure 9–18c). The embryos pass through metamorphosis in the vocal sacks and emerge as fully developed froglets. Males are not alone in caring for eggs. The females of a group of tree frogs carry the eggs on their back, in an open oval depression, a closed pouch, or individual pockets (Figure 9–18e). The eggs develop into miniature frogs before they leave their mother's back. A similar specialization is seen in the completely aquatic Surinam toad, *Pipa pipa*. In the breeding season the skin of the female's back thickens and softens. In egg laying the male and female in amplexus swim in vertical loops in the water. On the upward part of the loop the female is above the male and releases a few eggs, which fall onto his ventral surface. He fertilizes them and, on the downward loop, presses them against the female's back. They sink into the soft skin and a cover forms over each egg, enclosing it in a small capsule (Figure 9–18f). The eggs develop through metamorphosis in the capsules.

Tadpoles of the two species of the Australian frog genus *Rheobatrachus* are carried in the stomach of the female frog. The female swallows eggs or newly hatched larvae and retains them in her stomach through metamorphosis. This behavior was first described in *Rheobatrachus silus* and is accompanied by extensive morphological and physiological modifications of the stomach. These changes include distension of the proximal portion of the stomach, separation of individual muscle cells from the surrounding connective tissue, and inhibition of hydrochloric acid secretion, perhaps by prostaglandin released by the tadpoles. In January 1984, a second species of gastric-brooding frog, *R. vitellinus,* was discovered in Queensland. Strangely, this species lacks the extensive structural changes in the stomach that characterize the gastric brooding of *R. silus.* The striking differences between the two species suggest the surprising possibility that this bizarre reproductive mode might have evolved independently. Both species of *Rheobatrachus* disappeared within a few years of their discovery, and there is no obvious explanation for the apparent extinction of these fascinating frogs. The repeated pattern of amphibian population declines, even for species from habitats that appear undisturbed, suggests that some of these declines may be produced by global-level effects of human activities.

Viviparity Only a few species of anurans are viviparous (Wake 1993). The five species in the African bufonid genus *Nectophrynoides* show a spectrum of reproductive modes. One species deposits eggs that are fertilized externally and hatch into aquatic tadpoles, two species produce young that are nourished by yolk, and the two remaining species have embryos that feed on secretions from the walls of the oviduct. The golden coquí (*Eleutherodactylus jasperi,* a Puerto Rican leptodactylid) also gives birth to fully formed young, but in this case the energy and nutrients come from the yolk of the egg. (The golden coquí is yet another species of amphibian that has vanished and is presumed to be extinct.)

The Ecology of Tadpoles

Although many species of frogs have evolved reproductive modes that bypass an aquatic larval stage, a life history that includes a tadpole has certain advantages. A tadpole is a completely different animal from an adult anuran, both morphologically and ecologically.

Tadpoles are as diverse in their morphological and ecological specializations as adult frogs, and they occupy nearly as great a range of habitats (Figure 9–19). Tadpoles that live in still water usually have ovoid bodies and tails with fins that are as large as the muscular part of the tail, whereas tadpoles that live in fast-flowing water have more streamlined bodies and smaller tail fins. Semiterrestrial tadpoles wiggle through mud and leaves and climb on damp rock faces; they are often dorsoventrally flattened and have little or no tail fin, and many tadpoles that live in bromeliads have a similar body form. Direct-developing tadpoles have large yolk supplies, and reduced mouthparts and tail fins. The mouthparts of tadpoles also show variation that is related to diet (Figure 9–20). Filter-feeding tadpoles that hover in midwater lack keratinized mouth parts, whereas species that graze from surfaces have small beaks that are often surrounded by rows of denticles. Predatory tadpoles have larger beaks that can bite pieces from other tadpoles. Funnel-mouthed, surface-feeding tadpoles have greatly expanded mouthparts that skim material from the surface of the water.

Tadpoles of most species of anurans are filter-feeding herbivores, whereas all adult anurans are carnivores that catch prey individually. Because of these differences, tadpoles can exploit resources that are not available to adult anurans. This advantage may be a factor that has led many species of frogs to retain the ancestral pattern of life history in which an aquatic larva matures into a terrestrial adult. Many aquatic habitats experience annual flushes of primary production, when nutrients washed into a pool by rain or melting snow stimulate the rapid growth of algae. The energy and nutrients in this algal bloom are transient resources that are available for a brief time to the organisms able to exploit them.

Tadpoles are excellent eating machines. All tadpoles extract suspended food particles from water, and feeding and ventilation of the gills are related activities. The stream of water that moves through the mouth and nares to ventilate the gills carries with it particles of food. As the stream of water passes through the branchial basket, small food particles are trapped in mucus secreted by epithelial cells. The mucus, carrying particles of food with it, is moved from the gill filters to the ciliary grooves on the margins of the roof of the pharynx and then transported posteriorly to the esophagus.

Although all tadpoles filter food particles from a stream of water that passes across the gills, the method used to put the food particles into suspension differs among species. Some tadpoles filter

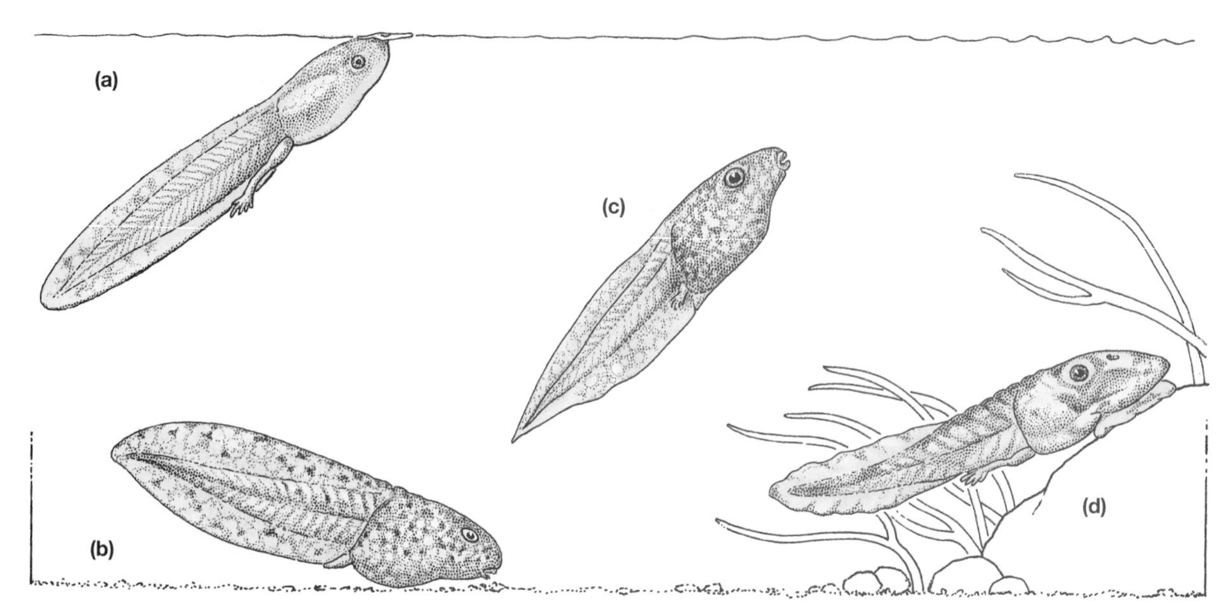

▲ Figure 9–19 Body forms of tadpoles. (a) *Megophrys minor,* Megophryidae. The mouthparts unfold into a platter over which water and particles of food on the surface are drawn into the mouth. (b) *Rana aurora,* Ranidae. A generalized feeder that nibbles and scrapes food from surfaces. (c) *Agalychnis callidryas,* Hylidae. A midwater suspension feeder shows the large fins and protruding eyes typical of midwater tadpoles. It maintains its position in the water column with rapid undulations of the end of its tail. (d) An unidentified species of *Nyctimystes,* Hylidae. A stream-dwelling tadpole that adheres to rocks in swiftly moving water with a sucker-like mouth while scraping algae and bacteria from the rocks. The low fins and powerful tail are characteristic of tadpoles living in swift water.

floating plankton from the water. Tadpoles of this type are represented in several families of anurans, especially the Pipidae and Hylidae, and usually hover in the water column. Midwater-feeding tadpoles are out in the open, where they are vulnerable to predators, and they show various characteristics that may reduce the risk of predation. Tadpoles of the African clawed frog, for example, are nearly transparent and they may be hard for predators to see. Some midwater tadpoles form schools that, like schools of fishes, may confuse a predator by presenting it with so many potential prey that it has difficulty concentrating its attack on one individual.

Many tadpoles are bottom-feeders that scrape bacteria and algae off the surfaces of rocks or the leaves of plants. The rasping action of keratinized mouthparts frees the material and allows it to be whirled into suspension in the water stream entering the mouth of a tadpole, and then filtered out by the branchial apparatus. Some bottom-feeding tadpoles like toads and spadefoot toads form dense aggregations that create currents to lift particles of food into suspension in the water. These aggregations may be groups of siblings. Toad (*Bufo americanus*) and cascade frog tadpoles (*Rana cascadae*) are able to distinguish siblings from nonsiblings, and they associate preferentially with siblings. They probably recognize siblings by scent. Toad tadpoles can distinguish full siblings (both parents the same) from maternal half-siblings (only the mother the same), and they can distinguish maternal half-siblings from paternal half-siblings. Prior experience and diet also play a role in kin recognition by tadpoles, and kin recognition might be an artifact of habitat selection.

Some tadpoles are carnivorous and feed on other tadpoles. The large Central American form *Leptodactylus pentadactylus* that preys on smaller species of anurans such as the túngara frog has a carnivorous tadpole that preys on eggs and tadpoles of other species of anurans. Predatory tadpoles have large mouths with a sharp, keratinized beak. Predatory individuals appear among the tadpoles of some species of anurans that are normally herbivorous. Some species of spadefoot toads in western North America

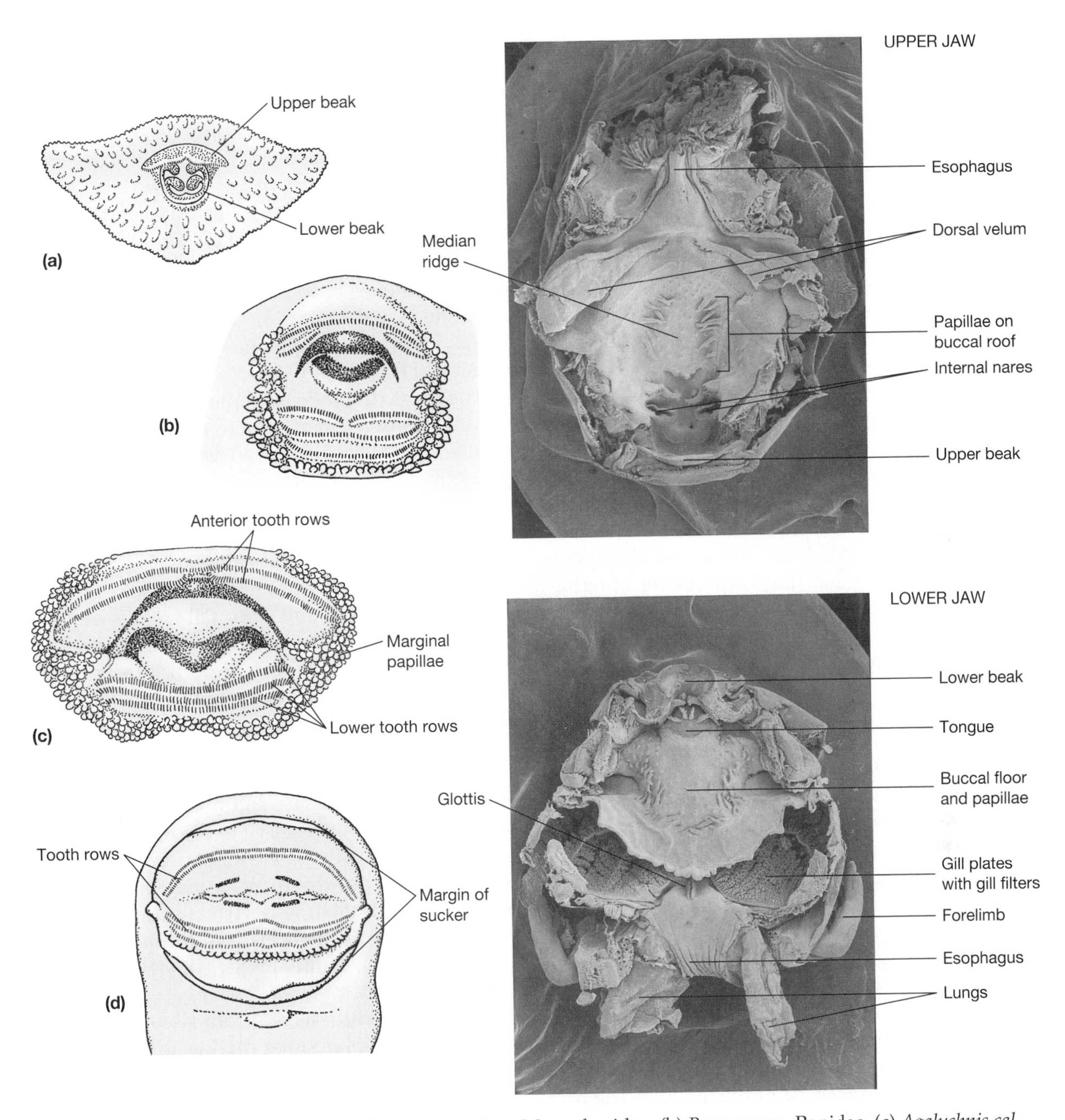

▲ Figure 9–20 Mouths of tadpoles. (a) *Megophrys minor*, Megophryidae. (b) *Rana aurora*, Ranidae. (c) *Agalychnis callidryas*, Hylidae. (d) *Nyctimystes* sp., Hylidae. (e) Scanning electron micrograph of the inside of the mouth and buccal region of a tadpole (*Alsodes monticola*, Leptodactylidae).

are famous for this phenomenon. Spadefoot tadpoles are normally herbivorous, but when tadpoles of the southern spadefoot toad, *Scaphiopus multiplicatus*, eat the freshwater shrimp that occur in some breeding ponds, they are transformed into the carnivorous morph. These carnivorous tadpoles have large heads and jaws and a powerful beak that allows them to bite off bits of flesh that are whirled into suspension and then filtered from the water stream. In addition to eating shrimp, they prey on other tadpoles.

In an Amazonian rain forest, tadpoles are by far the most important predators of frog eggs. In fact, egg predation decreases as the density of fish increases, apparently because the fish eat tadpoles that would otherwise eat frog eggs (Magnusson and Hero 1991). Carnivorous tadpoles are also found among some species of frogs that deposit their eggs or larvae in bromeliads. These relatively small reservoirs of water may have little food for tadpoles. It seems possible that the first tadpole to be placed in a bromeliad pool may feed largely on other frog eggs—either unfertilized eggs deliberately deposited by the mother of the tadpole, as is the case for the dart-poison frog *Dendrobates pumilio*, or fertilized eggs subsequently deposited by unsuspecting female frogs.

The feeding mechanisms that make tadpoles such effective collectors of food particles suspended in the water allow them to grow rapidly, but that growth contains the seeds of its own termination. As tadpoles grow bigger, they become less effective at gathering food because of the changing relationship between the size of food-gathering surfaces and the size of their bodies. The branchial surfaces that trap food particles are two-dimensional. Consequently, the food-collecting apparatus of a tadpole increases in size approximately as the square of the linear dimensions of the tadpole. However, the food the tadpole collects must nourish its entire body, and the volume of the body increases in proportion to the cube of the linear dimensions of the tadpole. The result of that relationship is a decreasing effectiveness of food collection as a tadpole grows; the body it must nourish increases in size faster than its food-collecting apparatus.

The morphological specializations of tadpoles are entirely different from those of adult frogs. The transition from tadpole to frog involves a complete metamorphosis in which tadpole structures are broken down and their chemical constituents are rebuilt into the structures of adult frogs.

9.3 Amphibian Metamorphosis

German biologist Friedrich Gudersnatch discovered the importance of thyroid hormones for amphibian metamorphosis quite by accident in the early twentieth century. Gudersnatch was able to induce rapid precocious metamorphosis in tadpoles by feeding them extracts of beef thyroid glands. Some details of the interaction of neurosecretions and endocrine gland hormones have been worked out, but no fully integrated explanation of the mechanisms of hormonal control of amphibian metamorphosis is yet possible (Hayes 1997).

The most dramatic example of metamorphosis is found among anurans, where almost every tadpole structure is altered. Anuran larval development is generally divided into three periods: (1) During premetamorphosis, tadpoles increase in size with little change in form; (2) in prometamorphosis the hind legs appear and growth of the body continues at a slower rate; and (3) during metamorphic climax the fore legs emerge and the tail regresses. These changes are stimulated by the actions of thyroxine, and production and release of thyroxine is controlled by a product of the pituitary gland, thyroid-stimulating hormone (TSH).

The action of thyroxine on larval tissues is both specific and local. In other words, it has a different effect in different tissues, and that effect is produced by the presence of thyroxine in the tissue; it does not depend on induction by neighboring tissues. The particular effect of thyroxine in a given tissue is genetically determined, and virtually every tissue of the body is involved (Table 9.5). In the liver, for example, thyroxine stimulates the enzymes responsible for the synthesis of urea (the urea cycle enzymes) and starts the synthesis of serum albumin. In the eye it induces the formation of rhodopsin. When thyroxine is administered to the striated muscles of a tadpole's developing leg, it stimulates growth; but when administered to the striated muscles of the tail, it stimulates the breakdown of tissue. When a larval salamander's tail is treated with thyroxine, only the tail fin disappears; but thyroxine causes the complete absorption of the tail of a frog tadpole.

Metamorphosis of salamanders is relatively undramatic compared with the process in anurans. Extensive changes occur at the molecular and tissue level in salamanders, but the loss of gills and absorption of the tail fin are the only obvious external

Table 9.5 Metamorphosis. Some of the morphological and physiological changes induced by thyroid hormones during amphibian metamorphosis.

Body form and structure
Formation of dermal glands
Restructuring of mouth and head
Intestinal regression and reorganization
Calcification of skeleton
Appendages
Degeneration of skin and muscle of tail
Growth of skin and muscle of limbs
Nervous system and sense organs
Increase in rhodopsin in retina
Growth of extrinsic eye muscles
Formation of nictitating membrane of the eye
Growth of cerebellum
Growth of preoptic nucleus of the hypothalamus
Respiratory and circulatory systems
Degeneration of the gill arches and gills
Degeneration of the operculum that covers the gills
Development of lungs
Shift from larval to adult hemoglobin
Organs
Pronephric resorption in the kidney
Induction of urea-cycle enzymes in the liver
Reduction and restructuring of the pancreas

changes. In contrast, the metamorphosis of a tadpole to a frog involves readily visible changes in almost every part of the body. The tail is absorbed and recycled into the production of adult structures. The small tadpole mouth that accommodated algae broadens into the huge mouth of an adult frog. The long tadpole gut, characteristic of herbivorous vertebrates, changes to the short gut of a carnivorous animal.

Metamorphic climax begins with the appearance of the forelimbs and ends with the disappearance of the tail. This is the most rapid part of metamorphosis, occupying only a few days after a larval period that lasts for weeks or months. One reason for the rapidity of metamorphic climax may lie in the vulnerability of larvae to predators during this period. A larva with legs and a tail is neither a good tadpole nor a good frog: The legs inhibit swimming and the tail interferes with jumping. As a result, predators are more successful at catching anurans during metamorphic climax than they are in prometamorphosis or following the completion of metamorphosis. Metamorphosing chorus frogs (*Pseudacris triseriata*) were most vulnerable to garter snakes when they had developed legs and still retained a tail. Both tadpoles (with a tail and no legs) and metamorphosed frogs (with legs and no tail) were more successful than the metamorphosing individuals at escaping from snakes (Figure 9–21). In water, the snakes captured 33 percent of the tadpoles offered, compared with 67 percent of the transforming frogs. On land the snakes captured 45 percent of the fully transformed frogs that were offered and 90 percent of the transforming ones. Life-history theory predicts that selection will act to shorten the periods in the lifetime of a species when it is most vulnerable to predation, and the speed of metamorphic climax may be a manifestation of that phenomenon.

9.4 Exchange of Water and Gases

Amphibians have a glandular skin that lacks external scales and is highly permeable to gases and water. Both the permeability and glandularity of the skin have been of major importance in shaping the ecology and evolution of amphibians. Mucus glands are distributed over the entire body surface and secrete mucopolysaccharide compounds. The primary function of the mucus is to keep the skin moist and permeable.

Percent of prey captured

In water 33%

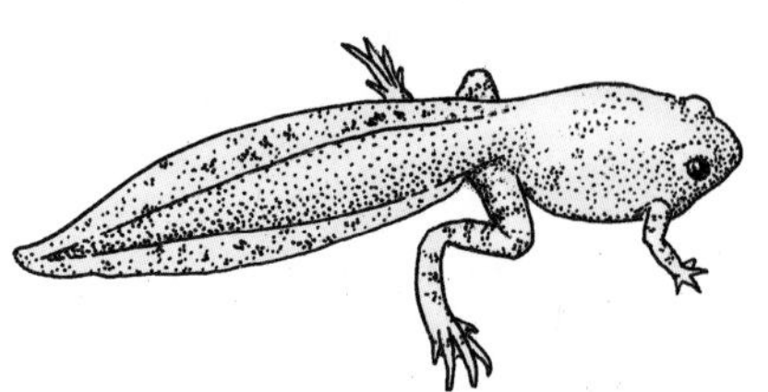

▲ Figure 9–21 Predation on tadpoles and frogs. Metamorphosing chorus frogs (*Pseudacris triseriata,* Hylidae) were more vulnerable to garter snakes than were tadpoles or fully transformed frogs.

For an amphibian, a dry skin means reduction in permeability. That, in turn, reduces oxygen uptake and the ability of the animal to use evaporative cooling to maintain its body temperature within equable limits. Experimentally produced interference with mucus gland secretion can lead to lethal overheating of frogs undergoing normal basking activity.

Both water and gases pass readily through amphibian skin. In biological systems, permeability to water is inseparable from permeability to gases, and amphibians depend on cutaneous respiration for a significant part of their gas exchange. Although the skin permits passive movement of water and gases, it controls the movement of other compounds. Sodium is actively transported from the outer surface to the inner, and urea is retained by the skin. These characteristics are important in the regulation of osmolal concentration and in facilitating uptake of water by terrestrial species.

Blood Flow in Larvae and Adults

Larval amphibians rely on their gills and skin for gas exchange, whereas adults of species that complete full metamorphosis lose their gills and develop lungs. Lungs develop at different larval stages in different families of amphibians, and as the lungs develop they are increasingly used for respiration. Late in their development tadpoles and partly metamorphosed froglets can be seen swimming to the surface to gulp air. As the gills lose their respiratory function, the carotid arches also change their roles. Arches 1 and 2 are lost early in embryonic development. Arches 3 through 5 supply blood to the gills, and arch 6 carries blood to the dorsal aorta via a connection called the ductus arteriosus (Figure 9–22a). At metamorphosis, arch 3 becomes the supply vessel for the internal carotid arteries (Figure 9–22b). Initially arches 4 and 5 supply blood to the dorsal aorta; but arch 5 is usually lost in anurans, so arch 4 becomes the main route by which blood from the heart enters the aorta. Arch 6 primarily supplies blood to the lungs and skin via the pulmocutaneous arteries.

Cutaneous Respiration

All amphibians rely on the skin surface for gas exchange, especially for the release of carbon dioxide. The balance between cutaneous and pulmonary uptake of oxygen varies among species, and within a species it depends on body temperature and how active the animal is. Amphibians show increasing reliance on the lungs for oxygen uptake as temperature and activity increase.

The heart of amphibians reflects the use of two respiratory surfaces in the patterns of blood flow

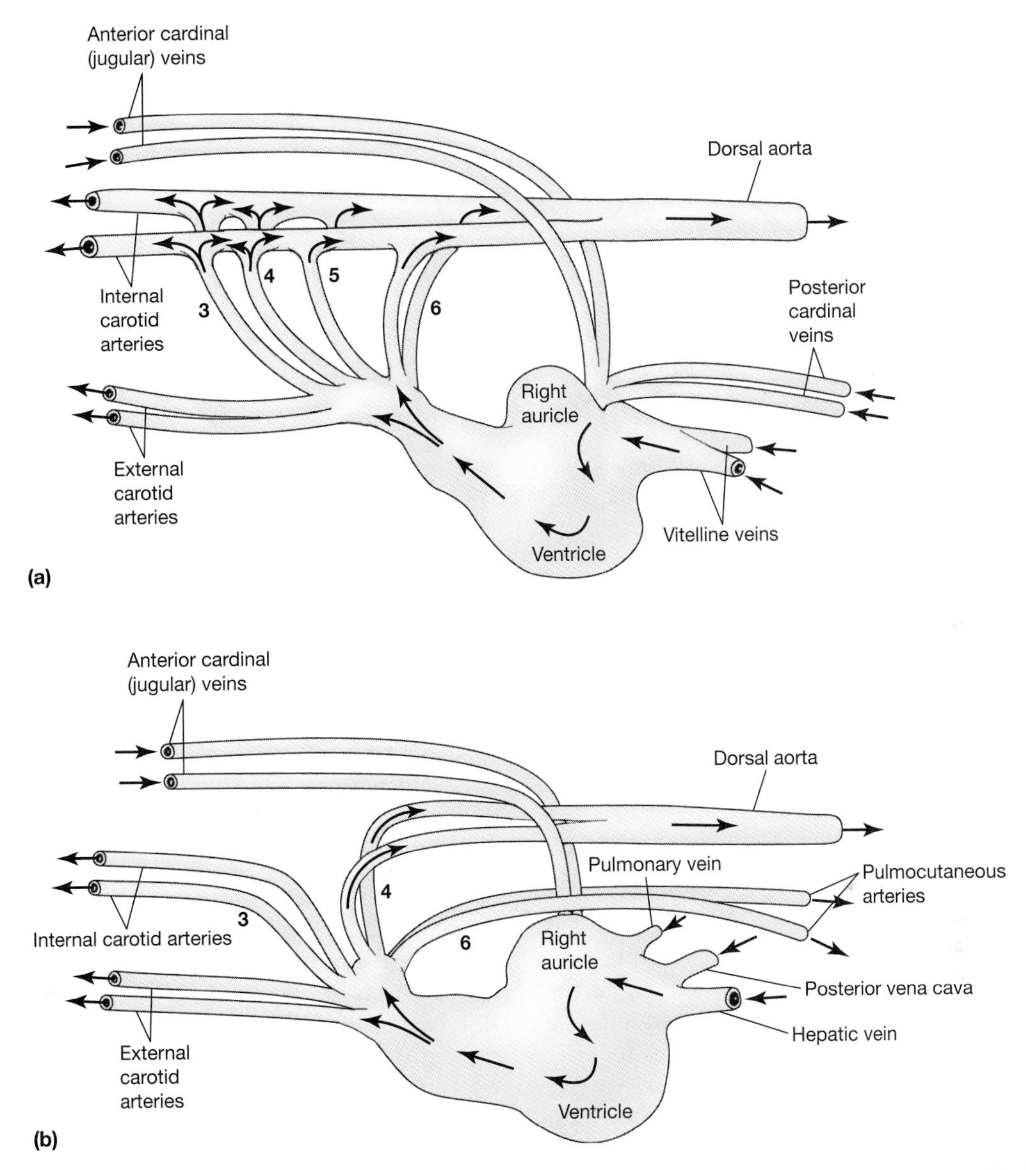

▲ Figure 9–22 Blood flow through the aortic arches of (a) a larval amphibian and (b) an adult without gills. The head is to the left. Arches 3, 4, and 5 carry blood to the gills of larvae, and arch 6 flows into the dorsal aorta. In adults arch 3 carries blood to the brain via the internal carotid arteries, and arch 6 sends blood to the pulmocutaenous arteries.

within the heart and the distribution of oxygenated and deoxygenated blood to different parts of the body. The anuran heart is the basis for the following description (Figure 9–23). The atrium of the heart is divided into right and left chambers, either anatomically by an interatrial septum or functionally by flow patterns. Blood from the systemic veins flows into the right side of the heart, and blood from the lungs flows into the left side. The ventricle shows a variable subdivision that correlates with the physiological importance of pulmonary respiration to the species. In frogs the ventricle is undivided, but the position within the ventricle of a particular parcel of blood before ventricular contraction appears to determine its fate on leaving the contracting ventricle. The short conus arteriosus contains a spiral valve of tissue that differentially guides the blood from the left and right sides of the ventricle to the aortic arches. The anatomical relationships within the heart are such that oxygen-rich blood returning to the

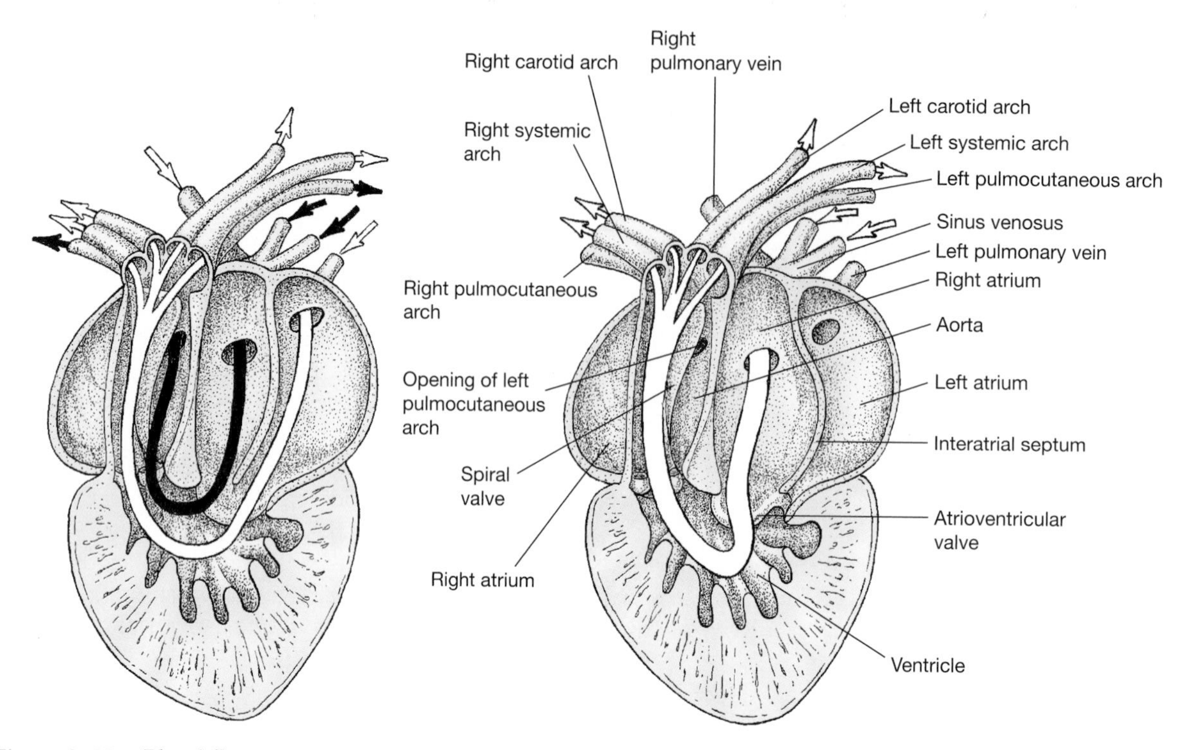

▲ Figure 9–23 Blood flow in the amphibian heart. Left, pattern of flow when lungs are being ventilated; right, flow when only cutaneous respiration is taking place. Dark arrows, blood with low oxygen content; light arrows, most highly oxygenated blood.

heart from the pulmonary veins enters the left atrium, which injects it on the left side of the common ventricle. The spongy muscular lumen of the ventricle minimizes the mixing of right- and left-side blood. Contraction of the ventricle tends to eject blood in laminar streams that spiral out of the pumping chamber, carrying the left-side blood into the ventral portion of the spirally divided conus. This half of the conus is the one from which the carotid (head-supplying) and systemic aortic arches arise.

Thus, when the lungs are actively ventilated, oxygen-rich blood returning from them to the heart is selectively distributed to the tissues of the head and body. Oxygen-poor venous blood entering the right atrium is directed into the dorsal half of the spiral-valved conus. It goes to the pulmocutaneous arch, destined for oxygenation in the lungs. However, when the skin is the primary site of gaseous exchange, as it is when a frog is underwater, the highest oxygen content is in the systemic veins that drain the skin. The lungs may actually be net users of oxygen, and because of vascular constriction little blood passes through the pulmonary circuit. Because the ventricle is undivided and the majority of the blood is arriving from the systemic circuit, the ventral section of the conus receives blood from an overflow of the right side of the ventricle. The scant left atrial supply to the ventricle also flows through the ventral conus. Thus the most oxygenated blood coming from the heart flows to the tissues of the head and body during this shift in primary respiratory surface, a phenomenon possible only because of the undivided ventricle. Variability of the cardiovascular output in amphibians is an essential part of their ability to use alternative respiratory surfaces effectively.

Permeability to Water

The internal osmolal pressure of amphibians is approximately two-thirds that characteristic of most other vertebrates. The primary reason for the dilute body fluids of amphibians is low sodium content—approximately 100 milliequivalents compared with 150 milliequivalents in other vertebrates. Amphibians can tolerate a doubling of the normal sodium concentration, whereas an increase from 150 milliequivalents to 170 milliequivalents is the maximum humans can tolerate.

Amphibians are most abundant in moist habitats, especially temperate and tropical forests, but a surprisingly large number of species live in dry regions. Anurans have been by far the most successful amphibian invaders of arid habitats. All but the harshest deserts have substantial anuran populations, and in different parts of the world, different families have converged on similar specializations. Avoiding the harsh conditions of the ground surface is the most common mechanism by which amphibians have managed to invade deserts and other arid habitats. Anurans and salamanders in deserts may spend 9 or 10 months of the year in moist retreat sites, sometimes more than a meter underground, emerging only during the rainy season and compressing feeding, growth, and reproduction into just a few months.

Many species of arboreal frogs have skins that are less permeable to water than the skin of terrestrial frogs, and a remarkable specialization is seen in a few tree frogs. The African rhacophorid *Chiromantis xerampelina* and the South American hylid *Phyllomedusa sauvagei* lose water through the skin at a rate only one-tenth that of most frogs. *Phyllomedusa* has been shown to achieve this low rate of evaporative water loss by using its legs to spread the lipid-containing secretions of dermal glands over its body surface in a complex sequence of wiping movements, but the basis for the impermeability of *Chiromantis* is not yet understood. These two frogs are unusual also because they excrete nitrogenous wastes as salts of uric acid rather than as urea (see Chapter 4). This uricotelism provides still more water conservation.

Behavioral Control of Evaporative Water Loss

For animals with skins as permeable as those of most amphibians, the main difference between rain forests and deserts may be how frequently they encounter a water shortage. The Puerto Rican coquí lives in wet tropical forests; nonetheless, it has elaborate behaviors that reduce evaporative water loss during its periods of activity. Male coquís emerge from their daytime retreat sites at dusk and move 1 or 2 meters to calling sites on leaves in the understory vegetation. They remain at their calling sites until shortly before dawn, when they return to their daytime retreats. The activities of the frogs vary from night to night, depending on whether it rained during the afternoon. On nights after a rainstorm, when the forest is wet, the coquís begin to vocalize soon after dusk and continue until about midnight, when they fall silent for several hours. They resume calling briefly just before dawn. When they are calling, coquís extend their legs and raise themselves off the surface of the leaf (Figure 9–24). In this position they lose water by evaporation from the entire body surface.

On dry nights the behavior of the frogs is quite different. The males move from their retreat sites to their calling stations, but they call only sporadically. Most of the time they rest in a water-conserving posture in which the body and chin are flattened against the leaf surface and the limbs are pressed against the body. A frog in this posture exposes only half its body surface to the air, thereby reducing its rate of evaporative water loss. The effectiveness of the postural adjustments is illustrated by the water losses of frogs in the forest at El Verde, Puerto Rico, on dry nights. Frogs in one test group were placed individually in small wire-mesh cages that were placed on leaf surfaces. A second group was composed of unrestrained frogs sitting on leaves. The caged frogs spent most of the night climbing around the cages trying to get out. This activity, like vocalization, exposed the entire body surface to the air, and the caged frogs had an evaporative water loss that averaged 27.5 percent of their initial body mass. In contrast, the unrestrained frogs adopted water-conserving postures and lost an average of only 8 percent of their initial body mass by evaporation.

Experiments showed that the jumping ability of coquís was not affected by an evaporative loss of as much as 10 percent of the initial body mass, but a loss of 20 percent or more substantially decreased the distance frogs could jump. Thus, coquís can use behavior to limit their evaporative water losses on dry nights to levels that do not affect their ability to escape from predators or to capture prey. Without those behaviors, however, they would probably lose enough water by evaporation to affect their survival.

Uptake and Storage of Water

The mechanisms that amphibians use for obtaining water in terrestrial environments have received less attention than those for retaining it. Amphibians do not drink water. Because of the permeability of their

▲ Figure 9–24 Male Puerto Rican coquí, *Eleutherodactylus coqui,* Leptodactylidae. (a) During vocalization nearly all the body surface is exposed to evaporation. (b) In the alert posture in which frogs wait to catch prey, most of the body surface is exposed. (c) In the water-conserving posture adopted on dry nights, half the body surface is protected from exposure.

skins, species that live in aquatic habitats face a continuous osmotic influx of water that they must balance by producing urine. The impressive adaptations of terrestrial amphibians are ones that facilitate rehydration from limited sources of water. One such specialization is the **pelvic patch.** This is an area of highly vascularized skin in the pelvic region that is responsible for a very large portion of an anuran's cutaneous water absorption. Toads that are dehydrated and completely immersed in water rehydrate only slightly faster than those placed in water just deep enough to wet the pelvic area. In arid regions, water may be available only as a thin layer of moisture on a rock, or as wet soil. The pelvic patch allows an anuran to absorb this water.

The urinary bladder plays an important role in the water relations of terrestrial amphibians, especially anurans. Amphibian kidneys produce urine that is hyposmolal to the blood, so the urine in the bladder is dilute. Amphibians can reabsorb water from urine to replace water they lose by evaporation, and terrestrial amphibians have larger bladders than aquatic species. Storage capacities of 20 to 30 percent of the body mass of the animal are common for terrestrial anurans, and some species have still larger bladders: The Australian desert frogs *Notaden nicholsi* and *Neobatrachus wilsmorei* can store urine equivalent to about 50 percent of their body mass, and a bladder volume of 78.9 percent of body mass has been reported for the Australian frog *Helioporus eyrei.*

Behavior is as important in facilitating water uptake as it is in reducing water loss. Leopard frogs, *Rana pipiens,* spend the summer activity season in grassy meadows where they have no access to ponds. The frogs spend the day in retreats they create by pushing vegetation aside to expose moist soil. In the retreats, the frogs rest with the pelvic patch in contact with the ground, and tests have shown that the frogs are able to absorb water from the soil. On nights when dew forms, many frogs move from their retreats and spend some hours in the early morning sitting on dew-covered grass before returning to their retreats. Leopard frogs show a daily pattern of water gain and loss during a period of several days when no rain falls: In the morning the frogs are sleek and glistening with moisture, and they have urine in their bladders. That observation indicates that in the morning the frogs have enough water to form urine. By evening the frogs have dry skins, and little urine in the bladder, suggesting that as they lost water by evaporation during the day they had reabsorbed water from the urine to maintain the water content of their tissues.

By the following morning the frogs have absorbed more water from dew and are sleek and well hydrated again. Net gains and losses of water are shown by daily fluctuations in body masses of the frogs; In the mornings they are as much as 4 or 5 percent heavier than their overall average mass, and in the evenings they are lighter than the average by a similar amount. Thus, these terrestrial frogs are able to balance their water budgets by absorbing water from moist soil and from dew to replace the water they lose by evaporation and in urine. As a result, they are independent of sources of water like ponds or streams and are able to colonize meadows and woods far from any permanent sources of water.

9.5 Poison Glands and Other Defense Mechanisms

The mucus that covers the skin of an amphibian has a variety of properties. In at least some species it has antibacterial activity, and a potent antibiotic that may have medical applications has been isolated from the skin of the African clawed frog. It is mucus that makes some amphibians slippery and hard for a predator to hold. Other species have mucus that is extremely adhesive. Many salamanders, for example, have a concentration of mucus glands on the dorsal surface of the tail. When one of these salamanders is attacked by a predator, it bends its tail forward and buffets its attacker. The sticky mucus causes debris to adhere to the predator's snout or beak, and with luck the attacker soon concentrates on cleaning itself, losing interest in the salamander. When the California slender salamander is seized by a garter snake, the salamander curls its tail around the snake's head and neck. This behavior makes the salamander hard for the snake to swallow, and also spreads sticky secretions on the snake's body. A small snake can find its body glued into a coil from which it is unable to escape.

Although secretions of the mucus glands of some species of amphibians are irritating or toxic to predators, an amphibian's primary chemical defense system is located in the poison glands (Figure 9–25). These glands are concentrated on the dorsal surfaces of the animal, and defense postures of both anurans and salamanders present the glandular areas to potential predators.

A great diversity of pharmacologically active substances has been found in the skins of amphibians. Some of these substances are extremely toxic; others are less toxic but capable of producing unpleasant sensations when a predator bites an amphibian. Biogenic amines such as serotonin and histamine, peptides such as bradykinin, and hemolytic proteins have been found in frogs and salamanders belonging to many families. Many of these substances, such as bufotoxin, physalaemin, and leptodactyline, are named for the animals in which they were discovered. Epibatidine, a toxic compound isolated from the skin of the Central American frog *Epipedobates tricolor,* is a potent painkiller. Unlike opioids such as morphine, epibatidine works by blocking a receptor for the neurotransmitter acetylcholine. An artificial form of epibatidine has been synthesized. Known as ABT-594, this compound lacks the toxic side effects of natural epibatidine and also lacks the addictive effect of opioids. It is currently undergoing safety trials in preparation for clinical tests of its effectiveness for relief of chronic pain in humans.

Cutaneous alkaloids are abundant and diverse among the dart-poison frogs, the family Dendrobatidae, of the New World tropics. Most of these frogs

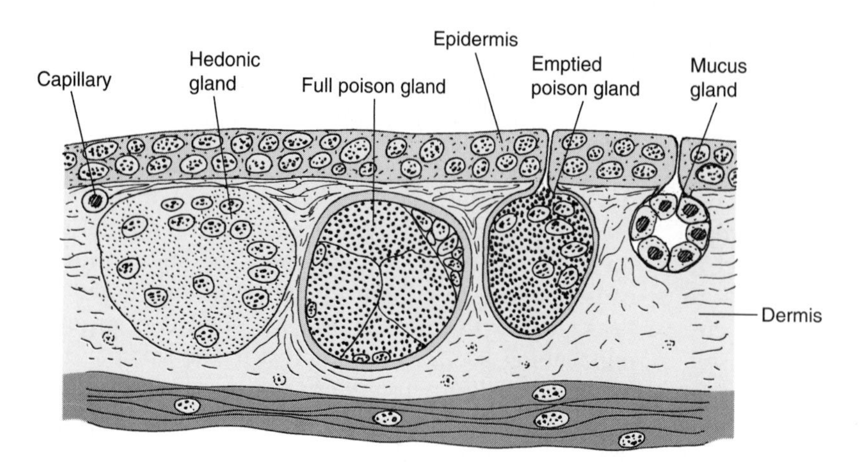

Figure 9–25 Amphibian skin. Cross section of skin from the base of the tail of a red-backed salamander, *Plethodon cinereus,* Plethodontidae. Three types of glands can be seen.

are brightly colored and move about on the ground surface in daylight, making no attempt at concealment. More than 200 new alkaloids have been described from dendrobatids, mostly species in the genera *Dendrobates, Minyobates, Epidobates,* and *Phyllobates.* Most of the alkaloids found in the skins of dart-poison frogs are similar to those found in ants, beetles, and millipedes that live in the leaf litter with the frogs, suggesting that frogs obtain alkaloids from their prey.

That hypothesis was tested by John Daly and his colleagues by raising juvenile Panamanian dart-poison frogs, *Dendrobates auratus,* under three dietary conditions (Daly et al. 2000). Dart-poison frogs live in the leaf litter on the forest floor, and leaf litter was used as the cage bedding for all experiments. Frogs with the most restricted diet were kept in a glass cage with carefully sealed joints and were fed only fruit flies. (Even these measures did not prevent a few tiny myrmicine ants from entering the cage during the experiment.) To ensure that the frogs in the glass cage had no natural prey, the leaf litter in their cage was frozen for two weeks before the experiment to kill the ants and other arthropods that live in it. Frogs receiving the other two diets were kept outdoors in screened cages with a mesh size large enough to allow fruit flies, ants, and other small arthropods to enter the cages. The frogs in one cage lived in leaf litter that had been frozen and ate primarily fruit flies that were attracted by bananas and fruit fly medium in the cage. Fruit flies were present in such abundance in this cage that they made up the bulk of the frogs' diet, but myrmicine ants also got into the cage. Frogs in the other screened cage lived in freshly gathered leaf litter that was replaced weekly; these frogs ate the ants and other arthropods that came in with the leaf litter. At the end of the experiment, analyses of the skins of the frogs showed that one frog from the glass cage had no detectable alkaloids and a second had had a trace amount of an alkaloid found in myrmicine ants—probably this frog had eaten the ants that had entered the cage despite the sealed joints. Frogs raised in the screened cage that was baited with bananas to attract fruit flies contained four alkaloids, all characteristic of myrmicine ants. Frogs in the cage that received fresh leaf litter had at least 16 different alkaloids in their skins that can be traced to millipedes, beetles, and ants. Wild-caught frogs from the area of the experiment had still more alkaloids in their skins—more than different 40 compounds. This experiment shows clearly that the frogs obtain toxic alkaloids from the prey that they eat, and that the more varied a frog's diet is, the more different alkaloids it contains. Furthermore, *Dendrobates auratus* from Cerro Ancón in Panama City had different alkaloids from those on Isla Taboga in the nearby Bay of Panamá, paralleling differences in the alkaloids found in ants at the two sites.

The name "dart-poison frogs" refers to the use by South American Indians of the toxins of some of these frogs to poison the tips of the blowgun darts used for hunting. The use of frogs in this manner appears to be limited to three species of *Phyllobates* that occur in western Colombia, although plant poisons like curare are used to poison blowgun darts in other parts of South America. A unique alkaloid, batrachotoxin, occurs in the genus *Phyllobates.* Batrachotoxin is a potent neurotoxin that prevents the closing of sodium channels in nerve and muscle cells, leading to irreversible depolarization and producing cardiac arrhythmias, fibrillation, and cardiac failure.

The bright yellow *Phyllobates terribilis* is the largest and most toxic species in the genus. The Emberá Choco Indians of Colombia use *Phyllobates terribilis* as a source of poison for their blowgun darts. The dart points are rubbed several times across the back of a frog, and set aside to dry. The Indians handle the frogs carefully, holding them with leaves—a wise precaution because batrachotoxin is exceedingly poisonous. A single frog may contain up to 1900 micrograms of batrachotoxin, and less than 200 micrograms is probably a lethal dose for a human if it enters the body through a cut. Batrachotoxin is also toxic when it is eaten. In fact, the investigators inadvertently caused the death of a dog and a chicken in the Indian village in which they were living when the animals got into garbage that included plastic bags in which the frogs had been carried. Cooking destroys the poison and makes prey killed by darts anointed with the skin secretions of *Phyllobates terribilis* safe to eat.

Skin secretions from another frog are used in the hunting magic of several Amazonian tribes. *Phyllomedusa bicolor*, a large, green hylid frog, secretes a variety of peptides, including a hitherto unknown compound that has been named adenoregulin (Daly et al. 1992). Mucus scraped from a frog's skin is dried and stored for later use. When it is mixed with saliva and rubbed into areas of freshly burned skin, it induces a feeling of illness, followed by listlessness, and finally a profound euphoria. Adenoregulin, the constituent of the mucus that is responsible for these effects, apparently enhances binding of adenosine and its analogs to A_1 receptors of the brain. These receptors are distributed throughout the brain, and a way to modify their function could contribute to treatments for a variety of central nervous system disorders, including depression, seizures, and loss of cognitive function in conditions such as Alzheimer's disease. Adenoregulin may be yet another example of a medically important compound from the rapidly dwindling tropical forests.

Many amphibians advertise their distasteful properties with conspicuous **aposematic** (warning) colors and behaviors. A predator that makes the mistake of seizing one is likely to spit it out because it is distasteful. The toxins in the skin may also induce vomiting that reinforces the unpleasant taste for the predator. Subsequently, the predator will remember its unpleasant experience and avoid the distinctly marked animal that produced it. Some toxic amphibians combine a cryptic dorsal color with an aposematic ventral pattern. Normally, the cryptic color conceals them from predators, but if they are attacked they adopt a posture that displays the brightly colored ventral surface (Figure 9–26).

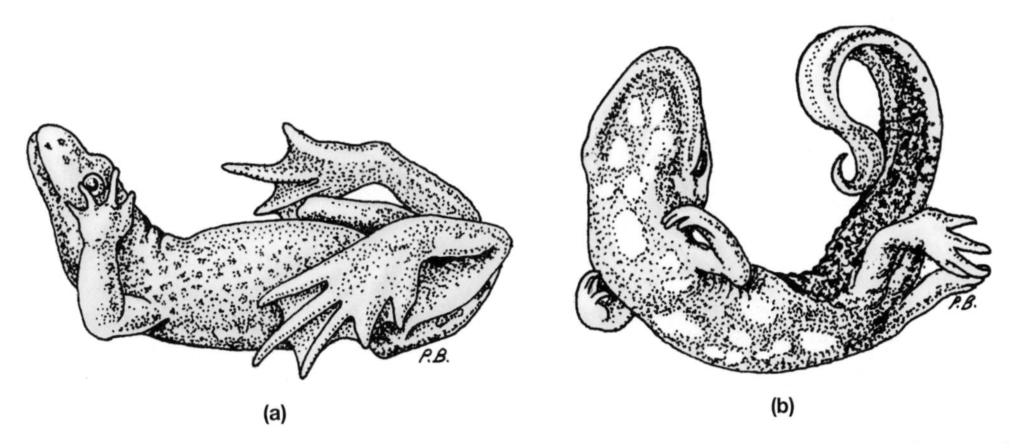

▲ Figure 9–26 Aposematic displays by amphibians. Brightly colored species of amphibians have displays that present colors as warnings that predators can learn to associate with the animals' toxic properties. (a) The European fire-bellied toad (*Bombina*, Bombinatoridae) has a cryptically colored dorsal surface and a brightly colored underside that is displayed in the *unken* reflex when the animal is attacked. (b) The Hong Kong newt (*Paramesotriton*, Salamandridae) has a brownish dorsal surface and a mottled red and black venter that is revealed by its aposematic display.

Some salamanders have a morphological specialization that enhances the defensive effects of their chemical secretions. The European salamander *Pleurodeles waltl* and two genera of Asian salamanders (*Echinotriton* and *Tylotriton*) have ribs that pierce the body wall when a predator seizes the salamander. You can imagine the shock for a predator that bites a salamander and finds its tongue and palate impaled by a dozen or more bony spikes! Even worse, the ribs penetrate poison glands as they emerge through the body wall and each rib carries a drop of poison into the wound.

Many anurans make long leaps to escape a predator, and others feign death. Some cryptically colored frogs extend their legs stiffly when they play dead. In this posture they look so much like the leaf litter on the ground that they may be hard for a visually oriented predator such as a bird to see. Very large frogs attack potential predators. They increase their apparent size by inflating the lungs, and hop toward the predator, often croaking loudly. That alone can be an unnerving experience, yet some of the carnivorous species such as the horned frogs of South America (*Ceratophrys*), which have recurved teeth on the maxillae and tooth-like serrations on the mandibles, also can inflict a painful bite.

Red efts are classic examples of aposematic animals (see the color insert). They are bright orange and are active during the day, making no effort to conceal themselves. Efts contain tetrodotoxin, which is a potent neurotoxin. Touching an eft to your lips produces an immediate unpleasant numbness and tingling sensation, and the behavior of animals that normally prey on salamanders indicates that it affects them the same way. As a result, an eft that is attacked by a predator is likely to be rejected before it is injured. After one or two such experiences, a predator will no longer attack efts. Support for the belief that this protection may operate in nature is provided by the observation that 4 of 11 wild-caught bluejays (*Cyanocitta cristata*) refused to attack the first red eft they were offered in a laboratory feeding trial (Tilley et al. 1982). That behavior suggests that those four birds had learned to avoid red efts before they were captured. The remaining seven birds attacked at least one eft, but dropped it immediately. After one or two experiences of this sort, the birds made retching movements at the sight of an eft and refused to attack.

Of course, aposematic colors and patterns work to deter predation only if a predator can see the aposematic signal. Nocturnal animals may have difficulty being conspicuous if they rely on visual signals. One species of dendrobatid frog apparently deters predators with a foul odor. The aptly named *Aromobates nocturnus* from the cloud forests of the Venezuelan Andes is the only nocturnal dendrobatid. It is an inconspicuous frog, about 5 cm long with a dark olive color. The frogs emit a foul, skunk-like odor when they are handled.

9.6 Mimicry

The existence of unpalatable animals that deter predators with aposematic colors and behaviors offers the opportunity for other species that lack noxious qualities to take advantage of predators that have learned by experience to avoid the aposematic species. In this phenomenon, known as **mimicry,** the mimic (a species that lacks noxious properties) resembles a noxious model, and that resemblance causes a third species—the dupe—to mistake the mimic for the model. Some of the best-known cases of mimicry among vertebrates involve salamanders. One that has been investigated involves two color morphs of the common red-backed salamander, *Plethodon cinereus.*

Red-backed salamanders normally have dark pigment on the sides of the body, but in some regions an erythristic (Greek *erythr* = red) color morph is found that lacks the dark pigmentation and has red-orange on the sides as well as on the back. These erythristic morphs resemble red efts, and could be mimics of efts. Red-backed salamanders are palatable to many predators, and mimicry of the noxious red efts might confer some degree of protection on individuals of the erythristic morph. That hypothesis was tested in a series of experiments (Brodie and Brodie 1980). Salamanders were put in leaf-filled trays from which they could not escape, and the trays were placed in a forest where birds were foraging. The birds learned to search through the leaves in the trays to find the salamanders. This is a very life-like situation for a test of mimicry because some species of birds are important predators of salamanders. For example, red-backed salamanders and dusky salamanders (*Desmognathus ochrophaeus*) made up 25 percent of the prey items fed to their young by hermit thrushes in western New York.

Three species of salamanders were used in the experiments, and the number of each species was adjusted to represent a hypothetical community of salamanders containing 40 percent dusky salamanders, 30 percent red efts, 24 percent striped red-backed salamanders, and 6 percent erythristic red-backed salamanders. The dusky salamanders are palatable to birds and are light brown; they do not resemble either efts or red-backed salamanders, and they served as a control in the experiment. The striped red-backed salamanders represent a second control: The hypothesis of mimicry of red efts by erythristic salamanders leads to the prediction that the striped salamanders, which do not look like efts, will be eaten by birds, whereas the erythristic salamanders, which are as palatable as the striped ones but which do look like the noxious efts, will not be eaten.

A predetermined number of each kind of salamander was put in the trays, and birds were allowed to forage for 2 hours. At the end of that time the salamanders that remained were counted. As expected, only 1 percent of the efts had been taken by birds, whereas 44 to 60 percent of the palatable salamanders had disappeared (Table 9.6). As predicted, the birds ate fewer of the erythristic form of the red-backed salamanders than they ate of the striped form.

These results show that the erythristic morph of the red-backed salamander does obtain some protection from avian predators as a result of its resemblance to the red eft. In this case the resemblance is visual, but mimicry can exist in any sensory mode to which a dupe is sensitive. Olfactory mimicry by amphibians might be effective against predators such as shrews and snakes, which rely on scent to find and identify prey. This possibility has scarcely been considered, but careful investigations may yield fascinating new examples.

Table 9.6 Different ial survival of salamanders exposed to foraging birds

Hypothesis: The erythristic morph of *Plethodon cinereus* is a mimic of the red eft.

Predictions:

1. Birds will not eat red efts because the efts are noxious.
2. Birds will readily eat dusky salamanders, which are palatable and not mimetic.
3. Birds will eat the striped *Plethodon*, which are also palatable and not mimetic.
4. Birds will mistake the erythristic *Plethodon* for efts and will not eat them.

Results

Percentage of Salamanders Gone from Trays

		Plethodon	
Red Efts	Dusky Salamanders	Striped	Erythristic
1.0	52.6	60.1	43.9

Interpretation

The predictions of the hypothesis were supported:

1. Birds did not eat the noxious red efts (prediction 1).
2. Birds did eat the palatable, nonmimetic dusky salamanders (prediction 2).
3. Birds ate the striped morph of *Plethodon* (prediction 3).
4. Birds ate significantly fewer of the mimetic morph of *Plethodon* than of the striped morph (prediction 4).

9.7 Why are Amphibians Vanishing?

Biologists from many countries met in England in 1989 at the First World Congress of Herpetology. In a week of formal presentations of scientific studies and in casual conversations at meals and in hallways, the participants discovered that an alarmingly large proportion of them knew of populations of amphibians that had once been abundant and now were rare, or even entirely gone. Events that had appeared to be isolated instances began to look like parts of a global pattern. As a result of that discovery, David Wake, of the University of California at Berkeley, persuaded the National Academy of Sciences to convene a meeting of biologists concerned about vanishing amphibians. Biologists from all over the world met at the West Coast center of the Academy in February 1990. All reported that populations of amphibians in their countries were disappearing, and often there was no apparent reason. Following that meeting, an international effort to identify the causes of amphibian declines was initiated by the Declining Amphibian Populations Task Force of the Species Survival Commission of the World Conservation Union (IUCN). In 1998 another meeting was convened, this time in Washington, D.C., that brought together authorities in disciplines ranging from herpetology and population biology through toxicology and infectious diseases to climate change and science policy. The conference concluded that "There is compelling evidence that, over the last 15 years, unusual and substantial declines have occurred in abundance and numbers of populations in globally distributed geographic regions" (Wake 1998).

The global decline of amphibian populations is alarming, especially because in many cases we have no idea *why* a species has disappeared from places in which it was formerly abundant. In some cases local events appear to provide an explanation. Habitat changes produced by logging are usually destructive to amphibians, for example, because frogs and salamanders depend on cool, moist microhabitats on the forest floor. When the forest canopy is removed, sunlight reaches the ground and conditions become too hot and dry for amphibians. Mining and extraction of oil also cause damage that can extend over large areas. The rock removed from mines often releases acid or toxic chemicals, cyanide used to extract gold from ore poisons surface water, oil wells spread toxic hydrocarbons, nitrates and nitrites drain from farmland and reach levels that cause deformities and death of larval amphibians—the list of abuses is nearly endless.

Some local causes of amphibian mortality are not only obvious, they are positively undignified. Federal land in the western United States is leased for grazing, and cattle drink from the ponds that are breeding sites for anurans. As the ponds shrink during the summer, they leave a band of mud that cattle cross when they come to drink. The deep hoof prints the cattle make can be death traps for newly metamorphosed frogs and toads that tumble in and cannot climb out. Even worse, a few juvenile anurans that have the bad luck to pass behind a cow at exactly the wrong moment are trapped and suffocated beneath a pile of fresh manure!

But the truly disturbing questions involve species such as the Costa Rica golden toad (*Bufo periglenes*) and other high-altitude frogs that live in habitats where there is no sign of local environmental damage. The global scope of the problem suggests that we should look for global explanations. Two factors that have probably contributed to some of these declines are acid precipitation and increased ultraviolet radiation.

Precipitation (rain, snow, and fog) over large parts of the world, especially the Northern Hemisphere, is at least a hundredfold more acidic than it would be if the water were in equilibrium with carbon dioxide in the air. The extra acidity is produced by nitric and sulfuric acids that form when water vapor combines with oxides of nitrogen and sulfur released by combustion of fossil fuels. Water in the breeding ponds of many amphibians in the Northern Hemisphere has become more acidic in the past 50 years, and this acidity has both direct and indirect effects on amphibian eggs and larvae. Embryos of many species of frogs and salamanders are killed or damaged at pH 5 or less (Figure 9–27). Larvae that hatch may be smaller than normal, and sometimes have strange lumps or kinks in their bodies. Spotted salamander larvae grow slowly in acid water because their prey-capture efforts are clumsy and they eat less than do larvae at higher pH (Preest 1993).

Another global phenomenon is an increase in the amount of ultraviolet radiation reaching the Earth's surface as a result of destruction of ozone in the stratosphere by chemical pollutants. The effects are most dramatic at the poles, and are spreading into lower latitudes in both hemispheres. Ultraviolet light, especially the 280- to 320-nanometer UV-B band, kills amphibian eggs and embryos (Blaustein et al. 1994). Only 50 to 60 percent of the eggs of the Cascade frog (*Rana cascadae*) and the western toad (*Bufo boreas*) in ponds in the Cascade Mountains of Oregon survived when they were exposed to incident sunlight, but when a filter that blocked UV-B was placed over the eggs, survival climbed to 70 to 85 percent.

On the other hand, some species of amphibians and some breeding sites do not appear to be affected, and the damage caused by UV-B radiation may depend on the interaction of several variables (Licht and Grant 1997). For example, the presence of overhanging vegetation or turbidity (suspended solids in the water of breeding ponds) may block UV-B before it reaches the eggs of amphibians. Similarly, some dissolved organic chemicals absorb ultraviolet light, and may protect amphibian eggs. Species of amphibians differ in their sensitivity to UV-B, and resistance may be correlated with the amount of the enzyme

DEVELOPMENT OF SPOTTED SALAMANDER EGGS

Normal

Abnormal

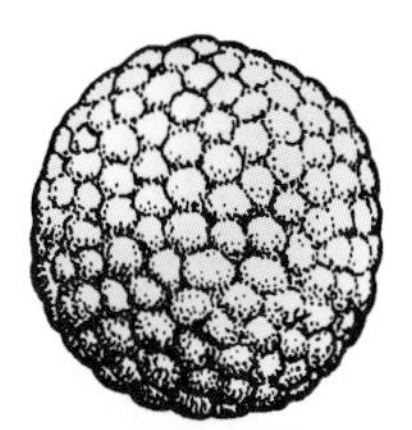

A. Cells divide evenly.

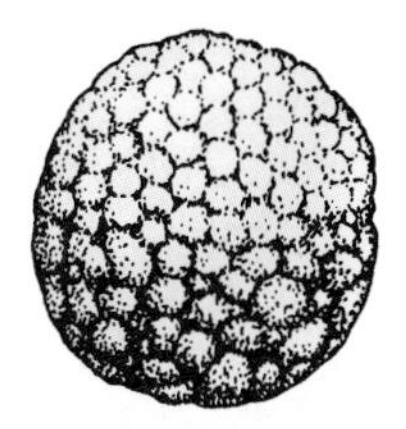

a. When exposed to highly acid conditions, cells divide unevenly.

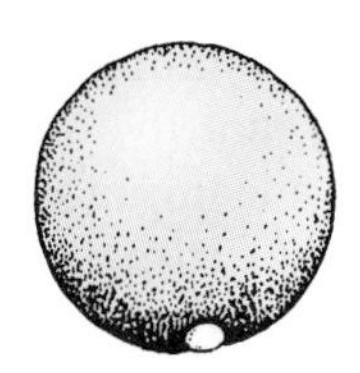

B. Yolk plug retracts.

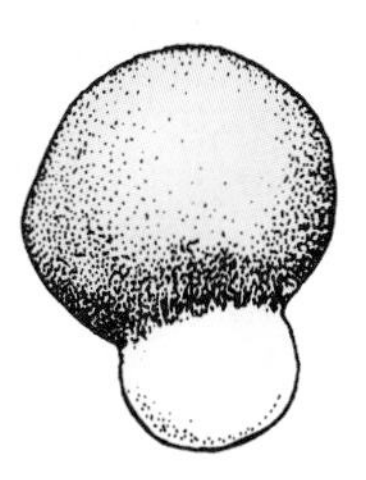

b. Yolk plug fails to retract fully and results in embryos shown in "c".

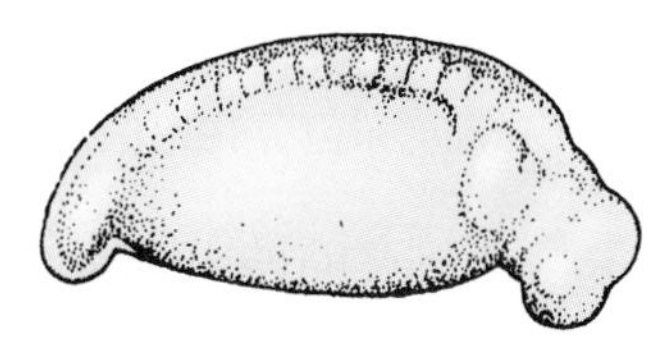

C. Embryo's development includes growth of its posterior form.

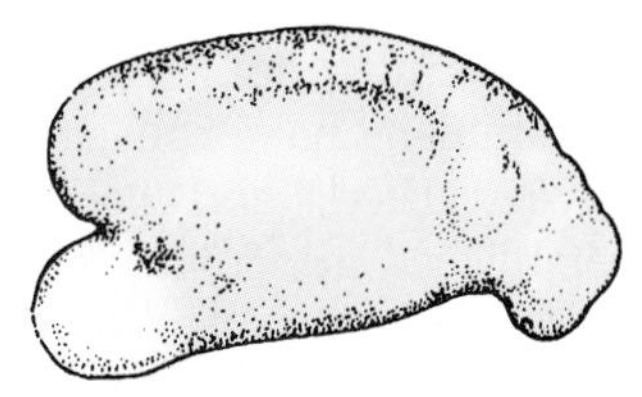

c. Posterior portion of embryo is deformed. Deformities like those seen in "a" through "c" are always lethal to an embryo.

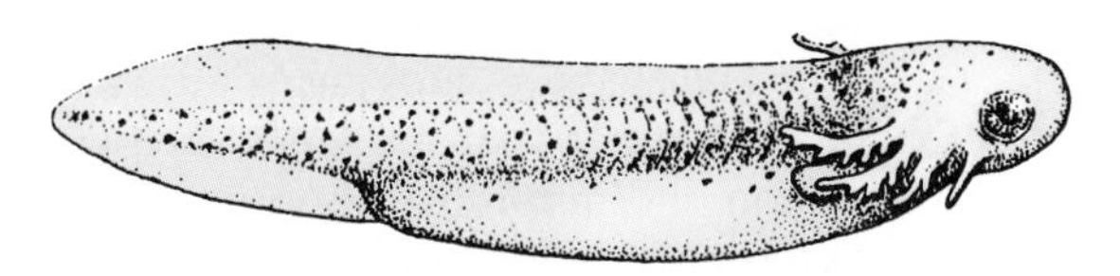

D. The gills and forelimbs develop as the embryo reaches a later stage.

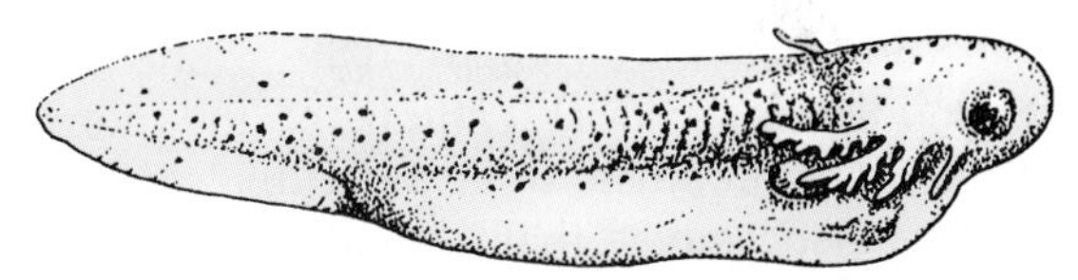

d. Conditions less acidic than those causing deformities in "a" through "c" produce damage at a later stage of development, including a swelling of the chest near the heart.

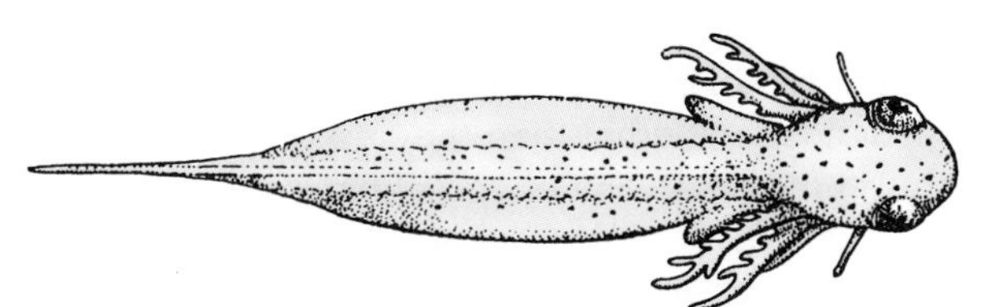

E. Gills become more lobed as the embryo nears hatching.

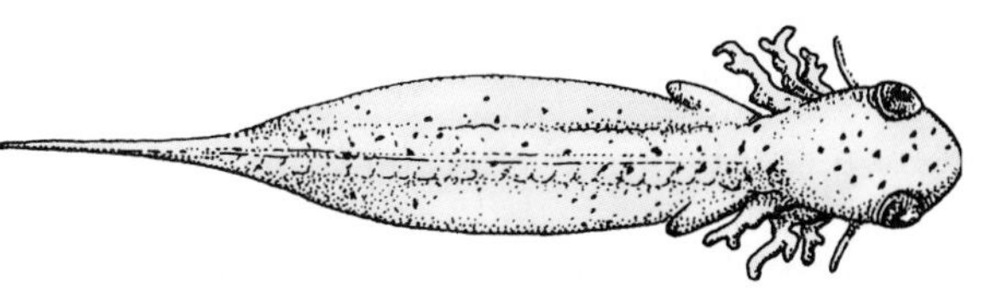

e. Conditions less acidic than those in "a" through "c" also produce stunted gills. Deformities in "d" and "e" are frequently lethal to the embryo.

▲ Figure 9–27 Effects of acidity on embryos of the spotted salamander, *Ambystoma maculatum,* (Ambystomatidae). Normal embryonic development is shown on the left and abnormalities observed in acid conditions on the right. Highly acidic conditions (pH values below about 5) kill embryos at early stages of development, whereas less acid conditions (between pH 5 and 6) produce abnormalities later in development. Above pH 6 embryos of spotted salamanders develop normally.

photolyase (an enzyme that repairs UV-induced damage to DNA) in the eggs of different species.

Sensitivity of amphibian eggs to UV-B may be contributing to the decline of some species. Because ozone depletion is a global phenomenon, this mechanism might account for the species that are declining in habitats that show no evidence of local environmental degradation. High-altitude species, which represent some of the most puzzling examples of declines, may be especially vulnerable to thinning of the ozone layer because the intensity of ultraviolet radiation normally increases with altitude. However, neither ozone depletion nor acid precipitation is yet severe at tropical latitudes, and other mechanisms are probably responsible for the disappearance of golden toads and other tropical frogs.

Recently attention has focused on the role of disease in the global decline of amphibians, and especially on two organisms that infect amphibians, iridoviruses and chytrid fungi. Disease-causing organisms and amphibians have coexisted as long as amphibians have existed, and some amphibian species have established apparently stable relationships with their pathogens. For example, iridoviruses infect pond-dwelling salamanders and appear to have shaped the biology of populations. Tiger salamanders (*Ambystoma tigrinum*) in Arizona have two larval forms, the normal morph and a cannibalistic morph with an enlarged head and jaws. Some populations consist entirely of the normal morph, whereas in others some individuals develop into cannibalistic larvae that eat individuals of the normal morph. A survey of the occurrence of normal and cannibal forms revealed that cannibals are absent from populations that are infected by iridoviruses, but occur regularly in populations without the viruses. Apparently in virus-free populations it is safe to eat other larvae, but when viruses are present a cannibal risks infection when it eats other larval salamanders.

The correspondence between the presence of iridoviruses and the life-history structure of salamander populations suggests that this host-pathogen association is ancient and that the salamander hosts and their iridovirus pathogens have evolved a stable relationship. This generalization probably applies to other amphibians that serve as hosts for iridoviruses, and it appears that iridoviruses cause fluctuations in amphibian populations but are not usually responsible for extinctions.

Chytrid fungi may be a new pathogen for amphibians, and they appear to be implicated in extinctions at sites as far apart as Central America and eastern Australia. Chytrids are a widespread group of soil fungi that normally infect algae, other fungi, plants, protozoans, and invertebrates. The first report of chytrids infecting vertebrates came in 1998, when they were identified as the probable cause of deaths of frogs in Australia and Central America. The following year chytrids were identified in dead frogs at the U.S. National Zoo. *Batrachochytrium,* the chytrid that infects amphibians, has motile reproductive zoospores that live in water and can penetrate the skin of an amphibian. When a zoospore enters an amphibian it matures to form branching structures that extend through the skin and a spherical reproductive body, the zoosporangium. A single zoospore matures in four or five days to a zoosporangium containing hundreds of zoospores that are released into the water. Chytrid infections do not appear to be lethal to tadpoles, but froglets die soon after metamorphosis, and adult frogs die when they are infected.

Amphibian declines have been studied especially well in Central America and eastern Australia, and chytrid infections are found in both areas. In Central America a wave of population declines and disappearances has been traced from Costa Rica into Panama during the period from 1988 through 1996 (Figure 9–28). The wave is moving southward at a rate of 50 to 100 km per year. In Australia a wave of chytrid infection and amphibian population decline may be moving northward though Queensland and the Northern Territories.

Why are amphibian populations suddenly succumbing to chytrid infections? Are chytrids new pathogens of amphibians, or has something happened to either amphibians or chytrids to make infections more lethal than they used to be? Studies of museum specimens suggest that chytrid infections may be a new phenomenon, at least in some areas. The changes in the skin can that accompany infections can be seen in museum specimens, and microscopic examination shows the distinctive fungal structures. Frogs collected twenty years ago show no sign of chytrids, suggesting that the fungus may have been introduced recently.

Frogs, also, may have changed in the recent past. Amphibians have immune systems that normally protect them from pathogens in the environment, but stress interferes with the immune system of ver-

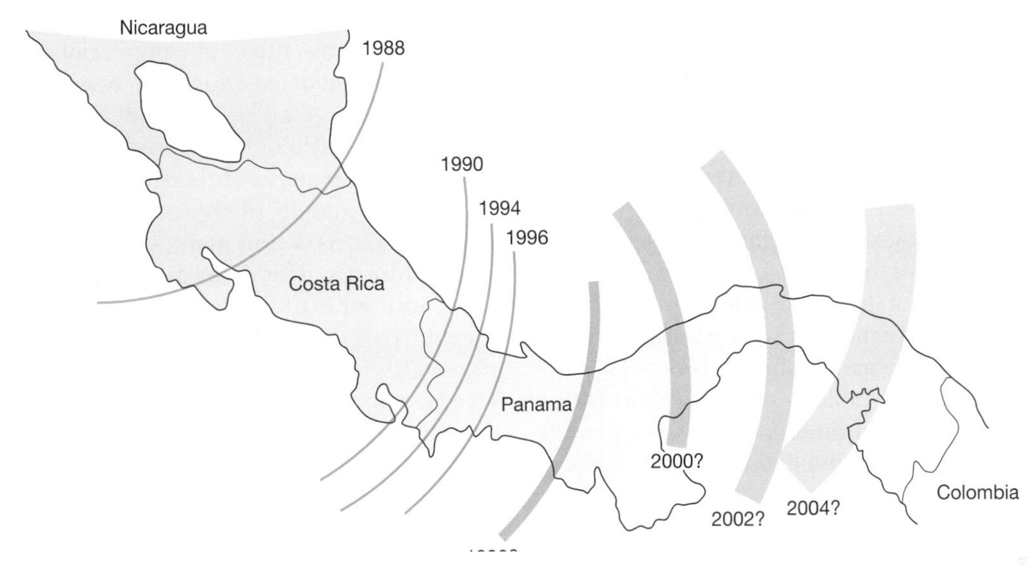

▲ Figure 9–28 North-to-south movement of infection by chytrid fungi and decline and disappearance of frog populations in Central America. A die-off occurred in 1988 in Monteverde, Costa Rica. In 1990, 1994, and 1996, disappearances were reported in sequence at Las Tablas, Costa Rica, and at Cerro Pando and Fortuna, Panama. Shaded arcs indicate anticipated progress of the infection at two-year intervals if it continues to move at 50 to 100 km/year.

tebrates. Amphibians that are stressed by acidity or ultraviolet radiation may be susceptible to diseases or parasites that they would ordinarily be able to resist. Interacting mechanisms of that sort are hard to identify, and the National Science Foundation is supporting a multidisciplinary study that combines the efforts of participants ranging from field biologists through experts in infectious disease to molecular biologists. This group is engaged in a multiyear program of field and laboratory studies to test hypotheses about the roles of iridoviruses and chytrid fungi in amphibian declines and the interactions of these factors with environmental changes and evolutionary changes in the disease organisms. Issues of conservation biology quickly transcend specific fields of study, and our best chance of finding solutions to the acute problems that affect many kinds of animals lies in pooling the techniques and perspectives of a broad spectrum of biological specialties.

Summary

Locomotor adaptations distinguish the lineages of amphibians. Salamanders (Urodela) usually have short, sturdy legs that are used with lateral undulation of the body in walking. Aquatic salamanders use lateral undulations of the body and tail to swim, and some specialized aquatic species are elongate and have very small legs. Frogs and toads (Anura) are characterized by specializations of the pelvis and hindlimbs that permit both legs to be used simultaneously to deliver a powerful thrust used both for jumping and for swimming. Many anurans walk quadrupedally when they move slowly, and some are agile climbers. The caecilians (Gymnophiona) are legless tropical amphibians; some are burrowers and others are aquatic.

The diversity of reproductive modes of amphibians exceeds that of any other group of vertebrates except the fishes. Fertilization is internal in derived salamanders, but most frogs rely on external fertilization. All caecilians have internal fertilization.

Many species of amphibians have aquatic larvae. Tadpoles, the aquatic larvae of anurans, are specialized for life in still or flowing water, and some species of frogs deposit their tadpoles in very specific sites, such as the pools of water that accumulate in the leaf axils of bromeliads or other plants. Specializations of tadpoles are entirely different from specializations of frogs, and metamorphosis causes changes in all parts of the body. Direct development that omits the larval stage is also widespread among anurans and is often combined with parental care of the eggs. Viviparity occurs in all three orders.

In many respects the biology of amphibians is determined by properties of their skin. Hedonic glands are key elements in reproductive behaviors, poison glands protect the animals against predators, and mucus glands keep the skin moist, facilitating gas exchange. Above all, the permeability of the skin to water limits most amphibians to microhabitats in which they can control water gain and loss. That sounds like a severe restriction, but in the proper microhabitat amphibians can utilize the permeability of their skin to achieve a remarkable degree of independence of standing water. Thus the picture that is sometimes presented of amphibians as animals barely hanging on as a sort of evolutionary oversight is misleading. Only a detailed examination of all facets of their biology can produce an accurate picture of amphibians as organisms.

An examination of that sort reinforces the view that the skin is a dominant structural characteristic of amphibians. This is true not only for the limitations and opportunities presented by the skin's permeability to water and gases, but also as a result of the intertwined functions of the skin glands in defensive and reproductive behaviors. The structure and function of the skin may be primary characteristics that have shaped the evolution and ecology of amphibians, and it may also be responsible for some aspects of their susceptibility to pollution. All over the world, populations of amphibians are disappearing at an alarming rate. Some of these extinctions may be caused by regional or global effects of human activities that are likely to affect other organisms as well.

Additional Readings

Alford, R. A. and S. J. Richards. 1999. Global amphibian declines: A problem in applied ecology. *Annual Review of Ecology and Systematics* 30:133–165.

Bannon, A. W., M. W. Decker, M. W. Holladay, P. Curzon, D. Donnelly-Roberts, P. S. Puttfarken, R. S. Bitner, A. Diaz, A. H. Dickenson, R. D. Porsolt, M. Williams, and S. P. Aneric. 1998. Broad-spectrum, non-opioid analgesic activity by selective modulation of neuronal nicotinic acetylcholine receptors. *Science* 279:77–81.

Blaustein, A. R. et al. 1994. UV repair and resistance to solar UV-B in amphibian eggs: a link to population declines. *Proceedings of the National Academy of Sciences, USA* 91:179–1795.

Blaustein, A. R. and D. B. Wake. 1995. The puzzle of declining amphibian populations. *Scientific American* 272(4):56–61.

Brodie, E. D., Jr., and E. D. Brodie III. 1980. Differential avoidance of mimetic salamanders by free-ranging birds. *Science* 208:181–182.

Daly, J. W. J. et al. 1992. Frog secretions and hunting magic in the upper Amazon: Identification of a peptide that interacts with an adenosine receptor. *Proceedings of the National Academy of Sciences, USA* 89:10960–10963.

Daly, J. W. et al. 2000. Arthropod-frog connection: Decahydroquinoline and pyrrolizidine alkaloids common to microsympatric myrmicine ants and dendrobatid frogs. *Journal of Chemical Ecology* 26:73–85.

Duellman, W. E. and L. Trueb. 1985. *The Biology of Amphibians.* New York: McGraw-Hill.

Halliday, T. R. 1990. The evolution of courtship behavior in newts and salamanders. *Advances in the Study of Behavior* 19:137–169.

Hayes, T. B. (Ed.). 1997. Amphibian metamorphosis: An integrative approach. *American Zoologist* 37:121–207.

Heyer, W. R. et al. (Ed.). 1993. *Measuring and Monitoring Biological Diversity: Standard Methods for Amphibians.* Washington, D.C.: Smithsonian Institution Press.

Jaeger, R. G. 1981. Dear enemy recognition and the costs of aggression between salamanders. *American Naturalist* 117:962–974.

Jaeger, R. G., K. C. B. Nishikawa, and D. E. Barnard. 1983. Foraging tactics of a terrestrial salamander: Costs of territorial defense. *Animal Behaviour* 31:191–198.

Kiesecker, J. M., A. R. Blaustein, and L. K. Belden. 2001. Complex causes of amphibian population declines. *Nature* 410:681–684.

Licht, L. E. and K. P. Grant. 1997. The effects of ultaviolet radiation on the biology of amphibians. *American Zoologist* 37:137–145.

Lips, K. R. 1999. Mass mortality and population declines of anurans at an upland sites in western Panamá. *Conservation Biology* 13:117–125.

Magnusson, W. E., and J.-M. Hero. 1991. Predation and evolution of complex ovoposition behaviour in Amazon rainforest frogs. *Oecologia* 86:310–318.

McDiarmid, R. W. and R. Altig. 1999. *Tadpoles: The Biology of Anuran Larvae.* Chicago: University of Chicago Press.

Phillips, K. 1994. *Tracking the Vanishing Frogs.* New York: St. Martin's Press.

Pough, F. H. 1988. Mimicry of vertebrates: Are the rules different? *American Naturalist* 131 (Supplement):S67–S102.

Pounds, J. A. 2001. Climate and amphibian declines. *Nature* 410:639–640.

Preest, M. R. 1993. Mechanism of growth rate reduction in acid-exposed larval salamanders, *Ambystoma maculatum. Physiological Zoology* 66:686–707.

Roth, G., and D. B. Wake. 1985. Trends in the functional morphology and sensorimotor control of feeding behavior in salamanders: An example of the role of internal dynamics in evolution. *Acta Biotheoretica* 34:175–192.

Ryan, M. J. 1985. *The Túngara Frog: A Study in Sexual Selection and Communication.* Chicago: University of Chicago Press.

Taigen, T. L., and K. D. Wells. 1985. Energetics of vocalization by an anuran amphibian (*Hyla versicolor*). *Journal of Comparative Physiology* B155:163–170.

Tilley, S. G., B. L. Lundrigan, and L. P. Brower. 1982. Erythrism and mimicry in the salamander *Plethodon cinereus. Herpetologica* 38: 409–417.

Tyler, M. J. (Ed.). 1983. *The Gastric Brooding Frog.* Beckenham, Kent, UK: Croom Helm.

Wake, D. B. 1998. Action on amphibians. *Trends in Ecology and Evolution* 13:379–380.

Wake, D. B., and S. B. Marks. 1993. Development and evolution of plethodontid salamanders: A review of prior studies and a prospectus for future research. *Herpetologica* 49: 194–203.

Wake, M. H. 1993. Evolution of oviductal gestation in amphibians. *Journal of Experimental Zoology* 266:394–413.

Wells, K. D., and T. L. Taigen. 1986. The effect of social interactions on calling energetics in the grey treefrog (*Hyla versicolor*). *Behavioral Ecology and Sociobiology* 19:9–18.

Wilkinson, J. W. (Ed.) *FROGLOG,* Bi-monthly newsletter of the Declining Amphibians Population Task Force <Daptf@open.ac.uk>.

Web Explorations

On-line resources for this chapter are on the World Wide Web at: http://www.prenhall.com/pough (click on the Table of Contents link and then select Chapter 9).

CHAPTER 10

Turtles

Turtles provide a contrast to amphibians in the relative lack of diversity in their life histories. All turtles lay eggs, and none exhibit parental care of the hatchlings. Turtles show morphological specializations associated with terrestrial, freshwater, and marine habitats, and marine turtles make long-distance migrations rivaling those of birds. Probably turtles and birds use many of the same navigation mechanisms to find their way. Most turtles are long-lived animals with relatively poor capacities for rapid population growth. Many, especially sea turtles and large tortoises, are endangered by human activities. Some efforts to conserve turtles have apparently been frustrated by a feature of the embryonic development of some species of turtles—the sex of an individual is determined by the temperature to which it is exposed in the nest. This experience emphasizes the critical importance of information about the basic biology of animals to successful conservation and management.

10.1 Everyone Recognizes a Turtle

Turtles found a successful approach to life in the Triassic and have scarcely changed since. The shell, which is the key to their success, has also limited the group's diversity (Figure 10–1). For obvious reasons, flying or gliding turtles have never existed, and even arboreality is only slightly developed. Shell morphology reflects the ecology of turtle species: The most terrestrial forms, the tortoises of the family Testudinidae, have high domed shells and elephant-like feet (Figure 10–1a). Smaller species of tortoises may show adaptations for burrowing. The gopher tortoises of North America are an example—their fore legs are flattened into scoops, and the dome of the shell is reduced. The Bolson tortoise of northern Mexico constructs burrows a meter or more deep and several meters long in the hard desert soil. These tortoises bask at the mouths of their burrows; when a predator appears, they throw themselves down the steep entrance tunnels of the burrows to escape, just as an aquatic turtle dives off a log. The pancake tortoise of Africa is a radical departure from the usual tortoise morphology (Figure 10–1b). The shell is flat and flexible because its ossification is much reduced. This turtle lives in rocky foothill regions and scrambles over the rocks with nearly as much agility as a lizard. When threatened by a predator, the pancake tortoise crawls into a rock crevice and uses its legs to wedge itself in place. The flexible shell presses against the overhanging rock and creates so much friction that it is almost impossible to pull the tortoise out.

Other terrestrial turtles have moderately domed **carapaces** (upper shells), like the box turtles of the family Emydidae (Figure 10–1c). This is one of several kinds of turtles that have evolved flexible regions in the **plastron** (lower shell), which allow the front and rear lobes to be pulled upward to close the openings of the shell. Aquatic turtles have low carapaces that offer little resistance to movement through water. The Emydidae and Bataguridae contain a large number of pond turtles (Figure 10–1d), including the painted turtles and the red-eared turtles often seen in pet stores and anatomy and physiology laboratory courses.

The snapping turtles (family Chelydridae) and the mud and musk turtles (family Kinosternidae) prowl along the bottoms of ponds and slow rivers and are not particularly streamlined (Figure 10–1f, g). The mud turtle has a hinged plastron, but the musk and snapping turtles have very reduced plastrons. They rely on strong jaws for protection. A reduction in the size of the plastron makes these species more agile than most turtles, and musk turtles may climb sev-

▲ Figure 10–1 Body forms of turtles: (a) Tortoise, *Testudo,* Testudinidae. (b) Pancake tortoise, *Malacochersus,* Testudinidae. (c) Terrestrial box *turtle, Terrapene,* Emydidae. (d) Pond turtle, *Trachemys,* Emydidae. (e) Soft-shelled turtle, *Aplone,* Trionychidae. (f) Mud turtle, *Kinosternon,* Kinosternidae. (g) Alligator snapping turtle, *Macroclemys,* Chelydridae. (h) African pond turtle, *Pelusios,* Pelomedusidae. (i) Australian snake-necked turtle, *Chelodina,* Chelidae. (j) South American matamata, *Chelys,* Chelidae. (k) Loggerhead sea turtle, *Caretta,* Cheloniidae. (l) Leatherback sea turtle, *Dermochelys,* Dermochelyidae.

eral feet into trees, probably to bask. If a turtle falls on your head while you are canoeing, it is probably a musk turtle.

The soft-shelled turtles (family Trionychidae) are fast swimmers (Figure 10–1e). The ossification of the shell is greatly reduced, lightening the animal, and the feet are large with extensive webbing. Soft-shelled turtles lie in ambush partly buried in debris on the bottom of the pond. Their long necks allow them to reach considerable distances to seize the invertebrates and small fishes on which they feed.

Extant turtles can be placed in 13 families (Table 10.1). The two lineages of living turtles can be traced through fossils to the Mesozoic. The **cryptodires** (*crypto* = hidden, *dire* = neck) retract the head into the shell by bending the neck in a vertical S-shape. The **pleurodires** (*pleuro* = side) retract the head by bending the neck horizontally. All the turtles discussed so far have been cryptodires, and these are the dominant group of turtles. Cryptodires are the only turtles now found in most of the Northern Hemisphere, and there are aquatic and terrestrial cryptodires in South America and terrestrial ones in Africa. Only Australia has no cryptodires. Pleurodires are now found only in the Southern Hemisphere, although they had worldwide distribution in the late Mesozoic and early Cenozoic. *Stupendemys,* a pleurodire from the Pliocene of Venezuela, had a carapace more than 2 meters long. All the living pleurodires are at least semiaquatic; but some fossil pleurodires had high, domed shells that suggest they may have been terrestrial. The most terrestrial of the living pleurodires are probably the African pond turtles (Figure

Table 10.1 Families of turtles

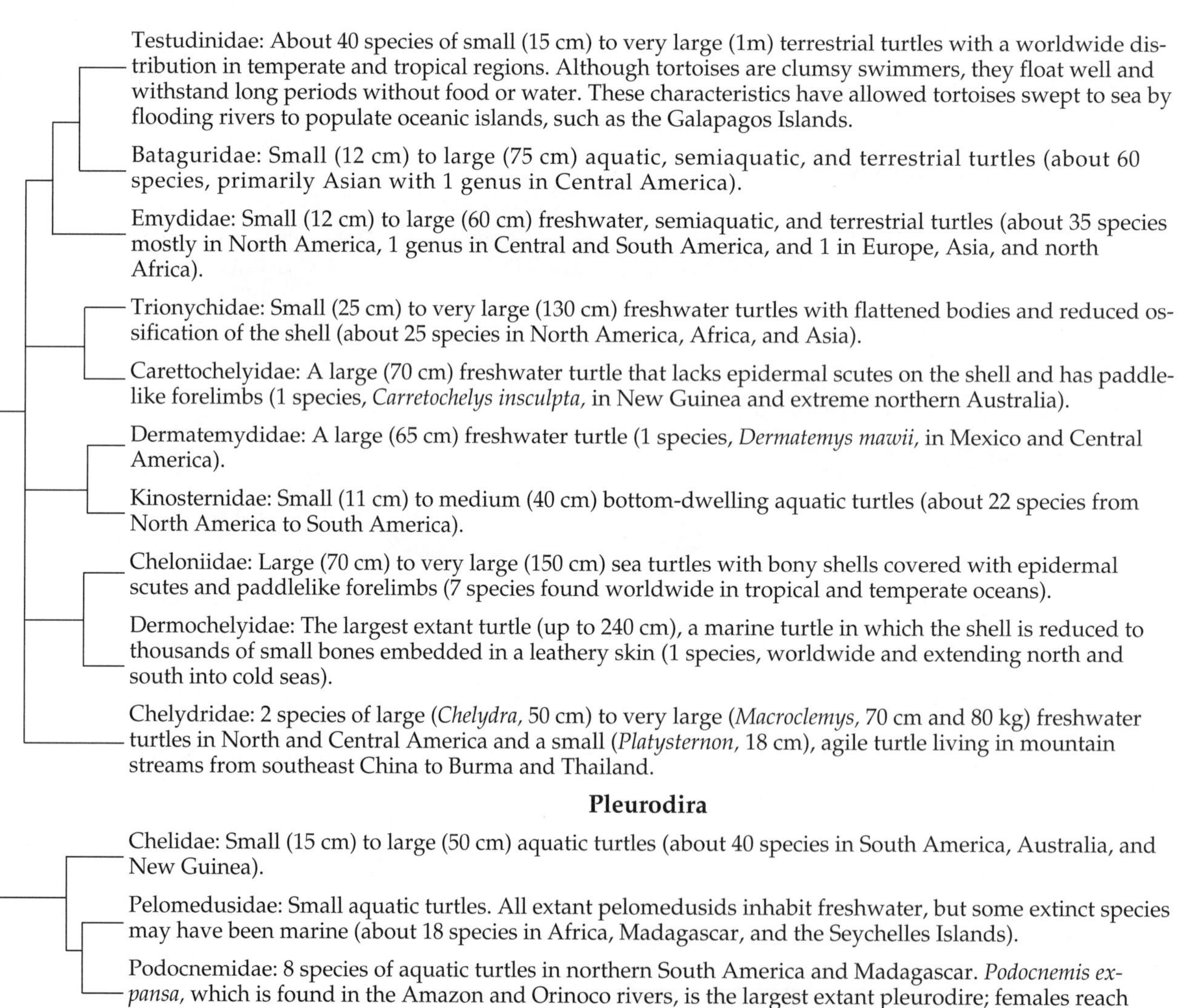

Cryptodira

Testudinidae: About 40 species of small (15 cm) to very large (1m) terrestrial turtles with a worldwide distribution in temperate and tropical regions. Although tortoises are clumsy swimmers, they float well and withstand long periods without food or water. These characteristics have allowed tortoises swept to sea by flooding rivers to populate oceanic islands, such as the Galapagos Islands.

Bataguridae: Small (12 cm) to large (75 cm) aquatic, semiaquatic, and terrestrial turtles (about 60 species, primarily Asian with 1 genus in Central America).

Emydidae: Small (12 cm) to large (60 cm) freshwater, semiaquatic, and terrestrial turtles (about 35 species mostly in North America, 1 genus in Central and South America, and 1 in Europe, Asia, and north Africa).

Trionychidae: Small (25 cm) to very large (130 cm) freshwater turtles with flattened bodies and reduced ossification of the shell (about 25 species in North America, Africa, and Asia).

Carettochelyidae: A large (70 cm) freshwater turtle that lacks epidermal scutes on the shell and has paddlelike forelimbs (1 species, *Carretochelys insculpta,* in New Guinea and extreme northern Australia).

Dermatemydidae: A large (65 cm) freshwater turtle (1 species, *Dermatemys mawii,* in Mexico and Central America).

Kinosternidae: Small (11 cm) to medium (40 cm) bottom-dwelling aquatic turtles (about 22 species from North America to South America).

Cheloniidae: Large (70 cm) to very large (150 cm) sea turtles with bony shells covered with epidermal scutes and paddlelike forelimbs (7 species found worldwide in tropical and temperate oceans).

Dermochelyidae: The largest extant turtle (up to 240 cm), a marine turtle in which the shell is reduced to thousands of small bones embedded in a leathery skin (1 species, worldwide and extending north and south into cold seas).

Chelydridae: 2 species of large (*Chelydra,* 50 cm) to very large (*Macroclemys,* 70 cm and 80 kg) freshwater turtles in North and Central America and a small (*Platysternon,* 18 cm), agile turtle living in mountain streams from southeast China to Burma and Thailand.

Pleurodira

Chelidae: Small (15 cm) to large (50 cm) aquatic turtles (about 40 species in South America, Australia, and New Guinea).

Pelomedusidae: Small aquatic turtles. All extant pelomedusids inhabit freshwater, but some extinct species may have been marine (about 18 species in Africa, Madagascar, and the Seychelles Islands).

Podocnemidae: 8 species of aquatic turtles in northern South America and Madagascar. *Podocnemis expansa,* which is found in the Amazon and Orinoco rivers, is the largest extant pleurodire; females reach shell lengths of 90 cm. The extinct *Stupendemys* (from Late Tertiary of Venezuela) was over 2m long.

10–1h), which readily move overland from one pond to another.

The snake-necked pleurodiran turtles (family Chelidae) are found in South America, Australia, and New Guinea (Figure 10–1i). As their name implies, they have long, slender necks. In some species the length of the neck is considerably greater than that of the body. These forms feed on fishes that they catch with a sudden dart of the head. Other snake-necked turtles have much shorter necks. Some of these feed on mollusks and have enlarged palatal surfaces used to crush shells. The same specialization for feeding on mollusks is seen in certain cryptodiran turtles.

An unusual feeding method among turtles is found in a pleurodire, the matamata of South America (Figure 10–1j). Large matamatas reach shell lengths of 40 cm. They are bizarre-looking animals. The shell and head are broad and flattened, and numerous flaps of skin project from the sides of the head and the broad neck. To these are added trailing bits of adhering algae. The effect is exceedingly cryptic. It is hard to recognize a matamata as a turtle even in an aquarium, and it is practically invisible against the mud and debris of a river bottom. In addition to obscuring the shape of the turtle, the flaps of skin on the head are sensitive to minute vibrations in water caused by the passage of a fish. When it senses the presence of prey, the matamata abruptly opens its mouth and expands its throat. Water rushes in, carrying the prey with it, and the matamata closes its mouth, expels the water, and swallows the prey. The matamata lacks the horny beak that other turtles use for seizing prey or biting off pieces of plants.

Marine turtles are cryptodires. The families Cheloniidae and Dermochelyidae show more extensive specialization for aquatic life than any freshwater turtle. All have the forelimbs modified as flippers. Cheloniids retain epidermal scutes on the shell (Figure 10–1k). The largest of the sea turtles of the family Cheloniidae is the loggerhead, which once reached weights exceeding 400 kg. The largest marine turtle, the leatherback, reaches shell lengths of more than 2m and weights in excess of 600 kg (Figure 10–11). The dermal ossification has been reduced to bony platelets embedded in connective tissue. This is a pelagic turtle that ranges far from land, and it has a wider geographic distribution than any other ectothermal amniote. Leatherback turtles penetrate far into cool temperate seas, and have been recorded in the Atlantic from Newfoundland to Argentina and in the Pacific from Japan to Tasmania. Leatherback turtles dive to depths of more than 1000m. One dive that drove the depth recorder off scale is estimated to have reached 1200m, which exceeds the deepest dive recorded for a sperm whale (1140m). Leatherback turtles feed largely on jellyfishes, whereas the smaller hawksbill sea turtles eat sponges that are defended by spicules of silica (glass) as well as a variety of chemicals (including alkaloids and terpenes) that are toxic to most vertebrates. Green turtles (*Chelonia mydas*) are the only herbivorous marine turtles.

10.2 But What Is a Turtle? Phylogenetic Relationships of Turtles

Turtles show a combination of ancestral features and highly specialized characters that are not shared with any other group of vertebrates, and their phylogenetic affinities are not fully understood. The turtle lineage probably originated among the early amniotes of the Late Carboniferous (Figure 10–2). Like those animals, turtles have anapsid skulls, but the shells and postcranial skeletons of turtles are unique. One view emphasizes the importance of the anapsid skull and places the affinities of turtles among the parareptiles that were discussed in Chapter 8 (Reisz and Laurin 1991, Lee 1993). A radically different opinion regards the anapsid skull of turtles as being a secondarily derived feature and places turtles among the diapsids (Rieppel and de Braga 1996). At the moment, the case for a diapsid origin appears strongest, but a lively debate is in progress and the issue is far from resolved.

The earliest turtles are found in Late Triassic deposits in Germany, Thailand, and Argentina. These animals had nearly all the specialized characteristics of derived turtles, and shed no light on the phylogenetic affinities of the group. *Proganochelys*, from Triassic deposits in Germany, was nearly a meter long (larger than most living turtles) and had a high, arched shell. The marginal teeth had been lost and the maxilla, premaxilla, and dentary bones were probably covered with a horny beak as they are in derived turtles. The skull of *Proganochelys* retained the supratemporal and lacrimal bones and the lacrimal duct, and the palate had rows of denticles; all these structures have been lost by derived turtles. The plastron of *Proganochelys* also contained some bones that have been lost by derived turtles, and the

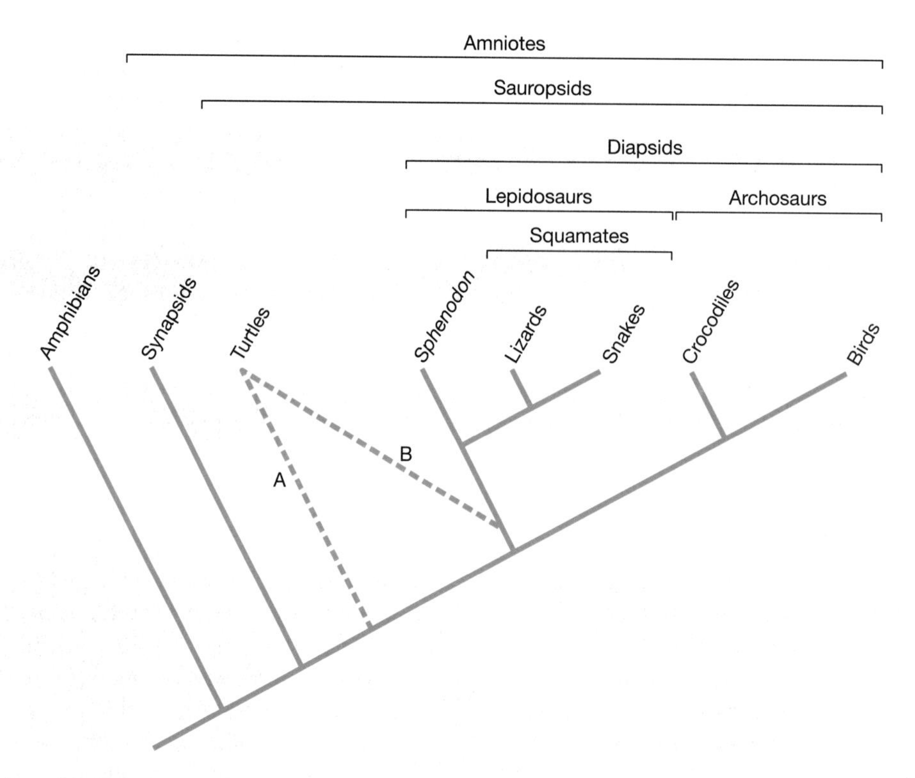

▲ Figure 10–2 Simplified cladogram of tetrapods. The two competing hypotheses of the relationships of turtles are shown: an origin from parareptiles (A) or from within diapsid reptiles (B).

vertebrae of the neck lack specializations that would have allowed the head to be retracted into the shell. *Palaeochersis* (Late Triassic of Argentina) and *Australochelys* (Early Jurassic of South Africa) are probably the sister group of later turtles, including Cryptodira + Pleurodira.

Turtles with neck vertebrae specialized for retraction are not known before the Cretaceous, but differences in the skulls and shells allow the pleurodiran and cryptodiran lineages to be traced back to the Late Triassic (shells of pleurodires) and Late Jurassic (skulls of cryptodires). The otic capsules of all turtles beyond the proganochelids are enlarged, and the jaw adductor muscles bend posteriorly over the otic capsule (Figure 10–3). The muscles pass over a pulley-like structure, the **trochlear process.** In cryptodires the trochlear process is formed by the anterior surface of the otic capsule itself, whereas in pleurodires it is formed by a lateral process of the pterygoid. Fusion of the pelvic girdle to the carapace and plastron further distinguishes pleurodires from cryptodires, which have a suture attaching the shell to the girdle. The beginnings of these changes are seen in *Australochelys.*

10.3 Turtle Structure and Functions

Turtles are among the most bizarre vertebrates. Covered in bone, with the limbs inside the ribs, and with horny beaks instead of teeth—if turtles had become extinct at the end of the Mesozoic, they would rival dinosaurs in their novelty. However, because they survived they are regarded as commonplace, and are often used in comparative anatomy courses to represent primitive amniotes (inappropriately, because they are so specialized).

Shell and Skeleton

The shell is the most distinctive feature of a turtle (Figure 10–4). The carapace is composed of dermal bone that typically grows from 59 separate centers of ossification. Eight plates along the dorsal midline form the neural series and are fused to the neural arches of the vertebrae. Lateral to the neural bones are eight paired costal bones, which are fused to the broadened ribs. The ribs of turtles are unique among tetrapods in being external to the girdles. Eleven pairs of peripheral bones, plus two unpaired bones

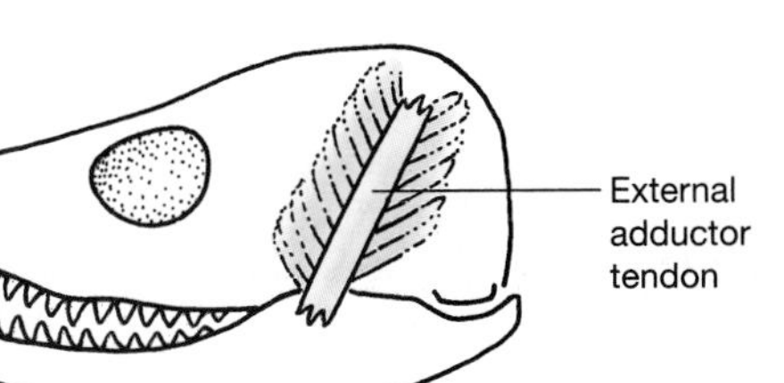

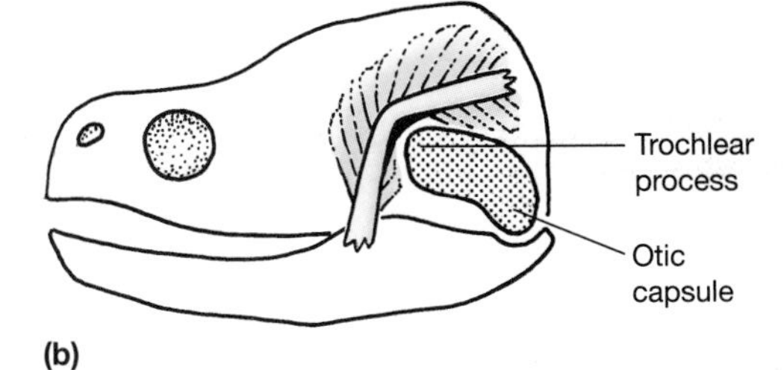

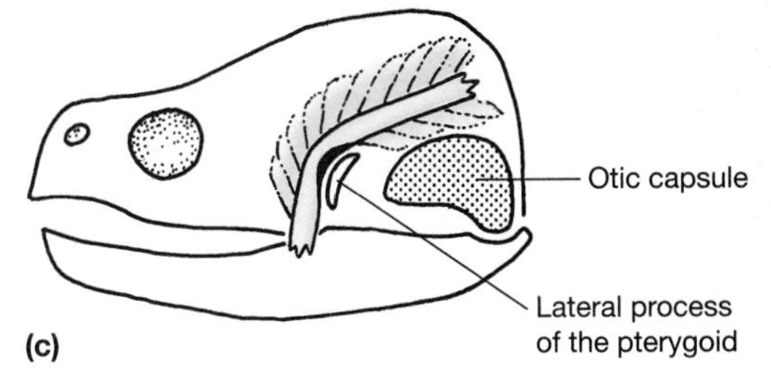

▲ Figure 10–3 Position of the external adductor tendon. (a) In the ancestral (parareptilian) condition. (b) In cryptodiran turtles. (c) In pleurodiran turtles.

in the dorsal midline, form the margin of the carapace. The plastron is formed largely from dermal ossifications, but the entoplastron is derived from the interclavicle, and the paired epiplastra anterior to it are derived from the clavicles. Processes from the hypoplastron fuse with the first and fifth pleurals, forming a rigid connection between the plastron and carapace.

The bones of the carapace are covered by horny scutes of epidermal origin that do not coincide in number or position with the underlying bones. The carapace has a row of five central scutes, bordered on each side by four lateral scutes. Eleven marginal scutes on each side turn under the edge of the carapace. The plastron is covered by a series of six paired scutes.

Flexible areas, called hinges, are present in the shells of many turtles. The most familiar examples are the North American and Asian box turtles (*Terrapene* and *Cuora*), in which a hinge between the hyoplastral and hypoplastral bones allows the anterior and posterior lobes of the plastron to be raised to close off the front and rear openings of the shell. Mud turtles (*Kinosternon*) have two hinges in the plastron; the anterior hinge runs between the epiplastra and the entoplastron (which is triangular in kinosternid turtles rather than diamond shaped), and the posterior hinge is between the hypoplastron and xiphiplastron. In the pleurodiran turtle *Pelusios,* a hinge runs between the mesoplastron and the hypoplastron. Some species of tortoises have plastral hinges; in *Testudo* the hinge lies between the hyoplastron and xiphiplastra, as it does in *Kinosternon;* but in another genus of tortoise, *Pyxis,* the hinge is anterior and involves a break across the entoplastron. The African forest tortoises (*Kinixys*) have a hinge on the posterior part of the carapace. The margins of the epidermal shields and the dermal bones of the carapace are aligned, and the hinge runs between the second and third pleural scutes and the fourth and fifth costals. The presence of hinges is sexually dimorphic in some species of tortoises. The erratic phylogenetic occurrence of kinetic shells and differences among related species indicate that shell kinesis has evolved many times in turtles.

Asymmetry of the paired epidermal scutes is quite common among turtles, and modifications of the bony structure of the shell are seen in some families. Soft-shelled turtles lack peripheral ossifications and epidermal scutes. The distal ends of the broadened ribs are embedded in flexible connective tissue, and the carapace and plastron are covered with skin. The New Guinea river turtle (*Carretochelys*) is also covered by skin instead of scutes, but in this species the peripheral bones are present and the edge of the shell is stiff. The leatherback sea turtle (*Dermochelys*) has a carapace formed of cartilage with thousands of small polygonal bones embedded in it, and the plastral bones are reduced to a thin rim around the edge of the plastron. The neural and costal ossifications of the pancake tortoise (*Malacochersus*) are greatly reduced, but the epidermal plates are well developed.

Extant turtles have only 18 presacral vertebrae, 10 in the trunk and 8 in the neck. The centra of the trunk vertebrae are elongated and lie beneath the dermal bones in the dorsal midline of the shell. The centra are constricted in their centers and fused to each other. The neural arches in the anterior two-thirds of the trunk lie between the centra as a result

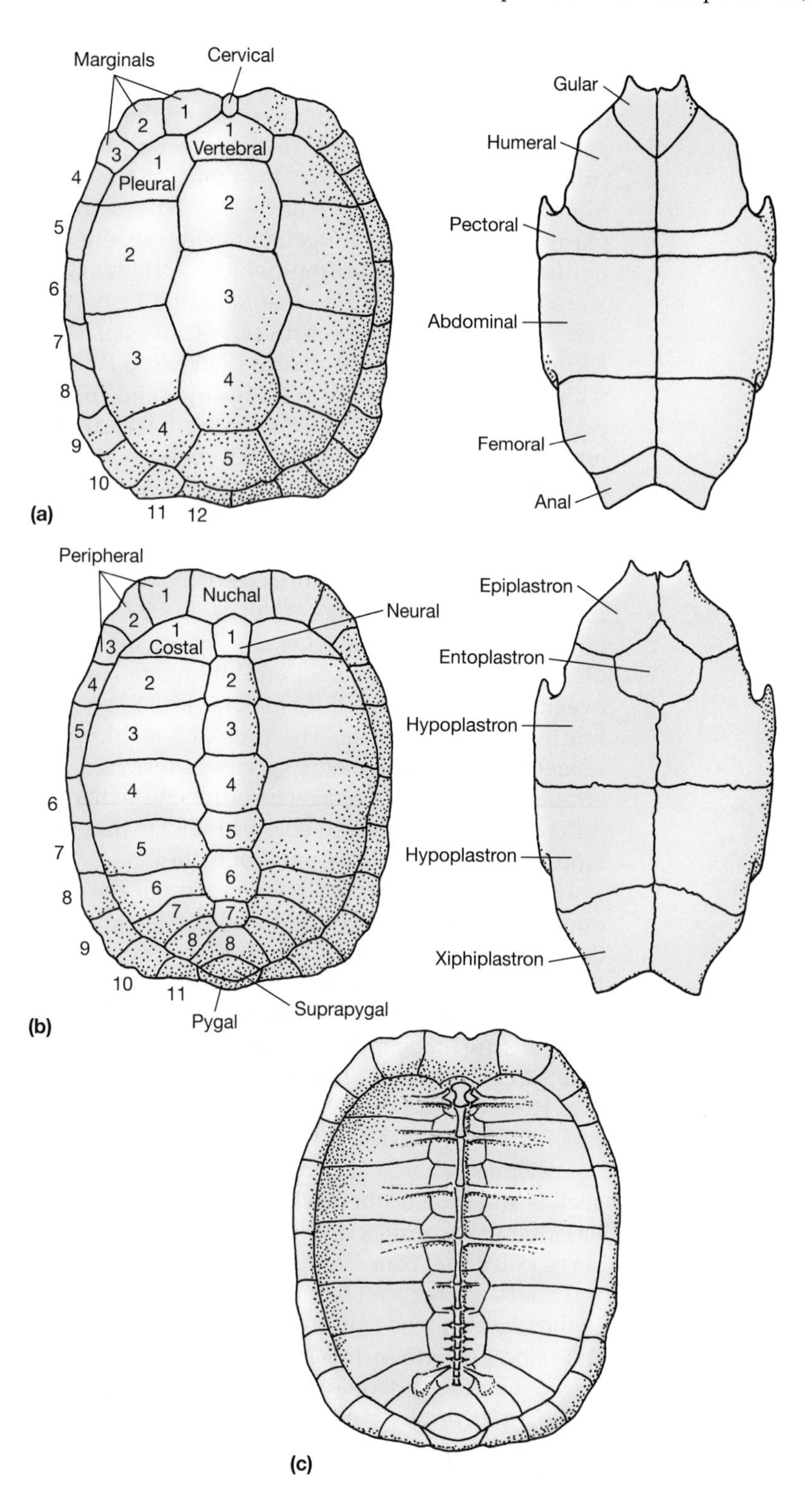

Figure 10–4 Shell and vertebral column of a turtle. (a) Epidermal scutes of the carapace (left) and plastron (right). (b) Dermal bones of the carapace (left) and plastron (right). (c) Vertebral column of a turtle, seen from the inside of the carapace. Note that anteriorly the ribs articulate with two vertebral centra.

of anterior displacement, and the spinal nerves exit near the middle of the preceding centrum. The ribs are also shifted anteriorly; they articulate with the anterior part of the neurocentral boundary, and in the anterior part of the trunk, where the shift is most pronounced, the ribs extend onto the preceding vertebra.

Cryptodires have two sacral vertebrae (the 19th and 20th vertebrae) with broadened ribs that meet the ilia of the pelvis. Pleurodires have the pelvic girdle firmly fused to the dermal carapace by the ilia dorsally and by the pubic and ischial bones ventrally, and the sacral region of the vertebral column is less distinct. The ribs on the 17th, 18th, 19th, and sometimes the 20th vertebrae are fused to the centra and end on the ilia or the ilium-carapacial junction.

The cervical vertebrae of cryptodires have articulations that permit the S-shaped bend used to retract the head into the shell. Specialized condyles (ginglymes) permit vertical rotation. This type of rotation, ginglymoidy, is peculiar to cryptodires, but the anatomical details vary within the group. In most families the hinge is formed by two successive ginglymoidal joints between the 6th and 7th and the 7th and 8th cervical vertebrae. The lateral bending of the necks of pleurodiran turtles is accomplished by ball-and-socket or cylindrical joints between adjacent cervical vertebrae.

The Heart

The circulatory systems of tetrapods can be viewed as consisting of two circuits: The systemic circuit carries oxygenated blood from the heart to the head, trunk, and appendages, whereas the pulmonary circuit carries deoxygenated blood from the heart to the lungs. The blood pressure in the systemic circuit is higher than the pressure in the pulmonary circuit, and the two circuits operate in series. That is, blood flows from the heart through the lungs, back to the heart, and then to the body. The morphology of the hearts of birds and mammals makes this sequential flow obligatory, but the hearts of turtles, lepidosaurs, and amphibians have the ability to shift blood between the pulmonary and systemic circuits.

This flexibility in the route of blood flow can be accomplished because the ventricular chambers in the hearts of turtles and lepidosaurs are in anatomical continuity, instead of being completely divided by a septum like the ventricles of birds and mammals. The pattern of blood flow can best be explained by considering the morphology of the heart and how intracardiac pressure changes during a heartbeat. Figure 10–5 shows a schematic view of the heart of a turtle. The left and right atria are completely separate, and three subcompartments can be distinguished in the ventricle. A muscular ridge in the core of the heart divides the ventricle into two spaces, the **cavum pulmonale** and the **cavum venosum**. The muscular ridge is not fused to the wall of the ventricle, and thus the cavum pulmonale and the cavum venosum are only partly separated. A third subcompartment of the ventricle, the **cavum arteriosum**, is located dorsal to the cavum pulmonale and cavum venosum. The cavum arteriosum communicates with the cavum venosum through an intraventricular canal. The pulmonary artery opens from the cavum pulmonale, and the left and right aortic arches open from the cavum venosum.

The right atrium receives deoxygenated blood from the body via the sinus venosus and empties into the cavum venosum, and the left atrium receives oxygenated blood from the lungs and empties into the cavum arteriosum. The atria are separated from the ventricle by flap-like atrioventricular valves that open as the atria contract, and then close as the ventricle contracts, preventing blood from being forced back into the atria. The anatomical arrangement of the connections between the atria, their valves, and the three subcompartments of the ventricle is crucial, because it is those connections that allow pressure differentials to direct the flow of blood and to prevent mixing of oxygenated and deoxygenated blood.

When the atria contract the atrioventricular valves open, allowing blood to flow into the ventricle. Blood from the right atrium flows into the cavum venosum, and blood from the left atrium flows into the cavum arteriosum. At this stage in the heartbeat, the large median flaps of the valve between the right atrium and the cavum venosum are pressed against the opening of the intraventricular canal, sealing it off from the cavum venosum. As a result, the oxygenated blood from the left atrium is confined to the cavum arteriosum. Deoxygenated blood from the right atrium fills the cavum venosum and then continues over the muscular ridge into the cavum pulmonale.

When the ventricle contracts, blood pressure inside the heart increases. Ejection of blood from the heart into the pulmonary circuit precedes flow into the systemic circuit because resistance is lower in the

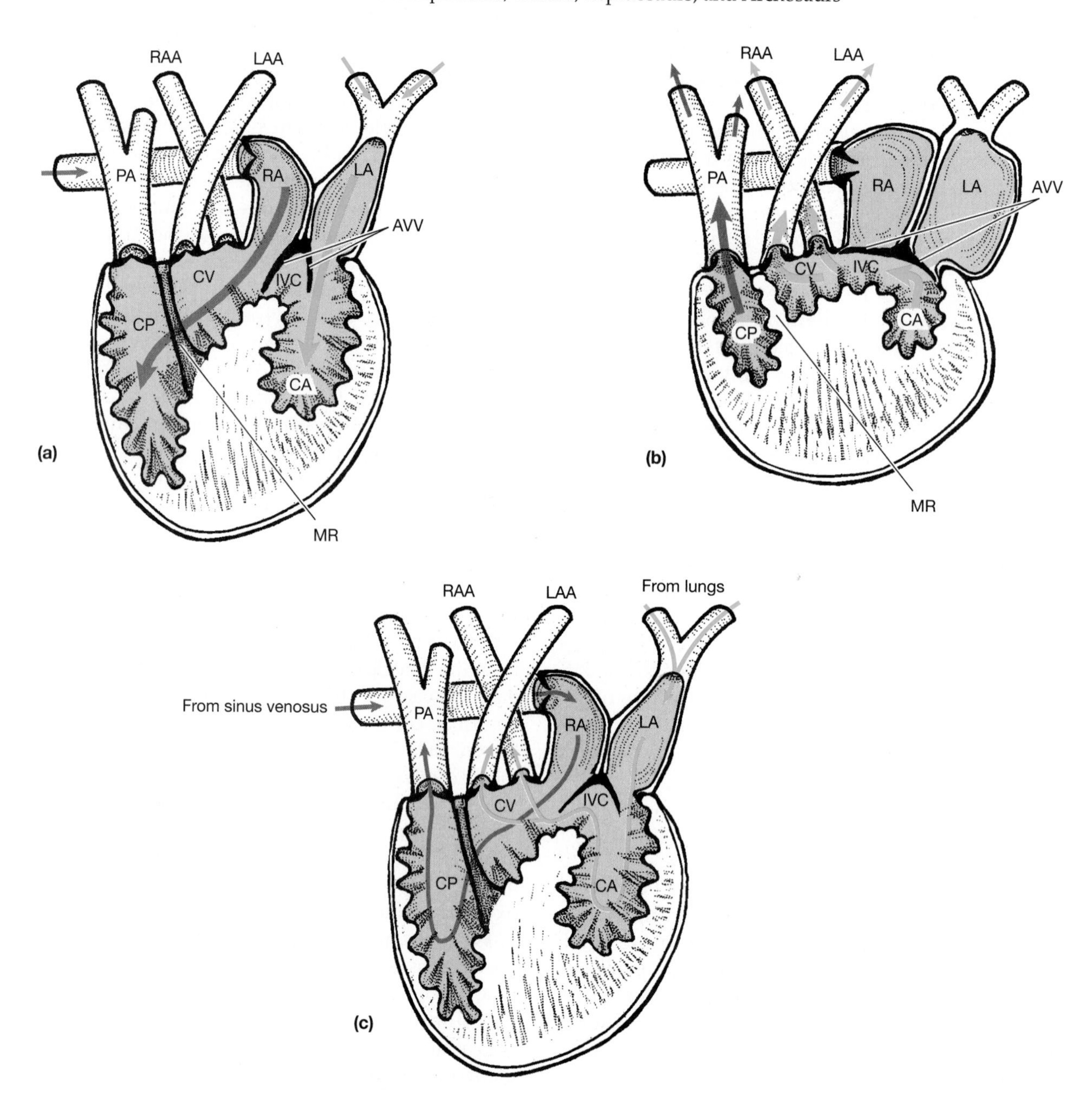

▲ Figure 10–5 Blood flow in the heart of a turtle. (a) As the atria contract, oxygenated blood (blue arrows) from the left atrium (LA) enters the cavum arteriosum (CA) while deoxygenated blood (gray arrows) from the right atrium (RA) first enters the cavum venosum (CV) and then crosses the muscular ridge (MR) and enters the cavum pulmonale (CP). The atrioventricular valve (AVV) blocks the intraventricular canal (IVC) and prevents mixing of oxygenated and deoxygenated blood. (b) As the ventricle contracts, the deoxygenated blood in the cavum pulmonale is expelled through the pulmonary arteries (PA); the AVV closes, no longer obstructing the IVC; and the oxygenated blood in the cavum arteriosum is forced into the cavum venosum and expelled through the right and left aortic arches (RAA and LAA). The adpression of the wall of the ventricle to the muscular ridge prevents mixing of deoxygenated and oxygenated blood. (c) Summary of the pattern of blood flow through the heart of a turtle.

pulmonary circuit. As deoxygenated blood flows out of the cavum pulmonale into the pulmonary artery, the displacement of blood from the cavum venosum across the muscular ridge into the cavum pulmonale continues. As the ventricle shortens during contraction, the muscular ridge comes into contact with the wall of the ventricle and closes off the passage for blood between the cavum venosum and cavum pulmonale.

Simultaneously, blood pressure inside the ventricle increases, and the flaps of the right atrioventricular valve are forced into the closed position, preventing backflow of blood from the cavum venosum into the atrium. When the valve closes, it no longer blocks the intraventricular canal. Oxygenated blood from the cavum arteriosum can now flow through the intraventricular canal and into the cavum venosum. At this stage in the heartbeat, the wall of the ventricle is pressed firmly against the muscular ridge, separating the oxygenated blood in the cavum venosum from the deoxygenated blood in the cavum pulmonale.

As pressure in the ventricle continues to rise, oxygenated blood in the cavum venosum is ejected into the aortic arches. This system effectively prevents mixing of oxygenated and deoxygenated blood in the heart, despite the absence of a permanent morphological separation of the two circuits.

Respiration

Primitive amniotes probably used movements of the rib cage to draw air into the lungs and to force it out, and lizards still employ that mechanism. The fusion of the ribs of turtles with their rigid shells makes that method of breathing impossible. Only the openings at the anterior and posterior ends of the shell contain flexible tissues. The lungs of a turtle, which are large, are attached to the carapace dorsally and laterally. Ventrally, the lungs are attached to a sheet of nonmuscular connective tissue that is itself attached to the viscera (Figure 10–6). The weight of the viscera keeps this diaphragmatic sheet stretched downward.

Turtles produce changes in pressure in the lungs by contracting muscles that force the viscera upward, compressing the lungs and expelling air, followed by contracting other muscles that increase the volume of the visceral cavity, allowing the viscera to settle downward. Because the viscera are attached to the diaphragmatic sheet, which in turn is attached to the lungs, the downward movement of the viscera

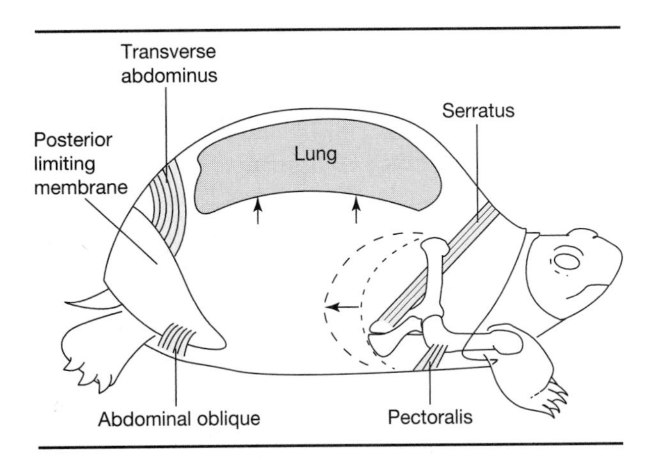

▲ Figure 10–6 Schematic view of the lungs and respiratory movements of a tortoise.

expands the lungs, drawing in air. In turtles both inhalation and exhalation require muscular activity. The viscera are forced upward against the lungs by the contraction of the transverse abdominus muscle posteriorly and the pectoralis muscle anteriorly. The transverse abdominus inserts on the cup-shaped connective tissue, the posterior limiting membrane, that closes off the posterior opening of the visceral cavity. Contraction of the transverse abdominus flattens the cup inward, thereby reducing the volume of the visceral cavity. The pectoralis draws the shoulder girdle back into the shell, further reducing the volume of the visceral cavity.

The inspiratory muscles are the abdominal oblique, which originates near the posterior margin of the plastron and inserts on the external side of the posterior limiting membrane, and the serratus, which originates near the anterior edge of the carapace and inserts on the pectoral girdle. Contraction of the abdominal oblique pulls the posterior limiting membrane outward, and contraction of the serratus rotates the pectoral girdle outward. Both of these movements increase the volume of the visceral cavity, allowing the viscera to settle back downward and causing the lungs to expand. The in-and-out movements of the forelimbs and the soft tissue at the rear of the shell during breathing are conspicuous.

The basic problems of respiring within a rigid shell are the same for most turtles, but the mechanisms show some variation. For example, aquatic turtles can use the hydrostatic pressure of water to help move air in and out of the lungs. In addition, many aquatic turtles are able to absorb oxygen and release carbon dioxide to the water. The pharynx and cloaca appear to be the major sites of aquatic gas exchange. In 1860, in *Contributions to the Natural History of the U.S.A.*, Louis Agassiz pointed out that the pharynx of soft-shelled turtles contains fringe-like processes and suggested that these structures are used for underwater respiration. Subsequent study has shown that when soft-shelled turtles are confined under water, they use movements of the hyoid apparatus to draw water in and out of the pharynx, and that pharyngeal respiration accounts for most of the oxygen absorbed from the water. The Australian turtle *Rheodytes leukops* uses cloacal respiration. Its cloacal orifice is as much as 30 millimeters in diameter, and the turtle holds it open. Large bursae (sacks) open from the wall of the cloaca, and the bursae have a well-vascularized lining with numerous projections (villi). The turtle pumps water in and out of the bursae at rates of 15 to 60 times per minute. Captive turtles rarely surface to breathe, and experiments have shown that the rate of oxygen uptake through the cloacal bursae is very high.

Patterns of Circulation and Respiration

The morphological complexity of the hearts of turtles and of squamates allows them to adjust blood flow through the pulmonary and systemic circuits to meet short-term changes in respiratory requirements. The key to these adjustments is changing pressures in the systemic and pulmonary circuits.

Recall that in the turtle heart, deoxygenated blood from the right atrium normally flows from the cavum venosum across the muscular ridge and into the cavum pulmonale. The blood pressure inside the heart increases as the ventricle contracts, and blood is first ejected into the pulmonary artery because the resistance to flow in the pulmonary circuit is normally less than the resistance in the systemic circuit. However, the resistance to blood flow in the pulmonary circuit can be increased by muscles that narrow the diameter of blood vessels. When this happens, the delicate balance of pressure in the heart that maintained the separation of oxygenated and deoxygenated blood is changed. When the resistance of the pulmonary circuit is essentially the same as that of the systemic circuit, blood flows out of the cavum pulmonale and cavum venosum at the same time, and some deoxygenated blood bypasses the lungs and flows into the systemic circuit (Figure 10–7). This process is called a **right-to-left intracardiac shunt.** *Right-to-left* refers to the shift of deoxygenated blood from the pulmonary circuit into the systemic circuit, and *intracardiac* means that it occurs in the heart rather than by flow between the major blood vessels.

Why would it be useful to divert deoxygenated blood from the lungs into the systemic circulation? The ability to make this shunt is not unique to turtles—it occurs also among lizards and snakes and in crocodilians. The heart morphology of lizards is very like that of turtles, and the same mechanism of changing pressures in the pulmonary and systemic circuits is used to achieve an intracardiac shunt. Crocodilians have hearts in which the ventricle is permanently divided into right and left halves by a septum, and they employ a different mechanism to achieve a right-to-left shunt.

The most general function for blood shunts may lie in the ability they provide to match patterns of lung ventilation and pulmonary gas flow (Wang et al. 1997). Lizards, snakes, crocodilians, and turtles normally breathe intermittently, and periods of lung ventilation alternate with periods of **apnea** (no breathing). A mathematical model indicates that a combination of right-to-left and left-to-right shunts could stabilize oxygen concentration in blood during alternating periods of apnea and breathing.

Another function of intracardiac shunts may be to reduce blood flow to the lungs during breath-holding to permit more effective use of the oxygen stored in the lungs. Diving is one situation in which reptiles hold their breath. Many reptiles are excellent divers, and even terrestrial and arboreal forms such as green iguanas may dive into water to escape predators, but that is not the only situation in which defensive behaviors interfere with breathing. Turtles are particularly prone to periods of breath-holding because their method of lung ventilation means they cannot breathe when they withdraw their heads and legs into their shells.

10.4 Ecology and Behavior of Turtles

Turtles are long-lived animals. Even small species like the painted turtle (*Chrysemys picta*) do not mature until they are 7 or 8 years old, and they may live to be 14 or older. Larger species of turtles live longer. Estimates of centuries for the life spans of tortoises

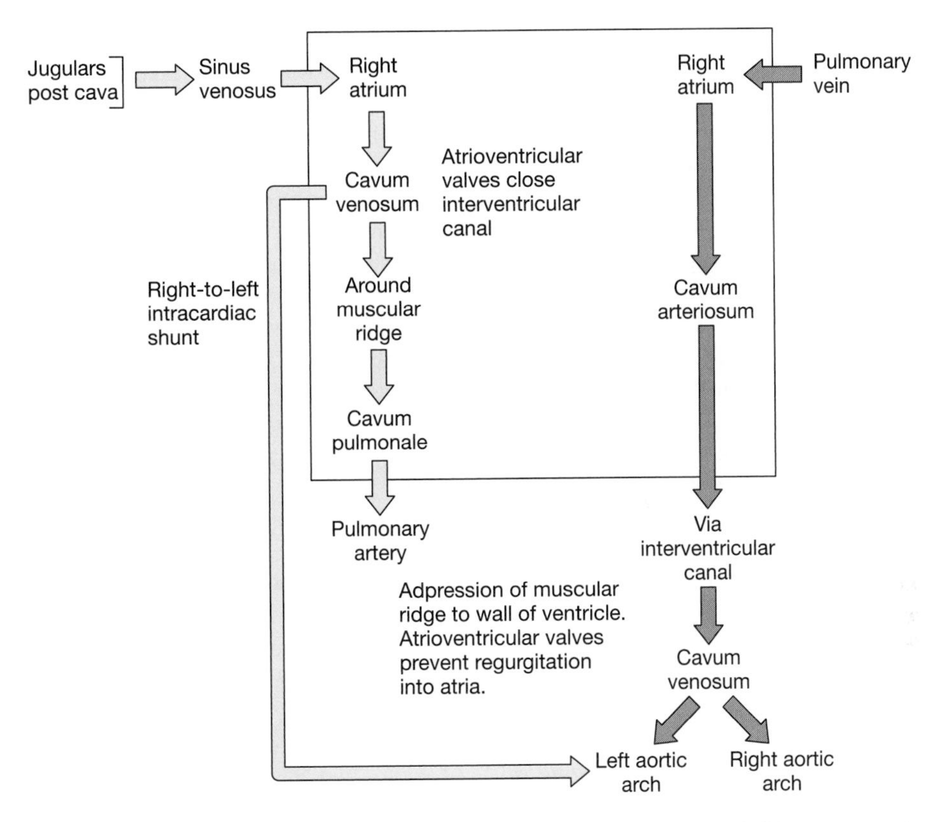

▲ Figure 10–7 Diagram of the right-to-left shunt of blood in the heart of a turtle. Light blue arrows show deoxygenated blood, and dark blue arrows show oxygenated blood. The box encloses the cycle of events during normal blood flow (compare with Figure 10–4). When resistance in the pulmonary circuit increases, some deoxygenated blood from the cavum venosum flows into the left aortic arch instead of into the pulmonary arteries.

are exaggerated, but large tortoises and sea turtles may live as long as humans, and even box turtles may live over 50 years. These longevities make the life histories of turtles hard to study. Furthermore, a long lifetime is usually associated with a low replacement rate of individuals in the population, and species with those characteristics are at risk of extinction when hunting or habitat destruction reduces their numbers. Conservation efforts for sea turtles and tortoises are especially important areas of concern.

Temperature Regulations and Body Size of Turtles

Turtles can achieve a considerable degree of stability in body temperature by regulating their exchange of heat energy with the environment. Turtles basking on a log in a pond are a familiar sight in many parts of the world, because few pond turtles are large enough to maintain body temperatures higher than the temperature of the water surrounding them. Emerging from the water to bask is the only way most pond turtles can raise their body temperatures to speed digestion, growth, and the production of eggs. In addition, aerial basking may help aquatic turtles to rid themselves of algae and leeches. Exposure to ultraviolet light may be involved in activating vitamin D, which is involved in controlling calcium deposition in their bones and shell. A few turtles are quite arboreal; these turtles have small plastrons that allow considerable freedom of movement for the limbs. The big-headed turtle (*Platysternon megacephalum*) from southeast Asia, lives in fast-flowing streams at high altitudes and is said to climb on rocks and trees to bask. In North America musk turtles (*Sternotherus*) bask on overhanging branches and drop into the water when they are disturbed.

Small terrestrial turtles, such as box turtles and small species of tortoises, can thermoregulate by moving between sun and shade. Small tortoises warm and cool quite rapidly, and they appear to behave very much like other small reptiles in selecting suitable microclimates for thermoregulation. Familiarity with a home range may assist this type of thermoregulation. A study conducted in Italy compared the thermoregulation of resident Hermann's tortoises (animals living in their own home ranges) with individuals that were brought to the study site and tested before they had learned their way around (Chelazzi and Calzolai 1986). The resident tortoises warmed faster and maintained more stable shell temperatures than did the strangers.

Turtles are unusual among reptiles in having a substantial number of species that reach large body sizes. The bulk of a large tortoise provides considerable thermal inertia, and large species like the Galápagos and Aldabra tortoises heat and cool slowly. The giant tortoises of Aldabra Atoll (*Geochelone gigantea*), which weigh 60 kilograms or more, allow their body temperatures to rise to 32 to 33°C on sunny days and cool to 28 to 30°C overnight.

Large body size slows the rate of heating and cooling, but it can make temperature regulation more difficult. A small turtle can find shade beside a bush or even a clump of grass, but a giant tortoise needs a bigger object—a tree, for example. In open, sunny habitats overheating can be a problem for giant tortoises. The difficulty is particularly acute for some tortoises on Grande Terre, an island in the Indian Ocean (Swingland and Frazier 1979, Swingland and Lessells 1979). During the rainy season each year some of the turtles on the island move from the center of the island to the coast. This movement has direct benefits, because the migrant turtles gain access to a seasonal flush of plant growth on the coast. As a result of the extra food, migrant females are able to lay more eggs than females that remain inland. There are risks to migrating, however, because shade is limited on the coast and the rainy season is the hottest time of the year. Tortoises on the coast must limit their activity to the vicinity of patches of shade, which may be no more than a single tree in the midst of a grassy plain. During the morning tortoises forage on the plain, but as their body temperatures rise they move back toward the shade of the tree. As the day grows hotter, tortoises try to get into the deepest shade, and the biggest individuals do this most successfully. As the big tortoises (which are mostly males) push their way into the shade, they force smaller individuals (most of which are females) out into the sun and some of these tortoises die of overheating.

Marine turtles are large enough to achieve a considerable degree of endothermy (Spotila and Standora 1985, Paladino et al. 1990). A body temperature of 37°C was recorded by telemetry from a green turtle swimming in water that was 20°C. The leatherback turtle is the largest living turtle; adults may weigh up to 1000 kilograms. Leatherbacks range far from warm equatorial regions, and in the summer can be found off the coasts of New England and Nova Scotia in water as cool as 8 to 15°C. Body temperatures of these turtles appear to be 18°C or more above water temperatures, and a countercurrent arrangement of blood vessels in the flippers may contribute to retaining heat produced by muscular activity.

Social Behavior and Courtship

Tactile, visual, and olfactory signals are employed by turtles during social interactions. Many pond turtles have distinctive stripes of color on their heads, necks, and forelimbs and on their hindlimbs and tail. These patterns are used by herpetologists to distinguish the species, and they may be species-isolating mechanisms for the turtles as well. During the mating season, male pond turtles swim in pursuit of other turtles, and the color and pattern on the posterior limbs may enable males to identify females of their own species. At a later stage of courtship, when the male turtle swims backward in front of the female and vibrates his claws against the sides of her head, both sexes can see the patterns on their partner's head, neck, and forelimbs (Figure 10–8).

Among terrestrial turtles, the behavior of tortoises is best known. Many tortoises vocalize during courtship; the sounds they produce have been described as grunts, moans, and bellows. The frequencies of the calls that have been measured range from 500 to 2500 hertz. Some tortoises have glands that become enlarged during the breeding season and appear to produce pheromones. The secretion of the subdentary gland found on the underside of the jaw of tortoises in the North American genus *Gopherus* appears to identify both the species and the sex of an

▲ Figure 10–8 Social behavior of turtles. (a) Male painted turtle (*Chrysemys picta*) courting a female by vibrating the elongated claws of his forefeet against the sides of her head. (b) The head-raising dominance posture of a Galápagos tortoise, *Geochelone*. (This behavior can sometimes be elicited by crouching in front of a male tortoise and raising your arm.)

individual. During courtship, males and females of the Florida gopher tortoise rub their subdentary gland across one or both forelimbs, and then extend the limbs toward the other individual, which may sniff at them. Males also sniff the cloacal region of other tortoises, and male tortoises of some species trail females for days during the breeding season. Fecal pellets may be territorial markers; fresh fecal pellets from a dominant male tortoise have been reported to cause dispersal of conspecifics.

Tactile signals used by tortoises include biting, ramming, and hooking. These behaviors are used primarily by males, and they are employed against other males and also against females. Bites are usually directed at the head or limbs, whereas ramming and hooking are directed against the shell. The epiplastral region is used for ramming, and in some species the epiplastral bones of males are elongated and project forward beneath the neck. A tortoise about to ram another individual raises itself on its legs, rocks backward, and then plunges forward, hitting the shell of the other individual with a thump that can be heard from a distance of 100m in large species. During hooking the epiplastral projections are placed under the shell of an adversary, and the aggressor lifts the front end of its shell and walks forward. The combination of lifting and pushing hustles the adversary along and may even overturn it.

Movements of the head appear to act as social signals for tortoises, and elevating the head is a signal of dominance in some species. Herds of tortoises have social hierarchies that are determined largely by aggressive encounters. Ramming, biting, and hooking are employed in these encounters, and the larger individual is usually the winner—although experience may play some role. These social hierarchies are expressed in the priority of different individuals in access to food or forage areas, mates, and resting sites. Dominance relationships also appear to be involved in determining the sequence in which individual tortoises move from one place to another. The social structure of a herd of tortoises can be a nuisance for zookeepers trying to move the animals from an outdoor pen into an enclosure for the night, because the tortoises resist moving out of their proper rank sequence.

Nesting Behavior

All turtles are oviparous. Female turtles use their hindlimbs to excavate a nest in sand or soil, and deposit a clutch that ranges from 4 or 5 eggs for small species to more than 100 eggs for the largest sea turtles. Turtles in the families Cheloniidae, Dermochelyidae, and Chelydridae lay eggs with soft, flexible shells, as do most species in the families Bataguridae, Emydidae, and Pelomedusidae. The eggs of turtles in the families Carettochelyidae, Chelidae, Kinosternidae, Testudinidae, and Trionychidae have rigid shells. Embryonic development typically requires 40 to 60 days, and in general soft-shelled eggs develop more rapidly than hard-shelled eggs. Some species of turtles lay their eggs in late summer or fall, and the eggs have a diapause (a period of arrested embryonic development) during the winter and resume development when temperatures rise in the spring. The Australian pleurodire *Chelodina rugosa* lays eggs underwater in temporary ponds, but embryonic development does not begin until the ponds dry out and the eggs are exposed to air (Kennett et al. 1998).

Environmental Effects on Egg Development

Temperature, wetness, and the concentrations of oxygen and carbon dioxide can have profound effects on the embryonic development of turtles. The temperature of a nest affects the rate of embryonic development, and excessively high or low temperatures can be lethal. The discovery that the sex of some reptiles is determined by the temperature they experienced during embryonic development has important implications for understanding patterns of life history as well as conservation of these species. Temperature-dependent sex determination is widespread among turtles, apparently universal among crocodilians, and is known for tuatara and a few species of lizards. The effect of temperature on sex determination is correlated with sexual size dimorphism of adults—high incubation temperatures produce the larger sex, which is usually females for turtles. The switch from one sex to the other occurs within a span of 3 or 4°C (Figure 10–9). Male crocodilians, tuatara, and lizards are usually larger than females, and in these groups high temperatures during embryonic development produce males. Temperatures of natural nests are not completely stable, of course. There is some daily temperature variation superimposed on a seasonal cycle of changing environmental temperatures. The middle third of embryonic development is the critical period for sex determination; the sex of the embryos depends on the temperatures they experience during those few weeks. When eggs are exposed to a daily temperature cycle, the high point of the cycle is most critical for sex determination.

Because of the narrowness of the thermal windows involved in sex determination and the variation that exists in environmental temperatures, both sexes are produced under field conditions, but not necessarily in the same nests. A nest site may be cooler in late summer when it is shaded by vegetation than it was early in the spring when it was exposed to the sun. Thus, eggs laid early in the season would produce females, whereas eggs deposited in the same place later in the year would produce males. Temperature can also differ between the top and the bottom of a nest. For example, temperatures averaged from 33 to 35°C in the top center of dry nests of American alligators in marshes, and males hatched from eggs in this area. At the bottom and sides of the same nests, average temperatures were from 30 to 32°C, and eggs from those regions hatched into females.

Temperature-dependent sex determination has important implications for efforts to conserve endangered species. Nests of the American alligator in marshes are cooler than those built on levees; females hatch from marsh nests and males from nests built on levees. The sex ratio of hatchlings from 8000 eggs collected from natural nests was five males to one female, reflecting the relative abundance of levee and marsh nests. It is not clear whether this large imbalance of the sexes represents a normal condition for alligators, or if it reflects a recent change in the availability of nest sites as a result of construction of levees.

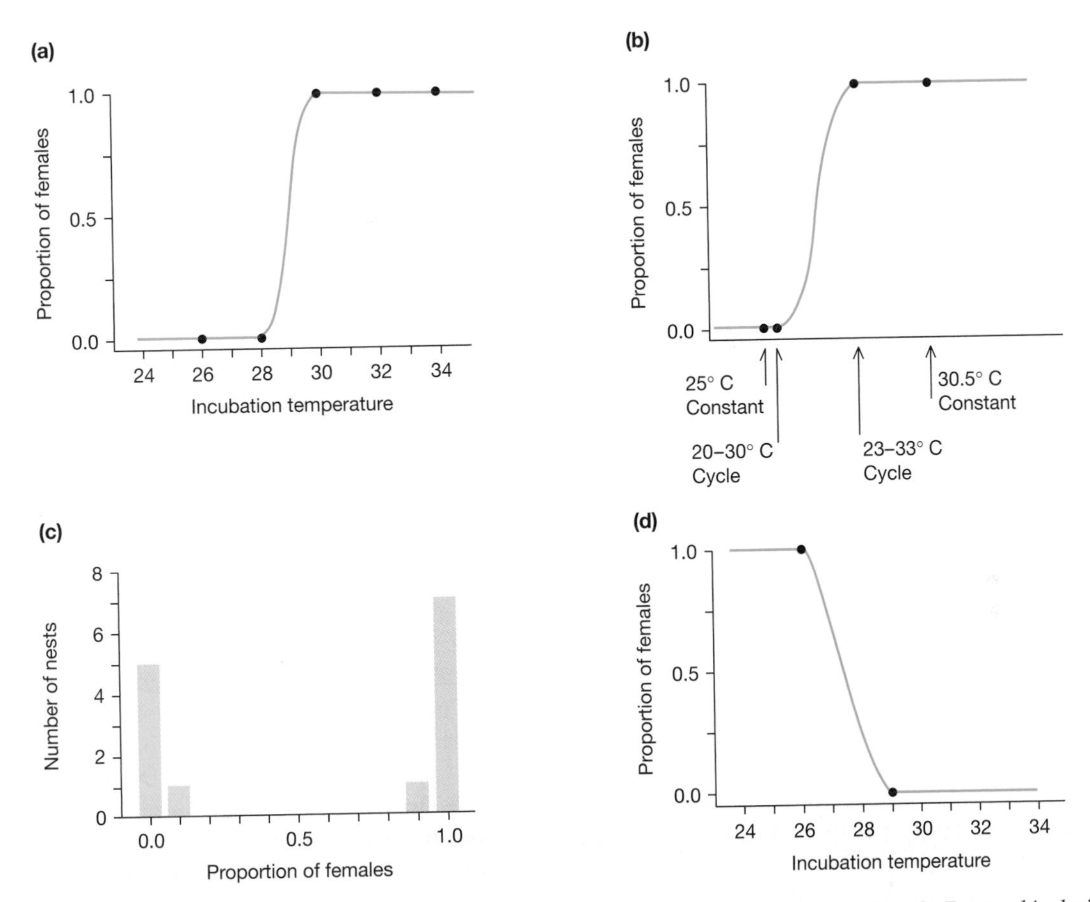

▲ Figure 10–9 Temperature-dependent sex determination. (a) Eggs of the European pond turtle *Emys orbicularis* hatch into males when they are incubated at 26 or 28°C and into females at 30°C or above. (b) The North American map turtle *Graptemys ouachitensis* shows the same pattern. A temperature that cycles between 20 and 30°C produces males, whereas a temperature cycle of 23 to 33°C produces females. (c) Natural nests of map turtles produce predominantly males or females, depending on the nest temperature. (d) Eggs of the lizard *Agama agama* also show temperature-dependent sex determination, but the male- and female-determining temperatures are the opposite of those in turtles—for the lizard, low temperatures produce females and high temperatures produce males.

Some conservation efforts have been confounded by temperature-dependent sex determination in sea turtles. A number of programs have depended on collecting eggs from natural nests and incubating them under controlled conditions. Unfortunately, these unnaturally uniform conditions of incubation can result in producing hatchlings of only one sex.

The amount of moisture in the soil surrounding a turtle nest is another important variable during embryonic development of the eggs. The wetness of a nest interacts with temperature in sex determination as well as influencing the rate of embryonic development and the size and vigor of the hatchlings produced. Dry substrates induced the development of some female painted turtles at low temperatures (26.5 and 27°C) that would normally have produced only males. The wetness of the substrate did not affect the sex of turtles from eggs incubated at 30.5 and 32°C: All the hatchlings from these eggs were females, as would be expected on the basis of temperature-dependent sex determination alone.

Wet incubation conditions produce larger hatchlings than do dry conditions, apparently because water is needed for metabolism of the yolk. When water is limited, turtles hatch early and at smaller body sizes, and their guts contain a quantity of yolk that was not used during embryonic development. Hatchlings from nests under wetter conditions are larger and contain less unmetabolized yolk. The large hatchlings that emerge from moist nests are able to run and swim faster than hatchlings from drier nests, and as a result they may be more successful at escaping from predators and at catching food.

Hatching and the Behavior of Baby Turtles

Turtles are self-sufficient at hatching, but in some instances interactions among the young may be essential to allow them to escape the nest. Sea-turtle nests are quite deep; the eggs may be buried 50 cm beneath the sand, and the hatchling turtles must struggle upward through the sand to the surface. After several weeks of incubation the eggs all hatch within a period of a few hours, and a hundred or so baby turtles find themselves in a small chamber at the bottom of the nest hole. Spontaneous activity by a few individuals sets the whole group into motion, crawling over and under one another. The turtles at the top of the pile loosen sand from the roof of the chamber as they scramble about, and the sand filters down through the mass of baby turtles to the bottom of the chamber.

Periods of a few minutes of frantic activity are interspersed by periods of rest, possibly because the turtles' exertions reduce the concentration of oxygen in the nest and they must wait for more oxygen to diffuse into the nest from the surrounding sand. Gradually, the entire group of turtles moves upward through the sand as a unit until it reaches the surface. As the baby turtles approach the surface, high sand temperatures probably inhibit further activity, and they wait a few centimeters below the surface until night when a decline in temperature triggers emergence. All the babies emerge from a nest in a very brief period, and all the babies in different nests that are ready to emerge on a given night leave their nests at almost the same time, probably because their behavior is cued by temperature. The result is the sudden appearance of hundreds or even thousands of baby turtles on the beach, each one crawling toward the ocean as fast as it can.

Simultaneous emergence is an important feature of the reproduction of sea turtles, because the babies suffer severe mortality crossing the few meters of beach and surf. Terrestrial predators—crabs, foxes, raccoons, and other predators—gather at the turtles' breeding beaches at hatching time and await their appearance. Some of the predators come from distant places to prey on the baby turtles. In the surf, sharks and bony fishes patrol the beach. Few, if any, baby turtles would get past that gauntlet were it not for the simultaneous emergence that brings all the babies out at once and temporarily swamps the predators.

Turtles exhibit no parental care, and the long period of embryonic development renders their nests vulnerable to predators. Females of many species of turtles scrape the ground in a wide area around the nest when they have finished burying their eggs. This behavior may make it harder for predators to identify the exact location of the nest. Major breeding sites of sea turtles are often on islands that lack mammalian predators that could excavate the nests. Another important feature of a nesting beach is provision of suitable conditions for the hatchling turtles. Newly hatched sea turtles are small animals; they weigh 25 to 50 grams, which is less than 0.05 percent of the body mass of an adult sea turtle. The enormous disparity in body size of hatchling and adult turtles is probably accompanied by equally great differences in their ecological requirements and their swimming abilities. Many of the major sea-turtle nesting areas are upcurrent from the feeding grounds, and that location may allow currents to carry the baby turtles from the breeding beaches to the feeding grounds.

We know even less about the biology of baby sea turtles than we do about the adults. Where the turtles go in the period following hatching has been a long-standing puzzle in the life cycle of sea turtles. For example, green turtles hatch in the late summer at Tortuguero on the Caribbean coast of Costa Rica. The turtles disappear from sight as soon as they are at sea, and they are not seen again until they weigh 4 or 5 kg. Apparently, they spend the intervening period floating in ocean currents. Material drifting on the surface of the sea accumulates in areas where currents converge, forming drift lines of flotsam that include sargassum (a brown algae) and the vertebrate and invertebrate fauna associated with it. These drift lines are probably important resources for juvenile sea turtles.

Navigation and Migration

Pond turtles and terrestrial turtles usually lay their eggs in nests that they construct within their home ranges. The mechanisms of orientation that they use to find nesting areas are probably the same ones they use to find their way among foraging and resting areas. Familiarity with local landmarks is an effective method of navigation for these turtles, and they may also use the sun for orientation. Sea turtles have a more difficult time, partly because the open ocean lacks conspicuous landmarks, and also because feeding and nesting areas are often separated by hundreds or thousands of kilometers. Most sea turtles are carnivorous. The leatherback turtle feeds on jellyfishes, ridley and loggerhead turtles eat crabs and other benthic invertebrates, and the hawksbill turtle uses its beak to scrape encrusting organisms (sponges, tunicates, bryozoans, mollusks, and algae) from reefs. Juvenile green turtles are carnivorous, but the adults feed on vegetation, particularly turtle grass (*Thalassia testudinium*), which grows in shallow water on protected shorelines in the tropics. The areas that provide food for sea turtles often lack the characteristics needed for successful nesting, and many sea turtles move long distances between their feeding grounds and their breeding areas.

The ability of sea turtles to navigate over thousands of kilometers of ocean and find their way to nesting beaches that may be no more than tiny coves on a small island is astonishing. The migrations of sea turtles, especially the green turtle, have been studied for decades. Turtles captured at breeding sites in the Caribbean and Atlantic Oceans have been individually marked with metal tags since 1956, and tag returns from turtle catchers and fishing boats have allowed the major patterns of population movements to be established (Figure 10–10).

Four major nesting sites of green turtles have been identified in the Caribbean and South Atlantic: one at Tortuguero on the coast of Costa Rica, one on Aves Island in the eastern Caribbean, one on the

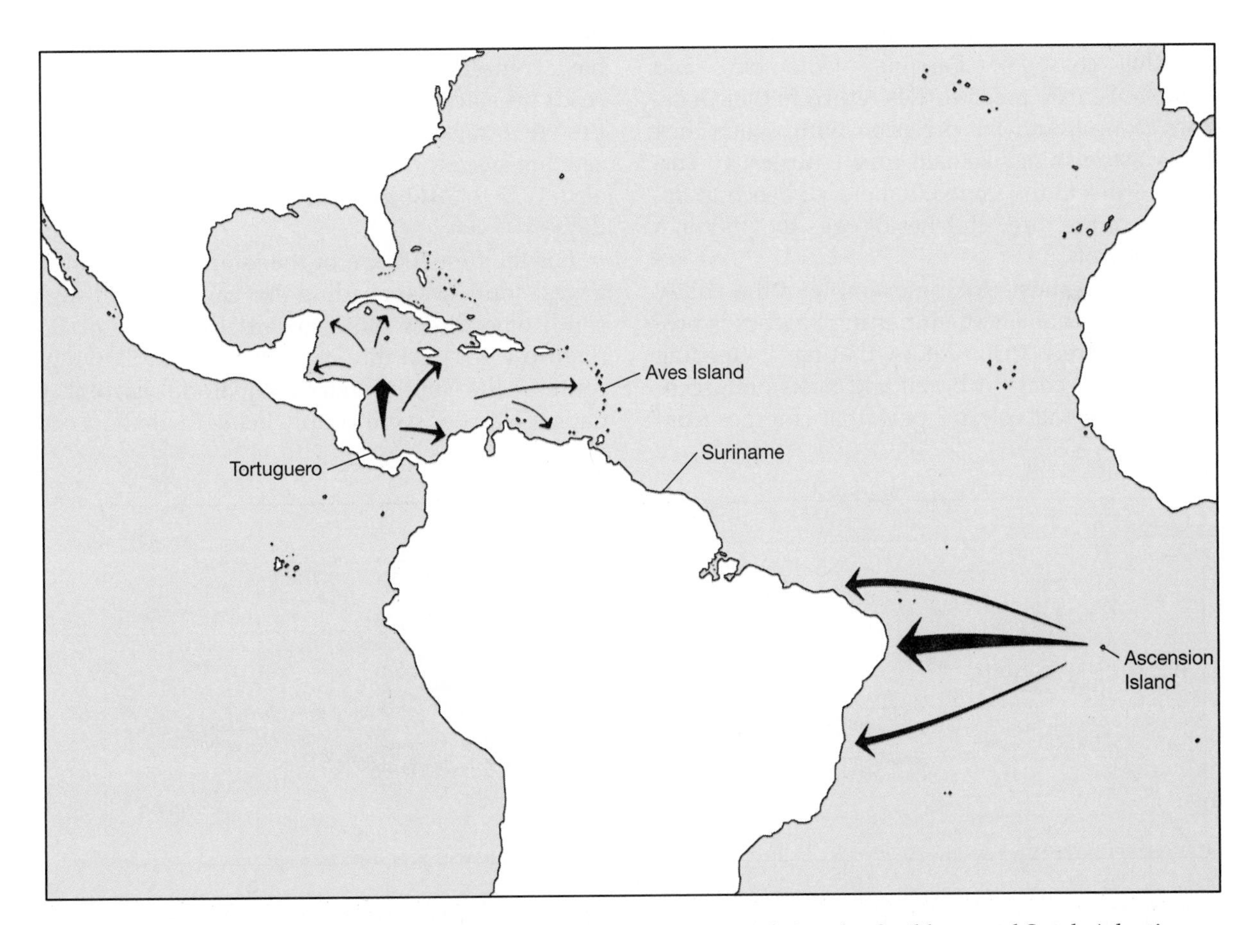

▲ Figure 10–10 Migratory movements of green turtles (*Chelonia mydas*) in the Caribbean and South Atlantic. The population that nests on beaches in the Caribbean is drawn from feeding grounds in the Caribbean and Gulf of Mexico. The turtles that nest on Ascension Island feed along the coast of northern South America.

coast of Suriname, and one on Ascension Island between South America and Africa. Male and female green turtles congregate at these nesting grounds during the nesting season. The male turtles remain offshore, where they court and mate with females, and the female turtles come ashore to lay eggs on the beaches. A typical female green turtle at Tortuguero produces three clutches of eggs about 12 days apart. About a third of the female turtles in the Tortuguero population nest in alternate years, and the remaining two-thirds of the turtles follow a 3-year breeding cycle. The coast at Tortuguero lacks the beds of turtle grass on which green turtles feed, and the turtles come to Tortuguero only for nesting. In the intervals between breeding periods, the turtles disperse around the Caribbean. The largest part of the Tortuguero population spreads northward along the coast of Central America. The Miskito Bank off the northern coast of Nicaragua appears to be the main feeding ground for the Tortuguero colony. A smaller number of turtles from Tortuguero swim south along the coast of Panama, Colombia, and Venezuela. Female green turtles return to their natal beaches to nest, and the precision with which they home is astonishing. Female green turtles at Tortuguero return to the same kilometer of beach to deposit each of the three clutches of eggs they lay in a breeding season.

Probably the most striking example of the ability of sea turtles to home to their nesting beaches is provided by the green turtle colony that has its feeding grounds on the coast of Brazil and nests on Ascension Island, a small volcanic peak that emerges from the ocean. The island is 2200 km east of Brazil and less than 20 km in diameter—a tiny target in the vastness of the South Atlantic.

How do turtles find their way across thousands of kilometers of ocean? Other animals use a variety of cues for navigation, and sea turtles probably do so as well. Chemosensory information may be one important component of their navigation. The South Atlantic equatorial current flows westward, washing past Ascension Island and continuing toward Brazil. Newly hatched green turtles may drift with the current from their natal beaches to the adult feeding grounds, and the odor plume of the island may help to guide female turtles back to the island to nest. That is, a female turtle leaving the coast of Brazil may swim upstream in the South Atlantic equatorial current (i.e., up the odor gradient) to locate Ascension Island.

It is impractical to locate female turtles off the coast of Brazil as they are about to begin their journey to the island, but it is easy to find turtles that have completed nesting at Ascension Island and are ready to start back to Brazil. Five female green turtles were tracked on their return trip using the Argos satellite system (Luschi et al. 1998). The turtles traveled 1777 to 2342 km and reached Brazil in 33 to 74 days.

For the first 500 km of the journey, they followed a west-south-west heading that carried them slightly south of a direct route toward the bulge of Brazil (Figure 10–11). At this stage they were following the route of the South Atlantic equatorial current. Perhaps the turtles were simply being carried off course

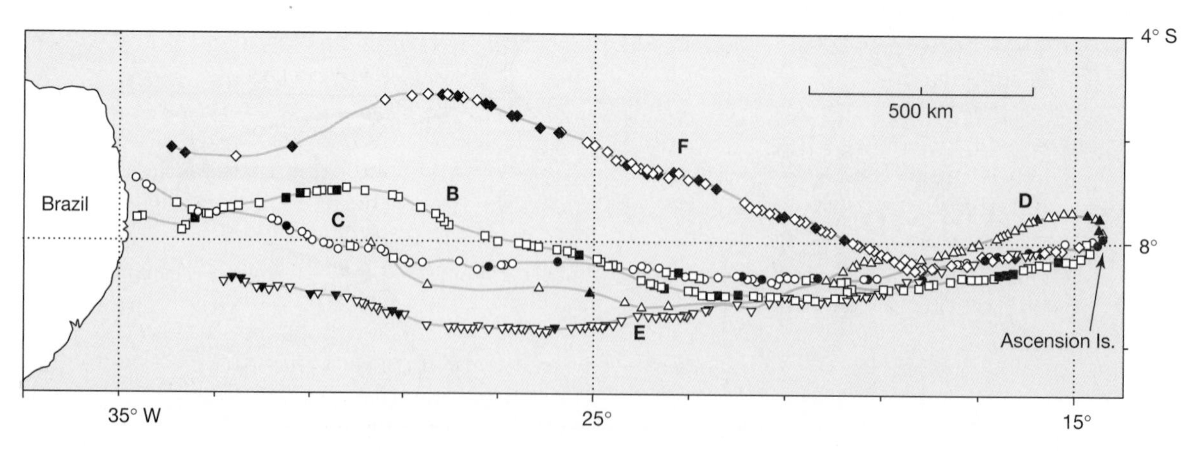

▲ Figure 10–11 Migratory routes of five green turtles (*Chelonia mydas*) tracked by satellite from Ascension Island to Brazil.

by the current, but they may have been using the same guidance system that they rely on for their outward journey—that is, staying within the plume of the island's scent. Even though the current carries them slightly south of a direct route to Brazil, they may save energy and move faster by initially staying in the current.

If they remained in the current for the entire trip they would be carried too far south, so the turtles make a midcourse correction. The new course heads west-north-west on a nearly direct route to the bulge on the coast of Brazil. The shift in direction might be triggered by the waning strength of the scent plume. The turtles spend more than 90 percent of the journey underwater, suggesting that they may be sampling the plume in three dimensions.

A study of navigation by hatchling loggerhead turtles showed that they used at least three cues for orientation: light, wave direction, and magnetism (Lohmann 1991). These stimuli play sequential roles in the turtles' behavior. When they emerge from their nests, the hatchlings crawl toward the brightest light they see. Normally the sky at night is lighter over the ocean than over land, and this behavior brings them to the water's edge. (Shopping centers, street lights, even porch lights on beachfront houses can confuse these and other species of sea turtles and lead them inland, where they are crushed on roads or die in the sun the next day.)

In the ocean, the loggerhead hatchlings swim into the waves. This response moves them away from shore and ultimately into the Gulf Stream. They drift with the current along the coast of the United States, and then eastward across the Atlantic. Off the coast of Portugal, the Gulf Stream divides into two branches. One turns north toward England, and the other swings south past the bulge of Africa and eventually back westward across the Atlantic. It's essential for the baby turtles to turn right at Portugal; if they fail to make that turn they are swept past England into the chilly North Atlantic, where they perish. If they do turn southward off the coast of Portugal, they are eventually carried back to the coast of tropical America—a round-trip that takes 5 to 7 years.

Magnetic orientation appears to tell the turtles when to turn right to catch the current that will carry them to the South Atlantic. We usually think of the Earth's magnetic field as providing two-dimensional information—north-south and east-west—but it's more complicated than that. The field loops out of the north and south magnetic poles of the Earth. At the equator the field is essentially parallel to the Earth's surface (in other words, it forms an angle of 0 degrees), and at the poles it intersects the surface at an angle of 90 degrees. Thus, the three-dimensional orientation of the Earth's magnetic field provides directional information (which way is magnetic north?) and information about latitude (what is the angle at which the magnetic field intersects the Earth's surface?).

When loggerheads in a pool on land that had no waves were exposed to an artificial magnetic field at the 57-degree angle of intersection with the Earth's surface that is characteristic of Florida, they swam toward artificial east—even when the magnetic field was changed 180 degrees, so that the direction they thought was east was actually west. That is, they were able to use a compass sense to determine direction. But that wasn't all they could do. When the angle of intersection of the artificial magnetic field was increased to 60 degrees, as if they were further north than they really were, the turtles turned south. A 60-degree angle of intersection corresponds to the latitude where the Gulf Stream forks off the coast of Portugal, and where the turtles must turn south to reach the South Atlantic. Thus, it appears that loggerhead turtles can use magnetic sensitivity to recognize both direction and latitude.

10.5 Conservation of Turtles

Many species of turtles have slow rates of growth and require long periods to reach maturity. These are characteristics that predispose a species to the risk of extinction when changing conditions increase the mortality of adults or drastically reduce recruitment of juveniles into the population. The plight of large tortoises and sea turtles is particularly severe, partly because these species are among the largest and slowest growing of turtles, and also because other aspects of their biology expose them to additional risk (Figure 10–12). The conservation of tortoises and sea turtles is a subject of active international concern and has led to the founding of a new journal, *Chelonian Conservation and Biology*.

The largest living tortoises are found on the Galápagos and Aldabra Islands. The relative isolation of these small and (for humans) inhospitable land-

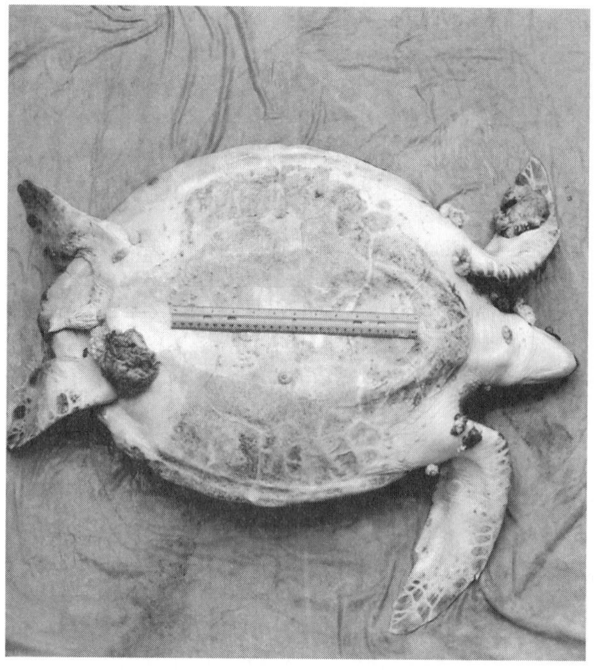

(a)

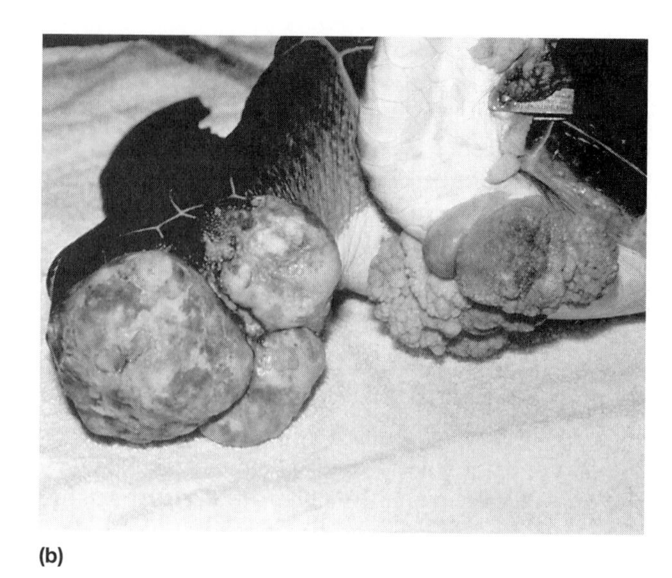

(b)

▲ Figure 10–12. A green turtle with cutaneous papillomas. (a) These tumors, which are probably caused by a virus, have been found on most species of sea turtles and in most parts of the world (Herbst 1994). The tumors grow to more than 30 cm in diameter and can appear on any skin-covered surface. (b) They are especially common on the conjuctiva of the eyes, and may grow over the cornea. Tumors were not recorded on green turtles in the Indian River Lagoon, Florida, until 1982, and by 1994 approximately 50 percent of the green turtles were affected. The first record of papilloma on green turtles in Kaneohe Bay, Hawaii, was in 1958. Since 1989 the incidence has ranged from 49 to 92 percent. The tumors can be lethal, and their increased frequency is an ominous development for species that were already endangered (Herbst 1994).

masses has probably been an important factor in the survival of tortoises. Human colonization of the islands has brought with it domestic animals such as goats and donkeys that compete with tortoises for the limited quantities of vegetation to be found in these arid habitats, and dogs, cats, and rats that prey on tortoise eggs and on baby tortoises.

The limited geographic range of a tortoise that occurs only on a single island makes it vulnerable to extinction. In 1985 and again in 1994 brush fires on the island of Isabela in the Galápagos Archipelago threatened the 20 surviving individuals of *Geochelone guntheri,* and emphasized the advantage of moving some or all of the turtles to the breeding station operated by the Charles Darwin Research Station on Santa Cruz Island. This station has a successful record of breeding and releasing another species of Galápagos tortoise, *Geochelone nigra hoodensis,* which is native to Española Island. In the early 1960s only 14 individuals of this form could be located. All were adults and apparently had not bred successfully for many years. All of the tortoises were moved to the research station, and the first babies were produced in 1971. On March 24, 2000, the one-thousandth captive-bred tortoise was released on Española. This success story shows that carefully controlled captive breeding and release programs can be an effective method of conservation for endangered species of turtles. But these programs also have inherent risks (Box 10-1).

Entire turtle faunas are threatened in some areas. Nearly all species of turtles in Southeast Asia are now at risk because of economic and political changes in the region (van Dijk et al. 2000). Turtles have traditionally been used in China for food and for their supposed medicinal benefits. A two-day survey in just two Chinese food markets found an estimated 10,000 turtles for sale. If the turnover time

BOX 10-1 SICK TURTLES

The desert tortoise (*Gopherus agassizi*) is one of the largest terrestrial turtles in North America. Its geographic range includes the southwestern corner of Utah, the southwestern third of Arizona, and adjacent parts of Nevada and California, and it extends southward into Mexico.

Populations of desert tortoises have declined since the 1950s as human activity has intruded on the desert habitat. Between 1979 and 1989, most tortoise populations in the Mohave and Colorado deserts decreased by 30 to 70 percent. The situation has become even more grave with the appearance of upper respiratory tract disease (URTD), which attacks desert tortoises, often with fatal results. Infected turtles first have a runny nose, which becomes progressively worse until the turtles exude foam from their nostrils, wheeze when they breathe, cease feeding, become listless, and ultimately die (Figure 10–13). In 1988 tortoises in the Desert Tortoise Natural Area in Kern County, California, first showed symptoms of URTD (Jacobson et al. 1991). In 1989, 627 dead tortoises were found and 43 percent of the live tortoises in the Natural Area showed symptoms of URTD.

▲ Figure 10–13 A gopher tortoise with a runny nose. Nasal discharge and swollen eyes are signs that this tortoise is infected with the *Mycoplasma* that causes upper respiratory tract disease.

A large variety of bacteria were cultured from the nasal passages of the sick turtles, including *Mycoplasma*, which has subsequently been shown to be the cause of the disease (Jacobson et al. 1995). Desert tortoises are popular pets in the desert southwest, and a high proportion of pet turtles are infected by *Mycoplasma*. The infection may have been introduced to the Desert Tortoise Natural Area when pet tortoises were released, and its spread may have been accelerated by the poor physical condition of the wild tortoises that resulted from habitat degradation and a prolonged drought. *Mycoplasma* infections are notoriously difficult to cure. Captive tortoises can be treated with a combination of antibiotics, but there is no practical treatment for wild tortoises.

URTD has now been reported in a population of the gopher tortoise (*Gopherus polyphenus*) on Sanibel Island off the coast of Florida. Again, captive tortoises appear to have introduced the infection into a wild population: Until 1978 tortoises used in tortoise races in Fort Myers were released on Sanibel Island, and infected tortoises from the races may have carried *Mycoplasma* with them.

These examples emphasize the risk of releasing animals that have been held in captivity into wild populations. Captive breeding programs must take extraordinary measures to ensure that the animals to be released are quarantined in a facility that is isolated from other animals. A breeding colony should be self-contained; once it is established, no outside animals should be introduced, and no equipment or containers should be moved in or out. Even the clothing of animal caretakers can carry pathogens, and a dressing room must be provided so caretakers can wash and change their clothes when they enter or leave. These precautions are time-consuming and expensive, but neglecting them can be disastrous.

is estimated conservatively to be a week, those two markets would consume more than a quarter of a million turtles annually. When that rate is extrapolated to all of the markets in China, the estimate rises to 12 million turtles sold annually in China alone (Altherr and Freyer 2000).

Turtles are long-lived animals with low reproductive rates, and those are exactly the wrong characteristics to withstand heavy predation. Very little is known about the natural history of Chinese turtles. In fact, some species such as *Cuora mccordi* are known scientifically only from specimens purchased in markets—wild populations have never been described. These species may never be known in the wild; specimens have not been seen in markets for several years and the species may be extinct.

As populations of turtles in China have been depleted, turtles have increasingly been imported from other countries in Southeast Asia. China is a huge country and movement within the country is restricted, especially for foreigners, so complete data are not available. Records for Hong Kong show that importation of turtles for food rose from 139,200 kg in 1977 to 1,800,024 kg in just the first 10 months of 1994. Vietnam is estimated to send between 1600 and 16,000 kg of turtles to China *every day!* No effective control of this trade is in place. Species protected by national laws and international conventions are included, and turtles are collected from nature reserves.

Although China is considered the biggest black hole for turtles (Behler 2000), it is by no means alone. Madagascar is home to endangered species of many other animals, including tortoises. Although these species are protected by international treaty, they are smuggled out of the country by the score for sale as pets in Japan, Europe, and North America where they fetch high prices. In the United States our own protected species are collected and sold illegally as pets, and the most endangered species command the highest prices. In addition, all species of turtles face threats ranging from death on roads and loss of habitat to being shot for target practice as they bask on logs. It is no exaggeration to say that "turtles are in terrible trouble" (Rhodin 2000). The conservation crisis for turtles is a severe as that facing amphibians.

Summary

The earliest turtles known, fossils from the Triassic, have nearly all the features of derived turtles. The first Triassic forms were not able to withdraw their heads into the shell, but this ability appeared in the two major lineages of living turtles, which were established by the Late Triassic. The cryptodiran turtles retract the head with a vertical flexion of the neck vertebrae, whereas the pleurodires use a sideward bend.

Turtles are among the most morphologically specialized vertebrates. The shell is formed of dermal bone that is fused to the vertebral column and ribs. In most turtles the dermal shell is overlain by a horny layer of epidermal scutes. The limb girdles are inside the rib cage. Breathing presents special difficulties for an animal that is encased in a rigid shell: Exhalation is accomplished by muscles that squeeze the viscera against the lungs, and inhalation is accomplished by muscles that increase the volume of the visceral cavity, thereby allowing the lungs to expand. The heart of turtles (and of lepidosaurs as well) is able to shift blood between the pulmonary and systemic circuits in response to the changing requirements of gas exchange and thermoregulation.

The social behavior of turtles includes visual, tactile, and olfactory signals used in courtship. Dominance hierarchies shape the feeding, resting, and mating behaviors of some of the large species of tortoises. All species of turtles lay eggs, and none provides parental care to their young. Coordinated activity by hatchling sea turtles may be necessary to enable them to dig themselves out of the nest, and simultaneous emergence of baby sea turtles from their nests helps them to evade predators as they rush down the beach into the ocean. Sea turtles migrate tens, hundreds, and even thousands of kilometers between their feeding areas and their nesting beaches, and use a large variety of cues for navigation.

The life history of many turtles makes them vulnerable to extinction. Slow rates of growth and long periods required to reach maturity are characteristic of turtles in general and of large species of turtles in particular. Turtles cannot withstand commercial exploitation of the sort that has occurred historically on many oceanic islands and is currently in progress in parts of Asia.

Additional Readings

Altherr, A. and D. Freyer. 2000. Asian turtles are threatened by extinction. *Turtle and Tortoise Newsletter.* No. 1:7–11.

Behler, J. 2000. Letter from the IUCN tortoise and freshwater turtle specialist group. *Turtle and Tortoise Newsletter.* No. 1:4–5.

Berry, K. 1989. *Gopherus agassizi,* desert tortoise. Pages 5–7 in The conservation biology of tortoises. *Occasional Papers of the IUCN Species Survival Commission,* No. 5, edited by I. R. Swingland and M. W. Klemens. IUCN, Gland, Switzerland.

Burke, V. J., N. B. Frazer, and J. W. Gibbons. 1993. Conservation of turtles: The chelonian dilemma. Pages 35–38 in *Proceedings of the 13th Annual Symposium on Sea Turtle Biology and Conservation.* Jekyll Island, Ga.: U.S. Department of Commerce, National Oceanic and Atmospheric Administration.

Chelazzi, G., and R. Calzolai. 1986. Thermal benefits from familiarity with the environment in a reptile. *Oecologia* 68:557–558.

Congdon, J. D., A. E. Dunham, and R. C. Van Loben Sels. 1994. Demographics of common snapping turtles (*Chelydra serpentina*): Implications for conservation and management of long-lived organisms. *American Zoologist* 34:397–408.

Ewert, M. A., and C. E. Nelson. 1991. Sex determination in turtles: Diverse patterns and some possible adaptive values. *Copeia* 1991:50–69.

Gibbons, J. W. (Ed.). 1990. *Life History and Ecology of the Slider Turtle.* Washington, D.C.: Smithsonian Institution Press.

Herbst, L. H. 1994. Fibropapillomatosis of marine turtles. *Annual Review of Fish Diseases* 4:389–425.

Hicks, J. W. et al. 1996. The mechanism of cardiac shunting in reptiles: A new synthesis. *Journal of Experimental Biology* 199:1435–1446.

Jacobson, E. R. 1993. Implications of infectious diseases for captive propagation and introduction programs of threatened/endangered reptiles. *Journal of Zoo and Wildlife Medicine* 24:245–255.

Jacobson, E. R. et al. 1995. Mycoplasmosis and the desert tortoise (*Gopherus agassizii*) in Las Vegas Valley, Nevada. *Chelonian Conservation and Biology* 1:279–284.

Jacobson, E. R., et al. 1991. Chronic upper respiratory disease of free-ranging desert tortoises (*Xerobates agassizi*). *Journal of Wildlife Disease* 27:296–316.

Kennett, R. K. Christian, and G. Bedford. 1998. Underwater nesting by the Australian freshwater turtle *Chelodina rugosa:* Effects of prolonged immersion and eggshell thickness on incubation period, egg survivorship, and hatchling size. *Canadian Journal of Zoology* 76:1–5.

Klemens, M. W. (Ed.) 2000. *Turtle Conservation.* Washington, D.C.: Smithsonian Institution Press.

Lee, M. S. Y. 1993. The origin of the turtle body plan: bridging a famous morphological gap. *Science* 261:1716–1720.

Lohmann, K. J. 1991. Magnetic orientation by hatchling loggerhead sea turtles. *Journal of Experimental Biology* 155:37–49.

Luschi, P. et al. 1998. The navigational feats of green sea turtles migrating from Ascension Island investigated by satellite telemetry. *Proceedings of the Royal Society* (London) Series B. 265:2279–2284.

Lutz, P. L., and J. A. Musick (Eds.). 1997. *The Biology of Sea Turtles.* Boca Raton, Fl.: CRC Press.

Merlen, G. 1999. *Restoring the Tortoise Dynasty: The Decline and Recovery of the Galapagos Giant Tortoise.* Quito, Ecuador: Charles Darwin Foundation.

Miller, K., G. C. Packard, and M. J. Packard. 1987. Hydric conditions during incubation influence locomotor performance of hatchling snapping turtles. *Journal of Experimental Biology* 127:401–412.

Morreale, S. J. et al. 1996. Migration corridor for sea turtles. *Nature* 384:319–320.

Packard, G. C., and M. J. Packard. 1988. Physiological ecology of reptile eggs. In C. Gans and R. B. Huey (Eds.), *Biology of the Reptilia,* Vol. 16. Philadelphia, Pa.: Alan Liss, pp. 523–605.

Paladino, F. V., M. P. O'Connor, and J. R. Spotila. 1990. Metabolism of leatherback turtles, gigantothermy, and thermoregulation of dinosaurs. *Nature* 344:858–860.

Rhodin, A. G. J. 2000. Publishers Editorial: Turtle Survival Crisis. *Turtle and Tortoise Newsletter.* No. 1:2–3.

Rieppel, O., and M. de Braga. 1996. Turtles as diapsid reptiles. *Nature* 384:453–455.

Rieppel, O. R. and R. Reisz. 1999. The origin and early evolution of turtles. *Annual Review of Ecology & Systematics* 30:1–22.

Reisz, R. R., and M. Laurin. 1991. *Owenetta* and the origin of turtles. *Nature* 349:324–326.

Swingland, I. R., and J. G. Frazier. 1979. The conflict between feeding and overheating in the Aldabran giant tortoise. Pages 611–615 in *A Handbook on Biotelemetry and Radio Tracking,* edited by C. J. Amlaner Jr., and D. W. MacDonald. Pergamon, Oxford, UK.

Swingland, I. R., and M. W. Klemens (Eds.). 1989. The conservation biology of tortoises. *Occasional Papers of the IUCN Species Survival Commission,* No. 5. Gland, Switzerland: IUCN.

Swingland, I. R., and C. M. Lessells. 1979. The natural regulation of giant tortoise populations on Aldabra Atoll. Movement polymorphism, reproductive success and mortality. *Journal of Animal Ecology* 48:639–654.

van Dijk, P. P., B. L. Stuart, and A. G. J. Rhodin (eds). 2000. Asian turtle trade. *Chelonian Research Monographs,* No. 2, Chelonian Research Foundation, Lunenburg, MA.

Wang, T., E. H. Krosniunas, and J. W. Hicks. 1997. The role of cardiac shunts in the regulation of arterial blood gases. *American Zoologist* 37:12–22.

Web Explorations

On-line resources for this chapter are on the World Wide Web at: **http://www.prenhall.com/pough** (click on the Table of Contents link and then select Chapter 10).

CHAPTER

11 The Lepidosaurs: Tuatara, Lizards, and Snakes

We note in our discussion of the fauna of the Mesozoic in Chapter 14 that many of the animals living then would not look strange in the modern world. Mammals, birds, and lepidosaurs (tuatara, lizards, and snakes) all originated and diversified in the Mesozoic, and their overshadowing by nonavian dinosaurs is as much a matter of our perception as of biological reality. Lizards and especially snakes are elements of the diapsid reptilian lineage that have had their greatest flowering in the Cenozoic, and the extant species of lepidosaurs outnumber the species of mammals.

Some aspects of the biology of lepidosaurs may give us an impression of the ancestral way of life of amniotic tetrapods, although many of the structural characteristics of lizards and snakes are derived. One derived characteristic of lizards is determinate growth; that is, increase in body size stops when the growth centers of the long bones ossify. This mechanism sets an upper limit to the size of individuals of a species, and may be related to the specialization of most lizards as predators of insects. The predatory behavior of lizards ranges from sitting in one place and ambushing prey to seeking food by traversing a home range in an active, purposeful way. Broad aspects of the biology of lizards are correlated with these foraging modes, including morphology, exercise physiology, reproductive mode, defense against predators, and social behavior. The anatomical specializations of snakes are associated with their elongate body form and include modifications of the jaws and skull that allow them to subdue and swallow large prey.

11.1 The Lepidosaurs

Lepidosaurs are the largest group of nonavian reptiles, containing more than 4000 species of lizards and 2700 species of snakes in addition to the two species of tuatara (Table 11.1). Lepidosaurs are predominantly terrestrial tetrapods with some secondarily aquatic species, especially among snakes. The skin of lepidosaurs is covered by scales and relatively impermeable to water. The outer layer of the epidermis is shed at intervals. Tuatara and most lizards have four limbs, but reduction or complete loss of limbs is widespread among lizards, and all snakes are limbless. Lepidosaurs have a transverse cloacal slit, rather than the longitudinal slit that characterizes other tetrapods.

Lepidosaurs are the sister lineage of archosaurs (crocodilians and birds). Within the Lepidosauria, the Sphenodontidae (tuatara) is the sister group of Squamata (lizards and snakes). Within the squamates, lizards can be distinguished from snakes in colloquial terms but not phylogenetically, because snakes are derived from lizards. Thus "lizards" is a paraphyletic group (i.e., one that does not include all the descendants of a common ancestor). Nonetheless, lizards and snakes are distinct in many aspects of their ecology and behavior, and a colloquial separation is useful in discussing them.

11.2 Radiation of Sphenodontids and the Biology of Tuatara

The sphenodontids were a diverse group in the Mesozoic. Triassic forms were small, with body lengths of only 15 to 35 centimeters. Most Triassic sphenodontids had teeth that were fused to the top edges of the jaw bones (**acrodont**) like the teeth of

Table 11.1 Lineages of extant lepidosaurs

Sphenodontidae: The sister group of squamates. Two species of tuatara, now restricted to islands off the coast of New Zealand.

Iguania

Iguanidae: More than 800 species of lizards, primarily found in the New World but with representatives on the islands of Fiji and Madagascar. See Table 11.2.

"Agamidae:" About 300 species of small to large (10 cm to 1m) lizards that extend through the Middle East into Africa and along the Indoaustralian archipelago into Australia.

Chamaeleonidae: 130 species of primarily arboreal lizards, but including a few grassland and terrestrial species, found in Africa and Madagascar and extending into southern Spain and along the west coast of the Mediterranean. Chameleons have the laterally flattened bodies characteristic of many arboreal lizards, and additional specializations including zygodactyl feet, prehensile tails, eyes that swivel to provide a 360-degree field of view, and a projectile tongue. The leaf chameleons (*Brookesia, Rhampholeon*) are as small as 25 mm, whereas some species of *Chamaeleo* grow to more than 60 cm. (Note that the family name and the genus are spelled with an *ae,* but the common name *chameleon* is spelled with an *e*.)

Scleroglossa

Gekkonidae: About 870 species of geckos with modified scales (setae) on the bottoms of the toes that allow them to climb vertical surfaces and even to hang by a single toe. The geographic distribution of gekkonids includes every continent except Antarctica. Extant species range in size from very small (30 mm) to medium size (30 cm). Eublepharines are about 25 species of small to medium-size terrestrial geckos, lacking the modifications of the toes that allow gekkonids to climb. Their eyes have moveable lids. Their geographic distribution is disjunct, extending from the southwestern United States through Central America and including species in Africa and Asia. Pygopodines are 34 species of elongate, nearly limbless terrestrial lizards found only in Australia.

Amphisbaenidae: 133 species of round-headed, spade-snouted, and keel-headed amphisbaenians found in the West Indies, South America, sub-Saharan Africa, and around the Mediterranean Sea.

? Trogonophiidae: 6 species of round-headed amphisbaenians that use an oscillating movement of the head in digging. Trogonophiids are found in north Africa and the Middle East.

Rhineuridae: 1 species, a spade-snouted amphisbaenian, found in Florida.

Bipedidae: 3 species of round-headed amphisbaenians (*Bipes*) found in Mexico. This lineage is unique among amphisbaenians in retaining well-developed forelimbs that are used to help penetrate the soil surface. Once underground, *Bipes* use the head for burrowing.

? Dibamidae: 2 genera of small to medium (5–25 cm) limbless, burrowing lizards. *Dibamus* (9 species) is found in Indomalaysia, whereas *Anelytropsis* (1 species) is from Mexico.

? Serpentes (see Table 11.3).

Gymnophthalmidae: About 145 species of small (less than 6 cm) lizards that live in the leaf litter of Neotropical forests. Limb reduction is widespread in this lineage.

Teiidae: About 105 species of active terrestrial lizards ranging from near the Canadian border in North America to central Argentina, and including the West Indies. Teids range from small insectivorous species of *Ameiva, Cnemidophorus,* and *Kentropyx* to species such as the tegus (*Tupinambis*) and the caiman lizard (*Dracena*), which can grow to a meter or more.

Lacertidae: About 215 species of small to medium-size terrestrial lizards from the Old World with a range of body forms very like those seen among teiids. Lacertids are found over all of Europe, Africa, and Asia. Like the teiids, lacertids are mostly terrestrial, and include some large predators as well as a great number of smaller species.

Xantusiidae: 17 species of tiny to small (3–10 cm), secretive lizards from southwestern North America, Central America, and Cuba. All have the eyelids fused into a transparent scale that covers the eye and gives the group its common name, spectacled lizards.

Table 11.1 Lineages of extant lepidosaurs *(continued)*

Scincidae: Some 1100 species make this one of the most species-rich lineages of lepidosaurs. Skinks occur on all continents except Antarctica. Nearly all are very small to medium-size (5–20 cm) terrestrial lizards, and many show limb reduction. Most are insectivorous, but some of the large (40 cm) species of the Australian blue-tongued skinks *(Tiliqua)* consume plant material, and the Solomon Islands arboreal skink (*Corucia zebrata*) has specializations of the gut that promote fermentative digestion of plant material by symbiotic microorganisms.

Cordylidae: 42 species of small to medium-size terrestrial or rock-dwelling lizards from sub-Saharan Africa. *Cordylus* is a rock dweller and has the dorsoventrally flattened body typical of lizards that seek shelter in crevices. It is heavily armored, and many species have exceedingly sharp spines along the body margins and the tail. Skull kinesis is discussed in the text in the context of feeding mechanisms employed by lizards, but *Cordylus* uses skull kinesis in a different way: By contracting its jaw muscles, *Cordylus* raises the braincase relative to the lower jaw, and it uses this mechanism to wedge itself into crevices. *Platysaurus* is a small rock dweller without the heavy body armor of *Cordylus,* and *Chamaesaura* is an elongate, nearly limbless lizard. Most cordylids are viviparous, except for *Platysaurus,* which lays eggs.

Gerrhosauridae: The 30 species of gerrhosaurids are found in sub-Saharan Africa and on Madagascar. This lineage shows parallels to the cordylids: *Gerrhosaurus* has heavy body armor, and *Tetradactylus* is elongate and has reduced limbs. *Angolosaurus skoogi* lives in shifting sand dunes in the Namib Desert, where it eats seeds that blow in from outside the dunes and buries itself in sand to avoid extreme temperatures. Gerrhosaurids are egg layers.

Anguidae: The 102 species of anguids occur in North and South America and in Europe, the Middle East, and southern China. All anguids have body armor and a fold along the side of the body that allows the trunk to expand and contract as they breathe. Most are terrestrial, foraging in leaf litter, and four genera are legless: *Anniella* (California and Baja California) is a small species (10–20 cm) that spends the day underground and emerges at night to hunt on the surface. *Anguis* (Europe), *Ophisaurus* (North America, Europe, and Asia), and *Ophioides* (South America) are limbless surface dwellers that forage in leaf litter and dense vegetation. *Anguis* is small, but *Ophisaurus* grows to a trunk length of 0.5m, with a tail at least as long as its trunk.

Xenosauridae: This small family includes 4 species of *Xenosaurus,* medium-size terrestrial lizards that live in the moist leaf litter of high-altitude cloud forests in Mexico.

Shinisauridae: A single species of *Shinisaurus,* a semiaquatic lizard from China that lives along stream banks and forages in the water, catching fishes and aquatic invertebrates.

Varanidae: The 40 species in this group range in size from *Varanus brevicauda,* which has a total length of only 10 cm, to the Komodo dragon, *V. komodoensis,* which can exceed 3m in length and may weigh 75 kg. All monitors are active predators, and the larger species patrol a large home range searching for food in holes and beneath rocks and logs. Some species are moderately arboreal. Varanids occur in Africa, Asia, and the East Indies, but about half the species are limited to Australia.

Lanthanotidae: The single species of *Lanthanotus,* the Bornean earless monitor, is an elongate lizard with a trunk 15–20 cm long. It is secretive and semiaquatic, spending the day in a burrow and foraging at night on both land and water.

Helodermatidae: The 2 species of helodermatids are the only poisonous lizards. These are large (25- to 40-cm trunk length), heavy-bodied lizards found in the southwestern United States and Mexico.

extant tuatara (*Sphenodon*), but others had teeth attached to the inner sides of the jaw bones (**pleurodont**), like those of some lizards. The tooth structure of Triassic sphenodontids suggests that the lineage included both insectivores and herbivores. Some sphenodontids from the Jurassic and Cretaceous were small terrestrial and arboreal forms, and others were marine animals as much as 1.5 meters long.

The two species of *Sphenodon,* known as tuatara, are the only extant sphenodontids. (*Tuatara* is a Maori word meaning "spines on the back," and no *s* is added to form the plural.) Tuatara formerly inhabited the North and South islands of New Zealand, but the advent of humans and their associates (cats, dogs, rats, sheep, and goats) exterminated tuatara on the mainland. Now populations are found on only about 30 small islands off the coast.

Tuatara have been fully protected in New Zealand since 1895, but only one species, *Sphenodon punctatus,* was recognized. In fact there is a second species of tuatara, *S. guentheri,* which was described in 1877. Because *S. guentheri* is much less common than *S. punctatus,* it was overlooked when laws protecting tuatara were written. As a result, *S. guentheri* did not receive the special protection that it needed. The only surviving population of *S. guentheri* is a group of fewer than 300 adults living on 1.7 hectares of scrub on the top of North Brother Island. These animals were regarded as not very important from a conservation perspective compared with the large populations of *S. punctatus* on some of the other islands. The presence of a lighthouse probably saved these tuatara: It was staffed until 1990 by resident keepers who deterred illegal landings and poaching. The East Island population of *S. guentheri* (the only other population of the species) became extinct during this century. This example illustrates the crucial role that taxonomy plays in conservation—a species must be recognized before it can be protected.

Adult tuatara are about 60 cm long. They are nocturnal, and in the cool, foggy nights that characterize their island habitats they cannot raise their body temperatures during activity by basking in the sun as lizards do. Body temperatures from 6 to 16°C have been reported for active tuatara, and these are low compared with most lizards. During the day, tuatara do bask in the sun and raise their body temperatures to 28°C or higher. Tuatara feed largely on invertebrates, with an occasional frog, lizard, or seabird for variety. The jaws and teeth of tuatara produce a shearing effect during chewing: The upper jaw contains two rows of teeth, one on the maxilla and the other on the palatine bones. The teeth of the lower jaw fit between the two rows of upper teeth, and the lower jaw closes with an initial vertical movement, followed by an anterior sliding movement. As the lower jaw slides, the food item is bent or sheared between the triangular cusps on the teeth of the upper and lower jaws.

Tuatara live in burrows that they may share with nesting seabirds. The burrows are spaced at intervals of 2 to 3m in dense colonies, and both male and female tuatara are territorial. They use vocalizations, behavioral displays, and color change in their social interactions.

The ecology of tuatara rests to a large extent on exploitation of the resources provided by colonies of seabirds. Tuatara feed on the birds, which are most vulnerable to predation at night. In addition, the quantities of guano produced by the birds, scraps of the food they bring to their nestlings, and the bodies of dead nestlings attract huge numbers of arthropods that are eaten by tuatara. These arthropods are largely nocturnal and must be hunted when they are active. Thus the crepuscular (occurring at dusk and dawn) activity of tuatara and the low body temperatures resulting from being active at those times of day are probably specializations that stem from the association of tuatara with colonies of nesting seabirds. This pattern of behavior and thermoregulation probably does not represent the ancestral condition even for sphenodontids, and there is no reason to interpret it as being ancestral for lepidosaurs or diapsids.

11.3 Radiation of Squamates

Determinate growth may be the most significant derived character of squamates. Growth occurs as cells proliferate in the cartilaginous epiphyseal plates at the ends of long bones, continues while the epiphyseal plates are composed of cartilage, and stops completely when the epiphyses fuse to the shafts of the bones, obliterating the cartilaginous plates. Determinate growth of this sort is characteristic of squamates (and also of birds and mammals). Crocodilians and turtles continue to grow all through their lives, although the growth rates of adults are much slower than those of juveniles. Development of determinate growth in squamates may initially have been associated with the insectivorous diet that researchers believe was characteristic of early lepidosaurs. Lizard-size animals can prey on insects without requiring the sorts of morphological or ecological specializations that are necessary for large insect-eating vertebrates such as mammalian anteaters.

The fossil record of lizards is largely incomplete through the middle of the Mesozoic, but Late Jurassic deposits in China and Europe include members of most lineages of extant lizards. The major groups of lizards had probably diverged by the end of the Jurassic.

The phylogenetic relationships of squamates are well understood in general, but disagreement surrounds some details. Two major lineages are recognized, Iguania and the Scleroglossa. Iguanians include the sister lineages "Agamidae" (which is not considered monophyletic) and Chamaeleonidae, as well as the large family Iguanidae (Table 11.2).

Scleroglossans include two large families (geckos and skinks) and several smaller families. The amphisbaenians (specialized burrowing lizards) and Serpentes (snakes) are nested within the Scleroglossa. Both groups have morphological and ecological specializations that make it convenient to discuss them individually, but their evolutionary status as lineages within the scleroglossan squamates must be remembered.

Lizards

The approximately 4000 species of lizards range in size from diminutive geckos only 3 cm long to the Komodo monitor lizard, which is 3m long at matu-

Table 11.2 Lineages of Iguanid lizards

Corytophaninae: 9 species of small to medium-size (10–20 cm) arboreal Neotropical lizards. They have long tails, laterally flattened bodies, and some (*Corytophanes, Laemanctus*) have crests on their heads that may make it hard for predators to recognize them as lizards when they are seen as silhouettes against a patch of light.

Crotaphytinae: The collared and leopard lizards; 12 species of medium-size (10–15 cm) North American desert lizards that prey on other lizards.

Hoplocercinae: 12 species of medium-size (up to 16 cm) terrestrial lizards with a geographic range extending from Panama to Brazil.

Iguaninae: 34 species of large (to >1.5m) terrestrial and arboreal herbivorous lizards. The Neotropical green iguana (*Iguana iguana*) is the most familiar member of the family, which also includes the black and ground iguanas of Central America and the West Indies, the Galápagos marine and land iguanas, and the Fijian iguanas. The North American chuckwalla (*Sauromalus*) is an iguanid. This rock-dwelling lizard is dorsoventrally flattened, and seeks shelter from predators in rock crevices, where it inflates its lungs to wedge itself in place.

Oplurinae: 7 species of medium-size (20–40 cm), rock-dwelling and arboreal lizards from Madagascar.

Phrynosomatinae: About 125 species of North and Central American lizards. The group takes its name from the horned lizards (*Phrynosoma*), but more than half the species are in the genus *Sceloporus* (spiny swifts), which range from southern Canada to Panama. Most phrynosomatids are terrestrial, and some are specialized to enter rock crevices (dorsoventrally flattened) or to live on loose sand (fringes of scales on the toes, valvular nostrils that exclude sand, earless or with skin folds and elongated scales that prevent sand grains from entering the ears). Some species of *Sceloporus* and *Urosaurus* are arboreal.

Polychrotinae: More than 440 species of primarily small (10 cm) South American lizards, nearly 400 of them in the genus *Anolis. Anolis carolinensis,* the green anole, has a geographic range that extends northward to North Carolina. Most species of polychrotids live in trees, bushes, or grass clumps.

"Tropidurinae:" More than 200 species of small to medium-size (10–20 cm) South American lizards that occupy regions from sea level to the high Andes, including deserts, rain forests, and grasslands.

rity and weighs some 75 kilograms. A reconstruction of the skeleton of a fossil monitor lizard, *Megalania prisca,* from the Pleistocene of Australia, is 5.5m long, and in life the lizard may have weighed more than 1000 kg. About 80 percent of extant lizards weigh less than 20 grams as adults and are insectivorous. Spiny swifts and japalures (Figure 11–1a and b) are examples of these small, generalized insectivores. Other small lizards have specialized diets: The North American horned lizards and the Australian thorny devil (Figure 11–1e and f) feed on ants. Most geckos (Figure 11–1g) are nocturnal, and many species are closely associated with human habitations.

Lizards are adaptable animals that have occupied habitats ranging from swamp to desert and even above the timberline in some mountains. Many species are arboreal. The most specialized of these are frequently laterally flattened, and they often have peculiar projections from the skull and back that help to obscure their outline. The Old World chameleons (Chamaeleonidae) are the most specialized arboreal lizards (Figure 11–1h). Their **zygodactylous** (Greek *zygo* = joined and *dactyl* = digit) feet grasp branches firmly, and additional security is provided by a prehensile tail. The tongue and hyoid apparatus are specialized, and the tongue can be projected forward more than a body's length to capture insects that adhere to its sticky tip. This feeding mechanism requires good eyesight, especially the ability to gauge distances accurately so that the correct trajectory can be employed. The chameleon's eyes are elevated in small cones and can move independently. When the lizard is at rest the eyes swivel back and forth, viewing its surroundings. When it spots an insect, the lizard fixes both eyes on it and cautiously stalks to within shooting range.

Most large lizards are herbivores. Many iguanas (family Iguanidae) are arboreal inhabitants of the tropics of Central and South America. Large terrestrial iguanas occur on islands in the West Indies and the Galápagos Islands, probably because the absence of predators has allowed them to spend a large part of their time on the ground. Smaller terrestrial herbivores like the black iguanas (Figure 11–1c) live on the mainland of Mexico and Central America, and still smaller relatives such as the chuckwallas and desert iguanas range as far north as the western United States. Many species of lizards live on beaches, but few extant species actually enter the water. The marine iguana of the Galápagos Islands is an exception. The feeding habits of the marine iguana are unique. It feeds on seaweed, diving 10m or more to browse on algae growing below the tide mark.

An exception to the rule of herbivorous diets for large lizards is found in the monitor lizards (family Varanidae). Varanids are active predators that feed on a variety of vertebrate and invertebrate animals, including birds and mammals (Figure 11–1j). Few lizards are capable of capturing and subduing such prey, but varanids have morphological and physiological characteristics that make them effective predators of vertebrates. The Komodo monitor lizard is capable of killing adult water buffalo, but its normal prey is deer and feral goats. Large monitor lizards were widely distributed on the islands between Australia and Indonesia during the Pleistocene and may have preyed on pygmy elephants that also lived on the islands.

The hunting methods of the Komodo monitor are very similar to those employed by mammalian carnivores, showing that a simple brain is capable of complex behavior and learning. In the late morning a Komodo monitor waits in ambush beside the trails deer use to move from the hilltops, where they rest during the morning, to the valleys, where they sleep during the afternoon. The lizards, familiar with trails used by the deer, often wait where several deer trails converge. If no deer pass the lizard's ambush it moves into the valleys, systematically stalking the thickets where deer are likely to be found. This purposeful hunting behavior, which demonstrates familiarity with the behavior of prey and with local geography, is in strong contrast to the opportunistic seizure of prey that characterizes the behavior of many lizards; but it is very like the hunting behavior of some snakes (Greene 1997).

Limb reduction has evolved repeatedly among lizards, perhaps as many as 62 times, and every continent has one or more families with legless, or nearly legless, species (Figure 11–1i). Leglessness in lizards is usually associated with life in dense grass or shrubbery in which a slim, elongate body can maneuver more easily than a short one with functional legs. Some legless lizards crawl into small openings among rocks and under logs, and a few are subterranean.

The amphisbaenians include about 140 species of extremely fossorial lizards (Latin *fossor* = a digger), which have specializations that are different from those of other squamates (Figure 11–2). The earliest amphisbaenian known is a fossil from the Late Cre-

▲ Figure 11–1 Variation in body form among lizards. Small, generalized insectivores: (a) spiny swift, *Sceloporus* (Phrynosomatinae, Iguanidae); (b) japalure, *Calotes* ("Agamidae"). Herbivores: (c) black iguana, *Ctenosaura* (Iguaninae, Iguanidae); (d) mastigure, *Uromastyx* ("Agamidae"). Ant specialists: (e) horned lizard, *Phrynosoma* (Phrynosomatinae, Iguanidae); (f) spiny devil, *Moloch* ("Agamidae"). Nocturnal lizard: (g) Tokay gecko, *Gekko* (Gekkonidae). Arboreal lizard: (h) African chameleon, *Chamaeleo* (Chamaeleonidae). Legless lizard: (i) North American glass lizard, *Ophisaurus* (Anguidae). Large predator: (j) monitor lizard, *Varanus* (Varanidae).

taceous. Most amphisbaenians are legless, but the three species in the Mexican genus *Bipes* have well-developed fore legs that they use to assist entry into the soil but not for burrowing underground. The skulls of amphisbaenians are used for tunneling, and they are rigidly constructed. The dental structure is also distinctive: Amphisbaenians possess a single median tooth in the upper jaw—a feature unique to this group of vertebrates. The median tooth is part of a specialized dental battery that makes amphisbaenians formidable predators, capable of subduing a wide variety of invertebrates and small vertebrates. The upper tooth fits into the space between two teeth in the lower jaw and forms a set of nippers that can bite out a piece of tissue from a prey item too large for the mouth to engulf as a whole.

The skin of amphisbaenians also is distinctive. The **annuli** (rings) that pass around the circumference of the body are readily apparent from external examination, and dissection shows that the integument is nearly free of connections to the trunk. Thus, it forms a tube within which the amphisbaenian's body can slide forward or backward. The separation of trunk and skin is employed during the concertina locomotion that all amphisbaenians use underground. Integumentary muscles run longitudinally from annulus to annulus. The skin over this area of muscular contraction is then telescoped and buckles outward, anchoring that part of the amphisbaenian against the walls of its tunnel. Next, contraction of muscles that pass anteriorly from the vertebrae and ribs to the skin slide the trunk forward within the

▲ Figure 11–1 *Continued*

tube of integument. Amphisbaenians can move backward along their tunnels with the same mechanism by contracting muscles that pass posteriorly from the ribs to the skin. (The name amphisbaenian is derived from the Greek roots [*amphi* = double, *baen* = walk] in reference to the ability of amphisbaenians to move forward and backward with equal facility.) A similar type of rectilinear (straight-line) locomotion is used by some heavy-bodied snakes. But the telescoping ability of the skin of snakes is generally restricted to the lateroventral portions of the body, whereas in amphisbaenians the skin is loose around the entire circumference of the body.

The burrowing habits of amphisbaenians make them difficult to study, but three major functional categories can be recognized. Some species have blunt heads; the rest have either vertically keeled or horizontally spade-shaped snouts. Blunt-snouted forms burrow by ramming their heads into the soil to compact it (Figure 11–2b). Sometimes an oscilla-

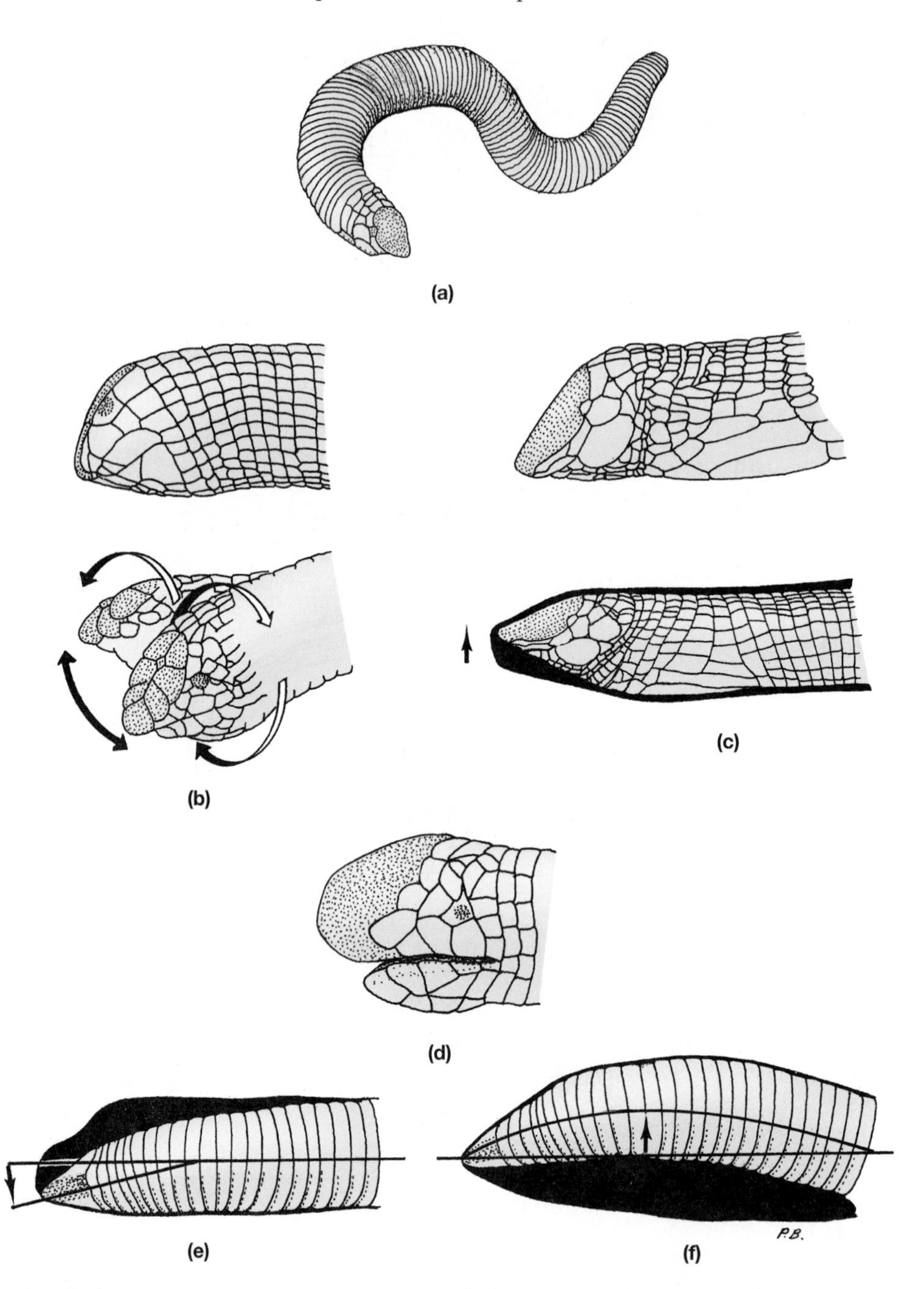

▲ Figure 11–2 Amphisbaenians. (a) *Monopeltis* from Africa. (b) Blunt-snouted (*Agamodon*, from Africa). (c) Shovel-snouted (*Rhineura* from Florida). (d) Wedge-snouted (*Anops*, Brazil). Widening the tunnel with movements of the head in loose soil (e) or with the anterior part of the body in dense soil (f).

tory rotation of the head with its heavily keratinized scales is used to shave material from the face of the tunnel. Shovel-snouted amphisbaenians ram the end of the tunnel, then lift the head to compact soil into the roof (Figure 11–2c). Wedge-snouted forms ram the snout into the end of the tunnel and then use the snout or the side of the neck to compress the material into the walls of the tunnel (Figure 11–2d, e, f). In parts of Africa representatives of the three types occur together and share the subsoil habitat. The unspecialized blunt-headed forms live near the surface where the soil is relatively easy to tunnel through, and the specialized forms live in deeper, more compact soil. The geographic range of unspecialized forms is greater than that of specialized ones, and in areas where only a single species of amphisbaenian occurs, it is a blunt-headed species.

The relationship between unspecialized and specialized burrowers is puzzling. One would expect that the specialized forms with their more elaborate methods of burrowing would replace the unspecialized ones, but this has not happened. The explanation may lie in the conflicting selective forces on the snout. On one hand, it is important to have a snout that will burrow through soil; but on the other hand, it is important to have a mouth capable of tackling a wide variety of prey. The specializations of the snout that make it an effective structure for burrowing appear to reduce its effectiveness for feeding. The blunt-headed amphisbaenians may be able to eat a wider variety of prey than the specialized forms can. Thus, in loose soil where it is easy to burrow, the blunt-headed forms may have an advantage. Only in soil too compact for a blunt-headed form to penetrate might the specialized forms find the balance of selective forces shifted in their favor.

Snakes

The 2700 species of snakes range in size from diminutive burrowing species, which feed on termites and grow to only 10 cm, to the large constrictors, which approach 10m in length (Table 11.3). The Scolecophidia includes three families of small burrowing snakes with shiny scales and reduced eyes. Traces of the pelvic girdle remain in most species, but the braincase is snake-like. Burrowing snakes in the families Aniliidae and Uropeltidae use their heads to dig through soil, and the bones of their skulls are solidly united. The sole xenopeltid, the sunbeam snake of southeastern Asia, is a ground-dwelling species that takes its common name from its highly iridescent scales. Boa constrictors (Boinae) are mostly New World snakes, whereas pythons (Pythoninae) are found in the Old World. The anaconda, a semiaquatic species of boa from South America, is considered the largest extant species of snake—it probably approaches a length of 10m—and the reticulated python of southeast Asia is nearly as large. Not all boas and pythons are large, however; some secretive and fossorial species are considerably less than 1m long as adults. The wart snakes in the family Acrochordidae are entirely aquatic; they lack the enlarged ventral scales that characterize most terrestrial snakes, and they have difficulty moving on land.

The Colubroidea includes most of the extant species of snakes, and the family Colubridae alone contains two-thirds of the extant species. The diversity of the group makes characterization difficult. Colubroids have lost all traces of the pelvic girdle, they have only a single carotid artery, and the skull is very kinetic. Many colubroid snakes are venomous, and snakes in the families Elapidae and Viperidae have hollow fangs at the front of the mouth that inject extremely toxic venom into their prey.

The body form of even a generalized snake such as the milk snake (Figure 11–3a) is so specialized that little further morphological specialization is associated with different habits or habitats. King snakes and milk snakes are constrictors and they crawl slowly, poking their heads under leaf litter and into holes that might shelter prey. Chemosensation is an important means of detecting prey for these snakes. Snakes have forked tongues, with widely separated tips that can move independently. When the tongue is projected, the tips are waved in the air or touched to the ground. Then the tongue is retracted, and chemical stimuli are transferred to the paired vomeronasal organs. The forked shape of the tongue of snakes (which is seen also among the Amphisbaenia, Lacertiformes, and Varanoidea) may allow them to detect gradients of chemical stimuli and localize objects.

Nonconstrictors such as the whip snakes and racers (Figure 11–3b) move quickly and are visually oriented. They forage by crawling rapidly, frequently raising the head to look around. Many arboreal snakes are extremely elongated and frequently have large eyes (Figure 11–3c). Their length distributes

Table 11.3 Lineages of extant snakes (Serpentes)

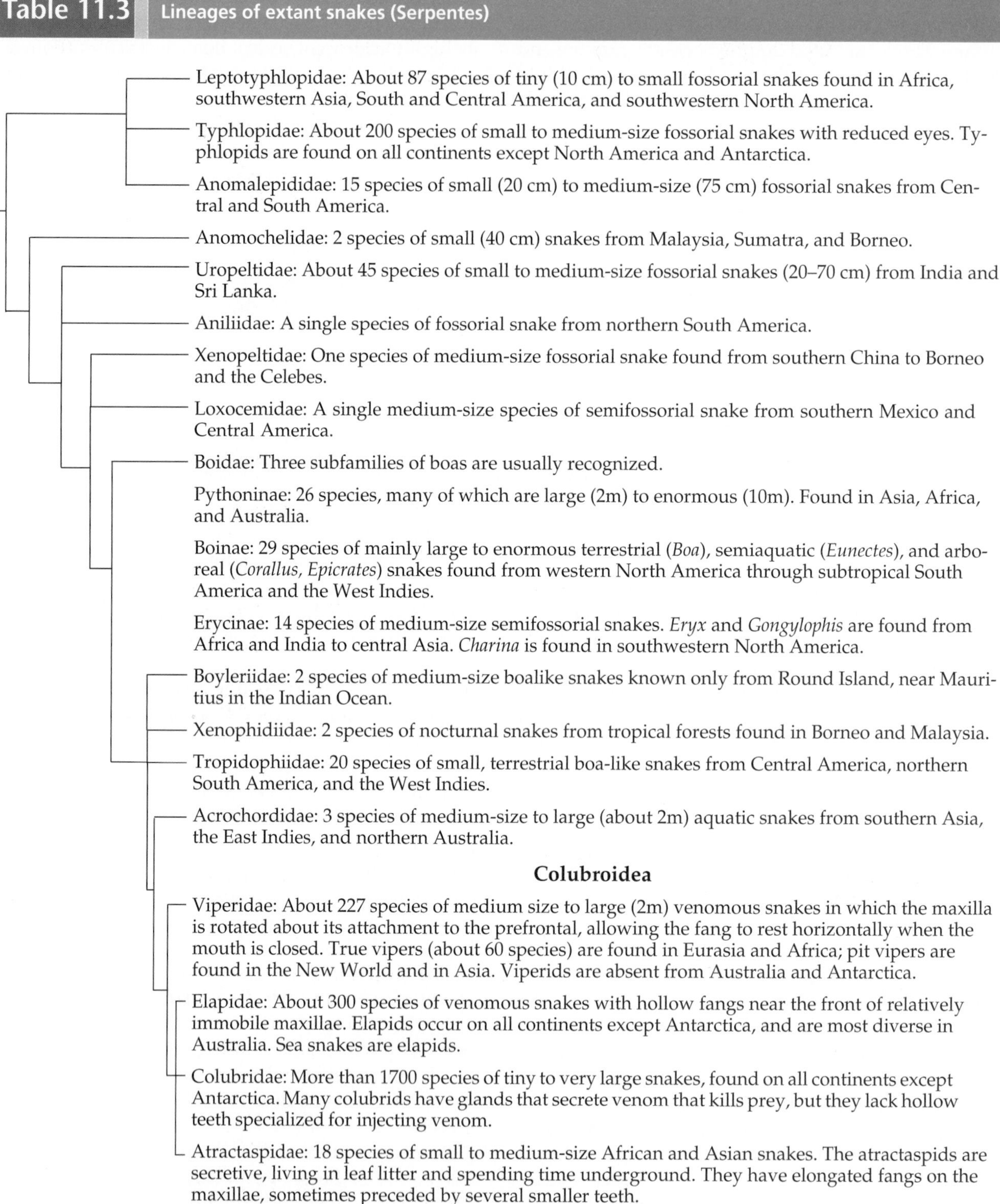

Leptotyphlopidae: About 87 species of tiny (10 cm) to small fossorial snakes found in Africa, southwestern Asia, South and Central America, and southwestern North America.

Typhlopidae: About 200 species of small to medium-size fossorial snakes with reduced eyes. Typhlopids are found on all continents except North America and Antarctica.

Anomalepididae: 15 species of small (20 cm) to medium-size (75 cm) fossorial snakes from Central and South America.

Anomochelidae: 2 species of small (40 cm) snakes from Malaysia, Sumatra, and Borneo.

Uropeltidae: About 45 species of small to medium-size fossorial snakes (20–70 cm) from India and Sri Lanka.

Aniliidae: A single species of fossorial snake from northern South America.

Xenopeltidae: One species of medium-size fossorial snake found from southern China to Borneo and the Celebes.

Loxocemidae: A single medium-size species of semifossorial snake from southern Mexico and Central America.

Boidae: Three subfamilies of boas are usually recognized.

Pythoninae: 26 species, many of which are large (2m) to enormous (10m). Found in Asia, Africa, and Australia.

Boinae: 29 species of mainly large to enormous terrestrial (*Boa*), semiaquatic (*Eunectes*), and arboreal (*Corallus, Epicrates*) snakes found from western North America through subtropical South America and the West Indies.

Erycinae: 14 species of medium-size semifossorial snakes. *Eryx* and *Gongylophis* are found from Africa and India to central Asia. *Charina* is found in southwestern North America.

Boyleriidae: 2 species of medium-size boalike snakes known only from Round Island, near Mauritius in the Indian Ocean.

Xenophidiidae: 2 species of nocturnal snakes from tropical forests found in Borneo and Malaysia.

Tropidophiidae: 20 species of small, terrestrial boa-like snakes from Central America, northern South America, and the West Indies.

Acrochordidae: 3 species of medium-size to large (about 2m) aquatic snakes from southern Asia, the East Indies, and northern Australia.

Colubroidea

Viperidae: About 227 species of medium size to large (2m) venomous snakes in which the maxilla is rotated about its attachment to the prefrontal, allowing the fang to rest horizontally when the mouth is closed. True vipers (about 60 species) are found in Eurasia and Africa; pit vipers are found in the New World and in Asia. Viperids are absent from Australia and Antarctica.

Elapidae: About 300 species of venomous snakes with hollow fangs near the front of relatively immobile maxillae. Elapids occur on all continents except Antarctica, and are most diverse in Australia. Sea snakes are elapids.

Colubridae: More than 1700 species of tiny to very large snakes, found on all continents except Antarctica. Many colubrids have glands that secrete venom that kills prey, but they lack hollow teeth specialized for injecting venom.

Atractaspidae: 18 species of small to medium-size African and Asian snakes. The atractaspids are secretive, living in leaf litter and spending time underground. They have elongated fangs on the maxillae, sometimes preceded by several smaller teeth.

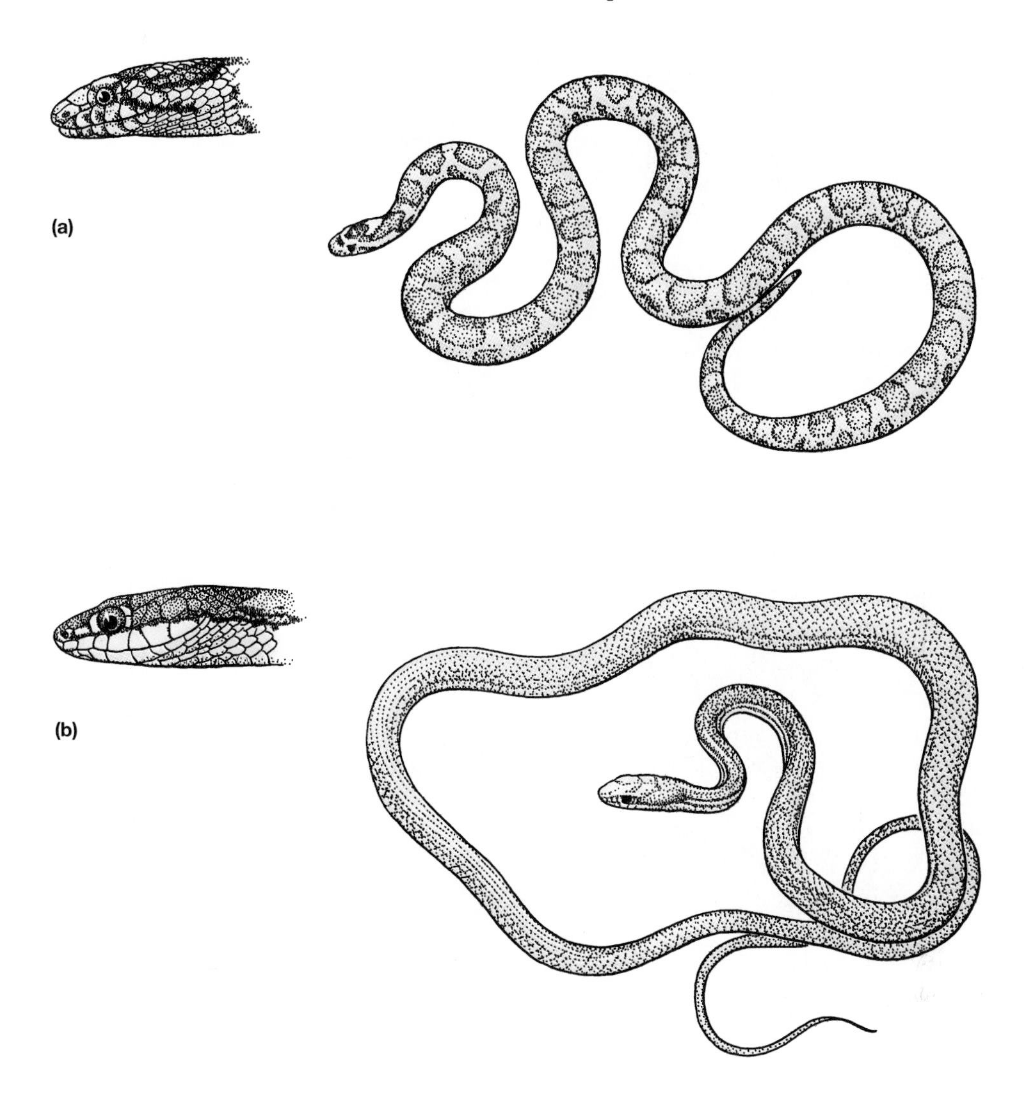

▲ Figure 11–3 Body forms of snakes. Slow-moving constrictors, such as the milk snake (a) *Lampropeltis,* are relatively short and stout. Active, visually oriented snakes such as racers, (b) *Masticophis,* are longer and faster moving. Arboreal snakes such as the parrot snake, (c) *Leptophis,* are still more elongate and can follow their prey out among the twigs at the ends of branches. Burrowing snakes such as the blind snakes, (d) *Typhlops,* have small rounded or pointed heads with little distinction between head and neck, short tails, and smooth, often shiny scales. Their eyes are greatly reduced in size. Vipers, especially the African vipers like the puff adder, (e) *Bitis arietans,* have large heads and stout bodies that accommodate large prey. Sea snakes, such as (f) *Laticauda,* have a tail that is flattened from side to side and valves that close the nostrils when they dive.

Figure continues on next page

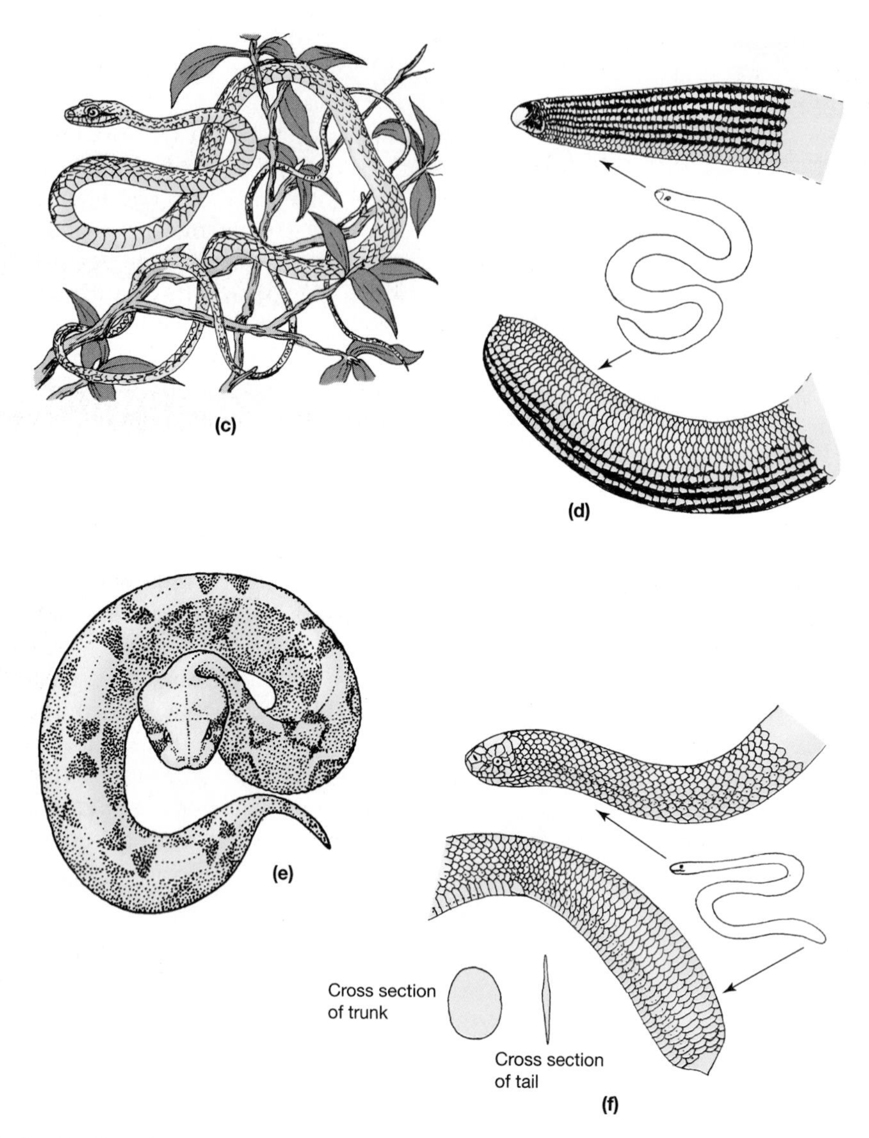

▲ Figure 11–3 *Continued*

their weight and allows them to crawl over even small twigs without breaking them. Burrowing snakes, at the opposite extreme of snake body form, are short and have blunt heads and very small eyes (Figure 11–3d). The head shape assists in penetrating soil, and a short body and tail create less friction in a burrow than would the same mass in an elongate body. Vipers, especially forms like the African puff adder (Figure 11–3e), are heavy-bodied with broad heads.

The sea snakes (Figure 11–3f) are derived from terrestrial elapids. Sea snakes are characterized by extreme morphological specialization for aquatic life: The tail is laterally flattened into an oar, the large ventral scales are reduced or absent in most species, and the nostrils are located dorsally on the snout and have valves that exclude water. The lung extends back to the cloaca and apparently has a hydrostatic role in adjusting buoyancy during diving as well as a respiratory function. Oxygen uptake through the skin during diving has been demonstrated in sea snakes. *Laticauda* are less specialized than other sea snakes, and may represent a separate radiation into the marine habitat. They retain enlarged ventral scales and emerge onto land to bask and to lay eggs. The other sea snakes are so specialized for marine life that they are helpless on land, and these species are viviparous.

The locomotor specializations of snakes reflect differences in their morphology associated with different predatory modes (discussed in the following section) and the properties of the substrates on which they move. In **lateral undulation** (also called serpentine locomotion; Figure 11–4a) the body is thrown into a series of curves. The curves may be irregular, as shown in the illustration of a snake crawling across a board dotted with fixed pegs. Each curve presses backward; the pegs against which the snake is exerting force are shown in solid color. The lines numbered 1 to 7 are at 3-inch intervals, and the position of the snake at intervals of 1 second is shown.

Rectilinear locomotion (Figure 11–4b) is used primarily by heavy-bodied snakes. Alternate sections of the ventral integument are lifted clear of the ground and pulled forward by muscles that originate on the ribs and insert on the ventral scales. The intervening sections of the body rest on the ground and support the snake's body. Waves of contraction pass from anterior to posterior, and the snake moves in a straight line. Rectilinear locomotion is slow, but it is effective even when there are no surface irregularities strong enough to resist the sideward force exerted by serpentine locomotion. Because the snake moves slowly and in a straight line it is inconspicuous, and rectilinear locomotion is used by some snakes when stalking prey.

Concertina locomotion (Figure 11–4c) is used in narrow passages such as rodent burrows that do not provide space for the broad curves of serpentine locomotion. A snake anchors the posterior part of its body by pressing several loops against the walls of the burrow and extends the front part of its body. When the snake is fully extended, it forms new loops anteriorly and anchors itself with these while it draws the rear end of its body forward.

Sidewinding locomotion (Figure 11–4d) is used primarily by snakes that live in deserts where windblown sand provides a substrate that slips away during serpentine locomotion. A sidewinding snake raises its body in loops, resting its weight on two or three points that are the only body parts in contact with the ground. The loops are swung forward through the air and placed on the ground, the points of contact moving smoothly along the body. Force is exerted downward; the lateral component of the force is so small that the snake does not slip sideward. This downward force is shown by imprints of the ventral scales in the tracks. Because the snake's body is extended nearly perpendicular to its line of

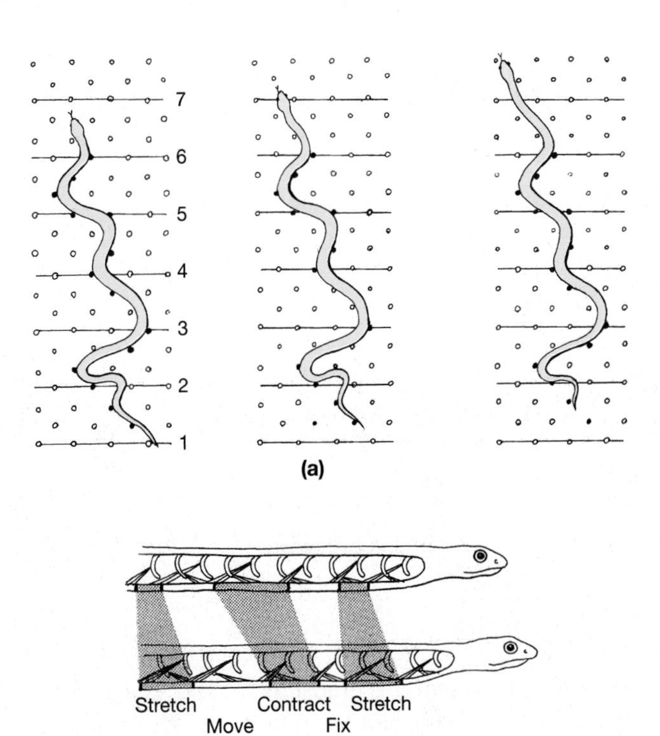

▶ Figure 11–4 Locomotion of snakes. (a) Lateral undulation. (b) Rectilinear. (c) Concertina. (d) Sidewinding.

Figure continues on next page

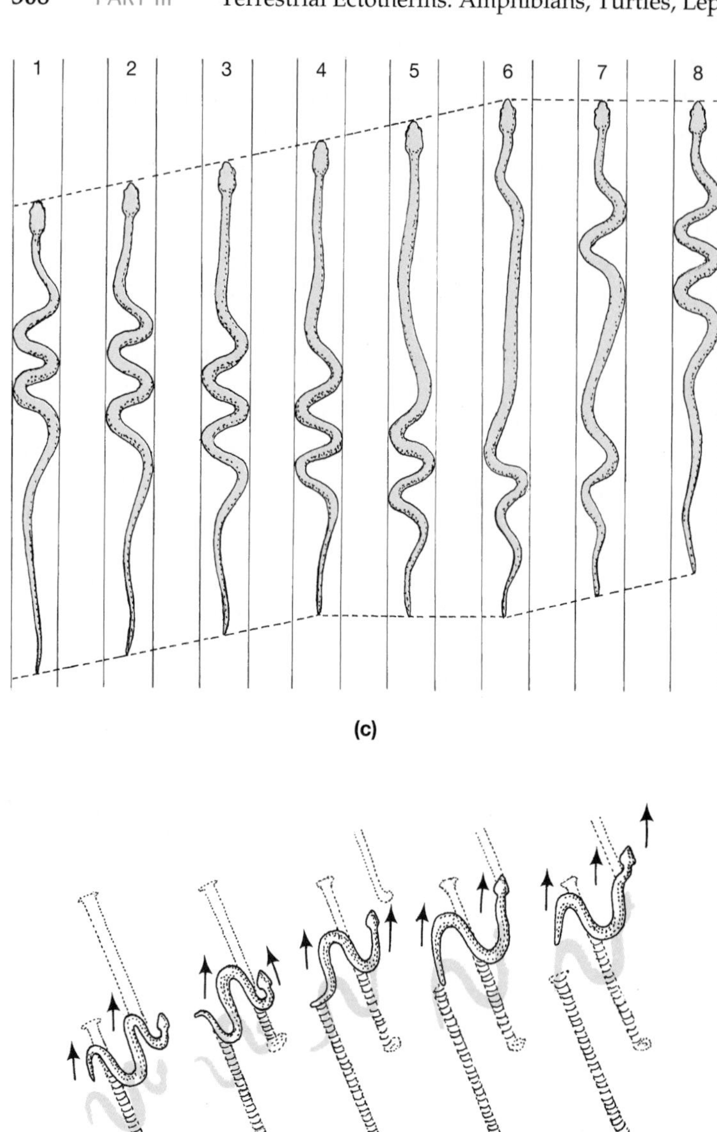

(c)

◀ Figure 11–4 *Continued*

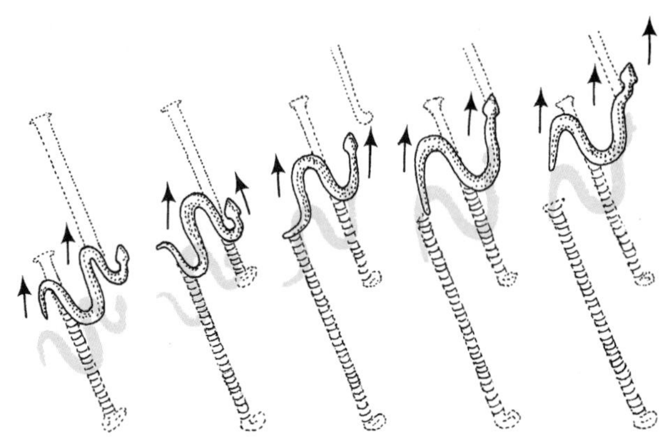

(d)

travel, sidewinding is an effective means of locomotion only for small snakes that live in habitats with few plants or other obstacles.

Snake skeletons are delicate structures that do not fossilize readily. In most cases we have only vertebrae, and little information has been gained from the fossil record about the origin of snakes. The earliest fossils known are from Cretaceous deposits and seem to be related to boas. Colubrid snakes are first known from the Oligocene, and elapids and viperids appeared during the Miocene.

The specializations of snakes compared with legless lizards appear to reflect two selective pressures—locomotion and predation. Elongation of the body is characteristic of snakes. The reduction in body diameter associated with elongation has been accompanied by some rearrangement of the internal anatomy of snakes. The left lung is reduced or entirely absent, the gallbladder is posterior to the liver, the right kidney is anterior to the left, and the gonads may show similar displacement.

Legless lizards face problems in swallowing prey. The primary difficulty is not the loss of limbs, because few lizards use the legs to seize or manipulate food. The difficulty stems from the elongation that is such a widespread characteristic of legless forms. As the body lengthens, the mass is redistributed into a tube with a smaller diameter. As the mouth gets smaller, the maximum diameter of the prey that can be swallowed also decreases, and an elongate animal is faced with the difficulty of feeding a large body through a small mouth. Most legless lizards are limited to eating relatively small prey, whereas snakes have morphological specializations that permit them to engulf prey considerably larger than the body diameter (see the following section). This difference may be one element in the great evolutionary success of snakes in contrast to the limited success of legless lizards and amphisbaenians.

11.4 Ecology and Behavior of Squamates

The past quarter-century has seen an enormous increase in the number and quality of field studies of the ecology and behavior of snakes and lizards. Studies of lizards have been particularly fruitful, in large measure because many species of lizards are conspicuous and active during the day. These diurnal species dominate the literature—much less is known about species with cryptic habits. The discussions in this section rely on studies of particular species, and it is important to remember that no one species or family is representative of lizards or snakes as a group.

Foraging and Feeding

The methods that snakes and lizards use to find, capture, subdue, and swallow prey are diverse; and they are important in determining the interactions among species in a community. Astonishing specializations have evolved: blunt-headed snakes with long lower jaws that can reach into a shell to winkle out a snail, nearly toothless snakes that swallow bird eggs intact and then slice them open with sharp ventral processes (hypapophyses) on the neck vertebrae, and chameleons that project their tongues to capture insects or small vertebrates on the sticky tips are only a sample of the diversity of feeding specializations of squamates.

Many of the feeding specializations of squamates are related to changes in the structure of the skull and jaws. The most conspicuous of these is the loss of the lower temporal bar and the quadratojugal bone that formed part of that bar (Figure 11–5). This modification is part of a suite of structural changes in the skull that contribute to the development of a considerable degree of kinesis. The living tuatara show the ancestral condition for squamates, with the quadratojugal linking the jugal and the quadrate bones to form a complete lower temporal arch. (This fully diapsid condition is not characteristic of all sphenodontids, however; some of the Mesozoic forms did not have a complete lower temporal arch.)

Early lizards are not well known. The fossil genera *Paliguana* and *Palaeagama* from the Late Permian and Early Triassic of South Africa are probably not true lizards, but they do show changes in the structure of the skull that probably parallel the changes that occurred in early squamates. The gap between the quadrate and jugal widened, and the complexly interdigitating suture between the frontal and parietal bones on the roof of the skull became straighter and more like a hinge. Additional areas of flexion evolved at the front and rear of the skull and in the lower jaw of some lizards. These changes were accompanied by the development of a flexible connection at the articulation of the quadrate bone with the squamosal, which provided some mobility to the quadrate. This condition, known as streptostyly, increases the force the pterygoideus muscle can exert when the jaws are nearly closed (Smith 1980).

In snakes the flexibility of the skull was increased still further by loss of the second temporal bar, which was formed by a connection between the postorbital and squamosal bones. A further increase in the flexibility of the joints between other bones in the palate and the roof of the skull produced the extreme flexibility of snake skulls. The third group of squamates has a completely different sort of skull specialization. The amphisbaenians are small, legless, burrowing animals. They use their heads as rams to construct tunnels in the soil, and the skull is heavy

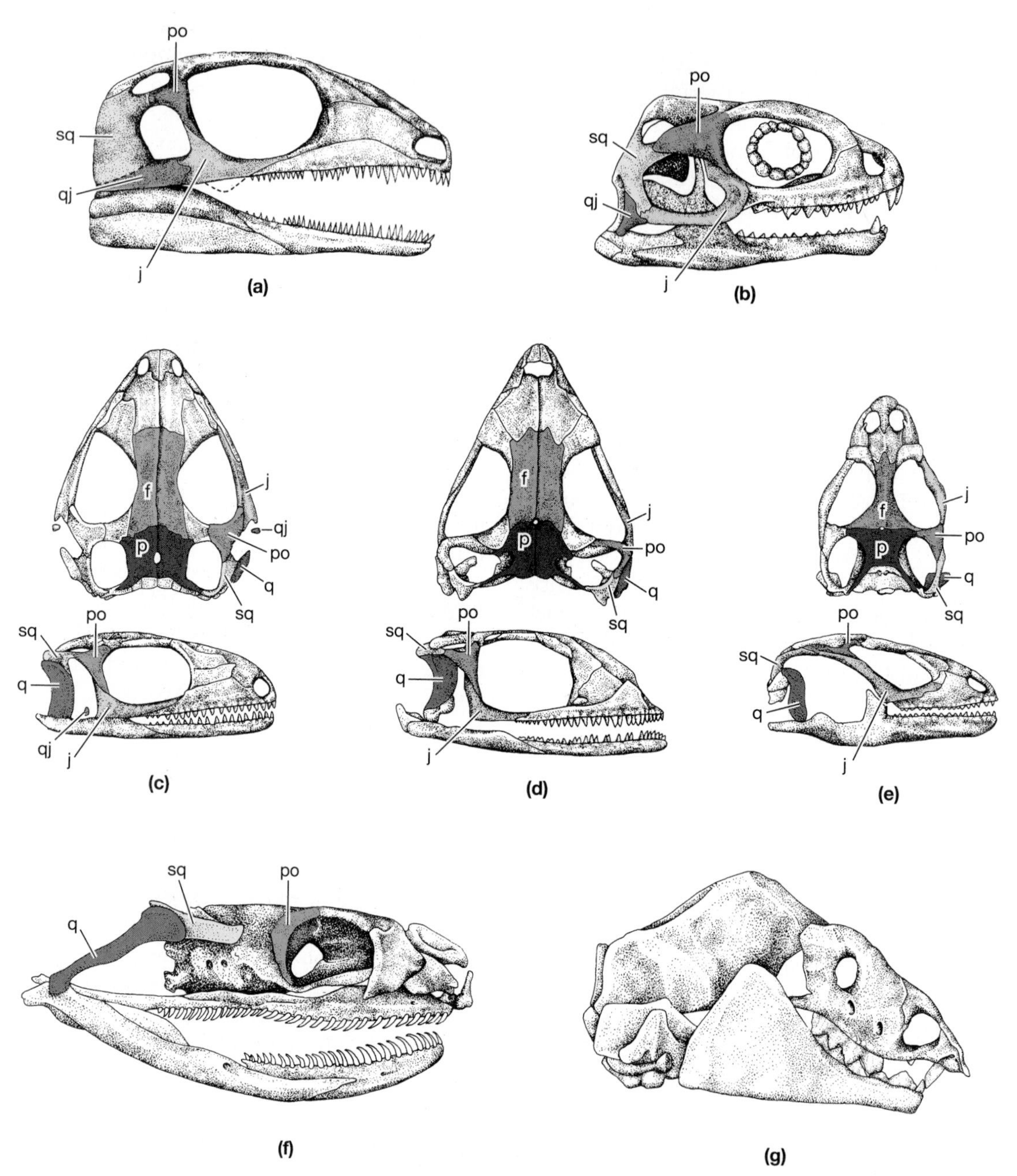

▲ Figure 11–5 Modifications of the diapsid skull among lepidosauromorphs. Fully diapsid forms like the Permian *Petrolacosaurus* (a) retain two complete arches of bone that define the temporal fenestrae. This condition is seen in living tuatara, *Sphenodon* (b). Lizards have achieved a kinetic skull by developing a gap between the quadrate and quadratojugal and by simplifying the suture butween the frontal and parietal bones, as shown by the modern collared lizard *Crotaphytus* (e). Probable transitional stages allowing increasing skull kinesis that occurred in a nonsquamate lineage are illustrated by *Paliguana* (c) and *Palaeagama* (d). In snakes (f) skull kinesis is further increased by loss of the upper temporal arch. Amphisbaenians (g), which use their heads for burrowing through soil, have specialized akinetic skulls: f, frontal; j, jugal; p, parietal; po, postorbital; q, quadrate; qj, quadratojugal; sq, squamosal.

with rigid joints between the bones. Their specialized dentition allows them to bite small pieces out of large prey.

Feeding Specializations of Snakes The entire skull of a snake is much more flexible than the skull of a lizard. In popular literature snakes are sometimes described as "unhinging" their jaws during feeding. That's careless writing and rather silly—unhinged jaws would merely flap back and forth. What those authors are trying to say is that snakes have extremely kinetic skulls that allow extensive movement of the jaws. A snake skull contains eight links, with joints between them that permit rotation (Figure 11–6). This number of links gives a staggering degree of complexity to the movements of the snake skull, and to make things more complicated, the linkage is paired—each side of the head acts independently. Furthermore, the pterygoquadrate ligament and quadrato-supratemporal ties are flexible. When they are under tension they are rigid, but when they are relaxed they permit sideward movement as well as rotation. All of this results in a considerable degree of three-dimensional movement in a snake's skull.

The mandibles of lizards are joined at the front of the mouth in a bony symphysis, but in snakes the mandibles are attached only by muscles and skin so they can spread sideward and move forward or back independently. Loosely connected mandibles and flexible skin in the chin and throat allow the jaw tips to spread, so that the widest part of the prey passes ventral to the articulation of the jaw with the skull.

Swallowing movements take place slowly enough to be observed easily (Figure 11–7). A snake usually swallows prey head first, perhaps because that approach presses the limbs against the body, out of the snake's way. Small prey may be swallowed tail first or even sideward. The mandibular and pterygoid teeth of one side of the head are anchored in the prey and the head is rotated to advance the opposite jaw as the mandible is protracted and grips the prey ventrally. As this process is repeated the snake draws the prey item into its mouth. Once the prey has reached the esophagus, it is forced toward the stomach by contraction of the snake's neck muscles. Usually the neck is bent sharply to push the prey along.

Most species of snakes seize prey and swallow it as it struggles. The risk of damage to the snake

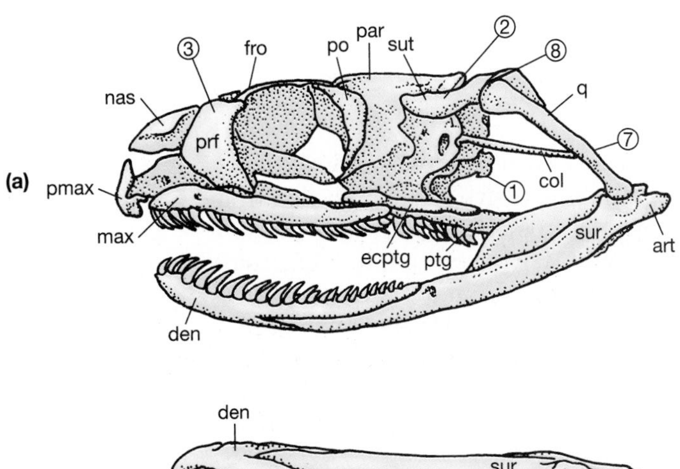

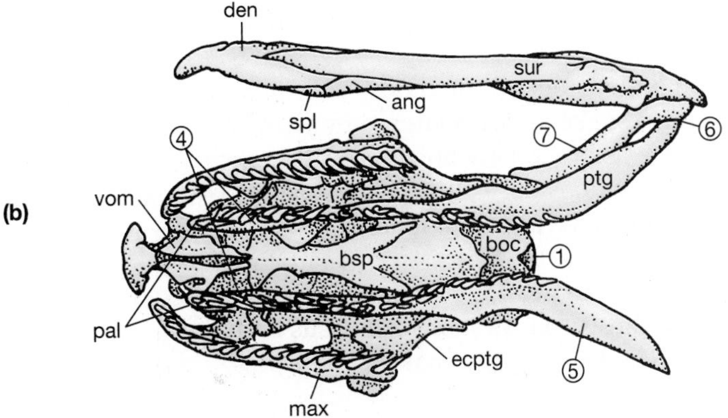

▶ Figure 11–6 Skull of a snake. (a) Lateral and (b) ventral views. A snake skull contains eight movable links: (1) braincase; (2) supratemporal; (3) prefrontal; (4) palatine; (5) pterygoid; (6) pterygoquadrate ligament; (7) quadrate; (8) quadrato-supratemporal tie. Legend: ang, angular; art, articular; boc, basioccipital; bsp, basisphenoid; col, columella; den, dentary; ecptg, ectopterygoid; fro, frontal; max, maxilla; nas, nasal; pal, palatine; par, parietal; pmax, premaxilla; po, postorbital; prf, prefrontal; ptg, pterygoid; q, quadrate; spl, spenial; sur, surangular; sut, supratemporal; vom, vomer.

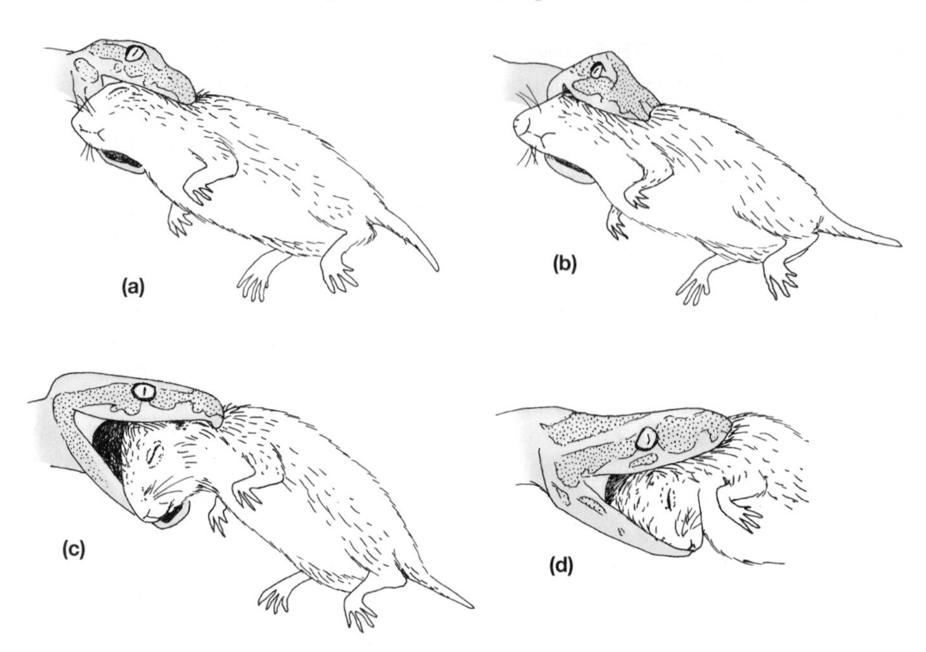

▲ Figure 11–7 Jaw movements during feeding. Snakes use a combination of head movements and protraction and retraction of the jaws to swallow prey. (a) Prey grasped by left and right jaws at the beginning of the swallowing process. (b) The upper and lower jaws on the right side have been protracted, disengaging the teeth from the prey. (c) The head is rotated counterclockwise, moving the right upper and lower jaws over the prey. The recurved teeth slide over the prey like the runners of a sled. (d) The upper and lower jaw on the right side are retracted, embedding the teeth in the prey and drawing it into the mouth. Notice that the entire head of the prey has been engulfed by this movement. Next the left upper and lower jaws will be advanced over the prey by clockwise rotation of the head. The swallowing process continues with alternating left and right movements until the entire body of the prey has passed through the snake's jaws.

during this process is a real one, and various features of snake anatomy seem to give some protection from struggling prey. The frontal and parietal bones of a snake's skull extend downward, entirely enclosing the brain and shielding it from the protesting kicks of prey being swallowed. Possibly the kinds of prey that can be attacked by snakes without a specialized feeding mechanism are limited by the snake's ability to swallow the prey without being injured in the process.

Constriction and venom are predatory specializations that permit a snake to tackle large prey with little risk of injury to itself. Constriction is characteristic of the boas and pythons as well as some colubrid snakes. Despite travelers' tales of animals crushed to jelly by a python's coils, the process of constriction involves very little pressure. A constrictor seizes prey with its jaws and throws one or more coils of its body about the prey. The loops of the snake's body press against adjacent loops, and friction prevents the prey from forcing the loops open. Each time the prey exhales, the snake takes up the slack by tightening the loops slightly. Two hypotheses have been proposed to explain the cause of death from constriction. The traditional view holds that prey suffocates because it cannot expand its thorax to inhale. Another possibility is that the increased internal pressure interferes with, and eventually stops, the heart (Hardy 1994).

Snakes that constrict their prey must be able to throw the body into several loops of small diameter to wrap around the prey. Constrictors achieve these small loops by having short vertebrae and short trunk muscles that span only a few vertebrae from the point of origin to the point of insertion. Contraction of these muscles produces sharp bends in the trunk that allow constrictors to press tightly against their prey. However, the trunk muscles of snakes are also used for locomotion, and the short muscles of constrictors produce several small-radius curves along the length of the snake's body. That morphology limits the speed with which constrictors can move, because rapid locomotion by snakes is accomplished by throwing the body into two or three broad loops. This is the pattern seen in fast-moving species such as whip snakes, racers, and mambas. The muscles that produce these broad loops are long, spanning many vertebrae, and the vertebrae are longer than those of constrictors.

In North America fast-moving snakes (colubroids) first appear in the fossil record during the Miocene, a time when grasslands were expanding. Constrictors, largely erycines, predominated in the snake fauna of the early Miocene, but by the end of that epoch the snake fauna was composed primarily of colubroids. Fast-moving colubroid snakes may have had an advantage over slow-moving boids in the more open habitats that developed during the Miocene, and that radiation of colubroids may have involved a complex interaction between locomotion and feeding. Rodents were probably the most abundant prey available to snakes, and rodents are dangerous animals for a snake to swallow while they are alive and able to bite and scratch. Constriction provided a relatively safe way for boids to kill rodents, but the long vertebrae and long trunk muscles that allowed colubroids to move rapidly through the open habitats of the later Miocene would have prevented them from using constriction to kill their prey.

Early colubroids may have used venom to immobilize prey. Duvernoy's gland, found in the upper jaw of many extant colubrid snakes, is homologous to the venom glands of viperids and elapids and produces a toxic secretion that immobilizes prey. (Some extant colubrids have venom that is dangerously toxic, even to animals as large as humans.) Thus, the evolution of venom that could kill prey may have been a key feature that allowed Miocene colubroid snakes to dispense with constriction and become morphologically specialized for rapid locomotion in open habitats. The presence of Duvernoy's gland appears to be an ancestral character for colubroid snakes, as this hypothesis predicts. Some colubrids, including the rat snakes (*Elaphe*), gopher snakes (*Pituophis*), and king snakes (*Lampropeltis*), have lost the venom-producing capacity of the Duvernoy's gland, and these are the groups in which constriction has been secondarily developed as a method of killing prey.

In this context, the front-fanged venomous snakes (Elapidae and Viperidae) are not a new development, but instead represent alternative specializations of an ancestral venom delivery system. Given the ancestral nature of venom for colubroid snakes, you might expect that different specializations for venom delivery would be represented in the extant snake fauna, as indeed they are. A variety of snakes have enlarged teeth (fangs) on the maxillae. Three categories of venomous snakes are recognized (Figure 11–8): opisthoglyphous, proteroglyphous, and solenoglyphous. This classification is descriptive, and represents convergent evolution by different phylogenetic lineages.

Opisthoglyphous (Greek *opistho* = behind and *glyph* = hollowed) snakes have one or more enlarged teeth near the rear of the maxilla with smaller teeth in front. In some forms the fangs are solid; in others there is a groove on the surface of the fang that may help to conduct saliva into the wound. Several African and Asian opisthoglyphs can deliver a dangerous or even lethal bite to large animals, including humans, but their primary prey is lizards or birds, which are often held in the mouth until they stop struggling and are then swallowed.

Proteroglyphous snakes (Greek *proto* = first) include the cobras, mambas, coral snakes, and sea snakes in the Elapidae. The hollow fangs of the proteroglyphous snakes are located at the front of the maxilla, and there are often several small, solid teeth behind the fangs. The fangs are permanently erect and relatively short.

Solenoglyphous (Greek *solen* = pipe) snakes include the pit vipers of the New World and the true vipers of the Old World. In these snakes the hollow fangs are the only teeth on the maxillae, which rotate so that the fangs are folded against the roof of the mouth when the jaws are closed. This folding mechanism permits solenoglyphous snakes to have long fangs that inject venom deep into the tissues of the prey. The venom, a complex mixture of enzymes and other substances (Table 11.4), first kills the prey and then speeds its digestion after it has been swallowed.

Snakes that can inject a disabling dose of venom into their prey have evolved a very safe prey-catching method. A constrictor is in contact with its prey while it is dying and runs some risk of injury from the prey's struggles. A solenoglyphous snake needs only

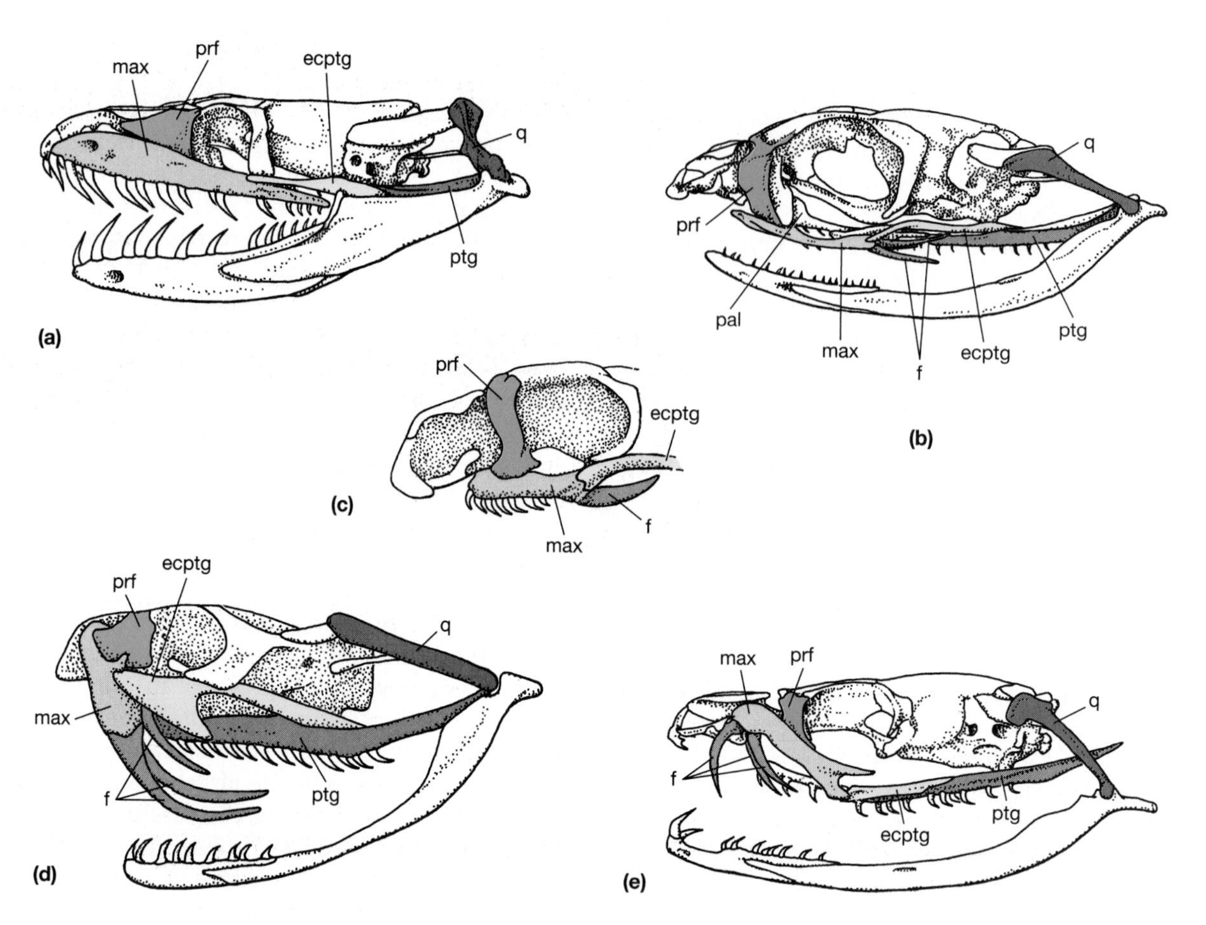

▲ Figure 11–8 Dentition of snakes. (a) Aglyphous (without fangs), African python, *Python sebae.* (b, c) Opisthoglyphous (fangs in the rear of the maxilla), African boomslang, *Dispholidus typus,* and Central American false viper, *Xenodon rhabdocephalus.* (d) Solenoglyphous (fangs on a rotating maxilla), African puff adder, *Bitis arietans.* (e) Proteroglyphous (permanently erect fangs at the front of the maxilla), African green mamba, *Dendroaspis jamesoni.* The fangs of solenoglyphs (d) are erected by an anterior movement of the pterygoid that is transmitted through the ectopterygoid and palatine to the maxilla, causing it to rotate about its articulation with the prefrontal, thereby erecting the fang. Some opisthoglyphs, especially *Xenodon* (c), have the same mechanism of fang erection. Legend: ecptg, ectopterygoid; f, fang; fro, frontal; max, maxilla; pal, palatine; par, parietal; pmax, premaxilla; prf, prefrontal; ptg, pterygoid; q, quadrate; sut, supratemporal.

to inject venom and allow the prey to run off to die. Later the snake can follow the scent trail of the prey to find its corpse. This is the prey-capture pattern of most vipers, and experiments have shown that a viper can distinguish the scent trail of a mouse it has bitten from trails left by uninjured mice.

Several features of the body form of vipers allow them to eat larger prey in relation to their own body size than can most nonvenomous snakes. Many vipers, including rattlesnakes, the jumping viper, and the African puff adder and Gaboon viper, are very stout snakes. The triangular head shape that is usually associated with vipers is a result of the outward extension of the rear of the skull, especially the quadrate bones. The wide-spreading quadrates allow bulky objects to pass through the mouth, and even a large meal makes little bulge in the stout body and thus does not interfere with locomotion. Vipers have specialized as relatively sedentary predators that wait in ambush and can prey even on quite large animals. The other family of terrestrial venomous snakes—the cobras, mambas, and their relatives—are primarily slim-bodied snakes that actively search for prey.

Foraging Behavior and Energetics of Lizards The activity patterns of lizards span a range from ex-

Table 11.4 Components of the venoms of squamates

Compound	*Occurrence*	*Effect*
Proteinases	All venomous squamates, espically vipers	Tissue destruction
Hyaluronidase	All venomous squamates	Increases tissue permeability, hastening the spread of other constituents of venom through the tissues
L-amino acid oxidase	All venomous squamates	Attacks a wide variety of substrates and causes great tissue destruction
Cholinesterase	High in elapids, may be present in sea snakes, low in vipers	Unknown; it is not responsible for the neurotixic effects of elapid venom
Phospholipases	All venomous squamates	Destroys cell membranes
Phosphatases	All venomous squamates	Breaks down high-energy compounds such as ATP, preventing cells from repairing damage
Basic polypeptides	Elapids and sea snakes	Blocks neuromuscular transmission

tremely sedentary species that spend hours in one place to species that are in nearly constant motion. Field observations of the tropidurid lizard *Leiocephalus schreibersi* and the teiid *Ameiva chrysolaema* in the Dominican Republic revealed two extremes of behavior. *Leiocephalus* rested on an elevated perch from sunrise to sunset and was motionless for more than 99 percent of the day. Its only movements consisted of short, rapid dashes to capture insects or to chase away other lizards. These periods of activity never lasted longer than 2 seconds, and the frequency of movements averaged 9.6 per hour. In contrast, *Ameiva* were active for only 4 or 5 hours in the middle of the day, but they were moving more than 70 percent of that time, and their velocity averaged one body length every 2 to 5 seconds.

The same difference in behavior was seen in a laboratory test of spontaneous activity: *Ameiva* was more than 20 times as active as *Leiocephalus.* In fact, the teiids were as active in exploring their surroundings as small mammals tested in the same apparatus. A xantusiid lizard tested in the laboratory apparatus had a pattern of spontaneous activity that fell approximately midway between that of the teiid and the tropidurid. Thus, a spectrum of spontaneous locomotor activity is apparent in lizards, extending from species that are nearly motionless through species that move at intermediate rates to species that are as active as mammals.

For convenience the extremes of the spectrum are frequently called sit-and-wait predators and widely foraging predators, respectively, and the intermediate condition has been called a cruising forager. Other field studies have shown that this spectrum of locomotor behaviors is widespread in lizard faunas. In North America, for example, spiny swifts (*Sceloporus*) are sit-and-wait predators, many skinks (*Eumeces*) appear to be cruising foragers, and whiptail lizards (*Cnemidophorus*) are widely foraging predators. The ancestral locomotor pattern for lizards may have been that of a cruising forager, and both sit-and-wait predation and active foraging may represent derived conditions. (A spectrum of foraging modes is not unique to lizards; it probably applies to nearly all kinds of mobile animals, including fishes, mammals, birds, frogs, insects, and zooplankton.)

The ecological, morphological, and behavioral characteristics that are correlated with the foraging modes of different species of lizards appear to define many aspects of the biology of these animals. For example, sit-and-wait predators and widely foraging predators consume different kinds of prey and fall victim to different kinds of predators. They have different social systems, probably emphasize different sensory modes, and differ in some aspects of their reproduction and life history.

These generalizations are summarized in Table 11.5 and are discussed in the following sections.

Table 11.5 Ecological and behavioral characteristics associated with the foraging modes of lizards. Foraging modes are presented as a continuum from sit-and-wait predators to widely foraging predators. In most cases data are available only for species at the extremes of the continuum. See the text for details.

	Foraging Mode		
Character	*Sit-And-Wait*	*Cruising Forager*	*Widely Foraging*
Foraging behavior			
Movements/hour	Few	Intermediate	Many
Speed of movement	Low	Intermediate	Fast
Sensory modes	Vision	Vision and olfaction	Vision and olfaction
Exploratory behavior	Low	Intermediate	High
Types of prey	Mobile, large	Intermediate	Sedentary, often small
Predators			
Risk of predation	Low	?	Higher
Types of predators	Widely foraging	?	Sit-and-wait and widely foraging
Body form			
Trunk	Stocky	Intermediate?	Elongate
Tail	Often short	?	Often long
Physiological characteristics			
Endurance	Limited	?	High
Sprint speed	High	?	Intermediate to low
Aerobic metabolic capacity	Low	?	High
Anaerobic metabolic capacity	High	?	Low
Heart mass	Small	?	Large
Hematocrit	Low	?	High
Energetics			
Daily energy expenditure	Low	?	Higher
Daily energy intake	Low	?	Higher
Social behavior			
Size of home range	Small	Intermediate	Large
Social system	Territorial	?	Not territorial
Reproduction			
Mass of clutch (eggs or embryos) relative to mass of adult	High	?	Low

However, a weakness of this analysis must be emphasized: Sit-and-wait species of lizards (at least, the ones that have been studied most) are primarily iguanians, whereas widely foraging species are mostly scleroglossans. That phylogenetic split raises the question of whether the differences we see between sit-and-wait and widely foraging lizards are really the consequences of the differences in foraging behavior, or if they are ancestral characteristics of iguanian versus scleroglossan lizards. If the latter is the case, the association with different foraging modes may be misleading. In either case, however, the model presented in Table 11.4 provides a useful integration of a large quantity of information about the biology of lizards; it represents a hypothesis that will be modified as more information becomes available.

Lizards with different foraging modes use different methods to detect prey: Sit-and-wait lizards normally remain in one spot from which they can survey a broad area. These motionless lizards detect the movement of an insect visually and capture it with a quick dash from their observation site. Sit-and-wait lizards may be most successful in detecting and capturing relatively large insects like beetles and grasshoppers. Active foragers spend most of their time on the ground surface, moving steadily and poking their snouts under fallen leaves and into crevices in the ground. These lizards apparently rely largely on chemical cues to detect insects, and they probably seek out local concentrations of patchily distributed prey such as termites. Widely foraging species of lizards appear to eat more small insects than do lizards that are sit-and-wait predators. Thus, the different foraging behaviors of lizards lead to differences in their diets, even when the two kinds of lizards occur in the same habitat.

The different foraging modes also have different consequences for the exposure of lizards to their own predators. A lizard that spends 99 percent of its time resting motionless is relatively inconspicuous, whereas a lizard that spends most of its time moving is easily seen. Sit-and-wait lizards are probably most likely to be discovered and captured by predators that are active searchers, whereas widely foraging lizards are likely to be caught by sit-and-wait predators. Because of this difference, foraging modes may alternate at successive levels in the food chain: Insects that move about may be captured by lizards that are sit-and-wait predators, and those lizards may be eaten by widely foraging predators. Insects that are sedentary are more likely to be discovered by a widely foraging lizard, and that lizard may be picked off by a sit-and-wait predator.

The body forms of sit-and-wait lizard predators may reflect selective pressures different from those that act on widely foraging species. Sit-and-wait lizards are often stout bodied, short tailed, and cryptically colored. Many of these species have dorsal patterns formed by blotches of different colors that probably obscure the outlines of the body as the lizard rests motionless on a rock or tree trunk. Widely foraging species of lizards are usually slim and elongate with long tails, and they often have patterns of stripes that may produce optical illusions as they move. However, one predator-avoidance mechanism, the ability to break off the tail when it is seized by a predator (**autotomy**), does not differ among lizards with different foraging modes (Box 11-1 & Figure 11–9).

What physiological characteristics are necessary to support different foraging modes? The energy requirements of a dash that lasts for only a second or two are quite different from those of locomotion that is sustained nearly continuously for several hours. Sit-and-wait and widely foraging species of lizards differ in their relative emphasis on the two metabolic pathways that provide adenosine triphosphate (ATP) for activity and in how long that activity can be sustained. Sit-and-wait lizards move in brief spurts, and they rely largely on anaerobic metabolism to sustain their movements. Anaerobic metabolism uses glycogen stored in the muscles as a metabolic substrate and produces lactic acid as its end product. It is a way to synthesize ATP quickly (because the glycogen is already in the muscles), but it is not good for sustained activity because the glycogen is quickly exhausted and lactic acid inhibits cellular metabolism. Lizards that rely on anaerobic metabolism can make brief sprints, but become exhausted when they are forced to run continuously. In contrast, aerobic metabolism uses glucose that is carried to the muscles by the circulatory system as a metabolic substrate, and it produces carbon dioxide and water as end products. Aerobic exercise can continue for long periods because the circulatory system brings more glucose and carries carbon dioxide away. As a result, widely foraging species can sustain activity for long periods without exhaustion.

The differences in exercise physiology are associated with differences in the oxygen transport systems of the lizards: Widely foraging species of

BOX 11-1 Caudal Autotomy—Your Tail or Your Life

Autotomy (self-amputation) of appendages is a common predator-escape mechanism among invertebrates and vertebrates. The tail is the only appendage that vertebrates are known to autotomize, and the capacity for caudal autotomy is developed to some degree among salamanders, tuatara, lizards, and a few amphisbaenians, snakes, and rodents (Arnold 1988). In most cases autotomy is followed by regeneration of a new tail.

The caudal autotomy of squamates occurs at distinctive fracture planes that are found in all but the four to nine anteriormost caudal vertebrae. The caudal muscles are segmental, and pointed processes from adjacent segments interdigitate. The caudal arteries have sphincter muscles just anterior to each fracture site, and the veins have valves. Autotomy appears to be an active process that requires contraction of the caudal muscles, bending the tail sharply to one side and initiating separation. The vertebral centrum ruptures, and the processes of the caudal muscles separate. The arterial sphincter muscles contract and the venous valves close, preventing loss of blood. An autotomized tail twitches rapidly for several minutes, and its violent writhing can distract the attention of a predator while the lizard itself scurries to safety (Figure 11–9). The tails of some juvenile skinks are

▲ Figure 11–9 Autotomy as a defensive mechanism. The freshly autotomized tail of a skink writhes and jerks, engaging the attention of a predatory king snake while the lizard escapes. In this sequence of photographs, a king snake seizes a skink by the tail (a). The skink autotomizes its tail and runs off (b), leaving the snake struggling to swallow the wriggling tail (c).

BOX 11.1 CAUDAL AUTOTOMY—YOUR TAIL OR YOUR LIFE (*CONTINUED*)

bright blue, and in experiments lizards with these colorful tails were more effective at using autotomy to escape from predatory snakes than were lizards with tails that had been painted black. Of course, an autotomized tail receives no blood flow and its muscular activity is sustained by anaerobic metabolism. The anaerobic metabolic capacity of lizard tail muscles appears to be substantially greater than that of limb muscles.

The point of autotomy is normally as far posterior on the tail as possible. When the tail of a lizard is seized with forceps, autotomy usually occurs through the plane of the vertebra immediately anterior to the point at which the tail was being held, thereby minimizing the amount of tail lost. When the tail is regenerated the vertebrae are replaced by a rod of cartilage that does not contain fracture planes. Consequently, future autotomy must occur anterior to the regenerated portion of the tail. Some geckos adjust the point of autotomy according to their body temperature: Autotomy occurs closer to the body when the lizard is cold than when it is warm. When a lizard is cold it cannot run as fast as when it's warm, and perhaps the longer segment of tail left behind by a cold lizard occupies the predator's attention long enough to allow the lizard to reach safety.

Leaving your tail in the grasp of a predator is certainly better than being eaten, but it is not free of costs. Because the tail acts as a signal of status among some species of lizards, a lizard that autotomizes a large part of its tail may fall to a lower rank in the dominance hierarchy. Losing the tail also affects the energy balance of a lizard: For example, the rate of growth of juvenile lizards that have autotomized their tails is reduced while their tails are being regenerated. Many lizards store fat in the tail, and the females mobilize this energy while they are depositing yolk in eggs. Sixty percent of the total fat storage was located in the tails of female geckos (*Coleonyx brevis*). Autotomy of the tail by gravid female geckos resulted in their producing smaller clutches of eggs than lizards with tails. Some lizards, especially the small North American skink *Scincella lateralis*, eat their autotomized tails if they can, thereby recovering the lost energy.

---

lizards have larger hearts and more red blood cells in their blood than do sit-and-wait species. As a result, each beat of the heart pumps more blood, and that blood carries more oxygen to the tissues of a widely foraging species of lizard than a sit-and-wait species.

Sustained locomotion is probably not important to a sit-and-wait lizard that makes short dashes to capture prey or to escape from predators, but sprint speed might be vitally important in both these activities. Speed may be relatively unimportant to a widely foraging lizard that methodically looks for prey under leaves and in cavities and can flee from bush to bush to confuse a predator. As you might predict from these considerations, sit-and-wait lizards generally have high sprint speeds and low endurance, whereas widely foraging species usually have lower sprint speeds and greater endurance.

The continuous locomotion of widely foraging species of lizards is energetically expensive. Measurements of energy expenditure of lizards in the Kalahari showed that the daily energy expenditure of a widely foraging species averaged 150 percent of that of a sit-and-wait species. However, the energy that the widely foraging species invested in foraging was more than repaid by its greater success in finding prey. The daily food intake of the widely foraging species was 200 percent that of the sit-and-wait predator. As a result, the widely foraging species had more energy available to use for growth and reproduction than did the sit-and-wait species, despite the additional cost of its mode of foraging.

Social Behavior

Squamates employ a variety of visual, auditory, chemical, and tactile signals in the behaviors they use to maintain territories and to choose mates. Iguanians use mainly visual signals, whereas scleroglossans (including snakes) use pheromones extensively. The various sensory modalities employed by animals have biased the amount of information we have about the behaviors of different species. Because humans are primarily visually oriented, we perceive the visual displays of other animals quite readily. The auditory sensitivity of humans is also acute, and we can detect and recognize vocal signals that are used by other species. However, the olfactory sensitivity of humans is low and we lack a well-developed vomeronasal system, so we are unable to perceive most chemical signals used by squamates. One result of our sensory biases has been a concentration of behavioral studies on organisms that use visual signals. Because of this focus, the extensive repertoires of visual displays of iguanian lizards figure largely in the literature of behavioral ecology, but much less is known about the chemical and tactile signals that are probably important for other lizards and for snakes.

The social behaviors of squamates appear to be limited in comparison with those of crocodilians, but many species show dominance hierarchies or territoriality. The signals used in agonistic encounters between individuals are often similar to those used for species and sex recognition during courtship. Parental attendance at a nest during the incubation period of eggs occurs among squamates, but extended parental care of the young is unknown.

Iguanian lizards employ primarily visual displays during social interactions. The polychrotine genus *Anolis* includes some 400 species of small to medium-size lizards that occur primarily in tropical America. Male *Anolis* have gular fans, areas of skin beneath the chin that can be distended by the hyoid apparatus during visual displays (see color insert). The brightly colored scales and skin of the gular fans of many species of *Anolis* are conspicuous signaling devices, and they are used in conjunction with movements of head and body.

Figure 11–10 shows the colors of the gular fans of eight species of *Anolis* that occur in Costa Rica. Since no two species have the same combination of colors on their gular fans, it is possible to identify a species solely by seeing the colors it displays. Each species also has a behavioral display that consists of raising the body by straightening the fore legs (called a push-up), bobbing the head, and extending and contracting the gular fan. The combination of these three sorts of movements allows a complex display. The three movements can be represented graphically by an upper line that shows the movements of the body and head and a lower line that shows the expansion and contraction of the gular fan. This representation is called a **display action pattern.** No two display action patterns are the same, so it would be possible to identify any of the eight species of *Anolis* by seeing its display action pattern.

The behaviors that territorial lizards use for species and sex recognition during courtship are very much like those employed in territorial defense—push-ups, head bobs, and displays of the gular fan. A territorial male lizard is likely to challenge any lizard in its territory, and the progress of the interaction depends on the response it receives. An aggressive response indicates that the intruder is a male and stimulates the territorial male to defend its territory, whereas a passive response from the intruder identifies a female and stimulates the territorial male to initiate courtship. These behaviors are illustrated by the displays of a male *Anolis carolinensis* shown in Figures 11–11 and 11–12. The first response of a territorial lizard to an intruder is the assertion-challenge display shown in Figure 11–11a. The dewlap is extended, and the lizard bobs at the intruder. The nuchal (neck) and dorsal crests are also slightly raised, and a black spot appears behind the eye. The next stage depends on the intruder's sex and its response to the territorial male's challenge (Figure 11–12). If the intruder is a male and does not retreat from the initial challenge, both males become more aggressive. During aggressive posturing (Figure 11–11b) the males orient laterally to each other, the nuchal and dorsal crests are fully erected, the body is compressed laterally, the black spot behind the eye darkens, and the throat is swelled. All these postural changes make the lizards appear larger and presumably more formidable to the opponent. If the intruder is a receptive female, the territorial male initiates courtship (Figure 11–11c).

The differences in color and movement that characterize the dewlaps and display action patterns of *Anolis* are conspicuous to human observers, but do the lizards also rely on them for species identification? Indirect evidence suggests that the lizards probably do use gular fan color and display action patterns for species identification. For example, examination of communities of *Anolis* containing many species shows that the differences in colors of gular fans and in display action patterns are greatest for those species that encounter one another most frequently.

Pheromonal communication probably occurs in several lineages of lizards, primarily scleroglossans, although chemical cues may be more important for some iguanians than has been realized. Territorial male *Sceloporus* and other phrynosomatid lizards rub

▲ Figure 11–10 Species-typical displays of *Anolis* lizards. Eight species of *Anolis* from Costa Rica can be separated into three groups based on the size and color pattern of their gular fans. *Simple* fans are unicolored, *compound* fans are bicolored, and *complex* fans are bicolored and very large. Display action patterns for each species are graphed beneath the lizard drawings. The horizontal axis is time (the duration of these displays is about 10 seconds), and the vertical axis is vertical height. Solid line shows movements of the head; dashed line indicates extension of the gular fan.

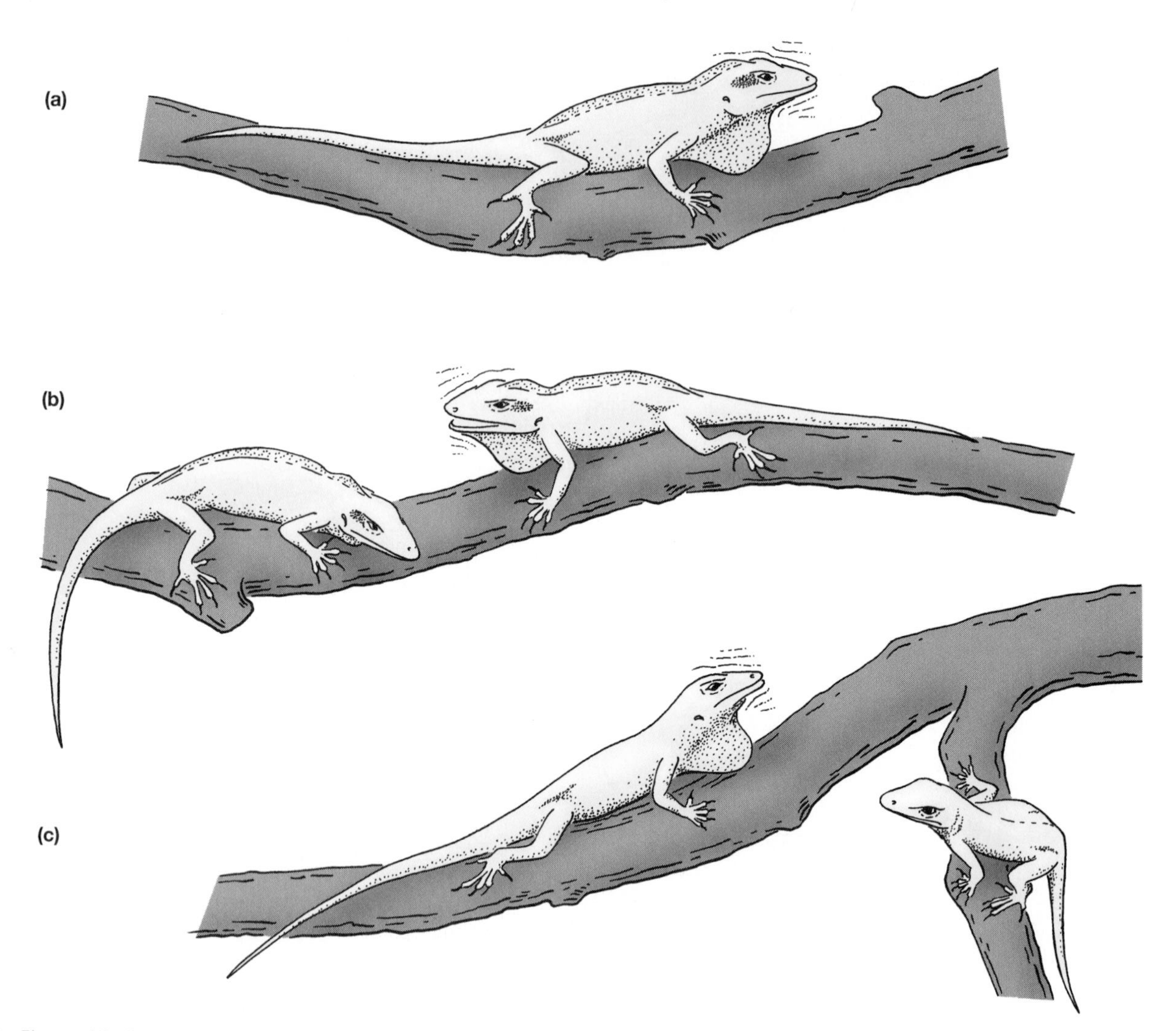

▲ Figure 11–11 Displays by a male *Anolis carolinensis.* (a) Assertion-challenge display. (b) Aggressive posturing between males. (c) Courtship. Note the extension of the dewlap, the species-typical head bob, and the absence of the dorsal and nuchal crests and the eyespot.

secretions from their femoral glands on objects in their territories. These secretions contain protein and sometimes lipids. Exploratory behavior by lizards, including iguanians, involves touching the tongue to the substrate, and the vomeronasal organ may detect pheromones in the femoral gland secretions. In addition, the secretions absorb light strongly in the ultraviolet portion of the spectrum. Some lizards are sensitive to ultraviolet light, and the femoral gland secretions rubbed onto rocks and branches may be both visual and olfactory signals.

Territoriality, the relative importance of vision compared with olfaction, and foraging behavior appear to be broadly correlated among lizards. The elevated perches from which sit-and-wait predators survey their home ranges allow them to see both intruders and prey, and they dash from the perch to repel an intruder or to catch an insect. In contrast, widely foraging lizards are almost entirely nonterritorial, and olfaction is as important as vision in their foraging behavior. These lizards spend most of their time on the ground, where their field of vision is lim-

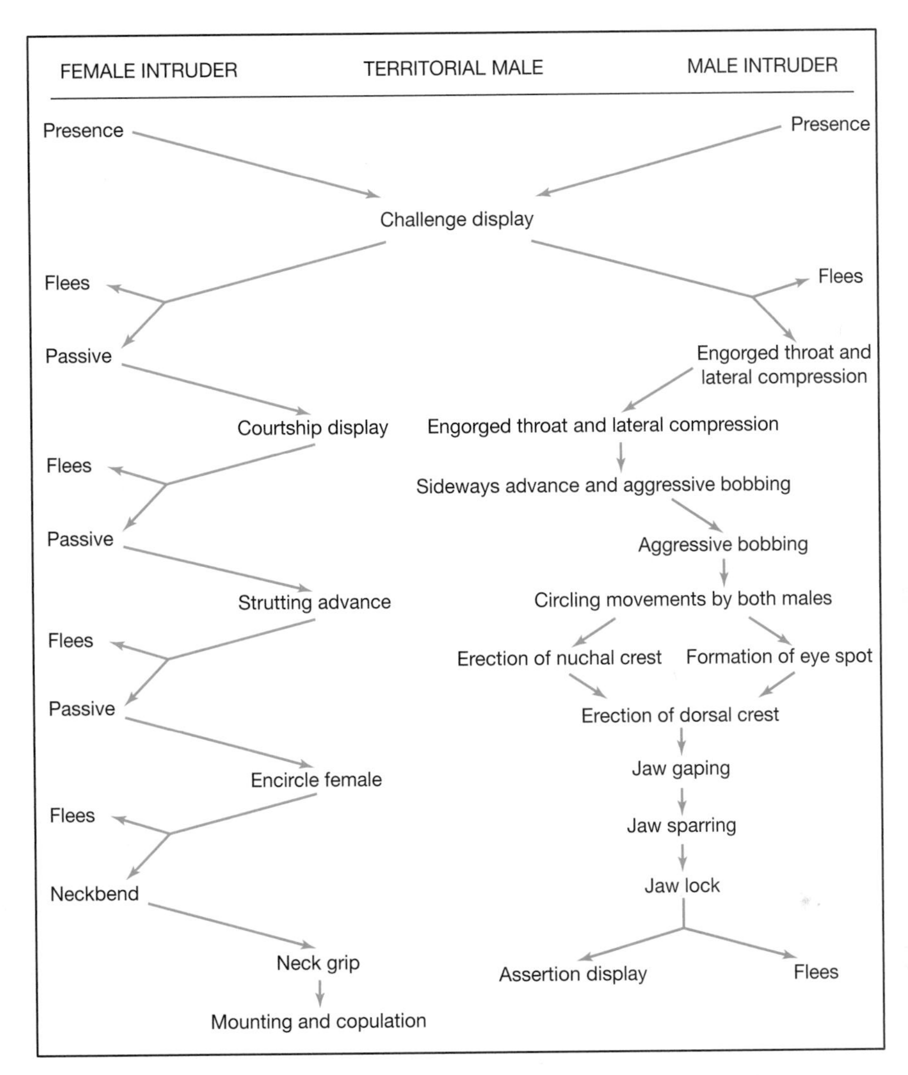

▲ Figure 11–12 Normal sequence of behaviors for a territorial male *Anolis carolinensis* confronting an intruding male or female anole. A territorial male challenges any intruder, and the response of the intruder determines the subsequent behavior of the territorial male.

ited and they probably have little opportunity to detect intruders.

Reproduction

Squamates show a range of reproductive modes from oviparity (development occurs outside the female's body and is supported entirely by the yolk—i.e., lecithotrophy) to viviparity (eggs are retained in the oviducts and development is supported by transfer of nutrients from the mother to the fetuses—matrotrophy). Intermediate conditions include retention of the eggs for a time after they have been fertilized, and the production of precocial young that were nourished primarily by material in the yolk. Oviparity is assumed to be the ancestral condition, and viviparity has evolved at least 45 times among lizards and 35 times among snakes. Viviparous squamates have specialized chorioallantoic placentae; in the Brazilian skink, *Mabuya heathi,* more than 99 percent of the mass of the fetus results from transport of nutrients across the placenta.

Oviparity A female lizard has only a certain amount of energy to devote to reproduction, and that energy can be used to produce a few large eggs (i.e., eggs with a large quantity of yolk) or several smaller eggs. All else being equal, large eggs produce large hatchlings and small eggs produce small hatchlings—so the question is, are large hatchlings more likely to survive than small hatchlings? And if large hatchlings do survive better, is the difference in survival great enough to make up for producing fewer individual hatchlings?

This question has been addressed in studies of the side-blotched lizard, *Uta stansburiana* (Sinervo et al. 1992). Female side-blotched lizards produce several clutches annually and mature in one year. The lizards normally lay small eggs in the first clutch and larger eggs in subsequent clutches. Does that shift in egg size reflect a survival advantage for large hatchlings later in the season?

To answer that question, Sinervo and his colleagues had to compare the survival of large and small hatchlings at different times of year. But the natural variation in the size of eggs is small, and it would be hard to detect differences in survival, even though a very small difference could have a significant effect in evolutionary time. To increase his chances of seeing an effect of hatchling size on survival, Sinervo created artificially small hatchlings by using a hypodermic needle and syringe to remove yolk from some eggs. In a different experiment (Sinervo and Licht 1991), he devised a method to produce giant hatchlings by surgically removing some ova from the ovary before yolk had been deposited. In that situation, the total amount of yolk available was divided among fewer eggs, and each egg was larger than normal. These methods allowed Sinervo to produce hatchlings that ranged from 50 percent to 150 percent of the normal size (Figure 11–13).

Studies at two field sites showed that female *Uta stansburiana* could maximize the number of surviving offspring from their first clutches by producing small eggs, whereas larger eggs would maximize survival of offspring in later clutches. That change in optimum egg size as the summer progresses probably reflects a changing ecological setting for the hatchlings. Hatchlings from the first clutch of eggs emerge into a world with fewer hatchling *Uta*—only their siblings and hatchlings from other first clutches compete with them for living space. In that situation, large body size may not confer an advantage, so a female can achieve the maximum number of surviving young by producing a large clutch of small eggs. Subsequent clutches don't have it so easy—they must establish home ranges in places that already have resident juvenile *Uta* from earlier clutches. In that situation, large hatchlings may be more likely to insert themselves successfully into the existing social system, and females achieve the maximum number of surviving young by producing a smaller clutch that contains large eggs.

Viviparity Viviparity is usually a high-investment reproductive strategy. Females of viviparous

◀ Figure 11–13 Hatchling side-blotched lizards, *Uta stansburiana.* The normal size for hatchlings (center) compared with gigantized (left) and miniaturized (right) individuals.

squamates generally produce relatively small numbers of large young, although there are exceptions to that generalization. Viviparity is not evenly distributed among lineages of squamates. Nearly half the origins of viviparity in the group have occurred in the family Scincidae, whereas viviparity is unknown in teiid lizards and occurs in only two genera of lacertids. Viviparity has advantages and disadvantages as a mode of reproduction. The most commonly cited benefit of viviparity is the opportunity it provides for a female snake or lizard to use her own thermoregulatory behavior to control the temperature of the embryos during development. This hypothesis is appealing in an ecological context, because a relatively short period of retention of the eggs by the female might substantially reduce the total amount of time required for development, especially in a cold climate.

Viviparity potentially lowers reproductive output because a female who is retaining one clutch of eggs cannot produce another. Lizards in warm habitats may produce more than one clutch of eggs in a season, but that is not possible for a viviparous species because development takes too long. In a cold climate lizards are not able to produce more than one clutch of eggs in a breeding season anyway, and viviparity would not reduce the annual reproductive output of a female lizard. Phylogenetic analyses of the origins of viviparity suggest that it has evolved most often in cold climates, as this hypothesis predicts, but other origins appear to have taken place in warm climates, and more than one situation favoring viviparity among squamates appears likely.

Viviparity has other costs. The agility of a female lizard is substantially reduced when her embryos are large. Experiments have shown that pregnant female lizards cannot run as fast as nonpregnant females, and that snakes find it easier to capture pregnant lizards than nonpregnant ones. Females of some species of lizards become secretive when they are pregnant, perhaps in response to their vulnerability to predation. They reduce their activity and spend more time in hiding places. This behavioral adjustment may contribute to the reduction in body temperature seen in pregnant females of some species of lizards, and it probably reduces their rate of prey capture as well.

In general, large species of squamates produce more eggs or fetuses than do small species, and within one species large individuals often have more offspring in a clutch than do small individuals. Both phylogenetic and ecological constraints play a role in determining the number of young produced, however. All geckos have a clutch size of either one or two eggs, and all *Anolis* produce only one egg at a time. Lizards with stout bodies usually have clutches that are a greater percentage of the mother's body mass than do lizards with slim bodies. The division between stout and slim bodies approximately parallels the division between sit-and-wait predators and widely foraging predators. It is tempting to infer that a lizard that moves about in search of prey finds a bulky clutch of eggs more hindrance than a lizard that spends 99 percent of its time resting motionless. However, because some of the divisions among modes of predatory behavior, body form, and relative clutch mass also correspond to the phylogenetic division between iguanian and scincomorph lizards, it is not possible to decide which characteristics are ancestral and which may be derived.

Parthenogenesis All-female (**parthenogenetic**) species of squamates have been identified in six families of lizards and one snake. The phenomenon is particularly widespread in the teiids (especially *Cnemidophorus*) and lacertids (*Lacerta*) and occurs in several species of geckos. Parthenogenetic species are known or suspected to occur among chameleons, agamids, xantusiids, and typhlopids. However, parthenogenesis is probably more widespread among squamates than this list indicates because parthenogenetic species are not conspicuously different from bisexual species. Parthenogenetic species are usually detected when a study undertaken for an entirely different purpose reveals that a species contains no males. Confirmation of parthenogenesis can be obtained by obtaining fertile eggs from females raised in isolation, or by making reciprocal skin grafts between individuals. Individuals of bisexual species usually reject tissues transplanted from another individual because genetic differences between them lead to immune reactions. Parthenogenetic species, however, produce progeny that are genetically identical to the mother, so no immune reaction occurs and grafted tissue is retained.

The chromosomes of lizards have allowed the events that produced some parthenogenetic species to be deciphered. Many parthenoforms appear to have had their origin as interspecific hybrids. These hybrids are diploid ($2n$) with one set of chromosomes from each parental species. For example, the diploid parthenogenetic whiptail lizard, *Cnemidophorus tesselatus*, is the product of hybridization between the bisexual diploid species *C. tigris* and *C. septemvittatus* (Figure 11–14). Some parthenogenetic species are triploids (3*n*). These forms are usually the result of a backcross of a diploid parthenogenetic individual to a male of one of its bisexual parental species or, less commonly, the result of hybridization of a diploid parthenogenetic species with a male of a bisexual species different from its parental species. A parthenogenetic triploid form of *C. tesselatus* is apparently the result of a cross between the parthenogenetic diploid *C. tesselatus* and the bisexual diploid species *C. sexlineatus.*

It is common to find the two bisexual parental species and a parthenogenetic species living in overlapping habitats. Parthenogenetic species of *Cnemidophorus* often occur in habitats like the floodplains of rivers that are subject to frequent disruption. Disturbance of the habitat may bring together closely related bisexual species, fostering the hybridization that is the first step in establishing a parthenogenetic species. Once a parthenogenetic species has become established, its reproductive potential is twice that of a bisexual species because every individual of a parthenogenetic species is capable of producing young. Thus, when a flood or other disaster wipes out most of the lizards, a parthenogenetic species can repopulate a habitat faster than a bisexual species.

Parental Care

Parental care has been recorded for more than 100 species of squamates. A few species of snakes and a larger number of lizards remain with the eggs or nest site. Some female skinks remove dead eggs from the clutch. Some species of pythons brood their eggs: The female coils tightly around the eggs, and in some species muscular contractions of the female's body produce sufficient heat to raise the temperature of the eggs to about 30°C, which is substantially above air temperature. One unconfirmed report exists of baby pythons returning at night to their empty eggshells, where their mother coiled around them and kept them warm. Little interaction between adult and juvenile squamates has been documented. In captivity female prehensile-tailed skinks (*Corucia zebrata*) have been reported to nudge their young toward the food dish, as if teaching them to eat. Prehensile-tailed skinks, which occur only on the Solomon Islands, are herbivorous and viviparous.

Free-ranging baby green iguanas have a tenuous social cohesion that persists for several months after they hatch. The small iguanas move away from the nesting area in groups that may include individuals from several different nests. One lizard may lead the way, looking back as if to see that others are following. The same individual may return later and recruit another group of juveniles. During the first 3 weeks after they hatch, juvenile iguanas move up into the forest canopy and are seen in close association with adults. During this time the hatchlings probably ingest feces from the adults, thereby inoculating their guts with the symbiotic microbes that facilitate digestion of plant material. After their fourth week of life, the hatchlings move down from the forest canopy into low vegetation, where they continue to be found in loosely knit groups of two to six or more individuals that move, feed, and sleep together. This association might provide some protection from predators, and if the hatchlings continue to eat fecal material, it is another opportunity to ensure that each lizard has received its full complement of gut microorganisms.

11.5 Thermoregulation

From the time of Aristotle onward, lizards, snakes, and amphibians have paradoxically been called cold blooded while they were thought to be able to tolerate extremeiy high temperatures. Salamanders frequently seek shelter in logs, and when a log is put on a fire, the salamanders it contains may come rushing out. Observations of this phenomenon gave rise to the belief that salamanders live in fire. In the first part of the twentieth century biologists were using similar lines of reasoning. In the desert lizards often sit on rocks. If you approach a lizard it runs away but the rock stays put, and touching the rock shows that it is painfully hot. Clearly, the reasoning went, the lizard must have been equally hot. Biologists marveled at the heat tolerance of lizards, and statements to this effect are found in authoritative textbooks of the period.

A study of thermoregulation of lizards by Raymond Cowles and Charles Bogert (1944) demonstrated the falsity of earlier observations and conclusions. Cowles and Bogert showed that reptiles can regulate their body temperatures with considerable precision, and that the level at which the tem-

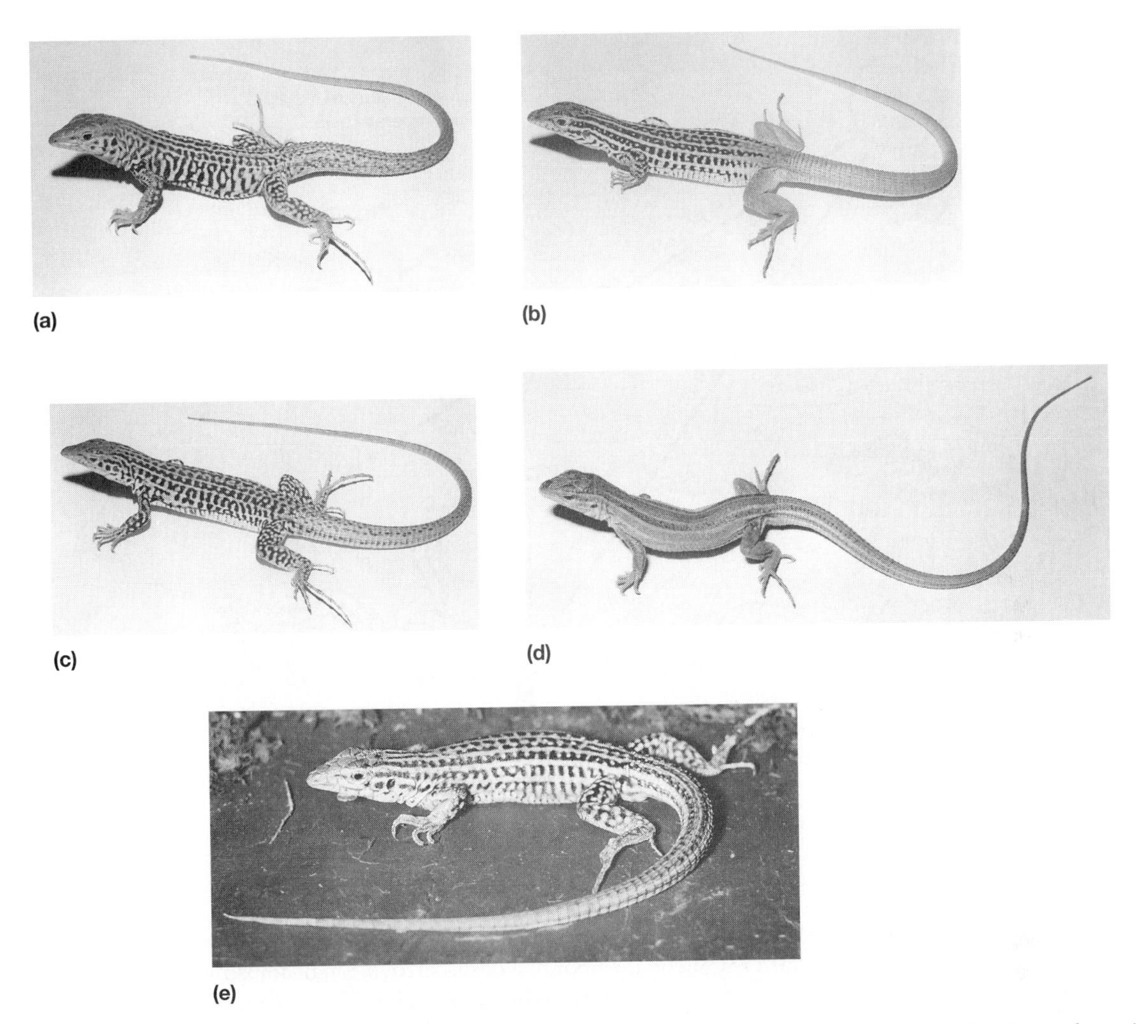

▲ Figure 11–14 The apparent sequence of crosses leading to the formation of diploid and triploid unisexual species of *Cnemidophorus*. Hybridization of the bisexual diploid species (a) *C. tigris* and (b) *C. septemvittatus* produced a unisexual diploid form with half of its genetic complement derived from each of the parental species (an allodiploid). This parthenogenetic form is called *C. tesselatus* (c). Hybridization between a diploid *C. tesselatus* and a male of the bisexual species *C. sexlineatus* (d) produced a unisexual triploid form with its genetic complement derived from three different parental species (an allotriploid). This parthenogenetic triploid form is also called *C. tesselatus* (e). Thus *C. tesselatus* consists of clones of both diploid and triploid lineages, although taxonomists soon will probably treat these two forms as separate species.

perature is regulated is characteristic of a species. The implications of this discovery for the biology of amphibians and reptiles are still being explored.

Energy Exchange between an Organism and Its Environment

A brief discussion of the pathways by which thermal energy is exchanged between a living organism and its environment is necessary to understand the thermoregulatory mechanisms employed by terrestrial animals. An organism can gain or lose energy by several pathways, and by adjusting the relative flow through various pathways an animal can warm, cool, or maintain a stable body temperature.

Figure 11–15 illustrates pathways of thermal energy exchange. Solar energy can reach an animal in several ways. Direct **solar radiation** impinges on an animal when it is standing in a sunny spot. In addition, solar energy is reflected from clouds and dust

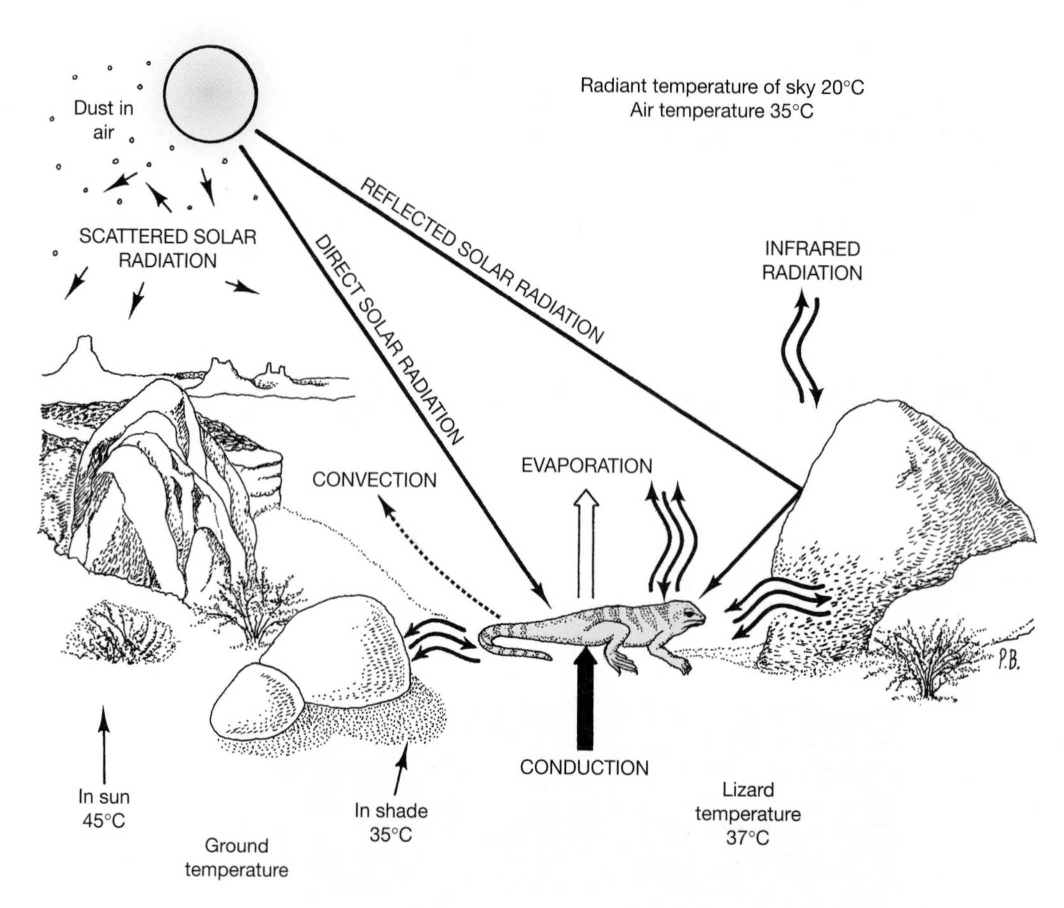

▲ Figure 11–15 Energy is exchanged between a terrestrial organism and its environment by several pathways. These are illustrated in simplified form by a lizard resting on the floor of a desert arroyo. Small adjustments of posture or position can change the magnitude of the various routes of energy exchange and give a lizard considerable control over its body temperature.

particles in the atmosphere, and from other objects in the environment, and reaches the animal by these circuitous routes. The wavelength distribution of the energy in all these routes is the same—the portion of the solar spectrum that penetrates the Earth's atmosphere. About half this energy is contained in the visible wavelengths of the solar spectrum (400 to 700 nanometers) and most of the rest is in the infrared region of the spectrum (> 700 nm).

Energy exchange in the **infrared** is an important part of the radiative heat balance. All objects, animate or inanimate, radiate energy at wavelengths determined by their absolute temperatures. Objects in the temperature range of animals and the Earth's surface (roughly –20 to +50°C) radiate in the infrared portion of the spectrum. Animals continuously radiate heat to the environment and receive infrared radiation from the environment. Thus, infrared radiation can lead to either heat gain or loss, depending on the relative temperature of the animal's body surface and the environmental surfaces as well as on the radiation characteristics of the surfaces themselves. In Figure 11–15, the lizard is cooler than the sunlit rock in front of it and receives more energy from the rock than it loses to the rock. However, the lizard is warmer than the shaded side of the rock behind it and has a net loss of energy in that exchange. The radiative temperature of the clear sky is about 20°C, so the lizard loses energy by radiation to the sky.

Heat is exchanged between objects in the environment and the air via **convection.** If an animal's surface temperature is higher than air temperature, convection leads to heat loss; if the air is warmer than the animal, convection is a route of heat gain. In still air, convective heat exchange is accomplished by convective currents formed by local heating; but in moving air, forced convection replaces natural convection and the rate of heat exchange is greatly increased. In the example shown, the lizard is warmer than the air and loses heat by convection.

Conductive heat exchange resembles convection in that its direction depends on the relative temperatures of the animal and environment. Conductive loss occurs between the body and the substrate where they are in contact. It can be modified by changing the surface area of the animal in contact with the substrate and by changing the rate of heat conduction in the parts of the animal's body that are in contact with the substrate. In this example the lizard gains heat by conduction from the warm ground.

Evaporation of water occurs from the body surface and from the pulmonary system. Each gram of water evaporated represents a loss of about 2450 joules (the exact value changes slightly with temperature). Evaporation of water transfers heat from the animal to the environment, and thus represents a loss of heat. The inverse situation, condensation of water vapor on an animal, would produce heat gain, but it rarely occurs under natural conditions.

Metabolic heat production is the final pathway by which an animal can gain heat. Among ectotherms metabolic heat gain is usually trivial in relation to the heat derived directly or indirectly from solar energy. There are a few exceptions to that generalization, and some of them are discussed later. Endotherms, by definition, derive most of their heat energy from metabolism; but their routes of energy exchange with the environment are the same as those of ectotherms and must be balanced to maintain a stable body temperature.

Behavioral Control of Body Temperatures by Ectotherms

The behavioral mechanisms involved in ectothermal temperature regulation are quite straightforward and are employed by insects, birds, and mammals (including humans) as well as by ectothermal vertebrates. Lizards, especially desert species, are particularly good at behavioral thermoregulation. Movement back and forth between sunlight and shade is the most obvious thermoregulatory mechanism they use. Early in the morning or on a cool day, lizards bask in the sun, whereas in the middle of a hot day they retreat to shade and make only brief excursions into the sun. Sheltered or exposed microhabitats may be sought out. In the morning when a lizard is attempting to raise its body temperature, it is likely to be in a spot protected from the wind. Later in the day when it is getting too hot, the lizard may climb into a bush or onto a rock outcrop where it is exposed to the breeze and its convective heat loss is increased.

An animal can alter the amount of solar radiation it absorbs by changing its body orientation to the sun, its body contour, and its skin color. Lizards use all of these mechanisms. An animal oriented perpendicular to the sun's rays intercepts the maximum amount of solar radiation, and one oriented parallel to the sun's rays intercepts minimum radiation. Lizards adjust their orientation to control heat gained by direct solar radiation. Many lizards can spread or fold their ribs to change the shape of the trunk. When the body is oriented perpendicular to the sun's rays and the ribs are spread, the surface area exposed to the sun is maximized and heat gain is increased. Compressing the ribs decreases the surface exposed to the sun and can be combined with orientation parallel to the rays to minimize heat gain. Horned lizards provide a good example of this type of control. If the surface area that a horned lizard exposes to the sun directly overhead when the lizard sits flat on the ground with its ribs held in a resting position is considered to be 100 percent, the maximum surface area the lizard can expose by orientation and change in body contour is 173 percent and the minimum is 28 percent. That is, the lizard can change its radiant heat gain more than sixfold solely by changing its position and body shape.

Color change can further increase a lizard's control of radiative exchange (see the color insert). Lizards darken by dispersing melanin in melanophore cells in the skin, and they lighten by drawing the melanin into the base of the melanophores. The lightness of a lizard affects the amount of solar radiation it absorbs in the visible part of the spectrum, and changes in heating rate (in the darkest color phase compared with the lightest) are from 10 to 75 percent.

Lizards can achieve a remarkable independence of air temperature due to their thermoregulatory capacities. Lizards occur above the timberline in many mountain ranges, and during their periods of activity on sunny days they are capable of maintaining body temperatures 30°C or more above air temperature. While air temperatures are near freezing, these lizards scamper about with body temperatures as high as those species that inhabit lowland deserts.

The repertoire of thermoregulatory mechanisms seen in lizards is greater than that of many other ectothermal vertebrates. Turtles, for example, cannot change their body contour or color, and their behavioral thermoregulation is limited to movements between sunlight and shade and in and out of water. Crocodilians are very like turtles, although young individuals may be able to make minor changes in body contour and color. Most snakes cannot change color, but some rattlesnakes lighten and darken as they warm and cool.

During the active parts of their day, desert lizards maintain their body temperatures in a zone called the **activity temperature range.** This is the region of temperature in which a lizard carries out its full repertoire of activities—feeding, courtship, territorial defense, and so on. For many species of desert lizards the activity temperature range is as narrow as 4°C, but for other ectotherms it may be as broad as 10°C. Different species of lizards have different activity temperature ranges. The thermoregulatory activities of a lizard are directed toward keeping it within its activity temperature range, but the precise temperature it maintains within this range depends on a variety of internal and external conditions. For example, many ectotherms maintain higher body temperatures when they are digesting food than when they are fasting. Female lizards when they are carrying young may maintain different body temperatures than at other times, and ectotherms with experimentally induced bacterial infections show a fever that is achieved by maintaining a higher-than-normal body temperature by behavioral means.

Physiological Control of the Rate of Change of Body Temperature

A new dimension was added to studies of ectothermal thermoregulation in the 1960s by the discovery that ectotherms can use physiological mechanisms to adjust their rate of temperature change. The original observations showed that several different kinds of large lizards were able to heat faster than they cooled when exposed to the same differential between body and ambient temperatures. Subsequent studies by other investigators extended these observations to turtles and snakes. From the animal's viewpoint, heating rapidly and cooling slowly prolongs the time it can spend in the normal activity range.

The basis of this control of heating and cooling rates lies in changes in peripheral circulation. Heating the skin of a lizard causes a localized vasodilation of dermal blood vessels in the warm area. Dilation of the blood vessels, in turn, increases the blood flow through them, and the blood is warmed in the skin and carries the heat into the core of the body. Thus, in the morning, when a cold lizard orients its body perpendicular to the sun's rays and the sunlight warms its back, local vasodilation in that region speeds heat transfer to the rest of the body.

The same mechanism can be used to avoid overheating. The Galápagos marine iguana is a good example. Marine iguanas live on the bare lava flows on the coasts of the islands. In midday, beneath the equatorial sun, the black lava becomes extremely hot—uncomfortably if not lethally hot for a lizard. Retreat to shade of the scanty vegetation or into rock cracks would be one way the iguanas could avoid overheating, but the males are territorial and those behaviors would mean abandoning their territories and probably having to fight for them again later in the day. Instead, the marine iguana stays where it is, using physiological control of circulation and the cool breeze blowing off the ocean to form a heat shunt that absorbs solar energy on the dorsal surface, carries it through the body, and dumps it out the ventral surface.

The process is as follows: In the morning the lizard is chilled from the preceding night and basks to bring its body temperature to the normal activity range. When its temperature reaches this level the lizard uses postural adjustments to slow the increase in body temperature, finally facing directly into the sun to minimize its heat load. In this posture the forepart of the body is held off the ground (Figure 11–16). The ventral surface is exposed to the cool wind blowing off the ocean, and a patch of lava under the animal is shaded by its body. This lava is soon cooled by the wind. Local vasodilation is produced by warming the blood vessels; it does not matter whether the heat comes from the outside (from the sun) or from inside (from warm blood). Warm blood circulating from the core of the body to the ventral skin warms it and produces vasodilation, increasing the flow to the ventral surface. The lizard's ventral skin is cooler than the rest of its body—it is shaded and cooled by the wind, and it loses heat by radiation to the cool lava in the shade created by the lizard's body. In this way the same mechanism that earlier in the day allowed the lizard to warm rapidly is converted to a regulated heat shunt that rapidly transports solar energy from the dorsal to the ventral surface and keeps the lizard from overheating. In combination with postural adjustments and other behavioral mechanisms, such as the choice of a site where the breeze is strong, these physiological adjustments allow a male iguana to remain on station in its territory all day.

Activity Temperature Ranges

The extensive repertoire of thermoregulatory mechanisms employed by ectotherms allows many species of lizards and snakes to keep body temperature within a range of a few degrees during the active

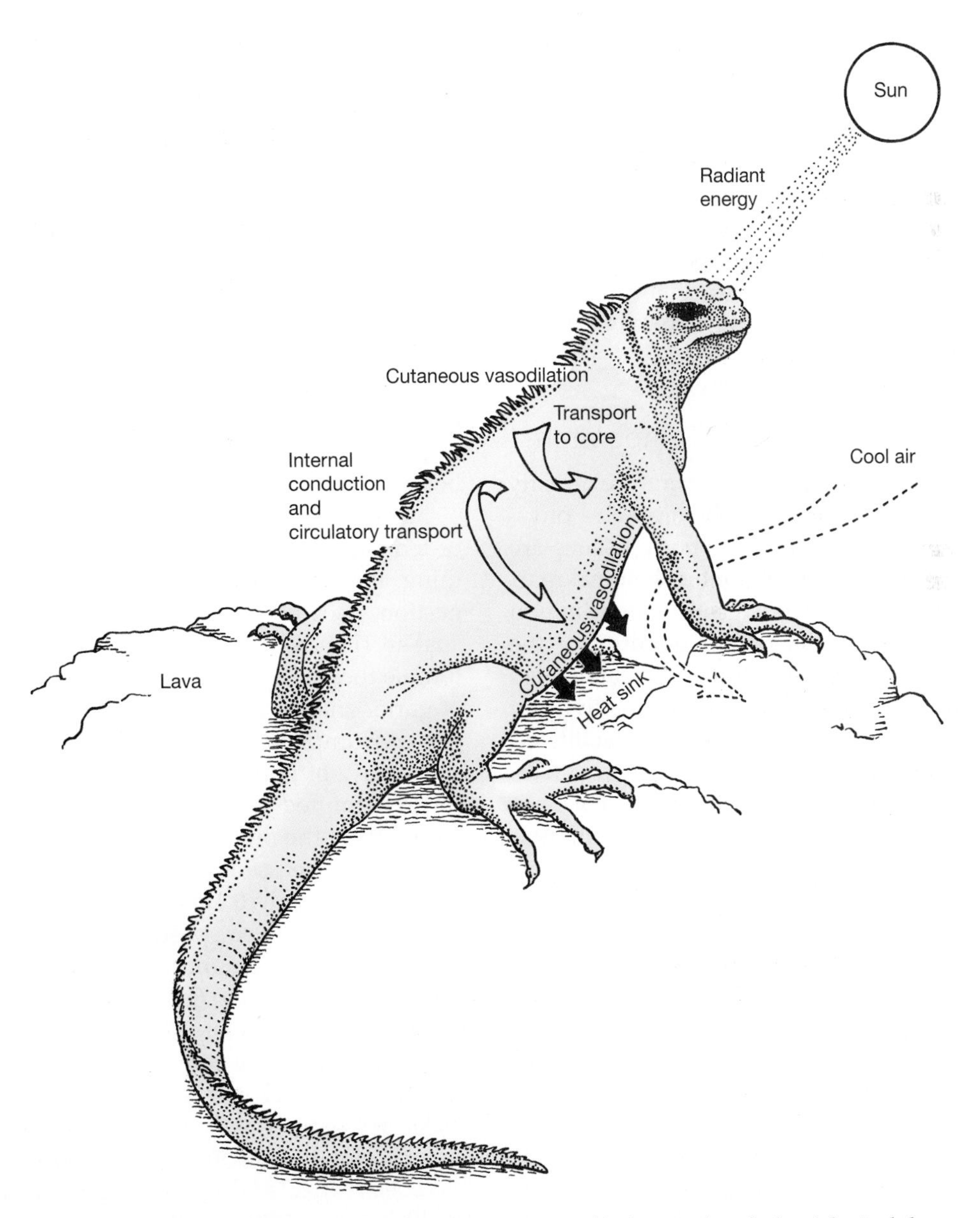

▲ Figure 11–16 The Galápagos marine iguana uses a combination of behavioral and physiological thermoregulatory mechanisms to shunt heat absorbed by its dorsal surface out its ventral surface.

part of their day. Many species of lizards have body temperatures between 33 and 38°C while they are active (the **activity temperature range**), and snakes often have body temperatures between 28 and 34°C.

These activity temperature ranges have been the focus of much research: Field observations show that thermoregulatory activities may occupy a considerable portion of an animal's time. Less obvious, but just as important, are the constraints that the need for thermoregulation sets on other aspects of the behavior and ecology of squamates. For example, some species of lizards and snakes are excluded from certain habitats because it is impossible to thermoregulate. In temperate regions the activity season lasts only during the months when it is warm and sunny enough to permit thermoregulation; at other times of the year, snakes and lizards hibernate. Even during the activity season, time spent on thermoregulation may not be available for other activities. Avery (1976) proposed that lizards in temperate regions show less extensive social behavior than do tropical lizards because thermoregulatory behavior in cool climates requires so much time.

Organismal Performance and Temperature

Minimizing variation in body temperature greatly simplifies the coordination of biochemical and physiological processes. An organism's body tissues are the site of a tremendous variety of biochemical reactions, proceeding simultaneously and depending on one another to provide the proper quantity of the proper substrates at the proper time for reaction sequences. Each reaction has a different sensitivity to temperature, and regulation is greatly facilitated when temperature variation is limited. Thus, coordination of internal processes may be a major benefit of thermoregulation for squamates. If the temperature stability that a snake or lizard achieves by thermoregulation is important to its physiology and biochemistry, you might expect that the internal economy of an animal functions best within its activity temperature range, and that is often the case. Examples of physiological processes that work best at temperatures within the activity range can be found at the molecular, tissue, system, and whole-animal levels of organization.

Temperature profoundly affects squamates' ability to carry out different activities. But not all activities are affected in parallel ways, and behavioral shifts (that is, qualitative rather than quantitative changes) are seen in some cases. Furthermore, the relationship between body temperature and physiological processes is a two-way interaction. Physiological capacities change in response to changes in body temperature, but under some conditions the sequence of cause and effect is reversed, and squamates manipulate their body temperatures in response to internal conditions such as feeding status or pregnancy.

The wandering garter snake (*Thamnophis elegans vagrans*) provides examples of the effects of body temperature on a variety of physiological and behavioral functions (Stevenson et al. 1985). Wandering garter snakes are diurnal, semiaquatic inhabitants of lakeshores and stream banks in western North America. They hunt for prey on land and in water, and feed primarily on fishes and amphibians. Chemosensation, accomplished by flicking the tongue, is an important mode of prey detection for snakes. Scent molecules are transferred from the tips of the forked tongue to the epithelium of the vomeronasal organ in the roof of the mouth. Garter snakes are diurnal; they spend the night in shelters, where their body temperatures fall to ambient levels (4 to 18°C), and emerge in the morning to bask. During activity on sunny days, the snakes maintain body temperatures between 28 and 32°C.

Stevenson and his associates measured the effect of temperature on the speed of crawling and swimming, the frequency of tongue flicks, the rate of digestion, and the rate of oxygen consumption of the snakes (Figure 11–17). Crawling, swimming, and tongue flicking are elements of the foraging behavior of garter snakes, and the rates of digestion and oxygen consumption are involved in energy utilization. The ability of garter snakes to crawl and swim was severely limited at the low temperatures they experience during the night when they are inactive. At 5°C snakes often refused to crawl, and at 10°C they were able to crawl only 0.1m per second and could swim only 0.25m/sec. The speed of both types of locomotion increased at higher temperatures. Swimming speed peaked near 0.6m/sec at 25 and 30°C, and crawling speed increased to an average of 0.8m/sec at 35°C. The rate of tongue flicking increased from less than 0.5 flick per second at 10°C to about 1.5 flicks per second at 30°C. The rate of digestion increased slowly from 10 to 20°C and more than doubled between 20 and 25°C. It did not increase further at 30°C, and dropped slightly at higher temperatures. The rate of oxygen consumption rose as tem-

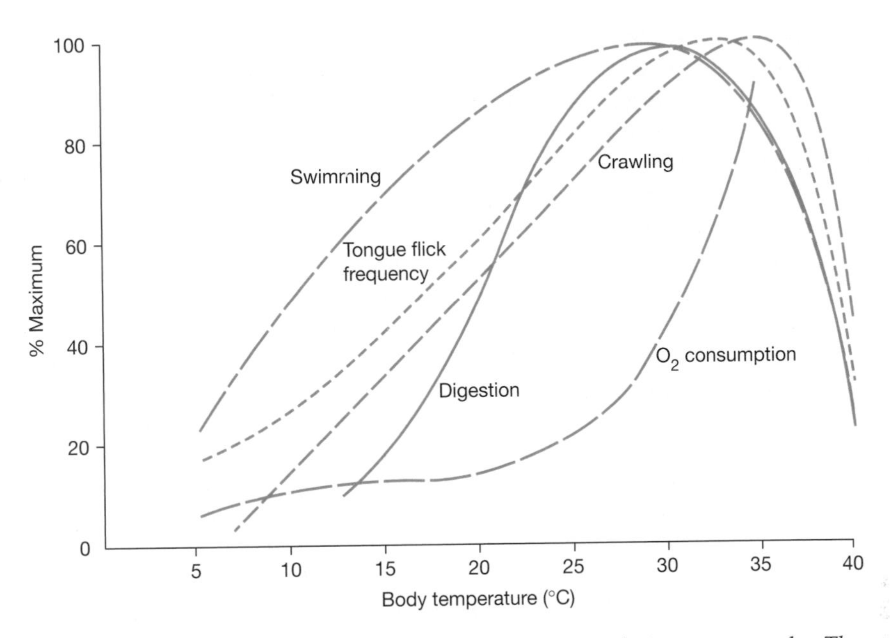

▲ Figure 11–17 Effect of temperature on performance. The ability of a wandering garter snake, *Thamnophis elegans vagrans*, to perform many activities essential to survival depends on its body temperature. The vertical axis shows the percentage of maximum performance achieved at each temperature.

perature increased from 20 to 35°C, which was the highest temperature tested because higher body temperatures would have been injurious.

All five measures of performance by garter snakes increased with increasing temperature, but the responses to temperature were not identical. For example, swimming speed did not increase substantially above 20°C, whereas crawling speed continued to increase up to 35°C. The rate of digestion peaked at 25 to 30°C and then declined, but the rate of oxygen consumption increased steadily to 35°C. More striking than the differences among the functions, however, is the apparent convergence of maximum performance for all the functions on temperatures between 28.5 and 35°C. This range of temperatures is close to the body temperatures of active snakes in the field on sunny days (28 to 32°C). Anywhere within that range of body temperatures, snakes would be able to crawl, swim, and tongue-flick at rates that are at least 95 percent of their maximum rates.

The relationship between the body temperatures of active garter snakes and the temperature sensitivity of various behavioral and physiological functions reported by Stevenson and his colleagues is probably common for squamates. That is, in most cases the body temperatures they maintain during activity are the temperatures that maximize organismal performance. However, at least two types of variation complicate the picture of squamate thermoregulation: changes in behavior that accompany changes in body temperature and changes in thermoregulation in response to the physiological status of an animal.

Behavioral Changes

A change in the body temperature of a squamate may be accompanied by a qualitative change in behavior instead of by the graded levels of performance shown by garter snakes. For example, *Agama savignyi* is an agamid lizard that lives in desert areas of the Middle East. It shows a pronounced temperature sensitivity of sprint speed: At a body temperature of 18°C it can run only 1m/sec, but at 34°C it runs about 3m/sec. *Agama savignyi* lives in open habitats where it may be some distance from shelters that could provide protection from predators. Clearly, the lizards are better able to run to a shelter when they are warm than when they are cool, and they display two types of defensive behavior depending on their body temperature. At body temperatures between 18 and 26°C most *A. savignyi* do not try to run from a predator; instead, they leap at the predator and try to bite. At body temperatures of 30°C or above, however, most lizards run away. This sort of qualitative shift in behavior at different body temperatures may be a widespread response among squamates to the effects of body temperature on their ability to carry out certain activities.

Effects of Nutritional Status and Bacterial Infections on Temperature Regulation

Several internal states of squamates and other ectotherms influence body temperature. A thermophilic (Greek *thermo* = heat and *philo* = loving) response after feeding is widespread: Individuals with food in the gut maintain higher body temperatures than do individuals without food. A higher body temperature accelerates digestion and increases digestive efficiency and water uptake, so a warm animal digests its food more rapidly and assimilates a higher proportion of the energy and water present in the food. Conversely, fasting animals regulate their body temperatures at low levels that reduce their metabolic rates and conserve their stored energy.

Behavioral fever is another common response of ectotherms. Individuals infected by bacteria change their thermoregulatory behavior and maintain body temperatures several degrees higher than those of uninfected controls. These behavioral fevers have been demonstrated in arthropods, fishes, frogs, salamanders, turtles, and lizards. The release of prostaglandin E_1, which acts on thermoregulatory centers of the anterior hypothalamus, appears to be the immediate cause of both the behavioral fevers of ectotherms and the physiological fevers of endotherms. Survival is enhanced by fever, apparently because bacterial growth is limited by a reduction in the availability of iron at higher temperatures.

Reproductive Status

Pregnancy affects thermoregulation by squamates. The rate of embryonic development of squamates is strongly affected by temperature, and a major advantage of viviparity is thought to be the opportunity it gives the mother to control the temperature of embryos during development. The body temperatures of female squamates during pregnancy may be different from the temperatures they would normally maintain. For example, pregnant female spiny swifts (*Sceloporus jarrovi*) had an average body temperature of 32.0°C, whereas male lizards in the same habitat had an average body temperature of 34.5°C (Beuchat 1986). The female lizards changed their thermoregulatory behavior after they had given birth, and the average body temperature of postparturient female lizards was 34.5°C, like that of the males. The low body temperatures of pregnant lizards were unexpected because there are several reasons why giving birth as early in the year as possible would be advantageous for the lizards.

That line of reasoning suggests that female lizards should maintain higher-than-normal body temperatures during pregnancy, or at least they should not reduce their body temperatures. Contrary to that prediction, the body temperatures maintained by pregnant squamates appear to converge toward 32°C whether the normal body temperature for the species is higher or lower. If the body temperature of pregnant female lizards is a compromise between the thermal requirements of the mother and the best temperature for embryonic development, this convergence might indicate that temperatures near 32°C are particularly favorable for embryonic development (Beuchat and Ellner 1987).

11.6 Temperature and Ecology of Squamates

Squamates, especially lizards, are capable of very precise thermoregulation, and microhabitats at which particular body temperatures can be maintained may be one of the dimensions that define the ecological niches of lizards. The five most common species of *Anolis* on Cuba partition the habitat in several ways (Figure 11–18). First, they divide the habitat along the continuum, from sunny to shady: Two species (*A. lucius* and *A. allogus*) occur in deep shade in forests, one (*A. homolechis*) in partial shade in clearings and at the forest edge, and two (*A. allisoni* and *A. sagrei*) in full sun. Within habitats in the sun-shade continuum, the lizards are separated by the substrates they use as perch sites. In the forest *A. lucius* perches on large trees up to 4m above the ground, whereas *A. allogus* rests on small trees within 2m of the ground. *A. homolechis*, which does not share its habitat with another common species of *Anolis*, perches on both large and small trees. In open habitats *A. allisoni* perches more than 2m above the

Color change is a temperature-regulating mechanism used by lizards such as the desert iguana (*Dipsosaurus dorsalis*). When they first emerge in the morning, desert iguanas are dark (upper left). By the time it has reached its activity temperature, a lizard has turned light (upper right). This color change reduces heat gained from the sun by 23 percent.

Luminescent bacteria in the light organs of fishes emit light as a by-product of their metabolism. (Lower left) A black dragonfish, *Idiacanthus*. A long barbel on the chin bears a luminous lure that is believed to attract prey close enough to be engulfed by the enormous jaws lined with sharp teeth. (Lower right) The flashlight fish, *Photoblepharon*, has a light-emitting organ under each eye. The fish can cover the organ with a pigmented shutter to conceal the light, or open the shutter to reveal it. It uses the light organ in social interactions with other flashlight fish, and in a blink-and-run defense to startle and confuse predators.

Photographs: Top left © Tom McHugh; Top right © Bucky Reeves; Bottom © Norbert Wu, DRK Inc.

Three species of salamanders form a mimicry complex in eastern North America. The red eft (*Notophthalmus viridescens*, top left) and red salamander (*Psuedotriton ruber*, top right) have skin toxins that deter predators. The red-backed salamander (*Plethodon cinereus*, middle left) is not protected by toxins, but predators confuse the erythristic form of that species (middle right) with the toxic species. The experiment described in the text used the mountain dusky salamander (*Desmognathus ochrophaeus*, bottom left) as a palatable control.

Photographs: red eft © Michael Lustbader; red salamander © Alan Blank, National Audubon Society; red-backed salamander, erythristic red-backed salamander, and mountain dusky salamander © F. H. Pough.

The gular fans of lizards are used in social displays. Color, size, and shape identify the species and sex of an individual. (All the lizards in these photographs are males.) (Top left) Carolina anole, *Anolis carolinensis*, from Florida. (Top right) knight anole, *Anolis equestris*, from Cuba. (Bottom left) *Anolis grahami* from Jamaica. (Bottom right) *Anolis chrysolepis* from Brazil.

Photographs: top left © J. H. Robinson; top right © Cosmos Blank, National Audubon Society; bottom left © Fred McConnaughey; bottom right © Jany Sauvanet.

Extinct

Golden Toad

Ivory-billed woodpecker

Extinct in the Wild

Przewalski's horse

Black-footed ferret

Critically Endangered

Northern hairy nosed wombat

Vulnerable

Great white shark

Examples of recently extinct vertebrates and species facing different levels of threat defined by the Species Survival Commission of the IUCN.

Photographs: golden toad © Michael Fogden; ivory-billed woodpecker © Tom McHugh, Przewalski's horse © Stephen J. Krasemann; black-footed ferret © Steve Kaufman; northern hairy nosed wombat © Dave Watts; great white shark © Jeffrey L. Rotman

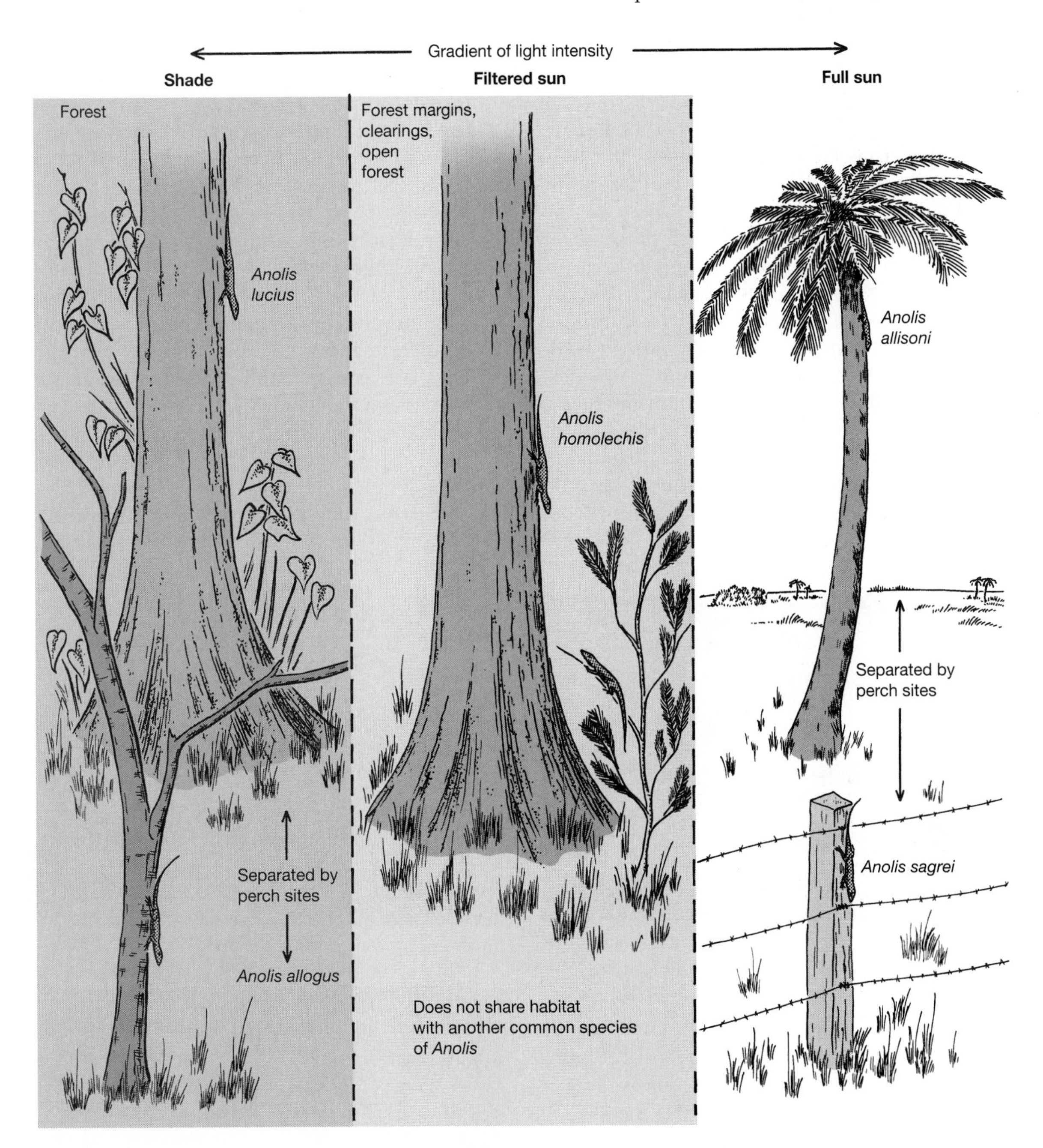

▲ Figure 11–18 Habitat partitioning by Cuban species of *Anolis*. See text for explanation.

ground on tree trunks and houses, and *A. sagrei* perches below 2m on bushes and fenceposts.

Some species of lizards do not thermoregulate. Lizards that live beneath the tree canopy in tropical forests often have body temperatures very close to air temperature (that is, they are thermally passive), whereas species that live in open habitats thermoregulate more precisely. The relative ease of thermoregulation in different habitats may be an important factor in determining whether a species of lizard thermoregulates or allows its temperature to vary with ambient temperature.

The distribution of sunny areas is one factor that determines the ease of thermoregulation. Sunlight penetrates the canopy of a forest in small patches that move across the forest floor as the Sun moves across the sky. These patches of sun are the only sources of solar radiation for lizards that live at or near the forest floor, and the patches may be too sparsely distributed or too transient to be used for thermoregulation. In open habitats sunlight is readily available, and thermoregulation is easier. The difference in thermoregulatory behavior of lizards in open and shaded habitats can be seen even in comparisons of different populations within a species. For example, *Anolis sagrei* occurs in both open and forest habitats on Abaco Island in the Caribbean. Lizards in open habitats bask and maintain body temperatures between 32 and 35°C from about 8:30 in the morning through about 5:00 in the afternoon. Lizards in the forest do not bask, and their body temperatures vary from a low of 24°C to a high of 28°C over the same period.

The task of integrating thermoregulatory behavior with foraging is relatively simple for sit-and-wait foragers such as *Anolis.* These lizards can readily change their balance of heat gain and loss by making small movements in and out of shade, or between calm and breezy perch sites, while continuing to scan their surroundings for prey. Widely foraging species may have more difficulty integrating thermoregulation and predation. They are continuously moving between sun and shade and in and out of the wind, and their body temperatures are affected by their foraging activity. These lizards sometimes have to stop foraging to thermoregulate, resuming foraging only when they have warmed or cooled enough to return to their activity temperature range.

(a)

(b)

(c)

▲ Figure 11–19 Three sympatric species of *Ameiva* from Costa Rica. (a) *Ameiva leptophrys,* which has an average adult mass of 83g. (b) *Ameiva festiva,* average adult mass 32g. (c) *Ameiva quadrilineata,* average adult mass 10g.

Body size is yet another variable that can affect thermoregulation. An example of the interaction of body size, thermoregulation, and foraging behavior is provided by three species of teiid lizards (*Ameiva*) in Costa Rica. *Ameiva* are widely foraging predators that move through the habitat, pushing their snouts beneath fallen leaves and into holes. Three species of *Ameiva* occur together on the Osa Peninsula of Costa Rica in a habitat that extends from full sun (a roadside) to deep shade (forest). The largest of the three species, *A. leptophrys*, has an average body mass of 83g; the middle species, *A. festiva*, weighs 32g; and the smallest, *A. quadrilineata*, weighs 10g (Figure 11–19). The three species forage in different parts of the habitat: *A. quadrilineata* spends most of its time in the short vegetation at the edge of the road, *A. festiva* is found on the bank beside the road, and *A. leptophrys* forages primarily beneath the forest canopy (Figure 11–20). The different foraging sites of the three species may reflect differences in thermoregulation that result from the variation in body size.

The thermoregulatory behavior of the three species is the same: A lizard basks in the sunlight until its body temperature rises to 39 to 40°C, then moves through the mosaic of sun and shade as it searches for food. The body temperature of the lizard drops as it forages, and a lizard ceases foraging and resumes basking when its body temperature has fallen to 35°C. Thus the time that a lizard can forage depends on how long it takes for its body temperature to cool from 39 or 40°C to 35°C. The rate of cooling for *Ameiva* in the shade is inversely proportional to the body size of the three species: *A. quadrilineata* cools in 4 minutes, *A. festiva* in 6 minutes, and *A. leptophrys* in 11 minutes (Figure 11–21). That relationship appears to explain part of the microhabitat separation of the three species: The smallest species, *A. quadrilineata*, cools so rapidly that it may not be able to forage effectively in shady microhabitats, whereas *A. leptophrys* cools slowly and can forage in the shade beneath the forest canopy. The species of intermediate body size, *A. festiva*, uses the habitat with an intermediate amount of shade.

The slow rate of cooling of *A. leptophrys* may explain why it is able to forage in the shade. But field observations indicate that its foraging is actually re-

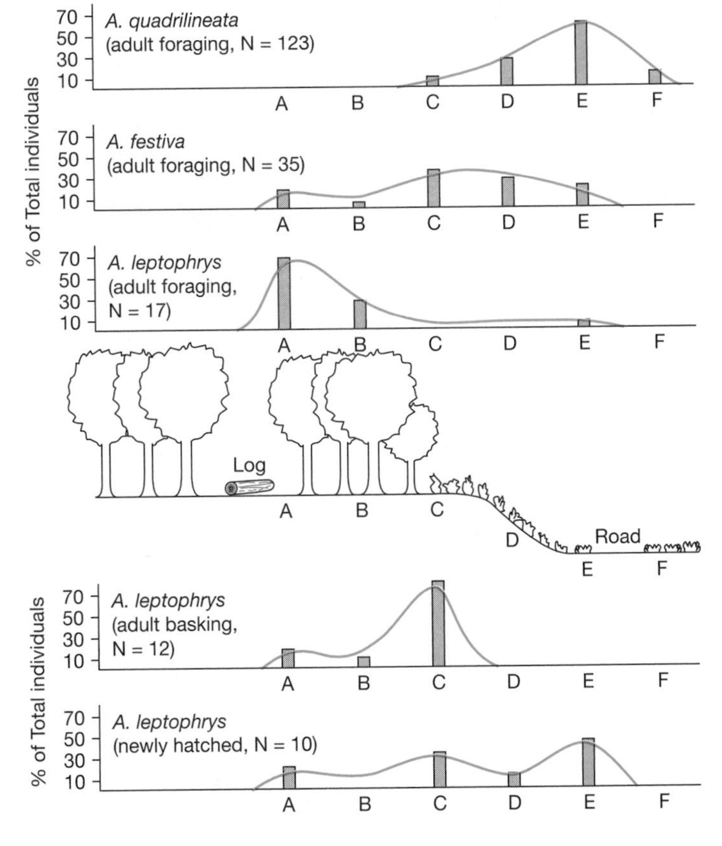

◀ Figure 11–20 Foraging sites of three species of *Ameiva* in Costa Rica. The histograms show the number of individuals of the three species seen in each of six locations: (a) a small clearing in the forest; (b) immediately inside the forest edge; (c) outside the edge of the forest; (d) midway between the edge of the forest and open area; (e) low vegetation beside a road; (f) low vegetation in a large open area without trees.

stricted to shade; it emerges from the forest only to bask. Does some environmental factor prevent *A. leptophrys* from foraging in the sun? The answer to that question may lie in the way the body temperatures of the three species increase when they are in open microhabitats. Body size profoundly affects the equilibrium temperature of an organism in the sun. A lizard warms by absorbing solar radiation, and as it gets warmer, its heat loss by convection, evaporation, and reradiation also increases. When the rate of heat loss equals the rate of heat gain, the body temperature does not increase further. Large lizards reach that equilibrium at higher body temperatures than do small ones. Computer simulations of the heating rates of the three *Ameiva* in sun showed that *A. quadrilineata* and *A. festiva* would reach equilibrium at body temperatures of 37 to 40°C, but *A. leptophrys* would continue to heat until its body temperature reached a lethal 45°C (Figure 11–21). This analysis suggests that *A. leptophrys* would die of heat stress if it spent more than a few minutes in a sunny microhabitat, but that the two smaller species of *Ameiva* would not have that problem.

Thus, as a result of the biophysics of heat exchange, the large body size of *A. leptophrys* apparently allows it to forage in shaded habitats (because it cools slowly), but prevents it from foraging in sunny habitats (because it would overheat). Field observations of the foraging behavior of hatchling *A. leptophrys* emphasize the importance of heat exchange in the foraging behavior of these lizards. Hatchling *A. leptophrys* forage in open habitats like *A. festiva* and *A. quadrilineata* rather than under the forest canopy like adult *A. leptophrys*. That is, the juveniles of the large species of *Ameiva* behave like adults of the smaller species, probably because of the importance of body size and heat exchange in determining the microhabitats in which lizards can thermoregulate.

The difference in the use of various microhabitats by these three species of lizards looks, at first glance, like an example of habitat partitioning in response to interspecific competition for food. That is, because all three species eat the same sort of prey, they could be expected to concentrate their foraging efforts in different microhabitats to reduce competition. How-

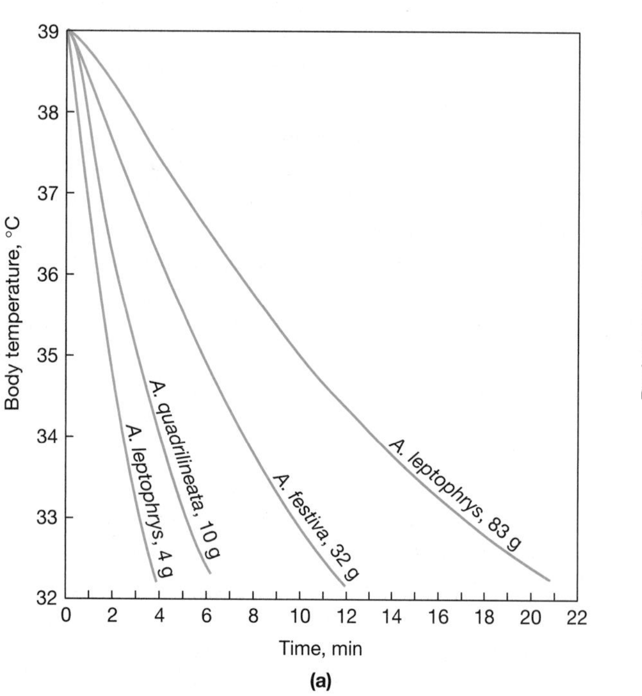

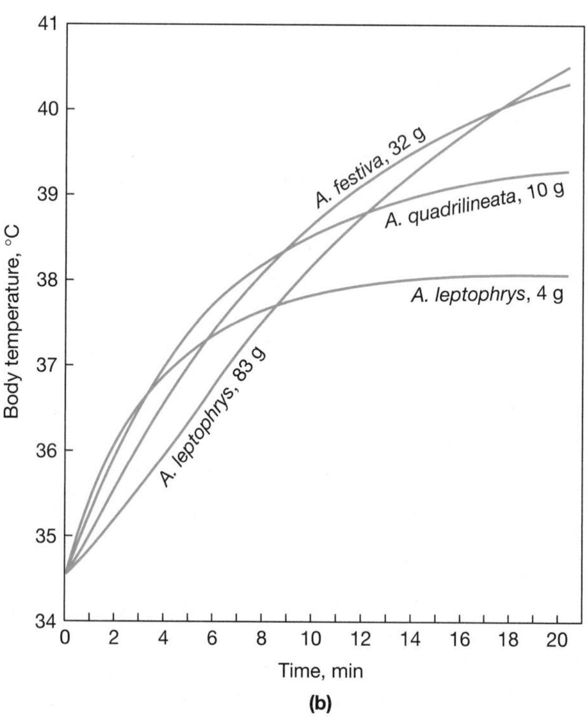

▲ Figure 11–21 Cooling and heating rates of the three sympatric species of *Ameiva*. The largest species, *A. leptophrys*, heats and cools more slowly than the smaller species. If it remained in the sun, its body temperature would rise above 40°C. The smaller species heat and cool more rapidly than the large species and reach temperature equilibrium at lower body temperatures.

ever, this analysis of the thermal requirements of the lizards suggests that interspecific competition for food is, at most, a secondary factor.

If competition for food were important, we would not expect to find hatchling *A. leptophrys* foraging in the same microhabitat as adult *A. quadrilineata*, because the similarity in size of the two lizards would intensify competition for food. The hypothesis that competition for food determines the microhabitat distribution of the animals predicts that the forms most similar in body size should be mostly widely separated in the habitat. In contrast, the hypothesis that energy exchange with the environment is critical in determining where a lizard can forage predicts that species of similar size will live in the same habitat, and that is approximately the pattern seen. Apparently, the physical environment (radiant energy) is more important than the biological environment (interspecific competition for food) in determining the microhabitat distributions of these lizards. That conclusion reflects the broad-scale ecological significance of the morphological and physiological differences between ectotherms and endotherms, a theme that is developed in the next chapter.

Summary

The extant lepidosaurs include the squamates (lizards and snakes) and their sister group, the Sphenodontidae. The lepidosaurs, with more than 6800 species, form the second largest group of extant tetrapods. The two species of tuatara of the New Zealand region are the sole extant sphenodontids. They are lizard-like animals, about 60 cm long, with a dentition and jaw mechanism that give a shearing bite. Sphenodontids were diverse in the Mesozoic and included terrestrial insectivorous and herbivorous species as well as a marine form.

Lizards range in size from tiny geckos only 3 cm long to the Komodo monitor lizard, which reaches a length of 3m. The Iguania is composed of stout-bodied lizards with study legs and considerable diversity of body form. Most Scleroglossa are elongate, and leglessness has developed independently many times within this lineage.

Differences in ecology and behavior parallel the phylogenetic divisions: Many iguanians are sit-and-wait predators that maintain territories and detect prey and intruders by vision. Iguanians often employ colors and patterns in visual displays during courtship and territorial defense. Many scleroglossan lizards are widely foraging predators that detect prey by olfaction and do not maintain territories. Pheromones are important in the social behaviors of many of these lizards.

Amphisbaenians are specialized burrowing lizards. Their skulls are solid structures that they use for tunneling through soil. Many amphisbaenians have blunt heads, and others have vertically keeled or horizontally spade-shaped snouts. The dentition of amphisbaenians appears to be specialized for nipping small pieces from prey too large to be swallowed whole. The skin of amphisbaenians is loosely attached to the trunk, and amphisbaenians slide backward or forward inside the tube of their skin as they move through tunnels with concertina locomotion.

Snakes are derived from scleroglossan lizards. Repackaging the body mass of a vertebrate into a serpentine form has been accompanied by specializations of the mechanisms of locomotion (serpentine, rectilinear, concertina, and sidewinding), prey capture (constriction and the use of venom), and swallowing (a highly kinetic skull).

Many squamates have complex social behaviors associated with territoriality and courtship, but parental care is only slightly developed. Fertilization is internal, and viviparity has evolved 80 or more times among squamates. Thermoregulation is another important behavior of squamates, and various activities are influenced by body temperature. The ecological niches of some lizards may be defined in part by the microhabitats needed to maintain particular body temperatures. Feeding status, pregnancy, and bacterial infections can change the thermoregulatory behavior of squamates, causing an affected individual to maintain a higher or lower body temperature than it otherwise would.

Additional Readings

Andrews, R. M., and B. R. Rose. 1994. Evolution of viviparity: Constraints on egg retention. *Physiological Zoology* 67:1006–1024.

Arnold, E. N. 1988. Caudal autotomy as a defense. In C. Gans and R. B. Huey (Eds.), *Biology of the Reptilia,* vol. 16. New York: Liss, pp. 235–273.

Auffenberg, W. 1981. *The Behavioral Ecology of the Komodo Monitor.* Gainesville: University Presses of Florida.

Avery, R. A. 1976. Thermoregulation, metabolism, and social behaviour in Lacertidae. In A. d'A. Bellairs and C. B. Cox (Eds.), *Morphology and Biology of Reptiles.* London: Academic, 245–259.

Beuchat, C. A. 1986. Reproductive influence on the thermoregulatory behavior of a live-bearing lizard. *Copeia* 1986:971–979.

Beuchat, C. A., and S. Ellner. 1987. A quantitative test of life history theory: Thermoregulation by a viviparous lizard. *Ecological Monographs* 57:45–60.

Cowles, R. B., and C. M. Bogert. 1944. A preliminary study of the thermal requirements of desert reptiles. *Bulletin of the American Museum of Natural History* 83:261–296.

Daugherty, C. H. et al., 1990. Neglected taxonomy and continuing extinctions of tuatara (*Sphenodon*). *Nature* 347:177–179.

Greene, H. 1997. *Snakes.* Berkeley: University of California Press.

Hardy, D. L., Sr. 1994. A re-evaluation of suffocation as the cause of death during constriction. *Herpetological Review* 25:45–47.

Huey, R. B., and A. F. Bennett. 1986. A comparative approach to field and laboratory studies in evolutionary biology. In M. E. Feder and G. V. Lauder (Eds.), *Predator-Prey Relationships,* University of Chicago Press, pp. 82–98.

Huey, R. B., and E. R. Pianka. 1981. Ecological consequences of foraging mode. *Ecology* 62:991–999.

Mason, R. T. 1992. Reptilian pheromones. In C. Gans and D. Crews (Eds.), *Biology of the Reptilia,* vol. 18. Chicago: University of Chicago Press, pp. 114–228.

Shine, R. 1985. The evolution of viviparity in reptiles: An ecological analysis. In C. Gans and F. Billett (Eds.), *Biology of the Reptilia,* vol. 15. New York: Wiley, pp. 606–694.

Shine, R. 1988. Parental care in reptiles. In C. Cans and R. B. Huey (Eds.), *Biology of the Reptilia,* vol. 16. New York: Liss, pp. 275–329.

Sinervo, B. 1994. Experimental manipulations of clutch size and offspring size in lizards: Mechanistic, evolutionary, and conservation considerations. In J. B. Murphy, J. T. Collins, and K. Adler (Eds.), Oxford, U.K.: *Captive Management and Conservation of Amphibians and Reptiles,* Contributions to Herpetology, Society for the Study of Amphibians and Reptiles.

Sinervo, B., P. Doughty, R. B. Huey, and K. Zamudio. 1992. Allometric engineering: A causal analysis of natural selection on offspring size. *Science* 258:1927–1930.

Sinervo, B., and P. Licht. 1991. Hormonal and physiological control of clutch size, egg size, and egg shape in side-blotched lizards (*Uta stansburiana*): Constraints on the evolution of life histories. *Journal of Experimental Zoology* 257:252–264.

Stevenson, R. D., C. R. Peterson, and J. S. Tsuji. 1985. The thermal dependence of locomotion, tongue flicking, digestion, and oxygen consumption in the wandering garter snake. *Physiological Zoology* 58:46–57.

Web Explorations

On-line resources for this chapter are on the World Wide Web at: http://www.prenhall.com/pough (click on the Table of Contents link and then select Chapter 11).

CHAPTER

12

Ectothermy: A Low-Cost Approach to Life

Ectothermy is an ancestral character of vertebrates, but like many ancestral characters it is just as effective as its derived counterpart, endothermy. Furthermore, the mechanisms of ectothermal thermoregulation are as complex and specialized as those of endothermy. Similarly, excretion of nitrogenous wastes in the form of uric acid (uricotely) is an ancestral character of the sauropsid lineage, but it has some advantages that are especially visible among desert-dwelling reptiles. Here we consider the consequences of ectothermy and uricotely in shaping broader aspects of the life-style of fishes, amphibians, and reptiles. The general conclusion from this examination is that success in difficult environments is as likely to reflect the ancestral features of a group as its derived characters.

12.1 Vertebrates and Their Environments

Vertebrates manage to live in the most unlikely places. Amphibians live in deserts where rain falls only a few times a year, and several years may pass with no rainfall at all. Lizards live on mountains at altitudes above 4000 meters where the temperature falls below freezing nearly every night of the year and does not rise much above freezing during the day.

Of course, vertebrates do not seek out only inhospitable places to live—birds, lizards, mammals, and even amphibians can be found on the beaches at Malibu (sometimes running between the feet of surfers), and fishes cruise the shore. However, even this apparently benign environment is harsh for some animals. Examining the ways that vertebrates live in extreme environments has provided much information about how they function as organisms; that is, how morphology, physiology, ecology, and behavior interact.

In some cases elegant adaptations allow specialized vertebrates to colonize demanding habitats. More common and more impressive than these specializations, however, is the realization of how minor are the modifications of the ancestral vertebrate body plan that allow animals to endure environmental temperatures from –70 to +70°C, or water conditions ranging from complete immersion in water to complete independence of liquid water.

The Importance of Ancestral Characters

No obvious differences distinguish animals from vastly different habitats—an Arctic fox looks very much like a desert fox, and a lizard from the Andes Mountains looks like one from the Atacama Desert. The adaptability of vertebrates lies in the combination of minor modifications of their ecology, behavior, morphology, and physiology. A view that integrates these elements shows the startling beauty of organismal function of vertebrates.

Water and Salt Balance

All living representatives of the sauropsid lineage, including turtles and birds, are uricotelic—that is, they excrete nitrogenous wastes, primarily in the form of uric acid. Indeed, uric acid and its salts account for 80 to 90 percent of urinary nitrogen in most species (Table 12.1).

The kidneys of sauropsids lack the long loops of Henle that allow mammals to reduce the volume of urine and raise its osmotic concentration to several

Table 12.1 Distribution of nitrogenous end products among sauropsids

	Total Urinary Nitrogen (%)		
Group	*Ammonia*	*Urea*	*Salts of Uric Acid*
Lepidosaurs			
Tuatara	3–4	10–28	65–80
Lizards and snakes	Small	0–8	90–98
Archosaurs			
Crocodilians	25	0–5	70
Birds	6–17	5–10	60–82
Turtles			
Aquatic	4–44	45–95	1–24
Desert	3–8	15–50	20–50

times the osmotic concentration of the blood plasma. Urine from the kidneys of sauropsids consists of a solution of uric acid and ions that has the same osmotic concentration as the blood plasma, or even a slightly lower concentration. If sauropsids depended solely on the urine-concentrating capacity of their kidneys, they would excrete all their body water in urine. This is where uricotely is important, because uric acid differs from urea in being only slightly soluble in water. Because it has such low solubility, it precipitates from a dilute solution. This is what happens when urine enters the cloaca or bladder. (Many sauropsids lack a urinary bladder entirely; others have an ephemeral bladder that is lost shortly after they hatch; and some sauropsids have a functional bladder throughout life.) The dissolved uric acid combines with ions in the urine and precipitates as a light-colored mass that includes sodium, potassium, and ammonium salts of uric acid as well as other ions held by complex physical forces. When the uric acid and ions precipitate from solution, the urine becomes less concentrated. In effect, water is released and reabsorbed into the blood. In this respect, excretion of nitrogenous wastes as uric acid is even more economical of water than is excretion of urea, because the water used to produce urine is reabsorbed and reused.

Water is not the only substance that is reabsorbed from the cloaca, however. Many sauropsids also reabsorb sodium ions and return them to the bloodstream. At first glance, that seems a remarkably inefficient thing to do. After all, energy was used to create the blood pressure that forced the ions through the walls of the glomerulus into the urine in the first place, and now more energy is being used in the cloaca or bladder to run the active transport system that returns the ions to the blood. The animal has used two energy-consuming processes and it is back where it started, with an excess of sodium and potassium ions in the blood. Why do that?

The solution to the paradox lies in a third water-conserving mechanism that is present in many sauropsids, salt-secreting glands that provide an extrarenal (= in addition to the kidney) pathway that disposes of salt with less water than the urine. Salt glands are widespread among lizards, and in all cases it is the lateral nasal glands that excrete salt. The secretions of the glands empty into the nasal passages, and a lizard expels them by sneezing or by shaking its head. In birds, also, the lateral nasal gland has become specialized for salt excretion. The glands are situated in or around the orbit, usually above the eye. Marine birds (pelicans, albatrosses, penguins) have well-developed salt glands, as do many freshwater birds (ducks, loons, grebes), shorebirds (plovers, sandpipers), storks, flamingos, carnivorous birds (hawks, eagles, vultures), upland game birds, the ostrich, and the roadrunner. Depressions in the supraorbital region of the skull of the extinct aquatic birds *Hesperornis* and *Ichthyornis* suggest that salt glands were present in these forms as well.

In sea snakes (Elapidae) and elephant-trunk snakes (Acrochordidae), the posterior sublingual gland secretes a salty fluid into the tongue sheath, from which it is expelled when the tongue is extended. In some species of homalopsines (a group of aquatic colubrid snakes from the Indoaustralian region) the premaxillary gland secretes salt. Salt-secreting glands on the dorsal surface of the tongue have been identified in several species of crocodiles, in a caiman, and in the American alligator.

Finally, in sea turtles and in the diamondback terrapin, a turtle that inhabits estuaries, the lacrimal glands are greatly enlarged (in some species each gland is larger than the turtle's brain) and secrete a salty fluid around the orbits of the eyes. Photographs of nesting sea turtles frequently show clear paths streaked by tears through the sand that adheres to the turtle's head. Those tears are the secretions of the salt glands. The huge glands leave an imprint on the skull, and the oldest sea turtle known, *Santanachelys gaffneyi* from the Early Cretaceous, clearly had salt-secreting glands. Unlike the situation in lizards, however, salt glands are uncommon among turtles. Terrestrial turtles, even those that live in deserts, do not have salt glands.

The diversity of glands involved in salt excretion among sauropsids indicates that this specialization has evolved independently in various groups. At least five different glands are used for salt secretion by diapsids, indicating that a salt gland is not an ancestral character for the group. Furthermore, although lizards and birds are not very closely related, both use the lateral nasal gland for salt secretion whereas crocodilians, which are closer to birds than to lizards, use lingual glands for salt secretion. Thus salt glands have evolved repeatedly among sauropsids, perhaps in response to the water-conserving opportunities offered by uricotely, which is an ancestral character of the sauropsid lineage.

Despite their different origins and locations, the functional properties of salt glands are quite similar. They secrete fluid containing primarily sodium or potassium cations and chloride or bicarbonate anions in high concentrations (Table 12.2). Sodium is the predominant cation in the salt-gland secretions of marine vertebrates, and potassium is present in the secretions of terrestrial lizards, especially herbivorous species such as the desert iguana. Chloride is the major anion, and herbivorous lizards may also excrete bicarbonate ions.

The total osmolal concentration of the salt gland secretion may reach 2000 mmoles • kg^{-1}—more than six times the osmolal concentration of urine that can

Table 12.2 Salt gland secretions from sauropsids

	Ion Concentration (mmole • kg^{-1})		
Species and Condition	*Na^+*	*K^+*	*Cl^-*
Lizards			
Desert iguana (*Dipsosaurus dorsalis*), estimated field conditions	180	1700	1000
Fringe-toed lizard (*Uma scoparia*), estimated field conditions	639	734	465
Snakes			
Sea snake (*Pelamis platurus*), salt loaded	620	28	635
Homalopsine snake (*Cerberus rhynchops*), salt loaded	414	56	—
Crocodilian			
Saltwater crocodile (*Crocodylus porosus*), natural diet	663	21	632
Birds			
Blackfooted albatross (*Diomeda nigripes*), salt loaded	800–900	—	—
Herring gull (*Larus argentatus*), salt loaded	718	24	—
Turtles			
Loggerhead sea turtle (*Caretta caretta*), seawater	732–878	18–31	810–992
Diamondback terrapin (*Malaclemys terrapin*), seawater	322–908	26–40	—

be produced by the kidney. This efficiency of excretion is the explanation of the paradox of active uptake of salt from the urine. As ions are actively reabsorbed, water follows passively, so an animal recovers both water and ions from the urine. The ions can then be excreted via the salt gland at much higher concentrations, with a proportional reduction in the amount of water needed to dispose of the salt. Thus, by investing energy in recovering ions from urine, sauropsids with salt glands can conserve water by excreting ions through the more efficient extrarenal route.

12.2 Heat—Ectotherms in Deserts

Deserts can be produced by various combinations of topography, air movements, and ocean currents. But whatever their cause, deserts have in common a scarcity of liquid water. A desert is defined as a region in which the potential loss of water (via evaporation and transpiration of water by plants) exceeds the input of water via precipitation. Dryness is at the root of many features of deserts that make them difficult places for vertebrates to live. The dry air characteristic of most deserts seldom contains enough moisture to form clouds that would block solar radiation during the day or radiative cooling at night. As a result, the daily temperature excursion in deserts is large compared with that of more humid areas. Scarcity of water is reflected by sparse plant life and a correspondingly low primary productivity in desert communities. Food shortages may be chronic and are exacerbated by seasonal shortages and unpredictable years of low production when the usual pattern of rainfall does not develop.

Not all deserts are hot; indeed, some are distinctly cold—most of Antarctica and the region of Canada around Hudson Bay and the Arctic Ocean are deserts. The low-latitude deserts north and south of the equator are hot deserts, however, and it is the combination of heat and dryness in low-latitude deserts that create the most difficult problems.

The scarcity of rain contributes to the low primary production of deserts and means that sources of liquid water for drinking are usually unavailable to small animals that cannot travel long distances. These animals obtain water from the plants or animals they eat, but plants and insects have ion balances that differ from those of vertebrates. In particular, potassium is found in higher concentrations in plants and insects than it is in vertebrate tissues, and excreting the excess potassium can be difficult if water is too scarce to waste in the production of large quantities of urine.

The low metabolic rates of ectotherms alleviate some of the difficulty caused by scarcity of food and water, but many desert ectotherms must temporarily extend the limits within which they regulate body temperatures or body fluid concentrations, become inactive for large portions of the year, or adopt a combination of these responses. Tortoises, lizards, and anurans from deserts illustrate these phenomena.

Terrestrial Ectotherms

Terrestrial habitats in deserts are often harsh—hot by day and cold at night. Solar radiation is intense, and air does not conduct heat rapidly. As a result, a sunlit patch of ground can be lethally hot, whereas a shaded area just a few centimeters away can be substantially cooler. Underground retreats offer protection from both heat and cold. The annual temperature extremes at the surface of the ground in the Mohave Desert extend from a low that is below freezing to a maximum above 50°C, but just 1 meter below the surface of the ground the annual range is only from 10 to 25°C. Desert animals rely on the temperature differences between sun and shade and between the surface and underground burrows to escape both hot and cold.

The Desert Tortoise The largest ectothermal vertebrates in the deserts of North America are tortoises. The Bolson tortoise (*Gopherus flavomarginatus*) of northern Mexico probably once reached a shell length of a meter, although predation by humans has apparently prevented any tortoise in recent times from living long enough to grow that large. The desert tortoise (*G. agassizii*) of the southwestern United States is smaller than the Bolson tortoise, but it is still an impressively large turtle (Figure 12–1). Adults can reach shell lengths approaching 50 cm and may weigh 5 kilograms or more. A study of the annual water, salt, and energy budgets of desert tortoises in Nevada shows the difficulties they face in that desert habitat (Nagy and Medica 1986).

Desert tortoises construct shallow burrows that they use as daily retreat sites during the summer and deeper burrows for hibernation in winter. The tortoises in the study area emerged from hibernation in spring, and aboveground activity extended through the summer until they began hibernation again in

(a)

(b)

▲ Figure 12–1 The desert tortoise, *Gopherus agassizii*. (a) An adult tortoise. (b) A tortoise entering its burrow.

November. Doubly labeled water was used to measure the energy expenditure of free-ranging tortoises (Box 12-1).

Figure 12–3 shows the annual cycle of time spent above ground and in burrows by the tortoises, and the annual cycles of energy, water, and salt balance. A positive balance means that the animal shows a net gain, whereas a negative balance represents a net loss. Positive energy and water balances indicate that conditions are good for the tortoises, but a positive salt balance means that ions are accumulating in the body fluids faster than they can be excreted. That situation indicates a relaxation of homeostasis and is probably stressful for the tortoises. The figure shows that the animals were often in negative balance for water or energy, and they accumulated salt during much of the year. Examination of the behavior and dietary habits of the tortoises through the year shows what was happening.

After they emerged from hibernation in the spring, the tortoises were active for about 3 hours every fourth day; the rest of the time they spent in their burrows. From March through May the tortoises were eating annual plants that had sprouted after the winter rains. They obtained large amounts of water and potassium from this diet, and their water and salt balances were positive. Desert tortoises lack salt glands, and their kidneys cannot produce concentrated urine, so they have no way to excrete the salt without losing a substantial amount of water in the process. Instead they retain the salt, and the osmolality of the tortoises' body fluids increased by 20 percent during the spring. This increased concentration shows that they were osmotically stressed due to the high concentrations of potassium in their food. Furthermore, the energy content of the plants was not great enough to balance the metabolic energy expenditure of the tortoises, and they were in negative energy balance. During this period the tortoises were using stored energy by metabolizing their body tissues.

As ambient temperatures increased from late May through early July, the tortoises shortened their daily activity periods to about 1 hour every sixth day. The rest of the time the tortoises spent estivating in shallow burrows. The annual plants died, and the tortoises shifted to eating grass and achieved positive energy balances. They stored this extra energy as new body tissue. The dry grass contained little water, however, and the tortoises were in negative water balance. The osmolal concentrations of their body fluids remained at the high levels they had reached earlier in the year.

In mid-July thunderstorms dropped rain on the study site, and most of the tortoises emerged from estivation. They drank water from natural basins, and some of the tortoises constructed basins by scratching shallow depressions in the ground that trapped rainwater. The tortoises drank large quantities of water (nearly 20 percent of their body mass) and voided the contents of their urinary bladders. The osmolal concentrations of their body fluids and urine decreased as they moved into positive water balance and excreted the excess salts they had accumulated when water was scarce. The behavior of the tortoises changed after the rain: They fed every 2 or 3

BOX 12–1 Doubly Labeled Water

The technique of doubly labeling water is widely employed in studies of the energy consumption of wild vertebrates, because it is the only method of measuring metabolism without restraining the animal in some way. The method measures carbon dioxide production, which in turn can be used to estimate oxygen consumption. "Doubly labeled" refers to water that carries isotopic forms of oxygen and hydrogen: The oxygen atom has been replaced with its stable isotope oxygen-18 (^{18}O), and one hydrogen atom (H) has been replaced by its radioactive form, tritium (^{3}H). (Doubly labeled water is prepared by mixing appropriate quantities of water containing the hydrogen isotope [^{3}HHO] with water containing the oxygen isotope [$H_2{}^{18}O$]. When this mixture is diluted by the body water of an animal, the concentrations of the isotopes are so low that few individual water molecules contain both isotopes.) A measured amount of doubly labeled water is injected into an animal, and allowed to equilibrate with the body water for several hours (Figure 12–2a), and then a small blood sample is withdrawn. The concentrations of tritium and oxygen-18 in the blood are measured, and the total volume of body water can be calculated from the dilution of the doubly labeled water that was injected.

After that first blood sample has been withdrawn, the animal is released and recaptured at intervals of several days or weeks. Each time the animal is recaptured, a blood sample is taken and the concentrations of tritium and oxygen-18 are measured. The calculation of the amount of carbon dioxide produced by the animal is based on the difference in the rates of loss of tritium and oxygen-18. Tritium, which behaves chemically like hydrogen, is lost as water—that is, as ^{3}HHO—whereas oxygen-18 is lost both as water ($H_2{}^{18}O$) and as carbon dioxide ($C^{18}OO$). Thus, the decline in the concentration of tritium in the blood of the animal is a measure of the rate of water loss, and the decline in the concentration of oxygen-18 is a measure of the rates of loss of carbon dioxide and water (Figure 12–2b). The difference between decrease in concentration of tritium and oxygen-18, therefore, is the rate of loss of carbon dioxide, and this is proportional to the rate of oxygen consumption. A full description of the use of doubly labeled water for metabolic studies can be found in Nagy (1983).

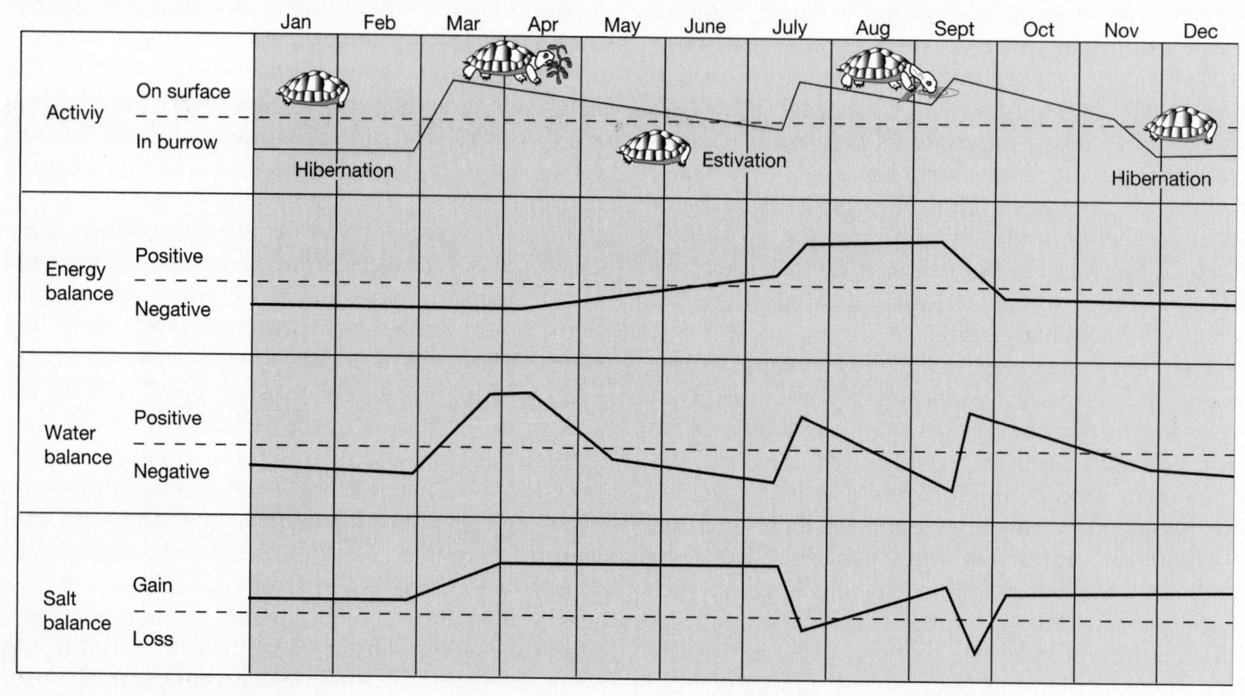

▲ Figure 12–2 Doubly labeled water. (a) The reaction of carbon dioxide and water to produce carbonic acid, and the subsequent reconversion of carbonic acid to water and carbon dioxide produce an equilibrium of ^{18}O between H_2O and CO_2 when water labeled with the isotope is injected into a vertebrate. (b) Differential washout of hydrogen and oxygen isotopes in the body water of an animal that has been injected with doubly labeled water.

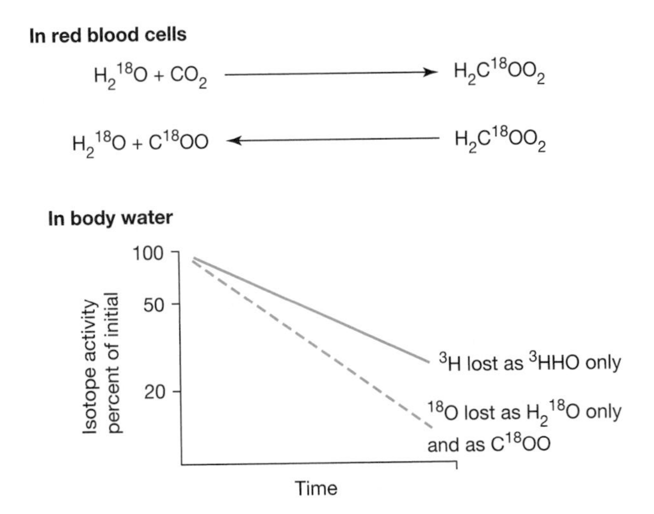

▲ Figure 12–3 Annual cycle of desert tortoises. See text for details.

days and often spent their periods of inactivity above ground instead of in their burrows.

August was dry, and the tortoises lost body water and accumulated salts as they fed on dry grass. They were in positive energy balance, however, and their body tissue mass increased. More thunderstorms in September allowed the tortoises to drink again and to excrete the excess salts they had been accumulating. Seedlings sprouted after the rain, and in late September the tortoises started to eat them.

In October and November the tortoises continued to feed on freshly sprouted green vegetation; but low temperatures reduced their activity, and they were in slightly negative energy balance. Salts accumulated and the osmolal concentrations of the body fluids increased slightly. In November the tortoises entered hibernation. Hibernating tortoises had low metabolic rates and lost water and body tissue mass slowly. When they emerged from hibernation the following spring they weighed only a little less than they had in the fall. Over the entire year, the tortoises increased their body tissues by more than 25 percent, and balanced their water and salt budgets, but they did this by tolerating severe imbalances in their energy, water, and salt relations for periods that extended for several months at a time.

The Chuckwalla The ability to tolerate physiological imbalances is an important aspect of the ability of ectothermal vertebrates to occupy habitats where seasonal shortages of food or water occur. The chuckwalla (*Sauromalus obesus*) is an herbivorous iguanid lizard that lives in the rocky foothills of desert mountain ranges (Figure 12–4). The annual cycle of the chuckwallas, like that of the desert tortoises, is molded by the availability of water. The lizards face many of the same problems that the tortoises encounter, but their responses are different. The lizards have nasal glands that allow them to excrete salt at high concentrations, and they do not drink rainwater but instead depend on water they obtain from the plants they eat.

▶ Figure 12–4 Chuckwalla, *Sauromalus obesus*.

Two categories of water are available to an animal from the food it eats: free water and metabolic water. Free water corresponds to the water content of the food, that is, molecules of water (H_2O) that are absorbed across the wall of the intestine. Metabolic water is a by-product of the cellular reactions of metabolism. Protons are combined with oxygen during aerobic metabolism, yielding a molecule of water for every two protons. The amount of metabolic water produced can be substantial; more than a gram of water is released by metabolism of a gram of fat (Table 12.3). For animals like the chuckwalla that do not drink liquid water, free water and metabolic water are the only routes of water gain that can replace the water lost by evaporation and excretion.

Chuckwallas were studied at Black Mountain in the Mohave Desert of California (Nagy 1972, 1973). They spent the winter hibernating in rock crevices and emerged from hibernation in April. Individual lizards spent about 8 hours a day on the surface in April and early May (Figure 12–5). By the middle of May, air temperatures were rising above 40°C and the chuckwallas retreated into rock crevices for about 2 hours during the hottest part of the day, emerging again in the afternoon. At this time of year annual plants that sprouted after the winter rains supplied both water and nourishment, and the chuckwallas gained weight rapidly. The average increase in body mass between April and mid-May was 18 percent (Figure 12–6). The water content of the chuckwallas increased faster than the total body mass, indicating that they were storing excess water.

By early June the annual plants had withered, and the chuckwallas were feeding on perennial plants that contained less water and more ions than the annual plants. Both the body masses and the water contents of the lizards declined. The activity of the lizards decreased in June and July: Individual lizards emerged in the morning or in the afternoon, but not at both times. In late June the chuckwallas reduced their feeding activity, and in July they stopped eating altogether. They spent most of the day in the rock crevices, emerging only in the late afternoon to bask for an hour or so every second or third day. From late May through autumn the chuckwallas lost water and body mass steadily, and in October they weighed an average of 37 percent *less* than they had in April when they emerged from hibernation.

Table 12.3 Quantity of water produced by metabolism of different substrates

Compound	*Grams of Water/Gram of Compound*
Carbohydrate	0.556
Fat	1.071
Protein	0.396 when urea is the end product; 0.499 when uric acid is the end product

The water budget of a chuckwalla weighing 200 grams is shown in Table 12.4. In early May the annual plants it is eating contain more than 2.5g of free water per gram of dry plant material, and the lizard shows a positive water balance, gaining about 0.8g of water per day. By late May, when the plants have withered, their free water content has dropped to just under a gram of water per gram of dry plant matter, and the chuckwalla is losing about 0.8g of water per day. The rate of water loss falls to 0.5g per day when the lizard stops eating.

Evaporation from the respiratory surfaces and from the skin accounts for about 61 percent of the total water loss of a chuckwalla. When the lizards stop eating, they also become inactive and spend most of the day in rock crevices. The body temperatures of inactive chuckwallas are lower than the temperatures of lizards on the surface. Because of their low body temperatures, the inactive chuckwallas have lower rates of metabolism. They breathe more slowly and lose less water from their respiratory passages. Also, the humidity is higher in the rock crevices than it is on the surface of the desert, and this reduction in the humidity gradient between the animal and the air further reduces evaporation. Most of the remaining water loss by a chuckwalla occurs in the feces (31 percent) and urine (8 percent). When a lizard stops eating, it also stops producing feces and reduces the amount of urine it must excrete. The combination of these effects reduces the daily water loss of a chuckwalla by almost 90 percent.

The food plants were always hyperosmolal to the body fluids of the lizards and had high concentrations of potassium. Despite this dietary salt load, osmotic concentrations of the chuckwallas' body fluids did not show the variation seen in tortoises, because the lizards' nasal salt glands were able to excrete ions at high concentrations. The concentration of potassium ions in the salt-gland secretions was nearly 10 times their concentration in urine. The formation of potassium salts of uric acid was the second major route of potassium excretion by the lizards,

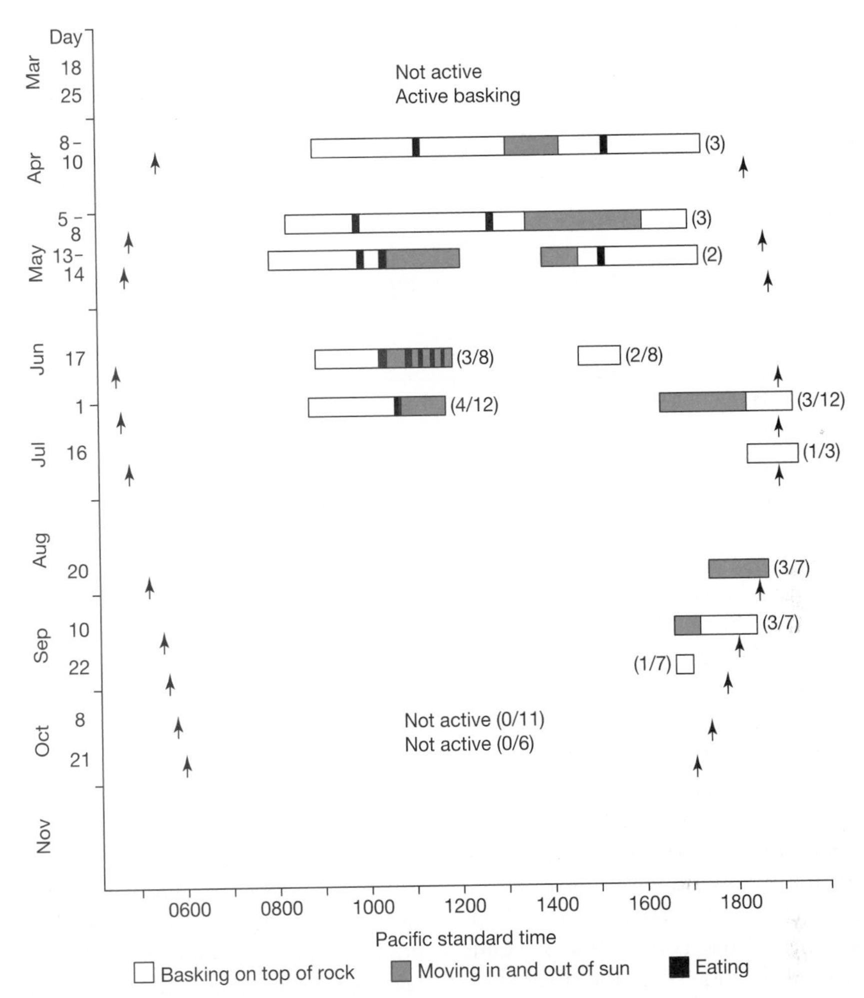

▲ Figure 12–5 Daily behavior patterns in chuckwallas through their activity season. Arrows indicate sunrise and sunset. Numbers in parentheses for April and May indicate the number of animals whose behavior was recorded. Thereafter the fraction in parentheses indicates the number of lizards active and observed out of the number known to be present.

and was nearly as important in the overall salt balance as nasal secretion. The chuckwallas would not have been able to balance their salt budgets without the two extrarenal routes of ion excretion, but with them they were able to maintain stable osmolal concentrations.

Both the chuckwallas and tortoises illustrate the interaction of behavior and physiology in responding to the characteristics of their desert habitats. The tortoises lack salt-secreting glands and store the salt they ingest, tolerating increased body fluid concentrations until a rainstorm allows them to drink water and excrete the excess salt. Some of the tortoises constructed basins that collected rainwater that they drank, whereas other tortoises took advantage of natural puddles. The chuckwallas were able to stabilize their body fluid concentrations by using their nasal glands to excrete excess salt, but they did not take advantage of rainfall to replenish their water stores. Instead they became inactive, reducing their rates of water loss by almost 90 percent, and relying on energy stores and metabolic water production to see them through the period of drought.

Conditions for the chuckwallas were poor at the Black Mountain site during Nagy's study. Only 5 cm of rain had fallen during the preceding winter, and that low rainfall probably contributed to the early withering of the annual plants that forced the chuck-

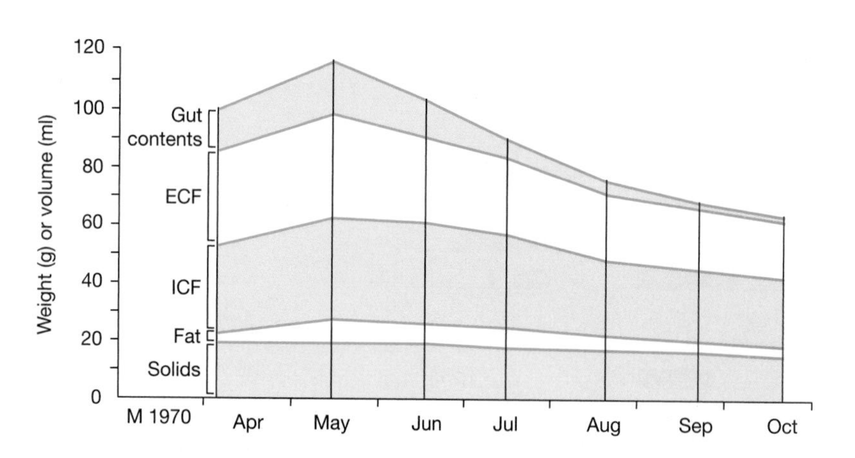

◄ Figure 12–6 Seasonal changes in body composition of chuckwallas. The total body water is composed of the extracellular fluid (ECF; blood plasma, urine, and water in lymph sacs) and the intracellular fluid (ICF; the water inside cells).

wallas to cease activity early in the summer. Unpredictable rainfall is a characteristic of deserts, however, and the animals living there must be able to adjust to the consequences. Rainfall records from the weather station closest to Black Mountain showed that in 5 of the previous 10 years, the annual total rainfall was about 5 cm. Thus the year of the study was not unusually harsh; conditions are sometimes even worse—only 2 cm of rain fell during the winter after the study. However, conditions in the desert are sometimes good. Fifteen centimeters of rain fell in the winter of 1968, and vegetation remained green and lush all through the following summer and fall. Chuckwallas and tortoises live for decades, and their responses to the boom-or-bust conditions of their harsh environments must be viewed in the context of their long life spans. A temporary relaxation of the limits of homeostasis in bad years is an effective trade-off for survival that allows the animals to exploit the abundant resources of good years.

Table 12.4 Seasonal changes in the water balance of a 200-gram chuckwalla

	Early May	*Late May*	*September*
Food intake (g dry mass/day)	2.60	2.86	0.00
Water content of food (g/g dry mass)	2.53	0.96	—
Water gain (g/day)			
Free water	6.56	2.74	0.0
Metabolic water	0.68	0.68	0.20
Total water gain	7.24	3.41	0.20
Water loss (g/day)	6.41	4.26	0.52
Net water flux (g/day)	+0.81	-0.84	-0.32

Desert Amphibians Permeable skins and high rates of water loss are characteristics that would seem to make amphibians unlikely inhabitants of deserts, but certain species are abundant in desert habitats. Most remarkably, these animals succeed in living in the desert *because* of their permeable skins, not despite them. Anurans are the most common desert amphibians, but tiger salamanders are found in the deserts of North America and several species of plethodontid salamanders occupy seasonally dry habitats in California.

The spadefoot toads are the most thoroughly studied desert anurans (Figure 12–7). They inhabit the desert regions of North America—including the edges of the Algodones Sand Dunes in southern California, where the average annual precipitation is only 6 cm and in some years no rain falls at all. An analysis of the mechanisms that allow an amphibian to exist in a habitat like that must include consideration of both water loss and gain. The skin of desert amphibians is as permeable to water as that of species from moist regions. A desert anuran must control its water loss behaviorally by its choice of sheltered microhabitats free from solar radiation and wind movement. Different species of anurans utilize different microhabitats—a hollow in the bank of a desert wash, the burrow of a ground squirrel or kangaroo rat, or a burrow the anuran excavates for itself. All these places are cooler and wetter than exposed ground.

Desert anurans spend extended periods underground, emerging on the surface only when conditions are favorable. Spadefoot toads construct burrows about 60 cm deep, filling the shaft with dirt and leaving a small chamber at the bottom, which they occupy. In southern Arizona the spadefoots construct these burrows in September, at the end of

▶ Figure 12–7 A desert spadefoot toad, *Scaphiopus multiplicatus.*

the summer rainy season, and remain in them until the rains resume the following July.

At the end of the rainy season when the spadefoots first bury themselves, the soil is relatively moist. The water tension created by the normal osmolal concentration of a spadefoot's body fluids establishes a gradient favoring movement of water from the soil into the toad. In this situation, a buried spadefoot can absorb water from the soil just as the roots of plants do. With a supply of water available, a spadefoot toad can afford to release urine to dispose of its nitrogenous wastes.

As time passes, the soil moisture content decreases and the soil moisture potential becomes more negative, until it equals the water potential of the spadefoot. At this point there is no longer a gradient allowing movement of water into the toad. When its source of new water is cut off, a spadefoot stops excreting urine and instead retains urea in its body, increasing the osmotic pressure of its body fluids. Osmotic concentrations as high as 600 mmole • $[kg\ H_2O]^{-1}$ have been recorded in spadefoot toads emerging from burial at the end of the dry season. The low water potential produced by the high osmolal concentration of the spadefoot's body fluids may reduce the water gradient between the animal and the air in its underground chamber, so that evaporative water loss is reduced. Sufficiently high internal concentrations should create potentials that would allow spadefoot toads to absorb water from even very dry soil.

The ability to continue to draw water from soil enables a spadefoot toad to remain buried for 9 or 10 months without access to liquid water. In this situation its permeable skin is not a handicap to the spadefoot—it is an essential feature of the toad's biology. If the spadefoot had an impermeable skin, or if it formed an impermeable cocoon as some other amphibians do, water would not be able to move from the soil into the animal. Instead, the spadefoot would have to depend on the water contained in its body when it was buried. Under those circumstances spadefoot toads would probably not be able to invade the desert, because their initial water content would not see them through a 9-month dry season.

A different pattern of adaptation to arid conditions is seen in a few tree frogs. The African rhacophorid *Chiromantis xerampelina* and the South American hylid *Phyllomedusa sauvagei* lose water through the skin at a rate only one-tenth that of most frogs. *Phyllomedusa* has been shown to achieve this low rate of evaporative water loss by using its legs to spread the lipid-containing secretions of dermal glands over its body surface in a complex sequence of wiping movements. These two frogs are unusual also because they are uricotelic rather than ureotelic. Their uricotelism provides still more water conservation.

Aquatic Ectotherms in the Desert

An aquatic habitat in a desert sounds like a contradiction. But in fact several different kinds of aquatic habitats are found in deserts, and some of them have distinctive faunas of fishes or amphibians. Temporary pools that are formed by heavy rains and last for only a few weeks are the breeding sites for most species of desert amphibians. The ephemeral nature of these habitats puts a premium on rapid development. Spadefoot toads, for example, can grow from an egg to metamorphosis in 2 or 3 weeks. Fishes require permanent water, and desert fishes are found in rivers, springs, and desert lakes.

Aquatic habitats in the desert have the same temperature extremes as terrestrial habitats, but there are some differences that are important to the animals living in them. Water has a much higher heat capacity than does air, so water temperature changes more slowly than does air temperature. Temperature changes from day to night are usually smaller in a pool of water than on land, but pools of water in the desert can reach lethally hot temperatures by the end of summer. Water conducts heat faster than does air, and the high heat capacity and conductance of water ensure that the body temperatures of small aquatic organisms are always close to water temperature. Aquatic organisms can regulate their body temperature by moving between areas of different water temperature.

Tadpoles The ability of aquatic animals to select the most favorable temperatures available is shown by observations of tadpoles of the foothill yellow-legged frog, *Rana boylii*. The tadpoles were in a small cove with a maximum depth of about 40 cm and a shallow area that was only 10 cm deep, and they moved about in this area as temperature changed (Figure 12–8). At night (2100 hours) all the tadpoles were in the deepest part of the cove where the temperature was warmest, and they remained there until morning. During the morning the sun warmed the water in the pool, and the shallow area warmed fastest. Between 0900 and 1000 hours the tadpoles moved from the deep water into the shallow area, and by 1100 hours all of the tadpoles were in the shallows. By midday the shallow water had became too warm, and all of the tadpoles moved back into the deep part of the pool, which remained cool. The shallow parts of the cove cooled rapidly in the late afternoon. Water temperature was essentially the same everywhere, and the tadpoles were distributed all over the cove. As temperature continued to fall, the tadpoles moved back into the deepest part of the pool for the night.

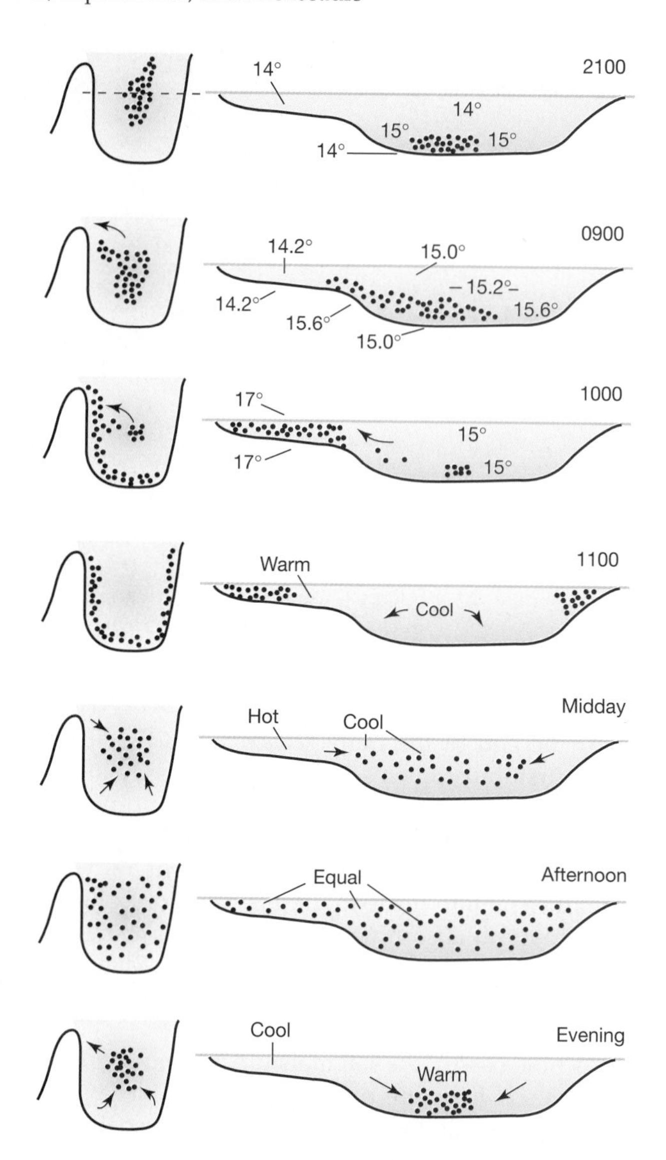

▲ Figure 12–8 Temperature selection by tadpoles. Tadpoles of *Rana boylii* change positions in a pool of water in response to temperature changes during the day. The left column shows the view looking down on the pool, and the right column shows a cross section of the pool at the position indicated by the dashed line in the top drawing.

Desert Fishes Permanent aquatic habitats in the desert may have populations of fishes, but relatively few species can tolerate the high temperatures and high salinities that characterize many of these bodies of water. Pupfish (Cyprinodontidae), minnows (Cyprinidae), and cichlids (Cichlidae) are the groups most often found in deserts. Desert lakes and springs rarely contain more than five species of fishes, and many habitats are occupied by a single species.

The problems that desert fishes encounter depend on the habitat in which they occur. Springs, such as those at Ash Meadows in Death Valley, may be relatively large or very small. Big Spring is 15m in diameter and 9m deep, whereas Mexican Spring is less than 2m in diameter and only 2 to 5 cm deep. The water in a spring usually has the same temperature and salinity year-round, but some springs are 30°C or above and others are near 20°C.

Ponds and lakes often show more seasonal variation in temperature and salinity. Some of these bodies of water are the remnants of enormous lakes that filled the desert valleys during the Pleistocene. Pyramid Lake in Nevada, for example, is a remnant of Lake Lahonton, which once covered more than 20,000 km^2 in California and Nevada. Former shorelines of Lake Lahonton are marked by three terraces that are now 34, 100, and 163 meters above the present lake level. Other remnants of Lake Lahonton include Honey Lake in California and Walker Lake and the Carson and Humboldt Sinks in Nevada. Similar Pleistocene lake terraces can be found in the Sahara Desert and in the deserts of the Middle East. As the Pleistocene lakes shrank, once widespread populations of fishes were isolated in the remnant bodies of water. (The Salton Sea of California has a different origin from the other lakes of the Southwest. It lies in the basin of Pleistocene Lake LaConte, but the Salton Sea was formed in 1905 when the Colorado River broke through the banks of an irrigation system and formed a lake 25m deep and more than 1300 km^2 in area.)

High temperatures, salinity, and variation in oxygen concentration make desert lakes difficult environments for fishes. Pupfish (species of *Cyprinodon*) are widespread in these lakes and cope well with both heat and salinity. At Quitobaquito Springs in Organ Pipe National Monument, for example, desert pupfishes (*Cyprinodon macularis*) are found in shallow water at temperatures of 40 to 41°C (which is only 2 or 3°C below the lethal temperature for the fish) rather than in water of 30°C that is a few meters away. Other species of *Cyprinodon* have been found voluntarily inhabiting water with temperatures of 43 to 44°C.

Desert lakes are often characterized by high salinities as well as by high temperatures. Juvenile *Cyprinodon* can tolerate salt concentrations at least up to 90g of dissolved salt/liter of water, which is approximately three times the concentration of seawater. Young pupfishes and large adults are slightly less tolerant of salinity than are juveniles, and eggs will not develop in salinities greater than 70 g/liter. A mosaic of salinities is often available to pupfishes because the density of water increases as the salinity increases, and water masses with different salinities tend not to mix. Thus, a fish can select areas of different salinity within a habitat just as it can select areas with different temperatures.

Streams are a third habitat for desert fishes. Stream flow varies through the year, and in dry seasons surface water in intermittent streams is reduced to isolated pools separated by dry areas of the stream bed. In the rainy season, however, the stream may flow continuously and even flood. Populations of fishes in desert streams typically wax and wane in response to changes in water flow. Salt Creek in Death Valley is inhabited by Devil's Hole pupfishes (*Cyprinodon salinus*). The population increases 100-fold as water from winter rains in the Mohave Desert flows down the stream in the winter and spring, and then crashes in summer and autumn as the stream dries. Some desert fishes, such as pupfishes and the longfin dace (*Agosia chrysogaster*), penetrate far into temporary aquatic habitats during periods of heavy rain and runoff. When desert streams are close to drying up, evaporation may consume the entire flow of water during the day, leaving only a damp streambed. Most species of fishes die under these conditions, but longfin dace survive beneath water-saturated mats of algae. Stream flow resumes at night when the temperature falls, and the dace emerge from beneath the algae, swimming about and feeding in a few millimeters of water. The fishes can survive in these conditions for several weeks, and if rain refills the stream they are able to return to areas of permanent water.

12.3 Cold—Ectotherms in Subzero Conditions

Temperatures drop below freezing in the habitats of many vertebrates on a seasonal basis, and some animals at high altitudes may experience freezing temperatures on a daily basis for a substantial part of the year. Birds and mammals respond to cold by increasing metabolic heat production and insulation, but ectotherms do not have those options. Instead, ectotherms show one of two responses—they avoid freezing by supercooling or synthesizing antifreeze compounds, or they tolerate freezing and thawing by using mechanisms that prevent damage to cells and tissues.

Cold Fishes

The temperature at which water freezes is affected by its osmolal concentration: Pure water freezes at 0°C, and increasing the osmolal concentration lowers the freezing point. Body fluid concentrations of marine fishes are 300 to 400 mmole • $[kg\ H_2O]^{-1}$, whereas seawater has a concentration near 1000 mmole • $[kg\ H_2O]^{-1}$. The osmolal concentrations of the body fluids of marine fishes correspond to freezing points of –0.6 to 0.8°C, and the freezing point of seawater is –1.86°C. The temperature of Arctic and Antarctic seas falls to –1.8°C in winter, yet the fishes swim in this water without freezing.

An early study of freezing avoidance of fishes was conducted by P. F. Scholander and his colleagues in Hebron Fjord in Labrador (Scholander et al. 1957). In summer the temperature of the surface water at Hebron Fjord is above freezing, but the water at the bottom of the fjord is –1.73°C (Figure 12–9). In winter the surface temperature of the water falls to –1.73°C, like the bottom temperature. Several species of fishes live in the fjord, and some are bottom dwellers, whereas others live near the surface. These two zones present different problems to the fishes. The temperature near the bottom of the fjord is always below freezing, but ice is not present because ice is lighter than water and remains at the surface. Surface-dwelling fishes live in water temperatures that rise well above freezing in the summer and drop below freezing in winter, and they are also in the presence of ice.

The body fluids of bottom-dwelling fish in Hebron Fjord have freezing points of –0.8°C year-round. Because the body temperatures of these fishes are –1.73°C, the fishes are supercooled. That is, the water in their bodies is in the liquid state even though it is below its freezing point. When water freezes, the water molecules become oriented in a crystal lattice. The process of crystallization is accelerated by nucleating agents that hold water molecules in the proper spatial orientation for freezing, and in the absence of nucleating agents pure water can remain liquid at –20°C. In the laboratory the fishes from the bottom of Hebron Fjord can be supercooled to –1.73°C without freezing; but if they are touched with a piece of ice, which serves as a nucleating agent, they freeze immediately. At the bottom of the fjord there is no ice, and the bottom-dwelling fishes exist year-round in a supercooled state.

What about the fishes in the surface waters? They do encounter ice in winter when the water temperature is below the osmotically determined freezing point of their body fluids, and supercooling would not be effective in that situation. Instead, the surface-dwelling fishes synthesize antifreeze substances in winter that lower the freezing point of their body fluids to approximately the freezing point of the seawater in which they swim.

Antifreeze compounds are widely developed among vertebrates (and also among invertebrates and plants). Marine fishes have two categories of organic molecules that protect against freezing: glycoproteins with molecular weights of 2,600 to 33,000, and polypeptides and small proteins, with molecular weights of 3,300 to 13,000 (Hew et al. 1986, Davies et al. 1988). These compounds are extremely effective in preventing freezing. For example, the blood plasma of the Antarctic fish *Trematomus borchgrevinki* contains a glycoprotein that is several hundred times more effective than salt (sodium chloride) in lowering the freezing point. The glycoprotein is absorbed onto the surface of ice crystals and hinders their growth by preventing water molecules from assuming the proper orientation to join the ice crystal lattice.

Chilled Lizards

Some terrestrial ectotherms rely on supercooling. Mountain-dwelling lizards such as Yarrow's spiny lizard (*Sceloporus jarrovi*), which lives at altitudes up to 3000m in western North America, are exposed to temperatures below freezing on cold nights, but sunny days permit thermoregulation and activity. These animals have osmolal concentrations of 300 mmole • $H_2O]^{-1}$, which correspond to freezing points of –0.6°C, but they withstand substantially lower temperatures before they freeze. For example, spiny lizards supercooled to an average temperature of –5.5°C before they froze. At –3°C the lizards had not frozen after 30 hours. The lizards spent the nights in rock crevices that were 5 to 6°C warmer than the air temperature, and the combination of this protection and their ability to supercool was usually sufficient to allow the lizards to survive. However, a few individuals at the highest altitudes were found frozen in their rock crevices during most winters.

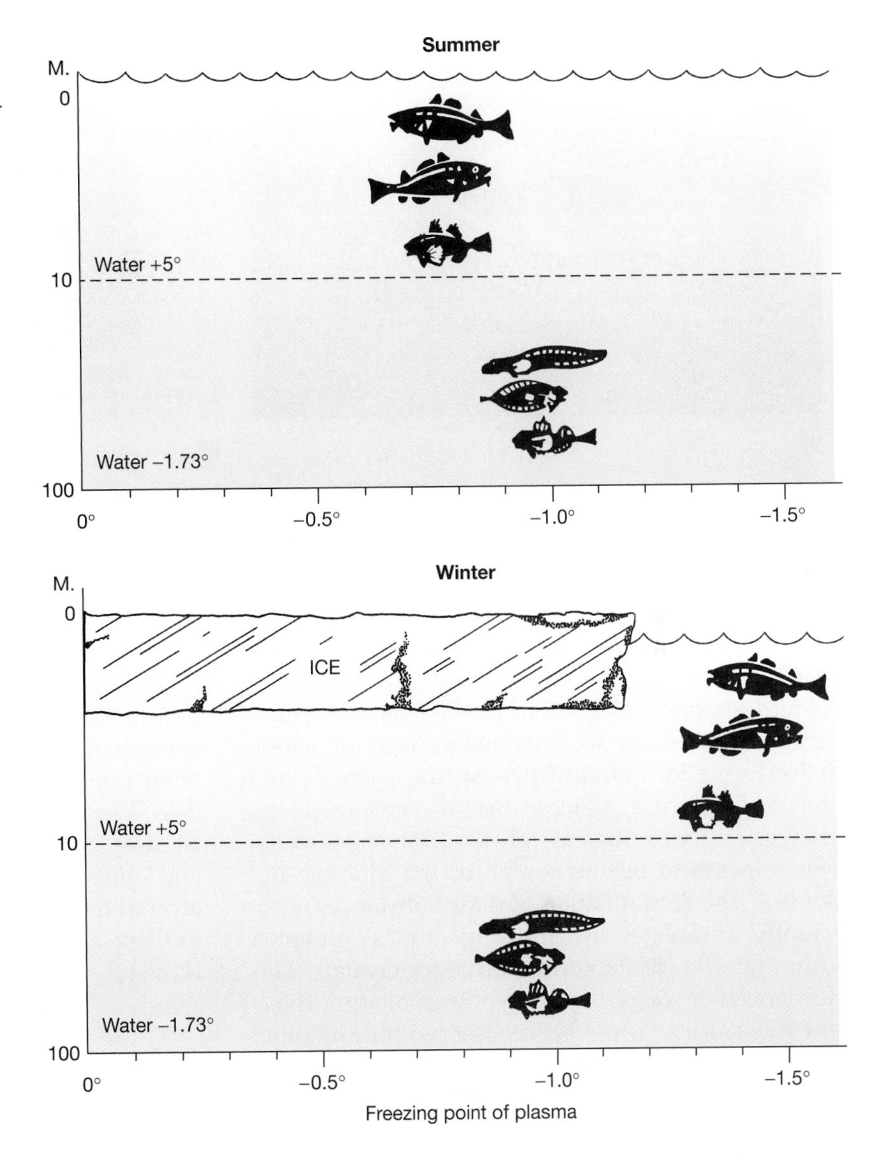

▶ Figure 12–9 Water temperatures and distribution of fishes in Hebron Sound in summer and winter. The freezing point of the fishes' blood plasma is indicated by the position of the symbol on the horizontal axis. Shallow-water fishes show a decrease of 0.7°C in the freezing point of their blood in winter, whereas deep-water fishes have the same freezing point all year.

Frozen Frogs

Terrestrial amphibians that spend the winter in hibernation show at least two categories of responses to low temperatures. One group, which includes salamanders, toads, and aquatic frogs, buries deeply in the soil or hibernates in the mud at the bottom of ponds. These animals apparently are not exposed to temperatures below the freezing point of their body fluids. As far as we know, they have no antifreeze substances and no capacity to tolerate freezing. However, other amphibians apparently hibernate close to the soil surface, and these animals are exposed to temperatures below their freezing points. Unlike fishes and lizards, these amphibians freeze at low temperatures, but they are not killed by freezing (Figure 12–10). These species can remain frozen at −3°C for several weeks, and they tolerate repeated bouts of freezing and thawing without damage. However, temperatures below −10°C are lethal.

Tolerance of freezing refers to the formation of ice crystals in the extracellular body fluids; freezing of the fluids inside the cells is apparently lethal. Thus, freeze tolerance involves mechanisms that control the distribution of ice, water, and solutes in the bodies of animals (Storey 1986). The ice content of frozen frogs is usually in the range of 34 to 48 percent. Freezing of more than 65 percent of the body water

(a)

(b)

▲ Figure 12–10 Wood frog (*Rana sylvatica*). (a) At normal temperature. (b) Frozen.

appears to cause irreversible damage, probably because too much water has been removed from the cells.

Freeze-tolerant frogs accumulate low-molecular-weight substances in the cells that prevent intracellular ice formation. Wood frogs, spring peepers, and chorus frogs use glucose as a cryoprotectant, whereas gray tree frogs use glycerol. Glycogen in the liver appears to be the source of the glucose and glycerol. The accumulation of these substances is apparently stimulated by freezing, and is initiated within minutes of the formation of ice crystals. This mechanism of triggering the synthesis of cryoprotectant substances has not been observed for any other vertebrates or for insects.

Frozen frogs are, of course, motionless. Breathing stops, the heartbeat is exceedingly slow and irregular or may cease entirely, and blood does not circulate through frozen tissues. Nonetheless, the cells are not frozen; they have a low level of metabolic activity that is maintained by anaerobic metabolism. The glycogen content of frozen muscle and kidney cells decreases, and concentrations of lactic acid and alanine (two end products of anaerobic metabolism) increase.

The ecological significance of freeze tolerance in some species of amphibians is unclear. The four species of frogs so far identified as being freeze tolerant all breed relatively early in the spring, and shallow hibernation may be associated with early emergence in the spring. Being among the first individuals to arrive at the breeding pond may increase the chances for a male of obtaining a mate, and it gives larvae the longest possible time for development and metamorphosis, but it also entails risks. Frequently, frogs and salamanders move across snow banks to reach the breeding ponds, and they enter ponds that are still partly covered with ice. A cold snap can lead to the entire surface of the pond freezing again, trapping some animals under the ice and others in shallow retreats under logs and rocks around the pond. Freeze tolerance may be important to these animals even after their winter hibernation is over.

12.4 The Role of Ectothermal Tetrapods in Terrestrial Ecosystems

Life as an animal is costly. In thermodynamic terms, an animal lives by breaking chemical bonds that were formed by a plant (if the animal is an herbivore) or by another animal (if it's a carnivore) and using the energy from those bonds to sustain its own activities. Vertebrates are particularly expensive animals because in general vertebrates are larger and more mobile than invertebrates. Big animals require more energy (i.e., food) than small ones, and active animals use more energy than sedentary ones.

In addition to body size and activity, an animal's method of temperature regulation (ectothermy, endothermy, or a combination of the two mechanisms) is a key factor in determining how much energy it uses, and therefore how much food it requires. Because ectotherms rely on external sources of energy to raise their body temperatures to the level needed for activity, whereas endotherms use heat generated internally by metabolism, ectotherms use substantially less energy than do endotherms. The metabolic rates (i.e., rates at which energy is used) of terrestrial ectotherms are only 10 to 14 percent of the metabolic rates of birds and mammals of the same body size. The lower energy requirements of ectotherms mean that they need less food than would an endotherm of the same body size.

Body size is another major difference between ectotherms and endotherms that relates directly to their mode of temperature regulation and affects their roles in terrestrial ecosystems. Ectotherms are smaller than endotherms, partly because the energetic cost of endothermy is very high at small body sizes. As body mass decreases, the mass-specific cost of living (energy per gram) for an endotherm increases rapidly, becoming nearly infinite at very small body sizes (Figure 12–11). This is a finite world, and infinite energy requirements are just not feasible. Thus energy requirements, among other factors, apparently set a lower limit to the body size possible for an endotherm.

The mass-specific energy requirements of ectotherms also increase at small body sizes, but because the energy requirements of ectotherms are about one-tenth those of endotherms of the same body size, an ectotherm can be about an order of magnitude smaller than an endotherm. A mouse-size mammal weighs about 20g, and few adult birds and mammals have body masses less than 10g. The very smallest species of birds and mammals weigh about 3g, but many ectotherms are only one-tenth that size (0.3g). Amphibians are especially small—20 percent of the species of salamanders and 17 percent of the species of anurans have adult body masses less than 1g, and 65 percent of salamanders and 50 percent of anurans are smaller than 5g. Squamates are generally larger than amphibians, but 18 percent of the species of lizards and 2 percent of snakes weigh less than 1g. Even the largest amphibians and squamates weigh substantially less than 100kg, whereas more than 5 percent of the species of extant mammals weigh 100kg or more (Figure 12–12).

Body shape is another aspect of vertebrate body form in which ectothermy allows more flexibility than does endothermy. An animal exchanges heat with the environment through its body surface, and the surface area of the body in relation to the mass of the body (the surface/mass ratio) is one factor that determines how rapidly heat is gained or lost. Small animals have higher surface/mass ratios than large ones, and that is why endothermy becomes increasingly expensive at progressively smaller body

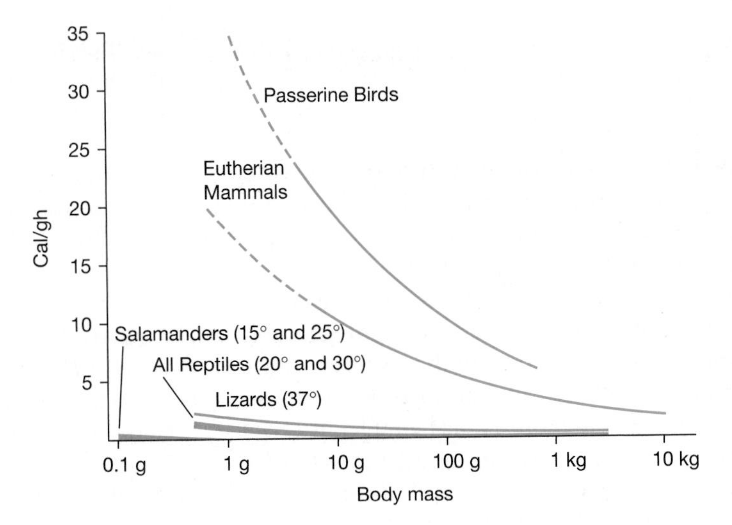

▶ Figure 12–11 Resting metabolic rate as a function of body size for terrestrial vertebrates. (This semilogarithmic presentation emphasizes the dramatic increase in mass-specific metabolic rate at small body sizes—compare it with the log-log graph in Figure 4–13.) Metabolic rates for salamanders are shown at 15 and 25°C as the lower and upper limits of the darkened area, and data for all reptiles combined are shown at 20 and 39°C. The curve for anurans falls within the all-reptiles area, and the relationship for nonpasserine birds is similar to that for eutherian mammals. Dotted portions of the lines for birds and mammals show hypothetical extensions into body sizes below the minimum for adults of most species of birds and mammals.

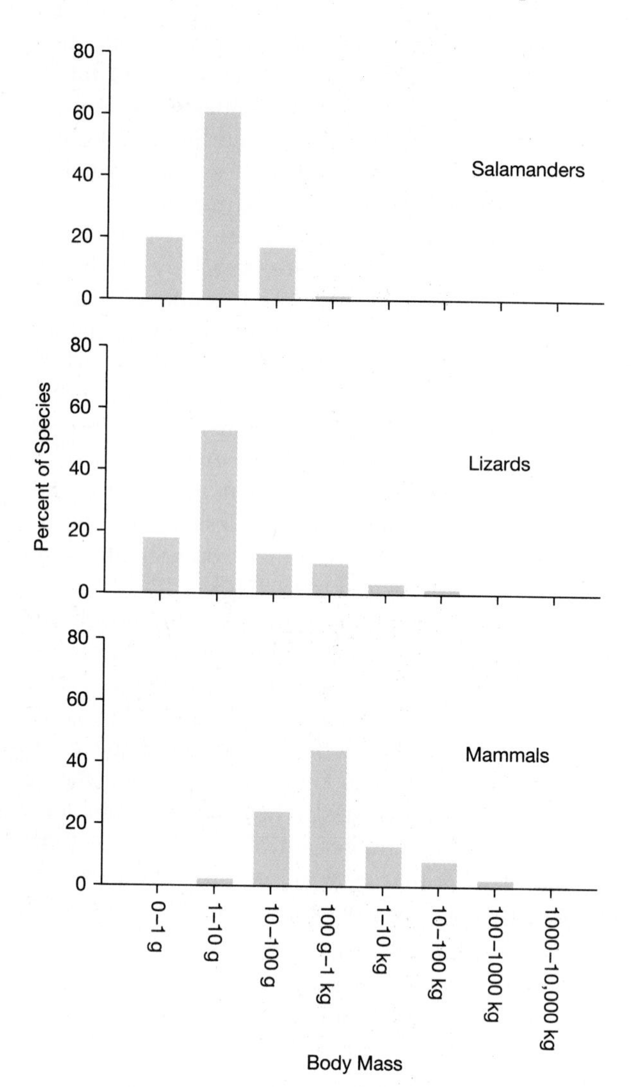

▲ Figure 12–12 Adult body masses of amphibians reptiles, and mammals.

sizes. Metabolic rates of small endotherms must be high enough to balance the high rates of heat loss across their large body surface areas.

Similarly, body shapes that increase surface/mass ratios have an energy cost that makes them disadvantageous for endotherms. There are no highly elongate endotherms, whereas elongate body forms are widespread among fishes (true eels, moray eels, pipefishes, barracudas, and many more), amphibians (all caecilians and most salamanders, especially the limbless aquatic forms), and reptiles (many lizards and all snakes). Dorsoventral or lateral flattening is another shape that increases surface/mass ratio. There are no flat endotherms, but flat fishes are common (dorsoventral flattening—skates, rays, flounders; lateral flattening—many coral-reef and freshwater fishes). Some reptiles are also flat (dorsoventral flattening—aquatic turtles, especially soft-shelled turtles, horned lizards; lateral flattening—many arboreal lizards, especially chameleons). Small body sizes and specialized body forms allow ectotherms to fill ecological niches that are not available to endotherms.

The amount of energy required by ectotherms and endotherms is not the only important difference between them; equally significant is what they do with that energy once they have it. Endotherms expend more than 90 percent of the energy they take in to produce heat to maintain their high body temperatures. Less than 10 percent—often as little as 1 percent—of the energy a bird or mammal assimilates is available for net conversion (that is, increasing the species' biomass by growth of an individual or production of young). Ectotherms do not rely on metabolic heat. The solar energy they use to warm their bodies is free in the sense that it is not drawn from their food. Thus, most of the energy they ingest is converted into the biomass of their species. Values of net conversion for amphibians and reptiles are between 30 and 90 percent (Table 12.5).

Due to that difference in how energy is used, a given amount of chemical energy invested in an ectotherm produces a much larger biomass return than it would have from an endotherm. A study of salamanders in the Hubbard Brook Experimental Forest in New Hampshire showed that, although their energy consumption was only 20 percent that of the birds or small mammals in the watershed, their conversion efficiency was so great that the annual increment of salamander biomass was equal to that of birds or small mammals. Similar comparisons can be made among lizards and rodents in deserts.

Small amphibians and squamates occupy key positions in the energy flow through an ecosystem. Because these animals are so small, they can capture tiny insects and arachnids that are too small to be eaten by birds and mammals. Because they are ectotherms, they are efficient at converting the energy in the food they eat into their own body tissues. As a result, the small ectothermal vertebrates in terrestrial ecosystems can be viewed as repackaging energy into a form that avian and mammalian predators can exploit. In other words, when a shrew or a bird searches for a meal in the Hubbard Brook Forest, the most abundant vertebrate prey it will find is salamanders. In this context, frogs, salamanders, lizards, and snakes occupy a position in terrestrial ecosys-

Table 12.5 Efficiency of biomass conversion by ectotherms and endotherms. These are net conversion eficiencies calculated as (energy converted/energy assimilated) ×100

Ectotherms *Species*	*Efficiency(%)*	*Endotherms* *Species*	*Efficiency(%)*
Red-backed salamander (*Plethodon cinereus*)	48	Kangaroo rat (*Dipodomys merriami*)	0.8
Mountain salamander (*Desmognathus ochrophaeus*)	76–98	Field mouse (*Peromyscus polionotus*)	1.8
Panamanian anole (*Anolis limifrons*)	23–28	Meadow vole (*Microtus pennsylvanicus*)	3.0
Side-blotched lizard (*Uta stansburiana*)	18–25	Red squirrel (*Tamiasciurius hudsonicus*)	1.3
Hognose snake (*Heterodon contortrix*)	81	Least weasel (*Mustela rixosa*)	2.3
Python (*Python curtus*)	6–33	Savanna sparrow (*Passericulus sandwichensis*)	1.1
Adder (*Vipera berus*)	49	Marsh wren (*Telmatodytes palustris*)	0.5
Average of 12 species	50	Average of 19 species	1.4

tems that is important both quantitatively (in the sense that they constitute a substantial energy resource) and qualitatively (in that ectotherms exploit food resources that are not available to endotherms).

In a very real sense, small ectotherms can be thought of as living in a different world from that of endotherms. As we saw in the case of the three species of *Ameiva* lizards in Costa Rica, interactions with the physical world may be more important in shaping the ecology and behavior of small ectotherms than are biological interactions such as competition. In some cases these small vertebrates may have their primary predatory and competitive interactions with insects and arachnids rather than with other vertebrates. For example, orb-web spiders and *Anolis* lizards on some Caribbean islands are linked by both predation (adult lizards eat spiders, and spiders may eat hatchling lizards) and competition (lizards and spiders eat many of the same kinds of insects). The competitive relationship between these distantly related animals was demonstrated by experiments. When lizards were removed from experimental plots, the abundance of insect prey increased, and the spiders consumed more prey and survived longer than in control plots with lizards present (Schoener and Spiller 1987).

Ectothermy and endothermy thus represent fundamentally different approaches to the life of a terrestrial vertebrate. An appreciation of ectotherms and endotherms as animals requires understanding the functional consequences of the differences between them. Ectothermy is an ancestral character of vertebrates, but it is a very effective way of life in modern ecosystems.

Summary

All living organisms are mosaics of ancestral and derived characters, interacting to make an organism and functional entity. Uricotely is an ancestral character of sauropsids that is retained in extant forms. Because uric acid precipitates from solution, it releases water that can be absorbed back into the blood. Thus sauropsids can excrete nitrogen with little loss of water despite the inability of their kidneys to produce urine that is more concentrated than their blood. Extrarenal salt-secreting glands have evolved independently at least five times among sauropsids, and these glands can produce salt solutions that are more than five times as concentrated as the blood, providing additional economy in the animal's water balance.

Ectothermy is another ancestral character that shapes the lives of vertebrates. Ectotherms do not use chemical energy from the food they eat to maintain high body temperatures. The results of that ancestral vertebrate characteristic are far-reaching for extant ectothermal vertebrates, and ectotherms and endotherms represent quite different approaches to vertebrate life.

Because of their low energy requirements, ectotherms can colonize habitats in which energy is in short supply. Ectotherms are able to extend some of their limits of homeostasis to tolerate high or low body temperatures and high or low body-water contents when doing so allows them to survive in difficult conditions.

When food is available, ectotherms are efficient at converting the energy it contains into their own body tissues for growth or reproduction. Net conversion efficiencies of ectotherms average 50 percent of the energy assimilated compared with an average of 1.4 percent for endotherms.

Ectotherms can be smaller than endotherms because their mass-specific energy requirements are low, and many ectotherms weigh less than a gram, whereas most endotherms weigh more than 10g. Because of this difference in body size, many small ectotherms—such as salamanders, frogs, and lizards—eat prey that is too small to be consumed by endotherms. The efficiency of energy conversion by ectotherms and their small body sizes lead to a distinctive role in modern ecosystems, one that is in many respects quite different from that of terrestrial ectotherms. Understanding these differences is an important part of understanding the organismal biology of terrestrial ectothermal vertebrates.

Additional Readings

Brattstrom, B. H. 1962. Thermal control of aggregation behavior in tadpoles. *Herpetologica* 18:38–46.

Davies, P. L., C. L. Hew, and G. L. Fletcher. 1988. Fish antifreeze proteins: Physiology and evolutionary biology. *Canadian Journal of Zoology* 66:2611–2617.

Hazard, L. C. 2001. Ion secretion by salt glands of desert iguanas (*Dipsosaurus dorsalis*). *Physiological and Biochemical Zoology* 74:22–31.

Hew, C. L., G. K. Scott, and P. L. Davies. 1986. Molecular biology of antifreeze. In H. C. Heller, X. J. Musacchia, and L. C. H. Wang (Eds.), *Living in the Cold: Physiological and Biochemical Adaptation*. New York: Elsevier, pp. 117–123.

Nagy, K. A. 1972. Water and electrolyte budgets of a free-living desert lizard, *Sauromalus obsesus*. *Journal of Comparative Physiology* 79:39–62.

Nagy, K. A. 1973. Behavior, diet and reproduction in a desert lizard, *Sauromalus obesus*. *Copeia* 1973:93–102.

Nagy, K. A. 1983. The doubly labeled water ($^{3}HH^{18}O$) method: A guide to its use. *UCLA Publication 12–1417*. Los Angeles: University of California.

Nagy, K. A., and P. A. Medica. 1986. Physiological ecology of desert tortoises in southern Nevada. *Herpetologica* 42:73–92.

Pough, F. H. 1980. The advantages of ectothermy for tetrapods. *The American Naturalist* 115:92–112.

Pough, F. H. 1983. Amphibians and reptiles as low-energy systems. In W. P. Aspey and S. I. Lustick (Eds.), *Behavioral Energetics: Vertebrate Costs of Survival*. Columbus: Ohio State University Press, pp. 141–188.

Schoener, T. W., and D. A. Spiller. 1987. Effect of lizards on spider populations: Manipulative reconstruction of a natural experiment. *Science* 236:949–952.

Scholander, P. F. et al. 1957. Supercooling and osmoregulation in Arctic fish. *Journal of Cellular and Comparative Physiology* 49:5–24.

Storey, K. B. 1986. Freeze tolerance in vertebrates: Biochemical adaptation of terrestrially hibernating frogs. In H. C. Heller, X. J. Musacchia, and L. C. H. Wang (Eds.), *Living in the Cold: Physiological and Biochemical Adaptations*. New York: Elsevier, pp. 131–138.

Web Explorations

On-line resources for this chapter are on the World Wide Web at: http://www.prenhall.com/pough (click on the Table of Contents link and then select Chapter 12).

CHAPTER

13

Geography and Ecology of the Mesozoic

Pangaea reached its greatest development during the early Mesozoic—the Earth's entire land surface had coalesced into a single continent that stretched from the South Pole to the North Pole. Early Mesozoic faunas and floras showed some regional differentiation due to climate, but would have had no geographic barriers to dispersal. With the breakup of Pangaea in the Jurassic and Cretaceous, floras and faunas were isolated and became distinct in different parts of the world.

Many new types of insects appeared in the Mesozoic, including social insects such as bees, ants, and termites. An important floral change was the appearance and rapid radiation of the **angiosperms** (flowering seed plants) in the Cretaceous. A major turnover occurred in terrestrial vertebrates at the end of the Triassic, with the large-animal fauna, including a diversity of mammal-like reptiles, being replaced by dinosaurs. Jurassic dinosaurs added a new ecological niche to the ecosystem—gargantuan herbivorous sauropods like *Apatosaurus* (better known by its old name, *Brontosaurus*). In the Cretaceous more advanced herbivorous dinosaurs appeared, the hadrosaurs (duck-billed dinosaurs) and ceratopsians (horned dinosaurs). These forms had jaws and teeth specialized for processing particular kinds of vegetation.

Other new vertebrates appearing in the Mesozoic include mammals, birds, and modern types of amphibians and reptiles. The Mesozoic was also the time of a great radiation of marine reptiles, none of which survived the end of the Cretaceous except for turtles. The ecological roles played by Mesozoic marine tetrapods were reinvented in the Cenozoic by marine mammals such as whales and seals.

13.1 Mesozoic Continental Geography

When the Triassic began, the entire land area of Earth was concentrated in the supercontinent Pangaea, which straddled the equator. The maximum development of Pangaea occurred at approximately the Middle/Late Triassic boundary. The southern part of Pangaea, areas that now form Antarctica and Australia, was close to the South Pole, and the northern part of modern Eurasia was within the Arctic Circle (Figure 13–1a).

The fragmentation of Pangaea began in the Jurassic with the separation of Laurasia and Gondwana by a westward extension of the Tethys Sea. Laurasia rotated away from the other continents, ripping North America from its connection with South America and increasing the size of the newly formed Atlantic Ocean (Figure 13–1b). Epicontinental seas were more extensive than they had been in the Triassic.

Separation of the continents, as well as rotation of the northern continents, continued throughout the Mesozoic. Epicontinental seas became even more extensive. By the Late Cretaceous the continents were approaching their current positions, although India was still close to Africa, and Australia, New Zealand, and New Guinea were still well south of their present-day positions (Figure 13–1c).

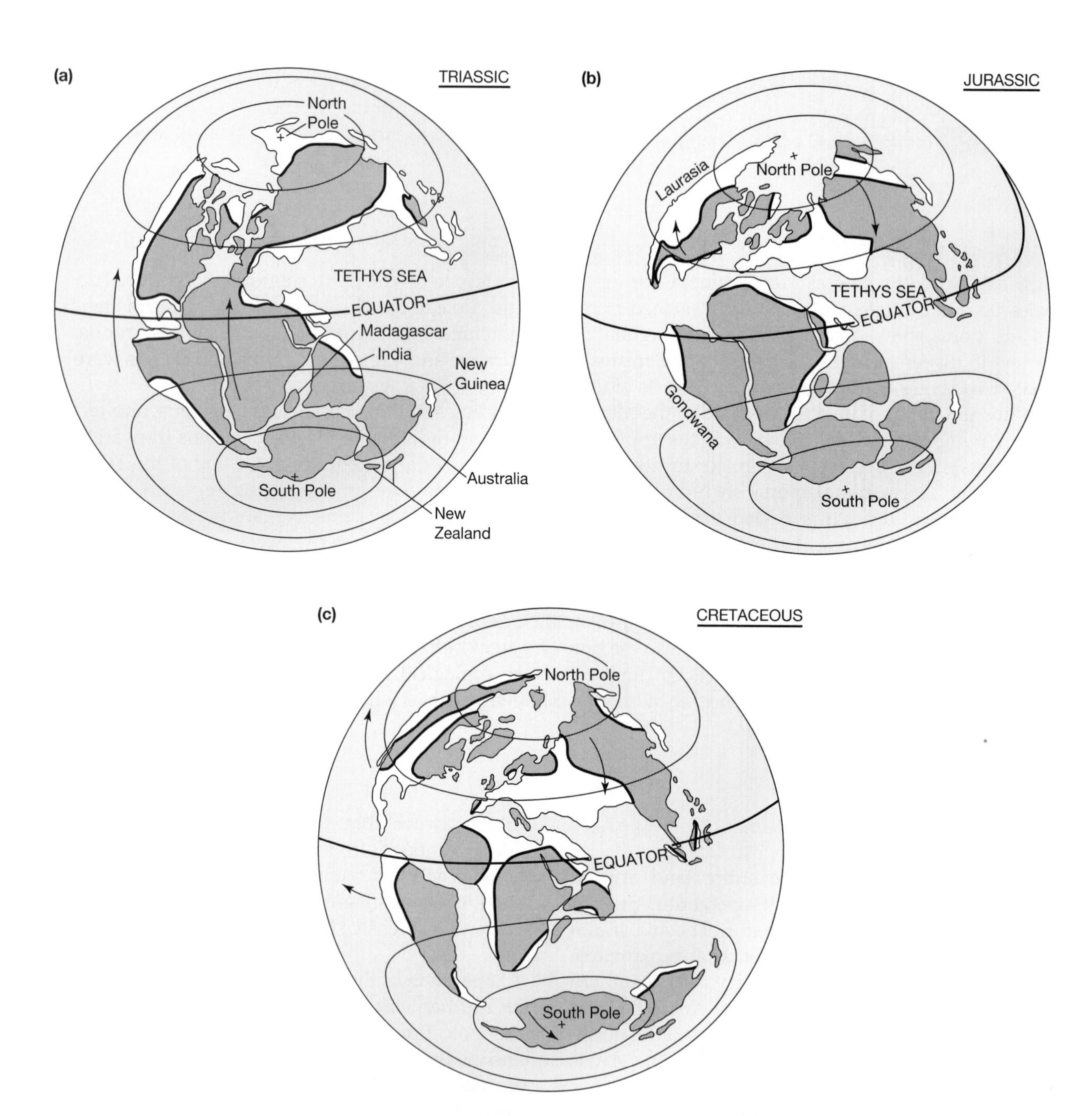

▲ Figure 13–1 Location of continental blocks in the Mesozoic. Continental land areas are shaded, and epicontinental seas are unshaded areas at the edges of the modern continents. Arrows indicate the direction of continental rotation.

13.2 Mesozoic Terrestrial Ecosystems

The Mesozoic was marked by a series of large-scale changes in flora and fauna beginning with the Permo-Triassic (Permian-Triassic) extinction event, the repercussions of which extended over a period of about 25 million years, and culminating in the Cretaceous-Tertiary transition. Terrestrial ecosystems had achieved an essentially modern food web by the end of the Permian. Plants grew in communities that contained mixtures of species. Herbivorous insects and vertebrates consumed these plants, and carnivorous invertebrates and vertebrates preyed on the herbivores. The kinds of plants and animals in early Mesozoic ecosystems were quite different from those in modern ecosystems, however.

The Permo-Triassic extinctions left a fairly impoverished terrestrial fauna. Insects lost a considerable amount of diversity, especially among plant-sucking forms, but new herbivorous forms appeared rapidly in the Triassic, including stick insects. Early Triassic vertebrate faunas were dominated by the herbivorous dicynodont mammal-like reptile *Lystrosaurus* as well as cynodont mammal-like reptiles and early archosaurs ("thecodonts"), diapsid reptiles that would eventually give rise to the dinosaurs. Later in the Triassic, the climate in the places where tetrapods were found shifted from warm and moist to hot and dry, and the archosaurs increased in diversity and faunal predominance. Triassic faunas still contained a diversity of cynodonts—both herbivorous and carnivorous—dicynodonts, and rhyncosaurs (large herbivorous diapsid reptiles) (see Figure 13–2).

Triassic vegetation included familiar modern groups of gymnosperms such as conifers (pines and other cone-bearing trees), ginkgophytes (relatives of the living ginkgo), and cycads and seed ferns. Other plants included ferns, tree ferns, and horsetails (Figure 13–2). While many of these plants exist today, they are not nearly as common and widespread as they were in the Mesozoic. Although Pangaea was a single continent in the Triassic, Triassic floras and faunas show regional characters that probably reflect differences in patterns of rainfall and seasonal temperature extremes in areas far from the sea. Conifers ranged in form from small bushes to tall forest trees, and many appear to have lived in relatively dry habitats. Cycads, too, appear to have lived in drier regions.

Triassic herbivorous vertebrates ranged in body mass between 10 and 1000 kilograms. This size distribution suggests that the major vegetational landscape may have been open woodland, because dense forest supports primarily small-bodied herbivores. All of these herbivores (rhyncosaurs, dicynodonts, gomphodont cynodonts, and aetosaur archosaurs) were generalized browsers that would have foraged within a meter of the ground. Higher-level browsers were not present until the evolution of prosauropod dinosaurs in the latest Triassic.

There were virtually no arboreal herbivorous vertebrates, although some sphenodontid reptiles and early mammals (haramyids) from the end of the Triassic might have been arboreal herbivores. Only when mammals occupied this habitat in the Cenozoic was there a significant radiation of canopy-dwelling herbivorous vertebrates. However, there were plenty of herbivorous insects in the Triassic, and along with them several arboreal insectivorous diapsid reptiles (including gliding forms).

During the Late Triassic there was a shift in vegetational communities, especially in Gondwana where the floras dominated by the seed fern *Dicroidium* were replaced by ones dominated by conifers. This floral turnover was matched by a major faunal turnover; the original diversity of large Triassic tetrapods became extinct, and was replaced by a variety of dinosaurs (Benton 1983). Mammals also made a first appearance at this time, as did other modern groups such as sphenodontids (the sister group of lizards), turtles, and crocodiles. Extinct groups such as pterosaurs (flying reptiles) appeared, along with various marine reptiles including ichthyosaurs and plesiosaurs.

The conifer- and fern-dominated vegetation of the Late Triassic continued into the Jurassic, though the seed ferns were then reduced in diversity. The floras of Laurasia and Gondwana were more similar to each other in the Jurassic than they were in the Triassic. Several new kinds of insects appeared in the Jurassic, including beetles, thrips, and a variety of bugs (hemipterans).

Jurassic ecosystems were dominated by the radiation of prosauropod and sauropod dinosaurs, the sauropods first appearing in the Middle Jurassic. Other Jurassic dinosaurs included the low-browsing stegosaurs and the carnivorous allosaurids. Lizards and the modern groups of amphibians first appeared in the Jurassic, and the earliest bird, *Archaeopteryx*, is known from the Late Jurassic. Sauropods were gigantic animals, with body masses ranging from 10,000 to 50,000 kg, and they may have influenced patterns of

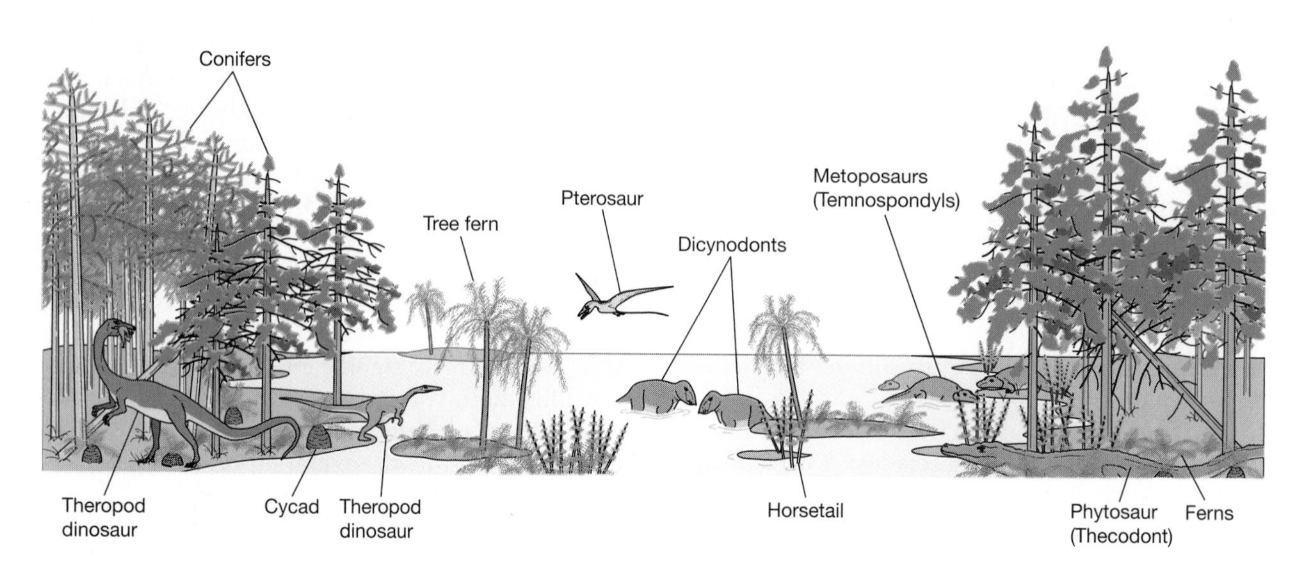

▲ Figure 13–2 Reconstruction of a scene from the early Late Triassic of New Mexico.

vegetational growth and structure as elephants do today. It is not clear how the vegetation of the Jurassic was able to support such a great biomass of herbivores. However, atmospheric carbon dioxide levels were exceptionally high during the Jurassic, and this may have allowed high levels of plant productivity.

The vegetation of the Early Cretaceous was similar to that of the Late Jurassic. However, by the Late Cretaceous the pattern of global vegetation was quite different. Flowering plants (angiosperms) appear in the fossil record in the Early Cretaceous, although extrapolations of the time of origin of angiosperms based on analysis of DNA suggest that they might have originated as early as the late Paleozoic (Martin et al. 1989). The angiosperms were the first plants to be pollinated by insects. Insects appearing in the Early Cretaceous include butterflies, aphids, short-horned grasshoppers, and gall wasps. In the Late Cretaceous a variety of social insects made their first appearance, including termites, ants, and hive-forming bees.

Angiosperms were initially located predominantly near the equator, but by the middle of the Cretaceous they were well established in the middle latitudes. However, they were never a common component of the high-latitude floras during the Mesozoic. By the end of the Cretaceous angiosperms formed 50 to 80 percent of the plants in many fossil assemblages, but their role in the vegetational landscape was different from that of today. The large trees were still the conifers. Angiosperms were smaller trees that took over the lower-level ground cover from ferns and cycads. These low-level angiosperms were all herbaceous plants; grasses, which are such a dominant feature of landscapes today, did not appear until well into the Cenozoic.

Different types of herbivorous dinosaurs appeared in the Cretaceous. The most important of these were the hadrosaurs and the ceratopsians—both low-level feeders with complex, shearing cheek teeth, probably for dealing with tough vegetation. With the breakup of Pangaea the tetrapod faunas started to show some distinct regional differences. The more derived tenanurine theropod dinosaurs became the dominant carnivores in Laurasia. In South America the more primitive ceratosaurian theropods were dominant, and tetanurans were rare. Sauropods appear to have continued as the dominant herbivores in South America and probably in Africa. In contrast, hadrosaurs and ceratopsians dominated the faunas of Late Cretaceous in the Northern Hemisphere. The hadrosaurs had rows of specialized teeth in the jaws for chewing vegetation, and the ceratopsians had turtle-like horny beaks that may have been used to tear open the tough outer covering of plants such as cycads. Both kinds of dinosaurs probably lived in migratory or nomadic herds that may have numbered in the thousands of individuals.

Other later Cretaceous faunal changes include the evolution of mammals with complex (tribosphenic) molars by the Early Cretaceous. By the mid Cretaceous the first mammals belonging to the three modern groups appeared—monotremes, marsupials, and eutherians. There was a considerable radiation of birds during the Cretaceous, but few of these forms were closely related to the modern birds that diversified in the Cenozoic. Snakes and modern types of crocodiles are also first known from the Cretaceous.

Mesozoic Climates

Mesozoic climates were equable worldwide, and there were no polar ice caps at any time during the era. Large temnospondyls (aquatic nonamniote tetrapods) are found in Triassic deposits from Australia, Antarctica, Greenland, and Spitzbergen, and coal deposits in both the Northern and Southern Hemispheres point to moist climates. In contrast, low and middle latitudes were probably dry—either seasonally or year-round—until the Late Cretaceous and early Cenozoic, when the presence of coal deposits in middle latitudes indicates the presence of swamps and suggests that the climate had become wetter. These dry lower latitudes had a type of vegetation different from the equatorial vegetation of today: During the Mesozoic there were no tropical rain forests like those of the present day. The Cretaceous plant record also suggests a highly equable world, with temperate forests extending into the polar regions, although a reduction in the diversity of plants at higher latitudes at the end of the Cretaceous signaled an episode of worldwide cooling.

Mesozoic Extinctions

The first round of Mesozoic extinctions was at the end of the Triassic, and had considerable effects on both terrestrial vertebrates and marine invertebrates. The fauna of the Triassic, which was essentially a holdover from the late Paleozoic, was replaced by forms that would dominate the Mesozoic, such as the dinosaurs; 18 families of tetrapods became extinct at this time. The Triassic extinctions coincided with the initial breakup of Pangaea, although there is also some evidence of a meteorite impact at this time. Lesser extinctions of tetrapods occurred in the Early Triassic and in the Late Jurassic, and both of these terrestrial extinction events were paralleled in the marine realm.

The extinctions at the end of the Cretaceous are famous as the ones that killed off the dinosaurs, even though the magnitude of the effect on the global fauna was nowhere near as large as the late Paleozoic (end Permian) extinctions. Many hypotheses have been proposed to explain why dinosaurs became extinct, ranging from the extraterrestrial (their gonads were zapped by radiation from an exploding supernova) to the ridiculous (constipation caused by angiosperms; see Benton 1990). It must be remembered that dinosaurs were not the only animals that suffered extinctions at this time. Thirty-six (40 percent) of the tetrapod families were extinct by the end of the Cretaceous, including not only all dinosaurs (except birds, which as dinosaur descendants count as dinosaurs, strictly speaking) but also all pterosaurs and marine reptiles. Birds and mammals suffered lesser extinctions, and insects were among the few groups that survived the Cretaceous relatively intact. Any explanation for the demise of the dinosaurs must also account for the disappearances of a large variety of other vertebrates and invertebrates, both on land and in the sea, as well as for the survival of many other lineages. The possible role of an asteroid impact in the end Cretaceous extinctions is discussed in Chapter 14.

Additional Readings

Benton, M. J. 1983. Dinosaur success in the Triassic: A noncompetitive evolutionary model. *Quarterly Review of Biology* 58:29–55.

Benton, M. J. 1989. Patterns of evolution and extinction in vertebrates. In K. C. Allen and D. E. G. Briggs (Eds.), *Evolution and the Fossil Record*. London: Belhaven Press, pp. 218–241.

Benton, M. J. 1990. Scientific methodologies in collision: The history of the study of the extinction of the dinosaurs. *Evolutionary Biology* 24:371–400.

Hallam, A. 1994. *An Outline of Phanerozoic Biogeography*. Oxford: Oxford University Press.

Martin, W., A. Gierl, and H. Saedler. 1989. Molecular evidence for pre-Cretaceous angiosperm origins. *Nature* 339:46–48.

Sepkoski, J. J. Jr. 1992. Phylogenetic and ecologic patterns in the Phanerozoic history of marine biodiversity. In N. Eldredge (Ed.), *Systematics, Ecology, and the Biodiversity Crisis.* New York: Columbia University Press, pp. 77–100.

Wing, S. L., and H.-D. Sues (rapporteurs). 1992. Mesozoic and early Cenozoic terrestrial ecosystems. In A. K. Behrensmeyer et al. (Eds.), *Terrestrial Ecosystems through Time.* Chicago: University of Chicago Press, pp. 327–416.

Web Explorations

On-line resources for this chapter are on the World Wide Web at: http://www.prenhall.com/pough (click on the Table of Contents link and then select Chapter 13).

CHAPTER

14

Mesozoic Diapsids: Dinosaurs, Crocodilians, and Others

The Diapsida is the most diverse lineage of amniotic vertebrates. The most spectacular diapsids are the dinosaurs, but the lineage also gave rise to most species of extant terrestrial vertebrates. Crocodilians and birds are diapsids, as are the squamates (lizards and snakes). A variety of lesser-known forms—including pterosaurs, ichthyosaurs, plesiosaurs and placodonts—fills the roster of Mesozoic diapsids.

The dinosaur fauna of the Mesozoic was unlike anything that has existed before or since. (As we noted in Chapter 1, birds are part of the dinosaur lineage. Because there is no cladistically correct term that refers only to dinosaurs minus birds, for convenience we will use dinosaurs in this sense.) Many dinosaurs were enormous. It is difficult to recreate the details of the lives they led because we have no living models of truly large terrestrial vertebrates. Even elephants are only as large as a medium-size dinosaur.

A second group of diapsids, the lepidosauromorphs, radiated into a variety of secondarily aquatic marine animals in the Mesozoic (ichthyosaurs, plesiosaurs, and placodonts), and then had another radiation, which produced the extant tuatara and squamates.

The remarkable success of large diapsids in the Mesozoic ended at the close of that era with the extinction of the dinosaurs. That mass extinction has attracted more than its share of attention because dinosaurs have great popular appeal; but as mass extinctions go, it was modest. However, it does provide a good opportunity to consider in detail the merits of two types of explanations of mass extinctions, the gradualism versus the catastrophism schools of thought.

14.1 Mesozoic Fauna

The Mesozoic era, frequently called the Age of Reptiles, extended for some 180 million years from the close of the Paleozoic 245 million years ago to the beginning of the Cenozoic only 65 million years ago. Through this vast period evolved a worldwide fauna that diversified and radiated into most of the adaptive zones occupied by all the terrestrial vertebrates living today and some that no longer exist (for example, the enormous herbivorous and carnivorous tetrapods called dinosaurs). Although the dinosaurs are the most familiar representatives of the Age of Reptiles, they are only one of many groups.

Inevitably such a huge number of animals is complicated and confusing, not only on first acquaintance, but even after study. Parallel and convergent evolution was widespread in Mesozoic tetrapods. Long-snouted fish eaters evolved repeatedly, as did heavily armored quadrupeds and highly specialized marine forms. A trend to bipedalism was general, and a secondary reversion to quadrupedal locomotion is seen in many forms. Knowledge of phylogenetic relationships is in a state of flux, and the scheme outlined in Figure 14–1 will undoubtedly need revision as additional material is analyzed. Current views of the ecology of dinosaurs are likewise undergoing a radical revision.

This chapter begins with a brief review of the phylogenetic relationships of Mesozoic tetrapods and some aspects of their functional morphology and major evolutionary trends. Crocodilians and

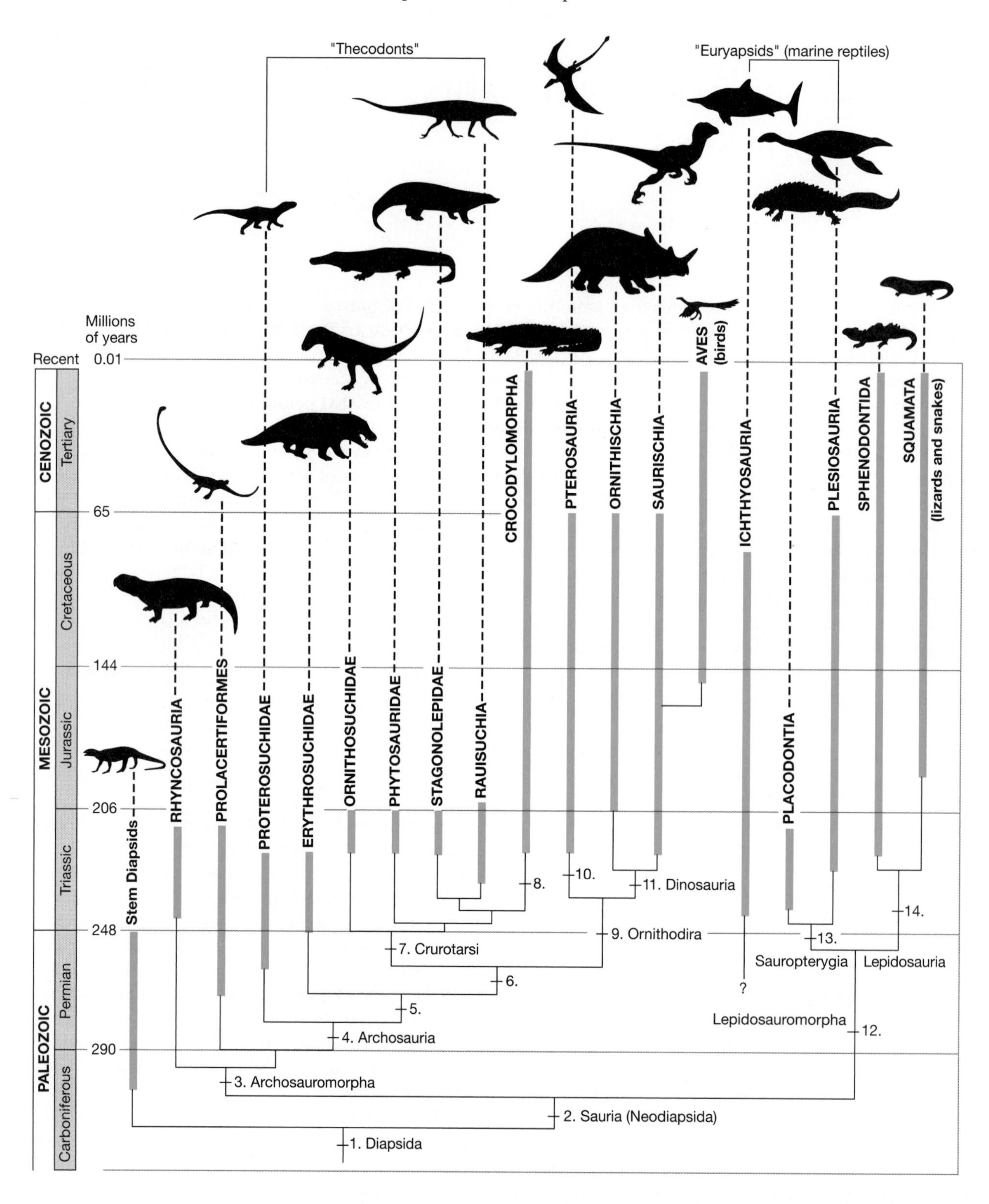

"Thecodonts"
"Euryapsids" (marine reptiles)
Millions of years
Recent
0.01
65
144
206
248
290
CENOZOIC
Tertiary
MESOZOIC
Cretaceous
Jurassic
Triassic
PALEOZOIC
Permian
Carboniferous
Stem Diapsids
RHYNCOSAURIA
PROLACERTIFORMES
PROTEROSUCHIDAE
ERYTHROSUCHIDAE
ORNITHOSUCHIDAE
PHYTOSAURIDAE
STAGONOLEPIDAE
RAUISUCHIA
CROCODYLOMORPHA
PTEROSAURIA
ORNITHISCHIA
SAURISCHIA
AVES (birds)
ICHTHYOSAURIA
PLACODONTIA
PLESIOSAURIA
SPHENODONTIDA
SQUAMATA (lizards and snakes)
1. Diapsida
2. Sauria (Neodiapsida)
3. Archosauromorpha
4. Archosauria
5.
6.
7. Crurotarsi
8.
9. Ornithodira
10.
11. Dinosauria
12.
13.
14.
Lepidosauromorpha
Sauropterygia
Lepidosauria
?

birds are the living animals most closely related to the Mesozoic forms, and the phylogenetic relationship of crocodilians and birds allows us to make inferences about the extinct ineages. Following a consideration of some aspects of the biology of dinosaurs, we consider their disappearance at the end of the Cretaceous.

14.2 Phylogenetic Relationships among Diapsids

Our understanding of the phylogenetic relationships of several groups of Mesozoic tetrapods has changed in the past decade. Many of these forms (thecodonts, crocodilians, pterosaurs, dinosaurs, squamates, and rhynchosaurs) had skulls with two temporal openings (a diapsid skull), and the Diapsida is considered a monophyletic lineage that includes most major groups of Mesozoic tetrapods (dinosaurs) as well as the living crocodilians, birds, tuatara (*Sphenodon*), and squamates (lizards, snakes, and amphisbaenians) (Figures 14–1 and 14–2).

The name *diapsid* means two arches and refers to the presence of an upper and a lower fenestra in the temporal region of the skull (see Chapter 15). More distinctive than the openings themselves is the morphology of the bones that form the arch separating them. The upper temporal arch is composed of a three-pronged postorbital bone and a three-pronged squamosal. The lower temporal arch is formed by the jugal and quadratojugal bones. The lower arch has been lost repeatedly

Legend: **1.** Diapsida—Skull with upper and lower temporal fenestrae, upper temporal arch formed by triradiate postorbital and triradiate squamosal, suborbital fenestra, ossified sternum, complex ankle joint between tibia and astragalus, first metatarsal less than half the length of the fourth metatarsal. **2.** Sauria (Neodiapsida)—Anterior process of squamosal narrow, squamosal mainly restricted to top of skull, tabular absent, stapes slender, cleithrum absent, fifth metatarsal hooked, trunk ribs mostly single-headed. **3.** Archosauromorpha—Cervical ribs with two heads, various features of limbs including concave-convex articulation between astragalus and calcaneum. **4.** Archosauria—Presence of an antorbital fenestra, orbit shaped like an inverted triangle, teeth laterally compressed with serrations. **5.** Pubis and ilium elongated, fourth trochanter on femur. **6.** Crown group Archosauria—Parietal foraman absent, no palatal teeth on pterygoid, palatine or vomer. **7.** Crurotarsi—Ankle (tarsus) in which the astragalus forms a distinct peg that fits into a deep socket on the calcaneum, and characters of the cervical ribs and the humerus. **8.** Crocodylomorpha—Secondary palate present and includes at least the maxillae. **9.** Ornithodira—Anterior cervical vertebrae longer than mid-dorsals, interclavicles absent, clavicles reduced or absent, tibia longer than femur, calcaneal tuber rudimentary or absent, metatarsals bunched together and 2–4 elongated. **10.** Pterosauria—Hand with three short fingers and elongate fourth finger supporting wing membrane, pteroid bone in wrist, short trunk, short pelvis with prepubic bones. **11.** Dinosauria—S-shaped swanlike neck, forelimb less than half the length of hindlimb, hand digit 4 reduced, and other characteristics of the palate, pectoral and pelvic girdles, hand, hindlimb, and foot. **12.** Lepidosauromorpha—Postfrontal enters border of upper temporal fenestra, supratemporal absent, teeth absent on lateral pterygoid flanges, characteristics of the vertebrae, ribs, and sternal plates. **13.** Sauropterygia—Elongation of postorbital region of skull, enlargement of upper temporal fenestra, elongate and robust mandibular symphysis, curved humerus, equal length of radius and ulna. **14.** Lepidosauria—Determinant growth with epiphyses on the articulating surfaces of the long bones, postparietal and tabular absent, fused astragalus and calcaneum, and other characteristics of the skull, pelvis, and feet. **15.** Squamata—Loss of lower temporal bar (including loss of quadratojugal), highly kinetic skull with reduction or loss of squamosal, nasals reduced, plus other characteristics of the palate and skull roof, vertebrae, ribs, pectoral girdle, and humerus.

◀ Figure 14–1 Phylogenetic relationships of the Diapsida. This diagram shows the probable relationships among the major groups of diapsids. (Turtles, which may be diapsids, are not included.) Dotted lines show interrelationships only; they do not indicate times of divergence nor the unrecorded presence of taxa in the fossil record. The numbers indicate derived characters that distinguish the lineages.

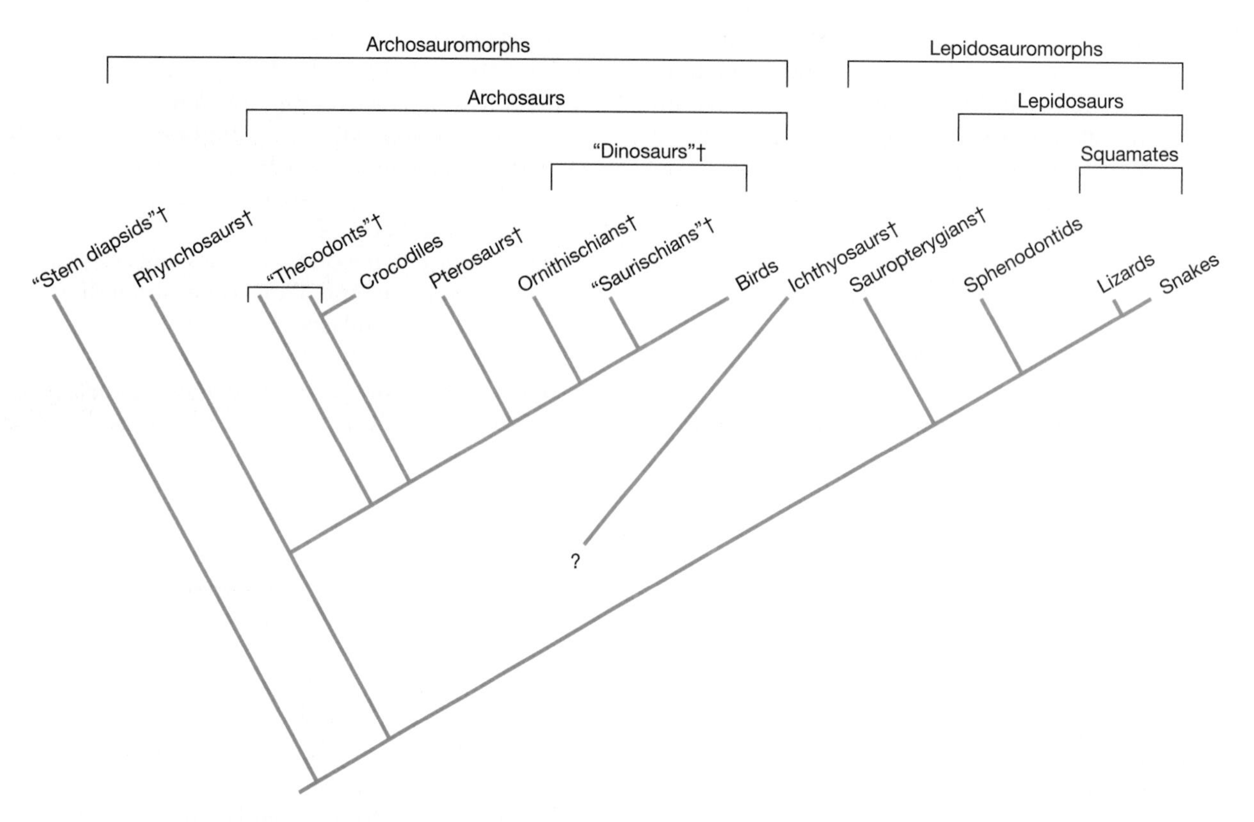

▲ Figure 14–2 Simplified cladogram of the Diapsida. Quotation marks indicate paraphyletic groups. A dagger (†) indicates an extinct taxon.

in the radiation of diapsids, and the upper arch is also missing in some forms. Living lizards and snakes clearly show the importance of those modifications of the skull in permitting increased skull flexion (kinesis) during feeding, and the same significance may attach to loss of the arches in some extinct forms. In addition to the two temporal fenestrae, derived diapsids have a suborbital fenestra on each side of the head anterior to the eye, and the presence of this fenestra modifies the relationships among the bones of the palate and the side of the skull.

The earliest diapsid known is *Petrolacosaurus,* from Late Carboniferous deposits in Kansas. It is a moderately small animal, 60 to 70 centimeters from snout to tail tip, with a long neck, large eyes, and long limbs (Figure 14–3). It gives the impression of having been an agile terrestrial animal that may have fed on large insects and other arthropods. The derived diapsids can be split into two groups, the **Archosauromorpha** and the **Lepidosauromorpha** (Figure 14–1). The archosauromorphs include crocodilians and birds, the extinct pterosaurs, dinosaurs, and several Late Permian and Triassic forms. The lepidosauromorphs include the tuatara (Sphenodontidae) and squamates as well as their extinct relatives. In addition, four groups of specialized marine tetrapods (the placodonts, plesiosaurs, ichthyosaurs, and *Hupehsuchus*) are tentatively considered to be lepidosauromorphs. The skulls of these animals have a dorsal temporal opening, but lack a lower temporal fenestra, and the postorbital and squamosal bones do not have the three-pronged shape characteristic of diapsids. However, these patterns are within the range of modifications of the basic diapsid skull that is seen among other members of the clade.

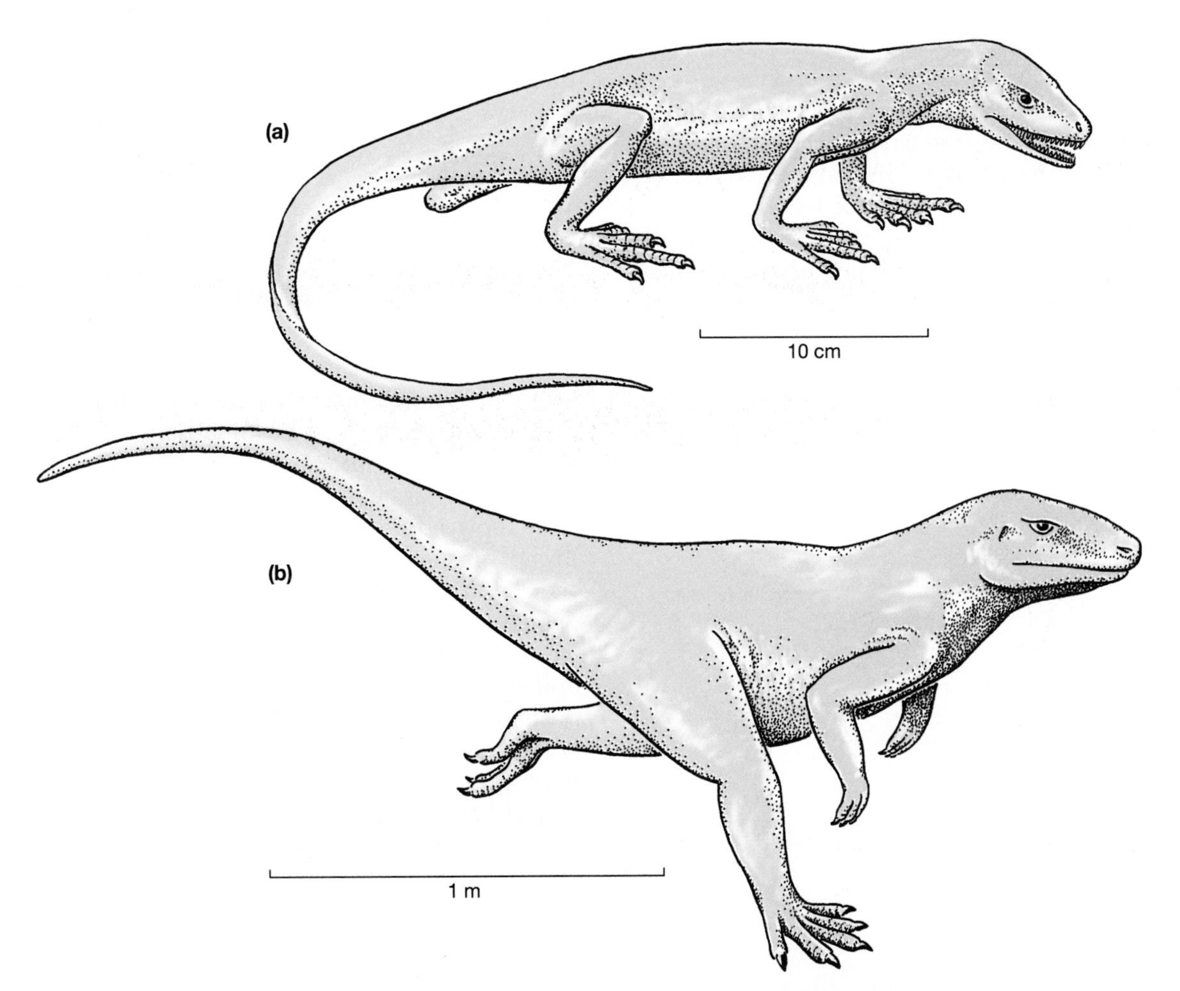

▲ Figure 14–3 Early diapsids. (a) *Petrolacosaurus* from the Late Carboniferous had forelimbs and hindlimbs that were about the same length. (b) *Euparkeria,* an early archosaur, had hindlimbs that were much longer than its forelimbs and was probably bipedal.

14.3 The Archosauromorpha

The Archosauromorpha includes the most familiar of the Mesozoic diapsids, the dinosaurs, as well as crocodilians, pterosaurs, birds, and some less familiar forms. The archosauromorphs are distinguished by several characteristics of the skull and axial skeleton. The proterosuchians (probably not a monophyletic group) include the Triassic forms known as thecodontians in earlier classifications. *Proterosuchus* was a quadrupedal, lizard-shaped carnivore, 2 or 3 meters long, from South Africa. Related forms are known from deposits in China, Bengal, Eurasia, Australia, and Antarctica. *Erythrosuchus,* another Triassic quadruped, was twice the size of *Proterosuchus* and massively built, whereas *Euparkeria* was a lightly built animal about 150 cm long. Its hindlimbs were half again as long as the forelimbs, suggesting that it was capable of bipedal locomotion (Figure 14–3).

14.4 Archosauria

The archosaurs are the animals most frequently associated with the great radiation of tetrapods in the Mesozoic. Dinosaurs and pterosaurs are distinctive components of many Mesozoic faunas, and other less familiar archosaurs were also abundant. The archosaurs are distinguished by the presence of an antorbital (in front of the eye) fenestra. The skull is deep, the orbit of the eye is shaped like an inverted triangle rather than being circular, and the teeth are laterally compressed (Figure 14–4a). A trend toward bipedalism was widespread but not universal among archosaurs, and the ventral side of the femur shaft had a distinctive area with a rough surface, the fourth trochanter Figure 14–4b). The powerful caudiofemoral muscle originated on the base of the tail and inserted on the trochanter. When the muscle contracted, it retracted (i. e., pulled back on) the thigh.

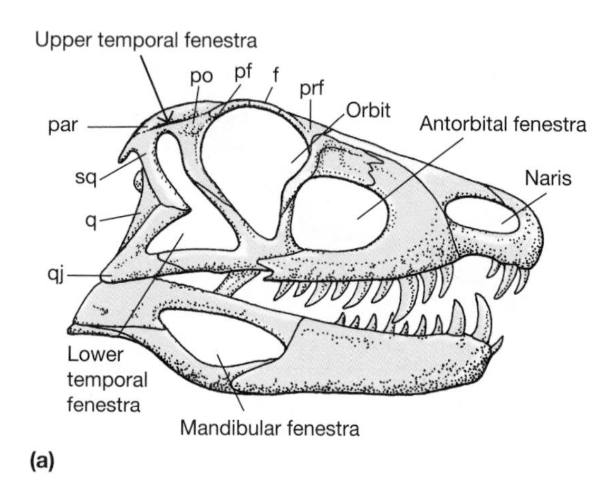

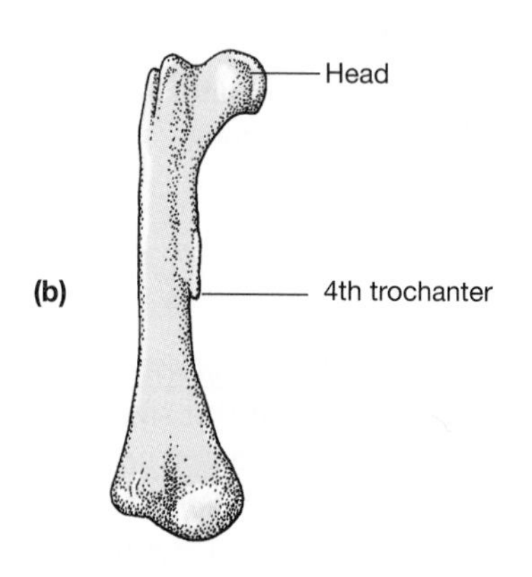

Legend: f, frontal; par, parietal; pf, postfrontal; po, postorbital; prf, prefrontal; sq, squamosal; q, quadrate; qj, quadratojugal.

▲ Figure 14–4 Morphological features of diapsids. (a) Skull of *Ornithosuchus* showing the characteristic features of archosaurs: two temporal arches, a keyhole-shaped orbit, and an antorbital fenestra. (b) Femur of *Thescelosaurus* showing the fourth trochanter.

Crurotarsi

The archosaur stock gave rise to two lineages of aquatic fish eaters, the phytosaurs and crocodilians. The phytosaurs were the earlier radiation, and during the Late Triassic they were abundant and important elements of the shoreline fauna. In contrast to crocodilians, in which the nostrils are at the tip of the snout and a secondary palate separates the nasal passages from the mouth, phytosaur nostrils were located on an elevation just anterior to the eyes. True crocodilians appeared in the Triassic and seem to have replaced phytosaurs by the end of that period. In most respects crocodilians conform closely to the skeletal structure of archosaurs, but the skull and pelvis are specialized. Crocodilians retained the nostrils at the tip of the snout and developed a secondary palate that carries the air passages posteriorly to the rear of the mouth. A flap of tissue arising from the base of the tongue can form a watertight seal between the mouth and throat. Thus, a crocodilian can breathe while only its nostrils are exposed without inhaling water. The increasing involvement of the premaxilla, the maxillae, and the pterygoids in the secondary palate can be traced from Mesozoic crocodilians to modern forms.

Modern crocodilians are semiaquatic animals, but Triassic crocodilians were terrestrial. They were thin, slender animals about the size of a large cat, and give the impression of having been active hunters that probably preyed on smaller diapsids. Traces of this terrestrial origin persist in living crocodilians. They have well-developed limbs, and some species make extensive overland movements. Crocodilians can gallop, moving the limbs from their normal laterally extended posture to a nearly vertical position beneath the body.

The Cretaceous was the high point in crocodilian evolution. The extension of warm climates to land areas that are now in cool temperate climate zones favored both diversity and large size. *Simosuchus clarki*, a fossil crocodilian recently discovered in Madagascar, was about a meter long. Unlike all other known crocodilians, *Simosuchus* had a short, blunt snout and teeth with multiple cusps, and unlike most crocodilians it may have been herbivorous. *Deinosuchus* (Greek *deino* = terrible and *suchus*

= crocodile) from the Cretaceous of Texas had a skull that was nearly 2m long. If this crocodilian had the same body proportions as extant forms, it would have had a total length of 12 to 15m and might well have preyed on dinosaurs.

Enormous crocodilians persisted long after dinosaurs disappeared. A skull of the Miocene crocodile *Purussaurus brasiliensis* found in the Amazon Basin in 1986 is 1.5m long. If that animal had the same proportions as an alligator, it would have had a total length of 11 to 12m and have stood 2.5m tall—that is, the height of the ceiling in most houses. An isolated lower jaw in the paleontology museum at the Universidade do Acre is 30 cm longer than the jaw of the complete skull, and may have come from an animal 13 to 14m long. These crocodilians would have been as large as *Tyrannosaurus rex*.

A heavy, laterally flattened tail propels the body of a crocodilian in water, and the legs are held against the sides of the body. In the Late Jurassic, a lineage of specialized marine crocodiles enjoyed brief success. These metriorhynchids had long skulls with pointed snouts. They lacked the dermal body armor typical of most crocodilians. They had a lobed tail very like that of the early ichthyosaurs, with the vertebral column turned downward into the lower lobe and the upper lobe supported by stiff tissue. The feet were paddle-like.

Biology of Extant Crocodilians In many respects crocodilians are the living archosaurs most like Mesozoic forms, and we will consider them first to set a context for understanding dinosaurs. Only 21 species of crocodiles now survive. Most are found in the tropics or subtropics, but three species have ranges that extend into the temperate zone. Systematists divide living crocodilians into three families: The Alligatoridae includes the two species of living alligators and the caimans (Figure 14–5). Except for the Chinese alligator, the Alligatoridae is solely a New World group. The American alligator is found in the Gulf coast states, and several species of caimans range from Mexico to South America and through the Caribbean. Alligators and caimans are freshwater forms, whereas the Crocodylidae in-

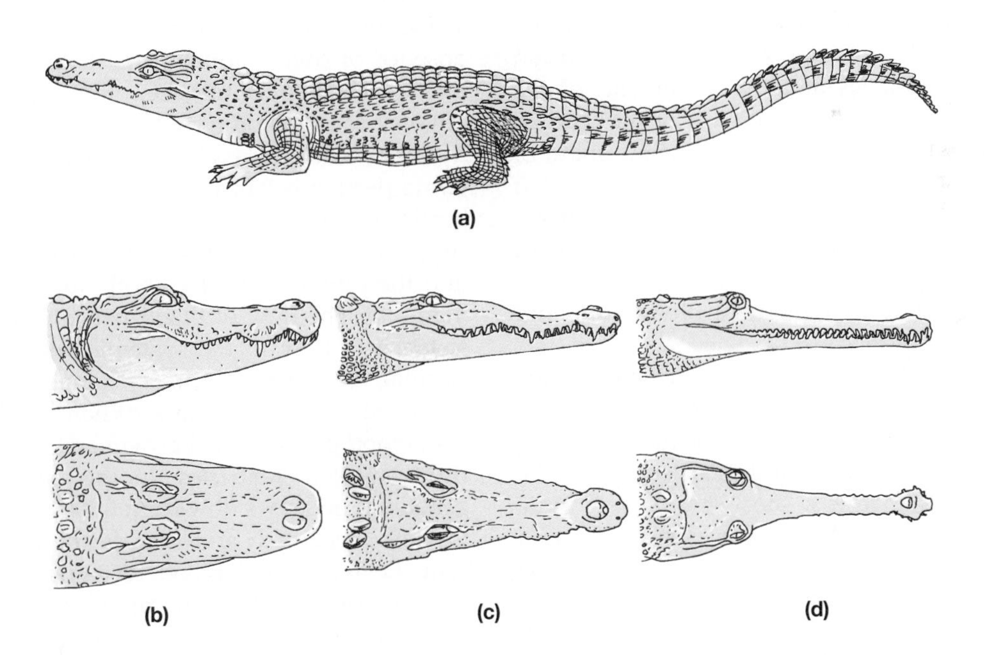

▲ Figure 14–5 Crocodilians. Modern crocodilians differ little from one another or from Late Mesozoic forms. The greatest interspecific variation in living crocodilians is seen in the head shape. Alligators and caimans are broad-snouted forms with varied diets. Crocodiles include a range of snout widths. The widest crocodile snouts are almost as broad as those of most alligators and caimans, and these species of crocodilians have varied diets that include turtles, fishes, and terrestrial animals. Other crocodiles have very narrow snouts, and these species are primarily fish eaters. (a) Cuban crocodile. (b) Chinese alligator. (c) American crocodile. (d) Gharial.

cludes species such as the saltwater crocodile that inhabits estuaries, mangrove swamps, and the lower regions of large rivers. This species occurs widely in the Indo-Pacific region and penetrates the Indo-Australian archipelago to northern Australia. In the New World, the American crocodile is quite at home in the sea, and occurs in coastal regions from the southern tip of Florida through the Caribbean to northern South America.

The saltwater crocodile is probably the largest living species of crocodilian. Until recently, adults may have reached lengths of 7m. Crocodilians grow slowly once they reach maturity, and it takes a long time to attain large size. In the face of intensive hunting in the past two centuries, few crocodilians now attain the sizes they are genetically capable of reaching. Not all crocodilians are giants; several diminutive species live in small bodies of water. The dwarf caiman of South America and the dwarf crocodile of Africa are about a meter long as adults and live in swift-flowing streams.

The third family of crocodilians, the Gavialidae, contains only a single species—the gharial, which once lived in large rivers from northern India to Burma. This species has the narrowest snout of any crocodilian; the mandibular symphysis (the fusion between the mandibles at the anterior end of the lower jaw) extends back to the level of the 23rd or 24th tooth in the lower jaw. A very narrow snout of this sort is a specialization for feeding on fish that are caught with a sudden sideward jerk of the head. We have already called attention to the evolution of similar skull shapes in a variety of Mesozoic marine tetrapods, including trematosaurs, phytosaurs, and the short-necked plesiosaurs.

The muscle that opens a crocodilian's mouth (the depressor mandibulae) runs from the rear of the skull to a retroarticular process on the mandible (i.e., an extension of the bone beyond its articulation). The depressor mandibulae is a short muscle with little mechanical advantage. A person can readily hold a crocodilian's mouth closed, as viewers can see on television nature shows almost daily. The jaw-closing muscles, in contrast, are very powerful. Broad-snouted crocodilians feed on adult turtles that they crush in their jaws.

Some crocodilians, such as the Nile crocodile, wait in ambush at the water's edge and attack large mammals like antelope and zebra when they come to drink. In tropical areas of Australia, warnings are posted beside rivers and lakes to alert people to the very real danger of attack by crocodiles. After seizing an animal, a crocodilian drags it under water to drown. When the prey is dead the crocodilian bites off large pieces and swallows them whole. Sometimes crocodilians wedge a dead animal into a tangle of submerged branches or roots to hold it as the crocodilian pulls chunks loose. Alternatively, crocodilians can use the inertia of a large prey item to pull off pieces: The crocodilian bites the prey and then rotates rapidly around its own long axis, tearing loose the portion it is holding. Some crocodilians leave large prey items to decompose for a few days until they can be dismembered easily.

Living crocodilians are ectotherms, and small individuals bask in the sunlight to raise their body temperatures. A basking crocodilian can increase its rate of heating by using a right-to-left intracardiac blood shunt to increase blood flow in the peripheral circulation, just as lizards do. However, the structure of the crocodilian heart is different from that of the squamate and turtle heart, and the intracardiac blood shunt is achieved in a different way.

Unlike turtles and squamates, crocodilians have a four-chambered heart—that is, the ventricle is divided into left and right halves by a septum. (It is the absence of that septum in the hearts of squamates and turtles that permits them to use pressure differentials to shift blood from the pulmonary [right] side of the ventricle across the muscular ridge to the systemic [left] side.) In the crocodilian heart, the right aortic arch opens from the left ventricle and receives oxygenated blood (Figure 14–6). The left aortic arch and the pulmonary artery both open from the right ventricle. The flow of blood is controlled by the resistance to flow in the systemic versus pulmonary circuits, and the pressures change depending on what the alligator is doing. When an alligator is at rest, blood pressure is approximately the same in the right and left ventricles. In this situation, deoxygenated blood does flow from the right ventricle into the left aortic arch and then posteriorly to the viscera. Deoxygenated blood contains hydrogen ions that are produced when carbon dioxide combines with the bicarbonate buffering system of the blood. The hydrogen ions that enter the left aortic arch may be used for the secretion of hydrochloric acid in the stomach during digestion. Note that the *right* aortic arch supplies blood to the head, so even in this situation the brain receives only oxygenated blood.

A different pattern of blood flow occurs when the alligator is active and pressure in the left ventricle

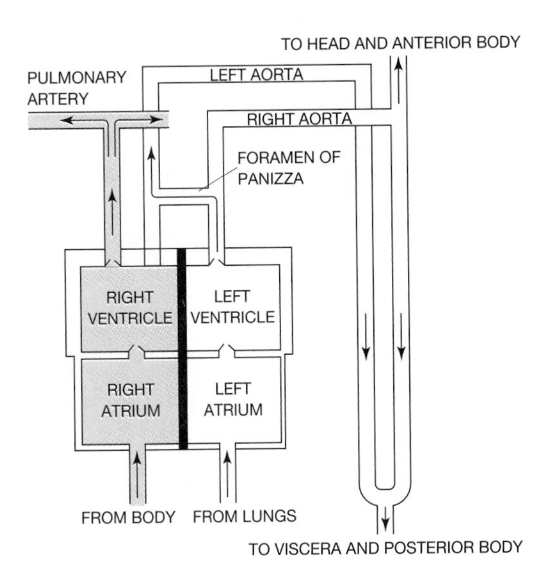

▲ Figure 14–6 The relationship of the heart and major vessels of a crocodilian. The right aortic arch opens from the left ventricle and receives oxygenated blood, which flows to both the anterior and posterior parts of the body. The left aortic arch opens from the right ventricle. When pressure in the right ventricle equals or exceeds pressure in the left, the atrioventricular valve opens and deoxygenated blood flows into the left aorta, which carries blood only to the posterior part of the body. When pressure in the left ventricle exceeds pressure in the right ventricle, the right atrioventricular valve is held shut, and oxygenated blood flows via the foramen of Panizza into the left aortic arch.

rises above that of the right ventricle. The left and right aortic arches are connected via the foramen of Panizza. When pressure in the right aortic arch exceeds that in the left, blood flows through this passage from the right aortic arch into the left. The increased pressure in the left aortic arch holds the ventricular valve closed, preventing entry of deoxygenated blood from the right ventricle. Thus, during activity both aortic arches receive oxygenated blood.

A third pattern of blood flow probably occurs when an alligator dives and is holding its breath. In this situation blood vessels in the pulmonary circuit are constricted and the pressure in the right ventricle rises to match the pressure in the left ventricle. Under these conditions a right-to-left shunt forms and a substantial volume of deoxygenated blood flows into the left aortic arch. The same right-to-left shunt probably occurs when a cold crocodilian is basking in the sun. Crocodilians (at least small individuals—no one has tested this with adult crocodilians!) can increase their rate of heating by increasing blood flow through the limbs. The legs have a large surface area in relation to their volume, and blood flowing through the legs rapidly transfers heat to the core of the body. Increasing the resistance to flow in the pulmonary circuit to produce a right-to-left shunt of blood in the heart is probably how crocodilians increase systemic blood flow to warm rapidly in the sun.

Crocodilians and Birds as Models for Dinosaurs One strength of the cladistic method of determining evolutionary relationships is the emphasis it places on shared derived characteristics of related organisms. Usually these are morphological characters, and they are used to draw inferences about phylogenetic relationships, but the process can be used in other ways. For example, if a phylogeny can be established by using morphological characters, other characteristics—ecological, behavioral, and physiological—can be superimposed on the phylogeny, and their evolution can be interpreted in an evolutionary context.

Morphological features common to crocodilians and birds place them in the archosaur lineage, and this relationship can be combined with information about the social behavior of the two groups to answer questions about the probable behavior of dinosaurs. For example, what sorts of parental care and vocal behavior did dinosaurs display? This question can be addressed by analyzing the phylogenetic history of parental care and vocalization in the archosaur lineage.

The extensive parental care provided by many birds has long been known, partly because most birds are conspicuous animals that are relatively easy to study. Vocalizations are important in the parental care of birds. The chicks cluck within the eggs as they near hatching, chirp as they beg for food, and shriek when they feel threatened. Adult birds of some species give warning calls that cause their young to freeze and remain motionless until the danger has passed.

Parental care and vocal communication with their young by crocodilians is less well known than that of birds, but it appears to be as extensive. All crocodilians probably protect their nests, and elaborate parental care has been described for some species. Vocal communication between juveniles and adults begins before eggs hatch and continues after the babies are out of the nest (Lang 1986).

Baby crocodilians begin to vocalize before they have fully emerged from the eggs, and these vocalizations are loud enough to be heard some distance away. The calls of the babies stimulate one or both parents to excavate the nest, using their feet and jaws to pull away vegetation or soil (Figure 14–7). For example, the female American alligator bites chunks of vegetation out of her nest to release the young when they start to vocalize. Then she picks up the babies in her mouth and carries them—one or two at a time—to water, where she releases them. This process is repeated until all hatchlings have been carried from the nest to the water. The parents of some species of crocodilians gently break the eggshells with their teeth to help the young escape. The sight of a crocodile, with jaws that could crush the leg of a zebra, delicately cracking the shell of an egg little larger than a hen's egg and releasing the hatchling unharmed is truly remarkable.

Young crocodilians stay near their mother for a considerable period—2 years for the American alligator, 3 years for the spectacled caiman of South America—and may feed on small pieces of food the female drops in the process of eating. Like many birds, baby crocodilians are capable of catching their own food shortly after they hatch and are not dependent on their parents for nutrition.

Adult crocodilians, like birds, are vocal archosaurs (Lang 1989). Head-slapping, tail-slapping, and a variety of vocalizations are used by male crocodilians during courtship and territorial displays. American alligators live in dense swamps, and males defend territories that are often out of sight of other males and females. A male's territory is often

(a)

(b)

▲ Figure 14–7 Parental care by the mugger crocodile, *Crocodylus palustris.* The numbered tag on the head allows individuals to be recognized. (a) Male parent picking a hatchling. (b, c) Male parent carrying hatchlings to the water, 9m away, where the mother was waiting.

centered on a deep hole it has dug that retains water when the swamp dries, and alligator holes are an important refuge for other animals in times of drought. The roar of a male alligator resounds through the swamp and announces his presence to other alligators up to 200m away. Females also roar, but only male alligators engage in a sub-audible vocalization (i.e., in a frequency below the range of human hearing) that causes drops of water to dance on the surface and travels for some distance underwater.

Frightened crocodilian hatchlings emit a distress squeak that stimulates adult male and female crocodilians to come to their defense. When staff members at a crocodile farm in Papua New Guinea rescued a hatchling New Guinea crocodile that had strayed from the pond, its distress call brought 20 adult crocodiles surging toward the hatchling (and the staff members!). The dominant male head-slapped the water repeatedly, and then charged into the chain link fence where the staff members were standing while the females swam about, gave deep guttural calls, and head-slapped the water.

Nesting behaviors and parental care of crocodilians overlap those of many birds. For example, the bush turkeys (megapods) of Australia bury their eggs in piles of soil and vegetation in craters they excavate in the ground and release their young at the end of incubation. The young disperse as soon as they emerge from the nest. The young of many birds—including familiar species such as ducks, chickens, and quail—are well developed at hatching (precocial) and able to find their own food. In these birds, as in crocodilians, the important function of parental care appears to be protecting the nests and young.

The most parsimonious explanation of the presence of well-developed vocalizations, elaborate nesting construction and guarding, and care of young in crocodilians and birds is that these behaviors were present in the common ancestor of these two groups. In other words, parental care of young and vocalization for social communication appear to be ancestral characters of the archosaur lineage, at least at the level of crocodilians. If that is the case, dinosaurs would have inherited parental care and vocal communication as a part of their ancestral behavioral repertoire, and we would expect them to have exhibited the behaviors seen in birds and crocodilians. Behavior is difficult to decipher from the fossil record, but evidence is accumulating to suggest that dinosaurs did indeed engage in parental care, vocal communication, and other forms of complex social behaviors.

Pterosauria

The archosaurs gave rise to two independent radiations of fliers. The birds are one of these lineages, and their similarity to other archosaurs is so striking that if they had disappeared at the end of the Mesozoic, they would be considered no more than another group of highly specialized archosaurs. The other lineage of flying archosaurs were the pterosaurs of the Late Triassic to Cretaceous (Figure 14–8). They ranged from the sparrow-size *Pterodactylus* to *Quetzalcoatlus,* with a wingspan of 13m. The wing formation of pterosaurs was entirely different from that of birds. The fourth finger of pterosaurs was elongate and supported a membrane of skin anchored to the side of the body and perhaps to the hind leg. A small splint-like bone was attached to the front edge of the carpus and probably supported a membrane that ran forward to the neck. The early rhamphorhynchoid pterosaurs had a long tail with an expanded portion on the end that was presumably used for steering; the later pterodactyloids lacked a tail.

Flight is a demanding means of locomotion for a vertebrate, and it is not surprising that pterosaurs and birds show a high degree of convergent evolution. The long bones of pterosaurs were hollow, as they are in birds and many other archosaurs, reducing weight with little loss of strength. The sternum, to which the powerful flight muscles attach, was well developed in pterosaurs, although it lacked the keel seen in birds. The eyes were large, and casts of the brain cavities of pterosaurs show that parts of the brain associated with vision were large and olfactory areas were small, as they are in birds. The cerebellum, which is concerned with balance and coordination of movement, was large in proportion to other parts of the brain, also as in birds.

Some pterosaurs lost their teeth and evolved a bird-like beak. Others had sharp, conical teeth in blunt skulls reminiscent of those of bats. Some pterosaurs with elongate skulls and stout, sharp teeth may have caught fish or small tetrapods. *Pterodaustro* had an enormously long snout lined with a comblike array of fine teeth that may have been used for sieving small aquatic organisms. *Dsungaripterus* had long jaws that met at the tips like a pair of

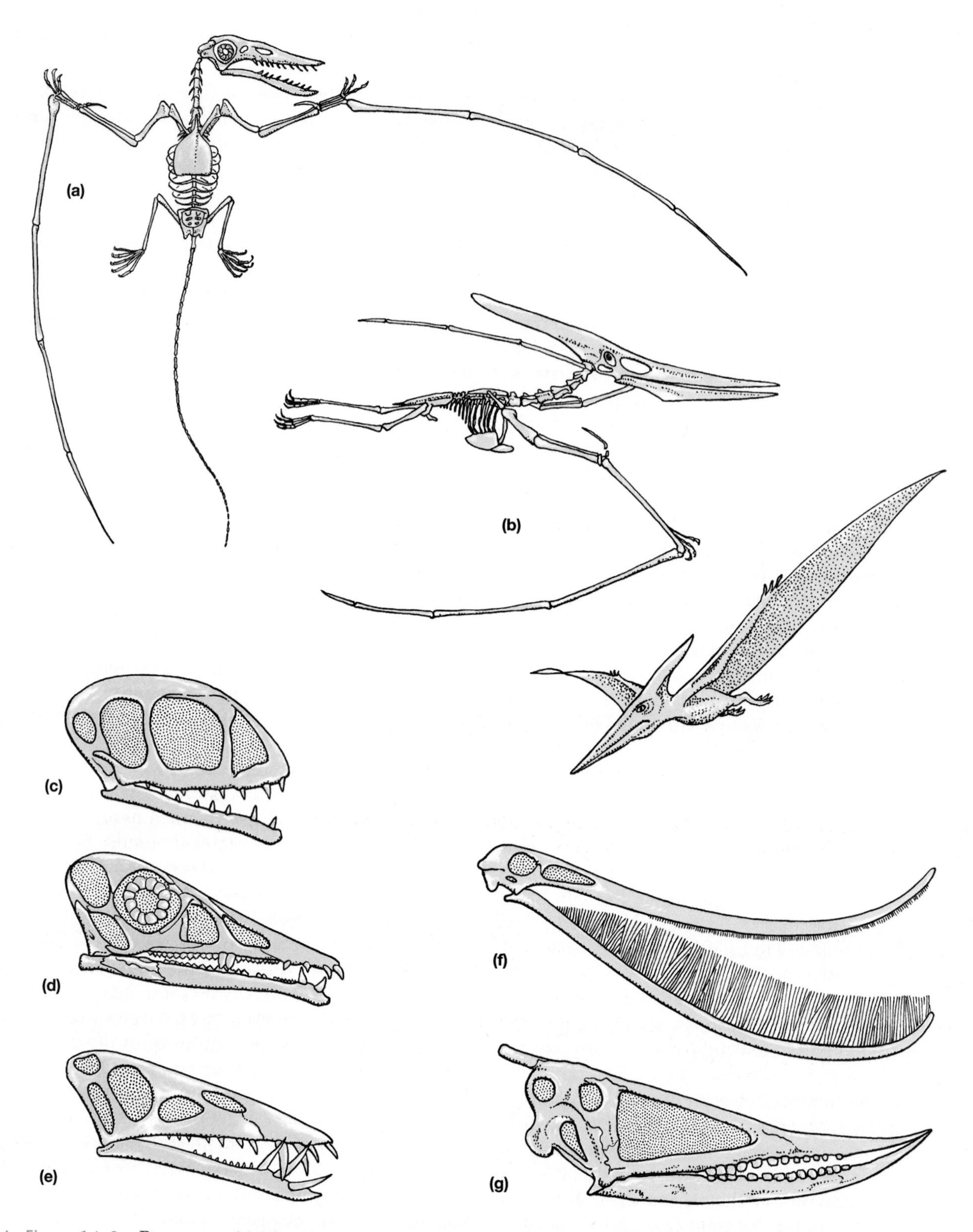

▲ Figure 14–8 Pterosaurs. (a) *Rhamphorhynchus* from the Jurassic. (b) *Pteranodon* from the Cretaceous. The skulls of pterosaurs suggest dietary specializations. (c) *Anurognathus* may have been insectivorous. (d) *Eudimorphodon* may have eaten small vertebrates. (e) *Dorygnathus* may have been a fish eater. (f) *Pterodaustro* had a comb-like array of teeth that may have been used to sieve plankton. (g) *Dsungaripterus* may have pulled mollusks from rocks with a horny beak and then crushed them with its molariform teeth.

forceps. The tips of the jaws were probably covered with a horny beak, and blunt teeth occupied the rear of the jaw. These animals may have plucked snails from rocks with their beaks and then crushed them with their broad teeth.

The flight capacities of pterosaurs have long been debated, and most hypotheses about their ecology have been based on the assumption that they were weak fliers. That assumption has led to suggestions of restrictions of activities and habitats of pterosaurs that seem unlikely for a group of animals that was clearly diverse and successful through much of the Mesozoic. An aerodynamic analysis suggests that the flying abilities of pterosaurs have been underestimated (Hazlehurst and Rayner 1992). This view suggests that small pterosaurs were slow, maneuverable fliers like bats. The large pterosaurs appear to have been specialized for soaring like frigate birds and some vultures.

Speculations about the flying abilities of pterosaurs depend on what assumptions are made about the shape of the wing. It extended outward to the tips of the fourth finger, but where was it attached to the body? Did it stop at the waist, or did it extend onto the hindlimbs as the wing does in bats? The wing structure may have varied among pterosaurs. A fossil of *Pterodactylus* shows the wing attached at least to the thigh, whereas an extremely well-preserved fossil of *Sordes pilosus* from Jurassic sediments in Kazakhstan shows that the hind legs were involved in the flight structures. The wing of this species attached along the outside of the hindlimb to the ankle, and another flight membrane—the uropatagium—stretched between the hind legs and was controlled by the fifth toe. This degree of involvement of the hindlimbs with the wings would have limited their role in terrestrial locomotion (as is the case for bats), and *Sordes* may have been a clumsy walker on flat surfaces but a good climber on rocks and trees.

Fossils of pterosaurs have not clearly shown the three-dimensional structure of the hind foot, and reconstructions of their locomotion on the ground have been speculative. A fossil of *Dimorphodon,* a primitive pterosaur discovered in Mexico, shows details of foot structure that were previously unknown. The way the foot bones articulate shows that *Dimorphodon* placed its foot flat on the ground during locomotion, as do birds, rather than walking on the toes as bipedal dinosaurs did (Clark et al. 1998). This interpretation is consistent with fossil tracks believed to represent pterosaurs, which show impressions of the entire hind foot and marks of the front feet as well.

14.5 Dinosaurs

By far the most generally known of the archosaurs are the Ornithischia and Saurischia. These groups are linked in popular terminology as dinosaurs, but they are independent radiations and differ in the specializations they developed. The groups had a common ancestor that was bipedal and both evolved some secondarily quadrupedal forms. Initially most of our information about dinosaurs came from North America and Europe, but in the past 20 years new discoveries in South America, Asia, and Africa have revealed details of a worldwide dinosaur fauna in the Mesozoic. The pattern that is emerging suggests that early in the Mesozoic, when the continents were still broadly connected, dinosaur faunas were quite similar. By the Cretaceous the continents had moved almost to their current positions, meaning they were more widely separated, and regional differences in dinosaur faunas were more pronounced.

Many morphological trends that can be traced in archosaur evolution appear to be associated with increased locomotor efficiency. The two most important developments were movement of the legs under the body and a widespread tendency toward bipedalism. Early archosauromorphs had a sprawling posture like that of many living amphibians and squamates. The humerus and femur were held out horizontally from the body, and the elbow and knee were bent at a right angle. Derived archosaurs have legs that are held vertically beneath the body.

Among early tetrapods, muscles originating on the pubis and inserting on the femur protract the leg (move it forward), muscles originating on the ischium abduct the femur (move it toward the midline of the body), and muscles originating on the tail retract the femur (move it posteriorly). The ancestral tetrapod pelvis, little changed from *Ichthyostega* through early archosauromorphs, was platelike (Figure 14–9a). The ilium articulated with one or two sacral vertebrae, and the pubis and ischium did not extend far anterior or posterior to the socket for articulation with the femur (acetabulum). The pubofemoral and ischiofemoral muscles extended ventrally from the pelvis to insert

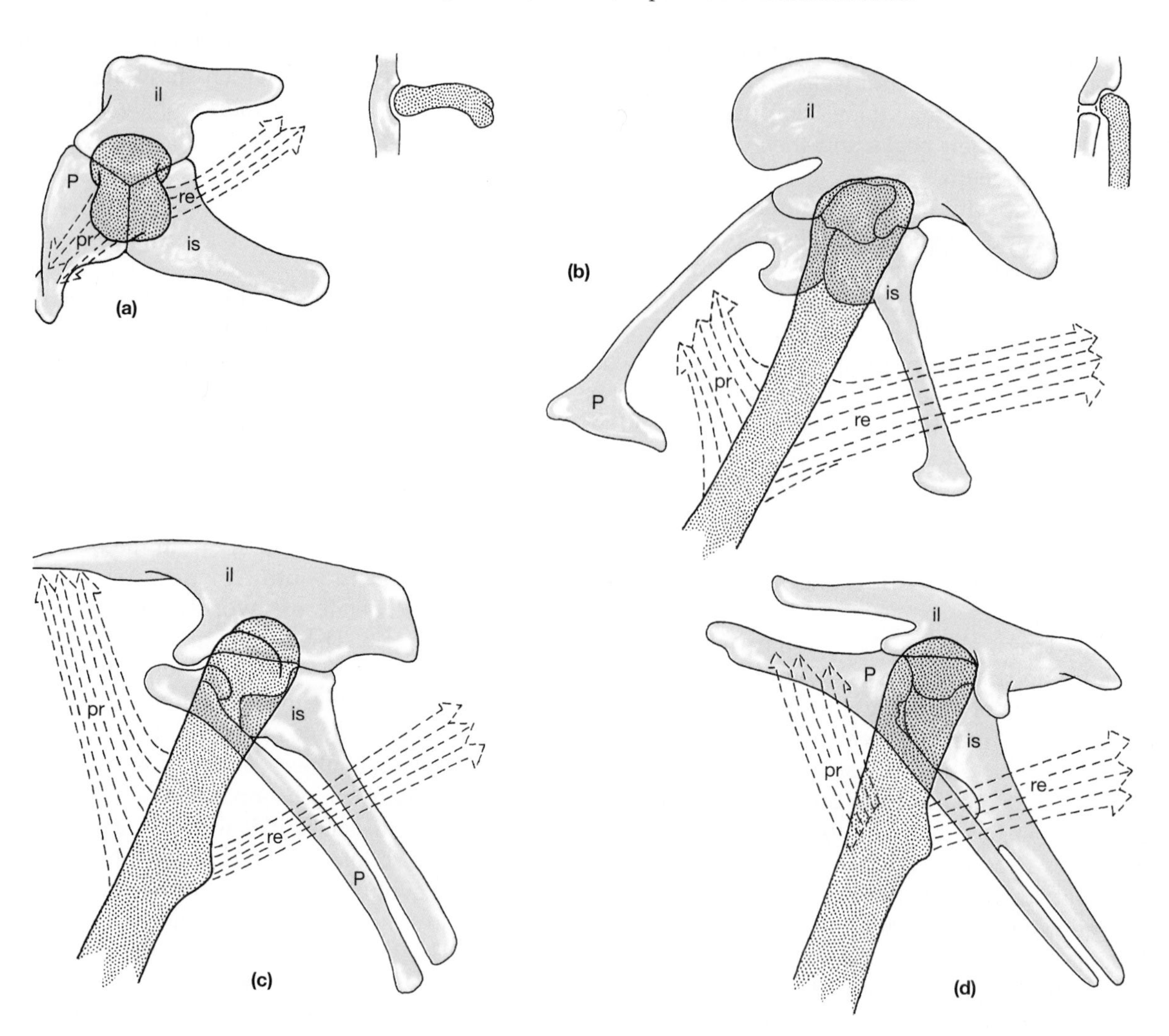

▲ Figure 14–9 Functional aspects of the pelvises of archosaurs. Pelvic morphology of an early archosaur (a, *Euparkeria*), a saurischian dinosaur (b, *Ceratosaurus*), and two ornithischian dinosaurs (c, *Scelidosaurus;* d, *Thescelosaurus*). The presumed action of femoral protractor muscles (pr) and retractors (re) is shown by arrows. Insets show an anterior view of the articulation of the femur with the pelvis. p, pubis; il, ilium; is, ischium.

on the femur. (The downward force of their contraction was countered by iliofemoral muscles that ran from the ilium to the dorsal surface of the femur.) As long as the femur projected horizontally from the body, this system was effective. The pubofemoral and caudofemoral muscles were long enough to swing the femur through a large arc relative to the ground. As the legs were held more nearly under the body, the pubofemoral muscles became less effective. As the femur rotated closer to the pubis, the sites of muscle origin and insertion moved closer together, and the muscles themselves became shorter. A muscle's maximum contraction is about 30 percent of its resting length; thus the shorter muscles would have been unable to swing the femur through an arc large enough for effective locomotion had there not been changes in the pelvis associated with the evolution of bipedalism.

The bipedal ornithischian and saurischian dinosaurs carried the legs completely under the body and show associated changes in pelvic structure. The two groups attained the same mechanically advantageous result in different ways. In quadrupedal saurischians, the pubis and ischium both became elongated and the pubis was rotated anteriorly, so that the pubofemoral muscles ran back from the pubis to the femur and were able to protract it (Figure 14–9b). The pubis of early ornithischians did not project anteriorly. Instead the ilium was elongated anteriorly, and it appears likely that the femoral protractors originated on the anterior part of the ilium, from which they ran posteriorly to the femur. This condition is seen in the pelvis of ornithischians such as *Scelidosaurus* (Figure 14–9c), and appears to be maintained in the ankylosaurs, a group of derived quadrupedal ornithischians. Other ornithischians developed an anterior projection of the pubis that ran parallel to and projected beyond the anterior part of the ilium (Figure 14–9d). This development occurred in both bipedal and quadrupedal lineages and provided a still more anterior origin for protractor muscles.

The trend toward bipedalism was important in opening new adaptive zones to archosaurs. A fully quadrupedal animal uses its forelimbs for walking, and any changes in limb morphology must be compatible with that function. As animals become increasingly bipedal, the importance of the forelimbs for locomotion decreases and the scope of specialized functions that can develop increases. Many of the smaller carnivorous dinosaurs that were fully bipedal used their forelimbs to seize prey. Specialization of forelimbs as wings occurred twice among diapsids, once in the evolution of birds and once in pterosaurs.

Bipedal animals have hind legs that are considerably longer than their front legs, and the degree of disproportion between hind legs and fore legs is assumed to reflect the extent of bipedalism in a given species. The quadrupedal dinosaurs had longer hind legs than front legs, and this condition is thought by most paleontologists to indicate that they were secondarily quadrupedal, having evolved from bipedal ancestors.

14.6 The Ornithischian Dinosaurs

Differences in the structure of the pelvis of saurischian and ornithischian dinosaurs indicate an early separation of the two groups. However, ornithischians and saurischians show parallels in body form that probably reflect the mechanical problems of being very large terrestrial animals. Ornithischians were herbivorous and radiated into considerably more diverse morphological forms than did the herbivorous sauropod saurischians. All ornithischians probably had cheeks, and horny beaks rather than teeth at the front of the mouth. The larger ornithischians were not as bipedal as some large theropod saurischians, and the forelimbs were never greatly reduced.

Three groups of ornithischian dinosaurs can be distinguished (Figures 14–10 and 14–11).

- Thyreophora—The armored dinosaurs. Quadrupedal forms including stegosaurs (forms with a double row of plates or spines on the back and tail) and ankylosaurs (heavily armored forms, some with club-like tails).
- Ornithopoda—Bipedal forms, including the duck-billed dinosaurs.
- Marginocephalia—The pachycephalosaurs (bipedal dinosaurs with enormously thick skulls) and ceratopsians (the quadrupedal horned dinosaurs).

Thyreophora—Stegosaurs and Ankylosaurs

The stegosaurs were a group of quadrupedal herbivorous ornithischians that were most abundant in the Late Jurassic, although some forms persisted to the end of the Cretaceous. *Stegosaurus,* a large form from the Jurassic of western North America, is the most familiar of these dinosaurs. It was up to 6m long, and its front legs were much shorter than its hind legs (Figure 14–12). A double series of leaf-shaped plates were probably set alternately on the left and right sides of the vertebral column. Two pairs of spikes on the tail made it a formidable weapon. *Kentrosaurus,* an African species that was contemporaneous with *Stegosaurus,* was smaller (2.5m), and had a series of seven pairs of spikes that started near the middle of the trunk and extended down the tail. Anterior to these spikes were about seven pairs of plates similar to those of *Stegosaurus,* but smaller.

The function of the plates of *Stegosaurus* has been a matter of contention for decades. Originally they were assumed to have provided protection from

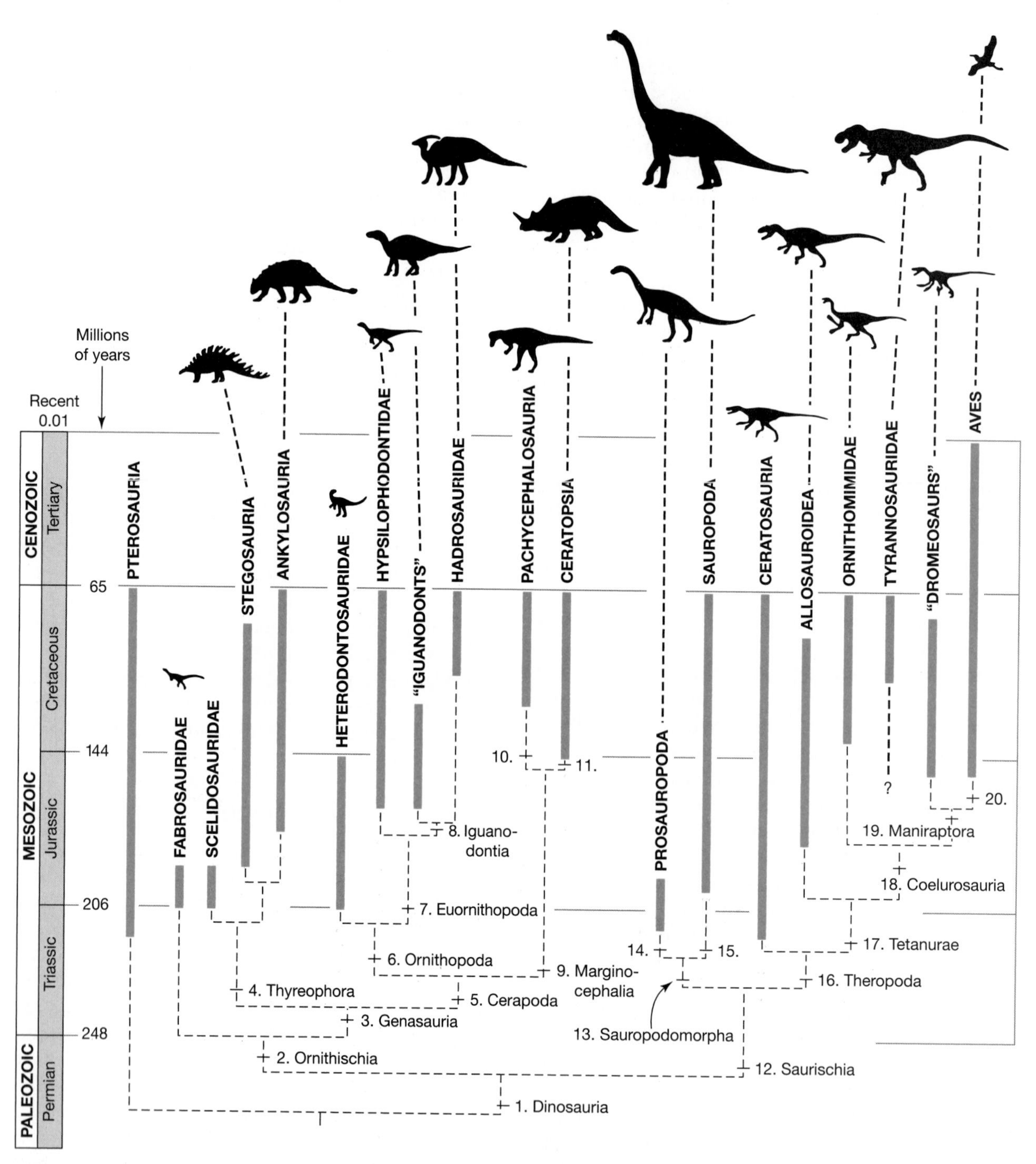

Legend: 1. Dinosauria—S-shaped swanlike neck, forelimb less than half the length of hindlimb, hand digit 4 reduced, and other characteristics of the palate, pectoral and pelvic girdles, hand, hindlimb, and foot. **2.** Ornithischia—Cheek teeth with low subtriangular crowns, reduced antorbital opening, predentary bone, toothless and roughened tip of snout, jaw joint set below level of upper tooth row, at least five sacral vertebrae, ossified tendons above sacral region, pelvis with pubis directed backward, small prepubic process on pubis. **3.** Genasauria—Muscular cheeks, reduction in size of mandibular foraman. **4.** Thyreophora—Characters of the orbit of the eye, rows of keeled scutes on the dorsal body surface. **5.** Cerapoda—Five or fewer premaxillary teeth, a diastema between premaxillary and maxillary teeth, characters of the pelvis. **6.** Ornithopoda—Premaxillary tooth row offset ventrally compared to max-

predators, and some reconstructions have shown the plates lying flat against the sides of the body as shields. A defensive function is not very convincing, however. Whether the plates were erect or flat, they left large areas on the sides of the body and the belly unprotected. Another idea is that the plates were used for heat exchange. Examination shows that the plates were extensively vascularized and could have carried a large flow of blood to be warmed or cooled according to the needs of the animal. *Kentrosaurus,* the African counterpart of *Stegosaurus,* had much smaller dorsal plates than *Stegosaurus;* the plates on *Kentrosaurus* extended only from the neck to the middle of the trunk. Posteriorly a double row of spikes extended down the tail. These spikes appear to have had a primarily defensive function rather than a thermoregulatory one. It is frustrating not to be able to compare the thermoregulatory behaviors of the two kinds of stegosaurs in a controlled experiment.

The short front legs of stegosaurs kept their heads close to the ground, and their heavy bodies do not give the impression that they were able to stand upright on their hind legs to feed on trees as ornithopods and perhaps sauropods did. Stegosaurs may have browsed on ferns, cycads, and other low-growing plants. The skull was surprisingly small for such a large animal, and had the familiar horny beak at the front of the jaws. The teeth show none of the specializations seen in hadrosaurs or ceratopsians, and the coronoid process of the lower jaw is not well developed. Unlike hadrosaurs and ceratopsians, which appear to have been able to grind or cut plant material into small pieces that could be digested efficiently, stegosaurs may have eaten large quantities of food without much chewing and used stones in a muscular gizzard to pulverize plant material.

The ankylosaurs are a group of heavily armored dinosaurs that are found in Jurassic and Cretaceous deposits in North America and Eurasia. Ankylosaurs

illary tooth row, jaw joint set well below level of tooth rows by ventral extension of quadrate. **7.** Euornithopoda—Absence of prominent boss in cheek region; **8.** Iguanodontia—Premaxillary teeth absent, external naris enlarged, wrist bones fused, also characters of the tooth surfaces and tooth enamel, lower jaw and skull. **9.** Marginocephalia—A shelf formed by the parietals and squamosals extends over the occiput, characters of the snout and pelvis. **10.** Pachycephalosauria—Thickened skull roof (frontals and parietals), other characters of the skull, dorsal vertebrae, forelimbs, and pelvis. **11.** Ceratopsia—Head triangular in dorsal view; tall, narrow anterior beak; jugals flare beyond the skull roof; deep, transversely arched palate, immobile mandibular symphysis. **12.** Saurischia—Construction of the snout including subnarial foramen, extension of the temporal musculature onto the frontal bones, elongation of the neck, modifications of the articulations between vertebrae, and modifications of the hand including a large thumb. **13.** Sauropodomorpha—Relatively small skull, anterior end of premaxilla deflected, teeth with serrated crowns, at least 10 cervical vertebrae forming an elongated neck. **14.** Prosauropoda—Elongation of claw on digit 1 of hand. **15.** Sauropoda—Four or more sacral vertebrae, straight femur with lesser trochanter reduced or absent. **16.** Theropoda—Articulation in middle of the lower jaw, construction of bones of the skull roof and palate, fenestra in the maxilla, characters of the vertebrae and neural arches, and transverse processes lacking posterior to a transition point in the middle of the tail, hand with elongated digits 1–3 armed with highly recurved claws, fibula and tibia closely adpressed, foot long and narrow with fifth metatarsal reduced to a splint, thin-walled (hollow) long bones. **17.** Tetanurae—Large fenestra posteriorly located in the maxilla, large fang-like teeth absent from dentary, maxillary tooth row ends anterior to orbit of eye, transition point in tail is farther anterior than in other theropods, expanded distal portion of pubis, characters of the foot. **18.** Coelurosauria—Fenestra in roof of mouth, characters of cervical vertebrae and ribs, furcula (wishbone) formed by fused clavicles, fused bony sternal plates, elongate forelimb and hand, characters of the foot. **19.** Maniraptora—Prefrontals reduced or absent, characters of the vertebrae, transition point in tail vertebrae close to base of tail, characteristics of the feet and pelvis. **20.** Aves—Progressive loss of teeth on maxilla and dentary, well-developed bill, feathers, characteristics of skull, jaws, vertebrae, and axial and appendicular skeleton.

◀ Figure 14–10 Phylogenetic relationships of the Dinosauria. This diagram shows the probable relationships among the major groups of dinosaurs, including birds. Dotted lines show interelationships only; they do not indicate times of divergence nor the unrecorded presence of taxa in the fossil record. The numbers indicate derived characters that distinguish the lineages.

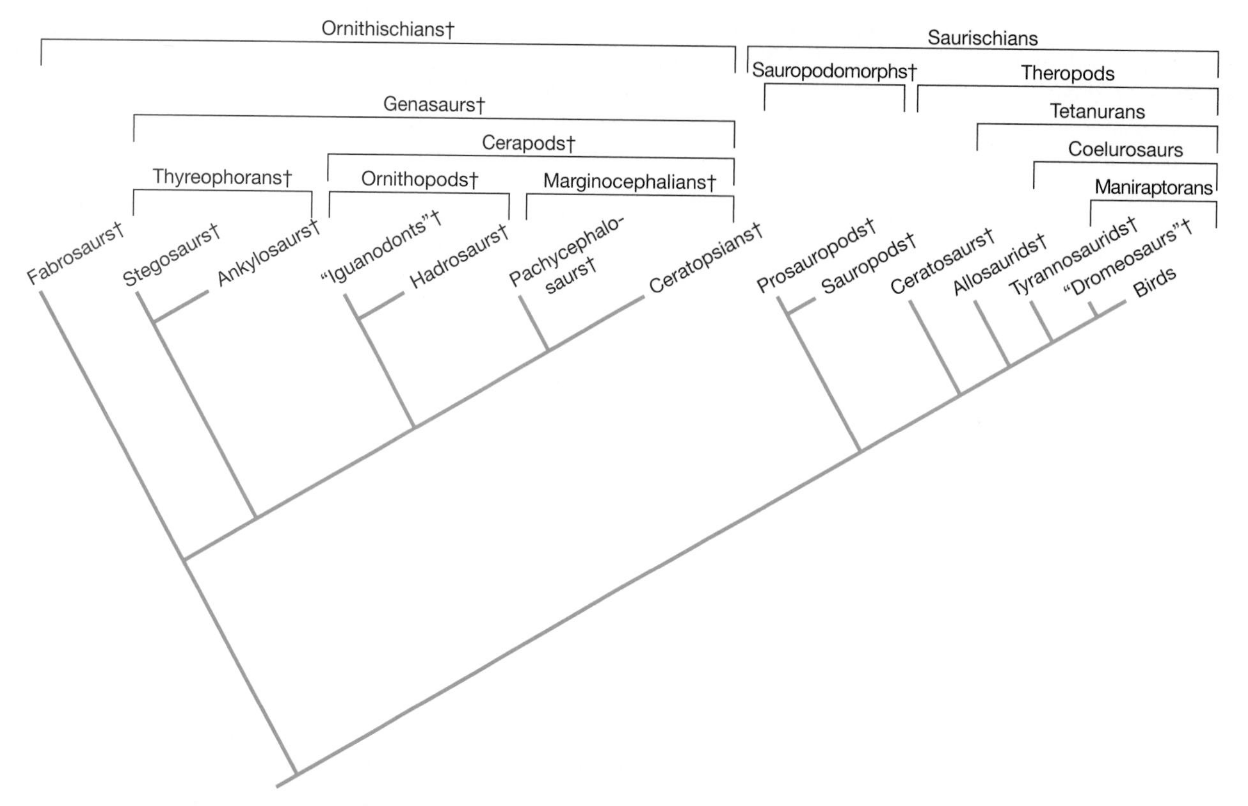

▲ Figure 14–11 Simplified cladogram of the Dinosauria. Quotation marks indicate paraphyletic groups. A dagger (†) indicates an extinct taxon.

were quadrupedal ornithischians that ranged from 2 to 6m in length. They had short legs and broad bodies, with **osteoderms** (bones embedded in the skin) that were fused together on the neck, back, hips, and tail to produce large shield-like pieces. Bony plates also covered the skull and jaws, and even the eyelids of *Euoplocephalus* had bony armor. Ankylosaurs had short tails, and some species had a lump of bone at the end of the tail that could apparently be swung like a club. The posteriormost caudal vertebrae of these club-tailed forms have elongated neural and hemal arches that touch or overlap the arches on adjacent vertebrae, and ossified tendons running down both sides of the vertebrae. Contraction of the muscles that inserted on these tendons probably pulled the posterior caudal vertebrae together to form a stiff handle for swinging the club head at the end of the tail. The tail of these animals resembles nothing so much as an enormous medieval mace. Other species had spines projecting from the back and sides of the body, and ankylosaurs must have been difficult animals for tyrannosaurids to attack. The brains of ankylosaurs appear to have had large olfactory stalks leading to complex nasal passages that probably had sheets of bone supporting an epithelium with chemosensory cells. If this interpretation is correct, ankylosaurs may have had a keen sense of smell.

Ornithopods

Ornithopods from the Early Jurassic, the heterodontosaurids and related groups, were mostly small (1 to 3m long) and bipedal. They had four toes on the hind feet and, unlike the bipedal saurischians, retained five toes on the front feet. Their cheek teeth were specialized for grinding plant material. Some heterodontosaurids had sharp tusks that may have been better developed in males than in females. The Cretaceous hypsilophodontids had horny beaks with which they may have cropped plant material that was subsequently ground by the high-crowned cheek teeth that gave the family its name (high-ridged tooth).

▶ Figure 14–12 Quadrupedal ornithischians. (a) *Stegosaurus* and (b) *Kentrosaurus* were stegosaurs. (c) *Euoplocephalus*, an ankylosaur. (d) *Styracosaurus*, a ceratopsian.

The first dinosaur fossil to be recognized as such was an ornithopod, *Iguanodon*, found in Cretaceous sediments in England (Figure 14–13a). Specimens have subsequently been found in Europe and Mongolia, and related forms have been discovered in Africa and Australia. *Iguanodon* reached lengths of 10m, although most specimens are smaller. Iguanodontids from the Early Cretaceous had large heads and elongated snouts that ended in broad, toothless beaks. Their teeth, which were in the rear of the jaws, were laterally flattened and had serrated edges, very like the lateral teeth of living herbivorous lizards like *Iguana*.

The first digit on each front foot of derived ornithopods was modified as a spine that projected upward. These spines show a striking resemblance to spines on the front feet of some frogs that are used as defensive weapons and during intraspecific encounters. *Ouranosaurus*, an ornithopod known from the early Middle Cretaceous of Africa, had a large sail that was supported by elongated neural spines on the vertebrae of the trunk and tail (Figure 14–13d).

Hadrosaurs The derived ornithopods include several specialized forms of hadrosaurs (duck-billed dinosaurs). Hadrosaurs were the last group of ornithopods to evolve, appearing in the middle of the Cretaceous. As their name implies, some duck-billed dinosaurs had flat snouts with a duck-like bill (Figure 14–13b). These were large animals, some reaching lengths of over 10m and weights greater than 10,000 kg. The anterior part of the jaws was toothless, but a remarkable battery of teeth occupied the rear of the jaws. On each side of the upper and lower jaws were four tooth rows, each containing about 40 teeth packed closely side by side to form a massive tooth plate. Several sets of replacement teeth lay beneath those in use, so a hadrosaur had several thousand teeth in its mouth, of which several hundred

▲ Figure 14–13 Bipedal ornithischians. (a) *Iguanodon.* (b) *Hadrosaurus.* (c) *Pachycephalosaurus.* (d) *Ouranosaurus.*

were in use simultaneously. Fossilized stomach contents of hadrosaurs consist of pine needles and twigs, seeds, and fruits of terrestrial plants.

The rise of the hadrosaurs was approximately coincident with a change in the terrestrial flora during the Middle Cretaceous. The bennettitaleans and seed ferns that had spread during the Triassic now were replaced by flowering plants (angiosperms). Simultaneous with the burgeoning of the angiosperms and hadrosaurian dinosaurs was a decline in the enormous sauropod dinosaurs such as *Diplodocus* and *Brachiosaurus.* Those lineages were most diverse in the Late Jurassic and Early Cretaceous, and only a few forms persisted after the middle of the Cretaceous.

Three kinds of hadrosaurs are distinguished: flat-headed, solid-crested, and hollow-crested (Figure 14–14). In the flat-headed forms (hadrosaurines) the nasal bones are not especially enlarged, although the nasal region may have been covered by folds of

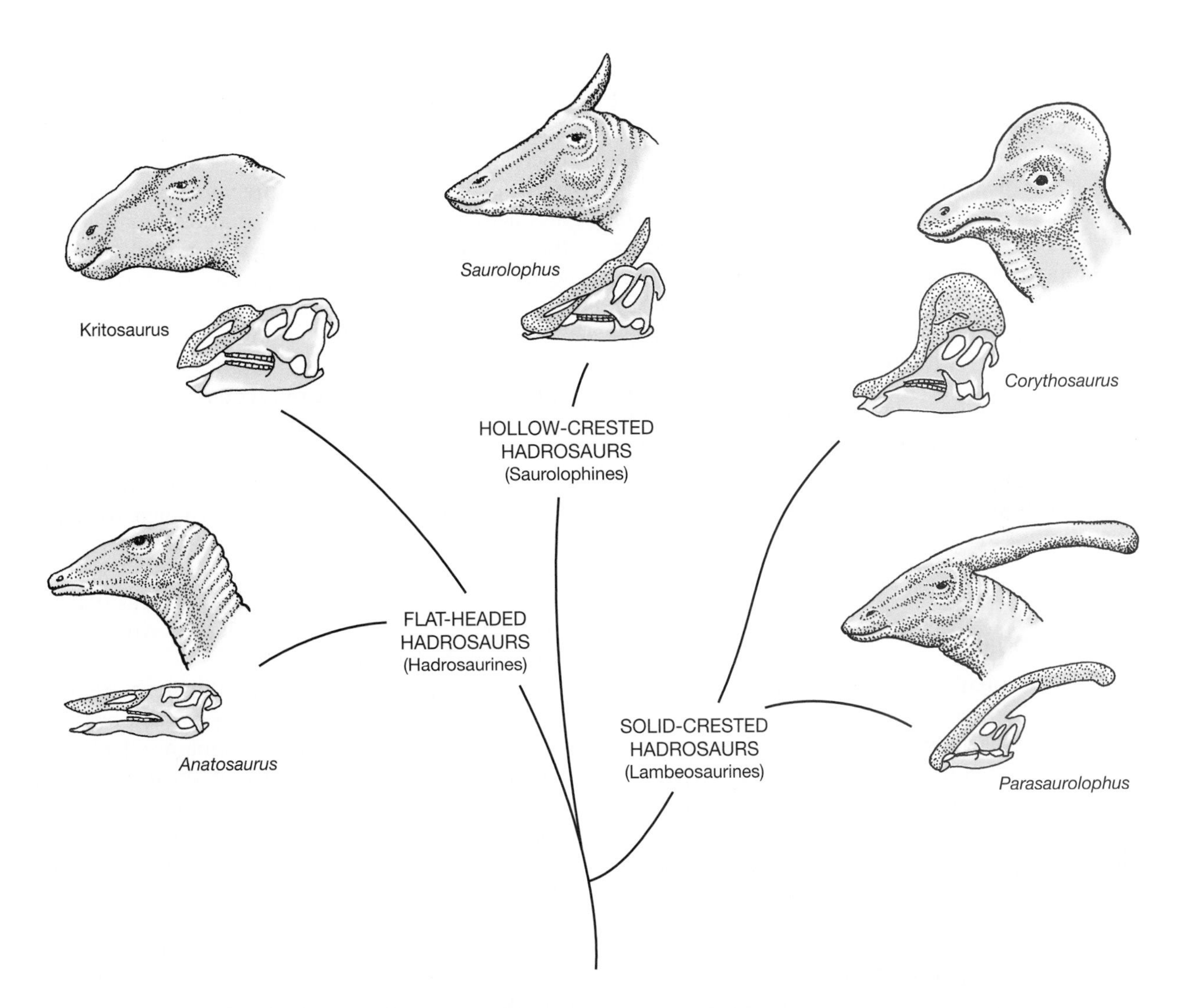

▲ Figure 14–14 Hadrosaurs. The bizarre development of the nasal and maxillary bones of some hadrosaurs gave their heads a superficially antelope-like appearance. In the flat-headed and solid-crested forms the nasal passages ran directly from the external nares to the mouth. In the hollow-crested forms the premaxillary and nasal bones contributed to the formation of the crests, and the nasal passages were diverted up and back through the crests before they reached the internal nares.

flesh. In the solid-crested forms (saurolophines) the nasal and frontal bones grew upward, meeting in a spike that projected over the skull roof. In the hollow-crested forms (lambeosaurines) a similar projection was formed by the premaxillary and nasal bones. In *Corythosaurus* those bones formed a helmet-like crest that covered the top of the skull, whereas in *Parasaurolophus* a long, curved structure extended over the shoulders. Although the crests of the lambeosaurines contained only bone, the nasal passages ran through the crests of the saurolophines. Inspired air traveled a circuitous route from the external nares through the crest to the internal nares, which were located in the palate just anterior to the eyes.

Perhaps these bizarre structures were associated with species-specific visual displays and vocalizations. The crests might have supported a frill attached to the neck, which could have been used in behavioral displays analogous to the displays of many living lizards that have similar frills. Possibly in the noncrested forms the nasal regions were covered by extensive folds of fleshy tissue that could be inflated by closing the nasal valves. Analogous structures can be found in the inflatable proboscises of elephant seals and hooded seals. The inflated structures of seals are resonating chambers used to produce vocalizations. The size and shape of the nasal cavities of lambeosaurine hadrosaurs suggest that adults produced low-frequency sounds, but juveniles would have had higher-pitched vocalizations.

Marginocephalia

The Marginocephalia, two lineages of highly specialized Late Cretaceous herbivores, were the last groups of ornithischians to appear.

Pachycephalosaurs The pachycephalosaurs are among the most bizarre ornithischians known (Figure 14–13c). The postcranial skeleton conforms to the bipedal pattern seen in ornithopods, but on the head an enormous bony dome thickens the skull roof. The bone is as much as 25 cm thick in a skull only 60 cm long. The angle of the occipital condyle indicates that the head was held so that the axis of the neck extended directly through the dome. The trunk vertebrae have articulations and ossified tendons that appear to have stiffened the vertebral column and resisted twisting. The pelvis was attached to at least six, and possibly to eight, vertebrae.

The thickened skull roof and the features of the postcranial skeleton have led some paleontologists to suggest that pachycephalosaurs used their heads like battering rams, perhaps for defense against carnivorous dinosaurs, or perhaps for intraspecific combat. An analogy has been drawn with goats and especially mountain sheep, in which males and females use head-to-head butting in social interactions. However, sheep and goats have horns that absorb some of the impact, and they have protective air sacs at the front of the brain. Pachycephalosaurs had neither of these specializations, although stretchable ligaments in the neck may have helped to absorb the shock of impact. The Galápagos marine iguana may be a better model for the behavior of pachycephalosaurs. These lizards have blunt heads with spikes very like miniature versions of the heads of pachycephalosaurs. Male marine iguanas press their heads together, twisting and wrestling during territorial disputes. Perhaps pachycephalosaurs used their bony heads in the same way.

Ceratopsians The most diverse marginocephalians, the horned dinosaurs or ceratopsians, appeared in the Early Cretaceous. By this time the easy transit from one continent to another that had characterized much of the Mesozoic was coming to an end. Early ceratopsians are found in western North America and eastern Asia (which were connected across the Bering Sea), but they were apparently excluded from the rest of the Northern Hemisphere by shallow epicontinental seas that covered the central parts of both North America and Eurasia in the late Mesozoic.

The distinctive features of ceratopsians are found in the frill over the neck, which is formed by an enlargement of the parietal and squamosal bones, a parrot-like beak, and a battery of shearing teeth in each jaw. The earliest ceratopsians were the psittacosaurs from Asia. These bipedal dinosaurs had no trace of a frill, but they did have a horny beak that covered a rostral bone at the front of the upper jaw. (The rostral bone is a distinctive feature of ceratopsian dinosaurs.) *Protoceratops,* one of the earliest quadrupedal ceratopsians, had developed a modest frill that extended backward over the neck and formed the origin for powerful jaw-closing muscles that extended anteriorly through slits at the rear of the skull and inserted on the coronoid process of the lower jaw. The teeth were arranged in batteries in each jaw, somewhat like those of hadrosaurs, but with an important difference. The teeth of ceratopsians formed a series of knife-like edges rather than a solid surface like hadrosaur teeth. The feeding method of ceratopsians seems likely to have consisted of shearing vegetation into short lengths, rather than crushing it as hadrosaurs did.

Protoceratops had a simple frill, unadorned by spikes, and also lacked a nasal horn. Derived ceratopsians had both of these elaborations. Two groups can be distinguished: In the short-frilled ceratopsians (*Monoclonius, Styracosaurus,* and others) the frill extended backward over the neck, whereas in the long-frilled forms (*Chasmosaurus, Pentaceratops,* and others) the frill extended past the shoulders. Both short- and long-frilled ceratopsians had nasal and brow horns developed to varying degrees. Probably the initial stages in the evolution of the frill involved jaw mechanics and the importance of strong jaw muscles. Even in *Protoceratops,* however, males had larger frills than females; this sexual dimorphism suggests that frills played a role in the social behavior of ceratopsians. Furthermore, the nasal and brow horns would have been formidable weapons for defense and for intraspecific combat. An analogy to the horns of antelope or the antlers of deer, which function in both defense and social behavior, seems appropriate for ceratopsians.

Social Behavior of Ornithischians

The morphological diversity of the ornithischian dinosaurs suggests that they were equally diverse in behavior and ecology. Social interactions based on visual displays and vocalizations may have been well developed among hadrosaurs, and pachycephalosaurs may have engaged in shoving contests like those of ungulate mammals. Individual interactions of these sorts may have been integrated into group behaviors. Fossilized eggs of dinosaurs provide information about nesting behaviors (Box 14-1). Evidence of parental care may be revealed by a nest of 15 baby hadrosaurs (*Maiasaura*) in the Late Cretaceous Two Medicine Formation in Montana. The babies were about a meter long—approximately twice the size of other hatchlings found in the same area—indicating that the group had remained together after they hatched. The teeth of the baby hadrosaurs showed that they had been feeding; some teeth were worn down to one-quarter of their original length. The object presumed to be a nest was a mound 3m in diameter and 1.5m high with a saucer-shaped depression in the center. Such a large structure would have made the babies conspicuous to predators, and it seems likely that a parent remained with the young. (*Maiasaura* can be translated as "good mother reptile.") The morphology of the inner ears of lambeosaurs suggests that adults would have been able to hear the high-pitched vocalizations of juveniles, strengthening the inference of parental care. The association between adults and young appears to have lasted for a considerable time. Fossils suggest that *Maiasauria* and the lambeosaur *Hypacrosaurus* grew to one-quarter of adult size before they left the nesting grounds, and a species of hypsilophodontid found at the same site grew to half its adult size. Both vocal communication and prolonged association of parent dinosaurs with their young are plausible in light of the behaviors known for crocodilians.

Additional fossils in the same formation suggest that the area contained nest sites of other species of hadrosaurs and of ceratopsians as well. Eggshells and baby dinosaurs are abundant in the Two Medicine Formation but rare in adjacent sediments. A similar concentration of conspicuous nests, eggs, and juveniles of the small ceratopsian *Protoceratops* discovered in Mongolia suggests parental care in this species as well.

Protection of nests and juveniles from predators is a plausible reason for parental care, and Katherine Troyer has suggested a second possible explanation for an association between juveniles and adults of herbivorous dinosaurs such as the hadrosaurs and ceratopsoids (Troyer 1984). Working with iguanas, which are herbivorous lizards, she showed that hatchlings must ingest the feces of adult iguanas (or of other hatchlings that have themselves ingested feces from adults) to inoculate their guts with the symbiotic microorganisms that permit them to digest plant material. The microorganisms responsible for the fermentation of plant material are anaerobic and soon die when feces are exposed to air. Thus, a close association between hatchlings and adults is necessary to achieve the transfer of the microorganisms. Herbivorous dinosaurs probably relied on fermentation of plant material just as most extant herbivorous vertebrates do, and some contact between juveniles and adults would have been neces-

BOX 14-1 Dinosaur Eggs and Nests

Fossils of dinosaur eggs have been found in Late Cretaceous deposits in Mongolia, China, France, India, and the United States. Most of these fossils are fragments of eggshells, but intact eggs containing embryos have been discovered. The fossil of an adult dinosaur has also been found, apparently brooding a clutch of eggs (Figure 14–15).

Concentrations of nests and eggs ascribed to sauropods in Cretaceous deposits in southern France suggest that these animals had well-defined nesting grounds. Eggs thought to be those of the large sauropod *Hypselosaurus priscus* are found in association with fossilized vegetation similar to that used by alligators to construct their nests. The orientation of the nests suggests that each female dinosaur probably deposited about 50 eggs. The eggs had an average volume of 1.9 liters, about 40 times the volume of a chicken egg. Fifty of these eggs together would weigh about 100 kg, or 1 percent of the estimated body mass of the mother. Crocodilians and large turtles have egg outputs that vary from 1 to 10 percent of the adult body mass, so an estimate of 1 percent for *Hypselosaurus* seems reasonable. The eggs might have been deposited in small groups instead of all together, because 50 eggs in one

(a)

(b)

(c)

▲ Figure 14–15 Dinosaur nests and eggs. (a) An adult oviraptor on a nest of eggs. This reconstruction is based on a fossil of an oviraptor found in the Gobi Desert. The adult was apparently brooding its eggs when it was buried by a giant sandstorm. (b) Fossilized nest of an oviraptorid dinosaur. (c) The fossilized skeleton of an embryo of an oviraptorid dinosaur (AMNHK17088). This is the first embryo of a carnivorous dinosaur ever found.

BOX 14-1 Dinosaur Eggs and Nests (*continued*)

clutch would have consumed oxygen faster than it could diffuse through the walls of the nest (Seymour 1979).

Two patterns of egg laying can be distinguished (Mikhailov 1997). Sauropod dinosaurs laid eggs in nests dug into the soil, much like those of extant turtles. In contrast, ornithischian and theropod dinosaurs laid eggs in an excavation that might have been filled with rotting vegetation to provide both heat and moisture for the eggs. (This method of egg incubation is used by many of the living crocodilians.) Nests of *Protoceratops* fall in this second category—the 30 to 35 eggs in each nest are arranged in concentric circles with their blunt ends up. Eggs of *Orodromeus makelai,* a hypsilophodontid, are also oriented vertically with the blunt end up, but are arranged in a spiral within the circular nest.

The fossil of a theropod dinosaur that apparently died while attending a nest of eggs was discovered in the Gobi Desert in 1923, but its significance was not recognized until 70 years later. The eggs, which were about 12 cm long and 6 cm in diameter, were thought to have been deposited by the small ceratopsian *Protoceratops andrewsi* because adults of that species were by far the most abundant dinosaurs at the site. The theropod was assumed to have been robbing the nest, and was given the name *Oviraptor philoceratops,* which means "egg seizer, lover of ceratops." In 1993, paleontologists from the American Museum of Natural History, the Mongolian Academy of Sciences, and the Mongolian Museum of Natural History discovered a fossilized embryo in an egg identical to the supposed *Protoceratops* eggs. To their surprise, the embryo was an *Oviraptor* nearly ready to hatch. This discovery suggests that the adult *Oviraptor* probably died while resting on its own nest.

sary to inoculate hatchling dinosaurs with the symbiotic microorganisms.

14.7 The Saurischian Dinosaurs

Two groups of saurischian dinosaurs are distinguished, the **Sauropodomorpha** and the **Theropoda** (see Figure 14–12). Sauropodomorphs, all of which are now extinct, were primarily quadrupedal herbivores, whereas theropods, which include the extant birds, are bipdal carnivores. Ten shared derived characters unite saurischians (Gauthier 1986); the most obvious is elongation of the mobile, S-shaped neck. This character distinguishes birds among living amniotes. Other bird-like characters of saurischians are found in modifications of the hand, skull, and postcranial skeleton.

Sauropodomorph Dinosaurs

The earliest sauropodomorph dinosaurs were the prosauropods, a group that was abundant and diverse in the Late Triassic and Early Jurassic. Three types of prosauropods are known, differing in size and tooth structure. The anchisaurids ranged in size from *Anchisaurus* (2.5m) to *Plateosaurus* (6m). The anchisaurids had long necks and small heads (Figure 14–16a), and the teeth of the best-known forms had large serrations. Modern herbivorous lizards (iguanas) have teeth with very much the same form, and anchisaurids were probably herbivorous. Supporting this view is the presence of **gastroliths** (Greek *gastro* = stomach and *lith* = stone) associated with some fossil prosauropods. These stones were probably swallowed by the dinosaurs and lodged in a muscular gizzard, where they assisted in grinding plant material to a pulp that could be digested more readily; some birds use gastroliths in this manner. Prosauropods appear to have had cheeks that retained food in the mouth as it was processed by the teeth. The earliest prosauropods were small, lightly built, and bipedal. Later forms were larger and heavier. Their body proportions suggest that they could stand vertically on their hind legs, but they probably used a quadrupedal posture most of the time.

Derived prosauropods, such as the melanorosaurids, were larger than the early prosauropods (*Riojasaurus* from the Late Triassic of Argentina was 11m long). No skulls of melanorosaurids have been found, so the structure of their teeth is unknown. The long, slender neck of *Riojasaurus* suggests that the head was small, like that of early prosauropods. The yunnanosaurids of China were smaller than the

▲ Figure 14–16 Sauropodomorph dinosaurs. (a) *Plateosaurus.* (b) *Camarasaurus.* (c) *Diplodocus.*

melanorosaurids and more lightly built, and they differed from the earlier prosauropods in having teeth shaped like flattened cylinders with a chisel-shaped tip. This is the tooth structure seen in the giant sauropod dinosaurs, and it is quite distinct from that of the laterally flattened, serrated teeth of early prosauropods, such as *Plateosaurus.*

The long necks of all the prosauropods suggest that they were able to browse on plant material at heights up to several meters above the ground. The ability to reach tall plants might have been a significant advantage during the shift from the low-growing *Dicroidium* flora to the taller bennettitaleans and conifers that occurred in the Late Triassic.

The derived sauropods of the Jurassic and Cretaceous were enormous quadrupedal herbivores. Most fossils consist of fragmentary material, and nearly complete skeletons are known for only about five of the nearly 90 genera that have been named. The sauropods were the largest terrestrial vertebrates that have ever existed. The largest of them may have reached body lengths exceeding 30m and weights of 100,000 kg. (For comparison, a large African elephant is about 5m long and weighs up to 5000 kg.)

Two major types of giant sauropods can be distinguished, the diplodocoids and the camarasauroids and brachiosaurids. Camarasauroids had necks with only 12 vertebrae. Their tails were shorter than those of diplodocoids (about 50 vertebrae) and lacked the whip-like extension that was characteristic of the diplodocoids (Figure 14–16b). The forelimbs of camarasauroids were relatively long, and the vertebral column was nearly horizontal. Brachiosaurids had still longer front legs, and the trunk sloped steeply downward from the shoulders to the hips. Camarasauroids and brachiosaurids had compact skulls with stout jaws and large, chisel-shaped teeth. The teeth of *Camarasaurus* and *Brachiosaurus* show evidence of heavy wear, suggesting that they fed on abrasive material.

The diplodocoids include *Apatosaurus* (formerly known as *Brontosaurus*) and *Diplodocus* (Figure 14–16c). These animals had long necks (15 cervical vertebrae) and long tails (up to 80 caudal vertebrae) that ended in a thin whiplash. Their front legs were relatively short, and the trunk slanted upward from the shoulders to the hips. Their skulls were elongate, teeth were limited to the front of the mouth, and the modest development of the lower jaw bones suggests that the jaw muscles were not particularly powerful.

Sauropods dominated terrestrial habitats in the Late Jurassic and Early Cretaceous, evolving with a flora of conifers, ginkos, ferns, cycads, and horsetails. Later in the Cretaceous, when angiosperms became the dominant land plants, sauropods were replaced in North America and western Asia by derived ornithischians. Sauropods persisted on the southern continents, however, and sauropods from South America may have reinvaded the southern part of North America at the end of the Cretaceous.

Biology of Sauropods

Both diplodocoids and camarasauroids were enormously heavy, and their vertebrae show features that helped the spinal column to withstand the stresses to which it was subjected (Figure 14–17). The vertebrae themselves were massive, and the neural arches were well developed. Strong ligaments transmitted forces from one arch to adjacent ones to help equalize stress. The head and tail were cantilevered from the body, supported by a heavy spinal ligament. The sides of the neural arches and centra had hollows in them, possibly occupied by air sacs in life, that reduced the mass of the bones with little reduction in strength. The feet of these forms were elephant-like, and fossilized tracks indicate that the hind legs bore about two-thirds of the body weight. Some trackways show no tail marks, suggesting that the tails were carried in the air, not dragged along the ground in the manner shown in older illustrations of these dinosaurs.

From the earliest discovery of these fossils, paleontologists doubted that such massive animals could

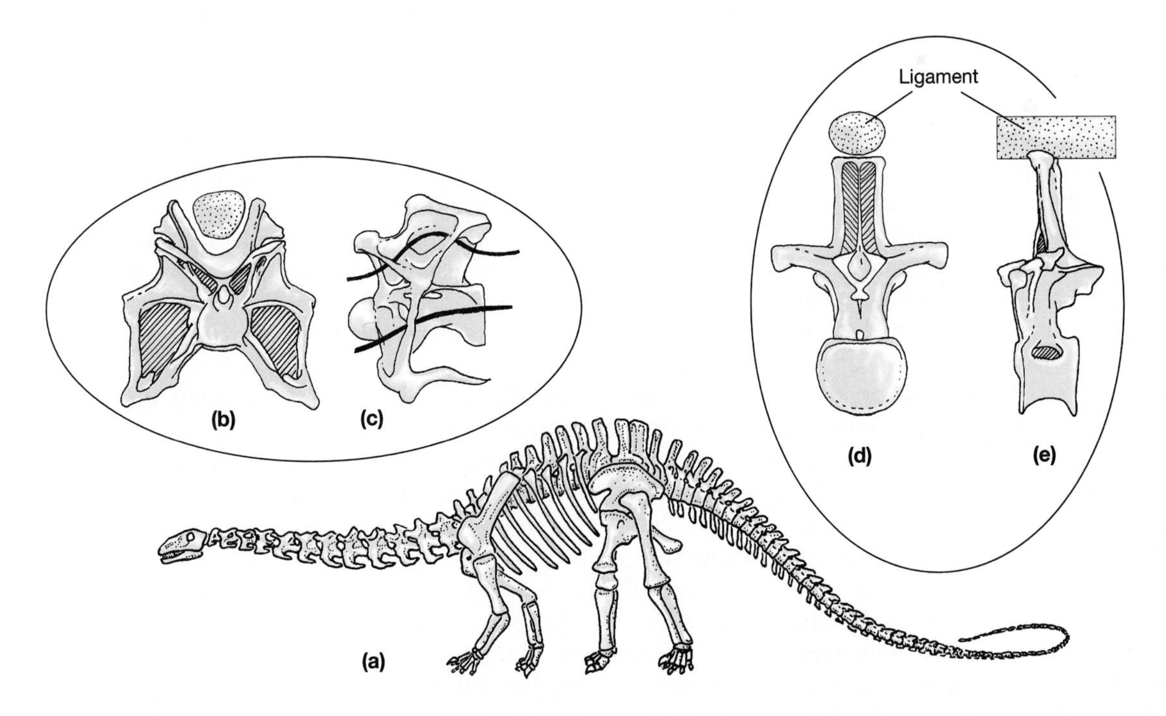

▲ Figure 14–17 Structural features of sauropods. The skeletons of large diplodocid sauropods like *Apatosaurus* (a) combined lightness with strength. Vertebrae from the dorsal region: (b) posterior view, c, lateral view, and neck (d, anterior view; e, lateral view) show the bony arches that acted like flying buttresses on a large building. (The black ribbons in [c] indicate the position of the arches.)

have walked on land, instead believing they must have been limited to a semiaquatic life in swamps. Mechanical analysis of sauropod skeletons does not support that conclusion, however. The skeletons of the large sauropods clearly reveal selective forces favoring a combination of strength with light weight. The arches of the vertebrae acted like flying buttresses on a large building, while the V-shaped neural spines of diplodocoids held a massive and possibly elastic ligament that helped to support the head and tail. In cross section the trunk was deep, shaped like the body of a terrestrial animal such as an elephant rather than rounded like that of the aquatic hippopotamus. The tails of sauropods are not laterally flattened like tails used for swimming. Instead they are round in cross section and, in diplodocoids, terminate in a long, thin whiplash. These structures look like counterweights and defensive weapons.

Fossil trackways of sauropods indicate that the legs were held under the body; the tracks of the left and right feet are only a single foot width apart. Analysis of the limb bones suggests that they were held straight in an elephant-like pose and moved fore and aft parallel to the midline of the body. This is what would be expected on mechanical grounds, because no other leg morphology is possible for a very large animal. Bone is far less resistant to bending forces exerted across its long axis than it is to compressional forces exerted parallel to the axis. As an animal's body increases in size, its mass grows as the cube of its linear dimensions; but the cross-sectional area of the bones increases as the square of their linear dimensions (see Chapter 8). The strength of bone is roughly proportional to its cross-sectional area. As a result, when the body size of an animal increases, the strength of the skeleton increases more slowly than the stress to which it is subjected. One evolutionary response to this relationship is disproportionate increase in the diameter of bones—an elephant skeleton is proportionally larger than a mouse skeleton. Another response is to transform bending forces to compression forces by bringing the legs more directly under the body. In a large animal, such as an elephant or a sauropod dinosaur, not only are the legs held under the body, but the knee joint tends to be locked as the animal walks. This morphology produces the straight-legged locomotion familiar in elephants; sauropods probably walked with an elephant-like gait, holding their heads and tails in the air.

Sauropod teeth are sometimes described as being small and weak. Certainly they were small in proportion to the size of the body, as was the entire skull. In absolute terms, however, they were neither small nor weak. They were larger than the teeth of browsing mammals, and there is no reason to believe that plant material was tougher in the Mesozoic than it is today. There were no flat (molariform) teeth to crush the ingested plant material. This function may have been served by gastroliths, and the breakdown of plant material may have been aided by symbiotic microorganisms as in modern ungulates.

The fossilized stomach contents of a sauropod dinosaur, found in Jurassic sediments in Utah, includes pieces of twigs and small branches about 2.5 cm long and 1 cm in diameter. The fragmented and shredded character of the woody material indicates that even without molariform teeth, the sauropod could crush its food. This discovery appears to confirm the view of sauropod ecology that was developed by studying the skeleton and analyzing plant fossils found in association with sauropod fossils. Sauropods probably occupied open country with an undergrowth of ferns and cycads and an upper story of conifers. They were preyed on by the large theropod carnivores and sought escape in flight or defended themselves by whipping their tails.

The long necks of sauropods are a puzzle. The conventional view is that they grazed from tree tops, perhaps even standing on their hind legs by using the tail as a counterweight. It may be significant that Mesozoic conifers bore branches only near the tops of the trees, far out of reach of any but a very long-necked dinosaur. On the other hand, two lines of reasoning argue against the idea that sauropods had giraffe-like feeding habits. An analysis of the joints between vertebrae in the necks of *Diplodocus* and *Apatosaurus* suggests that the necks were less flexible than had been assumed (Stevens and Parrish 1999). If this interpretation is correct, sauropods may have swept the head horizontally and vertically through limited arcs, covering a large volume of feeding space without having to move because their necks were so long.

Another mechanical problem that sauropods would have faced while feeding from treetops was the difficulty of pumping blood to a head that might have been as much as 20m above the ground and 6 or 7m above the level of the heart (Lillywhite 1991, Seymour and Lillywhite 2000). Blood is mostly water, and water is heavy. When their heads were raised to browse on trees, the tallest sauropods would have required ventricular blood pressures exceeding 500 millimeters of mercury to overcome the hydrostatic pressure of a 7-meter column of blood between the heart and the brain. A column of blood extending to a head 20m above the ground could have produced blood pressures as great as 1000 mm of mercury in the vessels of the legs and feet of a large sauropod. Pressures that high would have tended to force water across the walls of the capillaries, causing the legs and feet to swell. These problems would not have occurred if sauropods fed with their heads close to the level of their hearts.

Sauropods might have been an important force shaping the landscape and preventing ecological succession from transforming open countryside to dense forest. As such, their presence could have been important in creating and preserving suitable habitat for other species, much as elephants do in African savannas today. Like many herbivorous lizards, the sauropods may have been omnivorous and opportunistic in their feeding, taking whatever was readily available, including carrion. The fossilized stomach contents mentioned previously contain traces of bone as well as a tooth from the contemporary carnivorous dinosaur *Allosaurus.*

Sauropods lacked frills and other sexually dimorphic display structures of the sort seen among ornithishian dinosaurs, but that does not mean that social behavior was absent. After all, modern crocodilians are not sexually dimorphic or ornamented with frills, and they have elaborate social behaviors. The possibility of parental care and social behavior among sauropods is raised by the discovery of five baby prosauropod dinosaurs (*Mussaurus*) with the remains of two eggs in a nest in Late Triassic deposits in Argentina.

Fossil trackways reveal a few details from which glimpses of sauropod behavior can be reconstructed (Thulborn 1990). A famous trackway found in Texas—parts of which are now on display at the University of Texas at Austin and in the American Museum of Natural History in New York City—shows the footprints of a sauropod that was apparently being trailed by a large theropod, which was a few steps behind and slightly to the left. The theropod tracks duplicate several small changes in direction by the sauropod, and the rhythm of the therapod's stride was adjusted to match that of the sauropod. Mammalian predators such as lions make similar adjustments to match the stride of their prey before they attack. A drag mark made by the sauropod's right rear foot and two consecutive marks of the theropod's right foot (i.e., a hop) may even mark the point of an attack (Thomas and Farlow 1997).

Evidence of possible herd behavior by sauropods may be revealed by a series of tracks found in Early Cretaceous sediments at Davenport Ranch in Texas. These reveal the passage of 23 apatosaur-like dinosaurs some 120 million years ago. A group of individuals moving in the same direction at the same time would be remarkable for most living diapsids, but the apatosaur tracks suggest that this is what happened. Furthermore, the tracks may show that the herd moved in a structured fashion with the young animals in the middle, surrounded by adults.

Theropod Dinosaurs

Theropod dinosaurs included three general types of animals: large predators that probably attacked large prey using their jaws as weapons (ceratosaurs, allosaurs, and tryannosaurs), fast-moving predators that seized small prey with their forelimbs (ornithomimids), and fast-moving predators that attacked prey larger than themselves with a huge claw on the hind foot (dromeosaurs) (Figure 14–18).

Large Theropods Large carnivores are the impressive dinosaurs that form the centerpieces of paleontological displays in many museums. Increasing body size of theropods through the Mesozoic paralleled a similar size increase in the herbivorous saurischians and ornithischians that were their prey. *Dilophosaurus,* from the Early to Middle Jurassic of North America, is named for the paired bony crests on its head (Greek *di* = double, *loph* = crest, and *saurus* = lizard). Its jaws were slender and appear too weak to withstand the strain of attacking large prey. Although it was large (6m long), *Dilophosaurus* may have been a scavenger. *Ceratosaurus* of the Late Jurassic was also about 6m long, but had a heavier skull and jaws than *Dilophosaurus.* The head was large in proportion to the body, and the long teeth were fearsome weapons. Its front feet had four fingers with large claws. Allosaurs, which were contemporaries of the ceratosaurs, were larger—up to 12m long. Allosaurs had only three claws on the front feet. The Late Cretaceous tyrannosaurids such as *Tarbosaurus* and *Tyrannosaurus* were still longer, up to 15m long, and stood some 6m high. The front legs of the most specialized of these giants seem absurdly small; they were too short even to reach the mouth and had only two small fingers on each hand. Instead of relying on the forelimbs to capture prey as ornithomimids prob-

▲ Figure 14–18 Theropod dinosaurs. (a) *Coelophysis,* a Triassic ceratosaur. (b) *Struthiomimus,* a Cretaceous ornithomimid. (c) *Tyrannosaurus,* a large Cretaceous theropod. (d) *Deinonychus,* a Cretaceous deinonychosaur.

ably did, large theropods appear to have concentrated their weapons in the skull. The size of the head increased relative to the body, and the neck shortened. The head was lightened by the elaboration of antorbital and mandibular fenestrae, reducing the skull to a series of bony arches and providing maximum strength for a given weight.

The teeth of large therapods were as much as 15 cm long, dagger-shaped with serrated edges and driven by powerful jaw muscles. Marks from the teeth of predatory dinosaurs are sometimes found on fossilized dinosaur bones, and these records of prehistoric predation provide a way to estimate the force of a dinosaur's bite. The pelvis of a horned dinosaur (*Triceratops*) found in Montana bears dozens of bite marks from a *Tryannosaurus rex,* some as deep as 11.5 mm. A fossilized *Tryannosaurus* tooth was used to make an indentation that deep in the pelvis of a cow (Erickson et al. 1996). The force required to make the marks on the *Triceratops* pelvis were estimated to range from 6,410 to 14,400 N. These values exceed the force that can be exerted by several extant predators (dog, wolf, lion, shark). Interestingly, an alligator was the only predator tested that could deliver a bite as powerful as that of the *Tyrannosaurus,* and the jaws and teeth of alligators have many of the same structural characters as the jaws and teeth of *Tryannosaurus.*

Other experimental studies that used fossilized tyrannosaur teeth to bite meat showed that the serrations increased the cutting effect only slightly, but they trapped and retained meat fibers. This debris would have supported the growth of bacteria, and a tyrannosaur bite would almost surely have become infected. Perhaps tyrannosaurs did not necessarily kill large prey such as sauropods in the initial attack, but relied on infection to weaken the victim and make it susceptible to a subsequent attack. Bacteria on the teeth and claws of the largest extant lizard, the Komodo dragon, are thought to play exactly this role when the lizards prey on deer.

A coprolite (fossilized dung) the size of a loaf of bread from Saskatchewan, Canada, is believed to have been deposited by a *Tryannosaurus rex.* It contains bone fragments from a juvenile ornithischian, possibly the head frill of a *Triceratops* (Chin et al. 1998). The shattered bone in the coprolite suggests that tyrannosaurs repeatedly bit down on food in the mouth before they swallowed it. This feeding behavior is different from that of extant crocodilians, which swallow large mouthfuls of food without processing it.

Small theropods Many of the smaller theropods are found among the coelurosaurs, a mainly Cretaceous group that includes birds and all the theropods more closely related to birds than to allosaurs. Several characters of living birds are seen in coelurosaurs (Gauthier 1986). The most interesting characters from the perspective of the origin of birds include features usually thought to be associated with powered flight, especially a fused bony sternum and a furcula (wishbone) formed by fusion of the clavicles. The widespread occurrence of a furcula among nonflying relatives of birds shows that the original function of the furcula did not involve flight. Thus, the important role the furcula plays in flight by extant birds has evolved secondarily.

Small theropods were also found among other lineages. *Coelophysis* (Figure 14–18a), a Late Triassic form, is a ceratosaur about 3m in total length. *Coelophysis* was probably an active, cursorial predator of other small dinosaurs, lizards, and insects. A lineage of ostrich-like dinosaurs occurred in the Late Cretaceous. *Ornithomimus* was ostrich-like in size, shape, and probably in ecology as well. It had a small skull on a long neck, and its toothless jaws were covered with a horny bill. The forelimbs were longer than those of *Coelophysis,* and only three digits were developed on the hands. The inner digit was opposable and the wrist was flexible, making the hand an effective organ for capturing small prey. Like ostriches, *Ornithomimus* was probably omnivorous and fed on fruits, insects, small vertebrates, and eggs. Quite possibly it lived in herds, as do ostriches, and its long legs suggest that it inhabited open regions rather than forests.

Apparently not all ornithomimids preyed on small animals. A fossil from the Gobi Desert, *Deinocheirus* (terrible hand), had fingers more than 60 cm long that appear to have been used for grasping and dismembering large prey. The proportions of the hands and arms are like those of coelurosaurs. If this theropod had the same body proportions as other coelurosaurs, it may have been more than 7.5m tall—exceeding *Tyrannosaurus rex,* previously the tallest theropod known.

Dromeosaurs *Deinonychus* was unearthed by an expedition from Yale University in Early Cretaceous sediments in Montana (Figure 14–18d). It is a small theropod, a little over 2m long. Its distinctive features are the claw on the second toe of the hind foot and the tail. In other theropods the hind feet are clearly specialized for bipedal locomotion, very similar to bird feet. In these forms the third toe is the largest, the second and fourth are smaller, and the

▲ Figure 14–19 The foot of *Deinonychus,* showing the enlarged claw.

fifth has sometimes disappeared entirely. The first toe is turned backward, as in birds, to provide support behind the axis of the leg. The second toe of dromeosaurs, especially the claw on that toe, is enlarged (Figure 14–19). In its normal position the claw was apparently held off the ground and it could be bent upward even farther.

It seems likely that dromeosaurs used these claws in hunting, disemboweling prey with a kick. The structure of the tail was equally remarkable. The caudal vertebrae were surrounded by bony rods that were extensions of the prezygapophyses (dorsally) and hemal arches (ventrally) that ran forward about 10 vertebrae from their place of origin. Contraction of muscles at the base of the tail would be transmitted through these bony rods, drawing the vertebrae together and making the tail a rigid structure that could be used as a counterbalance or swung like a heavy stick. Possibly the tail was part of the armament of *Deinonychus,* used to knock prey to the ground where it could be kicked, and it may have served as a counterweight for balance as *Deinonychus* made sharp turns. Dromeosaurs probably relied on fleetness of foot to capture active prey. The discovery of five *Deinonychus* skeletons in close association with the skeleton of *Tenontosaurus,* an ornithischian three times their size, might indicate that *Deinonychus* hunted in packs (Maxwell 1999). Deinonychosaurs probably used their clawed forefeet to seize prey and then slashed at it with the sickle-like claws on the hind feet. This tactic appears to be illustrated by a remarkable discovery in Mongolia of a dromeosaur called *Velociraptor.* It was preserved in combat with a *Protoceratops,* its hands grasping the head of its prey and its enormous claw embedded in the midsection of the *Protoceratops.*

The discovery of *Deinonychus* stimulated a reexamination of fossils of several other genera of small theropod dinosaurs from the Cretaceous, including *Dromeosaurus* and *Velociraptor.* All these forms have an enlarged claw on the second toe of the hind foot, and they are now grouped with *Deinonychus* and birds in the Maniraptora. *Deinonychus*-like claws 35 cm long were discovered in Early Cretaceous sediments in Utah during autumn 1991. The claws were probably from a previously unknown theropod (nicknamed "super-slasher" by paleontologists) that was nearly as large as a *Tyrannosaurus rex* and had the speed, agility, and predatory behavior of *Deinonychus* (Figure 14–20).

▶ Figure 14–20 Two large deinonychosaurs ("super-slashers") attacking a brachiosaurid.

14.8 Dinosaur Soft Parts and Temperature Regulation

Fossils usually preserve only the bony parts of animals, because soft tissues decay too quickly for minerals to replace organic molecules. Physiological characteristics, too, are not directly visible in fossils and must be inferred from anatomy.

Dinosaur Organs

Soft tissues do fossilize under some conditions. When an animal becomes mummified (i.e., dries out before it decays) the skin and even internal structures may subsequently be replaced by minerals, creating a fossil of the mummy. Fossils of mummified hadrosaurs show that the skin was covered with a pavement of small and large polygonal tubercules. Many species show traces of a frill of skin extending along the back from the neck to the tail. Impressions of the skin of ceratopsians show plates, up to 5.5 cm in diameter, on a background of smaller structures. Dermal armor (osteoderms) fossilizes more readily than skin, and many sauropods were covered by a mosaic of dermal plates 10 to 12 cm in diameter distributed among small ossicles only 6 to 7 mm in diameter. Only among the dinosaurs most closely related to birds is there evidence of feather-like structures. Two dinosaurs with feathers have recently been unearthed from fossil deposits in China (see Chapter 15).

Internal organs are sometimes preserved. A specimen of *Sinosauropteryx* (a small carnivorous theropod from China) contains the remains of a lizard that was the dinosaur's last meal; another specimen contains two shelled eggs. These eggs are probably not the remains of a meal, because they are located far from the stomach. The report of a fossilized heart of a *Thescelosaurus* (an ornithopod) caused considerable excitement. It was interpreted as having four chambers (as would be expected from the structure of crocodilian hearts) but with a single aorta like that of birds instead of the double aorta characteristic of crocodilians (Fisher et al. 2000). Other paleontologists doubt that the structure is really a fossilized heart; instead it appears to be a concretion (Dalton 2000). Concretions are deposits of minerals. They often have strange shapes and sometimes look as if they had a biological origin, but they are purely geological.

Body Temperatures of Dinosaurs

The cladistic approach that was helpful for inferences about the social behavior of dinosaurs is less effective in deciding what thermoregulatory mechanisms they may have employed. Ectothermy is an ancestral characteristic in the archosaur lineage, and crocodilians are ectotherms. Clearly, the endothermy seen in birds is derived, but when did it appear? It could have been at any point between crocodilians and birds. That is, endothermy might be characteristic of pterosaurs + dinosaurs + birds, or of dinosaurs + birds, or it might be limited to birds. All the lineages between crocodilians and birds are extinct, so we cannot draw any direct evidence from them—paleophysiology is a speculative subject.

Much of the controversy about the temperature relations of dinosaurs stems from failure to distinguish between *homeothermy* and *endothermy.* Homeothermy means only that the body temperature of an organism is reasonably stable, whereas endothermy and ectothermy are mechanisms of temperature control. Ectothermy means that the primary source of heat for raising the body temperature comes from outside the body, whereas endothermy means that the primary source of heat is within the body—the animal's own metabolism. Either ectothermy or endothermy can produce homeothermy. Furthermore, ectothermy and endothermy lie at the ends of a spectrum, and many living animals occupy positions between those extremes.

The diversity of dinosaurs and the enormous size of many species complicate discussions of their thermoregulatory mechanisms. The great ecological and phylogenetic diversity of dinosaurs has largely been overlooked in the debate about their thermoregulatory mechanisms. Not all kinds of evidence apply to all species, and there were probably substantial differences in the biology of different lineages. Quite apart from any other consideration, the difference in body size of an *Ornithomimus* and a *Diplodocus* would make them very different kinds of animals.

Another factor is that we have no terrestrial animals to compare to dinosaurs. Even elephants (which weigh up to 5000 kg) are smaller than any sauropod, and the largest living reptiles (leatherback sea turtles and saltwater crocodiles) weigh about 1000 kg. Thus we have no models of dinosaurs to work with, and must extrapolate observations of living animals far beyond the size ranges for which we have measurements.

Gigantothermy

What can elephants and leatherback turtles tell us about the metabolism of dinosaurs? First, they tell us that when you compare animals of large body size, there is very little difference between the metabolic rates of endotherms and ectotherms. The metabolic rate of a 4000-kg elephant is 0.07 liters of oxygen per kilogram per hour, whereas the metabolic rate for a 400-kg leatherback turtle is 0.06 liters of oxygen per kg/hr. Thus, the metabolic distinction between endotherms and ectotherms disappears at large body sizes, and discussions of whether large dinosaurs were endotherms or ectotherms are meaningless.

Large animals have stable body temperatures simply because they are large. Leatherback turtles employ a form of thermoregulation that has been called gigantothermy (Paladino et al. 1990). The large body mass of these animals has two important consequences for temperature regulation. First, it produces enormous thermal inertia, so body temperatures change slowly. Second, it allows an animal to change the effective thickness of its insulation by changing the distribution of blood flow to the surface of its body versus the core. When a leatherback is in cold water, it can retain the heat produced by muscular activity in its body by restricting blood flow to the surface of the body and employing countercurrent heat exchangers in its limbs. In warmer water, when it needs to dissipate metabolic heat to avoid overheating, it can increase blood flow to the surface and bypass the countercurrent system in its limbs. These mechanisms, which allow leatherback turtles to make migratory journeys of 10,000 km between arctic oceans and the tropics, were certainly used by dinosaurs to stabilize their body temperatures.

Metabolic rates of leatherback turtles were used as a basis for computer simulations of body temperatures of dinosaurs (Spotila et al. 1991). Three metabolic rates were compared: (1) the standard resting metabolic rate of extant reptiles; (2) the measured resting rate for leatherback turtles, which is three times the standard reptilian rate; and (3) the metabolic rate of leatherback turtles while they are digging nests. Digging is hard work, and it raises the metabolic rate to about 10 times the standard reptilian rate. The model also incorporated two rates of blood-flow (low = resting rate, high = five times resting) and two blood flow patterns (low = flow to the body surface negligible, high = flow to the body surface three times greater than flow to the core).

Even a relatively small animal such as *Deinonychus* (which was about 1.5m tall) would have a body temperature 10 to 11°C above air temperature if it had the same metabolic rate as a nesting leatherback turtle, a low rate of blood flow, and low blood circulation to the surface of its body. Thus, the metabolic rates observed for large reptiles appear to be sufficient to allow even small dinosaurs to have body temperatures well above air temperature. Larger dinosaurs with metabolic rates higher than those of extant reptiles would have died from overheating.

Still more evidence pointing to a low metabolic rate for dinosaurs comes from studies of their nasal passages (Ruben et al. 1996). The lungs are delicate structures—the tissues separating air from blood must be thin to allow rapid exchange of carbon dioxide and oxygen. These delicate tissues would dry out if inspired air were not warmed to core body temperature and saturated with water vapor as it traveled through the nasal passages. The nasal passages of extant endotherms (both birds and mammals) contain respiratory turbinates, bony or cartilaginous structures covered with epithelium. These structures increase the surface area of the nasal passages, allowing inspired air to be humidified and warmed to core body temperature.

Ectotherms do not have respiratory turbinates, because the walls of the nasal passages alone provide enough surface area to accommodate the low breathing rates needed to satisfy their oxygen requirements. Ruben and his colleagues used computer-assisted tomography (CAT) scans of fossils to visualize the nasal passages of three kinds of theropod dinosaurs and one ornithischian. Respiratory turbinates are housed in an expansion of the nasal passages of mammals and birds. The nasal passages of reptiles, which lack respiratory turbinates, have smaller diameters than the nasal passages of birds and mammals of similar body size. The CAT scans revealed that the ornithischian and two of the theropods had narrow nasal passages like those of extant reptiles. The third theropod, *Dromeosaurus*, was like extant monitor lizards in having very short nasal passages that opened into the anterior part of the roof of the mouth. Thus, the nasal passages of all four dinosaurs resemble those of extant ectotherms.

14.9 Marine Diapsids—Placodonts, Plesiosaurs, and Ichthyosaurs

Several lineages of specialized marine tetrapods are tentatively grouped with the lepidosauromorphs (see Figure 14–1).

Placodonts

The Triassic placodonts were mollusk eaters specialized for crushing hard-shelled food rather than for rapid pursuit of prey. *Placodus* had large, flat maxillary teeth and a heavy palate with enormous teeth on the palatine bones (Figure 14–21a). The anterior teeth projected forward and might have been used to pull mussels or oysters off rocks. Another placodont, *Placochelys,* had a toothless beak instead of projecting teeth, but retained broad teeth in the rear of the mouth. In contrast, *Henodus* was almost toothless and may have crushed its food with horny plates like those of turtles. The similarity of some of these placodonts to turtles extended to their external appearance as well. *Henodus* and *Placochelys* had a body armor that was as extensive as that of turtles, but it was a mosaic of small polygonal dermal bones rather than large plates like those of turtles. In *Henodus* the dermal bones were apparently covered with epidermal scutes, as they are in turtles.

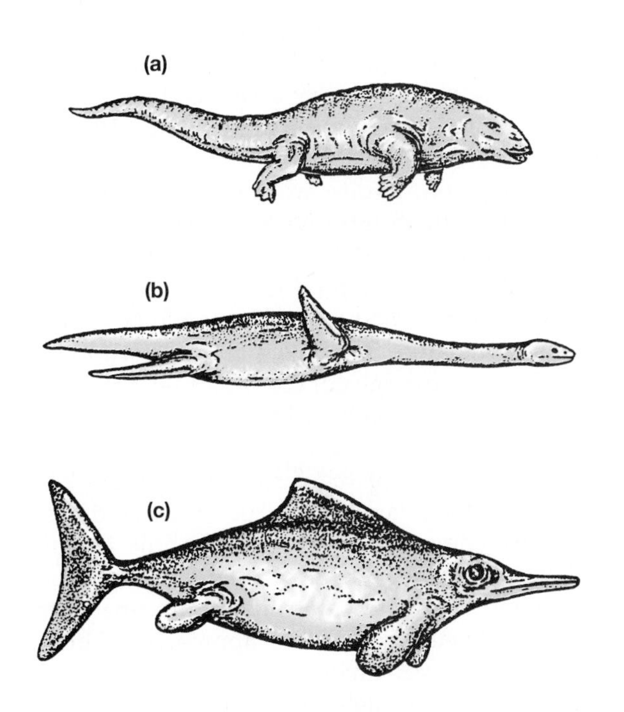

▲ Figure 14–21 Mesozoic aquatic reptiles. (a) The placodont *Placodus* from the Middle Triassic (approx. 1m long). (b) Late Jurassic plesiosaur *Cryptoclidus* (approx. 3m long). (c) Late Jurassic ichthyosaur *Opthalamosaurus* (approx. 2.5m long).

Plesiosaurs

The plesiosaurs appeared in the Late Triassic and persisted to the Cretaceous (Figure 14–21b). Two ecological types of plesiosaurs can be distinguished—one group is composed of long-necked animals with small heads, whereas the other contained short-necked animals with long skulls. Both had heavy, rigid trunks and appear to have rowed through the water with limbs that functioned like oars and may also have acted as hydrofoils, increasing the efficiency of swimming. Hyperphalangy, the addition of bones to the digits, increased the size of the paddles, and some plesiosaurs had as many as 17 phalanges per digit. In both types of plesiosaurs, the nostrils were located high on the head just in front of the eyes.

The long-necked plesiosaurs reached their zenith in *Elasmosaurus,* which lived in the Late Cretaceous. That form had 35 cervical vertebrae. The *Elasmosaurus* line is characterized by a progressive increase in the number of cervical vertebrae and a reduction in the size of the head. Not only did the number of cervical vertebrae increase, but individual vertebrae became longer. *Microcleidus* of the Middle Jurassic had 39 or 40 cervical vertebrae, and *Elasmosaurus* had 76. Even in the early forms the body was not well streamlined, and as the neck became longer, the streamlining became even poorer. The paddle size relative to the body size decreased from the Jurassic to the Cretaceous. Clearly, the *Elasmosaurus* line of plesiosaurs did not include rapid swimmers. They may have used an ambush strategy to capture fish (Massare 1987, 1988).

The short-necked plesiosaurs developed an increasingly streamlined body form. The neck became shorter and the paddles larger. These were probably speedier swimmers than the long-necked plesiosaurs, and they might have captured their prey by pursuit in the manner of seals and sea lions. The stomach region of the fossil of a short-necked plesiosaur from Japan contained a large number of jaws of ammonites, which were free-swimming cephalopods.

Ichthyosaurs

The ichthyosaurs were the most specialized of the aquatic tetrapods of the Mesozoic. In many aspects of their body form, they resemble porpoises (Figure 14–21c). Ichthyosaurs had a dorsal fin that was supported only by stiff tissue, not by bone, and the upper lobe of the caudal fin similarly lacked skeletal support. (The vertebral column of derived ichthyosaurs bent sharply downward into the ventral lobe of the caudal fin.) We know of the presence of these soft tissues because many ichthyosaur fossils in fine-grained sediments near Holzmoden in southern Germany contain an outline of the entire body preserved as a carbonaceous film.

Ichthyosaurs had both forelimbs and hindlimbs (unlike whales and dolphins, which retain only the forelimbs). The limbs of ichthyosaurs were modified into paddles by both hyperphalangy (as in plesiosaurs) and by hyperdactyly (the addition of extra digits). Fossil ichthyosaurs with embryos in the body cavity indicate that these animals gave birth to living young. One fossil appears to be an individual that died in the process of giving birth, with a young ichthyosaur emerging tail-first as do baby porpoises.

The stomach contents of ichthyosaurs, preserved in some specimens, include cephalopods, fishes, and an occasional pterosaur. Ichthyosaurs had large heads with long, pointed jaws that were armed with sharp teeth in most forms, although a few ichthyosaurs were toothless. Ichthyosaurs had very large eyeballs that were supported by a ring of sclerotic bones. *Ophthalmosaurus*, which had larger eyes than any other vertebrate, is believed to have hunted at great depth—500m or more—and detected light emitted by the photophores of its prey. Deep-diving animals risk caisson disease (the bends) if an emergency such as the need to avoid a predator forces them to rise rapidly to the surface. The two ichthyosaurs with the largest eyes (suggesting that they were the deepest-diving forms) were also the two that showed the highest incidence of caisson disease (Motani et al. 1998).

Triassic ichthyosaurs were elongate and poorly streamlined and may have used anguilliform locomotion. The greater streamlining of later forms may have been associated with the development of carangiform locomotion and rapid pursuit of prey like that of extant tunas. The Jurassic was the high point of ichthyosaur diversity. They were less abundant in the Early Cretaceous, and only a single genus remained in the Late Cretaceous. Ichthyosaurs became extinct before the end of the period. Even ichthyosaurs, which were the fastest swimmers among the Mesozoic marine reptiles, were probably not as swift as extant toothed whales, and the diversification of fast, agile fishes in the Cretaceous may have contributed to the decline of marine reptiles.

14.10 Terrestrial Vertebrates of the Late Mesozoic

There is a tendency to look on the Jurassic and Cretaceous as the Age of Dinosaurs and to forget that there was a very considerable nondinosaur terrestrial fauna as well as variations in the kinds of dinosaurs in different habitats. To a certain extent the dinosaurs do form a separate unit. The large theropods were probably the only animals capable of preying on adults of the large herbivores. Nonetheless, there must have been interactions between dinosaurs and nondinosaurs. As far as we know, all dinosaurs reproduced by laying eggs, and their eggs were a food source that could be exploited by small predators. The Nile monitor lizard today is a predator on the eggs of crocodiles, and it is likely that Cretaceous monitor lizards had a taste for dinosaur eggs.

Of course, one of the distinctive features of dinosaurs is their large size. The smallest dinosaur skeletons known indicate a total adult length of about half a meter. Even that is large in comparison with living squamates; most lizards are smaller than 20 cm, and only crocodilians approach the bulk of even moderate-size dinosaurs. There were, of course, adaptive zones available for smaller vertebrates in the Jurassic and Cretaceous. These were filled—as they are now—by sphenodontids, squamates, turtles, amphibians, and mammals. Among them were the animals that might have stolen eggs from nests and served as food for juvenile dinosaurs.

Although dinosaurs are so impressive and distinctive that we tend to think of them as inhabiting a world of their own, in reality they were part of an ecosystem including many vertebrates that would not appear strange to us today. All modern tetrapod groups—frogs, salamanders, lizards, snakes, turtles, crocodilians, birds, and mammals—arose in the Late Triassic or Early Jurassic, and increased in diversity through the Jurassic and Cretaceous. During the Late Cretaceous the diversity of these modern groups was approximately equal to that of dinosaurs and pterosaurs. In the latest Cretaceous the diversity of the modern groups rose dramatically and became twice that of the dinosaurs and pterosaurs, long before the terminal Cretaceous extinction event.

A striking feature of the Mesozoic world was the absence of large mammals, and the occupation of that adaptive zone by large archosaurs. Figure 14–22 shows the relative abundance of different groups of vertebrates at two Cretaceous fossil localities. The Lance locality appears to have been a wooded, swampy habitat with large streams and some ponds. The Bug Creek locality was probably downstream from such a swamp, in the delta of a major waterway. In both localities dinosaurs are a minor component of the community in their species diversity, although one dinosaur is the equivalent of a great many smaller animals in terms of biomass.

14.11 Late Cretaceous Extinctions

After thriving and dominating the terrestrial habitat for 150 million years, dinosaurs disappeared near the end of the Cretaceous. The extinctions that occurred at the Cretaceous-Tertiary transition are an example of a recurrent biological phenomenon—mass extinction. In fact, the Cretaceous-Tertiary extinction was relatively small as mass extinctions go, but it is the most widely known because the disappearance of dinosaurs has captured the attention of both scientists and the public.

Biologists have struggled for years to explain the rapid change in faunal composition that occurred at the end of the Mesozoic, and there are nearly as many different hypotheses as there are authors. The catastrophism school of thought maintains that the giant archosaurs were wiped out by some geological or cosmic disaster. The currently popular hypothesis in the catastrophism sweepstakes is the suggestion that the impact of a comet or an asteroid with Earth ignited worldwide firestorms or injected enough dust into the atmosphere to blot out the Sun, thereby suppressing photosynthesis and leading to the extinctions of animals.

Other paleontologists have argued that the extinctions we associate with the transition from the Cretaceous to the Tertiary were gradual, not abrupt, and

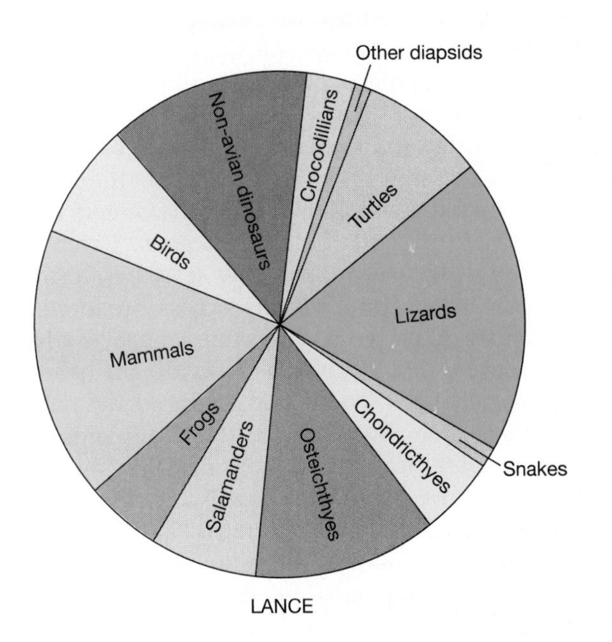

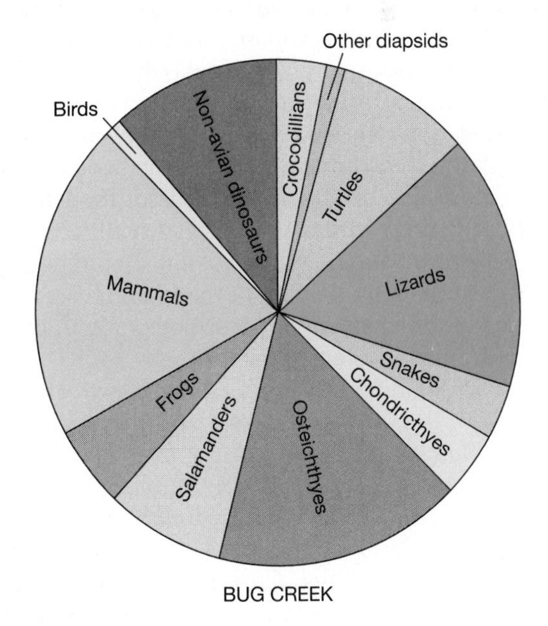

▲ Figure 14–22 Relative abundance of genera at Late Cretaceous fossil sites.

actually began millions of years before the end of the Cretaceous. Gradual extinctions could result from slow changes in climate or terrestrial habitats that might be consequences of continental drift and the accompanying changes in oceanic circulation.

Most paleontologists agree that an extraterrestrial impact did occur during the Late Cretaceous. A narrow band of rock containing a high concentration of iridium (known as an iridium anomaly) occurs in many parts of the world in sediments that were deposited in the Late Cretaceous. Iridium is normally a rare element in sedimentary rocks, but it occurs in high concentrations in the core of the Earth and in some volcanic deposits, and it is found in high concentrations in some extraterrestrial objects. Thus, a worldwide iridium anomaly suggests that some usual event—either a massive episode of volcanic activity or the impact of an extraterrestrial object—occurred in the very Late Cretaceous. Evidence suggesting that the event was an impact is provided by the presence of quartz crystals that show the effect of enormous force (known as shocked quartz) in Late Cretaceous. The enormous Chicxulub crater on the Yucatan coast of Mexico appears to be of the correct age (64.4 ± 0.5 million years) for this impact.

There is much less agreement about the consequences of the impact than about its occurrence. The two hypotheses for the extinction of dinosaurs—an extraterrestrial event versus normal ecological processes—make very different predictions about the time course of extinction. The enormous ocean waves, firestorms, and darkness that could have been produced by an impact are short-term events with durations measured in days to years. In contrast, changes in climates and habitats would probably require thousands or hundreds of thousands of years to cause extinctions. With such different predictions from the two hypotheses, it should be possible to test them by looking at the fossil record to see whether extinctions were abrupt or gradual, but it's not possible to make that analysis.

The major difficulty in deciding if extinctions at the end of the Cretaceous were abrupt or gradual is the problem of resolution. Dating events at the end of the Cretaceous involves large uncertainties: 1 to 2 million years for correlations of marine and terrestrial deposits and 0.5 to 4 million years for dates based on rates of decay of radioactive elements. Dates can help to determine the sequence of events but not their duration, so no direct test of the abruptness of the extinctions is possible. Additionally, dinosaur fossils are relatively rare, so it is difficult to distinguish statistically between a declining population prior to the K-T boundary and an artifact produced merely by the reduction of fossilized dinosaurs at that time.

Our inability to measure the duration of events applies to the fossil record of all kinds of life—marine invertebrates, terrestrial plants, and terrestrial vertebrates. A second difficulty is particularly acute for studies of dinosaurs. Only in western North America is there yet a fossil record of terrestrial vertebrates that extends from the Late Cretaceous across the Cretaceous-Tertiary boundary into the Paleocene. Thus, we are basing inferences about worldwide changes on just one area.

Hypotheses based on long-term changes in climate have been proposed repeatedly. The Mesozoic was apparently characterized by stable temperatures with little variation from day to night, from summer to winter, and from north to south. Geological evidence suggests that climates became cooler and more variable as the flow of warm water into the Arctic Ocean was reduced at the end of the Cretaceous, when the continents had nearly reached their present locations. Casper, Wyoming, is an important dinosaur fossil locality. A reconstruction of its Mesozoic temperature regime suggests that the long-term temperature extremes probably lay between 11 and 34°C, with a mean of 22°C. Clearly, the climate has changed since then: The present temperature extremes at Casper are –27 to +40°C, with a mean of 8°C.

In addition to these changes in climate, geological events at the end of the Cretaceous were changing the land surface and altering terrestrial habitats. A reduction in sea level of 150 to 200m drained the shallow inland seas that had filled the centers of the continents during the Late Cretaceous. Rivers cut their way down toward the new sea level, forming valleys instead of meandering across broad floodplains. Late Cretaceous dinosaurs appear to have been concentrated in river and floodplain habitats, and these changes would have reduced the habitat available to them. Furthermore, the floodplains were probably routes for north-south migration, and their interruption by valleys running east-west might have disrupted migratory patterns. The population sizes of large herbivorous dinosaurs and their theropod predators may have represented a delicate balance of predator-prey ratios and population densities. Perhaps changes in topography that restricted the movements of predators and their prey tilted the scales toward excessive predation or produced population densities that were too low to sustain normal social behavior and reproduction.

The second problem presented by the Cretaceous-Tertiary extinctions is explaining why some evolutionary lineages became extinct and others persisted. Nonavian dinosaurs disappeared, but turtles and crocodilians survived, as did birds and mammals. Nonavian dinosaurs were probably less diverse by the end of the Cretaceous than they once had been, but still they included many different kinds of animals—large and small, carnivorous and herbivorous, fast-moving and slow-moving. Some Cretaceous crocodilians were as big as the large theropods. Modern crocodilians are probably similar in their ecology and physiology to some carnivorous dinosaurs, turtles have similarities to herbivorous dinosaurs, and Mesozoic birds and mammals were basically like extant forms. What conditions would exterminate dinosaurs and allow crocodilians, turtles, birds, and mammals to survive?

Paleontologists who advocate the impact hypothesis of Cretaceous extinctions are beginning to address this question, but so far only at a scale that is too coarse to be convincing in an ecological context. For example, a recent review of the Cretaceous extinctions classified aquatic organisms (a group that includes turtles and crocodilians) as detritus feeders and attributed their survival to this characteristic (Fastovsky and Weishampel 1996). As we have seen, however, even the 21 extant species of crocodilians and 400 species of turtles are too diverse to fit such a superficial characterization, and the Cretaceous reptilian fauna was far more diverse.

The difficulty that proponents of the asteroid impact currently face in demonstrating a cause-and-effect relationship between the impact and the extinctions is a mismatch in the scale of the two events. An asteroid impact would have large-scale effects, whereas the Late Cretaceous extinctions appear to reflect fine-scale differences among lineages. Ascribing the late Mesozoic extinctions to an extraterrestrial impact is like using a sledge hammer to swat a housefly while sparing the fruit fly beside it.

The difficulty in understanding the Cretaceous-Tertiary extinctions is paralleled by our difficulty in understanding many of the extinctions that are occurring now. Frogs and other amphibians are declining on a worldwide basis, but no one cause for these declines has been identified. It seems most likely that multiple factors and interactions of these factors are responsible for the current wave of extinctions, and that multiple factors were involved in the Late Cretaceous extinctions as well.

Summary

The major groups of tetrapods in the Mesozoic were members of the diapsid (two arches) lineage. This group is distinguished particularly by the presence of two fenestrae in the temporal region of the skull that are defined by arches of bone. The archosauromorph lineage of diapsids contains the most familiar of the Mesozoic tetrapods, the dinosaurs. Two major groups of dinosaurs are distinguished, the Ornithischia and Saurischia.

The ornithischian dinosaurs were herbivorous, and had horny beaks on the snout and batteries of specialized teeth in the rear of the jaw. The ornithopods (duck-billed dinosaurs) and pachycephalosaurs (thick-headed dinosaurs) were bipedal, and the stegosaurs (plated dinosaurs), ceratopsians (horned dinosaurs), and ankylosaurians (armored dinosaurs) were quadrupedal.

The saurischians included the sauropod dinosaurs—enormous herbivorous quadrupedal forms like *Apatosaurus* (formerly *Brontosaurus*), *Diplodocus*, and *Brachiosaurus*—and the theropods, which were bipedal carnivores. Large theropods (of which *Tyrannosaurus rex* is the most familiar example) probably preyed on large sauropods. Other theropods were smaller: Ornithomimids were probably very like ostriches, and some had horny beaks and lacked teeth. Dromeosaurs were fast-running predators. Ornithomimids probably seized small prey with hands that had three fingers armed with claws, whereas dromeosaurs probably were able to prey on dinosaurs larger than themselves. They may have hunted in packs and used the enormous claw on the second toe to slash their prey. Birds had evolved by the Jurassic: *Archaeopteryx*, the earliest bird known, is very like small theropods.

Although the saurischians and ornithischians represent independent radiations and had different ecological specializations, they share many morphological features that appear to reflect the mechanical constraints of being very large terrestrial animals.

The phylogenetic relationship of crocodilians and birds allows us to draw inferences about some aspects of the biology of dinosaurs. Characters that are shared by crocodilians and birds are probably ancestral for dinosaurs. Social behavior, vocalization, and parental care are the norm for crocodilians and birds, and increasing evidence suggests that dinosaurs, too, had elaborate social behavior and vocalizations, and that at least some species cared for the young.

The Mesozoic was marked by a series of faunal changes. The extinctions at the end of the Cretaceous were the most dramatic of these, partly because the dinosaur fauna of the time was so spectacular, but they are not unique. In the Middle Cretaceous the diversity of sauropod dinosaurs declined while the ornithischians burgeoned. This faunal change was contemporaneous with the appearance of flowering plants (angiosperms). In this context of shifting faunas and floras throughout the Mesozoic, explanations of the Late Cretaceous extinctions that depend on worldwide catastrophes are less persuasive than hypotheses based on the gradual changes in sea level, topography, and climate that are shown in the fossil record of the late Mesozoic.

Additional Readings

Alexander, R. M. 1989. *Dynamics of Dinosaurs and Other Extinct Giants.* New York: Columbia University Press.

Archibald, J. D. 1989. The demise of the dinosaurs and the rise of the mammals. In K. Padian and D. J. Chure (Eds.), *The Age of Dinosaurs.* Knoxville: The Paleontological Society, University of Tennessee Press, pp. 48–57.

Carroll, R. L. 1987. *Vertebrate Paleontology and Evolution.* New York: Freeman.

Charig, A. 1972. The evolution of the archosaur pelvis and hindlimb: An explanation in functional terms. In K. A. Joysey and T. S. Kemp (Eds.), *Studies in Vertebrate Evolution.* Piscataway, N.J.: Winchester.

Chin, K., et al. 1998. A king-size theropod coprolite. *Nature* 393:680–682.

Clark, J. M., et al. Foot posture in a primitive pterosaur. *Nature* 391:886–889.

Dalton, R. 2000. Doubts grow over discovery of fossilized "dinosaur heart." *Nature* 407:275–276.

Dodson, P. 1993. Comparative craniology of the Ceratopsia. *American Journal of Science* 293-A:200–234.

Erickson, G. M., et al. 1996. Bite-force estimation for *Tyrannosaurus rex* from tooth marks on bones. *Nature* 382:706–708.

Fastovsky, D. E., and D. B. Weishampel. 1996. *The Evolution and Extinction of the Dinosaurs.* Cambridge, U.K.: Cambridge University Press.

Fisher, P. E., et al. 2000. Cardiovascular evidence for an intermediate or higher metabolic rate in an ornithischian dinosaur. *Science* 288:503–505.

Gauthier, J. 1986. Saurischian monophyly and the origin of birds. In K. Padian (Ed.), *The Origin of Birds and the Evolution of Flight, Memoirs of the California Academy of Sciences,* no. 8, pp. 1–55.

Hazlehurst, G. A., and J. M. V. Rayner. 1992. Flight characteristics of Triassic and Jurassic Pterosauria: An appraisal based in wing shape. *Paleobiology* 18:447–463.

Horner, J. R. 1984. The nesting behavior of dinosaurs. *Scientific American* 241 (4):130–137.

Jones, D. R., and G. Shelton. 1993. The physiology of the alligator heart: Left aortic flow patterns and right-to-left shunts. *Journal of Experimental Biology* 176:247–269.

Kirkland, J. I. 1994. Predation of dinosaur nests by terrestrial crocodilians. In K. Carpenter, K. F. Hirsch, and J. R. Horner (Eds.), *Dinosaur Eggs and Babies,* Cambridge, U.K.: Cambridge University Press, pp. 124–133.

Lang, J. W. 1986. Male parental care in mugger crocodiles. *National Geographic Research* 2:519–525.

Lang, J. W. 1989. Social behavior. In C. A. Ross (Ed.), *Crocodiles and Alligators,* New York: Facts on File, pp. 102–117

Lillywhite, H. B. 1991. Sauropods and gravity. *Natural History* December 1991, p. 33.

Massare, J. A. 1987. Tooth morphology and prey preference of Mesozoic marine reptiles. *Journal of Vertebrate Paleontology* 7:121–137.

Massare, J. A. 1988. Swimming capabilities of Mesozic marine reptiles: Implications for methods of predation. *Paleobiology* 14:187–205.

Maxwell, D. 1999. Days of the deinos. *Natural History* 108 (10):60–65.

Mikhailov, K. E. 1997. Eggs, eggshells, and nests. In P. J. Currie and K. Padian (Ed.), *Encyclopedia of Dinosaurs,* San Diego: Academic Press, pp. 205–209.

Motani, R. B. M. Rothschild, and W. Wahl, Jr. 1998. Large eyeballs in diving ichthyosaurs. *Nature* 402:747.

Motani, R, 2000. Rulers of the Jurassic seas. *Scientific American* 283:52–59.

Montani, R. B. M. 2000. Rulers of the Jurassic seas. *Scientific American* 283:52–59.

Paladino, F. V., M. P. O'Connor, and J. R. Spotila. 1990. Metabolism of leatherback turtles, gigantothermy, and thermoregulation of dinosaurs. *Nature* 344:858–860.

Ruben, J. A., et al. 1996. The metabolic status of some Late Cretaceous dinosaurs. *Science* 273:1204–1207.

Ruben, J. A., et al. 1997. Lung structure and ventilation in theropod dinosaurs and early birds. *Science* 278:1267–1270.

Seymour, R. S. 1979. Dinosaur eggs: Gas conductance through the shell, water loss during incubation and clutch size. *Paleobiology* 5:1–11.

Seymour, R. S. and H. B. Lillywhite. 2000. Hearts, neck posture and metabolic intensity of sauropod dinosaurs. *Proceedings of the Royal Society of London,* Series B. 267:1883–1887.

Spotila, J. R., et al. 1991. Hot and cold running dinosaurs: Body size, metabolism and migration. *Modern Geology* 16:203–227.

Stevens, K. A. and J. M. Parrish. 1999. Neck posture and feeding habits of two Jurassic sauropod dinosaurs. *Science* 284:798-800.

Thomas, D. A., and J. O. Farlow. 1997. Tracking a dinosaur attack. *Scientific American* 277 (6):74–79.

Thulborn, T. 1990. *Dinosaur Tracks.* London: Chapman & Hall.

Troyer, K. 1984. Microbes, herbivory and the evolution of social behavior. *Journal of Theoretical Biology* 106:157–169.

Weishampel. D. B., D. Dodson, and H. Osmólska. 1990. *The Dinosauria.* Berkeley: University of California Press.

Web Explorations

On-line resources for this chapter are on the World Wide Web at: http://www.prenhall.com/pough (click on the Table of Contents link and then select Chapter 14).

PART IV

Endotherms: Birds and Mammals

Birds and mammals are the vertebrates with which people are most familiar, partly because many species are large and diurnal and partly because birds and mammals have colonized nearly every terrestrial habitat on Earth. The success of birds and mammals in many habitats is related to their endothermy, which allows them to be active at night and in cold weather. These are conditions in which terrestrial ectotherms find it difficult or impossible to thermoregulate and, consequently, most of them are inactive.

Flight dominates the biology of birds: Most of the morphological features of birds are directly or indirectly related to the requirements of flight, and many distinctive aspects of their behavior and ecology stem from the mobility that flight provides. Migration, for example, is a particularly avian characteristic because, of all the terrestrial vertebrates, birds are best able to move long distances.

No one feature of mammals characterizes the group and dominates its biology as flight does for birds, but sociality comes close. Many features of the biology of mammals are related to their interactions with other individuals of their species, ranging from the period of dependence of young on their mother to lifelong alliances between individuals that affect their social status and reproductive success in a group.

Humans differ from other vertebrates in the extent to which they have come to dominate all of the habitats of Earth, and in their effects, direct and indirect, on other vertebrates. In this portion of the book we explore the evolution of birds and mammals, the adaptive zones opened to them by their distinctive characteristics, the origin and radiation of humans as an example of vertebrate evolution, and the impact of humans on other species.

CHAPTER

15

The Evolution of Birds and the Origin of Flight

Modern birds are easy to recognize—they have feathers and most of them can fly. A third characteristic of birds is less visible, but equally important: Birds have high metabolic rates and high body temperatures; that is, they are endotherms. These three characters are closely linked in modern birds. A high metabolic rate fuels the powerful muscles needed for flight, and feathers provide lift, propulsion, and streamlining when a bird flies. The insulation provided by feathers allows metabolic heat produced by the high metabolic rate to maintain a high and stable body temperature. Insulation, heat production, and high body temperatures are essential components of endothermy as it is seen in modern birds and mammals. But both birds and mammals evolved from ectothermal ancestors that had low metabolic rates and lacked insulation. How did they get from that condition to what we see now? The multiple roles of feathers in the biology of extant birds require an analysis that integrates several perspectives.

15.1 The Evolution of Endothermy

The difference in the sources of heat used by ectotherms and endotherms creates a paradox when biologists try to understand how an evolutionary lineage shifts from ectothermy to endothermy. Ectotherms rely on heat from outside their bodies, and the major specializations of ectothermal thermoregulation facilitate exchange of heat with the environment. The body surfaces of ectotherms have little insulation, probably because insulation would interfere with the gain and loss of heat. Metabolic rates of ectotherms are low, and ectotherms normally do not obtain sufficient heat from metabolism to warm the body significantly. Thus the thermoregulatory mechanisms of ectotherms are based on low metabolic rates, little insulation, and rapid exchange of heat with the environment.

Endothermal thermoregulation has exactly the opposite characteristics. The high metabolic rates of endotherms produce large quantities of heat, and that heat is retained in their bodies due to the insulation provided by hair or feathers. Endothermal thermoregulation consists largely of adjusting the layer of insulation so that heat loss balances the heat produced by high rates of metabolism.

An evolutionary shift from ectothermy to endothermy appears to encounter a Catch-22 situation: A high metabolic rate is no use unless an animal already has insulation to retain metabolically produced heat, because without insulation the heat is rapidly lost to the environment. However, insulation serves no purpose for an animal without a high rate of metabolism, because there is little internally produced heat for the insulation to conserve. Indeed, insulation can be a handicap for an ectotherm, because it prevents it from warming up. Raymond Cowles demonstrated that fact in the 1930s when he made small fur coats for lizards and measured their rates of warming and cooling. The potential benefit of a fur coat for a lizard is, of course, that it will keep the lizard warm as the environment cools off. However, the lizards in Cowles's experiments never achieved the benefit of insulation, because when they were wearing fur coats they were unable to get warm in the first place.

Those well-dressed lizards illustrate the paradox of the evolution of endothermy: Insulation is ineffective without a high metabolic rate, and the heat produced by a high metabolic rate is wasted without insulation. By this line of reasoning, neither one of the two essential features of endothermy would be selectively advantageous for an ectotherm without the previous development of the other. So how did endothermy evolve?

The solution to the paradox lies in finding a way that either increased metabolism or increased insulation could be advantageous for some reason other than regulating body temperature. If there is a benefit to increasing metabolism, for example, a lineage of animals could develop higher metabolic rates for that reason. Then, when they had the metabolic capacity to warm their bodies, it would be advantageous to develop insulation to retain the heat. This scenario provides a plausible explanation for the evolution of high metabolic rates, insulation (feathers or hair) and finally endothermy in birds and mammals.

15.2 Activity and Metabolism

Probably increased locomotor activity was the critical selective factor for the first steps toward endothermy in both birds and mammals. Active predators (animals that chase their prey) rely on aerobic metabolism to sustain high levels of activity. Anaerobic metabolism, which uses glycogen stored in the muscles as a substrate, can fuel a short sprint to escape a predator, but it is not effective for sustained activity because the supply of glycogen is quickly exhausted.

Long-distance runners rely on aerobic metabolism, which is fueled by glucose and oxygen carried to the muscles by the circulatory system. Glucose comes from the liver and oxygen from the lungs, and aerobic running can be sustained for long periods. Thus, part of becoming a pursuit predator involves increasing the capacity of systems and structures that are associated with aerobic metabolic capacity—the lungs, the heart and circulatory system, and the mitochondria in the muscles.

For many different kinds of animals, the maximum aerobic metabolic rate they can achieve is about ten times their resting metabolic rate. Why this should be so is not fully understood—presumably it has something to do with the cost of maintaining lungs, heart, circulatory system, and mitochondria that are needed to produce ATP at high rates during activity. Whatever the underlying mechanism, however, the significance of the link between resting and maximum rates of aerobic metabolism is clear. If the ancestors of birds and mammals were evolving high maximum levels of aerobic metabolism to pursue prey, their resting rates of metabolism were also creeping upward.

The body forms of dromesosaur dinosaurs such as *Deinonychus* strongly suggest that they were fleet-footed predators that pursued their prey, and the anatomical characters of the synapsid precursors of mammals also point to their evolution as active predators (Chapter 17). Thus, increased metabolic rates would have been advantageous in both the bird and mammalian lineages for reasons that had nothing to do with temperature regulation.

Once resting metabolic rates had been raised by selection for high maximal aerobic metabolism, heat was being produced even when an animal was at rest and a layer of insulation would retain that heat and raise body temperature. Feathers and hair would thus be advantageous because the ancestors of birds and mammals already had high metabolic rates.

15.3 Birds as Feathered Dinosaurs

The hypothesis that birds are descended from dinosaurs cannot be tested directly because the transitional forms are long gone, but it fits the evidence available. In the case of birds, dromeosaurs were evolving increasingly bird-like characters, and the similarity of birds and dinosaurs has long been recognized. In the 1860s and 1870s Thomas Henry Huxley was an ardent advocate of that relationship, writing that birds are nothing more than "glorified reptiles." Huxley, in fact, was so impressed by their many similarities that he placed birds and reptiles together in his classification as the class Sauropsida. For most of the next century traditional systematics, with its emphasis on strict hierarchical categories, obscured that evolutionary relationship by placing reptiles and birds at the same taxonomic level (class Reptilia and class Aves). Cladistic systematics emphasizes monophyltic evolutionary lineages, and now birds are again seen as the most derived theropod dinosaurs. The similarities of birds and theropods include the following derived characters:

- Elongate, mobile S-shaped neck
- Tridactyl foot with digitigrade posture

- Intertarsal ankle joint
- Hollow, pneumatic bones

These are general theropod characters. When we look specifically at dromeosaurs, still more shared derived characters emerge (Figure 15–1). Some dromeosaurs, including *Velociraptor* of *Jurassic Park* fame, had a wrist structure that permitted them to turn their hands as they seized prey. This wrist motion is used in flapping flight. A more derived dromeosaur, *Unenlagia*, a 2-meter-long terrestrial predator from Patagonia, added a shoulder joint that allowed it to flap its arms up and down, which is another essential component of powered flight.

Most dramatic was the discovery in the late 1990s of two genera of dromeosaurs with feathers (Qiang et al. 1998). These fossils were found at a site in Liaoning Province in northeastern China that has produced spectacular material of Late Jurassic or Early Cretaceous age (Stokstad 2001). *Caudipteryx*, the first of the feathered dinosaurs to be discovered, and *Protoarchaeopteryx* have both down-like and vaned feathers—although both had relatively short arms and neither could fly. *Protoachaeopteryx* has down-like feathers on the body and tail and a fan of symmetrical vaned feathers on the tail. *Caudipteryx* has vaned feathers attached to the second finger of the hand, where remiges (primary feathers) are found in modern birds, and a tuft of vaned feathers on the tail.

The entire body of a newly discovered dromeosaur from Liaoning Province is covered with feather-like structures. On the tail the filaments making up the structures radiate from a single point, and on the arms the filaments radiate from a central stem as in the feathers of modern birds (Qiang et al. 2001).

The Liaoning fossils are extraordinarily well preserved, and the fine sediments reveal details that would not be visible in coarser rocks. The Liaoning deposit consists of layers of volcanic ash, siltstone, and clay. Volcanic eruptions to the west, in what is now Inner Mongolia, sent clouds of sulfurous gases and ash drifting eastward, killing whatever lay in

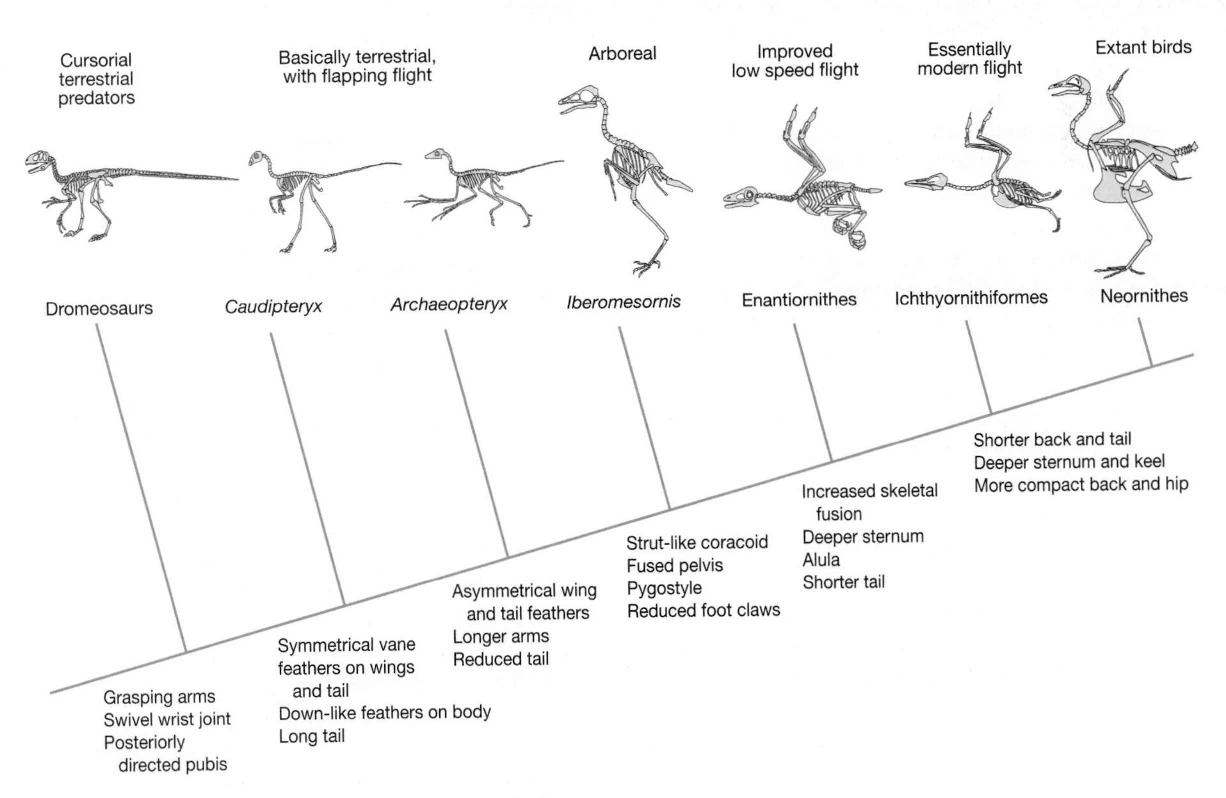

▲ Figure 15–1 Evolution of specializations for flight.

their path and immediately burying the corpses. Insects, frogs, fish lizards, crocodiles, mammals, and dinosaurs—the entire fauna of the region was entombed and preserved in exquisite detail.

But few deposits have fossils that show as much detail as those from Liaoning, and without that detail it is not possible to see the imprint of feathers. We cannot know how widespread feathers were among dromeosaurs because most fossil come from coarser rock that does not show fine detail.

The presence of two kinds of feathers in *Caudipteryx* and *Protoarchaeopteryx*, down-like feathers and vaned feathers, suggests that two selective processes may have been operating simultaneously. Down insulates birds, just as it insulates humans when it is stuffed into a down vest, and the presence of down is consistent with the hypothesis that the metabolic rates of dromeosaurs were high enough to produce heat that could be retained in the body by insulation. The filamentous body covering of *Sinosauropteryx* could also have been a layer of insulation.

The vaned feathers on the wings and tail of *Caudipteryx* and *Protoarchaeopteryx* would not have been useful for insulation. In modern birds vaned feathers provide lift, propulsion, and steering during flight, but neither of the fossil species was capable of flight. *Protoarchaeopteryx* had arms only half the length of modern birds, and *Caudipteryx* had even shorter arms. Furthermore, the vaned feathers of both species were symmetrical. Modern flying birds have asymmetric feathers, and that asymmetry allows feathers to produce lift and thrust. Probably the vaned wing and tail feathers of *Caudipteryx* and *Protoarchaeopteryx* were colorful and used mainly for social displays.

15.4 *Archaeopteryx* and the Origin of Flight

Archaeopteryx is the earliest bird known, and birds (Aviale) are defined as *Archaeopteryx* plus living birds (Neornithes) and all descendants of their most recent common ancestor. Fossils of *Archaeopteryx* from fine-grained sediments show imprints of feathers that were much better differentiated than the feathers of *Caudipteryx* and *Protoarchaeopteryx* (Figure 15–2). In addition to a presumed covering of body-contour feathers, *Archaeopteryx* had wing feathers that were differentiated into an outer series of primaries on the hand bones and an inner series of secondaries along the outer arm. This arrangement of flight feathers is essentially the same as that seen in extant birds. Furthermore, the flight feathers on the wings of *Archaeopteryx* have asymmetrical vanes like those of flying birds, suggesting that they had been shaped by aerodynamic forces associated with flapping flight.

The long tail of *Archaeopteryx* is like the tails of *Caudipteryx* and *Protoarchaeopteryx*, and no other bird retains a long tail. ("*Archaeoraptor*," a fossil from the Liaoning deposit that had feathers and a long tail, has been revealed as a hoax that was created by combining a partial fossil of a primitive bird with the tail of a dromeosaur. It had been purchased by a collector in America for $80,000, and with fossils commanding such prices the incentive to produce dramatic frauds is strong.) The rectrices (tail feathers) of *Archaeopteryx* are arranged in 15 pairs along the sides of the 6th through 20th caudal vertebrae.

Dromeosaurs, *Archaeopteryx*, and birds are so similar in their anatomy (Figure 15–3) that it is hard to imagine that birds did not evolve from theropod dinosaurs (Norell and Chiappe 1996, Padian and Chiappe 1998). The fossils we know do not form a series showing progressive change through time, however. *Archaeopteryx*, which is older than *Caudipteryx* and *Protoarchaeopteryx*, is also more bird-like. Because of this discordance, a few paleontologists still insist birds were derived from archosaur lineages that separated from the saurischian stock earlier than the theropods (Feduccia and Wild 1993, Feduccia 1996). It is more likely that *Archaeopteryx*, *Caudipteryx*, and *Protoarchaeopteryx* were all living fossils in their day, existing side by side with more derived forms—just as coelacanths, lungfish, and monotreme mammals (platypus and echindas) do today.

How Did Flight Evolve?

Birds, feathers, and flight are not synonymous. As we have seen, some dinosaurs had feathers, and flapping flight has evolved in three separate groups of vertebrates: pterosaurs, birds, and bats. The wings of these different vertebrates represent examples of convergent evolution, and the actual structural details of the wing design are quite different in the three groups. Only birds employ a complicated

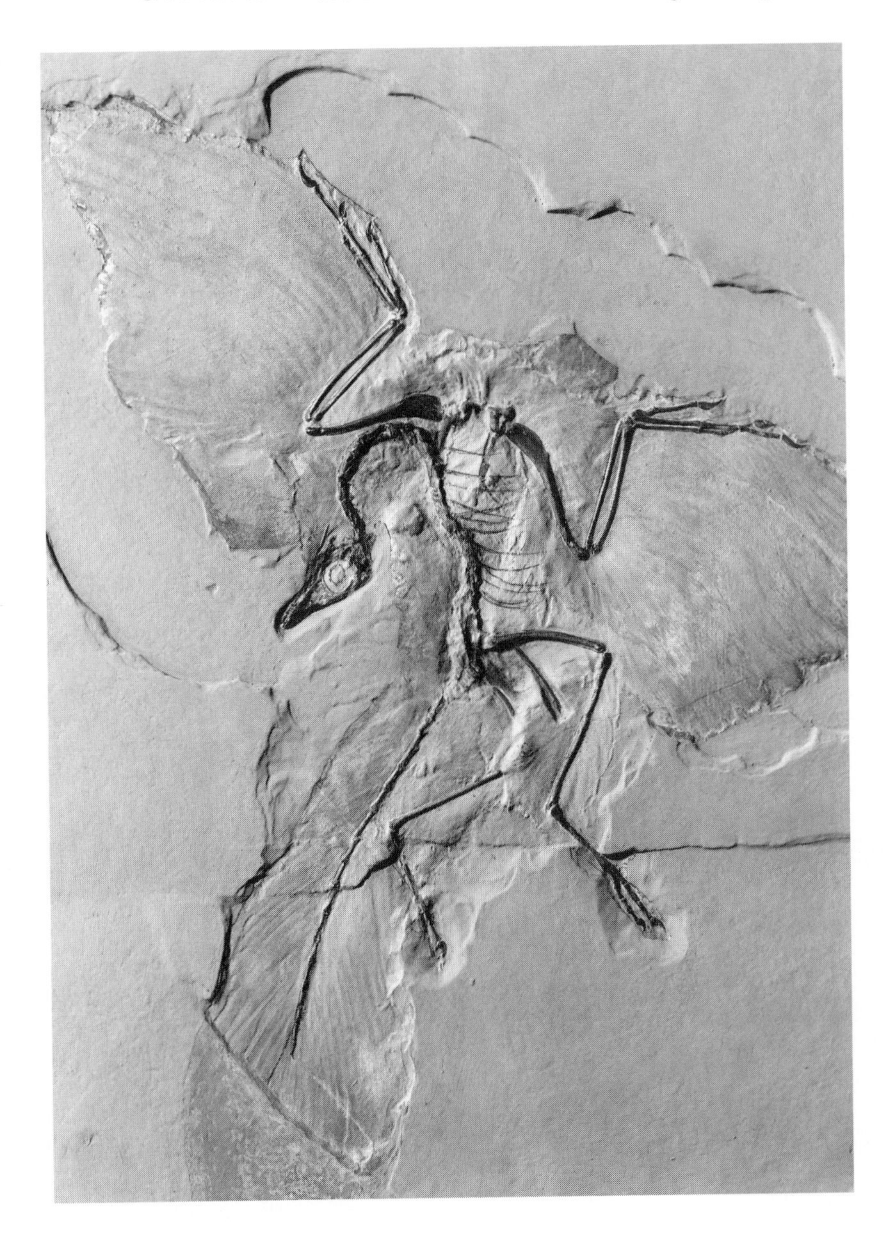

▶ Figure 15–2 *Archaeopteryx lithographica*.

series of overlapping feathers as the main wing surface.

What were the selective advantages for the evolution of wings and flight in the proavian ancestors of birds? Two competing hypotheses have existed for a century—the arboreal theory and the terrestrial theory (Figure 15–4). To examine these hypotheses, we need to give some further consideration to *Archaeopteryx*.

The arboreal theory of the origin of avian flight has long dominated the field (see Rayner 1988). According to this view, the ancestors of *Archaeopteryx* were tree climbers that jumped from branch to branch and from tree to tree much as some squirrels, lizards, and monkeys do. Under selective pressures favoring increased distance and accuracy of travel between trees, structures that provided some surface area for lift would be advantageous. *Draco* are arboreal lizards from the East Indies with extremely elongate ribs that support wings of skin on the sides of the trunk. They use their wings to glide from tree to tree. A flight starts with a dive from an elevated

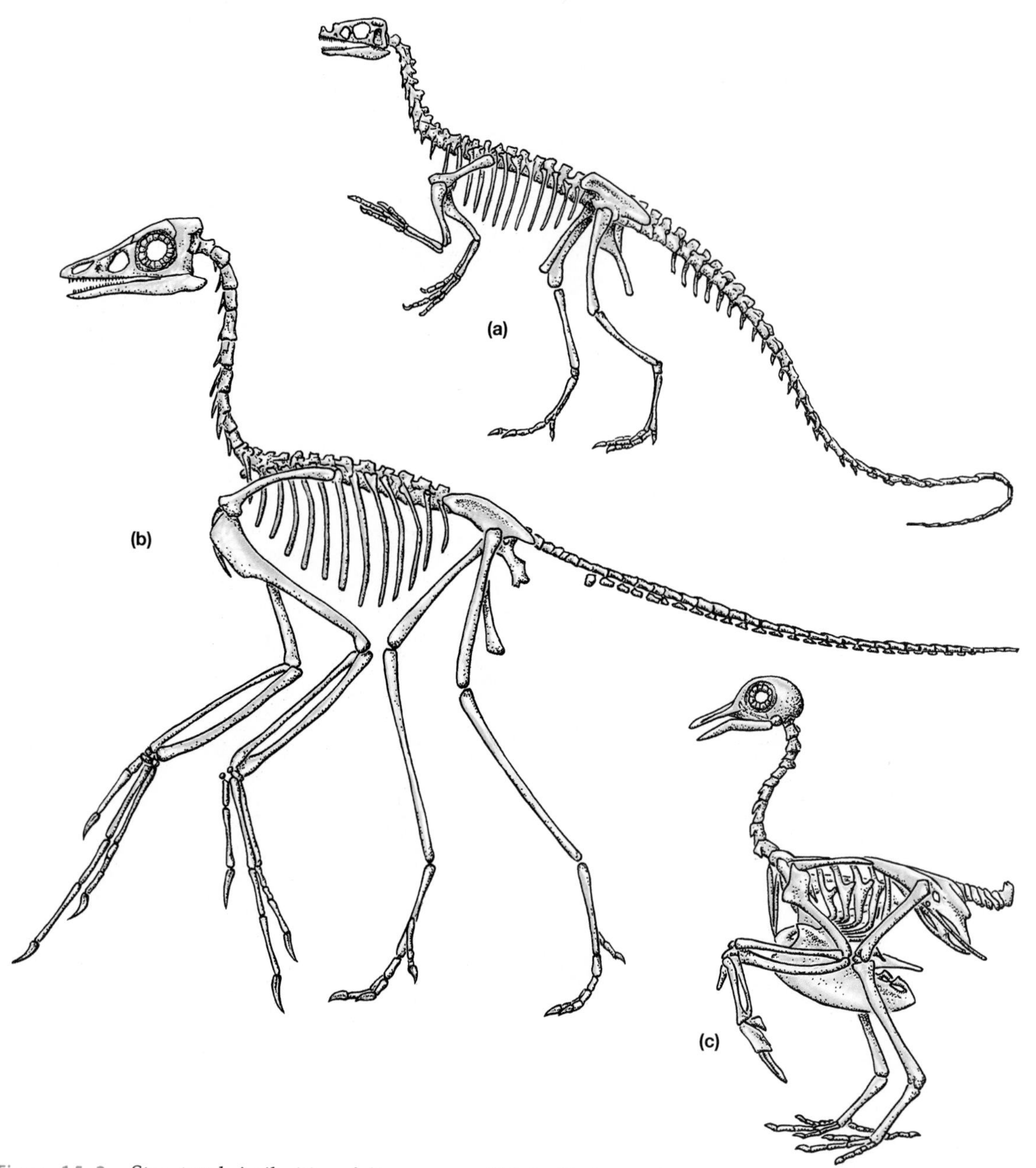

▲ Figure 15–3 Structural similarities of dinosaurs and birds. Skeleton of *Archaeopteryx* (b) compared with that of *Ornitholestes*, a dromeosaur (a), and a modern bird (c).

perch. The lizard descends at an angle of about 45 degrees, then levels out and uses the kinetic energy developed during the dive to glide nearly horizontally. A brief upward glide immediately precedes landing on another perch. Glides as long as 60 meters have been recorded with a loss of altitude of less than 2 meters. By this hypothesis, the evolution of flying forms would have passed from gliding stages through intermediate stages—such as *Archaeopteryx*, in which gliding was aided by weak flapping flight—to fully airborne flapping fliers.

On the other hand, given that the dromeosaur lineage consisted of bipedal, cursorial, terrestrial predators, is it necessary to invoke arboreal selection pressures for the evolution of avian flight? The from-the-ground-up theory postulates that flapping flight evolved directly from ground-dwelling, bipedal runners (Ostrom 1974).

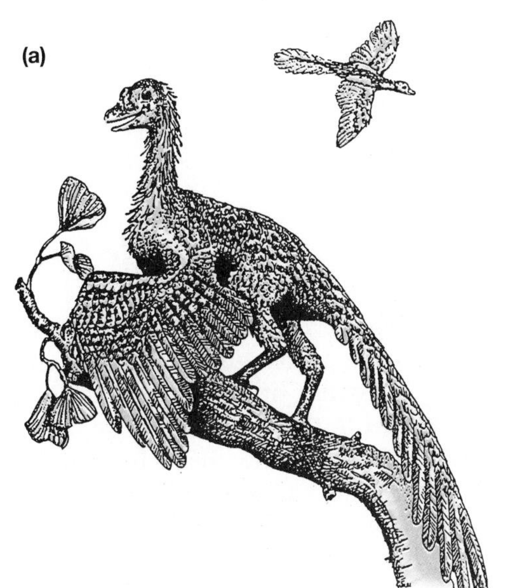

▲ Figure 15–4 Two reconstructions of *Archaeopteryx*. (a) The from-the-trees-down hypothesis, showing *Archaeopteryx* as an arboreal climber and (rear) glider. (b) The from-the-ground-up hypothesis, showing *Archaeopteryx* as a cursorial arboreal hunter.

According to the first version of this hypothesis (the cursorial theory), the proavian ancestors of birds were fast, bipedal runners that used their wings as planes to increase lift and lighten the load for running. In a later development, the wings were flapped as the animal ran to provide additional forward propulsion, much as a chicken flaps across the barnyard to escape from a dog. Finally, the pectoral muscles and flight feathers became sufficiently developed for full-powered flight through the air.

The cursorial theory in its original form failed as an explanation, because in mechanical terms flapping is not an effective mechanism to increase running speed. Maximum traction on the ground is required to achieve acceleration, and this traction can be provided only by solid contact of the feet with a firm substrate. Planing with the wings would have taken some of the body's weight off the feet and reduced traction. The amount of push that could be achieved with the small surfaces of the protowings probably would not even have compensated for the loss in force from the hind legs, much less increased acceleration.

A specimen of *Archaeopteryx* that had been misidentified as a coelurosaur for over 100 years revealed some previously unknown details of the hand and led John Ostrom to modify the cursorial theory. Some elements of the manus are extremely well preserved in this specimen and show the actual horny claws on digits 1 and 3. These claws look like the talons of a predatory bird.

The similarities in morphology between the hand, metacarpus, forearm, humerus, and pectoral apparatus of *Archaeopteryx* and several dromeosaurs may show that both the dromeosaurs and *Archaeopteryx* used their forelimbs to seize prey. The forelimb and shoulder of *Archaeopteryx* have not been structurally much modified from the skeletal condition of dromeosaurs, and *Archaeopteryx* differs from all other birds in lacking several features that are critical for powered flight—fused carpometacarpus, restricted wrist and elbow joints, modified coracoids, and a plate-like sternum with keel (Jenkins 1993). In fact, the only skeletal feature of *Archaeopteryx* that suggests flight is the well developed furcula (wishbone) that was present in dromeosaurs and is present in extant birds, although reduced or absent in flightless forms. Thus, the entire pectoral appendage (skeleton and muscles) of *Archaeopteryx* appears to have been as well adapted for predation as for flight. From these considerations, Ostrom postulated that the incipient wings of the proavian ancestors of *Archaeopteryx* evolved first as snares to trap insects or other prey against the ground or to bat them out of the air, so they could be grasped by the claws and teeth. The wings subsequently became modified into flapping appendages capable of subduing larger prey.

Aerodynamic models have suggested another refinement of the ground-up hypothesis—that evolution of the avian wing could have assisted horizontal jumps after prey. By spreading or moving its forelimbs, the proavian cursor could not only control pitch, roll, and yaw while leaping to catch a flying insect but also maintain its balance on landing.

How Well Could *Archaeopteryx* Fly?

Archaeopteryx was probably a late-surviving relict that was contemporaneous with more typically avian birds. It had teeth, claws on the fingers, and a long tail. If it did not have feathers, *Archaeopteryx* could readily be classified as a small dinosaur. However, several lines of evidence suggest that *Archaeopteryx* was capable of flight: Its skeletal proportions were similar to those of some extant flying birds, the number of primaries and secondaries were identical to those of extant birds, the asymmetry of its flight feathers is like that seen in extant flying birds, and the furcula was large. The seventh specimen of *Archaeopteryx* reveals a feature not visible in the fossils previously known—a rectangular sternum that was probably associated with strong flight muscles. On balance, these characteristics are consistent with the view that *Archaeopteryx* was a flying bird (Martin 1983b; Rayner 1988).

Unlike the proavian ancestors of birds, *Archaeopteryx* had wings that were large enough to contribute acceleration as it ran. *Archaeopteryx* probably took off by running and flapping its wings, just as many large birds do today. Once the animal was airborne, calculations of it's metabolic capacity suggest that an *Archaeopteryx* could fly at least 1.5 km at a velocity of 40 km/hr (Ruben 1991, 1993).

An animal that could take off from the ground and fly rapidly for several hundred meters would be able to escape predators or fly up into trees. Many living birds, including cursorial predators such as the North American roadrunner and the African secretary bird, use flight in exactly this way. Thus, *Archaeopteryx* can plausibly be interpreted as a ground-dwelling cursorial predator that could leap into the air to seize flying insects and fly rapidly to escape from its own predators. *Archaeopteryx* probably could not land in trees, because it retained a primitive foot structure that would not have allowed it to seize a branch. It seems likely that *Archaeopteryx* made running landings as chickens, quail, and pheasants do.

15.5 Early Birds

Our knowledge of the early evolution of birds after *Archaeopteryx* has increased enormously since 1980, due to the discovery of more fossil birds than had been unearthed in the entire previous century. Several important discoveries have been made at the Early Cretaceous Las Hoyas site in Spain. *Iberomesornis* had a pectoral girdle that was more derived than that of *Archaeopteryx*, with strut-like coracoids, a furcular process, and an ulna that was longer than the humerus. In addition, the tail was reduced to a series of free vertebrae and a pygostyle. The foot had curved claws and an opposable hallux (i.e., a rearward-pointing toe). This foot structure allows modern birds to perch, because as the legs bend a ligament curls the toes tightly around a branch. The presence of recurved claws and a reversed hallux in Early Cretaceous birds such as *Iberomesornis* suggests that they were able to land in trees.

A group of birds known as Enantiornithes ("opposite birds," because the metatarsals are fused differently from those of modern birds), first discovered in Argentina and now known from six continents, flourished between 140 million and 70 million years ago. The enantiornithine radiation, which includes most of the birds known from the Cretaceous, is a separate lineage from that of modern birds. *Sinornis* (Figure 15–5), from the Early Cretaceous of China, was the size of a sparrow and the smallest enantiornithine discovered so far. *Enantiornis* from Argentina was the size of a turkey vulture. Most enantiornithines were small to medium-size and probably lived in trees, but some had long legs and were probably wading birds, and others had powerful claws like modern hawks.

Enantiornithines continued the trends seen in *Iberomesornis*. *Sinornis* had a vertebral column that contained only 11 dorsal vertebrae, compared with 14 for *Archaeopteryx*. The short trunk and tail shifted the center of mass toward the forelimbs, as in extant birds, rather than toward the hindlimbs, as in cursorial terrestrial archosaurs. Many of the derived characters of enantiornithines are associated with flight.

▶ Figure 15–5 *Sinornis*, from Early Cretaceous lake bed deposits in China.

For example, the wrist could bend backward sharply, as in extant birds, so the wing could be tucked against the body. The sternum developed a deep keel. The alula, which increases maneuverability in slow-speed flight, is first seen in an enantiornithine from Las Hoyas, *Eoavulavis* (= dawn bird with an alula).

Gobipteryx, an enantiornithine known from skulls found in Late Cretaceous deposits in Mongolia, is unique among Cretaceous birds in lacking teeth. In the same deposits that yielded fossils of adult *Gobipteryx*, the Polish Mongolian Expedition found fossilized eggs, some containing well-preserved skeletons with skulls very like those of adult *Gobipteryx*. The skeletons of these embryos were well developed, and it is likely that the chicks were precocial at hatching (i.e., able to walk and to find their own food).

In addition to these fossils, an abundance of fossilized feather impressions and tracks indicates that by the Early Cretaceous, birds inhabited both the Northern and Southern hemispheres. The differences among the fossil species known—flying and flightless birds and a foot-propelled diver—combined with the wide geographic distribution of birds in the Cretaceous suggest that more of the evolution of birds took place in the Jurassic than has previously been appreciated. Unfortunately, the fossil record of birds from the Jurassic is poor. Avian evolution is an area of lively controversy, and views of the phylogenetic history of birds are likely to change substantially as additional material is studied.

Representatives of the lineage that includes modern birds (Ornithurae) are also known from the Cretaceous. The Ichthyornithiformes were flying birds with a well-developed keel on the sternum and short backs and tails, but they retained teeth. Several species of *Ichthyornis* have been named, mostly on the basis of differences in size. In general, ichthyornithiforms were the size of gulls and terns, and like these extant birds they may have flown far out to sea.

Another group of Late Cretaceous birds that lie in the lineage of modern birds, the Hesperornithiformes, were flightless, foot-propelled swimmers and divers. A fossil from England, *Enaliornis*, establishes the presence of hesperornithiforms in the Early Cretaceous, and they were a diverse group by the Late Cretaceous. The hesperornithiforms were medium-size to large flightless birds that were specialized for foot-propelled diving (Figure 15–6). The body and neck were elongate, the sternum lacked a keel, and the bones were not pneumatic. Teeth remained in the maxilla and dentary. Feathers preserved with two specimens of *Parahesperornis* were plumulaceous (down-like), and the birds may have had a furry appearance somewhat like that of the extant kiwi. The feet were placed far posteriorly on the body (a position characteristic of many foot-propelled diving birds), and the toes had lobes like those seen in extant grebes. The femur and tibiotarsus were locked in place and could not be rotated under the body. As a result, hesperornithiforms would not have been able to walk on land and probably pushed themselves along by sliding on their stomachs. Unlike modern diving birds, the hesperornithiforms had completely lost their wings. The ateral placement of the feet would have made it possible for hesperornithiforms to exert force directly backward during swimming and diving without an upward component that would have tended to drive them toward the surface. Pachyostosis (an increased density of bone) gave hesperornithiforms a high specific gravity that would have facilitated diving. Coprolites (fossilized feces) found in association with *Hesperornis* contain the remains of small fishes.

The Evolution of Derived Orders and Families of Birds

Modern birds, the Neornithes, began to diversify during the last part of the Cretaceous, replacing the Enantiornithes as the predominant avian forms. A major radiation of avian families occurred during the Cenozoic. The Eocene was the epoch of greatest diversification of birds. More surviving families arose in that period than at any other time. Most of these families consist of additional water birds and nonpasserine forest dwellers. A second radiation occurred in the Miocene and included a few additional families of water birds, but mostly land-dwelling passerines that were adapted to drier, less forested environments. Most families of birds had evolved by the end of the Miocene, and many still-existing genera and some species were present by the Pliocene. Birds apparently formed complex ecological communities by the middle of the Cenozoic.

Phylogeny of Extant Birds

Phylogenetic relationships among extant birds are poorly known and are the subject of continuing controversy. Several views of these relationships and the difficulties inherent in their study can be found in reviews by Cracraft (1986), Olson (1985), Sibley et al.

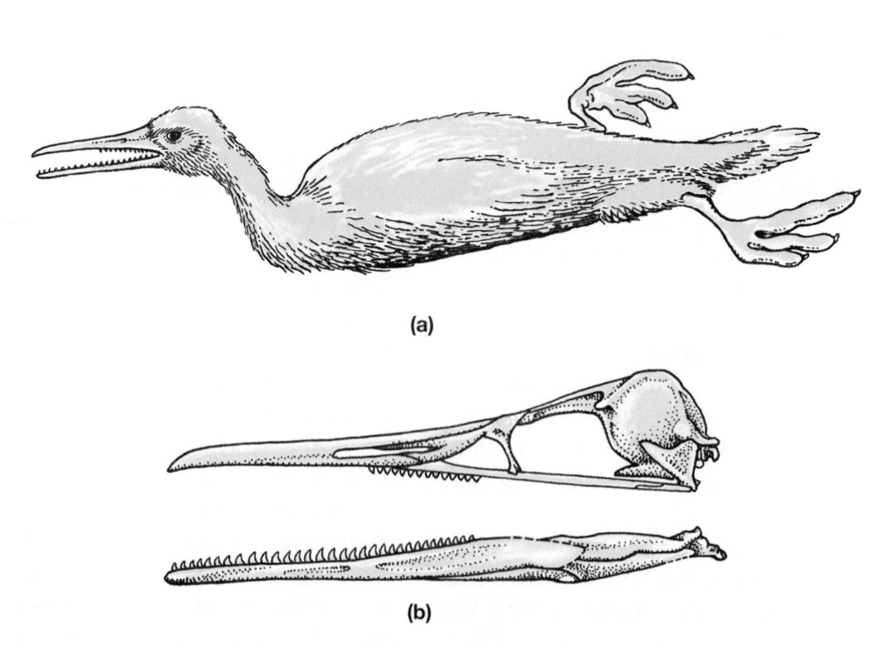

▲ Figure 15–6 The hesperornithiforms were flightless, toothed birds. (a) Restoration of *Hesperornis*. (b) Skull of *Parahesperornis*. Note the teeth in the maxilla and dentary bones.

(1988), Raikow (1985a), Houde (1986, 1987), Sibley and Ahlquist (1990), and Norrell and Clarke (2000). The most commonly used classification of birds derives from Storer (1971). A current hypothesis is presented in Table 15.1.

The lack of consensus about the phylogeny of extant birds makes it impossible to provide a cladogram that represents a widely accepted hypothesis of evolutionary relationships. Charles Sibley and his colleagues (Sibley et al. 1988, Sibley and Ahlquist

Table 15.1 Classification of extant birds to the level of orders. The geographic regions represent the distribution of the entire order; families within an order often have smaller distributions. The relationships of the flamingos and turacos are unknown.

Classification	*Approximate Number of*		*Geographic Region*
	Families	*Species*	
Tinamiformes (tinamous)	1	47	Neotropics
Struthioniformes (ostriches)	6	9	Neotropics, Africa, Australia, New Zealand, New Guinea
Galliformes (curassows, guans, chachalacas, megapodes, guineafowl, pheasants, quail, grouse, turkeys)	5	258	Worldwide
Anseriformes (screamers and waterfowl)	2	161	Worldwide
Passeriformes (perching birds, including the songbirds)			
Tyranni (broadbills, pittas, asities, New Zealand wrens, tyrant flycatchers, cotingas, manakins, woodcreepers, ovenbirds, antbirds, tapaculos)	15	1138	Pantropical
Passeres			
Crows and related forms	31	1113	Worldwide
Thrushes and related forms	7	611	Worldwide
Nuthatches, wrens, and related forms	14	1168	Worldwide
Larks, sparrows, finches, wood warblers, tanagers, blackbirds, and related forms	17	1651	Worldwide
Piciformes (jacamars, barbets, honeyguides, toucans, woodpeckers)	8	410	Worldwide
Coraciiformes (kingfishers, todies, motmots, bee-eaters, rollers, hoopoes, hornbills)	10	218	Worldwide
Trogoniformes (trogons, quetzals)	1	39	Pantropical except Australia
Coliiformes (mousebirds)	1	6	Africa
Apodiformes (swifts, hummingbirds)	3	422	Worldwide
Caprimulgiformes (nightjars, poorwills, frogmouths, oilbirds)	5	113	Worldwide
Strigiformes (owls)	2	186	Worldwide
Cuculiformes (cuckoos, hoatzin)	6	143	Worldwide
Psittaciformes (parrots)	3	358	Pantropical and Australia
Gruiformes (rails, coots, sungrebes, kagu, sunbittern, roatelos, buttonquails, cranes, limpkins, trumpeters, seriemas, bustards)	11	213	Worldwide
Charadriiformes (shorebirds, plovers, sandpipers, gulls, jaegers, skuas, skimmers, terns, auks, murres, puffins, sandgrouses)	19	366	Worldwide
Columbiformes (doves, pigeons)	1	310	Worldwide
Falconiformes (condors, hawks, eagles, kites, falcons, caracaras)	5	311	Worldwide
Ciconiiformes (herons, bitterns, whale-head storks, storks, ibises, spoonbills)	5	120	Worldwide
Pelecaniformes (tropic birds, boobies, gannets, cormorants, pelicans, frigatebirds)	6	67	Worldwide
Procellariiformes (albatrosses, shearwaters, petrels)	4	115	Worldwide
Podicipediformes (grebes)	1	21	Worldwide
Sphenisciformes (penguins)	1	17	Southern Hemisphere
Gaviiformes (loons)	1	5	New World, Eurasia
Phoenicopteriformes (flamingos)	1	5	Worldwide, except Australia
Musophagiformes (turacos)	1	23	Africa

1990) have presented a cladistic analysis based on comparisons of DNA. Their classification of passerine birds has been well received, but their analysis of the relationships of nonpasserine birds is more controversial. It provides a good example of some of the difficulties in reconstructing phylogeny solely by molecular methods.

One of the major controversies in bird phylogeny centers on the relationships of a group of flightless birds known as ratites. Extant ratites include ostriches (Africa), rheas (South America), emus and cassowaries (Australia), and kiwis (New Zealand). All these landmasses were part of the southern supercontinent Gondwana, and it has been suggested that the ratites arose from a single flightless ancestor that was widely distributed on Gondwana. By this hypothesis the ratites form a monophyletic group, and their current geographic distribution reflects the breakup of Gondwana in the late Mesozoic and early Cenozoic.

An alternative view proposes that the similarities of extant ratites are ancestral characters that were found in many lineages of birds. The ratites and another group of birds, the tinamous of Central and South America, share a paleognathous palatal structure. This palate is characterized by long prevomers that extend posteriorly to articulate with the palatines and pterygoids and by large basipterygoid processes that articulate with the pterygoids. The paleognathus palate is distinguished from the neognathus palate that is characteristic of other birds. Paleognathous palatal characters are seen in toothed birds such as *Hesperornis* and in the Cretaceous toothless bird *Gobipteryx*. These characters are also present in early stages of the embryonic development of neognathus birds. Thus the paleognathus condition of ratites and tinamous could have evolved by neoteny from a neognathus form. This possibility means that the paleognathus birds are not necessarily a monophyletic lineage. Furthermore, the discovery of birds related to ostriches in Paleocene and Eocene deposits in North America and Europe casts doubt on the assumption that the current distribution of ratites is the result of events associated with the breakup of Gondwana (Houde 1986).

The phylogeny and zoogeography of ratites are of considerable significance in current biochemical studies of the phylogeny of birds. In calibrating the DNA molecular clock, Sibley and Ahlquist assumed that the separation of the lineages of ratites was caused by the breakup of Gondwana. If the ratites represent two or more independent origins, or if the origins of some ratites were in the Northern Hemisphere rather than in Gondwana, the calibration of the DNA clock is erroneous. Conclusions about the times of divergence of other groups of birds that were based on that calibration are thus weakened.

15.6 Birds as Flying Machines

In many respects birds are variable: Beaks and feet are specialized for different modes of feeding and locomotion, the morphology of the intestinal tract is related to dietary habits, and wing shape reflects flight characteristics. Despite that variation, however, the morphology of birds is more uniform than that of mammals. Much of this uniformity is a result of the specialization of birds for flight.

Consider body size as an example: Flight imposes a maximum body size on birds. The muscle power required to take off increases by a factor of 2.25 for each doubling of body mass. That is, if species B weighs twice as much as species A, it will require 2.25 times as much power to fly at its minimum speed. If the proportion of the total body mass allocated to flight muscles is constant, the muscles of a large bird must work harder than the muscles of a small bird. In fact, the situation is still more complicated because the power output is a function of both muscular force and wing-beat frequency. Large birds have lower wing-beat frequencies than small birds, for mechanical and aerodynamic reasons. As a result, if species B weighs twice as much as species A, it will develop only 1.59 times as much power from its flight muscles, although it needs 2.25 times as much power to fly. Therefore, large birds require longer takeoff runs than small birds, and a bird could ultimately reach a body mass at which any further increase in size would move it into a realm in which its leg and flight muscles were not able to provide enough power to take off.

Calculations of this maximum size from aerodynamic principles suggest that it lies near 12 kg, and that estimate corresponds reasonably well with the observed body masses of birds. The mute swan weighs about 12 kg, and the trumpeter swan is about 17 kg. The largest flying bird known was a giant condor that had a wingspan estimated to be 7m and a possible body mass of 20 kg. Most large pterosaurs also had body masses estimated to be in the region of 20 kg.

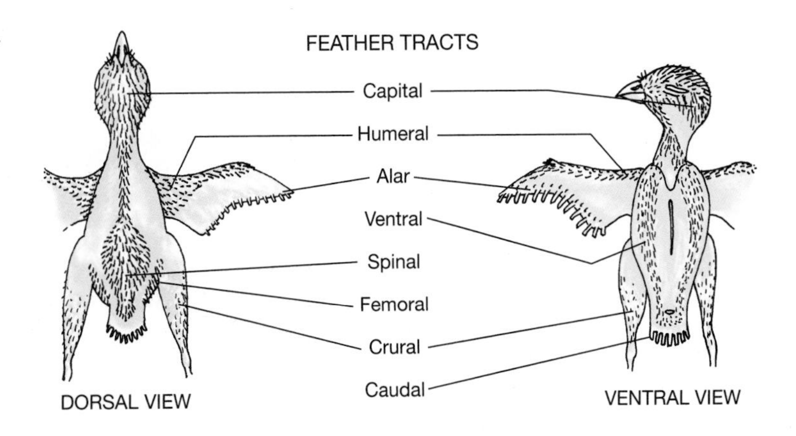

▶ Figure 15–7 Feather tracts of a typical songbird.

Flightless birds are spared the mechanical constraints associated with producing power for flight, but they still do not approach the body sizes of mammals. The largest extant bird is the flightless ostrich, which weighs about 150 kg, and the largest bird known, one of the extinct elephantbirds, weighed an estimated 450 kg. In contrast, the largest mammal, the blue whale, weighs more than 135,000 kg. If we restrict the comparison to quadrupedal, terrestrial mammals, the African elephant weighs up to 5000 kg.

The structural uniformity of birds is seen even more clearly if their body shapes are compared with those of other diapsids such as the dinosaurs. There are no quadrupedal birds, for example, nor any with horns or bony armor. Even those species of birds that have become secondarily flightless retain many ancestral characters—and the constraints associated with them. In this chapter we consider the body form and function of birds, especially in relation to the requirements of flight.

Feathers and Flight

Feathers develop from pits or follicles in the skin, generally arranged in tracts or **pterylae**, which are separated by patches of unfeathered skin, the **apteria** (Figure 15–7). Some species—such as ratites, penguins, and mousebirds—lack pterylae, and the feathers are uniformly distributed over the skin.

For all their structural complexity, feathers are remarkably simple and uniform in chemical composition. More than 90 percent of a feather consists of beta keratin, a protein related to the keratin that forms the scales of lepidosaurs. About 1 percent of a feather consists of lipids, about 8 percent is water, and the remaining fraction consists of small amounts of other proteins and pigments, such as melanin. The colors of feathers are produced by structural characters and pigments.

Basic Types of Feathers

Ornithologists usually distinguish five types of feathers: (1) contour feathers, including typical body feathers and the flight feathers (remiges and rectrices); (2) semiplumes; (3) down feathers of several sorts; (4) bristles; and (5) filoplumes. **Vaned feathers** (Figures 15–8 and 15–10b) include a short, tubular base, the **calamus**, which remains firmly implanted within the follicle until molt occurs. Distal to the calamus is a long, tapered **rachis**, which bears closely spaced side branches called **barbs**, the lowermost of which externally mark the division between calamus and rachis. The barbs on either side of the rachis constitute a surface called a **vane.** Vanes may be symmetrical or asymmetrical. The proximal portions of the vanes of a feather have a downy or plumulaceous texture, being soft, loose, and fluffy. This gives the plumage of a bird its excellent properties of thermal insulation. The more distal portions of the vanes have a pennaceous or sheet-like texture—firm, compact, and closely knit. This exposed part provides an airfoil, protects the downy undercoat, sheds water, reflects or absorbs solar radiation, and may have a role in visual or auditory communication. The barbules are structures that maintain the pennaceous character of the feather vanes. They are arranged in such a way that any physical disruption to the vane is easily corrected by the bird's preening behavior, in which the bird realigns the barbules by drawing its slightly separated bill over them.

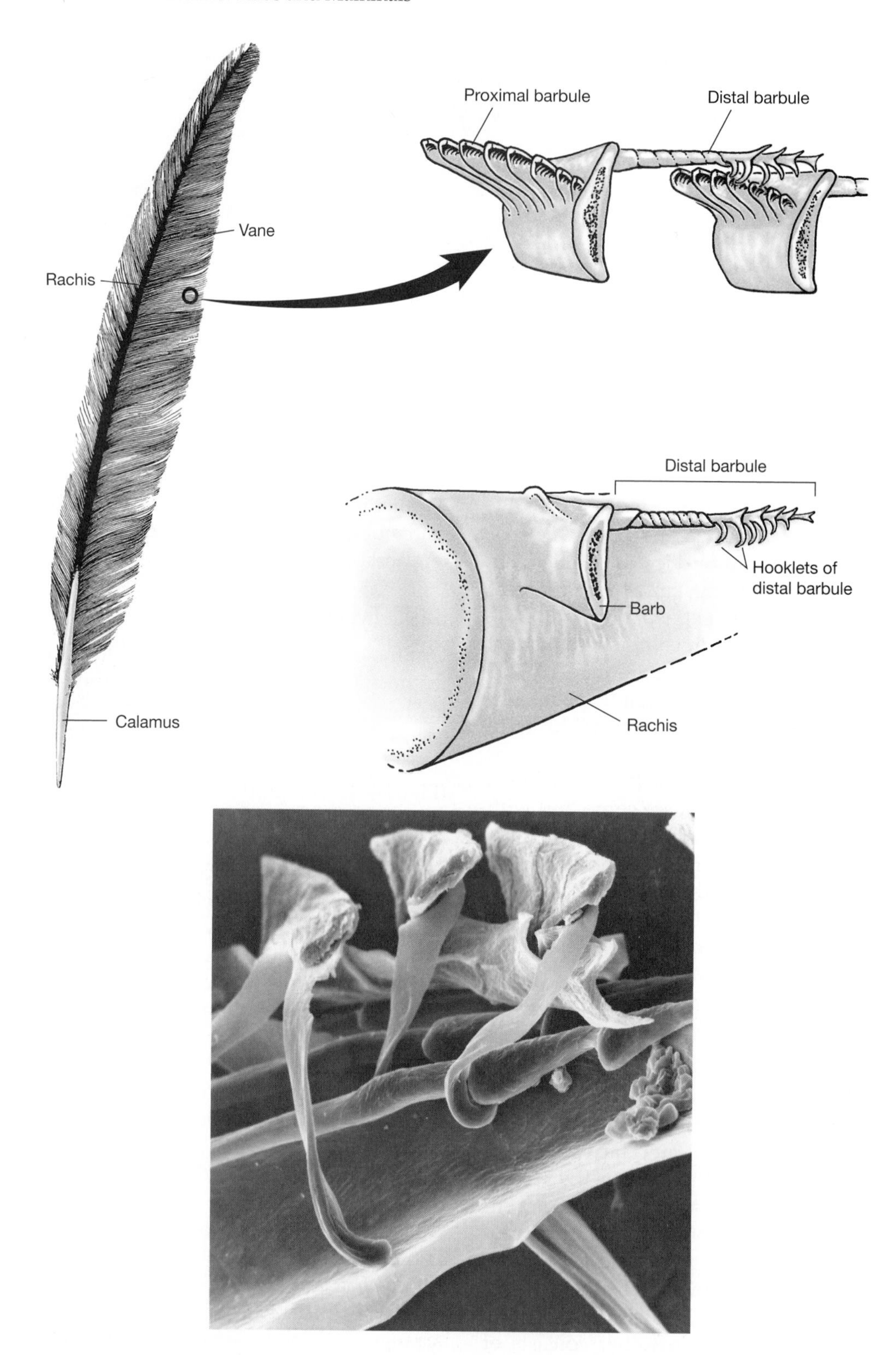

▲ Figure 15–8 Typical vaned feather (a wing quill) showing its main structural features. The inset and electron micrograph show details of the interlocking mechanism of the proximal and distal barbules.

▶ Figure 15–9 A red-tail hawk, showing the extreme development of slotting in the outer primaries.

The **remiges** (wing feathers, singular **remex**) and **rectrices** (tail feathers, singular **rectrix**) are large, stiff, mostly pennaceous contour feathers that are modified for flight. For example, the distal portions of the outer primaries of many species of birds are abruptly tapered or notched, so that when the wings are spread the tips of these primaries are separated by conspicuous gaps or slots (Figure 15–9). This condition reduces the drag on the wing and, in association with the marked asymmetry of the outer and inner vanes, allows the feather tips to twist as the wings are flapped and to act somewhat as individual propeller blades (see Figure 15–12).

Semiplumes are feathers intermediate in structure between contour feathers and down feathers. They combine a large rachis with entirely plumulaceous vanes and can be distinguished from down feathers in that the rachis is longer than the longest barb (Figure 15–10a). Semiplumes are mostly hidden beneath the contour feathers. They provide thermal insulation and help to fill out the contour of a bird's body.

Down feathers of various types are entirely plumulaceous feathers in which the rachis is shorter than the longest barb or entirely absent. Down feathers provide insulation for adult birds of all species. In addition, natal down, which is structurally simpler than adult down, provides an insulating covering on many birds at hatching or shortly thereafter. Natal downs usually precede the development of the first contour feathers, and down feathers are associated with apteria (the spaces between the contour feather tracts). Definitive downs are those that develop with the full body plumage. Uropygial gland downs are associated with the large sebaceous gland found at the base of the tail in most birds. The papilla of the gland usually bears a tuft of modified, brush-like down feathers that aid in transferring the oily secretion from the gland to the bill to provide waterproof dressing to the plumage.

Powder down feathers, which are difficult to classify by structural type, produce an extremely fine white powder composed of granules of keratin. The powder, which is shed into the general plumage, is nonwettable and is therefore assumed to provide another kind of waterproof dressing for the contour feathers. All birds have powder down, but it is best developed in herons.

Bristles are specialized feathers with a stiff rachis and barbs only on the proximal portion or none at all (Figure 15–10c). Bristles occur most commonly around the base of the bill, around the eyes, as eyelashes, and on the head or even on the toes of some birds. The distal rachis of most bristles is colored dark brown or black by melanin granules. The melanin not only colors the bristles but also adds to their strength, resistance to wear, and resistance to photochemical damage. Bristles and structurally intermediate feathers called semibristles screen out foreign particles from the nostrils and eyes of many birds; they also act as tactile sense organs and possibly as aids in the aerial capture of flying insects, as, for example, the long bristles at the edges of the jaws in nightjars and flycatchers.

Filoplumes are fine, hair-like feathers with a few short barbs or barbules at the tip (Figure 15–10d). In some birds, such as cormorants and bulbuls, the filoplumes grow out over the contour feathers and contribute to the external appearance of the plumage, but usually they are not exposed. Filoplumes are sensory structures that aid in the operation of other feathers. Filoplumes have numerous free nerve

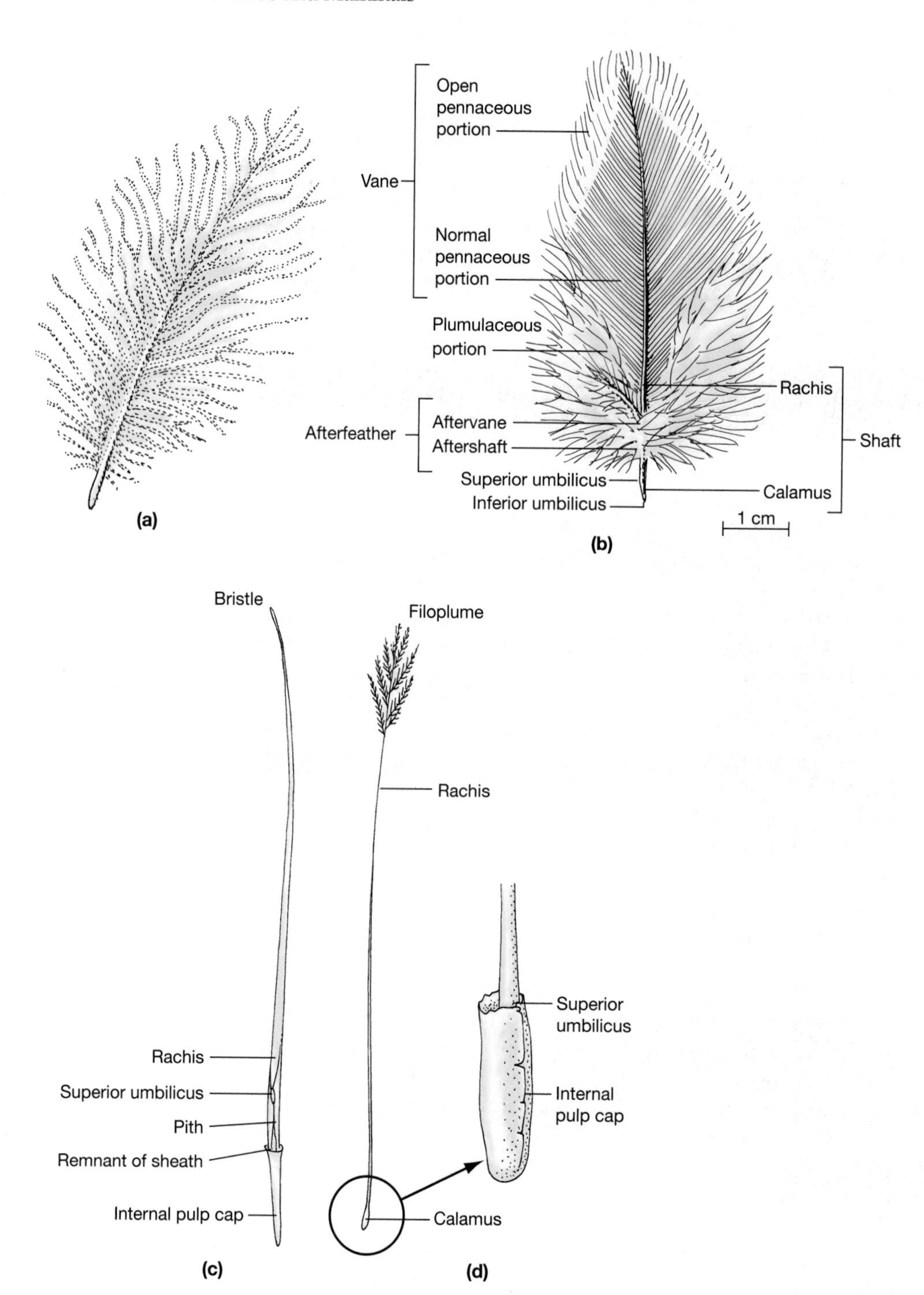

▲ Figure 15–10 Types of feathers. (a) Semiplume. (b) Body-contour feather. (c) Bristle. (d) Filoplume.

endings in their follicle walls, and these nerves connect to pressure and vibration receptors around the follicles. Apparently, the filoplumes transmit information about the position and movement of the contour feathers via these receptors. This sensory system probably plays a role in keeping the contour feathers in place and adjusting them properly for flight, insulation, bathing, or display.

Aerodynamics of the Avian Wing Compared with Fixed Airfoils

Unlike the fixed wings of an airplane, the wings of a bird function both as an airfoil (lifting surface) and as a propeller for forward motion. The avian wing is admirably suited for these functions, consisting of a light, flexible airfoil (Figure 15–11). The primaries, inserted on the hand bones, do most of the propelling when a bird flaps its wings, and the secondaries along the arm provide lift. Removal of flight feathers from the wings of doves and pigeons shows that when only a few of the primaries are pulled out, the bird's ability to fly is greatly altered, but a bird can still fly when as much as 55 percent of the total area of the secondaries has been removed.

A bird also has the ability to alter the area and shape of its wings and their positions with respect to the body. These changes in area and shape cause corresponding changes in velocity and lift that allow a bird to maneuver, change direction, land, and take

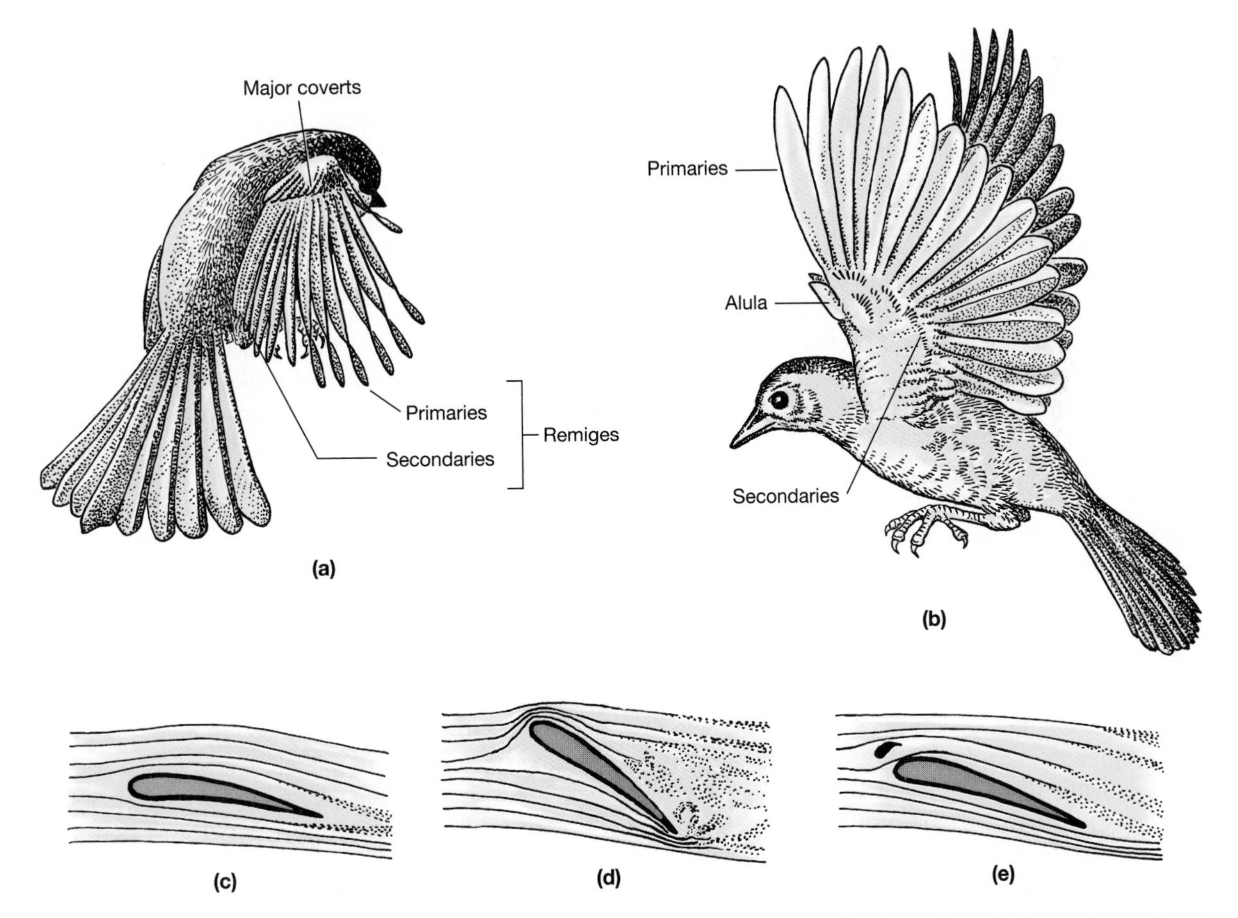

▲ Figure 15–11 The wing in flight. (a, b) Drawings from high-speed photographs show the twisting and opening of the primaries during flapping flight. (c–e) Airflow around a cambered airfoil. At a low angle of attack (c), the air streams smoothly over the upper surface of the wing and creates lift. When the angle of attack becomes steep (d), air passing over the wing becomes turbulent, decreasing lift enough to produce a stall. A wing slot formed by the alula (e) helps to prevent turbulence by directing a flow of rapidly moving air close to the upper surface of the wing.

off. Moreover, a bird's wing is not a solid structure like a conventional airfoil such as an airplane wing, but allows some air to flow through and between the feathers.

Obviously, the aerodynamic properties of a bird's wing in flight—even in nonflapping flight—are vastly more complex than those of a fixed wing on an airplane or glider. Nevertheless, it is instructive to consider a bird's wing in terms of the basic performance of a fixed airfoil. (See Norberg 1985 and Rayner 1988 for reviews.) Although a bird's wing actually moves forward through the air, it is easier to think of the wing as stationary with the air flowing past. The flow of air produces a force, which is usually called the **reaction**. It can be resolved into two components: the **lift**, which is a vertical force equal to or greater than the weight of the bird, and the **drag**, which is a backward force opposed to the bird's forward motion and to the movement of its wings through the air.

When the leading end of a symmetrically streamlined body cleaves the air, it thrusts the air equally upward and downward, reducing the air pressure equally on the dorsal and ventral surfaces. No lift results from such a condition. There are two ways to modify this system to generate lift. One is to increase the **angle of attack** of the airfoil, and the other is to bend its surface. Either change increases lift at the cost of increasing drag.

When the contour of the dorsal surface of the wing is convex and the ventral surface is concave (a **cambered airfoil**), the air pressure against the two surfaces is unequal because the air has to move farther and faster over the dorsal convex surface relative to the ventral concave surface (Figure 15–11c). The result is reduced pressure over the wing, or lift. When the lift equals or exceeds the bird's body weight, the bird becomes airborne. The camber of the wing varies in birds with different flight characteristics; it also changes along the length of the wing. Camber is greatest close to the body and decreases toward the wing tip. This change in camber is one reason that the proximal part of the wing generates greater lift than the distal part.

If the leading edge of the wing is tilted up so that the angle of attack is increased (Figure 15–11d), the result is increased lift—up to an angle of about 15 degrees, the **stalling angle**. This lift results more from a decrease in pressure over the dorsal surface than from an increase in pressure below the airfoil. If the smooth flow of air over the wing becomes disrupted, the airflow begins to separate from the wing because of the increased air turbulence over the wing. The wing is then stalled. Stalling can be prevented or delayed by the use of slots or auxiliary airfoils on the leading edge of the main wing. The slots help to restore a smooth flow of air over the wing at high angles of attack and at slow speeds. The bird's **alula** has this effect, particularly during landing or takeoff (Figure 15–11e). Also, the primaries act as a series of independent, overlapping airfoils, each tending to smooth out the flow of air over the one behind.

Another characteristic of an airfoil has to do with wing-tip vortexes—eddies of air resulting from outward flow of air from under the wing and inward flow from over it. This is **induced drag**. One way to reduce the effect of these wing-tip eddies and their drag is to lengthen the wing, so that the tip vortex disturbances are widely separated and there is proportionately more wing area where the air can flow smoothly. Another solution is to taper the wing, reducing its area at the wing tip where induced drag is greatest. The ratio of length to width is called the **aspect ratio**. Long, narrow wings have high aspect ratios and high lift-to-drag (L/D) ratios. High-performance sailplanes and albatrosses, for example, have aspect ratios of 18:1 and L/D ratios in the range of about 40:1.

Wing loading is another important consideration. This is the mass of the bird divided by the wing area. The lighter the loading, the less power is needed to sustain flight. Small birds usually have lighter wing loading than do large birds, but wing loading is also related to specializations for powered versus soaring flight. The comparisons in Table 15.2 illustrate both of these trends. Small species such as hummingbirds, barn swallows, and mourning doves have lighter wing loading than large species such as the peregrine, golden eagle, and mute swan; yet the 3-gram hummingbird, a powerful flier, has a heavier wing loading than the more buoyant, sometimes soaring barn swallow, which is more than five times heavier. Similarly, the rapid-stroking peregrine has a heavier wing loading than the larger, often soaring golden eagle.

Flapping Flight

Flapping flight is remarkable for its automatic, unlearned performance. A young bird on its maiden flight uses a form of locomotion so complex that it defies precise analysis in physical and aerodynamic

Table 15.2 Wing Loading of Some North American Birds

Species	*Body Mass (g)*	*Wing Area (cm^2)*	*(g/cm^2)*
Ruby-throated hummingbird	3.0	12.4	0.24
Barn swallow	17.0	118.5	0.14
Mourning dove	130.0	357.0	0.36
Peregrine falcon	1,222.5	1,342.0	0.91
Golden eagle	4,664.0	6,520.0	0.71
Mute swan	11,602.0	6,808.0	1.70

terms. The nestlings of some species of birds develop in confined spaces, such as burrows in the ground or cavities in tree trunks, in which it is impossible for them to spread their wings and practice flapping before they leave the nest. Despite this seeming handicap, many of them are capable of flying considerable distances on their first flights. Diving petrels may fly as far as 10 km the first time out of their burrows. On the other hand, young birds reared in open nests frequently flap their wings vigorously in the wind for several days before flying—especially large birds such as albatrosses, storks, vultures, and eagles. Such flapping may help to develop muscles, but it is unlikely that these birds are learning to fly. However, a bird's flying abilities do improve with practice for a period after it leaves the nest.

There are so many variables involved in flapping flight that it becomes difficult to understand exactly how it works. A beating wing is flexible and porous and yields to air pressure, unlike the fixed wing of an airplane. Its shape, wing loading, camber, sweepback, and the position of the individual feathers all change remarkably as a wing moves through its cycle of locomotion. This is a formidable list of variables, and it is no wonder that flapping flight has not yet fully yielded to explanation in aerodynamic terms. However, the general properties of a flapping wing can be described.

We can begin by considering the flapping cycle of a small bird in flight. A bird cannot continue to fly straight and level unless it can develop a force or thrust to balance the drag operating against forward momentum. The flapping of the wings, especially the wing tips (primaries), produces this thrust, whereas the inner wings (secondaries) are held more nearly stationary with respect to the body, and generate lift. It is easiest to consider the forces operating on the inner and outer wing separately. Most of the lift, and also most of the drag, comes from the forces acting on the inner wing and body of a flying bird. The forces on the wing tips derive from two motions that have to be added together. The tips are moving forward with the bird, but at the same time they are also moving downward relative to the bird. The wing tip would have a very large angle of attack and would stall if it were not flexible. As it is, the forces on the tip cause the individual primaries to twist as the wing is flapped downward (see Figure 15–11) and to produce the forces diagrammed in Figure 15–12.

The forces acting on the two parts of the wing combine to produce the conditions for equilibrium flight, shown. The positions of the inner and outer wings during the upstroke (dotted lines) and downstroke reveal that vertical motion is applied mostly at the wing tips (Figure 15–12a). Thus, the inner wing (the secondaries) acts as if the bird were gliding to generate the forces shown in Figure 15–12b, and the outer wing generates the force shown in part c. Most of the lift to counter the pull of gravity (M) is generated by the inner wing and body. Canting of the wing tip (the primaries) during the downstroke (Figure 15–12c) produces a resultant force (R) that is directed forward. The movement of the wing tip relative to the air is affected by the forward motion of the bird through the air (Figure 15–12d). As a result

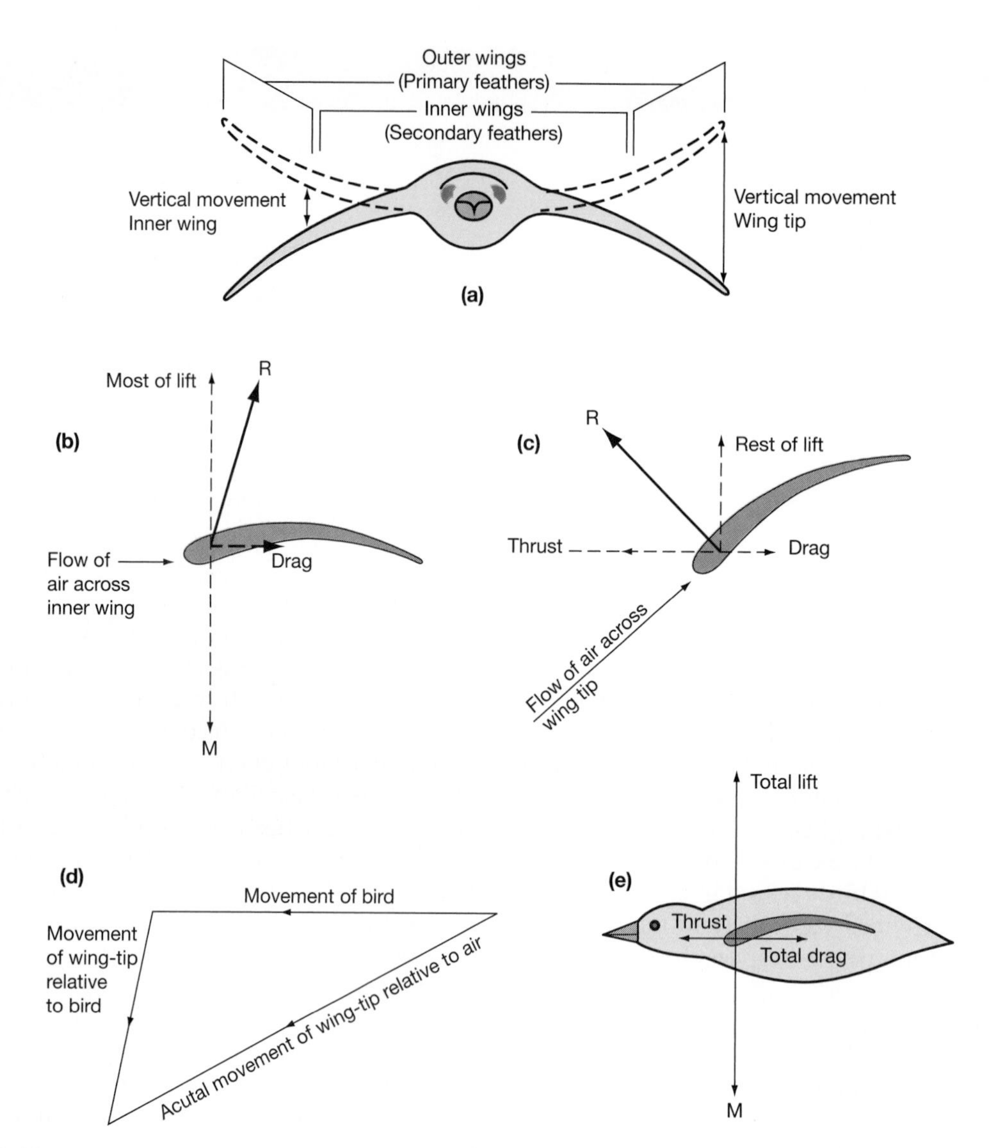

▲ Figure 15–12 Generalized diagrams of forces acting on the inner and outer wings and the body of a bird in flapping flight. M, gravitational force; R, resultant force. See text for explanation.

of this motion and the canting of the wing tips during the downstroke, the flow of air across the primaries is different from the flow across the secondaries and the body (Figure 15–12e). When flight speed through the air is constant, the forces acting on the inner wing and the body and on the outer wing combine to produce a set of summed vectors in which thrust exceeds total drag and lift at least equals the body mass.

As the wings move downward and forward on the downstroke, which is the power stroke, the trailing edges of the primaries bend upward under air pressure, and each feather acts as an individual propeller biting into the air and generating thrust. Contraction of the **pectoralis major**, the large breast muscle, produces the downstroke during level flapping flight. During this downbeat the thrust is greater than the total drag, and the bird accelerates. In small birds, the return stroke, which is upward and backward, provides little or no thrust and is mainly a passive recovery stroke. The bird decelerates during the recovery stroke.

For larger birds with slower wing actions, the time of the upstroke is too long to spend in a state of deceleration. A similar situation exists when any

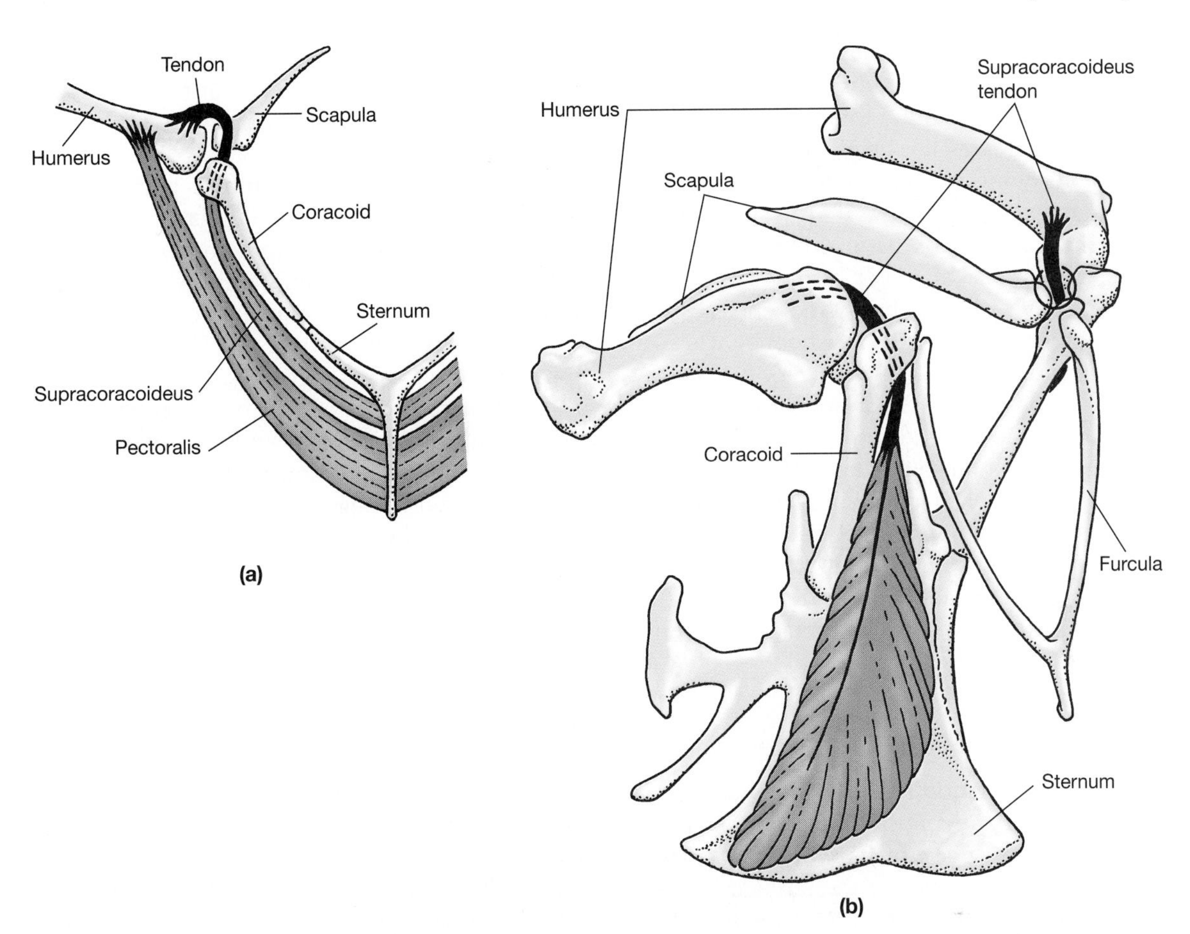

▲ Figure 15–13 Major flight muscles of birds. (a) Cross section through the sternum of a bird showing the relationships of the pectoralis major and supracoracoideus muscles. (b) Frontal and lateral view of the sternum and pectoral girdle of a bird showing the insertion of the supracoracoideus tendon through the foramen triosseum onto the dorsal head of the humerus. The foramen is formed by the articulation of the furcula, coracoid, and scapula.

bird takes off: It needs thrust on both the downstroke and the upstroke. Thrust on the upstroke is produced by bending the wings slightly at the wrists and elbow and by rotating the humerus upward and backward. This movement causes the upper surfaces of the twisted primaries to push against the air and to produce thrust as their lower surfaces did in the downstroke. In this type of flight the wing tip describes a rough figure eight through the air. As speed increases the figure-eight pattern is restricted to the wing tips.

A powered upstroke results mainly from contraction of the **supracoracoideus**, a deep muscle underlying the pectoralis major and attached directly to the keel of the sternum. It inserts on the dorsal head of the humerus by passing through the foramen triosseum, formed where the coracoid, furcula, and scapula join (Figure 15–13). In most species of birds the supracoracoideus is a relatively small, pale muscle with low myoglobin content, easily fatigued. In species that rely on a powered upstroke—for fast, steep takeoffs, for hovering, or for fast aerial pursuit—the supracoracoideus is relatively larger. The ratio of weights of the pectoralis major and the supracoracoideus is a good indication of a bird's reliance on a powered upstroke; such ratios vary from 3:1 to 20:1. The total weight of the flight muscles shows the extent to which a bird depends on powered flight. Strong fliers such as pigeons and falcons have breast muscles comprising more than

20 percent of body weight, whereas in some owls, which have very light wing loading, the flight muscles make up only 10 percent of total weight.

A flying bird increases its speed by increasing the amplitude of its wing beats, but the frequency of wing beats remains nearly constant at all speeds during level flight. Large birds have slower wing-beat frequencies than those of small birds, and strong fliers usually have slower beat frequencies than those of weak fliers. The furcula (wishbone) acts as a spring and a timing device. The frequency of respiration of many birds during flight appears to have a constant relationship to the wing-beat frequency. Some birds, especially those that have low wing-beat frequencies, breathe once per wing-beat cycle. In contrast, birds with high wing-beat frequencies have breathing cycles that span several wing beats. In general, inspiration appears to occur during or at the end of a wing upstroke, with expiration at the end of a downstroke.

Wing Structure and Flight Characteristics

An individual bird can modify the lift it generates by adjusting the camber and angle of attack of its wings, and by changing the speed of its wing beats. To a small degree a bird can change the surface area of the wing by changing the position of its feathers, but most change in wing structure occurs over evolutionary time. Wings may be large or small in relation to body size, resulting in light wing loading or heavy wing loading. They may be long and pointed, short and rounded, highly cambered or relatively flat, and the width and degree of slotting are additional important characteristics.

Depending on whether a bird is primarily a powered flier or a soaring form, the various segments of the wing (hand, forearm, upper arm) are lengthened to different degrees. Hummingbirds have very fast, powerful wing beats, requiring maximum propulsive force from the primaries. The hand bones of hummingbirds are longer than the forearm and upper arm combined. Most of the flight surface is formed by the primaries, and hummingbirds have only six or seven secondaries. Frigate birds are marine species with long, narrow wings specialized for powered flight as well as for gliding and soaring. All three segments of the forelimb are about equal in length. The soaring albatrosses have carried lengthening of the wing to the extreme found in birds: The humerus or upper arm is the longest segment, and there may be as many as 32 secondaries in the inner wing (Figure 15–14).

Ornithologists recognize four structural and functional types of wings (Figure 15–15). Seabirds, particularly those such as albatrosses and shearwaters that rely on **dynamic soaring**, have long, narrow, flat wings lacking slots in the outer primaries. Some albatrosses have aspect ratios of 18:1 and lift-to-drag ratios similar to those of high-performance sailplanes. Dynamic soaring is possible only where there is a pronounced vertical wind gradient, with

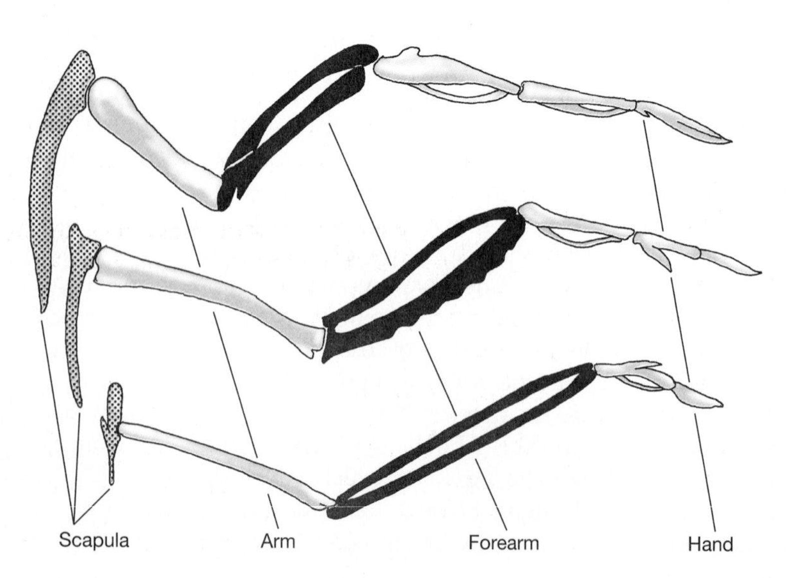

◀ Figure 15–14 Wing proportions. Comparison of the relative lengths of the proximal, middle, and distal elements of the wing bones of a hummingbird (top), frigate bird (middle), and albatross (bottom) drawn to the same size.

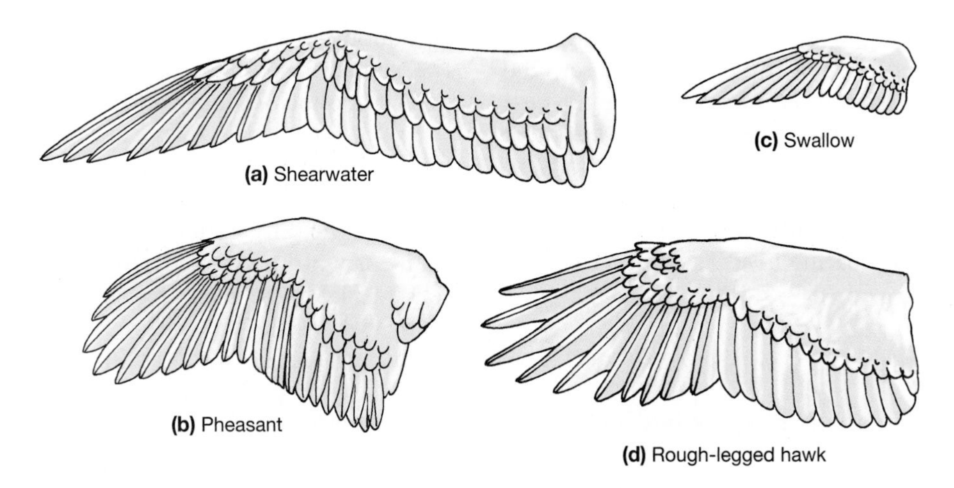

▲ Figure 15–15 Comparison of four basic types of bird wings. (a) Dynamic soaring. (b) Elliptical. (c) High aspect ratio. (d) High lift.

the lower 15 or so meters of air being slowed by friction against the ocean surface. Furthermore, dynamic soaring is feasible only in regions where winds are strong and persistent, such as in the latitudes of the Roaring Forties. This is where most albatrosses and shearwaters are found. Starting from the top of the wind gradient, an albatross glides downwind with great increase in ground speed (kinetic energy). Then, as it nears the surface, it turns and gains altitude while gliding into the wind. Because the bird flies into wind of increasing speed as it rises, its loss of airspeed is not as great as its loss of ground speed. Consequently it does not stall until it has mounted back to the top of the wind gradient, where the air velocity becomes stable. At that point the bird has converted much of its kinetic energy to potential energy, and it turns downwind to repeat the cycle.

Birds that live in forests and woodlands where they must maneuver around obstructions have **elliptical wings**. These wings have a low aspect ratio, tend to be highly cambered, and usually have a high degree of slotting in the outer primaries. These features are generally associated with rapid flapping, slow flight, and a high degree of maneuverability. Although some species with elliptical wings, notably upland gamebirds such as pheasants and grouse, have fast takeoff speeds, they maintain rapid flight only for short distances.

Many birds that are aerial foragers, make long migrations, or have a heavy wing loading that is related to some other aspect of their lives, such as diving, have **high aspect ratio** wings. These wings have a flat profile (little camber), and often lack slots in the outer primaries. In flight they show the swept-back attitude of jet fighter plane wings. All fast-flying birds have converged on this form.

The slotted **high-lift** wing is a fourth type. It is associated with static soaring typified by vultures, eagles, storks, and some other large birds. This wing has an intermediate aspect ratio between the elliptical wing and the high aspect ratio wing, a deep camber, and marked slotting in the primaries. When the bird is in flight, the tips of the primaries turn markedly upward under the influence of air pressure and body weight. Static soarers remain airborne mainly by seeking out and gliding in air masses that are rising at a rate faster than the bird's sinking speed. Hence, a light wing loading and maneuverability (slow forward speed and small turning radius) are advantageous. Broad wings provide the light wing loading, and the highly developed slotting enhances maneuverability by responding to changes in wind currents with changes in the positions of individual feathers instead of movements of the entire wing. A bird cannot soar in tight spirals at high speed, and flying slowly with enough lift to prevent stalling requires a high angle of attack. The deeply slotted primaries apparently make the combination of low speed and high lift possible. The distal, emarginated portion of each primary produces lift by acting as a separate high aspect ratio airfoil set at a high angle of attack. This design reduces induced drag.

In regions where topographic features and meteorological factors provide currents of rising air, static soaring is an energetically cheap mode of flight. By soaring rather than flapping, a large bird the size of a stork can decrease by a factor of 20 or more the energy required for flight per unit of time, whereas the saving is only one-tenth as much for a small bird such as a warbler. It is little wonder, then, that most large land birds perform their annual migrations by soaring and gliding as much of the time as possible, and some condors and vultures cover hundreds of kilometers each day soaring in search of food.

15.7 Body Form and Flight

Many aspects of the morphology of birds appear to have been molded by aerodynamic forces. (See Raikow 1985b and Rayner 1988 for reviews.) Feathers, for example, provide lift and streamlining during flight. Feathers are light, yet they are strong and resilient for their weight. Of course, flying is not the only function of feathers—they also provide the insulation necessary for endothermy, and their colors and shapes function in crypsis and display.

Structural modifications can be seen in several aspects of the structure of birds. The avian skeleton is not lighter in relation to the total body mass of a bird than is the skeleton of a mammal of similar size, but the distribution of mass is different. Many bones are air filled (pnuematized), and the skull is especially light, but the leg bones of birds are heavier than those of mammals. Thus, the total mass of the skeleton of a bird is similar to that of a mammal, but more of a bird's mass is concentrated in its hindlimbs.

Characteristics of some of the organs of birds reduce body mass. For example, birds lack urinary bladders, and most species have only one ovary (the left). The gonads of both male and female birds are usually small; they hypertrophy during the breeding season and regress when breeding has finished.

Power-producing features are equally important components of the ability of birds to fly. The pectoral muscles of a strong flier may account for 20 percent of the total body mass. The power output per unit mass of the pectoralis major of a turtledove during level flight has been estimated to be 10 to 20 times that of most mammalian muscles. Birds have large hearts and high rates of blood flow and complex lungs that use crosscurrent flows of air and blood to maximize gas exchange and to dissipate the heat produced by high levels of muscular activity during flight. The brains of birds are similar in size to the brains of rodents, and the forebrain and cerebellum are well developed. Birds rely heavily on visual information, and the optic lobes are especially large. The sense of smell is not well developed in most birds; the olfactory lobes are correspondingly small.

Streamlining

Birds are the only vertebrates that move fast enough in air for wind resistance and streamlining to be important factors in their lives. Many passerine birds are probably able to fly 50 km per hour or even faster when they must, although their normal cruising speeds are lower. Ducks and geese can fly at 80 or 90 km per hour, and peregrine falcons reach speeds as high as 200 km per hour when they dive on prey. Fast-flying birds have many of the same structural characters as those seen in fast-flying aircraft. Contour feathers make smooth junctions between the wings and the body and often between the head and body as well, eliminating sources of turbulence that would increase wind resistance. The feet are tucked close to the body during flight, further improving streamlining.

At the opposite extreme, some birds are slow fliers. Many of the long-legged, long-necked wading birds, such as spoonbills and flamingos, fall in this category. Their long legs trail behind them as they fly, and their necks are extended. They are far from streamlined, although they may be strong fliers.

Skeleton and Muscles

The hollow, air-filled (pneumatic) bones of birds (Figure 15–16) are probably an ancestral character of the archosaur lineage, not a derived character of birds. Not all birds have pneumatic bones. In general, pneumatization of bones is better developed in large birds than in small ones. Diving birds (penguins, grebes, loons) have little pneumaticity, and the bones of diving ducks are less pneumatic than those of nondivers.

The distribution of pneumaticity among the bones of the skeleton also varies. The skull is pneumatic in nearly all birds, although the kiwi, a flightless bird found in New Zealand, lacks air spaces in the skull. The sternum, pectoral girdle, and humerus are pneumatic and are part of a system of interconnected air sacs that allow a one-way flow of air through the lungs (discussed in the following section). Pneumaticity extends through the rest of the appendicular skeleton of some birds, even into the phalanges.

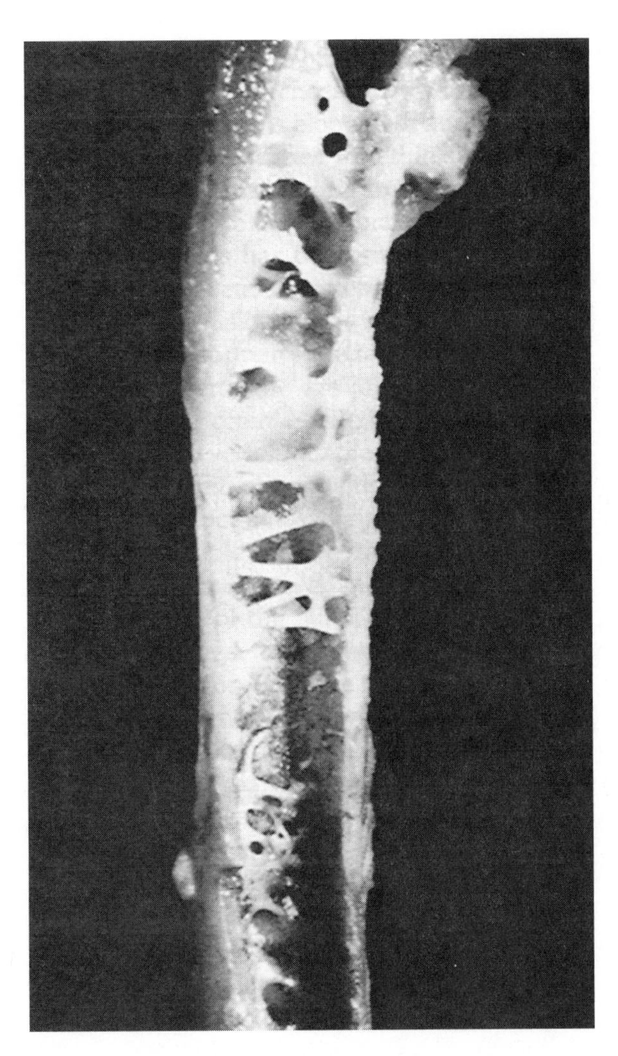

▲ Figure 15–16 Longitudinal slice of the humerus of a bird, showing the hollow core and reinforcing struts.

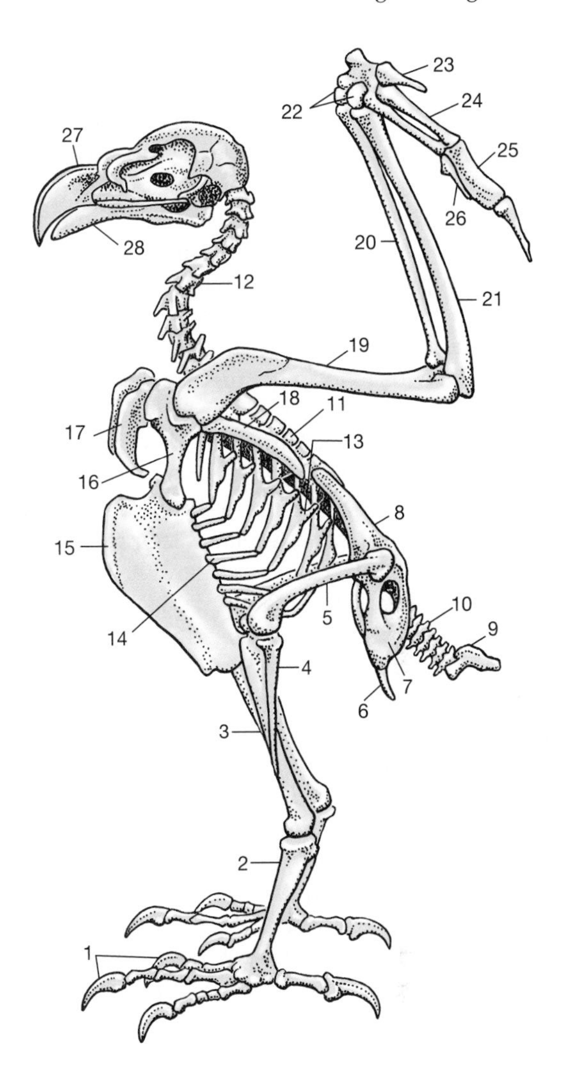

▲ Figure 15–17 Skeleton of an eagle. **1,** Toes; **2,** tarsometatarsus, formed by fusion of the distal tarsals to the metatarsals; **3** and **4,** tibiotarsus, formed by fusion of the calcaneum and astragalus to the end of the tibia **(3)** and fibula **(4)**; **5,** femur; **6,** pubis; **7,** ischium; **8,** ilium; **9,** pygostyle; **10,** caudal vertebrae; **11,** thoracic vertebrae; **12,** cervical vertebrae; **13,** uncinate process of rib; **14,** sternal rib; **15,** sternum; **16,** coracoid; **17,** furcula; **18,** scapula; **19,** humerus; **20,** radius; **21,** ulna; **22,** carpal bones; **23,** first digit; **24,** metacarpal; **25,** second digit; **26,** third digit; **27,** upper jaw; **28,** lower jaw.

Except for the specializations associated with flight, the skeleton of a bird is very much like that of a small dromeosaur (Figure 15–17). The pelvic girdle of birds is elongated, and the ischium and ilium have broadened into thin sheets that are firmly united with a **synsacrum,** which is formed by the fusion of 10 to 23 vertebrae. The long tail of ancestral diapsids has been shortened in birds to about five free caudal vertebrae and a **pygostyle** formed by the fusion of the remaining vertebrae. The pygostyle supports the tail feathers (rectrices). The thoracic vertebrae are joined by strong ligaments that are often ossified. The relatively immobile thoracic vertebrae, the synsacrum, and the pygostyle in combination with the elongated, roof-like pelvis produce a nearly rigid vertebral column. Flexion is possible only in the neck, at the joint between the thoracic vertebrae and the synsacrum, and at the base of the tail. The rigid trunk is balanced on the legs. The femur projects anteriorly, and its articulation with the tibiotarsus and fibula is close to the bird's center of gravity.

The wings are positioned above the center of gravity. The sternum is greatly enlarged compared with other vertebrates, and (except in flightless birds) it bears a keel from which the pectoralis and supracoracoideus muscles originate (Figure 15–13). Strong-flying birds have well-developed keels and large flight muscles. The scapula extends posteriorly above the ribs and is supported by the coracoid, which is fused ventrally to the sternum. Additional bracing is provided by the clavicles, which, in most birds, are fused at their distal ends to form the **furcula** (wishbone).

The relative size of the leg and flight muscles of birds is related to their primary mode of locomotion. Flight muscles comprise 25 to 35 percent of the total body mass of strong fliers such as hummingbirds and swallows. These species have small legs; the leg muscles account for as little as 2 percent of the body mass. Predatory birds such as hawks and owls use their legs to capture prey. In these species the flight muscles make up about 20 percent of the body mass and the limb muscles 10 percent. Swimming birds—ducks and grebes, for example—have an even division between limb and flight muscles; the combined mass of these muscles may be 30 to 60 percent of the total body mass. Birds such as rails, which are primarily terrestrial and run to escape from predators, have limb muscles that are larger than their flight muscles.

Muscle fiber types and metabolic pathways also distinguish running birds from fliers. The familiar distinction between the light meat and dark meat of a chicken reflects those differences. Fowl, especially domestic breeds, rarely fly; but they are capable of walking and running for long periods. The dark color of the leg muscles reveals the presence of myoglobin in the tissues and indicates a high capacity for aerobic metabolism in the limb muscles of these birds. The white muscles of the breast lack myoglobin and have little capacity for aerobic metabolism. The flights of fowl (including wild species such as pheasants, grouse, and quail) are of brief duration and are used primarily to evade predators. The bird uses an explosive takeoff, fueled by anaerobic metabolic pathways, followed by a long glide back to the ground. Birds that are capable of strong sustained flight have dark breast muscles with high aerobic metabolic capacities.

Heart, Lungs, and Gas Exchange

The heart of birds is composed of morphologically distinct left and right atria and ventricles. The separation of the ventricle into left and right halves is an archosaurian character that is seen in crocodilians as well as in birds. However, birds lack the capacity possessed by crocodilians to shunt blood between the pulmonary and systemic circulation. In birds, as in mammals, the blood must flow in series through the two circuits.

The respiratory system of birds is unique among extant vertebrates (Figure 15–18). The air sacs that occupy much of the dorsal part of the body and extend into the pneumatic spaces in many of the bones provide a system in which airflow through the lungs is unidirectional instead of tidal (in and out) as it is in mammalian lungs. The air sacs are poorly vascularized, and gas exchange occurs only in the parabronchial lung. The combined volume of the air sacs is about nine times the volume of the parabronchial lung itself. The flow of air through this extensive respiratory system may help to dissipate the heat generated by high levels of muscular activity during flight.

Air flows through the parabronchial lung in the same direction during both inspiration and exhalation (Figure 15–19). During inspiration the volume of the thorax increases, drawing air through the bronchus and into the posterior thoracic and abdom-

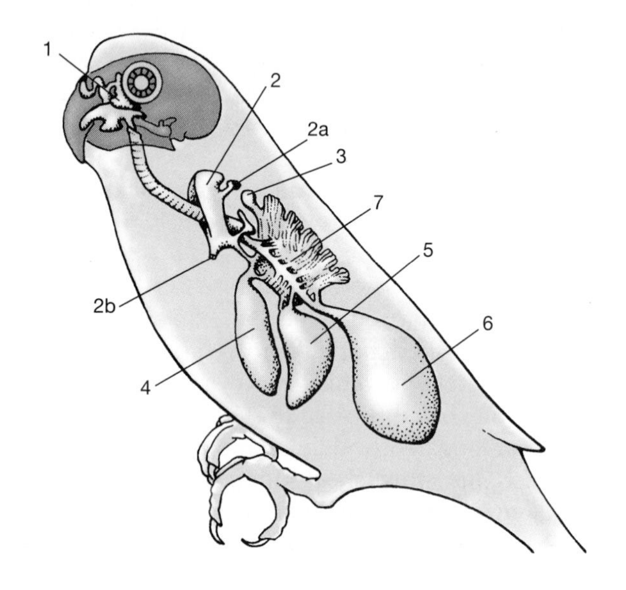

▲ Figure 15–18 The lung and air-sac system of the budgerigar. (Only the left side is shown.) **1,** Infraorbital sinus; **2,** clavicular air sac; **2a,** axillary diverticulum to the humerus; **2b,** sternal diverticulum; **3,** cervical air sac; **4,** cranial thoracic air sac; **5,** caudal thoracic air sac; **6,** abdominal air sacs; **7,** parabronchial lung.

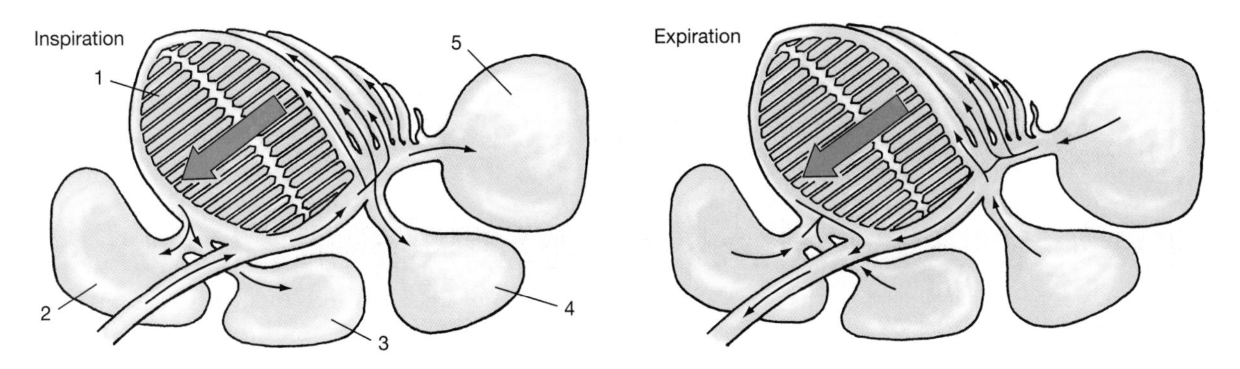

▲ Figure 15–19 Pattern of airflow during inspiration and expiration. Note that air flows through the parabronchial lung during both phases of the respiratory cycle. **1,** Parabronchial lung; **2,** clavicular air sac; **3,** cranial thoracic air sac; **4,** caudal thoracic air sac; **5,** abdominal air sacs.

inal air sacs and the parabronchial lung. Simultaneously, air from the parabronchial lung is drawn into the clavicular and anterior thoracic sacs. On expiration the volume of the thorax decreases, air from the posterior thoracic and abdominal sacs is forced into the parabronchial lung, and air from the clavicular and thoracic sacs is forced out through the bronchus.

Gas exchange takes place in a network of tiny, blind-end air capillaries that intertwine with equally fine blood capillaries. Airflow and blood flow pass in opposite directions, and the capillary pairs compose a crosscurrent exchange system. This arrangement is more effective at exchanging lung gases than is the mammalian lung, and the efficiency of gas exchange in the parabronchial lung may allow birds to breathe at higher altitudes than mammals (Box 15-1). Birds do sometimes penetrate to very high altitudes, either as residents or in flight. For example, radar tracking of migrating birds shows that they sometimes fly as high as 6500m, the alpine chough lives at altitudes around 8200m on Mount Everest, and bar-headed geese pass directly over the summit of the Himalayas at altitudes of 9200m during their migrations.

Vocalization is a second function of the respiratory systems of many vertebrates, including birds. However, the vocal organ of birds, the **syrinx** (plural **syringes**), is unique. The syrinx lies at the base of the trachea where the two bronchi diverge. Several of the cartilaginous rings that support the trachea and bronchi are modified and surrounded by muscles. Among birds, the anatomy of the syrinx varies—some are purely tracheal, some involve both the trachea and the bronchi, and others are purely bronchial. Two membranes are associated with the syrinx, one in the base of each bronchus. Songbirds have five to nine (usually seven) pairs of syringeal muscles, parrots and lyrebirds have three pairs, falcons have two, and most other birds have only a single pair of syringeal muscles. Storks, vultures, ratites, and most pelicans lack syringeal muscles.

The mechanism of avian song production is an area of controversy. One view holds that birds produce sound by vibrating the syringeal membranes, and they modulate sound by changing the vibration of those membranes (Greenewalt 1968, 1969). An alternative hypothesis maintains that pulsatile activity of the abdominal muscles and resonance in the trachea are important in modulating the amplitude and frequencies produced by the syringeal membranes (Gaunt et al. 1982, Nowicki 1987). A recent study suggests that it is not the syringeal membranes that vibrate, but rather the medial and lateral labia (Goller and Larsen 1997).

15.8 Feeding, Digestion, and Excretion

The digestive systems of birds show some differences from those of other vertebrates, although the distinctions are less pronounced than the differences in the respiratory systems. With the specialization of the forelimbs as wings, which largely precludes their having any role in prey capture, birds have concentrated their predatory mechanisms in their beaks and feet. The absence of teeth prevents birds from doing much processing of food in the mouth, and the gastric apparatus takes over some of that role.

BOX 15-1 HIGH-FLYING BIRDS

Did you know that birds regularly fly at altitudes higher than human mountain climbers can ascend without using auxiliary breathing apparatus? This amazing ability of birds to sustain activity at high altitudes is a result of the morphological characteristics of their pulmonary systems.

To fully appreciate the feats of these high-flying birds, we need to consider what factors are in play in the Earth's atmosphere. At the surface of the Earth, the atmosphere is most dense because the entire weight of the atmosphere is pressing down on it. At higher altitudes, the atmosphere becomes less and less dense.

At sea level, atmospheric pressure is 760 millimeters of mercury (760 torr in International System units). The composition of dry air by volume is 79.02 percent nitrogen and other inert gases, 20.94 percent oxygen, and 0.04 percent carbon dioxide. These gases contribute to the total atmospheric pressure in proportion to their abundance, so the contribution of oxygen is 20.94 percent of 760 torr, or 159.16 torr. The pressure exerted by an individual gas is called the partial pressure of that gas. The rate and direction of diffusion of gas between the air in the lungs and the blood in the pulmonary capillaries is determined by the difference in the partial

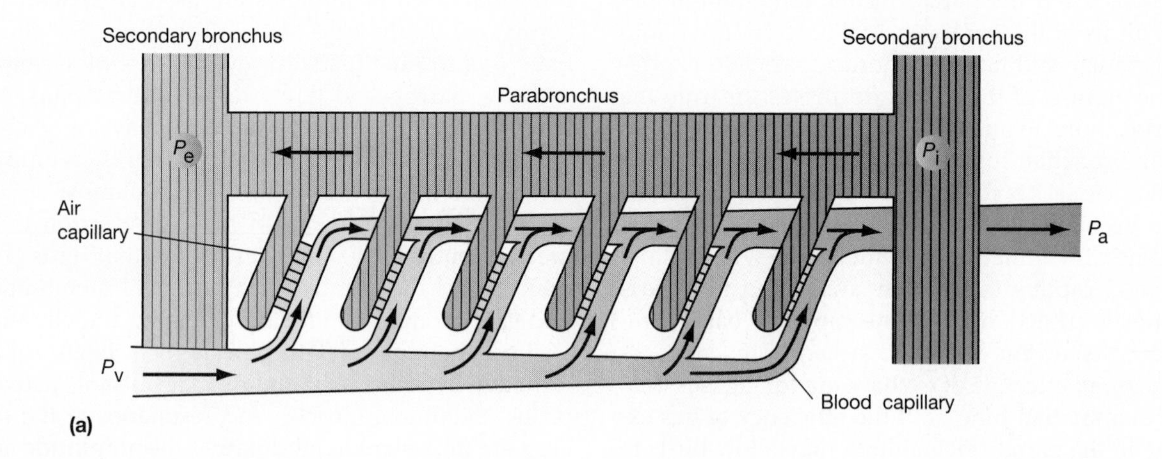

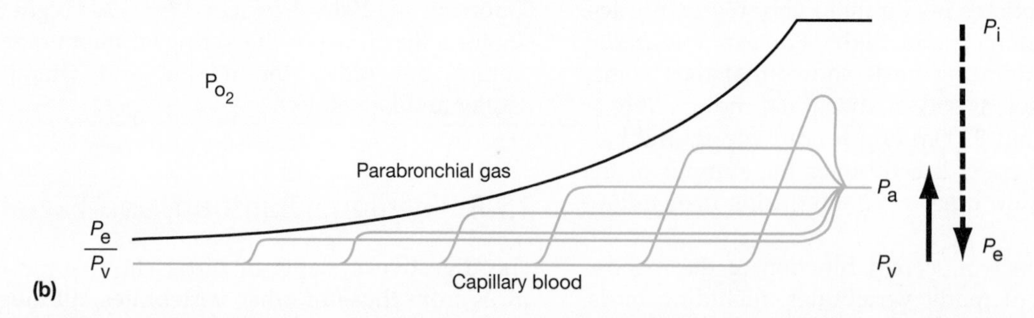

▲ Figure 15–20 Diagram of gas exchange in a cross-current lung. Air flows from right to left in this diagram and blood flows from left to right. P_e = oxygen pressure in the air exiting the parabronchus, P_v = oxygen pressure in the mixed venous blood entering the blood capillaries, P_a = oxygen pressure in the blood leaving the blood capillaries, P_i = oxygen pressure in the air entering the parabronchus. Top: General pattern of air and blood flow through the parabronchial lung. Bottom: Diagrammatic representation of cross-current gas exchange.

BOX 15-1 High-Flying Birds (continued)

pressures of the gas in the blood and in the lungs. Oxygen diffuses from air in the lungs into blood in the pulmonary capillaries because oxygen has a higher partial pressure in the air than in the blood, whereas carbon dioxide diffuses in the opposite direction because its partial pressure is higher in blood than in air.

At higher altitudes, the atmospheric pressure is lower. At 7700 meters, the atmospheric pressure is only 282 torr, and the partial pressure of oxygen in dry air is about 59 torr. Because of the low atmospheric pressure at this altitude, the driving force for diffusion of oxygen into the blood is small. (The actual pressure differential is reduced even below this figure because air in the lungs is saturated with water, and water vapor contributes to the total pressure in the lungs. The vapor pressure of water at 37°C is 47 torr. Thus, the partial pressure of oxygen in the lungs is 20.94 percent × (282 – 47 torr = 49 torr).

How does all this affect breathing? In mammals, it makes breathing at high altitudes difficult. The tidal ventilation pattern of the lungs of mammals means that the partial pressure of oxygen in the pulmonary capillaries can never be higher than the partial pressure of oxygen in the expired air. The best that a tidal ventilation system can accomplish is to equilibrate the partial pressures of oxygen in the pulmonary air and in the pulmonary circulation. In fact, failure to achieve complete mixing of the gas within the pulmonary system means that oxygen exchange falls short even of this equilibration, and blood leaves the lungs with a partial pressure of oxygen slightly lower than the partial pressure of oxygen in the exhaled air.

In birds, breathing is a different process. The cross-current blood flow system in their parabronchial lungs ensures that the gases in the air capillaries repeatedly encounter a new supply of deoxygenated blood (Figure 15–20, left side of diagram). When blood enters the system (on the left side of the diagram), it has the low oxygen pressure of mixed venous blood. The blood entering the leftmost capillary is exposed to air that has already had much of its oxygen removed farther upstream. Nonetheless, the low oxygen pressure of the mixed venous blood ensures that even in this part of the parabronchus, oxygen uptake can occur. Blood flowing through capillaries farther to the right in the diagram is exposed to higher partial pressures of oxygen in the parabronchial gas and takes up correspondingly more oxygen. The oxygen pressure of the blood that flows out of the lungs is the result of mixing of blood from all the capillaries. The oxygen pressure of the mixed arterial blood is higher than the partial pressure of oxygen in the exhaled air. As a result, birds are more effective than mammals at oxygenating their blood at high altitudes.

The Esophagus and Crop

Birds often gather more food than they can process in a short period, and the excess is held in the esophagus. Many birds have a **crop**, an enlarged portion of the esophagus that is specialized for temporary storage of food. The crop of some birds is a simple expansion of the esophagus, whereas in others it is a unilobed or bilobed structure (Figure 15–21). An additional function of the crop is transportation of food for nestlings. When the adult returns to the nest it regurgitates the material from the crop and feeds it to the young. In doves and pigeons, the crop of both sexes produces a nutritive fluid (crop milk) that is fed to the young. The milk is produced by fat-laden cells that detach from the squamous epithelium of the crop and are suspended in an aqueous fluid. Crop milk is rich in lipids and proteins, but contains no sugar. Its chemical composition is similar to that of mammalian milk, although it differs in containing intact cells. The proliferation of the crop epithelium and the formation of crop milk is stimulated by prolactin, as is the lactation of mammals. A nutritive fluid produced in the esophagus is fed to hatchlings by greater flamingos, the emperor penguin, and other species, especially those that eat seeds such as crossbills and the gray jay.

The hoatzin (*Opisthocomus hoazin*), a South American bird, is the only avian species known to employ foregut fermentation (Figure 15–21f). Hoatzins are herbivorous; green leaves make up more than

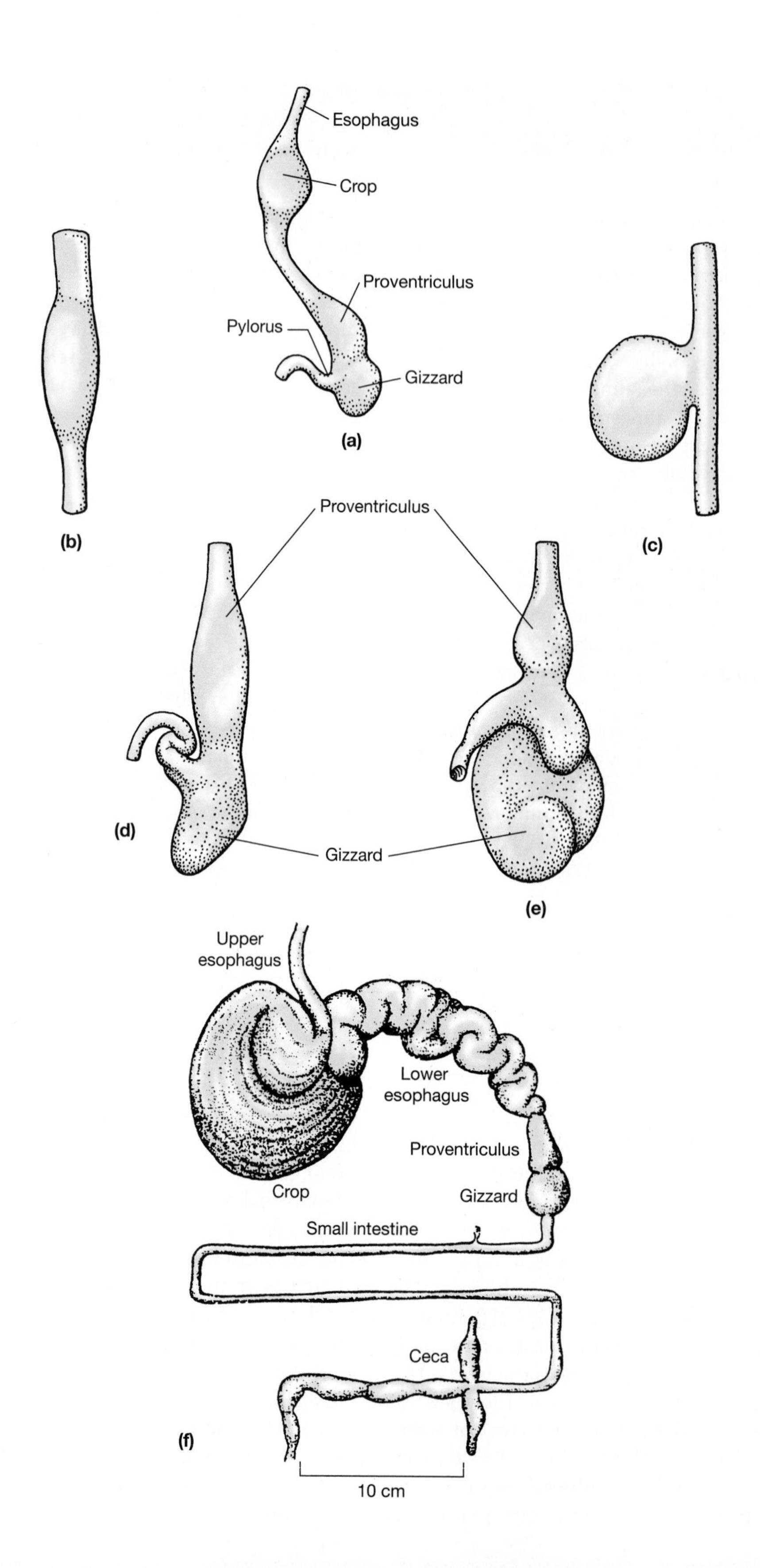
Esophagus
Crop
Proventriculus
Pylorus
Gizzard
(a)
(b)
(c)
Proventriculus
(d)
Gizzard
(e)
Upper
esophagus
Lower
esophagus
Proventriculus
Crop
Gizzard
Small intestine
Ceca
(f)
10 cm

80 percent of the diet. More than a century ago, naturalists observed that hoatzins smell like fresh cow manure, and a study of the crop and lower esophagus revealed that volatile fatty acids are the source of the odor (Grajal et al. 1989). Volatile fatty acids produced by the process of fermentation are absorbed by the gut. Bacteria and protozoa like those found in the rumen of cows break down the plant cell walls, and bacterial extracts from the hoatzin's crop are as effective as those from a cow's rumen in digesting plant material.

The Gastric Apparatus

The form of the stomachs of birds is related to their dietary habits. Carnivorous and piscivorous (fish-eating) birds need expansible storage areas to accommodate large volumes of soft food, whereas birds that eat insects or seeds require a muscular organ that can contribute to the mechanical breakdown of food. Usually, the gastric apparatus of birds consists of two relatively distinct chambers, an anterior glandular stomach (**proventriculus**) and a posterior muscular stomach (**gizzard**; see Figure 15–21a). The proventriculus contains glands that secrete acid and digestive enzymes. The proventriculus is especially large in species that swallow large items such as intact fruit (Figure 15–21d, e).

The gizzard has several functions, including food storage while the chemical digestion that was begun in the proventriculus continues, but its most important function is the mechanical processing of food. The thick, muscular walls of the gizzard squeeze the contents, and small stones that are held in the gizzards of many birds help to grind the food. In this sense the gizzard is performing the same function as a mammal's teeth. The pressure that can be exerted on food in the gizzard is intense. A turkey's gizzard can grind up two dozen walnuts in as little as 4 hours, and it can crack hickory nuts that require 50 to 150 kilograms of pressure to break.

The Intestine, Ceca, and Cloaca

The small intestine is the principal site of chemical digestion, as enzymes from the pancreas and intestine break down the food into small molecules that can be absorbed across the intestinal wall. The mucosa of the small intestine is modified into a series of folds, lamellae, and villi that increase its surface area. The large intestine is relatively short, usually less than 10 percent of the length of the small intestine. Passage of food through the intestines of birds is quite rapid: Transit times for carnivorous and fruit-eating species are in the range of a few minutes to a few hours. Passage of food is slower in herbivores and may require a full day. Birds generally have a pair of ceca at the junction of the small and large intestines. The ceca are small in carnivorous, insectivorous, and seed-eating species, but they are large in herbivorous and omnivorous species such as cranes, fowl, ducks, geese, and the ostrich. Symbiotic microorganisms in the ceca apparently ferment plant material.

The cloaca temporarily stores waste products while water is being reabsorbed. The precipitation of uric acid in the form of urate salts frees water from the urine, and this water is returned to the bloodstream. Species of birds that have salt-secreting glands can accomplish further conservation of water by reabsorbing some of the ions that are in solution in the cloaca and excreting them in more concentrated solutions through the salt glands (Chapter 4). The mixture of white urate salts and dark fecal material that is voided by birds is familiar to anyone who has washed an automobile.

15.9 The Hindlimbs and Locomotion

Unlike most tetrapods, birds usually are specialized for two or more different modes of locomotion: bipedal walking or swimming with the hindlimbs and flying with the forelimbs.

◀ Figure 15–21 Anterior digestive tract of birds. (a) The relationship among the parts. The relative sizes of the proventriculus and gizzard vary in relation to diet. Carnivorous and fish-eating birds like the great cormorant have a relatively small crop (b) and gizzard (c), whereas seed eaters and omnivores like the peafowl have a large crop (d) and muscular gizzard (e). (f) The digestive tract of the hoatzin. The muscular crop and the anterior esophagus are greatly enlarged. Cornified ridges on the inner surface of the crop probably grind the contents of the crop, reducing the particle size. This process is analagous to a cow chewing its cud, but has the advantage that mechanical processing and fermentation occur in the same structure.

Walking, Hopping, and Perching

Terrestrial locomotion may involve walking or running, supporting heavy bodies, hopping, perching, climbing, wading in shallow water, or supporting the body on insubstantial surfaces such as snow or floating vegetation. We will consider cursorial adaptations first, because birds evolved from bipedal dinosaurs and because the principles of cursorial adaptation have been well worked out in the quadrupedal mammals. Modifications usually associated with running in quadrupeds are (1) a progressive increase in the lengths of the distal limb elements relative to the proximal ones, (2) a decrease in the area of the foot surface that makes contact with the ground, and (3) a reduction in the number of toes. All three of these cursorial trends are expressed to varying degrees in running birds; however, problems of balance are more critical for bipeds than for quadrupeds, and these problems may have restricted the evolution among bipeds of some of the cursorial adaptations found in quadrupeds.

Because the center of gravity must lie over the feet of a biped to maintain balance, a reduction in the surface area in contact with the ground can be achieved only by some sacrifice in stability. No bird has reduced the length and number of toes in contact with the ground to the extent that the hoofed mammals have; but the large, fast-running ostrich has only two toes on each foot, and many other cursorial species have only the three forward-directed toes in contact with the ground (Figure 15–22).

Weight-bearing characteristics such as large, heavy leg bones arranged in vertical columns as supports of great body mass are well known in large mammals such as elephants. No surviving birds show specializations of this kind, but these features are seen in some of the large, flightless terrestrial birds of earlier times. The extinct elephantbirds (Aepyornithiformes) of Madagascar and moas (Dinornithiformes) of New Zealand were herbivores that evolved on oceanic islands in the absence of large carnivorous mammals and survived there until contact with humans in the post-Pleistocene period (Figure 15–23). The giant, carnivorous *Diatryma* (Gruiformes) and related species were successful continental forms in the Americas and Europe (see Chapter 16).

Hopping, a succession of jumps with the feet moving together, is a special form of locomotion found mostly in perching, arboreal birds. It is most highly developed in the passerines, and only a few nonpasserine birds regularly hop. Many passerines cannot walk, and hopping is their only mode of terrestrial locomotion. Some groups of passerines have a more terrestrial mode of existence; these birds are able to walk as well as hop (larks, pipits, starlings, and grackles are examples). The separation between walking and hopping passerines cuts across families. For example, in the family Corvidae, the ravens,

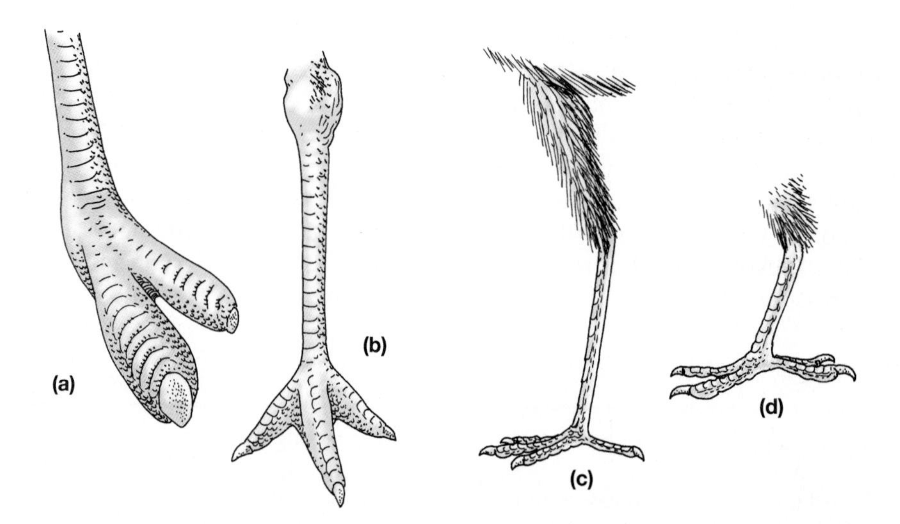

▲ Figure 15–22 Avian feet with various specializations for terrestrial locomotion. (a) Ostrich, with only two toes. (b) Rhea, with three toes. (c) Secretary bird, with a typical avian foot. (d) Roadrunner, with zygodactyl foot. (Not drawn to scale.)

▲ Figure 15–23 Large, flightless, terrestrial birds. (a) The extinct elephantbird of Madagascar and (b) the extinct predatory bird *Diatryma*, compared with (c) the cursorial ostrich.

crows, and rooks are walkers, whereas the jays and magpies are hoppers.

The most specialized avian foot for perching on branches is one in which all four toes are free and mobile, of moderate length, and with the hind toe well developed, lying in the same plane as the forward three, and opposable to them (**ansiodactyl**). Such a foot produces a firm grip and is highly developed in the passerine birds. The **zygodactylous** condition, with two toes forward and opposable to two extending backward is characteristic of birds such as parrots and woodpeckers, which climb or perch on vertical surfaces.

The tendons that flex the toes of a perching bird can lock the foot in a tight grip, so that the bird does not fall off its perch when it relaxes or goes to sleep. The plantar tendons, which insert on the individual phalanges of the toes, slip in grooves and sheaths that are positioned anterior to the knee joint and posterior to the ankle joint in such a way that the weight of the resting bird tightens the tendons and curls the toes around the perch. Thus, muscular contraction is not required to hold the toes closed. Furthermore, the tendons lying underneath the toe bones have hundreds of minute, rigid, hobnail-like projections that mesh with ridges on the inside surface of the surrounding tendon sheath. The projections and ridges lock the tendons in place in the sheaths and help to hold the toes in their grip around the branch.

Climbing

Birds climb on tree trunks or other vertical surfaces by using their feet, tails, beaks, and, rarely, their forelimbs. Several distantly related groups of birds have independently acquired specializations for climbing and foraging on vertical tree trunks. Species such as woodpeckers and woodcreepers, which use their toes as supports, begin foraging near the base of a tree trunk and work their way vertically upward, head first, clinging to the bark with strong feet on short legs. The tail is used as a prop to brace the body against the powerful pecking exertions of head and neck, and the pygostyle and free caudal vertebrae in these species are much enlarged and support strong, stiff tail feathers. A similar modification of the tail is found in certain swifts that perch on cave walls and inside chimneys.

Nuthatches and similarly modified birds climb on trunks and rock walls in both head-upward and head-downward directions while foraging, and in these species, which do not use their tails for support, the claw on the hallux is larger than those on the forward-directed toes, and is strongly curved.

Although a few nestling birds use their wings for support as they move around the nest, only the young hoatzin of South America has evolved a special modification of the forelimbs for climbing. Hoatzins take a long time to begin to fly—60 to 70 days. Until they can fly, they clamber through vegetation using large claws, on the first and second digits of the wing, that are moved by special muscles. Later in life the claws fall off, and the wing of the adult assumes a typical avian condition, but even the adults are weak fliers. They continue to use their wings to help in climbing among the dense branches of their tropical, swampy habitat. These characteristics of hoatzins may be related to their herbivorous diet. The large crop associated with foregut fermentation of plant material takes up a great deal of space in the trunk, and the sternum is reduced. As a result, there is less area for attachment of flight muscles than in other birds of the same size.

Swimming on the Surface

Although no birds have become fully aquatic like the ichthyosaurs and cetaceans, nearly 400 species are specialized for swimming. Nearly half of these aquatic species also dive and swim underwater.

Modifications of the hindlimbs are the most obvious avian specializations for swimming. Other changes include a wide body that increases stability in water, dense plumage that provides buoyancy and insulation, a large preen gland, producing oil that waterproofs the plumage, and structural modifications of the body feathers that retard penetration of water to the skin. The legs are at the rear of a bird's body, where the mass of leg muscles interferes least with streamlining and where the best control of steering can be achieved.

The feet of aquatic birds are either webbed or lobed (Figure 15–24). Webbing between the three forward toes (palmate webbing) has been independently acquired at least four times in the course of avian evolution. Totipalmate webbing of all four toes is found in pelicans and their relatives.

Lobes on the toes have evolved convergently in several phylogenetic lines of aquatic birds. There are two different types of lobed feet. Grebes are unique in that the lobes on the outer sides of the toes are rigid and do not fold back as the foot moves forward. A grebe rotates its foot 90 degrees so that the inner side points forward, the toes with their lobes slicing through the water like knife blades for an efficient recovery stroke with minimum drag. A simpler mechanism for the recovery stroke occurs in all the other lobe-footed swimmers, where the lobes are flaps that fold back against the toes during forward movement through water and flare open to present a maximum surface on the backward stroke.

Diving and Swimming Underwater

The transition from a surface-swimming bird to a subsurface swimmer has occurred in two fundamentally different ways: either by further specialization of a

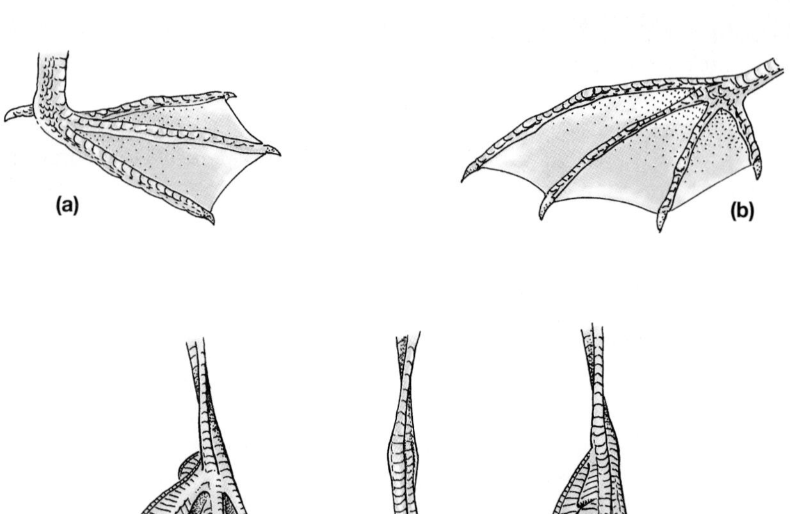

◀ Figure 15–24 Webbed and lobed feet of some aquatic birds. (a) Duck, showing partial webbing. (b) Cormorant, showing totipalmate webbing. The lobed foot of a grebe, showing how it is rotated during a stroke: (c) position of toes during backward power stroke in side view; (d) front view; (e) side view of the rotated foot during the forward recovery stroke.

hindlimb already adapted for swimming or by modification of the wing for use as a flipper under water. Highly specialized foot-propelled divers have evolved independently in grebes, cormorants, loons, and the extinct Hesperornithidae. All these families except the loons include some flightless forms (Figure 15–25). Wing-propelled divers have evolved in the Procellariiformes (the diving petrels), the Sphenisciformes (penguins), and the Charadriiformes (auks and related forms). Only among the waterfowl are there both foot-propelled and wing-propelled diving ducks, but none of these species is as highly modified for diving as specialists like the loons or auks. The water ouzels or dippers (*Cinclus*) are passerine birds that dive and swim underwater with great facility using their small, round wings, but they lack any other morphological specializations.

Other morphological, behavioral, and physiological modifications are important in the evolution of diving birds, such as buoyancy and oxidative metabolism under water. Diving birds overcome buoyancy by reducing the volume of their air sacs, having bones with reduced pneumaticity, expelling air from their plumage before submerging, and in the case of some penguins, by swallowing small stones that act as ballast. Like other diving air-breathers, birds constrict their peripheral blood flow, reduce the heart rate, and lower their metabolic rate while underwater. They also tend to have a large blood volume with a high oxygen-carrying capacity and muscles that are especially rich in myoglobin. They are able to tolerate high carbon dioxide levels in blood, and they can obtain considerable energy from anaerobic metabolism. These features allow diving times from 1 to 3 minutes, with a maximum recorded survival time of 15 minutes.

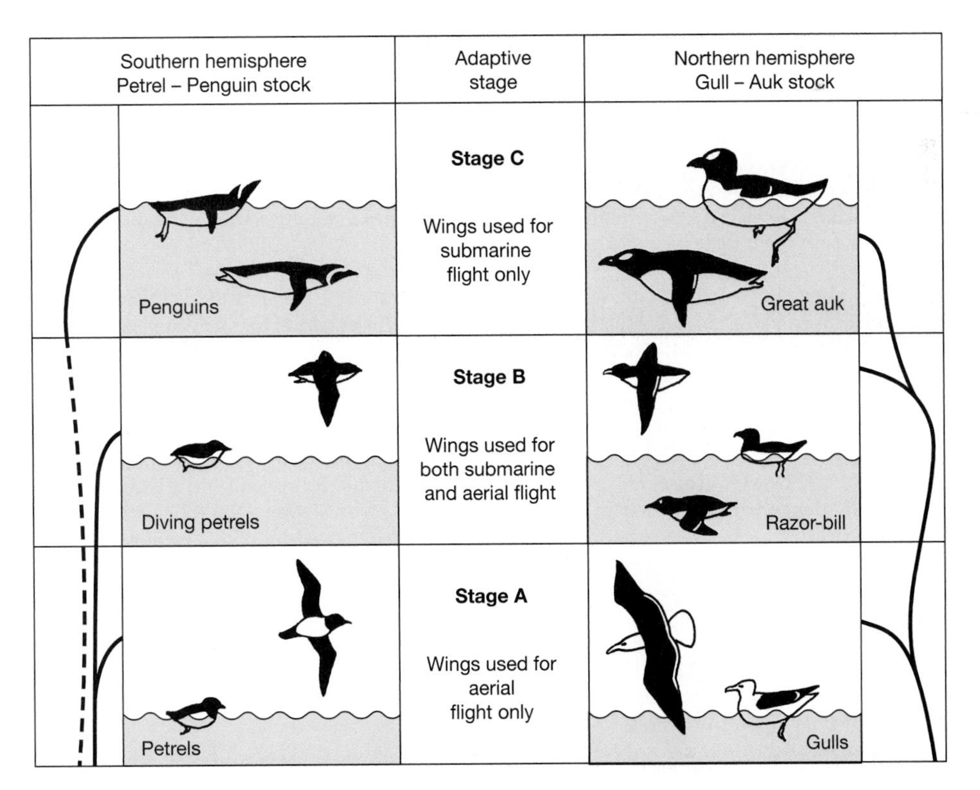

▲ Figure 15–25 Parallel evolution of swimming and diving birds in the Southern Hemisphere (Procellariiformes and Sphenisciformes) and the Northern Hemisphere (Charadriiformes).

Summary

Feathers are the distinguishing character of birds, and flight is the distinctive mode of avian locomotion. In many respects the morphology of birds is shaped by the demands of flight. Flapping flight is a more complicated process than flight with fixed wings like those of aircraft, but it can be understood in aerodynamic terms. Feathers compose the aerodynamic surfaces responsible for lift and propulsion during flight, and they also provide streamlining. There are many variations and specializations within the four basic types of wings, but in general high speed and elliptical wings are used in flapping flight, whereas high aspect ratio and high-lift wings are used for soaring and gliding. The hollow bones of birds, probably an ancestral character of the archosaur lineage, combine lightness and strength. The air sacs, some of which fill several of the pneumatic bones, create a one-way flow of air through the lung that enhances oxygen uptake. The air sacs probably also help to dissipate the heat produced by muscles during flight.

Some parts of the skeletons of birds are light in relation to the sizes of their bodies, and some of this lightness has been achieved by modification of the skull. Several skull bones that are separate in other diapsids are fused in birds, air sacs are extensively developed, and teeth are absent. The function of teeth in processing food has been taken over by the muscular gizzard, which is part of the stomach of birds. The gizzard is well developed in birds that eat hard items such as seeds; it may contain stones that probably assist in grinding food.

The basic design of the bird foot is very similar to the feet of dromeosaurs. Modifications of the feet are related to the life-style of a species—many flightless birds have reduced the number of toes, and aquatic birds have developed lobes on the toes—or webs between the toes—to increase the surface area acting on the water.

Additional Readings

Bang, B. G., and B. M. Wenzel. 1985. Nasal cavity and olfactory system. In A. S. King and J. McLelland (Eds.), *Form and Function in Birds,* vol. 3, London: Academic, pp. 195–225.

Chen, P., Z. Dong, and S. Zhen. 1998. An exceptionally well-preserved theropod dinosaur from the Yixian Formation of China. *Nature* 391:147–152.

Cracraft, J. A. 1986. The origin and early diversification of birds. *Paleobiology* 12:383–399.

Feduccia, A. 1996. *The Origin and Evolution of Birds.* New Haven, CT: Yale University Press.

Feduccia, A., and R. Wild. 1993. Birdlike characters in the Triassic archosaur *Megalancosaurus. Naturwissenshaften* 80: 564–566.

Gaunt, A. S., S. L. L. Gaunt, and R. M. Casey. 1982. Syringeal mechanics reassessed: Evidence from *Streptopelia. Auk* 99: 474–494.

Goller, F. and O. N. Larsen. 1997. A new mechanism of sound generation in songbirds. *Proceedings of the National Academy of Sciences, USA.* 94:14787–14791.

Grajal, A. S., et al. 1989. Foregut fermentation in the hoatzin, a Neotropical leaf-eating bird. *Science* 245:1236–1238.

Greenewalt, C. H. 1968. *Bird Song: Acoustics and Physiology.* Washington, D.C.: Smithsonian Institution Press.

Greenewalt, C. H. 1969. How birds sing. *Scientific American* 221:126–139.

Houde, P. 1986. Ostrich ancestors found in the Northern Hemisphere suggest new hypothesis of ratite origins. *Nature* 324:563–565.

Houde, P. 1987. Critical evaluation of DNA hybridization studies in avian systematics. *Auk* 104:17–32.

Jenkins, F. A., Jr. 1993. The evolution of the avian shoulder joint. *American Journal of Science* 293-A:253–267.

Maderson, P. F. A. and D. G. Homberger (Eds.). 2000. Evolutionary origin of feathers. *American Zoologist* 40:455–706.

Martin, L. D. 1983a. The origin and early radiation of birds. In A. H. Brush and G A. Clark, Jr. (Eds.). *Perspectives in Ornithology,* Cambridge, UK: Cambridge University Press, pp. 291–338.

Martin, L. D. 1983b. The origin of birds and of avian flight. In Richard F. Johnston (Ed.). *Current Ornithology,* vol. 1, New York, NY: Plenum, 105–129.

Mindell, D. P. (Ed.) 1997. *Avian Molecular Evolution and Systematics.* San Diego, CA: Academic Press.

Norberg, U. M. 1985. Flying, gliding, and soaring. In M. Hildebrand, et al. (Eds.), *Functional Vertebrate Morphology.* Cambridge: Harvard University Press.

Norell, M. A., and L. M. Chiappe. 1996. Flight from reason. (Review of *The Origin and Evolution of Birds* by Alan Feduccia). *Nature* 384:230.

Norell, M. A. and J. A. Clarke. 2001. Fossil that fills a critical gap in avian evolution. *Nature* 409: 181–184.

Nowicki, S. 1987. Vocal tract resonances in oscine bird sound production: evidence from birdsongs in a helium atmosphere. *Nature* 325:53–55.

Olson, S. L. 1985. The fossil record of birds. In D. S. Farner, J. R. King, and K. C. Parkes (Eds.). *Avian Biology*, vol. 8. Orlando, FL: Academic.

Ostrom, J. H. 1974. *Archaeopteryx* and the evolution of flight. *Quarterly Review of Biology* 49:27–47.

Padian, K. and L. Chiappe. 1998. The origin of birds and their flight. *Scientific American* 278(2):38–47.

Qiang, J., P. J. Currie, M. A. Norell, and J. Shu-An. 1998. Two feathered dinosaurs from northeastern China. *Nature* 395:753–761.

Qiang, J., M. A. Norrell, K.-Q. Gao, S.-A. Ji, and D. Ren. 2001. The distribution of integumentary structures in a feathered dinosaur. *Nature* 410:1084–1088.

Raikow, R. J. 1985a. Problems in avian classification. In Richard F. Johnston (Ed.). *Current Ornithology*, vol. 2, New York, NY: Plenum.

Raikow, R. J. 1985b. Locomotor system. In A. S. King and J. McLelland (Eds.), *Form and Function in Birds*, vol. 3. London: Academic, 57–147.

Rayner, J. M. V. 1988. Form and function in avian flight. In R. J. Johnston (Ed.), *Current Ornithology*, vol. 5. New York: Plenum, 1–66.

Ruben, J. 1991. Reptilian physiology and the flight capacity of *Archaeopteryx*. *Evolution* 45:1–17.

Ruben, J. 1993. Powered flight in *Archaeopteryx*: response to Speakman. *Evolution* 47:935–938.

Sibley, C. G., and J. E. Ahlquist. 1990. *Phylogeny and Classification of Birds*. New Haven, CT: Yale University Press.

Sibley, C. G., J. E. Ahlquist, and B. L. Monroe. 1988. A classification of the living birds of the world, based on DNA–DNA hybridization studies. *Auk* 105:409–423.

Stokstad, E. 2001. Exquisite Chinese fossils add new pages to book of life. *Science* 291:232–236.

Sues, H.-D. 2001. Ruffling feathers. *Nature* 410:1036–1037.

Web Explorations

On-line resources for this chapter are on the World Wide Web at: http://www.prenhall.com/pough. Click on the table of contents link and then select Chapter 15.

CHAPTER 16

The Ecology and Behavior of Birds

In the preceding chapter we examined the functional relationships among structural characteristics of birds and the physical and biological requirements of flight. In this chapter we consider some of the consequences of flight for the biology of birds. We have stressed mobility as a basic characteristic of vertebrates, and the ability to fly gives birds extraordinary mobility. In addition, birds are conspicuous animals. Many species are colorful and vocal, and they are active during the day so they are relatively easy to observe. As a result, studies of birds provide the basis for much of our understanding of the ecology and behavior of vertebrates.

Flight gives birds a degree of mobility that terrestrial animals lack. Birds can fly across oceans, and their respiratory systems are so effective that even high mountains are not barriers. The ability of birds to make long-distance movements is displayed most dramatically in their migrations. Even small species like hummingbirds travel thousands of kilometers between their summer and winter ranges. Birds use a variety of methods to navigate on these trips, including orienting by the Sun and the stars and probably using a magnetic sense. Migratory routes must change as environments change. The retreat of continental glaciers that occurred between 15,000 and 10,000 years ago caused continent-wide changes in vegetation, and some of the migratory routes birds follow now are no older than that.

A second characteristic of birds is diurnality; most species are active only by day. Most birds have excellent vision, and colors play important roles in their lives. Because humans, too, are diurnal and visually oriented, birds have been popular subjects for behavioral and ecological field studies. Two important areas of modern biology, optimal foraging theory and the behavioral ecology of mating systems, have drawn heavily on studies of birds for data that can be generalized to other vertebrates. Both types of studies emphasize the role of behavior in enhancing the fitness of individuals. The study of reproductive biology, in particular, includes examples in which individual fitness may be maximized by cooperating with other individuals. These altruistic behaviors are normally directed toward relatives of the individual engaging in the behavior.

16.1 Birds as Model Organisms

Birds might almost have been designed as the ideal vertebrate animals for biologists to study. They are diverse (about 9100 species), widespread, conspicuous, and largely diurnal. Most birds are visually oriented, and respond to stimuli such as colors, patterns, and movements that humans also are able to perceive. A worldwide corps of amateur ornithologists has helped to assemble the huge amount of information we have about the life history and population ecology of birds.

Studies of birds have contributed to our understanding of vertebrates, especially in areas such as ecology, morphology, and behavior (Konishi et al. 1989). These topics reveal fascinating examples of the interdependence of structure and function that emphasize the importance of broadly integrative studies in organismal biology. In this chapter we focus on aspects of feeding, reproduction, and behavior.

Divergence and Convergence in Feeding

Birds show morphological, physiological, and behavioral specializations associated with feeding on diverse sources of food. Modifications of the beak and tongue are often associated with dietary specializations.

Beaks and Tongues The presence of a horny beak in place of teeth is not unique to birds—we have noted the same phenomenon in turtles, rhynchosaurs, dinosaurs, pterosaurs, and the dicynodonts. But the diversity of beaks among birds is remarkable. The range of morphological specializations of beaks defies complete description, but some categories can be recognized (Figure 16–1).

Insectivorous birds such as warblers—which find their food on leaf surfaces—usually have short, thin, pointed bills that are adept at seizing insects. Aerial sweepers such as swifts, swallows, and nighthawks—which catch their prey on the wing—have short, weak beaks and a wide gape. Kinesis of the lower jaw substantially increases the gape of the nightjar; the distance between the two rami of the lower jaw increases from 12.5 millimeters when the mouth is closed to 40 millimeters when it is opened. Stiff feathers at the corners of the mouth further increase the insect-trapping area.

Many carnivorous birds, such as gulls, ravens, crows, and roadrunners, use their heavy pointed beaks to kill their prey. But most hawks, owls, and eagles kill prey with their talons and employ their beaks to tear off pieces small enough to swallow. Falcons stun prey with the impact of their dive, and then bite the neck of the prey to disarticulate the cervical vertebrae. The true falcons (the genus *Falco*) have a structure called a tomial tooth that aids in the process. The tomial tooth actually is a sharp projection from the upper bill that matches a corresponding notch on the bottom bill. Shrikes, a group of predatory passerine birds that employ neck biting to kill prey, also have a tomial tooth. Fish-eating birds such as cormorants and pelicans have beaks with a sharply hooked tip that is used to seize fish, while mergansers have long, narrow bills with a series of serrations along the sides of the beak, in addition to a hook at the tip. Darters and anhingas have harpoon-like bills that they use to impale fishes. The massive beak and jaw apparatus of the gigantic predatory bird *Diatryma* may have been used to kill and dismember prey (Box 16-1).

Spoonbills have flattened bills with broad tips, which they use to create currents that lift prey into the water column where it can be seized. Many

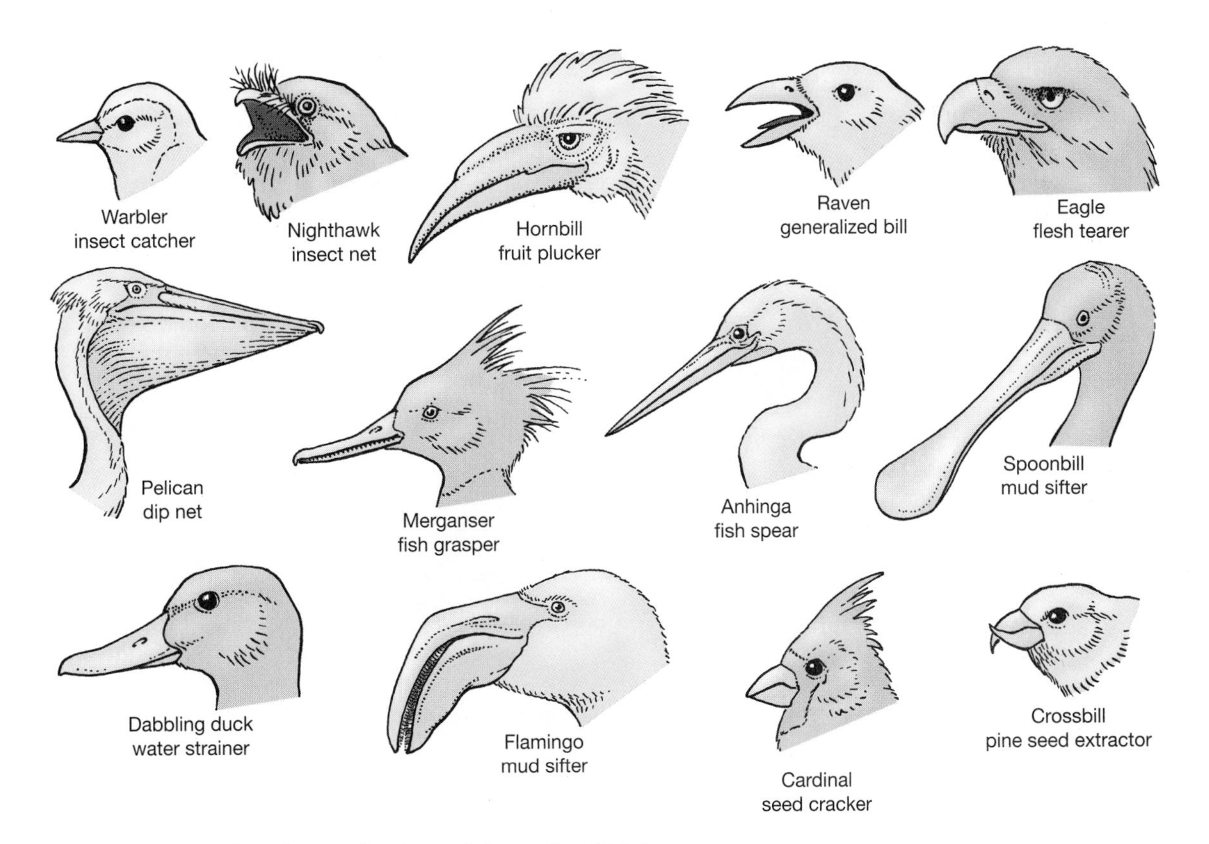

▲ Figure 16–1 Examples of specializations of the beaks of birds.

BOX 16-1 Giant Predatory Birds

We do not usually think of birds as being frightening, although some eagles have wingspans of 2 meters or more. However, when dinosaurs became extinct at the end of the Mesozoic the adaptive zone for bipedal carnivores was filled by giant flightless birds from the Paleocene until the Pleistocene (Marshall 1994).

The earliest of these dinosaur analogs were the diatrymas (terror cranes), which are known from Paleocene and Eocene deposits in North America and Europe. The diatrymas were 2m tall and had massive legs and toes with enormous claws. The head was huge—nearly as large as that of a horse—and had an enormous, hooked beak (Figure 16–2). A mechanical analysis of the feeding apparatus of *Diatryma* suggests that the birds were equipped to be ferocious predators: "Whatever *Diatryma* ate, it could bite it hard" (Witmer and Rose 1991). Indeed, the skull and jaws of *Diatryma* show thick mandibular rami and evidence of massive adductor muscles. These features are unlike those of any extant bird (perhaps fortunately for us); but they are seen in hyenas, where they are associated with the ability to crush bones. Perhaps *Diatryma*, like hyenas, was a scavenger as well as a predator and was able to crush bones to eat the marrow inside.

South America was isolated from the northern continents in the early Cenozoic, and another lineage of giant predatory birds, the phorusrachids, evolved on that continent. The phorusrachids, which were 1.5 to 2.5m tall, were more lightly built than the diatrymas and were probably faster runners. Like the diatrymas, the phorusrachids had huge beaks and powerful claws. Phorusrachids are known from South America from the Oligocene to the end of the Pliocene; they disappeared about the time of the great faunal interchange between North and South America. However, a phorusrachid is known from Pleistocene deposits in Florida, apparently representing a genus that moved north across the Central American land bridge.

▲ Figure 16–2 A terror crane, *Diatryma*, attacking an archaic ungulate (about the size of a German shepherd dog).

aquatic birds strain small crustaceans or plankton from water or mud with bills that incorporate some sort of filtering apparatus. Dabbling ducks have bills with horny lamellae that form crosswise ridges, and their tongues also have horny projections. The tongue and bill are densely invested with sensory corpuscles, forming a filter system that allows ducks to scoop up a billfull of water and mud, filter out the prey, and allow the debris to escape. Flamingos have a similar system. The bill of a flamingo is sharply bent, and the anterior part is held in a horizontal position when the flamingo lowers its head to feed. The flamingo's lower jaw is smaller than the upper jaw, and it is the upper jaw that vibrates rapidly up and down during feeding while the lower jaw remains motionless. (This reversal of the usual vertebrate pattern is made possible by the kinetic skull of birds.)

Seeds, containing the energy and nutrients that plants have invested in reproduction, are usually protected by hard coverings (husks) that must be removed before the nutritious contents can be eaten. Specialized seed-eating birds use one of two methods to husk seeds before swallowing them. One group holds the seed in its beak and slices it by making fore-and-aft movements of the lower jaw. Birds in the second group hold the seed against ridges on the palate and crack the husk by exerting an upward pressure with their robust lower jaw. After the husk has been opened, both kinds of birds use their tongues to remove the contents. Other birds have different specializations for eating seeds: Crossbills extract the seeds of conifers from between the scales of the cones, using the diverging tips of their bills to pry the scales apart and a prehensile tongue to capture the seed inside. Woodpeckers, nuthatches, and chickadees may wedge a nut or acorn into a hole in the bark of a tree and then hammer at it with their sharp bills until it cracks.

The production of fruits is a strategy of seed dispersal for plants, and birds are important dispersal agents. The bright colors of many fruits advertise their presence or ripeness and attract birds that eat them. The pulp surrounding the seed is digested during its passage through the bird's intestine, but the seeds are not damaged. The birds may discard the seed before eating the pulp, or the seeds may pass through the gut to be voided with the feces. In some cases the chemical or mechanical stress of its passage through the digestive system of a bird facilitates subsequent sprouting by the seed. Fruit-eating birds do not have to penetrate a hard covering, but the very size of the fruit may make it hard to swallow. Skull kinesis can be important for fruit eaters; the widening of the gape as the mouth opens may allow them to swallow large items.

The skulls of most birds consist of four bony units that can move in relation to each other. This skull kinesis is important in some aspects of feeding. The upper jaw flexes upward as the mouth is opened, and the lower jaw expands laterally at its articulation with the skull (Figure 16–3a). The flexion of the upper and lower jaws increases the bird's gape in both the vertical and horizontal planes and probably assists in swallowing large items. Many birds use their beaks to search for hidden food. Long-billed shorebirds probe in mud and sand to locate worms and crustaceans. These birds display a form of skull kinesis in which the flexible zone in the upper jaw has moved toward the tip of the beak, allowing the tip of the upper jaw to be lifted without opening the mouth (Figure 16–3b). This mechanism enables long-billed waders to grasp prey under the mud.

Tongues are an important part of the food-gathering apparatus of many birds. Woodpeckers drill holes into dead trees and then use their long tongues to investigate passageways made by wood-boring insects. The tongue of the green woodpecker, which extracts ants from their tunnels in the ground, extends four times the length of its beak. The hyoid bones that support the tongue are elongated and housed in a sheath of muscles that passes around the outside of the skull and rests in the nasal cavity (Figure 16–4). When the muscles of the sheath contract, the hyoid bones are squeezed around behind the skull, and the tongue is projected from the bird's mouth. The tip of a woodpecker tongue has barbs that impale insects and allow them to be pulled from their tunnels. Nectar-eating birds such as hummingbirds and sunbirds also have long tongues and a hyoid apparatus that wraps around the back of the skull. The tip of the tongue of nectar-eating birds is divided into a spray of hair-thin projections, and capillary force causes nectar to adhere to the tongue.

Dietary Specializations The morphology of birds' bills is usually closely correlated with methods of prey capture or dietary specialization. For example, the spoonbills are related to storks and herons. Like their relatives, spoonbills feed on aquatic animals including fishes, amphibians, and crayfishes, but they differ from storks and herons in bill structure and in feeding methods.

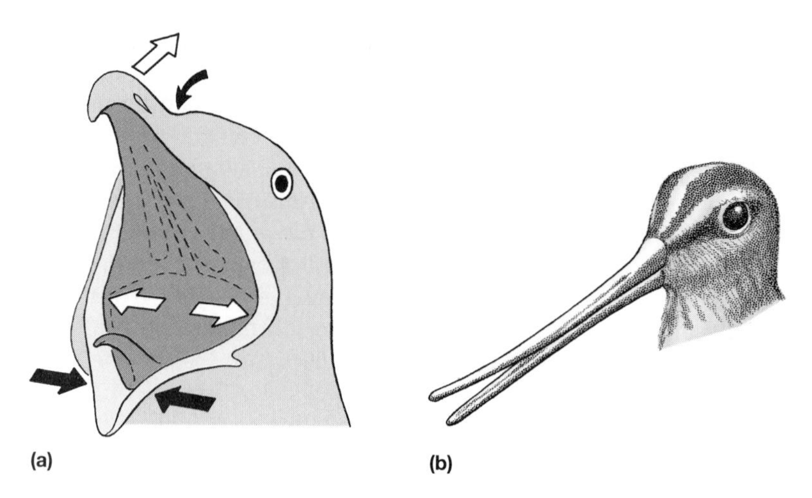

▲ Figure 16-3 Skull and jaw kinesis. (a) A yawning herring gull (*Larus argentatus*) shows the kinetic movements of the skull and jaws that occur during swallowing. White arrows show the positions of outward flexion; black arrows show inward flexion. (b) Long-billed wading birds that probe for worms and crustaceans in soft substrates can raise the tips of the upper bill without opening their mouth.

Storks and herons have long, pointed bills. They locate prey visually while wading in shallow water, and they seize prey items in their open bills. Because this method of hunting depends on being able to see individual prey items underwater, a heron may extend its wings, shading a patch of water to reduce the reflection from the water's surface. Spoonbills also hunt in this way, but they have a second method of capturing prey that is effective even in murky water (Weihs and Katzir 1994).

Spoonbills take their name from the shape of their beak, which is broadened from side to side, especially near the tip (Figure 16–5). The dorsal surface of the beak is curved, and the ventral surface is flat. In clear water a spoonbill wades forward slowly, seizing prey in its beak just as a heron does. In murky water, however, a spoonbill uses a different method of foraging, sweeping its beak through the water just a few centimeters above the bottom. The curved profile of the bill creates a vortex in the water that pulls small objects, including prey animals, off the bottom and up into the water column, where they are seized on the next sweep.

Bill morphology can affect the choice of food items on a very fine scale. The African black-bellied finch, *Pyrenestes ostrinus*, includes individuals with two different types of bills—large and small (Smith 1987). There is no sexual dimorphism in bill size, and no evidence of assortative mating. That is, a large-billed individual is equally likely to be male or female, and is equally likely to mate with an

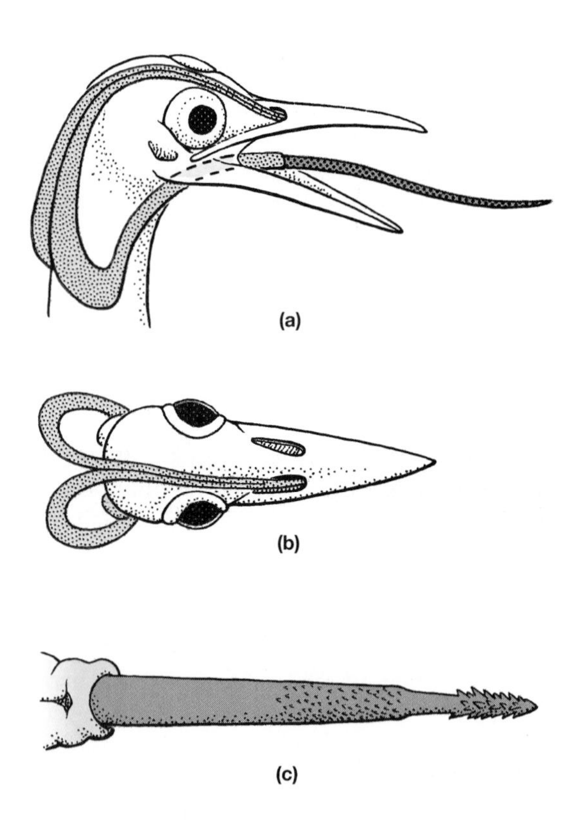

▲ Figure 16–4 Tongue projection mechanism of a woodpecker. The tongue itself is about the length of the bill, and it can be extended well beyond the tip of the beak by muscles that move the elongated branchial apparatus. The detail of the tongue shows the barbs on the tip that impale prey.

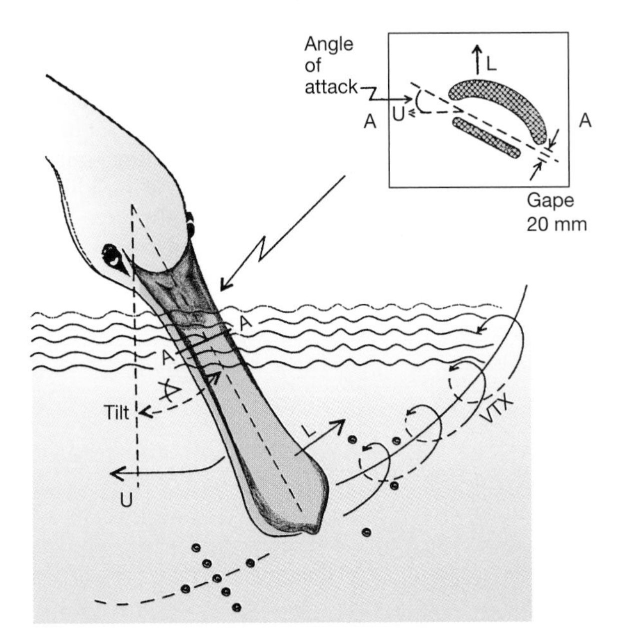

▲ Figure 16–5 Feeding mechanism of the spoonbill. When a spoonbill sweeps its beak through the water, the curved upper surface and flat lower surface create vortex currents (shown by spiraling arrows marked VTX) that lift prey from the bottom. In this drawing the bill is being swept from right to left (indicated by the arrow marked U). The sweeping motion generates a lift (L). The line A–A indicates the position of the cross section of the beak (shown in the inset).

individual of either bill morph. A mixed pair produces young with large and small bills, but no intermediates.

Finches are seed eaters, and bill size affects the ease with which they can eat different kinds of seeds. The African black-bellied finch eats the seeds of different species of sedges, and a range of seed sizes is available. Individuals of the large-billed morph are able to crack hard seeds more rapidly than small-billed individuals can, but small-billed birds consume soft seeds more rapidly than large-billed individuals can. These differences in seed handling time are reflected in the diets of the two morphs. Large-billed individuals are nearly three times as likely as small-billed birds to have hard seeds in their crop, and small-billed birds are three times as likely to have soft seeds (Figure 16–6). Thus, both bill morphs of *P. ostrinus* specialize on the sizes of seeds they can consume most effectively.

Foraging Behavior of Birds—The Concept of Optimization

Refined morphological specializations, such as the correspondence of beak shapes and diet, demonstrate such appropriate design that we are intuitively led to hypothesize the existence of similar perfection in physiological and behavioral adaptations to the environment. A general hypothesis is implicit in much of modern biology—that over evolutionary time, natural selection has favored organisms with genotypes providing them with characteristics that solve environmental problems in an optimal way for survival and reproduction. In the past quarter-century, this general hypothesis has seen vigorous testing and modification, in large part by ecologists working with **optimal foraging theory** (OFT) (Krebs et al. 1983, Stephens and Krebs 1987). Numerous other possible optimalities have been investigated, primarily those involving behavioral characteristics of animals, all borrowing from economics the procedures of cost-benefit analysis. However, the concept of optimization in biology remains controversial, and OFT is criticized by some ecologists as strongly as it is defended by others (Pierce and Ollason 1987, Stearns and Schmid–Hempel 1987).

The technique of OFT allows us to compare the predictions of a theoretical model to actual observations in the natural world. If the model and nature resemble one another, we have probably correctly considered the essential elements of the organism's adaptations in constructing the model; if they do not match well, the nature of the mismatch is a guide to other factors that must be considered. OFT has been used to investigate what animals feed on, where they go to feed, and how they search for food. It is possible to describe the rules animals use to make decisions to eat one food item and ignore another, or to hunt here and not there.

Most food resources used by birds occur in patches, and the food supply within a patch generally diminishes with time (seeds become depleted or insects take evasive action after becoming aware of a foraging predator). A foraging bird must decide how long to feed in one patch before moving to the next, taking into account that no food intake at all occurs when it is traveling between patches.

Theoretically a forager should behave as though it estimates the average rate of its food intake in the environment in which it is feeding and compares its current rate of food intake with the average, includ-

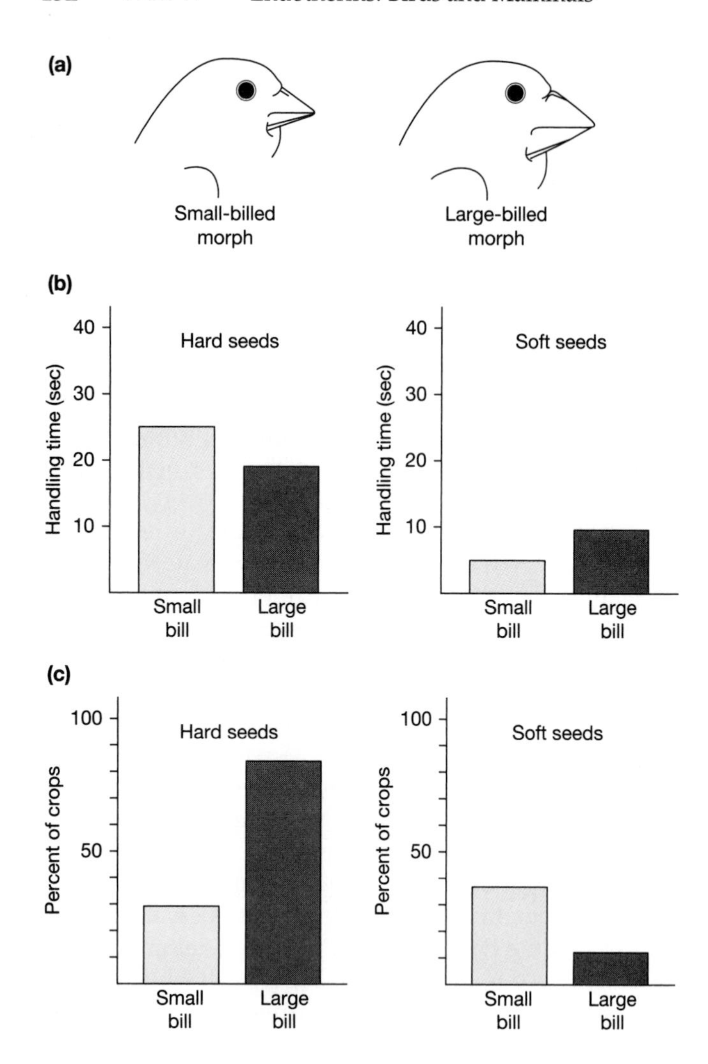

◀ Figure 16–6 Dimorphism in bill size and diet of the African black-bellied seed eater. (a) The two bill morphs, large and small. (b) The small-billed morph takes longer than the large-billed morph to eat hard seeds, but eats soft seeds faster than the large-billed morph can. (c) The small-billed morph is most likely to have soft seeds in its crop, whereas the large-billed morph is most likely to have hard seeds.

ing the cost of moving from one patch to another. When the rate of food capture in a patch falls to the environmental average, it is time to switch. This hypothesis is known as the **marginal value theorem** (Figure 16–7). The graph shows the rate of prey capture for a predator foraging in patches where the prey density is high or low. As the bird continues to forage in a patch, the rate of prey capture decreases. After a period of feeding the rate of prey capture in the patch will have decreased until it equals the average capture rate for the entire habitat, shown by the dotted horizontal line. Optimal foraging theory predicts that a predator will move to a new patch when its capture rate has dropped to the average for the habitat. The graph shows that a predator behaving in that manner will forage longer in a patch where prey is abundant than in a patch where prey is scarce, thereby maximizing its energy intake during the time it spends foraging.

Most field studies of this hypothesis have been unsuccessful because the researcher has little way of knowing the food abundance in a patch. On the other hand, laboratory tests of foraging have been rewarding. By creating patches with known prey numbers and varying not only those numbers but the time required to change patches, researchers have shown that there is a remarkable correspondence between the predictions of the model and the behavior of foraging birds.

What food to eat depends on the net energy content of the food after the energy costs of search and handling (the energy required to husk or kill) have been subtracted. OFT assumes that it is advantageous for a predator to maximize its energy intake

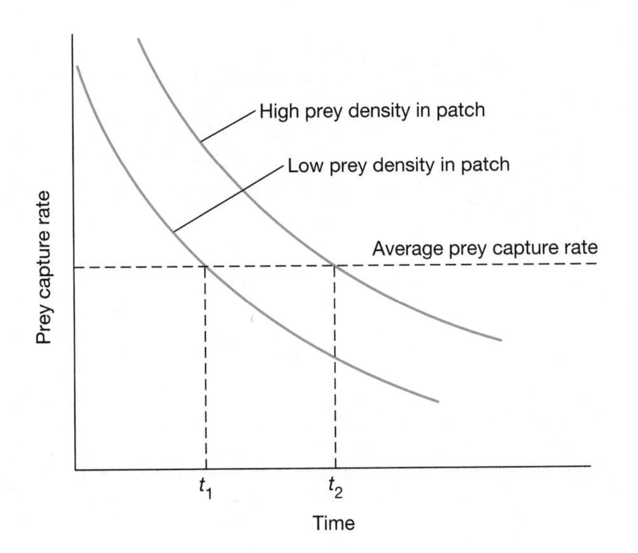

▲ Figure 16–7 The marginal value theorem of optimal foraging. As a forager captures prey, it reduces the density of prey. To maximize its rate of prey capture, a forager should move to a new patch when the rate of prey capture in its current patch falls to the average rate for the habitat. Thus, a forager will spend longer in patches that initially have high prey density than in those with low initial prey density.

per unit of time (Figure 16–8). The rate of energy intake for prey of different sizes depends on both the energy content of the prey and the time it takes to handle (subdue and swallow) the prey item. Large prey individuals contain more energy than small individuals, but they also require longer handling times. The graph shows this relationship for a wagtail eating dungfly larvae. Larvae that are 7 mm long provide the greatest energy return per second of handling time. Because food patches in nature are composed of several types of edible material, a foraging organism must decide whether to eat an item or ignore it. Optimal foraging theory predicts that an item with low net energy should be rejected if the predator could expect to find an energetically superior tidbit soon after rejecting the lower-quality item. Even if a feeding patch is rich in inferior food items, theoretically they should be completely ignored if there is a greater net energy gain from rejecting them in favor of some higher-quality item also in the patch. However, when high-quality items are scarce, the wise forager eats what is at its bill.

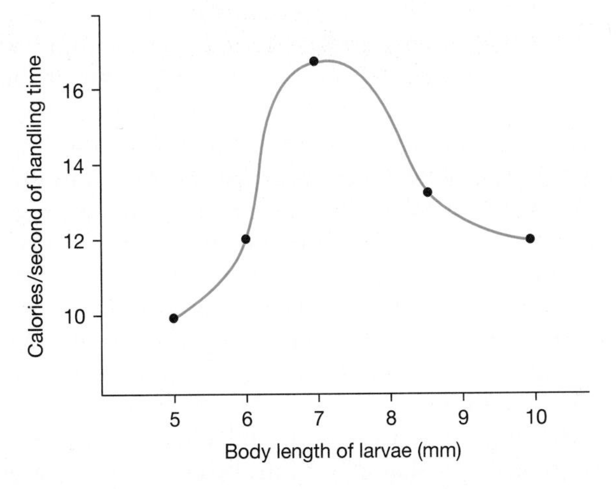

▲ Figure 16–8 Rate of energy intake for prey of different sizes. In this example, prey of intermediate size yield the greatest energy intake per second of handling time.

These predictions of what prey should be eaten and what should be ignored are excellent beginnings for laboratory experiments. They have also been tested successfully in the field. J. D. Goss-Custard (1977) studied the European redshank (*Tringa totanus*). These sandpipers feed on polychaete worms by probing in estuarine intertidal mud with their long bills. Such habitats often contain large numbers of several species of these worms, but little else that appeals to a redshank. Goss-Custard was able to study redshanks where only two species of worm occurred in significant numbers—a large species with a patchy distribution and a ubiquitous small species. As food items for the redshank, these two species seem to differ only in the greater quantity of nutritious flesh in the larger species. The rate at which large worms were captured increased as the density of large worms in a patch increased (Figure 16–9a). When the rate of capture of large worms was low, the birds ate many small worms; but they ignored small worms when they were able to capture large worms (Figure 16–9b). That is, they ignored less-profitable prey items (small worms) when more-profitable prey items (large worms) were readily available. As the rate of capture of large worms increased, the redshanks became more and more selective no matter what the density of small worms was—as the theoretical predictions indicate they should.

Not all studies of the prey optimization of foragers have matched theoretical predictions. Several of these inconsistencies can be resolved by recognizing that energy content is not the only value of food. Nutritional characteristics also affect the value of a

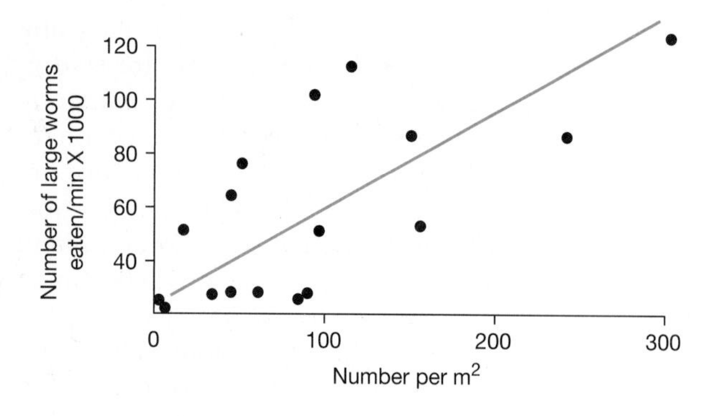

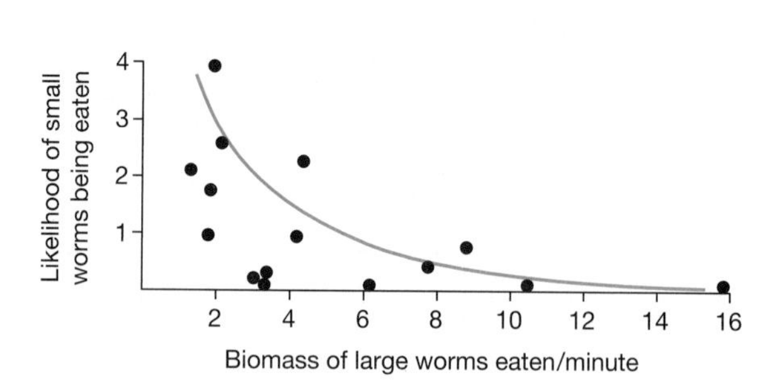

▲ Figure 16–9 Optimal foraging in action. Redshanks feeding on polychaete worms illustrate the behavior expected from an optimally foraging predator.

given food. Trace nutrients or the relative proportions of carbohydrate, fat, and protein may add to the value of a food item, or toxins may detract from its value.

Perhaps the greatest value of OFT has been in sharpening our perspective of the trade-offs necessary in adaptation to the complex environments in which animals live. However, many ecologists share the skepticism about OFT expressed by Pierce and Ollason (1987). Few behaviors match our simplistic models; animals must make compromises among simultaneous and conflicting demands. Adaptations also may be greatly affected by phylogenetic constraints. These complications must be remembered when the results of OFT studies appear to demonstrate that animals are behaving optimally.

16.2 The Sensory Systems

A flying bird moves rapidly through three-dimensional space and requires a continuous flow of sensory information about its position and the presence of obstacles in its path. Vision and hearing are the senses best suited to provide this sort of information on a rapidly renewed basis, and birds have well-developed visual and auditory systems. Their predominance is reflected in the brain: The optic lobes are large and the midbrain tectum is an important area for processing visual and auditory information. Olfaction is relatively unimportant for most birds and the olfactory lobes are small. The cerebellum, which coordinates body movements, is large. The cerebrum is less developed in birds than it is in mammals and is dominated by the corpus striatum. Many aspects of the behavior of birds are relatively stereotyped in comparison with the more plastic behavioral responses of mammals, and that difference may reflect the greater development of the neocortex of mammals.

Vision

The eyes of birds are large—so large that the brain is displaced dorsally and caudally, and in many species the eyeballs meet in the midline of the skull. The eyes of some hawks, eagles, and owls are as large as the eyes of humans. In its basic structure the eye of a bird is like that of any other vertebrate, but the shape varies from a flattened sphere to something approaching a tube (Figure 16–10). An analysis of the optical characteristics of birds' eyes suggests that these differences are primarily the result of fitting large eyes into small skulls. The eyes of a starling are small enough to be contained within the skull, whereas the eyes of an owl bulge out of the skull. An owl would require an enormous, unwieldy head to accommodate a flat eye like that of a starling. The tubular shape of the owl's eye allows it to fit into a reasonably sized skull.

Although the basic structures of vertebrate eyes are similar, the methods of focusing vary (Figure 16–10). Mammalian eyes have spherical lenses that account for most of the bending of light rays that focuses them on the retina. The focus is adjusted to accommodate nearby or distant objects by contracting or relaxing muscles within the ciliary body that change the lens shape. In birds, both the cornea and the lens contribute to focus, and accommodation is produced by changing the curvature of both structures. Muscles within the ciliary body change the

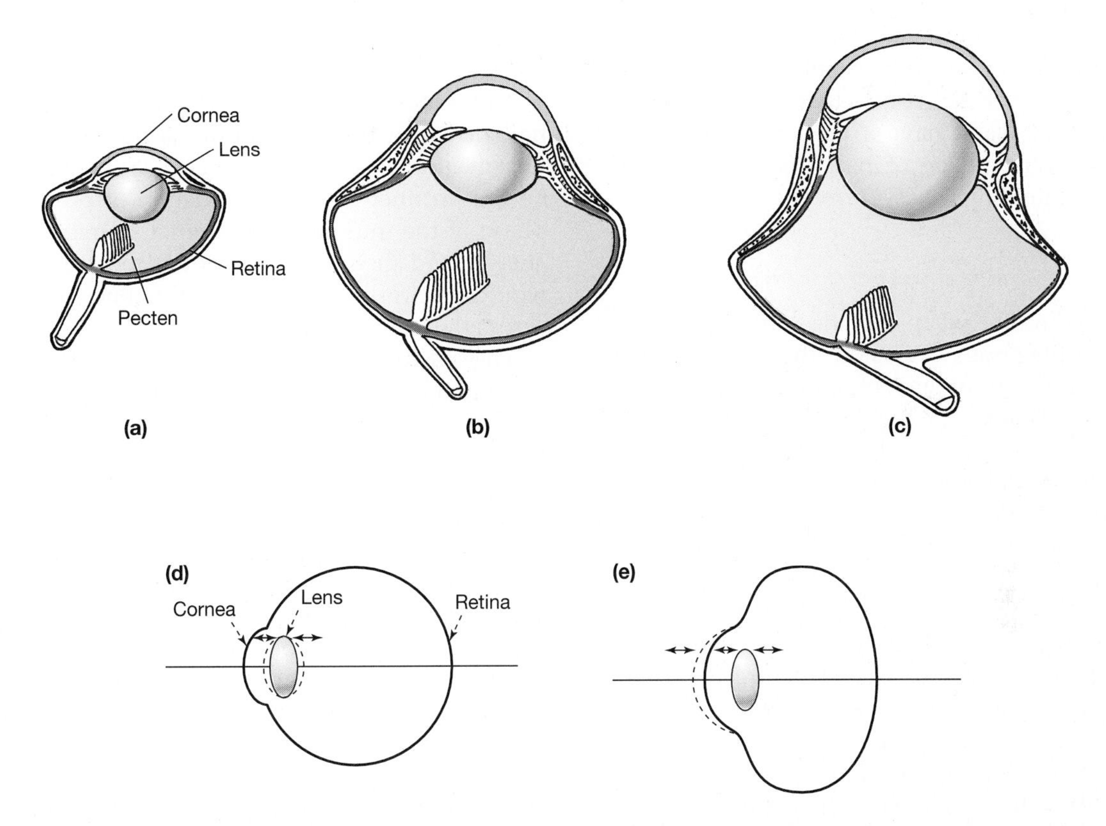

▲ Figure 16–10 The eyes of birds. (a–c) Variation in the shape of the eye of birds. (a) Flat, typical of most birds. (b) Globular, found in most falcons. (c) Tubular, characteristic of owls and some eagles. (d, e) Accommodation mechanisms of vertebrate eyes. (d) Mammalian eye: Contraction and relaxation of muscles associated with the ciliary bodies change the shape of the lens—more spherical for close vision, flatter for distant objects. (e) Bird eye: In addition to the change in shape of the lens, a second set of muscles anchored around the corneal margin change the curvature of the cornea.

lens shape, and a second set of muscles that are associated with the cornea change its curvature.

The pecten is a conspicuous structure in the eye of birds (Figure 16–10a). It is formed exclusively of blood capillaries surrounded by pigmented tissue and covered by a membrane; it lacks muscles and nerves. The pecten arises from the retina at the point where the nerve fibers from the ganglion cells of the retina join to form the optic nerve. In some species of birds the pecten is small, but in other species the pecten extends so far into the vitreous humor of the eye that it almost touches the lens. The function of the pecten remains uncertain after 200 years of debate. Proposed functions include reduction of glare, a mirror to reflect objects above the bird, production of a stroboscopic effect, and a visual reference point—but none of these seems very likely. The pecten is formed largely by blood capillaries, and its highly vascularized structure suggests that it may provide nutrition for the retinal cells and help to remove metabolic waste products that accumulate in the vitreous humor.

Oil droplets are found primarily in the cone cells of the avian retina, as they are in other sauropsids. The droplets range in color from red through orange and yellow to green; red droplets occur only in birds and turtles. Different-colored oil droplets are associated with various types of cone cells, and these cells are in different areas of the retina. For example, the central dorsal part of the retina of a pigeon has a predominance of photoreceptors with red droplets, while the ventral and lateral parts of the retina have cells with yellow droplets. The oil droplets are filters, absorbing some wavelengths of light and transmitting others. The function of the oil droplets is unclear; it is certainly complex, because the various colors of droplets are combined with different kinds of photoreceptor cells and different visual pigments. Birds like gulls, terns, gannets, and kingfishers that must see through the surface of water have a preponderance of red droplets. Aerial hawkers of insects (swifts, swallows) have predominantly yellow droplets.

Hearing

In birds, as in other diapsids, the columella (stapes) and its cartilaginous extension, the extracolumella, transmit vibrations of the tympanum to the oval window of the inner ear. The cochlea of birds has the same basic structure as that of other vertebrates, but it appears to be specialized for fine distinctions of the frequency and temporal pattern of sound. The cochlea of a bird is about one-tenth the length of the cochlea of a mammal, but it has about 10 times as many hair cells per unit of length. The space above the basilar membrane (the scala vestibuli) is nearly filled in birds by a folded, glandular tissue. This structure may damp sound waves, allowing the ear of a bird to respond very rapidly to changes in sounds.

The opening of the external auditory meatus is small, only a few millimeters in diameter in most birds, and are covered with feathers that may ensure laminar airflow across the opening during flight. The columellar muscle inserts on the columella. On contraction this muscle draws the columella away from the oval window, decreasing the sound energy transmitted to the inner ear. This mechanism may protect the ear against the noise of wind during flight, and it may also help to tune the auditory system. In starlings contraction of the columellar muscle increases the effect of the middle ear as a filter for sounds of different frequencies. The sounds that most readily pass the filter are those in the range of greatest auditory sensitivity, which in most birds corresponds to the frequency range of their own vocalizations.

Localization of sounds in space can be difficult for small animals such as birds. Large animals localize the source of sounds by comparing the time of arrival, intensity, or phase of a sound in their left and right ears; but none of these methods is very effective when the distance between the ears is small. The pneumatic construction of the skulls of birds may allow them to use sound that is transmitted through the air-filled passages between the middle ears on the two sides of the head to increase their directional sensitivity. If this is true, internally transmitted sound would pass from the middle ear on one side to the middle ear on the other side and reach the *inner* surface of the contralateral tympanum. There it would interact with the sound arriving on the external surface of the tympanum via the external auditory meatus. The vibration of each tympanum would be the product of the combination of pressure and phase of the internal and external sources of sound energy, and the magnitude of the cochlear response would be proportional to the difference in pressure across the tympanic membrane.

The sensitivity of the auditory system of birds is approximately the same as that of humans, despite the small size of birds' ears. Most birds have tympanic membranes that are large in relation to the head size. A large tympanic membrane enhances auditory sensitivity, and owls (which have especially sensitive hearing) have the largest tympani relative to their head size among birds. Sound pressures are amplified during transmission from the tympanum to the oval window of the cochlea because the area of the oval window is smaller than the area of the tympanum. The reduction ratio for birds ranges from 11 to 40. High ratios suggest sensitive hearing, and the highest values are found in owls; songbirds have intermediate ratios (20 to 30). (The ratio is 36 for cats and 21 for humans.) The inward movement of the tympanum as sound waves strike it is opposed by air pressure within the middle ear, and birds show a variety of features that reduce the resistance of the middle ear. The middle ear is continuous with the dorsal, rostral, and caudal air cavities in the pneumatic skulls of birds. In addition to potentially allowing sound waves to be transmitted to the contralateral ear, these interconnections increase the volume of the middle ear and reduce its stiffness, thereby allowing the tympanum to respond to faint sounds.

Owls are acoustically the most sensitive of birds. At frequencies up to 10 kilohertz (10,000 cycles per second), the auditory sensitivity of an owl is as great as that of a cat. Owls have large tympanic membranes, large cochleae, and well-developed auditory centers in the brain. Some owls are diurnal, others crepuscular (active at dawn and dusk), and some are entirely nocturnal. In an experimental test of their capacities for acoustic orientation, barn owls were able to seize mice in total darkness. If the mice towed a piece of paper across the floor behind them, the owls struck the rustling paper instead of the mouse, showing that sound was the cue they were using.

A distinctive feature of many owls is the facial ruff, which is formed by stiff feathers. The ruff acts as a parabolic sound reflector, focusing sounds with frequencies above 5 kHz on the external auditory meatus and amplifying them by 10 decibels. The ruffs of some owls are asymmetric, and that asymmetry appears to enhance the ability of owls to locate prey. When the ruffs were removed from the barn owls in Konishi's experiments, the owls made large errors in finding targets.

Asymmetry of the aural system of owls goes beyond the feathered ruff. The skull itself is markedly asymmetric in many owls (Figure 16–11), and these are the species with the greatest auditory sensitivity (Norberg 1977). The asymmetry ends at the external auditory meatus; the middle and inner ears of owls are bilaterally symmetric. The asymmetry of the external ear openings of owls assists with localization of prey in the horizontal and vertical axes (Norberg 1978). The time at which sounds are received by the two ears can indicate the horizontal direction of the source, and owls are capable of detecting differences of a few hundredths of a millisecond in the arrival times of a sound at the left and right ears. The vertical direction of a sound source can be determined by the differential sensitivity of the two ears to sounds coming from above and below the level of the owl. The soft tissue and feathers surrounding the face of an owl produce a situation of reversed asymmetry in which the actual directional sensitivity of the ears is the reverse of what would be expected from examination of the skull. The left ear, which opens low on the side of the head, is most sensitive to sounds originating *abov e* the level of the owl's head, and the right ear is sensitive to sounds coming from *below* the level of the head. This auditory asymmetry applies only to sounds with frequencies above 6000 hertz; sensitivity to lower frequencies is bilaterally symmetrical.

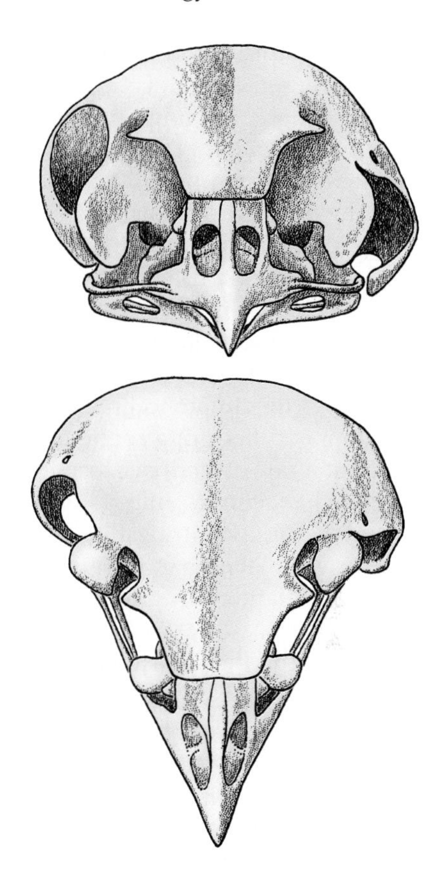

▲ Figure 16–11 The skulls of many owls show pronounced asymmetry in the position of the external auditory meatus that assists in localization of sound.

Olfaction

The sense of smell is well developed in some birds and poorly developed in others. The size of the olfactory bulbs is a rough indication of the sensitivity of the olfactory system. Relatively large bulbs are found in ground-nesting and colonial-nesting species, species that are associated with water, and carnivorous and piscivorous species of birds. Some birds use scent to locate prey. The kiwi, for example, has nostrils at the tip of its long bill and finds earthworms underground by smelling them. Turkey vultures follow airborne odors of carrion to the vicinity of a carcass, which they then locate by sight. Sponges soaked in fish oil and placed on floating buoys attracted shearwaters, fulmars, albatrosses, and petrels even in the dark.

Olfaction probably plays a role in the orientation and navigation abilities of some birds. The tube-nosed seabirds (albatrosses, shearwaters, fulmars, and petrels) nest on islands, and when they return from foraging at sea they approach the islands from downwind. Homing pigeons use olfaction (as well as other mechanisms) to navigate (Ioalè et al. 1990). The well-developed nasal bulbs of colonial-nesting species of birds suggest the possibility that olfaction is used for social functions such as recognition of individuals, but this has never been demonstrated.

Other Senses

Birds use a variety of cues for navigation, and many of these are extensions of the senses we have discussed into regions beyond the sensitivity of mammals. For example, birds can detect polarized light when it falls on the area of the retina that normally receives skylight, but other parts of the retina cannot detect polarization. Ultraviolet light (light with a wavelength of less than 400 nanometers) is not detected by most mammals, but pigeons are more sensitive to ultraviolet light in the region of 350 nm than they are to light at any wavelength in the visible part of the spectrum.

Some birds are sensitive to very small differences in air pressure. Experiments showed that pigeons were able to detect the difference in air pressure between the ceiling and floor of a room, and a pigeon flying upward at a rate of 4 cm per second would be able to detect a change in its altitude of 4 mm. This sensitivity may be useful during flight, and it may also play a role in allowing birds to anticipate changes in weather patterns that are important in the timing of migration.

Another kind of sensory information that may tell birds about the weather is infrasound—very-low-frequency sounds (less than 10 cycles per second) that are produced by large-scale movements of air. Thunderstorms, winds blowing across valleys, and other geophysical events produce infrasound that is propagated over thousands of miles. The lower-frequency limit of human hearing is around 20 Hz, but birds can hear well into the region of infrasound. Pigeons, for example, can detect frequencies as low as 0.05 Hz (3 cycles per *minute*).

The Earth's magnetic field provides a cue that could be used for orientation if an animal were able to perceive it, and considerable evidence suggests that birds can detect magnetic fields. The orientation of several species of birds during their migratory periods can be adjusted in predictable ways by placing them in an artificial magnetic field. The mechanism by which these birds detect the magnetic field is not known, but deposits of magnetite that have been found in the heads of pigeons may be involved (Wiltschko and Wiltschko 1988).

16.3 Social Behavior and Reproduction

Vision and hearing are the major sensory modes of birds, as they are of humans. One result of this correspondence has been the important role played by birds in behavioral studies. Most birds are active during the day and are thus relatively easy to observe. A tremendous amount of information has been accumulated about the behavior of birds under natural condition. This background has contributed to the design of experimental studies in the field and in the laboratory.

The activities associated with reproduction are among the most complex and conspicuous behaviors of birds. Much of our understanding of the evolution and function of the mating systems of vertebrates is derived from studies of birds. Classic work in avian ethology, such as Konrad Lorenz's studies of imprinting and Niko Tinbergen's demonstration of innate responses of birds to specific visual stimuli, has formed a basis for current studies of behavioral ecology.

Colors and Patterns

The colors of feathers are determined by a combination of pigments and structural characteristics, and the effects can be spectacular. Colors and patterns play important roles in the social behavior of birds and in their ability to avoid detection by predators. In addition, dark melanin pigments strengthen feathers and are often found in areas of high wear, such as the wing tips of seabirds.

Three types of pigments are widespread in birds. Dark colors are produced by melanin—eumelanins produce black, gray, and dark brown, whereas phaeomelanins are responsible for reddish brown and tan shades. Carotenoid pigments are responsible for most red, orange, and yellow colors. Birds obtain these pigments from their diet, and in some cases the intensity of color can be used to gauge the fitness of a prospective mate. Porphyrins are metal-containing compounds chemically similar to the pigments in hemoglobin and liver bile. Ultraviolet light causes porphyrins to emit a red fluorescence. Porphyrins are destroyed by prolonged exposure to sunlight, so they are most conspicuous in new plumage.

Structural colors result from tiny particles of melanin in the cells on the surface of feather barbs, which reflect specific wavelengths of light. Blue is produced by very small particles that reflect the shortest wavelengths, whereas some greens are produced by slightly larger particles of melanin. Structural colors can be combined with pigments—green parakeets combine a structural blue with a yellow carotenoid, and blue parakeets have a gene that blocks formation of the carotenoid. That gene is a simple recessive, so two blue parakeets can produce only blue offspring.

Iridescent colors, such as those on the heads and throats of hummingbirds and in the eye-shaped patterns on a peacock's tail, result from interference of light waves reflected from the outer and inner surfaces of hollow structures. The hue of the iridescent color depends on the distance between the reflecting surfaces, and the intensity of the color depends on how many layers of reflective structures the feathers contain. Hummingbirds, for example, have 7 to 15 layers of hollow melanin platelets in the barbules of the feathers. Perception of interference colors depends on the angle of view—color is visible when light is being reflected toward the eye of a viewer, but the feathers appear black when viewed from a different angle. As a result, iridescent colors flash on and off as a bird changes its position.

Vocalization and Visual Displays

Birds use colors, postures, and vocalizations for species, sex, and individual identification. Studies of birdsong have contributed greatly to our understanding of communication by vertebrates, and important general concepts such as species specificity in signals and innate predisposition to learning were first developed in studies of birdsong. Studies of the neural basis of song are leading to a close integration of behavior and neurobiology. (See Konishi 1985 for a review.)

Birdsong has a specific meaning that is distinct from a birdcall. The song is usually the longest and most complex vocalization produced by a bird. In many species songs are produced only by mature males, and only during the breeding season. Song is a learned behavior that is controlled by a series of song control regions (SCRs) in the brain. During the period of song learning, which occurs early in life, new neurons are produced. These neurons connect a part of the SCR associated with song learning to a region that sends impulses to nerve cells that in turn control the vocal muscles. Thus, song learning and song production are closely linked in male birds.

The SCRs are under hormonal control, and in many species of birds the SCRs of males are larger than those of females and have more and larger neurons and longer dendritic processes. The vocal behavior of female birds varies greatly across taxonomic groups: In some species females produce only simple calls, whereas in other species the females engage with males in complex song duets. The SCRs of females of the latter species are very similar in size to those of males (Table 16.1). The function of the SCR in female birds of species in which females do not vocalize has been unclear, but recent experiments suggest that it plays a role in species recognition. When the SCR of female canaries was inactivated, the birds no longer distinguished the vocalizations of male canaries from those of sparrows.

A birdsong consists of a series of notes with intervals of silence between them. Changes in frequency (frequency modulation) are conspicuous components of the songs of many birds, and the avian ear may be very good at detecting rapid changes in frequency. Birds often have more than one song type, and some species may have repertoires of several hundred songs.

Birdsongs identify the particular species of bird that is singing, and they often show regional dialects. These dialects are transmitted from generation to generation as young birds learn the songs of their parents and neighbors. In the indigo bunting, one of the best-studied species, song dialects that were characteristic of small areas persisted up to 15 years, which is substantially longer than the life of

Table 16.1 Sexual dimorphism in the song control regions of the brains of birds. The average ratio of the volumes of five SCRs in males compared with females (male:female) parallels the difference in the sizes of the song repertoires of males and females.

	Zebra Finch	*Canary*	*Chat*	*Bay Wren*	*Buff-Breasted Wren*
SCR volume ratio	4.0:1.0	3.1:1.0	2.3:1.0	1.3:1.0	1.3:1.0
Song repertoire	Males only	Males very much greater than females	Males much greater than females	Males the same as females	Males the same as females

an individual bird. Bird songs also show individual variation that allows birds to recognize the songs of residents of adjacent territories and to distinguish the songs of these neighbors from those of intruders. Male hooded warblers remember the songs of neighboring males—and recognize them as individuals when they return to their breeding sites in North America after spending the winter in Central America.

The songs of male birds identify their species, sex, and occupancy of a territory. Territorial males respond to playbacks of the songs of other males with vocalizations, aggressive displays, and even attacks on the speaker. These behaviors repel intruders, and broadcasting recorded songs in a territory from which a territorial male has been removed delays the occupation of the vacant territory by a new male.

Visual displays are frequently associated with songs; for example, a particular body posture that displays colored feathers may accompany singing. Male birds are often more brightly colored than females and have feathers that have become modified as the result of sexual selection. In this process, females mate preferentially with males that have certain physical characteristics. Because of that response by females, those physical characteristics contribute to the reproductive fitness of males, even though they may have no useful function in any other aspect of the animal's ecology or behavior. The colorful speculum on the secondaries of male ducks, the red epaulets on the wings of male red-winged blackbirds, the red crowns on kinglets, and the elaborate tails of male peacocks are familiar examples of specialized areas of plumage that are involved in sexual behavior and display.

Conspicuous or aerodynamically cumbersome feathers can make a male bird vulnerable to capture by visually guided predators, and the bright colors and special adornments of the breeding season are often discarded for a more sober, even cryptic, appearance during the rest of the year. Thus the male African standard-wing nightjar has specially elongated and flagged second primaries that are used in flight displays during courtship, but these feathers probably slow the male's flight and make it easier for an aerial predator to capture him (Figure 16–12). As soon as courtship is over, the male bites off the projecting parts of the feathers, leaving the stubs in the wings. The pattern of molting is so arranged that the primaries are not replaced until just before the next breeding season. This pattern of molting differs from that of all other caprimulgids and from that of female standard-wing nightjars. The usual pattern for caprimulgids is to begin molt in the spring with the outermost primary, and to move sequentially through the primaries to the tenth—and the female standard-wing nightjar follows this pattern. In the case of the males, however, molt begins in the center of the wing with the fifth and sixth primaries and ends with the tenth and the stump of the second. That sequence leaves just enough time for the second primary to grow to its full length at the beginning of the next courtship season.

The bright colors and other adornments of male birds may be an indication of good nutritional status, of resistance to parasites, or of the ability to evade predators. If these visual signals really do correlate with the quality of a male, they could provide a basis for females to evaluate the merits of several potential mates. This hypothesis, which is referred to as "truth in advertising," has been tested for several species of birds and seems to apply in most cases. The tail of a male peacock (*Pavo cristatus*) is a classic example of a sexually selected trait. Female peafowl mate preferentially with males that have long tails with many eyes. Cutting out some of the eyespots from a male peacock's tail reduces its success in attracting females. A study of peafowl showed that chicks sired by males whose tails have many eyes grew faster than chicks sired by males with smaller tails and had higher survival under seminatural conditions (Petrie 1994).

It has been assumed that preferred males produce better chicks because those males are genetically superior, but a study of mallard ducks suggested an additional possibility. When female mallards were paired with either attractive or less attractive males, the offspring of attractive males had higher viability—as has been shown in other studies (Cunningham and Russell 2000). But in this study the viability of the hatchlings was correlated with the size of the eggs they hatched from, and size of eggs is controlled by the female. Apparently female ducks that were paired with attractive males invested more energy in their eggs than did females paired with less attractive males. Thus, at least some of the advantage enjoyed by offspring of attractive males may result from behavior by the female.

Some species of birds use brightly colored objects to attract females; male bowerbirds, for example, decorate their bowers with feathers from other birds, shells, or shiny bits of glass and metal. A particularly

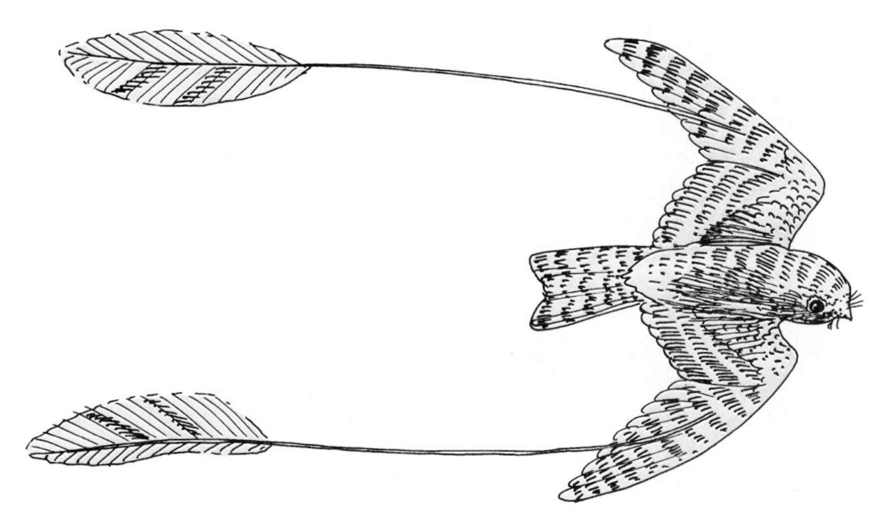

▲ Figure 16–12 Male standard-wing nightjar. The elongated second primaries are used in an aerial courtship display.

dramatic example of this behavior is provided by Archbold's bowerbird (*Archboldia papuensis*) of New Guinea (Frith and Frith 1990). Blue is an especially popular color for bowerbird ornaments, and male *Archboldia* collect the display plumes from the male King of Saxony bird of paradise (*Pteridophora alberti*). During the mating season, the male bird of paradise grows a single long feather from above each eye. These plumes look like thin wires with squares of blue plastic fastened to them at intervals. The Friths found that the bowers of several *Archboldia* were decorated with three to six *Pteridophora* plumes, which occupied a central position in the bower mat. When the Friths moved the plumes to the edges of the bower, the bower owner promptly returned them to their conspicuous location.

Vocalizations are not the only sounds that birds use in courtship; nonvocal sounds are produced by the feathers of some species. The drumming of male grouse in the spring is a familiar example of a nonvocal sound that plays a role in courtship. Sounds are often produced as a by-product of the beating of a bird's wings in flight, and only slight modification in the shapes of primaries or tail feathers is needed to produce the characteristic whistling and buzzing sounds made by certain kinds of ducks, bustards, and hummingbirds when they fly. Such sounds may be used in territorial advertisement or as individual location signals among birds flying at night or in heavy fog. Other species of birds have undergone more specific modification of their flight feathers to produce sounds used in displays. Among the tropical American manakins, researchers find not only narrowed and stiffened primaries involved in the production of sounds during displays, but also secondaries with thickened, clublike shafts that apparently act like castanets to produce clicks when the wings are moving (Figure 16–13). Other species, including goatsuckers, owls, doves, and larks, clap their wings together in flight, producing characteristic sounds associated with courtship or territorial defense.

Mating Systems and Parental Investment

The mating systems of vertebrates are believed to reflect the distribution of food, breeding sites, and potential mates. These resources affect individuals of the two sexes differently, and some types of resource distributions give one sex the opportunity to achieve multiple matings by controlling access to the resources. Studies of birds have contributed greatly to the development of theories of sexual strategies.

The energy cost of reproduction for males of most species of vertebrates is probably lower than the cost for females. Sperm are small and cheap to produce compared with eggs, which must supply the nutrients required for embryonic development. Furthermore, a male does not necessarily have any commitment beyond insemination, whereas a female must at least carry the eggs until they are deposited, and often is involved in brooding eggs and caring for the young as well. Courtship and territorial displays may be energetically expensive; but even if

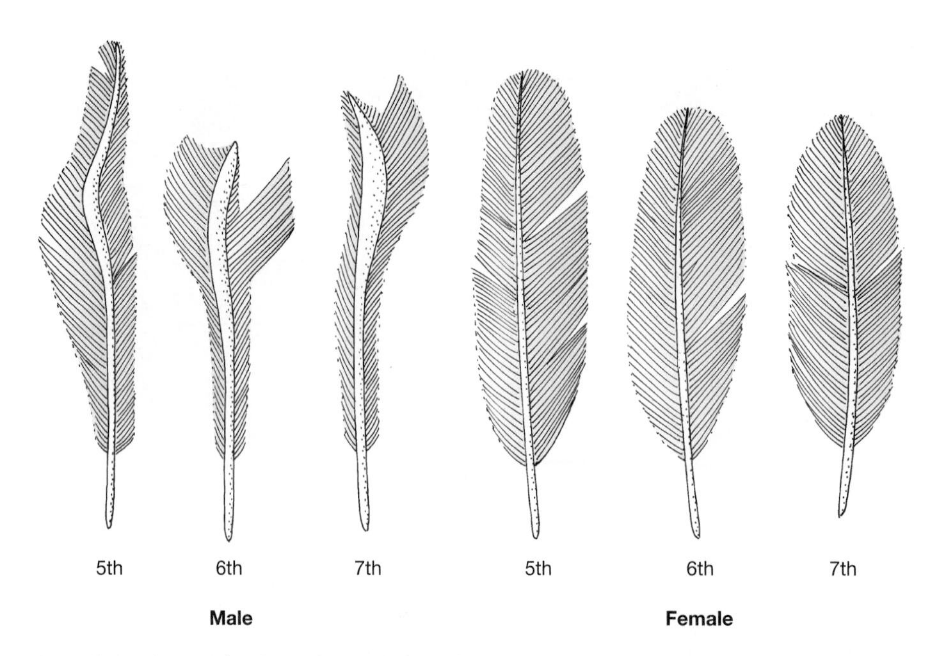

▲ Figure 16–13 Sound-producing feathers. Secondaries of the male (left) and female (right) South American manakin. The shafts of the male's feathers are thickened and produce sounds when the wings are moved in display.

that is generally true, most male vertebrates can potentially mate more often than females. A male is ready to mate again very soon after inseminating a female, but a female must ovulate and deposit yolk for a new clutch of eggs before she can attempt a second mating. Because of this disparity in the costs of reproduction for males and females, the routes to maximizing reproductive success may be different for the two sexes. For males the most productive strategy may be to mate with as many females as possible. A female may maximize her success by devoting time to careful choice of the best male as well as to care of her young.

The extent to which males can achieve multiple matings depends on ecological conditions—especially on the availability of food and nest sites, which are the resources most needed by females. Theoretically, a male could increase his opportunities to mate by defending these resources, excluding other males, and mating with all the females in the area. His ability to do that will depend on the spatial distribution of resources. If food and nest sites are more or less evenly distributed through the habitat, it is unlikely that a male could control a large enough area to monopolize many females. Under those conditions, all males will have access to the resources and to the females. On the other hand, if resources are clumped in space with barren areas between the patches, the females will be forced to aggregate in the resource patches. It will then be possible for a male to monopolize several females by defending a patch. Males that are able to defend good patches should attract more females than males defending patches of lower quality.

The temporal distribution of breeding females is also important in determining the potential for a male to achieve multiple matings. A male can mate with only one female at a time, so an important consideration is the number of receptive females relative to the number of breeding males at any moment. If the actual ratio of males to females in a population is 1:1, and if all the females became receptive at the same time, there would be one receptive female for each breeding male. In that situation, males have little opportunity to monopolize a large number of females; thus variance in mating success among males will be low. That is, most males will mate with one female, and relatively few males will have no matings or more than one mating. On the other hand, if females become receptive over a period of weeks, the number of breeding males at any given time will be larger than the number of females. In that situation it is possible for one male to mate with many females over the course of the breeding season. As the competition between males for mates increases, the variation in reproductive success of individual males increases, and sexual selection is likely to become more intense. These are the elements that contribute to determining the mating strategies of males and females.

Social vertebrates exhibit one of two broad categories of mating systems—monogamy or polygamy. Monogamy (Greek *mono* = one and *gamy* = marriage) refers to a pair bond between a single male and a single female. The pairing may last for part of a breeding season, an entire season, or for a lifetime. Polygamy (Greek *poly* = many) refers to a situation in which an individual has more than one mate in a breeding season.

Polygamy can be exhibited by males, females, or both sexes. In polygyny (Greek *gyn* = female), a male mates with more than one female; in polyandry (Greek *andr* = male), a female mates with two or more males. Promiscuity is a mixture of polygamy and polyandry in which both males and females mate with several different individuals.

Monogamy is a widespread social system of birds. Both parents in monogamous mating systems usually participate in caring for the young. Among species of birds that produce altricial young (which require extensive parental care), 93 percent are monogamous—compared with 83 percent among species with precocial young. Monogamy does not necessarily mean fidelity, however. Genetic studies of monogamous birds have shown that extra-pair copulation (mating with a bird other than the partner) is common, and may be the rule rather than the exception. Thus, some of the eggs in a nest may well have been fertilized by a male other than the partner of the female who deposited them, and he may have fertilized eggs that are being incubated in other nests.

Promiscuity is the second most common mating system for birds, accounting for 6 percent of the extant species of birds. Two percent of the species of birds are polygynous, and only 0.4 percent are polyandrous. Despite their relative rarity, promiscuous, polygynous, and polyandrous species of birds have been extensively studied. These unusual mating systems can reveal much about the mechanisms of sexual selection and evolution.

Monogamy Monogamy occurs in so many species of birds in so many different ecological conditions that no one mechanism is likely to explain its prevalence. Resource distribution and the degree of parental care appear to be two factors that are frequently important in monogamy. When nest sites and food are evenly distributed through a habitat, a male or female cannot control access to these resources. If neither sex has the opportunity to monopolize additional members of the opposite sex by controlling resources, monogamy is the reproductive strategy that maximizes the fitness of individuals. When the territory quality of one male is much like that of all other males, a female can probably maximize her reproductive success by pairing with an unmated male. Perhaps a more important incentive for monogamy for many species of birds is the need for attendance by both parents to raise a brood to fledging (i.e., leaving the nest). Dramatic examples include situations in which continuous nest attendance by one parent is necessary to protect the eggs or chicks from predators while the other parent forages for food. This situation is commonly observed in seabirds, which nest in dense colonies that sometimes include mixtures of two or more species. In the absence of an attending parent, neighbors raid the nest and kill the eggs or chicks. The male and female alternate periods of nest attendance and foraging, and some species engage in elaborate displays when the parents switch duties (Figure 16–14). A third situation that could make monogamy advantageous would be a sex ratio that deviates widely from 1:1. When one sex is in short supply, individuals of that sex may be the resource that individuals of the other sex defend. For example, female ducks suffer higher mortality than males, and the adult sex ratios of ducks are biased toward males. As a result of the shortage of female ducks, competition between males for mates is intense. A male duck pairs with a female several months before the breeding season begins and defends her against other males.

Polygyny When an individual male can control or gain access to several females, the male can increase his reproductive success by mating with more than one female. In **resource defense polygyny,** males control access to females by monopolizing critical resources—such as nest sites or food—that have patchy distributions. A male that stakes out a territory in a high-quality patch can attract many females. For this system to work, a female must benefit from mating with a male that already has one mate. That is, the reproductive fitness of a female must be greater as a secondary mate on a high-quality territory than it would be as a primary mate on a territory of lower quality. Red-winged blackbirds are a

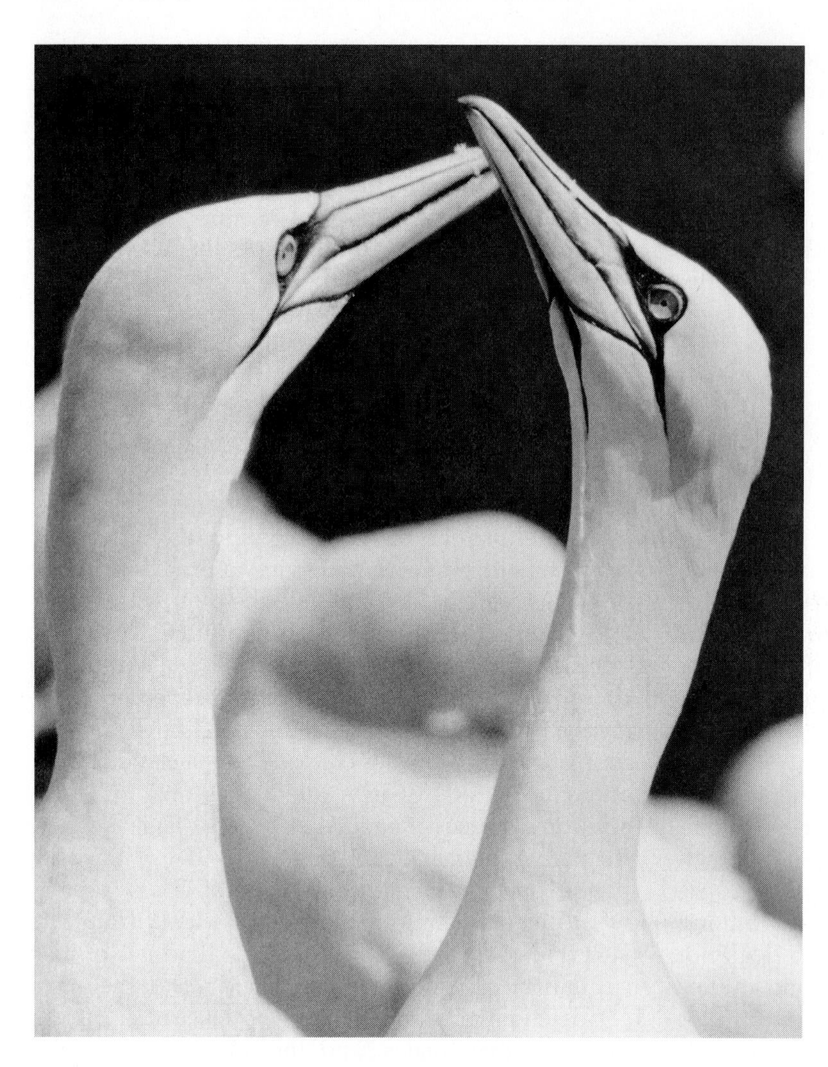

◀ Figure 16–14 Nest exchange display of northern gannets.

familiar example of resource defense polygyny. Male blackbirds arrive at their marshy breeding areas before females and compete for territories. When the females arrive, they have a choice among a variety of territories of different quality, each defended by a male blackbird. A female should choose to mate polygynously if the difference in quality of the territories is large enough so that she will raise more young than she would by mating monogamously with a male on a poorer territory.

In **male dominance polygyny,** males are not defending females, nor are they defending a resource that females require. Instead, males compete for females by establishing patterns of dominance or by demonstrating their quality through displays. This type of reproductive system is typical of situations in which the male is not involved in parental care and there is no potential for controlling resources or mates. Birds with precocial young in rich habitats often show male dominance polygyny, although this mating system is not limited to such species. The sizes of male territories in male dominance polygyny and the degree of aggregation of males are not set directly by the resources in the habitat, as is the case with resource defense polygyny. Instead, the distribution of males is determined by the sizes of the home ranges of females. Aggregations of many males in a small area are called **leks**. The prairie chicken of western North America is a well-studied lekking species. During the breeding season, male prairie chickens congregate in traditional lek sites. Each male occupies a small territory (from 13 to 100 square meters in area) in the lek, and within this territory performs a courtship display that includes elaborate postures (Figure 16–15). Two colorful sacs that are outgrowths from the esophagus are filled with air and project through the breast feathers. Air is expelled from these sacs with a popping sound. Females visit the leks and copulate with a single male. The central sites appear to be the most favored; and the 10 percent of males that occupy the most central sites obtain 75 percent of the matings.

Male Incubation and Polyandry Incubation and care of the young by both parents is considered to be the ancestral condition for birds, and it occurs in the vast majority of the extant species of birds. However, biparental care is not always required, and situations in which it is possible for one parent to care for the eggs and young allow the development of the polygynous mating strategies we have discussed. Males of polygynous species of birds do not participate in parental care, and all the tasks of incubation and rearing the young are performed by the female. Less commonly it is the female that is emancipated and the male that assumes parental responsibilities. The rarity of this situation probably reflects the relative parental investments of the two sexes. A male bird can leave a newly laid clutch of eggs and breed again as soon as he locates a receptive female, whereas a female who leaves her nest must ovulate and yolk a new clutch of eggs before she can obtain a second breeding. Thus, a female has more to lose by abandoning her eggs than does a male.

Despite the imbalance of energy investment in reproduction by males and females, there are a few situations in which the ability of a female bird to increase her fitness by multiple breedings equals or exceeds that of males, and these are the cases in which polygamy is balanced between males and females (promiscuity) or favors females (polyandry). Several species of charadriiforms show this pattern of breeding, including jacanas, stints, sandpipers, and phalaropes.

In polyandrous mating systems females control or gain access to multiple males. **Resource defense polyandry**, like its counterpart resource defense, polygyny, is based on the ability of one sex to control access to a resource that is critical for the other sex. This pattern of breeding among birds seems to be typical of situations in which the cost of each reproductive effort for the female is low (because food is abundant and a clutch contains only a few small eggs) and the probability of successful fledging is small. Spotted sandpipers (*Actitis macularia*) provide an example of this mating strategy. Predation on sandpiper nests is high, and the resource that female sandpipers control is replacement clutches for males that have lost their clutches to predators. Male spotted sandpipers form territories and incubate the

▶ Figure 16–15 Male greater prairie chicken displaying on a lek.

eggs. A female spotted sandpiper mates with a male and remains in his territory at least until she has laid three eggs. After that she may move away to breed with other males, leaving parental care to the male, or she may remain in the territory. If she remains she may or may not participate in parental care. A female that has moved away and bred with other males may return and breed again with the original male if their first clutch is destroyed.

The conspicuousness of birds and the relative ease with which they can be studied has made them a mainstay of sociobiological research. Our current understanding of vertebrate behavior owes much to research focusing on the diversity of avian mating systems, and on the correlations between ecological conditions and certain types of mating systems. Recent work has begun to emphasize the roles of individual experience and of lability of breeding systems. If environmental conditions determine the relative advantages of different mating systems, how should organisms respond to variation in these conditions? One possible response is a flexible mating system that responds in ecological time to changes in ecological conditions. Investigation of the short-term causes and consequences of variation in avian mating systems is emerging as an area of increasing importance for both ornithologists and behaviorists.

Oviparity, Nesting, and Brooding Eggs

Elaborate and diverse behaviors are associated with egg laying and parental care. Nest preparation by birds runs the gamut from nothing more than the fairy tern's selection of a branch on which to balance its egg, to the weaver bird's building of multiroom communal nests to be used by generation after generation. Incubation provides heat for the development of eggs, and the presence of a parent is a deterrent to many predators. However, some birds leave their eggs for periods of days while they forage, and brood parasites deposit their eggs in the nests of other species of birds and play no role in brooding or rearing their young.

Oviparity Although birds have a great diversity of mating strategies, their mode of reproduction is limited to laying eggs. No other group of vertebrates that contains such a large number of species is exclusively oviparous. Why is this true of birds?

Constraints imposed on birds by their specializations for flight are often invoked to explain their failure to evolve viviparity. Those arguments are not particularly convincing when observing bats, which have successfully combined flight and viviparity. Furthermore, flightlessness has evolved in at least 15 families of birds, but none of these flightless species has evolved viviparity.

Oviparity is presumed to be the ancestral reproductive mode for sauropsids. It is retained by both extant groups of archosaurs, the crocodilians and the birds. However, viviparity evolved in ichthyosaurs and has evolved nearly 100 times in the other major lineage of extant sauropsids, the lepidosaurs, so the capacity for viviparity is clearly present in some sauropsids. A key element in the evolution of viviparity among lizards and snakes appears to be the retention of eggs in the oviducts of the female for some period before they are deposited. This situation occurs when the benefits of egg retention outweigh its costs. For example, the high incidence of viviparity among snakes and lizards in cold climates may be related to the ability of a female ectotherm to speed embryonic development by thermoregulation. A lizard that basks in the sun can raise the temperature of eggs retained in her body, but after depositing the eggs in a nest, she has no control over their temperature and rate of development. Birds are endotherms and brood their eggs, thereby controlling their temperature after the eggs are laid. Thus egg retention provides no thermoregulatory advantage for a bird.

Broad aspects of the biology of birds may create an unfavorable balance of costs and benefits of egg retention. It is thus unlikely that any lineage of birds would take the first step in an evolutionary process that has repeatedly led to viviparity among snakes and lizards. If birds are viewed as being specialized for the production of one relatively large egg at a time and for complex egg incubation and parental care, the potential advantages of egg retention are greatly diminished, and the costs of decreased fecundity and increased risk of maternal mortality are increased (Blackburn and Evans 1986). Perhaps it is this balance of costs and benefits rather than any single factor that is responsible for the retention of the ancestral reproductive mode by all extant birds. The same line of reasoning probably can be applied to crocodilians, which construct nests and care for their young, and it can be extended with caution to speculations about the reproductive mode of dinosaurs.

Nesting Construction of nests is an important aspect of avian reproduction. Nests protect the eggs

▲ Figure 16–16 Diversity of bird nests. Some nests are no more than shallow depressions; other birds build elaborate structures. The piping plover (a), like many shorebirds, lays its eggs in a depression scraped in the soil. The bald eagle (b) constructs an elaborate nest that is used year after year. Coots (c) build floating nests, and the Australian mallee fowl (d) scrapes together a pile of sand and buries its eggs. Heat from the sun warms the eggs, and the male mallee fowl adds and removes sand to keep the temperature stable.

not only from such physical stresses as heat, cold, and rain but also from predators. Bird nests range from shallow holes in the ground to enormous structures that represent the combined efforts of hundreds of individuals over many generations (Figure 16–16). The nests of passerines are usually cup-shaped structures composed of plant materials that are woven together. Swifts use sticky secretions from buccal glands to cement material together to form nests, and grebes, which are marsh-dwelling birds, build floating nests from the buoyant stems of aquatic plants.

Most birds nest individually, but some lineages are exceptions: Only 16 percent of passerines nest in colonies, but 98 percent of seabirds are colonial nesters (Wittenberger and Hunt 1985, Kharitonov and Siegel-Causey 1988). Nesting colonies of some species of penguins, petrels, gannets, gulls, terns, and auks contain hundreds of thousands of individuals. Colonies are smaller in most other groups of birds; colonies of herons, storks, doves, swifts, and passerines contain 10–50 nests. Colonial nesting offers advantages and disadvantages. A colony is a concentration of potential prey that may attract predators, but the density of nesting birds may provide a degree of protection. In many colonies the nests are located two neck lengths apart, and an intruder is menaced from all sides by snapping beaks. Centrally placed nests may be better protected against predators than nests on the periphery of the colony.

Mixed colonies of two or more species of seabirds occur, but at least some of these may be transitional situations in which one species is in the process of displacing the other. For example, on the eastern coast of North America the greater black-backed gull (*Larus marinus*) is extending its range southward, apparently in response to the abundance of food available in garbage dumps. As it moves south, it is invading the breeding colonies of herring gulls (*Larus argentatus*). Great black-backed gulls, which are larger than herring gulls and breed earlier in the year, appear to be displacing herring gulls from some of their traditional breeding sites.

Incubation The megapodes, known as mound birds, bury their eggs in sand or soil and rely on heat from the sun or rotting vegetation for incubation. The Egyptian plover also buries its eggs in sand; but all other birds are believed to brood their eggs using metabolic heat. Some species of birds begin incubation as soon as the first egg is laid, and others wait until the clutch is complete. Though starting incubation immediately may protect the eggs, it means that the first eggs in the clutch hatch while the eggs that were deposited later are still developing. This forces the parents to divide their time between incubation and gathering food for the hatchlings. Furthermore, the eggs that hatch last produce young that are smaller than their older nestmates, and these young probably have less chance of surviving to fledge. Most passerines, as well as ducks, geese, and fowl, do not begin incubation until the next-to-last or last egg has been laid.

Prolactin, secreted by the pituitary gland, suppresses ovulation and induces brooding behavior, at least in those species of birds that wait until a clutch is complete to begin incubation. The insulating properties of feathers that are so important a feature of the thermoregulation of birds become a handicap during brooding, when the parent must transfer metabolic heat from its own body to the eggs. Prolactin plus estrogen or androgen stimulates the formation of brood patches in female and male birds, respectively. These brood patches are areas of bare skin on the ventral surface of a bird. The feathers are lost from the brood patch, and blood vessels proliferate in the dermis, which may double in thickness and give the skin a spongy texture. Not all birds develop brood patches, and in some species only the female has a brood patch, although the male may share in incubating the eggs. Ducks and geese create brood patches by plucking the down feathers from their breasts; they use the feathers to line their nests. Some penguins lay a single egg that they hold on top of their feet and cover with a fold of skin from the belly, thus enveloping the egg.

The temperature of eggs during brooding is usually maintained within the range 33 to 37°C, though some eggs can withstand periods of cooling when the parent is off the nest. Tube-nosed seabirds (Procellariiformes) are known for the long periods that adults spend away from the nest during foraging. Fork-tailed storm petrels (*Oceanodroma furcata*) lay a single egg in a burrow or rock crevice. Both parents participate in incubation, but the adults forage over vast distances—both parents may be absent from the nest for periods of hours or even several days at a time. The mean period of parental absence was 11 days (during an incubation period that averaged 50 days) for storm petrels studied in Alaska, and eggs were exposed to ambient temperatures of 10°C while the parents were away. Experimental studies showed that storm petrel eggs were not damaged by being cooled to 10°C every 4 days (Vleck and Kenagy 1980). The pattern of development of chilled eggs was like that of eggs incubated continuously at 34°C, except that each day of chilling added about one day to the total time required for the eggs to hatch.

Parent birds turn the eggs as often as several times an hour during incubation, moving individual eggs back and forth between the center and edge of the clutch. Temperature varies in a nest, and shifting the eggs about may ensure that they all experience about the same average temperature. In addition, turning the eggs may help to prevent premature fusion of the chorioallantoic membrane with the inner shell membrane. Embryos attain a stable orientation during incubation—that is, the same side is usually uppermost. This position is dictated by the asymmetric distribution of embryonic mass, and the process of turning the egg allows the embryo to assume its equilibrium position. Apparently this mechanism assures that when the chorioallantoic and inner shell membranes fuse (approximately midway through incubation), the embryo is in a position that will facilitate hatching.

Incubation periods are as short as 10 to 12 days for some species and as long as 60 to 80 days for others. In general, large species of birds have longer incubation periods than small species, but ecological factors also contribute to determining the length of the incubation period. The effect of parental absence in slowing development has already been mentioned, and the amount of time a parent is absent may depend on its foraging success. A high risk of predation may favor rapid development of the eggs. Among tropical tanagers, species that build open-topped nests near the ground are probably more vulnerable to predators than are species that build similar nests farther off the ground. The incubation periods of species that nest near the ground are short (11 to 13 days) compared to those of species that build nests at greater heights (14 to 20 days). Species of tropical tanagers with roofed-over nests have still longer incubation periods—17 to 24 days.

Egg Biology The inorganic part of eggshells contains about 98 percent crystalline calcite, $CaCO_3$, and the embryo obtains about 80 percent of its calcium from the eggshell. An organic matrix of protein and mucopolysaccharides is distributed through the shell and may serve as a support structure for the growth of calcite crystals. Eggshell formation begins in the isthmus of the oviduct. Two shell membranes are secreted to enclose the yolk and albumen, and carbohydrate and water are added to the albumen by a process that involves active transport of sodium across the wall of the oviduct followed by osmotic flow of water. The increased volume of the egg contents at this stage appears to stretch the egg membranes taut. Organic granules are attached to the egg membrane, and these **mammillary bodies** appear to be the sites of the first formation of calcite crystals (Figure 16–17). Some crystals grow downward from the mammillary bodies and fuse to the egg membranes, and other crystals grow away from the membrane to form cones. The cones grow vertically and expand horizontally, fusing with crystals from adjacent cones to form the palisade layer. Changes in the ionic composition of the fluid surrounding the egg during shell formation lead to an

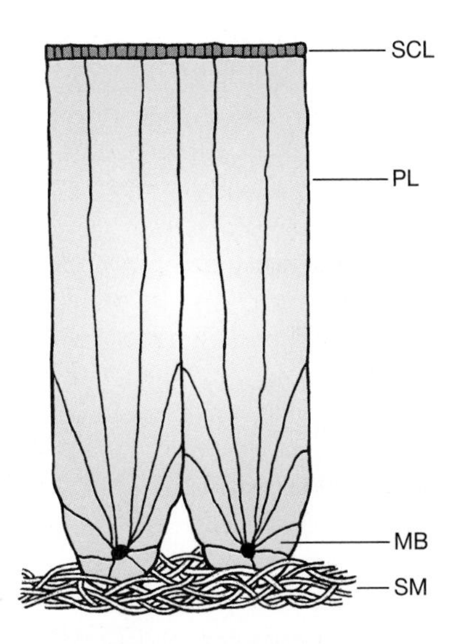

▲ Figure 16–17 Diagram of the crystal structure of an avian eggshell. Crystallization begins at the mammillary bodies (MB). Crystals grow into the outer shell membrane (SM) and upward to form the palisade layer (PL). Changes in the chemical composition of the fluid surrounding the growing eggshell are probably responsible for the change in crystal form in the surface crystalline layer (SCL).

increase in the concentrations of magnesium and phosphorus and a change in the pattern of crystallization in the surface layers of the shell.

The eggshell is penetrated by an array of pores that allow oxygen to diffuse into the egg and carbon dioxide and water to diffuse out (Figure 16–18). Pores occur at the junction of three calcite cones, but only 1 percent or less of those junctions form pores; the rest are fused shut. Pores occupy about 0.02 percent of the surface of an eggshell. The morphology of the pores varies in different species of birds: Some pores are straight tubes, whereas others are branched. The pore openings on the eggshell's surface may be occluded to varying degrees with organic or crystalline material.

Water evaporates from an egg during development, and the loss of water creates an air cell at the blunt end of the egg. The embryo penetrates the membranes of this air cell with its beak 1 or 2 days before hatching begins, and ventilation of the lungs begins to replace the chorioallantoic membrane in gas exchange. Pipping, the formation of the first cracks on the surface of the eggshell, follows about half a day after penetration of the air cell, and actual emergence begins half a day later. Shortly before hatching, the chick develops a horny projection on

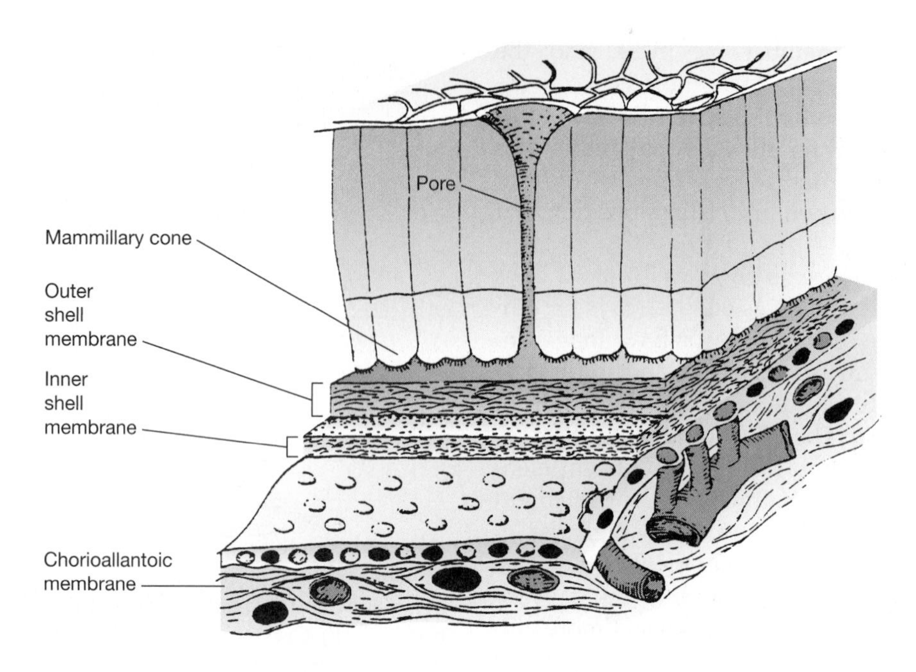

▲ Figure 16–18 A diagram of the structure of an eggshell. Pore canals penetrate the calcified region, allowing oxygen to enter and carbon dioxide and water to leave the egg. The gases are transported to and from the embryo via blood vessels in the chorioallantoic membrane.

its upper jaw. This structure is called the egg tooth, and it is used in conjunction with a hypertrophied muscle on the back of the neck (the hatching muscle) to thrust vigorously against the shell. The egg tooth and hatching muscle disappear soon after the chick has hatched. In those species of birds that delay the start of incubation until all the eggs have been laid, an entire clutch nears hatching simultaneously. Hatching may be synchronized by clicking sounds that accompany breathing within the egg, and both acceleration and retardation of individual eggs may be involved. A low-frequency sound produced early in respiration, before the clicking phase is reached, appears to retard the start of clicking by advanced embryos. That is, the advanced embryos do not begin clicking while other embryos are still producing low-frequency sounds. Subsequently, clicking sounds or vocalizations from advanced embryos appear to accelerate late embryos. Both effects were demonstrated by Vince (1969) in experiments with bobwhite quail eggs. She found that she could accelerate the hatching of a late egg by 14 hours when she paired it with an early egg that had started incubation 24 hours sooner, and the presence of the late egg delayed hatching of the early egg by 7 hours.

Parental Care

In the archosaur lineage, the ancestral form of reproduction appears to entail the depositing of eggs in a well-defined nest site, attendance at the nest by one or both parents, hatching of precocial young, and a period of association between the young and one or both parents. All the crocodilians that have been studied conform to this pattern, and evidence is increasing that at least some dinosaurs remained with their nests and young.

Extant birds follow these ancestral patterns, but not all species produce precocial young. Instead, hatchling birds show a spectrum of maturity that extends from precocial young that are feathered and self-sufficient from the moment of hatching to altricial forms that are naked and entirely dependent on their parents for food and thermoregulation (Figure 16–19 and Table 16.2). The most precocial birds at hatching are the megapodes of Australia and the East Indies, which also show the most ancestral form of nesting—burying their eggs in nests made from mounds of soil and vegetation very much like those of crocodilians. Newly hatched megapodes scramble to the surface already feathered and capable of flight. Most precocial birds are covered with down at hatching and can walk, but are not able to fly.

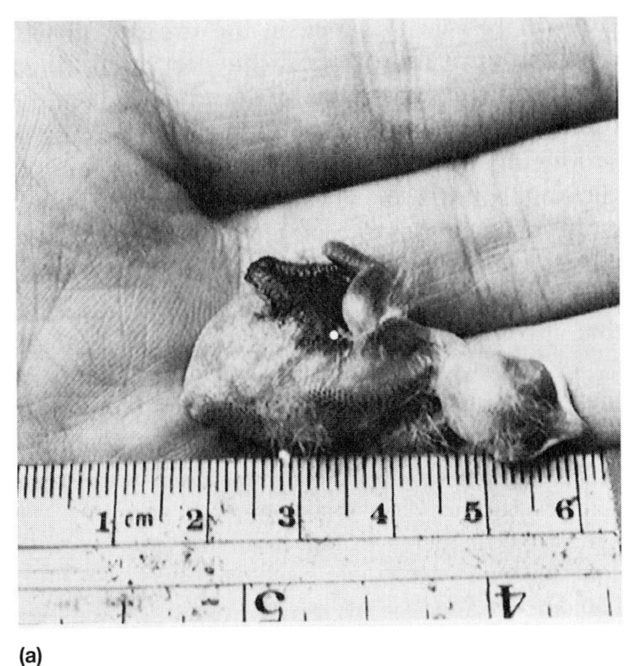

(a)

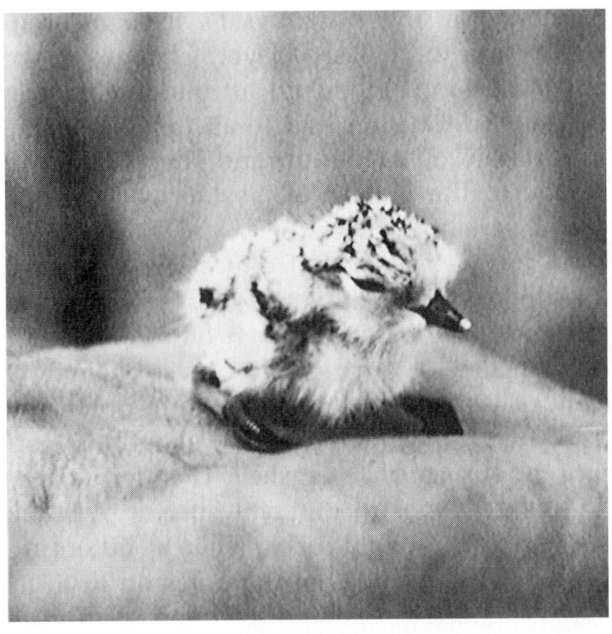

(b)

▲ Figure 16–19 Altricial and precocial chicks. Altricial chicks (a) such as that of the tree swallow (*Tachycinta bicolor*) are entirely naked when they hatch and unable even to stand up. Precocial species (b) such as the snowy plover (*Charadius alexandrinus*) are covered with down when they hatch and can stand erect and even walk. The plover in this photograph has just hatched; it retains the egg tooth on the tip of its bill, whereas the tree swallow chick is 5 days old. The dark color on the leg of the swallow is applied by a researcher to identify individual hatchlings for a study of parental care.

Table 16.2 Maturity of birds at hatching

Precocial: Eyes open, covered with feathers or down, leave nest after 1 or 2 days

1. Independent of parents: megapodes
2. Follow parents, but find their own food: ducks, shorebirds
3. Follow parents and are shown food: quail, chickens
4. Follow parents and are fed by them: grebes, rails

Semiprecocial: Eyes open, covered with down, able to walk but remain at nest and are fed by parents: gulls, terns

Semialtricial: Covered with down, unable to leave nest, fed by parents

1. Eyes open: herons, hawks
2. Eyes closed: owls

Altricial: Eyes closed, little or no down, unable to leave nest, fed by parents: passerines

The distinction between precocial and altricial birds extends back to differences in the amount of yolk originally in the eggs; it also includes differences in the relative development of organs and muscles at hatching, and the rates of growth after hatching (Table 16.3). Robert Ricklefs (1979) has proposed that the physical maturity of tissues at hatching, especially skeletal muscles, can be used to subdivide the growth patterns of birds. Mature tissues may not be capable of rapid growth, and these mature tissues may set the limits to the growth of other tissues. Ricklefs emphasized the relationship between mode of development and food supply, and David Winkler and Jeffrey Walters (1983) have pointed to a strong phylogenetic influence on development modes. Most differences in developmental mode occur between orders, not within them, even when species within an order differ substantially in their ecology.

Upon hatching, altricial young are guarded and fed by one or both parents. Adults of some species of birds carry food to nestlings in their beaks, but many species swallow food and later regurgitate it to feed

Table 16.3 Comparison of altricial and precocial birds

Amount of yolk in eggs	precocial > altricial
Amount of yolk remaining at hatching	precocial > altricial
Size of eyes and brain	precocial > altricial
Development of muscles	precocial > altricial
Size of gut	altricial > precocial
Rate of growth after hatching	altricial > precocial

the young. Hatchling altricial birds respond to any disturbance that might signal the arrival of a parent at the nest by gaping their mouths widely. The sight of an open mouth appears to stimulate a parent bird to feed it, and the young of many altricial birds have brightly colored mouth linings. Ploceid finches have covered nests, and the mouths of the nestlings of some species are said to have luminous spots that have been likened to beacons showing the parents where to deposit food in the gloom of the nest.

The duration of parental care is variable. The young of small passerines leave the nest about 2 weeks after hatching and are cared for by their parents for an additional one to three weeks. Larger species of birds, such as the tawny owl, spend a month in the nest and receive parental care for an additional 3 months after they have fledged; young wandering albatrosses require a year to become independent of their parents.

Nest Helpers

A peculiar feature of the reproductive biology of more than 200 species of birds is the existence of nest helpers, which feed and care for offspring that are not their own. Most examples of nest helpers occur in Australia or the tropics; only a few are known from Europe or North America. The mating systems of species with helpers vary from monogamy (the most common situation) through species with multiple breeders of either sex. Most species that have nest helpers are territorial, but some are colonial. Helper systems are characterized by regular involvement of the helpers in feeding and care of the young. Helpers defer their own breeding for one or more years while they assist in raising the offspring of other birds.

The peculiarity of helper systems lies in the expenditure of time and energy by helpers in caring for young that are not genetically their own. This altruistic behavior would appear to reduce the fitness of helpers, and that paradox has stimulated many studies. The concept of **kin selection** has contributed substantially to understanding helper systems. Stated in simplified form, this hypothesis proposes that an individual can increase its own fitness by providing assistance to a related individual because relatives share alleles of common descent, and these alleles (not individuals) are the units of inheritance. Thus, the **inclusive fitness** of an individual consists of (1) its own reproductive success, plus (2) the additional reproductive success of relatives that results from the altruistic behavior of the individual multiplied by the fraction of alleles shared with each relative, minus (3) any decrease in the reproductive success of the individual that results from its altruistic behavior.

The hypothesis of kin selection predicts that nest helpers will be related to the individuals they help, and this is often the case. For example, almost 50 percent of the cases of nest helpers among Florida scrub jays involved birds helping their own parents, and 25 percent involved birds helping a parent and a step-parent. Birds helping their own siblings accounted for 20 percent of the nest helpers, and less than 5 percent of the examples involved helping entirely unrelated individuals (Woolfenden 1981). However, this pattern is not universal; other examples of nest helpers involve more complicated genetic relationships among the participants, including cases where the helpers are unrelated to the individuals they help.

Nest helpers do help; nests with helpers almost always fledge more young than nests without helpers. Thus, kin selection could produce some benefit to the helpers, but why would the helpers not increase their fitness still more by breeding themselves? In other words, why *do* helpers help? Also, why do helpers *ever* help nonrelatives? Several general hypotheses have been proposed:

- A shortage of breeding territories, nest sites, or potential mates may make it difficult for young birds to breed. Helping to raise younger siblings may be the best way to mark time until an opportunity to breed presents itself.

- Becoming a nest helper may be a way to gain access to a territory and, eventually, to a mate.
- Some components of parental care are learned by experience. Birds that act as helpers for one or more breeding seasons may fledge more young when they do reproduce as a result of the experience they have gained.

These hypotheses are not mutually exclusive; they all may apply to some species. The last two hypotheses suggest that some advantage could be gained from helping even unrelated individuals.

16.4 Imprinting and Learning

The process known as imprinting has played a prominent role in studies of bird behavior. Imprinting is a special kind of learning that occurs only during a restricted period in ontogeny, called the critical period. Once imprinting is established, it is permanent and cannot be reversed. A flock of geese were imprinted on the famous ethologist Konrad Lorenz. The geese followed Lorenz around as if he were a mother goose.

The young of precocial bird species learn the characteristics that identify their parents in the hours immediately after hatching. Young ducks, for example, will imprint on an object that moves and makes a noise. Normally this object would be a parent, but in experimental situations young ducks will imprint on other animals (including humans) or on inanimate objects such as a ticking clock that is trundled along on a cart.

Most birds learn their own species' song early in life by hearing a parent bird sing. Studies of zebra finches show that the song-learning interval corresponds to a period during which new neurons are rapidly incorporated into the song control region of the brain. The images and vocalizations that birds learn early in life form the basis for their social and reproductive behavior as adults. Birds that are cross-fostered by adults of a different species, particularly males, subsequently attempt to mate with the foster parent's species, and birds that are hand-reared by humans identify their keepers as sexual partners when they are adults.

16.5 Migration and Navigation

The mobility that characterizes vertebrates is perhaps most clearly demonstrated in their movements over enormous distances. These displacements, which may cover half the globe, require both endurance and the ability to navigate. Other vertebrates migrate, even over enormous distances, but migration is best known among birds.

Migratory Movements of Birds

Migration is a widespread phenomenon among birds. About 40 percent of the bird species in the Palearctic are migratory, and an estimated total of some 5 billion birds migrate from the Palearctic every year. Migrations often involve movements over thousands of kilometers, especially in the case of birds nesting in northern latitudes, some marine mammals, sea turtles, and fishes. Short-tailed shearwaters, for example, make an annual migration between their breeding range in southern Australia and the North Pacific that requires a round-trip of more than 30,000 kilometers (Figure 16–20).

Data accumulated from recaptures of banded birds have established the origins, destinations, and migratory pathways for many species. A remarkable feature of most of these migrations is their relatively recent origin: Migrations are responses to seasonal changes in the availability of resources. These variations in the resource base are, in turn, the result of seasonal cycles in the climate—and worldwide patterns of climate have changed frequently during the past 2 million years. Many of the areas used as summer ranges by migratory birds were ice covered during glacial maxima and some current avian migratory patterns are relatively new.

Many birds return each year to the same migratory stopover sites, just as they may return to the same breeding and wintering sites year after year. Migrating birds may be concentrated at high densities at certain points along their traditional migratory routes. For example, species that follow a coastal route may be funneled to small points of land—such as Cape May, New Jersey—from which they must initiate long overwater flights. At these stopovers, migrating birds must find food and water to replenish their stores before venturing over the sea, and they must avoid the predators congregating at these sites. Development of coastal areas for human use has destroyed many important resting

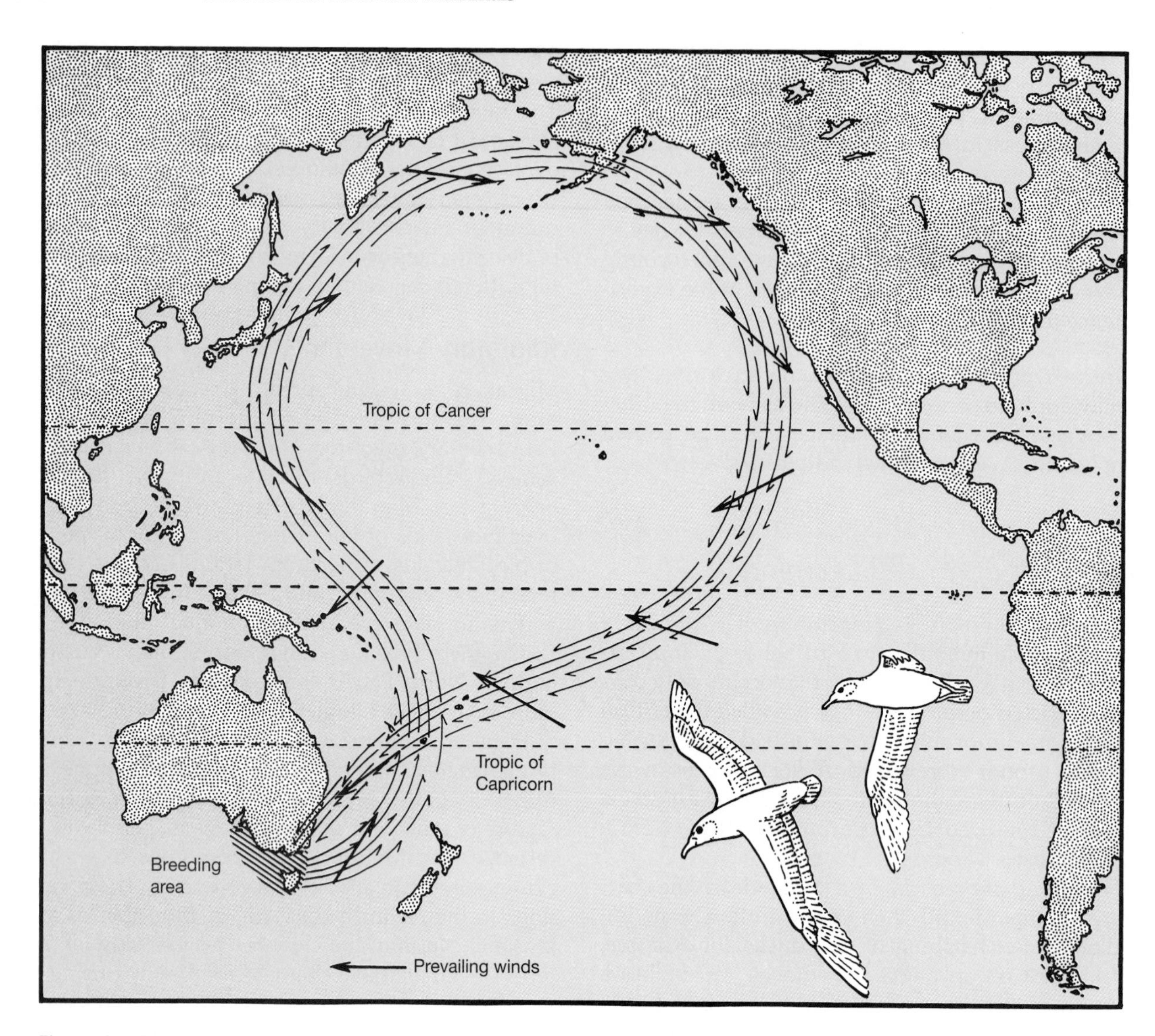

▲ Figure 16–20 The migratory path of the short-tailed shearwater. As this species migrates from its Australian breeding area to its northern range, it takes advantage of prevailing winds in the Pacific region to reduce the energy cost of migration.

and refueling stations for migratory birds. The destruction of coastal wetlands has caused serious problems for migratory birds on a worldwide basis. Loss of migratory stopover sites may remove a critical resource from a population at a particularly stressful stage in its life cycle.

Advantages of Migration The high energy costs of migration must be offset by energy gained as a result of moving to a different habitat. The normal food sources for some species of birds are unavailable in the winter, and the benefits of migration for those species are starkly clear. Other species may save energy mainly by avoiding the temperature stress of northern winters. In other cases the main advantage of migration may come from breeding in high latitudes in the summer where resources are abundant and long days provide more time to forage than the birds would have if they remained closer to the equator.

Physiological Preparation for Migration Migration is the result of a complex sequence of events that integrate the physiology and behavior of birds. Fat is the principal energy store for migratory birds, and birds undergo a period of heavy feeding and premigratory fattening (**Zugdisposition**, migratory preparation) in which fat deposits in the body cavity and subcutaneous tissue increase tenfold, ultimately reaching 20 to 50 percent of the nonfat body mass. Fat is metabolized rapidly when migration begins, and many birds migrate at night and eat during the day. Even diurnal migrants divide the day into periods of migratory flight (usually early in the day) and periods of feeding. In addition, pauses of several days to replenish fat stores are a normal part of migration. *Zugdisposition* is followed by **Zugstimmung** (migratory mood), in which the bird undertakes and maintains migratory flight. In caged birds, which are prevented from migrating, this condition results in the well-known phenomenon of **Zugunruhe** (migratory restlessness).

Preparation for migration must be integrated with environmental conditions, and this coordination appears to be accomplished by the interaction of internal rhythms with an external stimulus. Day length is the most important cue for *Zugdisposition* and *Zugstimmung* for birds in north temperate regions. Northward migration in spring is induced by increasing day length (Figure 16–21). The direction in which migratory birds orient during *Zugunruhe* depends on their physiological condition. In this experiment, the photoperiod was manipulated to bring one group of indigo buntings into their autumn migratory condition at the same time that a second group of birds was in its spring migratory condition. When the birds were tested under an artificial planetarium sky, the birds in the spring migratory condition oriented primarily in a northeasterly direction (Figure 16–21b), whereas birds in the fall migratory condition oriented in a southerly direction (Figure 16–21c). Dark circles show the mean nightly headings, pooled for several observations, for each of six birds in the spring migratory condition and five birds in the fall condition.

After breeding, many species of birds enter a refractory period in which they are unresponsive to long day lengths, their gonads regress to the nonbreeding condition, and they molt. Decreasing photoperiods in autumn may accelerate the southward migration. Many birds are again refractory to photoperiod stimulation for several weeks after the

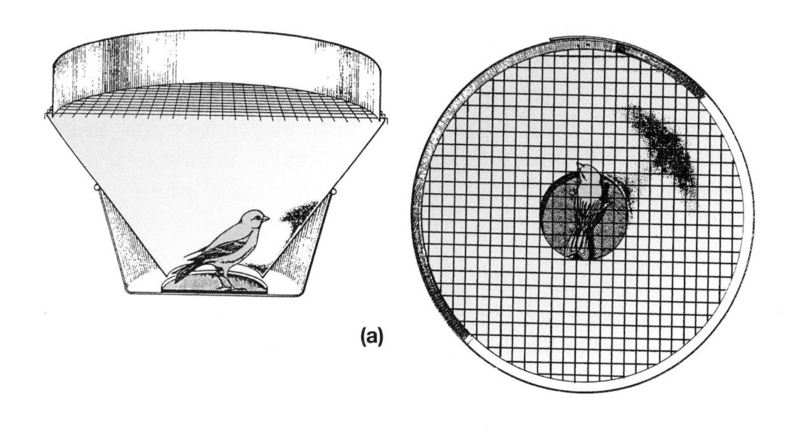

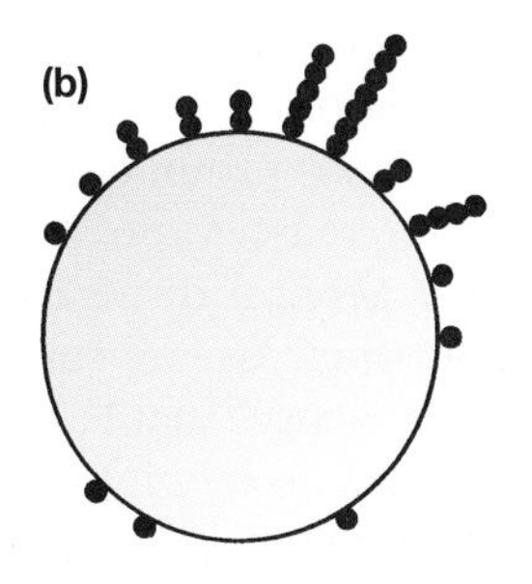

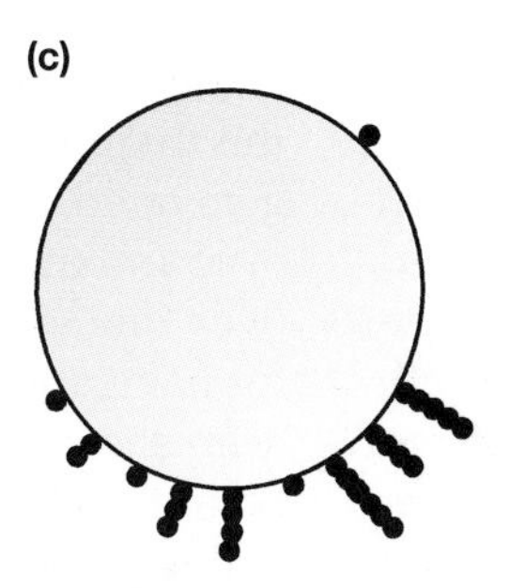

◀ Figure 16–21 Direction in which migratory birds orient during *Zugunruhe*. (a) The birds were tested in circular cages that allow a view of the sky. The bird stands on an ink pad, and each time it hops onto the sloping wall it leaves a mark on the blotting paper that lines the cage. (b) In spring the birds oriented toward the north, and (c) in autumn toward the south.

autumnal migration and require an interval of several weeks of short photoperiods before they can again be stimulated by long photoperiods.

Underlying the responses of birds to changes in day length is an endogenous (internal) rhythm. This circannual (about a year) cycle can be demonstrated by keeping birds under constant conditions. Fat deposition and migratory restlessness coincide in most species and alternate with gonadal development and molt, as they do in wild birds. When the rhythms are free running (that is, when they are not cued by external stimuli), they vary between 7 and 15 months. In other words, the birds' internal clocks continue to run, but in the absence of the cue normally provided by changing day length, the internal rhythms drift away from precise correspondence with the seasons.

Orientation and Navigation

The seasonal migrations of vertebrates that cover thousands of kilometers and, especially, their ability to return regularly to the same locations, year after year, pose an additional question—how do they find their way? Various hypotheses propose mechanisms to explain how animals navigate on these journeys. The explanations fall into two general categories: (1) Long-distance migration is an extension of the tendency to explore territory beyond the local home range, learning to recognize landmarks as one goes along; or (2) the ability to home through unfamiliar territory results from an internal navigation system. There are no clear answers, but experiments reveal not only that many vertebrates can find their way home when they are displaced but also that several different mechanisms and sensory modalities are involved. A general review of bird migration and navigation can be found in Alerstam (1990).

The homing pigeon has become a favorite experimental animal for studies of navigation. For as long as people have raised and raced pigeons, we have known that birds released in unfamiliar territory vanish from sight flying in a straight line, usually in the direction of home. How do pigeons accomplish this feat? There is no complete answer yet, but experiments have shown that navigation by homing pigeons (and presumably by other vertebrates as well) is complex and is based on a variety of sensory cues. On sunny days pigeons vanish toward home and return rapidly to their lofts. On overcast days vanishing bearings are less precise, and birds more often get lost. These observations led to the idea that pigeons use the Sun as a compass.

Of course, the Sun's position in the sky changes from dawn to dusk. That means a bird must know what time of day it is in order to use the Sun to tell direction, and its time-keeping ability requires some sort of internal clock. If that hypothesis is correct, it should be possible to fool a bird by shifting its clock forward or backward. For example, if lights are turned on in the pigeon loft 6 hours before sunrise every morning for about 5 days, the birds will become accustomed to that artificial sunrise. At any time during the day, they will act as if the time is 6 hours later than it actually is. When these birds are released they will be directed by the Sun, but their internal clocks will be wrong by 6 hours. Due to that error, the birds should fly off on courses that are displaced by 90° from the correct course for home. Clock-shifted pigeons react differently under sunny and cloudy skies, indicating that pigeons have at least two mechanisms for navigation (Figure 16–22). Each dot in these plots shows the direction in which a pigeon vanished from sight when it was released in the center of the large circle. The home loft is straight up in each diagram. The solid bar extending outward from the center of each circle is the average vector sum of all the individual vanishing points. When they are able to see the Sun, control birds that have been kept on the normal photoperiod orient predominantly in the direction of home (Figure 16–22a). Birds that have had their photoperiods shifted 6 hours fast disappear on bearings westward of the true direction of home (Figure 16–22b). However, when the Sun is obscured by clouds, the birds cannot use it for navigation and must rely on other mechanisms. Under these conditions both control and clock-shifted birds orient correctly toward home (Figure 16–22c, d).

Polarized light is another cue vertebrates use to determine directions. In addition, some vertebrates have been shown to detect ultraviolet light and to sense extremely low-frequency sounds, well below the frequencies humans can hear. Those sounds are generated by ocean waves and air masses moving over mountains and can signal a general direction over thousands of kilometers, but their use as cues for navigation remains obscure. The senses that have been shown to be involved in navigation by pigeons do not end here. Pigeons can also navigate by recognizing airborne odors as they pass over the terrain. Even magnetism is implicated: On cloudy days pi-

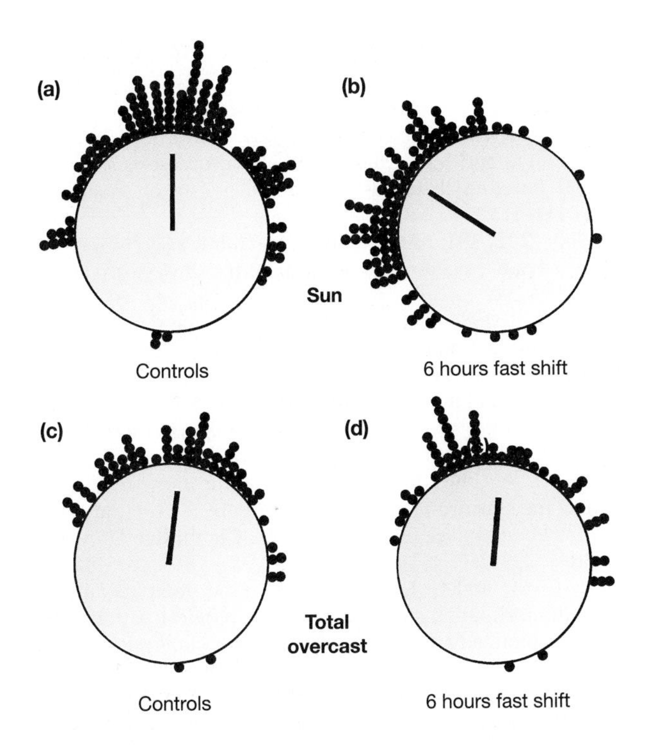

▲ Figure 16–22 Orientation of clock-shifted pigeons under sunny and cloudy skies. The line shows the average direction chosen by the birds.

geons wearing small magnets on their heads have their ability to navigate disrupted, but on sunny days magnets have no effect. When skies are clear, pigeons apparently rely on their Sun compass, ignoring magnetic cues.

Results of this sort are being obtained with other vertebrates as well, and lead to the general conclusion that a great deal of redundancy is built into navigation systems. Apparently, there is a hierarchy of useful cues. For example, a bird that relies on the Sun and polarized light to navigate on clear days could switch to magnetic direction sensing on heavily overcast days. For both conditions, the bird might use local odors and recognition of landmarks as it approaches home.

Many birds migrate only at night. Under these conditions a magnetic sense of direction might be important. Several species of nocturnally migrating birds use star patterns for navigation. Apparently, each bird fixes on the pattern of particular stars and uses their motion in the night sky to determine a compass direction. As in Sun compass navigation, an internal clock is required for this sort of celestial navigation, and artificially changing the time setting of the internal clock produces predictable changes in the direction in which a bird orients.

Despite numerous studies, the complexities of navigation mechanisms of vertebrates are far from being fully understood, and much controversy surrounds some hypotheses. The built-in redundancy of the systems makes it difficult to devise experiments that isolate one mechanism. When it is deprived of the use of one sensory mode, an animal is likely to have one or several others it can use instead. This redundancy itself, and the remarkable sophistication with which many vertebrates navigate, show the importance of migration in their lives.

Summary

The ecology and behavior of birds are directly influenced by their ability to fly. The mobility of birds allows them to exploit food supplies that have patchy distributions in time and space and to feed and reproduce in different areas. Birds are diurnal animals with excellent vision, and color and pattern are important elements of their social behavior.

Many of the complex social behaviors of birds are associated with reproduction, and birds have contributed greatly to our understanding of the relationship between ecological factors and the mating systems of vertebrates. All birds are oviparous, perhaps because egg retention is the first step in the evolution of viviparity and the specializations of the avian way of life do not make egg retention advantageous for birds. Monogamy, with both parents caring for the young, is the most common mating system for birds; but polygyny (a male mating with several females) and polyandry (a female mating with several males) also occur.

Migration is the most dramatic manifestation of the mobility of birds, and some species travel tens of thousands of kilometers in a year. Migrating birds use a variety of cues for navigation, including the position of the Sun, polarized light, the Earth's magnetic field, and infrasound.

Additional Readings

Alerstam, T. 1990. *Bird Migration*. Cambridge, U.K.: Cambridge University Press.

Blackburn, D. G., and H. E. Evans. 1986. Why are there no viviparous birds? *American Naturalist* 128:165–190.

Carey, C. 1983. Structure and function of avian eggs. In Richard F. Johnston (Ed.), *Current Ornithology*, vol. 1. New York: Plenum, 69–103.

Collias, N. E., and E. C. Collias. 1984. *Nest Building and Bird Behavior*. Princeton, N.J.: Princeton University Press.

Cunningham, E. J. A., and A. F. Russell. 2000. Egg investment is influenced by male attractiveness in the mallard. *Nature* 404:74–77.

Frith, C. B., and D. W. Frith. 1990. Archbold's bowerbird *Archboldia papuensis* (Ptilonorhynchidae) uses plumes from King of Saxony bird of paradise *Pteridophora alberti* (Paradisaeidae) as bower decoration. *Emu* 90:136–137.

Goss-Custard, J. D. 1977. Optimal foraging and size selection of worms by redshank, *Tringa totanus*. *Animal Behaviour* 25:10–19.

Ioalè, P., M. Nozzolini, and F. Papi. 1990. Homing pigeons do extract directional information from olfactory stimuli. *Behavioral Ecology and Sociobiology* 26:301–306.

Kharitonov, S. P., and D. Siegel–Causey. 1988. Colony formation in seabirds. In R. F. Johnston (Ed.), *Current Ornithology*, vol. 5. New York: Plenum, 223–272.

Konishi, M. 1985. Birdsong: from behavior to neuron. *Annual Review of Neurosciences* 8:125–170.

Konishi, M., et al. 1989. Contributions of bird studies to biology. *Science* 246:465–472.

Krebs, J. R., D. W. Stephens, and W. J. Southerland. 1983. Perspectives in optimal foraging. In A. H. Brush and G. A. Clark, Jr. (Eds.), *Perspectives in Ornithology*, Cambridge, U.K.: Cambridge University Press, 165–216.

Marshall, L. G. 1994. Terror birds of South America. *Scientific American* 270(2):90–95.

Norberg, R. Å. 1977. Occurrence and independent evolution of bilateral ear asymmetry in owls and implications for owl taxonomy. *Philosophical Transactions of the Royal Society of London* B 280:375–408.

Norberg, R. Å. 1978. Skull asymmetry, ear structure and function, and auditory localization in Tengmalm's owl, *Aeogilius funereus* (Linné). *Philosophical Transactions of the Royal Society of London* B 282:325–410.

Petrie, M. 1994. Improved growth and survival of offspring of peacocks with more elaborate trains. *Nature* 371:598–599.

Pierce, G. J., and J. G. Ollason. 1987. Eight reasons why optimal foraging theory is a complete waste of time. *Oikos* 49:111–118.

Ricklefs, R. E. 1979. Adaptation, constraint, and compromise in avian postnatal development. *Biological Review* 54:269–290.

Smith, T. B. 1987. Bill size polymorphism and intraspecific niche utilization in an African finch. *Nature* 329:717–719.

Stearns, S. C., and P. Schmid–Hempel. 1987. Evolutionary insights should not be wasted. *Oikos* 49: 118–125.

Stephens, D. W., and J. R. Krebs. 1987. *Foraging Theory*. Princeton, N.J.: Princeton University Press.

Vince, M. A. 1969. Embryonic communication, respiration, and the synchronization of hatching. In R. A. Hinde (Ed.), *Bird Vocalizations*. Cambridge, U.K.: Cambridge University Press, 233–260.

Vleck, C. M., and G. J. Kenagy. 1980. Embryonic metabolism of the fork-tailed storm petrel: Physiological patterns during prolonged and interrupted incubation. *Physiological Zoology* 53:32–42.

Weihs, D., and G. Katzir. 1994. Bill sweeping in the spoonbill, *Platalea leucordia:* evidence for a hydrodynamic function. *Animal Behaviour* 47:649–654.

Wiltschko, W., and R. Wiltschko. 1988. Magnetic orientation in birds. Pages 67–121 in *Current Ornithology*, volume 5, edited by R. F. Johnston. Plenum, New York, NY.

Winkler, D. W., and J. R. Walters. 1983. The determination of clutch size in precocial birds. In R. F. Johnston (Ed.), *Current Ornithology*, vol. 1, 33–68.

Witmer, L. M., and K. D. Rose. 1991. Biomechanics of the jaw apparatus of the gigantic Eocene bird *Diatryma:* implications for diet and mode of life. *Paleobiology* 17:95–120.

Wittenberger, J. F., and G. L. Hunt, Jr. 1985. The adaptive significance of coloniality in birds. In D. S. Farner, J. R. King, and K. C. Parkes (Eds.), *Avian Biology*, vol. 8, Orlando, Fla.: Academic, 1–78.

Woolfenden, G. E. 1981. Selfish behavior by Florida scrub jay helpers. In R. D. Alexander and D. W. Tinkle (Eds.), *Natural Selection and Social Behavior*. New York: Chiron, 257–260.

Web Explorations

On-line resources for this chapter are on the World Wide Web at: http://www.prenhall.com/pough (click on the Table of Contents link and then select Chapter 16).

CHAPTER

17

The Synapsida and the Evolution of Mammals

We must backtrack to the end of the Paleozoic (Chapters 7 and 8) to find the origin of the final lineage of vertebrates, the synapsids. The synapsids actually had their first radiation (pelycosaurs) and a significant portion of their second radiation (therapsids) in the Paleozoic, before the radiations of the diapsids we have already discussed. The third radiation of the synapsid lineage (mammals) did not reach its peak until the Cenozoic. Nonetheless, through the late Paleozoic and early Mesozoic the synapsid lineage was becoming increasingly mammal-like, and mammals and dinosaurs both appeared on the scene in the Late Triassic.

A notable feature within the evolution of the synapsid lineage is the acquisition of endothermy. It is gratifying to see in the fossil record just the sorts of structural change in the skeleton of synapsids that might be expected to accompany an increasingly active form of locomotion and foraging. The three groups of extant mammals—monotremes, marsupials, and eutherians (= placental mammals)—had evolved by the late Mesozoic, and they were accompanied by several groups of mammals that are now extinct.

17.1 The Origin of Synapsids

Synapsids include mammals and their extinct predecessors, commonly called mammal-like reptiles (see Figures 17–1 and 17–2). The term *synapsid* is often used incorrectly to refer to just the extinct nonmammalian forms, but it in fact includes all amniotes descended from a common ancestor with the synapsid type of temporal fenestration (see Chapter 8). *Mammal-like reptile* is an appealing term for the ancestors of mammals, yet it is a misleading one, and is no longer used in technical publications. The term is inaccurate in part because, as we saw in Chapter 8, mammals are not the descendants of any animals closely related to modern reptiles, and to think of them as some sort of large, peculiar lizard-like beast would indeed be inaccurate. Moreover, since the "mammal-like reptiles" contain the ancestry of mammals, they are a paraphyletic assemblage rather than a true evolutionary group. In this chapter we will stick with the more correct, if less euphonious, term *nonmammalian synapsid*.

The synapsid lineage was the first group of amniotes to radiate widely in terrestrial habitats. During the Late Carboniferous and the entire Permian, synapsids were the most abundant terrestrial vertebrates, and from the Early Permian into the Triassic they were the top carnivores in the food web. Most synapsids were medium to large-size animals, weighing between 10 and 200 kilograms, with a few weighing as much as half a ton or more (e.g., the dinocephalian *Moschops*). Most of the synapsid lineages disappeared by the Late Triassic. The surviving forms (represented only by mammals past the Early Cretaceous) were considerably smaller, mostly under a kilogram in body mass.

Synapsids are distinguished from other amniotes by the synapsid type of temporal fenestra, plus a few other skull features. Some features of synapsids represent the primitive condition in comparison to other amniotes (reptiles and birds). These include a glandular skin without the type of hard beta keratin typical of reptiles, the inability to excrete uric acid (suggesting perhaps an ovoviparous mode of

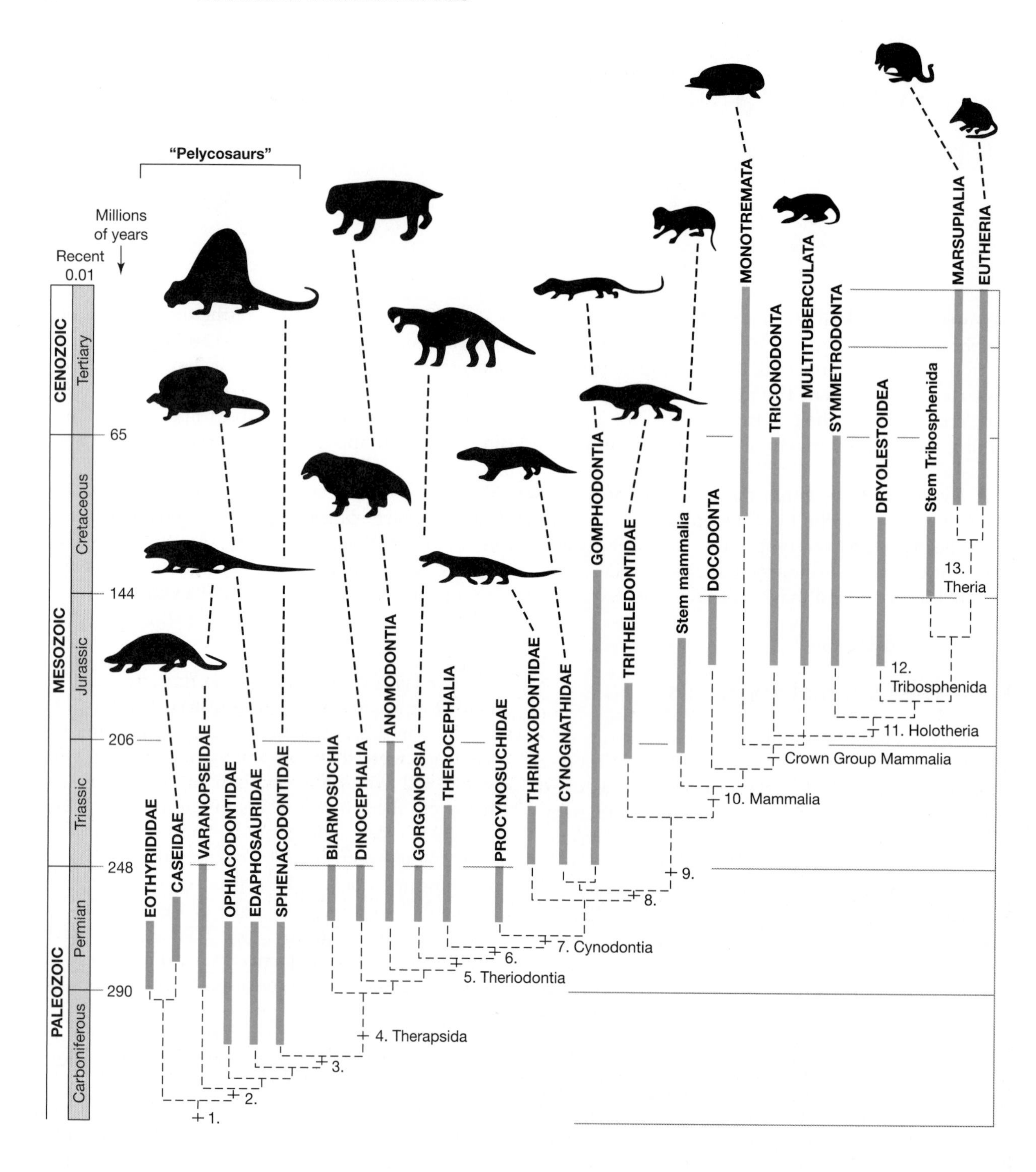

Legend: 1. Synapsida—Lower temporal fenestra present. **2.** Eupelycosauria—Snout deeper than it is wide, frontal bone forming a large portion of the margin of the orbit. **3.** Sphenacodontia—A reflected lamina on the angular bone, retroarticular process of the articular bone turned downward. **4.** Therapsida—Temporal fenestra enlarged, upper canine plus the bone containing it (the maxilla) enlarged, limb bones more slender, limbs held more underneath body (indicated by inturned heads of femur and humerus), shorter feet. **5.** Theriodontia—Coronoid process

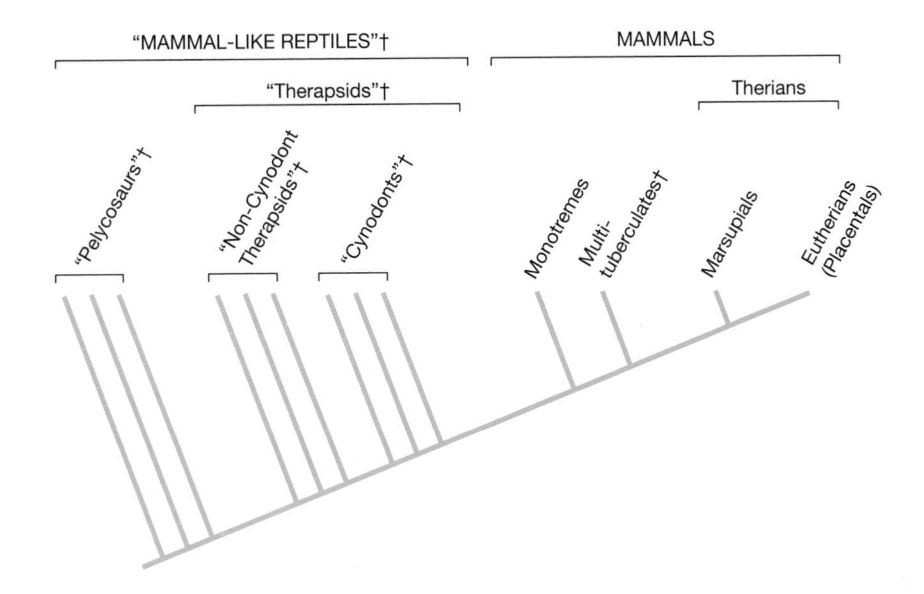

▲ Figure 17–2 Simplified cladogram of synapsids. Quotation marks indicate paraphyletic groups. A dagger (†) indicates an extinct taxon.

on dentary, flatter skull with wider snout. **6.** Eutheriodontia—Temporal fossa completely open dorsally. **7.** Cynodontia—Postcanine teeth with anterior and posterior accessory cusps and small cusps on the inner side, partial bony secondary palate, masseteric fossa on dentary and bowing out of zygomatic arch (evidence for the presence of a masseter muscle), lumbar ribs reduced or lost, distinct calcaneal heel. **8.** Eucynodontia—Dentary greatly enlarged, postdentary bones reduced, phalangeal formula of 2-3-3-3-3. **9.** Prismatic enamel, loss of postorbital bar, at least incipient contact between dentary and squamosal bones. **10.** Mammalia—Dentary-squamosal jaw articulation, double-rooted, precisely occluding postcanine teeth, and specializations of the portion of the skull housing the inner ear. ("Stem Mammalia" is a paraphyletic assemblage of Late Triassic and Jurassic genera, including [listed from more primitive to more derived] *Adelobasileus, Sinoconodon, Megazostrodon*, and *Morganucodon*.) **11.** Holotheria—Reversed triangles molar pattern. **12.** Tribosphenida—Tribosphenic molars (see Chapter 19). ("Stem Tribosphenida" is a paraphyletic assemblage of Early Cretaceous genera, including [listed from more primitive to more derived] *Aegialodon, Pappotherium*, and *Holoclemensia*, plus others. The somewhat more primitive *Vincelestes* might also be included in this grouping.) **13.** Theria—Details of braincase structure, and many features of the soft anatomy.

◀ Figure 17–1 Phylogenetic relationships of the Synapsida. This diagram shows the probable relationships among the major groups of synapsids. The dotted lines indicate interrelationships only, and are not indicative of times of divergence or of the presence of the taxon unrecorded in the fossil record. The numbers indicate derived characters that distinguish the lineages.

reproduction), and the absence of good color vision (suggesting perhaps a more nocturnal habit). The more primitive skin type suggests that synapsids never had the type of scaly covering typical of modern reptiles. Changes in the structure of the skull and skeleton of nonmammalian synapsids, and their probable relation to metabolic status and the evolution of the mammalian condition, are described later in Figure 17–6.

17.2 Diversity of Nonmammalian Synapsids

The two major groups of nonmammalian synapsids were the pelycosaurs and the therapsids (see Figure 17–3); Table 17.1 lists the major subgroups within each. Pelycosaurs, the more primitive of the two groups, were found in the Northern Hemisphere (Laurasia) and were predominantly Early Permian in age. The more derived therapsids were found predominantly in the Southern Hemisphere (Gondwana). They range in age from the Late Permian to the Early Cretaceous, but were predominantly known from the Late Permian to Early Triassic.

Pelycosaurs—Primitive Nonmammalian Synapsids

Pelycosaurs are known as the sailbacks of the late Paleozoic, although only a minority of them actually possessed sails. The best-known pelycosaur is probably *Dimetrodon* (Figures 17–3a), an animal frequently mislabeled as a dinosaur in children's books and packages of plastic toys. Pelycosaurs contain the ancestry of the more derived synapsids, including the mammals. Thus they represent a paraphyletic assemblage, and in theory should be termed something cumbersome like "nontherapsid synapsids." However, they were a unique group while they lived in the Late Carboniferous and the Permian, with morphological and ecological features that set them apart from the more derived therapsids. Pelycosaurs were basically generalized amniotes, albeit with some of their own specializations, and none showed any evidence of increased metabolic rate.

Most pelycosaurs were generalized carnivores. Some (the ophiacodontids) had long snouts and multiple pointed teeth (Figure 17–4a), and were semiaquatic fish-eaters. The caseids (Figure 17–3b) and edaphosaurids were herbivores, as we can determine by their blunt, peg-like teeth. Both forms had expanded rib cages, indicating that they had the large guts typical of herbivores, and heads that look surprisingly small in comparison to their barrel-shaped bodies. The most derived pelycosaurs were the sphenacodonts such as *Dimetrodon*—mainly large, carnivorous forms with large, sharp teeth (Figure 17–4b). An enlarged canine-like tooth in the maxillary bone is a key feature linking sphenacodonts to more derived synapsids. Additional derived features include an arched palate, which was the first step toward the development of a separation of the mouth and nasal passages seen in some therapsids and in mammals. Sphenacodonts also possess a structure in the lower jaw, called the reflected lamina of the angular. This is a flange of the angular bone toward the back of the jaw that juts out and backward, creating a little pocket of space between it and the main body of the jaw (see Figure 17–4b). The reflected lamina became more prominent in therapsids and assumed an important role in the evolution of the mammalian middle ear (see Box 17-1).

A remarkable feature of some edaphosaurids and sphenacodontids was the elongation of the neural spines of the trunk region into the well-known pelycosaur sail (Figure 17–3a). These sails must have been evolved independently in the two groups, because they are absent from more primitive forms in each group and differ in structure—sphenacodontid spines are smooth and edaphosaurid spines had horizontal projections. Marks of blood vessels on the spines indicate that the tissue they supported was heavily vascularized. These vessels, combined with evidence of spines that have been broken and then healed in place, suggest that in life connective tissue and a web of skin covered the spines. There is no evidence of any degree of sexual dimorphism in the possession of these sails, so it is unlikely that they were devices evolved primarily for display (as, for example, antelope horns).

Such a large increase in the surface area of an animal would affect its heat exchange with the environment, and it seems likely that the sails were temperature-regulating devices. In the morning a *Dimetrodon* could orient its body perpendicular to the Sun's rays and allow a large volume of blood to flow through the sail, where the blood would be warmed by the Sun and the heat carried back into the animal's body. When a *Dimetrodon* was warm

▲ Figure 17–3 Diversity of nonmammalian synapsids. (a) The sphenacodontid pelycosaur *Dimetrodon*. (b) The caseid pelycosaur *Cotylorhyncus*. (c) The dinocephalian therapsid *Moschops*. (d) The gorgonopsid therapsid *Lycaenops*. (e) The cynodont *Probelesodon*.

Table 17.1 Major Groups Of Nonmammalian Synapsids

Pelycosaurs

Eothyrididae: The most primitive pelycosaurs, small (cat-size) and probably insectivorous. Early Permian of North America (e.g., *Eothyris*).

Caseidae: Large (pig-size) herbivorous forms. Middle Permian of North America and Europe (e.g., *Caseia, Cotylorhynchus* [Figure 17–3b]).

Varanopseidae: Generalized, medium-size forms. Early Permian of North America (e.g., *Varanops*).

Ophiacodontidae: The earliest known pelycosaurs (but not the most primitive). Medium-size, with long, slender heads, reflecting semiaquatic fish-eating habits. Late Carboniferous to Early Permian of North America and Europe (e.g., *Ophiacodon* [Figure 17–4a]).

Edaphosauridae: Large-size herbivores, some with a sail. Early Permian of North America (e.g., *Edaphosaurus*).

Sphenacodontidae: Large-size carnivores, some with a sail. Early Permian of North America and Europe (e.g., *Haptodus, Dimetrodon* [Figure 17–3a, 17–4b]).

Noncynodont Therapsids

Biarmosuchia: The most primitive therapsids, medium-size (dog-size) carnivores. Late Permian of Eastern Europe (e.g., *Biarmosuchus*).

Dinocephalia: Medium to large-size (cow-size) carnivores and herbivores. Some large herbivores had thickened skulls, possibly for head-butting in intraspecific combat. Late Permian of Eastern Europe and South Africa (e.g., *Titanophoneus, Moschops* [Figure 17–3c]).

Anomondontia: Small (rabbit-size) to large-size herbivores, the most diverse of the therapsids. Includes the dicynodonts, which retained only the upper canines and substituted the rest of the dentition with a turtle-like horny beak. Includes burrowing and semiaquatic forms. Late Permian to Late Triassic worldwide (including Antarctica), except for Australia (e.g., *Dicyonodon* [Figure 17–4c], *Lystrosaurus* [Figure 17–5]).

Gorgonopsia: Medium- to large-size carnivores. Late Permian of Eastern Europe and South Africa (e.g., *Scymnognathus* [Figure 17–4d], *Lycaenops* [Figure 17–3d]).

Therocephalia: Small- to medium-size carnivores and insectivores. Paralleled cynodonts in acquisition of secondary palate, complex postcanine teeth, and evidence for nasal turbinate bones. Late Permian to mid Triassic of Eastern Europe and South and East Africa (e.g., *Pristerognathus*).

Cynodont Therapsids

Procynosuchidae: The most primitive cynodonts, medium-size (rabbit-size) carnivores and insectivores. Late Permian of Eastern Europe and South and East Africa (e.g., *Procynosuchus*).

Thrinaxodontidae: Medium-size (rabbit-size) carnivores and insectivores. Early Triassic of Eastern Europe, South Africa, South America and Antarctica (e.g., *Thrinaxodon* [Figure 17–7b, 17–8]).

Cynognathidae: Medium- to large-size (dog-size) carnivores. Early Triassic of South Africa and South America (e.g., *Cynognathus* [Figure 17–4e]).

Gomphodontia: Small- (rodent-size) to medium-size (dog-size) herbivores. Include the large diademodontids (e.g., *Diademodon*, Early Triassic of Africa and East Asia) and transversodontids (e.g., *Transversodon*, Middle Triassic of South and East Africa, South America, East Asia), and the small-size tritylodontids (e.g., *Tritylodon, Oligokyphus* [Figure 17–4f]). Late Triassic to Middle Jurassic of North America, Europe, East Asia and South Africa, plus Early Cretaceous of Russia. Tritylodonts were rodent-like forms that paralleled the mammalian condition of the postcranial skeleton. Some researchers consider them as separate from the other gomphodonts, and as the sister group to mammals.

Chiniquodontidae: Small-size carnivores and insectivores, closely related to Tritheledontidae (not included in Figure 17–1). Middle to Late Triassic of North and South America (e.g., *Probelesodon* [Figure 17–3e, 17–7c]).

Tritheledontidae: Small-size carnivores and insectivores. Approached mammalian condition in form of jaw joint and postcranial skeleton. Late Triassic of South America to Early Jurassic of South Africa (e.g., *Diarthrognathus*).

Large, medium, and small refer to size within a particular group. A medium-size noncynodont therapsid is much larger than a medium-size cynodont.

enough, blood flow through the sail could be restricted, and the heat would be retained within the body.

Therapsids—More Derived Nonmammalian Synapsids

From the Middle Permian to the Late Triassic, there was a flourishing fauna of more derived synapsids that are grouped under the general name of therapsids. (Note that the strictly correct name here should be "nonmammalian therapsids," because these animals include the ancestors of mammals. The term *therapsid* thus includes not only these beasts but also mammals.) Therapsids all had modifications suggesting an increase in metabolic rate over the more primitive pelycosaurs (see Figure 17–6); thus they are often portrayed as possessing hair. They were all fairly heavy-bodied, large-headed, stumpy-legged forms (Figures 17–3c, and 17–5). The image of this body form combined with incipient hairiness prompted the cartoonist Larry Gonick (1990) to declare them "too ugly to survive."

Therapsid Diversity Like the earlier pelycosaurs, therapsids radiated into herbivorous and carnivorous forms. Some of the herbivorous therapsids were large, heavy-bodied, slow-moving animals (Figures 17–3c, 17–5), and some of them may have congregated in herds as ungulates (hoofed mammals) do today. Other herbivorous forms were small and superficially rather like rodents. The carnivorous therapsids included large, ferocious-looking animals (Figure 17–3d) that may have played an ecological role similar to that of big cats today, smaller ones (Figure 17–3e) that may have been more fox-like, and rabbit-size forms that were probably insectivorous. There were numerous different lineages of therapsids. One of them—the cynodonts—we hold as especially important, because this was the group that gave rise to mammals. While cynodonts were highly derived therapsids, other therapsid lineages could be considered equally derived in their own particular fashion (e.g., the dicynodonts, with their specializations for herbivory).

Therapsids were known from a wider biogeographical range than pelycosaurs. Pelycosaurs have been found primarily in deposits in North America and in Russia, parts of the world that were near the paleoequator at that time (Chapter 7). Late Permian therapsids were found farther afield, becoming particularly common in deposits in southern Africa, a region that was under glaciation in the Carboniferous. Latest Permian and Triassic therapsids were known from all over Gondwana, as well as from the Northern Hemisphere. The radiation of therapsids into more southern, colder climes has been held as one piece of evidence suggesting they had a higher metabolic rate than the pelycosaurs. There also appears to be a profound difference in the types of terrestrial ecosystems inhabited by therapsids. The pelycosaur-dominated communities of the Late Carboniferous and Early Permian were still rooted in aquatic ecosystems, while the later therapsid-dominated faunas had terrestrial plants as their trophic base.

The therapsids were the dominant large land mammals in the Late Permian, sharing center stage only with the large herbivorous parareptilian pareiasaurs. The transition from the Permian to the Triassic was a period of great extinction. From the diversity of Permian therapsids, only the tusked dicynodonts (specialized herbivores) and the therocephalians and the cynodonts (derived carnivores) survived and diversified in the Early Triassic. But other vertebrate groups also diversified in the Triassic, most notably the diapsid reptiles (see Chapter 14). Therapsids became an increasingly minor component of the terrestrial fauna during the Triassic, with near extinction at its end. Only three groups of therapsids—the true mammals and the cynodont tritylodontids and tritheledontids, all small-size forms, survived into the Jurassic.

Derived Features of Therapsids Primitive therapsids differed from sphenacodontid pelycosaurs in several ways that appear to be related to increasing specialization as predators. The larger temporal fenestra provided space for the origin of large external adductor muscles on the skull roof, the upper canines were longer, and the entire skull was more rigid than it was in pelycosaurs. The only differentiation of the dentition in sphenacodonts was a large canine tooth (Figure 17–4b), but the dentition of therapsids showed a distinct differentiation into incisors, canines, and postcanine teeth (Figure 17–4d, e). The postcranial skeleton of therapsids also showed changes from the pelycosaur condition—the pectoral and pelvic girdles were less massive, and the limbs more slender. The shoulder joint appears to have allowed more freedom of movement of the forelimb (see Figure 17–6).

More derived therapsids include the dicynodonts (Anomodontia) and the theriodonts (gorgonopsids,

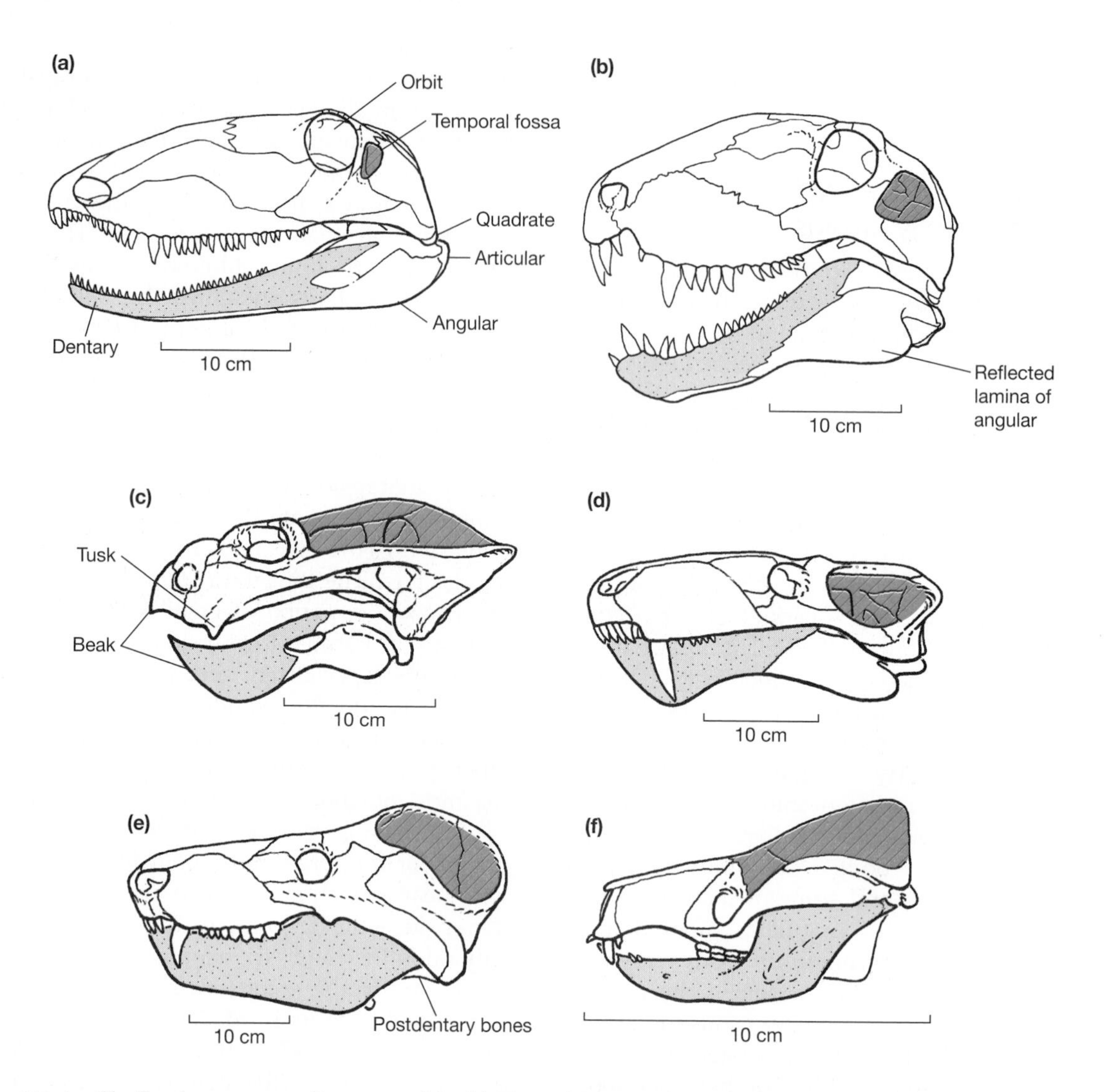

▲ Figure 17–4 Skulls of nonmammalian synapsids. (a) The ophiacodontid pelycosaur *Ophiacodon*. (b) The sphenacodontid pelycosaur *Dimetrodon*. (c) The dicynodont therapsid *Dicynodon*. (d) The gorgonopsid therapsid *Scymnognathus*. (e) The cynodont *Cynognathus*. (f) The tritylodontid cynodont *Oligokyphus*. Light-blue stippling = the dentary bone, dark blue cross-hatching = the temporal fossa.

therocephalians and cynodonts). Dicynodonts dominated terrestrial diversity in the Late Permian and diversified into several different ecological types, including burrowers and semiaquatic forms such as *Lystrosaurus* (Figure 17–5). A distinctive feature of dicynodonts was the extreme specialization of the skull for a herbivorous diet. In most forms all of the marginal teeth were lost, except for the upper canines, which were retained as a pair of tusks (Figure 17–4c); the jaws were covered with a horny beak like that of turtles. A beak has some advantages over teeth as a structure for grinding plant material. Unlike a row of teeth, a horny beak can provide a continuous cutting surface, and it can be replaced continuously as it is worn away. The structure of the jaw articulation of dicynodonts permitted extensive fore-and-aft movement of the lower jaw, shredding the food between the two cutting plates.

▲ Figure 17–5 A reconstruction of the Early Triassic dicynodont therapsid *Lystrosaurus*.

Theriodonts, the major predators of the Late Permian and Early Triassic, were characterized by the development of the **coronoid process** of the dentary, a vertical flange near the back of the lower jaw that provided additional area for the insertion of the adductor musculature (see Figure 17–4d–f). This process would provide a lever arm for the action of the jaw closing muscles, perhaps important in these initially carnivorous therapsids. Accompanying changes in the skull roof opened up the temporal fossa completely in the therocephalians and cynodonts, providing more space for the origination of these jaw muscles.

Cynodonts appeared in the Late Permian, had their heyday in the Triassic, and were mostly extinct by the end of that period. However, some lineages persisted into the Jurassic (tritheledontids) and even into the Early Cretaceous (tritylodontids). Of course, it could be observed that cynodonts have persisted as mammals to the present day, and so these extinct forms should more properly be termed nonmammalian cynodonts. A marked reduction in body size characterized the cynodont lineages, especially the carnivorous forms. Some early cynodonts were the size of large dogs, but by the Middle Triassic the carnivorous cynodonts were only about the size of a rabbit. The earliest mammals of the latest Triassic were less than 100 millimeters, about the size of a shrew.

Cynodonts had a variety of derived features, making them more mammal-like than other therapsids (see Figure 17–6). All cynodonts had multicusped cheek teeth (that is, with small accessory cusps anterior and posterior to the main cusp). The herbivorous gomphodonts had expanded teeth, with blunt cusps forming opposing grinding and cutting surfaces on upper and lower teeth (Figure 17–4f). The lower jaw was apparently pulled backward into occlusion with the upper jaw, similar to the mode of jaw movement in dicynodonts. The sculptured surfaces of the bones forming the upper and lower jaws suggest the presence of blood vessels and skin glands, with the inference that cynodonts (and also therocephalians) had a mammal-like muzzle and lips, perhaps with whiskers. An enlarged infraorbital foramen—the hole under the eye through which the sensory nerves from the snout pass back to the brain—suggests a highly innervated face, and so also supports the notion of a more mobile, sensitive muzzle. Cynodonts and therocephalians also had evidence of turbinates, scroll-like bones in the nasal passages that warm the incoming air and help to prevent respiratory water loss.

17.3 Evolutionary Trends in Synapsids

The synapsid lineage crossed a physiological boundary as the animals moved from ectothermy to endothermy, and this change was accompanied by changes in ecology and behavior. Physiology, ecology, and behavior do not fossilize directly; but some of the changes that were occurring in metabolism, ventilation, and locomotion can be traced indirectly through changes in the skull and the axial skeleton.

Skeletal Modifications and Their Relationship with Metabolic Rate

Numerous changes in the skull and skeleton of synapsids can be observed over evolutionary time. One way to interpret these changes is to map them onto a phylogeny that considers the mammalian condition as the inevitable evolutionary outcome, and to congratulate the nonmammalian synapsids for achieving ever more mammal-like status. However, there is no simple foresight in evolution: The changes in the skeleton that were occurring were not simply directed toward some mammalian condition, but rather reflected an increasing level of metabolic activity, adaptive for these animals in their own right. Some of the most obvious changes are shown in Figure 17–6.

Metabolic rate does not directly show on the skeleton, but features associated with a higher metabolic rate may indeed be apparent. Animals with high metabolic rates require greater amounts of food and oxygen per day, so any changes suggesting improvements in the rate of feeding or respiration may indicate metabolic rate increase. Only animals with high metabolic rates are capable of sustained activity. Thus indicators of greater levels of activity may also reflect higher metabolic rates.

Numbers of the features listed here refer to those in Figure 17–6.

1. *Size of the temporal fenestra*—A larger fenestra indicates a greater volume of jaw musculature, and hence implies more food eaten per day. Small in pelycosaurs, it suggests a fairly small size of jaw-closing musculature. An increasingly larger opening in more derived synapsids, along with an increasing tendency to enclose the braincase with dermal bone, results in a distinct **temporal fossa** for the origin of a larger volume of jaw musculature (see also Figure 17–4). The external adductor (jaw-closing) muscle now passes through the fenestra to originate from the lateral surface of the skull roof. In the mammal the fossa is enlarged further with the loss of the postorbital bar, so that the orbit is confluent with the temporal fossa.
2. *Condition of the lower temporal bar*—A bar bowed out from the skull behind the orbit indicates the presence of a **masseter muscle**, originating from this bony bar and inserting on the lower jaw, again suggesting more food processing (see Figure 17–7). The temporal bar lies very close to the upper border of the lower jaw in pelycosaurs and noncynodont therapsids, indicating that no part of the jaw musculature could have inserted on the outside of the lower jaw. In cynodonts and mammals the bar is bowed outward, forming the **zygomatic arch** and indicating the presence of a masseter muscle. A corresponding fossa on the dentary in these animals also indicates the presence of this muscle.
3. *Lower jaw and jaw joint*—Changes reflect the increased compromise between food processing and hearing in synapsids, as explained later. The lower jaw in pelycosaurs resembles the general amniote condition. The tooth-bearing portion, the **dentary**, takes up only about half of the jaw. By the level of cynodonts, the dentary has greatly expanded, and the postdentary bones have been reduced in size (see also Figure 17–4). In mammals the dentary now forms a new jaw joint with the skull.
4. *Teeth*—Greater specialization of the dentition reflects an increased emphasis on food processing.

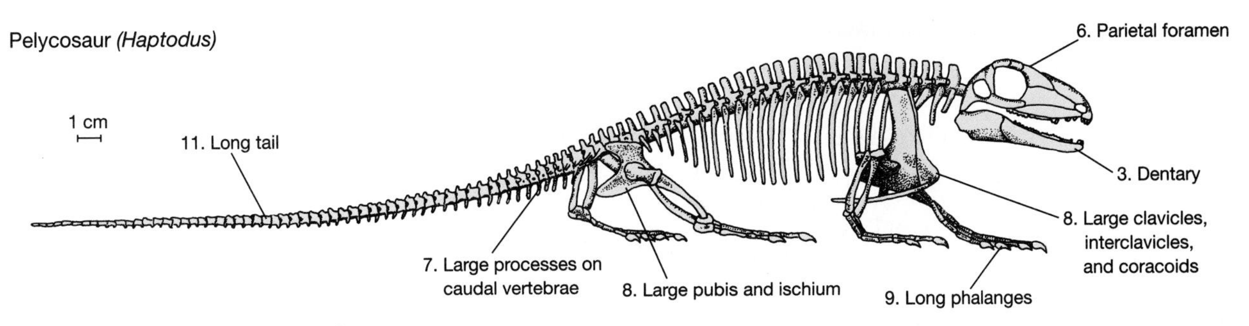

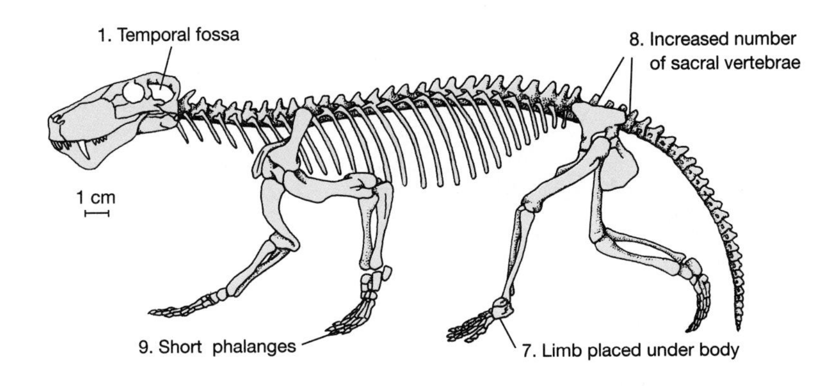

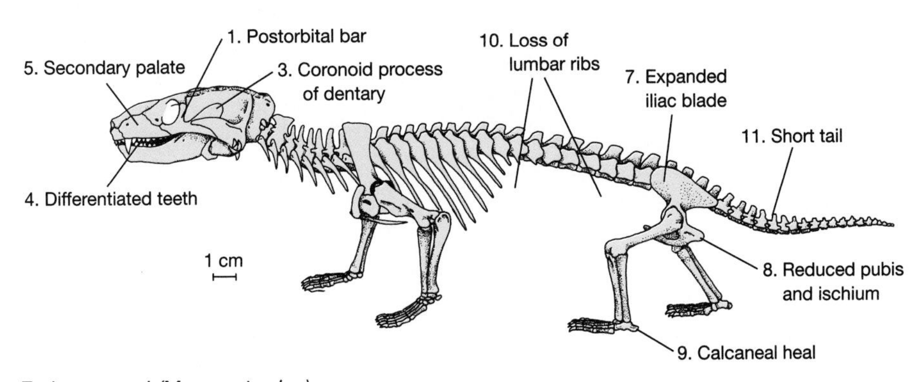

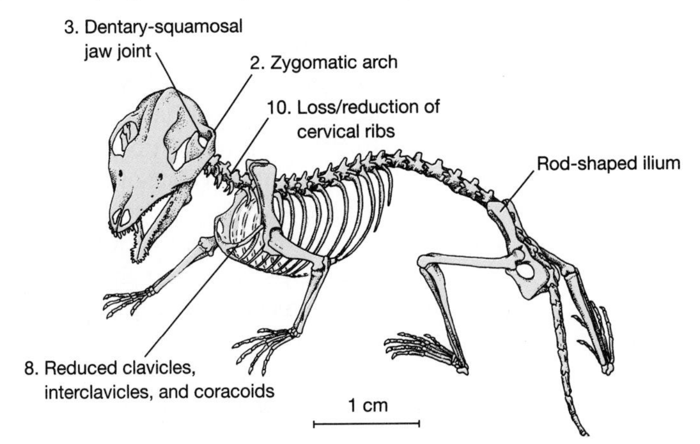

▲ Figure 17–6 Skeletal features of synapsids.

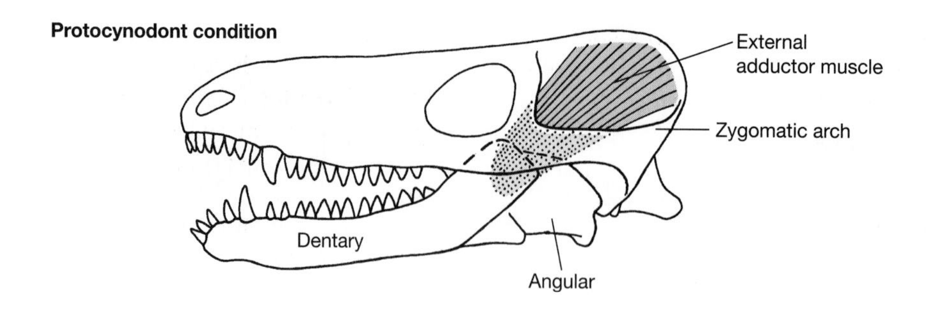

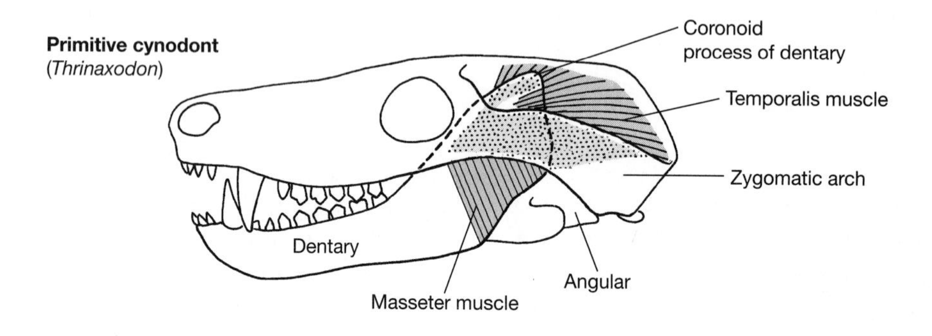

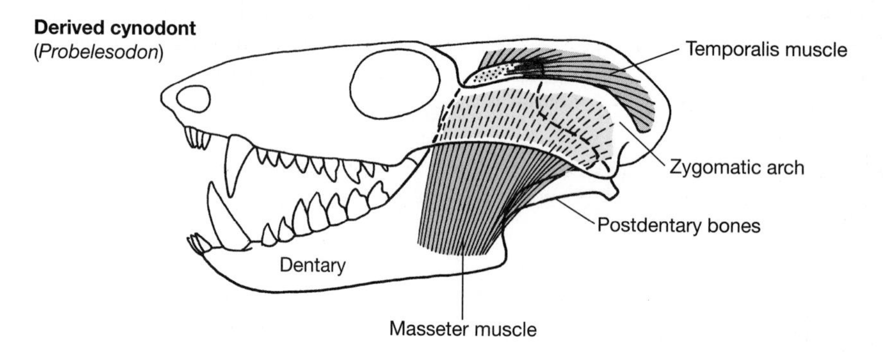

▲ Figure 17–7 Cynodont skulls and jaw musculature.

The teeth of pelycosaurs are **homodont**; that is, they are virtually all the same size and shape, with no evidence of regionalization of function. In more derived synapsids the teeth become increasingly **heterodont**; that is, differentiated in size, form, and function, although actual mastication with precise occlusion of the cheek teeth is a mammalian feature. All therapsids have the teeth differentiated into incisors, canines, and postcanine teeth, and in cynodonts these postcanine teeth may be complex. Mammals have reduced the number of tooth replacements to two (**diphyodonty**), have the postcanine teeth further differentiated into premolars (replaced) and molars (not replaced), and have the lower teeth set closer together than the uppers—indicative of chewing with a rotary jaw motion.

5. *Development of a secondary palate*—A secondary palate allows breathing and eating at the same time, and also helps to bolster the skull against stresses from increased amounts of food processing. No secondary palate is apparent in pelycosaurs. An incipient, incomplete one is present in some noncynodont therapsids, and a

complete one is present in derived cynodonts and in mammals (dicynodont therapsids evolved a secondary palate convergently). Mammals also have merged the originally double nasal opening into a single median one, probably reflecting an increase in size of the nasal passages and a higher rate of respiration.

6. *Presence of a parietal foramen*—A hole in the skull for the pineal eye reflects control of temperature regulation by behavioral means. It is present in pelycosaurs and in most therapsids, lost within cynodonts (and, convergently, in some therocephalians).
7. *Position of the limbs*—Limbs placed more underneath the body (upright posture) are reflective of a higher level of activity, resolving the conflict between running and breathing (see later discussion). Pelycosaurs have the sprawling limb posture typical of primitive amniotes. All therapsids show some degree of development of an upright posture. Also in therapsids evidence can be seen, in the expanded iliac blade and the development of the greater trochanter of the femur, of a switch to a gluteal type of hindlimb musculature typical of mammals (see Chapter 19). In pelycosaurs, large processes on the caudal vertebrae indicate the rentention of the more primitive amniote method of limb retraction using the caudofemoralis muscle.
8. *Shape of the limb girdles*—The primitive amniote condition reflects a sprawling posture, with large ventral components to the limb girdles. With a more upright posture, more of the weight passes directly through the limbs, and the supportive undercarriage of the limb girdles can be reduced. Pelycosaurs have large clavicles, interclavicles and coracoids in the pectoral girdle, and large pubes and ischia in the pelvic girdle. Therapsids have more lightly built girdles with the reduction of the ventral elements and the expansion of the dorsal ones, such as the iliac blade. Therapsids also show an increase in the number of sacral vertebrae from the two or three of pelycosaurs to four or more. Mammals have a very reduced pubis and a rod-shaped ilium, probably reflecting a change in muscle positioning and muscle forces generated with the change in the vertebral column to allow dorsoventral flexion (see item 10).
9. *Shape of the feet*—Long-toed feet indicate feet used more as holdfasts, typical of a sprawling gait. Short-toed feet indicate feet used more as levers, with a more upright posture. Pelycosaurs have long fingers and toes. All therapsids and mammals have shorter feet, with derived cynodonts and mammals having pattern of two segments on the first digit (thumb or big toe), and three on the other four fingers or toes (count this on your own hands) A distinct **calcaneal heel** is seen in cynodonts and mammals, providing a lever arm for a greater degree of push-off from the gastrocnemius (calf) muscle. Mammals also have an opposable big toe.
10. *Form of the vertebral column*—The loss of the lumbar ribs suggests the presence of a diaphragm, indicating a higher rate of respiration. Lumbar ribs are present in pelycosaurs and noncynodont therapsids, are absent from mammals, and are reduced or absent in cynodonts. With the complete loss of the lumbar ribs, mammals have evolved distinctive differences between the thoracic and lumbar vertebrae, indicative of the mammalian mode of dorsoventral flexion, although the distinctive bounding gait of modern mammals is probably limited to therians (see Chapters 19 and 20). Mammals have also reduced or lost the ribs on the cervical vertebrae, and restricted the number of neck vertebrae to seven. Seven cervical vertebrae are also seen in cynodonts and most other therapsids.
11. *Tail*—A long, heavy tail, as in pelycosaurs and the primitive amniote condition, reflects locomotion conducted primarily by axial movements. A shorter tail, as in most therapsids and mammals, reflects a more upright posture where limb propulsion is more evident than axial flexion. A more slender tail base in these animals is also indicative of the switch in limb retractor muscles (see item 7).

Evolution of Jaws and Ears

The original synapsid condition of the lower jaw was for a tooth-bearing dentary occupying the anterior jaw half, with a variety of bones (known collectively as **postdentary bones**) forming the posterior half, as seen in other tetrapods and bony fishes. The jaw articulation was between the **articular** in the lower jaw and the **quadrate** in the skull. A trend within cynodonts was to enlarge the size of the dentary and decrease the size of the postdentary bones. This trend was probably related to the increase in the volume of jaw adductor musculature, which inserted onto the dentary (Figure 7–7). In the most derived cynodonts a **condylar process** of the dentary grew backward, and eventually contacted the squamosal bone of the skull. In mammals and some very derived cynodonts, this contact between the dentary and the squamosal formed a new jaw joint, the **dentary-squamosal** jaw joint. In these derived cynodonts, and in the earliest mammals, this new jaw joint coexisted with the old one; but in later mammals the dentary-squamosal jaw joint is the sole one. The bones forming the old jaw joint are now part of the middle ear (Box 17-1).

The bare facts of the transition have been known for a century or so, but the evolutionary interpretation of these facts has changed over time. Originally it was assumed that the lizard-like middle ear, with the stapes alone forming the auditory ossicle, was the primitive condition for all tetrapods, and hence the one possessed by the early synapsids. We now have good evidence that an enclosed middle ear evolved separately in modern amphibians and amniotes, and probably at least three times convergently within amniotes (see Chapter 8). But before scientists had this information, they assumed that with the transition to mammals, an originally single-boned middle ear was transformed into a three-boned one. (This evolutionary scenario also carried the tacit assumption that the mammalian condition of both ear bones and jaw articulation was somehow superior to that of other tetrapods.) The change in jaw morphology was not assumed to be directly connected with the change in the ears. Rather, it was assumed that once a new jaw joint was formed, the leftover bones became incorporated into the middle ear.

On reconsideration, there are some problems with this story. Even if it were true that the mammalian type of jaw joint *was* somehow superior, how could researchers explain the millions of years of cynodont evolution during which the dentary was enlarging prior to contacting the skull? Did those earlier cynodonts know that their descendants would one day make a better jaw joint out of this bone? Why is the original jaw articulation an inherently weak one? It appears to work well enough in other vertebrates (no one has ever accused *Tyrannosaurus rex* of having a weak jaw). And why disrupt a perfectly functional middle ear to insert some extra bones that happened to be available? Even if a three-boned middle ear were ultimately superior, there would be a period of adjustment to a new condition that would probably be less effective than the previous condition.

These questions were addressed, and solved, by Edgar Allin (1975). He came up with the suggestion (revolutionary at the time, but in retrospect merely sensible) that the chain of bones that make up the middle ear of mammals had been used for hearing all along in synapsids (at least since the level of derived pelycosaurs; see Box 17-1). Using the same set of bones for two different functions, as a jaw joint and as a hearing device, might seem like a rather clumsy, makeshift arrangement in comparison with that seen in modern-day animals. However, this mode of hearing would probably be adequate for the early synapsids which, as ectotherms with low food intakes, did not use their jaws a lot in feeding.

But a problem would arise in the more derived synapsids, which had a higher metabolic rate. A higher metabolic rate would need to be fueled by a greater amount of food intake, and so would require a greater volume of jaw musculature and greater use of the jaw in feeding. This would lead to greater stresses on the jaw joint, which would be incompatible with the role of these bones in hearing. Allin interpreted the evolutionary history of the jaw in cynodonts as representing an *evolutionary conflict* between the functions of feeding and hearing.

An initial evolutionary step would be to enlarge the dentary bone and transfer all the jaw-muscle insertions to this bone. This change would isolate the auditory postdentary bones somewhat from the feeding apparatus, allowing them to become smaller as befits auditory ossicles. Note that the reason earlier researchers had interpreted the jaw of synapsids as weak is that the postdentary bones remained loose and wobbly; further, they did not fully fuse with one another and with the dentary as in many other tetrapods. However, given Allin's interpretation, there was a very good reason for these bones *not* to fuse up: If they had done so, they would have compromised their role as vibrating auditory ossicles. Figure 17–9 illustrates the possible appearance of a cynodont with this type of ear, with an eardrum held in the lower jaw.

BOX 17-1 Evolution of the Mammalian Middle Ear

More than a century ago, embryological studies demonstrated that the malleus and incus of the middle ear of mammals were homologous with the articular and quadrate bones that formed the ancestral jaw joint of gnathostomes. More recently the transition has been traced in fossils, from its beginning in basal synapsids through therapsids to early mammals. Nonmammalian synapsids have a jaw joint formed from the quadrate (in the skull) and the articular (in the lower jaw), as in other jawed vertebrates. The first indication of a mammalian type of middle ear is seen in the sphenacodontid pelycosaurs (Figure 17–4b). These animals have a structure called the "reflected lamina" on the angular bone of the lower jaw, which is believed, at least in later synapsids, to have housed a tympanum (eardrum). In therapsids the stapes, which acted as a skull brace in primitive amniotes, now has an articulation with the quadrate, as in the primitive gnathostome condition of a hyomandibula (= stapes)-quadrate contact. The quadrate-stapes contact in therapsids resembles the incus-stapes articulation of mammals, suggesting that they were using their jaw joint to hear with as well as to form the jaw hinge.

The transition from the nonmammalian to the mammalian condition can be visualized by comparing the posterior half of the skull of *Thrinaxodon*, a fairly primitive cynodont (Figure 17–8a), with that of *Didelphis*, the Virginia opossum (Figure 17–8b). The reflected lamina was probably the principal support of the tympanic membrane, and vibrations were transmitted to the stapes via the articular and quadrate. The mammalian jaw joint is a new structure, formed by the dentary of the lower jaw and squamosal bone of the skull. The tympanum and middle ear ossicles lie behind the jaw joint and are much reduced in size, but they retain the same relation to each other as they had in cynodonts. The lower jaw of a fetal mammal viewed from the medial side shows that the angular (tympanic), articular (malleus), and quadrate (incus) develop in the same positions they have in the cynodont skull (Figure 17–8c). The homologue of the angular, the tympanic bone, supports the tympanum of mammals. The retroarticular process of the articular bone is the manubrium of the malleus, and the ancestral jaw joint persists as the articulation between the malleus and incus.

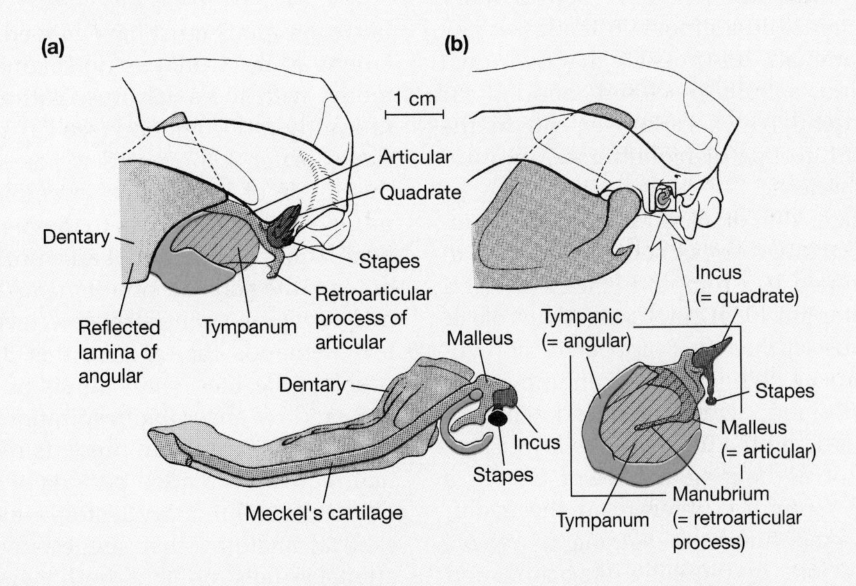

▲ Figure 17–8 Anatomy of the back of the skull and the middle ear bones of synapsids. (a) *Thrinaxodon*, a cynodont. (b) *Didelphis*, the Virginia opossum. (c) Embryonic mammal.

▲ Figure 17–9 Reconstruction of the head of the primitive cynodont *Thrinaxodon*, showing the probable appearance of the ear region.

In progressively derived cynodonts, the dentary increased further in size. Finally, the condylar process of the dentary was large enough to contact the squamosal in the skull and to form a new jaw joint, freeing up the original jaw joint from its old function and allowing it to be devoted entirely to the function of hearing. A classic example of an evolutionary intermediate is provided by the tritheledontid *Diarthrognathus*. This animal gets its name from its double jaw joint (Greek: *di* = two, *arthro* = joint, and *gnath* = jaw). It has both the ancestral articular-quadrate joint and, next to it, the mammalian dentary-squamosal articulation. These postdentary bones are still retained in a groove in the lower jaw in the earliest mammals. It is possible that their final incorporation into a fully enclosed middle ear occurred convergently, to a certain extent, in the different mammal groups of monotremes, multituberculates, and therians.

Since 1975, when Allin devised the preceding evolutionary scenario, other researchers have added to the story and refined it. Crompton (Crompton and Parker 1978, Crompton 1985) emphasized the role of the masseter muscle that first appeared in cynodonts. In mammals the masseter can move the lower jaw laterally and, acting in concert with the pterygoideus muscle on the inside of the lower jaw, can be thought of as holding the lower jaw in a supportive sling. Crompton interpreted the evolution of the masseter muscle as helping to relieve stresses at the jaw joint by this sling-like action, and so helping to resolve the conflict between jaw use and hearing.

Bramble (1978) pointed out that while a fully formed masseter might act in this fashion, a small, incipient one at its evolutionary appearance would not perform a sling-like role. He suggested that a small, forwardly directed masseter would counteract the backward pull of the enlarged temporalis, which otherwise might act to dislocate the jaw when the animal bit down hard on something. He also interpreted a preadaptive role for the condylar process of the dentary that eventually forms the new jaw joint. This process increases in size in cynodonts, but its initial function prior to the formation of a new jaw joint had never been deciphered. Bramble's biomechanical analysis of jaw function interpreted the initial function of this structure as preventing dislocation between the dentary and the postdentary bones during biting.

Then, in 1996, Tim Rowe considered that the final separation of the middle ear bones from the lower jaw in true mammals is correlated with an increase in the size of the mammalian forebrain. He proposed that the expansion of the skull during embryonic development to accommodate the expansion of the brain would dislocate these bones from their original position, and thus provide the initial condition for their subsequent enclosure in a discrete middle-ear cavity distinct from the lower jaw.

Locomotion and the Evolution of Upright Posture

Early tetrapods must have moved with lateral undulations of the trunk, as do salamanders and lizards today, with the axial musculature contributing significantly to locomotion (see Chapter 8). David Carrier (Carrier 1987) noted a potential problem with this mode of locomotion in amniotes. The hypaxial muscles the animal uses to move itself also aid in rib ventilation, meaning that an animal like a lizard cannot breathe and run simultaneously. For an ectothermal tetrapod, with fairly low levels of activity and low demands for oxygen, this does not present a problem. Besides, short bursts of speed are initially powered by anaerobic respiration in all vertebrates, even in us. A problem presents itself only if the animal requires extended periods of sustained locomotion. But evolutionary history—as exemplified by a skeletal anatomy that suggests active, fast-running animals—tells us that both synapsid and archosaurian lineages did include fast-moving animals. That is, with increasing levels of activity, there developed an *evolutionary conflict* between running and breathing, similar to the conflict that we have just considered between feeding and hearing.

Figure 17–10 illustrates the problem. In addition to the conflict in muscle use, the side-to-side bending of the lizard's rib cage pushes the air from one lung into the other, interfering with airflow in and out of the mouth. The immediate solution to this problem is to devise a means of locomotion that allows the trunk to be held rigid and unbending, involving instead the primary use of the limbs for propulsion. Therapsids solved this problem by adopting an upright posture, bringing their limbs more underneath the trunk. With the limbs in this position, their use in propulsion will have less of a bending effect on the trunk. A different way of solving this problem was adopted by the archosaurian amniotes with the evolution of bipedality, as seen in dinosaurs and birds. Bipedal locomotion involves only the hindlimbs, without movements of the trunk. However, among synapsids only a few derived mammals developed bipedality.

Further refinements came with true mammals, which had the capacity for dorsoventral flexion of the vertebral column—possible only after the lumbar ribs had been lost or reduced. This loss of ribs, first seen in cynodonts, may have been associated with the acquisition of a muscular diaphragm. Living mammals use the diaphragm, in addition to the ribs, to inhale air into the lungs. Experiments in which the diaphragm was immobilized showed that it is especially important in obtaining additional oxygen during activity (Ruben et al. 1987).

The bounding gait of derived, therian mammals allows respiration and locomotion to work together in a *synergistic* fashion, rather than in a conflicting one. The inertial movements of the viscera (especially the liver), modulated by contractions of the diaphragm, help to force air in and out of the lungs with each bounding stride (Figure 17–10; Bramble and Carrier 1983). Humans have little direct experience of this basic mammalian condition: Our bipedal locomotion has led to a complete decoupling of locomotion and ventilation, just as in archosaurs. But an

▲ Figure 17–10 The effect of axial bending on lung volume of a running lizard (top view) and a galloping dog (side view). The bending axis of the lizard's thorax is between the right and left lungs. As the lizard bends laterally, the lung on the concave (left) side is compressed and air pressure in that lung increases (shown by +), while air pressure in the lung on the convex (right) side is reduced (shown by –). Air may be pumped between the lungs (arrow), but little or no air moves in or out of the animal. In contrast, the bending axis of a galloping mammal's thorax is dorsal to the lungs. As the vertebral column bends, volume of the thoracic cavity decreases; pressure in both lungs rises (shown by +), pushing air out of the lungs (arrow). When the vertebral column straightens, volume of the thoracic cavity increases, pressure in the lungs falls (shown by –), and air is pulled into the lungs (arrow).

appreciation of this condition helps to explain certain features of locomotion in quadrupedal mammals. For example, a galloping horse increases its speed by increasing its stride length rather than its stride frequency, thus avoiding interference with the rhythm of lung ventilation.

Evolution of the Nasal and Palatal Regions

We have been considering various skeletal features in synapsids that might be clues to their metabolic status. John Ruben and his co-workers (Ruben 1995, Hillenius 1992) have suggested that one structure that might be considered as a sure sign of endothermy is the presence of nasal turbinate bones—thin, scroll-like bones—seen today in the living endothermal vertebrates, birds and mammals (see Chapter 15). The maxilloturbinates, in particular, are important in conserving water during respiration; the naso- and ethmoturbinates are more concerned with olfaction (Figure 17–11).

High rates of ventilation, as demanded by an endothermal metabolism, present a problem for air-breathing vertebrates. The tissues of the lungs where gas exchange takes place are extremely thin, because gas must be exchanged very rapidly via diffusion; these tissues are therefore susceptible to drying. To minimize water loss, the inhaled air must be warmed to lung temperature and saturated with water vapor before reaching the gas-exchange surfaces. This process occurs as the air passes over warm, moist surfaces in the nasal passages. During exhalation water condenses on the turbinates—which are relatively cooler than the air expired from the lungs—thus conserving water. The higher the rate of ventilation, the more elaborate are the structures a vertebrate needs to conserve respiratory water loss.

Lizards have low rates of ventilation, and the simple walls of their nasal passages are adequate to treat the air before it reaches their lungs. Birds and mammals have much higher rates of ventilation than lizards, and require structures that increase the surface area of the nasal passages. Although these fragile turbinate bones do not usually fossilize, the ridges where they attach in the nasal passages *do* preserve, at least in synapsids. Those ridges are found both in therocephalians and cynodonts, suggesting mammal-like rates of respiration in these animals.

There are other fundamental differences between living mammals and reptiles in the mouth and pharyngeal regions. The secondary palate of mammals, seen also in cynodonts, completely separates the nasal passages from the mouth. This structure indicates a higher metabolic rate in a variety of ways; as previously discussed, it reflects both increased oral food processing and the need to eat and breathe at the same time.

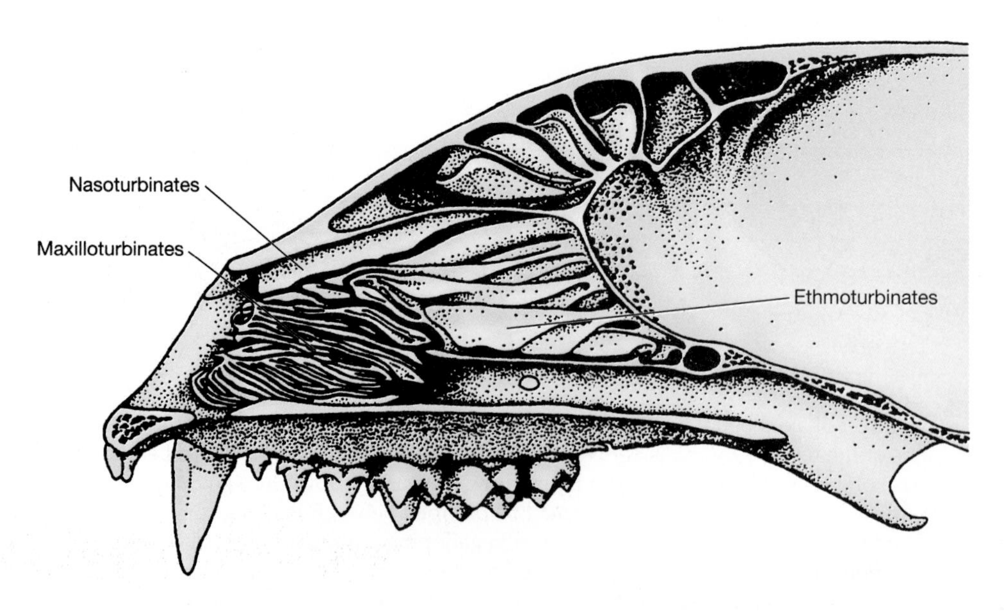

▲ Figure 17–11 Longitudinal section through the snout of a raccoon, showing the nasal passages and turbinate bones.

17.4 The First Mammals

Living mammals today are characterized by two salient features: hair and mammary glands (Latin *mamma* = a teat). Neither of these features is directly preserved in the fossil record, although we will propose that we can infer the point in synapsid evolution at which these features were acquired. Another feature typical of most living mammals is viviparity—giving birth to young rather than laying eggs. However, we know that this was not a feature of the earliest mammals because some living primitive mammals, the **monotremes** (the platypus and echidna of Australia and New Guinea), still lay eggs. How, then, can we define where in synapsid evolutionary history we should start to apply the term *mammal*?

We have already argued that at least some cynodonts had higher metabolic rates than the other therapsids; and if they were indeed endothermal, it is possible they had hair for insulation. However, some unexpected bony evidence allows some inferences about when an insulating coat of fur first evolved. Modern mammals have a gland associated with the eye, called the **Harderian gland.** This gland secretes an oily substance that travels down the nasolacrimal duct to the nose, where it can be wiped off on the paws. This secretion is then used for preening the fur, providing a lipid barrier to cold and wetness, probably an essential feature for the insulating fur of a small mammal.

The presence of a Harderian gland is reflected in the skull, and the fossil record shows that this gland was not present in any cynodont. But it was present in *Morganucodon,* one of the earliest known mammals (Hillenius 2000). This suggests that a hairy coat was a feature of the first mammals. Cynodonts may have had mammal-like metabolic rates, and may have had some type of hair (perhaps sensory whiskers). Perhaps the need to evolve fur as insulation was not really pressing until the very small size of the earliest mammals, which would have been more subject to heat loss due to the relatively greater body surface area (Ruben and Jones 2000).

Traditionally the acquisition of a dentary-squamosal jaw joint was used to define the first fossil mammal, because only mammals possess this type of jaw joint today. Later evidence revealed that some animals we would still classify as derived cynodonts, such as *Diarthrognathus,* also possessed a dentary-squamosal jaw articulation while still retaining the old quadrate-articular one. On the other hand, some of the earliest mammals also retained this old jaw joint. In practice, it has always been pretty easy to determine an early mammal. They are tiny, a couple of orders of magnitude smaller than cynodonts (a body mass of less than 100 grams—shrew size). In contrast, the smallest cynodonts would have weighed about a kilogram—the size of a rabbit—except for some smaller tritylodontids that were around the size of a mouse or a rat. But did this transition to a very small body size really signify a biological shift into a different type of adaptive zone that we can clearly signify as mammalian?

The traditional point of mammalian transition is marked not only by this jaw joint but also by some derived features of the skull, reflecting enlargement of the brain and inner ear regions, as well as by postcanine teeth with divided roots. Other workers (e.g., Rowe 1988) would prefer to limit the group Mammalia to those animals bracketed by the interrelationships of surviving mammals (i.e., to the "Crown Group Mammalia" in Figure 17–1). However, we will argue later for the traditional division between mammals and cynodonts to represent some real biological transitions, including the evolution of lactation and suckling.

Features of the Earliest Mammals

The oldest well-known mammals include *Morganucodon* (also known as *Eozostrodon*) from the earliest Jurassic of Wales (Figure 17–12), and *Megazostrodon* from South Africa (Figure 17–6d). A somewhat older possible mammal is *Adelobasileus* (about 225 million years old), known only from an isolated braincase

▲ Figure 17–12 Reconstruction of the Early Jurassic mammal *Morganucodon.*

from the Late Triassic of Texas. Another Early Jurassic mammal is the Chinese *Sinoconodon*, known from a skull that appears to be much more primitive than that of *Morganucodon*. Most of our information about mammals comes from their teeth. At their small size, the fragile mammal bones do not easily preserve; but the harder, enamel-containing teeth are more likely to fossilize. Fortunately, mammalian teeth turn out to be very informative about their owners' life-style.

Mammals are diphyodont; that is, they have only two sets of replacing teeth (like our milk teeth and our permanent teeth) and the molars are not replaced at all, but erupt fairly late in life. Mammals also have molars with precise occlusion that is produced by an interlocking arrangement of the upper and lower teeth. We can determine this in fossils because the teeth show distinct wear facets, produced from abrasion between teeth and showing that the teeth always meet in the same manner. Precise occlusion makes it possible for the cusps on the teeth to cut up food very thoroughly, creating a large surface area for digestive enzymes to act on, and thereby promoting rapid digestion. Only mammals **masticate** (thoroughly chew) their food in this fashion. The cheek teeth of mammals also are set in an alternating fashion, such that each upper tooth occludes with two lower teeth and each lower tooth occludes with two upper teeth. In addition, the lower jaws of mammals are closer together than are the upper jaws, a condition termed **anisognathy**.

In the anisognathous condition, the lower jaw must be moved sideways (outward) before the lower teeth can contact the uppers; since the two sides of the jaws are connected, only one side can be used for chewing at any one time. The mammalian condition, where the jaws are moved in a rotary fashion, contrasts with the simple up-and-down movement of the jaws in reptiles, where the upper and lower teeth are the same distance apart (**isognathy**; Figure 17–13).

Morganucodon and similar early mammals had wear on their teeth indicating precise occlusion, and they also had anisognathus jaws that would mandate sideways movement of the lower jaw to occlude the teeth. Thus we can infer that they used the basic mammalian pattern of jaw movement. Because mammals have only two sets of teeth per lifetime, their teeth must be durable; teeth with durable, prismatic enamel are a distinctive mammalian feature.

We can also deduce some features of the soft-tissue anatomy of the earliest mammals by comparing the monotremes with more derived living mammals,

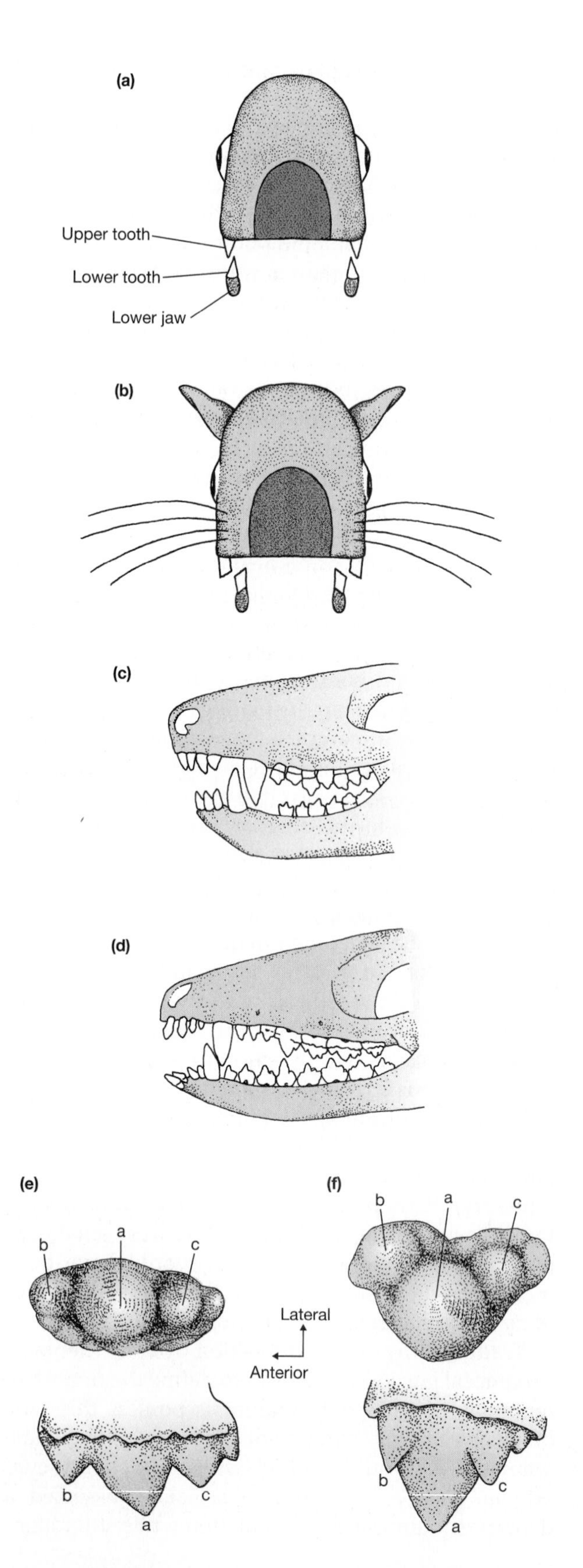

the **therians** (marsupials and eutherians). Monotremes have a more primitive ear than therians. The cochlea in their inner ear is not highly coiled (we can actually see that this was also the case in preserved braincases of the earliest mammals), and they lack a pinna, or external ear. Thus it is likely that the earliest mammals were also pinna-less, even though this makes them look rather nonmammalian in reconstruction (Figure 17–12). Although monotremes produce milk, their mammary glands lack nipples, so these were also probably lacking in the earliest mammals. Monotremes, although endothermal, have a metabolic rate lower than that of therians, lack effective sweat glands, and are not good at evaporative cooling: early mammals were probably similar.

However, monotremes are not good examples for deductions about the behavior and ecology of the earliest mammals. Monotremes are relatively large mammals, and they are specialized in their habits (Chapter 19). The earliest mammals were small and, judging by their teeth, insectivorous. Like small, insectivorous marsupials and eutherians today, they were probably nocturnal and solitary in their behavior, with the mother-infant bond as the only strong social bond. We do know from preserved braincases that early mammals had large olfactory lobes, indicating the importance of the sense of smell, and proportionally larger cerebral hemispheres than even the derived cynodonts. Early mammals may have been capable of much more sophisticated sensory processing than were cynodonts.

The Evolution of Lactation and Suckling

Researchers have two main questions about the evolution of the typically mammalian features of lactation and suckling. First, when did providing the young with milk originate, and did it coincide with the point that we consider as the cynodont-mammal transition? Second, how did milk and the mammary gland itself evolve?

Lactation Caroline Pond (1977) has argued that precise occlusion and diphyodonty indicate the presence of lactation in early mammals. She points out that precise occlusion would not work with the multiply-erupting system of tooth replacement (polyphyodonty) seen in living reptiles and in nonmammalian synapsids, because only a fully erupted tooth occludes properly with its counterpart in the other half of the jaw. This would be especially true for the tall, pointy teeth of early mammals (Figure 17–13), where a small amount of malocclusion could break the tooth. Precise monitoring of occlusion is less important in mammals with flatter teeth, such as humans.

Thus diphyodonty must have *preceded* precise occlusion in evolution. But an animal could reduce its number of sets of teeth only if it was fed milk during its early life. If the newborn were fed only on liquid food, the jaw could grow while it had no need of teeth, and permanent teeth could erupt in a near adult-size jaw. Thus, if an animal has precise occlusion and diphyodonty, it must first have evolved lactation.

Pond's hypothesis allows us to infer that the earliest mammals lactated, but had they inherited this feature from the cynodonts? In the next section we discuss the evolution of suckling, and conclude that cynodont babies would not have been able to suckle. This does not mean that cynodonts did not have parental care. Obviously, parental care would have to be in place before lactation could evolve. Parental care would also be essential for an endotherm, to ensure the proper thermal environment for juveniles and the developing young.

◀ Figure 17–13 Occlusion and molar form in cynodonts and early mammals. (a) Cross-sectional view through the muzzle of a cynodont. Lower teeth and upper teeth are the same distance apart (isognathy), and jaw movement is a simple vertical up-and-down motion. (b) Cross-sectional view through the muzzle of a mammal. Lower teeth are closer together than upper ones, because of a narrower lower jaw (anisognathy); jaw movement is rotatory, with chewing on only one side of the jaw at any one time. (c) Side view of the jaws of a cynodont (*Thrinaxodon*). The postcanine teeth all look similar (and are replaced continually), and are arranged so each tooth in the upper jaw lines up with one in the lower jaw. (d). Side view of the jaws of a mammal (*Morganucodon*). Postcanine teeth are now divided into simple premolars (replaced once) and more complex molars (not replaced); upper and lower molars are offset so that they interdigitate on occlusion (each tooth in the upper jaw meshes with two in the lower jaw, and vice versa). (e) Schematic upper molar of *Morganucodon* in occlusal ("tooth's eye") and lateral views, representing the original mammalian pattern (similar to that of cynodonts). The three main cusps are set in a straight line. (f) Schematic upper molar of *Kuehneotherium* in occlusal and lateral view, representing the more derived holotherian pattern. Arrangement of the cusps is triangular, with the major cusp set more medially (in the uppers, more laterally in the lowers), to effect the "reversed triangles" pattern of occlusion.

Thus we have some fairly strong evidence to suggest that lactation appeared with the origin of mammals. But how did the whole system evolve in the first place? Mammary glands cannot be specifically derived from any of the specialized skin glands (e.g., sweat glands) seen in living mammals. However, they do appear to be rather closely linked to the type of glands associated with hair (sebaceous glands) (Blackburn 1991). Thus they probably had the primitive ability to secrete small amounts of organic materials, as do sebaceous glands in mammals today. One suggestion for the evolution of lactation is that these glands originally secreted substances (aggregating pheromones) signaling the offspring to recognize their mother, and to congregate with her (Figure 17–3e; see also Duvall 1986).

An evolutionary scenario related specifically to the properties of milk was advanced by Blackburn and colleagues (1989). As they noted, all milk contains proteins that are related to the lysozyme enzymes that attack bacteria; even human milk has antimicrobial properties. The researchers suggested that the original use of milk might have been for protection of the eggs in a nest against microorganisms. Once a secretion of this type had evolved, any evolutionary change to a more copious, more nutritive secretion accidentally ingested by the young could only have been of benefit. This proto milk would initially have supplemented the egg yolk, and then later supplanted it.

What is the evolutionary advantage of lactation for mammals? Lactation allows the production of offspring to be separated from seasonal food supply. Unlike birds, which must lay eggs only when there is the appropriate food supply for the fledglings (spring and summer), mammals can store food as fat and convert it into milk at a later date. Provision of food in this manner by the mother alone also means that she does not have to be dependent on paternal care to rear her young. Finally, lactation makes viviparity less strenuous on the mother, because the young can be born at a relatively undeveloped stage and cared for outside of the uterus.

Suckling The ability to suckle is a unique mammalian feature. Mammals can form fleshy seals against the bony hard palate with the tongue and with the epiglottis, effectively isolating the functions of breathing and swallowing (Figure 17–14). Mammals make use of these seals to enable suckling on the nipple without ceasing to breathe through the nose. Adult humans have lost the more posterior seal that allows this action, with the movement of the larynx more ventrally in early childhood. This makes us more liable to choke on our food, and also enables us to breathe voluntarily through the mouth as well through the nose. But we retain the more anterior seal: This is what stops you from swallowing water when you gargle.

This mammalian pharyngeal anatomy is also important for our mode of swallowing a discrete, chewed bolus of food, rather than shoving large items wholesale down the gullet (like a snake swallowing a mouse). Changes in the bony anatomy of the palate and surrounding areas suggest that these functions came into place only with the most derived cynodonts, suggesting that this type of swallowing, and the capacity for suckling, are fundamentally mammalian attributes. The use of these seals in feeding is important for baleen whales, which take in a large mouthful of water and then strain it out through the baleen plates to filter out the plankton. The absence of these pharyngeal seals in other tetrapods may explain why Mesozoic marine reptiles, such as ichthyosaurs, never adopted this whale-like mode of filter feeding (Collin and Janis 1997). Filter-feeding fish, such as the whale shark, entrap food particles on the gills as water passes through, so they do not need these types of mouth seals.

Another characteristic feature of mammals is the possession of facial muscles, which are lacking in other vertebrates (Figure 17–15). These muscles make possible our varieties of facial expressions, but they were probably first evolved in the context of suckling, for mobile lips and cheeks that enable the young to suck. The facial muscles are thought to be homologous with the neck constrictor muscles (*constrictor colli*) of other amniotes, because both types of muscles are innervated by cranial nerve seven. There is some evidence that the acquisition of facial muscles may have occurred somewhat differently in various mammal lineages, because monotremes extend a different portion of the *constrictor colli* muscle into the face during development than do therians, and they lack mobile lips (although this could be a secondary feature, related to the development of a beak).

There is a great deal of difference in the elaboration of these muscles in different mammals. Most mammals do not have highly expressive faces. However, it is not just primates that are capable of

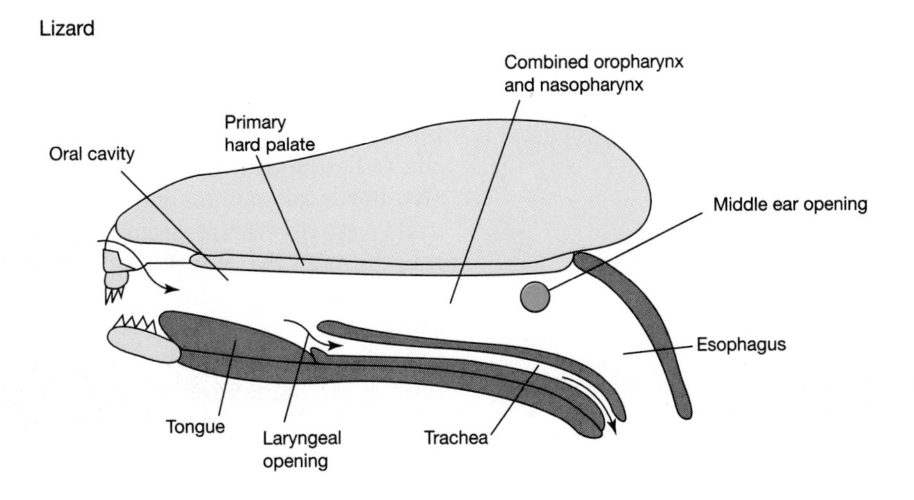

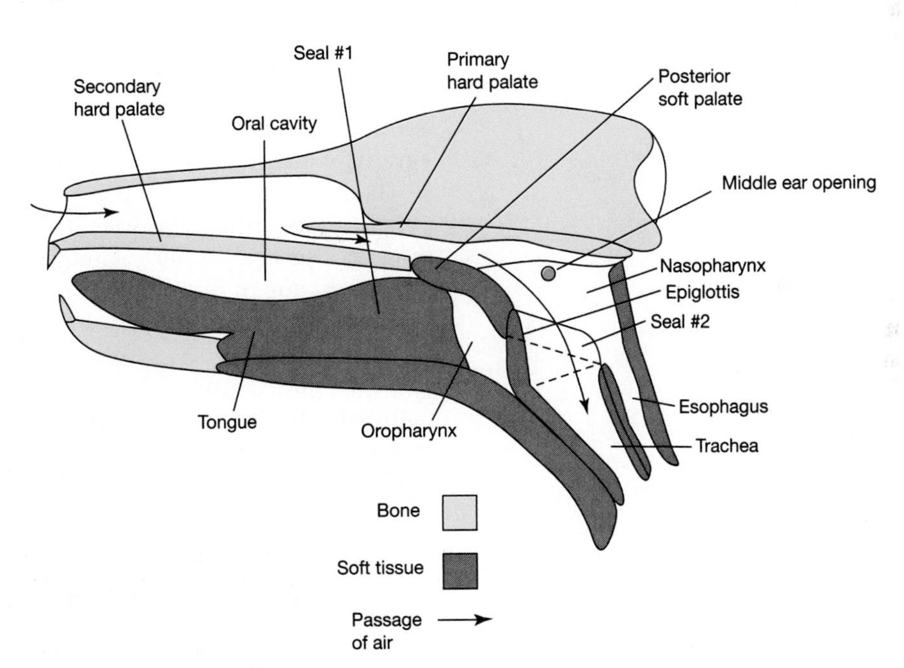

▲ Figure 17–14 Longitudinal-sectional views of the oral and pharyngeal regions.

facial expressions. Horses use their lips in feeding, and are capable of quite a wide variety of expressions, whereas cows use their tongues and are poker-faced. This is why Mr. Ed (the television talking horse) seems plausible to us, whereas a cow could never play that role. All mammals with well-developed facial muscles display similar expressions for similar emotions; for example, an angry human snarls in a similar way to an angry dog, or even an angry horse. This similarity of expression is perhaps surprising, since different mammals must have elaborated their facial muscles independently from ancestors with little capacity for facial expression.

Lizard (reptile)—no muscles of facial expression

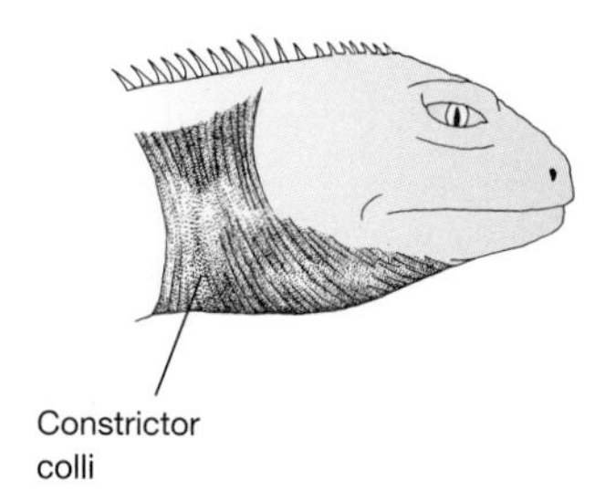

Rodent (mammal)—moderate development of muscles of facial expression

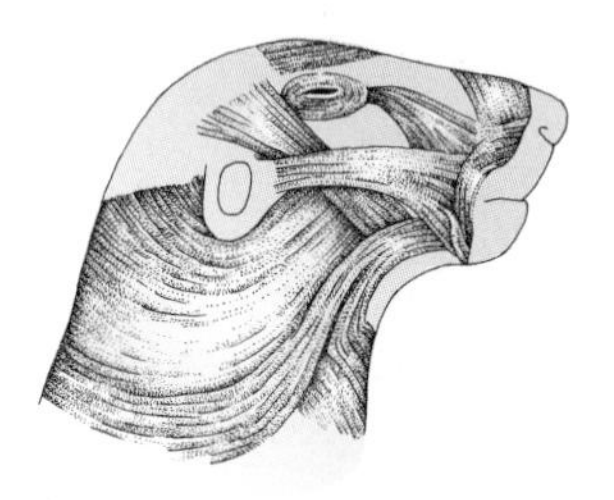

Primate (mammal)—extensive development of muscles of facial expression

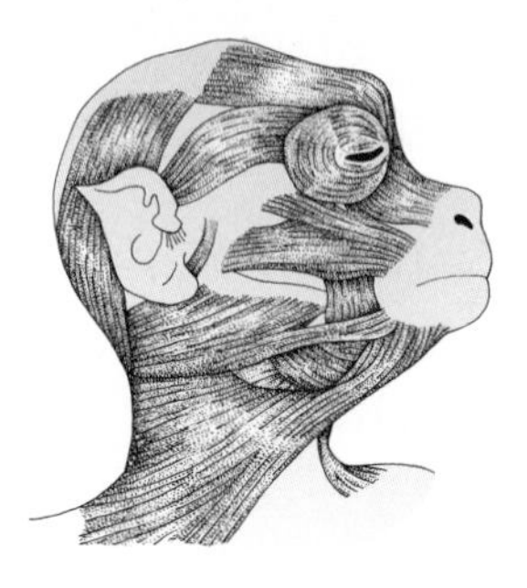

▲ Figure 17–15 Muscles of facial expression.

The Radiation of Mesozoic Mammals

The Cenozoic is usually termed the Age of Mammals. While it is true that mammals were small and basically shrew-like until the end of the Cretaceous, the radiation of Mesozoic mammals still represents two-thirds of mammalian history. Mesozoic mammals were diverse taxonomically, but fairly homogenous in adaptive morphotype. They were all fairly small: Many were shrew-size, and none was much bigger than a present-day opossum (weighing 2 to 3 kg). Their teeth reveal mainly insectivorous diets with some omnivory, but these mammals were too small to be predatory carnivores or herbivores with a highly fibrous diet.

Although it is unlikely there was ever direct competition between mammals and dinosaurs (scenarios of mammals causing dinosaur extinction by eating their eggs notwithstanding), it is true that mammals did not diversify into larger-bodied forms with more varied diets until the dinosaurs' extinction. The presence of dinosaurs must in some way have been preventing the radiation of mammals into a broader variety of adaptive niches.

There were two major periods of mammalian diversification during the Mesozoic. The first, spanning the Jurassic to Early Cretaceous, produced an early radiation of forms that in the main did not survive past the Mesozoic: morganucodonts, docodonts, triconodonts, symmetrodonts, dryolestids, and the like (Figure 17–1). Although these early mammals are fascinating to students of early mammalian history, we will not discuss them in more detail here. (For an extensive review, see Lillegraven et al. 1979.)

The second radiation, which got underway in the Early Cretaceous, was composed of more derived mammals, including the first therians, and is discussed more extensively in Chapter 19. Changes in other aspects of the terrestrial ecosystem were also apparent at this time. The Early Cretaceous marks the time of the initial radiation of the angiosperms, the flowering plants (Chapter 13). Among other tetrapods, the Cretaceous dinosaur faunas were distinctively different from earlier ones, and snakes made their fist appearance, possibly in association with a diversity of small mammals for them to feed on.

Summary

The synapsid lineage is characterized by a single lower temporal fenestra on each side of the skull. The first synapsids were the pelycosaurs of the Late Carboniferous and Early Permian, including the familiar sailbacks. Many pelycosaurs were large animals; some may have weighed as much as 250 kg. The most derived pelycosaurs, the sphenacodontids, had several more mammal-like features, including the beginnings of a three-boned middle ear.

The earliest therapsids were more derived in possessing an upright posture, limbs more designed for propulsion, and a greater volume of jaw musculature. The most derived therapsids, the Triassic cynodonts, had many features indicative of endothermy, including turbinate bones in the nose, a secondary palate, and a reduction of the rib cage suggesting the presence of a diaphragm. Many features of the cynodont skull that changed over its history can be understood in the context of an evolutionary conflict between chewing and hearing, because some of the bones of the mammalian middle ear formed the original synapsid jaw joint.

The first mammals of the latest Triassic had teeth that were replaced only once and that precisely interlocked to chew food. Their skeletons suggest that by this time they had developed the typically mammalian bounding gait. A substantial reduction in body size accompanied the evolution of the first mammals—they were only 100 millimeters long and probably weighed less than 50 grams. We can deduce quite a lot about the probable biology of these early mammals, not just from the skeletal remains but also from considering the biology of the living monotremes and primitive therians. By inference from patterns of tooth replacement and the presence of a Harderian gland, we can conclude that both lactation and a fur coat evolved with the earliest mammals. Milk itself may have been evolved for its antimicrobial properties in the nests of oviparous mammals.

The radiation of Mesozoic mammals was mainly one of small insectivorous or omnivorous forms. The modern groups of mammals—monotremes and therians—can trace their origin back to the Early Cretaceous, a time of evolutionary turnover not only in mammals but also in other tetrapods as well as in plants. Mammals did not diversify larger, more specialized forms until the Cenozoic, after the extinction of the dinosaurs.

Additional Readings

Allin, E. F. 1975. Evolution of the mammalian middle ear. *Journal of Morphology* 147:403–437.

Blackburn, D. G. 1991. Evolutionary origins of the mammary gland. *Mammal Review* 21:81–96.

Blackburn, D. G., V. Hayssen, and C. J. Murphy. 1989. The origins of lactation and the evolution of milk: A review with new hypotheses. *Mammal Review* 19:1–26.

Bramble, D. M. 1978. Origin of the mammalian feeding complex: Models and mechanisms. *Paleobiology* 4:271–301.

Bramble, D. M., and D. R. Carrier 1983. Running and breathing in mammals. *Science* 262:235–240.

Carrier, D. R. 1987. The evolution of locomotor stamina in tetrapods: Circumventing a mechanical constraint. *Paleobiology* 13:326–341.

Collin, R. and C. M. Janis 1997. Morphological constraints on tetrapod feeding mechanisms. Why were there no suspension-feeding marine reptiles? In J. M. Callaway and E. L. Nicholls (Eds.), *Ancient Marine Reptiles.* New York: Academic Press, 451–466.

Crompton, A. W. 1995. Masticatory function in nonmammalian cynodonts and early mammals. In J. J. Thomason (Ed.), *Functional Morphology in Vertebrate Paleontology.* Cambridge, U.K.: Cambridge University Press, 55–75.

Crompton, A. W., and P. Parker. 1978. Evolution of the mammalian masticatory apparatus. *American Scientist* 66: 192–210.

Duvall, D. 1986. A new question of pheromones: Aspects of possible chemical signaling and reception in the mammal-like reptiles. In N. Hotton III, et al. (Eds.), *The Ecology and Biology of Mammal-Like Reptiles,* Washington, D.C.: Smithsonian Institution Press, 219–238.

Gonick, L. 1990. *The Cartoon History of the Universe,* volumes 1–7. New York: Doubleday.

Hillenius, W. J. 1992. The evolution of nasal turbinates and mammalian endothermy. *Paleobiology* 18:17–29.

Hillenius, W. J. 2000. The septomaxilla of nonmammalian synapsids: soft tissue correlates and a new functional interpretation. *Journal of Morphology* 245:207–229.

Hopson, J. A. 1994. Synapsid evolution and the radiation of non-eutherian mammals. In D. R. Prothero and R. M. Schoch (Eds.), *Major Features of Vertebrate Evolution,* Short Courses in Paleontology no. 7. Knoxville, Tn.: Paleontological Society and University of Tennessee Press, 190–219.

Hotton, N. III. 1991. The nature and diversity of synapsids: Prologue to the origin of mammals. In H.-P. Schultze and L. Trueb (Eds.), *Origins of the Higher Groups of Tetrapods,* Ithaca, N.Y.: Cornell University Press, 598–634.

Hotton, N. III, et al. (Eds.). 1986. *The Ecology and Biology of Mammal-like Reptiles,* Washington, D.C.: Smithsonian Institution Press.

Kemp, T. S. 1982. *Mammal-like Reptiles and the Origin of Mammals.* London: Academic.

King, G. 1990. *The Dicynodonts: A Study in Paleobiology*. London: Chapman & Hall.
Lillegraven, J. A., Z. Kielan–Jaworowska, and W. A. Clemens. 1979. *Mesozoic Mammals: The First Two-Thirds of Mammalian History*. Berkeley: University of California Press.
Pond, C. M. 1977. The significance of lactation in the evolution of mammals. *Evolution* 31:177–199.
Rowe, T. 1988. Definition, diagnosis, and origin of Mammalia. *Journal of Vertebrate Paleontology* 8:241–264.
Rowe, T. 1996. Coevolution of the mammalian middle ear and neocortex. *Science* 273:651–654.
Ruben, J. A. 1995. The evolution of endothermy in mammals and birds: From physiology to fossils. *Annual Review of Physiology* 57:69–95.
Ruben, J. A., A. F. Bennett, and F. L. Hisaw. 1987. Selective factors in the origin of the mammalian diaphragm. *Paleobiology* 13:54–59.
Ruben, J. A., and T. D. Jones. 2000. Selective factors associated with the origin of fur and feathers. *American Zoologist* 40:585–596.
Smith, K. K. 1992. The evolution of the mammalian pharynx. *Zoological Journal of the Linnean Society* 104:313–349.

Web Explorations

On-line resources for this chapter are on the World Wide Web at: http:/www.prenhall.com/pough (click on the Table of Contents link and then select Chapter 17).

CHAPTER

18

Geography and Ecology of the Cenozoic

The role of Earth's history in shaping the evolution of vertebrates is difficult to overestimate. The positions of continents have affected climates and the ability of vertebrates to migrate from region to region. The geographical continuity of Pangaea in the late Paleozoic and early Mesozoic allowed tetrapods to migrate freely across continents, and the faunas were fairly similar in composition across the globe. By the late Mesozoic, however, Pangaea no longer existed as a single entity. Epicontinental seas extended across the centers of North America and Eurasia, and the southern continents were separating from the northern continents and from one another. This isolation of different continental blocks resulted in the isolation of their tetrapod faunas, and limited the possibility for migration. As a result, tetrapod faunas became progressively more different on different continents. Distinct regional differences became apparent in the dinosaur faunas of the late Mesozoic, and they have been a prominent feature of Cenozoic mammalian faunas.

Continental drift in the late Mesozoic and early Cenozoic moved the northern continents that had formed the old Laurasia (North America and Eurasia) from their early Mesozoic near-equatorial position into higher latitudes. The greater latitudinal distribution of continents, plus changes in the patterns of ocean currents that were the result of continental movements, caused cooling trends in high northern and southern latitudes during the late Cenozoic. With the formation of the Arctic ice cap some 5 million years ago, this cooling led to a series of ice ages that began in the Pleistocene and continue to the present. We are currently living in a relatively ice-free interglacial period. Both the fragmentation of the landmasses and the changes in climate during the Cenozoic have been important factors in the evolution of mammals.

18.1 Cenozoic Continental Geography

The breakup of Pangaea in the Jurassic began with the movement of North America that opened the ancestral Atlantic Ocean. Rifts formed in Gondwana, and India moved northward on its separate oceanic plate (Figure 18–1), eventually to collide with Eurasia. The collision of the Indian and Eurasian plates in the mid-Cenozoic produced the Himalayas, the highest mountain range in today's world.

South America, Antarctica, and Australia separated from Africa during the Cretaceous, but maintained connections into the early Cenozoic (Figure 18–1). In the middle to late Eocene, Australia separated from Antarctica and, like India, drifted northward. (Note that New Guinea, today an island north of Australia that is separated from the mainland continent by a shallow sea, is actually a part of the Australian continental block. It was in direct land contact with Australia for much of Earth's history—this is why the faunas and of the two regions are so similar.) Intermittent land connections between South America and Antarctica were retained until the middle Cenozoic via the Scotia Island arc, and mammalian faunas from Antarctica in the Late Eocene bear a strong resemblance to those from South America. New Zealand separated from Australia sometime in the middle Mesozoic, apparently too early in the history of mammalian diversification to carry an endemic component of mammals. Although New Zealand has many other endemic tetrapods—most notably the primitive diapsid reptile *Sphenodon* (the tuatara)—its only endemic mammal is a bat, which evidently reached the island by flying

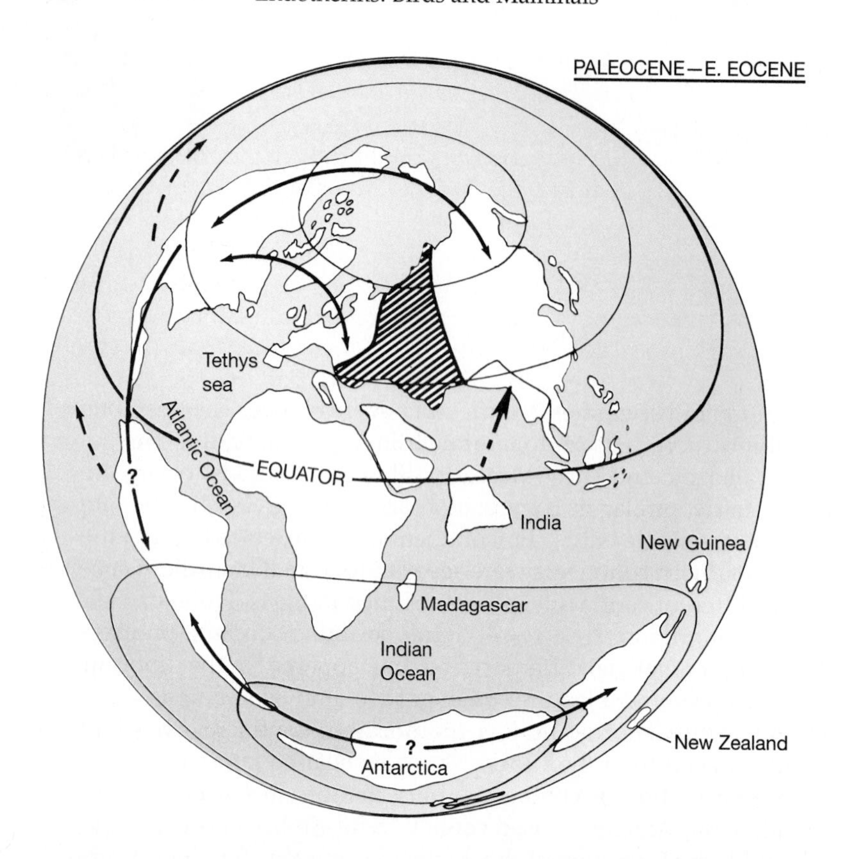

Figure 18–1 Continental positions in the late Paleocene and early Eocene. An epicontinental sea, the Turgai Straits (crosshatching), extended across Eurasia. Dashed arrows show the direction of continental drift, and solid arrows indicate major land bridges mentioned in the text.

from Australia. Other mammals that are found in New Zealand today, such as possums, were brought in by humans.

In the Northern Hemisphere a land bridge between Alaska and Siberia—the trans-Bering Bridge—broadly connected North America to Asia at high but relatively ice-free latitudes (Figure 18–1). Intermittent connections also persisted between eastern North America and Europe via Greenland and Scandinavia from the Cretaceous to the early Eocene. However, migrations of mammals between North America and Europe via this route appear to have been prominent only during the early Eocene. The connection apparently broke during the Eocene, and later migrations between North America and the Old World occurred via the Bering land bridge linking Siberia and Alaska. Other tectonic movements also influenced Cenozoic climates. For example, the uplift of mountain ranges such as the Rockies, the Andes, and the Himalayas in the middle Cenozoic cast rain shadows along their western flanks, resulting in the replacement of woodlands by grasslands.

18.2 Cenozoic Terrestrial Ecosystems

The evolution of animals and plants during the Cenozoic is closely connected with climatic changes, which in turn are related to changes in the positions of continents. The key to understanding changes in climatic conditions lies in knowing that the major landmasses were moving *away* from the equatorial region and toward the poles. The passage of landmasses over the polar region—Antarctica in the early Cenozoic and Greenland in the later Cenozoic—allowed the formation of polar ice caps, culminating in the periods of glaciation (ice ages) of the Pleistocene. The story of Cenozoic terrestrial ecosystems is also the story of the temperate regions of the higher latitudes, which became cooler and drier with accompanying changes in vegetation (for example, the replacement of lush tropical-like forests with woodland and grassland).

The radiation of modern types of mammals is a prominent feature of the Cenozoic, following their origin in the latest Triassic, and their persistence through the Mesozoic at small body size and low levels of morphological diversity. Thus the Cenozoic is commonly known as the Age of Mammals, even though the time elapsed since the start of the Cenozoic represents only about a third of the time spanned by the total history of mammals (and even though the most diverse vertebrates of the Cenozoic, in terms of numbers of species, were the teleost fishes). The radiation of larger mammals is almost certainly related to the extinction of the dinosaurs, which left the world free of large tetrapods and provided a window of opportunity for other groups.

In the warm world of the early Cenozoic, tropical-like forests were found in high latitudes—even within the Arctic Circle. Apparently the vegetation of these polar regions could withstand six months of near darkness. The vegetation of polar regions in the early Cenozoic was an exceptionally broad-leaved type, unknown in the world today. The broad leaves presumably allowed the trees to obtain the maximum amount of sunlight possible during the short summer and under the dim conditions of winter. A fossil assemblage from the early Eocene of Ellesmere Island (within the Canadian Arctic Circle) shows the presence of reptiles such as turtles and crocodiles, as well as of mammals resembling (although not closely related to) the tree-dwelling primates and flying lemurs of present-day Southeast Asia. Figure 18–2 illustrates the subtropical conditions seen in Europe in the early Eocene.

Many of the early Cenozoic lineages of mammals are now extinct. These groups are often called archaic mammals, a rather pejorative term that really refers to our own perspective from the comfort of the Recent. Archaic mammals were mainly small- to medium-size generalized forms with a predominance of arboreal types. Larger, specialized predators, and herbivores with teeth suggesting a high-fiber herbivorous diet, did not appear until the late Paleocene. Members of most present-day orders did not make their first appearance until the Eocene; but there are some notable exceptions, most importantly the orders Carnivora and Xenarthra (first known from the early Paleocene) and Rodentia (first known from the late Paleocene).

It appears that browsing by herbivorous dinosaurs kept forests at bay during the Cretaceous, much as large herbivorous mammals such as elephants maintain savanna habitats today. Because forests dominated terrestrial habitats after the dinosaurs' extinction, most of the niches available for early Cenozoic mammals were arboreal ones. It was not until the climatic changes of the Eocene produced more open habitats that larger-size mammals began to radiate into the terrestrial niches they occupy today.

Modern types of birds also diversified at the start of the Cenozoic, including large terrestrial birds—both carnivorous forms (now all extinct) and herbivores like the present-day ostrich. Many of the present-day groups of lizards had their origins in the latest Cretaceous or early Cenozoic, while the modern groups of turtles and amphibians were largely established in the Cretaceous. Modern types of freshwater crocodiles were the only ones to survive into the Cenozoic; there was a modest radiation of terrestrial crocodiles in the Southern Hemisphere during the Paleocene and Eocene. Champsosaurs (crocodile-like in form but only very distantly related to crocodilians) were also prominent freshwater aquatic predators in the early Cenozoic. Among the insects, modern butterflies and moths first appeared in the middle Eocene.

The later Cenozoic world was in general cooler and dryer than that of the early Cenozoic. Changes in terrestrial ecosystems reflected these climatic changes. Temperate forests and woodland replaced the tropical-like vegetation of the higher latitudes, and tropical forests were confined to equatorial regions. Extensive grasslands first appeared in the Miocene, forming swathes of savanna (grassland with scattered trees) across North America, southern South America, and central Asia. In the Pliocene or Pleistocene, savannas appeared in more tropical areas such as East Africa, and the more temperate grasslands turned into treeless prairie or steppe. New types of vegetation also appeared in the Plio-Pleistocene: tundra and taiga in the Arctic regions and deserts in the tropical and temperate regions.

The radiation of mammals in the late Cenozoic reflected these vegetational changes. Large grazing mammals such as horses, antelope, rhinos, and elephants evolved along with the emerging grasslands—and with them the carnivores that preyed on them, such as large cats and dogs. Some small mammals also diversified, most notably modern types of rodents such as rats and mice. This diversification of small mammals may explain the concurrent late Cenozoic diversification of modern types of snakes. Late Cenozoic lizards included some very large

varanoids, including not only the largest lizard known today, the Komodo dragon, but also a cow-size predator, *Megalania*, in the Pleistocene of Australia. Many modern types of birds first appeared in the late Cenozoic, including the passerines (song birds) and modern types of birds of prey such as eagles, hawks and vultures. The more open habitats of the late Cenozoic also favored the diversification of social insects that live in grasslands, such as ants and termites.

18.3 Cenozoic Climates

The Cenozoic was in general a cooler and dryer time than the Mesozoic, although the early part (until the mid-Eocene, around 50 million years ago) was a period of relative warmth. The later Cenozoic was characterized by the buildup of ice at the poles, culminating in the Ice Age at the start of the Pleistocene, two million years ago.

Tertiary Climates

The warm and humid conditions typical of the Jurassic and Early to Middle Cretaceous changed toward the end of the Cretaceous, when widespread moderate cooling took place. Yet the world of the early Cenozoic still reflected the hothouse world of the Mesozoic. There was an increase in atmospheric carbon dioxide at the start of the Eocene, which would have resulted in warming via a greenhouse effect. The peak of climatic warming in the higher latitudes occurred around the early-middle Eocene boundary (Figure 18–3). From this high point, the higher-latitude regions started to cool, with a rather precipitous drop in mean annual temperature in the latest Eocene, plunging the Earth into the start of the colder world of the later Cenozoic (Retallack 2001).

What caused this dramatic change in the Earth's temperatures? A primary reason was probably a reverse greenhouse effect related to sharply falling levels of atmospheric carbon dioxide. Additionally, cold polar-bottom water was formed over the poles when Australia broke away from Antarctica and Greenland broke away from Norway. Ocean circulation carried the cold water toward the equator, cooling the temperate latitudes. The Antarctic ice cap probably formed by the end of the Eocene, although the Arctic ice cap did not form until some 30 million years later.

Following a rather cool Oligocene, temperatures started to rise in higher latitudes in the early Miocene, reaching a second, less dramatic peak in the middle Miocene. The Miocene world was in general drier than the Eocene, and the combination of warmth and dryness promoted the spread of grasslands. Miocene vegetational changes, and the expansion of grass at the expense of other plants, may also have been related to further decreasing global levels of atmospheric carbon dioxide. Grass plants are capable of evolving a different biochemical pathway for photosynthesis (the C_4 pathway), which is more efficient in conditions of low carbon dioxide. Most other plants cannot use this pathway.

The early Miocene warming may be due to the opening of Drake's Passage between Antarctica and South America, which isolated the cold polar water around Antarctica. However, expansion of the Antarctic ice cap in the late Miocene once again brought cooling to the higher latitudes, a trend that has persisted to the present day with occasional remissions. The closing of the Isthmus of Panama that joined North and South America in the Pliocene also profoundly affected global climate. Warm waters could no longer circle the equator, and it is during this time that the Gulf Stream was formed, carrying warm water from the subtropical American regions to western Europe. Elsewhere in the world, disruption of oceanic circulation resulted in cooling and led to the formation of the Arctic ice cap.

Note that changes in the diversity of mammals paralleled changing temperatures at higher latitudes throughout the Cenozoic, with peaks seen in the mid-Eocene and mid-Miocene (Figure 18–3). This pattern makes sense when we think of today's world, in which a much greater diversity of mammals exists in

◄ Figure 18–2 A reconstruction of a scene from the early Eocene of Europe showing a tropical-like rainforest. The trees include sequoia, pine, birch, palmetto, swamp cypress, and tree ferns. In the foreground are cycads and magnolia. The birds are ibises, with a *Diatryma* (an extinct flightless predator) in the background. The hippo-like mammals are *Coryphodon*, belonging to the extinct ungulate-like order Dinocerata. An early primate (*Cantius*, an omomyid) climbs up a liana vine. Crouching among the ferms is an oxyaenid *Palaeonictis*, a cat-like predator belonging to the extinct order Creodonta. Its potential prey are the early artiodactyls, *Diacodexis*, in the foreground.

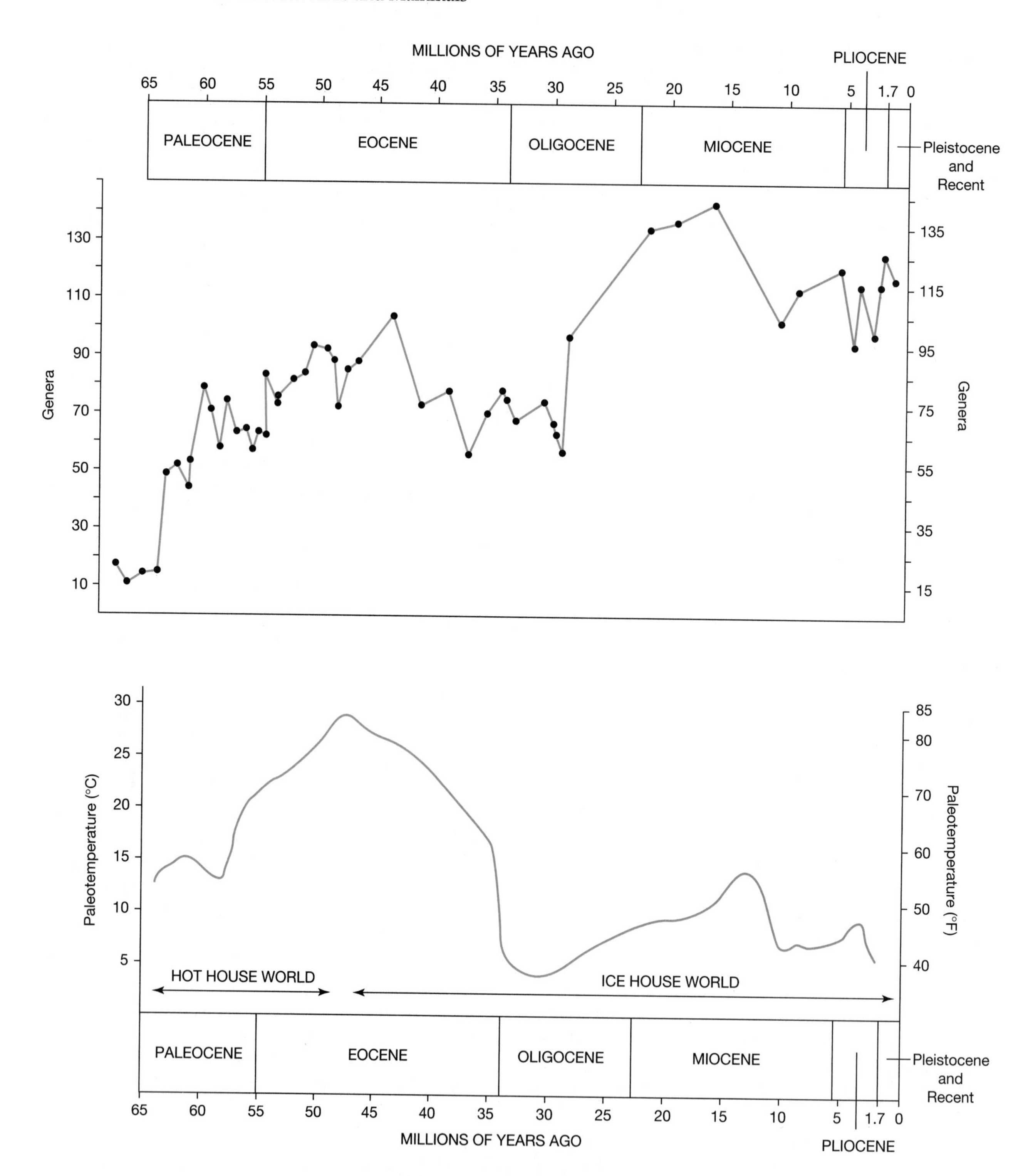

▲ Figure 18–3 Numbers of mammalian genera in North America and mean annual paleotemperatures in the North Sea during the Cenozoic. Note that peaks in total genera and in temperature substantially coincide.

the tropical forests than in the temperate woodlands. It may seem paradoxical that the diversity appears to be greater at the mid-Miocene peak than at the warmer mid-Eocene peak. Several phenomena probably contribute to this paradox. Some of these factors may be artifacts of the fossil record, whereas others probably represent real ecological differences. First there is the pull of the Recent. That is, the nearer we get to the present day, the less likely it is that fossils have been eroded and lost forever, so the *apparent* diversity of fossils in young deposits is greater than in old ones. Second, there were more large mammals in the Miocene than there had been in the Eocene, and large teeth and bones have a greater chance of preservation than small ones—again biasing the fossil record. Finally, the changing Cenozoic climates had produced a fractionation of habitat types by the Miocene. Tropical forests were still found at the equator, but other kinds of habitats (such as woodland and grassland) were found at higher latitudes, and these diverse habitats probably really did support a great diversity of mammals.

The Pleistocene Ice Ages

The extensive episodic continental glaciers that characterize the Pleistocene were events that had been absent from the world since the Paleozoic. These ice ages had an important influence not only on Cenozoic mammal evolution in general but also on our own evolution—and even on our present civilizations. The world can still be considered to be in the grip of an ice age, but right now we inhabit a warmer, interglacial, period.

For example, currently the volume of ice on Earth constitutes about 26 million cubic kilometers. During glacial episodes in the Pleistocene, there was as much as 77 million km^3 of ice, perhaps even more. An enormous volume of water still is locked up in glaciers and polar ice caps. The melting of the glaciers of the last glacial episode at the end of the Pleistocene, around 10 thousand years ago, caused the sea level to rise by about 140 meters (almost half the height of the Empire State Building) comparable to its present relatively stable condition. If the present-day glaciers were to melt, sea level would rise at least another 50 meters, submerging most of the world's coastal cities. Entire countries such as Bangladesh and some Pacific Island nations would be completely submerged. No wonder present-day environmentalists are concerned about the possibilities of global warming caused by human activities!

Today glaciers cover 10 percent of the Earth's land surface, mostly in polar regions but also on high mountains. At times in the Pleistocene, an ice mass that was probably between 3 and 4 km thick covered as much as 30 percent of the land, and extended southward in North America to 38° N latitude (southern Illinois; Figure 18–4). A similar ice sheet covered northern Europe. However, much of Alaska, Siberia, and Beringia (Pleistocene land that is now underwater between Alaska and northeast Asia) were free of ice, and housed a biome known as the Mammoth Steppe or the steppe-tundra that is unknown today. It was obviously much more productive than present-day high-latitude habitats, because it contained a mammalian faunal assemblage that rivaled the diversity of modern African savanna faunas. This fauna combined mammals now absent from higher latitudes, such as lions and rhinos, with animals that persist in Arctic latitudes today, such as reindeer and musk ox.

These continental glaciers advanced and retreated several times during the Pleistocene. (The Southern Hemisphere was less affected because at that time the southern continental landmasses were further from the poles than the northern ones, as they are today.) There were four major episodes of glaciation, as shown in Figure 18–4, but we now know that many (perhaps 20 or more) minor ones were interspersed among these major ones.

Continental glaciation had a greater effect on world climates than just ice covering the high latitudes. Popular books depict mammoths struggling to free themselves from ice, but glaciers advance slowly enough for animals to migrate toward the equator—although problems may occur if routes are blocked by mountain ranges or seaways. This would have been more of a problem for Eurasian mammals than for North American mammals. In Eurasia the major mountain ranges (the Alps and the Himalayas) run in an east-to-west direction, while in North America the mountain ranges (the Rockies and the Appalachians) run from north to south. However, Eurasian animals fleeing colder climates had the advantage of broad connections between temperate and tropical zones, both in Asia and in Africa. In contrast, North American animals would have to traverse the extremely narrow Isthmus of Panama to reach the more tropical areas of South

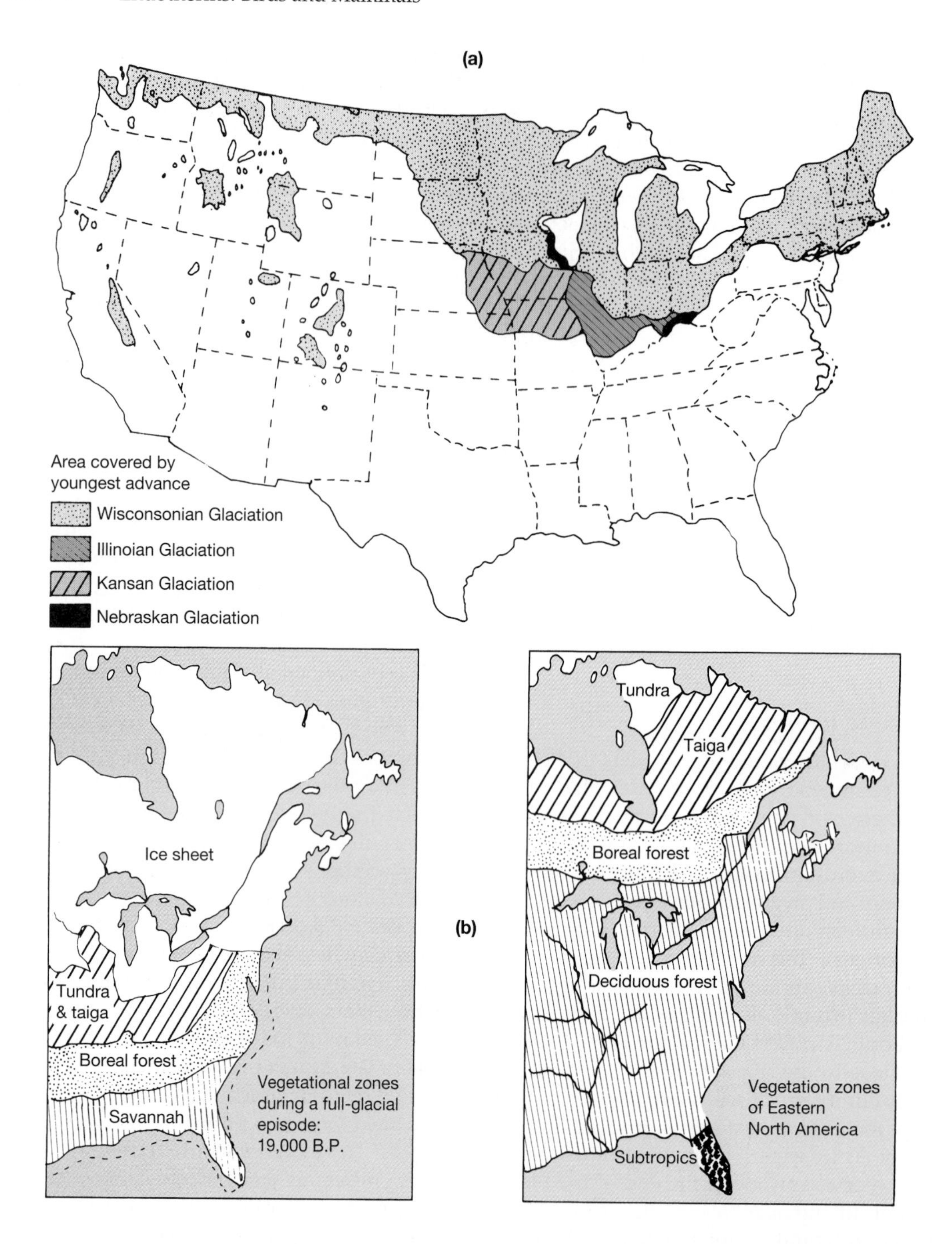

▲ Figure 18–4 Pleistocene glaciation. (a) Extent of glaciation in North America. Several advances of the continental glacier are shown. In the West, montane glaciers advanced and retreated. The major episodes of glaciation, named in North America for the southern extent of the ice sheets, are listed in order from youngest (the Wisconsin, commencing approximately 100,000 years ago) to the oldest (the Nebraskan, commencing approximately 1.30 million years ago). (b) Effect of glacial periods on the position of biomes in North America. Biomes are shifted southward during periods of glacial advance (left), and extend northward during interglacial periods like the present (right). Note that in addition to shifting biomes, entire types of habitats may appear and disappear. There is no savanna habitat today in the southeastern United States, and there was no subtropical habitat in Florida during the last full glacial episode.

America. This may have limited the migration of certain types of mammals (see Chapter 19).

Drying of the ice-free portions of the Earth due to the volume of water tied up in glaciers was at least as important for terrestrial ecosystems as the glaciers themselves. Many of the equatorial areas that today are covered by lowland rain forests were then much dryer, even arid. Today's relatively mild interglacial period is apparently colder and drier than other interglacial periods in the Pleistocene. For example, during other interglacial periods hippos were found in what is now the Sahara Desert, and hippos were also found in England.

With each glacial episode the forests in the Amazon and Congo basins contracted into several isolated refugia separated by savannas. During interglacial periods, the forests again spread over the basins, and the savannas contracted and were fractionated. Due to these alternating expansions and contractions, plant and animal populations were repeatedly isolated and remixed. During isolations, one species might be separated into several populations. The effect of glaciation on equatorial aridity, and the resultant contraction of moist habitats, may explain the high species diversity that now exists in these areas.

What caused these episodes of glaciation? A longstanding theory suggests that the amount of solar radiation impinging on the Earth (solar insolation) varies enough to affect the Earth's climate. In the 1930s the Yugoslavian astronomer Milutin Milankovitch proposed that episodes of glaciation are initiated by the fortuitous (or perhaps unfortunate) combination of several small variations of the passage of the Earth's orbit around the Sun, and the position of the Earth relative to the Sun. Three cycles interact here, each with its own characteristic periodicity (time elapsed between extremes of the cycle): (1) the Earth's elliptical orbit around the Sun (with a periodicity of 100,000 years); (2) the tilt of the Earth's rotational axis (with a periodicity of 40,000 years); and (3) the precession (wobble) of Earth's rotational axis (with a periodicity of 26,000 years).

Each of these orbital properties produces different effects. Change in tilt and precession modify the distribution of sunlight with respect to season and latitude, but not total global insolation, whereas changes in the Earth's orbit result in minute changes in global insolation. Normally these properties are cycling out of phase, like discordant keys played on a piano, but every so often they line up together like notes making a chord. Milankovitch suggested that the critical factor leading to a glacial episode is a change in the amount of summer insolation at high latitudes. It appears that glacial episodes get their start not from the world as a whole getting colder year round, but from cool summers that prevent the melting of winter ice. In contrast, the winters during glacial periods may have been warmer than those of the present day.

It is important to realize that these Milankovitch cycles must have been in existence throughout Earth's history. However, it was only after the formation of the Arctic ice cap in the Pleistocene (possibly as early as the Pliocene) that there existed enough buildup of polar ice to plunge the Northern Hemisphere into an ice age.

18.4 Cenozoic Extinctions

The best-known extinction of the Cenozoic is probably the one at the end of the Pleistocene, although this was by no means the extinction of greatest overall magnitude. The Pleistocene extinction appears dramatic because of the extinction of the megafauna, the diversity of large mammals (i.e., those over 20 kg in body mass). This included many very large mammals that are now totally extinct, such as glyptodonts and ground sloths in North and South America, mammoths in Holarctica and Africa, and diprotodontids in Australia. It also included many larger and exotic forms of more familiar mammals, such as the saber-toothed cats of Holarctica and Africa; the Irish elk; cave bears and woolly rhinos of Eurasia; and the huge kangaroos, wombats, and echidnas in Australia. Large terrestrial birds also suffered in these extinctions—including herbivores such as the moas of New Zealand and the elephant bird of Madagascar as well as carnivores such as the New World phorusrachiforms.

There is much debate about the cause of these extinctions. The main extinctions occurred at the end of the last glacial period, some 10,000 years ago. (Surprisingly enough, animals appear to be more vulnerable to extinction when the climate changes from glacial to interglacial, rather than the other way around, probably because the former event appears to occur with greater rapidity.) Thus climatic change would be an obvious explanation. However, many scientists have noted that it is only the last glacial period, rather than any of the previous ones, that brought extinctions of such magnitude. This observation suggests that part if not all of the blame for megafaunal extinctions should be placed on the spread of modern humans and modern hunting techniques, which was concurrent with that time period (see Chapter 24).

About 30 percent of mammalian genera became extinct at the end of the Pleistocene. That is approximately the magnitude of the other two major Cenozoic extinctions (late Eocene and late Miocene). However, the preceding two extinctions differ in several critical ways from the Pleistocene. The late Eocene extinctions were associated with the dramatic fall in higher-latitude temperatures (Figure 18–3). Higher-latitude forests turned to temperate woodlands, with the accompanying disappearance of mammals adapted to these tropical-like forests. This included not only a diversity of archaic mammals but also some early more modern types, such as higher-latitude primates and early horses. The early Cenozoic diversity of amphibians and reptiles in higher latitudes was also greatly reduced during the late Eocene.

The late Miocene extinctions were associated again not only with falling higher-latitude temperatures but also with global drying. The major extinctions were of browsing mammals (including a large diversity of browsing horses), which suffered habitat loss as the savanna woodlands turned into open grasslands and prairie. North America was especially hard hit by the climatic events of the late Miocene, because of its relatively high latitudinal position and the fact that animals could not migrate to more tropical areas in South America before the formation of the Isthmus of Panama.

Most significantly for the overhunting hypotheses, mammals of all body sizes (not just large ones) were affected in both the Eocene and Miocene. Other organisms, both terrestrial and marine, also experienced profound extinctions. The late Pleistocene extinction affected primarily large mammals and birds.

Additional Readings

Cerling, T. E. et al. 1997. Global vegetation change through the Miocene/Pliocene boundary. *Nature* 389:153–158.

Guthrie, R. D. 1990. *Frozen Fauna of the Mammoth Steppe.* Chicago: University of Chicago Press.

Imbrie, J., and A. Berger (Eds.). 1984. *Milankovitch and Climate Change.* Amsterdam: Elsevier.

Janis, C. M. 1993. Tertiary mammal evolution in the context of changing climates, vegetation, and tectonic events. *Annual Review of Ecology and Systematics* 24:467–500.

Kerr, R. A. 1987. Ocean drilling details steps to an icy world. *Science* 236:912–913.

Kürschner, W. M. 2001. Leaf sensor for CO_2 in deep time. *Nature* 411:247–248.

Martin, P. S., and R. G. Klein. 1984. *Quaternary Extinctions.* Tucson: University of Arizona Press.

Retallack, G. J. 2001. A 300-million-year record of atmospheric carbon dioxide from fossil plant cuticles. *Nature* 411: 287–290.

Terborgh, J. 1992. *Diversity and the Tropical Rainforest.* New York: Scientific American Library.

Zachos, J., M. Pagani, L. Sloan, E. Thomas, and K. Billups. 2001. Trends, rhythms, and aberrations in global climate 65 Ma to present. *Science* 292:686–693.

Web Explorations

On-line resources for this chapter are on the World Wide Web at: http://www/prenhall.com/pough (click on the Table of Contents link and then select Chapter 18).

CHAPTER

19

Mammalian Characteristics and Diversity

Cenozoic mammals are a highly diverse group of organisms, adapted to a wide variety of life styles and displaying a large amount of ecomorphological diversity. In addition to the familiar features of lactation and hair, mammals have other derived characters that set them apart from other vertebrates. However, in some respects mammals are more primitive than other amniotes (reptiles and birds). Although most mammals alive today are eutherians (placental mammals), an understanding of mammal diversity and specializations requires an understanding of how eutherians differ from marsupials, as well as how therians (marsupials and eutherians) differ from monotremes. Much of the diversity seen among Cenozoic therian mammals reflects the isolation of different groups of mammals on different continental landmasses; many of these ecomorphological types evolved convergently on different continents. The changing climates of the higher latitudes during the course of the Cenozoic also resulted in a wider diversity of mammals adapted to new habitats, such as grasslands, and changes in sea level and continental positions resulted in various mammalian intercontinental migrations

19.1 Features Shared by All Mammals

Mammals are perceived as the dominant terrestrial animals of the Cenozoic, but their species diversity (about 4500 species) is less than half that of birds (about 9100 species), and considerably less than that of lepidosaurian reptiles (about 6800 species). Mammal species diversity is in fact less than that of amphibians (about 4900 species), making them the smallest lineage of tetrapods.

Mammals do, of course, include the largest living terrestrial and aquatic vertebrates (the blue whale, at around 120 tons, is the largest animal ever known). And perhaps mammals have cornered the market on morphological diversity; no other vertebrate taxon has forms as different from each other as a whale is from a bat. Even among strictly terrestrial forms there is a tremendous morphological difference between, for example, a mole and a giraffe. But it is prudent to consider that mammals do not rise above other vertebrates when some measures of evolutionary success, such as species diversity, are considered. Figure 19–1 illustrates the diversity of the major extant groups of therian mammals (i.e., excluding monotremes).

Lactation

The females of all mammalian species lactate, feeding their young by producing milk. Mammary glands are entirely absent from the males of marsupials, but they are present and potentially functional in male monotremes and eutherians. There have been examples of human males producing milk under certain circumstances, and recently a species of fruit bat has been identified in which the male actually produces milk (Francis et al. 1994). It has long been a mystery why male mammals retain mammary glands but do not lactate (see Daly 1979); indeed, breast cancer affects human males as well as females.

Although all mammals lactate, only therians have nipples so that the young can suck directly from the breast rather than from the mother's fur. Mammary hairs, probably primitive mammalian features, are present in monotremes and marsupials, and the mammae develop from areola patches confined to

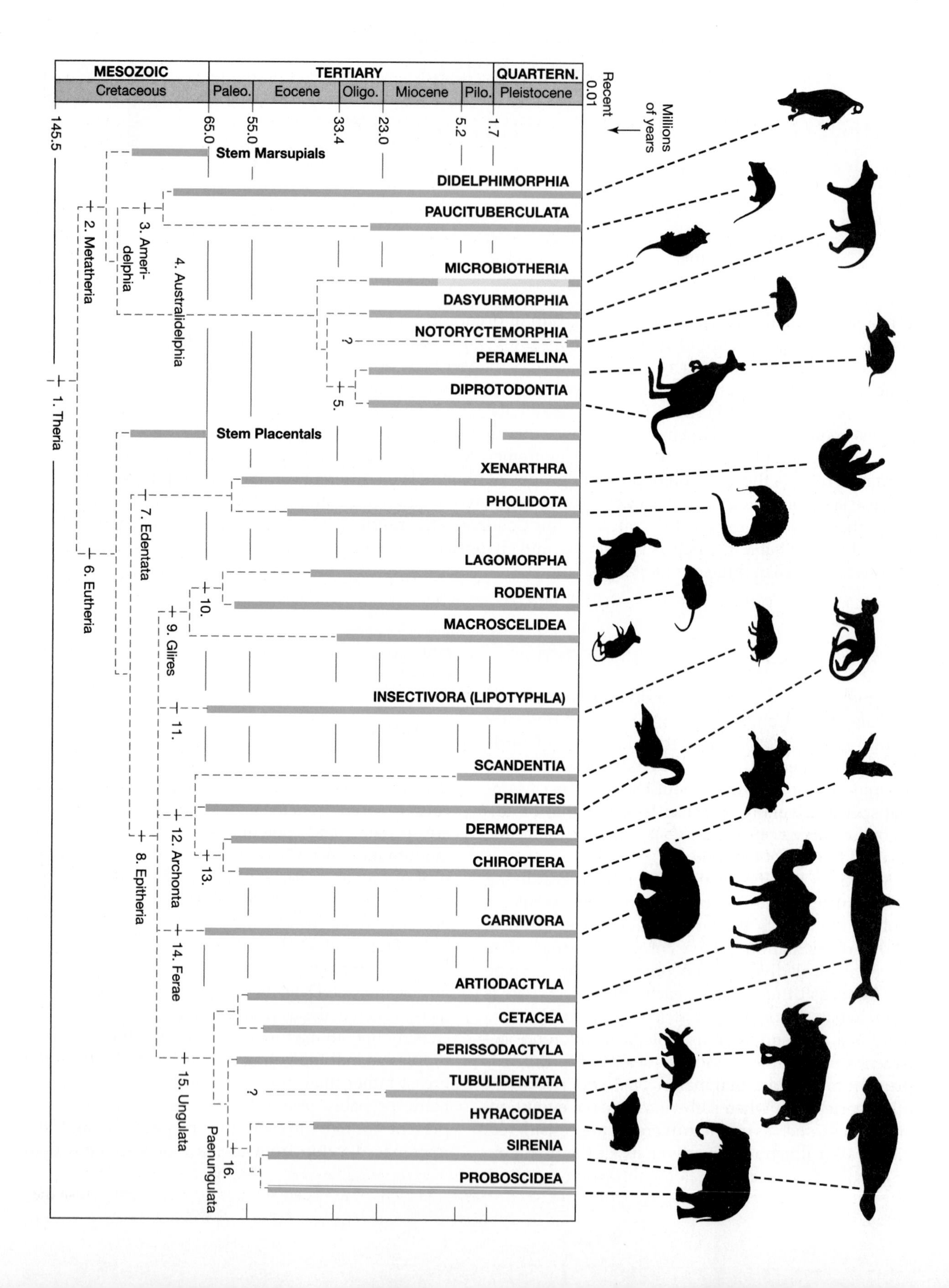

MESOZOIC
TERTIARY
QUARTERN.
Cretaceous
Paleo.
Eocene
Oligo.
Miocene
Pilo.
Pleistocene
Recent
0.01
Millions of years
145.5
65.0
55.0
33.4
23.0
5.2
1.7
Stem Marsupials
DIDELPHIMORPHIA
PAUCITUBERCULATA
MICROBIOTHERIA
DASYURMORPHIA
NOTORYCTEMORPHIA
PERAMELINA
DIPROTODONTIA
Stem Placentals
XENARTHRA
PHOLIDOTA
LAGOMORPHA
RODENTIA
MACROSCELIDEA
INSECTIVORA (LIPOTYPHLA)
SCANDENTIA
PRIMATES
DERMOPTERA
CHIROPTERA
CARNIVORA
ARTIODACTYLA
CETACEA
PERISSODACTYLA
TUBULIDENTATA
HYRACOIDEA
SIRENIA
PROBOSCIDEA
1. Theria
2. Metatheria
3. Ameri-delphia
4. Australidelphia
5.
6. Eutheria
7. Edentata
8. Epitheria
9. Glires
10.
11.
12. Archonta
13.
14. Ferae
15. Ungulata
16. Paenungulata
?
?

the abdominal region. In eutherians these mammary hairs are absent, and the mammae develop from mammary lines that form along the entire length of the abdomen.

Skeletomuscular System

In Chapter 17 we discussed how the mammalian skeletal system was gradually evolved within the synapsid lineage. Here we will focus on those features that are uniquely mammalian in comparison with other extant amniotes. **Epiphyses** (growth areas) on the long bones reflect the mammalian feature of determinate growth. The epiphyses are the ends of the bones that are separated from the shaft (the diaphysis) by a zone of growth cartilage in immature mammals. At maturity the bone fuses up in this area, and the epiphyses no longer appear as distinct structures separate from the rest of the bone.

Cranial Features In the mammalian skull the dermal bones that originally formed the skull roof have grown down around the brain and completely enclose the braincase (Figure 19–2a). The bones that form the lower border of the synapsid temporal opening are bowed out into a zygomatic arch, which we commonly refer to as the cheek bone. This is the bony bar that you can feel just below your eyes.

The mammalian jaw is formed from a single bone, the dentary, in contrast to the multiboned jaw of all other jawed vertebrates. The dentary articulates directly with the squamosal bone, a dermal bone in the skull. The jaw articulation of other jawed vertebrates is formed by the quadrate in the skull and the articular in the lower jaw. In mammals these bones have joined the stapes, resulting in the unique mammalian three-boned middle ear.

The cranial muscles of mammals are distinctly different from those of other amniotes. The main jaw adductor muscle is divided into two: a temporalis that runs from the top of the head to the coronoid process of the dentary, and a masseter that runs from the zygomatic arch to the outside of the lower jaw (Figure 19–2b). A muscle that inserts on the outside of the lower jaw in this fashion is a uniquely mammalian feature, and this is the muscle that enables mammals to move their jaw sideways during chewing. The main jaw-opening muscle of therian mammals is the digastric. This muscle has a complex

Legend: 1. Theria—Mammary glands with nipples, viviparity with loss of eggshell, digastric muscle used in jaw opening, anal and urogenital openings separate in adults, spiraled cochlea, scapula with supraspinous fossa, and numerous features of skull and dentition. **2.** Metatheria—Dentition essentially monophyodont (P3 is only tooth replaced), development of chorioallantoic membrane suppressed, pseudovaginal canal present at parturition, and various detailed features of skull, dentition (inc. upper molars with wide stylar shelves), and ankle joint. **3.** Ameridelphia—Sperm paired in epididymis. **4.** Australidelphia—Details of dentition and ankle joint (also supported by biochemical data). **5.** Syndactylous foot (digits 2 and 3 enclosed in a common skin sheath), W-shaped outer border (ectoloph) of upper molars. **6.** Eutheria—Egg shell membrane lost, intrauterine gestation prolonged with suppression of estrous cycle, corpus callosum connects cerebral hemispheres, ureters pass lateral to Mullerian ducts, fusion of Mullerian ducts into a median vagina, penis simple (not bifid at tip), plus details of dentition (inc. upper molars with narrow stylar shelves). **7.** Edentata—Details of skull anatomy, sacrum strongly fused to pelvis, tooth development suppressed with loss of anterior teeth and enamel poorly developed or absent. **8.** Epitheria—Stirrup-shaped stapes, plus some skull features. **9.** Glires—Details of skull anatomy. **10.** Details of skull anatomy, first pair of incisors lost, second pair of incisors large and evergrowing, details of placenta formation. **11.** Insectivora—details of skull anatomy, simplification of the hindgut with the loss of the cecum, reduction of pubic symphysis in pelvis. **12.** Archonta—Pendulous penis, plus details of ankle structure. **13.** Marked elongation of forelimbs, flight membrane between fingers, plus details of skull anatomy. **14.** Ferae (also includes the extinct order Creodonta)—Restriction of carnassial shearing teeth to the posterior part of the dentition, bony laminar separating cerebrum from cerebellum in brain, plus details of ankle joint. **15.** Ungulata—Bunodont dentition, reduced canines, astragalus with short robust head. **16.** Paenungulata—Styloglossus tongue muscle bifurcate, details of structure of skull, wrist bones, and placenta.

◀ Figure 19–1 Phylogenetic relationships of extant mammalian orders. This diagram shows the probable relationships among living therian mammals. Dotted lines show interrelationships only, and are not indicative of the times of divergence of or the unrecorded presence of taxa in the fossil record. Numbers indicate derived characters that distinguish the lineages.

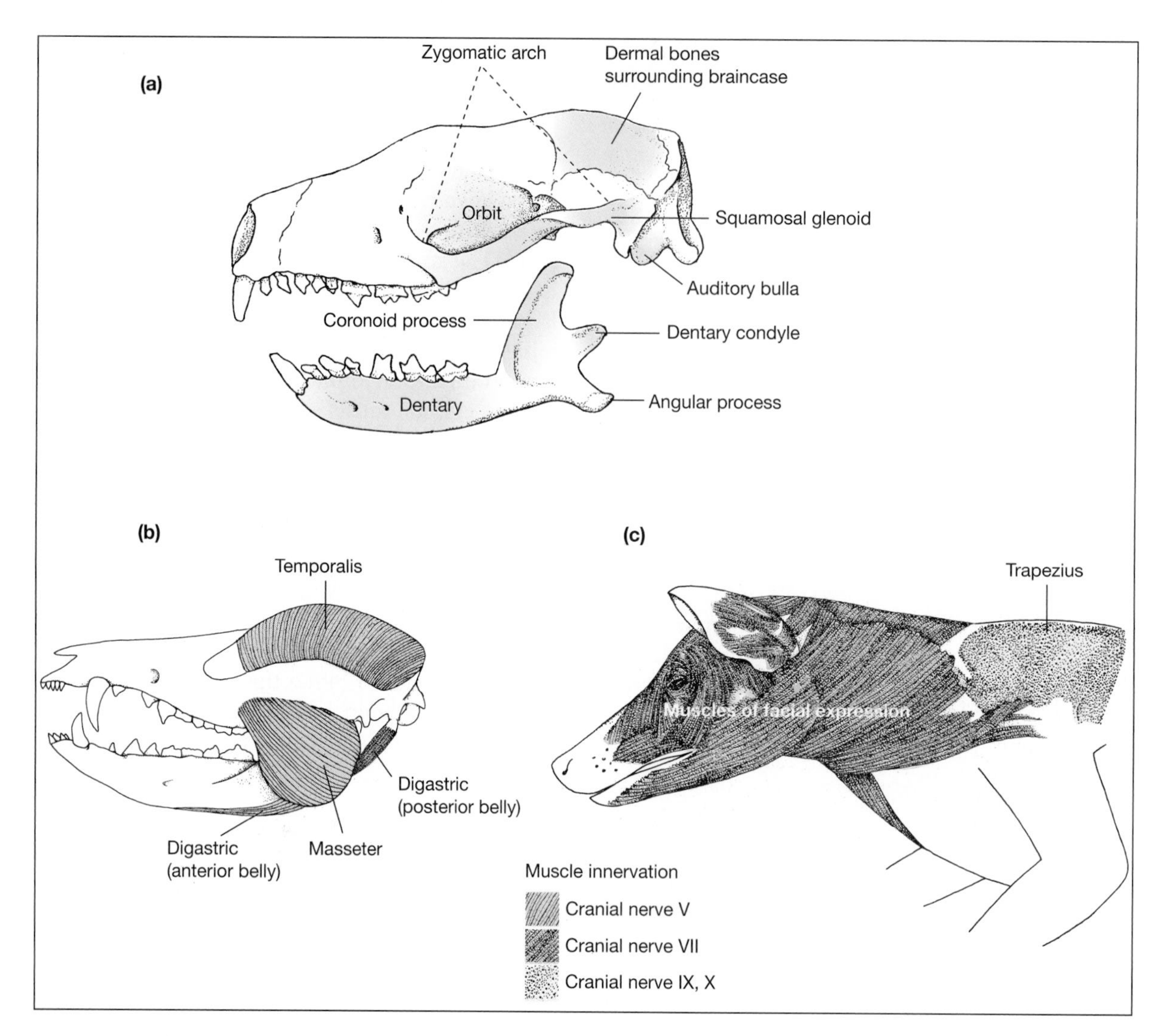

▲ Figure 19–2 Cranial anatomy of mammals. (a) Skull of a generalized mammal (a hedgehog): the condyle on the dentary fits into the glenoid on the squamosal to form the jaw joint. (b) The muscles of mastication. (c) Superficial view of cranial muscles showing overlying facial musculature.

evolutionary history. The back half is innervated by cranial nerve seven (as is the jaw-opening muscle of other tetrapods), while the front half is innervated by cranial nerve five. Mammals are also unique in having muscles of facial expression that are derived from neck muscles of reptiles and innervated by cranial nerve seven (Figure 19–2c).

The dentition of mammals is divided into several types of teeth (heterodont): incisors, canines, premolars, and molars. Most mammals have two sets of dentitions in their lifetime (diphyodonty). The first set—or the milk teeth—consists of incisors, canines, and premolars only, although the form of these premolars may be like that of the adult molars. The permanent, adult dentition consists of the second set of the original teeth with the addition of the later-erupting molars. (Our last molars are known as wisdom teeth, so called because they erupt at the age—late teens—by which we supposedly have attained wisdom.) Mammals are the only animals that masticate (chew the food) and swallow a discrete bolus of food. Therian mammals have unique types of molars, called **tribosphenic molars** (Box 19-1).

Postcranial Features Unlike the sprawling posture of extant reptiles, mammals have an upright posture with the limbs positioned underneath the body (Figure 19–4). However, the highly upright posture of familiar mammals such as cats, dogs, and

BOX 9-1 The Evolution of Tribosphenic Molars

Very early in mammalian history, as early as the earliest Jurassic, there was a difference observable in the types of molar teeth and patterns of occlusion. The primitive type, exemplified by the basal mammal *Morganucodon*, is to have the three main molar cusps in a more-or-less straight line. The more derived type, exemplified by the contemporaneous symmetrodont *Kuehneotherium*, is to shift the principal, middle cusp so that the teeth assume a triangular form in occlusal (food's-eye) view (Figure 19–3). The apex of the triangle formed by the upper tooth points inward, while that of the lower tooth points outward, forming an intermeshing relationship called reversed triangle occlusion. The upper triangle is known as the trigon, and the lower one the trigonid. The longer sides of these triangular teeth result in a greater amount of available area for shearing action. The teeth also interdigitate in a more complex fashion than do the teeth of more primitive mammals.

A further dental complication seen in later, more derived mammals is the development of the tribosphenic molar. The group of mammals possessing tribosphenic molars is termed the Tribosphenida, and includes the Theria (marsupials and eutherians). This type of molar adds a new cusp, the protocone, to the trigon in the uppers, which occludes against a basined addition to the lowers called the talonid (Figure 19–3). The tribosphenic molar adds the function of crushing and punching to the original tooth, which acted mainly to cut and shear. The possession of this tooth presumably reflects a greater diversity of dietary items taken.

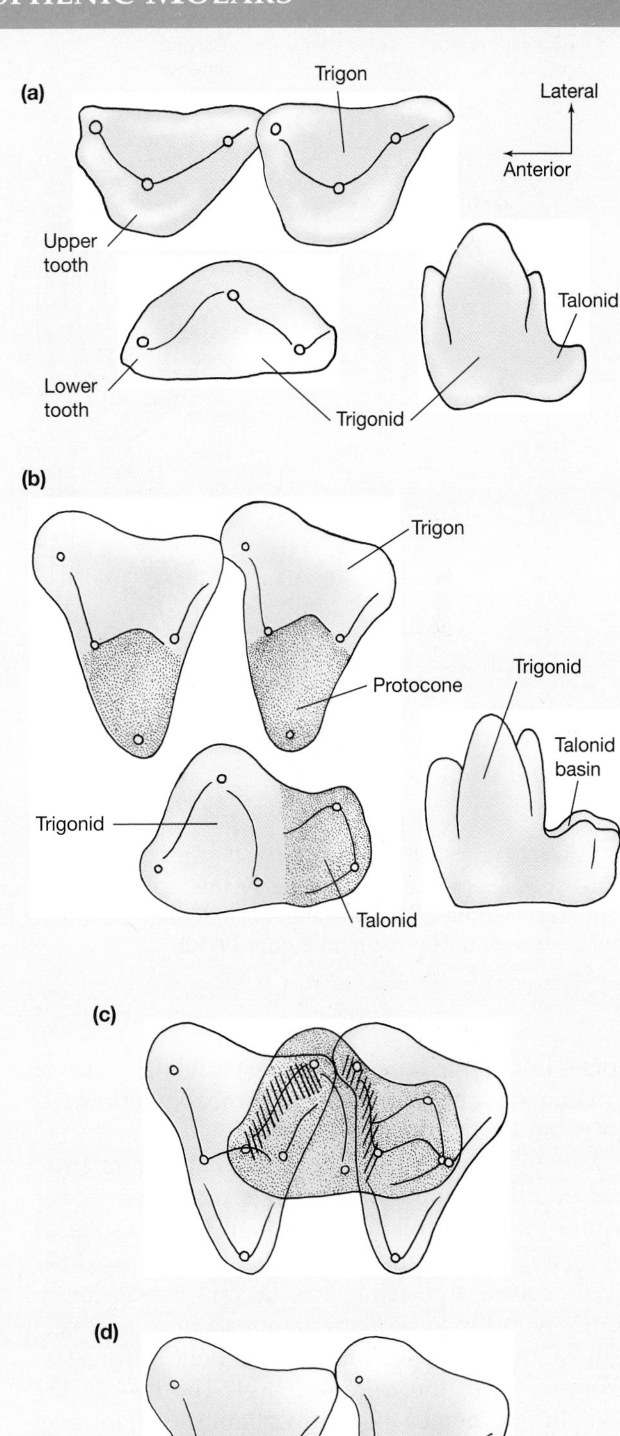

▶ Figure 19–3 Evolution of mammalian molars. (a) Schematic occlusal view of reversed triangle molars of a nontherian holotherian mammal (e.g., *Kuehneotherium*). The lower molar is also illustrated in side view. (b) Similar view of the tribosphenic molars of a therian. The new portions (the protocone in the uppers and the basined talonid in the lowers) are shaded. Parts (c) and (d) show the action of the lower molars in occlusion with the uppers: The lower molar is shaded, and areas of tooth contact are crosshatched. (c) Initial contact between the teeth at the start of occlusion. The trigonid cusps produce a shearing action along side the cusps at the back of the trigon of the anterior upper tooth and at the front of the trigon of the posterior upper tooth. (d) Mortar-and-pestle action of the molars at the end of the occlusal power stroke. The protocone (= the pestle of the combination) fits into the basin formed between the cusps in the trigonid (the mortar).

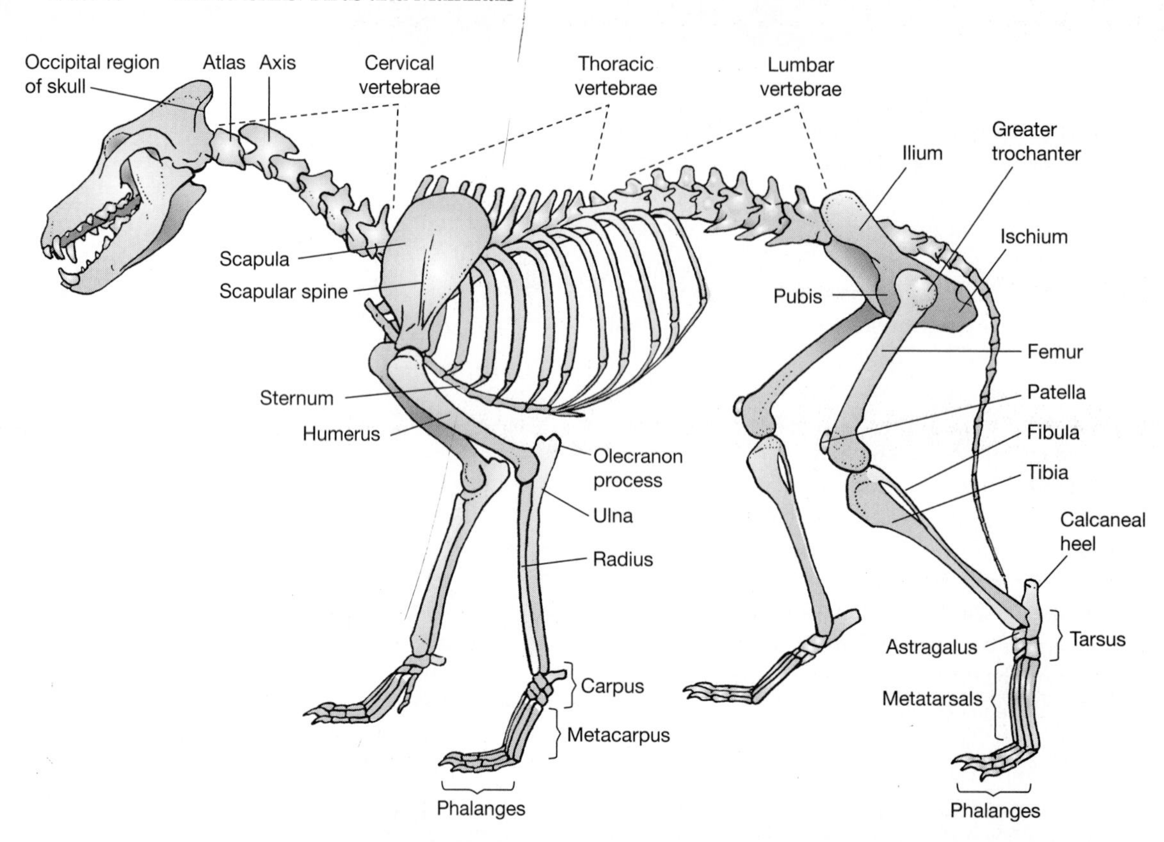

▲ Figure 19-4 Skeleton of a dire wolf (extinct Pleistocene wolf). The long neural spines of the thoracic vertebrae serve as the attachment area of the nuchal ligament, which runs from the back of the head and helps to hold the head up. This animal is standing on the phalanges in a digitigrade form of foot posture, in contrast with the more primitive plantigrade foot posture of the opossum in Figure 19–12b.

horses is a derived one; the semi-sprawling stance of a mammal such as the opossum probably represents the primitive mammalian condition.

Mammals have an ankle joint that differs from that of other amniotes. The site of movement is not within the bones of the ankle joint (the mesotarsal joint); it is between the tibia and one of the proximal ankle bones, the astragalus (a crurotarsal joint). Along with this new joint, mammals have a projection of the other proximal ankle bone, the calcaneum, to form the calcaneal heel. The heel is the point of insertion of the gastrocnemius (calf) muscle. In the pelvic girdle the ilium is rod-shaped and directed forward, and the pubis and ischium are short, in contrast with the more plate-like pubis and ischium of reptiles. The femur has a distinct trochanter on the proximal lateral side (= the greater trochanter) for the attachment of the gluteal muscles, which are now the major retractors of the hind limbs. It is the gluteals that give mammals their characteristic rounded rear ends.

With a very few exceptions, all mammals have seven cervical (neck) vertebrae. (Manatees and one type of tree sloth have six cervical vertebrae, and the giraffe actually has eight [Solounias 1999].) They also have a uniquely specialized atlas-axis complex of the first two cervical vertebrae. Mammals can rotate their head on their neck in two places: not only in the more general up-and-down fashion, at the joint between the head and the atlas, but also in the more derived side-to-side fashion, at the joint between the atlas and the axis. Mammals have restricted the ribs to the anterior (thoracic) trunk vertebrae. The lumbar ribs now have zygaphophyseal connections that allow for dorsoventral flexion; they also have large transverse processes for the attachment of the longissimus dorsi (one of the epaxial muscles) that produces this movement during locomotion. The

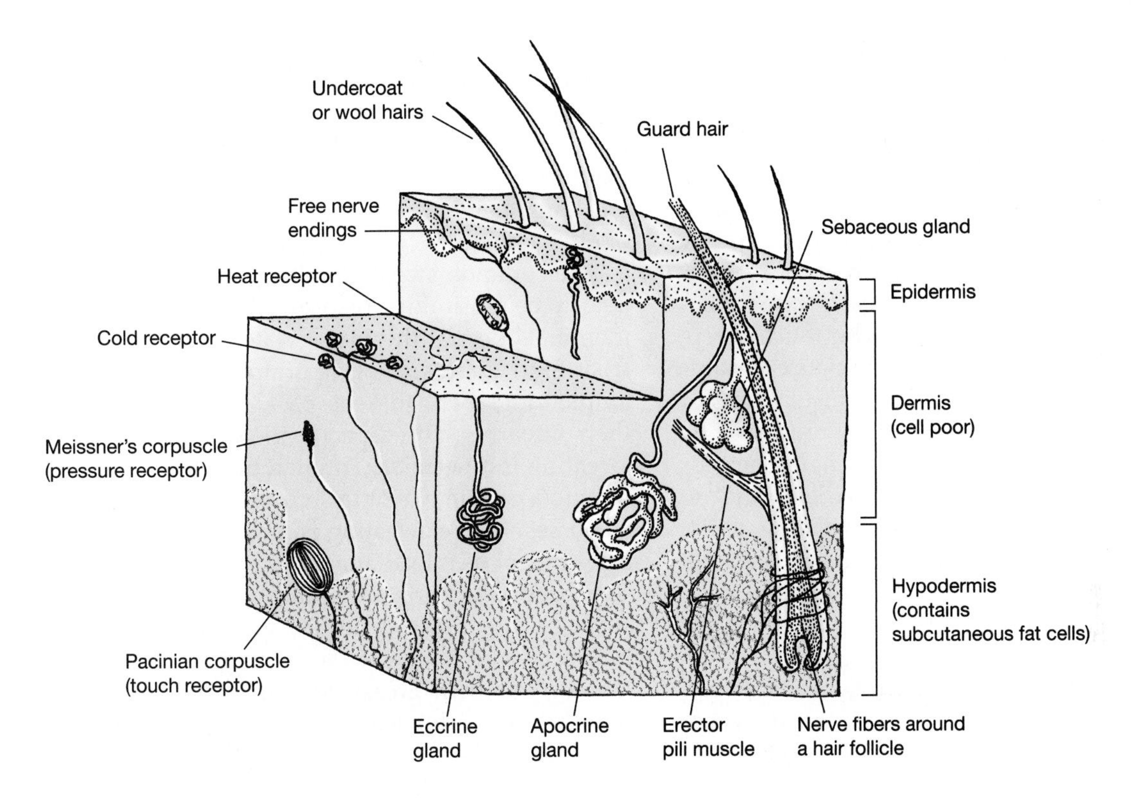

▲ Figure 19–5 Structure of mammalian skin.

capacity to twist the spine in a dorsoventral fashion in mammals may relate to the ability of mammals to lie down on their sides, something that other vertebrates cannot do easily. This ability may have been important in the evolution of suckling, with ventrally placed nipples.

The Integument

In many ways the outside covering of mammals is the key to their unique way of life. We have emphasized that endothermy is an energetically expensive process, and much of the ability of mammals to live in harsh climates is attributable to properties of their integument. The variety of mammalian integuments is enormous. Some small rodents have a delicate epidermis only a few cells thick. Human epidermis varies from a few dozen cells thick over much of the body to over a hundred cells thick on the palms and soles. Elephants, rhinoceroses, hippopotamuses, and tapirs were once classified together as pachyderms (Greek: *pachy* = thick and *derm* = skin) because their epidermis is several hundreds of cells thick. The texture of the external surface of the epidermis varies from smooth (in fur-covered skins and the hairless skin of cetaceans) to rough, dry, and crinkled (many hairless terrestrial mammals). The tail of opossums and many rodents is covered by epidermal scales similar to those of lizards but lacking the hard beta keratin found in birds and reptiles.

Figure 19–5 illustrates the typical structure of mammalian skin. Note that while mammalian skin is like that of other vertebrates in basic form, with epidermal, dermal, and hypodermal layers, there are also unique components. Mammalian skin has hair, lubricant- and oil-producing sebaceous glands, and apocrine and eccrine glands that secrete volatile substances, water, and ions. Typically mammalian structures derived from the keratinous layer of the epidermis include nails, claws, hoofs, and horns. Sensory nerve endings include free nerve endings (probably pain receptors), beaded nerve nets around blood vessels, Meissner's corpuscles (touch receptors), Pacinian corpuscles (pressure receptors), nerve terminals around hair follicles, and warmth and cold receptors. Vascular plexuses (intertwined blood vessels) of the skin are the basis for countercurrent blood flow.

Hair Hair has a variety of functions including camouflage, communication, and sensation via **vibrissae** (= whiskers). Vibrissae grow on the muzzle, around the eyes, or on the lower legs; touch receptors are associated with these specialized hairs. However, the basic function of hair is insulation. Fur consists of closely placed hairs, often produced by multiple hair shafts arising from a single complex root. Its insulating effect depends on its ability to trap air within the fur coat, and its insulating ability is proportional to the length of the hairs. The erector pili muscles that attach midway along the hair shaft (see Figure 19–5) pull the hairs erect to deepen the layer of trapped air.

Cold stimulates a general contraction of the erector pili via the sympathetic nerves, as do other stressful conditions such as fear and anger. Hair erection increases the depth of the **pelage** (the hairy covering), thus trapping a larger volume of air. A curious side effect noticeable in near-naked mammals such as humans is the dimples (goose pimples) on the skin's surface over the insertion of contracted erector pili muscles. Hair erections can serve for communication as well as for thermoregulation; mammals can use them to send a warning of fear or anger (as seen in the display of a puffed-up cat or the raised hackles of a dog).

Prominent features of the pelage of extant mammals are its growth, replacement, color, and mobility. A hair is composed of keratin, and it grows from a deep invagination of the germinal layer of the epidermis called the hair follicle (Figure 19–5). The color of hair depends on the quality and quantity of melanin injected into the forming hair by melanocytes at the base of the hair follicle. The color patterns of mammals are built up by the colors of individual hairs. Because exposed hair is nonliving, it wears and bleaches. Replacement occurs by growth of an individual hair or by **molting,** in which old hairs are replaced. Most mammals have pelage that grows and rests in seasonal phases; molting usually occurs only once or twice a year.

Glandular Structures Secretory structures of the skin develop from the epidermis. There are three major types of skin glands in vertebrates: **sebaceous, apocrine**, and **eccrine** glands. Except for the eccrine glands, skin glands are associated with hair follicles, and the secretion in all of them is under neural and hormonal control. A full component of these skin glands is found in monotremes as well as in therians, and thus they may be assumed to be a basic feature of all mammals.

The common sweat glands of humans do not appear to be a primitive mammalian feature, and most mammals do not thermoregulate by secreting fluid from skin glands for evaporative cooling. For example, dogs pant to keep cool, and kangaroos lick their forearms. The capacity to cool off by copious sweating is a peculiarly human trait, seen elsewhere only to a certain extent in some large ungulates such as horses, where sweating is largely associated with activity, not heat load alone.

Sebaceous glands are found over the entire body surface. They produce an oily secretion, sebum, that lubricates and waterproofs the hair and skin. Sheep lanolin, our own greasy hair, and the grease spots that the family dog leaves on the wallpaper where it curls up in the corner are all sebaceous secretions. Apocrine glands have a restricted distribution in most mammals, and their secretions appear to be used in chemical communication. In humans apocrine glands are found in the armpit and pubic regions—these are the secretions that we usually try to mask with deodorant. In some other mammals, such as large ungulates, these glands are scattered over the body surface and may be used in evaporative cooling.

Many mammals have specialized scent glands that are modified sebaceous or apocrine glands. Sebaceous glands secrete a viscous substance, usually employed to mark objects, while apocrine glands produce volatile substances that may be released into the air as well as placed on objects. Scent marking is used to indicate the identity of the marker and to define territories. Scent glands are usually placed on areas of the body that can be easily applied to objects, such as the face, chin, or feet. Domestic cats often rub their face and chin to mark objects, including their owners. Many carnivorans have anal glands so that scents can be deposited along with the urine and feces, and anal scent glands (apocrine glands) are a well-known feature of skunks. There are also apocrine glands in the ear that produce ear wax.

Eccrine glands produce a secretion that is mainly watery, with little organic content. In most mammals eccrine glands are restricted to the soles of the feet, prehensile tails, and other areas that contact environmental surfaces, where they improve adhesion or enhance tactile perception. Eccrine glands are found over the body surface only in primates and especially in humans, where they secrete copious amounts of fluid for evaporative cooling. Profuse thermoregulatory sweating has evolved convergently within mammals, because eccrine glands function as sweat glands in humans while ungulates sweat via apocrine glands. In humans sweat glands may act in conjunction with nearby apocrine glands, contributing to odor production under conditions of stress and excitement.

Mammary glands have a more complex, branching structure than other skin glands. They have several basic features in common with apocrine and sebaceous glands: structure, body distribution, and chemical composition of secretion. The evolution of mammary glands may have occurred with the formation of a new type of skin gland that contained properties of both apocrine and sebaceous glands; though they resemble the other two types of glands, mammary glands cannot be fully homologised with either one alone (Blackburn 1991).

Claws, Nails, and Hooves Some integumentary appendages are involved in locomotion, offense, defense, or display. Claws, nails, and hooves are accumulations of keratin that protect the terminal phalanx of the digits (Figure 19–6). Permanently extended claws are the primitive condition; the retractable claws of cat-like carnivores are derived. The hoof gives ungulates a small foot (which is mechanically advantageous for a running animal) that is solid enough the bear the animal's weight. The fingernails of humans and other primates are a simpler structure than either the retractable claw or the hoof, but were derived from ancestral claws.

Internal Anatomy

Mammals have numerous differences from other amniotes in their internal anatomy and physiology. Some of these relate to their endothermic metabolism; similar systems have evolved convergently in the other endothermic vertebrates, birds. Others are uniquely mammalian and reflect their evolutionary history.

Adipose Tissue Both birds and mammals differ from other tetrapods in their distribution of adipose tissue. Rather than concentrations of **adipocytes** (fat storage cells) in discrete fat bodies, these cells are scattered around the bird or mammalian body as subcutaneous fat, as fat associated with various internal organs (e.g., heart, intestines, kidneys), as deposits in skeletal muscles, and as cushioning for joints. Mammalian adipose tissue (white fat) is not simply inert material used only as an energy store when fasting or as an insulating layer of blubber in marine mammals, as has often been assumed. More recent studies have revealed that adipocytes secrete a wide variety of messenger molecules that coordinate important metabolic processes. Some small depots have site-specific properties that equip them to interact locally with the immune system and possibly other organs (see Pond, 1998).

Mammals also have a unique type of adipose tissue—brown fat. This tissue is specially adapted to generate heat and can break down lipids or glucose to generate energy as heat at a rate up to ten times that of the muscles. Brown fat is especially prominent in newborn mammals, where it is important for general thermoregulation, and in the adults of hibernating species, where it is important for rewarming the body in coming out of hibernation.

Cardiovascular System The heart of mammals differs from that of ectothermic amniotes in having a complete ventricular septum and only a single systemic arch (Figure 19–7), although the original double arch is apparent in development. A similar condition is seen in birds, but this condition clearly arose convergently in the two groups because it is the left systemic arch that is retained (as the single **aorta**) in mammals, and the right arch in birds.

The historical double condition of the single mammalian aorta can be further observed in the detailed vascular anatomy; a dead-end portion of the right systemic arch is retained to serve as the point of origin of the right subclavian artery, which leads to the right arm. This anatomy suggests that mammals never had the completely double system of aortic arches seen in reptiles (see Chapter 10). The implication of this would be that much of the complexity of the mammalian heart—in the division of ventricles and separation of the original ventral aorta trunk into systemic and pulmonary systems—occurred convergently with the condition in other amniotes.

Mammals also differ from other vertebrates in the form of their erythrocytes (red blood cells), which lack nuclei in the mature condition. Additionally, while monotremes retain a small sinus venosus as a distinct chamber, therians have incorporated this

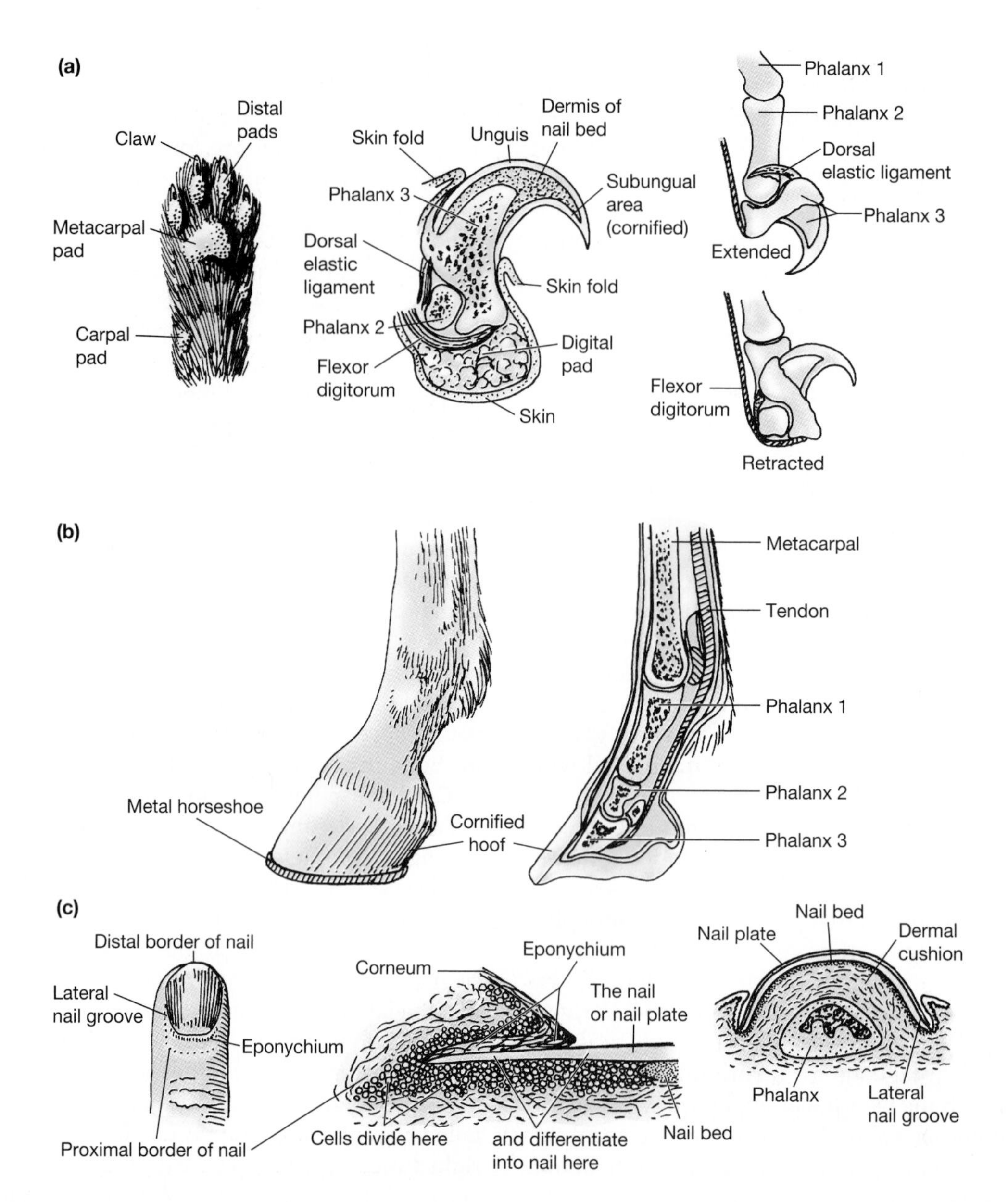

▲ Figure 19–6 Skin appendages associated with terminal phalanges. (a) Retractable claws. Left: Hair and thick epidermal pads associated with the base of the claws. Center: Cross section of a claw showing its close relationships with the blood vessels, dermis, and bone of the third (terminal) phalanx. Right: Claw retraction mechanism characteristic of cats. (b) The hoof of a horse. Left: Normal appearance of the hoof of a shod horse. (Horseshoes are devices used to minimize wear of the hoof on unnaturally [human produced] hard and abrasive surfaces.) Right: Longitudinal section of lower foot showing relationship of phalanges to hoof. (c) The human nail. Left: Distinct regions on a nail correspond to the regional specializations of the epidermis associated with the nail (Center). Right: A cross section of the end of a finger shows the close association of the nail with the dermis and terminal phalanx of the digit.

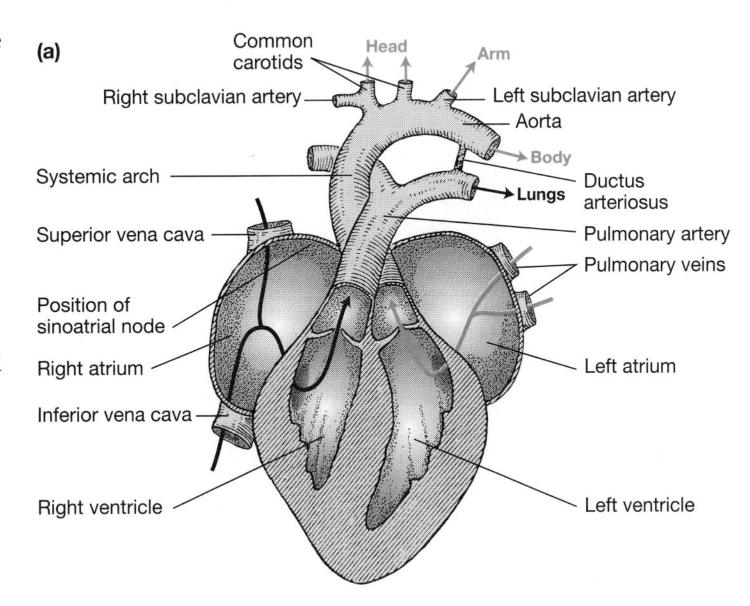

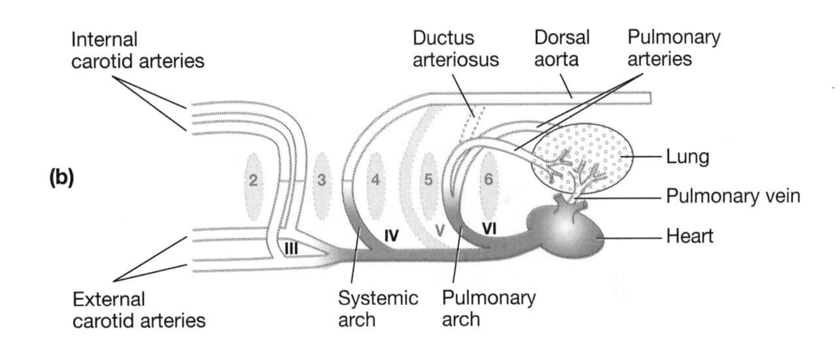

▶ Figure 19–7 (a) Diagrammatic view of the mammalian heart. Black arrows indicate the passage of deoxygenated blood, blue arrows the passage of oxygenated blood.
(b) Derived condition of the aortic arches in a mammal. (This is a schematic drawing intended to show relationships with the more primitive vertebrate condition shown in Figure 8–10.) The ductus arteriosus (= ligamentum arteriosum in the adult condition) is the remains of the dorsal part of the pulmonary arch; it is patent and functional in the amniote fetus, where it is used as a bypass shunt for the lungs. Arabic numbers: gill slits/visceral pouches. Roman numerals: aortic arches. Numberings represent presumptive primitive condition for vertebrates.

structure into the right atrium as the sinoatrial node, which now acts as the heart's pacemaker.

Respiratory System Mammals have large, lobed lungs with a spongy appearance due to the presence of a finely branching system of bronchioles in each lung, terminating in tiny thin-walled, blind-ending chambers (the sites of gas exchange) called **alveoli**. They also have a muscular sheet, the diaphragm, which aids the ribs in inspiration and divides the original pleuroperitoneal cavity into (1) **peritoneal cavity** surrounding the viscera and (2) paired **pleural cavities** surrounding the lungs.

Urogenital System Many sauropsids lose the bladder, in association with excreting a semisolid paste of uric acid. In contrast, all mammals retain the bladder and excrete relatively dilute urine. Mammalian kidney function also differs from that of other vertebrates. Mammals have entirely lost the renal portal system seen in other vertebrates, which supplies venous blood to the kidney in addition to the arterial blood supplied by the renal artery. Mammals also have a new portion of the kidney tubule called the loop of Henle, correlating with their ability to excrete urine that has a higher concentration of salt than the body fluids. Marine birds and reptiles have various types of salt glands in their heads to excrete excess salt, but marine mammals can handle this salt excretion via the kidney.

Therian mammals differ from other vertebrates in various ways. In most vertebrates the urinary, reproductive, and alimentary systems reach the outside via a single common opening, the **cloaca**, while in therians the cloaca is replaced by separate openings for the urogenital and alimentary systems. The testes are placed in a scrotum outside of the body in the males (in most, but not all species; see Chapter 20), and the penis is used for urination as well as for the passage of sperm, with the ducts leaving the testes and the bladder combined into a single **urethra** (Figure 19–8). The clitoris in the female is the homolog of the penis, but is not used to pass urine.

A bifid (forked) glans of the penis is seen in monotremes and in most marsupials, as opposed to the single glans of eutherians. Some male eutherians have a bone in their penis, the os penis or **bacculum**—females may also have a corresponding structure called the os clitoris. This structure is seen among certain species in primates, rodents, insectivorans, carnivorans, and bats.

Most female mammals conjoin the urethra and the vagina into a single **urogenital** sinus leading to the outside (Figure 19–8), while primates and some rodents are unique in having separate openings for the urinary and genital systems. Note that in the more usual mammalian condition the clitoris is within the urogenital sinus, where it receives direct stimulation from the penis during copulation, a rather more practical arrangement than in humans. Primates also are unusual in having a pendulous penis that cannot be retracted into the body. Most male mammals extend the penis from an external sheath (normally the only visible portion) only during urination and copulation, again a seemingly more practical design. The anatomy of primates may be related to the relatively high position of the body wall. In most mammals the body wall encloses a greater portion of the femur (thigh bone) than it does in primates, providing more intra-abdominal space for luxuries such as a retractable penis.

Sex Determination and Sex Chromosomes

In all vertebrates the gonad is initially indifferent and capable of producing either an ovary or a testis. Individuals with both types of gonads present and functional are called **hermaphrodites**. Hermaphroditism is common in nonamniotic vertebrates, especially fishes, but is virtually absent among amniotes. The gender of most mammals is generally obvious from birth, expressed by genitalia and other secondary sex characteristics, but in other vertebrates external genitalia and sexually dimorphic structures are absent or not expressed until maturity is achieved.

Sex determination in mammals is by distinctive sex chromosomes, the X and Y chromosomes: females have two X chromosomes and males have an XY combination. Sex chromosomes are a derived feature among vertebrates. Birds also have sex chromosomes, but here the chromosomal designation is different: ZZ produces a male and ZW a female. This difference alone is an indication that the sex chromosomes evolved independently in birds and mammals. Some reptiles have sex chromosomes, but others do not—in these species sex is determined by environmental temperature during development (Chapter 11).

It appears that a gene located on the Y chromosome in mammals initiates male gonadal development, and female gonadal development results from its absence. Once a gonad has had its primary sex declared as female or male, the sex hormone estrogen or testosterone is produced. These hormones affect development of the secondary sex characters. In humans the genitalia, breasts, hair patterns, and differential growth patterns are secondary sex characteristics. Horns, antlers, and dimorphic color patterns are familiar differences that we associate with gender in other mammals.

Sensory Systems

The sensory systems of mammals differ from those of other tetrapods in various ways. Mammals have exceptionally large brains among vertebrates, and their brain evolved along a somewhat independent pathway to that of other amniotes. In their sensory systems mammals are more reliant on hearing and olfaction than most other tetrapods, and they are less reliant on vision.

The Brain The enlarged portion of the cerebral hemispheres of mammals, the neocortex or **neopallium**, is formed in a unique fashion. The dorsal part of the cortex is enlarged to form the neopallium (whereas sauropsids enlarge the lateral portion), and it has a complex, laminated structure. In more derived mammals the neopallium dominates the entire forebrain and becomes highly folded, which greatly increases its surface area.

Other unique features of the mammalian brain include divided optic lobes in the midbrain, an infolded cerebellum, and a large representation of the area for cranial nerve seven in the brain, which is associated with development of the facial musculature. In addition to the anterior commissure that links the hemispheres in all amniotes, eutherians have a new nerve tract linking the two cerebral hemispheres—the corpus callosum.

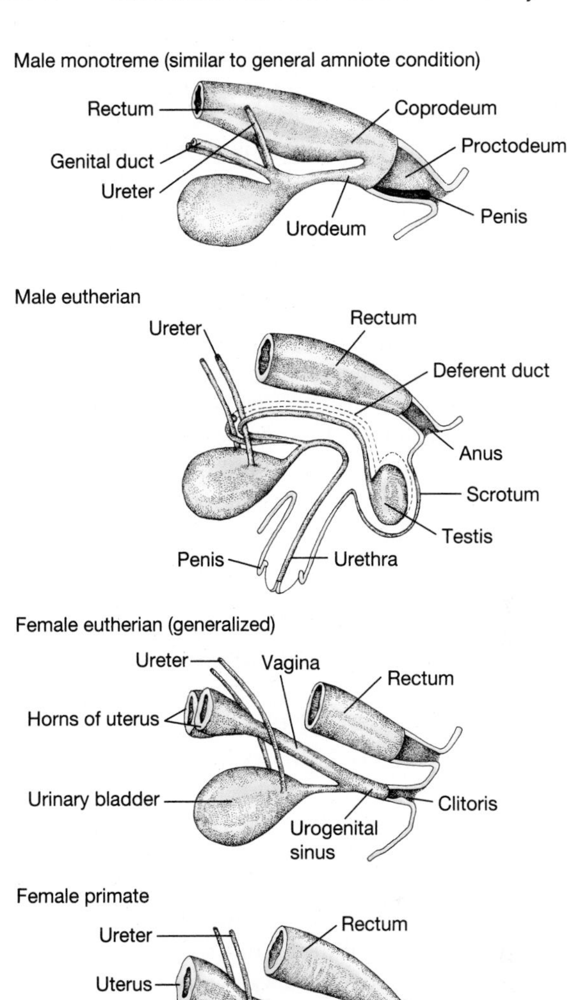

▶ Figure 19-8 Anatomy of the urogenital ducts in mammals. The head is to the left.

Olfaction The keen sense of smell of most mammals is probably related to their primarily nocturnal behavior. The olfactory receptors are located in specialized epithelium on the scroll-like nasoturbinal and ethmoturbinal bones in the nose. The olfactory bulb is a prominent portion of the brain in many mammals, but primates have a relatively small olfactory portion of the brain and a poor sense of smell, probably in association with their diurnal habits. The sense of smell is also reduced or completely absent in whales, in association with their aquatic existence.

Vision Mammals evolved as nocturnal animals, and visual sensitivity (forming images in dim light) was more important than visual acuity (forming sharp images). Mammals have retinas composed primarily of rod cells, which have high sensitivity to light but are relatively poor at acute vision. Lizards and birds have high visual acuity because of their nearly pure cone retinas. Mammals approach this acuity only in one small region of the retina, the all-cone **fovea**. In addition to providing high visual acuity, cones are the basis for color vision. Cones have relatively low sensitivity, however; they are at least two orders of magnitude less sensitive to light than rods, and they do not work at night and in other dim-light conditions. Thus, while reptiles and birds have superior color vision, mammals have superior night vision.

Most mammals also have a well-developed **tapetum lucidum**, a reflective layer behind the retina that reflects light back through the retina, thus providing a second chance for a photon of light to stimulate a retinal receptor cell. The tapetum causes the eye shine that you see when you point a flashlight at a dog or cat. The tapetum is best developed in nocturnal mammals, and is lost in the diurnal anthropoid primates, including humans.

Hearing Mammals have a middle ear that is more complex than that of other tetrapods. It contains a chain of three bones (stapes, malleus, and incus) rather than just a single bone. Mammals have a greater acuity of hearing than other tetrapods; however, while a chain of ear ossicles seems to have a mechanical advantage over a single one—boosting the force of vibration on the inner ear—having three bones (rather than one) in the middle ear was evidently an evolutionary accident rather than the result of selection for increased auditory capacity (see Chapter 17).

Several other features of therian mammals also contribute to increased auditory acuity. These include a long cochlea capable of a greater degree of pitch discrimination. The cochlea of mammals is so long that it must be coiled to fit inside the otic capsule (Figure 19–9). In addition, the external ear, or **pinna**, helps to determine sound direction. The pinna, in combination with the narrowing of the external auditory meatus of mammals, concentrates sound from the relatively large area encompassed by the external opening of the pinna to the small, thin, tympanic membrane. The pinna is unique to mammals, although it has a feathery analog in certain owls (see Chapter 16). Most mammals can move their pinnae to pick up sound, although anthropoid primates lack this capacity. The auditory sensitivity of a terrestrial mammal is reduced if the pinnae are removed. Aquatic mammals use entirely different systems to hear underwater, and have reduced or lost the pinnae. Cetaceans, for example, use the lower jaw to channel sound waves to the inner ear.

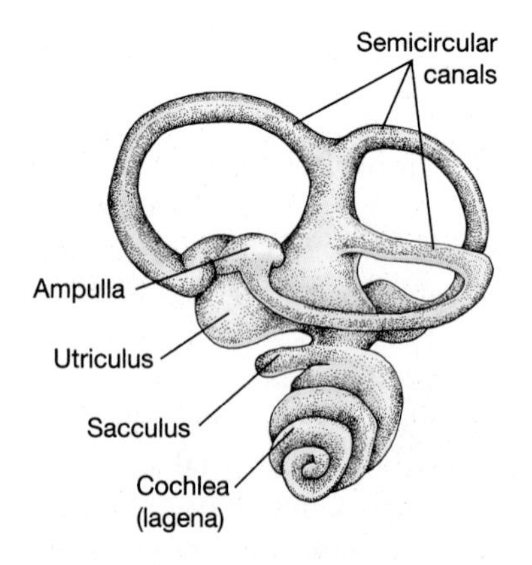

▲ Figure 19–9 Design of the vestibular apparatus in therian mammals, with the cochlea (= lagena) now lengthened and coiled.

19.2 Major Lineages of Mammals

Traditionally, the class Mammalia has been divided into three subclasses: **Allotheria** (multituberculates, now extinct), **Prototheria** (monotremes), and **Theria** (marsupials, infraclass **Metatheria**, and eutherians [= placentals], infraclass **Eutheria**). This classification does not really take into account the large diversity of Mesozoic mammals (see Chapter 17). Many of the Mesozoic groups were originally allied with the monotremes, although more recent work has shown that monotremes are more closely related to therians than was previously thought. However, these three groups do reflect basic divisions in body plans between those mammals surviving into the Cenozoic.

Multituberculates and monotremes were originally seen as being very distantly related to therians. Living monotremes retain some primitive anatomical features in addition to their egg-laying habits. Until quite recently they were considered to be derived from a different group of Late Triassic stem mammals from the therians, and they were once even considered to be an independent radiation from the cynodonts. The teeth of multituberculates are so different from those of other early mammals that they, too, were thought to have a very early offshoot from the main lineage.

Living monotremes are toothless as adults, so the characters of dental anatomy (on which much of mammalian phylogeny is based) cannot easily be used to evaluate their relationships. However, teeth are found in juvenile platypuses, and teeth are known from fossil forms. A single jaw of a fossil form (*Steropodon*) from the Early Cretaceous of Australia showed fully formed teeth that were surprisingly clearly triangular in shape. The shape of these teeth suggests that monotremes are at least more closely related to therians than some of the Mesozoic mammals (see Figure 17–1).

Multituberculates—"Rodents of the Mesozoic"

Multituberculates were a very long-lived group, known from the Late Jurassic to the early Cenozoic (late Eocene). They are known almost entirely from the Northern Hemisphere: Only recently has a Gondwanan radiation been recognized in South America and Madagascar. Multituberculates probably occupied an adaptive niche similar to that of rodents, small terrestrial and semiarboreal omnivores. Their very narrow pelvis suggests that they did not lay eggs, but may have given birth to immature young.

Some multituberculates were rather squirrel-like: The structure of their ankle bones shows that they could rotate their foot backward to descend trees head-first, like a squirrel, and their caudal vertebrae indicate a prehensile tail (Figure 19–10). These squirrel-like multituberculates also had an enlarged lower posterior premolar that formed a shearing blade, perhaps used to open hard seeds (Figure 19–11a). Other multituberculates were more terrestrial, and rather like a ground hog or a wombat. The extinction of multituberculates in the Eocene was probably due to competition with the rodents, which first appeared in the late Paleocene.

▶ Figure 19–10 The early Cenozoic multituberculate *Ptilodus*.

Multituberculates get their name from their molars, which are broad, multicusped (multituberculed) teeth specialized for grinding rather than shearing. Wear on the teeth indicates that multituberculates moved their lower jaw backward while bringing their teeth into occlusion. Multituberculate teeth and jaw movements were rather similar to those of the tritylodontids, the lineage of cynodonts that survived into the Jurassic, and competition between multituberculates and tritylodontids may have contributed to the trytilodontids' demise. The teeth and jaw movements of multituberculates were also similar to those of rodents, except rodents move their lower jaws *forward* into occlusion (Figure 19–11).

The position of the multituberculates within the Mammalia is controversial. Multituberculates have a rather derived, therian-like shoulder girdle (see Figure 19–12), which originally was considered an independent evolution within this group. This feature is now considered to link them even more closely to therians than are monotremes (as shown in Figure 17–1).

In contrast, Zofia Kielan-Jaworowska (1997), who has spent a lifetime studying multituberculates, considers them to be very primitive, especially in the nature of their brains (as determined from casts of the inside of the skull). If this were true, then the derived nature of the shoulder girdle must represent convergence between multituberculates and therians. The phylogenetic position of multituberculates is an area of active debate and controversy; better fossil material of Jurassic multituberculates would greatly help to resolve their relationships.

Differences between Therians and Nontherians

Therians are distinguished from monotremes by derived features. The most obvious one is giving birth to young rather than laying eggs. The differences in

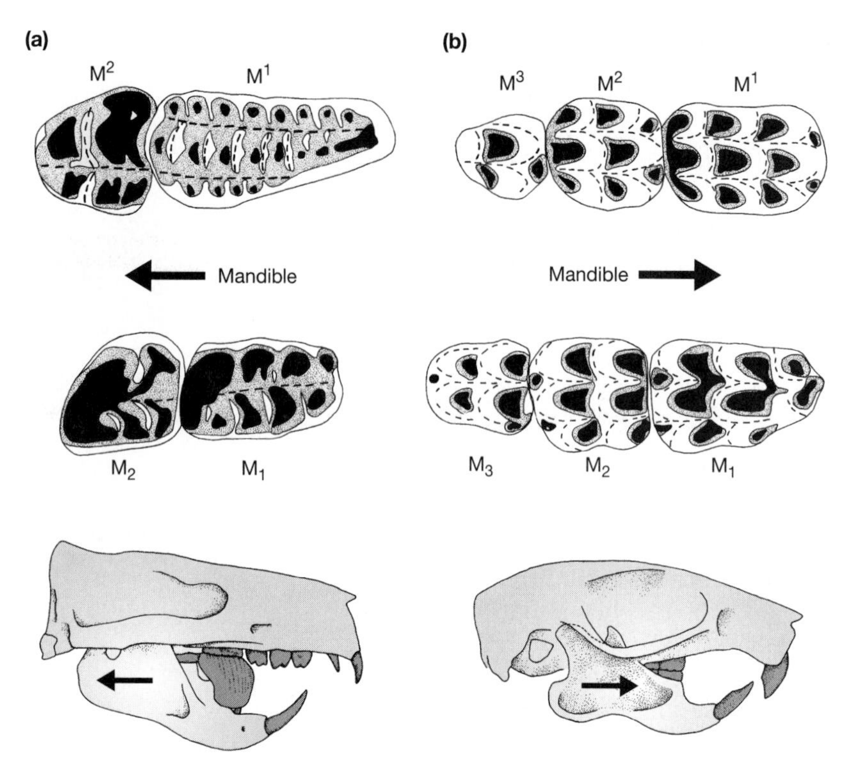

▲ Figure 19–11 Comparison of the skulls and jaws of (a) a multituberculate (*Ptilodus*) and (b) a rodent (*Hapalomys*). The cusps on the multituberculate's molars are concave anteriorly on the uppers and concave posteriorly on the lowers, indicating retraction of the mandible during chewing. The cusps on the rodent's molars are concave posteriorly on the uppers and concave anteriorly on the lowers, indicating protraction of the mandible during chewing. Black areas are exposed dentine, blue areas are worn enamel, clear areas are unworn enamel, and dashed lines are valleys between cusps.

reproduction between different types of mammals are considered in Chapter 20. Therians also have mammae with nipples, a cochlea in the inner ear with at least two and a half coils, an external ear, and tribosphenic molars (see Box 19-1).

Therians also have distinctive features of the skull and skeleton. Therians have completely lost the sclerotic rings around the eyes. Monotremes retain sclerotic cartilages, although they do not ossify to form a bony ring as seen in many other amniotes, including nonmammalian synapsids. Monotremes retain the septomaxilla bone in the skull that is lost in almost all therians. In the postcranial skeleton, therians have completely lost ribs on the cervical vertebrae; these are retained in monotremes. While all mammals have a specialized ankle joint, known as a crurotarsal joint, a more derived hinge joint between the tibia and the astragalus in the ankle, with complete superposition of the astragalus on the calcaneum, is seen only in therians (see Figure 19–12). This ankle joint probably made improved forms of locomotion possible, such as running and hopping.

The therian shoulder girdle is also extremely derived. While monotremes have the derived mammalian form of pelvis, their shoulder girdle is more reminiscent of the typically reptilian condition. Therians have lost the ventral elements of the shoulder girdle, the coracoid and interclavicle bones, and have expanded the scapula with the addition of a scapular spine. This spine is actually the old anterior border of the scapula, and the new portion in therians is the supraspinous fossa in front of the spine. The clavicle (collar bone) is retained in most therians, but is lost in many running-adapted eutherians (e.g., dogs and horses).

This reduction of the ventral elements of the shoulder girdle allows the scapula to move as an independent limb segment around its dorsal border, adding to the length of the stride during locomotion. Additionally the hypaxial muscles have become modified to

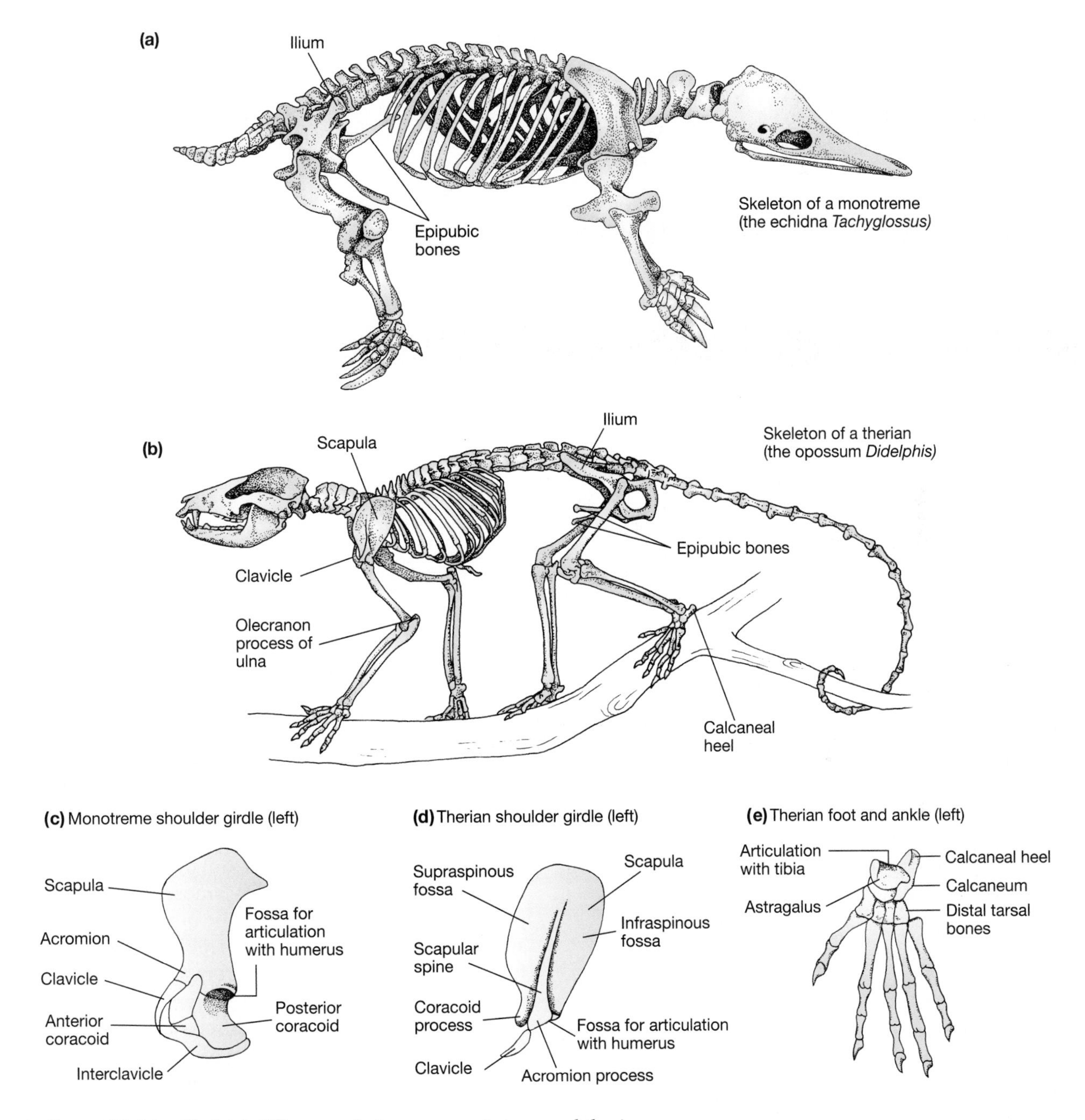

▲ Figure 19–12 Skeletal differences between monotremes and therians.

hold the limb girdle in a muscular **scapular sling** (Figure 19–13). This muscle arrangement probably also aids with scapular mobility during locomotion and cushions the impact of body weight landing on the front limbs during bounding. The therian mammalian type of bounding gait (see Chapters 8 and 20) may be dependent on this modification of the shoulder girdle anatomy and musculature.

The shoulder musculature has also been reorganized to reflect the change in scapular function (Figure 19–13). A muscle that originally protracted the humerus on the shoulder girdle, the supracora-

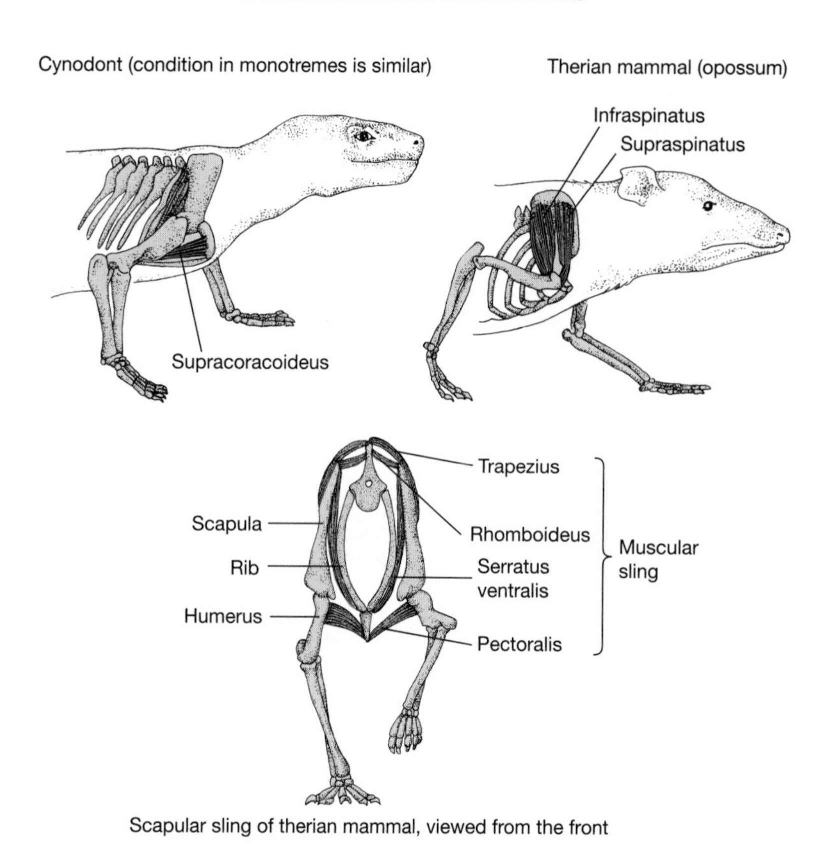

◄ Figure 19–13 Specializations of shoulder girdle musculature.

coideus, has been modified into two new muscles—the infraspinatus and supraspinatus—which may help to stabilize the limb on the shoulder girdle during bounding. (The supracoracoideus became modified in a different fashion in birds to form a wing elevator; see Chapter 15.)

Differences between Marsupials and Eutherians

Although there are numerous differences between marsupials and eutherians in their reproductive biology, as will be discussed in Chapter 20, there are few major differences in their anatomy.

Features in the skull and dentition distinguish marsupials from eutherians, although not all of these features apply to all marsupials. Marsupials characteristically have an inflected angle to the dentary bone (at the pterygoideus muscle insertion) that is lacking in eutherians, and their nasal bones abut the frontal bones with a flared, diamond shape in contrast to the rectangular shape of the eutherian nasals (see Figure 19–14). Eutherians may also have an elaboration of bone around the ear region, the **auditory bulla,** that may increase auditory acuity. Marsupials usually lack a bulla, or they may possess a small one formed from a different bone from that of eutherians. Herbivorous eutherians may have a bar of bone behind the orbit called the postorbital bar, but this is never seen in marsupials. During ontogeny, most eutherians replace all their teeth except for the molars, while marsupials replace only the last premolar (although this feature cannot be determined from an adult skull).

Marsupials also differ from eutherians in their **dental formula,** or numbers of different types of teeth. The primitive dental formula of eutherians (maximum number of teeth usually seen in eutherians) is

$$I\,3/3 \quad C\,1/1 \quad P\,4/4 \quad M\,3/3$$

This notation indicates three upper and three lower incisors on each side (a total of 12 incisors), one upper and lower canine (a total of 4), four upper and lower premolars (a total of 16), and three upper and lower molars (a total of 12). Many eutherians have fewer teeth than this, but they rarely have more. Humans have lost a pair of incisors and two pairs of premolars from each side, so we have a total

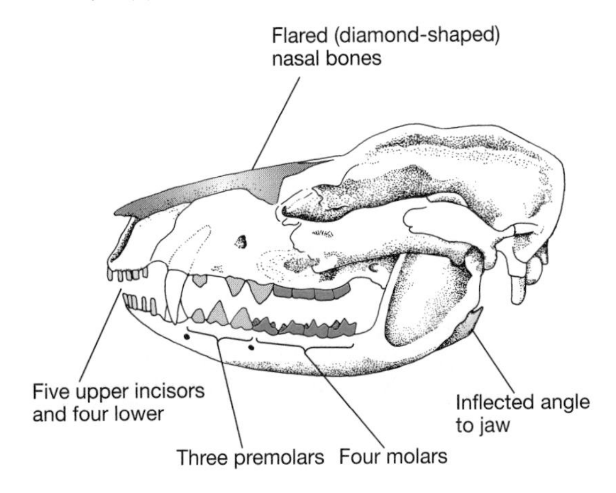

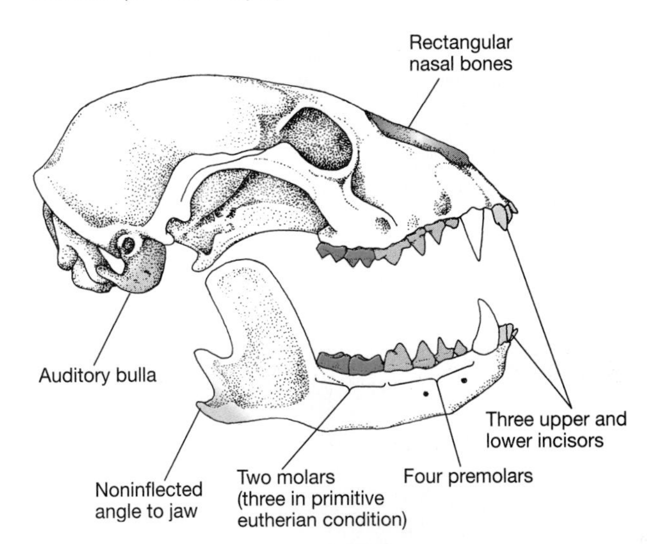

▲ Figure 19–14 Skull differences between marsupials and eutherians.

of only 32 teeth instead of the primitive eutherian component of 44. A few mammals with highly specialized diets, such as armadillos and dolphins, have more teeth than the standard eutherian formula.

Marsupials have a greater number of incisors and molars than eutherians, and fewer premolars. The standard (maximum) marsupial dental formula is

$$I\,5/4 \quad C\,1/1 \quad P\,3/3 \quad M\,4/4$$

The postcranial skeleton of Marsupials can be distinguished from the skeleton of eutherians primarily by the **epipubic bones** that project forward from the pubis (see Figure 19–12). (There are differences in other areas, such as the form of the ankle bones, which would be apparent only to a specialist.) It once was thought that epipubic bones were a unique feature that supported the pouch; but it is now known that epipubic bones are a primitive mammalian feature, retained even in a few very early eutherians (Novacek et al. 1997).

The loss of epipubic bones in most eutherians probably relates to the fact that they are rigid components of the abdominal wall that would interfere with the expansion of the abdomen during pregnancy. The original function of these bones in mammals appears now to be the insertion of various muscles (e.g., the rectus abdominus) that would have been affected by the reduction of the large pubo-ischiatic plate of the cynodont pelvic girdle (see Chapter 17).

Mammalian Ordinal Diversity

The orders of living therian mammals are listed in Tables 21.1 and 21.2, and one current hypothesis of interrelationships of these orders is shown in Figure 19–1. Controversies in the interrelationships among the various eutherian mammal orders are discussed later.

Monotremes Monotremes are grouped in the infraorder Ornithodelphia (Greek *ornitho* = bird and *delphy* = womb, referring to the single functional oviduct in the platypus and many birds; contrast with the *didelphid* opossums), and in the order Monotremata (Greek *mono* = one and *trema* = hole, referring to the cloaca—the single opening of the excretory and reproductive tracts). There are two families. The Ornithorhynchidae (Greek *rhynchus* = beak) contains the platypus, a semiaquatic animal that feeds on aquatic invertebrates in the streams of eastern Australia. The family Tachyglossidae (Greek *tachy* = fast and *glossa* = tongue) contains two types of echidnas, the short-nosed echidna of Australia, which eats mainly ants and termites, and the long-nosed echidna of New Guinea, which includes earthworms in its diet (see Figure 19–15).

Monotremes may originally have had a wide distribution over Gondwana; a platypus is known from the Paleocene of South America. Recently a greater diversity of small-sized teeth belonging to Cretaceous Australian monotremes has been discovered, suggesting a previously unsuspected Mesozoic diversity of these animals (Flannery et al. 1995). How-

(a)

(b)

(c)

▲ Figure 19-15 Diversity of living monotremes (drawn approximately to scale; [a] is the size of a large cat). (a) The platypus, *Ornithorhynchus anatinus.* (b) The short-nosed echidna (Australia), *Tachyglossus aculeatus.* (c) The long-nosed echidna (New Guinea), *Zaglossus bruijni.*

ever, the few known Cenozoic monotreme fossils represent only echidnas or platypuses.

The rather sprawling stance of monotremes, reminiscent of reptiles, may reflect specializations for swimming or digging rather than a truly primitive condition. Monotremes also have their own unique specializations. Both platypuses and echidnas lack teeth as adults and have a leathery (rather than horny) bill or beak. This beak contains receptors that sense electromagnetic signals from the muscles of other animals and are used for sensing prey underwater or in a termite nest. Male platypuses have a spur on the hind leg attached to a venom gland, which is used to poison rivals or predators. Monotremes have lost the lacrimals and frontal bones in the skull, which are retained in other mammals.

Marsupials Marsupials traditionally have been considered as a single order, **Marsupialia**. More recent studies suggest they can be divided into at least four lineages that are equivalent in morphological and genetic diversity to eutherian orders. One current scheme of marsupial classification is shown in Table 19.1, and Figure 19–16 illustrates a diversity of living marsupials.

There is a fundamental split in the interrelationships of living marsupials into the Ameridelphia of the New World and the Australidelphia of (mainly) Australia. A third group, now extinct and possibly more primitive than either of these living groups, was the Deltatheroidea of the Late Cretaceous of Asia.

The ameridelphian order Didelphimorphia includes didelphoids (opossums plus related extinct forms, such as the Northern Hemisphere marsupials of the early Cenozoic), and some extinct Cretaceous forms. Present-day opossums are quite a diverse group of small to medium-size marsupials, mainly arboreal or semi-arboreal omnivores, including animals such as the herbivorous woolly opossum and the otter-like yapok. A much greater diversity of didelphoids existed in the Tertiary, including jerboa-like hopping forms and mole-like forms. Other ameridelphians are the caenolestids or rat opossums (Figure 19–16b), small, terrestrial, shrew-like forms, and the extinct borhyaenoids, a large Cenozoic radiation of carnivorous forms. Borhyaenoids included ferret-like, dog-like, and bear-like forms, and even a Plio-Pleistocene saber-tooth parallel, *Thylacosmilus*.

The one remaining type of South American marsupial is the monito del monte (Figure 19–16c), a tiny mouse-like animal living in the montane forests of Chile and Argentina. Molecular studies have shown this animal to be related to the Australian marsupials, and it may be a distant relict of the stock of originally South American marsupials that migrated across Antarctica to Australia in the early Cenozoic.

The Australian Australidelphia fall into three major orders. The dasyurids are carnivorous forms: the marsupial cats and marsupial mice (which

Table 19.1 Classification of extant marsupial orders and approximate numbers of families and species. Note that different researchers may recognize different numbers of families within an order. (See Figure 19–19 for details of geographic regions.)

Classification	*Families/Species*	*Major Examples*
Cohort Ameridelphia		
Didelphimorphia	1/77	Opossums; 20 g to 6 kg; Neotropical region (plus one North American species).
Paucituberculata	1/5	Caenolestids or rat opossums; 15 to 40g; Neotropical region.
Cohort Australidelphia		
Microbiotheria	1/1	The monito del monte; ~25 g; Neotropical region.
Dasyuromorphia	3/60	Marsupial mice, native cats, Tasmanian devil, Tasmanian wolf (thylacine), marsupial anteater (numbat); 5g to 20 kg; Australian region.
Notoryctemorphia	1/1	Marsupial mole; 50g; Australian region.
Peramelina	2/21	Bandicoots and bilbies; 100g to 5 kg; Australian region.
Diprotodontia	9/110	Possums, flying phalangers, cuscuses, honey possum (noolbender), koala, wombats, potoroos, wallabies, kangaroos. 12g to 90 kg; Australian region; possums and wallabies introduced into New Zealand by humans.

would be better termed marsupial shrews, since they are carnivorous and insectivorous rather than omnivorous). Some larger dasyurids include the Tasmanian devil (Figure 19–16d), and the Tasmanian (marsupial) wolf or thylacine (Figure 19–20). The thylacine has reportedly been extinct since the 1930s, but occasional claims for its continued existence come from purported sightings or footprints. Both these Tasmanian animals were known from mainland Australia before the arrival of humans and their eutherian true dogs, the dingoes. The marsupial anteater (the numbat; Figure 19–20) is related to the dasyurids, and the marsupial mole (Figure 19–16e) may be a more distant relation.

The **peramelids** include the bandicoots and bilbies (Figure 19–16f). These animals look rather like rabbits, but they are insectivorous rather than herbivorous. Peramelids share with the final group of marsupials (the diprotodontians) a condition of the hind feet known as syndactyly, in which the second and third toes are reduced in size and enclosed within the same skin membrane so that they appear to be a single toe. The syndactylous toes are used for grooming.

The largest group of marsupials is the **diprotodontians**. Diprotodontians get their name from their modified lower incisors, which project forward rather like the incisors of rodents (Greek *di* = two, *proto* = first, and *dont* = tooth). This lineage includes herbivorous and omnivorous forms today, although the recently extinct marsupial lion *Thylacoleo* appears to represent a return to carnivory from an herbivorous ancestry. *Thylacoleo* shows some interesting specializations: Coming from a herbivorous ancestry, it had lost its canines but modified the incisors into canine-like teeth.

The three major radiations within the diprotodontians are the phalangeroids, phascolarctoids, and macropodoids. Phalangeroids represent an arboreal radiation of rather primate-like animals, also including gliders. They comprise six families, including possums, phalangers (Figure 19–20), ring tails, cuscuses, and the diminutive honey possum or noolbenger (Figure 19–16g)—the only nectar-eating mammal that is not a bat. Phascolarctoids or vombatiformes include the arboreal koala (Figure 19–16h) and the terrestrial, burrowing wombats (Figure 19–20). Extinct phascolarctoids include the

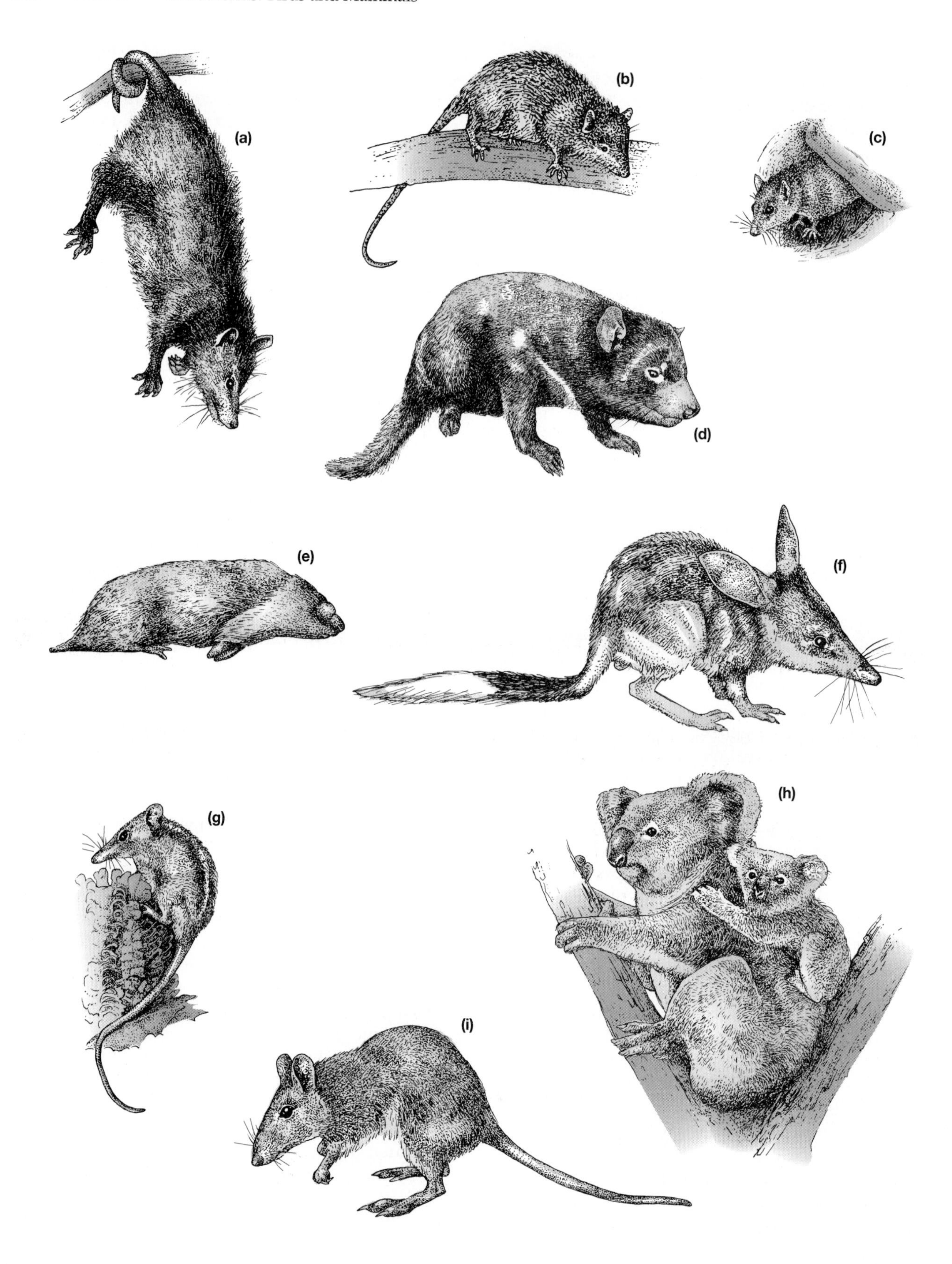
(a)
(b)
(c)
(d)
(e)
(f)
(g)
(h)
(i)

bison-size diprotodontids that looked like giant wombats and grazed on the Plio-Pleistocene Australian savannas. Macropodoids include the small, omnivorous rat kangaroos or potaroos (Figure 19–16i), and the larger, herbivorous true kangaroos (including wallabies and tree kangaroos). The largest kangaroos today have a body mass of about 90 kg, but larger ones (up to twice that size) existed in the Pleistocene, including a radiation of one-toed, short-faced browsing sthenurine kangaroos, now all extinct. Sthenurines, with their short faces and stout forearms, could be the source of legends of giant rabbits in Australia's interior. Perhaps, as desert forms, they also had large ears for cooling.

Eutherians Eutherian mammals can be grouped into a number of distinct taxa, but their interrelationships remain obscure, suggesting that the diversification of these groups from an ancestral stock occurred very rapidly. Figure 19–17 illustrates a diversity of eutherian mammals.

The phylogeny of eutherian mammals shown in Figures 19–1 and 19–18a and in Table 19.2 was devised by Malcolm McKenna (McKenna 1975), who grouped mammalian orders into more inclusive grandorders. The most primitive group of eutherian mammals is the **edentates** (order Xenarthra): sloths (Figure 19–17a), anteaters, and armadillos. *Edentate* means "without teeth," and while only anteaters are completely toothless, all of these animals have simplified their dental pattern. The pangolins or scaly anteaters (order Pholidota) may be related to the edentates.

Insectivorans (= members of the order Insectivora) were often considered to be the basal stock from which other eutherians were derived. However, although modern insectivores such as shrews may superficially resemble ancestral eutherian mammals, they are not closely related. The Insectivora includes true shrews, moles, hedgehogs, and the Madagascan tenrecs (Figure 19–17b).

Tree shrews were once thought to be primitive primates. This is now no longer believed to be true, but they are grouped with the primates, bats, and dermopterans (flying lemurs; Figure 19–17e) in the grandorder Archonta. Elephant shrews (Figure 19–17c), so named because of their trunk-like nose, are grouped in the Glires along with rodents and rabbits.

The Ferae includes the extant Carnivora, and the extinct Creodonta, whose members took the role of the large-size predators in the early Cenozoic. Members of the Carnivora (= **carnivorans**) are distinguished by the possession of a pair of specialized shearing teeth, or **carnassials**, formed from the upper last (fourth) premolar and the lower first molar. Carnivorans include not only specialized carnivores such as cats but also secondarily herbivorous forms, such as the panda bear, and one of the three living groups of secondarily aquatic mammals, the **pinnipeds** (seals, sea lions, and walruses), that are related to bears.

The largest grandorder, both in terms of included orders and the size of its members, is the **Ungulata**, or hoofed mammals (hooves are terminal appendages made out of keratin; see Figure 19–6). The common living ungulates are the **artiodactyls** (even-toed ungulates; Figure 19–17h) and the **perissodactyls** (odd-toed ungulates; Figure 19–17g). Some archaic ungulates in the early Cenozoic were also omnivorous, or even carnivorous; in South America there were several orders that are now entirely extinct. A surprise in the reexamination of mammalian interrelationships over the past couple of decades has been the realization that whales and dolphins (order **Cetacea**) are ungulates taxonomically, related to the artiodactyls.

A final grouping of the ungulates is the African-originating paenungulates. This includes the rodent-like **hyraxes** (Figure 19–17i, also known as the conies of the Bible), the aquatic **sirenians** (sea cows and dugongs), and the **proboscideans** (elephants and extinct relatives). The embryos of elephants show evidence of an aquatic ancestry in the structure of their kidneys (Gaeth et al. 1999), suggesting that the grouping of sirenians, desmostylians (semiaquatic

◀ Figure 19–16 Diversity of living marsupials (drawn approximately to scale; [d] is the size of a large cat). (a) The common North American opossum, *Didelphis virginiana* (Didelphidae: Didelphimorphia). (b) The shrew opossum, *Lestoros inca* (Caenolestidae: Paucituberculata). (c) The monito del monte, *Dromiciops australis* (Microbiotheriidae: Microbiotheria). (d) The Tasmanian devil, *Sarcophilus harrisii* (Dasyuridae: Dasyuromorphia). (e) The marsupial mole, *Notoryctes typhlops* (Notoryctidae: Notoryctemorphia). (f) The bilbey, or rabbit-eared bandicoot, *Macrotis lagotis* (Thylacomyidae: Peramelina). (g) The honey possum, *Tarsipes rostratus* (Tarsipedidae, Phalangeroidea, Diprotodontia). (h) The koala, *Phascolarctos cinereus* (Phascolarctidae, Phascolartoidea, Diprotodontia). (i) The long-nosed potoroo (rat kangaroo), *Potorous tridactylus* (Macropodidae, Macropodoidea, Diprotodontia).

(a)
(b)
(c)
(d)
(e)
(f)
(h)
(g)
(i)

extinct forms), and proboscideans may have been an originally aquatic paenungulate stock. Proboscideans would then represent a secondary return to a more terrestrial existence. However, popular speculation that the elephant's trunk was originally a snorkel is mistaken; the skulls of early fossil proboscideans show that these animals lacked trunks.

Controversies in Eutherian Origin and Interrelationships

The Eutheria are a diverse group that appears to have had a very rapid evolutionary radiation. With such rapid evolutionary events it is often hard to determine interrelationships easily, and the interrelationships of the eutherian orders have been a subject of debate for decades. More recently, the use of molecular techniques in systematics has led to a confrontation between molecular biologists and the more traditional morphologists. This confrontation includes not only the interrelationships of the eutherian orders but also their time of divergence from each other.

Morphological versus Molecular Phylogenies The classification of eutherian mammals just discussed was based on morphological characters. More recently phylogenies have been generated using information from various molecules. Figure 19–18 shows the phylogeny of eutherians determined from morphology (a simplified version of the phylogeny in Figure 19–1) and contrasts it with a consensus of several different phylogenetic trees derived from molecular analyses (see Waddell et al. 1999, Madsen et al. 2001, Murphy et al. 2001).

Several features of the phylogenies are the same, but there are some interesting differences. One is the grouping of the endemic African orders as the Afrotheria (van Dijk et al. 2001). This includes not only the aardvark plus paenungulate grouping but also the elephant shrews and, among the insectivorans, the tenrecs (now found only on Madagascar) and the golden moles (placed in a separate family from the true moles of the Northern Hemisphere). The sister taxon to this African grouping is the South American edentates, suggesting a basic split at the base of the eutherian radiation into a southern, Gondwanan grouping and a northern, Laurasian one.

Morphologists are largely intrigued by the possibility of the Afrotheria, because it agrees with the early biogeographic separation of Africa and does not conflict with the fossil record in other respects. However, many morphologists are less happy about another grouping of eutherians by the molecular biologists—that of placing the whales within the order Artiodactyla, as the sister group to hippos.

To people familiar only with living animals, the concept of a whale-hippo link is rather appealing. Both are large, aquatic animals with several features in common, such as the ability to suckle their young underwater. However, knowledge of the fossil record makes this alliance much less likely. We have an almost perfect fossil record sequence from the Eocene site of early whale evolution—in Pakistan along the shores of the ancient Tethys Sea. This record not only shows a progression among early whales from a more terrestrial to a more aquatic existence but also illustrates intermediate steps between early whales and the archaic carnivorous mesonychid ungulates. Additionally, no aquatic hippo-like animals are known until about 10 million years ago, and their probable ancestors were terrestrial. Indeed, the surviving pygmy hippo is a predominantly terrestrial beast. This casts doubt on a common aquatic ancestor for whales and hippos.

How can such conflicts between molecules and morphology be resolved? Maureen O'Leary performed an interesting computer simulation (O'Leary and Geisler 1999). As long as she included only information from living animals, O'Leary could generate a cladogram based on morphological information that resembled the one obtained from molecular information, placing whales with hippos. But when she added data from extinct animals, the cladogram shifted to place the whales with the mesonychids, outside of the Artiodactyla. The implication is that partial evidence can bias results, and that the

◀ Figure 19-17 Diversity of living eutherians (not drawn to scale). (a) Two-toed sloth, *Choloepus didactylus* (Megalonychidae: Xenarthra); cat-size. (b) Common tenrec, *Tenrec ecaudatus* (Tenrecidae: Insectivora); rat-size. (c) Golden-rumped elephant shrew, *Rhynchocyon chrysopygus* (Macroscelidida: Macroscelidea); rat-size. (d) Naked mole rat, *Heterocephalus glaber* (Bathyergidae: Rodentia); mouse-size. (e) Flying lemur, *Cynocephalus volans* (Cynocephalidae: Dermoptera); cat-size. (f) The spotted hyena, *Crocuta crocuta* (Hyaenidae: Carnivora); wolf-size. (g) Asiatic tapir, *Tapirus indicus* (Tapiridae: Perissodactyla); pony-size. (h) African water chevrotain, *Hyemoschus aquaticus* (Tragulidae: Artiodactyla); rabbit-size. (i) The rock hyrax, *Procavia capensis* (Procaviidae: Hyracoidea); rabbit-size.

Table 19.2 Classification of extant eutherian orders and approximate numbers of families and species. Note that different researchers may recognize different numbers of families within an order. The geographic regions represent the distribution of the entire orders (and for the present day only); families within an order often have smaller distributions. (See Figure 19–19 for details of biogeographic regions.)

Order	*Families/Species*	*Major Examples*
		Grandorder Edentata
Xenarthra	3/30	Anteaters, sloths, armadillos; 20g to 33 kg; Neotropical region (plus some armadillos in southern USA).
Pholidota	1/7	Pangolins (scaly anteaters); 2 to 33 kg; Ethiopian and Oriental regions.
		Grandorder Glires
Lagomorpha	2/69	Rabbits, hares, pikas; 180g to 7 kg; worldwide except Antarctica, introduced in Australia by humans.
Rodentia	29/1814	Rats, mice, squirrels, guinea pigs, capybara; 7g to over 50 kg; worldwide except Antarctica.
Macroscelidea	1/15	Elephant shrews; 25 to 500g; Ethiopian region with one species in Morocco and Algeria.
		Grandorder Insectivora
Insectivora (= Lipotyphla)	6/390	Hedgehogs, moles, shrews, tenrecs; 2g to 1 kg; worldwide except Australia and Antarctica (although only a single species of shrew is known from South America, a Pleistocene immigrant).
		Grandorder Archonta
Scandentia	1/16	Tree shrews; 400g; Oriental region.
Primates	9/235	Lemurs, monkeys, apes, humans; 85g to over 275 kg; primarily Oriental, Ethiopian, and Neotropical regions, humans are now worldwide.
Dermoptera	1/2	Flying lemurs; 1 to 2 kg; Oriental region.
Chiroptera	15/986	Bats; 4g to 1.4 kg; worldwide (including New Zealand) except Antarctica.
		Grandorder Ferae
Carnivora	12/274	Dogs, bears, raccoons, weasels, hyaenas, cats, sea lions, walruses, seals (these last three are often assigned to the suborder Pinnipedia); 70g to 760 kg, some marine forms over 100 kg; worldwide.
		Grandorder Ungulata
Artiodactyla	10/213	Even-toed ungulates: Swine, hippopotamuses, camelids, deer, giraffe, antelope, sheep, cattle; 2 to 2500 kg; worldwide except Antarctica (introduced into Australia and New Zealand by humans).
Cetacea	9/80	Porpoises, dolphins, sperm whales, baleen whales; 20 to 120,000 kg; worldwide in oceans and in some rivers and lakes in Asia, South America, northern America, and Eurasia.
Perissodactyla	3/17	Odd-toed ungulates: Horses, tapirs, rhinoceroses; 150 to 3600 kg; Worldwide except Antarctica (horses introduced by humans into North America and Australia).

Table 19.2 Classification of extant eutherian orders and approximate numbers of families and species (*continued*)

Order	*Families/Species*	*Major Examples*
Tubulidentata	1/1	Aardvark; 64 kg; Ethiopian region.
Hyracoidea	1/7	Hyraxes (= conies or dassies); 4 kg; Ethiopian region and Asia Minor.
Proboscidea	1/2	Elephants and fossil relatives; 4500 to 7000 kg; Ethiopian and Oriental regions.
Sirenia	2/4	Dugongs, manatees; 140 to over 1000 kg; coastal waters and estuaries of all tropical and subtropical oceans except the eastern Pacific (in the Atlantic drainage they enter rivers).

inability to obtain data from extinct species may present serious problems in constructing cladograms based solely on molecular information.

Timing of Eutherian Origin The radiation of modern eutherians has long been assumed to be correlated with the disappearance of the dinosaurs, and there is no fossil evidence of members of extant orders until the Cenozoic. Blair Hedges claimed a much earlier time for the initial radiation, using molecular information and employing the notion of a constant rate of molecular change (the so-called molecular clock). He estimated that modern eutherians first arose 120 million years ago, in the late Early Cretaceous, and concluded that their divergence was related to the splitting up of the supercontinent Pangaea (Hedges et al. 1996, Kumar and Hedges 1998). However, this flatly contradicts the fossil-record evidence. While it is true that the fossil record is incomplete, Hedges' scheme would imply that the hidden fossil record for modern eutherians during the time of the dinosaurs was as long as the known Cenozoic record following dinosaur extinction.

This proposed early divergence could possibly represent genetic divergence only among animals that still looked pretty similar during the Mesozoic, with changes in morphology not occurring until the Cenozoic. An alternative interpretation is that rapid morphological evolution at the start of the Cenozoic resulted in the speeding up of the rate of molecular evolution, biasing molecular clock estimations. Additionally, the fossil record for the Cretaceous is actually rather good. Many types of mammals are preserved, but none apparently belonging to the modern eutherian orders. The fossil record would have to be around 100 times worse than it actually is for it to be statistically likely that modern eutherians were present during most of the Cretaceous but not preserved (Foote et al. 1999, Benton 1999).

19.3 Cenozoic Mammal Evolution

During the early Cenozoic the continents had not moved as far from their equatorial positions as they are today, but the major landmasses were more isolated than at present (see Chapter 18). Australia had broken free from the other southern continents by the mid-Eocene, but North and South America did not come into contact until the Pliocene. India collided with Asia in the Miocene and formed the Himalayan mountain ranges. Africa made contact with Eurasia in the late Oligocene or early Miocene, closing off the original east-west expanse of the Tethys Sea to form the now-enclosed Mediterranean basin.

Biogeography of Cenozoic Mammals

The radiation of mammals occurred during this fractionation of the continental masses, when different stocks became isolated on different continents. This separation of ancestral stocks from one another by the Earth's physical processes, rather than by their own movements, is termed **vicariance**. Some of the difference in the distribution of present-day mammals (for example, the isolation of the monotremes in Australia and New Guinea) results from **vicariance biogeography**, that is, animals and plants being carried passively on moving landmasses.

Other patterns of mammal distribution can be explained by **dispersal**, which reflects movements of the animals themselves, usually by the spread of populations rather than the long-distance

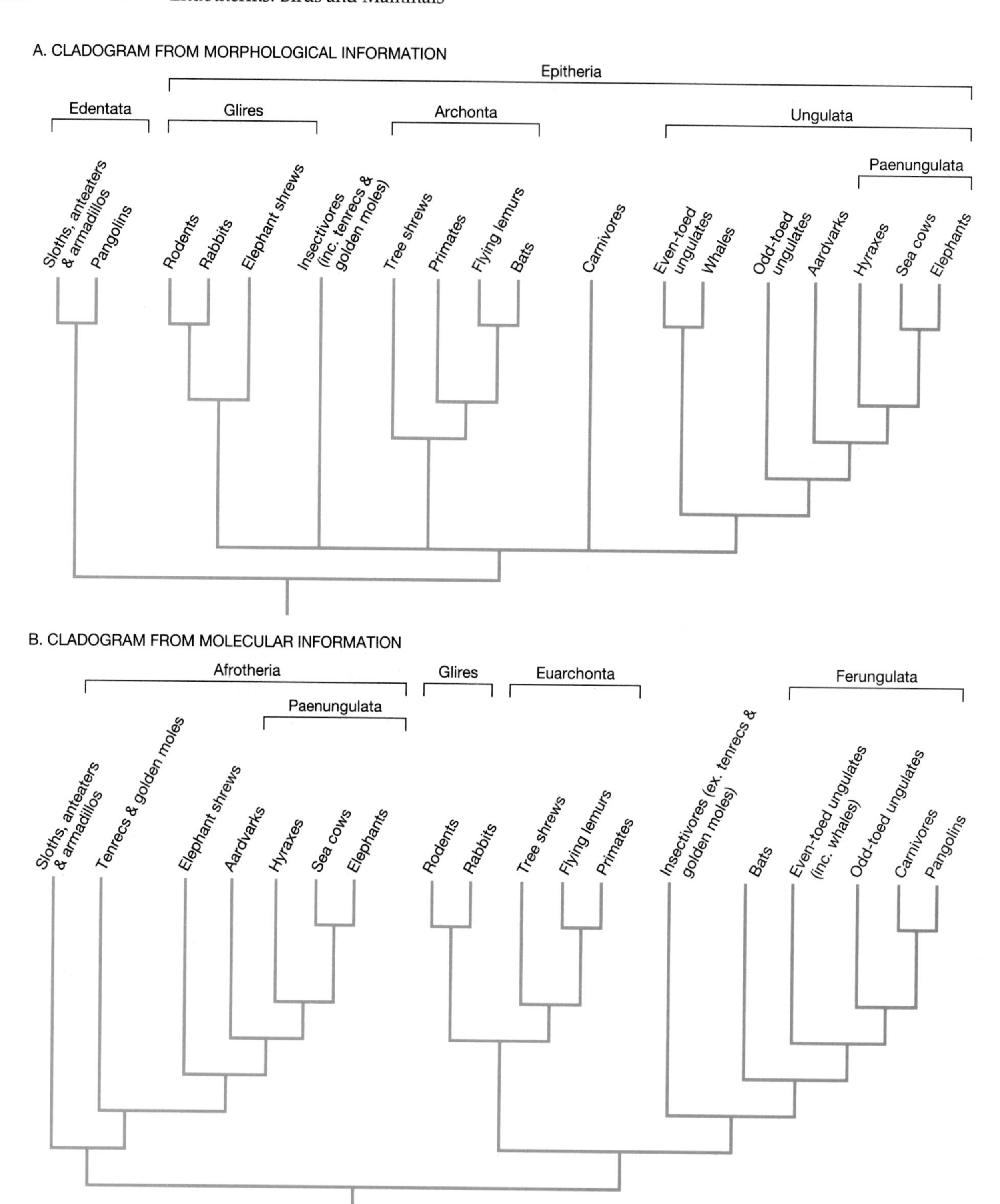

▲ Figure 19–18 Contrasting phylogenies of the interrelationships among orders of eutherian mammals.

movements of individual animals. The presence of marsupials in Australia apparently represents dispersal from South America via Antarctica in the early Cenozoic. The subsequent isolation of Australia may have allowed marsupials to diversify on that continent. Dispersal can produce extinctions as well as radiations. For example, when the Turgai Straits that had separated Europe and Asia during the Paleocene and Eocene dried up in the early Oligocene, mammals from Asia flooded into Europe, and some uniquely European mammals (mainly archaic types) became extinct. This episode of extinctions was so dramatic that it is known as the Grand Coupure (the Great Dying).

Today the mammals of the Northern Hemisphere (Holarctica) and of Africa, Madagascar, South America, and Australia are strikingly different from one another. These geographic groupings fall into three major faunal provinces: (1) Laurasian fauna, including Holarctica, Africa, and Madagascar, consisting almost exclusively of eutherians; (2) South American, or New World tropical fauna, containing a mixture of eutherians and marsupials; and (3) Australian fauna, containing monotremes, marsupials, and a sprinkling of eutherians.

However, the fauna of Africa was once even more distinct from the fauna of the rest of Holarctica—before Africa collided with Eurasia in the Oligocene/Miocene. For example, hyraxes, which now survive only as small rodent-like forms (Figure 19–17i), assumed ecomorphological roles taken today by antelope and pigs. Similarly, the South American fauna was more distinct from that of Holarctica before South America was connected with North America in the Pliocene. Even the North American fauna was much more distinct from that of the rest of Holarctica for most of the Cenozoic than it is today. Much of the present North American fauna (e.g., deer, bison, and rodents such as voles) emigrated from Eurasia in the Pliocene and Pleistocene via the Bering land bridge. Today only Madagascar and Australia retain distinctly different faunas.

Convergent Evolution of Mammalian Ecomorphological Types

The term *ecomorphology* describes the way in which an animal's form (its morphology) is adapted to its behavior in its environment (its ecology). The separate evolution of different basic mammalian stocks on different continents strikingly illustrates the convergent evolution of ecomorphs in Cenozoic mammals.

Figure 19–19 illustrates examples of convergence. Specialized running herbivores (Figure 19–19a–d) include the extinct litopterns of South America, the Holarctic artiodactyl antelope of Eurasian origin, and the perissodactyl horses (most of the evolution of horses took place in North America). These ungulates are strikingly similar in many respects (see also Figure 20–13). Even the Australian marsupials such as the kangaroo demonstrate convergences in jaws, teeth, and feeding behavior to eutherian herbivores. Convergent evolution also took place among large-bodied, slower-moving herbivores (Figure 19–19e–h), such as the Holarctic rhinoceroses (perissodactyls), the elephants, which originated in Africa, the extinct notoungulates of South America, and the extinct marsupial diprotodontids of Australia.

Carnivorous mammals show similar convergences (Figure 19–19i–j). The true wolf of Holarctica, the recently extinct marsupial wolf or thylacine of Australia, and the extinct marsupial borhyaenids of South America have similar body morphologies and tooth form, although only the true wolves have the long legs that distinguish fast-running predators. Similar types of intercontinental convergences, again involving both marsupials and eutherians, occurred with a carnivorous ecomorph that does not exist today—the saber-toothed cat-like predator.

Mammals specialized for feeding on ants and termites (myrmecophagy; Figure 19–19l–o) include the giant anteater in South America, the aardvark in Africa, the pangolin in tropical Asia and Africa, and the spiny anteater (a monotreme) in Australia. Ants and termites are social insects that employ group defense. They often build impressive earthen nests containing thousands of individuals, some of which are strong-jawed soldiers. The convergent specializations of these unrelated or distantly related mammals include a reduction in the number and size of teeth, changes in jaw and skull shape and strength, and forelimbs modified for digging (see Chapter 20). Some mammals with a carnivorous ancestry have adopted a myrmecophagous life style and show similar, though less extensive, specializations. These species include a marsupial anteater, the Australian numbat (Figure 19–20), an African hyena, the aardwolf, and a Madagascan viverrid, the fanalouc.

Other examples of convergent evolution of ecomorphs abound, such as burrowing moles or mole-

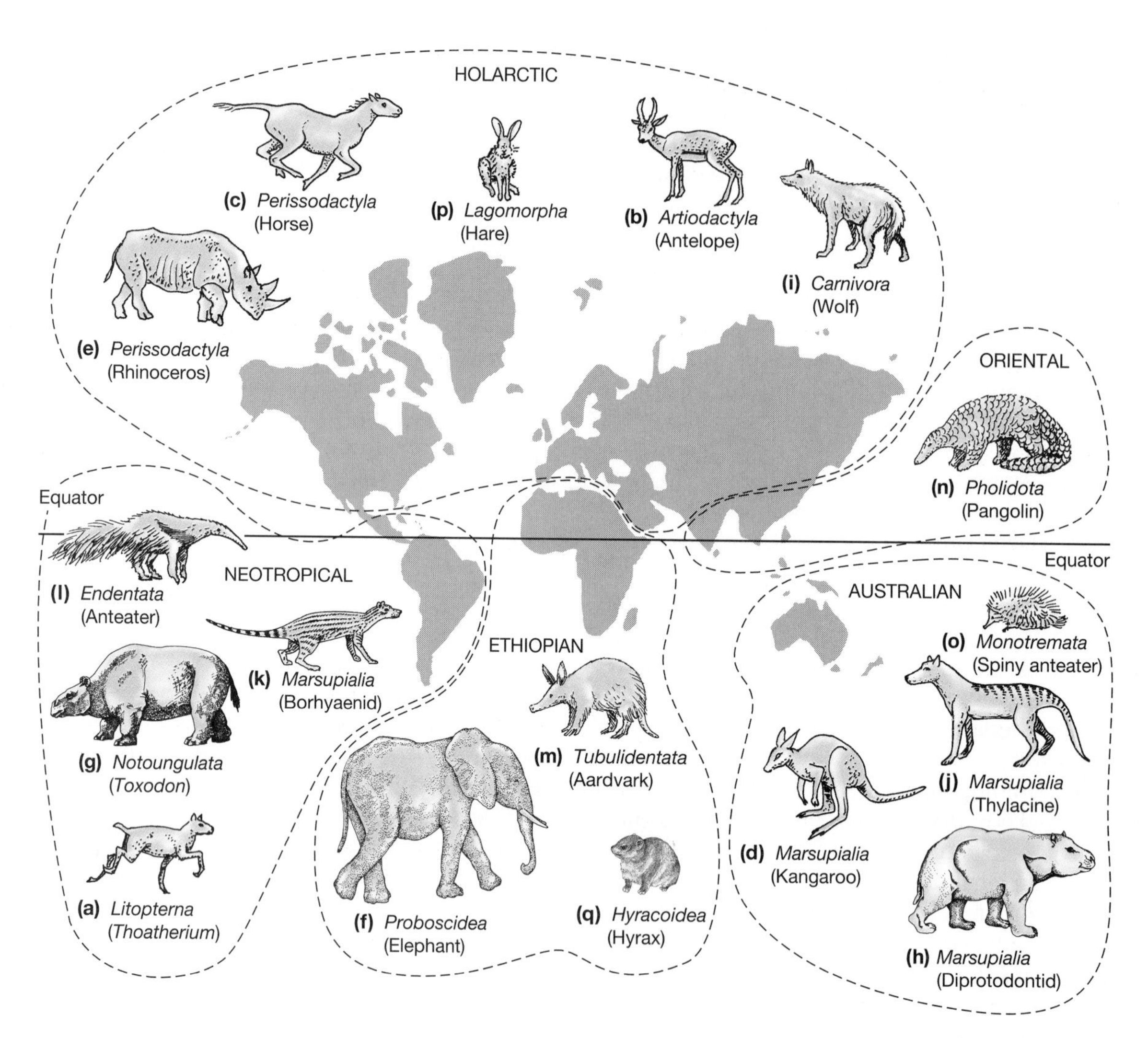

▲ Figure 19–19 Radiation and convergence of mammals evolving in isolation during the Cenozoic. Mammals are grouped with landmasses on which they probably originated. Major biogeographical regions of the world today are illustrated. Current convention, reflecting the historical patterns of colonization by Europeans, is to call American continents the New World and other continents (including Australia) the Old World. The northern continents, North America and northern Eurasia, can be grouped together as Holarctica; they can be further subdivided into the Nearctic (North America, including Greenland), and the Palaearctic (Europe and northern Asia, including Asia Minor). India and southeast Asia fall within the tropics, and together are termed the Oriental region. Africa (including Madagascar) forms the Ethiopian or African region. The Ethiopian and Oriental regions are sometimes grouped together as the Old World Tropics, or Paleotropics. South and Central America form the Neotropical region, or New World Tropics. Australia and associated islands (including New Guinea, Tasmania, and New Zealand) make up the Australian region.

like animals and gliders. Mole-like ecomorphs are discussed in Chapter 20. Gliding ecomorphs include the flying squirrels of the Northern Hemisphere and the marsupial flying phalangers in Australia (see Figure 19–20), as well as the flying lemurs (not true primates) of southeast Asia (Figure 19–17e) and a completely different type of flying squirrel in Africa. Sometimes similar forms evolved under what appear to be much less isolated conditions. Thus hares and rabbits (Figure 19–19p) evolved in Holarctica where other small herbivores, the related rodents, also occurred. The rodent-like animals of Africa are the hyraxes (Figure 19–19q), which had a much greater diversity of sizes and body forms in the past.

Distribution of Early Cenozoic Mammals

At the start of the Cenozoic all mammals were small and fairly unspecialized. The marsupials appear to have been omnivorous and arboreal, like modern opossums, while the eutherians were mostly shrew-like terrestrial primitive insectivorous forms, or archaic ungulates.

The origin of the marsupials appears to have been in North America, while that of the eutherians was in Asia. The original stock of North American marsupials has been extinct for at least 30 million years. The present-day North American native marsupial, the common opossum *Didelphis virginiana*, is a Pleistocene immigrant from South America. Opossums are still moving north and were first recorded in Canada in the 1950s.

Multituberculates were primarily Northern Hemisphere residents, but a distinct southern branch, the gondwanatheres, has now been identified from the Cretaceous of South America, Africa, and Madagascar. Monotremes were probably originally Australian in origin, but a fossil monotreme was found in Patagonia, at the tip of South America (Pascual et al. 1992). It is not clear if this animal represents dispersal from Australia (across Antarctica) or the remnants of an earlier, unrecorded, Gondwanan fauna that spread across the southern continents.

The more derived mammals with tribosphenic molars (including therians) were probably originally a Northern Hemisphere radiation, perhaps originating after North America and Eurasia separated from Gondwana. The northern tribosphenic mammals are first known from the late Early Cretaceous. Some mammals with molars that have been thought to be tribosphenic in shape (known only from lower molars, in fact) are known from southern continents: *Ambondro* from the Middle Jurassic of Madagascar (predating the northern mammals by 25 million years) and *Ausktribosphenos* from the Early Cretaceous of Australia. However, more recent research suggests that these southern mammals may have evolved their molars convergently with the northern tribosphenic mammals, representing a southern branch of mammalian evolution that may be ancestral to monotremes (Luo et al. 2001).

Marsupials are considered the quintessential Australian mammals, but they did not reach that continent until the early Cenozoic. Their origin was in North America and their initial Cenozoic diversification in South America. During the Cenozoic marsupials dispersed not only across the southern continents but also across the northern ones. Marsupials are known from the Eocene and/or Oligocene of Europe, Asia, and northern Africa, but apparently they were never particularly numerous or diverse there. Their mid-Cenozoic extinction in that part of the world may be just part of the extinction of many archaic mammals taking place at that time. There is no need to invoke a complex scenario of competition with eutherians.

Cenozoic Mammals of the Southern Continents

The mammals of South America differ from those of the Northern Hemisphere, but less so than do the Australian mammals. These two continents provide particularly clear examples of the effects of biogeographical isolation, as does the island of Madagascar.

Cenozoic Isolation of Australia The mammalian fauna of Australia has always been composed almost entirely of marsupials and monotremes. The earliest Australian marsupial fossils are of early Eocene age, and there are abundant remains from the early Miocene onward. South America, Antarctica, and Australia were still close together in the early Cenozoic. Australia was probably populated by marsupials that moved from South America across Antarctica, which was warm and moist until about 35 million years ago. Marsupial fossils are known from the Eocene of western Antarctica, and in fact the major barrier to dispersal to Australia was probably crossing the midcontinental mountain ridge between western and eastern Antarctica.

Once marsupials reached Australia, they enjoyed the advantages of long-term isolation, evolving to fill

▲ Figure 19–20 Convergences in body form and habits between eutherian and marsupial mammals.

a variety of niches with food habits ranging from complete herbivory to carnivory. Figure 19–20 illustrates some Australian marsupials and notes their convergences with northern eutherians. Note that the possums and phalangers are a diverse radiation of arboreal mammals that are convergent not only with squirrels, but with primates.

A particularly striking example of convergence exists between the aye-aye of Madagascar (a lemur, a type of primitive primate; see Chapter 23) and the striped possum of the York Peninsula of northeastern Australia. Both pry wood-boring insects out from under tree bark, and both have chisel-like incisors and have elongated one of their fingers to act as a probe. Certain insectivorans (apatemyids) from the early Cenozoic of Holarctica had similar specializations, but went extinct in the early Miocene when woodpeckers evolved—birds that also feed on wood-boring insects. Madagascar and Australia are the only places in the world today that lack woodpeckers, which suggests that the woodpeckers elsewhere outcompeted the wood-boring mammals.

The early Eocene fauna of Australian mammals revealed a remarkable discovery: In addition to several marsupial species, it preserved the molar of a eutherian that was similar to archaic ungulates known from the early Cenozoic of North America (Godthelp et al. 1992). This fossil suggests that marsupials did not make the journey across Antarctica alone; they were accompanied by at least one type of eutherian. This eutherian evidently became extinct, because other terrestrial eutherians are unknown in Australia until the late Miocene. This animal rather turns the tables on the supposed marsupial inferiority in the presence of eutherians (see Chapter 20). The only other eutherians known in the early Cenozoic of Australia are bats. Their affinities appear to be with the bats of Asia, which suggests that they flew across to Australia via quite a different route.

Rodents originally arrived in Australia in the late Miocene, probably via dispersal along the island chain between Southeast Asia and New Guinea. Australian rodents are an interesting endemic radiation today; they are related to the mouse/rat group of Eurasian rodents, but have evolved into unique Australian forms such as the small jerboa-like hopping mice and the large (otter-size) water rats. True mice and rats arrived later, in the Pleistocene. However, these rodents apparently had surprisingly little overall effect on the Australian marsupials. A far greater threat has been the prehistoric invasion by humans and dogs (dingoes), and the recent introduction of domestic mammals such as foxes, rabbits, and cats. The Tasmanian wolf and Tasmanian devil, both carnivorous marsupials, were known from mainland Australia prior to the arrival of dingoes. Today numerous other mainland forms are threatened or endangered by humans and their introduction of exotic species.

Madagascar Madagascar is an island off the coast off East Africa. It apparently separated from Africa in the mid-Mesozoic, and its present-day mammalian fauna represents subsequent immigration. But the source of these immigrants, whether from the African mainland or from Asia (the source of the present-day indigenous people of Madagascar), remains in debate. The best-known endemic mammals of Madagascar are the lemurs, a radiation of primitive primates known nowhere else in the world (see Chapter 23). Diverse as today's lemurs are, their diversity was much greater in the recent past. As recently as a few thousand years ago, much larger (up to gorilla-size) lemurs—both terrestrial and arboreal—paralleled the radiation of the great apes among the anthropoid primates. The extinction of these giant lemurs appears to be related to the arrival of humans a couple of thousand years ago, probably due to habitat destruction rather than to hunting.

The other native Madagascan mammals belong to the orders Carnivora and Insectivora, and are composed of forms unknown from elsewhere in the world today. The native insectivorans are the tenrecs, many resembling large shrews (Figure 19–17b). Tenrecs also have diversified into hedgehog-like spiny forms, mole-like burrowers, and otter-like swimmers. The carnivorans are viverrids, related to the African and Asian mainland civets and genets. The most remarkable of these is the fossa (*Cryptoprocta ferox*), an animal a little larger than a big house cat. In the absence of true cats, the fossa has evolved to become a cat-like predator. Although it is more stockily built than true cats, with shorter, more muscular legs that reflect its arboreal habits, its skull is virtually indistinguishable from that of a cat.

South American Mammals and the Great American Interchange From the Late Jurassic until the late Cenozoic, South America was isolated from North America by a seaway between Panama and the northwestern corner of South America. In the Pliocene, about 2.5 million years ago, the Panamanian land bridge was established between North and South America. Animals from the two continents were free to mix for the first time in more than 100 million years (Marshall et al. 1982, Marshall 1988). Faunal interchange by island disperal or rafting commenced in the late Miocene as the two American landmasses drew nearer to each other. This event, the Great American Interchange (GAI), is a spectacular example of the effects of dispersal and faunal intermingling between two previously separated landmasses (see Figure 19–21 for the cast of characters involved).

Superficially the mammals moving from North America to South America appear to have been more numerous and to have fared better than those moving in the opposite direction. For many years, the interchange was viewed as an example of the

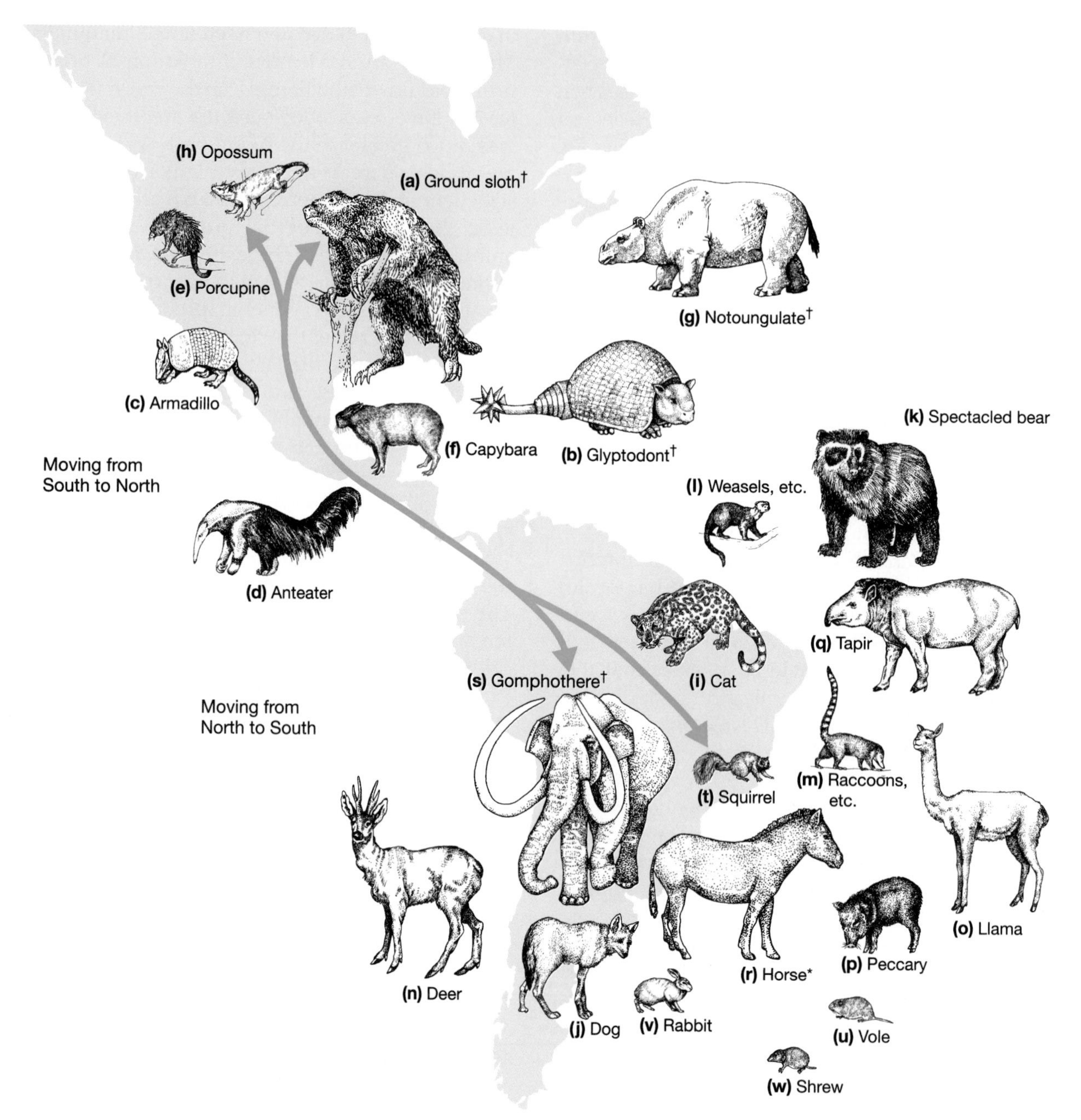

▲ Figure 19–21 Mammalian taxa involved in the Great American Interchange. A dagger (†) indicates taxa that are now totally extinct; an asterisk (*) indicates taxa now extinct in that area.

competitive superiority of Northern Hemisphere mammals. However, with reconsideration of the available evidence, a different interpretation is preferred today. Before discussing the GAI in more detail, first we need to consider the diversity of mammals that inhabited South America before the interchange.

Three major groups of mammals can be distinguished in South America prior to the GAI. In fact, much of the original diversity of the South American mammal fauna was extinct by the time of the GAI, and competition with northern immigrants was not responsible for their demise.

- *Early inhabitants, known from the Paleocene, evolving in situ or originating from North America; marsupials, edentates, and archaic ungulates.* Equally important were the groups of North American mammals that failed to colonize South America at this time, notably insectivorans, carnivorans, and rodents. We have no explanation of why these groups did not succeed in populating South America, but their absence is important because it left adaptive zones open, and marsupials radiated into them during the isolation of South America in the early Cenozoic.

Edentates were known only from South America prior to the Pliocene. The past diversity of edentates was considerably greater than that of today, including the armadillo-related glyptodonts, cow-size beasts encased in a turtle-like carapace of dermal bone (Figure 19–21b), and ground sloths, ranging from the size of a large dog to the size of a rhino (Figure 19–21a).

The endemic South American ungulates, now all extinct, radiated into several orders. The ones surviving at the time of the GAI were the litopterns and the notoungulates. Litopterns were gracile, cursorial forms, and were either pony-like (Figure 19–19a), or larger and camel-like. Notoungulates were of stockier build, and mainly diverged into smaller rodent-like forms, or larger, rhino-like forms (Figure 19–21g).

- *Late Eocene or early Oligocene colonizers, probably arriving by rafting from Africa.* These include the caviomorph rodents (e.g., guinea pigs, agoutis, and capybaras) and the South American monkeys. The caviomorphs diversified in the Miocene and Pliocene to include the largest rodents that have ever lived. *Telicomys*, a dinomyid (= terrible mouse), was the size of a small rhinoceros, and even today the pig-size capybara (Figure 19–21f) is extremely large for a rodent.
- *Late Miocene arrivals, arriving from North America via an island arc linking the two continents.* These include raccoons (moving from North to South) and some small types of ground sloths (moving from South to North).

What happened when the establishment of the Panamanian land bridge in the Pliocene allowed these two very different faunas to mix? Some animals from North America moved southward, and fewer South American forms moved northward (see Figure 19–21). (This discrepancy in the numbers of immigrants disappears when we consider that North America has a greater land area than South America, and hence a greater diversity of mammalian taxa to donate.) On each continent the newcomers and the native fauna appeared to coexist; for the most part the immigrants enriched the existing fauna rather than displacing it. However, a later disparity exists. Although Pleistocene extinctions affected the largest mammals on both continents, the southern immigrants to North America were more profoundly affected than North American forms in South America.

Today about half the generic diversity of South American mammals consists of forms with a North American origin, although some notable northern immigrants to South America are now extinct there—for example, horses and the elephant-related gomphotheres (Figure 19–21s). The southern species that persist in North America are mostly confined to Central America (e.g., capybaras) and the southern United States (e.g., armadillos). Opossums and porcupines are the exceptions that remained successful in northern North America.

The key to understanding the apparent greater success of the immigrants to South America lies in understanding biogeography. The equator and much of tropical America lies within South America. At times of climatic stress, such as during glacial periods, South America retains more equable habitats than North America, and fewer extinctions would be expected. Vegetational changes associated with climatic changes obviously also played a role. It is thought that the Isthmus of Panama had some savanna-like habitats linking North and South America during the Pliocene, acting as a corridor allowing for the movement south of mammals like horses and deer and the movement north of mammals like glyptodonts and ground sloths.

By the Pleistocene the Central American corridor was evidently closed to migration of savanna-adapted mammals, possibly because of encroachment by tropical forests. Mammals that arrived in North America only in the Pleistocene, such as bison and mammoths, never reached South America. The glyptodonts and ground sloths would also have been unable to return to South America. When the North American savannas disappeared at the end of the Pleistocene, these southern-originating edentate mammals were unable to adapt to the cooler prairies that housed the bison, and so went extinct.

A final point of consideration in the GAI is that counts of who moved where—and when—usually consider only species known from the fossil record (as in Figure 19–21). Because tropical habitats rarely preserve fossils, we have little information about the fossil history of Central America. Yet the large diversity of opossum-like marsupials, edentates, monkeys, and caviomorph rodents in Central America today can only have come from South America. Some other Central American mammals that are traditionally considered as northern taxa, such as cats (ocelots and pumas) and ungulates (brocket deer and tapirs), may also have evolved into their present form in South America and reimmigrated back into North America. Because of our present-day political boundaries, we often forget that Central America is biologically part of North America, not South America. A proper tally of the immigrants from South America to Central America is necessary before we can write the final chapter on the GAI.

Summary

All mammals share uniquely derived features. Lactation is the most obvious mammalian feature, and all mammals have hair and a variety of skin glands used for hair lubrication and waterproofing, olfactory communication, and thermoregulation. Mammary glands evolved from these types of skin glands.

Other unique mammalian features include the following. In the skull: a single-boned lower jaw (the dentary) and a three-boned middle ear; a heterodont, diphyodont dentition; a masseter muscle for mastication and muscles of facial expression. In the skeleton: determinate growth and long bone with epiphyses; usually only seven cervical vertebrae; thoracic ribs only in the trunk, and lumbar vertebrae capable of dorsoventral flexion; upright posture with a crurotarsal ankle joint, a calcaneal heel, and gluteal muscles as the main limb retractors. In the internal anatomy: fat is stored subcutaneously and in association with major organs; large lungs with branching bronchioles ending in alveoli; a heart with a complete ventricular septum and a single (left) systemic arch; red blood cells that lack nuclei; loss of the renal portal system; and kidney tubules with loops of Henle. In the nervous system: a brain with a large neopallium; a good sense of olfaction and hearing, with an enlarged cochlea in the inner ear, but poor vision and limited color vision in most mammals. Mammals also have a XY chromosomal system of sex determination.

The major groups of living mammals are monotremes and therians (marsupials and eutherians). Therians are not only more derived than monotremes in their mode of reproduction (viviparity as opposed to egg-laying), but they are more derived in several features of the skeleton and musculature, especially the anatomy of the shoulder girdle, which may allow for a greater range of locomotory behaviors. Therians also have tribosphenic molars, pinnae (external ears), a cochlea with two and a half coils, and separated openings for the urogenital and alimentary ducts, with the loss of the cloaca. Only a few dental and skeletal differences distinguish marsupials and eutherians, most notably the loss of the epipubic bones in eutherians. Eutherians are by far the most diverse group of living mammals, and they are found worldwide. Controversy exists about how the different eutherian orders are related to one another, and about their time of origination and diversification.

The diversity of Cenozoic mammals can be best understood in the context of changing patterns of biogeography. Distribution of some mammalian groups reflects vicariance, as in the monotremes of Australia. Other patterns of distribution reflect dispersal, such as the migration of marsupials to Australia from South America. Isolation of the various continents has resulted in unique mammalian faunas in Australia, Madagascar, and South America. The South American fauna was even more different from the rest of the world until the formation of the Isthmus of Panama around 2 million years ago.

Additional Readings

Archer, M., T. F. Flannery, A. Ritchie, and R. E. Molnar. 1985. First Mesozoic mammal from Australia—an early Cretaceous monotreme. *Nature* 318:363–366.

Benton, M. J. 1999. Early origins of modern birds and mammals: Molecules vs. morphology. *BioEssays* 21:1043-1051.

Blackburn, D. G. 1991. Evolutionary origins of the mammary gland. *Mammal Review* 21:81–96.

Daly, M. 1979. Why don't male mammals lactate? *Journal of Theoretical Biology* 78:325–345.

Flannery, T. F., et al. 1995. A new family of monotremes from the Cretaceous of Australia. *Nature* 377:418–420.

Foote, M., et al. 1999. Evolutionary and preservational constraints on the origins of major biologic groups: Limiting divergence times of eutherian mammals. *Science* 283:1310–1314.

Francis, C. M., et al. 1994. Lactation in male fruit bats. *Nature* 367:691–692.

Gaeth, A. P., R. V. Short, and M. B. Renfree. 1999. The developing renal, reproductive, and respiratory systems of the African elephant suggest an aquatic ancestry. *Proceedings of the National Academy of Sciences* 96:5555–5559.

Godthelp, H., M. Archer, and R. Cifelli. 1992. Earliest known Australian Tertiary mammal fauna. *Nature* 356:514–516.

Hedges, S. B., et al. 1996. Continental breakup and the ordinal diversification of birds and mammals. *Nature* 381:226–229.

Jenkins, F. A., Jr., and D. W. Krause. 1983. Adaptations for climbing in North American multituberculates (Mammalia). *Science* 220:712–715.

Kielan-Jaworowska, Z. 1997. Characters of multituberculates neglected in phylogenetic analyses of early mammals. *Lethaia* 29:249–266.

Krause, D. W. 1982. Jaw movement, dental function, and diet in the Paleocene multituberculate *Ptilodus*. *Paleobiology* 8:265–281.

Krause, D. W. 1986. Competitive exclusion and taxonomic displacement in the fossil record: The case of rodents and multituberculates in North America. *Contributions to Geology*, University of Wyoming, Special Paper 3:95–117.

Krause, D. W., et al. 1997. Cosmopolitanism among Late Cretaceous mammals. *Nature* 390:504–507.

Kumar, S., and S. B. Hedges. 1998. A molecular timescale for vertebrate evolution. *Nature* 392:917–919.

Luo, Z.-X., R. L. Cifelli, and Z. Kielan-Jaworowska. 2001. Dual origin of tribosphenic mammals. *Nature* 409:53–57.

Madsen, O. et al. 2001. Parallel adaptive radiations in two major clades of placental mammals. *Nature* 409:610–614.

Marshall, L. G. 1988. Land mammals and the Great American Interchange. *American Scientist* 76:380–388.

Marshall, L. G., S. D. Webb, J. J. Sepkoski, Jr. and D. M. Raup. 1982. Mammalian evolution and the great American interchange. *Science* 215:1351–1357.

McKenna, M. C. 1975. Toward a phylogenetic classification of the Mammalia. In P. Luckett and F. S. Szalay (Eds.), *Phylogeny of the Primates*. New York: Plenum Press, 21–46.

Murphy, W. J. et al. 2001. Molecular phylogenetics and the origins of placental mammals. *Nature* 409:614–628.

Novacek, M. J. 1992. Mammalian phylogeny: Shaking the tree. *Nature* 356:121–125.

Novacek, M. J. 1994. The radiation of placental mammals. In D. R. Prothero and R. M. Schoch (Eds.), *Major Features of Vertebrate Evolution*. Short Courses in Paleontology no. 7, The Paleontological Society. Knoxville: University of Tennessee Press, 220–237.

Novacek, M. J., et al. 1997. Epipubic bones in eutherian mammals from the Late Cretaceous of Mongolia. *Nature* 389:483–486.

O'Leary, M. A., and J. H. Geisler. 1999. The position of Cetacea within Mammalia: Phylogenetic analysis of morphological data from extinct and extant taxa. *Systematic Biology* 48:455–490.

Pascual, R., et al. 1992. First discovery of monotremes in South America. *Nature* 256:704–706.

Pond, C. M. 1998. *The Fats of Life*. Cambridge: Cambridge University Press.

Rich, P. V., and G. F. van Tets. 1985. *Kadimakara, Extinct Vertebrates of Australia.* Lilydale, Austr.: Pioneer Design Studio.

Rich, T. H., et al. 1997. A tribosphenic mammal from the Mesozoic of Australia. *Science* 278:1438–1442.

Rowe, T. 1988. Definition, diagnosis, and origin of Mammalia. *Journal of Vertebrate Paleontology* 8:241–264.

Solounias, N. 1999. The remarkable anatomy of the giraffe's neck. *Journal of Zoology* 247:257–268.

Stehli, F. S., and S. D. Webb (Eds.). 1985. *The Great American Interchange*. New York: Plenum.

van Dijk, M. A. M. et al. 2001. Protein sequence signatures support the African clade of mammals. *Proceedings of the National Academy of Sciences, USA* 98:188–193.

Vaughan, T. A., J. A. Ryan, and N. J. Czaplewski. 2000. *Mammalogy*, 4th. ed. Philadelphia: Saunders.

Vrba, E. S. 1992. Mammals as a key to evolutionary theory. *Journal of Mammalogy* 73:1–28.

Waddell, P. J., N. Okada, and M. Hasegawa. 1999. Towards resolving the interordinal relationships of placental mammals. *Symposium on the Origin of Mammalian Orders, Systematic Biology* 48:1–5. (See also other papers in this volume.)

Webb, S. D. 1991. Ecogeography and the Great American Interchange. *Paleobiology* 17:266–280.

Web Explorations

On-line resources for this chapter are on the World Wide Web at: http://www.prenhall.com/pough (click on the Table of Contents link and then select Chapter 19).

CHAPTER 20 Mammalian Specializations

Cenozoic mammals are a highly diverse group of organisms, adapted to a wide variety of life styles and displaying broad ecomorphological diversity. The three major types of living mammals—monotremes, marsupials, and eutherians—can be distinguished by profound differences in their reproduction. Differences in morphological specializations can also be seen in the skull and teeth for feeding, in the postcranial skeleton for locomotion, and in the brain and sense organs.

20.1 Mammalian Reproduction

The mode of reproduction is the major, and most obvious, difference between the three major groups of living mammals. While all mammals lactate and care for their young, monotremes are unique among living mammals in laying eggs. Among the therians, marsupials and eutherians have profound differences in their relative lengths of gestation; marsupials are familiar to many people as pouched mammals, the pouch housing the young that would still be inside the uterus of a comparable eutherian.

All mammals grow from an initial embryonic ball of cells (the **blastocyst**) that forms both the embryo and the **trophoblast.** The trophoblast is a differentiation of extraembryonic tissue specialized for obtaining nutrition in the uterus, for producing hormones to signal the state of pregnancy to the mother, and (in therians) for helping the embryo implant to the uterine wall. All mammals have a trophoblast, but the distinction between the inner cell layers of the blastocyst (forming the embryo) and outer cell layers (forming the trophoblast) is more distinct in eutherians than in monotremes and marsupials (Tyndale-Biscoe and Renfree 1987).

Additionally, all mammals have a glandular uterine epithelium (**endometrium**), which secretes materials that nourish embryos in the uterus, and a corpus luteum (plural = corpora lutea, Latin *corpus* = body and *lut* = yellowish), a hormone-secreting structure formed in the ovary from the follicular cells remaining after the egg is shed. Hormones secreted by the corpus luteum are essential for the establishment and at least the initial maintenance of pregnancy, although only in eutherians is there feedback from the placenta to maintain the life of the corpus luteum.

The egg-laying mode of monotremes is most likely the primitive mammalian condition, because egg-laying is primitive for amniotes. With the evolution of viviparity the uterine glands that add the shell and other egg components are lost. Thus it seems highly likely that it would be difficult or impossible to return to oviparity once a lineage has become dedicated to viviparity.

Reproductive Mode of Monotremes

The reproductive tract of monotremes retains the primitive amniote condition. The two oviducts remain separate, and do not fuse in development except at the base, where they join with the urethra from the bladder to form the **urogenital sinus** (see Figure 20–4a, later). The oviducts swell to form a uterus, where the fertilized egg is retained (only the left oviduct is functional in the platypus). In all mammals the eggs are fertilized in the anterior portion of the oviduct, the **fallopian tube,** before they enter the uterus. The ovaries of monotremes are bigger than those of therians, and monotremes provide the embryo with a large amount of yolk. However, the eggs of monotremes are much smaller at ovulation than those of reptiles or birds of similar body size.

The amount of yolk is not sufficient to sustain the embryo until hatching, and the eggs are retained in the uterus, where they are nourished by maternal secretions and increase in size before the shell is secreted (matrotrophy). (The eggshell is leathery, like that of some lizards, rather than rigid like the calcareous eggshells of birds.) Hatchling monotremes have a bird-like egg tooth (or caruncle) that they use to open the shell.

Monotremes lay one or two eggs, and the young hatch relatively rapidly after the egg is laid (after only 12 days of incubation in the case of the platypus). The young are hatched at an undeveloped, almost embryonic stage (see Figure 20–1a), and brooding by the mother continues for a further 16 weeks. The platypus usually lays its eggs in a burrow, whereas echidnas keep their eggs in a ventral pouch that resembles the pouch of marsupials but is probably not homologous with it. All monotremes have a low reproductive rate—no more than once a year.

Reproductive Mode of Therians

All therians have a placenta, which is formed from the extraembryonic membranes of the fetus. Marsupials and eutherians are often thought to differ in their type of placentation, and it is sometimes stated that marsupials lack a placenta. However, there is more similarity in placentation between these two types of mammals than commonly believed.

All therians have an initial **choriovitelline** placenta, developed from the yolk sac (the extraembryonic membrane common to all vertebates). Eutherians also have a later-developing **chorioallantoic** placenta, developed from the combination of the

▶ Figure 20–1 Mammalian neonates.

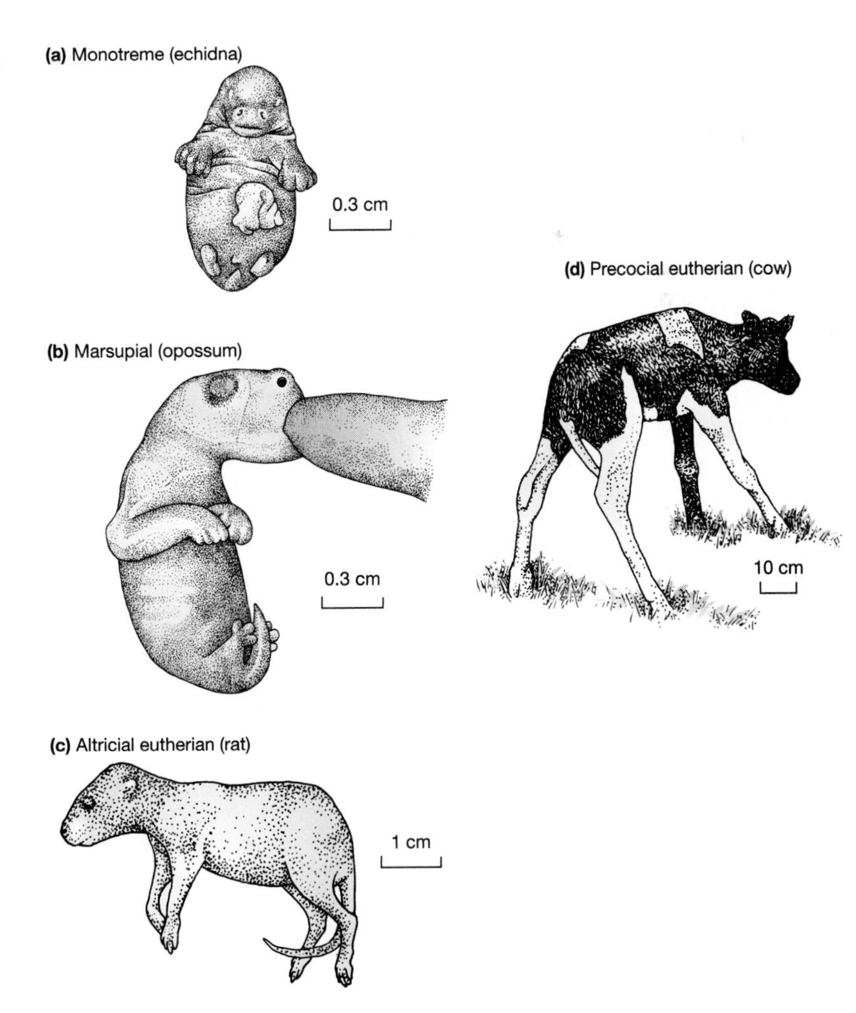

chorionic and allantoic amniote membranes (extraembryonic membranes seen only in amniotes). The chorioallantoic membrane of echidnas also grows within the egg after laying. Some eutherian species retain the choriovitelline placenta even after the chorioallantoic placenta has developed (Figure 20–2). The placenta of some eutherians (e.g., anthropoid primates and some rodents) penetrates deeply into the uterine lining, so much so that only a layer or two of tissues separates the fetal and maternal blood systems. Six distinct layers separate fetal and maternal blood in most eutherians, however. There is much variety in the form of the mammalian placenta as well as the types of placentation (see Mossman 1987 for details).

Most marsupials have only a choriovitelline placenta during their short gestation, but some marsupials show a transitory chorioallantoic placenta near the end of gestation (see Figure 20–3). The chorioallantoic placenta is best developed in bandicoots, where it is vascularized and invasive, but a less-developed chorioallantoic placenta is also seen koalas and wombats. The lack of a chorioallantoic placenta in most marsupials actually represents *suppression* of the chorioallantoic membrane.

In all therians, the ureters draining the kidney enter the base of the bladder rather than the cloaca or urogenital sinus, as in other animals. In females, the oviducts are fused in the midline anterior to the urogenital sinus to form a uterus. However, both these features evidently evolved convergently in marsupials and eutherians. It may seem an odd feature of design that the ureters would not enter the bladder in a vertebrate. However, while some type of tube draining the kidney is a feature of all vertebrates, a bladder is a tetrapod invention that forms as an outpocketing of the cloaca. Backwashing of the urine from the cloaca into the bladder may be a good enough design for most vertebrates; indeed, the bladder is lost in a number of nonmammalian amniotes, including many lepidosaurs and most birds. However, in a viviparous mammal urine might enter the uterus containing the developing young, and repositioning of the ureters may have been adaptive at this evolutionary juncture (Renfree 1993).

In all therians **parturition** (giving birth) and lactation are under the control of hormones produced by the pituitary and hypothalamus, often with complicated feedback control loops. Hormone production by the embryo and the placenta also plays a key role. In litter-producing mammals the young may be born in a single litter (**semelpary**), or in several litters spread over the mother's lifetime (**iteropary**). Single young are seen in both large and small mammals, but few larger mammals have large litters.

Embryonic diapause (maintaining the embryo in a state of suspended animation prior to implantation) is an important reproductive feature of some therians. This capacity enables the mother to space successive litters and to separate the time of mating and fertilization from the start of gestation. Thus, diapause allows mating and the birth of young to occur at optimal times of the year. Embryonic diapause is particularly well developed in kangaroos (see Figure 20–5c), and has often been perceived as a derived marsupial feature. However, embryonic diapause occurs in a wide variety of eutherians, including carnivorans, rodents, bats, edentates and at least one artiodactyl, the roe deer (Renfree and Shaw 2000).

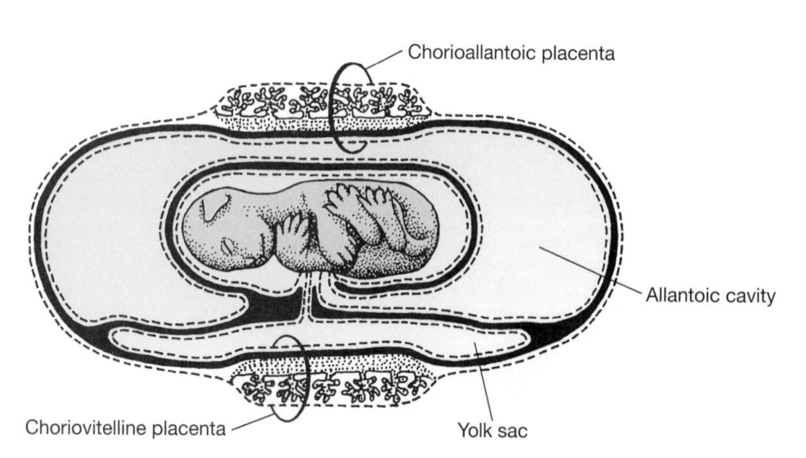

◀ Figure 20–2 Two types of mammalian placental structures as seen in a transitional stage of an implanted embryo of a cat. Both a choriovitelline placenta and a chorioallantoic placenta are present at this stage. The chorioallantoic placenta grows outward and takes over the function of the earlier-forming choriovitelline placenta.

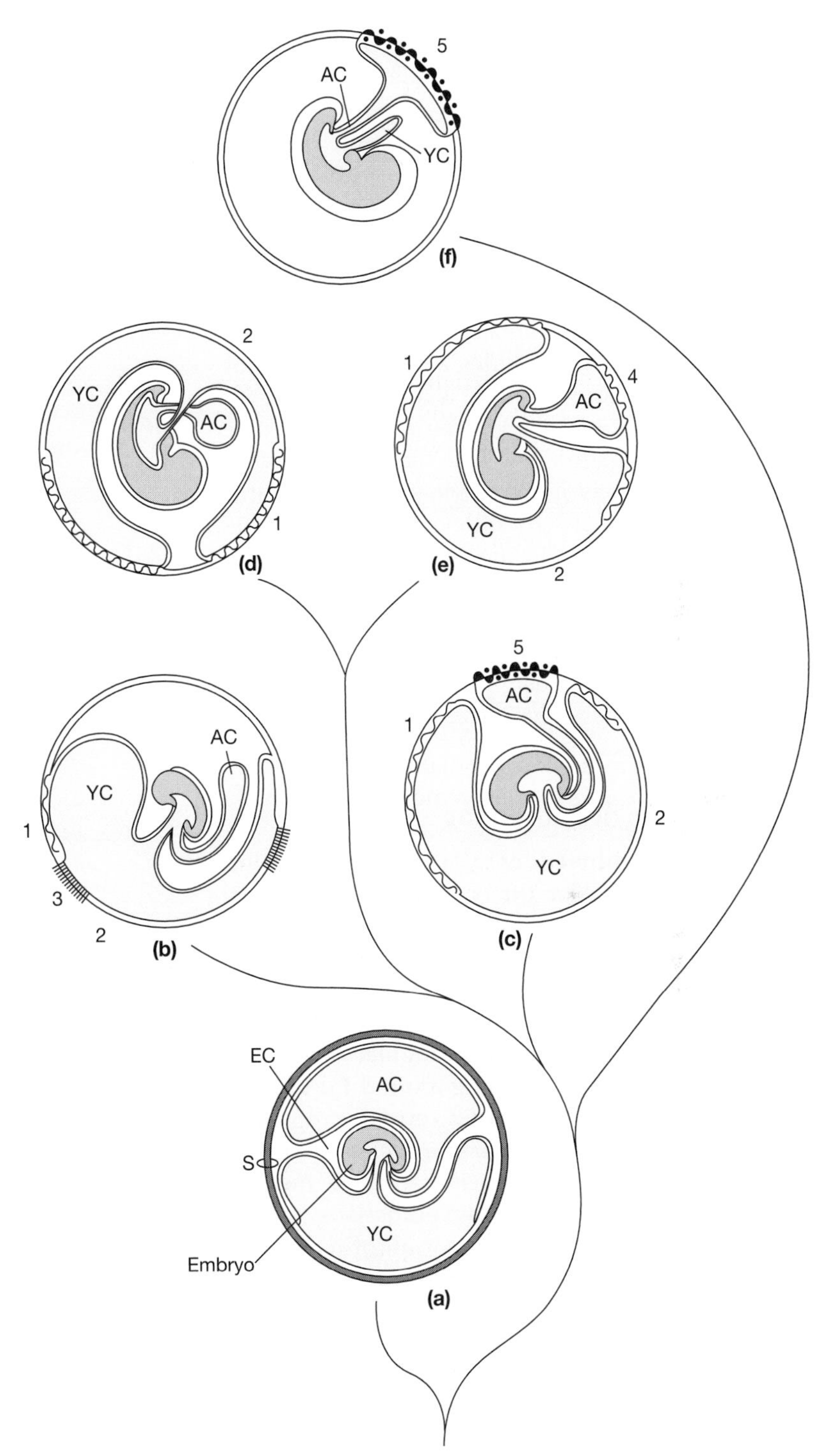

▶ Figure 20–3 Types of placentation in marsupials and eutherians. (a) Egg-laying monotreme. (b) Dasyurid—the allantois reaches the chorion and then retreats from it without forming a placental structure. (c) Bandicoot—a complex chorioallantoic placenta is formed at the close of gestation and the choriovitelline placenta remains functional until the young are born. (d) Possums and kangaroos—the allantois may grow to a large size but remains enshrouded in the folds of the yolk sac wall. (e) Koala and wombat—the allantois reaches the chorion, forming an apposed chorioallantoic placenta. (f) Eutherian—choriovitelline placenta is short-lived, and a complex chorioallantoic placenta is the functional one for most of the gestation. Key: 1 = vascular choriovitelline placenta; 2 = nonvascular choriovitelline placenta; 3 = syncytialized choriovitelline placenta; 4 = apposed (nonvascular) chorioallantoic placenta; 5 = syncytialized (vascular) chorioallantoic placenta; AC = allantoic cavity; EC = extraembryonic coelom; S = shell; YC = yolk sac cavity.

Position of the Testes

Monotremes are like other amniotes in having testes that are retained within the abdomen. Some therians, both marsupials and eutherians, also retain the testes in the abdomen, either in the original position high within the body, or partially descended and housed subcutaneously at the base of the abdomen. But the testes of most therian mammals descend into a scrotum during development.

The value of a scrotum is obscure: The traditional idea is that a scrotum provides a cool environment for sperm production, but there is no simple correlation between core body temperature and testicular position among mammals. The scrotum probably evolved convergently in marsupials and eutherians, because the control of scrotal development is different in the two groups. In marsupials testicular descent is under direct genetic control, whereas in eutherians it is hormonally determined. Additionally, the scrotum is in front of the penis in marsupials and behind the penis in most eutherians (although there are some exceptions; e.g., rabbits have a prepenile scrotum).

Specializations of Eutherians

In eutherians the ureters pass laterally around the developing reproductive ducts to enter the bladder. This arrangement allows the oviducts of females to fuse in the midline anterior to the urogenital sinus for much of their length (Figure 20–4d). In males this anatomical arrangement results in the **vasa deferentia** (the male reproductive tracts, singular = vas deferens Latin *vas* = vessel) looping around the ureters in their passage from the scrotum to the urogenital sinus (Figure 20–4e). All eutherians have a single, midline vagina, but not all have a single median uterus as seen in humans. Most eutherians have a uterus that is bipartite for some or all of its length, and a bipartite uterus may sometimes occur as a developmental abnormality in humans.

In most eutherians the urogenital sinus and the alimentary canal have separate openings with a distinct external space between them, the perineum. Edentates and some insectivorans are the exceptions here, with a more marsupial-like condition of more closely appositioned urogenital and alimentary openings (often incorrectly called a cloaca). The separation of female urogenital openings into distinct external urethral and vaginal openings is seen only in primates and some rodents.

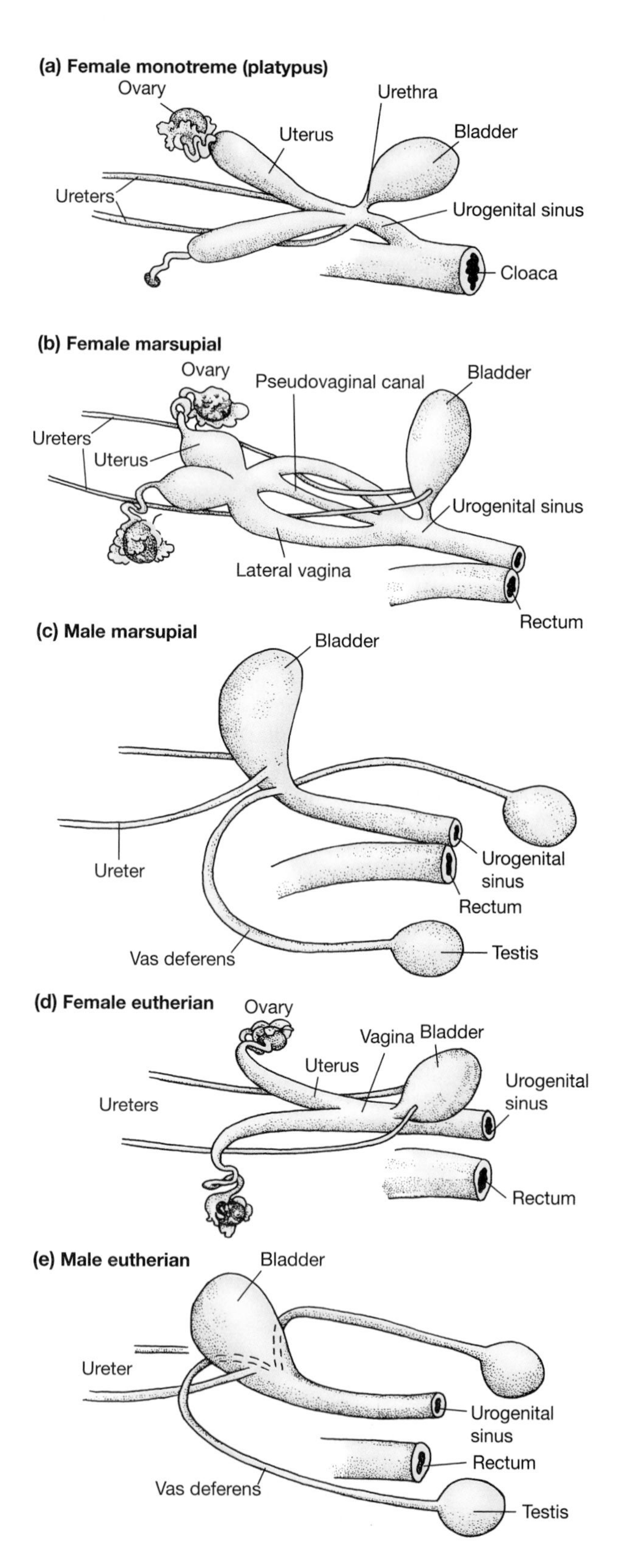

▲ Figure 20–4 Mammalian reproductive tracts.

The corpus luteum (or the corpora lutea, if more than one egg has been shed) is maintained for a longer period than one **estrous cycle** (the normal reproductive cycle of growth, maturation, and release of an egg) by hormones secreted by the pituitary and/or the placenta. (In humans and some other anthropoid primates the estrous cycle is called the **menstrual cycle,** since it is associated with a periodic shedding of the uterine endometrial lining not seen in other mammals.) The maintenance of the corpus luteum allows eutherians to retain the young in the uterus for a period greater than a single estrous cycle, in contrast to the usual marsupial condition. The length of gestation is correlated with body size; larger eutherians have longer gestation periods.

Some eutherians are born in a highly **altricial** state (i.e., poorly developed, as in many rodents and insectivorans), in which they are only slightly more developed than some marsupial young (Figure 20–1c). Others are born in more developed stages, extending to a highly **precocial** state (i.e., well developed, as in most ungulates) in which the young can run within a few hours of birth (Figure 20–1d). All eutherians, however precocial, still require a period of lactation for the transfer of essential antibodies from the mother as well as for nutrition. However, the period of lactation in most eutherians is relatively short in comparison with other mammals—usually shorter than the period of gestation.

In all mammals, larger-bodied species tend to have fewer young than smaller ones. In eutherians a distinct difference exists between large carnivores and large herbivores. While almost all ungulates have only a single, precocial young (or at the most, twins), large carnivores such as bears and lions have several altricial young per litter.

Specializations of Marsupials

In marsupials the ureters pass medial to the developing reproductive ducts to enter the bladder. This arrangement prevents the oviducts of the females from fusing in the midline, at least posteriorly, and means that the vasa deferentia of the males do not have to loop around the ureters (Figure 20–4c). The female reproductive tract consists of two lateral vaginae that unite anteriorly, from which point the two separate uteri diverge. The lateral vaginae are for the passage of sperm only. Birth of the young is through a midline structure, the median vagina or **pseudovaginal canal,** which develops at the first parturition (Figure 20–4b).

Hormonal feedback from the embryonic trophoblast to the pituitary and hypothalamus alerts the mother to the state of pregnancy and influences the secretory activity of the uterus. Marsupials do not maintain the corpus luteum, and the young are ejected at the end of the estrous cycle in most species (the swamp wallaby has a gestation period that is longer than the estrous cycle). The length of gestation in marsupials is relatively independent of body size (although the total time taken to rear the young is not). In contrast to eutherians, marsupials retain direct evidence of their oviparous ancestry; a transient shell membrane appears, and some **neonates** (newborns) have an egg tooth.

Marsupial gestational development is very different from that of both monotremes and eutherians. It clearly represents the derived condition, because both eutherians and monotremes are more similar to other amniotes. Marsupial neonates have well-developed forelimbs in comparison with other altricial neonates (Figure 20–1b), their lungs are relatively large at birth, and various aspects of their cranial development are also speeded up. Development of the jaws, secondary palate, facial muscles, and tongue is advanced (while that of the central nervous system is retarded), so that the newborn marsupial can attach itself to a nipple and begin suckling.

When young marsupials are born, they must make their way from the vagina to attach themselves to a nipple to complete their development. Most, but not all, marsupials enclose the nipples within a pouch. A pouch is absent in some dasyurids (marsupial mice, etc.) and some didelphids (opossums).

The general mode is the one seen in macropodids (kangaroos, etc.), in which the young climbs up to the pouch unaided. The mother adopts a distinct sitting birth posture (see Figure 20–5a, b) and licks a path from the vagina to the pouch, but does not otherwise aid the young in its journey. Marsupial neonates are evidently equipped with an instinct to climb upward. Some dasyurids and didelphids have even more altricial young than kangaroos. These babies are ejected directly into the pouch (or into the mammary area of pouchless species) at birth. The newborn young of these species are passive, unlike newborn kangaroos.

The time that young marsupials spend developing while attached to the nipple greatly exceeds the length of gestation. Lactation also continues for some time after the young have become sufficiently mature to detach from the nipple. This is when we

(a) Birth posture of red kangaroo

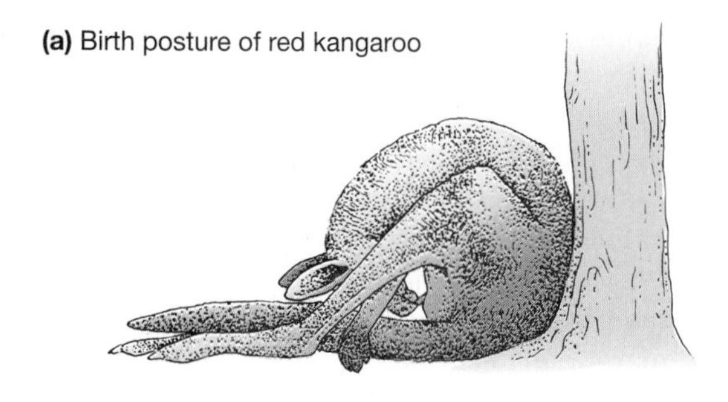

(b) Birth posture of grey kangaroo

(c) Red kangaroo with 3 different young at 3 different developmental stages

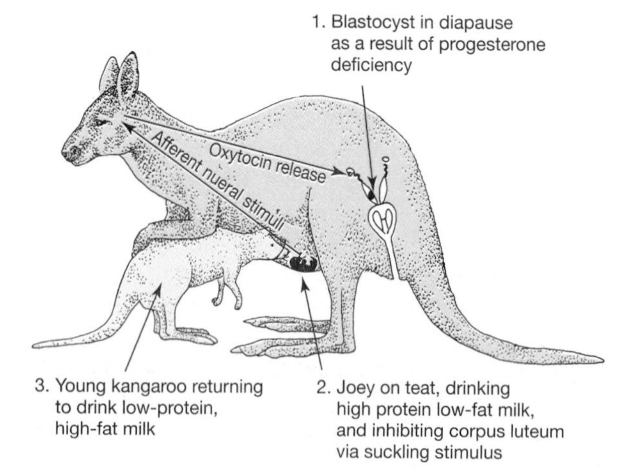

▲ Figure 20–5 Birth posture and embryonic diapause in kangaroos.

typically see the pouch young or young-at-foot, hopping in and out of the pouch.

Although the composition of the milk varies little during pregnancy in eutherians, there is a marked variation in marsupials as well as in monotremes. The first milk is dilute and protein-rich, while the later milk is more concentrated and richer in fats. Concurrent asynchronous lactation has been observed in some kangaroos; that is, an immature pouch young is attached to one nipple while a more mature, independent pouch young drinks from another nipple, and the mother produces different kinds of milk at the two nipples (see Figure 20–5c). The composition of milk is probably determined by how long each nipple has been producing milk.

The Primitive Therian Condition

Given that the young of both monotremes and marsupials are highly altricial, this is probably the primitive condition for mammals in general and for all therian mammals in particular. Altricial young may be adaptive for small, endothermal vertebrates. These young are essentially endothermal as neonates (newborns), with correspondingly low metabolic rates. Parental brooding keeps them warm and, because of their low metabolic rates, the newborns convert a high proportion of the food provided by the parents into body tissue. Of course, since the production of such altricial young, whether from eggs or by live birth, is dependent on the existence of lactation and parental care, altricial young are likely to be a derived mammalian feature rather than one inherited from cynodonts (see Chapter 17).

The reproductive mode of the earliest therians is not clear. The specializations of the reproductive anatomy of marsupials and eutherians cannot be derived from each other; they must represent separate evolutionary histories in each group. Possibly the adoption of viviparity in the common ancestor of marsupials and eutherians stimulated the migration of the ureters to the base of the bladder separately in the two types of mammals, with different consequences for the subsequent possible morphologies of the reproductive tracts. Alternatively, viviparity may have been evolved independently in the two groups.

The early therians of the Cretaceous period were small animals, most less than a kilogram in body mass. They are likely to have had the life-history features shared by extant small living therians: short life span, several litters produced in rapid succession (or a single large litter), and a short gestation period. Some fetal eutherians show evidence of mouth seals—tissue that develops around the lateral margins of the mouth of neonatal marsupials to aid in attachment to the nipples. This feature might suggest that attachment to a nipple is a primitive therian feature. However, a pouch is definitely a derived feature of marsupials. Indeed, a pouch may even have been evolved independently on different occasions within marsupials, because it is absent in many primitive marsupials and differs in form in different derived lineages.

20.2 Some Extreme Eutherian Reproductive Specializations

Numerous reproductive specializations exist among living eutherians. Some the most bizarre are those seen in the naked mole rat (*Heterocephalus glaber;* see Figure 19–17d) and the spotted hyena (*Crocuta crocuta;* see Figure 19–17f).

Naked mole rats are found in arid areas in sub-Saharan Africa and live as underground burrowers feeding on plant roots and tubers. They are **eusocial,** a type of social system otherwise seen only in social insects such as ants, termites, and bees. Like these insects, animals within a colony are closely related but, unlike the insects, they are all diploid in chromosome number.

Mole rats live in colonies of up to 40 individuals with only one breeding female, the queen. The queen somehow hormonally suppresses breeding by the other female members of the colony, and produces one to four litters per year with up to two dozen young in each litter. Other colony members are divided into three social castes: the smaller frequent workers that engage in cooperative burrowing and feed the community; the infrequent workers (also small) whose role is similar, but who appear to do only about 25 percent as much work as the frequent workers; and the larger nonworkers that only care for the young. Males and females are equally represented in these castes.

All the males in a colony produce sperm, but only the nonworkers are large enough to copulate successfully with the queen. If the queen dies, one of the faster-growing female infrequent workers may become the new queen. The adaptive advantages of this social structure may relate to sharing energy costs in conditions of sparse food supply and long periods of seasonal aridity.

Spotted hyenas, found in the African savannas, are the only hyenas that regularly hunt in packs, and they also have females with masculinized genitalia. The clitoris is so enlarged that it resembles a male penis (complete with the capacity for erection), and the labia are fused to form a structure resembling a scrotum. For many years it was mistakenly thought that these hyenas were hermaphrodites.

Various adaptive hypotheses have been proposed for these features, but it is unlikely that the appearance of the external genitalia themselves is adaptive. Rather, it appears that high levels of male hormones (androgens such as testosterone) are advantageous for females because they cause aggressive behavior. This could be useful in an ecological situation (perhaps recently encountered in evolution) where food is short and more aggressive females and their offspring obtain more than their share of food during communal predation. These high levels of male hormones have the side effect of masculinizing the females' genitalia during development.

Although hyena females may use their genitalia for behavioral display, there are clearly many disadvantages to this condition. All the functions of the original urogenital sinus must now be transmitted through this penis-like structure; thus female hyenas not only urinate through the enlarged clitoris, they also copulate and give birth through this structure. Perhaps unsurprisingly there is a high incidence of mortality among females giving birth for the first time.

20.3 Are Eutherians Reproductively Superior to Marsupials?

It used to be considered that the marsupial mode of reproduction was inferior to that of eutherians. This conclusion was based primarily on the reasoning that marsupials had been unable to compete with eutherians except in Australia, where they had evolved in isolation (but see Chapter 19). It was claimed that marsupials were unable to maintain the young in the uterus past one estrous cycle because the mother was unable to recognize the condition of pregnancy due to the lack of hormonal feedback between the developing young and the maternal brain. But we now know that hormonal feedback of this nature does, in fact, exist in marsupials even though they lack the extended gestation of eutherians.

It has also been argued that the lack of midline fusion of the oviducts in marsupials would make it impossible to carry large young to full term. However, many eutherians have uteri that are completely separate without evidence of midline fusion. Separated uteri are seen, for example, in cows, which no one would consider to give birth to small young. It seems more likely that, rather than the marsupial reproductive mode being a primitive therian condition or an inferior version of the eutherian condition, marsupials and eutherians evolved different but essentially equal reproductive strategies. Marsupials extended postgestation lactation, whereas eutherians emphasized intrauterine life.

In the past couple of decades an alternative viewpoint of marsupial reproduction has developed; that is, that marsupial reproduction may be superior under some conditions (e.g., Kirsch 1977, Parker 1977). For example, kangaroos eject pouch young while fleeing predators (actually, humans chasing them in cars; there are no native pursuit predators in Australia). It has been suggested that in this situation the marsupial mother, freed from the burden of carrying her young, could escape to breed again, whereas in the case of a pregnant eutherian both mother and young would perish.

An alternative marsupial advantage relates to the fact that it takes a marsupial longer to produce fully independent young than a eutherian of similar body size. (This is partly because the transfer of nutrients via the milk is less efficient than across the placenta.) Both marsupial and eutherian mothers invest similar amounts of resources in their young, but the time from conception to weaning is half again as long for a marsupial as for a eutherian. It has been argued that the slower rate of resource investment in marsupials means that, should the young die, the marsupial has invested less than the eutherian and so has lost less energy overall.

How would this work? Imagine that a marsupial and a eutherian both conceive a litter in March, and then experience a sudden dry spell in June, resulting in the death of the young. The eutherian, which had been investing more food to develop the litter more rapidly, might be worse off than the marsupial, which had not devoted so many resources to its young. At this point, the marsupial might have enough stored energy left to conceive again immediately, whereas the eutherian would not. This feature of marsupial reproduction might be adaptive in arid, unpredictable climates such as those in Australia, where droughts may occur regularly.

These proposals and other similar ideas seem plausible, but they have a critical flaw in that they consider only present-day mammals. In seeking evolutionary reasons for the development of marsupial or eutherian reproductive strategies, we must consider the animals and conditions that existed at the time of divergence, back in the Early Cretaceous. The ability to dump the pouch young if pursued is unlikely to be adaptive for a small mammal having but a single litter per lifetime (as seen in many extant dasyurids). Likewise, harsh, arid climates were not a feature of the Cretaceous. Australia did not develop its arid interior until the late Cenozoic, and marsupials did not even reach Australia until the early Cenozoic. Both reproductive strategies probably represent perfectly good means of achieving the same goal for the small, primitive therians of the Early Cretaceous.

Is there any feature of marsupial reproduction that would make their potential adaptive diversity different from eutherians at the time in the Cenozoic when mammals became larger and diversified into a greater variety of ecomorphological types? The lack of marsupial species diversity (about 6 percent that of eutherians) has been cited as an example of evolutionary inferiority. But due to accidents of history, marsupials have had less land area to evolve on during the later Cenozoic than eutherians, and so would be expected to have less species diversity.

One ecomorphological mode is probably impossible for a marsupial to achieve—fully aquatic life, such as a whale. A fully aquatic marsupial could not carry altricial young in a pouch, because they would be unable to breathe air. One semiaquatic marsupial, the South American yapok or water opossum, seals up its pouch during short underwater forays; but in general marsupials have avoided aquatic situations. It is also hard to imagine a marsupial giving birth under water in the fashion of a whale—the tiny neonates would probably be swept away by currents before they could reach the pouch.

True flight has not been evolved by any marsupial, although there are several species of gliding possums. Because flight has evolved only once among eutherians, however, the absence of flight in marsupials could just be a matter of chance. In contrast, eutherians have evolved semiaquatic and fully aquatic forms numerous times in parallel, and the relative absence of this mode of life among marsupials may represent a genuine constraint.

The mode of marsupial birth might limit the size of the adults. Today there are no marsupials with a body mass greater than about 90 kg; in the past only a very few were larger than this. The biggest were the bison-size diprotodontids—no marsupial ever reached the size of a rhino or an elephant. The birth posture (see Figure 20–5) might be difficult to adopt for a large animal, and the neonate would have a much greater absolute distance to climb to reach the nipples.

It has also been suggested that the need for marsupials to have well-developed forelimbs with large claws at birth in order to climb up to the pouch has somehow constrained the possibilities of later morphological specializations of those limbs, such as the formation of flippers or wings. The need for well-developed forelimbs at birth might also limit the type of cursorial (running-adapted) morphologies seen in ungulate eutherian mammals, in which the number of digits has been reduced to one or two, as in the hooves of antelope and horses. This might explain why the marsupial ungulate-equivalents, the kangaroos, hop with specialized hind legs but retain rather unspecialized fore legs—in contrast to the running eutherians, which specialize both fore and hind legs. Hopping in modern kangaroos is an exceedingly efficient mode of locomotion, and so it has been supposed that the evolution of hopping related to selection for saving locomotor energy. But hopping is not so efficient in smaller mammals such as jerboas (rodents), and since the early-appearing hopping marsupials were small animals, some other reason must be sought for its initial evolution.

Although the marsupial mode of reproduction may not be adaptively inferior overall, it remains true that eutherians have a faster reproductive rate than marsupials, especially at smaller body sizes. Certainly one reason that feral eutherians are diversifying at the expense of the native Australian marsupials is because they can reproduce faster and more often.

20.4 Specializations for Feeding

The mastication of food is exceedingly important for mammals. As endotherms they need large quantities of food, and they rely on mastication to reduce the particle size of food to speed the digestive processes. Birds manage to be endothermal without teeth; many of them have a gizzard within the digestive tract to grind up their food. Other amniotes (for example, some dinosaurs) show evidence of mastication. But all mammals masticate their food, and all mammals have a distinct swallowing reflex whereby they ingest a discrete bolus of finely chewed food. The mammalian tongue, important in both oral food processing and in swallowing, has a unique system of intrinsic muscles. The muscular cheeks of mammals are derived from a unique set of facial muscles (Chapter 19).

Dentition

Mammalian teeth are formed like those of other vertebrates—they have a **thecodont** type of rooting in the jaw bone, where the teeth are set in sockets and held in place by periodontal ligaments. A typical mammalian molar is shown in Figure 20–6a. Eutherian mammals have a maximum (in each jaw half, upper or lower) of three incisors, one canine, four premolars, and three molars (see Chapter 19). Mammals usually use their incisors to seize food, canines to stab prey, premolars to pierce and crush food, and molars to break down the food into fine particles by mastication. Therians have tribosphenic molars that originally had a triangular (tritubercular) crown (Figure 20–6d).

The incisors of mammals that gnaw, such as rodents and rabbits, may be enormously enlarged and grow continuously throughout life. Rodent incisors have enamel only on the anterior surface. Because the enamel is the hardest part of the tooth, it wears more slowly than the dentine behind it, producing a self-sharpening chisel edge (see Figure 20–6b, c).

Canines are often lost in herbivores, which have no need to subdue their food items, but they may be retained in modified form. The tusks of pigs and walruses are modified canines (but the tusks of elephants are modified incisors). Ruminant artiodactyls incorporate the lower canine into the incisor row, so that they appear to have a total of eight lower incisors (the upper incisors are replaced by a horny pad). Male horses may have small, apparently functionless canines, but these are rarely seen in females. Upper canines are generally larger in male primates than in females, even slightly so in humans. Hornless ruminants, such as the mouse deer (Figure 19–17h), may retain large upper canines in the male for fighting and display.

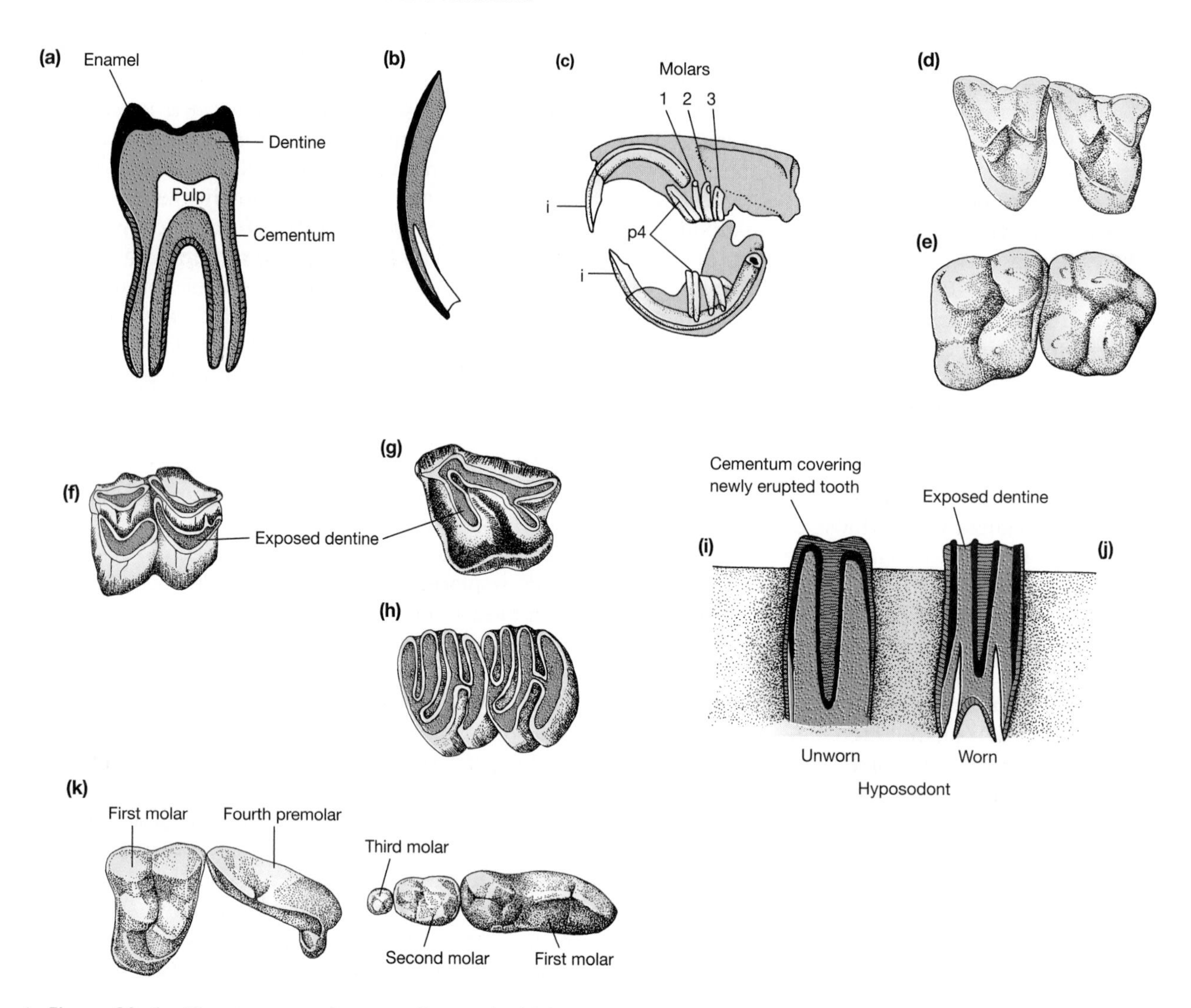

▲ Figure 20–6 The structure of mammalian teeth. (a) Sectioned mammalian molar showing general dental form. (b) Sectioned rodent ever-growing incisor, showing enamel on anterior surface only. (c) Sectioned rodent skull, showing ever-growing incisors, and ever-growing (hypselodont) cheek teeth. (d) Tritubercular upper molar, as found in the primitive therian mammal condition. (e) Bunodont, quadritubercular upper molars, as found in omnivorous mammals. (f) Selenodont upper molar of a deer. (g) Lophodont upper molar of a rhino. (h) Multilophed upper molars of a rodent. (i) Sectioned unworn hypsodont molar, showing covering of cementum. (j) Worn hypsodont molar, showing sharp enamel ridges interspersed with softer areas of dentine and cementum. (k) Carnivore carnassial shearing teeth (in a coyote).

The molars and premolars are usually different in form. Primitively, premolars are single-cusped slicing or puncturing teeth, whereas molars are three-cusped (or more) teeth for more thorough food processing. In many herbivores the entire postcanine tooth row is used for general mastication, and the premolars have been modified to resemble the molars (**molarized**). This condition can be seen in ruminant artiodactyls, and especially in horses where all the postcanine teeth appear identical.

Omnivorous and fruit-eating mammals have reduced the originally pointed cusps of their molars to rounded, flattened structures, suitable for crushing and pulping. They have also added a fourth cusp to the upper molars and increased the size of the talonid basin in the lower molars, so that these teeth now appear square (quadritubercular) rather than triangular. These teeth are called **bunodont** (Greek *buno* = a hill or mound and *dont* = tooth) in reference to the rounded cusps (Figure 20–6e). We have bunodont molars, as do most primates and other omnivores such as pigs and raccoons.

Herbivorous mammals evolved from omnivorous ancestors. Herbivores need the molars to grind up flat, fibrous food rather than to pulp and crush the food in the fashion of the bunodont molars of omnivores. In herbivores the simple cusps of the bunodont tooth have been run together into ridges, or **lophs.** These lophed teeth work best when the enamel has been worn off the top of the ridges to reveal the underlying dentine. In these partially worn teeth, each ridge then consists of a pair of sharp enamel blades lying on either side of the intervening dentine (see Figure 20–6f–j). When these teeth occlude and the jaws move sideways the food encounters multiple sets of flat, shearing blades, in a manner analogous to cheese moving over a cheese grater.

Different lineages of herbivorous mammals evolved lophed teeth convergently, and we use different terminologies to describe them. The teeth of perissodactyls (and also hyraxes) have straight lophs that run predominantly in a lateral to medial direction across the teeth; these teeth are called **lophodont** (Figure 20–6g). The teeth of ruminant artiodactyls have crescentic lophs that run in a predominantly anterior to posterior direction across the teeth; these teeth are called **selenodont** (Figure 20–6f).There is probably little difference in the function of lophodont and selenodont teeth; their difference in form reflects merely difference in ancestry. Among other herbivorous mammals, kangaroos and rabbits have a form of lophodont molars; koalas a form of selenodont molars; and wombats, rodents, warthogs, and elephants a highly complex form of molars termed multilophed or lamellar (Figure 20–6h).

All herbivorous mammals face a similar problem with their teeth, that of dental durability. Durability of the dentition is not an issue for most vertebrates, which continually replace their teeth. But all mammals have inherited the condition of diphyodonty (a single set of replacement teeth) from their original ancestor. Diphodonty was probably essential for precise occlusion in early mammals (see discussion in Chapter 17), but it presents a problem in that the adult dentition must last a lifetime. Herbivores face a particular problem because vegetation is more abrasive than other forms of food. Grazers have especially high tooth wear, because grasses contain silica in the cell wall.

There are a variety of ways in which herbivorous mammals have made their dentition more durable. In the manatee (a sirenian) and in the nabarlek (a rock wallaby), the normal mode of bringing in the molars from the back of the jaw is tricked into making a continual supply of molars that replace the worn teeth, which then drop out of the front of the jaw. Elephants do something similar, although they have not added extra teeth. Instead they make each molar enormous, the size of the entire original tooth row, and bring in a total of six molars (three milk molars and three permanent molars) one at a time during an animal's lifetime. However, the most common mode of making the molars more durable is to make them high crowned or **hypsodont** (Greek: *hyps* = high; see Figure 20–6i, j). The opposite of hypsodont, the primitive mode for teeth, is low crowned or **brachydont** (Greek: *brachs* = short).

Hypsodont cheek teeth look like regular teeth when seen in the jaw, but the division at the base between the crown and the root is not visible, as with brachydont teeth. The crowns of hypsodont teeth have been extended into the depth of the jaw bones, and the skull may be highly modified to accommodate the teeth. This condition can be seen in the very deep lower jaw, the deep cheek region, and the posteriorly moved orbit of grazers (Figure 20–8d) in which these modifications provide room to accommodate the high-crowned cheek teeth.

Larger hypsodont mammals (i.e., ungulates) also extend the layer of **cementum,** a bone-like material that in most mammals covers only the root and the base of the crown (Figure 20–6a), so that it covers the entire tooth (Figure 20–6i). This is done for mechanical reasons; the high lophs of the teeth must be laid down during tooth development, before the tooth erupts. Without cementum acting as filler, the individual lophs would be like tall, free-standing blades once the tooth had erupted, liable to fracture. As the animal ages and wears down the teeth, the teeth push up from the base to provide a continuous occlusal surface, much as lead pushes out of a mechanical pencil.

Most hypsodont mammals have a finite amount of tooth crown. When the tooth is worn out, the animals can no longer eat. (Most mammals in the wild will have died of natural causes long before their teeth wear out, but domestic horses surviving into their twenties and thirties often must be fed soft food because they have virtually no molars left.) However, some mammals have molars in which the roots do not close and the tooth is functionally ever-growing or **hypselodont.** For a variety of reasons, this appears to be an option primarily for small mammals (Janis and Fortelius 1988). Hypselodont molars are seen most commonly in rabbits and some rodents. They are also present in wombats, dugongs, and sloths (in sloths it is the root that is continually growing, not the crown).

Carnivorous mammals characteristically have large canines for subduing their prey; they also have specialized shearing postcanine teeth. Eutherian mammals in the order Carnivora (carnivorans) have a pair of specialized teeth modified into a set of tightly shearing blades, the **carnassials** (Figure 20–6k), formed from the last premolar in the upper jaw and the first molar in the lower jaw. The extinct eutherian creodonts also had carnassials, but these were formed from different teeth farther back in the jaws. The marsupial wolf lacked true carnassials; instead, each molar was somewhat specialized into a blade-like shape but none was significantly larger than the others.

Craniodental Specializations of Mammals

The hedgehog (Figure 20–7a) shows the generalized insectivorous condition that can be taken to represent the primitive mammalian mode. The opossum (Figure 19–14) is also a good example of a generalized mammal. The molars of generalized mammals usually retain the primitive triangular shape and pointed individual cusps that are useful for puncturing insect cuticle.

Anteaters Mammals that specialize on ants and termites are termed **myrmecophageous** (Greek *myrme* = ant and *phago* = to eat). Such specializations have evolved convergently among eutherians in edentates (in both anteaters and armadillos), pangolins (scaly anteaters), and aardvarks. Less extensive modifications are seen among some carnivorans: the aardwolf (a hyena), the bat-eared fox, and the fanalouc (a Madagascan civet). The numbat, or marsupial anteater, is a somewhat specialized myrmecophage, while the echidna (a monotreme) is highly specialized for this dietary habit.

Myrmecophageous mammals tend to elongate the jaws and progressively reduce the teeth (shown in an extreme condition in the giant anteater, Figure 20–7b). In addition to long snouts, they also have enlarged salivary glands and highly elongated tongues. The more derived myrmecophageous

(a) Hedgehog

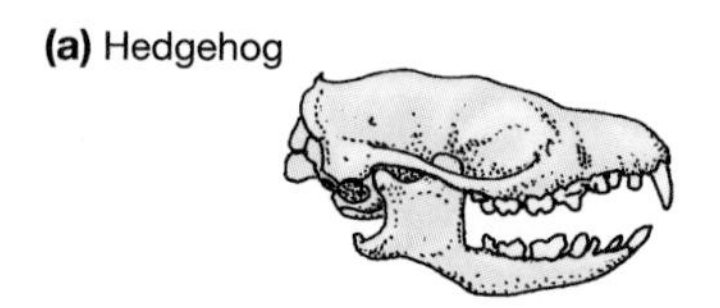

(b) Giant anteater

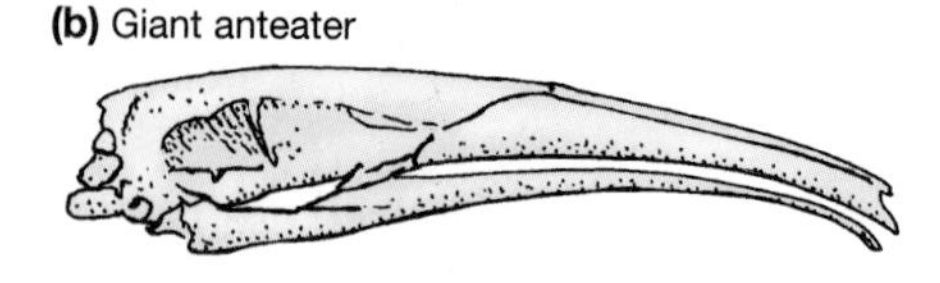

(c) Nectar-eating bat

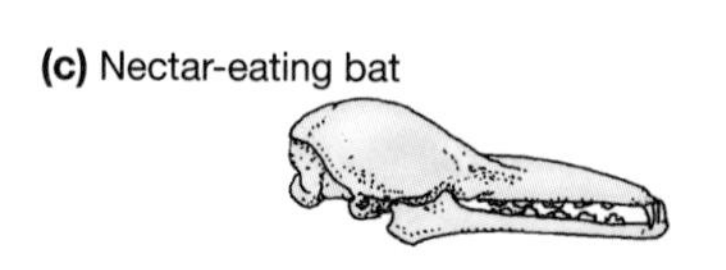

(d) Porpoise

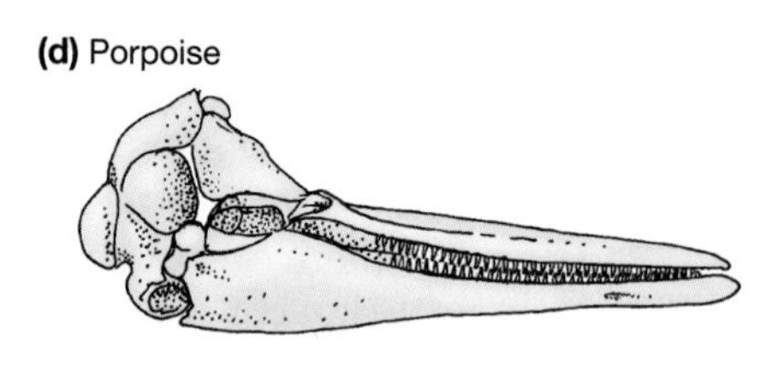

(e) Right whale

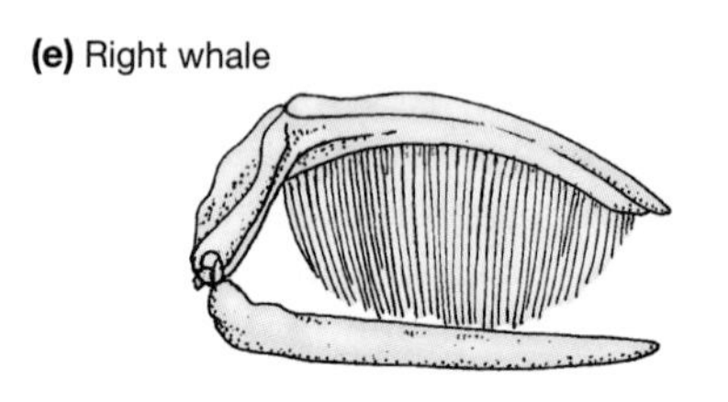

(f) Walrus

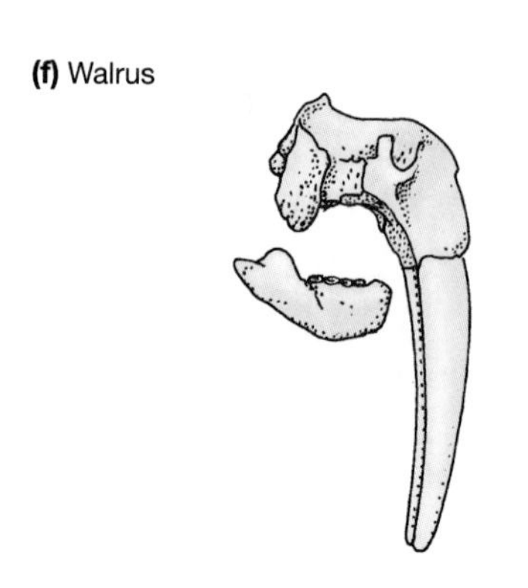

▲ Figure 20–7 Some feeding specializations of the teeth and skulls of mammals.

mammals usually specialize on ripping apart termite mounds and have digging specializations of the postcranial skeleton for this purpose. Reduction of the teeth is also seen in mammals that specialize on nectar (e.g., some bats, Figure 20–7c, and the honey possum).

Aquatic Feeders Aquatic fish-eating (**piscivorous**) or squid-eating mammals such as porpoises and dolphins (Figure 20–7d) have highly elongate jaws that have lost the anterior-most teeth. The remaining teeth are single-cusped and pointed in form, and they are greatly increased in number. Their skulls and teeth have become convergent with other piscivorous tetrapods such as ichthyosaurs (extinct marine reptiles) and crocodiles. Seals show some modifications in the same direction, but retain short jaws with a full complement of anterior teeth.

The baleen whales (mysticetes) have replaced their teeth with sheets of a fibrous, stiff, horn-like epidermal derivative known as **baleen,** which extends downward from the upper jaw (Figure 20–7e). These whales filter-feed, using the baleen to strain planktonic organisms from the water. The walrus (Figure 20–7f) feeds on mollusks, and its postcanine teeth are flat for crushing their shells (the enormous canine tusks are used mainly for display).

Differences between Carnivorous and Herbivorous Mammals The basic mode of mammalian mastication is best understood by considering the difference between carnivorous and herbivorous mammals (see Figure 20–8). All mammals use a combination of masseter, temporalis, and pterygoideus muscles to close the jaws, and the digastric (in therians) muscle to open the jaw. (Monotremes have a similar, but nonhomologous, jaw-opening muscle called the detrahens.) The relative sizes of the muscles used, especially in jaw closing, and the shape of the skulls reflect the different demands of masticating flesh and masticating vegetation. The temporalis has its greatest mechanical advantage at initiating jaw closure at moderate to large gapes, when the incisors and canines are likely to be used. Thus a large temporalis is typical of carnivores, which use a forceful bite with their canines to kill and subdue prey. The masseter is situated so that its line of action creates forces closer to the back of the tooth row, and it is also positioned to move the jaw sideways. Thus a large masseter is typical of herbivores, which grind their fibrous food using lateral jaw movements and complex molars.

Carnivorous mammals (and most generalized mammals, such as the opossum) have a fairly large temporalis muscle and a large coronoid process of the jaw for its insertion. The temporal fossa on the skull is also large for the origin of this muscle, and a sagittal crest along the midline of the skull may also increase this area of origin. The masseter muscle is of moderate size. The jaw joint is on the same level as the tooth row in the primitive mammalian condition (Figure 20–8a). This position of the jaw joint results in the teeth coming into contact sequentially as the jaw closes, like the blades of a pair of scissors—a design well suited for teeth that primarily cut and shear (see Figure 20–8b).

Carnivorous mammals have modifications of the jaw joint (the postglenoid process) to prevent the strong temporalis muscle from dislocating the lower jaw. They also tend to have a high occipital region (the back of the head), reflecting extensive musculature linking the head to the neck. The muscles are probably important for resisting struggling prey. The extinct saber-toothed carnivores had extremely high occipital regions.

The skulls of herbivorous mammals are more modified than those of carnivorous ones. Herbivores have skulls and teeth modified to grind up tough, resistant food; they also have modifications of the digestive system to aid in coping with a diet of plant matter. Plant food is much more abundant than animal food and does not need to be subdued, but the energy content of plant material is generally lower than that of animal tissues. The protein content of leaves and stems is usually low, and the protein is enclosed by a tough cell wall formed by **cellulose** (a complex carbohydrate). Thus the skulls and teeth of herbivorous mammals must adapted for grinding up large quantities of tough, fibrous material.

Herbivorous mammals have increased the size of the masseter and reduced the size of the temporalis in comparison to the primitive mammalian condition. A large masseter muscle is important for sideways jaw motion and grinding with the back teeth. This muscle morphology is reflected in the increased size of the angle of the lower jaw (for the insertion of the masseter) and the reduced size of the coronoid process and temporal fossa. The jaw joint has been shifted so that it is offset from the tooth row and is situated higher up on the skull (Figure 20–8c).

This offsetting of the jaw joint means that the part of the jaw between the tooth row and the joint can act as a sort of handle, increasing the moment arm

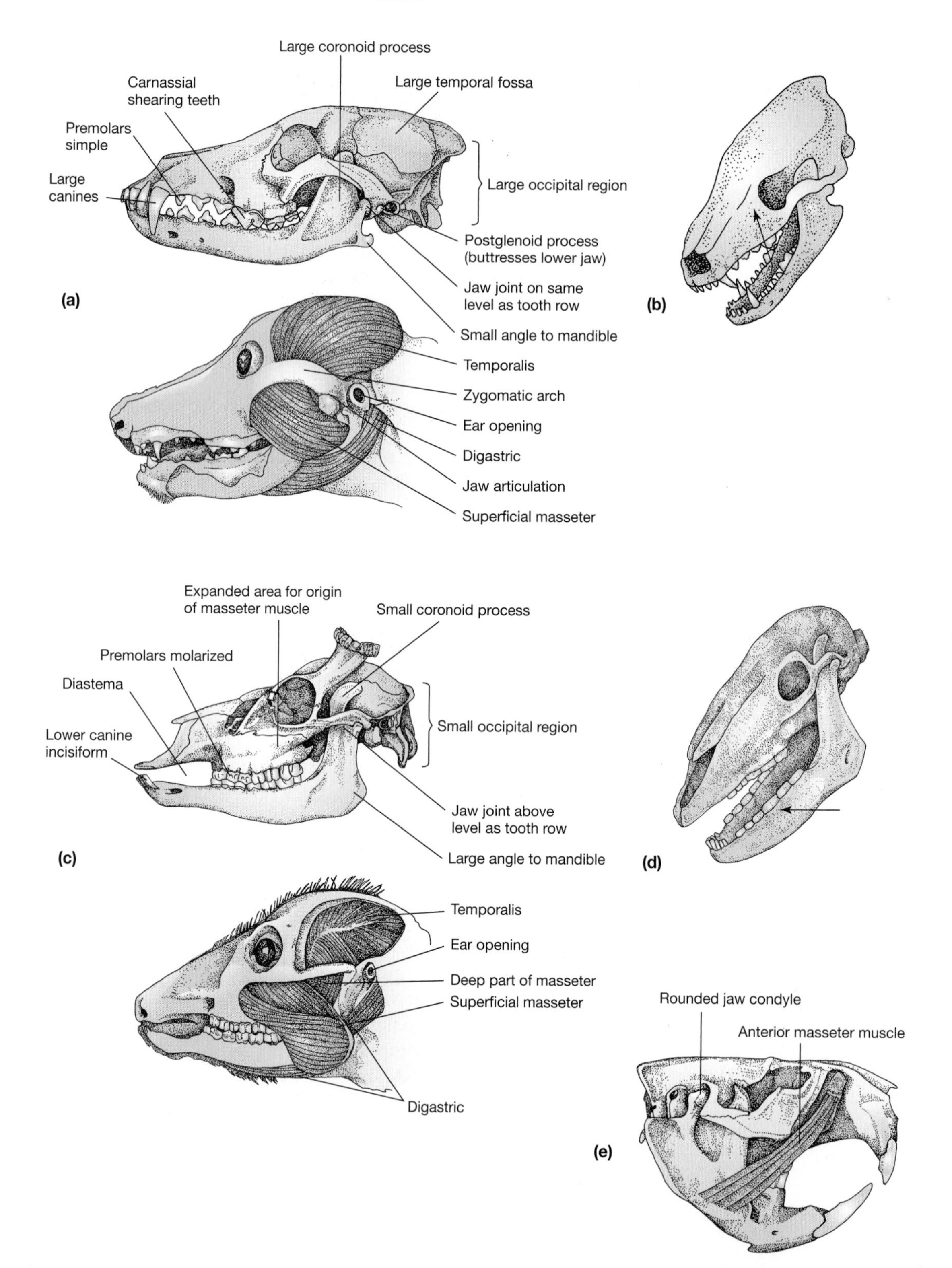

▲ Figure 20–8 Craniodental differences between carnivorous and herbivorous mammals. (a) Carnivore skull and musculature (a dog). (b) Action of carnivore jaws. (c) Herbivore skull and musculature (a deer). (d) Action of herbivore jaws. (e) Rodent skull and musculature (beaver).

for the action of the masseter muscle. This allows the teeth to be occluded simultaneously (rather than sequentially), much as the offset handle of a cooking spatula allows you to apply the entire blade of the spatula to the bottom of the frying pan while keeping your hand above the pan's rim (see Figure 20–8d). This simultaneous occlusion of the teeth is important for grinding the food. Some herbivorous dinosaurs also had offset jaw joints, but in this case the jaw joint was below the level of the tooth row. It is the offsetting of the jaw joint from the tooth row that is the important mechanical feature, not whether the joint is above or below the row.

Herbivores also usually have elongated snouts, resulting in a gap between the cheek teeth and the incisors called the diastema. (If horses did not have a diastema, we would not be able to put a bit in their mouths.) The actual function of the diastema is uncertain. It may allow extra space for the tongue to manipulate food, or it may just be a reflection of the elongation of the jaw for other reasons. A long jaw allows an animal to select food with its incisors without poking its eye on the vegetation; note that herbivores that can use their hands to select food, such as monkeys, rodents and even kangaroos, have shorter faces than most ungulates.

Many herbivorous eutherian mammals ossify the cartilaginous partition at the back of the orbit, to form a bony postorbital bar. This bar is probably important in absorbing stress from the jaws during the constant chewing of herbivores, thus protecting the braincase. Herbivores usually have a fairly low occipital region, because they do not need to have a carnivore-like attachment for muscles that help brace the head on the neck. An exception is pigs, which root with their snout and have extensive musculature linking the back of the head to the neck.

Many rodents have a highly specialized type of food processing. Their upper and lower tooth rows are the same distance apart—unlike the condition in most mammals, where the lower tooth rows are closer together than the uppers (see Chapter 19). This derived condition in rodents is combined with a rounded jaw condyle, which allows forward and backward jaw movement, and the insertion of a portion of the masseter muscle far forward on the skull, so that the lower jaw can be pulled forward into occlusion (Figure 20–8e). This jaw apparatus allows rodents to chew on both sides of the jaw at once, presumably a highly efficient mode of food processing. Note that this mode of chewing can be achieved only with flattened, lamellar teeth (Figure 20–6h), because the high ridges of other types of teeth would prevent this jaw motion. Elephants, which also possess lamellar teeth, chew in a somewhat similar manner; but they have obviously evolved this capacity convergently with the rodent condition, and the muscle arrangement is different.

Herbivores, Microbes, and the Ecology of Digestion

The specialized teeth of herbivores can rupture the cell walls of vegetation and expose the cell contents, but only special enzymes (**cellulases**) can digest the cellulose forming the cell wall and constituting a large percentage of the plant material. However, no multicellular animal has the ability to synthesize cellulase. Thus the efficient use of plants as food requires cellulase enzymes produced by microorganisms that live as symbionts in the guts of herbivorous animals. Several different types of mammals have independently evolved chambers within the digestive tract to house symbiotic microorganisms that convert the cellulose and lignin of plant cell walls into digestible nutrient.

The many separate evolutions of fermentative digestion among vertebrates have resulted in distinctly different solutions to the problems posed by plants as food. Horses and other perissodactyls are examples of **hindgut** (**monogastric**) fermenters. These have a simple stomach and have enlarged both the colon (the large intestine) and the cecum (Figure 20–9, left). Other hindgut fermenters include elephants, hyraxes, New World howler monkeys, wombats, koalas, rabbits, and many rodents. Some degree of hindgut fermentation is probably a primitive vertebrate character: it is known among birds, lizards, turtles, and fishes, as well as omnivorous and carnivorous mammals such as humans and dogs.

Cows and other ruminant artiodactyls are examples of **foregut** (**ruminant**) fermenters, in which the nonabsorptive forestomach is divided into three chambers that store and process food, followed by a fourth chamber in which digestion occurs (Figure 20–9, right). Ruminants are so called because they ruminate, or chew the cud (described later). Camels resemble other ruminants in many respects, but have only three-chambered stomachs (the omasum is lacking). A simpler type of foregut fermentation, without extensive stomach division or cud-chewing, is found in many other mammals, including Old

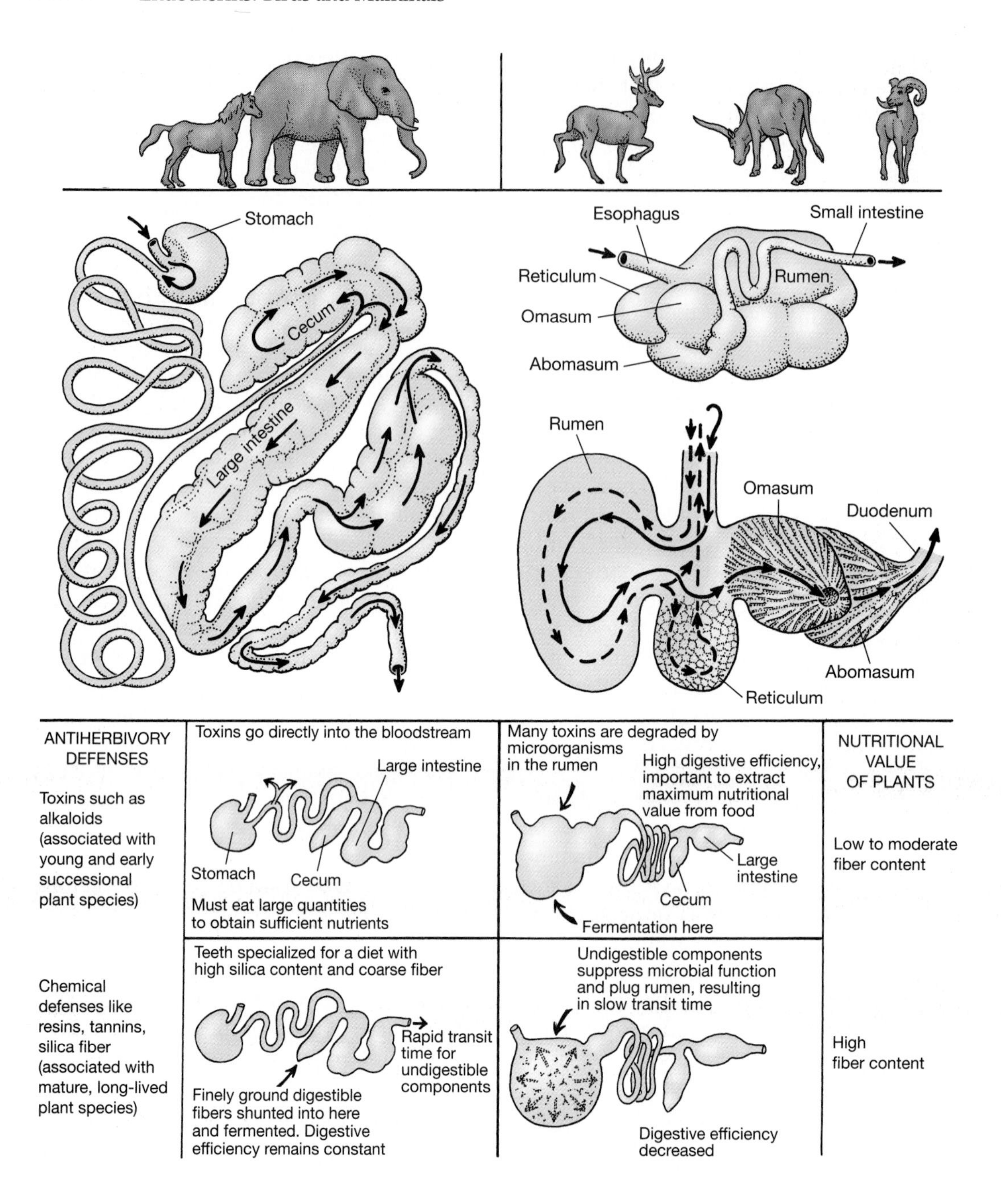

▲ Figure 20–9 Hindgut and ruminant foregut digestive systems. (Left) The hindgut fermenting system. Fermentation occurs in the enlarged cecum and colon (large intestine). (Right) The ruminant system, showing the four-chambered stomach.

World colobine monkeys, kangaroos, hippos, and some rodents.

Hindgut fermenters chew their food thoroughly, fracturing the plant cell walls with their teeth so that the cell contents are released. These cell contents are then processed and absorbed in the stomach and small intestine. The cellulose of the plant cell wall is not digested until it reaches the cecum and colon, where it is attacked by the symbiotic microorganisms. Cellulose is fermented to form substances known as volatile fatty acids, which are absorbed through the walls of the cecum and colon.

Some small hindgut fermenters, such as rabbits and rodents, ferment the food largely in the cecum and do not absorb much of the initial products of fermentation. Instead they rely on **coprophagy,** re-eating the first set of feces that are produced and recycling the nutrients. This digestive strategy would probably not work for a larger animal, which would not so easily be able to contort itself to ingest the feces as they emerge from the anus.

Ruminant foregut fermenters do not need to chew their food as thoroughly on initial mastication, since the cell walls will be chemically disrupted in the stomach. Ruminants do not have such extensive modifications of the skull and teeth as do hindgut fermenters. The food is initially retained in the two front chambers of the stomach, the **rumen** and the reticulum. Here microorganisms break down the cellulose, and the food is repeatedly regurgitated and rechewed (chewing the cud).

The limitation of passage from the reticulum to the third compartment, the omasum, is based on the particle size of the digested food. Thus this rechewing of the food acts in part to regulate the reduction of particle size and the passage of the food out of the fermentation chambers into the omasum (the function of which is not entirely clear), and the abomasum (the true stomach). As most or all of the cellulose has been broken down and absorbed before the true stomach, the digestive process of ruminants from this point on is like that of most other mammals. However, there may be a small amount of additional fermentation later on, in the moderately enlarged cecum.

Distinct advantages and disadvantages are associated with each of the two kinds of fermentative digestion. Foregut fermentation can be extremely efficient because the microorganisms attack the plant material *before* it reaches the small intestine, where most absorption takes place. In contrast, the food has already passed the small intestine of a hindgut fermenter before it is mixed with microorganisms in the cecum and colon, although if the teeth have done their work most of the cell contents have been digested and absorbed by this point. But this prestomach digestion can also be disadvantageous for ruminants. Not only is the cellulose fermented, but so are all the other food components, including free sugars and protein.

This protein fermentation actually ends up being beneficial to the ruminant, because they engage in a process termed **nitrogen cycling.** Microorganisms ferment the protein into ammonia, which is then taken via the circulation to the liver, where it is converted to urea. This urea is transported by the circulatory system to the rumen, where it is used by the microorganisms for their own growth. Surplus microorganisms overflow from the fermentation chambers into the abomasum, where they are digested. Thus all the protein that the ruminant ever actually gets to digest is microbial protein. An advantage of this system is that the microorganisms make all of the essential amino acids needed in the diet. Thus, a ruminant can be more limited in its selection of plant species than a hindgut fermenter, which must find all of its essential amino acids by eating a variety of plant sources.

An additional advantage of foregut fermentation is the role microorganisms in the rumen play in detoxifying chemical compounds that would be harmful to a vertebrate. A hindgut fermenter receives no such benefit and must absorb plant toxins into its bloodstream and transport them to its liver for detoxification.

On the other hand, a hindgut fermentation system processes material rapidly, whereas ruminants process food more slowly. Food moves through the gut of a horse in 30 to 45 hours, compared to 70 to 100 hours for a cow. Hindgut fermentation works well with food that has relatively high concentrations of fiber, because a large volume of food can be processed rapidly. The system is not efficient at extracting energy from the cellulose, but by processing a large volume of food rapidly, a horse can obtain a large quantity of energy from the cell contents in a short time.

In contrast, a ruminant foregut system is slow because food cannot pass out of the rumen until it has been ground into very fine particles. Ruminants do not do well on diets containing high levels of fiber, because this slows the passage rate of the food even further—the animal can literally starve to death with a stomach full of food. Ruminants are very efficient at extracting maximum amounts out of the cellulose of food of moderate fiber content, but they cannot process highly fibrous food, as can hindgut fermenters.

These differences in digestive physiology are reflected in the ecology of foregut and hindgut fermenters. Hindgut fermenters can survive on very low-quality food such as straw, as long as it is available in large quantities. Consequently, the feral horses in the American West can survive on land too poor for cattle to graze; these ruminants (cattle) are unable to process the low-quality food fast enough to subsist. In contrast, hindgut fermenters cannot survive so well in areas where the absolute quantity of food is limiting.

Ruminants are the main herbivores in places like the Arctic and deserts, where the food is of moderately good quality but severely limited in quantity; such food best supports an animal that can make the most efficient use of its dietary intake. Ruminants also have an advantage in desert conditions because of their nitrogen cycling. Because this cycle uses some of the waste urea from other sources, less urea needs to be excreted by the kidneys—thus, less water needs to be used to produce the urine. Ruminants are better able to go without water for a few days than are perissodactyls, and ruminants are more typically found in arid habitats. Ruminants may have become the dominant herbivores in the later Cenozoic because the increased seasonality resulted in vegetation where it was easier for a selective feeder to make a living.

20.5 Specializations for Locomotion

As we saw in Chapter 8, the specialized rapid mammalian gait is the bound, leaping from the hind feet to the fore feet with flexion of the backbone. The tree shrew illustrated in Figure 20–10 provides a good example of the skeletal structure of a small, generalized mammal adapted for a mixture of terrestrial and arboreal locomotion (**scansorial**), and shows how the bounding gait appears in a small mammal of this type. The earliest mammals had already lost the lumbar ribs, a condition inherited from their cynodont ancestors, and so potentially had the ability for this bounding type of motion. However, it seems likely that the type of bounding seen in the tree shrew was not evolved until the appearance of therians, with their distinctive modifications of the shoulder girdle (see Chapter 19).

Bounding is an asymmetric gait: That is, the movements of the two feet of a diagonal pair (fore and hind) are unevenly spaced in time. Cantering and galloping are modified versions of bounding employed by larger mammals. Other mammalian gaits, such as the walk and the trot, are symmetrical—the movements of the two feet of a pair are evenly spaced in time. (On the sound track for a Western a regular, *clip-clop-clip-clop* signifies a trotting horse, while an irregular *pattaboom-pattaboom* signifies a galloping one.) During bounding or galloping the hind limbs are used together to launch the animal into the air, and it lands on its front feet.

The tree shrew illustrates the mode in which the limbs and back are flexed during locomotion in small mammals. In larger mammals the limbs are held in a less flexed position and the back is less arched, due to the increased stress experienced on the bones by larger animals (Chapter 8). A mammal the size of a horse stands and moves with relatively stiff, straight legs, and gallops rather than bounds (see Figure 20–10c). In a mammal the size of an elephant, the legs are virtually pillar-like and the fastest gait is a rapid walk. Asymmetric gaits permit coupling of locomotion and respiration (the exact opposite of the problem initially facing early tetrapods in fast locomotion; see Chapter 17). As the animal leaps off the hind legs, the liver and other viscera are pushed backward, helping with the expansion of the thorax in inspiration. When the animal lands on its fore legs the viscera are pushed forward onto the diaphragm, helping with expiration (see Figure 17–10).

Specialized Forms of Locomotion

The basic form of mammalian locomotion is the bounding and scrambling of small, scansorial mammals such as tree shrews and squirrels. However, larger animals experience the world differently from smaller ones because of physical size and scaling effects, and larger mammals are usually modified for more specialized forms of locomotion (see Chapter 8). Figure 20–11 contrasts the specializations of running (**cursorial**) mammals and digging (**fossorial**) mammals. (Arboreal specializations of mammals can be seen in the primates discussed in Chapter 23.)

Cursorial Limb Morphology Cursorial mammals need to maximize their stride length and have thus elongated their legs. The cost of locomotion is in the number of strides it takes to travel a given distance, and a long-legged mammal will cover the distance in fewer strides than a shorter-legged one. Long legs also provide a long outlever arm for the major locomotor muscles, such as the triceps in the forelimb and the gastrocnemius in the hindlimb.

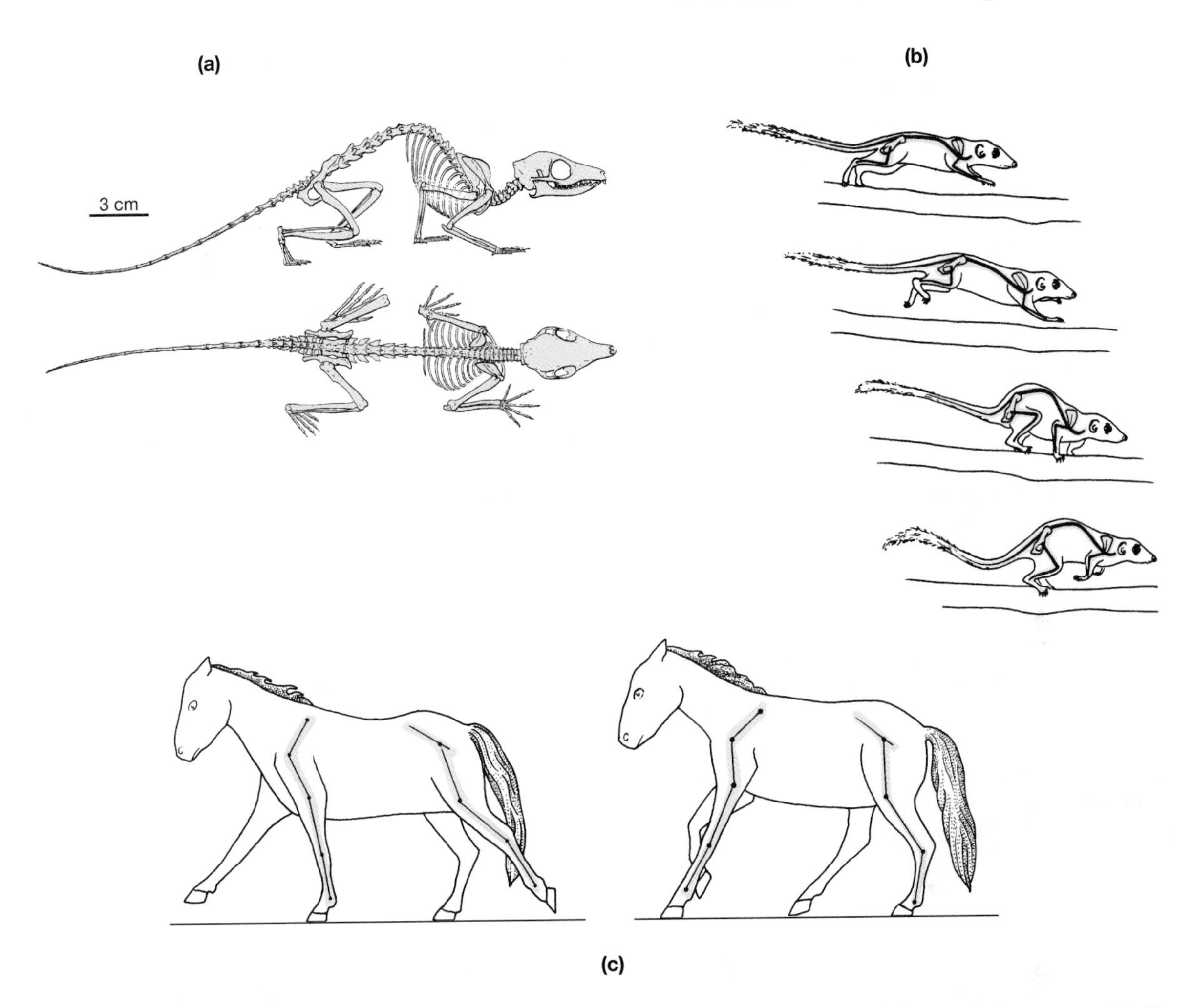

▲ Figure 20–10 Gait and locomotion in mammals. (a) Skeleton of tree shrew (*Tupaia glis*) in typical posture with flexed limbs and quadrupedal stance. (b) Sequential phases of the bounding run in a tree shrew. Note the relatively flexed limbs and mobile back. (c) Sequential phases of the gallop in a horse. Note the relatively straight limb angles and immobile back.

This arrangement favors speed of motion rather than power. Imagine using a ruler resting on a pencil to flip a coin at the ruler's tip into the air. You could flip the coin faster if most of the length of ruler was on the side of the coin, and only a short portion was on the other side of the pencil for you to use as a lever. This is similar to the limb setup in Figure 20–11b, left. Here the olecranon process of the ulna (the portion of the elbow onto which the triceps muscle attaches) is the short lever arm (the part of the ruler that you would push down on), and the whole limb from the elbow on down is the long lever arm (the part of the ruler on which the coin would be balanced).

Additionally, cursorial mammals have made their legs long by elongating only certain portions of the limb, not each segment. They elongate primarily the lower portions of the limb, the radius and ulna (in the forelimb) or the tibia and fibula (in the hindlimb) and the metapodials. The humerus and femur are not elongated, nor are the phalanges. Contrast this with our own long hindlimbs, where the femur is elongated but the whole foot (metapodials plus phalanges) is short. Along with this limb lengthening is a change in foot posture, from the primitive, flat on the ground **plantigrade** posture (as in ourselves) to a tip-toe **digitigrade** foot posture, as in dogs and cats (Figure 20–11a).

Ungulates go even further with this type of foot posture and adopt an **unguligrade** stance, which would be equivalent to a ballerina standing on point.

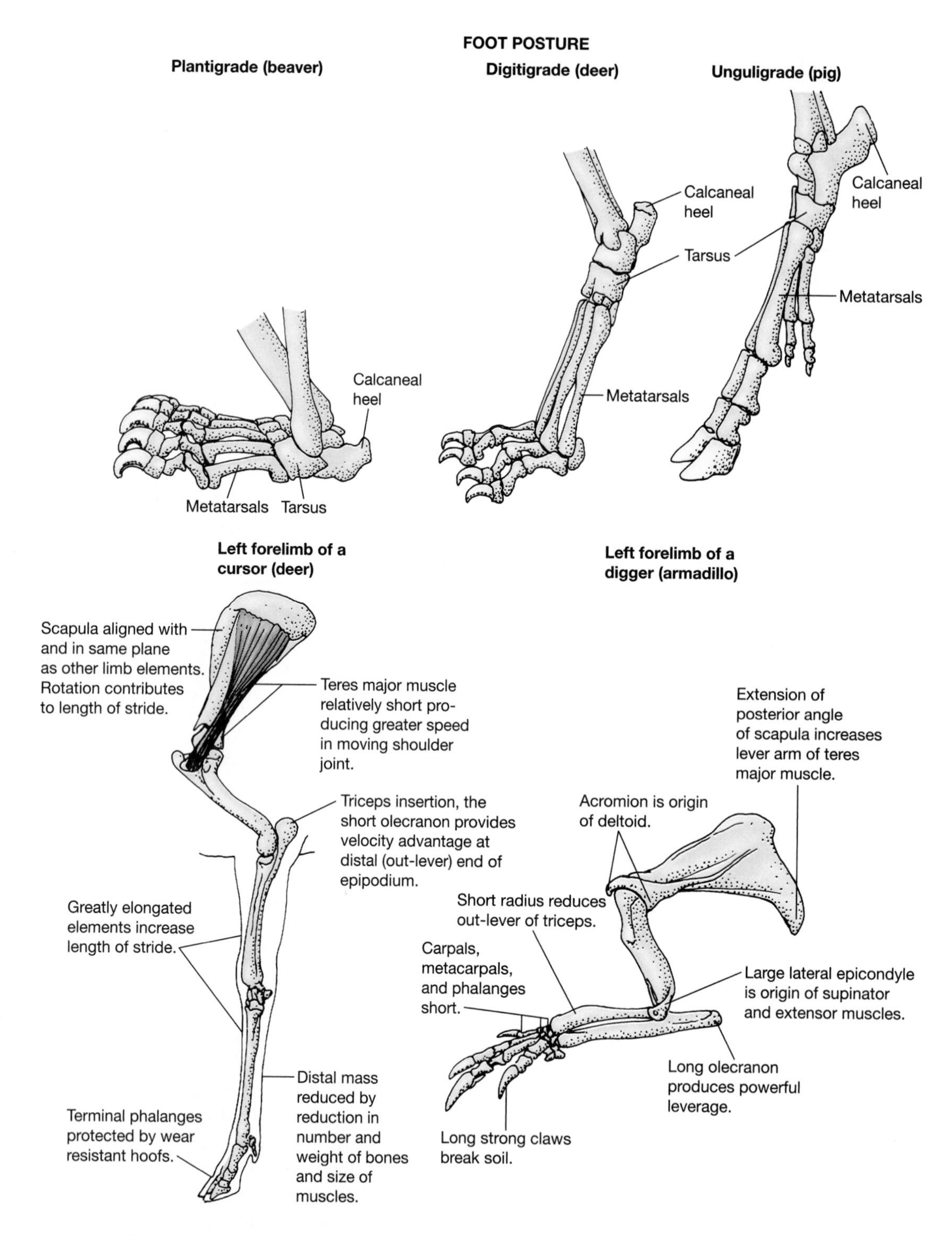

▲ Figure 20–11 Contrasting specializations of the limbs of mammals.

Lengthening the lower portion of the limb also allows the muscles to be limited to the proximal portion of the limb and attached to long elastic tendons that insert distally. There is almost no muscle in a horse's leg below the wrist (the horse's so-called knee joint) or the ankle (the hock). These tendons allow storage of elastic energy during locomotion and contribute to locomotor efficiency. Additionally, locating the muscles in the proximal part of the limb, close to the body, means that the lower part of the limb is lighter, and so less mass has to be accelerated distally with each stride.

The most obvious example of elastic storage in leg tendons is a kangaroo hopping, where the animal bounces on landing as if using a pogo stick. However, all cursorial animals rely on energy storage in tendons in gaits faster than a walk (see Figure 20–12), and extension of these tendons may be part of the evolutionary reason for limb elongation and changes in foot posture. Even humans rely on elastic energy storage in tendons, especially in the Achilles tendon that attaches the gastrocnemius muscle to the calcaneal heel. People who have damaged their Achilles tendons (a common sports injury) find running difficult or impossible.

Other cursorial modifications include ones that restrict the motion of the limb to a fore-and-aft plane, so that most of the thrust on the ground contributes to forward movement. The clavicle is reduced or lost, and the wrist and ankle bones are relatively immobile. (Note how easily you can turn your hand at the wrist so that your palm faces upward; a dog can't do that, and a horse has almost no ability to rotate this joint.)

The number of digits may also be reduced, perhaps to lighten the weight of the limb so that it can be accelerated and stopped more easily during striding (the higher position of the muscles also results in a lighter lower limb). Carnivores often lose digit 1, but otherwise compress the digits together rather than reduce them in number. Artiodactyls reduce or lose digits 1, 2, and 5, becoming effectively four-toed like a pig or two-toed like a deer (see Figure 20–11b, left). Perissodactyls lose digits 1 and 5, and reduce digits 2 and 4, becoming three-toed like a rhino, or single-toed like a horse (see Figure 20–13).

The evolution of cursorial specializations in mammals is an interesting story. No early Cenozoic (Paleocene) mammal showed cursorial specializations,

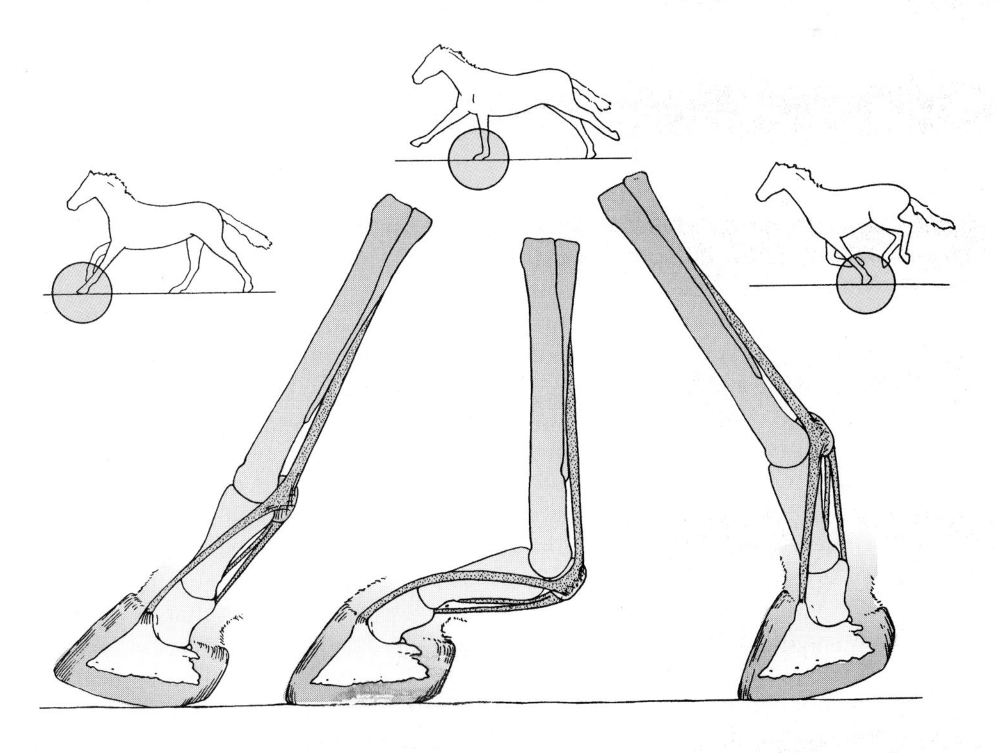

▲ Figure 20–12 Springing action of the energy-storing tendons (suspensory ligaments) in the foot of a horse.

and early members of the Carnivora and the ungulate orders of the Eocene were also fairly unspecialized. However, during the later Cenozoic we find these mammals becoming larger and developing a more cursorial form of limb morphology. Such cursorial specializations occurred convergently among many different ungulate lineages (see Box 20-1). As ancient ungulates and carnivorans were noncursorial, whereas their modern descendants do have cursorial specializations, it has long been assumed that these specializations, especially the evolution of longer legs, must have arisen in the context of predator-prey relationships. Longer legs would have given a carnivore a little more speed to pursue the herbivore, resulting in selection for ungulates with longer legs to make a faster escape.

Although most living carnivores are ambush predators that do not pursue their prey for more than a few yards, today we do have some pursuit predators, such as wolves and spotted hyenas, that hunt in groups. Cheetahs pursue their prey but are not pack hunters. Lions hunt in groups, but do not pursue their prey for a long distance. Only pursuit predators have highly cursorial limb morphologies. Think of the difference in limb length between a dog (long-limbed pursuit predator) and a cat (shorter-limbed ambush predator).

This evolutionary scenario of a coevolutionary arms race between predator and prey is an appealing one, but the fossil record does not support it. If coevolution were the driving force, ungulates and carnivores would have evolved their longer limbs at the same time, in lockstep fashion, but this is not the case. For example, ungulates living in North America evolved longer limbs and other cursorial specializations in the early Miocene, some 20 million years before the fossil record yields evidence of carnivores built like present-day pursuit predators in the Pliocene, around 5 million years ago.

Why, then, did ungulates evolve cursorial specializations if not to flee predators? It appears that all of the limb modifications that make a mammal a faster runner will also make it more efficient at slower gaits, such the trot. The early Miocene is the time when habitats in North America started to turn from productive woodlands to less productive grasslands, meaning that ungulates would have to forage further each day for food. Supporting evidence for this hypothesis comes from looking at ungulate evolution on other continents. In Eurasia, where the transition from woodlands to grasslands happened later in the Miocene, the cursorial specializations of ungulates also appeared later, contemporaneous with the change in vegetation. In South America, where the vegetation changed in the late Oligocene, cursorial ungulates appeared earlier, again reflecting the changing habitats. But carnivores with a pursuit type of cursorial morphology first appeared on all continents only five million years ago—15 to 25 million years later than their cursorial prey.

Thus ungulate cursorial specializations may have been evolved to allow foraging over long distances rather than for predator escape. Cursorial limb morphology might make fast running possible, but that is probably not why it was evolved in the first place. Cursorial specializations of carnivores may also have evolved to boost stamina at slower speeds, in the context of traversing a home range to obtain food, with the capacity for fast running being an unexpected benefit. Even the highly cursorial pack-hunting carnivores may have originally evolved their limb morphologies for stamina, perhaps following migratory herds in the highly seasonal habitats that first appeared in the Pliocene.

A final unresolved issue is why ungulates appear to be more specialized for cursorial locomotion than carnivores. A horse has legs that are more elongated, and a more derived foot stance than a dog; the horse also has lost all of the toes but one and has much more restricted movement at the limb joints. Yet horses are not significantly faster runners than dogs; if this were true, foxhunting on horseback with hounds would be impossible. Perhaps the limbs of carnivores are subject to multiple selection pressures because they must use their forelimbs to aid in killing and digging. One might expect that because carnivores had fewer limb modifications for fast running than ungulates, they would show more extensive modifications of other systems, such as the heart and the lungs; but fast-running ungulates also have enlarged hearts and lungs.

The answer may lie in the digestive physiology of ungulates. The gut and gut contents of an ungulate may comprise up to 40 percent of the total body mass (more in a ruminant than in a horse). Thus, for any given body weight, an ungulate has proportionally less muscle mass than a carnivore of the same size, because so much of its mass is composed of the gut. It may be that ungulates have to be more modified in the skeleton than carnivores to compensate for this difference in relative muscle mass in proportion to total body weight.

BOX 20-1 Convergence of Ungulate Specializations

Cursorially adapted ungulates with skull and dental specializations for grinding fibrous food evolved convergently many times, on several different continents. Among marsupials kangaroos also converge on ungulates in some features, although of course they are bipedal hoppers rather than quadrupdal runners.

Figure 20–13 shows convergence in the limbs and the teeth between horses, perissodactyls evolving in North America, and litopterns—an extinct order of ungulates only distant related to any living ungulate, evolving in South America. By the early Miocene (around 20 million years ago), both litopterns (*Diadiaphorus*) and horses (*Hypohippus, Miohippus,* and *Protohippus*) had evolved lophed teeth and fairly elongated limbs with a tridactyl foot (a reduction in the number of toes to a main one and two other accessory ones). However, the late Miocene *Thoatherium,* living around 10 million years ago, shows a precocious evolution of a fully monodactyl (one-toed) condition that was not attained by horses until genera such as *Equus* (the extant genus of horse) first appeared in the Pliocene, some 5 million years later.

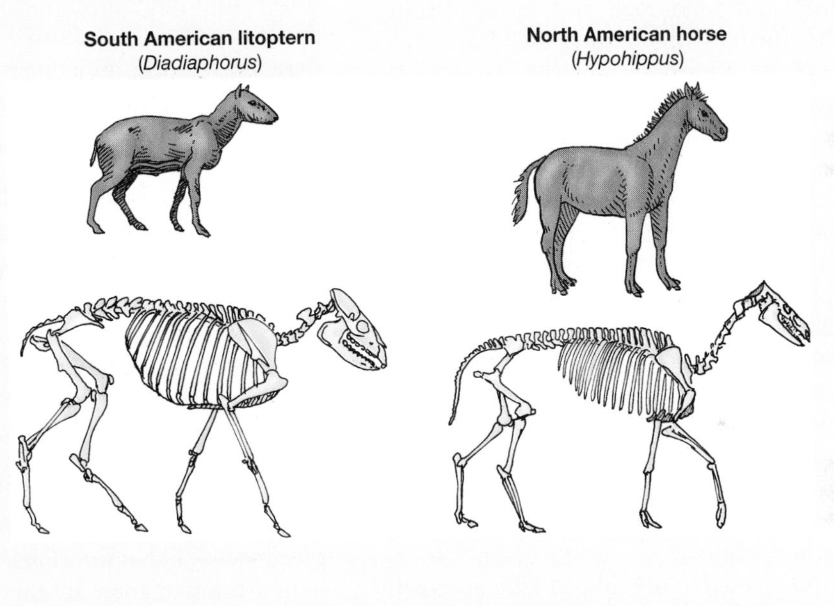

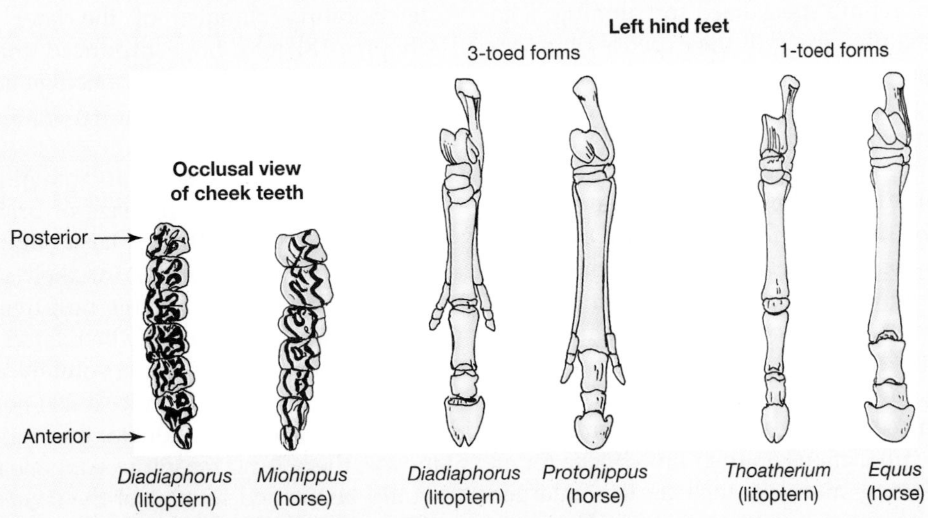

▲ Figure 20–13 Convergence between unrelated ungulates from different continents.

Fossorial Limb Morphology Mammals do several types of digging. The most common is scratch digging at the surface, as seen in anteaters and armadillos. This is the type of digging that a dog uses to bury a bone. Animals that are truly fossorial (i.e., burrowing under the ground surface) have a variety of specializations. No mammal is limbless and elongated like a burrowing lizard or caecelian, although mammals that follow their prey down burrows (such as ferrets and dachshunds) have elongated bodies and rather short legs.

Some subterranean diggers are called rapid scratch diggers. These animals, such as the African golden mole and the Australian marsupial mole, dig with both fore and hind feet, and move through sandy soil by pushing the grains aside and back without constructing an open burrow. True moles, living in more compact soil, rotate their forelimb to the earth rather than simply retracting it, a form of digging termed rotation thrust digging. Moles burrow just below the surface of the ground, seeking worms and insect larvae in the roots of plants, and push the soil upward as they tunnel. Finally, many rodents (gophers are an example) use their evergrowing incisors to dislodge soil rather than their limbs, which is termed chisel-tooth digging. Gophers have a pronounced diastema and can pull their lips together behind the incisors to prevent soil from entering the mouth. Gophers push the soil they excavate out onto the surface of the ground, forming the mounds that are a familiar feature of the landscape wherever gophers occur.

A limb specialized for digging is almost exactly the opposite of a limb specialized for running. Running limbs maximize speed at the expense of power, whereas digging limbs maximize power at the expense of speed. In this case the thought experiment with the ruler would be in using it to lever up a heavy object. Now you would want to put the short end of the ruler under the object, and have a long lever arm at the other end for you to push down on, as with a crow bar. Fossorial mammals achieve this mechanical design in the forelimb with a long olecranon process and a relatively short forearm (Figure 20–11, right).

Digging mammals retain all five digits, tipped with stout claws for breaking the substrate. They also have large projections on their limb bones for attachment of strong muscles, such as the enlarged acromion process on the scapula. Scratch diggers at the surface, such as anteaters, have a very stout pelvis, with many vertebrae involved in the sacrum, for bracing the hindlimb while digging with the forelimb. However, underground burrowers such as moles do not have this type of strengthened pelvis and sacrum.

20.6 Specializations of the Sensory Systems

The brain of living mammals is distinctive in the enlargement of the area of the telencephalon called the **neopallium.** As discussed in Chapter 19, mammals differ from other amniotes in their poor color vision, and most living mammals are much more dependent on hearing and olfaction than on vision, both for perceiving their environment and for communication.

Evolution of Mammalian Brain Size

Both mammals and birds have brains that are several times larger than those of ectothermal tetrapods. Some late Cenozoic mammals have evolved especially large brains, with this trend occurring in parallel among a variety of mammalian lineages. Although we tend to think of the tendency to evolve a large brain as a natural outcome of mammalian evolution, it is not at all clear precisely what the adaptive advantage has been.

Large brains do appear to be associated with complex social behavior. Pack-hunting carnivores are bigger-brained than solitary ones, for example. The deteriorating climates of the later Cenozoic (see Chapter 18) may have produced conditions where smarter mammals may have had an adaptive advantage, but many small-brained mammals survived these conditions as well.

Evaluating how brainy an animal is can be problematical. For a start, the size of brains scales with negative allometry; that is, larger animals have proportionally smaller brains for their size than small animals. The reason for this negative scaling is not known for certain, but may be related to the fact that bigger bodies don't need absolutely more brain tissue to operate. For example, if 200 nerve cells in the brain were needed to control the right hind leg of a mouse, there is no reason to suppose that more cells would be needed to control the right hind leg of an elephant. Of course, larger animals have brains that are absolutely larger than smaller ones; they are just not quite as large as one would predict for their size, all other things being equal. Comparisons of brain size often neglect differences in body size. For example, the brains of dolphins are just as big as human brains, but most dolphins are much bigger than humans and they would be expected to have larger brains.

To get around the effect of body size, the size of brains is usually estimated in relative terms. One such estimate is the **encephalization quotient (EQ),** a measure of the actual brain size in relation to the expected brain size for an mammal of that body mass (including all mammals, primates and others, in the equation of determining the average figure; see Jerison 1973). The average EQ of an extant nonprimate mammal is 0.5. Insectivores, rodents, and marsupials mostly have EQ values of less than 1, while primates and whales have values of greater than 1 (great apes have values of about 3, while humans have a value of about 8). Some carnivorans and ungulates (e.g., dogs, cats, horses, and elephants) also have EQ values of greater than 1.

Figure 20–14 shows a plot of the changes of EQs of ungulates over time, and plots for other mammalian orders would have similar shapes. Although average brain size has increased over time, and the largest brains known belong to living ungulates, some living ungulates have brains as small as some of the early Cenozoic forms. In other words, the range of brain sizes has increased rather than the brains of all mammals increasing in size.

Specializations for Color Vision

It is a common misconception that mammals other than primates are color-blind: Older books depict dogs and horses seeing the world as if on a black-and-white television set. While a retina comprised mainly of rods is probably the primitive mammalian condition—as seen in many extant small, nocturnal mammals such as insectivorans—all mammals, even monotremes, have some cones in their retinas. Thus it is likely that at least a rudimentary ability to distinguish color is present in all mammals, possibly inherited from their ancestry with other amniotes. Some diurnal mammals, such as squirrels, have retinas that are densely packed with cones. The ability to distinguish colors also is clearly present in many other mammals, including dogs, cats, and horses (Jacobs 1993), although nonprimate mammals may not attach the sensory significance to color that humans do.

Humans and some other anthropoid primates (apes and Old World monkeys) have three kinds of color receptor cones, each registering for a different waveband of light; thus we can perceive three primary colors (red, green, and blue). The refined color vision of anthropoid primates probably relates to their diurnal habits, and may compensate for their relatively poor sense of olfaction. In contrast most other mammals only have two color receptor cones—they see the world more like a person with red-green color blindness. Our sense of color vision pales beside that of birds, which have four kinds of color receptor cones and can also see farther into the ultraviolet than we can. Colorful as birds' plumage appears to us, we must see only a pallid subset of the colors with which birds perceive each another.

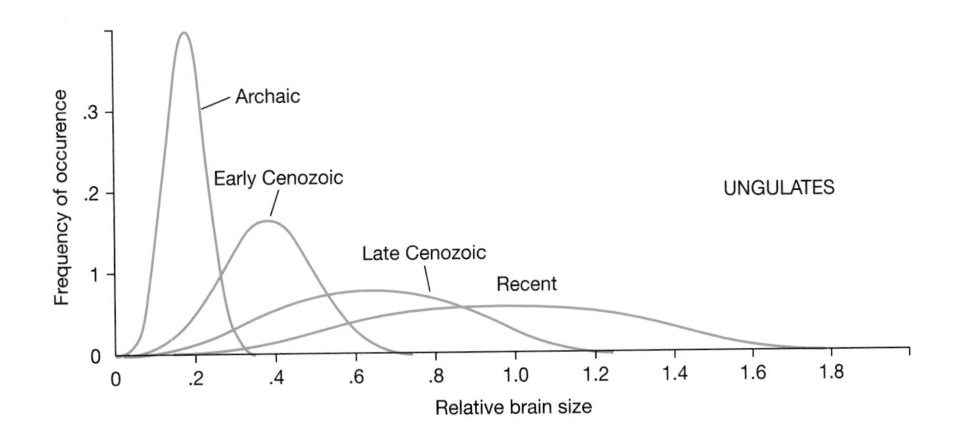

▲ Figure 20–14 Changes in the relative brain size of Northern Hemisphere ungulates during the Cenozoic.

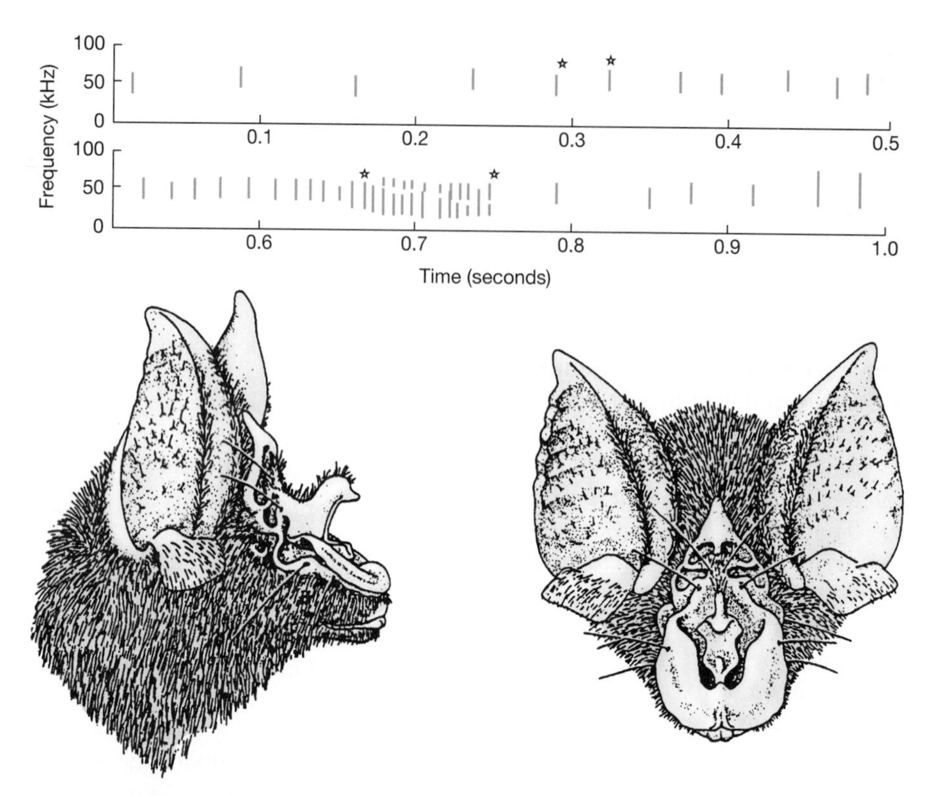

▲ Figure 20–15 Bat echolocation sounds and modifications of the nose to broadcast them. The sound spectrograph analysis is of the frequency modulated (FM) pulses emitted by the little brown bat, *Myotis lucifugus,* during an interception maneuver. Frequency in kilohertz is plotted against time during the continuous 1-second record. Filled stars indicate typical loud pulses near the time of detection of the target; open stars indicate the onset and completion of the terminal buzz just before capture of prey.

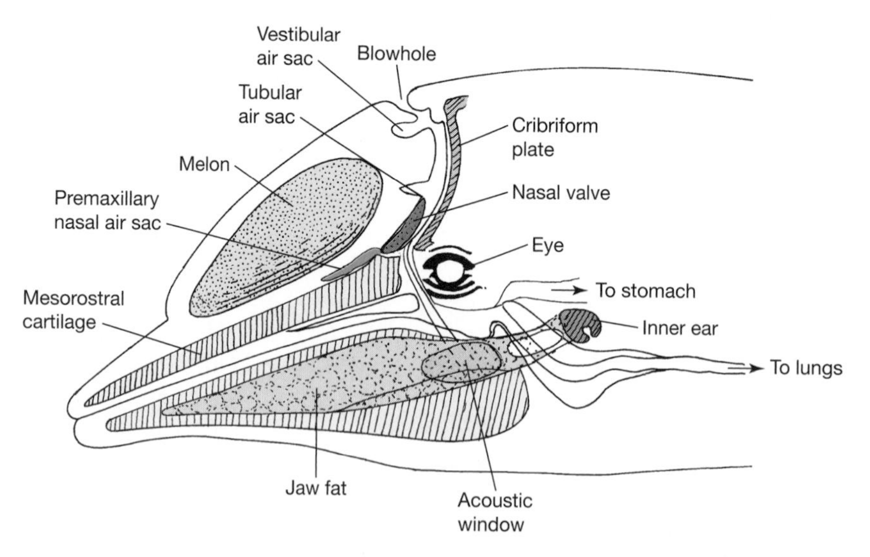

▲ Figure 20–16 Odontocete echolocation. Sound production and reception apparatus shown for the bottle-nosed dolphin (*Tursiops truncatus*). Sound produced by air shuttled between air sacs through the nasal valve is focused by the oil of the melon into a forwardly directed beam. Some sound may also be guided by the mesorostral cartilage. Returned sound is channeled through the mandibular (jaw) fat bodies and especially the acoustic window on the lower jaw to the otherwise isolated and fused middle and inner ears.

Specializations for Hearing

Audition has several advantages as a distance sense compared with vision and olfaction. Sound is not readily blocked by obstacles in the environment (as is light), and it is transmitted more directionally and much faster than odors. The therian mammalian ear shows the importance of audition, with the extensively coiled cochlea in the inner ear and the pinna or outer ear. The combined effect of these specialized structures is exceptional frequency discrimination, broad sensitivity to various intensities of sound, and precision of directionality. Cranial casts of early mammals show that auditory areas of the brain were large.

Bats and cetaceans rely on hearing as their primary distance sense for navigation and location of prey. An examination of **echolocation** illustrates how a sensory capacity that is ancestral for mammals can be elaborated under certain environmental conditions. Several mammals emit sounds above 20 kilohertz (20,000 cycles per second), called **ultrasound** because it is above the range of normal human sensitivity (the 10 octaves between 20 Hz and 20 kHz). Elephants emit sounds below the range of human hearing (**infrasound**) in connection with group movements and reproductive receptivity. When infant rodents stray outside their nest, they emit ultrasounds that stimulate adults to retrieve them. Some adult rodents, a few marsupials, dermopterans, pinnipeds, many insectivores, microchiropteran bats, and odontocete cetaceans (toothed whales) emit such sounds throughout their life as part of a sound-based distance-sensing system of echolocation.

The most thoroughly studied echolocating mammals are the microchiropteran bats and the toothed cetaceans (e.g., porpoises), which use this sensory modality to locate prey under conditions where vision would not be appropriate. (The types of whales whose songs you can buy on records are the filter-feeding baleen whales, which do not produce ultrasound, probably because they do not need to target moving prey.) Bats hunt for insects at night, and toothed whales hunt for fish and other marine animals in murky waters. The high-frequency sounds emitted bounce off surrounding objects, including prey items, and the sound received back in the animal's ear can translate these reflected sound waves into information about spatial relationships. Bats and cetaceans differ in their production and reception of ultrasound because they operate in different media. In some ways bats are less specialized than cetaceans, because air is the original medium for the sensory modality of hearing.

We ourselves have a certain capacity to use sound to form a picture of our surroundings. For example, you can normally tell by shaking a bottle of pills whether it is full or nearly empty, or whether it contains small or large pills, and your voice echoes in an empty room but not in a room full of furniture. Blind people, of course, have greatly increased their sensitivity to being able to use sound to determine spatial relationships in their environment.

Bats produce a stream of ultrasound from their larynx, which is enlarged but not greatly modified from the general mammalian condition. The sounds are emitted through the mouth or the nose, which often have highly complex folds and flaps to concentrate the sound, giving microchiropteran bats their typically gargoyle-like faces (see Figure 20–15). In contrast the megachiropteran fruit bats, which do not echolocate, have rather fox-like faces, and these bats are often known as flying foxes. (Ironically, it is these harmless fruit bats that usually play the role of microchiropteran vampire bats in horror movies, because of their large size and fox-like faces.) The external ears of microchiropteran bats are also large and complexly shaped to receive the ultrasound.

Cetaceans have a problem that bats do not; because sound travels faster in water than in air, sound waves for any given frequency are longer in water than in air. Short sound waves are needed for echolocation, and to produce short waves cetaceans emit clicks that are of much higher frequency than those emitted by bats. The ultrasound is not produced through the larynx, because there is only a limited amount of air in the lungs that cannot be refilled underwater. Instead, sound is produced in the nasal passages and focused through an oil-filled body on the forehead called the melon (see Figure 20–16).

There is also a problem with the reception of sound in water. The middle ear of mammals is designed to carry sound from an air-air interface at the eardrum. When you put the mammalian ear in water, sound bounces off the eardrum; to make things worse, sound readily enters the body everywhere else. That is why you can hear only a diffuse roaring sound when you are underwater and cannot localize the source of the sound. To solve these problems, cetaceans isolate the inner ear from the rest of the skull with special sound-absorbing tissues, and sound is transmitted to the inner ear directly via a fat body that runs alongside the lower jaw.

Summary

The major groups of living mammals can be distinguished by their mode of reproduction. Therians (marsupials and eutherians) are more derived than monotremes, giving birth to live young (viviparity) as opposed to laying eggs. Marsupials give birth to very immature young that complete their development attached to the nipples usually, but not always, enclosed in a pouch. Eutherians give birth to more mature young and have a shorter period of lactation than marsupials. The marsupial mode of reproduction has often been considered inferior to the eutherian one, but the differences may reflect only their separate evolutionary histories, with neither method being inherently superior to the other.

Cenozoic mammals have diversified into a variety of feeding types, reflected in different anatomies of their skulls and dentitions. The skulls and teeth of herbivores are in general more specialized than those of omnivores and carnivores because fibrous plant material is difficult to chew and abrades the teeth. Vegetation is also more difficult to digest than other diets, and many herbivores have evolved a symbiotic association with microorganisms in their guts, which ferment the plant fiber and aid in its chemical breakdown.

The teeth of herbivorous mammals are particularly specialized. In order to shred tough plant material the original cusps have been run together into cutting blades, or lophs. Also, herbivorous mammals have to ensure that their teeth last a lifetime, a problem all mammals share since they replace their teeth only once, and the molars are not replaced at all. The most common way is to make the teeth high-crowned, or hypsodont, although different solutions exist in some mammals, such as elephants and manatees. Carnivorous mammals usually have a pair of specialized blade-like cutting teeth, the carnassials.

The primitive mammalian mode of locomotion is probably some sort of bounding. With the radiation of larger mammals in the Cenozoic, more specialized types of locomotion evolved. Cursorial (running) mammals have elongated their legs and changed their foot posture, restricting the range of limb motion and sometimes reducing the number of digits. Although such specializations bestow the ability for fast running, and long limbs are assumed to have evolved in a coevolutionary fashion in predator and prey, ungulates in fact evolved their longer legs many millions of years before carnivores did. An alternative explanation for the evolution of cursorial specializations are for stamina during foraging. Fossorial limbs, designed for digging, are short and stout with heavy muscles and large claws. Fossorial specializations have evolved convergently in several lineages of mammals, including aardvarks, armadillos, moles, and rodents.

Many mammals—including ungulates, carnivorans, whales, and primates—have evolved larger brains over the course of the Cenozoic, but many small-brained mammals persist today. Contrary to the popular notion that nonprimate mammals are color-blind, some sort of color vision probably characterizes all mammals—but only Old World monkeys and apes share with humans the capacity for trichromatic color vision. Hearing was probably always a highly important mammalian sense, and the capacity to emit and perceive ultrasound (echolocation) may well be a primitive mammalian capacity. The sensory modality of echolocation is best developed in microchiropteran bats and in toothed whales such as porpoises.

Additional Readings

Alexander, R. M. 1982. *Locomotion of Animals.* Glasgow: Blackie.

Alexander, R. M. 1992. *Exploring Biomechanics: Animals in Motion.* New York: Scientific American Library.

Au, W. W. L. 1993. *The Sonar of Dolphins.* New York: Springer.

Biewener, A. A. 1989. Mammalian terrestrial locomotion and size. *BioScience* 39 (11):776–783.

Frank, L. G. 1996. Female masculinzation in the spotted hyena: Endocrinology, behavioral ecology, and evolution. In J. L. Gittleman (Ed.), *Carnivore Behavior, Ecology, and Evolution,* volume 2. Ithaca, N.Y.: Comstock Publishing Associates, 78–131.

Hildebrand, M. 1995. *Analysis of Vertebrate Structure,* 4th ed. New York: Wiley.

Jacobs, G. H. 1993. The distribution and nature of colour vision among the mammals. *Biological Reviews* 68:413–471.

Janis, C. M. 1976. The evolutionary strategy of the Equidae and the origins of rumen and cecal digestion. *Evolution* 30:757–774.

Janis, C. M., and M. Fortelius. 1988. On the means whereby mammals achieve increased functional durability of their dentitions, with special reference to limiting factors. *Biological Reviews* 63:197–230.

Janis, C. M. and P. B. Wilhelm. 1993. Were there mammalian pursuit predators in the Tertiary? Dances with wolf avatars. *Journal of Mammalian Evolution* 1:103–126.

Jarvis, J. U. M. 1981. Eusociality in a mammal: Cooperative breeding in the naked mole rat. *Science* 212:571–573.

Jerison, H. J. 1973. *Evolution of the Brain and Intelligence*. New York: Academic Press.

Kirsch, J. A. 1977. The six percent solution: Second thoughts on the adaptedness of the Marsupialia. *American Scientist* 65:276–288.

Lillegraven, J. A., et al. 1987. The origin of eutherian mammals. *Biological Journal of the Linnean Society* 32:281–336.

Mossman, H. W. 1987. *Vertebrate Fetal Membranes*. New Brunswick, N.J.: Rutgers University Press.

Norris, K. S. 1974. *The Porpoise Watcher*. Toronto: George J. McLeod.

Parker, P. 1977. An ecological comparison of marsupial and placental patterns of reproduction. In B. Stonehouse and D. Gilmore (Eds.), *The Biology of Marsupials*. London: MacMillan Press, 273–286.

Renfree, M. B. 1993. Ontogeny, genetic control, and phylogeny of female reproduction in monotreme and therian mammals. In F. S. Szalay, M. J. Novacek, and M. C. McKenna (Eds.), *Mammal Phylogeny, Mesozoic Differentiation, Multituberculates, Monotremes, Early Therians, and Marsupials*, New York: Springer Verlag, 4–20.

Renfree, M. B. and G. Shaw. 2000. Diapause. *Annual Review of Physiology* 62:353–375.

Renfree, M. B., and R. V. Short. 1988. Sex determination in marsupials: Evidence for a marsupial-eutherian dichotomy. *Philosophical Transactions of the Royal Society, London, B* 322:41–53.

Sharman, G. B. 1976. Evolution of viviparity in mammals. In C. R. Austin and R. V. Short (Eds.), *Reproduction in Mammals, 6: The Evolution of Reproduction*. Cambridge, U.K.: Cambridge University Press, 32–70.

Smith, K. K. 1997. Comparative patterns of craniofacial development in eutherian and metatherian mammals. *Evolution* 51:1663–1678.

Tyndale-Biscoe, C. H., and M. B. Renfree. 1987. *Reproductive Physiology of Marsupials*. Cambridge, U.K.: Cambridge University Press.

Vaughan, T. A., J. M. Ryan, and N. J. Czaplewski. 2000. *Mammalogy*, 4th ed. Philadelphia: Saunders.

Werdelin, L., and A. Nilsonne. 1999. The evolution of the scrotum and testicular descent in mammals: A phylogenetic view. *Journal of Theoretical Biology* 196:61–72.

Web Explorations

On-line resources for this chapter are on the World Wide Web at http://www.prenhall.com/pough (click on the Table of Contents link and then select Chapter 20).

CHAPTER 21

Endothermy: A High-Energy Approach to Life

Endothermy is a derived character of mammals and birds. The two lineages evolved endothermy independently, but the costs and benefits are the same for both. Active predation was probably the initial step toward endothermy (see Chapters 15 and 17). Sustaining activity requires the high levels of aerobic metabolism characteristic of endotherms. Although ectotherms can achieve high levels of activity for short periods, they rely on anaerobic metabolism to do this and become exhausted in a few minutes. Only endotherms can sustain activity at high levels for prolonged periods.

Once high rates of metabolism had evolved, covering the body with a layer of insulation would allow an animal to use metabolic heat production to maintain a high body temperature. Endothermy is a superb way to become relatively independent of many of the challenges of the physical environment, especially cold. Birds and mammals can live in the coldest habitats on Earth, provided that they can find enough food. That qualification expresses the major problem of endothermy: It is energetically expensive. Endotherms need a reliable supply of food and the conspicuous interactions of endotherms are often with their biological environment—predators, competitors, and prey—rather than with the physical environment as is often the case for ectotherms. Because energy intake and expenditure are important factors in the daily lives of endotherms, calculations of energy budgets can help us to understand the consequences of some kinds of behavior.

When all efforts at homeostasis are inadequate, endotherms have two more methods of dealing with harsh conditions: (1) Birds and large species of mammals can migrate to areas where conditions are more favorable. (2) Many species of small mammals and some birds can become torpid. This response, a temporary drop in body temperature, conserves energy and prolongs survival—at the cost of abandoning the benefits of homeothermy.

21.1 Endothermal Thermoregulation

Birds and mammals are endotherms. They regulate their high body temperatures by precisely balancing metabolic heat production with heat loss. An endotherm can change the intensity of its heat production by varying metabolic rate; it can change heat loss by varying insulation. In this way, an endotherm maintains a constant high body temperature by adjusting heat production to equal heat loss from its body under different environmental conditions.

Endotherms produce metabolic heat in several ways. Besides the obligatory heat production derived from the basal or resting metabolic rate, there is the heat increment of feeding, often called the **specific dynamic action** or **specific dynamic effect** of the food. This added heat production after eating apparently results from the energy used to assimilate molecules and synthesize protein, and it varies depending on the type of foodstuff being processed. It is highest for a meat diet and lowest for a carbohydrate diet.

Activity of skeletal muscle produces large amounts of heat. This is especially true during locomotion, which can result in a heat production exceeding the basal metabolic rate by 10- to 15-fold. This muscular heat can be advantageous for balancing heat loss in a cold environment. It can also be a problem requiring special mechanisms of dissipation in hot environments that approach or exceed the body temperature of the animal. Cheetahs, for example, show a rapid increase in body temperature when they chase prey, and it is usually overheating that causes a cheetah to break off a pursuit. **Shivering,** the generation of heat by muscle-fiber contractions in an asynchronous pattern that does not result in gross movement of the whole muscle, is an important mechanism of heat production.

Endotherms usually live under conditions in which ambient temperatures are lower than the regulated body temperatures of the animals themselves. Heat loss to the environment is thus a more usual circumstance than heat gain. Balancing heat production and heat loss is one of the most important regulatory functions of an endotherm.

Any material that traps air is an insulator against conductive heat transfer. Hair and feathers provide insulation by trapping air, and a mammal or bird can adjust the depth of the layer of trapped air by raising and lowering the hair or feathers. We humans have goose bumps on our arms and legs when we are cold because our few remaining hairs rise to a vertical position in an ancestral mammalian attempt to increase our insulation.

These physiological responses to temperature are controlled from neurons located in the hypothalamus of the brain. The hypothalamic thermostat controls temperature regulation by ectotherms, too, but in these animals changes in hypothalamic temperature initiate thermoregulatory behaviors (e.g., orienting to maximize heat loss or gain) rather than physiological processes.

Mechanisms of Endothermal Thermoregulation

Body temperature and metabolic rate must be considered simultaneously to understand how endotherms maintain their body temperatures at a stable level in the face of environmental temperatures that may range from –70 to +40°C. Most birds and mammals conform to the generalized diagram in Figure 21–1.

Each species of endotherm has a definable range of ambient temperatures over which the body temperature can be kept stable by using physiological and postural adjustments of heat loss and heat production. This ambient temperature range is called the **zone of tolerance.** Above this range the animal's ability to dissipate heat is inadequate; both the body temperature and metabolic rate increase as ambient temperature increases, until the animal dies. At ambient temperatures below the zone of tolerance, the animal's ability to generate heat to balance heat loss is exceeded, body temperature falls, metabolic rate declines, and cold death results. Large animals usually have lower values for the lower critical temperature and lower lethal temperature than small animals because heat is lost from the body surface, and large animals have smaller surface-to-mass ratios than those of small animals. Similarly, well-insulated species have lower values for the lower critical temperature and lower lethal temperature than those of poorly insulated ones, but thinly insulated species usually have higher values for the upper critical temperature than those of heavily insulated ones.

The **thermoneutral zone** (TNZ) is the range of ambient temperatures within which the metabolic rate of an endotherm is at its basal level and thermoregulation is accomplished by changing the rate of heat loss. The thermoneutral zone is also called the zone of physical thermoregulation because an animal adjusts its heat loss using processes such as fluffing or sleeking its hair or feathers, postural changes such as huddling or stretching out, and changes in blood flow (vasoconstriction or vasodilation) to exposed parts of the body (feet, legs, face).

The larger an animal is and the thicker its insulation, the lower the temperature it can withstand before physical processes become inadequate to balance its heat loss. The **lower critical temperature** is the point at which an animal must increase metabolic heat production to maintain a stable body temperature. In the **zone of chemical thermogenesis,** the metabolic rate increases as the ambient temperature falls. The quality of the insulation determines how much additional metabolic heat production is required to offset a change in ambient temperature. That is, well-insulated animals have relatively shallow slopes for the graph of increasing metabolism below the lower critical temperature, and poorly insulated animals have steeper slopes (see Figure 21–4). Many birds and mammals from the Arctic and Antarctic are so well insulated that they can withstand the lowest temperatures on Earth (about –70°C) by increasing their basal metabolism only threefold.

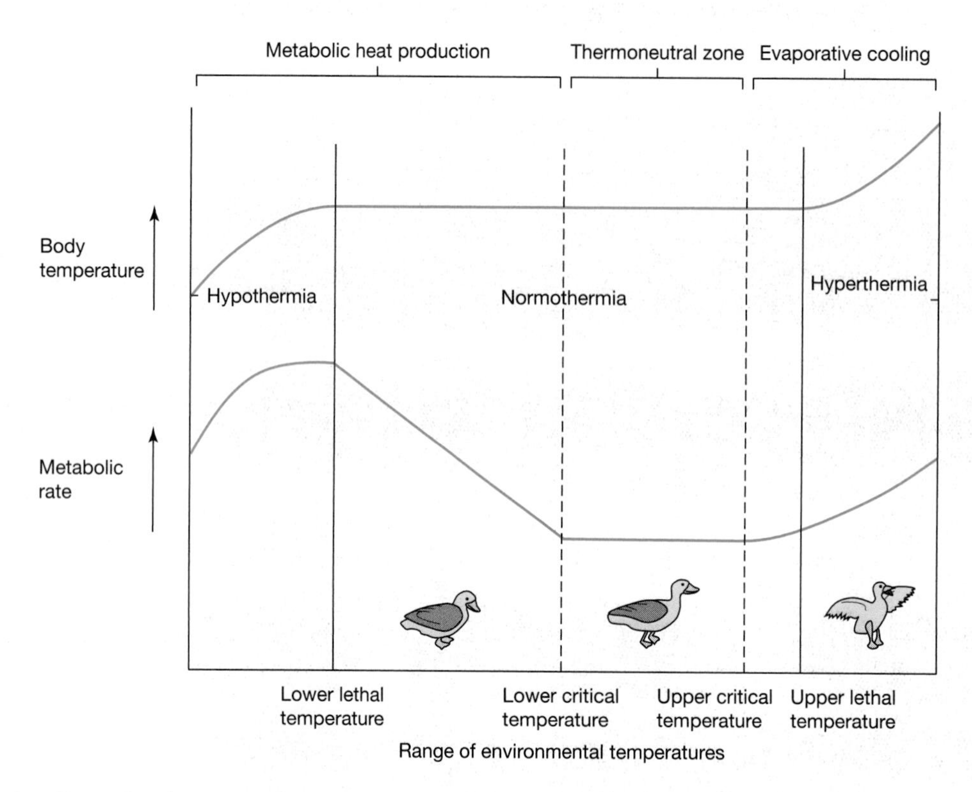

▲ Figure 21–1 Generalized pattern of changes in body temperature and metabolic heat production of an endothermic homeotherm in relation to environmental temperature. Normal body temperature varies somewhat for different mammals and birds.

Less well-insulated animals may be exposed to temperatures below their **lower lethal temperature.** At that point, metabolic heat production has reached its maximum rate and is still insufficient to balance heat loss to the environment. Under those conditions the body temperature falls, and the Q_{10} effect of temperature on chemical reactions causes the metabolic rate to fall as well. A positive-feedback condition is initiated in which falling body temperature reduces heat production, causing a further reduction in body temperature. Death from **hypothermia** (low body temperature) follows.

Endotherms are remarkably good at maintaining stable body temperatures in cool environments, but they have difficulty at high ambient temperatures. The **upper critical temperature** represents the point at which nonevaporative heat loss has been maximized by using all of the physical processes available to an animal—exposing the poorly insulated areas of the body and maximizing cutaneous blood flow. If these mechanisms are insufficient to balance heat gain, the only option vertebrates have is to use evaporation of water by panting, sweating, or gular flutter. The temperature range from the upper critical temperature to the upper lethal temperature is the **zone of evaporative cooling.** Some mammals sweat, a process in which water is released from sweat glands on the surface of the body. Evaporation of the sweat cools the body surface. Other animals pant, breathing rapidly and shallowly so that evaporation of water from the respiratory system provides a cooling effect. Many birds use a rapid fluttering movement of the gular region to evaporate water for thermoregulation. Panting and gular flutter require muscular activity, and some of the evaporative cooling they achieve is used to offset the increased metabolic heat production they require.

At the **upper lethal temperature,** evaporative cooling cannot balance the heat flow from a hot environment, and body temperature rises. The Q_{10} effect of temperature produces an increase in the rate of metabolism, and metabolic heat production raises the body temperature, increasing the metabolic rate still further. This process can lead to death from **hyperthermia** (high body temperature).

The difficulty that endotherms experience in regulating body temperature in high environmental temperatures may be one of the reasons that the body temperatures of most endotherms are in the range 35 to 40°C. Most habitats seldom have air temperatures that exceed 35°C. Even the tropics have average yearly temperatures below 30°C. Thus the high body temperatures maintained by endotherms ensure that in most situations the heat gradient is from animal to environment. Still higher body temperatures, around 50°C, for example, could ensure that mammals were always warmer than their environment. There are upper limits to the body temperatures that are feasible, however. Many proteins denature near 50°C. During heat stress, some birds and mammals may tolerate body temperatures of 45 to 46°C for a few hours; but only some bacteria, algae, and a few invertebrates exist at higher temperatures. This is another case in which the direction of vertebrate evolution has been established by a balance between biological needs and physiochemical realities.

Costs and Benefits of Endothermal Thermoregulation

Endothermy has both benefits and costs compared to ectothermy. On the positive side, endothermy allows birds and mammals to maintain high body temperatures when solar radiation is not available or is insufficient to warm them—at night, for example, or in the winter. The thermoregulatory capacities of birds and mammals are astonishing; they can live in the coldest places on Earth. On the negative side, endothermy is energetically expensive. We have pointed out that the metabolic rates of birds and mammals are nearly an order of magnitude greater than those of amphibians and reptiles. The energy to sustain those high metabolic rates comes from food, and endotherms need more food than do ectotherms.

Of course, a host of other differences distinguish the ecology and behavior of endotherms and ectotherms. These also can be considered costs or benefits of the different methods of thermoregulation. In this chapter we concentrate on how endotherms use energy, how they thermoregulate in cold and in hot environments, and how they control their water gains and losses. These topics are intimately related, because the high metabolic rates of endotherms are associated with more precise homeostatic control of their internal environment than is necessary for many ectotherms. For example, rising body temperature can be countered by evaporative cooling (sweating or panting), but too much evaporative cooling depletes water stores and leads to other problems. Body size plays a large role in determining the problems that endotherms encounter and the responses that are possible for them.

The same sorts of habitats that are challenging for ectotherms are also difficult for endotherms, although not always for the same reasons. Cold temperatures, for example, make ectothermal thermoregulation difficult or impossible and may present a risk of freezing. Endotherms have enough insulation and thermogenic (heat-producing) capacity to survive low temperatures, but they need a plentiful and regular supply of food to do that. Indeed, their high energy requirements appear to shape several aspects of the biology of endotherms, such as the relationship among body size, diet, and home range (discussed in the next chapter).

21.2 Energy Budgets of Vertebrates

An understanding of the costs of living can be obtained by constructing an energy budget for an animal. An energy budget, like a financial budget, shows income and expenditure but uses units of energy as currency. The energy costs of different activities can be evaluated by converting the energy intake (food) and loss (metabolism, feces, and urine) to a common unit of measurement. We will use kilojoules (kJ) for these calculations. (One kJ equals 0.24 kilocalories. The kilocalorie is the unit called Calories on labels showing nutritional content of packaged foods.) This quantitative approach to the study of ecology and behavior offers the promise of understanding some of the reasons why animals behave as they do.

Studies of vampire bats (*Desmodus rotundus*) by Brian McNab (1973) have revealed a clear-cut relationship between energy intake, energy expenditure, and the species' geographic range. These bats of the suborder Microchiroptera inhabit the Neotropics and are specialized to feed exclusively on blood. Their daily pattern of activity is simple: They spend about 22 hours in their caves, fly out at night to a feeding site, and return after they have fed. Typically, a vampire flies about 10 kilometers round-trip at 20 km per hour to find a meal. Thus, a bat spends half an hour per day in flight. The remaining time outside the cave may be spent in feeding.

The vampire's food is convenient for energetic calculations because of the relatively constant caloric content of blood. Here is the infomation needed to construct an energy budget:

- I = ingested energy (blood). The blood a bat drinks provides the energy needed for all of its life processes: maintenance, activity, growth, and reproduction.
- E = excreted energy. As in any animal, not all of the food ingested is digested and taken up by the bat. The energy contained in the feces and urine is lost.
- $I - E$ = assimilated energy. This is the energy actually taken into the bat's body.
- M = resting metabolism. This can be subdivided into M_i (metabolism while the bat is inside the cave) and M_o (nonflight metabolism while the bat is outside the cave).
- A = cost of activity, a half-hour of flight per day.
- B = biomass increase. This term is the profit a bat shows in its energy budget. It may be stored as fat or used for growth or for reproduction (production of gametes, growth of a fetus, or nursing a baby).

In its simplest form, the energy budget is

$$\text{energy in} = \text{energy out} \pm \text{biomass change}$$

The biomass term appears as ± because an animal metabolizes some body tissues when its energy expenditure exceeds its energy intake. (This is what every dieter hopes to do to lose weight.) Translating this general equation into the terms defined gives

$$I - E = M_i + M_o + A + B$$

All these terms can be measured and expressed as kilojoules per bat per day (kJ/bat · day). These calculations are based on McNab's studies and apply to a Brazilian vampire bat weighing 42g.

- **Ingested energy**—In a single feeding a vampire can consume 57 percent of its body mass in blood, which contains 4.6 kJ/g. Thus the ingested energy is

$$42\text{ g} \cdot 57\% \cdot 4.6\text{ kJ/g fluid blood} = 110.1\text{ kJ}$$

- **Excreted energy**—A vampire excretes 0.24g urea in the urine plus 0.95g of feces daily. Urea contains 10.5 kJ/g and the feces contain 23.8 kJ/g. Thus the excreted energy is

$$0.24\text{g urea} \cdot 10.5\text{ kJ/g} + 0.95\text{g feces} \cdot 23.8\text{ kJ/g} = 2.5\text{ kJ} + 22.6\text{ kJ} = 25.1\text{ kJ}$$

- **Assimilated energy**—The energy the bat actually assimilates from blood equals the energy ingested (110.1 kJ) minus that excreted (25.1 kJ). Thus a vampire's energy intake is 85.0 kJ/day:

$$110.1\text{ kJ} - 25.1\text{ kJ} = 85.0\text{ kJ}$$

- **Resting Metabolism**—In a tropical habitat, 20°C is a reasonable approximation of the temperature a bat experiences both inside and outside the cave. While at rest in the laboratory at 20°C, a vampire's metabolic rate is 3.8 cm^3 O_2/g · hr. The terms for metabolism can be calculated and converted to joules using the energy equivalent of oxygen (20.1 J/cm^3 O_2):

$$M_i = 42\text{ g} \cdot 3.8\text{ cm}^3\text{ O}_2/\text{g} \cdot \text{hr} \cdot 20.1\text{ J/cm}^3\text{ O}_2 \cdot 22\text{ hr/day} = 70.6\text{ kJ}$$

$$M_o = 42\text{ g} \cdot 3.8\text{ cm}^3\text{ O}_2/\text{g} \cdot \text{hr} \cdot 20.1\text{ J/cm}^3\text{ O}_2 \cdot 1.5\text{ hr/day} = 4.8\text{ kJ}$$

- **Activity**—The metabolism of a bat flying at 20 km/hr is 11.4 cm^3 O_2/g · hr. The cost of the round trip from the cave to the feeding site is

$$A = 42\text{ g} \cdot 11.4\text{ cm}^3\text{ O}_2/\text{g} \cdot \text{hr} \cdot 20.1\text{ J/cm}^3\text{ O}_2 \cdot 0.5\text{ hr} = 4.8\text{ kJ}$$

- **Biomass change**—The quantities calculated so far are fixed values that the bat cannot avoid except by changing its behavior. The biomass change is a variable value. If the assimilated energy is greater than the fixed costs, this energy profit can go to biomass increase. Fixed costs that exceed the assimilated energy are reflected as a loss of biomass. For the situation described there is an energy profit:

$$I - E = M_i + M_0 + A \pm B$$
$$110.1\ \text{kJ} - 25.1\ \text{kJ} = 70.6\ \text{kJ} + 4.8\ \text{kJ} + 4.8\ \text{kJ} \pm B$$
$$B = 4.8\ \text{kJ/bat} \cdot \text{day}$$

These calculations show that vampires can live and grow under the conditions assumed. What happens if we change some of the assumptions? McNab points out that the northern and southern limits of the geographic range of vampires conform closely to the winter isotherms of 10°C (Figure 21–2). That is, the minimum temperature outside the cave during the coldest month of the year is 10°C; the bats do not occur in regions where the minimum temperature is lower. Is this coincidence, or is 10°C the lowest temperature the bats can withstand? Calculating an energy budget for a vampire under these colder conditions provides an answer.

Caves have very stable temperatures that usually do not vary from summer to winter. We will assume that temperature remains constant at 20°C in the cave our imaginary vampires inhabit. In that case only the conditions a bat encounters outside the cave are altered. Because of limitations of stomach capacity, ingestion cannot increase beyond 57 percent of body mass, the value assumed in the previous calculation. Therefore, we need recalculate only M_o, A, and B.

- **Metabolism outside**—At 10°C a bat must increase its metabolic rate to maintain its body temperature, and laboratory measurements

▶ Figure 21–2 Geographic range of the vampire bat, *Desmodus rotundus*. The area in which vampire bats are found closely approximates the 10°C isotherm for the minimum average temperature during the coldest month of the year (dashed line) at the northern limit of its range (in Mexico) and the southern limit (in Uruguay, Argentina, and Chile). Positions of the 10°C isotherm and the bats' altitudinal range in the Andes Mountains are not known and are indicated by question marks.

indicate the resting metabolic rate increases to $6.3\ cm^3\ O_2/g \cdot hr$:

$$= 42\ g \cdot 6.3\ cm^3\ O_2/g \cdot hr \cdot 20.1\ J/cm^3\ O_2 \cdot 1.5\ hr$$

$$= 8.0\ kJ$$

- **Activity**—The cost of activity will not change, because the metabolic rate of the bat during flight ($11.4\ cm^3\ O_2/g \bullet hr$) is higher than the resting metabolic rate needed to keep it warm ($6.3\ cm^3\ O_2/g \bullet hr$). Only the term changes, increasing from 4.8 to 8.0 kJ, and the sum of the energy costs becomes 83.4 kJ/bat • day.

Because the assimilated energy remains at 85.1 kJ/bat day, only 1.8 kJ is available for biomass increase. The assumptions in these calculations introduce a degree of uncertainty, and probably 1.7 kJ is not statistically different from 0 kJ. Thus, at 10°C a bat uses all its energy staying alive. A vampire bat could survive under those conditions, but it would have no energy surplus for growth or reproduction. If the temperature outside the cave were lower than 10°C, the bat would have a negative energy balance and would lose body mass with each meal. The agreement between our calculations and the actual geographic distribution of the bats suggests that energy may be a significant factor in limiting their northward and southward spread.

Additional calculations reveal more about the selective forces that shape the lives of vampire bats. Although a bat's stomach can hold a volume of blood equal to 57 percent of its body mass, a bat cannot fly with that load. The maximum flight load is 43 percent of the body mass. Before it can take off to start the flight back to its cave, therefore, a bat must reduce the weight gained from its meal. Vampires do this by rapidly excreting water. Within 2 minutes after it begins to eat, a vampire starts to emit a stream of dilute urine. Experiments reveal that a vampire produces urine at a maximum rate of 0.24 ml/g of body mass per hour. Thus in the hour and a half the bat may spend in feeding, it could excrete as much as 15g of water—more than enough to allow it to fly (Busch 1988).

Although rapid excretion of water solves the bat's immediate problem, it introduces another. The bat is left with a stomach full of protein-rich food that will yield a large amount of urea. To excrete this urea, the bat needs water to form urine. By the time a vampire gets back to its cave, it is facing a water shortage instead of a water excess. Unlike many mammals, vampires seldom drink water, depending instead on blood for their water requirements. Like other mammals adapted to conditions of water scarcity, vampire bats have kidneys capable of producing very concentrated urine to conserve water. Because of its unusual ecology and behavior, a vampire bat can be considered to live in a desert of its own making in the midst of a tropical forest.

21.3 Endotherms in the Arctic

As the energy budget for the vampire bat revealed, most endotherms (especially small ones) expend most of the energy they consume just keeping themselves warm—even in the moderate conditions of tropical and subtropical climates. Nonetheless, endotherms have proven themselves very adaptable in extending their thermoregulatory responses to allow them to inhabit even Arctic and Antarctic regions. Not even small body size is an insuperable handicap to life in these areas: Small birds such as redpolls and chickadees weighing only 10g overwinter in central Alaska.

Aquatic life in cold regions places still more stress on an endotherm. Because of the high heat capacity and conductivity of water, an aquatic animal may lose heat at 50 to 100 times the rate it would if it were moving at the same speed through air. Even a small body of water is an infinite heat sink for an endotherm; all of the matter in its body could be converted to heat without appreciably raising the temperature of the water. How, then, do endotherms manage to exist in such stressful environments?

There are potentially two solutions to the problems of endothermal life in cold environments and the special problems of aquatic endotherms in particular. A stable body temperature could be achieved by increasing heat production or by decreasing heat loss. On closer examination the option of increasing heat production does not seem particularly attractive. Any significant increase in heat production would require an increase in food intake. This scheme poses obvious ecological difficulties in terrestrial Arctic and Antarctic habitats where primary production is extremely low, especially during the coldest parts of the year. For most polar animals the quantities of food necessary would probably not be available, and the metabolic rates of most polar endotherms are similar to those of related species from temperate regions.

Because they lack the option of increasing heat production significantly, conserving heat within the body is the primary thermoregulatory mechanism of polar endotherms. Insulative values of pelts from Arctic mammals are two to four times as great as those from tropical mammals. In Arctic species insulative value is closely related to fur length (Figure 21–3). Small species such as the least weasel and the lemming have fur only 1 to 1.5 centimeters long. Presumably, the thickness of their fur is limited because longer hair would interfere with the movement of their legs. Large mammals (caribou, polar and grizzly bears, dall sheep, and Arctic fox) have hair 3 to 7 cm long. There is no obvious reason why their hair could not be longer; apparently, further insulation is not needed. The insulative values of pelts of short-haired tropical mammals are similar to those measured for the same hair lengths in Arctic species. Long-haired tropical mammals, like the sloths, have less insulation than Arctic mammals with hair of similar length.

A comparison of the lower critical temperatures of tropical and Arctic mammals illustrates the effectiveness of the insulation provided (Figure 21–4). **Lower critical temperature** is the environmental temperature at which metabolic heat production must rise above its basal level to maintain a stable

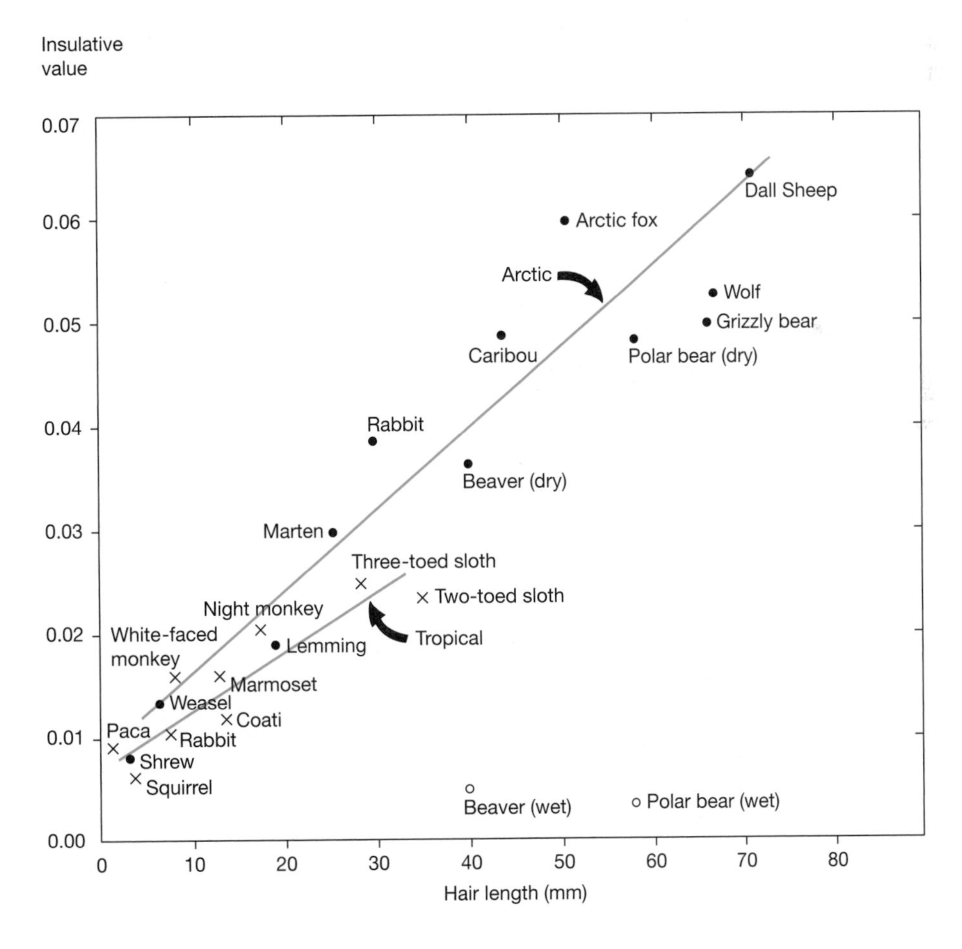

▲ Figure 21–3 Insulative values of the pelts of Arctic mammals. In air (dark circles) the insulation is proportional to the length of the hair. Pelts from tropical mammals (x) have approximately the same insulative value as those of Arctic mammals at short hair lengths, but long-haired tropical mammals like sloths have less insulation than Arctic mammals with hair of the same length. Immersion in water greatly reduces the insulative value of hair, even for such semiaquatic mammals as the beaver and polar bear (light circles).

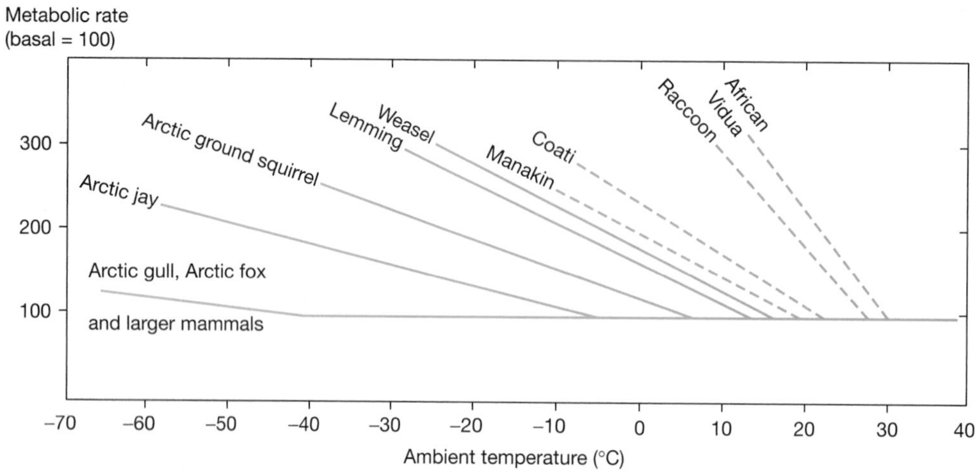

▲ Figure 21–4 Lower critical temperatures for birds and mammals. Solid lines, arctic birds and mammals; dashed lines, tropical birds and mammals. The basal metabolic rate for each species is considered to be 100 units to facilitate comparisons among species.

body temperature. Tropical mammals have lower critical temperatures between 20 and 30°C. As air temperatures fall below their lower critical temperatures, these animals are no longer in their thermoneutral zones and must increase their metabolic rates to maintain normal body temperatures. For example, a tropical raccoon has increased its metabolic rate approximately 50 percent above its standard level at an environmental temperature of 25°C.

Arctic mammals are much better insulated; a combination of insolation and relatively high resting metabolic rates can give even small species like the least weasel and the lemming lower critical temperatures that are between 10 and 20°C in still air. Larger mammals have thermoneutral zones that extend well below freezing. Because of their effective insulation, arctic birds and mammals can maintain resting metabolic rates at lower environmental temperatures than tropical species, and they show smaller increases in metabolism (i.e., flatter slopes) below the lower critical temperature. The Arctic fox, for example, has a lower critical temperature of –40°C, and at –70°C (approximately the lowest air temperature it ever encounters) has elevated its metabolic rate only 50 percent above its standard level. Under those conditions the fox is maintaining a body temperature approximately 110°C above air temperature. Arctic birds are equally impressive. The Arctic gray jay has a lower critical temperature below 0°C in still air and can withstand –70°C with a 150 percent increase in its metabolic rate. The Arctic glaucous gull, like the Arctic fox, has a lower critical temperature near –40°C and can withstand –70°C with a modest increase in metabolism.

Clearly, hair or feathers can provide superb insulation for terrestrial endotherms and for semiaquatic species that can leave the water to dry out. External coverings are of limited value to fully aquatic animals, however; when air trapped between hairs or feathers is displaced by water, most of the insulative value is lost. The insulation of beaver and polar bear hair falls almost to zero when it is wet through (Figure 21–3). In water, fat is a far more effective insulator than hair, and fully aquatic mammals have thick layers of blubber. This blubber forms the primary layer of insulation; skin temperature is nearly identical to water temperature, and there is a steep temperature gradient through the blubber so that its inner surface is at the animal's core body temperature.

The insulation provided by blubber is so effective that pinnipeds and cetaceans require special heat-dissipating mechanisms to avoid overheating when they engage in strenuous activity, or venture into warm water or onto land. This heat dissipation is achieved by shunting blood into capillary beds in the skin outside the blubber layer and into the flippers, which are not covered by blubber. Selective perfusion of these capillary beds enables a seal or porpoise to adjust its heat loss to balance its heat production. When it is necessary to conserve energy, a countercurrent heat exchange system in the blood vessels supplying the flippers is brought into operation; when excess heat is to be dumped, blood is shunted away from the countercurrent system into superficial veins.

The effectiveness of the insulation of marine mammals is graphically illustrated by the problems experienced by the northern fur seal (*Callorhinus ursinus*) during its breeding season. Northern fur seals are large animals; males attain body masses in excess of 250 kg. Unlike most pinnipeds, fur seals have a dense covering of fur, which is probably never wet through, as well as blubber. They are inhabitants of the North Pacific. For most of the year they are pelagic, but during summer they breed on the Pribilof Islands in the Bering Sea north of the Aleutian Islands. Male fur seals gather harems of females on the shore. There they must try to prevent the females from straying, chase away other males, and copulate with willing females.

George Bartholomew and his colleagues have studied both the behavior of the fur seals and their thermoregulation (Bartholomew and Wilke 1956). Summers in the Pribilof Islands (which are near 57° N latitude) are characterized by nearly constant overcast and air temperatures that rise only to 10°C during the day. These conditions are apparently close to the upper limits the seals can tolerate. Almost any activity on land makes them pant and raise their hind flippers (which are abundantly supplied with sweat glands), waving them about. If the Sun breaks through the clouds, activity suddenly diminishes—females stop moving about, males reduce harem-guarding activities and copulation, and the adult seals pant and wave their flippers. If the air temperature rises as high as 12°C, females, which never defend territories, begin to move into the water. Forced activity on land can produce lethal overheating.

At the time of the study, seal hunters herded the bachelor males from the area behind the harems inland preparatory to killing and skinning them. Bartholomew recorded one drive that took place in the early morning of a sunny day while the air temperature rose from 8.6°C at the start of the drive to 10.4°C by the end. In 90 minutes the seals were driven about 1 kilometer, with frequent pauses for rest.

> *The seals were panting heavily and frequently paused to wave their hind flippers in the air before they had been driven 150 yards from the rookery. By the time the drive was half finished most of the seals appeared badly tired and occasional animals were dropping out of the pods (groups of seals). In the last 200 yards of the drive and on the killing grounds there were found 16 "roadskins" (animals that had died of heat prostration) and in addition a number of others prostrated by overheating.*

The average body temperature of the roadskins of this drive was 42.2°C, which is 4.5°C above the 37.7°C mean body temperature of adults not under thermal stress.

Fur seals can withstand somewhat higher temperatures in water than they can in air because of the greater heat conduction of water, but they are not able to penetrate warm seas. Adult male fur seals apparently remain in the Bering Sea during their pelagic season. Young males and females migrate into the North Pacific, but they are not found in waters warmer than 14 to 15°C, and they are most abundant in water of 11°C. Their inability to regulate body temperature during sustained activity and their sensitivity to even low levels of solar radiation and moderate air temperatures probably restrict the location of potential breeding sites and their movements during their pelagic periods. Summers in the Pribilofs are barely cool enough to allow the seals to breed there. An increase in summer temperature associated with global warming might drive the seals from their traditional breeding grounds.

21.4 Migration to Avoid Difficult Conditions

Every environment has unfavorable aspects for some species, and these unfavorable conditions are often seasonal, especially in latitudes far from the equator. The primary cause of migrations is usually related to seasonal changes in climatic factors such as temperature or rainfall. In turn, these conditions influence food supply and the occurrence of suitable breeding conditions.

Long-distance migration is more feasible for birds and marine animals than for terrestrial species, partly because geographic barriers are less of a problem and partly because the energy cost of transport is less for swimming and flying than for walking (see Chapter 8). We can consider the costs and benefits of migrating by considering two kinds of animals that represent extremes of body size. The baleen whales are the largest animals that have ever lived, and hummingbirds are among the smallest vertebrate endotherms; yet both whales and hummingbirds migrate.

Whales

The annual cycle of events in the lives of the great baleen whales is particularly instructive in showing how migration relates to the use of energy and how it correlates with reproduction in the largest of all animals. Most baleen whales spend summers in polar or subpolar waters of either the Northern or the Southern Hemisphere, where they feed on krill or other crustaceans that are abundant in those cold, productive waters. For three or four months each year, a whale consumes a vast quantity of food that is converted into stored energy in the form of blubber and other kinds of fat. During this same time pregnant female whales nurture their unborn young, which may grow to one-third the length of their mothers before birth.

Near the end of summer the whales begin migrating toward tropical or subtropical waters where the females bear their young. The young grow rapidly on the rich milk provided by their mothers, and by spring the calves are mature enough to travel with their mothers back to Arctic or Antarctic waters. The calves are weaned about the time they arrive in their summer quarters. From a bioenergetic and trophic point of view, the remarkable feature of this annual migration is that virtually all of the energy required to fuel it comes from ravenous feeding and fattening during the three or four months spent in polar seas. Little or no feeding occurs during migration or during the winter period of calving and nursing. Energy for all these activities comes from the abundant stores of blubber and fat.

The gray whale (*Eschrichtius robustus*) of the Pacific Ocean has one of the longest and best-known migrations (Figure 21–5). The summer feeding waters are in the Bering Sea and the Chukchi Sea north of the Bering Strait in the Arctic Ocean. A small segment of the population moves down the coast of Asia to Korean waters at the end of the Arctic summer, but most gray whales follow the Pacific Coast of North America, moving south to Baja California and adjacent parts of western Mexico. They arrive in December or January, bear their young in shallow, warm waters, and then depart northward again in March. Some gray whales make an annual round-trip of at least 9000 km, and the adults eat little or nothing for the eight months they are away from their northern feeding grounds.

The amount of energy expended by a whale in this annual cycle is phenomenal. The basal metabolic

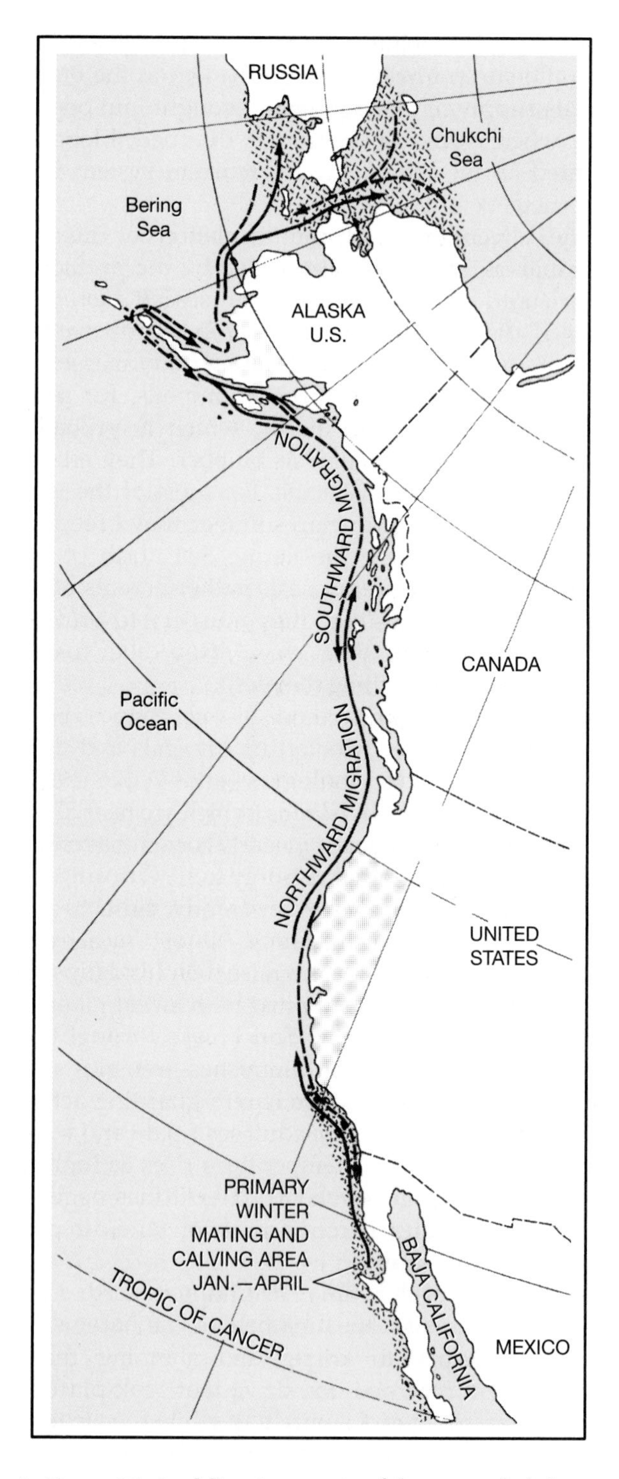

▲ Figure 21–5 Migratory route of the gray whale between the Arctic Circle and Baja California.

rate of a gray whale with a fat-free body mass of 50,000 kg is approximately 979,000 kJ per day. If the metabolic rate of a free-ranging whale, including the locomotion involved in feeding and migrating, is about three times the basal rate (a typical level of energy use for mammals), the whale's average daily energy expenditure is around 2,937,000 kJ. Body fat contains 38,500 kJ/kg, so the whale's daily energy expenditure is equivalent to metabolizing over 76 kg of blubber or fat per day. Assuming an energy content of 20,000 kJ/kg for krill and a 50 percent efficiency in converting the gross energy intake of food into biologically usable energy, the energy requirement for existence is equivalent to a daily intake of 294 kg of food.

In addition to satisfying its daily energy needs, a whale on the feeding grounds must accumulate a store of blubber. To live for 245 days without eating, the whale must metabolize a minimum of 18,375 kg of fat. Accumulating that amount of fat in 120 days of active feeding in Arctic waters at a conversion efficiency of 50 percent requires the consumption of 70,438 kg of krill, or 586 kg per day. The total food intake per day on the feeding grounds to accommodate the whale's daily metabolic needs plus energy storage for the migratory period would be not less than 294 + 586 = 880 kg of krill per day.

This is a minimum estimate for females, because the calculations do not include the energetic costs of the developing fetus or the cost of milk production. Nor do they include the cost of transporting 20,000 kg of fat through the water. But a large whale can do all this work and more without exhausting its insulative blanket of blubber because nearly half the total body mass of a large whale consists of blubber and other fats.

Why does a gray whale expend all this energy to migrate? The adult is too large and too well insulated ever to be stressed by the cold Arctic and subarctic waters, which do not vary much from 0°C from summer to winter. It seems strange for an adult whale to abandon an abundant source of food and go off on a forced starvation trek into warm waters that may cause stressful overheating. The advantage probably accrues to the newborn young, which, though relatively large, lacks an insulative layer of blubber. If the young whale were born in cold northern waters it would probably have to use a large fraction of its energy intake (milk produced from its mother's stored fat) to generate metabolic heat to regulate its body temperature. That energy could otherwise be used for rapid growth. Apparently it is more effective, and perhaps energetically more efficient, for the mother whale to migrate thousands of kilometers into warm waters to give birth and nurse in an environment where the young whale can invest most of its energy intake in rapid growth.

Hummingbirds

At the opposite end of the size range, hummingbirds are the smallest endotherms that migrate. Ornithologists have long been intrigued by the ability of the ruby-throated hummingbird (*Archilochus colubris*), which weighs only 3.5 to 4.5g, to make a nonstop flight of 800 km during migration across the Gulf of Mexico from Florida to the Yucatan Peninsula.

Like most migratory birds, ruby-throated hummingbirds store subcutaneous and body fat by feeding heavily prior to migration. A hummingbird with a lean mass of 2.5g can accumulate 2g of fat. Measurements of the energy consumption of a hummingbird hovering in the air in a respirometer chamber in the lab indicate an energy consumption of 2.89 to 3.10 kJ per hour. Hovering is energetically more expensive than forward flight, so these values represent the maximum energy used in migratory flight. Even so, 2 grams of fat produces enough energy to last for 24 to 26 hours of sustained flight. Hummingbirds fly about 40 km per hour, so crossing the Gulf of Mexico requires about 20 hours. Thus, by starting with a full store of fat, they have enough energy for the crossing with a reserve for unexpected contingencies such as a headwind that slows their progress. In fact, most migratory birds wait for weather conditions that will generate tailwinds before they begin their migratory flights, thereby further reducing the energy cost of migration.

21.5 Torpor as a Response to Low Temperatures and Limited Food

We have stressed the high energy cost of endothermy, because the need to collect and process enough food to supply that energy is a central factor in the lives of many endotherms. In extreme situations environmental conditions may combine to overpower a small endotherm's ability to process and transform enough chemical energy to sustain a high body temperature through certain critical phases of its life. For diurnally active birds, long cold nights during which there is no access to food can be lethal, especially if the bird has not been able to feed fully during the daytime. Cold winter seasons usually present a dual problem for resident endotherms—the need to maintain high body temperature when environmental temperatures are low despite the seasonal scarcity of food energy. In response to such problems, some birds and mammals have mechanisms that permit them to avoid the energetic costs of maintaining a high body temperature under unfavorable circumstances by entering a state of torpor (adaptive hypothermia). By entering torpor an endotherm is giving up many of the advantages of endothermy, but in exchange it realizes an enormous saving of both energy and water. Thus endotherms enter torpor only when they would face critical shortages of energy or water if they remained at normal body temperature.

Physiological Adjustments during Torpor

When an endotherm becomes torpid, profound changes occur in a variety of physiological functions. Although body temperatures may fall very low during torpor, temperature regulation does not entirely cease. In **deep torpor** an animal's body temperature drops to within 1°C or less of the ambient temperature, and in some cases (bats, for example) extended survival is possible at body temperatures just above the freezing point of the tissues. Arctic ground squirrels actually allow the temperature of parts of their bodies to supercool as low as –2.9°C. Oxidative metabolism and energy use are reduced to as little as one-twentieth of the rate at normal body temperatures. Respiration is slow, and the overall breathing rate can be less than one inspiration per minute. Heart rates are drastically reduced and blood flow to peripheral tissues is virtually shut down, as is blood flow posterior to the diaphragm. Most of the blood is retained in the body's core.

Deep torpor is a comatose condition, much more profound than the deepest sleep. Voluntary motor responses are reduced to sluggish postural changes, but some sensory perception of powerful auditory and tactile stimuli and ambient temperature changes is retained. Perhaps most dramatically, a torpid animal can arouse spontaneously from this state by endogenous heat production from brown fat that restores the high body temperature characteristic of a normally active endotherm. Some endotherms can rewarm under their own power from the lowest levels of torpor; others must warm passively with an increase in ambient temperature, until some threshold is reached at which arousal starts.

There are varying degrees of torpor, from the deepest states of hypothermia to the lower range of body temperatures reached by normally active endotherms during their daily cycles of activity and sleep. Nearly all birds and mammals, especially those with body masses under 1 kg, undergo **circadian temperature cycles.** These cycles vary from 1 to 5°C or more between the average high-temperature characteristic of the active phase of the daily cycle and the average low temperature characteristic of rest or sleep. Small birds (sunbirds, hummingbirds, chickadees) and small mammals (especially bats and rodents) may drop their body temperatures during quiescent periods from 8 to 15°C below their regulated temperature during activity. Even a bird as large as the turkey vulture (about 2.2 kg) regularly drops its body temperature at night. When all these different endothermic patterns are considered together, no really sharp distinction can be drawn between torpor and the basic daily cycle in body temperature that characterizes most small to medium-size endotherms.

Body Size and the Occurrence of Torpor

Species of endotherms capable of deep torpor are found in several groups of mammals and birds. The echidna, platypus, and several species of small marsupials display patterns of hypothermia. But the phenomenon is most diverse among eutherians, particularly among bats and rodents. Certain kinds of hypothermia have also been described for some insectivores, particularly the hedgehog, some primates, and some edentates. Deep torpor, contrary to popular notion, is not known for any of the carnivores, even though some of them den in the winter and remain inactive for long periods. Among birds, torpor occurs in some of the goatsuckers or nightjars and in hummingbirds, swifts, mousebirds, and some passerines (sunbirds, swallows, chickadees, and others). Other species, including larger ones like turkey vultures, show varying depths of hypothermia at rest or in sleep but are not in a semicomatose state of deep torpor.

The largest mammals that undergo deep torpor are marmots, which weigh about 5 kg, and torpor and body size are closely related. Torpor is not as advantageous for a large animal as for a small one. In the first place, the energetic cost of maintaining high body temperature is relatively greater for a small animal than for a large one and, as a consequence, a small animal has more to gain from becoming torpid. Second, a large animal cools off more slowly than a small animal so it does not lower its metabolic rate as rapidly.

Furthermore, large animals have more body tissue to rewarm on arousal, and their costs of arousal are correspondingly larger than those of small animals (Figure 21–6). An endotherm weighing a few grams, such as a little brown bat or a hummingbird, can warm up from torpor at the rate of about 1°C per minute, and be fully active within 30 minutes or less depending on the depth of hypothermia. A 100-gram hamster requires more than 2 hours to arouse, and a marmot takes many hours. Entrance into torpor is slower than arousal. Consequently, daily torpor is feasible only for very small endotherms; there would not be enough time for a large animal to enter and arouse from torpor during a 24-hour period. Moreover, the energy required to warm up a large mass is very great. A 4-gram hummingbird needs only 0.48 kilojoule to raise its body temperature from 10 to 40°C. That is 1/85 of the total daily energy expenditure of an active hummingbird in the wild. By contrast, a 200 kg bear would require 18,000 kJ to warm from 10 to 37°C, the equivalent of a full day's energy expenditure. The smaller potential saving and the greater cost of arousal make daily torpor impractical for any but small endotherms.

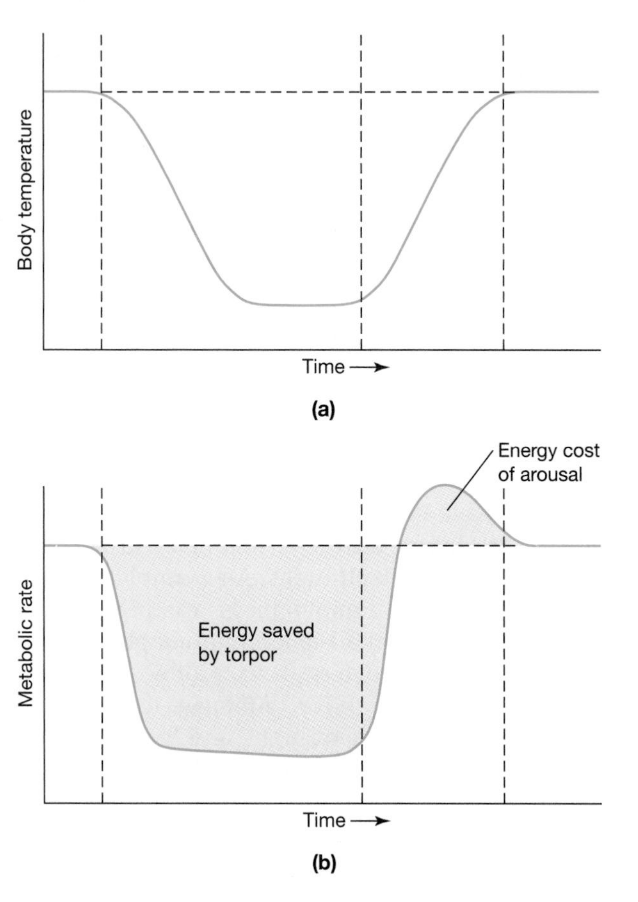

▲ Figure 21–6 Changes in body temperature and metabolic rate during torpor. A decrease in metabolic rate (shown in part b) precedes a fall in body temperature (part a) to a new set point. An increase in metabolism produces the heat needed to return to normal body temperatures; metabolic rate during arousal briefly overshoots resting rate.

Medium-size endotherms are not entirely excluded from the energetic savings of torpor, but the torpor must persist for a longer period to realize a saving. For example, ground squirrels and marmots enter prolonged torpor during the winter **hibernation** when food is scarce. They spend several days at very low body temperatures (in the region of 5°C), then arouse for a period before becoming torpid again. Still larger endotherms would have such large total costs of arousal (and would take so long to warm up) that torpor is not cost effective for them even on a seasonal basis. Bears in winter dormancy, for example, lower their body temperatures only about 5°C from normal levels, and metabolic rate decreases about 50 percent. Even this small drop, however, amounts to a large energy saving through the course of a winter for an animal as large as a bear. That small reduction in body temperature, combined with the large fat stores bears accumulate before retreating to their winter dens, is sufficient to carry them through the winter.

Energetic Aspects of Daily Torpor

Studies of daily torpor in birds have emphasized the flexibility of the response in relation to the energetic stress faced by individual birds. Susan Chaplin's work with chickadees provides an example (Chaplin 1974). These small (10 to 12g) passerine birds are winter residents in northern latitudes, where they regularly experience ambient temperatures that do not rise above freezing for days or weeks (Figure 21–7).

Figure 21–7 The black-capped chickadee, *Parus atricapillus*.

Chaplin found that in winter, chickadees around Ithaca, New York, allow their body temperatures to drop from the normal level of 40 to 42°C that is maintained during the day down to 29 to 30°C at night. This reduction in body temperature permits a 30 percent reduction in energy consumption. The chickadees rely primarily on fat stores they accumulate as they feed during the day to supply the energy needed to carry them through the following night. Thus the energy available to them and the energy they use at night can be estimated by measuring the fat content of birds as they go to roost in the evening and as they begin activity in the morning. Chaplin found that in the evening chickadees had an average of 0.80g of fat per bird. By morning the fat store had decreased to 0.24g. The fat metabolized during the night (0.56g per bird) corresponds to the metabolic rate expected for a bird that had allowed its body temperature to fall to 30°C.

Chaplin's calculations show that this torpor is necessary if the birds are to survive the night. It would require 0.92g of fat per bird to maintain a body temperature of 40°C through the night. That is more fat than the birds have when they go to roost in the evening. If they did not become torpid, they would starve before morning. Even with torpor, they use 70 percent of their fat reserve in one night. They do not have an energy supply to carry them far past sunrise, and chickadees are among the first birds to begin foraging in the morning. They also forage in weather so foul that other birds, which are not in such precarious energy balance, remain on their roosts. The chickadees must reestablish their fat stores each day if they are to survive the next night.

Hummingbirds, too, may depend on the energy they gather from nectar during the day to carry them through the following night. These very small birds (3 to 10g) have extremely high energy expenditures and yet are found during the summer in northern latitudes and at high altitudes. An example of the lability of torpor in hummingbirds was provided by studies of nesting broad-tailed hummingbirds at an altitude of 2900 meters near Gothic, Colorado (Calder and Booser 1973). Ambient temperatures drop nearly to freezing at night, and hummingbirds become torpid when energy is limiting. Calder and Booser were able to monitor the body temperatures of nesting birds by placing an imitation egg containing a temperature-measuring device in the nest. These temperature records showed that hummingbirds incubating eggs normally did not become torpid at night. The reduction of egg temperature that results from the parent bird's becoming torpid does not damage the eggs, but it slows development and delays hatching. Presumably, there are advantages to hatching the eggs as quickly as possible. As a re-

▲ Figure 21–8 Richardson's ground squirrel, *Spermophilus richardsonii.*

sult, brooding hummingbirds expend energy to keep themselves and their eggs warm through the night, provided that they have the energy stores necessary to maintain the high metabolic rates needed.

On some days bad weather interfered with foraging by the parent birds, so that they apparently went into the night with insufficient energy supplies to maintain normal body temperatures. In this situation the brooding hummingbirds did become torpid for part of the night. One bird that had experienced a 12 percent reduction in foraging time during the day became torpid for 2 hours, and a second that had lost 21 percent of its foraging was torpid for 3.5 hours. Torpor can thus be a flexible response that integrates the energy stores of a bird with environmental conditions and biological requirements such as brooding eggs.

Energetic Aspects of Prolonged Torpor

Hibernation is an effective method of conserving energy during long winters, but hibernating animals do not remain at low body temperatures for the whole winter. Periodic arousals are normal, and these arousals consume a large portion of the total amount of energy used by hibernating mammals. An example of the magnitude of the energy cost of arousal is provided by Lawrence Wang's study of the Richardson's ground squirrel (*Spermophilus richardsonii,* Figure 21–8) in Alberta, Canada (Wang 1978).

The activity season for ground squirrels in Alberta is short: They emerge from hibernation in mid-March, and adult squirrels reenter hibernation 4 months later, in mid-July. Juvenile squirrels begin hibernation in September. When the squirrels are active they have body temperatures of 37 to 38°C, and their temperatures fall as low as 3 to 4°C when they are torpid. Figure 21–9 shows the body temperature of a juvenile male ground squirrel from September through March; periods of torpor alternate with arousals all through the winter. Hibernation began in mid-September with short bouts of torpor followed by rewarming. At that time the temperature in the burrow was about 13°C. As the winter progressed and the temperature in the burrow fell, the intervals between arousals lengthened and the body temperature of the torpid animal declined. By late December the burrow temperature had dropped to 0°C, and the periods between arousals were 14 to 19 days. In late February the periods of torpor became shorter, and in early March the squirrel emerged from hibernation.

A torpor cycle consists of entry into torpor, a period of torpor, and an arousal (Figure 21–10). In this example entry into torpor began shortly after noon on February 16, and 24 hours later the body temper-

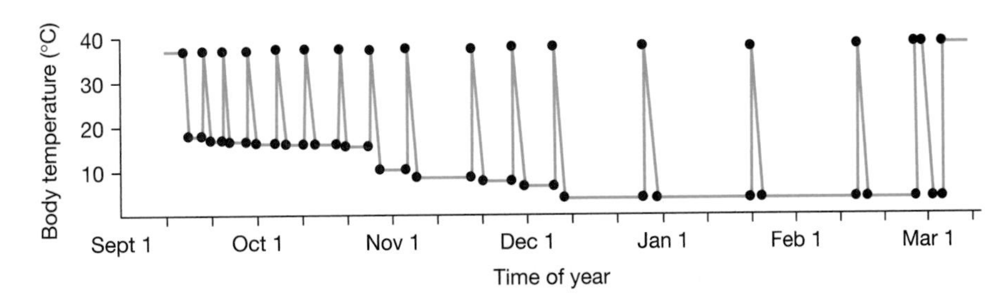

▲ Figure 21–9 Record of body temperature during a complete torpor season for a Richardson's ground squirrel.

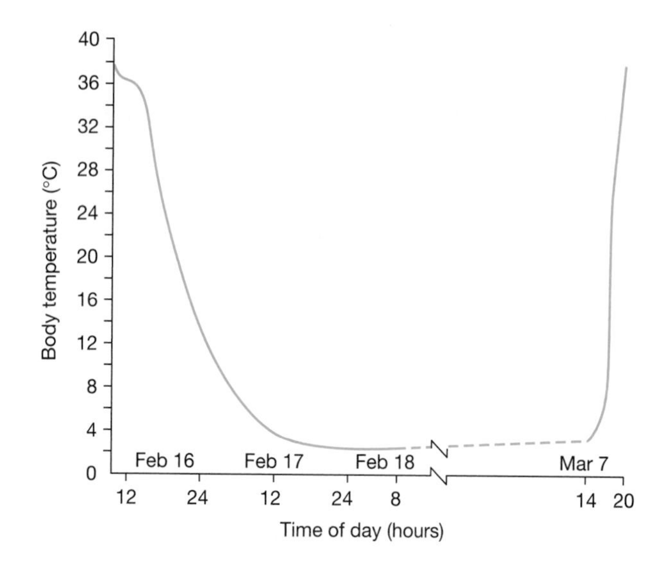

▲ Figure 21–10 Record of body temperature during a single torpor cycle for a Richardson's ground squirrel.

ature had stabilized at 3°C. This period of torpor lasted until late afternoon on March 7, when the squirrel started to arouse. In 3 hours the squirrel warmed from 3 to 37°C. It maintained that body temperature for 14 hours, and then began entry into torpor again.

These periods of arousal account for most of the energy used during hibernation (Table 21.1). The energy costs associated with arousal include the cost of warming from the hibernation temperature to 37°C, the cost of sustaining a body temperature of 37°C for several hours, and the metabolism above torpid levels as the body temperature slowly declines during reentry into torpor. For the entire hibernation season, the combined metabolic expenditures for those three phases of the torpor cycle account for an average of 83 percent of the total energy used by the squirrel.

Surprisingly, we have no clear understanding of why a hibernating ground squirrel undergoes these arousals that increase its total winter energy expenditure nearly fivefold. Ground squirrels do not store food in their burrows, so they are not using the periods of arousal to eat. They do urinate during arousal, and it is possible that some time at a high body temperature is necessary to carry out physiological or biochemical activities such as resynthesizing glycogen, redistributing ions, or synthesizing serotonin. Arousal may also allow a hibernating animal to determine when environmental conditions are suitable for emergence. Whatever their function, the arousals must be important because the squirrel pays a high energy price for them during a period of extreme energy conservation.

21.6 Endotherms in Deserts

Hot, dry areas place a more severe physiological demands on endotherms than do the polar conditions we have already discussed. Endotherms encounter two problems regulating body temperature in hot deserts. The first results from a reversal of the normal relationship of an animal to the environment. In most environments an endotherm's body temperature is warmer than the air temperature. In this situ-

Table 21.1 Use of Energy during Different Phases of the Torpor Cycle by Richardson's Ground Squirrel

	Percentage of Total Energy per Month			
Month	*Torpor*	*Warming*	*Intertorpor Homeothermy*	*Reentry*
July	8.5	17.2	56.5	17.8
September	19.2	15.2	49.9	15.7
November	20.8	23.1	43.1	13.0
January	24.8	24.1	40.0	11.1
March	3.3	14.0	76.4	6.3
Average for season	16.6	19.0	51.6	12.8

ation heat flow is from the animal to its environment, and thermoregulatory mechanisms achieve a stable body temperature by balancing heat production and heat loss. Very cold environments merely increase the gradient between an animal's body temperature and the environment. The example of Arctic foxes with lower critical temperatures of –40°C illustrates the success that endotherms have had in providing sufficient insulation to cope with enormous gradients between high core body temperatures and low environmental temperatures.

In a hot desert the gradient is not increased; it is reversed. Desert air temperatures can climb to 40 or 50°C during summer, and the ground temperature may exceed 60 or 70°C. Instead of losing heat to the environment, an animal is continually absorbing heat, and that heat plus metabolic heat must somehow be dissipated to maintain the animal's body temperature in the normal range. It can be a greater challenge for an endotherm to maintain its body temperature 10°C below the ambient temperature than to maintain it 100°C above ambient.

To make matters worse, water is scarce in deserts. Evaporative cooling is the major mechanism an endotherm uses to reduce its body temperature. The evaporation of water requires approximately 2400 kJ/kg. (The exact value varies slightly with temperature.) Thus, evaporation of a liter of water dissipates 2400 kJ, and evaporative cooling is an effective mechanism as long as an animal has an unlimited supply of water. In a hot desert, however, where thermal stress is greatest, water is a scarce commodity; its use must be carefully rationed. Calculations show, for example, that if a kangaroo rat were to venture out into the desert sun, it would have to evaporate 13 percent of its body water per hour to maintain a normal body temperature. Most mammals die when they have lost 10 to 20 percent of their body water, so it is obvious that under desert conditions, evaporative cooling is of limited utility except as a short-term response to a critical situation.

Nonetheless, diverse assemblages of birds and mammals inhabit deserts. The mechanisms they use are complex and involve combinations of ecological, behavioral, morphological, and physiological adjustments that act together to enhance the effectiveness of the entire system. The ancestral structures and physiological characteristics of birds and mammals are remarkably versatile. In many cases those characters alone are all an animal needs to confront conditions that seem extraordinarily harsh.

Kidneys and Water Conservation

Metabolism produces waste products that must be excreted. Nitrogenous wastes are a particular problem for desert animals because they are excreted in solution, so an animal must use some of its limited water supply to eliminate these chemical compounds.

Birds Birds have two advantages over mammals when it comes to eliminating nitrogenous wastes. Birds have the ancestral sauropsid uricotelic nitrogen pathway, which produces uric acid that combines with ions and precipitates as urate salts, thereby releasing water that can be reabsorbed and reused. Furthermore, some birds have an extrarenal route for salt excretion—glands in the nasal passages that secrete salt at concentrations far above the concentrating power of any kidney, avian or mammalian (see Table 21.2). Not all birds have salt glands, and they are not necessary for life in the desert, but they are an additional advantage for the desert species that do have them.

Mammals Mammals have no significant extrarenal routes of excretion. Except for trivial quantities of urea and ammonia secreted by sweat glands, mammals depend on the concentrating power of their kidney for nitrogen excretion, although ruminant artiodactyls recycle some urea in the rumen (see Chapter 20). Mammals can do this because the ancestral mammalian kidney is such an effective device for concentrating urine that very little additional specialization is needed for life in the desert. In fact, many mammals live most of their lives with limited access to water. The deciduous forests of northeastern North America, which are home to the white-footed deer mouse, receive a meter or more of rain and snow in a normal year. They are mesic (moist) environments, but that does not mean that a deer mouse has water to drink. During summer and fall the forest floor dries out between storms, and for days or weeks at a time the deer mouse has no water to drink within its home range. Because leaving its home range in an attempt to find water would be foolhardy, deer mice—and other small mammals—rely on the water in their food and the concentrating ability of their kidneys for survival. In this respect there is little difference between living in a deciduous forest and living in a desert, and rodents have colonized deserts around the world.

The mammalian kidney is a highly derived organ composed of millions of nephrons, the basic units of kidney structure that are recognizable in all

Table 21.2 Maximum urine concentrations of some tetrapods

Species	*Maximum Observed Urine Concentration (mmole • kg^{-1})*	*Approximate Urine: Plasma Concentration Ratio*
American alligator (*Alligator missisippiensis*)	312	0.95
Desert iguana (*Dipsosaurus dorsalis*)	300	0.95
Desert tortoise (*Gopherus agassizii*)	622	1.8
Pelican (*Pelecanus erythrorhynchos*)	700	2
House sparrow (*Passer domesticus*)	826	2.4
House finch (*Carpodacus mexicanus*)	850	2.4
Human (*Homo sapiens*)	1430	4
Bottlenose porpoise (*Tursiops truncatus*)	2658	7.5
Hill kangaroo (*Macropus robustus*)	2730	7.5
Camel (*Camelus dromedarius*)	2800	8
White rat (*Rattus norvegicus*)	2900	8.9
Marsupial mouse (*Dasycercus eristicauda*)	3231	10
Cat (*Felis domesticus*)	3250	9.9
Desert woodrat (*Neotoma lepida*)	4250	12
Vampire bat (*Desmodus rotundus*)	6250	20
Kangaroo rat (*Dipodomys merriami*)	6382	18
Australian hopping mouse (*Notomys alexis*)	9370	22

vertebrates (Figure 21–11). Each nephron consists of a glomerulus that filters the blood and a long tube in which the chemical composition of the filtrate is altered. A portion of this tube, the **loop of Henle,** is a derived character of mammals that is largely responsible for the ability of mammals to produce concentrated urine. The mammalian kidney is capable of producing urine more concentrated than that of any nonamniote—and in most cases, more concentrated than that of sauropsids as well (Table 21.2). Understanding how the mammalian kidney works is important for understanding how mammals can thrive in places that are seasonally or chronically short of water.

Urine is concentrated by removing water from an ultrafiltrate, leaving behind the wastes. Because cells are unable to transport water directly, they use osmotic gradients to manipulate the movement of water molecules. In addition, the cells lining the nephron actively reabsorb substances important to the body's economy from the ultrafiltrate and secrete toxic substances into it. The cells lining the nephron differ in permeability, molecular and ion transport activity, and reaction to the hormonal and osmotic environments in the surrounding body fluids.

The cells of the proximal convoluted tubule (PCT) have an enormous surface area produced by long, closely spaced microvilli, and the cells contain many mitochondria (Figure 21–11). These structural features reflect the function of the PCT in actively moving sodium from the lumen of the tubule to the peritubular space and capillaries; sodium transport is followed by passive movement of chloride to the peritubular space to neutralize electric charge (Figure 21–12). Water then flows osmotically in the same

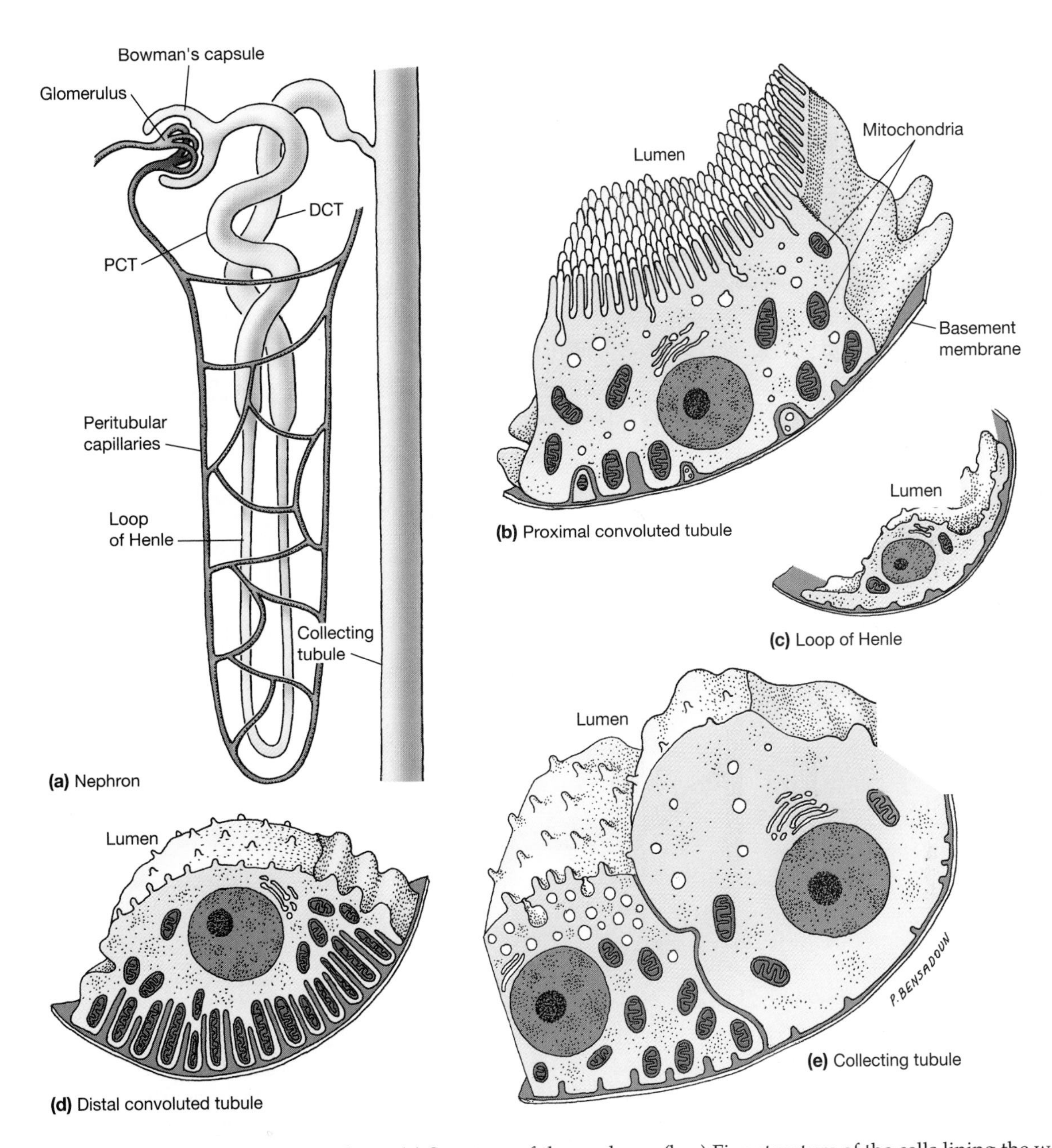

▲ Figure 21–11 The mammalian nephron. (a) Structure of the nephron. (b–e) Fine structure of the cells lining the walls of the nephron.

direction. Farther down the nephron, the cells of the thin segment of the loop of Henle are wafer-like and contain fewer mitochondria. The descending limb of the loop of Henle permits passive flow of sodium and water, and the ascending limb actively removes sodium from the ultrafiltrate. Finally, cells of the collecting tubule appear to be of two kinds. Most seem to be suited to the relatively impermeable state characteristic of periods of sufficient body water. Other cells are mitochondria-rich and have a greater surface area. They are probably the cells that respond to the presence of antidiuretic hormone (ADH) from the pituitary gland, triggered by insufficient body fluid. Under the influence of ADH, the collecting tubule actively exchanges ions, pumps urea, and becomes permeable to water, which flows

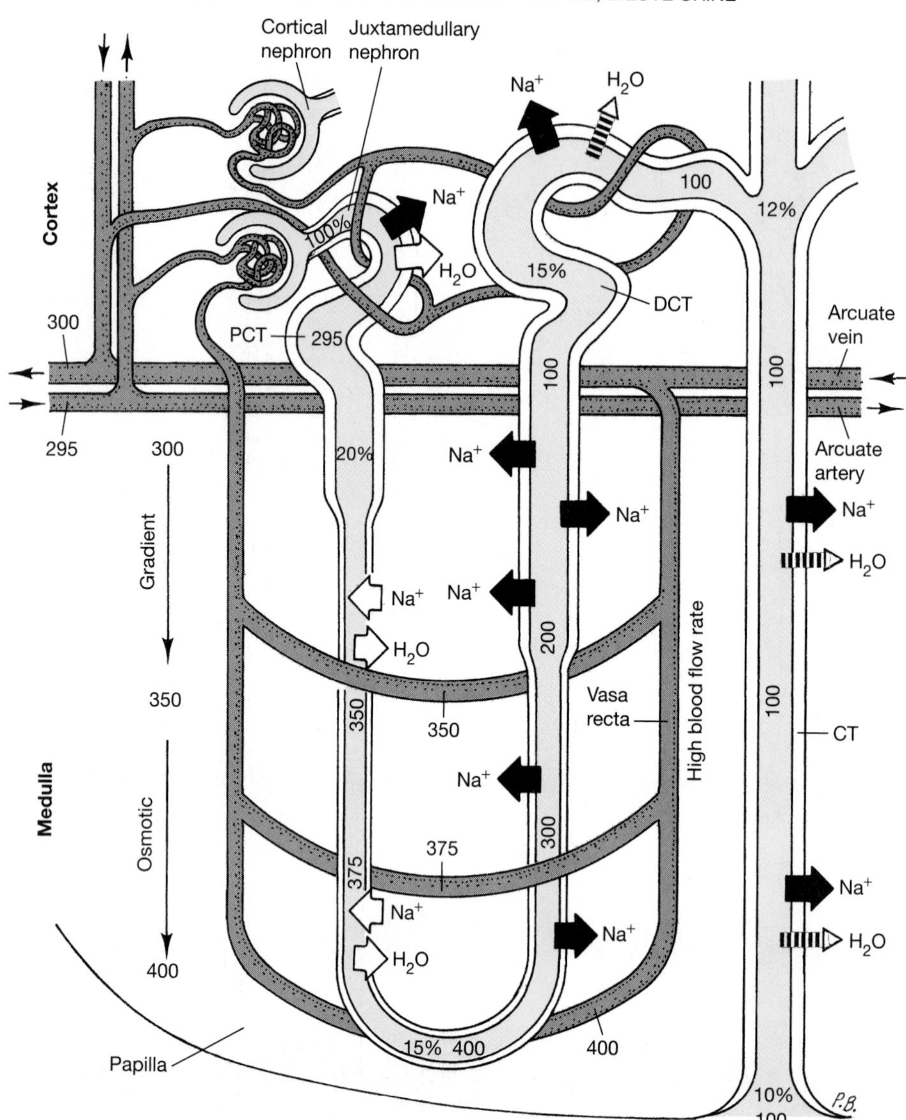

▲ Figure 21–12 Diagram showing how the mammalian kidney produces dilute urine when the body is hydrated and concentrated urine when the body is dehydrated. Black arrows indicate active transport and white arrows indicate passive flow. The numbers represent the approximate osmolality of the fluids in the indicated regions. Percentages are the volumes of the forming urine relative to the volume of the initial ultrafiltrate. (a) When blood osmolality drops below normal concentration (about 300 mmole · kg^{-1}), excess body water is excreted. (b) When osmolality rises above normal, water is conserved.

from the lumen of the tubule into the concentrated peritubular fluids.

The nephron's activity is a six-step process, each step localized in a region with special cell characteristics and distinctive variations in the osmotic environment. The first step is production of an ultrafiltrate at the glomerulus (Figure 21–12). The ultrafiltrate is isosmolal with blood plasma and resembles whole blood after the removal of (1) cellular elements, (2) substances with a molecular weight of 70,000 or greater (primarily proteins), and (3) substances with molecular weights between 15,000 and 70,000, depending on the shapes of the molecules. Humans produce about 120 milliliters of ultrafiltrate per minute—that is 170 liters (45 gallons) of glomerular filtrate per day! We excrete only about 1.5 liters of urine because the kidney reabsorbs more than 99 percent of the ultrafiltrate produced by the glomerulus.

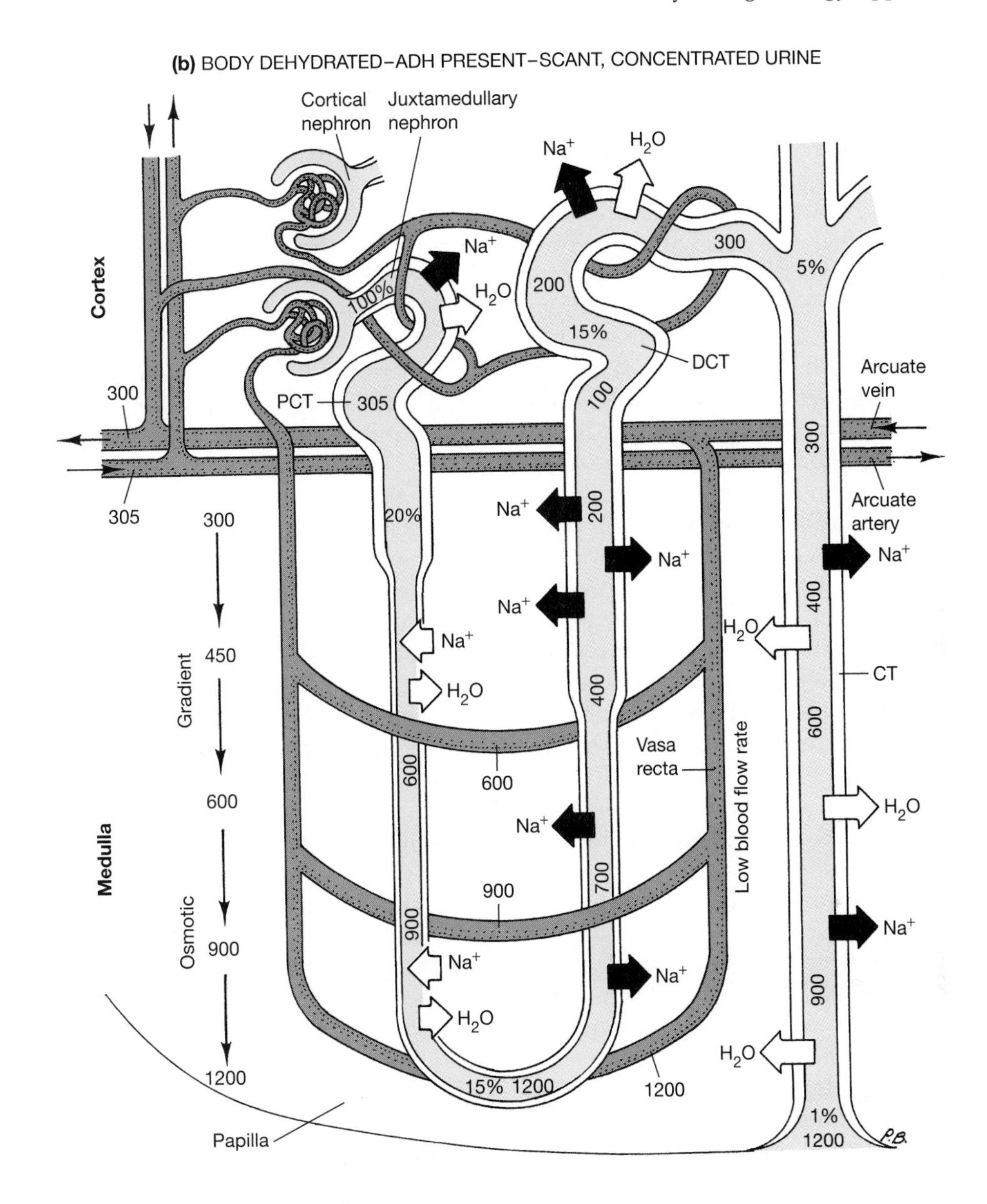

▲ Figure 21–12 *Continued*

The second step in the production of the urine is the action of the **proximal convoluted tubule** (PCT, Figure 21–12) in decreasing the volume of the ultrafiltrate. The PCT cells have greatly enlarged surfaces that actively transport sodium and perhaps chloride ions from the lumen of the tubules to the exterior of the nephron. Water flows osmotically through the PCT cells in response to the removal of sodium chloride. By this process about two-thirds of the salt is reabsorbed in the PCT, and the volume of the ultrafiltrate is reduced by the same amount. Although the urine is still very nearly isosmolal with blood at this stage, the substances contributing to the osmolality of the urine after it has passed through the PCT are at different concentrations than in the blood.

The next alteration occurs in the descending limb of the loop of Henle. The thin, smooth-surfaced cells freely permit diffusion of sodium and water. Because the descending limb of the loop passes through tissues of increasing osmolality as it plunges into the medulla, water is lost from the urine and it becomes more concentrated. In humans the osmolality of the fluid in the descending limb may reach 1200 mmoles · kg^{-1}. Other mammals can achieve considerably higher concentrations. By this mechanism the volume of the forming urine is reduced to 25 percent of the initial filtrate volume, but it is still large. In an adult human, for example, 25 to 40 liters of fluid per day reach this stage.

The fourth step takes place in the ascending limb of the loop of Henle, which has cells with numerous large, densely packed mitochondria. The ATP produced by these organelles is used to remove sodium from the forming urine. Because these cells are impermeable to water, the volume of urine does not decrease as sodium is removed, and because sodium was removed without loss of water, the urine is hyposmolal to the body fluids as it enters the next segment of the nephron. Although this sodium-pumping, water-impermeable, ascending limb does not concentrate or reduce the volume of the forming urine, it sets the stage for these important processes.

The very last portion of the nephron changes in physiological character, but the cells closely resemble those of the ascending loop of Henle. This region, the **distal convoluted tubule** (DCT), is permeable to water. The osmolality surrounding the DCT is that of the body fluids, and water in the entering hyposmolal fluid flows outward and equilibrates osmotically. This process reduces the fluid volume to 5 to 20 percent of the original ultrafiltrate.

The final touch in the formation of a scant, highly concentrated mammalian urine occurs in the **collecting tubules.** Like the descending limb of the loop of Henle, the collecting tubules course through tissues of increasing osmolality, which withdraw water from the urine. The significant phenomenon associated with the collecting tubule, and to a lesser extent with the DCT, is its conditional permeability to water. During excess fluid intake, the collecting tubule demonstrates low water permeability: Only half of the water entering it may be reabsorbed and the remainder excreted. In this way a copious, dilute urine can be produced. When a mammal is dehydrated the collecting tubules and the DCT become very permeable to water, and the final urine volume may be less than 1 percent of the original ultrafiltrate volume. In certain desert rodents so little water is contained in the urine that it crystallizes almost immediately on urination.

A polypeptide called **antidiuretic hormone** (ADH; also known as **vasopressin**) is produced by specialized neurons in the hypothalamus, stored in the posterior pituitary, and released into the circulation whenever blood osmolality is elevated or blood volume drops. ADH increases the permeability of the collecting tubule to water and facilitates water reabsorption to produce a scant, concentrated urine. Alcohol inhibits the release of ADH, inducing a copious urine flow, and this can result in dehydrated misery the morning after a drinking binge.

The key to concentrated urine production clearly depends on the passage of the loops of Henle and collecting tubules through tissues with increasing osmolality. These osmotic gradients are formed and maintained within the mammalian kidney as a result of its structure (Figure 21–13), which sets it apart from the kidneys of other vertebrates. Particularly important are the structural arrangements within the kidney medulla of the descending and ascending segments of the loop of Henle and its blood supply, the **vasa recta.** These elements create a series of parallel tubes with flow passing in opposite directions in adjacent vessels (countercurrent flow). As a result, sodium secreted from the ascending limb of the loop of Henle diffuses into the medullary tissues to increase their osmolality, and this excess salt is distributed by the countercurrent flow to create a steep osmotic gradient within the medulla (see Figure 21–12b). The final concentration of a mammal's urine is determined by the amount of sodium accumulated in the fluids of the medulla. Physiological alterations in the concentration in the medulla result primarily from the effect of ADH on the rate of blood flushing the medulla. When ADH is present, blood flow into the medulla is retarded and salt accumulates to create a steep osmotic gradient. Another hormone, aldosterone, from the adrenal gland increases the rate of sodium secretion into the medulla to promote an increase in medullary salt concentration.

In addition to these physiological means of concentrating urine, a variety of mammals have morphological alterations of the medulla. Most mammals have two types of nephrons: those with a cortical glomerulus and abbreviated loops of Henle that do not penetrate far into the medulla, and those with juxtamedullary glomeruli, deep within the cortex, with loops that penetrate as far as the papilla of the renal pyramid (Figure 21–13b). Obviously, the longer, deeper loops of Henle experience large osmotic gradients along their lengths. The flow of blood to these two populations of nephrons seems to be independently controlled. Juxtamedullary glomeruli are more active in regulating water excretion; cortical glomeruli function in ion regulation.

Form and function are intimately related in mammalian kidneys. A thick medulla corresponds to long renal pyramids, long loops of Henle, large osmotic

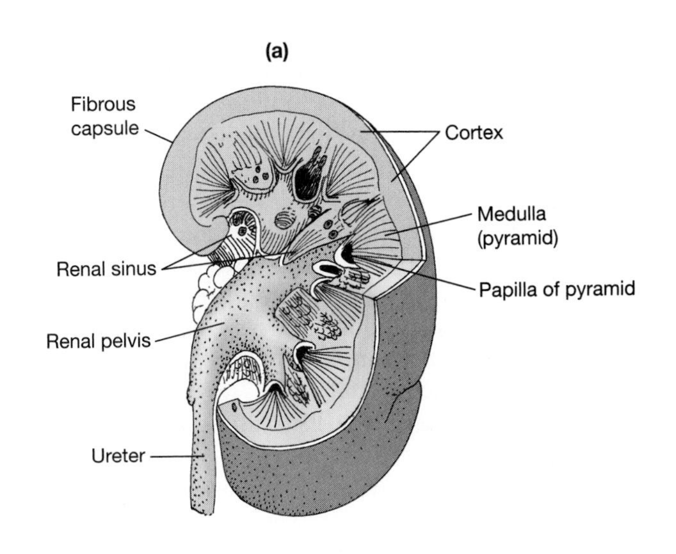

◀ Figure 21–13 Gross morphology of the mammalian kidney. (a) Structural divisions of the kidney and proximal end of the ureter. (b) Enlarged diagram of a section extending from the outer cortical surface of the apex of a renal pyramid, the renal papilla.

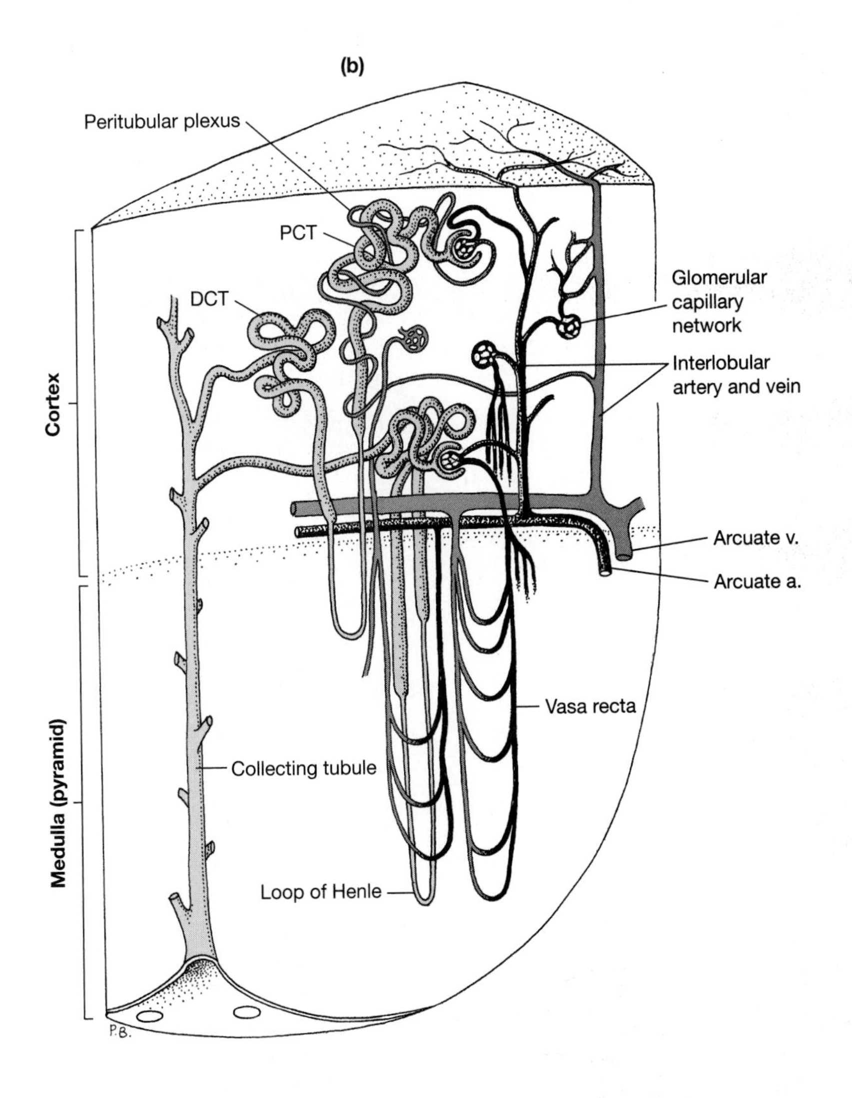

gradients, and great concentrating power. Maximum urine osmolalities of mammals are proportional to the relative medullary thickness of their kidneys (Figure 21–14). Some desert rodents have exceptionally long renal pyramids and urine concentrations that exceed 7000 mmole · kg^{-1}. (The relative medullary thickness of human kidneys is an unimpressive 3.0, and maximum urine concentration is only 1430 mmole · kg^{-1}.) Figure 21–14 shows a strong correlation between relative medullary thickness and maximum urine concentration. But substantial variation is apparent, indicating that other anatomical or physiological factors are involved. For mammals as a group, relative medullary thickness accounts for 59 percent of interspecific variation in maximum urine concentration (Beuchat 1990).

Strategies for Desert Survival

Deserts are harsh environments, but they contain a mosaic of microenvironments that animals can use to find the conditions they need. We can categorize three major classes of responses of endotherms to desert conditions, as follows:

- Relaxation of homeostasis—Some endotherms have relaxed the limits of homeostasis. They survive in deserts by tolerating greater-than-normal ranges of variation in characters such as body temperature or body water content.
- Avoidance—Some endotherms manage to avoid desert conditions by behavioral means. They live in deserts but are rarely exposed to the full rigors of desert life.
- Specializations—Physiological mechanisms such as torpor in response to shortages of food or water are used by some desert organisms.

These categories are not mutually exclusive; many desert animals combine elements of all three responses.

Relaxation of Homeostasis—Large Mammals in Hot Deserts Large animals, including humans, have specific advantages and disadvantages in

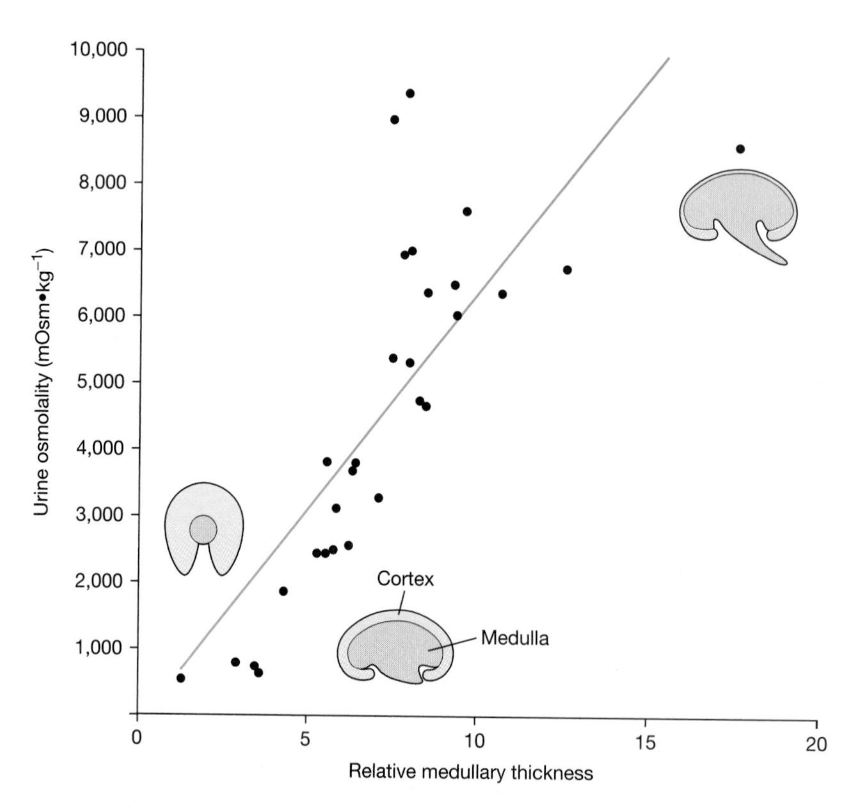

▲ Figure 21–14 Relationship of maximum urine concentration to relative medullary thickness for 29 species of rodents. The sketches illustrate kidneys with different relative medullary thicknesses.

desert life that are directly related to body size. A large animal has nowhere to hide from desert conditions. It is too big to burrow underground, and few deserts have vegetation large enough to provide useful shade to an animal much larger than a jackrabbit. On the other hand, large body size offers some options not available to smaller animals. Large animals are mobile and can travel long distances to find food or water, whereas small animals may be limited to home ranges only a few meters or tens of meters in diameter. Large animals have small surface/mass ratios and can be well insulated. Consequently, they absorb heat from the environment slowly. A large body mass gives an animal a large thermal inertia; that is, it can absorb a large amount of heat energy before its body temperature rises to dangerous levels.

The dromedary camel (*Camelus dromedarius*) is the classic large desert animal (Figure 21–15). There are authentic records of journeys in excess of 500 km, lasting two or three weeks, during which the camels did not have an opportunity to drink. Dromedaries make their longest trips in winter and spring, when air temperatures are relatively low and scattered rainstorms may have produced fresh vegetation that provides them a little food and water.

Camels are large animals—adult body masses of dromedary camels are 400 to 450 kg for females and up to 500 kg for males. The camel's adjustments to desert life are revealed by comparing the daily cycle of body temperature in a camel that receives water daily and one that has been deprived of water (Figure 21–16). The watered camel shows a small daily cycle of body temperature with a minimum of 36°C in the early morning and a maximum of 39°C in midafternoon. When a camel is deprived of water, the daily temperature variation triples. Body temperature is allowed to fall to 34.5°C at night and climbs to 40.5°C during the day.

The significance of this increased daily fluctuation in body temperature can be assessed in terms of the water that the camel would expend to prevent the 6°C rise by evaporative cooling. With a specific heat of 4.2 kJ/(kg · °C), a 6°C increase in body temperature for a 500-kg camel represents storage of 12,600 kJ of heat. Evaporation of a kilogram of water dissipates approximately 2400 kJ. Thus, a camel would have to evaporate slightly more than 5 liters of water to maintain a stable body temperature at the nighttime level, and it can conserve that water by tolerating hyperthermia during the day.

In addition to saving water that is not used for evaporative cooling, the camel receives an indirect benefit from hyperthermia via a reduction of energy flow from the air to the camel's body. As long as the camel's body temperature is below air temperature, a gradient exists that causes the camel to absorb heat from the air. At a body temperature of 40.5°C the camel's temperature is equal to that of the air for much of the day, and no net heat exchange takes place. Thus the camel saves an additional quantity of water by eliminating the temperature gradient between its

▶ Figure 21–15 Dromedary camels. In the heat of the day, most of these camels have faced into the Sun to reduce the amount of direct solar radiation they receive and are pressed against each other to reduce the heat they gain by convection and reradiation.

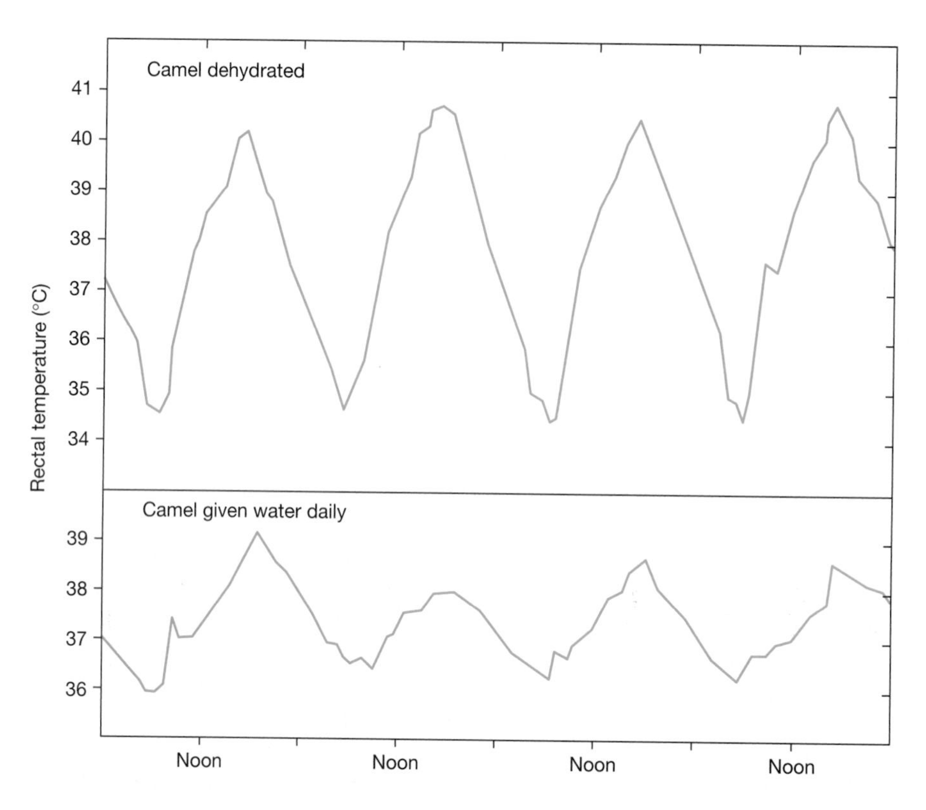

▲ Figure 21–16 Daily cycles of body temperature of camels. A dehydrated camel (top) relaxes its control of body temperature compared to a camel with daily access to water (bottom).

body and the air. The combined effect of these measures on water loss is illustrated by data from a young camel (Table 21.3). When deprived of water, the camel reduced its evaporative water loss by 64 percent and reduced its total daily water loss by half.

Behavioral mechanisms and the distribution of hair on the body aid dehydrated camels in reducing their heat load. In summer camels have hair 5 or 6 cm long on the back and up to 11 cm long over the hump. On the ventral surface and legs the hair is only 1.5 to 2 cm long. Early in the morning camels lie down on surfaces that have cooled overnight by radiation of heat to the night sky. The legs are tucked beneath the body and the ventral surface, with its short covering of hair, is placed in contact with the cool ground. In this position a camel exposes only its well-protected back and sides to the sun and places its lightly furred legs and ventral surface in contact with cool sand, which may be able to conduct away some body heat. Camels may assemble in small groups and lie pressed closely together through the day. Spending a day in the desert sun squashed between two sweaty camels may not be your idea of fun, but in this posture a camel reduces its heat gain because it keeps its sides in contact with other camels (both at about 40°C) instead of allowing solar radiation to raise the fur surface temperature to 70°C or above.

Despite their ability to reduce water loss and to tolerate dehydration, the time eventually comes when even camels must drink. These large, mobile animals can roam across the desert seeking patches of vegetation produced by local showers and move from one oasis to another, but when they drink they face a problem they share with other grazing animals: Water holes can be dangerous places. Predators frequently center their activities around water holes, where they are assured of water as well as a continuous supply of prey animals. Reducing the time spent drinking is one method of reducing the risk of predation, and camels

Table 21.3 Daily Water Loss of a 250-kg Camel

Condition	Water Loss (L/day) by Different Routes			
	Feces	*Urine*	*Evaporation*	*Total*
Drinking daily (8 days)	1.0	0.9	10.4	12.3
Not drinking (17 days)	0.8	1.4	3.7	5.9

can drink remarkable quantities of water in very short periods. A dehydrated camel can drink as much as 30 percent of its body mass in 10 minutes. (A very thirsty human can drink about 3 percent of body mass in the same time.)

The water a camel drinks is rapidly absorbed into its blood. The renal blood flow and glomerular filtration rate increase, and urine flow returns to normal within a half-hour of drinking. The urine changes from dark brown and syrupy to colorless and watery. Aldosterone stimulates sodium reabsorption, which helps to counteract the dilution of the blood by the water the camel has drunk. Nonetheless, dilution of the blood causes the red blood cells to swell as they absorb water by osmosis. Camel erythrocytes are resistant to this osmotic stress, but other desert ruminants have erythrocytes that would burst under these conditions. Bedouin goats, for example, have fragile erythrocytes, and the water a goat drinks is absorbed slowly from the rumen. Goats require two days to return to normal kidney function after dehydration.

Large African antelope, such as the 100-kg oryx (*Oryx beisa*) and 200-kg eland (*Taurotragus oryx*), use heat storage like the dromedary, but allow their body temperatures to rise considerably above the 40.5°C level recorded for the camel. Rectal temperatures of 45°C have been recorded for the oryx and 46.5°C for the Grant's gazelle. Body temperatures above 43°C rapidly produce brain damage in most mammals, but Grant's gazelles maintained rectal temperatures of 46.5°C for as long as 6 hours with no apparent ill effects. These antelope keep brain temperature below body temperature by using a countercurrent heat exchange to cool blood before it reaches the brain. The blood supply to the brain passes via the esternal carotid arteries, and at the base of the brain in these antelope the arteries break into a rete mirabile that lies in a venous sinus (Figure 21–17). The blood in the sinus is venous blood, returning from the walls of the nasal passages where it has been cooled by the evaporation of water. This chilled venous blood cools the warmer arterial blood before it reaches the brain. A mechanism of this sort is widespread among mammals.

Large animals illustrate one approach to desert life. Too large to escape the rigors of the environment, they survive by tolerating a temporary relaxation of homeostasis. Their success under the harsh conditions in which they live is the result of complex interactions between diverse aspects of their ecology, behavior, morphology, and physiology. Only when all of these features are viewed together does an accurate picture of an animal emerge.

Avoiding Desert Conditions The mobility of vertebrates allows them to escape some of the stresses of desert conditions, either by finding shelter (such as a burrow) or by moving to shade or water. In general larger animals can move farther than small animals but also need larger burrows, patches of shade, or quantities of water. Time of activity also provides options for mitigating the harshness of desert conditions—many desert animals are nocturnal, especially during the hottest parts of the year.

Rodents are the preeminent small mammals of arid regions. It is a commonplace observation that population densities of rodents are higher in deserts than in moist situations. Several ancestral features of rodent biology allow them to extend their geographic ranges into hot, arid regions. Among the most important of these characters are the normally nocturnal habits of many rodents and their practice of living in burrows. A burrow provides escape from the heat of a desert, giving an animal access to a

▶ Figure 21–17 The countercurrent heat-exchange mechanism that cools blood going to a gazelle's brain.

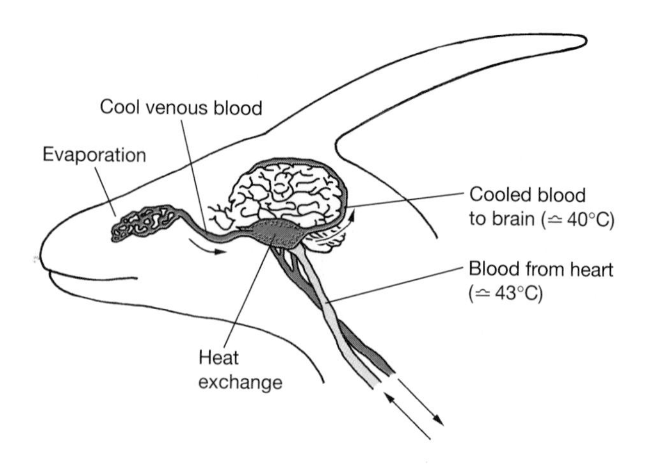

sheltered microenvironment while soil temperatures on the surface climb above lethal levels.

Kangaroo rats are among the most specialized desert rodents in North America; Merriam's kangaroo rat (*Dipodomys merriami*) occurs in desert habitats from central Mexico to northern Nevada. A population of this species lives in extremely harsh conditions in the Sonoran Desert of southwestern Arizona (Tracy 2000). During the summer daytime temperatures at the ground surface approach 70°C, and even a few minutes of exposure would be deadly. Kangaroo rats spend the day in burrows 1 to 1.5m underground, where air temperatures do not exceed 35°C even during the hottest parts of the year. In the evening, when the kangaroo rats emerge to forage, external air temperatures have fallen to about 35°C.

Not all rodents that live in deserts are nocturnal. Ground squirrels are diurnal and thus conspicuous inhabitants of deserts (Figure 21–18). They can be seen running frantically across the desert surface even in the middle of day. The almost frenetic activity of desert ground squirrels on intensely hot days is a result of the thermoregulatory problems that small animals experience under these conditions. Studies of the antelope ground squirrel (*Ammospermophilus leucurus*) at Deep Canyon—near Palm Springs, California—provide information about how the squirrels' behavior is affected by the heat load of the environment (Chappell and Bartholomew 1981a, b).

The heat on summer days at Deep Canyon is intense, and standard operative temperatures in the sun are as high as 70 to 75°C (Box 21-1). Standard operative temperature rises above the thermoneutral zone of ground squirrels within 2 hours after sunrise, and the squirrels are exposed to high heat loads for most of the day. They have a bimodal pattern of activity that peaks in midmorning and again in the late afternoon. Relatively few squirrels are active in the middle of the day. The body temperatures of antelope ground squirrels are labile, and body temperatures of individual squirrels vary as much as 7.5°C (from 36.1 to 43.6°C) during a day. The squirrels use this lability of body temperature to store heat during their periods of activity.

The high temperatures limit the time that squirrels can be active in the open to no more than 9 to 13 minutes. They sprint furiously from one patch of shade to the next, pausing only to seize food or to look for predators. The squirrels minimize exposure to the highest temperatures by running across open areas, and seek shade or their burrows to cool off. On a hot summer day a squirrel can maintain a body temperature below 43°C (the maximum temperature it can tolerate) only by retreating every few minutes to a burrow deeper than 60 cm, where the soil temperature is 30 to 32°C. The body temperature of an antelope ground squirrel shows a pattern of rapid oscillations, rising while the squirrel is in the sun and falling when it retreats to its burrow (Figure 21–19). Ground squirrels do not sweat or pant; instead, they use this combination of transient heat storage and passive cooling in a burrow to permit diurnal activity. The strategy the antelope ground squirrel uses is basically the same as that employed by a camel—saving water by allowing the body temperature to rise until the heat can be dissipated passively. The difference between the two animals is a consequence of their difference in body size: A camel weighs 500 kg and can store heat for an entire day

◀ Figure 21–18 The antelope ground squirrel, *Ammospermophilus leucurus*.

and cool off at night, whereas an antelope ground squirrel weighs about 100g and heats and cools many times in the course of a day.

The tails of many desert ground squirrels are wide and flat, and the ventral surfaces of the tails are usually white. The tail is held over the squirrel's back with its white ventral surface facing upward. In this position it acts as a parasol, shading the squirrel's body and reducing the standard operative temperature. The tail of the antelope ground squirrel is relatively short, extending only halfway up the back, but the shade it gives can reduce the standard operative temperature by as much as 6 to 8°C. The Cape ground squirrel (*Xenurus inauris*) of the Kalahari Desert has an especially long tail that can be extended forward nearly to the squirrel's head. Cape ground squirrels use their tails as parasols (Figure 21–21), and observations of the squirrels indicated that the shade may significantly extend their activity on hot days.

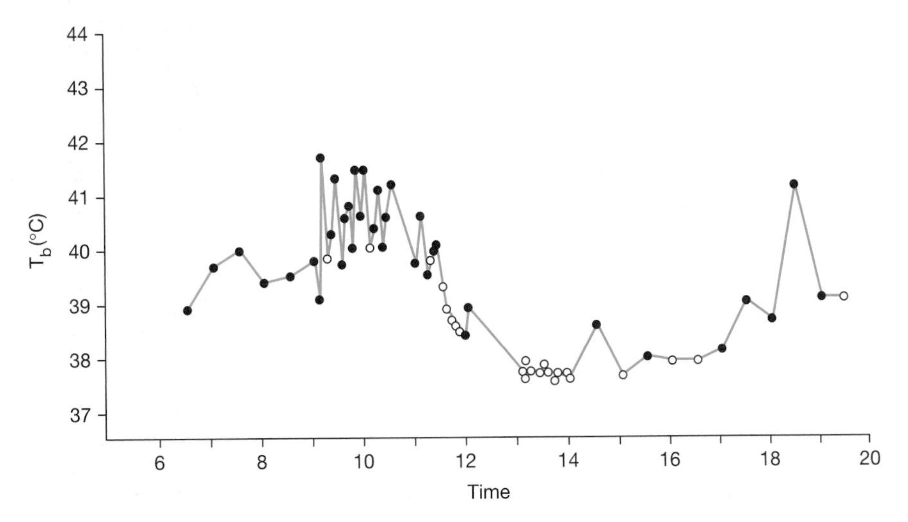

▲ Figure 21–19 **Short-term cycles of activity and body temperature of an antelope ground squirrel.** The squirrel warms up during periods of activity on the surface (dark circles), and cools off when it retreats to its burrow (light circles).

BOX 21-1 How Hot Is It?

Measurements of environmental temperatures figure largely in studies of the energetics of animals. But the actual process of making the measurements is complicated, and no one measurement is necessarily appropriate for all purposes. The exchange of energy between animals and their environments involves radiation, convection, conduction, and evaporation. In addition, metabolic heat production contributes significantly to the body temperatures of endotherms. The thermal environment of an animal is determined by all of the routes of heat exchange operating simultaneously, and its body temperature includes the effect of metabolism as well. The question *How hot is it?* translates to *What is the heat load for an animal in this environment?* Answering that question requires integrating all the routes of energy exchange to give one number that represents the environmental heat load. (It is easier to think of heat load as coming from a hot environment and being represented by a risk of overheating [hyperthermia], but the same reasoning applies to a cold environment. In that situation the problem is loss of heat and the risk of hypothermia.)

Physiological ecologists have developed several measurements of the environmental heat load on an organism, and Figure 21–20 illustrates four of these. The data come from a study of the thermoregulation of the antelope ground squirrel in a desert canyon in California.

The easiest measurement to make is the temperature of the air (T_a frequently called **ambient temperature**). At Deep Canyon, California, in June air temperature rises from about 25°C at dawn to a peak above 50°C in late afternoon, and then declines. Air temperature is a factor in conductive and convective heat exchange. An animal gains heat by conduction and convection when the air temperature is warmer than the animal's surface temperature and loses heat when the air is cooler. Conductive heat exchange is usually small, but convection can be an important component of the overall energy budget of an organism. However, the magnitude of convective heat exchange depends on wind speed as well as air temperature. Thus, measuring air temperature provides only part of the information needed to assess just one of the three important routes of heat exchange. Consequently, air temperature is not a very useful measure of heat load.

If air temperature is unsatisfactory as a measure of environmental heat load because it makes only a small contribution to the overall energy exchange, perhaps a measurement of the major source of heat is what is needed. **Solar insolation** in this arid habitat is the major source of heat, and the magnitude of the insolation (Q_r) can be measured with a device called a pyranometer. Solar insolation rises from 0 at dawn to about 900 watts per square meter in midday, and falls to zero at sunset. This measurement provides information about how much solar energy is available to heat an animal, but that is still only one component of the energy exchange that determines the heat load.

The **effective environmental temperature (T_e)** combines the effects of air temperature, ground temperature, solar insolation, and wind velocity. The effective environmental temperature is measured by making an exact copy of the animal (a manikin), equipping it with a temperature sensor such as a thermocouple, and

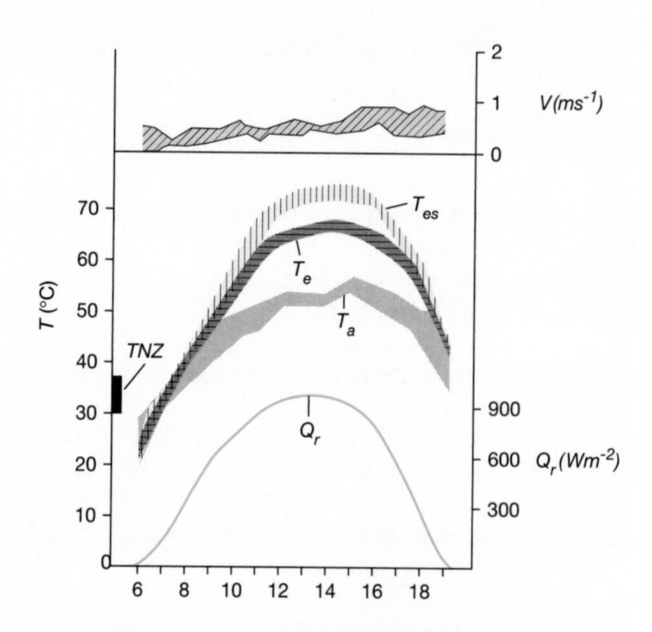

▲ Figure 21–20 Ground-level meterological conditions in open sunlit areas at Deep Canyon, California, during June. Wind velocity (V) in meters per second, solar insolation (Q_r) in watts per square meter, effective environmental temperature (T_e) and standard operative temperature (T_{es}) in °C. The thermoneutral zone (TNZ) of ground squirrels is indicated.

BOX 21-1 How Hot Is It? (*continued*)

putting the manikin in the same place in the habitat that the real animal occupies. Taxidermic mounts are frequently used as manikins: The pelt of an animal is stretched over a framework of wire or a hollow copper mold of the animal's body. Because the manikin has the same size, shape, color, and surface texture as the animal, it responds in the same way as the animal to solar insolation, infrared radiation, and convection. The equilibrium temperature of the manikin is the temperature that a metabolically inert animal would have as a result of the combination of radiative and convective heat exchange. At Deep Canyon the temperatures of manikins of antelope ground squirrels increased more rapidly than air temperatures and stabilized near 65°C from midmorning through late afternoon. The temperatures of the manikins were 15°C higher than air temperature, showing that the heat load experienced by the ground squirrels was much greater than that estimated from air temperature alone.

Because the manikin is a hollow shell—a pelt stretched over a supporting structure—it does not duplicate thermoregulatory processes that have important influences on the body temperature of a real animal. Metabolic heat production increases the body temperature of a ground squirrel, evaporative water loss lowers body temperature, and changes in peripheral circulation and raising or lowering the hair change the insulation. The effects of these factors can be incorporated mathematically if the appropriate values for metabolism, insulation, and whole-body conductance are known. The result of this calculation is the **standard operative temperature (T_{es})**. An explanation of how to calculate T_{es} can be found in Bakken (1980). For the ground squirrels in our example, the standard operative temperature was nearly 10°C higher than the effective temperature and about 25°C higher than the air temperature. For much of the day the T_{es} of a ground squirrel in the sun at Deep Canyon was 30°C or more above the squirrel's upper critical temperature of 43°C. Similar calculations can provide values for T_{es} in other microenvironments the squirrels might occupy—in the shade of a bush, for example, or in a burrow. This is the information needed to evaluate the squirrels' behavior to determine if their activities are limited by the need to avoid overheating.

Use of Torpor by Desert Rodents The significance of daily torpor as an energy conservation mechanism in small birds was illustrated earlier. Many desert rodents also have the ability to become torpid. In most cases the torpor can be induced by limiting the food available to an animal. When the food ration of the California pocket mouse (*Perognathus californicus*) is reduced slightly below its daily requirements, it enters torpor for a part of the day. In this species even a minimum period of torpor results in an energy saving. If a pocket mouse were to enter torpor and then immediately arouse, the process would take 2.9 hours. Calculations indicate that the overall energy expenditure during that period would be reduced 45 percent compared to the cost of maintaining a normal body temperature for the same period. In this animal, the briefest possible period of torpor gives an energetic saving, and the saving increases as the time spent in torpor is lengthened.

The duration of torpor is proportional to the severity of food deprivation for the pocket mouse. As its food ration is reduced, it spends more time each day in torpor and conserves more energy. Adjusting the time spent in torpor to match the availability of food may be a general phenomenon among seed-eating desert rodents. These animals appear to assess the rate at which they accumulate food supplies during foraging rather than their actual energy balance. Species that accumulate caches of food will enter torpor even with large quantities of stored food on hand if they are unable to add to their stores by continuing to forage. When seeds were deeply buried in the sand, and thus hard to find, pocket mice spent more time in torpor than they did when the same quantity of seed was close to the surface (Reichman and Brown 1979). This behavior is probably a response to the chronic food shortage that may face desert rodents because of the low primary productivity of desert communities and the effects of unpredictable variations from normal rainfall patterns, which may almost completely eliminate seed production by desert plants in dry years.

(a) (b)

▲ Figure 21–21 Cape ground squirrel (*Xenurus inauris*) using its tail as a parasol. (a) The erected tail shades the dorsal surface of the animal. (b) The tail is held over the back of a horizontal squirrel, shading its head and body.

Birds in Deserts Although birds are relatively small vertebrates, the problems they face in deserts are more like those experienced by camels and antelope than like those of small mammals. Birds are predominantly diurnal, and few seek shelter in burrows or crevices. Thus, like large mammals, they meet the rigors of deserts head on and face the antagonistic demands of thermoregulation in a hot environment and the need to conserve water.

Birds are much more mobile than mammals of the same body size. A kangaroo rat or ground squirrel is confined to a home range less than 100m in diameter, but it is quite possible for a desert bird with the same body size as those rodents to fly many kilometers to reach water. For example, mourning doves in the deserts of North America congregate at dawn at water holes, some individuals flying 60 km or more to reach them.

The normally high and labile body temperatures of birds give them another advantage that is not shared by mammals. With body temperatures normally around 40°C, birds face the problem of a reversed temperature gradient between their bodies and the environment for a shorter portion of each day than does a mammal. Furthermore, birds' body temperatures are normally variable, and birds tolerate moderate hyperthermia without apparent distress. These are all ancestral characters that are present in virtually all birds. Neither the body temperatures nor the lethal temperatures of desert birds are higher than those of related species from nondesert regions.

The mobility provided by flight does not extend to fledgling birds, and the most conspicuous adaptations of birds to desert conditions are those that ensure a supply of water for the young. Altricial fledglings, those that need to be fed by their parents after hatching, receive the water they need from their food. One pattern of adaptation in desert birds ensures that reproduction will occur at a time when succulent food is available for fledglings. In the arid central region of Australia, bird reproduction is precisely keyed to rainfall. The sight of rain is apparently sufficient to stimulate courtship, and mating and nest building commence within a few hours of the start of rain. This rapid response ensures that the baby birds will hatch in the flush of new vegetation and insect abundance stimulated by the rain.

A different approach, very like that of mammals, has been evolved by columbiform birds (pigeons and doves), which are widespread in arid regions. Fledglings are fed on pigeon's milk, a liquid substance produced by the crop under the stimulus of prolactin. The chemical composition of pigeon's milk is very similar to that of mammalian milk; it is primarily water plus protein and fat, and it simultaneously satisfies both the nutritional requirements and the water needs of the fledgling. This approach places the water stress on the adult, which must find enough water to produce milk as well as meet its own water requirements.

Seed-eating desert birds with precocial young, like the sandgrouse found in the deserts of Africa and the Near East, face particular problems in providing water for their young. Baby sandgrouse begin to find seeds for themselves within hours of hatching. But they are unable to fly to water holes as their parents do, and seeds do not provide the water they need. Instead, adult male sandgrouse transport water to their broods. The belly feathers of males have a unique structure in which the proximal portions of the barbules are coiled into helices. When the feather is wetted, the barbules uncoil and trap water. The feathers of male sandgrouse hold 15 to 20 times their weight of water, and the feathers of females hold 11 to 13 times their weight.

Male sandgrouse in the Kalahari Desert of southern Africa fly to water holes just after dawn and soak their belly feathers, absorbing 25 to 40 milliliters of water. Some of this water evaporates on the flight back to their nests, but calculations indicate that a male sandgrouse could fly 30 km and arrive with 10 to 28 ml of water still adhering to its feathers. As the male sandgrouse lands, the juveniles rush to him, and seizing the wet belly feathers in their beaks, strip the water from them with downward jerks of their heads. In a few minutes, the young birds have satisfied their thirst and the male rubs himself dry on the sand.

Summary

Endothermy is an energetically expensive way of life. It allows organisms considerable freedom from the physical environment, especially low temperatures, but it requires a large base of food resources to sustain high rates of metabolism. Endothermy is remarkably effective in cold environments; some species of birds and mammals can live in the coldest temperatures on Earth. The insulation provided by hair, feathers, or blubber is so good that little increase in metabolic heat production is needed to maintain body temperatures 100°C above ambient temperatures. In fact, some aquatic mammals, such as northern fur seals, are so well insulated that overheating is a problem when they are on land or in water warmer than 10 or 15°C.

Hot environments are more difficult for endotherms than cold environments, because endothermal thermoregulation balances internal heat production with heat loss to the environment. When the environment is hotter than the animal, movement of heat is reversed. Evaporative cooling is effective as a short-term response to overheating, but it depletes the body's store of water and creates new problems. Small animals, nocturnal rodents for example, can often avoid much of the daily heat load in hot environments by spending the day underground in burrows and emerging only at night. Larger animals have nowhere to hide and must meet the heat load head-on. Camels and other large mammals of desert regions relax their limits of homeostasis when confronted by the twin problems of high temperatures and water shortage: They allow their body temperatures to rise during the day and fall at night. This physiological tolerance is combined with behavioral and morphological characteristics that reduce the amount of heat that actually reaches their bodies from the environment.

Environments that are both hot and dry—deserts—pose a dual challenge. Animals must have a way of cooling themselves, but water is in short supply. Minimizing the water used to excrete metabolic wastes is an important consideration for desert animals, and birds and mammals achieve water conservation in different ways. Birds have the ancestral sauropsid character of excreting nitrogenous wastes as salts of uric acid; this process releases water as the urate salts precipitate from solution in the urine. Mammals have a unique structure in the kidney, the loop of Henle, that allows them to produce urine with high concentrations of urea and salts. The loop of Henle probably evolved early in the mammalian lineage as an alternative to the uricotelism of sauropsids.

Mobility is an important part of the response of large endotherms to both hot and cold environments: Seasonal movements away from unfavorable conditions (migration) or regular movements between scattered oases that provide water and shade are options available to medium-size or large mammals. The great mobility of birds makes migration feasible even for relatively small species.

When environmental rigors overwhelm the regulatory capacities of an endotherm and resources to sustain high rates of metabolism are unavailable, many small mammals (especially rodents) and some birds enter torpor, a state of adaptive hypothermia.

During torpor the body temperature is greatly reduced, and the animal becomes inert. Periods of torpor can be as brief as a few hours (nocturnal hypothermia is widespread) or can last for many weeks. Mammals that hibernate (enter torpor during winter) arouse at intervals of days or weeks, warming to their normal temperature for a few hours and then returning to a torpid condition. Torpor conserves energy at the cost of forfeiting the benefits of endothermy.

The most remarkable feature of the ability of birds and mammals to live in diverse climates is not the specializations of Arctic or desert animals, remarkable as they are, but the realization that only minor changes in the basic endothermal pattern are needed to permit existence over nearly the full range of environmental conditions on Earth.

Additional Readings

Aschoff, J. 1982. The circadian rhythm of body temperature as a function of body size. In C. R. Taylor, K. Johansen, and L. Bolis (Eds.), *A Companion to Animal Physiology.* Cambridge, U.K.: Cambridge University Press, 173–188.

Bakken, G. S. 1980. The use of standard operative temperature in the study of the thermal energetics of birds. *Physiological Zoology* 53:108–119.

Barnes, B. M. 1989. Freeze avoidance in a mammal: Body temperatures below 0°C in an Arctic hibernator. *Science* 244:1593–1595.

Bartholomew, G. A., and F. Wilke. 1956. Body temperature in the northern fur seal, *Callorhinus ursinus. Journal of Mammalogy* 37:327–337.

Beuchat, C. A. 1990. Body size, medullary thickness, and urine concentrating ability in mammals. *American Journal of Physiology* 258 (*Regulatory, Integrative, Comparative Physiology* 27): R298–R308.

Busch, C. 1988. Consumption of blood, renal function, and utilization of free water by the vampire bat, *Desmodus rotundus. Comparative Biochemistry and Physiology* 90A:141–146.

Calder, W. A. 1994. When do hummingbirds use torpor in nature? *Physiological Zoology* 67:1051–1076.

Calder, W. A., and J. Booser. 1973. Hypothermia of broad-tailed hummingbirds during incubation in nature with ecological correlations. *Science* 180:751–753.

Chaplin, S. B. 1974. Daily energetics of the black-capped chickadee, *Parus atricapillus,* in winter. *Journal of Comparative Physiology* 89:321–330.

Chappell, M. A., and G. A. Bartholomew. 1981a. Standard operative temperatures and thermal energetics of the antelope ground squirrel *Ammospermophilus leucurus. Physiological Zoology* 54:81–93.

Chappell, M. A., and G. A. Bartholomew. 1981b. Activity and thermoregulation of the antelope ground squirrel *Ammospermophilus leucurus* in winter and summer. *Physiological Zoology* 54:215–223.

Davenport, J. 1992. *Animal Life at Low Temperature.* New York: Chapman & Hall.

French, A. R. 1986. Patterns of thermoregulation during hibernation. In H. C. Heller, X. J. Musacchia, and L. C. H. Wang (Eds.), *Living in the Cold: Physiological and Biochemical Adaptations,* New York: Elsevier, 393–402.

Heller, H. C. (Ed.). 1987. Living in the cold. *Journal of Thermal Biology* 12(2) (The entire issue is devoted to this topic.)

McNab, B. K. 1973. Energetics and distribution of vampires. *Journal of Mammalogy* 54:131–144.

Reichman, O. J., and J. H. Brown. 1979. The use of torpor by *Perognathus amplus* in relation to resource distribution. *Journal of Mammalogy* 60:550–555.

Tracy, R. L. 2000. *Adaptive Variation in the Physiology of a Widely Distributed Mammal and Re-examination of the Bases for its Desert Survival.* Unpublished doctoral dissertation, Arizona State University.

Wang, L. C. H. 1978. Energetic and field aspects of mammalian torpor: The Richardson's ground squirrel. In L. C. H. Wang and J. W. Hudson (Eds.), *Strategies in Cold: Natural Torpidity and Thermogenesis.* New York: Academic, 109–145.

Web Explorations

On-line resources for this chapter are on the World Wide Web at: http://www.prenhall.com/pough (click on the Table of Contents link and then select Chapter 21).

CHAPTER

22

Body Size, Ecology, and Sociality of Mammals

We have noted the increase in brain size that has occurred during the evolution of mammals. We have further suggested that the origin of derived features of the mammalian brain might partly be sought in the nocturnal habits that are postulated for Mesozoic mammals. Relying on scent or hearing instead of vision to interpret their surroundings, ancestral mammals may have experienced selection for an increased ability to associate and compare stimuli received over intervals of time. The associative capacity resulting from changes in the brain during the evolution of mammals might also contribute to more complex behavior, and the association of mother and young during nursing could provide an opportunity to modify behavior by learning.

Indeed, social behaviors and interactions between individuals play a large role in the biology of mammals. These behaviors are modified by the environment, and clear-cut relationships between energy requirements, resource distribution, and social systems can often be demonstrated. In this chapter we consider some examples of those interactions that illustrate how complex is the evolution of mammalian social behavior. In addition, we consider the social behavior of several species of primates. The social behavior of many primates is elaborate, but not necessarily more complex than that of some other kinds of mammals, including canids. However, since primates have been the subjects of more field studies than have other mammals, we know a great deal about their social behavior, its consequences for the fitness of individuals, and even a little about the way some species of primates view their own social systems.

22.1 Social Behavior

Sociality means living in structured groups, and some form of group living is found among nearly all kinds of vertebrates. However, the greatest development of sociality is found among mammals. Much of the biology of mammals can best be understood in the context of what sorts of groups form, the advantages of group living for the individuals involved, and the behaviors that stabilize groups.

Mammals may be particularly social animals due to the interaction of several mammalian characteristics, no one of which is directly related to sociality, that in combination create conditions in which sociality is likely to evolve. Thus, the relatively large brains of mammals (which presumably facilitate complex behavior and learning), prolonged association of parents and young, and high metabolic rates and endothermy (with the resulting high resource requirements) may be viewed as conditions that are conducive to the development of interdependent social units.

Of course, not all mammals are social; in fact, there are more species of solitary mammals than social ones. Solitary and social species are known among marsupials and eutherians. Monotremes appear to be solitary (as are most eutherians of that body size), but the three living monotreme species are too small a sample to form a basis for speculations about the phylogenetic origins of sociality among mammals. Of course, the social behavior of mammals does not operate in a vacuum; it is only one part of the biology of a species. Social behavior interacts with other kinds of behavior (such as food gathering, predator avoidance, and reproduction), with the morphological and physiological characteristics of a species, and with the distribution of resources in the habitat. Our emphasis in this chapter is on those interactions, and we illustrate the interrelationships of behavior and ecology with examples drawn from both predators and their prey.

The social behavior of mammals is an area of active research, and our treatment is necessarily limited. Additional information can be found in reviews by Wilson (1975), Eisenberg (1981), Wittenberger (1981), Rubenstein and Wrangham (1986), Gittleman (1989), and Caro (1994). Some interesting examples of the costs and benefits of social behavior were briefly reviewed by Lewin (1987).

22.2 Population Structure and the Distribution of Resources

From an ecological perspective, the distribution of resources needed by a species is usually a major factor in determining its social structure. If resources are too limited to allow more than one individual of a species to inhabit an area, there is little chance of developing social groupings. Thus the distribution of resources in the habitat and the amount of space needed by an individual to fill its resource requirements are important factors influencing the sociality of mammals.

Most animals have a **home range,** an area within which they spend most of their time and find the food and shelter they need. Home ranges are not defended against the incursions of other individuals—an area that is defended is called a **territory.** The value of a home range probably lies in the familiarity of an individual animal with the locations of food and shelter. Many species of vertebrates employ a type of foraging known as traplining, in which they move over a regular route and visit specific places where food may be available. For example, a mountain lion may carefully approach a burrow where a marmot lives, beginning its stalk long before it can actually see whether the marmot is outside its burrow; a hummingbird may return to patches of flowers at intervals that match the rate at which nectar is renewed. This kind of behavior demonstrates a familiarity with the home range and with the resources likely to be available in particular places.

The **resource dispersion hypothesis** predicts that the size of the home range of an individual animal will depend primarily on two factors: the resource needs of the individual, and the distribution of resources in the environment. That is, individuals of species that require large quantities of a resource such as food should have larger home ranges than individuals of species that require less food. Similarly, the home ranges of individuals should be smaller in a rich environment than in one where resources are scarce. The resource dispersion hypothesis is a very general statement of an ecological principle. It applies equally well to any kind of animal and to any kind of resource. The resources usually considered are food, shelter, and access to mates. In Chapter 16 we considered the role of monopolization of resources by individuals in relation to the mating systems of birds, and here we discuss the role of resource dispersion in relation to home range size and sociality of mammals.

Body Size and Resource Needs

Studies of mammals have concentrated on food as the resource of paramount importance in determining the sizes of home ranges. The energy consumption of vertebrates increases in proportion to body mass, raised to a power that is usually between 0.75 and 1.0. If we assume that energy requirements determine home range size, we can predict that home range size will also increase in proportion to the 0.75 or 1.0 power of body mass. That prediction appears to be correct in general, but perhaps wrong in detail (Figure 22–1). That is, the sizes of the home ranges of mammals do increase with increasing body size, but the rates of increase (the slopes in Figure 22–1) are somewhat greater than expected. Home ranges appear to be proportional to body mass raised to powers between 0.75 and 2.0. This relationship between energy requirements and home range sizes suggests that energy needs are important in determining the size of the home range, but that additional factors are involved. One possibility is that the efficiency with which an animal can find and use resources decreases as the size of a home range increases. If that is true, the sizes of home ranges would be expected to increase with the body sizes of animals more rapidly than energy requirements increase with body size.

The failure of the resource dispersion hypothesis to predict the exact relationship between body size and the size of home ranges indicates that we have more to learn about how animals use the resources of their home ranges. So far we have been assuming that resources are distributed evenly throughout the

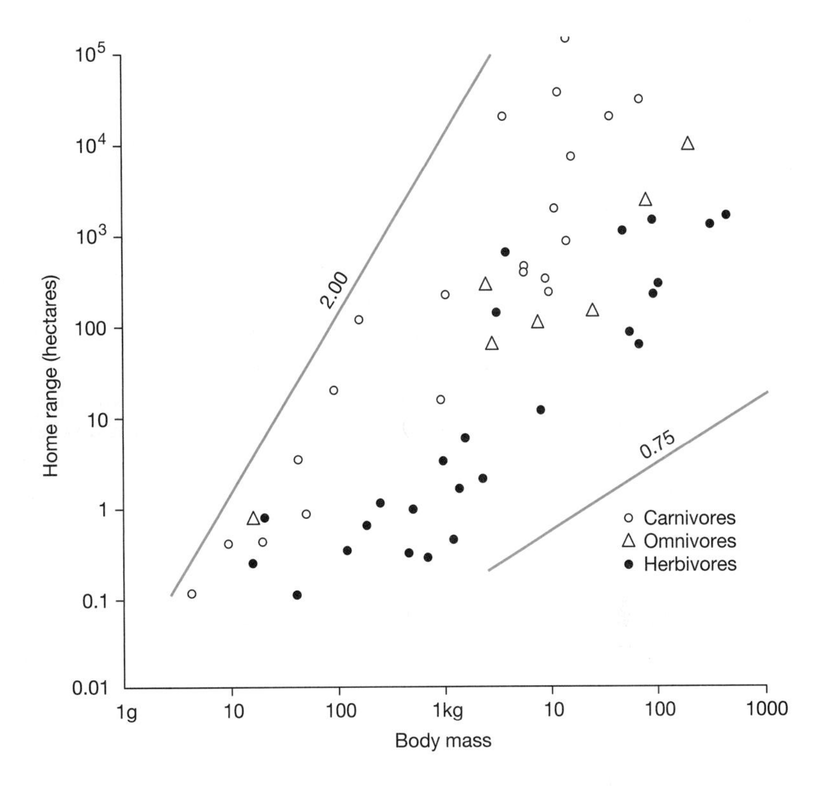

▶ Figure 22–1 Home range size of mammals as a function of body mass. The lines have slopes of 0.75 and 2.0, which illustrate allometric increases in home range size on this double logarithmic scale.

home range, but that assumption overlooks the structural complexity of most habitats. What insights can be obtained from a more realistic consideration of how mammals gather food?

The Availability of Resources

Three factors seem likely to be important in determining the availability of food to mammals: what they eat, whether their food is evenly dispersed through the habitat or is found in patches, and how they gather their food. We will consider examples of each of these factors.

Dietary Habits Figure 22–1 shows that home range size increases with body size, and that dietary habits also affect home range size. Herbivores have smaller home ranges than do omnivores of the same body size, and carnivores have larger home ranges than do herbivores or omnivores. For example, the home range of a red deer (an herbivore) that weighs 100 kilograms is approximately 100 hectares. A bear (an omnivore) of the same body size has a home range larger than 1000 hectares, and a tiger (a carnivore) has a home range of more than 10,000 hectares.

The relationship between home range size and dietary habits of mammals probably reflects the abundance of different kinds of food. The grasses and leaves eaten by some herbivores are nearly ubiquitous, and a small home range provides all the food an individual requires. The plant materials (seeds and fruit) eaten by omnivores are less abundant than leaves and grasses, and different species of plants produce seeds and fruit at different seasons. Thus, a large home range is probably necessary to provide the food resources needed by an individual omnivore. The vertebrates that are eaten by carnivores are still less abundant, and a correspondingly larger home range is apparently needed to ensure an adequate food supply.

Distribution of Resources We have continued to assume that resources are evenly distributed through the habitat, and that one part of a home range is equivalent to another part in terms of the availability of food. This assumption may be valid for some grazing and browsing herbivorous mammals, but it is clearly not true for mammals that seek out fruiting trees (which represent patches of food) or for any carnivorous mammal that preys on animals occurring in groups. The sizes of the home ranges of animals that use food occurring in patches should reflect the distribution of patches: Home ranges should be small if patches are abundant and large if patches are widely dispersed.

That relationship is well illustrated by the home ranges of Arctic foxes (*Alopex lagopus*) in Iceland (Hersteinsson and Macdonald 1982). The foxes live in social groups consisting of one male and two females plus the cubs from the current year. The home ranges of the individuals of a group overlap widely with each other, and there is very little overlap between the home ranges of members of different groups. The home ranges of the foxes are located along the coast and do not extend far into the uplands (Figure 22–2). Between 60 and 80 percent of the diet is composed of items the foxes find on the shore—carcasses of seabirds, seals, and fishes, and invertebrates from clumps of seaweed washed up on the beach. Little food is available for foxes in the uplands. The foxes concentrate their foraging on the beach during the 3 hours before low tide, which is the best time for beachcombing. The foxes approach the shore carefully, stalking along gullies. They creep out on the beach carefully, apparently looking for birds that are resting or feeding. If birds are present, the foxes stalk them. If no birds are present, the foxes search the beach for carrion.

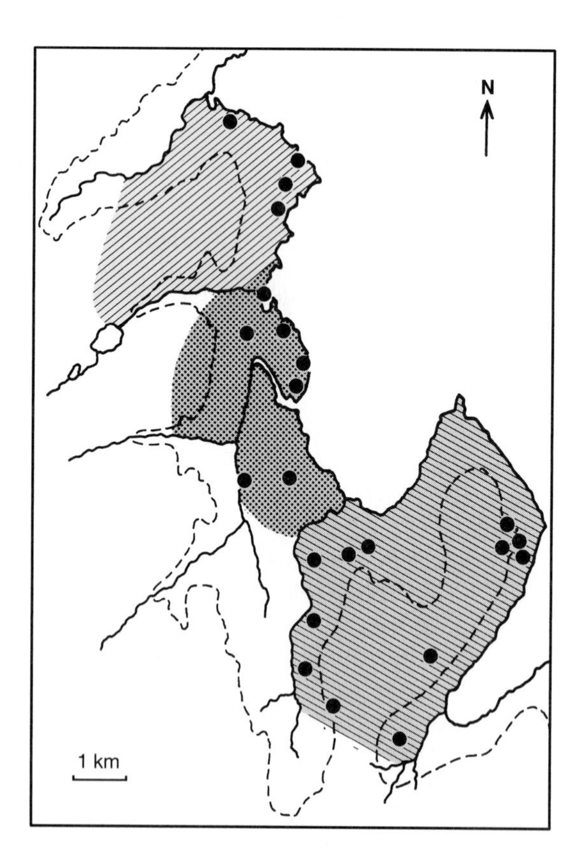

▲ Figure 22–2 Map of the territorial boundaries of three groups of Arctic foxes in Iceland. The 200-meter contour line is shown. Black dots mark the sites of dens used by the foxes.

The researchers studied three groups of foxes using radiotelemetry to follow the movements of individuals. The areas of the home ranges varied more than twofold, from 8.6 to 18.5 square kilometers (Table 22.1). The sizes of the home ranges were slightly more similar when only the coastline was considered: Each territory included between 5.4 and 10.5 km of coastline.

The coastline was, of course, the main source of the foxes' food, but not all areas of the coastline accumulated floating objects. The distribution of food on the beaches was patchy and depended on the directions of currents. As a result, some parts of the shore were more productive than others. The length of productive coastline occupied by each group of foxes was quite similar—from 5.4 to 6.0 km. Farmers in Iceland use driftwood to make fence posts, and the amount of driftwood that was harvested by the farmers from the coasts in the home ranges of the three groups of foxes varied only from 1800 to 2100 logs per year. Because both driftwood and carrion are moved by currents and deposited on the beaches, the harvest of driftwood by farmers probably reflects the harvest of carrion by the foxes. Thus, the home range sizes of the three groups appear to match the distribution of their most important food resource, and the productive areas of the home ranges of the three groups are very similar despite the more than twofold difference in total areas of their home ranges.

Group Size and Hunting Success It is readily apparent that the average home range size of a species of mammal can influence the social system of that species. Individuals of a species probably encounter each other frequently when home ranges are small, whereas individuals of species that roam over thousands of hectares may rarely meet. Thus the distribution of resources in relation to the resource needs of a species is one factor that can limit the degree to which social groupings can occur. However, sociality may influence resource distribution if groups of animals are able to exploit resources that are not available to single individuals.

Table 22.1 Home ranges of three groups of Arctic foxes in Iceland

	Group 1	*Group 2*	*Group 3*	*Average*
Total area (km^2)	10.3	8.6	18.5	12.5
Length of coastline (km)	5.6	5.4	10.5	7.2
Length of productive coastline (km)	5.6	5.4	6.0	5.7
Driftwood productivity (logs/year)	1800	1800	2100	1900

The influence of sociality on resource distribution may be seen among predatory animals that can hunt individually or in groups. Some species of prey are too large for an individual predator to attack, but are vulnerable to attack by a group of predators. For example, spotted hyenas (*Crocuta crocuta*) weigh about 50 kg. When hyenas hunt individually, they feed on Thomson's gazelles (*Gazella thomsoni*, 20 kg) and juvenile wildebeest (*Connochaetes taurinus*, about 30 kg; Figure 22–3). However, when hyenas hunt in packs, they feed on adult wildebeest (about 200 kg) and zebras (*Equus burchelli*, about 220 kg). Some species of prey have defenses that are effective against individual predators but less effective with groups of predators. For example, the success rate for solitary lions (*Panthera leo*) hunting zebras and wildebeest is only 15 percent, whereas lions hunting in groups of six to eight individuals are successful in up to 43 percent of their attacks. Groups of lions make multiple kills of wildebeest more than 30 percent of the time, but individual lions kill only a single wildebeest.

The relationship of sociality and body size of a predator to the size of its prey is shown in Figure 22–4: Social predators (defined as those that hunt in groups of eight to ten individuals) attack larger prey than weakly social predators (average group sizes of 1.6 to 3.1 individuals), and these weakly social predators attack larger prey than do solitary predators (average group sizes of 1.0 to 1.3 individuals). One consequence of sociality for predatory mammals thus appears to be an increase in the potential food resources of an environment: Individual predators may be able to extend the range of prey species they can attack by hunting in groups.

Of course, the major disadvantage to hunting in a group is that there are more mouths to feed when a kill is made. The food requirement of a group of predators is the sum of the individual requirements of the members of the group, and the per-capita amount of food obtained by hunting in a group would have to exceed that caught by a solitary hunter to make group hunting advantageous.

Packer and Ruttan (1988) have reviewed factors that could contribute to the evolution of cooperative hunting. The question is, do predators form groups *because* that allows them to hunt large prey, or *must* they hunt large prey because they live in groups for some other reason? A study of lions on the Serengeti Plains suggests that the second hypothesis is correct (Packer et al. 1990). Female lions are the only female felids to live in social groups. Female lions defend a group territory and protect their cubs from other groups of female lions. The high population densities that are characteristic of lions may have favored group defense of a territory. The presence of large prey makes it possible for lions to hunt in groups, but group hunting does not increase the amount of food available per lion—a lion hunting by herself can catch as much prey as her share of a group capture. So groups form because of the advantages they provide in the social structure of the population of lions on the Serengeti, and group hunting is a by-product of that social structure.

A similar interpretation has been suggested for the formation of groups of male cheetahs (Caro 1994). Male cheetahs may live alone or form permanent coalitions of two or three individuals that live and hunt together. In Caro's study these coalitions were often composed of littermates, and a coalition was more successful in occupying a territory than was a single male. Competition for territories was intense, and territorial disputes were an important

▲ Figure 22–3 Spotted hyenas. Spotted hyenas (a) may hunt individually or in packs. They prey on small animals like the Thomson's gazelle (b) when they hunt individually. When they hunt in packs, they attack larger prey such as the wildebeest (c) and zebras (d).

source of mortality for male cheetahs. Cheetahs hunting singly concentrated on small prey such as Thomson's gazelles, whereas coalitions attacked larger prey such as wildebeest. Overall foraging success increased with group size for male cheetahs, but Caro concluded that the benefit of a coalition in holding a territory and controlling access to females was probably more important than its effect on food intake.

22.3 Advantages of Sociality

The advantages of cooperative hunting may provide a basis for sociality among some carnivorous mammals. But sociality is not limited to species of mammals that hunt in groups, and the potential advantages of sociality are not confined to predatory behavior. Mammals may derive benefits from sociality in terms of avoiding predation and facilitating feeding and in reproduction and care of young.

Defenses against Predators

One probable advantage of sociality is the reduced risk of predation for an individual that is part of a group compared to the risk for a solitary individual. The benefits of sociality in avoiding predation take many forms. A group of animals may be more likely to detect the approach of a predator than an individual would be, simply because a group has more eyes, ears, and noses to keep watch. An individual in a group may be able to devote a larger proportion of

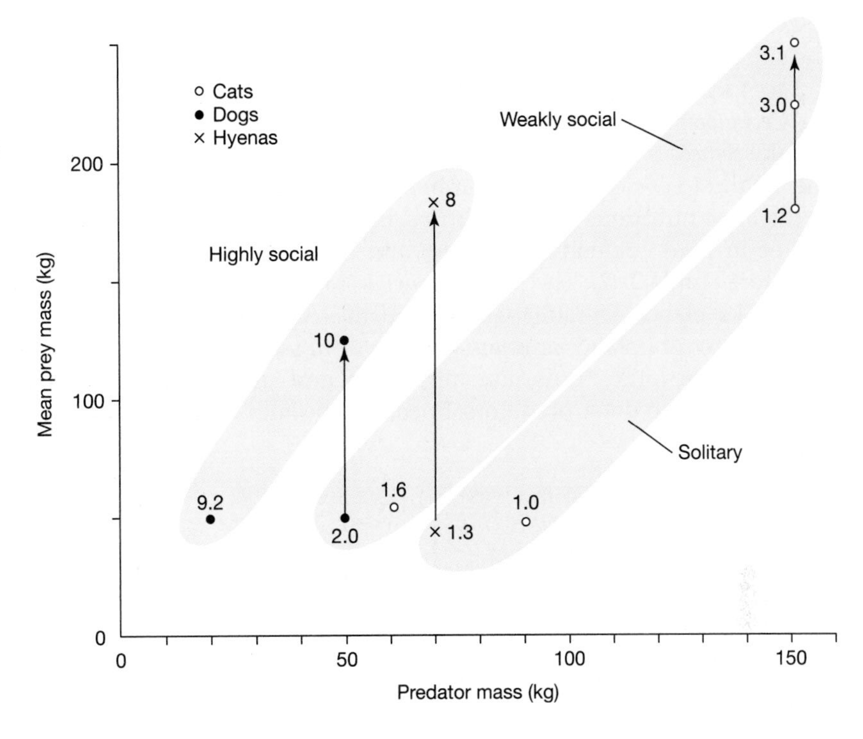

▶ Figure 22–4 Size of prey in relation to predator mass for solitary predators and for predators that hunt in small or large groups. The numbers are average group sizes; vertical lines connect points for species that hunt in groups of variable size.

its time to feeding and less to watching for predators than a solitary individual can. Mammals that live in groups generally occur in open habitats, whereas solitary species are usually found in forests; that relationship may partially reflect the antipredator aspects of group living. It is important to note that the benefits of sociality in predator avoidance result from the reduced risk of predation for an *individual* that is part of a group, not for the species as a whole.

Sociality and Reproduction

Groupings of animals are important factors in mating systems and in care of young. In Chapter 16 we described mating systems in the context of avian biology, and most aspects of that discussion apply equally well to the mating systems of mammals. In the next section we extend the analysis by considering the specific relationships among body size, habitat, diet, antipredator behavior, and mating systems of several species of African ungulates.

The extensive period of dependence of many young mammals on their parents provides a setting in which many benefits of sociality can be manifested. Maternal care of the offspring is universal among mammals, and males of many species also play a role in parental care. Group living provides opportunities for complex interactions among adults and juveniles that involve various sorts of **alloparental behavior** (care provided by an individual that is not a parent of the young receiving the care). Collaborative rearing of young of several mothers is characteristic of lions and of many canids. Frequently, nonbreeding individuals join the mothers in protecting and caring for the young. Among dwarf mongooses (*Helogale undulata*) this kind of behavior extends to the care of sick adults; similar behaviors are reported for mammals as diverse as elephants and cetaceans. Many social groups of mammals consist of related individuals, and these nonbreeding helpers may increase their inclusive fitness by assisting in rearing the offspring of their kin.

22.4 Body Size, Diet, and the Structure of Social Systems

The complex relationships among body size, sociality, and other aspects of the ecology and behavior of herbivorous mammals are illustrated by the variation in social systems of African antelopes (family

Bovidae) (Jarman 1974, Leuthold 1977, Estes 1991). The smallest species of antelopes have adult weights of 3 to 4 kg (the dik-diks, *Madoqua,* and some duikers, *Cephalophus*); one of the the largest (the African buffalo, *Syncerus caffer*) weighs 400 kg (Figure 22–5). The smallest species are forest animals that browse on the most nutritious parts of shrubs, live individually or in pairs, defend a territory, and hide from predators (Table 22.2). The largest species (including the 500 kg eland, *Taurotragus oryx,* and the African buffalo) are grassland animals that feed unselectively, live in large herds, are migratory, and use group defense to deter predators. Species with intermediate body sizes are also intermediate in these ecological and behavioral characteristics. It seems likely that the correlated variation in body size, ecology, and behavior among these antelopes reveals functional relationships among these aspects of their biology. How might such diverse features of mammalian biology interact?

An antelope's feeding habits appear to provide a key that can be used to understand other aspects of its ecology and behavior. The diets of different species are closely correlated with their body size and habitats. In turn, those relationships are important in setting group size. The size of a group deter-

(a)

(b)

(c)

▲ Figure 22–5 The Bovidae includes species with a wide range of adult body sizes. The dik-dik (a) is among the smallest species, the impala (b) is medium-size, and the African buffalo (c) is one of the largest.

Table 22.2 Correlations of the ecology and social systems of African ungulates

Diet Type	*Examples*	*Body Mass (kg)*	*Food Habits*	*Group Size*	*Mating System*	*Predator Avoidance*
I	Dik-dik, some duikers	3–20	Highly selective browsers	1 or 2	Stable pair, territorial	Hide
II	Thomson's gazelle, impala	20–100	Moderately selective browsers and grazers	2 to 100	Male territorial in breeding season, temporary harems	Flee
III	Wildebeest, hartebeest	100–200	Grazers, selective for growth stage	Large herds	Nomadic, temporary harems	Flee, hide in herd, threaten predator
IV	Eland, buffalo	300–900	Unselective browsers and grazers	Large herds	Male hierarchy	Group defense

mines the distribution of females in time and space, and this is a major factor in establishing the mating system used by males of a species. Group size also plays an important role in determining the appropriate antipredator tactics for a species. Mating systems and antipredator mechanisms are central factors in the social organization of a species.

Body Size and Food Habits

Antelope are ruminants, relying on symbiotic microorganisms in the rumen to convert cellulose from plants into compounds that can be absorbed by the vertebrate digestive system. The effectiveness of ruminant digestion is proportional to body size. This relationship exists because the volume of the rumen in species with different body sizes is proportional to body mass, whereas metabolic rates are proportional to the 0.75 power of body mass. The ecological consequence of this difference in allometric slopes is illustrated in Figure 22–6: A large ruminant has proportionately more capacity to process food than does a small ruminant. For animals of very small body size, the metabolic requirements become high in relation to the volume of rumen that is available to ferment plant material (see Chapter 20).

Because of this relationship, small ruminants must be more selective feeders than large ruminants. That is, a large ruminant has so much volume in its rumen that it can afford to eat large quantities of food of low nutritional value. It does not extract much energy from a unit volume of this food, but it is able to obtain its daily energy requirements by processing a large volume of food. Small ruminants, in contrast, must eat higher-quality food and rely upon obtaining more energy per unit volume from the smaller volume of food that they can fit into their

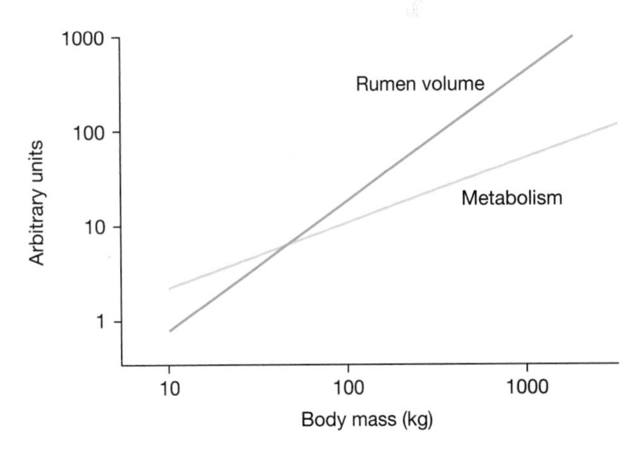

▲ Figure 22–6 Rumen volume and energy requirements in relation to body size. Rumen volume increases in proportion to body size (an allometric slope of 1), whereas energy requirements are proportional to metabolism (an allometric slope of 0.75). Thus large species are more effective ruminants than small species. Both axes are drawn with logarithmic scales, and the scale of the vertical axis is in arbitrary units.

rumen in a day. In fact, 40 kg is the approximate lower limit of body size at which an unselective ruminant can balance its energy budget; species larger than 40 kg can be unselective grazers, whereas smaller species must eat only the most nutritious parts of plants.

The species of antelopes in this example can be divided into four feeding categories:

- **Type I** species are selective browsers. They feed preferentially on certain species of plants, and they choose the parts of those plants that provide the highest-quality diet—new leaves (which have a higher nitrogen and lower fiber content than mature leaves) and fruit. Dik-diks and duikers fall in this category, and they have adult body masses between 3 and 20 kg. These animals show little sexual dimorphism in body size and appearance. The males have small horns, while the females are hornless or may also have small horns.
- **Type II** species are moderately selective grazers and browsers. They eat more parts of a plant than the type I species, and they may have seasonal changes in diet as they exploit the availability of fresh shoots or fruits on particular species of plants. Thomson's gazelle (*Gazella thomsoni*) and the impala (*Aepyceros melampus*) weigh 20 to 100 kg and have type II diets. These animals show substantial sexual dimorphism in body size, with the males being about a third again larger than the females in body mass. They are highly dimorphic in appearance: Males have large, elaborate horns and may have a different coat color from females. Females are either hornless, or have horns that are much smaller than those of males.
- **Type III** species are primarily grazers that are unselective for species of grass, but selective for the parts of the plant. That is, they eat the leaves and avoid the stems. Hence, they are selecting for a growth stage: They avoid grass that is too short, because that limits food intake. They also avoid grass that is too long, because it has too many stems that are low-quality food. Wildebeest (*Connochaetes taurinus*) and hartebeest (*Alcelaphus buselaphus*), which weigh about 200 kg, are type III feeders. Type III species show little sexual dimorphism in size or appearance, and the horns of females are nearly as large as those of males.
- **Type IV** species are very large and are unselective grazers and browsers. They eat all species of plants and all parts of the plant. Eland (*Taurotragus oryx*, 500 kg) and buffalo (*Syncerus caffer*, 400 kg) are type IV species. Males are substantially larger than females, but there is little dimorphism in horns.

Food Habits and Habitat

The food habits of different antelope species are important in determining what sorts of habitats provide the resources they need. Selective feeding operates at three levels: vegetation type, species and individual groups of plants, and parts of plants eaten. The type of vegetation present largely depends on the habitat—forests contain shrubs and bushes, whereas the plains are covered with grass. The resources needed by species with type I diets are found in forests where the presence of a diversity of species with different growth seasons ensures that new leaves and fruit will be available throughout the year. Species with type II diets are found in habitats that are a mosaic of woodland and grassland, and type III species (which are primarily grazers) are found in savanna and grassland areas. Species with type II and type III diets may move from place to place in response to patterns of rainfall. For example, wildebeest require grass that has put out fresh new growth, but that has not had time to mature. To find grass at this growth stage, wildebeest have extensive nomadic movements that follow the seasonal pattern of rain on the African plains.

Type IV feeders eat almost any kind of plant material, and they can find something edible in almost any habitat. They occupy a range of habitats, including grassland and brush, and do not have nomadic movements.

Habitat and Group Size

The habitats in which antelope feed and the types of food they eat set certain constraints on the sorts of social groupings that are possible. For example, species with type I diets live in forests and feed on scattered, distinct items. They eat an entire leaf or fruit at a bite, and they must move between bites. A type I feeder completely removes the items it eats, so it changes the distribution of resources in its habitat (Figure 22–7). A second individual cannot feed close

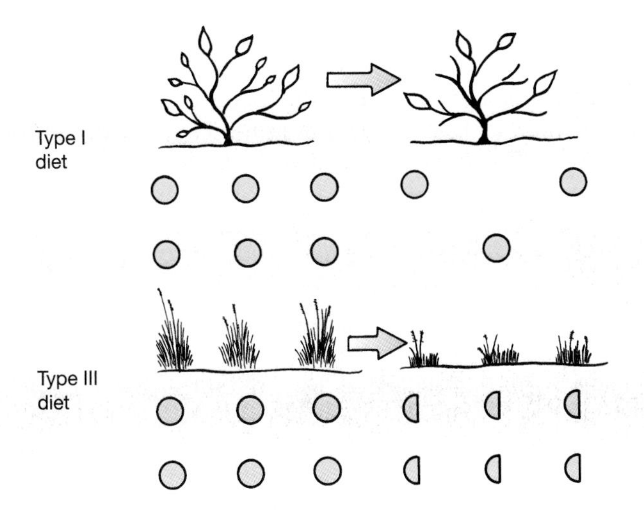

▲ Figure 22–7 The effect of feeding by a selective browser with a type I diet and a grazer with a type III diet. (a) The browser removes entire food items (new leaves or fruit), thereby changing the distribution of food in the habitat, as well as the abundance of food. (b) The grazer removes part of a grass clump, changing the abundance of food in the habitat but not its distribution.

behind the first, because the food resources of an individual bush or shrub are entirely consumed by the first individual to feed there. As a result, the feeding behavior of species with type I diets makes it impossible for a group of animals to feed together. If one individual attempts to follow behind another to feed, the second animal must search to find food items overlooked by the individual ahead; consequently, it falls behind. Alternatively, it can move aside from the path of the first animal to find an area that has not already been searched. In either case, small animals in dense vegetation rapidly lose track of each other, and no cohesive group structure is maintained.

Instead, type I species are solitary or occur in pairs, and the individuals of a pair are only loosely associated as they feed. A type I diet places a premium on familiarity with a home range, because a tree or bush is a patch of food that must be visited repeatedly to harvest fruit or new leaves as they appear.

Species of antelope with type II and type III diets are less selective than species with type I diets, and their feeding has less impact on the distribution of resources. These species do not remove all of the food resource in an area, and other individuals can feed nearby. Type III feeders, in particular, graze as they walk—taking a bite of grass, moving on a few steps, and taking another bite. This mode of feeding changes the abundance of food, but not its distribution in space; herds of wildebeest graze together, all moving in the same direction at the same speed and maintaining a cohesive group. Rainfall is a major determinant of the distribution of suitable food in the habitat of these species. The rainstorms that stimulate the growth of grass are erratic, and the patch sizes in which food resources occur are enormous—hundreds of square kilometers of new grass where rain fell are separated by hundreds of kilometers of old, dry grass that did not receive rain. Instead of having home ranges or territories, species with type II and type III diets move nomadically with the rains. Group sizes change as the distribution of resources changes, from half a dozen to 60 individuals for species with type II diets, and from herds of 300 or 400 to superherds of many thousands of individuals during the nomadic movements of wildebeests.

Species of antelope with type IV diets are so unselective in their choice of food that they can readily maintain large groups. Herds of buffalo may number in the hundreds. Because these species can eat almost any kind of vegetation, the distribution of resources does not change seasonally and the size of the herds is stable.

Group Size and Mating Systems

The mating systems used by African antelope are closely related to the size of their social groups and the distribution of food, because those are the major factors determining the distribution of females and the potential for males to obtain opportunities to mate by controlling resources that females need. The females of species with type I diets are dispersed because the distribution of resources in the habitat does not permit groups of individuals to form. A male of a type I species can defend food resources, but individuals must disperse through the territory to feed, and it is not feasible for a male to maintain a territory that attracts a group of females. Males of type I species pair with one female; the male defends its territory year-round, the pair bond with an individual female appears to be stable, and offspring are driven out of the territory as they mature.

A group of individuals of type II species contains several males and females. The evenly distributed nature of food of these species makes it difficult for a male to monopolize resources. Only some of the males are territorial, and even this territoriality is manifested for only part of the year. A territorial male tries to exclude other males from its territory and to gather groups of females. Exclusive mating rights are achieved by holding a patch of ground and containing females within it, driving them back if they try to leave. These species have no long-term association between a male and a particular female.

Type III species are nomadic, and males establish territories only when the herd is stationary. During these periods male wildebeest have access to groups of females within their territories, but the association between a male and females is broken when the herd moves on. However, mothers and their daughters maintain associations for 2 or 3 years. Unmated male wildebeest form bachelor herds with hierarchies, and individuals at the top of a bachelor hierarchy try to displace territorial males. If a territorial male is displaced, it joins a bachelor herd at the bottom of the hierarchy and must work its way up to the top before it can challenge another territorial male.

The social structure of buffalo (a type IV species) differs in two respects from that of wildebeests (type III): (1) Each herd includes many mature males that form a dominance hierarchy. Ability to attain dominance over other males is largely related to body size, and the males of type IV antelope grow throughout life. A mature male may thus be twice as heavy as a female. The individuals near the top of the hierarchy court receptive females, but no territoriality or harem formation is seen. (2) The female membership of the herd is fixed, and this situation results in a degree of genetic relationship among all the members of a herd of buffalo. That genetic relationship among individuals creates situations in which kin selection may be a factor in the social behavior of the African buffalo.

Mating Systems and Predator Avoidance

Prey species have various ways to avoid predators, but only some will work in a given situation. In general, a prey species can (1) avoid detection by a predator, (2) flee after it has been detected but before the predator attacks, (3) flee after the predator attacks, or (4) threaten or attack the predator. Body size, habitat, group size, and the mating system all contribute to determining the risk of predation faced by a species, as well as which predator avoidance methods are most effective.

Predators usually attack prey that are the same size as the predator or smaller. Thus, small species of prey animals potentially have more species of predators than do large species. Species of antelope with type I diets are small; consequently, they are at risk from many species of predators. Furthermore, small antelope may not be able to run fast enough to escape a predator after it has attacked. On the other hand, these small antelope live in dense habitats, where they are hard to see. They are cryptically colored and secretive, and they rely on being inconspicuous to avoid detection by predators. If they are pursued, they may be able to use their familiarity with the geography of their home range to avoid capture.

Groups of animals are more conspicuous to predators than are individuals, but groups also have more eyes to watch for the approach of a predator. Species of antelope with type II diets live in small groups in open habitats, where they can detect predators at a distance. These antelope avoid predators by fleeing either before or after the predator attacks (Box 22-1). Small predators may be attacked by the antelope, but usually only when a member of the group has been captured. This sort of defense is normally limited to a mother protecting her young; the rest of the group does not participate.

Species of antelope in the type III diet category are large enough to have relatively few predators, and in a group they may be formidable enough to scare off a predator. Wildebeest sometimes form a solid line and walk toward a predator: This behavior is effective in deterring even lions from attacking. Many predators of wildebeest focus their attacks on calves, and defense of a calf is usually undertaken only by the mother. Much of the antipredator behavior of wildebeest depends on the similarity of appearance of individuals in the herd to each other. Field observations have shown that the individuals in a group of animals that are distinctive in their markings or behavior are most likely to be singled out and captured by predators.

One of the unavoidable events that makes a female wildebeest distinctive is giving birth to a calf, and the reproductive biology of the species has specialized features that appear to minimize the risks associated with giving birth. For wildbeest, the breeding season and birth are highly synchronized. Mating occurs in a short interval; consequently, 80 percent of the births occur within a period of 2 or 3 weeks. Furthermore, nearly all of the births that will

BOX 22-10 ALTRUISM—OR TAUNTING?

A distinctive behavior—stotting, which consists of leaping vertically into the air—is used by some species of antelope when they are threatened by a predator. The function of stotting is unclear. It may be an alarm signal that alerts other individuals of the species to the presence of a predator, but the advantage to the individual that gives the warning is not clear. Altruistic behavior of this type is usually associated with kin selection; but the individuals in groups of antelope with type II diets are not closely related, and they do not show other types of altruistic behavior such as group defense of young. It has been suggested that some behaviors of prey species that had been considered to be altruistic alarm signals are really signals directed to the predator by fleet-footed prey.

Alarm signals are given by many other kinds of vertebrates. A familiar example is the white underside of the tail of deer (Figure 22–8). A deer that sees a predator at a distance does not immediately flee, but stands watching the predator. It may flick its tail up and down, exposing the white ventral surface in a series of flashes. European hares stand erect on their hind legs when a fox in the open approaches within 30 meters of the hare. In this posture, the hare is readily visible to the fox.

This kind of behavior is not limited to mammals. For example, several related species of fleet-footed lizards that live in open desert habitats have dorsal colors that blend with the substrate on which they live, and a pattern of white with black bars on the underside of the tail. These lizards stand poised for flight as a predator approaches, looking back over their shoulder at the predator. The tail is curled upward—exposing the contrasting black-and-white pattern on its ventral surface—and waved from side to side. Is it possible that in behaviors of this sort the prey animal is signaling to the predator that it has been detected, and to attack it will be unprofitable because the prey is ready to flee? Does the prey's behavior deter pursuit by the predator?

This intriguing hypothesis requires experimental test of the prediction that predators are less likely to attack prey individuals that engage in these behaviors than individuals that do not signal their awareness of the predator's approach. However, if the hypothesis can be supported by experiments, some puzzling examples of apparently altruistic signals can be reinterpreted as behavior that benefits the individual giving the signal (Hasson 1991).

▲ Figure 22–8 Alarm signals to conspecifics or signals to a predator? The white-tail deer displays the white underside of its tail when it detects a predator.

take place on a day occur in the morning in large aggregations of females, all giving birth at once. A female wildebeest that is slightly out of synchrony with other members of her group can interrupt delivery at any stage up to emergence of the calf's head in order to join the mass parturition. Presumably, this remarkable synchronization and control of parturition reflects the advantage of presenting predators with a homogeneous group of cows and calves rather than a group with only a few calves that could readily be singled out for attack.

Type IV species, such as buffalo, are formidable prey even for a pride of lions. They escape much potential predation simply because of their size. When buffalo are attacked, they engage in group defense; if a calf is captured, its distress cries bring many members of the group to its defense. This altruistic behavior probably represents kin selection, because the stability of the female membership of buffalo herds results in genetic relationships among the individuals.

Horns and Antlers as Social Signals

Horns and antlers are conspicuous features of the antelope we have been discussing, and they are characteristic of many large ungulates. Their primary roles appear to be social recognition, sexual display, and jousting between males, although they may also be used for defense. Figures 22–9 illustrates various types of mammalian cranial appendages, which is the collective term for horns and antlers.

Today all horned ungulates are found among ruminant artiodactyls (deer, giraffes, antelope, and other bovids), but in the past some other types of artiodactyls also had horns. Cervids (deer, caribou, and moose) have antlers. Both horns and antlers are outgrowths from the frontal bone of the skull. A mountain sheep illustrates the typical horn structure of ungulates (Figure 22–9a). The horn consists of a bony core covered by a sheath composed of keratin. The horn grows from its base, and the keratin portion of the horn extends well beyond the bony core. (This is why the tips of cows' horns can be blunted without causing pain to the animal—the keratin is dead material, like fingernails, hooves, and hair.)

Giraffids (giraffes and the okapi), too, have unusual horns called ossicones (Figure 22–9c). The bony core is not an outgrowth of the frontal bone; instead it is a separate bone that fuses with the frontal bone during development. Giraffes' horns are covered with skin rather than being formed by a sheath of keratin.

Caribou (cervids) have antlers rather than horns (Figure 22–9d). Antlers are confined to males of most species of cervids, but female reindeer and caribou have antlers. (Reindeer and caribou are the only species of cervids that form herds containing males and females.) Unlike horns, antlers are branched, consist only of bone, and usually are shed annually. As they grow, antlers are covered by a layer of highly vascularized skin (the velvet). When the antler is mature, blood flow to the velvet is cut off and the skin sloughs off to reveal the bony antler. (In Asia, velvet is reputed to impart virility. New Zealand has a major deer-farming industry, and dried velvet sells for thousands of dollars per kilo.)

Although the cranial appendages of ruminant artiodactyls appear rather similar, they are not homologous in their mode of growth. They appear to have evolved independently within different ruminant lineages. Modern rhinos are unlike other horned ungulates because their horns are formed entirely of keratin (the epidermal protein that forms hair and fingernails) and are found in both males and females (although some extinct rhinos had bony horns that were present in males only). Rhino horns are single (not paired) structures that form on the midline of the nose region (Figure 22–9b). In contrast, the horns of extant ruminants are paired and form above the eyes, although some fossil artiodactyls had single horns on the nose or on the back of the head, in addition to the paired horns over the eyes.

The evolution of ruminant horns appears to be tied in with their socioecology. The evolution of ruminant horns can be understood in the context of changing Cenozoic habitats, which in turn led to changes in diet, body size, behavior, and morphology.

The ancestors of horned ruminants first appeared in the fossil record in the late Oligocene, when they were small, hornless animals with teeth indicating a diet of fruit and young leaves (i.e., a type I diet). They would have appeared similar to present-day chevrotains (primitive ruminants) and duikers (antelope), which inhabit the tropical forests of Asia and Africa. By the early Miocene the Eurasian woodlands where these animals lived had become more seasonal and more open in structure. These changes in vegetation changed the availability of food resources. Ruminants responded by becoming somewhat larger (goat-size rather than rabbit-size) and evolving teeth more capable of eating fibrous vegetation such as mature leaves (a type II diet).

This new diet enabled the ruminants to adopt a new type of social behavior. The social behavior of the small, early ruminants was probably like that of the chevrotains: solitary or monogamous with an individual home range. Mature leaves are much more abundant and concentrated in space than new leaves and fruits or berries, which are widely dispersed in the environment. The new, larger ruminants that ate mature leaves could find their food in a smaller home range. With food concentrated in a smaller area, leaf-eating ungulates could become territorial, defending a territory large enough for several animals. This ecological strategy would not have been practical for smaller ungulates, because home ranges large enough to support several animals would be too big to patrol effectively.

Thus territorial ruminants move from a monogamous type of mating system, with only a single fe-

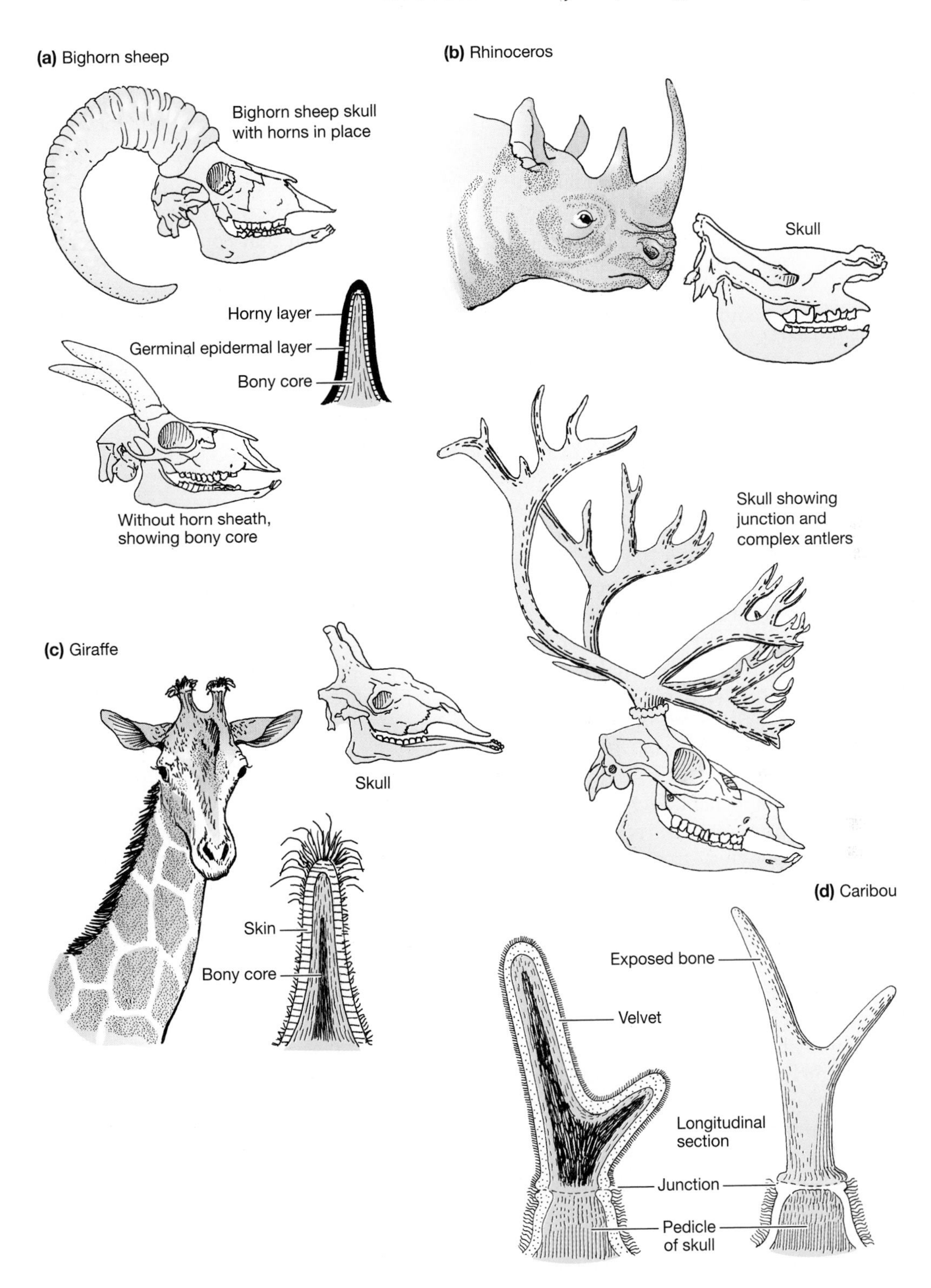

▲ Figure 22–9 Cranial appendages. (a) Horns of a bighorn sheep. (b) Rhinoceros horns. (c) Giraffe ossicones. (d) Antlers of a caribou.

male for each male, to a polygynous one with many potential mates for each male. In this situation some males could have greater reproductive success (i.e., mate with more females) than other males. Intense male-male competition promoted the evolution of horns or equivalent structures used for social displays. (This is because horns are used for ritual display and stylized combat, and they may actually reduce the incidence of injury during male-male interactions.)

That explanation is, of course, a historical speculation; we will never know for certain what actually happened. However, three lines of evidence are consistent with this interpretation—the timing of the evolution of horns in African and Eurasian ungulates, sexual dimorphism in the occurrence of horns and antlers, and the failure of North American ungulates to evolve horns.

Horns or their equivalents appeared in different families of ruminants in Africa and Eurasia at about the same time in the early Miocene. The evolution of horns was correlated with a change in habitat (seen from the plant fossil record) and an increase in body size (seen in the fossil record of the animals). Among present-day ruminants smaller, solitary forms are hornless. Larger forms, where the females are in groups and the males are territorial, have horns. African and Eurasian ungulates crossed this size threshold as the climate and habitat changed in the Miocene. Thus the evolution of horns correlates with a likely change in behavior, from solitary and monogamous to group-forming and polygamous. What we see today as an ecomorphological shift along a habitat gradient (forest to woodland) may be what happened in evolution over a temporal shift of changing habitats in the Miocene.

Horns appear to have evolved initially only in males. Fossils show that early members of all horned ruminant lineages included both horned and hornless individuals. Presumably, those with horns were males and those without horns were females. This sexual dimorphism suggests that horns were initially used in male-male interactions. If horns had originally been used for some activity that both sexes engage in, such as defense against predators, both sexes should have evolved them simultaneously.

Large, grazing ruminants, type III and IV species such as wildebeest and buffalo, now have home ranges too large to be defended as territories. These animals are no longer sexually dimorphic, although instead of the males loosing their horns the females have evolved horns as well. The females use these horns in competition with the males for feeding resources, now that they live with them year-round (Jarman 2000).

The failure of horns to evolve in North American ungulates such as camels and horses may be partially explained by the different pattern of vegetational change on that continent. Grasslands rapidly replaced forests in North America, without a persistent stage of open forests. In grassland habitats, camels and horses were unlikely to have passed through an evolutionary stage where territorial defense would have been a worthwhile ecological strategy. Perhaps this is why they never evolved the type of sexual dimorphism seen in antelopes. Additionally, hindgut fermenters like horses are less efficient feeders than ruminants, and so may always have required a home range area that is too large to defend as a territory. Apparently there has never been a suitable ecological niche for a unicorn.

Horses and camelids (camels and llamas) have a different type of social system, called harems. Both these types of ungulates form permanent associations of females and their young, usually accompanied by only a single male. Males that are not part of a harem association form bachelor herds. In this social system the male defends a group of females from other males, rather than defending a piece of real estate. The term *harem* conjures up visions of a male controlling and dominating a group of females, and that was the original interpretation applied by behavioral ecologists. However, more careful observation (and the presence of more female ecologists who have brought new perspectives to the field) has revealed that it is the bond between the females that is the basis of the harem. The females then allow a male to join their social grouping because he keeps other males from constantly pestering them and interfering with the time they can spend feeding.

22.5 Primate Societies

The phylogenetic relationship of humans to other primates has led some biologists to assume that these are the animals that should have the most elaborate social systems, and that studying the social systems of primates will provide information about the evolution of human behavior. Both assumptions are controversial: A growing base of information indicates not only that complex social systems exist among many kinds of vertebrates other than primates but also that interpreting primate behavior in the context of human evolution is fraught with difficulty and must be cautiously approached. Nonetheless, some primates do have elaborate and complex social systems, and more long-term research has focused on the social systems of primates than on any other vertebrates. A review of primate behavior emphasizes its variety and complexity and sets the stage for considering the evolution of humans in the next chapter.

The approximately 200 species of primates (Table 22.3) are ecologically diverse. They live in habitats ranging from lowland tropical rain forests, to semideserts, to northern areas that have cold, snowy winters. Some species are entirely arboreal, whereas others spend most of their time on the ground. Many are generalist omnivores that eat fruit, flowers, seeds, leaves, bulbs, insects, bird eggs, and small vertebrates. However, many of the colobus (*Colobus*) and howler monkeys (*Alouatta*) are specialized folivores (leaf eaters) with digestive tracts in which bacteria and protozoans ferment cellulose, and some of the small prosimians and callithricid monkeys are insectivores.

Social Systems of Primates

R. W. Wrangham (1982) proposed that the social systems of primates can best be classified based on the amount of movement of females occuring between groups (Table 22.4). Four categories can be defined on this basis.

Female transfer systems In species with this type of social organization, most females move away from the group in which they were born to join another social group. Because of this migration of females among groups, the females in a group are not closely related to each other. In contrast, males often remain with their natal groups, and associations of male kin may be important elements of the social behaviors of these species of primates. Male chimpanzees, for example, cooperate in defending their territories from invasion by neighboring males. Most species of primates with female transfer systems live in relatively small social groups.

Nonfemale transfer systems Most females of these species spend their entire lives in the group in which they were born. Social relations among the females in a group are complex and based on kinship. Males of these species emigrate from their natal group as adolescents, and they may continue to move among groups as adults. In some of these species a single male lives with a group of females until he is displaced by a new male. In other species several males may be part of the group and maintain an unstable dominance hierarchy among themselves. Cooperation by several adult males may allow them to resist challenges from younger, stronger males that they would not be able to subdue if they acted as individuals. Group size is usually larger for nonfemale transfer species than for species with female transfer.

Monogamous species A single male and female form a pair, sometimes accompanied by juvenile offspring. These species of primates show little sexual dimorphism, the sexes share parental care and territorial defense, and the offspring are expelled from the parents' territory during adolescence.

Solitary species These species live singly or as females with their infants and juvenile offspring. Male prosimians maintain territories that include the home ranges of several females and exclude other males from their territories, whereas male orangutans do not defend territories. Instead, they repulse other males when a female within the male's home range comes into estrus.

Ecology and Primate Social Systems

Three ecological factors appear to be particularly important in shaping the social systems of primates, as they are for other vertebrates:

- **Distribution of resources.** The defensibility of food resources appears to determine whether individuals will benefit from not attempting to defend resources, defending individual territories, or forming long-term relationships with other individuals and jointly defending resources.
- **Group size.** The distribution of food in time and in space may determine how large a group can be, and whether the group can remain stable or must break into smaller groups when food is scarce.
- **Predation.** The risk of predation may determine whether individuals can travel alone or require the protection of a group, whether the benefit of the additional protection provided by a large versus a small group outweighs the added competition among individuals in a large group, and whether the presence of males is needed to protect young.

Table 22.3 Social organization of extant primates

Taxon	*Social Organization*
Prosimii	
Lemuriformes	
Lemuroidea	Largely solitary or monogamous pairs
Aye-aye (*Daubentonia*, 1 species)	
Lemurs (*Lemur* and 9 other genera, 18 species)	
Indri (*Indri*, 1 species)	
Sifaka (*Propithecus*, 2 species)	
Lorisoidea	Largely solitary
Bushbabies (*Galago*, 8 species)	
Lorises (*Loris*, 1 species; *Nycticebus*, 2 species)	
Potto (*Perodicticus*, 1 species)	
Angwantibo (*Artocebus*, 1 species)	
Tarsiiformes	
Tarsiidae Tarsiers (*Tarsius*, 3 species)	Solitary or monogamous pairs
Anthropoidea	
Platyrrhini (New World monkeys)	
Ceboidae	
Cebidae	
Callithrichinae (marmosets and tamarins, 5 genera, 16 species)	Largely monogamous pairs
Cebinae (capuchins and squirrel monkeys, 2 genera, 5 species)	Monogamous pairs or small to large groups
Aotinae (owl monkey, 1 species)	
Atelidae	
Atelinae (howler and spider monkeys, 4 genera, 13 species)	
Pithecinae (sakis and ukaris, 3 genera, 8 species)	
Callicebinae (titis, 1 genus, 3 species)	
Catarrhini (Old World monkeys and apes)	
Cercopithecoidea	Mostly small to large groups
Cercopithecidae	
Cercopithecinae	
Vervet monkey, guenons, and others (*Cercopithecus*, 17 species)	
Mangabeys (*Cercocebus*, 5 species)	
Macaques (*Macaca*, 12 species)	
Baboons (*Papio*, 4 species; *Theropithecus*, 1 species)	
Colobinae	
Colobus monkeys (*Colobus*, 7 species)	
Langurs (*Nasalis, Presbytis, Pygathrix, Rhinopithecus*, 20 species)	
Hominoidea (apes and humans)	
Hylobatidae	
Gibbons (*Hylobates*, 9 species)	Monogamous pairs
"Pongidae"	
Orangutan (*Pongo*, 1 species)	Solitary
Gorilla (*Gorilla*, 1 species)	Small groups with a variable number of resident males
Chimpanzee (*Pan*, 2 species)	Closed social network containing several breeding males and females
Hominidae	
Human (*Homo*, 1 species)	Closed social network containing several breeding males and females

Table 22.4 Characteristics of the social systems of primates

System	*Group Size*	*Number of Males in Group*	*Male Behavior*	*Example*
Female transfer	Small	One or many	Territorial, harems, sometimes male kinship groups	Chimpanzee, gorilla, howler monkeys, hamadryas baboons, colobus monkeys, some langurs
Nonfemale transfer	Large	One or several	Male hierarchy, whole group (males and females) may exclude conspecifics from food sources	Most cercopithecines: yellow baboons, mangabeys, macaques, guenon monkeys
Monogamous	Male and female, plus juvenile offspring	One	Both sexes participate in territorial defense and parental care	Gibbons, marmosets, tamarins, indri, titis
Solitary	Individual, or female plus juvenile male offspring	—	Range of male overlaps ranges of several females	Bushbabies, tarsiers, lorises, orangutans

Behavioral Interactions

Life within a group of primates is a balance between competition and cooperation (Figure 22–10). Competition is manifested by aggression. Some aggression—for example, the defense of food, sleeping sites, or mates—is closely linked to resources. Other types of aggression involve establishing and maintaining dominance hierarchies, which can be an indi-

(a)

(b)

▲ Figure 22–10 Social behaviors of yellow baboons (*Papio cyanocephalus*). (a) A male friend grooming a female baboon in estrus. (b) Aggression among male baboons.

rect form of resource competition if high-ranking individuals have preferential access to resources.

Cooperation, too, is diverse. Grooming behavior in which one individual picks through the hair of another, removing ectoparasites and cleaning wounds, is the most common form of cooperation. Other types of cooperation include sharing food or feeding sites, collective defense against predators, collective defense of a territory or a resource within a home range, and formation of alliances between individuals. Two-way, three-way, and even more complex alliances that function during competition within a group are common among primates.

Kinship and the concept of inclusive fitness play important roles in the interpretation of primate social behavior. A behavior must not decrease the fitness of the individual exhibiting the behavior if it is to persist in the repertoire of a species. Because fitness is nearly impossible to demonstrate in wild populations, behaviorists normally search for effects that are likely to be correlated with fitness, such as access to females (for males), interbirth interval (for females), or the probability that offspring will survive to reproductive age. Behaviors that increase these measures are assumed to increase fitness. The behaviors may directly benefit the individual displaying the behavior (personal fitness), or they may be costly to the personal fitness of the individual but sufficiently beneficial to its close relatives to offset the cost to the individual (inclusive fitness).

Social Relationships among Primates

Four general types of relationships among individuals have been described in the social behavior of primates (For more details see Watts [1985], Richard [1985], Smuts et al. [1987], Dunbar [1988], Cheney and Seyfarth [1990], and Nishida [1990].)

Adult-Juvenile Associations Primates are born in a relatively helpless state compared to many mammals, and they depend on adults for unusually long periods. The relationship of a mother to her infant is variable within a species—some mothers are protective, whereas others are permissive. Permissive mothers often wean their offspring earlier than protective mothers and may have shorter intervals between the birth of successive offspring, although this relationship has not been observed in all species. The offspring of permissive mothers may suffer higher rates of mortality than the offspring of protective mothers, and the incompetence of some inexperienced mothers appears to lead to high mortality among firstborn offspring.

Older siblings often participate in grooming and carrying an infant, but they may also assault, pinch, and bite the infant while it is being fed or groomed by the mother. Allomaternal behavior provided by an adult female who is not the mother includes cuddling, grooming, carrying, and protecting an infant. Several factors seem to influence allomaternal behavior: Young infants are preferred to older ones, infants of high-ranking mothers receive more attention and less abuse than infants of low-ranking mothers, and siblings may participate more than unrelated females in allomaternal behavior. Males of the monogamous New World primates participate extensively in caring for infants, carrying them for much of the day and sharing food with them, whereas the relationships of males of Old World primates with infants are more often characterized by proximity and friendly contact than by care.

Female Kinship Bonds The females of some species of semiterrestrial Old World primates live in groups that include several males and females. This social organization is typical of yellow baboons *(Papio cynocephalus)*, several species of macaques *(Macaca)*, and vervet monkeys *(Cercopithecus aethiops)*. Females of these species remain for their entire lives in the troops in which they were born. Female kinship bonds and kin selection play important roles in the behaviors of females in nonfemale transfer systems because the females in a group are related to each other.

The females within a group form a dominance hierarchy and compete for positions in the hierarchy. Related females within a group are referred to as **matrilineages.** Females consistently support their female relatives during encounters with members of other matrilineages. The supportive relationship among females within a matrilineage is an important element of a group's social structure. For example, when their female kin are nearby, young animals can dominate older and larger opponents from subordinate matrilineages. Furthermore, high-ranking females retain their position in the hierarchy even when age or injury reduces their fighting ability. An adolescent female yellow baboon normally attains a rank in the group just below that of her mother, and this inheritance of status provides stability in the dominance relationships among the females of a group. However, the social rank of the matrilineage is not fixed: Low-ranking female yellow baboons, with their female kin, may challenge higher-ranking individuals; if they are successful, their entire matrilineage may rise in rank within the group.

Female kinship bonds are clearly important elements of the social structure of nonfemale transfer systems, but the exact contribution of the long-term relationships among females to the fitness of individual females is not clear. In some species high-ranking females are young when they first give birth and have short interbirth intervals and high infant survival, but those correlations are not present in all the species that have been studied. Furthermore, female kinship bonds are manifested weakly if at all in female transfer systems, which include most species of apes and many species of monkeys.

Male-Male Alliances Male primates in nonfemale transfer species often form dominance hierarchies, but male rank depends mainly on individual attributes and is therefore less stable than female dominance systems based on matrilineage. Young adult males, which are usually recent immigrants from another group, have the greatest fighting ability and usually achieve the highest rank. Some older males achieve stable alliances with each other that enable them to overpower younger and stronger rivals in competition for opportunities to court receptive females. These males probably achieve greater mating success by engaging in these reciprocal alliances than they would achieve on the basis of their individual ranks in the hierarchy.

Cooperative relationships among males are most common in female transfer systems, because the males of these species remain in their natal group. As a result, kin relationships exist among the males in a group. In red colobus monkeys (*Colobus badius*), for example, only males born in the group appear to be accepted by the adult male subgroup, and the membership of this subgroup can remain stable for years. Adult males spend much of their time in close proximity to each other and cooperate in aggression against males of a neighboring group. For example, Goodall (1986) reported the systematic killing of an entire group of male chimpanzees (*Pan troglodytes*) by the males of a neighboring group, which then took over the females in that community. Within a group, male chimpanzees spend more time together than do females, and they engage in a variety of cooperative behaviors, including greeting, grooming, and sharing meat. However, this apparent cooperation is simply a way of cementing relationships that are based on intense and sometimes violent competition over females.

Male-Female Friendships among Baboons Barbara Smuts's (1985) observations of a group of yellow baboons revealed that interactions between individual male and female baboons were not randomly distributed among members of the group. Instead, each female had one or two particular males called friends. Friends spent much time near each other and groomed each other often. These friendships lasted for months or years, including periods when the female was not sexually receptive because she was pregnant or nursing a baby. Male friends were solicitous of the welfare of their female friends and of their infants. Similar male-female friendships have been described in mountain gorillas *(Gorilla gorilla beringei),* gelada baboons *(Theropithecus gelada),* hamadryas baboons *(Papio hamadryas),* rhesus macaques *(Macaca mulatta),* and Japanese macaques *(Macaca fuscata).*

The advantage of these friendships for a female appears to lie in the protection that males provide to the females and their offspring from predators as well as from other members of the group. The advantage for a male of friendship with a female is less apparent. If the female's offspring had been sired by the male, protecting it would contribute to the male's fitness. However, in Smuts's study of yellow baboons, only half the friendships between males and infants involved relationships in which the male was the likely sire of the infant. The other friendships involved males that had never been seen mating with the mother of the infant. The advantage of friendship for males may depend on long-term associations with females. Smuts noted that males who participated in a friendship with a female had a significantly increased chance of mating with that female many months later, when she was again receptive.

How Do Primates Perceive Their Social Structure?

The preceding summary of primate social structures represents the results of tens of thousands of hours of observations of individual animals over periods of many years. Statistical analyses of interactions between individuals—grooming sessions, aggression, defense—reveal correlations associated with factors including age, personality, kinship, and rank. Do the animals themselves recognize those relationships?

That is a fascinating but difficult question, especially with studies of free-ranging animals. Observations are accumulating that suggest that primates probably do recognize different kinds of relationships among individuals. For example, when juvenile rhesus macaques are threatened by another monkey, they scream to solicit assistance from other individuals who are out of sight. The kind of scream they give varies depending on the intensity of the interaction (threat or actual attack) and the dominance rank and kinship of their opponent. Furthermore, a mother baboon appears able to interpret the screams of her juvenile and to respond more or less vigorously depending on the nature of the threat her offspring faces. When tape-recorded screams were played back to the mothers, the mothers responded most strongly to screams that were given during an attack by a higher-ranking opponent, less strongly to screams that were given in response to interactions with lower-ranking opponents, and least strongly to screams that were given in interactions with relatives.

In a similar experiment with free-ranging vervet monkeys, the screams of a juvenile were played back to three females, one of them the mother of the juvenile. The mother responded more strongly to the screams than did the other two monkeys, as might be expected if females can recognize the voices of their own offspring. However, the other two monkeys responded to the screams by looking toward the mother, suggesting that they were able not only to associate the screams with a particular juvenile but also to associate that juvenile with its mother.

Observations of redirected aggression also suggest that some primates classify other members of a group by matrilineage and friendships. When a baboon or macaque has been attacked and routed by a higher-ranking opponent, the victim frequently attacks a bystander who took no part in the original interaction. This behavior is known as redirected aggression—and the targets of redirected aggression are relatives or friends of the original opponent more frequently than would be expected by chance. Vervet monkeys show still more complex forms of redirected aggression: They are more likely to behave aggressively toward an individual when they have recently fought with one of that individual's close kin. Furthermore, an adult vervet is more likely to threaten a particular animal if that animal's kin and one of its own kin fought earlier that same day. This sort of feud is seen only among adult vervets, suggesting that it takes time for young animals to learn the complexities of the social relationships of a group.

These sorts of observations suggest that adult primates have a complex and detailed recognition of the genetic and social relationships of other individuals in their group. Furthermore, they may be able to recognize more abstract categories—such as *relative* versus *nonrelative, close relative* versus *distant relative,* or *strong friendship bond* versus *weak friendship bond*—that share similar characteristics independent of the particular individuals involved.

Summary

Sociality, the formation of structured groups, is a prominent characteristic of the behavior of many species of mammals. However, social behavior is only one aspect of the biology of a species, and social behaviors coexist with other aspects of behavior and ecology, including finding food and escaping from predators.

The size and geography of an animal's home range is related to the distribution and abundance of resources, the body size of the animal, and its feeding habits. Large species have larger home ranges than small species, and for any given body size the sizes of home ranges are in the order carnivores $>$ omnivores $>$ herbivores.

Social systems are related to the distribution of food resources and to the opportunities for an individual (usually, a male) to increase access to mates by controlling access to resources. Dietary habits, the structural habitat in which a species lives, and its means of avoiding predators are closely linked to body size and mating systems. These aspects of biology form a web of interactions, each influencing the others in complex ways.

The social systems of primates—especially cercopithecoid monkeys—have been the subjects of field studies, and more information about social behavior under field conditions is available for primates than for other mammals. The social systems of primates are complex but not unique among mammals. Some primates are solitary or monogamous; others live in groups and display behaviors that suggest not just recognition of other individuals but also recognition of the genetic and social relationships among other individuals. Studies of other kinds of mammals will probably reveal similar phenomena. Understanding the behavior of mammals requires a broad understanding of their ecology and evolutionary histories.

Additional Readings

Caro, T. M. 1994. *Cheetahs of the Serengeti Plains.* Chicago: University of Chicago Press.

Cheney, D., and R. Seyfarth. 1990. *How Monkeys See the World.* Chicago: University of Chicago Press.

Clutton-Brock, T. H., and P. H. Harvey. 1983. The functional significance of variation in body size among mammals. In J. F. Eisenberg and D. G. Kleiman (Eds.), *Advances in the Study of Mammalian Behavior,* Special Publication 7. Lawrence, Kans: The American Society of Mammalogists.

Dunbar, R. I. M. 1988. *Primate Social Systems.* Ithaca, N.Y.: Cornell University Press.

Eisenberg, J. F. 1981. *The Mammalian Radiations.* Chicago: University of Chicago Press.

Estes, R. D. 1991. *The Behavior Guide to African Mammals.* Berkeley: University of California Press.

Gittleman, J. L. (Ed.). 1989. *Carnivore Behavior, Ecology, and Evolution.* Ithaca, N.Y.: Cornell University Press.

Gittleman, J. L., and P. H. Harvey. 1982. Carnivore home range size, metabolic needs, and ecology. *Behavioral Ecology and Sociobiology* 10:57–63.

Goodall, J. 1986. *The Chimpanzees of Gombe.* Cambridge: Harvard University Press.

Hasson, O. 1991. Pursuit-deterrent signals: Communication between prey and predator. *Trends in Ecology and Evolution* 6:325–329.

Hersteinsson, P., and D. W. Macdonald. 1982. Some comparisons between red and Arctic foxes, *Vulpes vulpes* and *Alopex lagopus,* as revealed by radio tracking. In C. L. Cheesman and R. B. Mitson (Eds.), *Symposia of the Zoological Society of London,* no. 49. London: Academic, 259–289.

Janis, C. M. 1982. Evolution of horns in ungulates: Ecology and paleoecology. *Biological Reviews* 57:261–318.

Jarman, P. J. 1974. The social organization of antelope in relation to their ecology. *Behaviour* 58: 215–267.

Jarman, P. J. 2000. Dimorphism in social Artiodactyla: Selection upon females. In E. S. Vrba and G. B. Schaller (Eds.), *Antelope, Deer, and Relatives,* New Haven: Yale University Press, 171–179.

Kelt, D. A. and D. Van Vuren. 1999. Energetic constraints and the relationship between body size and home range area in mammals. *Ecology* 80:337–340.

Leuthold, W. 1977. *African Ungulates: A Comparative Review of Their Ethology and Behavioral Ecology.* New York: Springer.

Lewin, R. 1987. Social life: A question of costs and benefits. *Science* 236:775–777.

Macdonald, D. W. 1983. The ecology of carnivore social behavior. *Nature* 301:379–381.

McNab, B. K. 1983. Ecological and behavioral consequences of adaptation to various food resources. In Eisenberg and Kleiman (Eds.), *Advances in the Study of Mammalian Behavior,* 664–697.

Nicolson, N. A. 1987. Infants, mothers, and other females. In B. Smuts, D. Cheney, et al. (Eds.), *Primate Societies,* Chicago: University of Chicago Press, 330–342.

Nishida, T. (Ed.) 1990. *The Chimpanzees of the Mahale Mountains. Sexual and Life History Strategies.* Tokyo: Tokyo University Press.

Packer, C., and L. Ruttan. 1988. The evolution of cooperative hunting. *The American Naturalist* 132:159–198.

Packer, C., D. Scheel, and A. E. Pusey. 1990. Why lions form groups: Food is not enough. *American Naturalist* 136:1–19.

Richard, A. F. 1985. *Primates in Nature.* San Francisco: Freeman.

Rubenstein, D. I., and R. W. Wrangham (Eds.). 1986. *Ecological Aspects of Social Evolution: Birds and Mammals.* Princeton, N.J.: Princeton University Press.

Smuts, B. 1985. *Sex and Friendship in Baboons.* Hawthorne, N.Y.: Aldine.

Smuts, B., et al. (Eds). 1987. *Primate Societies.* Chicago: University of Chicago Press.

Van Soest, P. J. 1982. *Nutritional Ecology of the Ruminant.* Corvallis, Oreg.: O & B Books.

Watts, E. S. (Ed.). 1985. *Nonhuman Primate Models for Human Growth and Development.* New York: Liss.

Wilson, E. O. 1975. *Sociobiology.* Cambridge: Harvard University Press.

Wittenberger, J. F. 1981. *Animal Social Behavior.* Boston: Duxbury.

Wrangham, R. W. 1982. Mutualism, kinship, and social evolution. In King's College Sociobiology Group (Eds.), *Current Problems in Sociobiology.* Cambridge: Cambridge University Press.

Web Explorations

On-line resources for this chapter are on the World Wide Web at http://www.prenhall.com/pough. (Click on the Table of Contents link, and then select Chapter 22).

CHAPTER

23 Primate Evolution and the Emergence of Humans

Humans are primates, and complex social systems such as those discussed in Chapter 22 are an ancestral feature of the primate lineage. Primates have been a moderately successful group for most of the Cenozoic, although since the end of the Eocene they have been primarily confined to tropical latitudes (with the obvious exception of humans). Primates include not only the anthropoids, the group of apes and monkeys to which humans belong, but also the prosimians, animals such as bushbabies and lemurs. Molecular techniques of studying genetic relationships suggest that the separation of humans from the African great apes was more recent than had been inferred from anatomical studies, and that the closest extant relatives of humans are chimpanzees. Fossils of the hominid genus *Australopithecus*, the sister taxon to our own genus *Homo*, clearly show that bipedal walking arose before the acquisition of a large brain. *Homo* and *Australopithecus* lived together in Africa for over a million years, and the extinction of the australopithecines was probably related to climatic changes rather than to competition with our ancestors. The fossil record indicates that the first stone tools are at least 2.5 million years old. Tool cultures are clearly associated with the first appearance of the genus *Homo*, but may not be an exclusive characteristic of our genus.

23.1 Primate Origins and Diversification

Humans arose from arboreal ancestors that lived in early Cenozoic forests 65 million years ago. Our closest extant relatives are the chimpanzees and gorillas of Africa. These statements, which Darwin's Victorian contemporaries found so startling and upsetting, are commonplace observations now. In the following sections we summarize the current state of the evidence and inferences about human phylogeny and evolution.

Characteristics of Primates

Humans share many biological traits with the animals that are variously called apes, monkeys, and prosimians. Using the principle of homology (Chapter 1), comparative anatomy demonstrates that all of these mammals can be grouped in the order Primates (Latin *prima* = first). Obvious traits that characterize primates are listed in Table 23.1. It is important to understand that several of these characteristics can be identified as trends within the primate lineage and thus may not be apparent in all primitive members of the lineage. Nor are many of these characters unique to primates: for example, many mammals retain the clavicle, pigs have bunodont molars similar to those of primates, and many ungulates and kangaroos have only a single young per pregnancy.

Most of these traits have been attributed to an arboreal life. All the basic modifications of the appendages can be seen as contributing to arboreal locomotion, as can the stereoscopic depth perception that results from binocular vision and the enlarged brain that coordinates visual perception and locomotory response. Most primates are arboreal, but some have become secondarily terrestrial (baboons, for example), and humans are the most terrestrial of all. Even so, many of the traits that are most distinctively human have been said to derive from earlier arboreal specializations. However, arboreality cannot be the entire basis for these primate characteristics.

Squirrels provide a telling counterexample. They are fully arboreal and can match or exceed the climb-

Table 23.1 Characteristics of Primates

Retention of the clavicle (which is reduced or lost in many mammalian lineages) as a prominent element of the pectoral girdle.
A shoulder joint allowing a high degree of limb movement in all directions and an elbow joint permitting rotation of the forearm.
The general retention of five functional digits on the fore and hind limbs.
Enhanced mobility of the digits, especially the thumb and big toes, which are usually opposable to the other digits.
Claws modified into flattened and compressed nails.
Sensitive tactile pads developed on the distal ends of the digits.
A reduced snout and olfactory apparatus with most of the skull posterior to the orbits.
A reduction in the number of teeth compared to primitive mammals but with the retention of simple bunodont molar cusp patterns (see Chapter 20).
A complex visual apparatus with high acuity, color perception, and a trend toward development of forward-directed binocular eyes.
A large brain relative to body size, in which the cerebral cortex is particularly enlarged.
A trend toward derived fetal nourishment mechanisms.
Only two mammary glands (some exceptions).
Typically, only one young per pregnancy associated with prolonged infancy and pre-adulthood.
A trend toward holding the trunk of the body upright leading to facultative bipedalism.

ing skills of similar-size primates, but show few of the specializations seen in primates. Although squirrels do have a clavicle and good mobility of the shoulder and elbow joint, the range of movement is not as great as that of primates. The thumbs of squirrels are greatly reduced and nonopposable, and the remaining digits have long, sharp, recurved claws lacking tactile pads at their tips. Squirrels have large olfactory organs and snouts, laterally directed eyes, and no notable brain enlargement compared with fully terrestrial and diurnal rodents, although they do have good color vision. Squirrel life histories are typical of small mammals: adults have relatively short life spans and produce large litters of fast-growing young. Thus primate characteristics are not required by arboreality.

Evolutionary Trends and Diversity in Primates

Table 22.3 (in Chapter 22) presents a traditional classification of modern primates in the context of their social systems, and Figure 23–1 is a simplified representation of their interrelationships. It is generally agreed that the first primates evolved from a lineage of arboreal generalized mammals not unlike the present-day tree shrews of Southeast Asia (family Tupaiidae). Tree shrews have long bushy tails and relatively short limbs, appearing rather like squirrels (Figure 20–10). The skull and teeth show most of the unspecialized characteristics of generalized, primitive mammals, including a long snout, tritubercular cheek teeth, and laterally placed eyes. Tree shrews were once considered as true, primitive primates. Today they are acknowledged as close relatives within the grandorder Archonta (see Chapter 19), but are placed in their own order, Scandentia. Nonetheless, they offer a good extant anatomical model for the basal stock of our lineage.

Plesiadapiformes The first primate-like mammals (**plesiadapiforms**) (Greek *plesi* = near; Latin *adapi* = rabbit and *form* = shape) appear in the earliest Cenozoic. Plesiadapiforms were rather squirrel-like, and ranged from chipmunk-size to marmot-size. They included several different lineages, varying in their diets (as can be judged from their teeth) from generalized omnivores to insectivores and gum eaters. Plesiadapiforms were most diverse in the Paleocene of North America, although they were known from across the Northern Hemisphere. Their numbers declined in the Eocene, and they were extinct by the end of this epoch. The decline in plesiadapiform diversity coincides with the evolution and radiation of rodents in the late Paleocene. There may well have been competition between these two lineages, leading to the eventual extinction of the plesiadapiforms.

Plesiadapiforms are the sister group to true primates, and shared some derived features of the teeth and skeleton with them. They differed in their smaller brains, their longer snout, the lack of a postorbital bar, the lack of an opposable **hallux** (big toe), and the specialization of rodent-like incisors in some forms (see Figure 23–2a). Plesiadapiforms also apparently retained claws. In contrast, all true primates

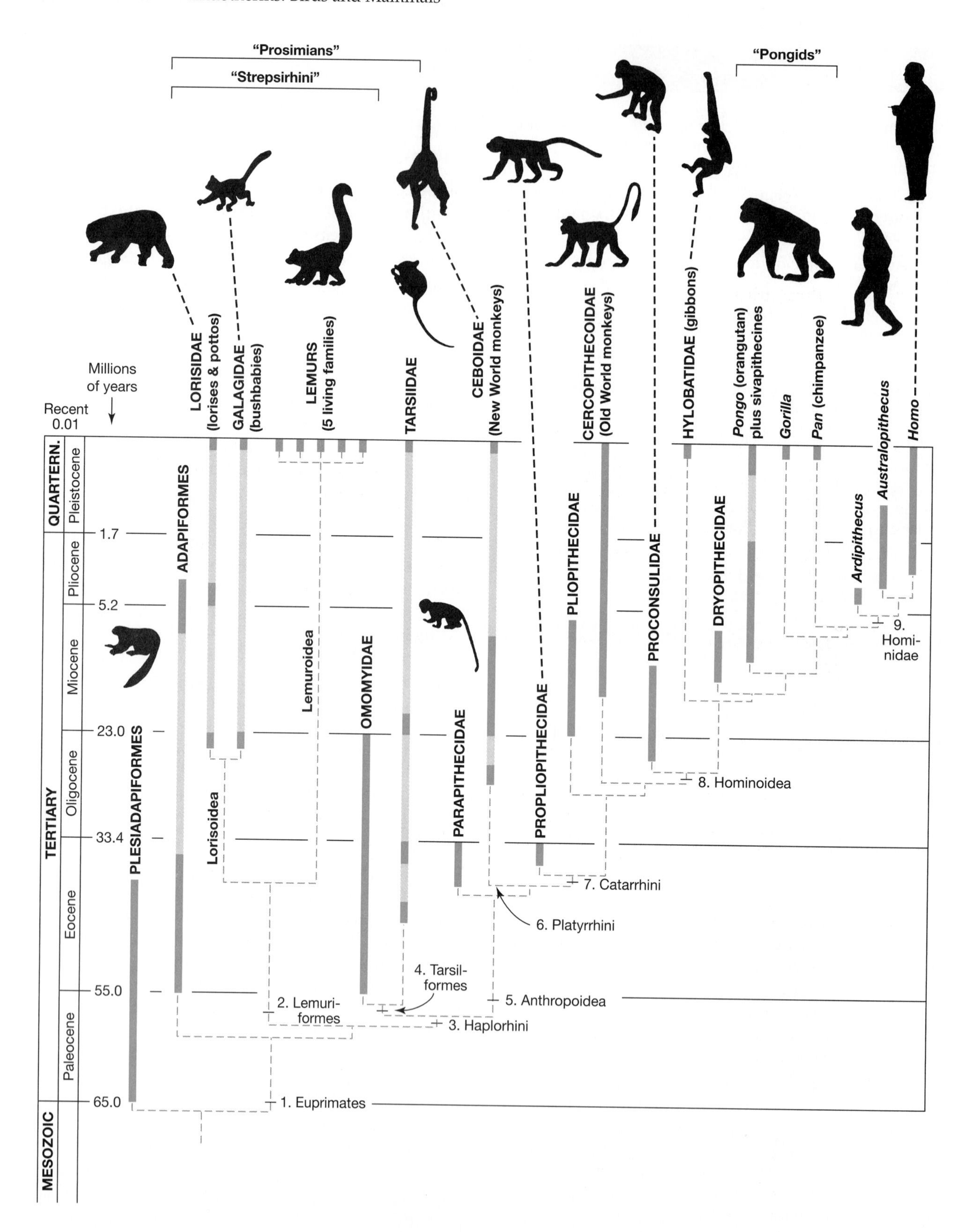

"Prosimians"
"Strepsirhini"
"Pongids"
Millions of years
Recent 0.01
LORISIDAE (lorises & pottos)
GALAGIDAE (bushbabies)
LEMURS (5 living families)
TARSIIDAE
CEBOIDAE (New World monkeys)
CERCOPITHECOIDAE (Old World monkeys)
HYLOBATIDAE (gibbons)
Pongo (orangutan) plus sivapithecines
Gorilla
Pan (chimpanzee)
Australopithecus
Homo
QUARTERN.
Pleistocene
TERTIARY
Pliocene
Miocene
Oligocene
Eocene
Paleocene
MESOZOIC
1.7
5.2
23.0
33.4
55.0
65.0
ADAPIFORMES
PLESIADAPIFORMES
Lorisoidea
Lemuroidea
OMOMYIDAE
PARAPITHECIDAE
PROPLIOPITHECIDAE
PLIOPITHECIDAE
PROCONSULIDAE
DRYOPITHECIDAE
Ardipithecus
9. Hominidae
8. Hominoidea
7. Catarrhini
6. Platyrrhini
4. Tarsilformes
5. Anthropoidea
2. Lemuriformes
3. Haplorhini
1. Euprimates

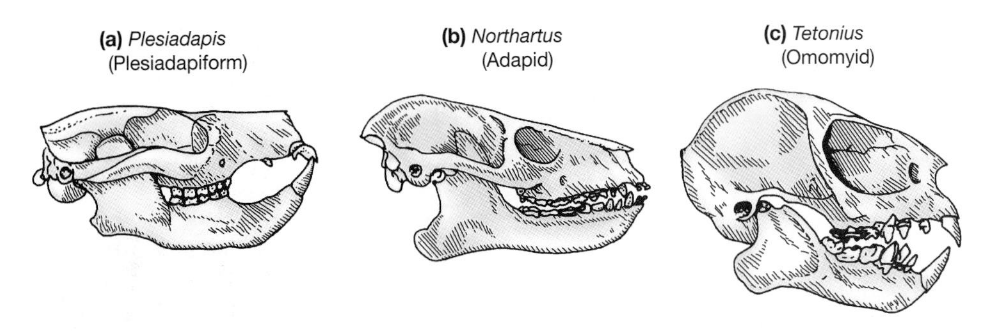

▲ Figure 23–2 Reconstruction of fossil skulls of some early primates and primate-like mammals (not to the same scale).

have nails—except for the marmosets of South America, which have secondarily reverted to having claws.

Prosimians The first true primates, or **Euprimates** (Greek *eu* = good), are known from the early Eocene of North America, Eurasia, and northern Africa. Isolated teeth from the late Paleocene of Morocco may belong to an even earlier primate. These early primates belong to a group that has traditionally been called **prosimians** (Greek *pro* = before; Latin *simi* = ape)—the bush babies of Africa; the lemurs of Madagascar; and the lorises, pottos, and tarsiers of Southeast Asia (Figure 23–3). Prosimians are in general small, nocturnal, long-snouted, and relatively small-brained, compared to the more derived **anthropoids** (Greek *anthrops* = man), or apes and monkeys. Prosimian diets are more generalized, and there are few specialized herbivores. However, the grouping of prosimians is a paraphyletic one, since several derived features (for example, a short snout with a dry nose, rather than a wet, dog-like one) indicate tarsiers are more closely related to the anthropoids than are the other prosimians (Figure 23–1). An alternative division of the primates is into the **Strepsirhini** (Greek *strepsi* = twisted and *rhin* = nose)—lemuroids and lorisoids—and **Haplorhini** (Greek *haplo* = simple)—tarsiers and anthropoids. The Strepsirhini is also paraphyletic, although living

Legend: 1. Euprimates—Cheek teeth bunodont; a nail (instead of a claw) always present in extant forms, at least on the pollex (thumb); postorbital bar present. **2.** Lemuriformes—Grooming claw present on second toe; lower front teeth modified into a tooth comb. **3.** Haplorhini [tarsiiformes plus anthropoids]—Cranium short; orbit and temporal fossa separated ventrally by a postorbital wall; dry nose and free (rather than tethered) upper lip. **4.** Tarsiiformes—Eyes greatly enlarged. **5.** Anthropoidea (monkeys and apes)—Fused frontal bones; fused mandibular symphysis; tubular external auditory meatus; lower molars increase in size posteriorly, the third only slightly larger than the second, all with five cusps, the hypoconulid small. **6.** Platyrrhini (New World monkeys)—Widely spaced and rounded nostrils; contact between jugal and parietal bones on lateral wall of skull behind orbit; first two lower molars lack hypoconulids. **7.** Catarrhini (Old World monkeys and apes)—Narrowly spaced nostrils; number of premolars reduced to two; contact between frontal and sphenoid bones in lateral wall of skull; tympanic bone extents laterally to form a tubular auditory meatus (ear tube). **8.** Hominoidea (apes and humans)—Lower molars with expanded talonid basin surrounded by five main cusps; broad palate and nasal regions; enlarged brain; broad thorax with dorsally positioned scapula; reduced lumbar region, with expanded sacrum and the absence of a tail. **9.** Hominidae (humans)—Relatively small incisors and canines; short snout; ventrally positioned foramen magnum; short, broad ilium; long legs in comparison with arms; big toe not opposable.

◀ Figure 23–1 Phylogenetic relationships of the primates. This diagram shows the probable relationships among the major groups of primates. Dotted lines show interrelationships only; they do not indicate the times of divergence nor the unrecorded presence of taxa in the fossil record. Light bars indicate ranges of time when the taxon is known to be present, but is unrecorded in the fossil record. Numbers indicate derived characters that distinguish the lineages.

▲ Figure 23–3 Diversity of living prosimians. (a) Ring-tailed lemur, *Lemur catta* (Lemuridae, Lemuroidea, Lemuriformes). (b) Indri, *Indri indri* (Indriidae, Lemuroidea, Lemuriformes). (c) Aye-aye, *Daubentonia madagascarensis* (Daubentoniidae, Lemuroidea, Lemuriformes). (d) Demidoff's bush baby, *Galagoides demidovii* (Galagidae, Lorisoidea, Lemuriformes). (e) Potto, *Perodicticus potto* (Lorisidae, Lorisoidea, Lemuriformes). (f) Tarsier, *Tarsius spectrum* (Tarsiidae, Tarsiformes).

strepsirhines form a monophyletic grouping, the lemuriformes (Figure 23–1).

Most Eocene prosimians were larger than the plesiadapiforms, with larger brains and more obviously specialized arboreal morphological features such as relatively longer, more slender limbs. They can be grouped into two main lineages: the larger **adapids**, which were long-snouted with teeth more specialized for herbivory, and the smaller **omomyids** (Greek *omo* = shoulder and *mys* = mouse), which were short-snouted with teeth more specialized for insectivory or gum eating (Figure 23–2). Judging by the size of the orbits, the adapids were probably diurnal, while the larger-eyed omomyids were probably nocturnal. Adapids are thought to be related to present-day lemurs and lorisoids, although they may actually be more primitive than other euprimates (see Figure 23–1); omomyids are closer to tarsiers.

The diversification of the adapids and omomyids throughout the Northern Hemisphere reflects the tropical-like climates of the higher latitudes during the earlier part of the Eocene (see Chapter 18). With the late Eocene climatic deterioration in the temperate latitudes, primates declined and eventually disappeared from areas outside of Africa and tropical Asia. The primitive types of prosimians were largely extinct by the end of the Eocene, although some specialized forms (sivaladapids) survived into the late Miocene and early Pliocene in Asia. Even today, almost all nonhuman primates are restricted to the tropics (Figure 23–4). This has been true for most of the later Cenozoic, barring some excursions of apes into more northern portions of Eurasia during the warming period in the late Miocene.

Present-day prosimians are a moderately diverse Old World tropical radiation, first known from the

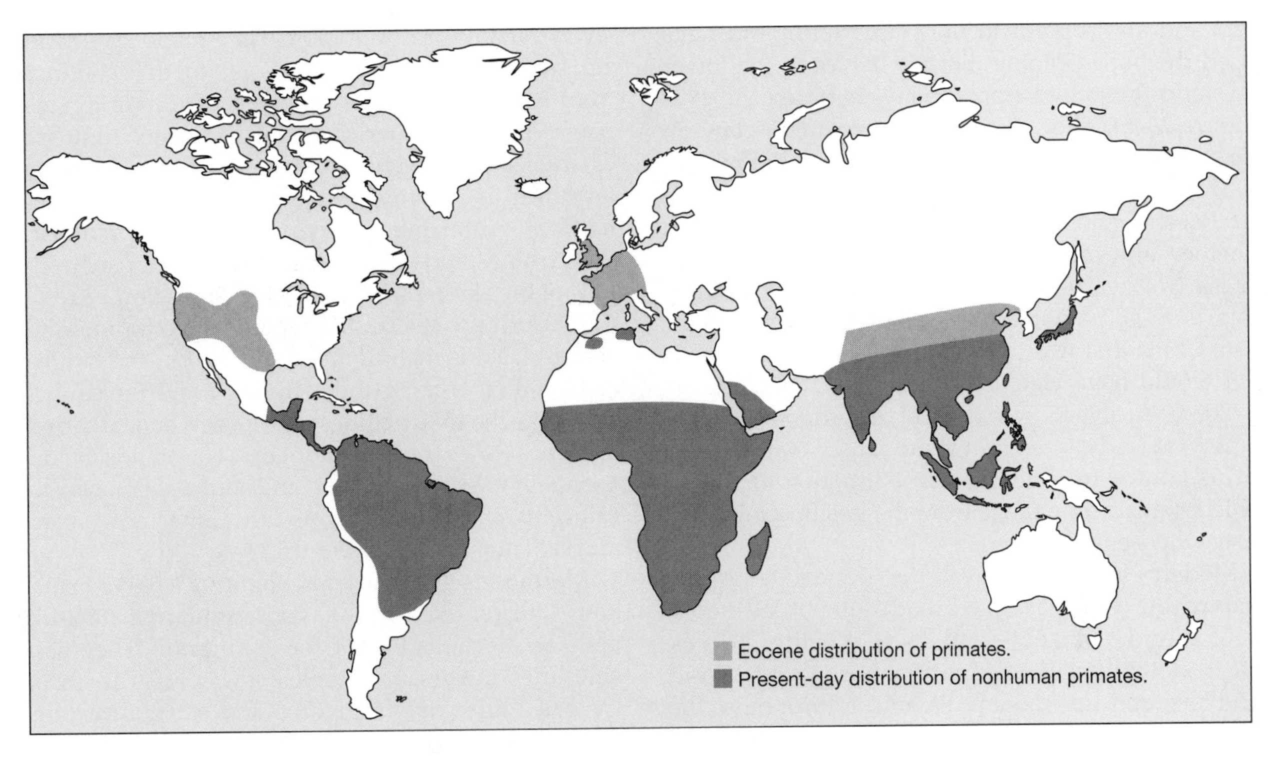

▲ Figure 23–4 Distribution map of primates in the Recent and Eocene times.

late Oligocene of Africa. The modern lineages are only sparsely known from the fossil record, probably because fossil preservation is rare in tropical forest habitats. The lemurs of the island of Madagascar have undergone an evolutionary diversification into five different families. These include some large (raccoon-size) diurnal specialized herbivores such as the indri (*Indri indri*), and the peculiar aye-aye (*Daubentonia madagascarensis*), which uses its specialized long middle finger to probe grubs out of tree bark (Figure 23–3c).

Until relatively recent times (only a couple of thousand years ago) there was a much greater variety of lemurs, including much larger forms. Arboreal forms resembled giant koalas and orangutans, and terrestrial forms resembled ground sloths and baboons. It seems that the lemurs, in isolation from the rest of the world, evolved their own version of primate diversity including parallels to the anthropoid apes. Unfortunately much of this diversity is now gone, probably because of the immigration of humans to the island. Many present-day lemurs are threatened with extinction, owing to further destruction of forest habitat (Jolly 1980).

Anthropoids Modern anthropoids are in general larger than prosimians, with larger brains housing relatively small olfactory lobes, and have a frugivorous (fruit-eating) or folivorous (leaf-eating) diet rather than an omnivorous or insectivorous one. They also are usually diurnal, with more complex social systems. In arboreal locomotion they are either arboreal quadrupedal above-branch climbers or suspensory under-branch climbers, whereas prosimians are usually clinging and leaping forms. Suspensory climbing, where the animals move relatively slowly by clinging underneath the branches, is not the same as **brachiation,** swinging rapidly from the underside of one branch to the underside of the next using the hands to grasp the branches. Suspensory climbing is general for large anthropoids, whereas brachiation is seen only in gibbons and spider monkeys. However, brachiation can be considered a specialized form of suspensory locomotion.

Modern anthropoids can be distinguished from prosimians by a variety of skull features that reflect a large brain size, and by a fibrous diet that requires extensive chewing. Anthropoids have a bony wall behind the orbit (as do tarsiers, a feature linking tarsiers and anthropoids in the Haplorhini), and have fused the bones joining the two halves of the lower jaws and the paired frontal bones between the eyes (Figure 23–5). They also lack the grooming claw on the second toe that is seen in modern prosimians.

The earliest anthropoids are known from the middle Eocene, and there is considerable debate as to whether anthropoids had their origin in Asia or Africa (Kay et al. 1997). The oldest known anthropoid is *Eosimias* (Greek *eos* = dawn), a tiny animal from China that was only about two inches in length and would have weighed about 10 grams (Gebo et al. 2000). A diverse radiation of anthropoids with a variety of body sizes (but none larger than a large cat) is known from the Fayum Formation of Egypt, which spans a time range from the late Eocene to the early Oligocene.

Modern anthropoids can be divided into the **Platyrrhini** (Greek *platy* = broad)—the broad-nosed New World monkeys—and the **Catarrhini** (Greek *cata* = downwards)—the narrow-nosed Old World monkeys and apes. Some Fayum anthropoids, the smaller (marmoset-size), monkey-like parapithecids, represent forms more primitive than any living anthropoid. Some others, such as the larger, more apelike propliopithecids, represent early catarrhines.

Platyrrhines, the New World monkeys or **ceboids** (Greek *cebus* = monkey), first appeared in South America in the Oligocene and have been an exclusively New World radiation. They must have rafted across the Atlantic Ocean to get to this continent from Africa: Rodents of probable African origin also reached South America about the same time (see Chapter 18). Platyrrhines are more primitive than catarrhines in retaining three premolars in each jaw half; all catarrhines have only two premolars. Platyrrhines and catarrhines also differ in some details of the skull, especially in the ear region.

Platyrrhines can be divided into the cebids and the atelids. Cebids include the cebines (e.g., the familiar capuchin or organ grinder monkey and the squirrel monkey), the callitrichines (marmosets and tamarins), and the aotines (the owl monkey). The atelids include the atelines (wooly, howler, and spider monkeys), the callicebines (titi monkeys), and the pithecines (uakaris and saki monkeys); see Figure 23–6.

Marmosets and tamarins (Figure 23–6a) are small and squirrel-like, and have secondarily claw-like nails on all digits except for the big toe. They have simplified molars, and while a few species are insectivorous, most are frugivores and eat gum exuded from trees. They are also unusual among anthropoid primates in producing twins. There is no anthropoid equivalent to the marmosets in the Old World, although perhaps the bush baby prosimians are analogous in certain respects.

The owl monkeys (aotines) are the only nocturnal anthropoids. The ateline monkeys (Figure 23–6c, d)

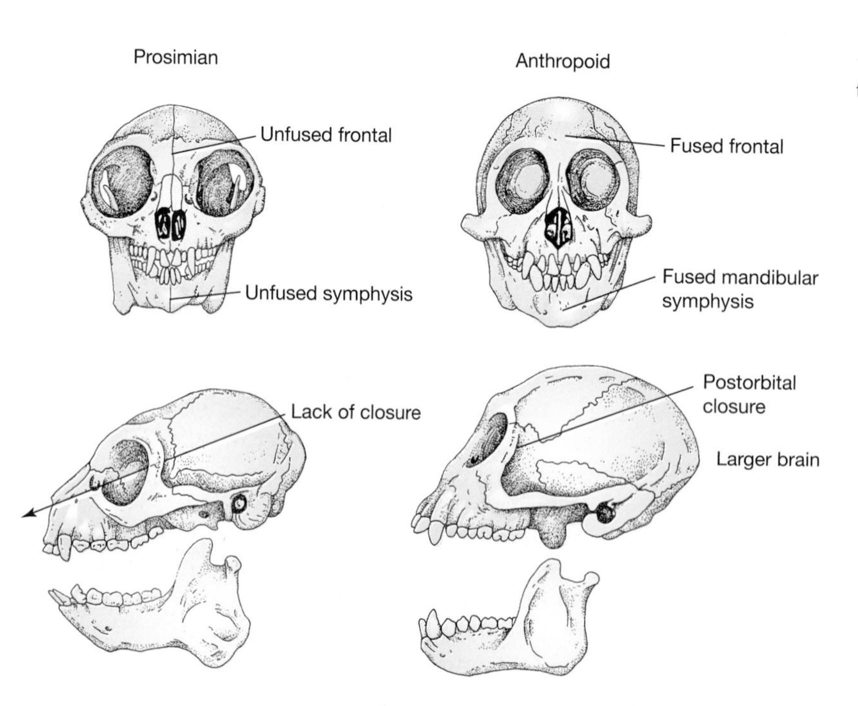

Figure 23–5 Cranial differences between prosimians and anthropoids.

▲ Figure 23–6 Diversity of living monkeys. (a) Pygmy marmoset, *Cebuella pygmaea* (Callitrichinae, Cebidae). (b) Squirrel monkey, *Saimiri sciureus* (Cebinae, Cebidae, Ceboidea). (c) Spider monkey, *Ateles paniscus* (Atelinae, Atelidae, Ceboidea). (d) Red howler monkey, *Alouatta seniculus* (Atelinae, Atelidae, Ceboidea). (e) Pig-tailed macaque, *Macaca nemistrina* (Cercopithecinae, Cercopithecidae, Cercopithecoidea). (f) Savanna baboon, *Papio anubis* (Cercopithecinae, Cercopithecidae, Cercopithecoidea). (g) Hanuman langur, *Presbytis vetulus* (Colobinae, Cercopithecidae, Cercopithecoidea). (h) Red colobus, *Piliocolobus badius* (Colobinae, Cercopithecidae, Cercopithecoidea).

are distinguished by having a prehensile tail and a specialized suspensory mode of arboreal locomotion, aided by the tail. The cebid radiation paralleled that of Old World monkeys to a certain extent, although there is no terrestrial radiation equivalent to that of baboons and macaques. Nor was there ever a cebid radiation equivalent to the anthropoid great apes.

The lack of ape-like forms among the cebids is surprising, considering that there was an evolutionary radiation of ape-like forms (now all extinct) among the Madagascan lemurs. Perhaps the extensive radiation of ground sloths in South America inhibited a terrestrial radiation among the primates. However, a striking parallel does exist between the spider monkey and the gibbon (which is usually considered a lesser ape). Both are specialized brachiators, with exceptionally long arms for swinging through the branches, and they have evolved a remarkable convergence in a wrist joint modification that allows exceptional hand rotation. Spider monkeys can be distinguished from gibbons chiefly by their use of a prehensile tail as a fifth limb during locomotion. (Gibbons, like all apes, lack a tail entirely.)

The catarrhines include the Old World monkeys, apes, and humans. We catarrhines have nostrils that are close together and open forward and downward, and we have a smaller bony nasal opening from the skull than is the case for platyrrhines. There is a trend toward large body size in our lineage; the great apes and humans are the largest living primates, rivaled only by some of the extinct lemurs. The tail is often short or absent, and prehensile tails have never evolved. The group consists of two clades: the Old World monkeys—**Cercopithecoidea** (Greek *cerco* = tail and *pithecus* = ape)—and the apes and humans—**Hominoidea** (Latin *homini* = man—the latter including the gibbons (Hylobatidae), and the great apes and humans. Humans, including extinct genera such as *Australopithecus*, are traditionally placed in their own family, **Hominidae** (see Figure 23–1).

Present-day Old World monkeys include two groups, **colobines** (Greek *colobo* = shortened) and **cercopithecines**. Colobines are found in both Africa and Asia, including colobus monkeys, langur monkeys, proboscis monkeys, and the golden monkey. They are more folivorous than cercopithecines, and have more lophed, higher-cusped molars and a complex forestomach for fermentation of plant fiber. Colobines are primarily arboreal animals, with a long tail and hind legs longer than the fore legs (Figure 23–6g, h).

Cercopithecines are predominately an African radiation, although the genus *Macaca* (macaques) occurs in Asia (including high-latitude places such as Japan and Tibet) and Europe (on Gibraltar, where it is known as the Barbary ape). Cercopithecines include macaques, mangabeys, baboons, guenons, and the patas monkey. They are more omnivorous or folivorous than colobines, as reflected in their broader incisors and their flatter, more bunodont molars. Cercopithecines are also more terrestrial, as reflected by their short tail and their equal-lengthed fore- and hindlimbs. They have cheek pouches for storing food, and a hand with a longer thumb and shorter fingers than colobines (Figure 23–6e, f).

The first Old World monkeys are known from the middle Miocene. These earliest monkeys, victoriapithecines, are more primitive than any known Old World monkey, and so form the sister group to the modern cercopithecoids. Monkeys are known from a slightly later date than the first true apes, the generalized proconsulids from the early Miocene of Africa. Monkeys are actually more derived than apes in certain respects: They have teeth that are more specialized for herbivory, some have gut specializations for the fermentation of cellulose, and they are more specialized for arboreality than the generalized Miocene apes.

The radiation of monkeys in the late Miocene and Pliocene coincided with the reduction in diversity of the earlier radiation of generalized apes and ape-like forms. Because we ourselves are apes, we often think of the monkeys as being the earlier, more primitive anthropoid radiation. But among the Old World anthropoids, the converse is actually true; apes were originally the more primitive, generalized forms. The radiation of cercopithecoid monkeys is more derived in many respects and ultimately more successful in terms of species diversity.

23.2 Origin and Evolution of the Hominoidea

Apes and humans are placed in the Hominoidea. Hominoids are distinguished morphologically from other recent anthropoids by a pronounced widening and dorsoventral flattening of the trunk relative to body length, so that the shoulders, thorax, and hips have become proportionately broader than in monkeys.

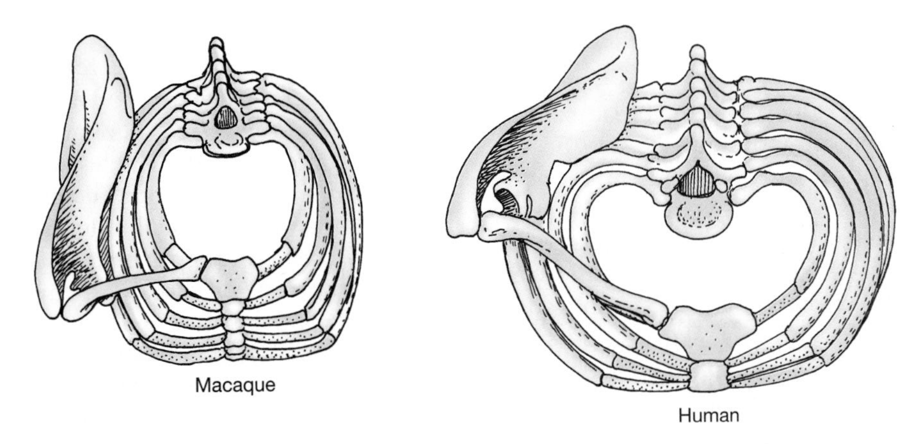

▲ Figure 23–7 Cephalic view (from the top with head and neck vertebrae removed) of chest and pectoral girdle (right half) of macaque and human. Note the broader chest and the dorsal position of the scapula in the hominoid; curvature of the ribs is also greater, with the vertebral column lying more in the middle of the rib cage, closer to the center of gravity. These features all make it easier for a hominoid to balance in an upright position—while monkeys must bend their knees and lean forward if balancing on their hind legs, to avoid tipping over backward.

The clavicles are elongated, the iliac blades of the pelvis are wide, and the sternum is a broad structure, the bony elements of which fuse soon after birth to form a single flat bone. The shoulder blades of hominoids lie over a broad, flattened back, in contrast to their lateral position next to a narrow chest in monkeys (Figure 23–7) and most other quadrupeds.

The pelvic and pectoral girdles of hominoids are relatively closer together than in other primates, because the lumbar region of the vertebral column is short. The caudal vertebrae have become reduced to vestiges in all Recent hominoids, and normally no free tail appears postnatally. Balance in a bipedal pose is assisted by the flat thorax, which places the center of gravity near the vertebral column. These and other anatomical specializations of the trunk are common to all hominoids and help to maintain the erect postures that these primates assume during sitting, vertical climbing, and walking bipedally (see Figure 23–8).

The skulls of hominoids also differ from those of other catarrhines in their extensive formation of sinuses—hollow, air-filled spaces lined with mucous membranes that develop between the outer and inner surfaces of skull bones. Chimpanzees, gorillas, and humans share the derived character of true frontal sinuses.

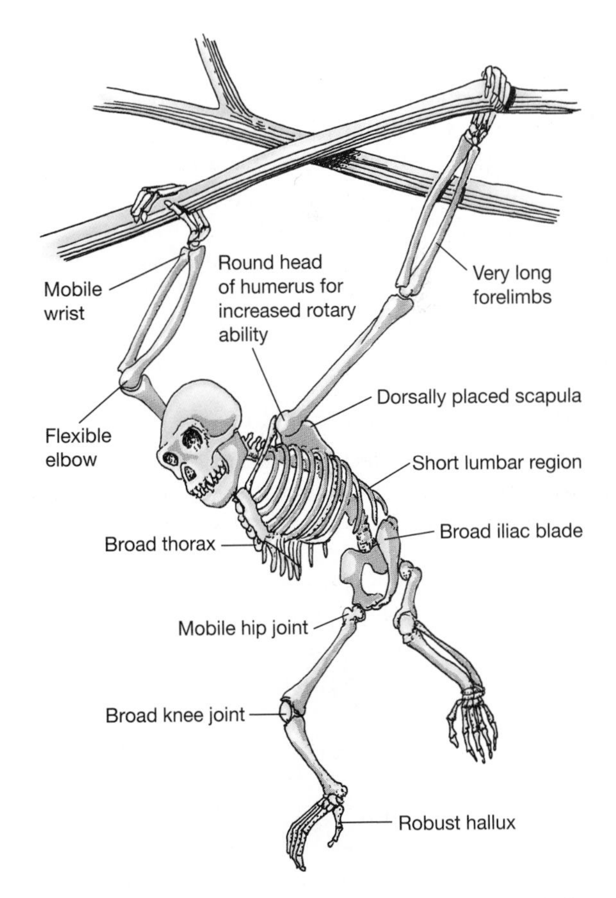

▲ Figure 23–8 Skeleton of generalized hominoid, showing morphological specializations for suspensory locomotion.

Diversity and Evolution of Nonhuman Hominoids

Primates that we can call apes in the broad sense of the word have been around since the late Eocene. However, primates that can be included in a monophyletic Hominoidea date only from the early Miocene, when the anthropoid lineage diverged into the hominoids and the cercopithecoids (Old World monkeys). Modern apes are a highly specialized radiation of large tropical animals. The Miocene radiation of apes was of more generalized animals, which also radiated into more temperate parts of the Old World.

Differences in the cheek teeth between hominoids and Old World monkeys are of paramount significance in identifying fossils, because many primate fossils are represented only by individual teeth. The Old World monkeys have lower molars (M1 and sometimes M2) with four cusps, one at each corner of a rectangular crown; the anterior pair of cusps, like the posterior pair, is connected by a ridge. Hominoid lower molars have crowns with five cusps, but in extant humans (although not most fossil relatives) this pattern is frequently obscured by crenulations or variations in the number of cusps. When five cusps do appear, the grooves between the cusps usually resemble the letter Y, with the open part of the Y embracing the hypoconid cusp (Figure 23–9). This molar pattern has persisted among hominoids for more than 20 million years.

Diversity of Present-day Apes

Present-day apes include the Asian gibbons (including siamangs) and orangutan as well as the African chimpanzees and gorilla (Figure 23–10). Traditionally, gibbons have been accorded their own family, Hylobatidae, and the other apes have been placed in the family "Pongidae." However, since it is now clear that pongids represent a paraphyletic grouping (see Figure 23–1), the official term, Pongidae, has been abandoned, although a term that encompasses the nonhuman great apes is still a useful one.

Nine species of gibbons (genus *Hylobates*) are found in Southeast Asia, both on the mainland (from India to China) and in the islands (Borneo, Sumatra, Java, and other nearby islands). They are the smallest apes, and they differ from other apes in their monogamous social system. Gibbons move through the trees most frequently by brachiation. They become entirely bipedal when moving on the ground, holding their arms outstretched for balance like a tightrope walker uses a pole.

There are two subspecies of the orangutan, *Pongo pygmaeus*; one lives on Borneo and one on Sumatra, although their range was greater in prehistoric times. Orangutans are around the same size as humans and are extremely sexually dimorphic, with males being twice the weight of females. Their behavior is fairly solitary, the main groups consisting of females and their offspring. Orangutans are arboreal but rarely swing by their arms, preferring slow quadrupedal climbing among the branches of trees, usually hanging below the branch (suspensory climbing).

Gorillas and chimpanzees live in the tropical forests of central Africa. Both are more terrestrial than gibbons and orangutans. On the ground they move quadrupedally by knuckle walking, a mode of locomotion in which they support themselves on the dorsal surface of digits three and four rather than

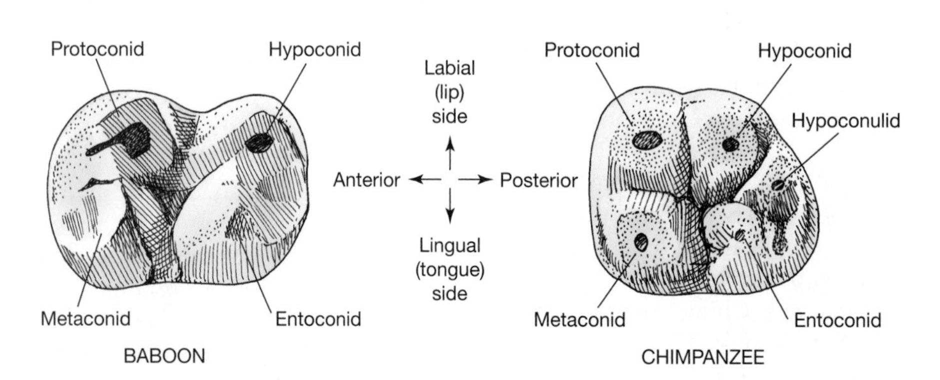

▲ Figure 23–9 Right lower molar patterns of Old World monkey (baboon) and hominoid (chimpanzee) compared.

▶ Figure 23–10 Diversity of living apes. (a) Siamang (a type of gibbon), *Hylobates syndactylus*. (b) Orangutan, *Pongo pygmaeus*. (c) Gorilla, *Gorilla gorilla*. (d) Common chimpanzee, *Pan troglodytes*.

placing their weight flat on the palm of the hand, as we do when we walk on all fours (see Figure 23–10c).

Three geographically isolated subspecies of gorilla (*Gorilla gorilla*) have been described: the western lowland gorilla, the eastern lowland gorilla, and the mountain gorilla. Gorillas are the largest and most terrestrial of the apes. Unlike the orangutan, they are highly social and live in groups. Like the orangutan, gorillas are highly sexually dimorphic in body size—the males may weigh up to 200 kg, twice the mass of the females—and they are the most folivorous of the apes.

There are two (possibly three) species of chimpanzees. The larger and more widely distributed common chimpanzee (*Pan troglodytes*) is known primarily from Central and East Africa. The western subspecies (*Pan troglodytes verus*), from Nigeria and the Cameroon, may be a separate species. There is also a smaller pygmy chimpanzee or bonobo (*Pan paniscus*), known from Central Africa south of the Zaire river, which lives in more forested habitats than other chimpanzees. Chimpanzees are more omnivorous than the more strictly herbivorous gorillas; they are also more arboreal, exhibiting a greater degree of suspensory locomotion. They are only moderately sexually dimorphic and, like gorillas, live in groups. The pygmy chimpanzee may be more closely related to humans than is the common chimpanzee (Zihlman et al. 1978).

Although all modern hominoids can stand erect and walk to some degree on their hind legs, only humans display an erect bipedal mode of striding locomotion involving a specialized structure of the pelvis and hind limbs, thereby freeing the forelimbs from obligatory functions of support, balance, or locomotion.

Diversity of Fossil Apes

Fossil primates commonly called apes are known from the late Eocene to early Oligocene Fayum Formation of Egypt, which appears to represent a tropical forest type of environment. These primates belong to the extinct family Propliopithecidae (Greek *plio* = more, referring here to the Pliocene epoch), and include the cat-size *Aegyptopithecus,* originally thought to be an early true ape. Propliopithecids are considered ape-like rather than monkey-like because they have low-crowned, bunodont cheek teeth, suggesting a diet composed of fruit and seeds but lacking in mature leaves. Their skeletal anatomy suggests that they were primarily arboreal quadrupeds, not terrestrial but also not highly specialized for climbing. However, these primates are more primitive than any living catarrhine; they predate the divergence of modern monkeys and apes. They thus cannot be classified as hominoids. Other ape-like primates that are not true hominoids are the pliopithecids, gibbon-like forms from the middle to late Miocene of Eurasia.

The first true hominoids were the early Miocene **proconsulids** of East Africa. They apparently occupied primarily forested habitats. Proconsulids were generalized quadrupedal arboreal animals, ranging from the size of a small monkey to the size of a female gorilla, with bunodont molars suggesting a frugivorous diet. The early Miocene *Morotopithecus* from Uganda is the first hominoid to have features of skeletal anatomy (the shoulder and vertebral column) that are shared by living apes and humans.

By the middle Miocene more derived hominoids had diversified into a variety of ecological types. Some remained in Africa, such as the genera *Afropithecus, Kenyapithecus,* and *Equatorius*. Others spread into Eurasia, following the general middle Miocene warming trend, and both cercopithecine and colobine monkeys are also known from Eurasia in the late Miocene and Pliocene. The later Cenozoic Eurasian hominoids include the **dryopithecids** (Greek *dryo* = tree) and **sivapithecids** (after the Hindu god *Shiva*), which are more closely related to the great apes and humans than are gibbons. Evidence from paleoclimatic studies and the nature of the other animals in the fauna suggests that these Miocene hominoids primarily occupied woodland or forest habitats.

Dryopithecids were probably more primitive than any living great ape. In contrast, the sivapithecids appear to be related to the living orangutan. The teeth and jaws of sivapithecids suggest that the animals were about a meter tall. Perhaps the best known of these fossil apes is *Sivapithecus*. This genus now includes the animal formerly known as *Ramapithecus* (Latin *rami* = branch), known from around 17 million years ago, which was originally considered an ancestral hominid. The dentition of these primates, with reduced canines and thick-enameled molars, indicate that they fed on material that required crushing (thick enamel) and grinding (no interlocking canines to inhibit rotary chewing movements), suggesting a diet of nuts and hard fruits. (Early hominids also have thick-enameled molars, which is one reason that *Ramapithecus* was originally thought to be a hominid.)

The orangutan lineage also includes the late Miocene to Pleistocene *Gigantopithecus* of Asia. The Pleistocene species represented the largest primate that has ever lived; at an estimated body mass of 300 kg, it would have been twice the size of an average gorilla. Some people have speculated that a surviving lineage of *Gigantopithecus* is behind the legends of the Yeti in Tibet, and the Bigfoot or Sasquach of northwestern North America. (Bigfoot, if it exists, could have reached North America by migrating across Beringia during the Pleistocene, as did so many other mammals.)

A few fossil apes, such as the European *Graecopithecus* (also known as *Ouranopithecus;* Greek *ourano* = heaven), which lived between 10 and 8 million years ago, appear to belong to the same clade as the great apes plus humans. It is notable that humans have retained the ancestral characteristics of our clade in numerous dental and associated cranial features, while the lineages of great apes have derived unique specializations. Of course, humans have also evolved their own specializations independently of the various great ape lineages. It remains important to note, however, that all of the living hominoids are derived in comparison with the Mio-Pliocene ape radiation. The fact that gibbons and orangutans are in a less derived position on the cladogram than are humans or African apes (Figure 23–1) does not imply that earlier apes looked like these modern forms.

23.3 Origin and Evolution of Humans

Humans have traditionally been placed in a separate family, called the Hominidae, which includes our own genus *Homo* and the related *Australopithecus* and *Ardipithecus*. However, this classification is more a reflection of the chauvinism of those species doing the classifying than any profound difference in anatomy between apes and humans. In fact, molecular evidence suggests a much closer relationship between apes and humans. If the two were any other types of mammals, they would undoubtedly be placed in the same family.

We now live in an unusual time for hominids, since it is only in the past 30,000 years that only a single species has been in existence. As recently as 50,000 years ago our species, *Homo sapiens*, shared the planet with *Homo neanderthalensis* and *Homo erectus*; throughout hominid history (with perhaps, the exception of the first half a million years) several species of hominids have coexisted (see Figure 23–13).

Anatomical differences between humans and apes appear in the skull and jaws, the trunk and pelvis, and to a lesser extent in the limbs. Humans have short jaws in association with the shortening of the entire muzzle, certain teeth are smaller than their counterparts in other apes, and the entire dentition is relatively uniform in size and shape. In particular, the canines lie on the same occlusal plane as the incisors and cheek teeth. Apes have a prominent diastema between the canines and incisors that accommodates occlusion of the jaws, whereas in humans all the teeth touch their adjacent members (Figure 23–11).

The jaws of an ape are rectangular or U-shaped. The four incisors lie at right angles to the canines and cheek teeth, which form nearly parallel lines along the jaw's two sides. Ancestors of the hominids, and fossil hominoids such as *Sivapithecus*, had an almost V-shaped jaw, and some of the earliest hominids had a U-shaped jaw. The jaw of members of the genus *Homo* is bow-shaped, with the teeth running in a curve that is widest at the back of the mouth. The human palate is prominently arched, whereas the ape palate is flatter between their parallel rows of cheek teeth.

Various evolutionary trends can be identified within the hominids. The points of articulation of the skull with the vertebral column (the occipital condyles), and the foramen magnum (for the passage of the spinal cord), shifted from the ancestral position at the rear of the braincase to a position under the braincase. This change balances the skull on top of the vertebral column, in association with an upright, vertical posture of the trunk. The braincase itself became greatly enlarged in association with an increase in forebrain size. By the end of the middle Pleistocene a prominent vertical forehead developed, in contrast to the sloping foreheads of the apes. The brow ridges and crests for muscular attachments on the skull became reduced in size, in association with the reduction in size of the muscles that once attached to them. The human nose became a more prominent feature of the face, with a distinct bridge and tip.

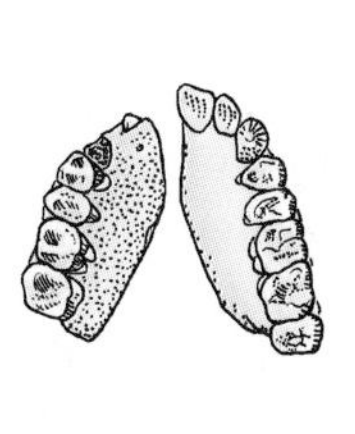

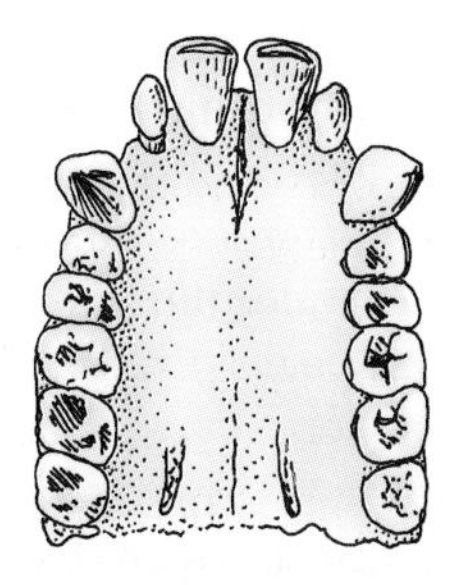

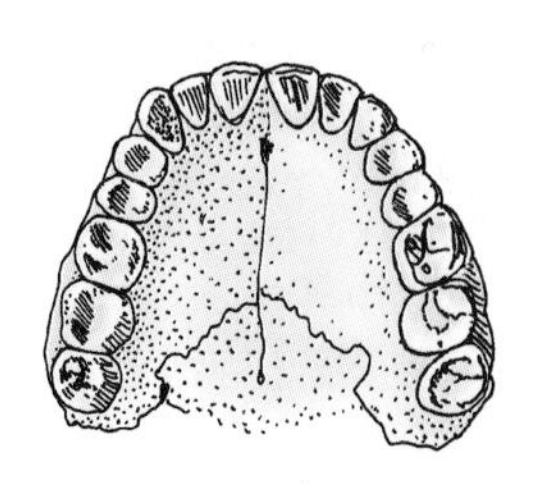

▶ Figure 23–11 Comparison of ape and human upper jaws.

Relationships within the Hominoidea

The primate phylogenetic tree (Figure 23–1) shows that the human, chimpanzee, and gorilla are more closely related to each other, in terms of recency of their common ancestor, than they are to orangutans or gibbons. These relationships among extant apes and humans were first determined on the basis of comparative anatomy by T. H. Huxley in 1863. They have gained additional support and clarity from recent biochemical, serological, and cytological comparisons as well as from the fossil record of earlier hominoids. Most of these studies, especially the molecular ones, support a close relationship between chimpanzees and humans, although some anatomical studies place chimpanzees and gorillas closer to each other than either is to humans.

The extant African apes have no known fossil record. At present we must depend on molecular studies to understand the history of our closest surviving relatives. This evidence indicates that gorillas separated from the chimpanzee and human common ancestor between 8 and 10 million years ago. The gorillas were subsequently isolated by unknown phenomena into eastern and western populations about 3 million years ago. Sometime in the interim, perhaps as recently as 6 million years ago, chimpanzee and human lineages separated. Two or possibly three species of chimpanzee exist today with geographic differences in genetic patterns similar to that found in gorillas. However, the differences in chimpanzee populations are thought to have occurred more recently, between about 2.5 and 1.6 million years ago.

A surprising conclusion from the emerging discipline of molecular evolution is that humans are more closely related to the extant great apes of Africa than either morphologists or paleontologists had thought. For example, hybridization of nonrepeated DNA sequences of humans and chimpanzees indicate 98 percent identity. Electrophoretic measurement of genetically controlled polymorphic proteins (allozymes) in the six extant genera of hominoids show that the degree of genetic difference among human, chimpanzee, gorilla, and orangutan is no greater than that observed among species in the same genus in many other taxa of vertebrates, including most mammals.

Calculations of divergence times are based on the concept of the molecular clock (the supposedly unvarying rate of generic change within a clade over evolutionary time). They suggest that hominids last shared a common ancestor with the African apes no more than 5 to 10 million years ago, in the late Miocene to early Pliocene. This is a much more recent split than most biologists have previously supposed. It indicates a phase of rapid evolutionary change in the lineage leading to the specialized apes, as well as in our own lineage.

Because of this close genetic relatedness between living hominoids, there has been a recent trend to readjust their higher-level classification. Many taxonomists now consider the family Hominidae, which we use here in the more traditional sense to include only the genera *Homo*, *Australopithecus*, and *Ardipithecus*, to include all of the great apes (i.e., those originally classified as "pongids"). Humans would now be reduced to the level of the subtribe Hominina, which also includes chimpanzees, within the subfamily Homininae, which also includes at least the gorilla (Delson and Tattersall 1997, Lewin 1998).

However, this classification is very new, and primate taxonomy nomenclature is in a state of flux. For example, some researchers consider gibbons to be within the family Hominidae, while others prefer to leave them as their own family Hylobatidae. Some researchers would place orangutans in the subfamily Homininae within the newly defined family Hominidae; others would accord them their own subfamily Ponginae. Because there is no one widely accepted classification reflecting these new views of primate relationships, we have adhered to the traditional classification (e.g., Fleagle 1998) in this presentation, acknowledging that it exaggerates the actual differences among lineages.

If the molecular time estimates are correct, a gap of only 1 to 5 million years separates the divergence of humans from chimpanzees from the first undoubted hominid fossils, which are first known from 4.4 million years ago. About 8 million years ago, at the beginning of an essentially blank 4-year period for hominoid fossils in East Africa, tectonic forces split the region almost in two along a north-to-south feature still dramatically obvious today, the Rift Valley. New mountain boundaries and resulting changes in the ecological landscape appear to have divided an as-yet-unknown ancestral population of hominoids. The western population survived in moist forested lowlands and became the ancestors of the chimpanzees. The eastern population lived in the rain shadow of the new mountains, where forests gave way to woodlands and woodlands to savannas, and became hominids. As conditions became dryer and more seasonal 2 to 3 million years ago, savannas became increasingly widespread in the rain shadow, and the spread of savanna environments is thought to have been critical in determining the course of human evolution.

Primitive Hominids

The earliest hominids are the **australopithecines** (Latin *australi* = southern). Despite their role in human ancestry, in terms of their biology they perhaps are best thought of as bipedal apes with a modified dentition. Microwear analysis of the teeth of australopithecines suggests that they were primarily fruit eaters, perhaps including some meat in the diet as do present-day chimpanzees. However, they were not carnivorous hunters, as has been proposed for early members of the genus *Homo*. Although isolated fragments as old as 5.6 million years are known from Africa, they are too small to yield much information. No interpretable hominid remains older than about 4.3 million years had been discovered until recently.

All australopithecines appear to have been capable of bipedal walking. Human-like footprints were found by Mary Leakey and her associates at Laetoli in Tanzania, in volcanic ash beds radiometrically dated between 3.6 and 3.8 million years in age, and were probably made by *Australopithecus afarensis*, which is known from fossils at the same site. Analysis of these footprints indicates that they do not differ substantially from modern human trails made on a similar substrate, demonstrating the antiquity of bipedalism in hominid ancestry, long before the acquisition of an enlarged brain. However, australopithecines have some ape-like features both of limb anatomy and of the semicircular canals in the inner ear (structures responsible for orientation and balance), suggesting the retention of ape-like orientation in an arboreal environment rather than a fully terrestrial life-style (Spoor et al. 1996). It appears that australopithecines were able to stand and walk bipedally to a certain extent, but still spent much of their time in the trees, and probably were not capable of sustained running.

The most completely known early australopithecine, and until very recently the earliest known hominid, is *Australopithecus afarensis*. This species is best known from a very substantial part of a single young adult female skeleton, known in popular literature by the nickname Lucy. Lucy was discovered by Donald Johanson in the Afar region, also in Ethiopia, not far from the Red Sea, in a deposit dated at 3.2 million years. Lucy is an astonishing specimen in several respects. She is the most complete pre-*Homo* hominid fossil ever found, consisting of more than 60 pieces of bone from the skull, lower jaw, arms, legs, pelvis, ribs, and vertebrae. Her overall body size was small. Young but fully grown when she died, Lucy was only about 1m tall and weighed perhaps 30 kg (see Figure 23–12). Other finds indicate that males of her species were larger, averaging 1.5m tall and weighing around 45 kg.

Lucy's hip and limbs clearly indicate that she stood erect and walked bipedally. However, detailed analysis of the fossils and reconstruction of musculature demonstrate a type of bipedalism somewhat different from modern humans, and retention of a partly arboreal anatomy (Berge 1994). Arboreality is also reflected in her hands and feet, with fingers and toes significantly longer and more curved than in modern humans. However, the hands of australopithecines were more human-like than those of fully arboreal apes such as gibbons and orangutans, and they lacked the highly robust fingers of the knuckle-walking chimpanzees and gorillas (Clarke 1999). Lucy and her kind, who lived between 3.6 and 2.9 million years ago, may have slept and taken refuge in trees, with bipedalism being important in the efficient harvesting of fruits in their open forest and woodland habitat.

However, it does seem that Lucy walked more like we do than the "caveman" image of walking with a stoop and bent hips and knees. That notion of early human walking has more basis in Hollywood than in science—it is someone's idea of an intermediate gait between chimpanzees and humans that probably never existed. Computer simulations and anatomical studies indicate that all hominids walked with an upright, straight-hipped, straight-kneed gait (Crompton et al. 1998). In fact, the virtual Lucy devised by Robin Crompton will inevitably tumble if you try to simulate her walking in caveman style.

Lucy's teeth and lower jaw are also clearly human-like. Her diet may have been rich in hard objects, probably fruits that would have been unevenly distributed in space and time. Unfortunately, we know little about the size and shape of Lucy's cranium. Other specimens of *A. afarensis* indicate a brain size of 380 to 450 cubic centimeters, quite close to that of modern chimpanzees and gorillas.

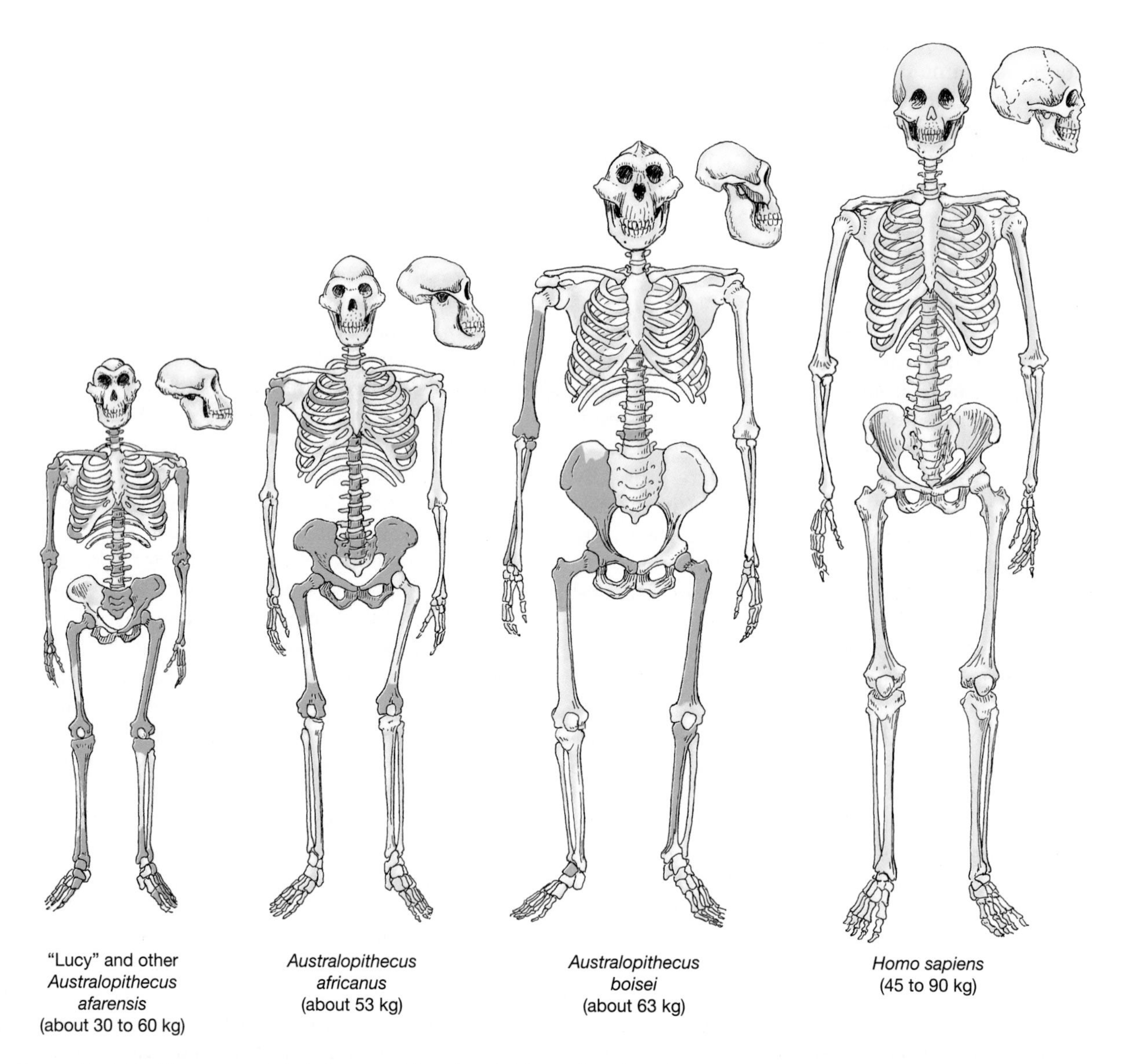

▲ Figure 23–12 Reconstructed skeletons of four hominid species, showing relative stature and stance. Darker portions of the postcranial skeletons of the fossils show the parts actually known from specimens. Although almost all australopithecine skulls are fragmentary, sufficient material exists to reconstruct entire skulls with fair certainty. Too little of the most ancient australopithecine (*Ardipithecus ramidus*) has been discovered to permit confident reconstruction. Note the shorter iliac blade, the less funnel-shaped rib cage, the relatively longer legs, and the shorter fingers and toes in *Homo*.

More recently, earlier and even more primitive hominids have been discovered. A new genus of fossil hominid based on 17 specimens (all but four of them teeth) represents the remains of a form that lies close to the divergence between our closest African ape relatives and the lineage to humans (White et al. 1994). *Ardipithecus ramidus* (formerly *Australopithecus ramidus*), as the fragments have been named, comes from Ethiopian sediments about 4.4 million years of age and indicative of a wooded habitat (WoldeGabriel et al. 1994).

Ardipithecus (Afar *Ardi* = ground floor) has many chimpanzee-like features, and lacks some of the derived traits shared by early species of *Australopithecus* and all later hominids. Tentative evaluation of the scant evidence, however, indicates that *Ardipithe-*

cus was bipedal, judging from the position of the foramen magnum. It also had a human-like rather than ape-like arm (especially the shoulder joint and elbow), and incisiform canines with less sexual dimorphism than modern African apes—traits that are hallmarks of the human lineage.

Several new species of *Australopithecus* have been found in the past decade, filling in more gaps in the story of human evolution. Figure 23–13 shows one current scheme of how the australopithecine species were related to each other and to our own genus *Homo*. The earliest known member of the genus *Australopithecus* is now *A. anamensis*, which ranges from 4.2 to 3.9 million years old and was found by Meave Leakey and Alan Walker in several sites in Kenya (Leakey et al., 1995). This hominid appears to be intermediate in anatomy between *Ardipithecus* and *Australopithecus afarensis*, with a less ape-like dentition than *Ardipithecus*, and with fragments of limb bones strongly suggesting bipedality. Its estimated body mass, around 50 kg, is greater than that of either of these other early hominids.

Australopithecus bahrelghazali is slightly younger than *A. anamensis* and contemporaneous with *A. afarensis* (Brunet et al. 1995). This hominid was found in Chad, in central Africa, in a fossil site with an interpreted environment of lakeside woodland. This is the first known australopithecine to be found west of the Rift Valley, suggesting that early hominids were more widespread in Africa—and more diverse in their habitats—than had previously been supposed.

A still younger species, *Australopithecus garhi*, is the newest hominid species to be named. Dated as 2.5 million years old, it is only slightly older than the earliest known specimen of the genus *Homo* (Asfaw et al. 1999). This species, known from a fragmentary skull from Ethiopia, has been hailed as the missing link between the genera *Australopithecus* and *Homo*. This skull was found near some fossilized butchered animal bones, leading to the speculation that *A. garhi* may have been the first species in our lineage to eat meat and use tools.

The youngest (i.e., latest surviving) of the early australopithecines, which are often termed gracile australopithecines because of their relatively small stature and delicate build, was *Australopithecus africanus*, which was known between 2.3 and 3 million years ago). Although *A. africanus* was fully bipedal, it also had very robust arm bones, suggesting that it may have spent even more time in trees than *A. afarensis*. This greater degree of arboreality implies that it may not be in the direct evolutionary line to *Homo*. However, *A. africanus* may have differed from *A. afarensis* in including more meat in its diet (Sponheimer and Lee-Thorp 1999).

There is controversy about the interrelationships of these early hominids. Some researchers would interpret them as an ancestor-descendant evolutionary line—with *Ardipithecus* giving rise to *Australopithecus anamensis*, which in turn gave rise to *A. afarensis*; and *A. afarensis* giving rise to both *Homo* and the later australopithecines. Others see a more bushy interrelationship between these taxa, as shown in Figure 23–13. Additional evidence of early diversity in the hominid lineage is provided by the description of a new genus, *Kenyanthropus* (Leakey et al. 2001).

The later australopithecines were distributed in East and South Africa from 2.5 to around 1.2 million years ago. Termed robust australopithecines, they are often placed in their own genus *Paranthropus*. These were relatively large, powerfully built forms with pronounced sagittal crests on the skulls, enormous grinding postcanine teeth, and the body proportions of a football player (although they were no more than about 1.5m in height). The three robust species—*Australopithecus robustus* of South Africa and *A. aethiopicus* and *A. boisei* of East Africa—were terrestrial, savanna-dwelling vegetarians, somewhat analogous in their dietary habits to the forest-inhabiting gorillas. In comparison with other australopithecines, their molars were big and exhibited heavy wear, suggesting a more coarse and fibrous diet. It is not known how these robust australopithecines are interrelated, or even whether they represent a single radiation from within the gracile australopithecines. The robust type of australopithecine may have arisen independently on more than one occasion.

The gracile australopithecines were primarily an early Pliocene radiation. In the late Pliocene, at around 2.5 million years ago, the hominid lineage split into two—one lineage leading to our direct ancestors, the genus *Homo*, and the other leading to the robust australopithecines. This particular time also appears to record global climatic changes. It was 2.5 million years ago when the Isthmus of Panama formed, linking North and South America. This event caused changes in oceanic currents, because there was no longer a direct passage between Atlantic and Pacific oceans; the modern Gulf Stream circulation probably dates from this time. There was significant global cooling, with the onset of Arctic glaciation. Paleontological evidence from both the

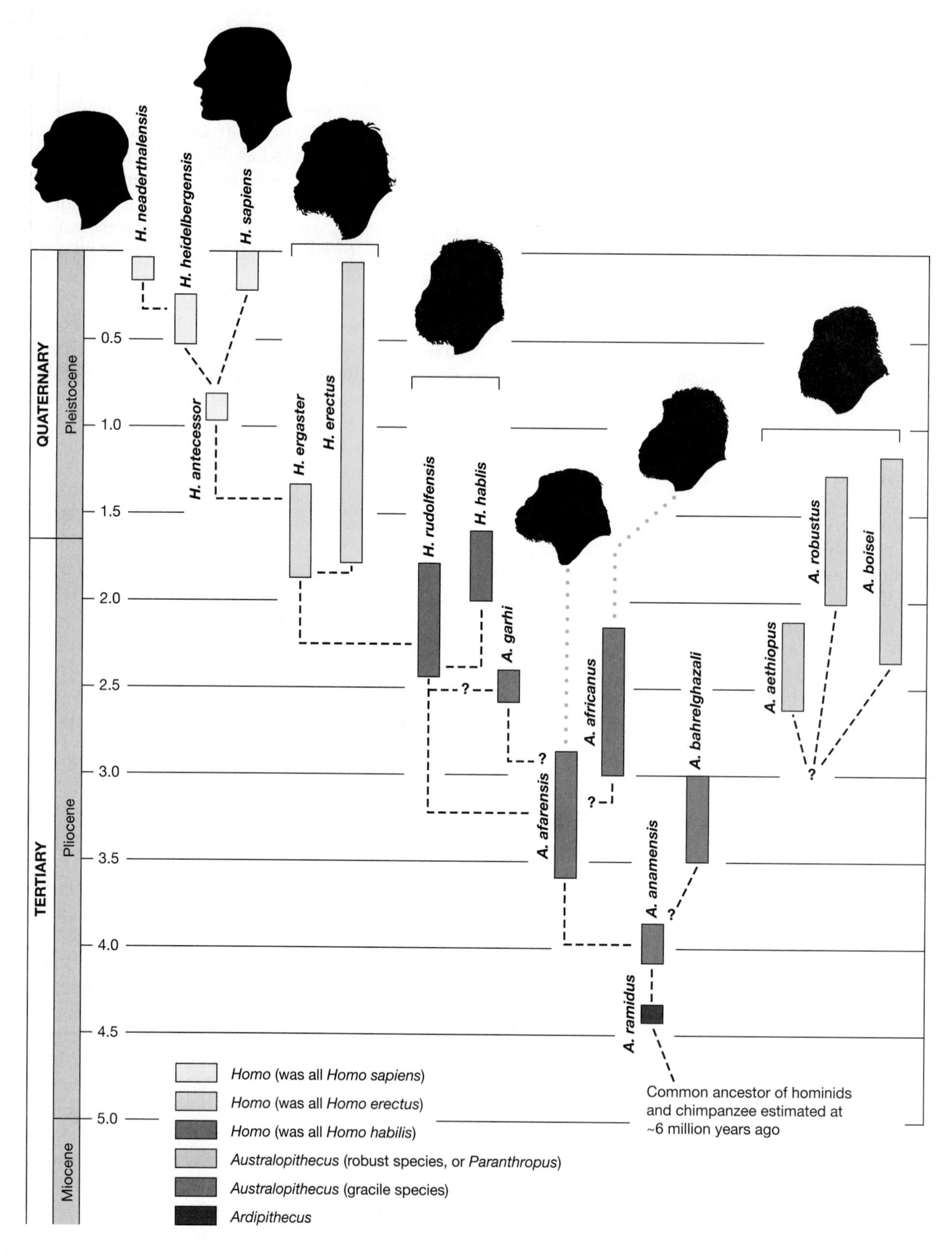

▲ Figure 23–13 A hypothesis of the phylogenetic relationships within the Hominidae.

flora and fauna of Africa suggests the climate changed at this time, to a drier, more savanna-like habitat; grazing antelopes increased in abundance, and new types of carnivores appeared. These global environmental changes may have been the trigger in changing the course of hominid evolution.

Thus during the late Pliocene, from 2.5 million years onward, there were two evolving lineages of hominids—early true humans and robust australopithecines. Further climatic cooling and drying resulted in the reduced abundance of robust australopithecines at the start of the Pleistocene, and their extinction later in the epoch. However, had the Pleistocene climatic changes been different—such as reverting to a wetter and warmer regime—it might have been our ancestors who became extinct and the robust australopithecines that survived.

Derived Hominids (the genus *Homo*)

The earliest species of *Homo,* originally all ascribed to *H. habilis* (= "handy man"), existed in East Africa from 2.4 to 1.6 million years ago. *Homo habilis* is best distinguished from australopithecines by its larger cranial capacity: between 500 and 750 cubic centimeters in contrast to the value of 380 to 450 cm^3 for *Australopithecus afarensis* (Falk 1985). *Homo habilis* also differs from *Australopithecus* in having a smaller face, and a smaller jaw and dentition, with smaller cheek teeth and larger front teeth.

The actual taxonomic status of these highly variable fossil remains of this early genus of *Homo* is much debated. Many researchers now split the original *H. habilis* into two species: *H. rudolfensis,* known from 2.4 to 1.8 million years ago, and *H. habilis* (redefined, known from 1.9 to 1.6 million years ago). Despite being older than *H. habilis, H. rudolfensis* was somewhat larger-brained, and apparently more closely related to later species of *Homo.*

Around 2 million years ago a new hominid appeared in the fossil record—*Homo erectus* (Latin *erect* = upright), originally described in the late nineteenth century as *Pithecanthropus erectus,* and known at that time as Java Man or Peking Man. *Homo erectus* probably originated in East Africa, where it coexisted for at least several hundred thousand years with two of the robust australopithecines. *H. erectus* was the first intercontinentally distributed hominid. It appears to have spread to Asia no later than 1.8 million years ago, and subsequently perhaps into Europe.

Nowadays the older African version of *H. erectus* is usually called *Homo ergaster,* with the name of *erectus* reserved for the Asian hominid. *Homo ergaster* is thought to be more closely related to later hominids, and it disappeared around 1.3 million years ago. *Homo erectus* survived for much longer. It was originally thought that the species disappeared between 200,000 and 300,000 years ago, but recent discoveries have shown that it survived for around a quarter of million years longer. Remains from Java have been dated to between 27 and 53 thousand years old (Swisher et al. 1996), making *H. erectus* a contemporary of the Neandertals and modern *H. sapiens.* We shall refer to *H. erectus* only, implying inclusion of both *H. erectus* and *H. rudolfensis* in our discussion of the characteristics of hominids at this evolutionary level.

Three characteristics are especially important features of *Homo erectus.* First, they were substantially larger than earlier hominids (up to 1.85m tall and weighing at least 65 kg—the same size as modern humans), with a major increase in female size that reduced sexual dimorphism. In earlier hominids males were apparently twice the size of females, whereas in *H. erectus* the males were only about 20 or 30 percent larger than the females, as in our own species. The reduction in sexual dimorphism in *H. erectus* and later species of *Homo* imply behavioral changes, from a ranging pattern in which females foraged in smaller territories than males and a probable polygynous mating system to more stable, mixed-sex groups with monogamous pair bonding. The substantial increase of body size of *H. erectus* may be directly related to a significant increase in group mobility and ultimately to intercontinental distribution.

Second, *H. erectus* also had a larger brain than earlier *Homo* species, with cranial capacities ranging from 775 to 1100 cm^3. However, because later species of *Homo* were also larger animals than earlier *Homo,* this difference in brain size in *H. erectus* mainly reflected larger overall body size.

Third, *Homo erectus* was the first hominid to have a human-like nose, with downward-facing nostrils. Its other facial features were primitive—a prognathous (projecting) face with relatively large teeth; almost no chin; a flat, sloping forehead; prominent, bony eyebrow ridges; and a broad, flat nose (Figure 23–14c). The skeletal proportions were similar to those of modern humans, although the femoral (thigh bone) head was relatively smaller; but

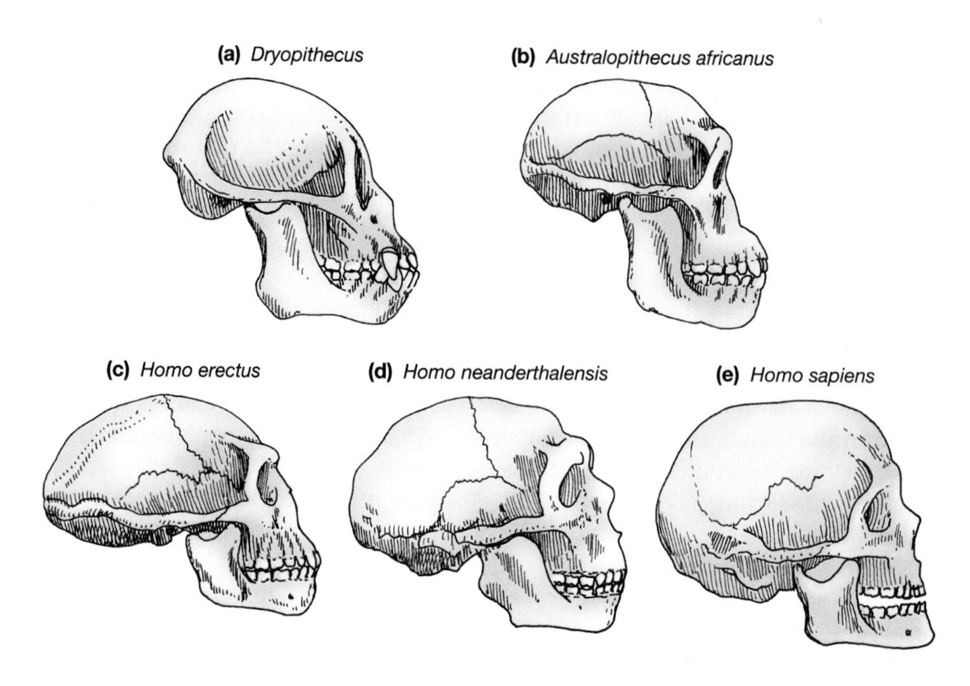

▲ Figure 23–14 Comparison of skulls of hominids (c–e) and more primitive hominoids (a, b).

the bones were more robust, suggesting a more muscular build.

Homo erectus represents a large change in the evolutionary history of humans. With this pair of species, a number of firsts were recorded in human prehistory. These included the first appearance of hominids outside of Africa, the first evidence for tool making and the use of fire, the first evidence of hunting and the use of home bases, and the first evidence of delayed tooth eruption—with relatively small teeth compared to body size.

The smaller teeth suggest that the cooking of food may have started at this time. The delayed tooth eruption suggests a human-like extended childhood, which would also imply a human-like extended life span. One explanation for this change in life history is that care of the young by grandmothers, freeing the mothers to reproduce again sooner, became an important feature in this stage of human evolution (O'Connell et al. 1999). This might also be associated with the uniquely human (among living mammals) attribute of female menopause, a postreproductive phase in an extended human life span. Other features of human reproductive behavior may have originated at this time, such as concealed estrus. Humans are unique among living mammals in the lack of obvious signaling of the female's ovulation; there are various speculations about the reasons for this phenomenon, including promotion of community stability in a mixed-sex but basically monogamous society (Diamond 1992).

Origin of Modern Humans

The species *Homo sapiens* (Latin *sapien* = wise), as originally defined, first appeared around 500,000 years ago—although a newly discovered cranium suggests that *H. sapiens* may be as old as one million years (Abbate et al., 1998). *H. sapiens* included not only the modern types of humans, which date from around 200,000 years ago, but also the Neandertals (also known as Neanderthals, named for the Neander valley in Germany), which sometimes have been classified as the subspecies *Homo sapiens neanderthalensis*. More recent evidence, from a variety of sources, has complicated this picture.

Specimens of *Homo sapiens* have a slightly larger brain, a thicker and more robust skull, larger teeth, and less prognathism (projection of the jaw beyond the plane of the upper face) than *Homo erectus*. Early, archaic types of *Homo sapiens*, known from both Africa and Europe between 500,000 and 300,000 years ago, have been accorded their own species, *H. heidelbergensis*. The term *H. sapiens* is now reserved for modern humans, and Neandertals are accorded their own species, *H. neanderthalensis*. Recent analysis of ancient DNA from the bone of a Neandertal showed considerable genetic difference from modern humans, suggesting that they were not directly ancestral to modern humans nor, perhaps, quite as closely related as once thought (Krings et al. 1997). Restudy of the morphology of Neandertals also shows their skulls to have many features that are distinctly different from ours (Tattersall and Schwartz 1998).

H. heidelbergensis was thought to have been ancestral to both modern humans and Neandertals, but it may actually be more closely related to the Neandertals than to modern humans (see Figure 23–13). Additionally, a new hominid found 800,000 years ago in Spain has a face very like that of a modern human. Researchers have named this hominid *Homo antecessor*, and have nominated it as the common ancestor of Neandertals and modern humans (Gibbons 1997).

There has been considerable debate about the origin of *Homo sapiens* (summarized in Lahr 1994). Some workers hold to a multiregional model of human evolution. They suggest that *H. sapiens* may have evolved independently in different areas, each from an already distinctive local population of *H. erectus*. Others hold to a single, African origin for modern *H. sapiens*, implying a second wave of migrations out of Africa, replacing *H. erectus* in Asia and the Neandertals in Europe. Analysis of both fossil and modern material indicates that there was most probably a single African origin for all *Homo sapiens*.

A single African origin is also supported by the so-called Mitochondrial Eve hypothesis. This theory is based on studies of mitochondrial DNA, which as part of the egg cytoplasm rather than the egg nucleus is passed to the offspring only from the mother. A 1987 study by Alan Wilson analyzed the mitochondrial DNA from 147 individuals representing a large diversity of the world's races. Wilson's analysis showed that the greatest diversity in living humans resides in Africa, leading to the conclusion of a single African origin (e.g., Wilson and Cann 1992). Wilson also concluded that human mitochondrial DNA can be traced to a single human female around 200,000 years ago, and that a bottleneck in human evolution existed then—with all living humans coming from this line of descent. (Studies based on divergence times for genes on the Y chromosome, which are passed only from father to son, have suggested for a similar time for the origin of modern humans—naturally, this proposal has been called the African Adam hypothesis.) It is interesting that these molecular studies, although they have met with some criticism, generally agree with the fossil record, which shows that modern *Homo sapiens* dates from around this time.

The Neandertals

The first recognized discovery of fossils of *Homo neanderthalensis* were from the Neander Valley in western Germany in 1856, and Neandertal features first appear in fossils about 150,000 years ago. Neandertals were 1.5 to 1.7m tall and had receding foreheads, prominent brow ridges, and weak chins. Their brains were as large or larger than those of modern-day *Homo sapiens*, but they were enlarged in a slightly different fashion. The Neandertal brain had a larger visual cortex area than ours, at the back of the head, while we have a larger, central, temporal region (see Figure 23–14d, c). The Neandertals were stone toolmakers, producing tools known as the Mousterian tool industry, with a well-organized society and advancing culture.

On the average, Neandertals appear to have been much stronger than extant humans. They had short limbs with prominent attachments for massive muscles. Another striking difference between Neandertals and extant humans is the size of facial features: Neandertals had thick brow ridges, large noses, and broad midface regions. They had well-developed incisor and canine teeth, and their facial features may have been associated with structural modifications that supported these large teeth. The front teeth of Neandertals typically show very heavy wear, sometimes down to the roots. Were Neandertals processing tough food between their front teeth, or perhaps chewing hides to soften them, as do some modern aboriginal peoples?

Neandertals probably hunted the wild horse, mammoth, bison, giant deer, and woolly rhinoceros from close range: Their Mousterian-style hunting tools were of the punching, stabbing, and hacking type. Although almost all skeletal remains of Neandertals show evidence of serious injury during life, one in five persons was over 50 years old at the time of death. It was not until after the Middle Ages that historic human populations again achieved this longevity. Whether the Neandertals had the capacity for complex speech remains controversial, but they were the first humans known to bury their dead, probably with considerable ritual. Of special importance are burials at Shanidar cave in Iraq that include a variety of plants recognized in modern times for their medicinal properties.

The emerging picture of *H. neanderthalensis* is quite different from that proposed for much of the time since their discovery in 1856, when they were portrayed as primitive savages. They now appear to have been adept hunters who probably lived in a complex society. The Neandertals and the remaining populations of *Homo erectus* vanished between 40,000 and 30,000 years ago, and 40,000 years ago marks the first appearance of modern *Homo sapiens* in Europe. There is much current debate about the role of *H. sapiens* in the disappearance of the other species: Did our species gradually outcompete the others in a noncombative fashion, or was there some type of direct conflict? Modern humans do not have the distinguishing DNA markers of Neandertals, suggesting that interbreeding between these two species did not take place.

Other Theories of Human Evolution

Anatomically modern humans, known in southern Europe as Cro-Magnon people, are distinguished from earlier specimens of *Homo sapiens* by suites of subtle skeletal features equivalent to those distinguishing modern races. It is with the Cro-Magnons that an unbroken trail of evidence and history leads to the present. One popular view of human evolution is that many of the features that distinguish humans from apes, especially in the skull and face, can be explained by the process of neoteny, or the evolution of a new species essentially from the juvenile form of its ancestors. People have noted that humans bear a greater resemblance to juvenile chimpanzees than they do to adults, especially in the flatter face and the higher forehead, and have speculated that the process of neoteny has been responsible.

There has been some rather flippant speculation that human features arose out of some process for the sexual selection of juvenile features; perhaps the obsession with the sexual appeal of youth is a property of our genus. However, this possibly appealing explanation has proven to be a rather superficial one. It has been shown that the specific features of humans that resemble those of juvenile apes come about by specific, new growth processes, not by delaying of growth as would be expected if neoteny were responsible.

Another theory of the uniqueness of humans is the Aquatic Ape hypothesis. This idea, first suggested by Sir Alistair Hardy in the 1960s and further developed by Elaine Morgan in subsequent decades (e.g., Morgan 1982), proposes that humans went through an aquatic phase in their evolution. This supposed aquatic phase would link many unique features of humans that otherwise seem to be inexplicable in the context of terrestrial evolution on the savannas: the deposition of subcutaneous body fat, the loss of body hair, the breathing control necessary for speech, human profligate loss of salt in the sweat with the proliferation of eccrine glands for evaporative cooling (see Chapter 19), and even aspects of bipedal posture. The Aquatic Ape hypothesis has been dismissed without examination by almost all anthropologists, who can rightly be criticized in this context for automatically discounting ideas written outside of the scientific tradition. Langdon (1977) and Douglas (2000) offer less polemic treatments of this hypothesis, acknowledging some valid points if not defending it in total.

23.4 Evolution of Human Characteristics—Bipedality, Larger Brains, and Language

Humans are classically distinguished from other primates by three derived features: a bipedal stance and mode of locomotion, an extremely enlarged brain, and the capacity for speech and language. Here we examine possible steps in the evolution of each of these key features.

Bipedality

The most radical changes in the hominid postcranial skeleton are associated with the assumption of a fully erect, bipedal stance in the genus *Homo*. Anatomical modifications include the S-shaped curvature of the vertebral column, the modification of the pelvis and position of the acetabulum (socket in the hip for the ball joint with the femur) in connection with upright bipedal locomotion, and the lengthening of the leg bones and their positioning as vertical columns directly under the head and trunk (see Figure 23–15). The secondary curve of the spine

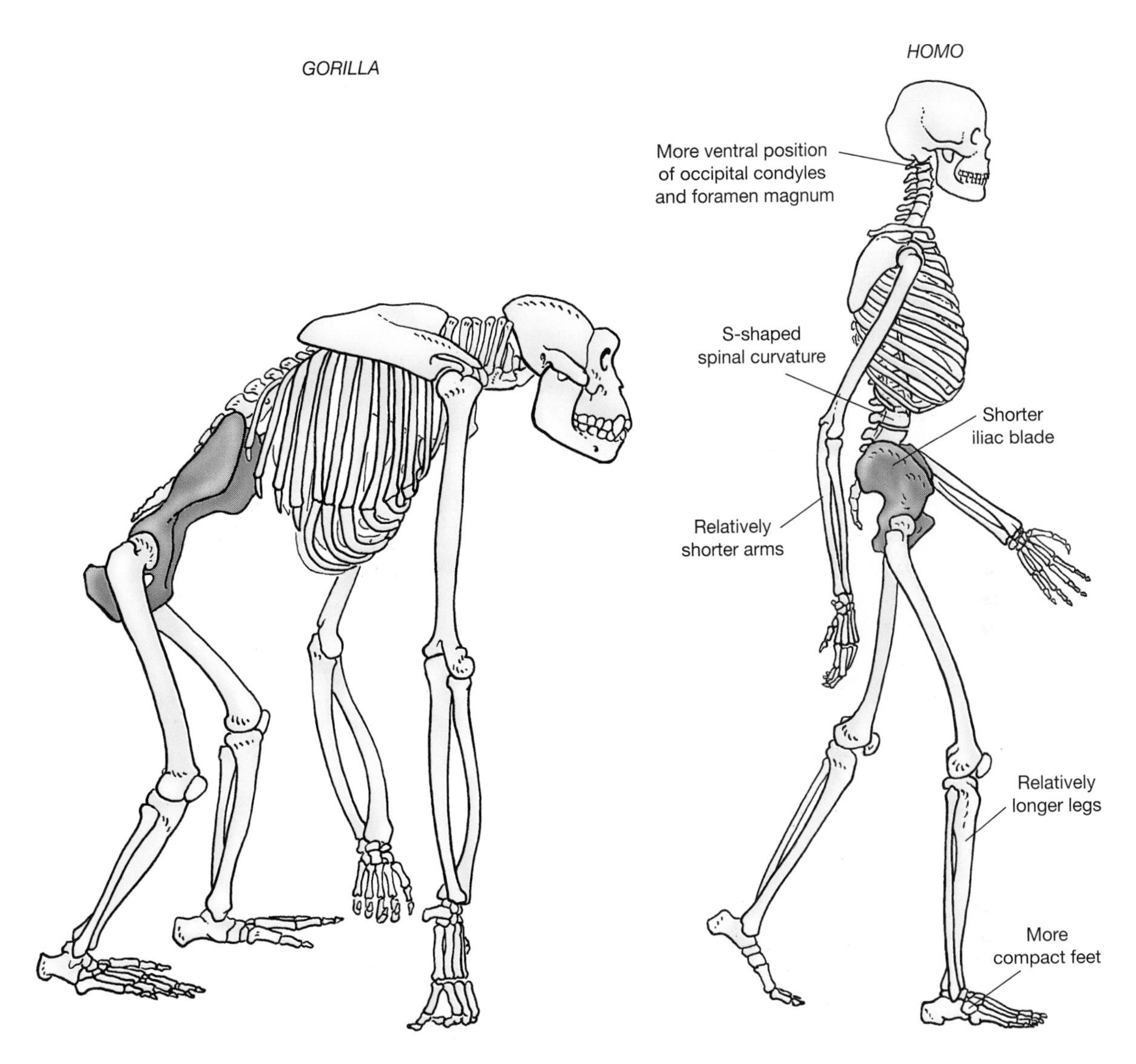

▲ Figure 23–15 Comparison of skeletons of gorilla and human. The pelvis is more darkly shaded.

in humans is a consequence of bipedal locomotion, and forms only when an infant learns to walk. We have by no means perfected our spines for the stresses of bipedal locomotion, which are quite different from those encountered by quadrupeds. One consequence of these stresses is the high incidence of lower back problems in modern humans.

Humans stand in a knock-kneed position, which allows us to walk with our feet placed on the midline and reduces rolling of the hips from side to side. This limb position leaves some tell-tale signatures on the femur (the thigh bone), both in the articulation with the hip and with the knee joint. This type of bony evidence can aid researchers in deducing whether fossil species were fully bipedal. An unfortunate consequence of this limb position is that humans, especially athletes, are rather prone to knee dislocations and torn knee ligaments. Because women have wider hips than men, their femurs are inclined toward the knees at an even more acute angle, and they are more prone to such knee injuries.

The feet of humans show drastic modification for bipedal, striding locomotion. The feet have become flattened except for a tarso-metatarsal arch with corresponding changes in the shapes and positions of the tarsals (ankle bones) and with close, parallel

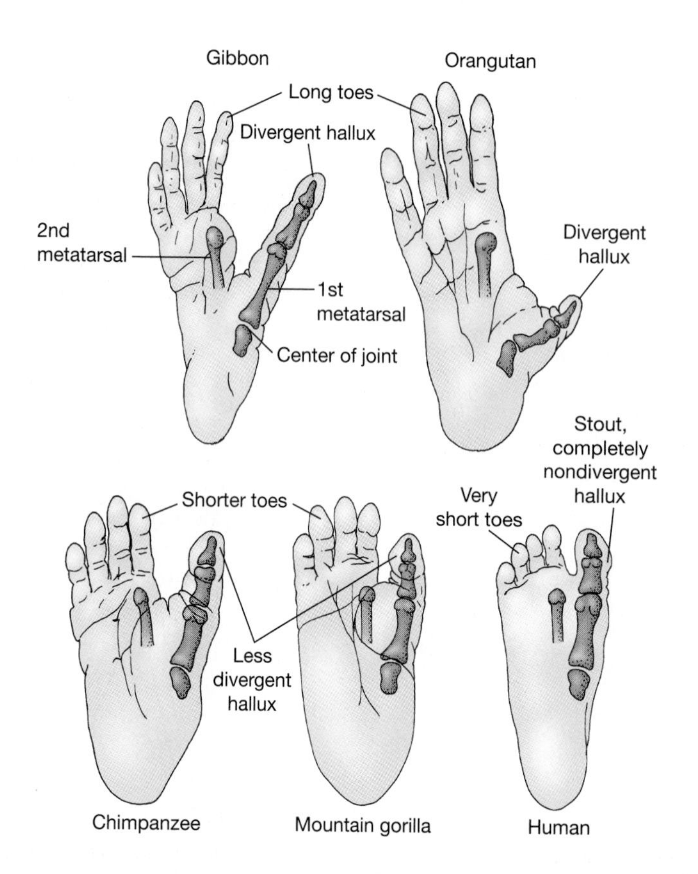

◀ Figure 23–16 Comparison of feet of extant hominoids, showing some skeletal parts (metatarsals for digits in I and II and phalanges of digit I) in relation to foot form.

alignment of all five metatarsals and digits. In addition the big toe is no longer opposable, as in apes and monkeys (Figure 23–16), although it may still have had some capacity to diverge from the rest of the toes in early hominids.

It is important to emphasize that among mammals the bipedal, striding mode of terrestrial locomotion is unique to the human line of evolution. Bipedality may have been a key change that made possible the evolution of other distinctively human traits, such as the perfection of tool-using hands, and thus indirectly probably stimulated the evolution of ever larger forebrains. What evolutionary events turned the human into an obligate terrestrial biped from ancestors that clearly were arboreal? How did our tool-using hands and our big brain with its associated implications for intelligence, derived social organization, and culture evolve?

There are almost as many hypotheses about the reasons for human bipedality as there are anthropologists. Among the suggested reasons are improved predator avoidance (being able to look over tall grass), freeing the hands for carrying objects (either for hunting or collecting foods), thermoregulation (an upright ape presents a smaller surface area to the Sun's rays), and efficiency of locomotion. An obvious problem with all such hypotheses is that they are difficult to test. Although humans are extremely efficient at bipedal locomotion, especially at walking, it seems unlikely that bipedality evolved specifically for efficient, striding locomotion. Other apes are not nearly as efficient as humans at bipedal walking, but this is the evolutionary condition from which human bipedality must have commenced. That is, a protohominid or early hominid must have walked bipedally in an inefficient way first, before selection could act on the skeleton.

The recent reinterpretation of the postcranial skeleton of the Miocene ape *Oreopithecus* (Greek *oreo* = mountain), known from the Italian area of Tuscany, suggests a parallel to the evolution of bipedality in humans (Köhler and Moyà-Solà 1997). This animal appears to have evolved a type of bipedality convergent with the condition in hominids: It had an S-shaped spine, a pelvic girdle rather like the condition in *Australopithecus*, and a knock-kneed angle to the femur. However, the foot differed markedly from that of humans. Rather than a compact foot, with a big toe aligned with the other toes (Figure 23–16), it had a widely divergent big toe—apparently providing a broad, tripodal base of support. This foot anatomy, combined with the short legs, indicates a slow, shuffling gait rather than human-like, efficient striding. It also suggests that at least in this species, bipedality must have evolved for some other reason besides efficient locomotion. It has been suggested that bipedality evolved in *Oreopithecus*, an animal living in an area that was then an island insulated from predators, to increase food-gathering efficiency. Interestingly enough, *Oreopithecus* also appears to have had a human-like, precision grip between thumb and forefinger (Moyà-Solà et al. 1999).

In some respects, human bipedality may not be such a mystery as it first appears. Because we are most closely related to chimpanzees and gorillas, there has been the tendency to take their feature of knuckle walking as the primitive condition for the common ancestor. But in fact protohominids were likely at least semiarboreal rather than specialized for terrestrial quadrupedal locomotion (Steudal 1996). However, there is current debate about the origins of knuckle walking in hominids. Some researchers claim that the wrist bones of gorilla and chimpanzee are sufficiently different to imply that knuckle-walking evolved in parallel in these two genera (Dainton and Macho 1999). Other researchers claim that some evidence from the wrist of *Australopithecus anamensis* shows evidence of a knuckle-walking ancestry, with the implication that this was the original gait of the common ancestor of gorillas, chimpanzees, and humans (Richmond and Strait 2000).

Other apes, such as gibbons and orangutans, tend to walk in a clumsy bipedal stance on the ground: Their upright trunk, adapted for arboreal locomotion, predisposes them to do this. A tendency to walk bipedally appears to be a primitive hominoid feature, further developed in *Oreopithecus* as well as in hominids (and perhaps also in *Gigantopithecus*, if the Bigfoot rumors are true). It may be that the chimpanzees and gorillas are the odd ones out among hominoids in walking quadrupedally on the ground.

Origin of Large Brains

Brain tissue is a metabolically expensive tissue both to grow and to maintain. Most of the growth of the brain occurs during embryonic development, and requires energy input from the mother. Thus, selective pressures for larger brains can be satisfied only in an environment that provides sufficient energy, especially to the pregnant or lactating female. The evolution of larger brains may have required increased foraging efficiency (partially achieved through larger female size and mobility), and high-quality foods in substantial quantities (partially achieved through the use of tools and fire).

Larger brains also would have required a change in life-history pattern that probably exaggerated the ancestral primate character of slow rates of pre- and postnatal development (thus lowering daily energy demands and also a females' lifetime reproductive output). The origin of bipedality may be linked to the origin of large brain size and a change in the human mode of reproduction (see Box 23-1).

Origin of Speech and Language

Although other animals can produce sounds, and many mammals communicate by using a specific vocabulary of sounds (as anyone who has kept domestic pets well knows), the use of a symbolic language is a uniquely human attribute. Although apes and chimpanzees have been taught to use some human words and form simple sentences, this is a long way from the complexity of human language. But where in human evolution did language evolve, and how can we tell this from the fossil record? The first evidence of human writing is only a few millennia old: obviously language evolved before this, but how long before?

One line of evidence comes from the study of brains. Although brains themselves do not fossilize, evidence can be obtained from **endocasts**, the impressions left by brains on the inside of the skull. One unique human brain feature associated with speech is an expansion on the left side of the brain called Broca's area. A similar area has been interpreted to be present in *Homo rudolfensis* and later species of *Homo* (Falk 1985). But does this mean that speech and language dates back to the start of the genus *Homo*? Certainly, trauma to this area results in damage to speech production and comprehension in

BOX 23-1 CORRELATION OF BIPEDALITY WITH INCREASED BRAIN SIZE IN HUMAN EVOLUTION

Steven Stanley (e.g., Stanley 1996) has noted that brain growth in humans is unlike that in other animals, including apes. Our large brain size is in large part due to a prolonged period of postnatal growth in the first year of life, continuing the rapid growth that occurs in utero. In contrast, brain growth in apes slows markedly after birth (Figure 23–17).

The obvious hypothesis for this pattern of brain growth is that humans already have large enough heads to cause a problem at birth for the mothers, and so further brain enlargement must occur after birth. However, this pattern of brain growth means that human babies are born in a particularly helpless condition, and require intensive care for the first year of life.

Stanley argues that this type of brain development must have postdated the evolution of full terrestriality, and the abandoning of arboreal habits. Australopithecine infants must have been like those of modern apes, mature enough to cling to their mother's fur soon after birth, as is essential in an arboreal habitat where the female would need to use her arms and hands for climbing. Only in a fully terrestrial hominid could a helpless infant be looked after properly.

Thus, whatever the reason for the evolution of the fully human type of bipedality, it must have preceded the evolution of the very large human brain. The transition to a prolonged postnatal period of brain growth probably occurred at around the evolutionary level of *Homo erectus*. Here there is dental evidence (delayed tooth eruption) for the type of prolonged childhood typical of humans, but not of other animals, including apes (see Figure 23–18).

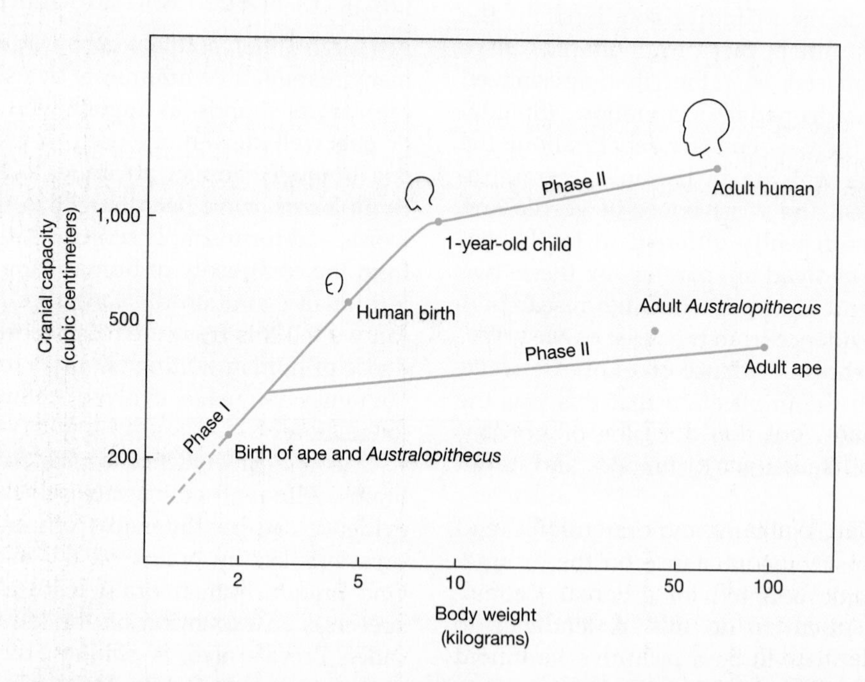

▲ Figure 23–17 Patterns of brain growth in humans and apes. The pattern for *Australopithecus* resembled that of a living ape, such as a chimpanzee, whereas in humans the continuation of the high prenatal rate of growth through the first year of life produces a much larger brain as an adult.

BOX 23-1 CORRELATION OF BIPEDALITY WITH INCREASED BRAIN SIZE IN HUMAN EVOLUTION (CONTINUED)

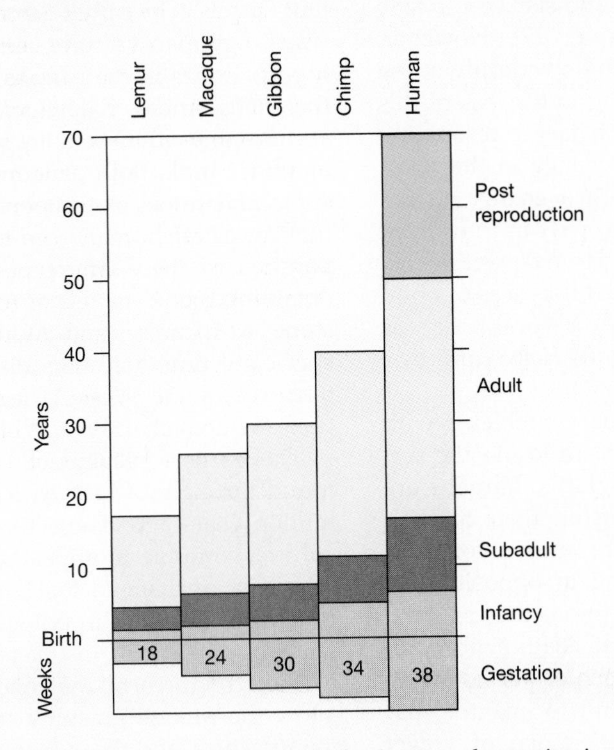

▲ Figure 23–18 Progressive changes in the length of life phases and gestation in primates. Note the proportional increase in the length of the postjuvenile stages as gestation increases. Only humans have a significant postreproductive stage, although the stage has also occasionally been observed in chimpanzees.

present-day humans, but this does not necessarily mean that human-like language had evolved by this point.

In contrast, Alan Walker argues that controlled speech would not have been possible until a later stage than *Homo ergaster* (Walker and Shipman 1996). In *H. ergaster* the spinal cord in the region of the thorax is much smaller than in modern humans. The implication is that *H. ergaster* lacked the capacity for the complex neural control of the intercostal muscles that allows modern humans to control breathing in such a way that we can talk coherently. Richard Kay (Kay et al. 1998) pointed out that the exit in the skull (the hypoglossal canal) for the nerve that innervates the tongue muscles (cranial nerve 12) was smaller in other hominids (and also smaller in chimpanzees and gorillas) than in the modern humans and Neandertals. This provides further evidence for the lack of human-like speech until relatively recently. (See also Cartmill, 1998, for a popular account of this and other hypotheses about the evolution of speech.)

Even if more derived *Homo* species had evolved the capacity for language, they would not have been able to produce the range of vowel sounds that we can produce until a change in the anatomy of the pharynx and vocal tract had taken place

(Lieberman 1984). The primitive position of the mammalian larynx is high up in the neck, right behind the base of the tongue. In this position the epiglottis can make a seal with the back of the pharynx, allowing for breathing and suckling to take place simultaneously (see Chapter 17). However, in modern humans the larynx shifts ventrally at the age of one to two years, resulting in the loss of this seal and the creation of a much larger resonating chamber for vocalization. This change in the vocal tract anatomy is associated with a change in the shape of the base of the skull, so we can infer from fossil skulls when the shift in larynx position occurred. Although there are some differences of opinion, a fully modern condition of the vocal tract was probably not a feature of the genus *Homo* until true *H. sapiens*.

There remain some evolutionary problems with the origin of this vocal tract trait. In losing the seal between the palate and the epiglottis, humans are especially vulnerable to choking on their food. It seems likely that this liability to choke would be a powerful selective force acting in opposition to repositioning the larynx. In addition, the death of infants from SIDS (sudden infant death syndrome) is associated with the developmental period when the larynx is in flux. It is difficult to imagine that the ability to produce a greater range of vowel sounds could counteract these antagonistic selective forces. Perhaps there was another, more immediately powerful reason for repositioning the larynx.

One advantage that repositioning the larynx affords us is the ability to voluntarily breathe through our mouths. Other mammals are unable to do this. One can subdue a rambunctious horse quite easily by clamping one's hand over its nose and blocking the nasal passages. Although dogs appear to mouth breathe, their panting is actually a means of ventilatory cooling: Air is passing through the nose and out of the mouth, not in and out of the lungs as it does when we pant.

Voluntary mouth breathing has an obvious importance in the production of speech. But perhaps a more important function of mouth breathing is apparent to anyone with a bad cold. Many of us would suffocate every winter if we were unable to breathe through our mouths. It is tempting to speculate that the human capacity for at least well-enunciated speech owes its existence to a prior encounter of our species with the common cold virus.

Origins of Human Technology and Culture

Humans are tool users and toolmakers *par excellence.* Some other animals also use tools to a limited extent, usually in rather stereotyped and instinctive ways. Egyptian vultures open ostrich eggs by picking up stones in their beaks and dropping them on the shells, and a Galápagos finch holds twigs and cactus spines in its beak to probe for insects in holes or under bark. Both baboons and chimpanzees use sticks and stones as weapons very much in the way that ancestral humans must have done, and chimpanzees use these same types of materials as tools in obtaining food. The use of roots, wooden clubs, and stones as hammers and anvils to crack five different species of nuts may be a culturally transmitted behavior in some West African chimpanzee populations (Whiten et al. 1999, Whiten and Boesch 2001).

The earliest recognized simple stone tools are found throughout East Africa and date to 2.5 to 2.7 million years ago. These tools, called the Oldowan culture, continue to appear in the geological record relatively unchanged for 1 million years. For some time most paleoanthropologists believed they were made by *Homo*, not *Australopithecus*. Careful analysis of the structures of the modern human hand that allow the precise grasping necessary for stone tool manufacture, and the evidence of these structures on the bones of the hand—especially the thumb—may lead to a very different conclusion (Susman 1994). Based on the thumb morphology data available, the Oldowan tools could have been made by either *Homo habilis* or *Australopithecus robustus*. However, the earlier gracile australopithecines did not have the hand bone characteristics of precision grip.

Stone tools no more sophisticated than the Oldowan culture are the only tools known earlier than about 1.4 million years ago. In fact, similar tools were made up to historical times by societies that retained stone-age technologies, such as the Australian Aborigines. Novel tools, especially cleavers and so-called hand axes, appear in East Africa about 1.4 million years ago. These later tools of *H. ergaster*, known as the Acheulean tool industry, changed very little in style for the next 1.2 million years—although the materials chosen changed, and later samples have a much higher proportion of small tools than do early examples. The lack of dramatic advance in tool manufacture over this immense period is surprising, especially in light of the spread of these tools to southwest Asia and western Europe.

Thus, one of the remarkable conclusions from paleoanthropological findings is that the use of tools, chipped stones, and possibly modified bone and antler precedes the origin of the big-brained *Homo sapiens* by at least 1.5 million years. The use of tools by ancestral hominids may have been a major factor in the evolution of the modern *Homo* type of cerebral cortex, although tool use is not necessarily associated with brain enlargement. Once the manufacture and use of tools began to increase fitness, selective pressures might have favored neural mechanisms promoting improved crafting and use of tools (Calvin 1994). In fact, the elaborate brain of *Homo sapiens* may be the consequence of culture as much as its cause.

By at least 750,000 years ago, *Homo erectus* was making advanced types of stone tools and had also learned to control and use, but perhaps not to make, fire. Actual hearths in caves are not widely recognized before 500,000 years ago. With fire humans could cook their food, increasing its digestibility and decreasing the chance of bacterial infection, and preserve meat for longer periods than it would remain usable in a raw state. They could also keep themselves warm in cold weather, ward off predators, and light up the dark to see, work, and socialize. The earliest evidence of wooden throwing spears, found in association with stone tools and the butchered remains of horses, is from 400,000 years ago in Germany (Thieme 1997).

It is often assumed that much of the evolution of human tool use and culture developed in the context of humans hunting other animals. The term "Man the Hunter" was coined back in the 1960s, although numerous researchers who happen to be women have noted that there are some inherent biases with the perception that human evolution has been shaped solely by male activities. However, a strong counterargument for the importance of the role of "Woman the Gatherer" has more recently emerged (e.g., Morbeck et al. 1997, Hrdy 1999).

Anthropologists have pointed out that much of our perception of human evolution as being an upward and onward quest has as much to do with the traditional western cultural myths of the "Hero's tale" as with any hard anthropological evidence (Landau 1984). More recent examination of the archeological evidence for hunting in early hominids, such as stone tools and bones with cut marks, has suggested that early humans were scavengers rather than hunters (e.g., Blumenschine and Cavallo 1992).

Neandertal people practiced ritual burial in Europe and the Near East at least 60,000 years ago, suggesting that religious beliefs had developed by that time. Not long afterward the cave bear became the focus of a cult in Europe. By 40,000 years ago, Cro-Magnon people began constructing their own dwellings and living in communities—things Neandertal peoples also probably did. The domestication of animals and plants, the development of agriculture, and the dawn of civilization were soon to follow.

Summary

Evidence for the origin of *Homo sapiens* comes from the Cenozoic fossil record of primates and from comparative study of living monkeys, apes, and human beings. The first primate-like mammals are known from the early Paleocene, but the earliest true primates are not known until the early Eocene. These were rather similar to the extant lemurs, and were initially known from North America as well as from the Old World. All were arboreal, and some possessed somewhat larger brains in relation to body size than other mammals of that time.

After the climatic deterioration of the late Eocene, primates were confined to the tropical regions until the middle Miocene. By the Oligocene, two other distinct groups of higher primates had evolved: the platyrrhine monkeys of the New World tropics and the Old World catarrhine monkeys and apes. The anthropoid apes and human-like species, including *Homo sapiens,* are grouped in the Hominoidea. Many morphological features distinguish the hominoids from other catarrhines, and enlargement of the brain has been a major evolutionary force molding the shape of the hominoid skull, especially in the later part of human evolution.

The first known hominoids occur in the early Miocene, around 25 million years ago. By the late Miocene hominoids had diversified into a number of environments and had spread widely over Africa, Europe, and Asia. A variety of hominid (human) fossils occur in late Pliocene and early Pleistocene deposits of Kenya and Ethiopia. The earliest hominids, with small cranial volumes and often heavy features of skull, tooth, and jaw, are the australopithecines are known from 4.4 to 1.2 million years ago. Australopithicines were apparently bipedal, but retained considerable arboreal habits.

The earliest member of the genus *Homo* dates from around 2.5 million years ago, concurrent with the earliest stone tools found in East Africa. *Homo* appears to be a primarily terrestrial genus, and this terrestriality may have been an important feature in the production of helpless infants whose brains continued to grow after birth. A global climatic change at around this time may have prompted the evolution both of *Homo* and of the robust australopithecines.

Homo erectus ranged across Africa and Eurasia from about 1.8 million to around 30,000 years ago. This hominid had a brain capacity approaching the lower range of *Homo sapiens,* made stone tools, and used fire. *Homo sapiens,* the only surviving species of the family Hominidae, came into existence around 200,000 years ago. By 40,000 to 60,000 years ago, some populations of *Homo* had a well-organized society with a rapidly developing culture, especially obvious in their use of stone tools. For much of human history several species of the genus *Homo* have existed in the world together. The present-day situation, with *Homo sapiens* as the sole existing hominid, is a highly unusual one.

Additional Readings

Abbate, E. et al. 1998. A one-million-year-old *Homo* cranium from the Danakil (Afar) depression of Eritrea. *Nature* 393:458–460.

Andrews, P. 1992. Evolution and environment in the Hominoidea. *Nature* 360:641–646.

Andrews, P., and L. Martin. 1991. Hominoid dietary evolution. *Philosophical Transactions of the Royal Society, London, Series B* 334:199–209.

Asfaw, B. et al. 1999. *Australopithecus garhi*: A new species of early hominid from Ethiopia. *Science* 284:629–635.

Benefit, B. R., and M. L. McCrossin. 1997. Earliest known Old World monkey skull. *Nature* 388:368–371.

Berge, C. 1994. How did the australopithecines walk? A biomechanical study of the hip and thigh of *Australopithecus afarensis*. *Journal of Human Evolution* 26:259–273.

Blumenschine, R. J., and J. A. Cavallo. 1992. Scavenging and human evolution. *Scientific American* 267(4):90–96.

Brunet, M., A. Beauvilain, Y. Coppens, E. Heintz, A. H. E. Montay, and D. Pilbeam. 1995. *Australopithecus bahrelghazali*, a new species of early hominid from Koro Toro region, Chad. *Comptes Rendues*, series II, 322:907–913.

Burenhult, G. (Ed.). 1993. *The First Humans (The Illustrated History of Humankind, Volume 1).* New York: Harper Collins.

Calvin, W. H. 1994. The emergence of intelligence. *Scientific American* 271(4):100–107.

Cartmill, M. 1998. The gift of the gab. *Discover* 19(11):56–64.

Ciochon, R. L., and R. S. Corruccini (Eds.). 1983. *New Interpretations of Ape and Human Ancestry*. New York: Plenum.

Clarke, R. J. 1999. Discovery of complete arm and hand of 3.3 million-year-old *Australopithecus* skeleton from Sterkfontein. *South African Journal of Science* 95:477–480.

Coppens, Y. 1994. East side story: The origin of humankind. *Scientific American* 270(5):88–95.

Crompton, R. H. et al. 1998. The mechanical effectiveness of erect and "bent-hip, bent-knee" bipedal walking in *Australopithecus afarensis*. *Journal of Human Evolution* 35:55–74.

Dainton, M., and G. A. Macho. 1999. Did knuckle walking evolve twice? *Journal of Human Evolution* 36:171–194.

Delson, E., and I. Tattersall. 1997. Primates, in *Encyclopedia of Human Biology*, vol. 7. New York: Academic Press, 93–104.

Diamond, J. 1992. *The Third Chimpanzee* New York: Harper Collins.

Douglas, K. 2000. Eve's watery origins. *New Scientist* 2266:29–33.

Falk, D. 1985. Hadar AL 162–28 endocast as evidence that brain enlargement preceded cortical reorganization in hominid evolution. *Nature* 313:45–47. (See also *Nature* 321:536–537, 1986.)

Fleagle, J. G. 1998. *Primate Adaptation and Evolution*, 2nd. ed., San Diego: Academic Press.

Gebo, D. L. et al. 1997. A hominoid genus from the early Miocene of Uganda. *Science* 276:401–404.

Gebo, D. L. et al. 2000. The oldest known anthropoid postcranial fossils and the early evolution of higher primates. *Nature* 404:276–278.

Gibbons, A. 1997. A new face for human ancestors. *Science* 276:1331–1333.

Hill, A. et al. 1992. Earliest *Homo*. *Nature* 355:719–722.
Hrdy, S. B. 1999. *Mother Nature*. New York: Pantheon.
Johanson, D. C., and M. A. Edey. 1981. *Lucy: The Beginnings of Humankind*. New York: Simon & Schuster.
Johanson, D. C., and B. Edgar. 1996. *From Lucy to Language*. New York: Simon and Schuster.
Jolly, A. 1980. *A World Like Our Own: Man and Nature in Madagascar*. New Haven: Yale University Press.
Jones, S., R. Martin, and D. Pilbeam. 1992. *The Cambridge Encyclopedia of Human Evolution*. Cambridge, U.K.: Cambridge University Press.
Kay, R. F., C. Ross, and B. A. Williams. 1997. Anthropoid origins. *Science* 275:797–804.
Kay, R. F., M. Cartmill, and M. Balow, 1998. The hypoglossal canal and the origin of human vocal behavior. *Proceedings of the National Academy of Sciences* 95:5417–5419.
Klein, R. G. 1989. *The Human Career*. Chicago: University of Chicago Press.
Köhler, M. and S. Moyà–Solà. 1997. Ape-like or hominid-like? The positional behavior of *Oreopithecus bambolii* reconsidered. *Proceedings of the National Academy of Sciences* 94:11747–11750.
Krings, M. et al. 1997. Neandertal DNA sequences and the origin of modern humans. *Cell* 90:19–30.
Lahr, M. M. 1994. The multiregional model of modern human origins: A reassessment of its morphological basis. *Journal of Human Evolution* 26:23–56.
Landau, M. 1984. Human evolution as a narrative. *American Scientist* 72:262–268.
Langdon, J. H. 1997. Umbrella hypotheses and parsimony in human evolution: A critique of the Aquatic Ape hypothesis. *Journal of Human Evolution* 33:479–494.
Larick, R. and R. L. Ciochon. 1996. The African emergence and early Asian dispersals of the genus *Homo*. *American Scientist* 84:538–551.
Leakey, M. G. et al. 1995. New four-million-year-old hominid species from Kanapoi and Allia Bay, Kenya. *Nature* 376:565–571.
Leakey, M. G. et al. 2001. New homonin genus from eastern Africa shows diverse middle Pliocene lineages. *Nature* 410:433–440.
Lemonick, M. D., and A. Dorfman. 1999. Up from the apes. *Time* 154(8):34–42.
Lewin, R. 1998. *Principles of Human Evolution: A Core Textbook*. Boston: Blackwell Scientific.
Lieberman, P. 1984. *The Biology and Evolution of Language*. Cambridge: Harvard University Press.
Maas, M. C., D. W. Krause, and S. G. Strait. 1988. The decline and extinction of plesiadapiforms (Mammalia: ?Primates) in North America: Displacement or replacement? *Paleobiology* 14:410–431.
McHenry, H. M. 1994. Behavioral ecology implications of early hominid body size. *Journal of Human Evolution* 27:77–87.
Morbeck, M. E., A. Galloway, and A. L. Zihlman (Eds.). 1997. *The Evolving Female*. Princeton, N.J.: Princeton University Press.
Morell, V. 1994. Will primate genetics split one gorilla into two? *Science* 265:1661.
Morgan, E. 1982. *The Aquatic Ape. A Theory of Human Evolution*. New York: Stein and Day.
Moyà-Solà, S., M. Köhler, and L. Rook. 1999. Evidence of hominid-like precision grip capability in the hand of the Miocene ape *Oreopithecus*. *Proceedings of the National Academy of Sciences* 96:313–317.
O'Connell, J. F., K. Hawkes, and N. G. Blurton Jones. 1999. Grandmothering and the evolution of *Homo erectus*. *Journal of Human Evolution* 36:461–485.
Richmond, B. G., and D. S. Strait. 2000. Evidence that humans evolved from a knuckle-walking ancestor. *Nature* 404:382–386.
Sponheimer, M., and J. Lee-Thorpe. 1999. Isotopic evidence for the diet of an early hominid, *Australopithecus africanus*. *Science* 283:368–370.
Spoor, F. et al. 1996. Evidence for a link between human semicircular canal size and bipedal behavior. *Evolutionary Anthropology* 30:183–187.
Stanley, S. M. 1996. *Children of the Ice Age*. New York: Harmony Books.
Steudal, K. 1996. Limb morphology, bipedal gait, and the energetics of hominid locomotion. *American Journal of Physical Anthropology* 99:345–355.
Stringer, C. B., and R. McKie. 1996. *African Exodus*. New York: Henry Holt.
Susman, R. L. 1994. Fossil evidence for early hominid tool use. *Science* 265:1570–1573.
Susman, R. L., and J. T. Stern. 1982. Functional morphology of *Homo habilis*. *Science* 217:931–934.
Swisher, C. C. III et al. 1996. Latest *Homo erectus* of Java: Potential contemporaneity with *Homo sapiens* in Southeast Asia. *Science* 274:1870–1874.
Szalay, F. S., and E. Delson. 1979. *Evolutionary History of the Primates*. New York: Academic.
Tattersall, I. 1993. *The Human Odyssey: Four Million Years of Human Evolution*. New Jersey: Prentice Hall.
Tattersall. I. 1995. *The Last Neanderthal*. New York: Macmillan.
Tattersall, I., and J. H. Schwartz. 1998. Morphology, paleoanthropology, and Neanderthals. *Anatomical Record (New Anat)* 259:113–117.
Thieme, H. 1997. Lower Palaeolithic hunting spears from Germany. *Nature* 385:807–810.
Vrba, E. S., E. H. Denton, and M. L. Prentice. 1989. Climatic influence on early hominid behavior. *Ossa* 14:127–156.
Walker, A., and P. Shipman. 1996. *The Wisdom of the Bones: In Search of Human Origins*. New York: Alfred A. Knopf.
White, T. D. 1980. Evolutionary implications of Pliocene hominid footprints. *Science* 208:175–176.
White, T. D., G. Suwa, and B. Asfaw. 1994. *Australopithecus ramidus*, a new species of early hominid from Aramis, Ethiopia. *Nature* 371:306–312.
Whiten, A. and C. Boesch. 2001. The cultures of chimpanzees. *Scientific American* 284(1):60–67.
Whiten, A. et al. 1999. Cultures in chimpanzees. *Nature* 399:682–685.
Wilson, A. C., and R. L. Cann. 1992. The recent African genesis of humans. *Scientific American* 266(4):68–73.
WoldeGabriel, G., T. D. White, and H. Buffetaut. 1994. Ecological and temporal placement of early Pliocene hominids at Aramis, Ethiopia. *Nature* 371:330–333.
Wood, B. 1987. Who is the "real" *Homo habilis*? *Nature* 327(6119):187–188.

Wood, B. 1992. Origin and evolution of the genus *Homo*. *Nature* 355:783–790.

Wood, B. 1994. The oldest hominid yet. *Nature* 371:280–281.

Zihlman, A. L. et al. 1978. Pygmy chimpanzee as possible prototype for the common ancestor of humans, chimpanzees and gorillas. *Nature* 275:744–746.

Web Explorations

On-line resources for this chapter are on the World Wide Web at: http://www.prenhall.com/pough (click on the Table of Contents link and then select Chapter 23).

CHAPTER

24

The Impact of Humans on Other Species of Vertebrates

A review of the biology of vertebrates must include consideration of the effect of the current dominance of one species, *Homo sapiens,* on other members of the vertebrate clade. Never before in the history of the Earth has a single species so profoundly affected the abundance, and even the prospects for survival, of other species.

Some of the influence of humans derives from the size of our population—more than 6 billion now and projected to reach 8 billion by 2028—and our worldwide geographic distribution. But other species of vertebrates match or surpass humans in numbers, and some at least come close to matching us in geographic distribution. Advances in technology that began with stone tools some 2.5 million years ago are what sets humans apart. Even primitive human societies use more resources than do other species of animals. In the aggregate, extraction and consumption of energy and other resources by human societies has worldwide impacts ranging from oil spills and contamination of soil and water with heavy metals to the loss of wild habitats to agriculture and urbanization. Per-capita energy use has nearly tripled since 1850, and most of the increase has occurred in the past few decades. The most highly industrialized societies have the greatest per-capita rates of consumption of energy and other resources. The United States, for example, has about 5 percent of the world's population and accounts for about 25 percent of the world's consumption of resources. The developed nations as a group have 20 percent of the world's population and are responsible for 60 percent of resource consumption. Societies with high rates of consumption have correspondingly high rates of production of greenhouse gases and waste products. Less technologically developed societies are not necessarily ecologically benign, however. They present special problems as growing human populations increasingly impinge on areas that have so far been relatively undisturbed. Wealthy nations at least have the resources to control pollution and protect habitats as national parks, monuments, and wildlife reserves. Poorer societies struggle daily to make ends meet and understandably regard conservation as a luxury beyond their reach. Perhaps the most insidious process affecting the relationship between humans and other vertebrates is the spread of Western cultural values that emphasize material possessions, thereby increasing demand for consumer products and expanding the geographic extent of high-impact societies.

This book has emphasized the fossil record of vertebrates and their characteristics as organisms, and both kinds of information are essential parts of efforts to conserve natural habitats and to protect endangered species. We tend to assume that the natural state of an environment is the way it was at the time of the first written record, but fossil evidence shows that humans had enormous impacts on vertebrate faunas long before writing was invented some 5000 years ago. Extinctions of vertebrates due to human activities began about 50,000 years ago in Australia, and they have accelerated ever since. The scientific study of vertebrates dates back only a few hundred years, but it, too, has accelerated rapidly. Information about the biology of vertebrates can be used to identify causes of population declines and extinctions and possibly to prevent some species that are currently endangered from becoming extinct.

24.1 Humans and the Pleistocene Extinctions

Starting from the appearance of the earliest vertebrates in the Late Cambrian or Ordovician, the diversity of vertebrates increased slowly through the Paleozoic and early Mesozoic, and then more rapidly during the past hundred million years. This overall increase has been interrupted by eight periods of extinction for aquatic vertebrates and six for terrestrial forms (Benton 1990). Extinction is as normal a part of evolution as species formation, and the duration of most species in the fossil record appears to be from 1 million to 10 million years. Periods of major extinction (a reduction in diversity of 10 percent or more) are associated with changes in climate and the consequent changes in vegetation. But that pattern of reasonably long-lived species and extinctions that were correlated with changes in climate and vegetation changed at about the time that humans became a dominant factor in many parts of the world (Wilson 1992, Steadman 1995).

For example, the number of genera of Cenozoic mammals reached a peak in the mid-Miocene and a second peak in the early Pleistocene. Only 60 percent of the known Pleistocene genera are living now, and the extinct forms include most species of large terrestrial mammals—those weighing more than 20 kg. These large mammals, plus some enormous species of birds and reptiles, are collectively called the Pleistocene megafauna. The term is most commonly applied to North American species such as ground sloths, mammoths, mastodons, and the giant Pleistocene beaver that was the size of a bear, but other continents also had megafaunas that became extinct in the Pleistocene.

The first humans to reach Australia were met by a megafauna that included representatives of four groups: marsupials, flightless birds, tortoises, and echidnas. The largest Australian land animals in the Pleistocene were several species of herbivorous mammals in the genus *Diprotodon,* the largest of which probably weighed about 2000 kg. The largest kangaroos weighed 200 kg, and one of them may have been carnivorous. The largest echidnas reached weights of 20 to 30 kg and were waist-high to an adult human. The flightless bird *Genyornis newtoni* was twice the height of a human, and the horned turtles in the family Meilanidae were nearly as large as Volkswagen Beetle cars. Perhaps the most dramatic species in the Australian megafauna was a monitor lizard at least 6 meters long (as large as a medium-size *Allosaurus*). The turtles, the monitor lizard, *Genyornis,* and all marsupial species weighing more than 100 kg became extinct in the late Pleistocene (Gillespie et al. 1978, Martin and Klein 1984, Flannery 1995, Miller et al. 1999).

What caused these extinctions of large animals in the late Pleistocene? Some researchers speculate that delayed effects of the Miocene climatic change and the Wisconsin glaciation, which reached its peak between 20,000 and 15,000 years ago, were the primary causes of the megafaunal extinctions. Others point out that the Wisconsin glaciation was preceded by 11 other glacial advances, some of which were more severe than the Wisconsin, and none of those glacial episodes was accompanied by widespread extinctions. The new element that entered the picture in the late Pleistocene was hunting humans, with their stone tools and social skills. The hypothesis that overhunting was the primary cause of the Pleistocene extinctions has been presented forcefully by Paul Martin at the University of Arizona (Martin and Klein 1984, Diamond 1989, Martin 1990, Stuart 1991). A third hypothesis also identifies humans as the cause of the extinctions, but proposes that the extinctions were caused by the introduction of new diseases and the use of fire to alter the habitat rather than by overhunting.

What types of information point to humans as direct or indirect causes of the Pleistocene extinctions? The most striking argument in favor of human causation is the apparent correspondence in the times of the arrival of humans on continents and islands and the extinction of the megafauna (Figure 24–1). The earliest extinctions appear to have occurred in Australia sometime between 50,000 and 40,000 years ago. In North and South America, at least eight species of large mammals survived until about 10,000 years ago. Humans colonized islands later than continents, and extinctions occurred between 10,000 and 4000 years ago on islands in the Mediterranean Sea, 4000 years ago on islands in the Arctic Ocean north of Russia, 2000 years ago on Madagascar, and only a few hundred years ago on islands in the Pacific Ocean.

These extinctions occurred either before or after the Wisconsin glaciation at times when climates were not changing rapidly. In addition, most of the species of birds and mammals that became extinct were large, whereas small species were far less af-

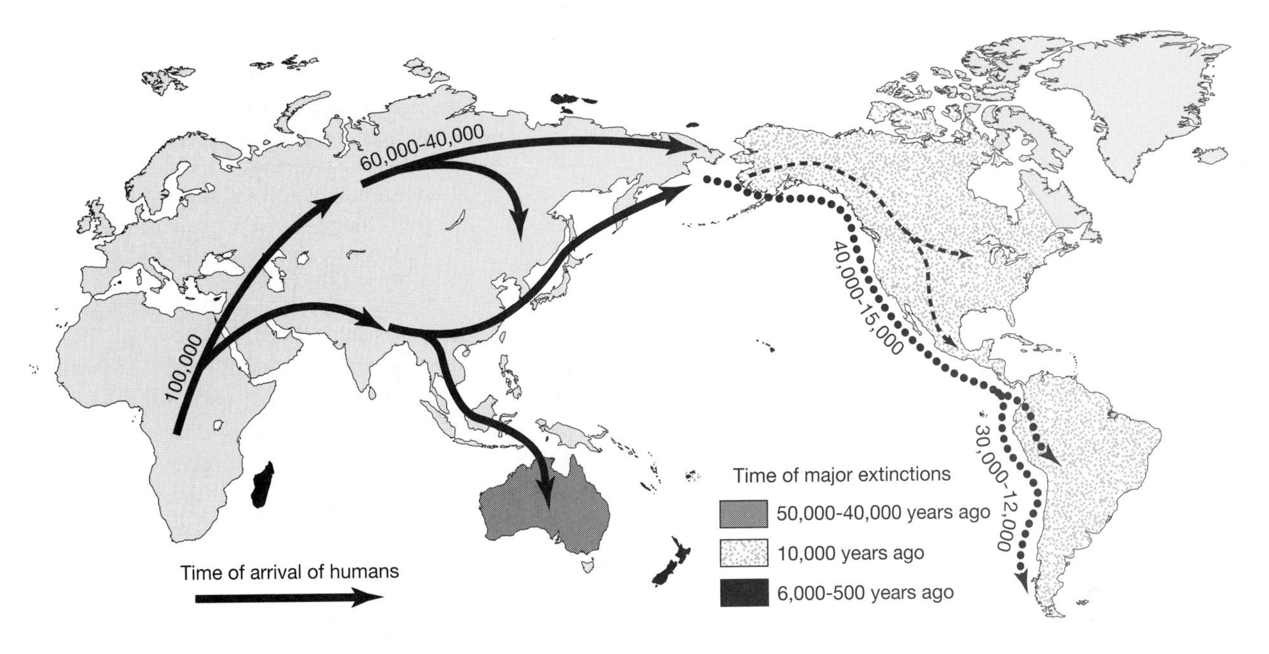

▲ Figure 24–1 Spread of human populations and extinction of native species of vertebrates.

fected. Taken together, these factors point to humans rather than to climate change as the primary reason for the extinctions. But how could a relatively small number of humans have continent-wide effects on faunas that had persisted through successive glacial advances and retreats?

Stone-Age Human Hunters

In North America high rates of mammalian extinction occurred primarily after the peaks of the Wisconsin glaciation (Figure 24–2). Mastodons, mammoths, and ground sloths all became extinct around 10,000 years ago. Martin contends that large mammals were hunted by invading nomadic humans, and some archaeological data support this view. Apparent kill sites have been found with the bones of many individuals of these prey species, sometimes with cut marks from stone tools on the bones or even with stone spear points embedded in the bones.

Some of the evidence for overhunting is dramatic. Eleven species of moa (giant flightless birds) existed on New Zealand when the Maori arrived in the late thirteenth century, and all appear to have been extinct within 100 years. The role of Maori hunters in the extinction of moa is amply documented. On the North Island of New Zealand a butchering site was discovered on the sand dunes at Koupokonui. The remains of hundreds of individuals of three species of moa were found in and around ovens. Uncooked moa heads and necks had been left in piles to rot, while the legs were roasted. At Wairau Bar on the South Island the ground is littered with the bones of moa, an estimated 9000 individuals plus 2400 eggs, and at Waitaki Mouth are the remains of an estimated 30,000 to 90,000 moa (Flannery 1995, Holdaway and Jacomb 2000).

Disease and Fire

Diseases transmitted to other species by humans and the animals associated with humans may also have played a role in megafaunal extinctions (MacPhee 1999, in press). Although there is no paleontological evidence of disease in the Pleistocene megafauna, emerging infectious diseases now threaten both biodiversity and human health (Daszak et al. 2000). Modern examples of transmission of disease from domestic to wild animals abound—within the past two decades, lions and wild dogs in Africa have been infected by canine distemper transmitted from

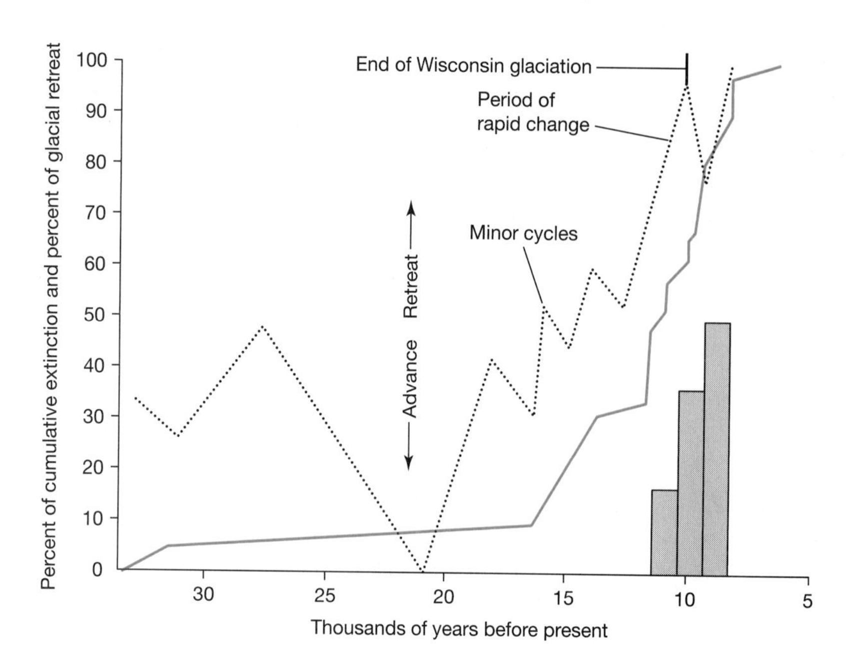

◀ Figure 24–2 Coincidence between the extinction of North American mammals, glacial retreat, and the population of humans. Solid line represents the cumulative percentage extinction of 40 mammalian species; dashed line is the percentage of withdrawal of the late Wisconsin glaciation; histograms are arbitrary units of relative abundance of evidence for the occurrence of humans in North America.

domestic dogs. Wild dogs (*Lycaon pictus*) are now nearly extinct in the Serengeti Plains. Less than a century ago, packs contained 100 or more animals, but today packs consist of only 10 or so adults. The population of wild dogs on the Serengeti is less than 60 animals, and the entire surviving population of the species is no more than 5000 individuals.

In general a newly introduced disease does not kill every individual of a previously healthy population. Some individuals survive, perhaps because they are resistant or maybe just because they are lucky enough to avoid infection. Nonetheless, a disease that drastically reduces the number of individuals of a species may start a process that leads to extinction, and this may be happening to some populations of African wild dogs (Gorman et al. 1998). The hunting method of wild dogs—prolonged pursuit of antelope until an individual is captured—is energetically expensive. A pack of wild dogs hunts cooperatively, with one individual taking up the chase as another tires. The drastic reduction that has occurred in pack size means that each dog must work harder. To make things worse, spotted hyenas (*Crocuta crocuta*) steal the kills made by wild dogs, and a small pack of dogs probably has more difficulty defending its kills from hyenas than a larger pack. Wild dogs normally hunt for about 3.5 hours per day, and calculations of the energy cost of hunting and the energy gained from prey show that they just meet their daily energy needs on this schedule. If hyenas steal some of the kills, the wild dogs must increase the time they spend hunting. A 10 percent loss of prey would force the wild dogs to double their hunting time, and a 25 percent loss would force them to spend 12 hours a day hunting. Wild dogs are already working at nearly their physiological limits when they hunt for 3.5 hours per day, and they probably cannot survive if they lose much food to hyenas. Thus, a drastic reduction in pack size sets the stage for a competitive interaction with hyenas, and this interaction could be the factor that drives a pack of wild dogs to extinction.

With modern humans came the use of fire to manipulate the habitat. *Genyornis newtoni* was a flightless bird that inhabited inland plains and some coastal areas of Australia when the first humans arrived about 50,000 years ago. Although *Genyornis* was a ponderous bird and was probably less fleet-footed than emus (*Dromaius novaehollandiae*), there is only one site known with evidence that humans hunted *Genyornis*. The reason that *Genyornis* became extinct and emus survived may lie in their feeding habits. Emus eat a wide variety of items including grasses, whereas the chemical composition of *Genyornis* eggshells suggests that they were browsers, eating leaves from shrubs. Wildfires were a part of the

Australian landscape long before humans arrived. Natural fires occurred during the dry season and did not recur until the vegetation in a burned area had regenerated to the point of creating enough fuel to sustain another fire. The early human inhabitants of Australia may have set fires at other times of the year and at shorter intervals than the natural fire cycle. A regime of more frequent burning would have converted the shrub lands that *Genyornis* depended on to the grasslands and spinifex that characterize the inland Australian plains today. Thus, habitat change produced by the new fire regime created by humans may have been responsible for the extinction of *Genyornis*. The large herbivorous mammals that became extinct in Australia were also browsers, and the same habitat changes may have been responsible for their disappearance (Miller et al. 1999, Flannery 1999).

The Role of Climate

Many researchers suggest that changing climates and their influence on habitat availability were as important as humans in producing extinctions. In this view, large mammals are more susceptible to environmental stress than small species. Severe restrictions of forage and available water and elimination of migratory corridors to more suitable habitats would affect large mammals, with their correspondingly large requirements, more than smaller mammals. Thus, large North American mammals may have been on the decline when intensive human hunting began, or changes in climate and habitat may have killed the last remaining individuals of a species.

Dale Guthrie (in Martin and Klein 1984) has pointed out that most climatic theories of Pleistocene extinction are too ambiguous and diffuse to account for the final extinction event, however effective they are at depicting environmental stress. In the general trend of replacement of closed-canopy forests by more open, grass-dominated steppes and savannas that began in the mid Cenozoic, Guthrie sees increasing seasonality as being of primary significance. The climatic trend throughout the Cenozoic was toward a shorter growing season at higher latitudes. At first this trend increased the mosaic patchwork of forest, scrub, and grassland over much of the world. This increasingly complex pattern of vegetation supported a large, diverse fauna.

Indeed, the spread of grasses promoted faunal diversification. Responding to the spread of grasslands, the generally large-bodied ungulates reached maximum diversity in the mid-Miocene, contributing significantly to the total number of mammalian genera known from that epoch. But as the trend continued toward seasonal extremes of temperature and aridity in the Pliocene, the diversity of vegetation declined. The flora over broad expanses of continents changed from a mosaic of plant associations to a latitudinal and altitudinal zonation of broad bands of plant communities with low diversity. The lower overall diversity of floras and greater regional homogeneity reached its maximum after the Wisconsin glacial advance. The reduction of plant diversity may have eliminated some herbivores, especially large browsers, and restricted all but a few species to limited ranges.

The postglacial period beginning 15,000 to 10,000 years ago appears to have been more stressful to large mammal faunas than the previous 2.5 million years (Stuart 1991). Shorter growing seasons mean less plant productivity to support herbivores, even in the restricted ranges where they found suitable floras. Kiltie (in Martin and Klein 1984) argues that the rapid advent of short growing seasons and heightened seasonality critically decreased the resources available to many species of large herbivorous mammals with long periods of growth to maturity and long gestation periods. These changes constituted environmental challenges to which the megafauna could not rapidly adapt physiologically or genetically.

These vegetation changes occurred in North America after the Wisconsin glaciation, but not necessarily on other continents, and large mammals persisted in some regions. In the New World, horses and (in North America) camels became extinct, but they survived in Eurasia and Africa. Others, like bison and deer, survived in the New World in spite of humans. The Egyptian civilization knew of mammoths, and other evidence indicates that the pharaohs and mammoths were contemporaries, although geographically widely separated (Rosen 1994). The Australian megafauna disappeared between 50,000 and 40,000 years ago, but the megafauna on New Zealand survived until a few hundred years ago. There is currently no final solution to the puzzle of Pleistocene extinctions, and there probably is no single cause. In some areas human predation appears to be a sufficient explanation, whereas in others the indirect effects of humans (disease and fire) may have been more important.

24.2 Humans and Recent Extinctions

As we move closer to the present, the role of humans as the major cause of extinction becomes unambiguous. The acres of moa bones found at butchering sites in New Zealand clearly demonstrate enormous predation by Maori hunters about 800 years ago. Excavation of fossils preserved in tube-like lava caves formed by volcanoes show that the Hawaiian Islands probably had more than 100 species of native birds when Polynesian colonists arrived about 300 A.D. (Olson and James 1991a,b). By the time European colonists reached Hawaii in the late eighteenth century, that number had been reduced by half.

The Age of Exploration, which began in the fifteenth century, brought sophisticated weapons and commercial trade to areas that had known only stone tools and hunter-gatherer economies. Ships sailed from Europe to all corners of the world, stopping en route to renew their supplies of food and water from oceanic islands. Not surprisingly, extinctions on islands began about two centuries before extinctions on continents (Figure 24–3). Notable examples include the dodo (*Didus ineptus*), a flightless bird related to pigeons that lived on the island of Mauritius in the Indian Ocean. The dodo was last seen alive in 1681. In the Hawaiian Islands about one-third of the native species of birds that were still surviving when Captain Cook arrived in the late eighteenth century are now extinct. Animal species continue to become extinct today (Figure 24–3), and a worldwide survey of extinctions since the start of European colonization reveals two trends: Island extinctions began almost two centuries earlier than continental extinctions, and both island and continental extinctions have increased rapidly from the early or mid-nineteenth century through the twentieth century (World Conservation Monitoring Centre 1992). More than 800 species have become extinct in the past 500 years, a rate of extinction that is 1000 to 10,000 times higher than it would be without the effect of humans. The year 2000 edition of the Red Book published by the World Conservation Union lists a total of 11,046 species of plants and animals currently at risk of extinction.

The World Conservation Monitoring Centre has summarized the best information available about the conservation status of animals. They place species in categories of risk by using criteria that focus on the absolute size of wild populations and changes in the populations in the past 10 years.

- **Extinct**—A species is Extinct when there is no reasonable doubt that the last individual has died. The ivory-billed woodpecker (*Campehphilus principalis*) was the largest species of woodpecker with a head-plus-body length up to 56 cm. (See the color insert.) It became extinct in North America in the 1970s. The last confirmed sighting of the species was in Ojito de Agua, Cuba, in 1987. The golden toad (*Bufo periglenes*) from the montane cloud forest of Costa Rica was first described in 1967 and had vanished by 1987. Climate change and infection by chytrid fungi have been proposed as the causes of its extinction.
- **Extinct in the wild**—A species is Extinct in the Wild when it is known to survive only in cultivation, in captivity, or as a naturalized population (or populations) well outside the past range. Przewalski's horse (*Equus przewalskii*) is the only true wild horse. The species formerly ranged through Mongolia, China, and Turkestan. About 200 horses remain, all in zoos. Captive-bred animals that were released in Mongolia have successfully reared foals, but conservation biologists fear that the unusually severe winter of 2000–2001 may have killed some or all of this group. The black-footed ferret (*Mustela nigripes*) was a resident of prairie dog towns in North America. The last wild population became extinct in 1981, and more than 200 captive-bred ferrets have been released since 1991.
- **Critically endangered**—A species is Critically Endangered when it is facing an extremely high risk of extinction in the wild in the immediate future. The northern hairy-nosed wombat (*Lasiorhinus krefftii*) is a burrowing marsupial weighing as much as 32 kg that lives in colonies, spending the day underground and emerging at night to feed. It once ranged over a large part of eastern Australia from New South Wales to Queensland. Its populations have been decimated by competition with rabbits, sheep, and cattle that were introduced by Europeans, and the remaining population of the species consists

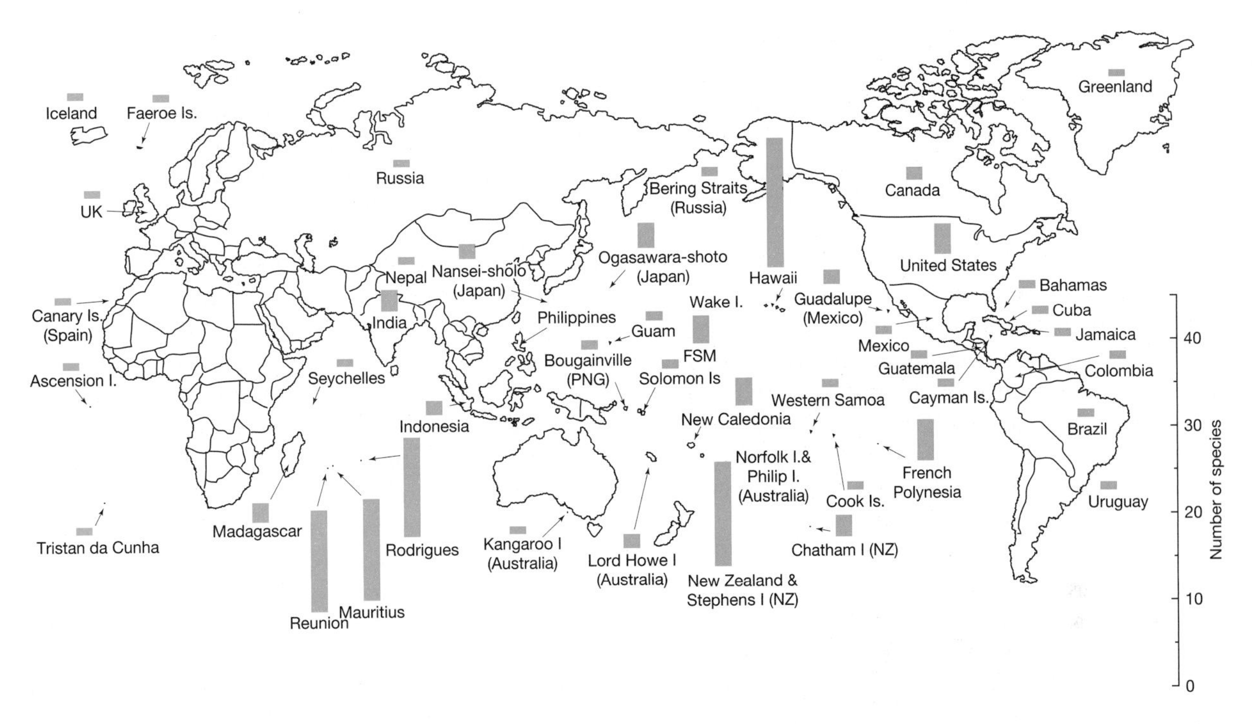

▲ Figure 24–3 The numbers of confirmed extinctions of species of birds since 1600. Islands have suffered more extinctions than continental areas.

of 65 animals in Epping Forest in Queensland. The common sturgeon (*Acipenser sturio*) is a large fish (up to 3m and 300 kg). Its historic range included the entire coastline of Europe from the North Cape to the Baltic, Mediterranean, and Black seas. The species is now extinct in some of its former spawning rivers, including the Elbe, Rhine, and Vistula. Breeding populations are restricted to a few European rivers—the Gironde in France, the Guadalquivir in Spain, and the lower Danube. Overharvesting is the primary cause of its decline. The flesh is prized, and the roe is used to make caviar—so gravid females are especially sought and are killed before they can reproduce.

- **Endangered**—A species is Endangered when it is facing a very high risk of extinction in the wild in the near future. Examples of endangered species include the giant panda (*Ailuropoda melanoleuca*), which was once distributed through Myanmar, northern Vietnam, and a large part of eastern and southern China. Wild pandas now occur only in fragmented populations in mountain ranges in western China; the total population is thought to be about 1200 animals. Attempts to breed giant pandas in captivity outside of China have been generally unsuccessful. Both the chimpanzee (*Pan troglodytes*) and the bonobo or pygmy chimpanzee (*Pan paniscus*) are endangered. Chimpanzees once occupied a broad area of equatorial Africa, whereas bonobos have always been limited to the Zaire Basin.
- **Vulnerable**—A species is Vulnerable when it is facing a high risk of extinction in the wild in the medium-term future. The great white shark (*Carcharodon carcharias*) has a worldwide distribution in warm and temperate seas. As a top predator, it has always had a low population density, and overfishing is considered a potential threat. Habitat destruction and poaching are major concerns in the case of the orangutan (*Pongo pgymaeus*), which used to be widespread in tropical and subtropical Asia and now occurs only on the islands of Borneo and Sumatra.

A summary of the status of vertebrates in different classes shows that a total of 1096 species of mammals (about 25 percent of all extant species) and 1107 species of birds (about 12 percent of extant species) are extinct, endangered, threatened, or vulnerable (Table 24.1). The percentages of species at risk in other groups appears to be lower, but that impression is misleading. For these groups of vertebrates there are simply no data available about the status of most species, so they could not be included in the analysis.

It is not easy to calculate the rate at which extinctions are occurring, and different assumptions can produce different values. Estimates of the number of species of birds that will be committed to extinction by the year 2015 range from 450 to 1350 (Heywood et al. 1994). (Committed to extinction means that a species' populations in the wild are no longer viable, and the species will inevitably become extinct unless major conservation actions reverse the current trend.) What is clear is that destruction of habitat is the major threat, affecting 60 percent of threatened species of birds and nearly 80 percent of threatened species of mammals (Figure 24–4).

24.3 Organismal Biology and Conservation

Much of the scientific information employed in conservation programs is organized by the Species Survival Commission (SSC) of IUCN—the World Conservation Union. The SSC is a worldwide network of 7000 wildlife biologists, wildlife veterinarians, zoo biologists, marine biologists, and academic scientists with specialized knowledge of groups of animals and plants. The commission publishes SSC Action Plans that describe the current situation for species or groups of species, identifying the major threats and proposing mechanisms to address those issues. More than 60 Action Plans have been published, and many are available online at the IUCN website.

Developing management plans for endangered species requires enormous amounts of information about the basic biology of the species concerned. Without a thorough understanding of how a species works, well-intentioned management efforts can be ineffective or even have negative effects on the species. In the following sections we describe several examples of situations in which an understanding of ecology, behavior, or physiology of a species is central to effective management.

What Is Critical in a Critical Habitat?

Federal law and some state laws require assessment of the habitat requirements of species that are considered at risk, and wildlife biologists may be charged with the responsibility of determining the critical habitat for a species. In biological terms the elements of its habitat that are critical for the success

Table 24.1 Numbers of species of vertebrates that are extinct, endangered, or vulnerable. World Conservation Monitoring Center 2001 <http://www.wcmc.uk.org/>.

Class	*Recently Extinct*	*Extinct in the Wild*	*Critically Endangered*	*Endangered*	*Vulnerable*	*Total*
Mammals	86	3	169	315	612	1096
Birds	104	4	168	235	704	1107
Reptiles	20	1	41	59	153	253
Amphibians	5	0	18	31	75	124
Sarcopterygians	0	0	0	1	0	1
Acrinopterygians	80	11	156	125	434	715
Elasmobranchs	0	0	1	7	7	15

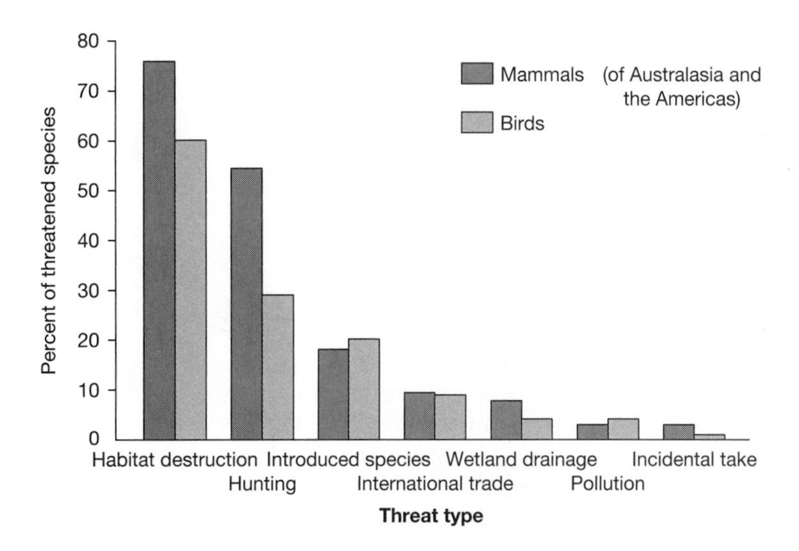

▶ Figure 24–4 The major threats affecting birds (on a worldwide basis) and mammals (Australasia and the Americas). Habitat destruction is the single most important threat for both kinds of animals, affecting 60 percent of birds and 76 percent of mammals. Hunting (for food and sport) is a greater threat to mammals than to birds. Introduced species may be predators or competitors, and *international trade* refers to commercial exploitation for fur, feathers, and the pet trade. *Incidental take* is the term used to designate accidental mortality, such as dolphins that are drowned by boats fishing for tuna.

of a species are likely to be subtle and complex, and the differences between legislative and biological perspectives can create tension and conflict. The problems of identifying and then protecting critical habitat are particularly severe for migratory species, for which even the most superficial definition of critical habitat must include both ends of their migratory paths and the areas in which they stop as they move back and forth. To complicate conservation efforts even more, migratory animals usually move through several national jurisdictions and may pass over or through international waters where no national jurisdiction exists. The legislative and diplomatic effort required to protect these species is enormous, and must be based on sound biological information.

Critical Habitat for Migratory Animals The migration of the gray whale (*Eschrichtius robustus*) from its summer feeding grounds north of the Arctic Circle to breeding sites in Ojo de Liebre, Bahia Magdalena, and San Ignacio Lagoons on the Pacific coast of Baja California was described in Chapter 21. The adult whales apparently make this 9000-km migration because newborn calves need the warm waters of the lagoons to maintain stable body temperatures (Figure 24–5). Thus, the lagoons are an essential part of the gray whale's habitat. More than 300 other species of animals also depend on the lagoons, and a thriving shellfishing industry provides livelihood for local communities. The United Nations has declared the lagoons a World Heritage Site, and both Ojo de Liebre and San Ignacio have been protected by Mexican legislation since 1972 because of the role they play in the life cycle of whales. In 1988 Mexico included the lagoons in the Vizcaino Biospheric Preserve.

Consequently, in 1994 when Salt Exporters, Inc.—a company jointly owned by Mitsubishi Corporation and the Mexican government—announced plans to build a plant to extract salt from seawater at San Ignacio Lagoon, the news was greeted with protests and demonstrations locally and by environmentalists around the world. The proposed plant, which would cover 300 square kilometers (116 square miles) in the heart of the lagoon system, would supply all of Japan's industrial salt needs for the manufacture of plastic and chlorine. The proposed plant would draw nearly 23,000 liters (6000 gallons) of water per minute from the lagoon and would have an enormous environmental impact. Salt Exporters already operates a salt plant at nearby Guerro Negro, and environmentalists blame that plant for the deaths of nearly 100 sea turtles by salt poisoning.

Environmental organizations around the world cooperated in opposing plans for the San Ignacio plant, sponsoring a letter-writing campaign that generated nearly a million letters of protest, promoting boycotts of Mitsubishi products by consumers and of Mitsubishi stock by mutual funds, and enlisting the help of governments to apply political pressure. In

Figure 24–5 Mother gray whale and calf in San Ignacio Lagoon.

March 2000, after five years of pressure, Mitsubishi and the Mexican government announced that they were abandoning the proposal to build the plant.

Birds are the best-known migratory animals, and migration is a central feature of the biology of many species. Loss of habitat at summer and winter ranges, and at stopping places along the migratory routes, appears to be contributing to declining populations of shore birds and songbirds that migrate between summer ranges in North America and winter ranges in the tropics.

About 350 species of songbirds occur in North America. About 250 of them spend their winters in the New World tropics, which extend from southern Mexico through Central and northern South America and into the West Indies. The remaining 100 species of North American songbirds are either year-round residents in northern habitats or migrate only short distances south of their summer ranges. Neotropical migrants compose the majority of species of songbirds in most habitats in North America. In most wooded areas, half of the breeding species are Neotropical migrants; in some northern regions of North America, more than 90 percent of the songbirds are migrants.

Bird watchers have a nationwide system of local bird censuses, and starting in the 1970s and 1980s these counts revealed dramatic decreases in the numbers of some Neotropical migrants such as wood thrushes (*Hylocichla mustelina*) and cerulean warblers (*Dendroica cerulea*) in certain areas—as much as 1 percent annually for the past 30 years. In contrast, populations of year-round residents and short-distance migrants, such as chickadees (*Parus atricapillus*) and northern cardinals (*Cardinalis cardinalis*), were stable or even increasing. The pattern of decline is not consistent across species or regions of North America, however. Populations of some species of Neotropical songbirds are declining, but others are increasing. Populations of songbirds in the forested areas of the eastern United States are declining, but those in the West are not.

This complex picture results from the variety of factors that affect migratory birds in their summer and winter ranges and in stopover points on their routes of migration. Changes in land use are probably the basis for declining bird populations, both directly and indirectly. Neotropical migrants breed in their summer ranges, and fragmentation of forests into smaller and smaller patches has reduced the total amount of breeding habitat available and changed the nature of the habitat that remains. Nests near the edges of woodlands generally suffer higher rates of failure than those nearer the center, and as forest patches grow smaller a greater proportion of the habitat is near an edge. Part of the increased rate of nest failure near the edges of woodlands may result from nest parasitism by brown-headed cowbirds (*Molothrus ater*), which are obligate nest parasites (Figure 24–6). Cowbirds do not build their own nests, but lay their eggs in the nests of other species; some 200 species of birds have been reported to be parasitized by cowbirds. A female cowbird lays 20 to 40 eggs, one or two in each nest, and a few pairs of cowbirds can parasitize all the nests in a small woodland. A female cowbird often removes an egg of the host species from a nest when she lays her own, and cowbird eggs develop and hatch more rapidly than the eggs of the host species, giving the cowbird nestling an advantage. Larger and pushier than the nestlings of the host species, cowbird nestlings take so much of the food their unwitting foster parents bring to the nest that the host bird's nestlings may starve.

Cowbird parasitism is insidious. When a nest with eggs or fledglings is lost to a predator, the parent birds usually build another nest and start a second clutch. They may succeed in reproducing that season despite the loss of their first clutch. In contrast, the parent birds that serve as hosts for the cowbirds do not distinguish the cowbird eggs or nestlings from their own. Thus, when a pair of birds has raised a cowbird to fledging, they behave as if they had successfully fledged their own young and do not nest again that year.

▲ Figure 24–6 Nest parasitism by brown-headed cowbirds. (a) A female cowbird about to deposit an egg in the nest of a blue grossbeak. (b) Two cowbird eggs in a robin's nest. The parent birds do not remove the cowbird eggs, even though they are speckled brown and white and look nothing like the plain blue robin egg. (c) An adult yellow warbler feeding a cowbird chick that is nearly as large as the foster parent.

Cowbirds are insectivorous and feed primarily in open fields. Because their movements are not limited by the need to care for their young, they can move long distances between the fields in which they feed and the woodlands in which they find the nests of other birds. Populations of cowbirds have increased dramatically as woodland habitats in North America have been replaced by open fields, and the small remaining woodlands offer songbirds few options for concealing their nests from cowbirds. Thus habitat fragmentation in the summer breeding range of Neotropical migrants appears to be responsible directly and indirectly for some of the decrease in their populations.

Habitat change in the winter range is probably also responsible for some population declines. The huge landmass of North America funnels down to a much smaller area in Central America, where many migrant birds overwinter. Competition for food and space in the small landmass of Central America may be one of the reasons that migrants do not breed in their winter ranges. Because the land area in the winter range is small, habitat changes caused by agricultural practices even on a relatively small scale could affect large numbers of birds. Coffee originated in Ethiopia and was brought to Latin America by Spanish colonists. Traditionally coffee (and also cacao, the source of chocolate) have been grown as understory trees, beneath a canopy of taller trees. This traditional method produces shade-grown coffee. Coffee and cacao plantations of this sort are similar in structure to natural forests, although they are less complex and have fewer species of trees (Figure 24–7). Traditional coffee and cacao plantations provide important habitats for birds. A survey of traditional coffee and cacao plantations in Chiapas, Mexico, by the Smithsonian Migratory Bird Center found that more than 150 species of birds live in them, including Neotropical migrants.

In the past 20 years, hybrid coffee trees that grow in full sun have replaced traditional plantations. Sun-grown coffee produces substantially greater yields than shade-grown, and its use has spread rapidly. Currently 17 percent of cropland planted in coffee in Mexico, 40 percent in Costa Rica, and 69 percent in Colombia has no shade canopy. Sun-grown coffee requires intensive cultivation, with heavy use of chemical fertilizers, insecticides, herbicides, and fungicides. This approach appears to lead to increased soil erosion, acidification, and higher amounts of toxic runoff. Furthermore, because sun-grown coffee plantations lack the complex structure of shade-grown coffee plantations, they do not provide habitats for forest birds. The diversity of birds plummets when a coffee plantation is converted from traditional shade-grown coffee to sun-grown. Studies in Colombia and Mexico found 94 to 97 percent fewer bird species in sun-grown coffee plantations than in shade-grown plantations.

Migratory birds depend on forested habitats in their winter ranges, but the land that can be set aside in parks and reserves in Central and northern South America is insufficient to maintain healthy bird populations. If migratory birds are to survive, human-dominated landscapes in the Neotropics must provide suitable habitats for them, such as traditional shade-grown coffee plantations. The problem is that sun-grown coffee produces a faster and larger return on a financial investment. Shade-grown coffee is environmentally friendly, but not economically attractive. To promote shade-grown coffee, the Smithsonian Migratory Bird Center has developed the Bird-Friendly™ label, which can be used by coffee plantations that meet specific criteria (Figure 24–8).

Subtle Elements of Critical Habitats The presence or absence of birds in a coffee plantation and dead sea turtles floating in the water near a salt plant are conspicuous indicators of habitat quality. They are easy to demonstrate and are effective rallying points for public opinion and political pressure. More complicated, and probably more widespread, are subtle interactions between elements of a habitat and the needs of a species. These interactions have the potential to determine the survival or extinction of populations of threatened species, but they can be detected only by careful study and insightful analysis of basic biological information. The desert tortoise *(Gopherus agassizii)* of North America provides an example of this sort of interaction (Oftedal and Allen 1996).

Desert tortoises are herbivores and, as we noted in Chapter 12, they lack salt-excreting glands. Plants contain higher concentrations of potassium than do animals, so herbivorous animals must excrete some of the potassium they get in their food. Desert tortoises excrete excess potassium along with nitrogenous wastes in the form of salts of uric acid. Because potassium excretion is chemically linked to nitrogen excretion, an excess of potassium in a tortoise's diet robs the tortoise of nitrogen. That is, a tortoise on a

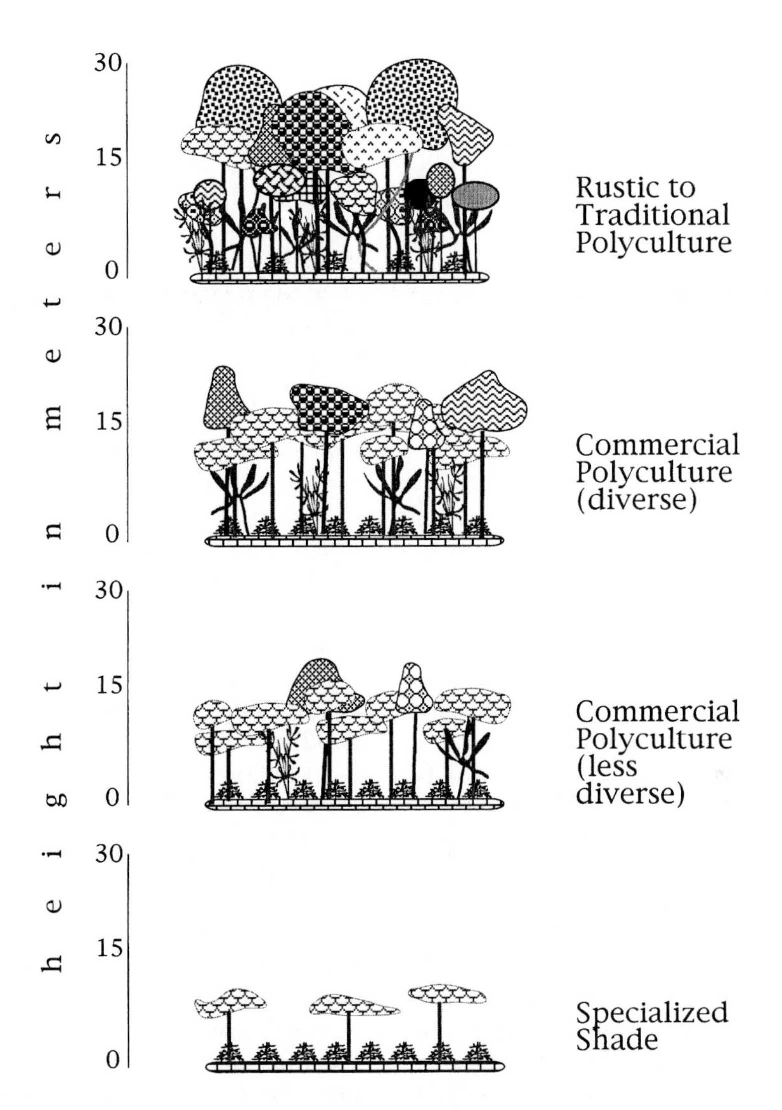

▲ Figure 24–7 Structural habitats in different kinds of coffee plantations.

high-potassium diet can use so much nitrogen getting rid of the excess potassium that it does not have enough nitrogen for protein synthesis. Even high-nitrogen diets (20 percent crude protein) do not allow young desert tortoises to grow when the diet also contains 3.8 percent potassium.

Plant species vary in potassium content. Feeding trials with tortoises have shown that when tortoises are given a choice of diets with a range of potassium contents, they select the one with the least potassium. When they are offered only high-potassium diets, they reduce the amount of food they eat. Wild tortoises may be able to survive only if plants with favorable ratios of nitrogen to potassium are available. Two factors are thought to have decreased the availability of these plants and jeopardized the

▲ Figure 24–8 The Bird-Friendly label for shade-grown coffee.

nutritional status of tortoises—the spread of nonnative plants in North American deserts and the introduction of livestock grazing in tortoise habitats.

In many parts of North America, including the southwestern deserts where desert tortoises occur, some native species of plants have been replaced by Eurasian plants that accompanied the spread of European settlers. There is concern—although no definitive data as yet—that the Eurasian plants may be less suitable for tortoises than the native plants they have replaced. In particular, if the Eurasian plants are higher in potassium or lower in nitrogen than native plants, tortoises could have difficulty obtaining and retaining the nitrogen they need for normal growth.

Desert tortoises evolved in habitats that contained rodents, rabbits, desert bighorn sheep, and pronghorn antelope—herbivores that may have competed with the tortoises for preferred food plants. The major new element in the landscape since the arrival of Europeans is ranchers who stock sheep and cattle at densities far above the capacity of the range to support them. The damage that livestock grazing causes on a landscape scale is obvious to any ecologist, although it is vehemently denied by lobbyists for the ranching industry at legislative and regulatory hearings. What is not clear is whether cattle or sheep actually compete with tortoises for preferred plant species, and whether the presence of these domestic animals affects the growth and survival of tortoises. Determining the impact of commercial grazing on tortoises and other native species will require the combined efforts of ecologists and botanists working in the field as well as nutritionists and physiologists in the laboratory.

Animal Behavior and Management Strategies

Animals have complex suites of behaviors that allow them to find appropriate food, evade predators, function in a group, and identify suitable mates. Some of these behaviors are innate; that is, an animal is born with the ability to produce the correct response to certain stimuli, but many behaviors are learned or honed by trial and error. Animals raised in captivity do not experience the same stimuli they would in their natural environment and may not be able to function effectively if they are released. As captive breeding programs are developed to restore species to the wild, increasing effort is being devoted to ensuring that the animals to be released are competent to survive in the wild (Shepherdson et al. 1998). In other cases, basic information about social behavior of a species has been used in management of wild populations (Clemmons and Buchholz 1997).

Imprinting, Learning, and the Captive Husbandry of Birds The process known as imprinting is an important feature of the behavior of birds and mammals. Imprinting is a special kind of learning that occurs only during a restricted time in ontogeny called the critical period. Once imprinting is established, it is permanent and cannot be reversed.

Colors, patterns, sound, and movement are the major stimuli for imprinting among birds, whereas scent is the most important stimulus for mammals. Two types of social imprinting can be distinguished: Filial imprinting is the process of learning to recognize the individual characteristics of the parents, whereas sexual imprinting refers to learning the characteristics of other members of the species. Filial imprinting is responsible for keeping the young with the mother after they move away from the nest or den.

Sexual imprinting during infancy is probably one of the mechanisms that allow an individual to seek a mate of its own species when it matures. In addition, sexual imprinting may also prevent an individual from mating with close relatives, thereby preventing inbreeding.

Birds and mammals normally imprint on their parents and siblings because those are the only objects in the den or nest that are emitting visual, auditory, and olfactory stimuli. In the absence of their parent, however, infants may imprint on inanimate objects or on members of another species, including humans.

The confusion of species-identification by birds that have imprinted on a foster parent or a keeper can be disastrous for programs in which endangered species are reared in captivity and then released. Young birds must recognize appropriate mates if they are to establish a breeding population, and captive-rearing programs go to great lengths to ensure that the young birds are properly imprinted.

For example, the entire population of California condors (*Gymnogyps californicus*) had dwindled to 27 birds in the mid-1980s. At that time all of the condors were captured and placed initially in two captive colonies, one at the Los Angeles Zoo and the other at the San Diego Wild Animal Park. Subsequently a third colony was established at the World Center for Birds of Prey in Boise, Idaho. The condors have reproduced successfully in captivity, and the total captive population now consists of 102 condors. As of January 2000 a total of 49 young condors had been released in California and Arizona, and the first eggs laid by released birds were found in March and April 2001. To ensure that hatchling condors do not imprint on humans during their critical period, they are reared in enclosed incubators and fed by a technician who inserts her hand into a rubber glove modeled to look like the head of an adult condor (Figure 24–9a).

Still more training may be necessary to produce captive-reared young that can survive after they have been released. A husbandry program for the northern bald ibis (*Geronticus eremita*) is an example of how complicated this process can be (Figure 24–9b). Within historic times the geographic range of the bald ibis extended from the Middle East and North Africa north to Switzerland and Germany, but the wild population has dwindled to fewer than 220 birds in a reserve in Morocco. Bald ibises flourish in captivity, and there are more than 700 captive individuals. This species seems ideal for reintroduction—it is prolific, and its disappearance from the wild seems to have been caused by human predation rather than by pollution or loss of habitat. Yet two attempts to establish populations by releasing captive-reared birds have failed.

The reason for the failures seems to have been the absence of normal social behaviors in the captive-reared birds. Bald ibises are social birds with extended parental care, and it seems that juveniles learn appropriate behaviors from adults. For some reason, this learning did not occur in captivity. An attempt is now under way to instruct young bald ibises in these social skills—human foster parents are hand-rearing the birds, teaching them to find their way to fields where they can forage, to recognize predators and other dangers such as automobiles, and to engage in mutual preening, which is an important social behavior.

Adolescent Elephants The populations of elephants in African game parks have increased substantially since the parks were established. This population growth testifies to the effectiveness of the parks in protecting the elephants, but it creates serious problems with overcrowding. As their populations increase, foraging elephants tear down mature trees to eat the upper branches and often move outside of the parks, where they raid gardens and orchards. To reduce these problems South Africa has developed a process of selective removal of elephants. Some individuals are trapped and moved to other parks, and others are killed.

Between 1981 and 1993 young male and female African elephants (*Loxodonta africana*) that had been orphaned when their mothers were killed were relocated from Kruger Park to Pilanesburg. There were no elephants already at Pilanesburg, and the orphans matured in the absence of older elephants. When the first elephant calf was born in 1989, the relocation program looked like a success story, except for one problem—the young males were attacking and killing white rhinoceros (*Ceratotherium simum*). By 1997 the elephants had killed more than 40 rhinoceros (Slotow et al. 2000).

The attacks on rhinoceros were made by male elephants when they were in musth. Musth is an annual period during which the level of testosterone circulating in the blood of male elephants rises dramatically. Elephants in musth adopt a distinctive posture that allows them to be recognized from a distance (Figure 24–10). Their temporal glands swell and secrete an oily material, and sexual and aggressive behavior increases. In natural populations, males first enter musth when they are 25 to 30 years old. The duration of musth increases as they get older—from few days to a few weeks for animals between 25 and 30 to 2 to 4 months for animals older than 40. Thus, the young males at Pilanesberg had developed musth earlier than would be expected, and their periods of musth

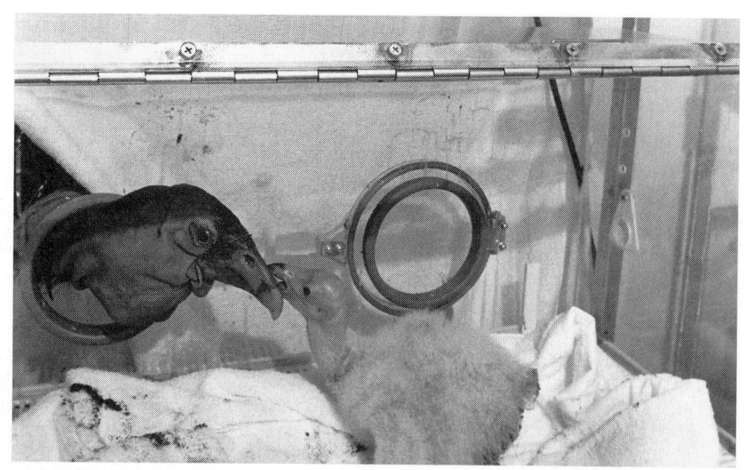

(a)

(b)

◄ Figure 24–9 Captive husbandry of endangered species of birds. (a) A hatchling California condor in its incubator with the model condor head used to feed it. (b) A bald ibis.

lasted from weeks to months, which was far longer than normal for such young males.

The African elephant is an endangered species, and the white rhinoceros is listed as vulnerable. Both species face sufficient risk without the added complication of having one species attacking and killing the other. The aggressive behavior of the young male elephants at Pilanesburg is not typical of African elephants, but it was not unique to this population. It occurred in some other populations, especially in Hluhluwe-Umfolozi Park, which is in northern KwaZulu Natal province. This was the second population to be established with orphans from Kruger Park, and the young male elephants at Hluhluwe-Umfolozi began to kill rhinoceros about two years after the behavior appeared in the Pilanesberg population. About 45 rhinos have been killed at Hluhluwe-Umfolozi, and smaller numbers have been killed in other areas.

The populations of elephants that are killing rhinoceros have one feature in common—all were established by moving only young animals to locations where they matured in the absence of adults. That

(a)
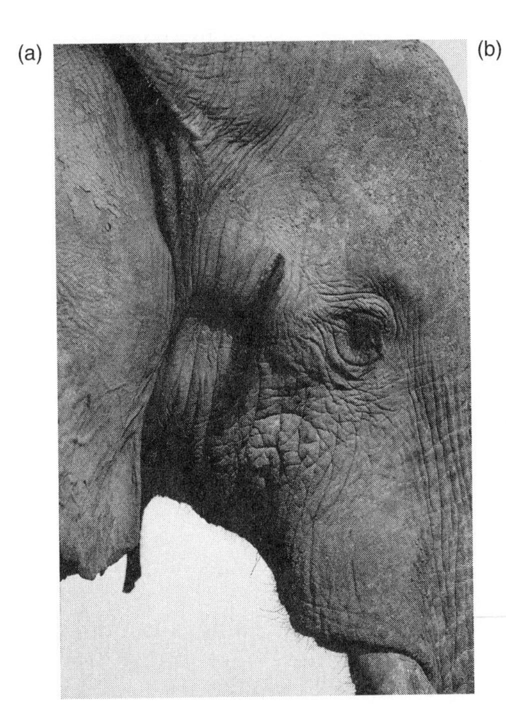
(b)

▲ Figure 24–10 A male African elephant in musth. (a) Secretion from the temporal gland of a male in musth. (b) A large bull sparring with a smaller male.

abnormal situation turned out to be the key to the early and prolonged musth and aggressive behavior of the young males. In populations of elephants that include older males, young males are unlikely to be in musth, and when they do enter musth the periods are short. Social interactions are responsible for this situation—young males in musth engage in aggressive interactions with the older bulls in musth, and the young males are defeated and driven away by the bulls. Within minutes to hours after being defeated in an aggressive encounter, a young male loses the signs of musth. Repeated interactions with older bulls drive young males out of musth, and the presence of older males in a population may delay the onset of musth in young males.

In 1998 six bull elephants were moved from Kruger Park to Pilanesberg, and their presence had a dramatic effect on the young males. The duration of musth in young males dropped sharply, in most cases falling from weeks to just a few days (Table 24.2). Gratifyingly, the young males also stopped killing rhinoceros.

All of the original elephant populations that were founded in South Africa during the early 1980s were composed of orphan animals, because the techniques needed to move large elephants did not exist at that time. As methods of handling wild elephants improved, larger animals could be transported, and by the late 1980s adult females were being moved. Rob Slotow and his colleagues were the first to move large adult males. They built a special trailer capable of moving bull elephants up to about 45 years old, animals that are as tall as 3.4m at the shoulder. In May 2000 they used this trailer to move 10 Kruger Park males to Hluhluwe-Umfolozi Park, hoping that adding them to the population would modify the behavior of young males as it had at Pilanesberg.

Sex Determination and Sea Turtle Conservation

Conservation of sea turtles provides special challenges. Some species range over thousands of square kilometers of ocean, including international waters and coastal areas that are under the jurisdictions of many different nations. For example, female leatherback turtles (*Dermochelys coriacea*) that had

Table 24.2 Periods of Musth in Young Elephants

Pilanesburg Male Identification Number	*Duration of Musth in days* — *Before introduction of older males*	*After introduction of older males*
18	>16, >14	6, 1
6	55, 40	3, 1, 12–26
22	90, >86	2, 53–77
5	134	76, 24

laid eggs on Playa Grande on the Pacific coast of Costa Rica were tracked by tethering radio transmitters to their shells. Signals from the transmitters were picked up by satellite when the turtles were on the surface, allowing the animals to be tracked across the open ocean (Morreale et al. 1996). From 1992 through 1995, seven turtles were tracked for periods of 29 to 87 days as they traveled over distances of 417 to 2780 km. Remarkably, the turtles all followed the same path to the southwest, passing by the Galápagos Islands and continuing into the Pacific Ocean. With the vast Pacific Ocean before them, the seven turtles remained in a corridor that was no more than 500 km wide. In the process, they passed from the territorial limits of Costa Rica into international waters, then into Ecuadorian waters around the Galápagos Islands and back into international waters. Green turtles breeding in the Caribbean may pass through the territorial waters of a half-dozen countries as they swim from their feeding grounds to the nesting beaches. As is the case with migratory birds and whales, movements between national jurisdictions and international waters add enormously to the problem of establishing and enforcing provisions to protect the turtles.

Protection of sea turtles presents biological challenges as well as legal ones. The extreme faithfulness that turtles show for a particular breeding site limits the amount of genetic variation at each site. For example, an examination of loggerhead turtles (*Caretta caretta*) in the Mediterranean Sea showed substantial genetic separation between turtles nesting at adjacent sites on the coast of Turkey (Schroth et al. 1996). Some of these breeding sites are being destroyed by real estate development. Because the turtles nesting at each site are genetically distinct from those at other sites, the loss of a single breeding site results in the loss of a portion of the genetic variation of the species. Thus, preserving the genetic diversity of the species as a whole depends on preserving the breeding sites of all the subpopulations.

The Kemp's Ridley Sea Turtle All seven species of sea turtles face threats, but the melancholy distinction of being the most endangered sea turtle goes to Kemp's ridley (*Lepidochelys kempi;* Figure 24–11). This species has only one major nesting site, a 14-mile stretch of beach on the coast near Rancho Nuevo in Tamaulipas, Mexico. Kemp's ridley nests by day, and once did so in enormous numbers. The influx of female turtles to the beach is called an *arribada* (Spanish for *arrival*), and a movie made in 1947 shows an *arribada* estimated to contain 47,000 turtles, all nesting at once. Although Kemp's ridley turtles had probably been an important source of food for inhabitants of the region since pre-Columbian times, the location of the nesting beach was unknown to the scientific community until 1966. By then the largest *arribadas* consisted of only 3000 to 5000 turtles, and their numbers have continued to decrease as adult females and eggs taken from the nesting beach have been used for food, and as adults and juveniles have drowned in fishing and shrimping nets in the Gulf of Mexico. By 1994, despite conservation efforts, the entire population probably contained fewer than 800 females (Shaver 1990).

▲ Figure 24–11 An adult female Kemp's ridley sea turtle.

From 1978 to 1988 the Mexican Instituto Nacional de Pesca worked with the U.S. Fish and Wildlife Service, the National Marine Fisheries Services, the National Park Service, and the Texas Parks and Wildlife Department to establish a second breeding population of Kemp's ridley at Padre Island National Seashore in Texas. Each year about 2000 eggs were collected in Mexico as they were laid and then shipped to Padre Island, where they were incubated and hatched. Additional eggs were incubated on the beach at Rancho Nuevo.

The goal of the project was to produce at least a 1:1 ratio of male and female hatchlings, and preferably a preponderance of females, to establish a population of turtles that would return to Padre Island to breed. Several individuals, including National Parks Service employee Donna Shaver, cooperated in a study of the sex ratio of the hatchling turtles that had been produced between 1978 and 1984. Embryos that had failed to hatch had been preserved for later examination, and the sex of the dead turtles could be determined by histological examination of their gonads. The results were discouraging (Figure 24–12). In three of the five years for which adequate samples were available, only one-third of the hatchlings were females, and the highest proportion of females ever achieved was only 50 percent (Shaver et al. 1988). Those ratios were far from the project's goal of at least a 1:1 sex ratio, and the deviation was in the wrong direction—an excess of males was being produced.

The phenomenon of temperature-dependent sex determination in turtles had been described in the biological literature some 30 years earlier, but its implications for conservation of sea turtles had not been appreciated (Mrosovsky and Yntema 1981). The Padre Island project, like many other sea turtle conservation projects around the world, incubated the eggs in moist sand in styrofoam boxes that were kept in a covered egg house on the beach. Shaver found that the temperature inside the boxes was slightly too cool, and this was why most of the eggs developed into male turtles.

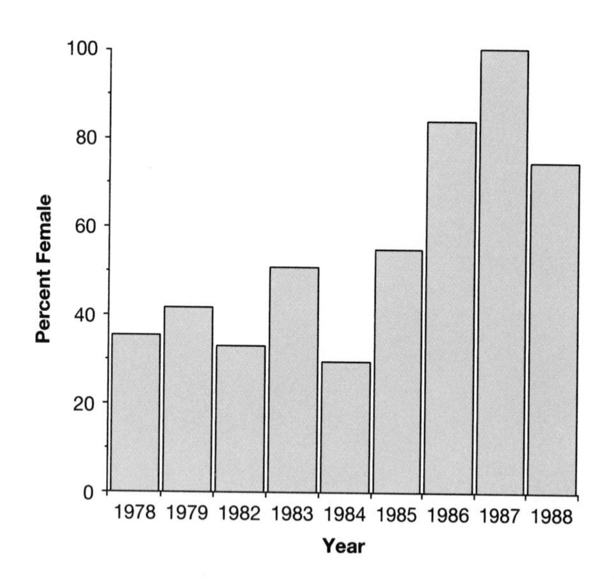

▲ Figure 24–12 Temperature-dependent sex determination and sea turtle conservation. Proportion of female hatchlings produced in the egg house at Padre Island National Seashore from 1978 through 1988. Note the increase in the proportion of females from 1985 onward, after incubation temperatures were raised. (No reliable samples are available for 1980 and 1981.)

From 1985 onward, temperatures in the egg houses at Rancho Nuevo and Padre Island were raised about 3°C. This small increase in incubation temperature was sufficient to shift sex determination in favor of females, and from 1985 through the end of the project in 1988 the proportion of female hatchlings increased dramatically (Figure 24–12). This example illustrates the importance of applying basic biological information about organisms to management and conservation programs.

Other Species of Sea Turtles Attempts to save sea turtles are in progress all over the world, sponsored by a variety of government agencies, private organizations, and dedicated individuals. The population of green turtles (*Chelonia mydas*) that nests on Tortuguero Beach on the coast of Costa Rica is the largest in the Caribbean. Recoveries of tagged animals show that turtles from this population disperse to feeding grounds throughout the Greater Caribbean. Following a period of heavy exploitation when nearly every turtle that came ashore to nest was captured, the turtles nesting on the beach have been protected and studied since the late 1950s (Bjorndal et al. 1999). The harvest of adult turtles has been controlled, and regular surveys have counted the number of females that emerge to nest. A graph of the number of nesting turtles shows an encouraging upward trend from 1971 through 1996 (Figure 24–13). Interpretation of this trend, and the encouragement that can be drawn from it, are limited by the long life spans and slow growth of sea turtles. Caribbean green turtles take several decades to reach maturity. Thus, changes in survival of nests and of juvenile turtles would not affect the numbers of adult females for many years. These data can be compared to observations of a star 25 light years away—what we see now is the condition of the star 25 years ago, and we have no way of knowing if it has increased or decreased in brightness since then.

Other questions arise from our inadequate knowledge of the biology of sea turtles. For example, is it a wise management practice to dig up nests of turtle eggs from the nesting beaches and incubate them in a protected area? Predation on eggs in natural nests can be high, but moving the eggs risks upsetting the normal sex ratio. Head-starting is another technique that has been used in sea turtle conservation projects. Baby turtles are kept in captivity for weeks or months and allowed to grow before they are released at sea. This practice avoids the very high losses of baby turtles to predators that occur when the newly hatched babies make their way down the beach and into the sea. However, a baby turtle may be imprinting on the characteristics of the beach and the adjacent water as it moves from nest to sea. If that is the case, head-starting may deprive the turtle of information it needs to return to the nesting beach as an adult. Furthermore, a head-starting project may do no more than feed the predators at the site of release if head-started turtles are not released carefully and in appropriate places. We simply cannot evaluate the effects of these manipulations, because we do not know enough about the biology of sea turtles.

These questions are a subset of a broader set of questions about the effectiveness of conservation methods (Dodd and Siegel 1991, Burke 1991, Reinert 1991). The problems cited by these authors emphasize the central role of information about all aspects of the biology of organisms in successful conservation plans. This sort of information is not easy to obtain for any species of organism, and turtles are more difficult to study than most animals. For example, a total of 22,255 yearling Kemp's ridley

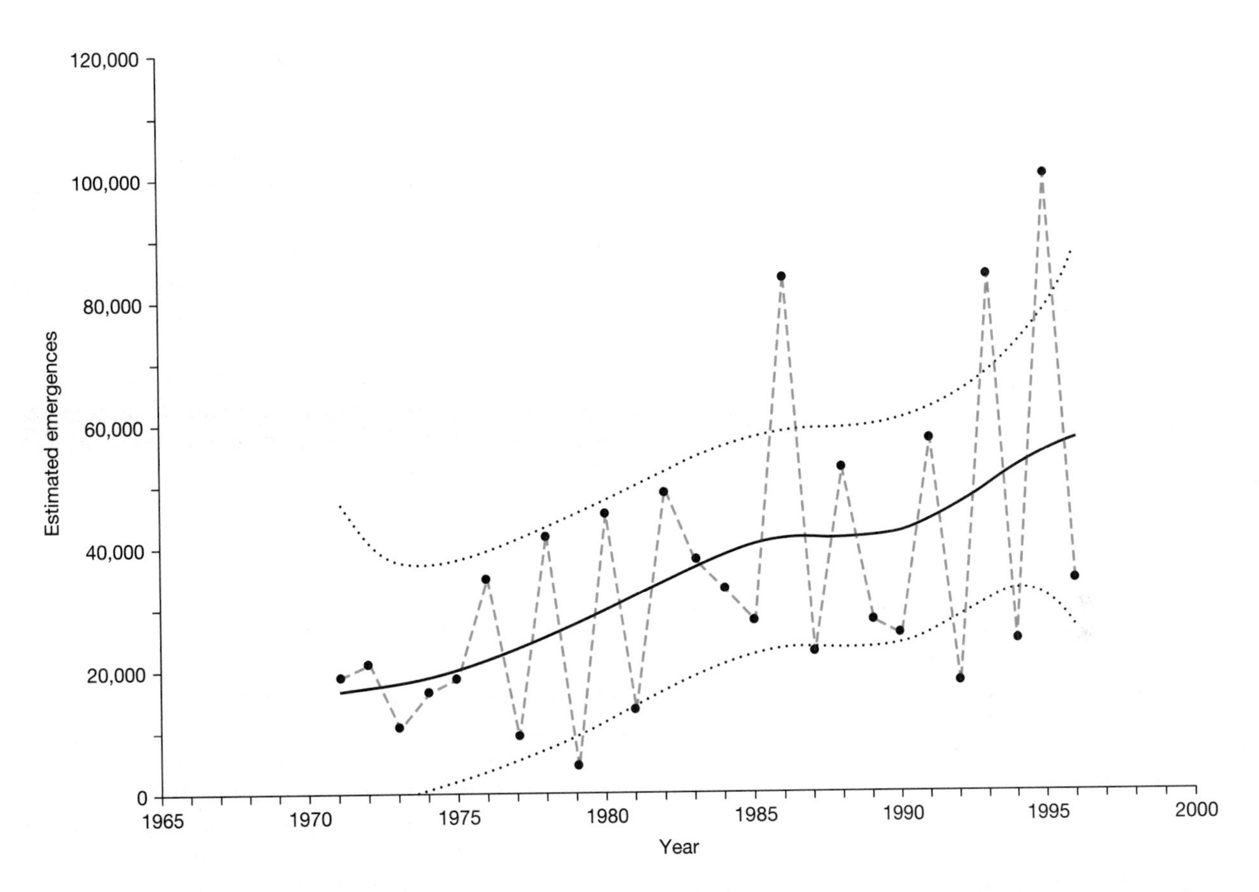

▲ Figure 24–13 Nesting emergences of female green sea turtles at Tortuguero, Costa Rica. The points are annual counts of turtles emerging to nest. The solid line is a statistical estimate of the number of turtles nesting annually and the dashed lines represented ± 2 standard errors of the estimate.

turtles were tagged and released by the Galveston Laboratory of the National Marine Fisheries Service during the head-start experiment, which lasted from 1978 to 1993 (Caillouet et al. 1995). During that period 805 tagged turtles were recaptured, and estimates of annual survival range from 10 to 50 percent. Kemp's ridley turtles are believed to mature in 10 years, and a mathematical calculation shows that an annual survival of 45 percent would be needed to produce one surviving turtle at age 10 from the average of 1437 yearlings that were released each year. It is not yet known whether enough Kemp's ridley turtles from the head-start project survived to maturity to affect the population, but the odds faced by an individual head-started turtle are daunting.

Global Issues in Conservation Biology

We have focused on problems faced by particular species and particular habitats, but global changes are a growing area of concern for conservation biology. Consider these examples of global problems that have impacts at the levels of individual organisms.

Global Climate Change Global warming seems to be a reality, although the causes are hotly debated. A report issued in 2001 by the United Nations Intergovernmental Panel on Climate Change predicted that the average air temperature of the Earth could increase by more than 5°C in the next century. Already indications of warming are apparent. Growing seasons in high latitudes begin earlier in the spring and extend later into the fall than they used to. A 90-year record shows that the growth rate of Alaskan white spruce has decreased as temperatures have increased, apparently because of temperature-induced drought stress (Barber et al. 2000). An analysis of 25 years of nesting records for birds in the United Kingdom reveals that the date of nesting is related to temperature, and birds are laying earlier as temperatures rise (Crick and Sparks 1999). The cues that migratory birds use to begin their northward journeys are not affected by global warming, and as a result birds may arrive at their breeding grounds after the optimum conditions for laying eggs and rearing young have passed (Both and Visser 2001). Anticipating the responses of organisms to global climate change is a crucial test case for evolutionary biology and ecology (Kareiva et al. 1993).

Acid Precipitation Rain and snow are acidified by sulfur- and nitrogen-containing gases produced when fossil fuels are burned. More than 90 percent of sulfur emissions and 95 percent of nitrogen emissions in North America are of human origin, primarily from burning coal and oil to generate electricity and from engines in vehicles. These gases mix with water in the atmosphere and produce rain and snow that contains sulfuric and nitric acids. The same processes occur in other industrialized countries, and the acids are carried by weather systems across national borders and even between continents. Nowhere on Earth is beyond the reach of atmospheric pollutants.

As acid precipitation falls, it changes the chemical composition of soil and water. Calcium is removed from soil and carried to streams and rivers and eventually to the sea. When the soil in the watershed of a river has been depleted of calcium, other ions are dissolved. Aluminum, which follows calcium and magnesium in order of solubility, is of particular concern because it is toxic to aquatic organisms. The effect of acid precipitation has been studied for several decades in Norway. Aluminum is the major cause of fish death in the southern part of Norway. Thirteen thousand fish populations were monitored from 1960 to 1990; 21 percent of them showed a reduction in size and an additional 19 percent disappeared entirely.

The loss of calcium from the soil has an impact on terrestrial animals as well. Some birds in European forests are producing eggs with thin, porous shells that break easily. Part of the problem is a lack of calcium in the birds' diets, because acid in rain and snow has dissolved calcium from the soil. These depleted soils no longer support the populations of snails that once supplied the calcium the birds need to form eggshells (Graveland et al. 1994).

Endocrine Disruptors A variety of chemicals mimic the effects of hormones and can disrupt normal functions of the endocrine systems of vertebrates. Some plants, for example, contain chemicals that affect the reproductive cycles of rodents feeding on them; these compounds are probably part of the plant's defense against herbivores. Synthetic chemicals that are released into the environment by human activities can disrupt the endocrine systems of vertebrates, including humans (Iguchi and Sato 2000).

Pesticides are a major source of endocrine-disrupting chemicals, as are plastics, lubricants, and solvents. Phthalates leach from plastics (including those used to make infant nursing bottles) and have been implicated in the high incidence of thelarche (premature breast development) in girls as young as two years in Puerto Rico.

Endocrine disruptors can alter sexual development when animals are exposed during embryonic development. Waste from a pulp paper mill released into the Fenholloway River in the panhandle of Florida contains androstenedione, the anabolic steroid made famous by baseball player Mark McGuire. In the river androstenedione has masculinized the mosquitofish (*Gambusia affinis*)—populations downstream from the mill contain only males (McNatt et al. 2000). This is the first example of an environmental androgen, a pollutant that masculinizes embryos. In contrast, environmental estrogens are disturbingly common. The list of vertebrates that have been feminized by environmental estrogens includes several species of fish (trout, carp, salmon, and sturgeon), amphibians (frogs and salamanders), reptiles (alligators and turtles), birds (gulls, herons, and birds of prey), and mammals (the Florida panther) (Guillette et al. 2000). In these cases males have smaller testes than normal and may be infertile. Humans are probably not immune from the effects of environmental estrogens. Between 1938 and 1991 the average sperm densities of human males in Europe and North America declined from 113 million per milliliter of seminal fluid to 66 million per ml, and the average volume of seminal fluid decreased from 3.40 mL to 2.75 ml. Feminization by environmental estrogens has been suggested as the cause of this phenomenon (Carlsen et al. 1992).

Charcteristics of Vulnerable Species

The proliferation of humans and their technology imperils the continued existence of nearly all species, including humans themselves, but some species have biological characteristics that make them especially vulnerable. We have already noted the early and rapid extinctions of species of birds and mammals that were restricted to islands, but there are many other factors that contribute to survival potential of a species (Table 24.3). A large carnivore that occupies a position at the top of the food chain, has a narrow habitat tolerance and a restricted geographic range that nonetheless crosses national boundaries or international waters, has a low reproductive rate, is hunted for sport, and is intolerant of humans clearly has a large number of risk factors.

The polar bear (*Ursus maritimus*) fits this description. Males are up to 3.4m from nose to tail and weigh more than 600 kg; females are somewhat smaller (up to 2.4m and 315 kg). Their circumpolar habitat includes parts of five countries and the Arctic Ocean. They are nomadic, traveling more than 8000 kilometers in a year, moving between national jurisdictions and international waters. A female polar bear reproduces only every 3 or 4 years and usually has one cub that remains with its mother for more than two years. Females reach reproductive maturity when they are 4 to 6 years old and have a life span of about 25 years in the wild.

The Arctic is a remote part of the Earth, far from industrialized areas, but that has not protected it from pollution. Winds and ocean currents carry acid precipitation and both inorganic and organic pollutants from the entire northern hemisphere to the Arctic. Polar bears are at the top of the Arctic food chain and ringed seals (*Phoca hispida*) are their preferred food. Ringed seals contain high levels of polychlorinated biphenyls (PCBs), chemicals that are used in electrical equipment. The concentration of PCBs in seal blubber from the Russian and European Arctic is higher than in seals from Greenland, Canada, and Alaska, and these high levels are attributed to the continued use of PCBs in Russian electrical equipment. Dioxin is another pollutant that is affecting the Arctic, and most of the dioxin contamination in the Canadian Arctic can be traced to facto-

Table 24.3 Characteristics of species that affect survival potential. The examples are used to illustrate specific characteristics; they are not meant to indicate that the species named are at a high or low risk in all respects.

Higher Risk	*Lower Risk*
Large body size (tiger)	Small body size (bobcat)
Carnivore (mountain lion)	Herbivore (white-tailed deer)
Small geographic range (Puerto Rican parrot)	Large geographic range (Canada goose)
Narrow habitat tolerance (orangutan)	Broad habitat tolerance (Virginia opossum)
Hunted for food or for the market where there is no effective legal protection (Asian box turtle)	Hunted for sport in managed game programs (mourning dove)
Geographic range crosses national boundaries or includes international waters (leatherback sea turtle)	Geographic range restricted to one or a few countries (American alligator)
Intolerant of the presence of humans (grizzly bear)	Tolerant of the presence of humans (black bear)
Low reproductive rate: long gestation period, small litter, slow maturity (giant panda)	High reproductive rate: short gestion period, large litter, rapid maturity (raccoon)
A predator high in the food chain that is subject to the effects of biological magnification of chemical pollutants (peregrine falcon)	Lower level predator that feeds on less contaminated prey (sparrow hawk)

ries and incinerators in the United States. PCBs and dioxin are endocrine disruptors, and they appear to be interfering with reproduction in polar bears (UNEP 2000).

By choosing characteristics from low end of the risk scale we can develop a composite picture of the typical wild animals of the late 21st Century. They will be small and either herbivores or predators no higher than midway along the food chain, have high reproductive potential and broad habitat requirements that include human altered landscapes. Species with these characteristics include rock doves (*Columba livia,* the pigeons familiar to city dwellers), house sparrows (*Passer domesticus*), Norway rats (*Rattus norvegicus*), and carp (*Cyprinus carpio*).

24.4 The Paradoxes of Conservation

Although conservation must address biological issues, conservation is not a purely biological issue. Human societies that burn fossil fuels and manufacture products will inevitably produce pollutants that travel far from their sources. It is relatively easy to believe that it is wrong for a multinational corporation to build a salt-extraction plant that will destroy a protected habitat, or for a paper mill to release pollutants to save itself the cost of removing them from its waste discharge. Is it equally certain that a subsistence-level farming family in a developing country should not disrupt the habitat of an endangered species to grow the food it needs to survive?

Rich and poor nations respond differently to the often-conflicting demands of earning a living versus conserving natural resources. A wealthy nation, especially a large country like the United States, can afford to set aside land to protect habitats and organisms; but that option is not always possible for a poor nation. Effective conservation efforts cannot focus only on biological questions. In the real world, conservation requires intricate balancing of biological, political, economic, and cultural values, and no one response is right for all species and all habitats. Some species need protected areas, but many of the most important areas of relatively undisturbed habitat are in countries where poverty and pressing social problems make complete protection an unrealistic goal. Conservation programs must address social and economic issues as well as biological ones if they are to succeed. The people living near parks and management areas must believe that protecting those habitats and the species they contain will contribute more to their own standard of living than they could gain by clearing the land and killing the animals.

Efforts of this sort are under way. The National Resources Defense Council is working with the villages around San Ignacio Lagoon to develop sustainable fisheries that are environmentally friendly. These fisheries also will provide an economic basis to make it easier to resist future proposals for industrial development that would damage the lagoon. Project Piaba in Brazil began as a study of fish diversity in the middle Rio Negro basin and has grown into a community-based interdisciplinary project that fosters ecologically sound economic development (Chao 2001). The Rio Negro is the primary source of ornamental aquarium fish; more than 20 million fish are exported annually, with a retail value of more than $100 million. If it were not for the fishery, the area would be developed for mining, forestry, and agriculture. Since sustaining the fishery requires sustaining the ecosystem, scientists working with Project Piaba are studying the structure and function of the aquatic systems. They are using that information to develop fishery management procedures to give the local people an incentive to preserve the integrity of the environment that supports the fishery.

The magnitude and complexity of the problems facing conservation biologists are daunting, and the scale on which remedial efforts must be attempted is nearly beyond comprehension. Allison Jolly (1985) has calculated that the total world population of all wild primates combined is less than that of any of the Earth's major cities. The entire extant populations of many species of primates are no larger than that of a small town. Conditions are equally critical for many other vertebrates. Two decades ago, Jolly (1980) expressed the problems we face as biologists today.

> This realization has been painful. It began for me in Madagascar, where the tragedy of forest felling, erosion, and desertification is a tragedy without villains. Malagasy peasant farmers are only trying to change wild environments to feed their own families, as mankind has done everywhere since the Neolithic Revolution. The realization grew in Mauritius, where I watched the world's last five echo parakeets land on one tree and knew they will soon be no more. It has

come through an equally painful intellectual change. I became a biologist through wonder at the diversity of nature. I became a field biologist because I preferred watching nature go its own way to messing it about with experiments. At last I understood that biology, as the study of nature apart from man, is a historical exercise. From the Neolithic Revolution to its logical sequels of twentieth-century population growth, biochemical engineering of life forms, and nuclear mutual assured destruction, the human mind has become the chief factor in biology. . . . the urgent need in [vertebrate] studies is conservation. It is sheer self-indulgence to write books to increase understanding if there will soon be nothing left to understand.

Reviews of the biodiversity crisis agree that an essential first step in coming to grips with the problem is a clear understanding of what species exist, where they are, and what are the critical elements in their survival. This is an enterprise that must enlist biologists from specialties as diverse as systematics, ecology, behavior, physiology, genetics, nutrition, and animal husbandry. There are so many species about which we know almost nothing, and there is so little time for us to learn.

Summary

The diversity of vertebrates has increased steadily (albeit with several episodes of extinction) for the past 500 million years, peaking in the mid-Miocene, about 15 million years ago. Much of the decline in diversity of vertebrates (and other forms of life) since then can be traced to the direct and indirect effects of humans on the other species with which we share the planet. Major threats to the continued survival of species of vertebrates include habitat destruction, pollution, and hunting. At the base of all these phenomena is the enormous increase in human population size. The world's human population is currently about 6 billion people, double what it was only 50 years ago. A quarter of a million humans are added to the population each day, a population the size of New York City is added each month, and nearly 100 million additional people demand resources each year. Consumption of resources and pollution of the enviromment increase at rates far greater than the rate of population growth. This differential is increasing as global communication (especially television) exposes the overwhelming majority of humans to a Western life-style, raising expectations and aspirations worldwide. Typically the use of resources in a modern technological society increases four to five times faster than population growth, and the release of pollutants rises in proportion to resource use.

Conservation efforts are complicated by political and economic issues, and strategies that are appropriate for developed countries may be impractical in developing nations. Programs that integrate the needs of humans and wildlife have the best chances for long-term success. Information about the basic biology of organisms plays an essential role in conservation by defining the critical elements of a habitat that must be preserved to ensure that a species can survive, by identifying sources of problems, and by guiding management of wild populations and reintroduction programs.

Additional Readings

Baptistotte, C., J. T. Scalfoni, and N. Mrosovksy. 1999. Male-producing thermal ecology of a southern loggerhead turtle nesting beach in Brazil: Implications for conservation. *Animal Conservation* 2:9–13.

Barber, V. A., G. P. Juday, and B. P. Finnay. 2000. Reduced growth of Alaska white spruce in the twentieth century from temperature-induced drought stress. *Nature* 405:668–672.

Benitez, J., and I. Perfecto. 1989. Efecto de diferentes tipos de manejo de cafe sobre las comunidades de hormigas. *Agroecologia Neotropical* 1(1):11–15.

Benton, M. J. 1990. Patterns of evolution and extinction in vertebrates. In K. Allen and D. Briggs (Eds.), *Evolution and the Fossil Record*. Washington, D.C.: Smithsonian Institution Press, 218–241.

Bjorndal, K. A, J. A. Wetherall, A. B. Bolten, J. A. Mortimer. 1999. Twenty-six years of green turtle nesting at Tortuguero, Costa Rica: An encouraging trend. *Conservation Biology* 13:126–134.

Both, C., and M. E. Visser. 2001. Adjustment to climate change is constrained by arrival date in a long-distance migrant bird. *Nature* 411:296–298.

Burke, R. L. 1991. Relocations, repatriations, and translocations of amphibians and reptiles: Taking a broader view. *Herpetologica* 47:350–357.

Caillouet, C. W. Jr., et al. 1995. Survival of head-started Kemp's ridley sea turtles (*Lepidoshelys kempii*) released into the Gulf of Mexico or adjacent bays. *Chelonian Conservation and Biology* 1:285–292.

Carlsen, E., et al. 1992. Evidence for decreasing quality of semen during the past 50 years. *British Medical Journal* 305:609–619.

Chao, N. L. 2001. Project Piaba. *Communiqué* (American Zoo and Aquarium Association), January 2001, p. 14.

Clemmons, J. R., and R. Buchholz (Eds.). 1997. *Behavioral Approaches to Conservation in the Wild.* Cambridge, U.K.: University Press.

Crick, H. P. Q., and T. H. Sparks. 1999. Climate change related to egg-laying trends. *Nature* 399:423–424.

Daszak, P., A. A. Cunningham, and A. D. Hyatt. 2000. Emerging infectious diseases of wildlife—threats to biodiversity and human health. *Science* 287:443–449.

Diamond, J. 1989. Quaternary megafaunal extinctions: Variations on a theme by Paganini. *Journal of Archaeological Science* 16:167–175

Dodd, C. K. Jr., and R. A. Siegel. 1991. Relocations, repatriation, and translocation of amphibians and reptiles: Are they conservation strategies that work? *Herpetologica* 47:336–350.

Ehrenfeld, D. 1981. Options and limitations in the conservation of sea turtles. In K. A. Bjorndal (Ed.), *Biology and Conservation of Sea Turtles,* Washington, D.C.: Smithsonian Institution Press, 457–463.

Flannery, T. F. 1995. *The Future Eaters.* New York: George Brazillier.

Flannery, T. F. 1999. Debating extinction. *Science* 283: 182–183.

Gillespie, R., et al. 1978. Lancefield swamp and the extinction of the Australian megafauna. *Science* 200:1044–1048.

Gorman, M. L., et al. 1998. High hunting costs make African wild dogs vulnerable to kleptoparasitism by hyaenas. *Nature* 391:479–481.

Graveland, J., et al. 1994. Poor reproduction in forest passerines from decline of snail abundance on acidified soils. *Nature* 368:446–448.

Guillette, L. J. Jr., et al. 2000. Alligators and endocrine disrupting contamintants: A current perspective. *American Zoologist* 40:438–452.

Heywood, V. H., et al. 1994. Uncertainties in extinction rates. *Nature* 368: 105.

Holdaway, R. N., and C. Jacomb. 2000. Rapid extinction of the moas (Aves: Dinornithiformes): Model, test, and implications. *Science* 287:2250–2254.

Iguchi, T., and T. Sato. 2000. Endocrine disruption and developmental abnormalities of female reproduction. *American Zoologist* 40:402–411.

Jolly, A. 1980. *A World Like Our Own: Man and Nature in Madagascar.* New Haven: Yale University Press.

Jolly, A. 1985. *The Evolution of Primate Behavior,* 2nd ed. New York: Macmillan.

Kareiva, P. M., J. G. Kingsolver, and R. B. Huey. 1993. *Biotic Interactions and Global Change.* Sunderland, Mass.: Sinauer Associates.

MacPhee, R. D. E. (Ed.). 1999. *Extinctions in Near Time: Causes, Contexts, and Consequences.* New York: Plenum.

MacPhee, R. D. E. (Ed.). In press. *Humans and Other Catastrophes: A New Look at Extinctions and the Extinction Process.* New York: Plenum.

Martin, P. S., 1990. 40,000 years of extinctions on the "planet of doom." *Palaeogeography, Palaeoclimatology, Palaeoecology* 82:187–201.

Martin, P. S., and R. G. Klein (Eds.). 1984. *Quaternary Extinctions, a Prehistoric Revolution.* Tucson: University of Arizona Press.

McNatt, H. B., R. A. Angus, and W. M. Howell. 2000. Effects of paper mill effluent on a population of eastern mosquitofish, *Gambusia holbrooki.* Center for Bioenvironmental Research at Tulane/Xavier Universities. (http://e.hormone.tulane.edu) 2000. October 15–18. New Orleans: Tulane University.

Miller, G. H., et al. 1999. Pleistocene extinction of *Genyornis newtoni:* Human impact on Australian fauna. *Science* 283:205–208.

Morreale, S. J., et al. 1996. Migration corridor for sea turtles. *Nature* 384:319–320.

Mrosovsky, N., and C. L. Yntema. 1981. Temperature dependence of sexual differentiation in sea turtles: Implications for conservation practices. In K. A. Bjorndal (Ed.), *Biology and Conservation of Sea Turtles,* Washington, D.C.: Smithsonian Institution Press, 271–280.

Oftedal, O. T., and M. E. Allen. 1996. Nutrition as a major facet of reptile conservation. *Zoo Biology* 15:491–497.

Olson, S. L., and H. F. James. 1991a. Descriptions of thirty-two new species of birds from the Hawaiian Islands. Part 1: Non-passeriformes. *Ornithological Monographs* 45:1–88.

Olson, S. L., and H. F. James. 1991b. Descriptions of thirty-two new species of birds from the Hawaiian Islands. Part 2: Passeriformes. *Ornithological Monographs* 46:1–88.

Owen-Smith, N. 1987. Pleistocene extinctions: The pivotal role of megaherbivores. *Paleobiology* 13:352–362.

Pimentel, D., et al. 1992. Conserving biological diversity in agricultural/forestry systems. *BioScience* 42:354–362.

Pritchard, P. C. H. 1980. The conservation of sea turtles: Practices and problems. *American Zoologist* 20:609–617.

Ramsey, S. L., and D. C. Houston. 1999. Do acid rain and calcium supply limit eggshell formation for blue tits *(Parus caeruleus)* in the U.K.? *Journal of Zoology* 247:121–125.

Reinert, H. K. 1991. Translocations as a conservation strategy for amphibians and reptiles: Some comments, concerns, and observations. *Herpetologica* 47:357–363.

Rosen, B. 1994. Mammoths in ancient Egypt? *Nature* 369: 364.

Schroth, W., B. Streit, and B. Schierwater. 1996. Evolutionary handicap for turtles. *Nature* 384:521–522.

Shaver, D. 1990. Kemp's ridley project at Padre Island enters a new phase. *Park Science* 10(1):12–13.

Shaver, D., et al. 1988. Styrofoam box and beach temperatures in relation to incubation and sex ratios of Kemp's ridley sea turtles. Pages 103–108 in *Proceedings of the Eighth Annual Workshop on Sea Turtle Conservation and Biology,* NOAA

Technical Memorandum NMFS-SEFC-214, 24–26 February 1988, Fort Fisher, N.C.

Shepherdson, D. J., J. D. Mellen. and M. Hutchins (Eds.) 1998. *Second Nature: Environmental Enrichment for Captive Animals.* Washington, D.C.: Smithsonian Institution Press.

Slotow, R., et al. 2000. Older bull elephants control young males. *Nature* 408:425–426.

Steadman, D. W. 1995. Prehistoric extinctions of Pacific island birds: Biodiversity meets zooarchaeology. *Science* 267:1123–1131.

Stuart, A. J. 1991. Mammalian extinctions in the Late Pleistocene of northern Eurasia and North America. *Biological Reviews* 66:453–562.

UNEP. 2000. *Geo-2000 Global Environment Outlook.* United Nations Environment Programme http://edscns16.cr.usgs.gov

Vitousek, P. M. 1994. Beyond global warming: Ecology and global change. *Ecology* 75:1861–1876.

Wilson, E. O. 1992. *The Diversity of Life.* New York: Norton.

World Conservation Monitoring Centre. 1992. *Global Biodiversity: Status of the Earth's Living Resources.* London, UK: Chapman & Hall.

Web Explorations

On-line resources for this chapter are on the World Wide Web at http://www/prenhall.com/pough (click on the Table of Contents link and then select Chapter 24).

Glossary

abduction Movement away from the midventral axis of the body. *See also* adduction.
acetabulum Socket in the pelvis that receives the head of the femur.
acid precipitation Rain and snow acidified by sulfur- and nitrogen-containing gases produced by burning fossil fuels.
acrodont Teeth fused to the top of the jawbones. *See also* pleurodont, thecodont.
activity temperature range The body temperatures maintained by an ectothermal animal when it is thermoregulating.
adduction Movement toward the midventral axis of the body. *See also* abduction.
adductor mandibularis The major muscle that closes the jaws.
adipocytes Fat storage cells.
advertisement call The vocalization of a male anuran used in courtship and territorial behavior.
aerobic metabolism Metabolic breakdown of carbohydrates in the presence of oxygen yielding carbon dioxide and water as the end products.
agnathans Jawless vertebrates. *See also* gnathostomes.
allantois The extra-embryonic membrane of amniotes that develops as an outgrowth of the hind gut.
allochthonous Originating somewhere other than the region where found.
alloparental behavior Care provided by an individual that is not a parent of the young animal receiving care.
allopatry Situation in which two or more populations or species occupy mutually exclusive, but often adjacent, geographic ranges.
allotherians The multituberculates, an extinct suborder of mammals.
altricial Helpless at birth or hatching, as in pigeons and cats, for example. *See also* precocial.
alula The tuft of feathers on the first digit of a bird's wing that reduces turbulence in airflow over the wing.
alveoli The sites of gas exchange in the lungs.
ambient temperature The temperature of the environment. Usually refers to air temperature.
amble A gait of tetrapods, a speeded-up walk with at least one foot on the ground and two or three feet off the ground at any one time.
ammocetes larva The larval form of lampreys.
ammonotelic Excreting nitrogenous wastes primarily as ammonia.
amnion The innermost extra-embryonic membrane of amniotes.
amniotes Those vertebrates whose embryos possess an amnion, chorion, and allantois (i.e., turtles, lepidosaurs, crocodilians, birds, and mammals) in addition to the yolk sac of all vertebrates.
amphicoelous (amphicelous) The condition in which the vertebral centrum has both the anterior and posterior surfaces concave.
amphioxus The lancelet, *Branchiostoma lanceolatum.*
amphistylic An upper jaw suspended from multiple sites of attachment to the skull.
amplexus Clasping of a female anuran by a male during mating. Axillary amplexus refers to the male clasping the female in the pectoral region; inguinal amplexus is clasping in the pelvic region.
ampullae of Lorenzini Electroreceptors found in the skin of the snout of chondrichthyans.
anadromous Migrating up a stream or river from a lake or ocean to spawn (of fishes). *See also* catadromous.
anaerobic metabolism Metabolic breakdown of carbohydrates in the absence of oxygen yielding lactic acid as an end product.
anapsid A skull lacking temporal fenestrae, or an animal with an anapsid skull. *See also* diapsid, synapsid.
anastomoses Direct connections between arterioles and venules that bypass the capillary system.
angiosperm Most advanced and recently evolved of the vascular plants, characterized by production of seeds enclosed in tissues derived from the ovary. The ovary and/or seed is eaten by many vertebrates, and the success of the angiosperms has had important consequences for the evolution of terrestrial vertebrates.
angle of attack The angle above the horizontal of the leading edge of a bird's wing or shark's fin.
anisodactyl The arrangement of toes seen in perching birds, with three in front opposed to one behind.
anisognathy The situation in which the tooth rows in the upper and lower jaws are not the same distance apart. *See also* isognathy.
annulus (plural *annuli*) Rings extending around a structure.
antidiuretic hormone, ADH (also known as vasopressin) A hypothalamic hormone that causes the kidney to conserve water by increasing the permeability of the collecting tubules.
apatite The mineral form of calcium phosphate found in bone.
aphotic "Without light" (e.g., in deep-sea habitats or caves).
apnea "Without breath" (i.e., holding the breath, as during diving).
apocrine gland Type of gland in which the apical part of the cell from which the secretion is released breaks down in the process of secretion. *See also* holocrine gland.
apomorphic A character that is changed from its preexisting (ancestral) condition, usually one that is onique to the taxon possessing it. *See also* autapomorphy.
apomorphy (Also known as a derived character) A character that has changed from its ancestral condition. *See also* synapomorphy.

aposematic Device (color, sound, behavior) used to advertise the noxious qualities of an organism.
appendicular Of the limbs (e.g., leg bones or muscles).
apteria Regions of skin without feathers.
arcade Curve or arch in a structure, such as the tooth row of humans.
archaic Of a form typical of an earlier evolutionary time.
archinephric duct The ancestral kidney drainage duct.
archipterygium Fin skeleton, as in a lungfish, consisting of symmetrically arranged rays that extend from a central skeletal axis.
aspect ratio The ratio of the length of a wing to its width. Long wings have high aspect ratios.
atrium (plural *atria*) An entrance. The atria of the heart receive blood from the sinus venosus or the veins.
auditory bulla An elaboration of bone around the ear region in mammals that may increase auditory acuity.
aural Of the external or internal ear or sense of hearing.
auricle (also called the pinna) The external flap of the mammalian ear.
australopithecines Extinct Pliocene/Pleistocene hominids, the sister group of *Homo.*
autapomorphy An attribute unique to one evolutionary lineage of organisms. *See also* apomorphy.
autochthonous Originating in the region where found.
autonomic nervous system The part of the peripheral nervous system that controls glands, smooth muscles, and internal organs and produces largely involuntary responses, including the sympathetic and parasympathetic portions in mammals.
autotomy The voluntary release of a portion of the body to escape a predator, as when a lizard loses its tail. Autotomized structures are subsequently regrown.
axial Within the trunk region.

bacculum (also known as *os penis*) A bone in the penis of some eutherian mammals.
baleen (also known as whalebone) Sheets of fibrous, horn-like epidermal tissue that extend downward from the upper jaw and are used for filter feeding by the baleen whales (mysticetes).
barb Side branch from the rachis of a feather.
benthic Living at the soil/water interface at the bottom of a body of water.
bilateral symmetry (bisymmetry) Characteristic of a body that can be divided into mirror-image halves.
binominal nomenclature The Linnean system that assigns generic and species names to organisms (e.g., *Homo sapiens* for modern humans).
biomass Living organic material in a habitat (available as food for other species).
biome Biogeographic region defined by a series of spatially interrelated and characteristic life forms (e.g., tundra, mesopelagic zone, tropical rain forest, coral reef).
bipedality Locomotion on two legs, as in humans.
birdsong The longest and most complex vocalization produced by a bird. The song identifies the species and in many species is produced only by mature males and only during the breeding season.
blastocyst (also known as blastula) The hollow ball of cells that forms during cleavage of a fertilized egg. *See also* trophoblast.
blubber An insulating layer of fat beneath the skin, typical of marine mammals.
bone A mineralized tissue that forms the skeleton of vertebrates. Bone is about 50 percent mineralized. Not to be confused with the anatomical structure called "a bone," which may be composed of bone tissue and/or cartilage tissue.
bound A gait of tetrapods consisting of jumping off the hind legs and landing on the front legs.
brachial Pertaining to the forelimb.
brachiation Locomotion by swinging from the underside of one branch to another.
brachydont Molar teeth with low crowns. *See also* hypsodont.
branchial Pertaining to the gills.
branchiomeric Referring to segmentation of structures associated with, or derived from, the ancestral pharyngeal arches. *see also* metameric.
branchiomeric muscles Muscles powering the visceral arches.
buccal pumping Drawing air or water in and out of the mouth region by raising and lowering the floor of the mouth.
bunodont Molar teeth with low, rounded cusps. *See also* lophodont, selenodont.

calamus The tubular base of a feather that remains in the follicle.
calcaneus (or calcaneum) The large metatarsal bone that forms the heel of mammals.
cambered airfoil A structure, such as the wing of a bird, that is convex dorsally and produces lift when air flows across it.
carapace Dorsal shell, as of a turtle.
carnassials Teeth of eutherian mammals in the order Carnivora that are specialized as shearing blades.
catadromous Migrating down a river or stream to a lake or ocean to spawn (of fishes). *See also* anadromous.
catastrophism Hypothesis of major evolutionary change as a result of unique catastrophic events of broad geographic and thus ecological effect.
cavum arteriosum, cavum pulmonale, cavum venosum The chambers formed during ventricular contraction in the hearts of turtles and lepidosaurs.
cellulase An enzyme that digests cellulose. No multicellular organism produces cellulase, and herbivorous vertebrates rely on unicellular symbionts living in their digestive tracts to produce cellulase.
cellulose A complex carbohydrate that forms the cell walls of plants.
cementum A bone-like substance that fastens the teeth in their sockets.
centrum (plural *centra*) Bony portion of a vertebra that surrounds the notochord.
cephalic Pertaining to the head.
ceratotrichia Keratin fibers that support the web of the fins of Chondrichthyes.
character/character state Any identifiable characteristic of an organism. Characters can be anatomical, behavioral, ecological, or physiological.
chloride cells Cells in the gills of fishes and skin of amphibians specialized to transport sodium and chloride ions.

choana (plural *choanae*) Internal nares.

chondrification Formation of cartilage.

chondrocranium (also known as the neurocranium) A structure that surrounds the brain. Initially formed of cartilage, it is replaced by endochondral bone in most bony fishes and tetrapods.

Chordata Phylum of animals characterized by having a notochord at some stage of life.

chorioallantoic placenta A placenta developed from the chorionic and allantoic extraembryonic membranes that replaces the choriovitelline placenta during the embryonic development of all eutherian mammals and some marsupials. *See also* placenta.

chorion The outermost extra-embryonic membrane of amniotes.

choriovitelline placenta A placenta developed from the yolk sac that is characteristic of all therian mammals during early development. *See also* placenta.

circadian "About a day." Circadian rhythms are cycles that have a period of approximately 24 hours.

clade Phylogenetic lineage originating from a common ancestral taxon and including all descendants. *See also* grade.

cladistic Pertaining to the branching sequences of phylogeny. *See also* phylogenetic systematics.

cladogram Branching diagram representing the hypothesized relationships of taxa.

cleidoic egg One independent of environment except for heat and gas (carbon dioxide, oxygen, water vapor) exchange. Characteristic of amniotes.

cline Change in a biological character along a geographic gradient.

cloaca A common opening of the reproductive and excretory tracts.

cochlea (also known as lagena in nonmammalian tetrapods) The portion of the inner ear that houses the hair cells.

coelom (celom) Body cavity, lined with tissue of mesodermal origin.

coevolution Complex biotic interaction through evolutionary time resulting in the adaptation of interacting species to unique features of the life histories of the other species in the system.

collagen A fibrous protein that contributes to many structures.

columella The single auditory ossicle of the middle ear of nonmammalian tetrapods, the stapes of mammals. Homologous to the hyomandibula of fishes.

concealed estrus Estrus that is not revealed by external signals, such as swelling of the genitalia. In the anthropoid lineage concealed estrus is a derived character of humans.

condyle A rounded articular surface of a bone.

cones Photoreceptor cells in the vertebrate retina that are differentially sensitive to light of different wavelengths and thus perceive color. *See also* double cone.

conodont Small spine-like or comb-like structures formed of apatite found in marine sediments from the Late Cambrian to the Late Triassic. They are considered to be tooth-like elements of an early vertebrate, the conodont animal.

conspecific Belonging to the same species as that under discussion. *See also* heterospecific.

continental drift The movement of continental blocks on the mantle of the Earth. *See also* plate tectonics.

conus arteriosus An elastic chamber in front of the ventricle of some gnathostomes.

convergent evolution Appearance of similar characters in widely separated evolutionary lineages (e.g., wings in bats and birds). *See also* parallel evolution.

coprolite Fossilized dung.

coprophagy Eating the first set of feces that are produced, thereby recycling nutrients that would otherwise be lost.

coronoid process A vertical flange near the rear of the dentary bone that increases the area for attachment of the temporalis muscle.

corpus luteum (plural *corpora lutea*) A hormone-secreting structure formed in the ovary from the follicular cells remaining after an egg is released.

cosmine Form of dentine containing branching canals characteristic of the cosmoid scales of sarcoptoygian fishes.

countercurrent exchange Fluid streams flowing in opposite directions in adjacent vessels to promote exchange of heat or dissolved substance.

countershaded Referring to a color pattern in which the aspect of the body that is more brightly lighted (normally, the dorsal surface) is darker colored than the less brightly illuminated surface. The effect of countershading is to make an animal harder to distinguish from its background.

cranial kinesis Movement within the skull or of the upper jaws independent of the skull.

cranial nerves The nerves that emerge directly from the brain; 10 pairs in the primitive vertebrate condition, 12 pairs in amniotes.

cranial Pertaining to the cranium or skull, a unique and unifying characteristic of all vertebrates.

Craniata Animals with a cranium. Hagfishes have a cranium but lack vertebrae and are sometimes classified in the Craniata but not in the Vertebrata.

crepuscular Animals that are active at dawn and dusk.

critical period The restricted time during the ontogeny of an individual when imprinting occurs.

critically endangered A species is considered critically endangered when it is facing an extremely high risk of extinction in the wild in the immediate future.

cryptodires Turtles that bend the neck in a vertical plane to retract the head into the shell.

cupula A cup-shaped, gelatinous secretion of a neuromast organ in which the kinocilium and microvilli are embedded.

cursorial Specialized for running.

dear enemy recognition A situation in which a territorial animal responds more strongly to strangers than to its neighbors from adjacent territories.

demersal More dense than water and therefore sinking, as in the eggs of many fishes and amphibians.

denticles Small tooth-like structures in the skin, as in shark skin.

dentine A mineralized tissue found in the teeth of extant vertebrates and the dermal armor of some primitive fishes. Dentine is about 90 percent mineralized.

derived character Also known as an apomorphy. A character that has changed from its ancestral condition. *See also* shared derived characters.

dermal bone A type of bone that forms within the skin.

dermatocranium Dermal bones that cover a portion of the skull.

dermis The deeper cell layer of vertebrate skin of mesodermal and neural crest origin. *See also* epidermis.

detritus Particulate organic matter that sinks to the bottom of a body of water.

deuterostomy Condition in which the embryonic blastopore forms the anus of the adult animal; characteristic of chordates. *See also* protostomy.

diaphysis The shaft of a bone. *See* epiphysis.

diapsid A skull with two temporal fenestrae or an animal with a diapsid skull. *See also* anapsid, synapsid.

digitigrade Standing with the heel off the ground and the toes flat on the ground, as in dogs. *See also* plantigrade, unguligrade.

diphyodonty One replacement of the dentition during an animal's lifetime, as in most mammals.

dispersal Movements of animal populations via movement of the animals themselves. *See also* vicariance biogeography.

distal Away from the body. *See also* proximal.

distal convoluted tubule Portion of a kidney nephron responsible for changing the concentration of the ultrafiltrate by actively transporting salt.

diurnal Animals that are active during the day.

double cone Type of retinal photoreceptor in which two cones share a single axon. *See also* cones.

drag Backward force opposed to forward motion.

dryopithecids Later Cenozoic Eurasian hominoids, more primitive than any living ape.

durophagous Feeding on hard material.

eccrine gland A gland in the mammalian skin that produces a watery secretion with little organic content, forms the sweat glands of humans.

echolocation Determining location in three-dimensional space by sending out a pulse of sound and listening to echoes returning from objects in the environment.

ecosystem Community of organisms and their entire physical environment.

ectoderm One of the embryonic germ layers, the outer layer of the embryo.

ectotherm An organism that relies on external sources of heat to raise its body temperature.

edentulous Lacking teeth.

elastin A fibrous protein that can stretch and recoil.

electrocytes Muscle cells modified to produce an electric discharge.

embryonic diapause Maintaining the embryo in a stage of arrested development before it implants on the wall of the uterus.

enamel A mineralized tissue found in the teeth of extant vertebrates and the dermal armor of some primitive fishes. Enamel is about 99 percent mineralized.

encephalization quotient The ratio of the actual brain size of a species compared to the brain size expected from its body size.

endangered A species is considered endangered when it is facing a very high risk of extinction in the wild in the near future.

endemism Property of being endemic (i.e., found only in a particular region).

endocasts Fossil impressions of the insides of body cavities.

endochondral bone A type of bone that forms in cartilage.

endocrine disruptor A natural or synthetic chemical that interferes with normal development by duplicating the physiological effect of a hormone.

endocrine glands Glands that discharge hormones into the blood.

endoderm Innermost of the germ-cell layers of late embryos.

endometrium The glandular uterine epithelium of mammals that secretes materials that nourish the embryo.

endotherm An organism that relies on internal (metabolic) heat to raise its body temperature.

eon The largest division of geologic time. The history of Earth has occupied three eons: the Archean, Proterozoic, and Phanerozoic.

epaxial Referring to muscles on the dorsal portion of the trunk. *See also* hypaxial.

epicercal (Also known as heterocercal.) A tail fin with the upper lobe larger than the lower lobe. *See also* hypocercal.

epicontinental sea (epeiric sea) Sea extending within the margin of a continent.

epidermis The superficial cell layer of vertebrate skin of ectodermal origin. *See also* dermis.

epigenetic Pertaining to an interaction of tissues during embryonic development that results in the formation of specific structures.

epiphysis 1. Pineal organ, an outgrowth of the roof of the diencephalon. 2. (plural, *epiphyses*) Accessory center of ossification at the ends of the long bones of mammals, birds, and some squamates. In mammals the epiphyses are the actual articulating ends of the long bones themselves, with the cartilaginous zone of growth between the epiphysis and diaphysis. When the ossifications of the shaft (diaphysis) and epiphysis meet, lengthwise growth of the shaft ceases. This process produces a determinate growth pattern.

epiphyte Plant that grows nonparasitically on another plant.

epipubic bones Bones in noneutherian mammals that project anteriorly from the pubis.

estivation (aestivation) Form of torpor, usually a response to high temperatures or scarcity of water.

estrous cycle The normal reproductive cycle of growth, maturation, and release of an egg.

estrus (oestrus) The periodic state of sexual excitement in the females of most mammals (but not humans) that immediately precedes ovulation and during which a female is most receptive to mating. Also known as heat.

estuarine Pertaining to, or formed in, a region where the fresh water of rivers mixes with the seawater of a coast.

euryhaline Capable of living in a wide range of salinities. *See also* stenohaline.

euryphagous Eating a wide range of food items; a food generalist. *See also* stenophagous.

eurythermal Capable of tolerating a wide range of temperatures. *See also* stenothermal.

eurytopic Capable of living in a broad range of habitats.

eusocial Applied to a species or group of animals that display all of the following characters: cooperation in caring for the young, reproductive division of labor, more or less sterile individuals aiding individuals engaged in reproduction, and overlap of two or more generations of life stages capable of contributing to colony labor.

eustachian tube Passage connecting the middle ear to the pharynx.
exocrine glands Glands that discharge through a duct into a cavity or onto the body surface.
explosive breeding A very short breeding season.
extinct A species is considered extinct when there is no reasonable doubt that the last individual has died.
extinct in the wild A species is considered extinct in the wild when it is known to survive only in cultivation, in captivity, or as a naturalized population (or populations) well outside the past range.
extra-pair copulation Mating with an individual other than the partner in a monogamous breeding system.
extraperitoneal Positioned in the body wall beneath the lining of the coelom (the peritoneum) in contrast to being suspended in the coelom by mesentaries.

fallopian tube The anterior portion of the oviduct where eggs are fertilized.
ferment To break down food in the absence of oxygen, as in the stomach of a ruminant.
fever An increase in body temperature in response to infection.
flow-through ventilation Flow of respiratory fluid (air or water) in one direction, as across the gills of a fish.
foramen (plural *foramina*) An opening in a bone (e.g., for the passage of nerves or blood vessels).
foregut fermenter A mammal in which fermentation of foodstuffs is carried out in a modified stomach (e.g., a cow).
fossorial Specialized for burrowing.
fovea centralis Area of the vertebrate retina containing only cone cells, where the most acute vision is achieved at high light intensities.
free nerve ending A sensory nerve ending in the skin that is believed to sense pain.
furcula Avian wishbone formed by the fusion of the two clavicles at their central ends.
fusiform Torpedo shaped.

gallop A gait of mammals that is a modified bound.
gametes Sex cells—that is, eggs (ova) and sperm.
gastrolith A stone swallowed to aid digestion by grinding food in the gizzard.
genetic fitness The contribution of one genotype to the next generation relative to the contributions of other genotypes.
genus A group of related species.
geosyncline Portion of the Earth's crust that has been subjected to downward warping. Sediments frequently accumulate in geosynclines.
gestation Period during which an embryo is developing in the reproductive tract of the mother.
gill arch Assemblage of tissues associated with a gill; the term may refer to the skeletal structure only or to the entire epithelial muscular and connective tissue complex.
gizzard The muscular stomach of birds and other archosaurs.
glomerulus A capillary tuft associated with a kidney nephron that produces an ultrafiltrate of the blood.
gnathostomes Jawed vertebrates. *See also* agnathans.
gonads The organs that produce gametes—ovaries in females, testes in males.
Gondwana Supercontinent that existed either independently or in close contact with all other major continental landmasses throughout vertebrate evolution until the middle of the Mesozoic and was composed of all the modern Southern Hemisphere continents plus the subcontinent of India.
grade A level of morphological organization achieved independently by different evolutionary lineages. *See also* clade.
Great American Interchange (GAI) Faunal interchange between North and South America when the Central American land bridge (the Isthmus of Panama) was formed about 2.5 million years ago.
gymnosperms Group of plants in which the seed is not contained in an ovary—conifers, cycads, and ginkos.

hallux The big toe.
Harderian gland A gland associated with the eye of modern mammals that secretes an oily substance used to preen the fur.
head-starting Rearing neonatal animals in captivity for a period before they are released in the wild.
hemal arch Structure formed by paired projections ventral to the vertebral centrum and enclosing caudal blood vessels.
hermaphroditic Having both male and female gonads.
heterocercal (also known as epicercal) A tail fin with the upper lobe larger than the lower lobe.
heterocoelus Having the articular surfaces of the vertebral centra saddle-shaped, as in modern birds.
heterodont A dentition with teeth of different sizes and shapes in different regions of the jaw. *See also* homodont.
heterospecific Belonging to a different species from that under discussion. *See also* conspecific.
heterosporous plants Plants with large and small spores; the smaller give rise to male gametophytes and the larger to female gametophytes (equivalent to protogymnosperms).
hibernation A period of torpor in the winter when food is scarce.
hindgut fermenter A mammal in which fermentation of foodstuffs is carried out in the intestine (e.g., a horse).
holocrine gland Type of gland in which the entire cell is destroyed with the discharge of its contents. *See also* apocrine gland.
home range The area within which an animal spends most of its time. A home range is not defended. *See also* territory.
homodont A dentition in which teeth do not vary in size or shape along the jaw. *See also* heterodont.
homologous Inherited via common ancestry.
homology The fundamental similarity of individual structures that belong to different species within a monophyletic group.
homoplasy Similarities that do not indicate common ancestry. Structures resulting from parallel and convergent evolution and evolutionary reversal are examples of homoplasy.
hormone A chemical messenger molecule carried in the blood from its site of release to its site of action.
HOX/homeobox genes A sequence of DNA that codes a special protein for regulating the genes that control devel-

opment by specifying embryonic cell division along the anterior-posterior axis of the body.

hydrofoil A water-planing surface, such as the pectoral fins of sharks.

hydrosphere Free liquid water of the Earth—oceans, lakes, rivers, and so on.

hyoid arch The second gill arch.

hypapophyses Sharp processes, on the ventral surface of the neck vertebrae of the egg-eating snake (*Dasypeltis*), that slice through the shell of an egg.

hypaxial Referring to muscles on the ventral portion of the trunk. *See also* epaxial.

hyperdactyly A condition in which the number of digits is increased above the usual tetrapod complement of five.

hyperosmolal Of greater osmotic activity.

hyperphalangy Increase in the number of bones (phalanges) in the digits.

hyperthermia High body temperature.

hypertrophy Increase in the size of a structure.

hypocercal A tail fin with the lower lobe larger than the upper lobe. *See also* epicercal.

hypophysis The pituitary gland.

hyposmolal Of lower osmotic activity.

hypothermia Low body temperature.

hypotremate Having the main gill openings on the ventral surface and beneath the pectoral fins, as in skates and rays. *See also* pleurotremate.

hypselodont Molar teeth with ever-growing crowns. *See also* brachydont.

hypsodont Molar teeth with high crowns.

imprinting A special kind of learning that occurs only during a restricted time (called the critical period) in the ontogeny of an individual.

inclusive fitness The sum of individual fitness plus the effect of kin selection.

index of refraction The amount of deflection of a ray of light as it passes from one medium into another.

individual fitness *See* genetic fitness.

infraorbital foramen A hole beneath the eye through which nerves and blood vessels pass to the muzzle.

infrared The portion of the electromagnetic spectrum with wavelengths from 750 nanometers (just beyond visible light) to 1 millimeter (just before microwave radiation). Often called thermal radiation or heat.

infrasound Sound frequencies below the range of human hearing, approximately 20 hertz.

ingroup The group of organisms being considered. *See also* outgroup.

inguinal Pertaining to the groin.

intercalary cartilage A cartilage lying between the last two bones in the toes of some tree frogs.

intercalary plates Extra elements in the vertebral column of elasmobranchs that protect the spinal cord and major blood vessels.

interspecific Pertaining to phenomena occurring between members of different species.

intraspecific Pertaining to phenomena occurring between members of the same species.

isognathy The situation in which the tooth rows in the upper and lower jaws are the same distance apart. *See also* anisognathy.

isohaline Of the same salt concentration.

isosmolal Of the same osmotic activity.

isostasy Condition of gravitational balance between segments of the Earth's crust or of return to balance after a disturbance.

isostatic movement Vertical displacement of the lithosphere due to changes in the mass over a point or region of the Earth.

isotherm Line on a map that connects points of equal temperature.

iteropary Producing several individual babies or litters of young during the lifetime of a female. *See also* semelpary.

Jacobson's organ Also known as vomeronasal organ. An olfactory organ in the roof of the mouth of tetrapods.

keratin A fibrous protein found only in vertebrates that forms epidermal structures such as hair, scales, feathers, and claws.

kin selection Favoring the perpetuation of one's own genes by helping relatives to reproduce.

kinocilium A sensory cell located in neuromast organs.

lactation Producing milk form mammary glands to nourish young; characteristic of mammals.

lagena (also known as cochlea in mammals) The portion of the inner ear that houses the hair cells.

lateral plate mesoderm The ventral part of the mesoderm, surrounding the gut.

lateral line system The sensory system on the body surface of aquatic vertebrates that detects water movement.

Laurentia Paleozoic continent that included most of present-day North America, Greenland, Scotland, and part of northwestern Asia.

lecithotrophy Embryonic development nourished by the yolk when eggs are retained with the reproductive tract until they hatch. *See also* ovoviviparity.

leptocephalus larva Specialized, transparent, ribbon-shaped larva of tarpons, true eels, and their relatives.

lift Vertical forced opposed to gravity.

Linnean system A system of naming living organisms developed by the Swedish naturalist Carl von Linné (Carolus Linnaeus) in the eighteenth century.

lithosphere Crust of the Earth.

littoral Pertaining to the shallow portion of a lake, sea, or ocean where rooted plants are capable of growing.

loop of Henle The portion of the renal tubule of mammals that extends into the medulla. Essential for establishing the concentration gradient that produces a small volume of highly concentrated urine.

lophodont Molar teeth with ridges (lophs) that run in a predominantly internal-external direction across the tooth. *See also* bunodont, selenodont.

lophophorate Pertaining to several kinds of marine animals that possess ciliated tentacles (lophophores) used to collect food (e.g., pterobranchs).

lower critical temperature The point at which an endotherm must increase its metabolic heat production to maintain a stable body temperature.

lower lethal temperature The temperature at which even maximum metabolic heat production is inadequate to maintain a stable body temperature.

mammary gland A gland, found in mammals, that secretes milk. Mammary glands have characters of both apocrine and eccrine glands.
mammilary bodies Organic granules attached to the egg membrane that are the sites of first formation of calcite crystals making up the egg shell.
mandibular arch The most anterior of the gill arches, forming the jaws of gnathostomes.
marginal value theorem The hypothesis that an animal stops foraging in a patch of food when the rate of energy intake falls to the average rate for the habitat as a whole.
marsupium An external pouch in which the young of marsupial mammals develop.
masseter muscle A jaw muscle of mammals originating from the zygomatic arch and inserting on the lower jaw.
masticate To chew thoroughly.
matrilineage Related females within a group that support each other in social interactions.
matrotrophy Embryonic development nourished by materials transferred from the maternal circulation. Placentrophic matrotrophy describes the situation in which a placenta is the site of transfer of nutrients and wastes between the embryo and maternal circulation.
megafauna Species of large terrestrial animals (mammals, birds, and reptiles) that became extinct when human populations expanded, mostly between 50,000 and 10,000 years ago.
Meissner's corpuscle A sensory nerve ending in the skin that is believed to sense touch.
melanocyte A pigment cell containing melanin.
meninges Sheets of tissue enclosing the central nervous system. In mammals these are the dura mater, arachnoid, and pia mater.
menstrual cycle The periodic shedding of the endometrial lining of the uterus; characteristic of humans and some other anthropoid primates.
mesenteries Membranous sheets derived from the mesoderm that envelop and suspend the viscera from the body wall within the coelom.
mesoblast Mesodermal cell.
mesoderm Central of three germ layers of late embryos.
metameric Pertaining to ancestral segmentation, used in reference to serially repeated units along the body axis.
metamorphic climax The period in the life of a tadpole that begins with appearance of the forelimbs and ends with disappearance of the tail.
metamorphosis The developmental transition from larval to adult body form.
microvilli Sensory cells located in neuromast organs.
mimicry A tripartite system in which one organism (the mimic) counterfeits the signal of a second organism (the model), thereby deceiving a third organism (the dupe). The signal can be any characteristic of the model that the dupe can perceive—color, pattern, scent, etc.
molting Replacement of old hairs or feathers with new ones.
monogamy A mating system based on a pair bond between a single male and female. *See also* polygamy.
monophyletic Having a single evolutionary origin. *See also* paraphyletic, polyphyletic.
monophyletic lineage A taxon composed of a common ancestor and all its descendants.
monophyly Relationship of two or more taxa having a common ancestor.
morph Genetically determined variant in a population.
morphotypic Referring to a type of classification based entirely on physical form.
musth An annual period of elevated testosterone levels in the blood of male elephants.
myomeres Blocks of striated muscle fiber arranged along both sides of the body, most obvious in fishes.
myrmecophagy Eating ants and termites.

naris (plural *nares*) The external opening of the nostril.
nasolabial groove A channel from the external naris to the lip found in plethodontid salamanders.
Neo-Darwinism *See* New Synthesis.
neonates Newborn individuals.
neopallium (also known as neocortex). The derived expanded portion of the mammalian cerebral cortex.
neoteny Retention of larval or embryonic characteristics past the time of reproductive maturity. *See also* paedomorphosis and progenesis.
nephron The basic functional unit of the kidney.
neural arch Dorsal projection from the vertebral centrum that, at its base, encloses the spinal cord.
neural crest A type of embryonic tissue unique to vertebrates that forms many structures, especially in the head region.
neurocranium (also known as chondrocranium) Portion of the head skeleton encasing the brain.
neuromast organs Clusters of sensory hair cells and associated structures on the surface of the head and body of aquatic vertebrates, usually enclosed within the lateral line system.
neuron The basic functional unit of the nervous system.
New Synthesis (also known as Neo-Darwinism). The combination of genetics and evolutionary biology developed in the early twentieth century.
niche The functional role of a species or other taxon in its environment—the ways in which it interacts with both the living and nonliving elements.
nocturnal Animals that are active at night.
notochord A dorsal stiffening rod that gives the phylum Chordata its name.

occipital Pertaining to the posterior part of the skull.
odontodes Small tooth-like elements in the skin. The original tooth-like components of primitive vertebrate dermal armor. The denticles of sharkskin are odontodes.
ontogentic Pertaining to the development of an individual.
ontogeny The development of an individual. *See also* phylogeny.
operculum Flap or plate of tissue covering the gills.
opistoglyphs Venomous snakes with enlarged teeth in the rear of the jaw; rear-fanged snakes.
optimal foraging theory The hypothesis that an animal adjusts its foraging behavior to maximize energy return per unit time.
orogeny Process of crustal uplift or mountain building.
osmosis Movement of water across a membrane from a region of high activity (low solute concentration) to a region of low activity (high solute concentration).
osseous Bony.
osteoderm A bone embedded in the skin; characteristic of crocodilians.

ostracoderm Armored jawless aquatic vertebrates known from the Ordovician to the Devonian.
otolith A mineralized structure in the inner ear of teleost fishes.
outgroup Group of organisms that is related to but removed from the group under study. One or more outgroups are examined to determine which character states are evolutionary novelties (apomorphies).
ovary The female gonad.
oviparity Depositing eggs that develop outside the body.
ovoviviparity Embryonic development nourished by the yolk when eggs are retained with the reproductive tract until they hatch. (Lecithotrophy is the preferred term for this type of embryonic nourishment.)

pachyostosis Increased density of bone; characteristic of diving animals.
Pacinian corpuscle A sensory nerve ending in the skin that is believed to sense pressure.
paedomorphosis Condition in which a larva becomes sexually mature without attaining the adult body form. Paedomorphosis may be achieved by neoteny or by progenesis.
palatoquadrate Upper jaw element of primitive fishes and Chondrichthyes, portions of which contribute to the palate, jaw articulation, and middle ear of other vertebrates.
pancreas A glandular outgrowth of the intestine that secretes digestive enzymes.
pancreatic islets (also known as the islets of Langerhans) Clusters of endocrine cells in the pancreas that secrete insulin.
Pangaea (Pangea) Single supercontinent that existed during the mid-Paleozoic and consisted of all modern continents apparently in direct physical contact with a minimum of isolating physical barriers.
parallel evolution Appearance of similar characters in lineages that have separated recently (e.g., long hind legs in hopping rodents from the North American and African deserts). *See also* convergent evolution.
paraphyletic Referring to a taxon that includes the common ancestor and some but not all of its descendants. *See also* monophyletic, polyphyletic.
parasympathetic nervous system The division of the autonomic nervous system that maintains normal body functions, such as digestion.
parsimonious In evolutionary biology, the hypothesis that requires the fewest changes from ancestral to derived character states.
parthenogenesis Reproduction by females without fertilization by males.
parturition Giving birth.
pectoralis major The large breast muscle that powers the downstroke of the wings in a bird.
pelage The hairy covering of a mammal.
pelagic Living in the open ocean.
pelvic patch A vascularized area in the pelvic region of anurans that is responsible for uptake of water.
pericardial cavity The portion of the coelom that surrounds the heart.
pericardium Thin sheets of lateral-plate mesoderm that line the pericardial cavity.
peritoneum Thin sheets of lateral-plate mesoderm that line the pleuroperitoneal cavity.
Phanerozoic Period of time (Eon) since the start of the Cambrian.
pharyngeal arches (also known as visceral skeleton) The gill supports between the pharyngeal gill slits.
pharyngeal slits Openings in the pharynx that were originally used to filter food particles from the water.
pharyngotremy Condition in which the pharyngeal walls are perforated by slit-like openings; found in chordates and hemichordates.
pharynx The throat region.
pheromone A chemical signal released by one individual that affects the behavior of other individuals of the species.
photophore Light-emitting organ.
phylogenetic Pertaining to the development of an evolutionary lineage. *See also* ontogenetic.
phylogenetic systematics (also known as cladistics). A classification system that is based on the branching sequences of evolution.
phylogeny The evolutionary development of a group. *See also* ontogeny.
physoclistic Lacking a connection from the gut to the swim bladder in adults (of fishes).
physostomous Having a connection between the swim bladder and gut in adults (of fishes).
piloerection Contraction of muscles attached to hair follicles resulting in the erection of the hair shafts.
pinna The external ear of mammals.
piscivorous Eating fish.
pitch Tilt up or down parallel to the long axis of the body. *See also* roll, yaw
placenta Extraembryonic tissue that obtains nutrients from the endometrium of the uterus and secretes hormones to signal the state of pregnancy to the mother. *See also* choriovitelline placenta and chorioallantoic placenta.
placentrophic matrotrophy Embryonic development nourished by materials transferred from the maternal circulation via a placenta.
placoid scale Primitive type of scale found in elasmobranchs and homologous with vertebrate teeth.
plantigrade Standing with the foot flat on the ground, as in humans. *See also* digitigrade, unguligrade.
plastron Ventral shell, as of a turtle.
plate tectonics Theory of Earth history in which the lithosphere is continually being generated from the underlying core at specific areas and reabsorbed into the core at others, resulting in a series of conveyor-like plates that carry the continents across the face of the Earth.
plesiomorphic Pertaining to the ancestral character from which an apomorphy is derived.
plesiomorphy An ancestral character (i.e., one that has not changed from its ancestral condition). *See also* symplesiomophy, synapomorphy.
pleurodires Turtles that bend the neck in a horizontal plane to retract the head into the shell.
pleurodont Teeth fused to the inner surface of the jaw bones. *See also* acrodont, thecodont.
pleuroperitoneal cavity The portion of the coelom that surrounds the viscera.

pleurotremate Having the main gill openings on sides of the body anterior to the pectoral fins, as in sharks. *See also* hypotremate.

polarity The direction of evolutionary change in a character.

polyandry A mating system in which a female mates with more than one male.

polygamy A mating system in which an individual has more than one mate in a breeding season. *See also* monogamy.

polygyny A mating system in which a male mates with more than one female.

polymorphism Simultaneous occurrence of two or more distinct phenotypes in a population.

polyphyletic Referring to a taxon that does not contain the most recent common ancestor of all the subordinate taxa of the taxon (i.e., not a true taxonomic unit but an assemblage of similar taxa such as "marine mammals"). *See also* monophyletic, paraphyletic.

portal system Portion of the venous system specialized for the transport of substances from the site of production to the site of action. A portal system begins and ends in capillary beds.

postzygapophysis Articulating surface on the posterior face of a vertebral neural arch. *See also* prezygapophysis.

precocial Well developed and capable of locomotion soon after birth or hatching (e.g., as in chickens and cows). *See also* altricial.

prezygapophysis Articulating surface on the anterior face of a vertebral neural arch. *See also* postzygapophysis.

progenesis Accelerated development of reproductive organs relative to somatic tissue, leading to paedomorphosis.

prolonged breeding A long breeding season.

proprioception The neural mechanism that senses the positions of the limbs in space. A derived character of tetrapods.

proteroglyphs Venomous snakes with permanently erect fangs at the front of the jaw (i.e., cobras and their relatives).

Proterozoic Later part of the Precambrian, from about 1.5 billion years ago until the beginning of the Cambrian 54 million years ago. *See also* Phanerozoic.

protostomy Condition in which the embryonic blastopore forms the mouth of the adult animal. *See also* deuterostomy.

prototherians The monotreme mammals.

protraction Movement away from the center of the body usually in a forward direction. *See also* retraction.

protrusible Capable of being moved away (protruded) from the body.

proventriculus The glandular stomach of birds.

proximal Close to the body. *See also* distal.

proximal convoluted tubule Portion of a kidney nephron responsible for changing the concentration of the ultrafiltrate by actively transporting salt.

pseudovaginal canal A midline structure in marsupials through which the young is born.

pterylae Tracts of follicles from which feathers grow.

pygostyle The fused caudal vertebrae of a bird that support the tail feathers.

rachis The central structure of a feather, from which barbs extend.

ram ventilation A respiratory current across the gills; created by swimming with the mouth open.

rectrices (singular *rectrix*) Tail feathers.

refugium Isolated area of habitat fragmented from a formerly more extensive biome.

regional heterothermy Maintaining different temperatures in different parts of the body.

remiges (singular *remex*) Wing feathers.

resource dispersion hypothesis The proposal that the size of an animal's home range will be determined by its needs for resources, such as food, and by the spatial distribution of resources in the environment.

rete mirabile "Marvelous net," a complex mass of intertwined capillaries specialized for exchange of heat and/or dissolved substances between countercurrent flows.

retraction Movement toward the center of the body or in a backward direction. *See also* protraction.

reversal Return to an ancestral feature (e.g., the streamlined body form of whales and porpoises).

ricochet A bipedal hopping gait, as in kangaroos and many rodents.

rod Photoreceptor cell in the vertebrate retina specialized to function effectively under conditions of dim light.

roll Rotate around the long axis of the body. *See also* pitch, yaw.

rostrum Snout; especially an extension anterior to the mouth.

ruminant An herbivorous mammal with a specialized stomach in which microorganisms ferment plant material.

scapulocoracoid cartilage In elasmobranchs and certain primitive gnathostomes, the single solid element of the pectoral girdle.

scutes Scales, especially broad or inflexible ones.

sebaceous gland A gland in mammal skin that secretes oily or waxy materials.

sebum An oily secretion produced by sebaceous glands.

secondary lamellae Projections from the gill filaments where gas exchange occurs.

selenodont Molar teeth with crescentic ridges or lophs rather than cusps that run in a predominantly anterior to posterior direction across the tooth. *See also* bunodont, lophodont.

semelpary Reproducing only once during the lifetime of a female. *See also* iteropary.

serial Repeated, as in the body segments of vertebrates.

shared derived characters Derived characters shared by two or more taxa. *See also* synapomorphy.

shivering Generation of heat by asynchronous contraction of muscle fibers.

sinus Open space in a duct or tubular system.

sinus venosus The posteriormost chamber of the heart of nonamniotes, and some reptiles, that receives blood from the systemic veins.

sister group Group of organisms most closely related to the study taxa, excluding their direct descendants.

sivapithecids Later Cenozoic Eurasian hominoids related to the extant orangutan.

sociality Living in structured groups.

solenoglyphs Venomous snakes with long fangs in the front of the jaw that are rotated when the mouth is open; vipers.

solute A substance dissolved in a liquid.

somatic nervous system The part of the peripheral nervous system that innervates structures derived from the somatic mesoderm controlling voluntary movements of skeletal muscles and returning sensations from the periphery.

somite Member of a series of paired segments of the embryonic dorsal mesoderm of vertebrates.

spawning The process by which fishes deposit and fertilize eggs.

species In biological time, groups of organisms that are reproductively separated from other groups. In evolutionary time, a lineage that follows its own evolutionary trajectory.

specific dynamic action Increased heat production associated with digesting food.

speciose Referring to a taxon that contains a large number of species.

spermatophore A packet of sperm transferred from male to female during mating by most species of salamanders.

splanchnocranium The visceral or pharyngeal skeleton associated with the gills.

spleen An organ in which blood cells are produced, stored, and broken down.

squamation Scaly covering of the body.

standard metabolic rate The rate of metabolism that sustains vital functions (respiration, blood flow, etc.) in an animal at rest.

stapes Called the *columella* in non-mammalian tetrapods. The single auditory ossicles of the middle ear of tetrapods other than mammals, part of ossicular chain of mammals. Homologous to the hyomandibula of fishes.

stenohaline Capable of living only within a narrow range of salinity of surrounding water; not capable of surviving a great change in salinity. *See also* euryhaline.

stenophagous Eating a narrow range of food items; a food specialist. *See also* euryphagous.

stenothermal Capable of living or being active in only a narrow range of temperatures. *See also* eurythermal.

stratigraphy Classification, correlation, and interpretation of stratified rocks.

stratum (plural *strata*) A layer of material.

supercooling Lowering the temperature of a fluid below its freezing point without initiating crystallization.

supracoracoideus The muscle that produces the powered upstroke of the wings of a bird.

surface-to-volume ratio The ratio of body surface area to body volume, often expressed as $cm^2 \bullet cm^{-3}$.

swim bladder (also known as gas bladder) A buoyancy structure of bony fishes. Usually filled with gas, but in coelacanths it is filled with fat.

symbiont An organism that lives with (usually inside or attached to) another organism, to their mutual benefit.

sympathetic nervous system The division of the autonomic nervous system that produces largely involuntary responses, which prepare the body for stressful or highly energetic situations.

sympatry Occurrence of two or more species in the same area.

symphysis A joint between bones formed by a pad or disk of fibrocartilage that allows a small degree of movement.

symplesiomorphy Character shared by a group of organisms that is found in their common ancestor (i.e., a primitive character). *See also* plesiomorphy.

synapomorphy Derived characters (apomorphies) shared by two or more taxa. *See also* plesiomorphy.

synapsid A skull with a single temporal fenestra or an animal with a synapsid skull. *See also* anapsid, diapsid.

synsacrum Fused vertebrae and ribs of birds that articulate with the pelvis.

syrinx The vocal organ of birds, lying at the base of the trachea.

tadpole The larval form of anurans.

talonid Basinlike heel on a lower molar tooth, found in therian mammals.

tapetum lucidum A reflective layer behind the retina that increases sensitivity in low light by directing light back through the retina.

tarsometatarsus Bone formed by fusion of the distal tarsal elements with the metatarsals in birds and some dinosaurs. *See also* tibiotarsus.

taxon Any scientifically recognized group of organisms united by common ancestry.

temperature-dependent sex determination Situation in which the sex of an individual is determined by the temperature it experienced during embryonic development. Universal among crocodilians, widespread among turtles, occasional among squamates.

temporal fossa An opening in bone of the temporal region of the skull that allows passage of jaw muscles from the skull to the lower jaw.

tentacle A sensory organ of caecilians that allows chemical substances to be transported from the surroundings to the vomeronasal organ.

territory An area that is defended against incursion by other individuals of the species. *See also* home range.

testis (plural *testes*) The male gonad.

tetrapods Terrestrial vertebrates descended from a four-legged ancestor.

thecodont Teeth set in sockets in the jaw bones. Also refers to a paraphyletic assemblage of basal, extinct archosaurian reptiles. *See also* acrodont, pleurodont.

therians Marsupial and eutherian mammals.

thermoneutral zone The range of ambient temperatures within which an endotherm can maintain a stable body temperature by changing the rate of heat loss to the environment.

thermophilic Favoring high temperatures.

thermoregulation Control of body temperature.

tibiotarsus Bone formed by fusion of the tibia and proximal tarsal elements in birds and some dinosaurs. *See also* tarsometatarsus.

tidal ventilation In-and-out flow of respiratory fluid, as in the lungs of a tetrapod.

torpor A period of inactivity accompanied by a reduction in the regulated body temperature.

tribosphenic molars Tooth form unique to therian mammals.

troglodyte Organism that lives in caves.

trophic Pertaining to feeding and nutrition.

trophoblast Embryonic tissue of mammals specialized for implanting the embryo on the wall of the uterus, obtaining nutrients from the mother, and secreting hormones to signal the state of pregnancy to the mother.
trot A gait of tetrapods in which diagonal pairs of limbs are moved together with a period of suspension between each pair of limb movements when all four feet are off the ground.
turbinates Scroll-like bones in the nasal passages covered by moist tissues that warm and humidify inspired air.
tympanic membrane or tympanum The eardrum.

ultrafiltrate A fluid produced in the glomerulus of a nephron; composed of blood with the cells and large molecules removed by filtration.
ultrasound Sound frequencies above the range of human hearing, approximately 20 kilohertz.
unguligrade Standing with only the tips of the toes on the ground, as in horses. *See also* digitigrade, plantigrade.
upper critical temperature The point at which an endotherm must initiate evaporative cooling to maintain a stable body temperature.
urea cycle The enzymatic pathway by which urea is synthesized from ammonia.
ureotelic Excreting nitrogenous wastes primarily as urea.
urethra The duct in amniotes that carries urine from the bladder to the outside. In male therian mammals, part of the urethra also carries sperm.
uricotelic Excreting nitrogenous wastes primarily as uric acid and its salts.
urogenital Pertaining to the organs, ducts, and structures of the excretory and reproductive systems.
urogenital sinus Combined opening of the urethra and vagina in most female mammals. (Primates and some rodents have separate openings.)
urostyle A solid rod formed by fused posterior vertebrae; found in anurans.

vacuoles Membrane-bound spaces within cells containing secretions, storage products, etc.
vane The surface on either side of the central structure of a feather (the rachis).
vas deferentia (singular *vas deferens*) The male reproductive tract of mammals.
vasa recta The blood vessels surrounding the loop of Henle.
vascular Relating to blood and blood vessels.
vascular plexus Intertwined blood vessels in the skin that are the basis for countercurrent blood flow.
vasodilation Expansion of blood vessels to increase blood flow to a region.
ventricle A chamber. The ventricle of the heart is the portion that applies force to eject blood from the heart.
Vertebrata Animals that have vertebrae.
vibrissae The sensory whiskers of mammals.
vicariance biogeography Animals and plants being carried passively on moving landmasses. *See also* dispersal.
viscera Internal organs suspended within the coelom.
visceral arches Gills and jaws.
visceral nervous system The part of the peripheral system that innervates portions of the body derived from the lateral plate mesoderm. Includes the autonomic nervous system and sensory nerves that relay information from the viscera and blood vessels. May also include the special branchial motor system of the cranial nerves, although this is now in dispute.
visceral skeleton Skeleton primitively associated with the pharyngeal arches, uniquely derived from the neural-crest cells and forming in mesoderm immediately adjacent to the endoderm lining the gut.
viviparity Giving birth to young as opposed to laying eggs.
vomeronasal organ (also known as Jacobson's organ) An olfactory organ in the roof of the mouth of tetrapods.
vulnerable A species is considered vulnerable when it is facing a high risk of extinction in the wild in the medium-term future.

Weberian apparatus A chain of small bones that conducts vibrations from the swim bladder to the inner ear of some bony fishes.
wet adhesion The process by which arboreal species of frogs stick to smooth surfaces, such as leaves.

yaw Swing from side to side relative to the long axis of the body. *See also* pitch, roll.

zone of chemical thermogenesis The range of temperatures within which an endotherm can maintain a stable body temperature via metabolic heat production.
zone of tolerance The range of ambient temperatures over which an endotherm can maintain a stable body temperature.
Zugdisposition Preparation for migration by accumulating fat.
Zugstimmung The condition in which a bird makes migratory flights.
Zugunruhe Restlessness of caged birds that are prevented from migrating.
zygapophysis Articular process of the neural arch of a vertebra. *See also* postzygapophysis and prezygapophysis.
zygodactylous Type of foot in which the toes are arranged in two opposable groups.
zygomatic arch A temporal bar that is bowed outward to accommodate a large masseter muscle in mammals.

Credits

CHAPTER 2 **Fig. 2–2** Illustration by Marianne Collins from Stephen Jay Gould, *Wonderful Life: The Burgess Shale and the Nature of History* (New York: Norton, 1989). Reproduced with the permission of Marianne Collins and W. W. Norton & Company, Inc., copyright © 1989 by Stephen Jay Gould. **Fig. 2–5** After E. S. Goodrich, *Studies on the Structure and Development of Vertebrates* (London, U.K.: Macmillan, 1930). **Fig. 2–6** Part (c) based on B. Stahl, *Vertebrate History* (New York: McGraw-Hill). Part (d) after K. V. Kardong, *Vertebrates—Comparative Anatomy, Function, Evolution,* 2nd ed. (Dubuque, IA: Wm. C. Brown Publishers, 1998). Reprinted by permission of the McGraw-Hill Companies. **Fig. 2–7** After J. G. Maisey, 1996, *Discovering Fossil Fishes* © 1996 Nevraumont Publishing Co., Inc. Fig. 7, p. 39. Reprinted by permission of Navraumont Publishing Co., Inc. **Fig. 2–8** After K. V. Kardong, *Vertebrates—Comparative Anatomy, Function, Evolution,* 2nd ed. (Dubuque, IA: Wm. C. Brown, 1998). Reprinted by permission of the McGraw-Hill Companies. **Fig. 2–9** After W. F. Walker Jr., and K. F. Liem, *Functional Anatomy of Vertebrates,* 2nd ed. (Fort Worth, TX: Saunders, 1994). Copyright © 1994 by Saunders College Publishing, reproduced by permission of the publisher. **Fig. 2–10** Part (a) after W. F. Walker Jr., and K. F. Liem 1994; Part (b) after L. B. Radinsky in M. H. Wake, *Hyman's Comparative Vertebrate Anatomy* (Chicago: University of Chicago Press, 1979). **Fig. 2–11** Modified from L. B. Radinsky, *The Evolution of Vertebrate Design* (Chicago, IL: University of Chicago Press, 1987). **Fig. 2–12** Modified from A. G. Kluge et al., *Chordate Structure and Function,* 2d ed. (New York: Macmillan, 1977), and Q. Bone and N. B. Marshall, *Biology of Fishes* (Glasgow, U.K.: Blackie, 1982). **Fig. 2–14** Part (b) after R. Lawson in M. H. Wake, *Hyman's Comparative Vertebrate Anatomy* (Chicago, IL: University of Chicago Press, 1979). **Fig. 2–16** After K. V. Kardong, *Vertebrates—Comparative Anatomy, Function, Evolution,* 2nd ed. (Dubuque, IA: Wm. C. Brown Publishers, 1998). Reprinted by permission of the McGraw-Hill Companies. **Fig. 2–17** After K. V. Kardong, *Vertebrates—Comparative Anatomy, Function, Evolution,* 2nd ed. (Dubuque, IA: Wm. C. Brown Publishers, 1998). Reprinted by permission of the McGraw-Hill Companies. **Fig. 2–18** After K. V. Kardong, 1998. Reprinted by permission of the McGraw-Hill Companies.

CHAPTER 3 **Fig. 3–1** Part (a) modified from D.-G. Shu et al., *Nature* 402 (1999): 42–46. Part (b) modified from D. Elliot, *Science* 237 (1987): 190–192. **Fig. 3–2** Part (a) modified from M. A. Purnell, *Lethaia* 28 (1995): 187–188. Part (b) modified from M. A. Purnell, *Lethaia* 27 (1994): 129–138. **Fig. 3–3** Based primarily on P. C. J. Donoghue et al., *Biological Reviews* 75 (2000): 191–251, with information from J. G. Maisey, *Cladistics* (1986): 201–256, and J. Mallatt and J. Sullivan, *Molecular Biology and Evolution* 15 (1998): 1706–1718. **Fig. 3–5** Modified from D. Jensen, *Scientific American* 214 (1996): [2]: 82–90. **Fig. 3–7** Modified after J. A. Moy–Thomas and R. S. Miles, *Paleozoic Fishes* (Philadelphia, PA: Saunders College Publishing, 1971). **Fig. 3–8** Modified primarily after J. A. Moy–Thomas and R. S. Miles, *Paleozoic Fishes* (Philadelphia, PA: Saunders College Publishing, 1971) and P. Janvier in J. Hanken and B. K. Hall, *The Skull,* vol. 2 (Chicago: University of Chicago Press, 1993). **Fig. 3–12** Drawing modified from one provided courtesy of Jon Mallatt. **Fig. 3–13** Based on G. V. Lauder and K. F. Liem, *Bulletin of the Museum of Comparative Zoology* 150: 95–197; B. G. Gardiner, *Bulletin of the British Museum (Natural History) Geology* 37 (1984): 173–427; J. G. Maisey, *Cladistics* 2(1986): 201–256; R. L. Carroll, *Vertebrate Paleontology and Evolution* (New York: W. H. Freeman, 1988); B. G. Gardiner and B. Schaeffer, *Zoological Journal of the Linnean Society* 97 (1989): 135–187; P. E. Olsen and A. R. McCune, *Journal of Vertebrate Paleontology* 11 (1991): 269–292; J. S. Nelson, *Fishes of the World* (New York: Wiley, 1994); and P. Janvier, *Early Vertebrates* (Oxford, U.K.: Clarendon, 1996). **Fig. 3–15** Modified after J. A. Moy–Thomas and R. S. Miles, *Paleozoic Fishes* (Philadelphia, PA: Saunders College Publishing, 1971). **Fig. 3–16** Modified after J. A. Moy–Thomas and R. S. Miles, *Paleozoic Fishes* (Philadelphia, PA: Saunders College Publishing, 1971). **Table 3.2** J. G. Maisey, 1986, *Cladistics* 2 (1986): 201–256, and P. Janvier, *Early Vertebrates* (Oxford, U.K.: Clarendon Press, 1996).

CHAPTER 4 **Fig. 4–1** Modified from G. M. Hughes, *Comparative Physiology of Vertebrate Respiration* (Cambridge: Harvard University Press, 1963). **Fig. 4–2** Modified after G. M. Hughes, *Comparative Physiology of Vertebrate Respiration* (Cambridge: Harvard University Press, 1963); M. Hildebrand, *Analysis of Vertebrate Structure,* 3d ed. (New York: Wiley); and P. B. Moyle, *Fish: An Enthusiast's Guide.* (Berkeley, CA: University of California Press, 1993). **Fig. 4–4** Modified after A. Flock in *Lateral Line Detectors,* P. Cahn (ed.) (Bloomington: Indiana University Press, 1967); and R. F. Hueter and P. W. Gilbert in *Underwater Naturalist,* S. H. Gruber (ed.), special double issue, 19(4) and 20(1), 1991: 48–55. **Fig. 4–5** Modified after E. Schwartz, in *Handbook of Sensory Physiology,* vol. 3, part 3, A. Fessard (ed.) (New York: Springer, 1974). **Fig. 4–6** Modified in part from J. Bastian, *Physics Today* 47(2) 1994: 30–37. **Fig. 4–8** Modified from A. J. Kalmijn in *Handbook of Sensory Physiology,* vol. 3, part 3, A. Fessard (ed.) (New York: Springer, 1974). **Fig. 4–10** Produced with the assistance of Carl Hopkins. **Fig. 4–14** Modified from B. Schmidt–Nielsen in *Nitrogen Metabolism and the Environment,* J. W. Campbell and L. Goldstein (eds.) (London, Academic Press, 1972), 79–103. **Fig. 4–15** Parts (b–c) from data in F. E. J. Fry, Animals in aquatic environments: fishes, pp. 715–728 in *Handbook of Physiology, Adaptation to the Environment,* D. B. Dill, E. F. Adolph, and C. G. Wilber (eds.). (Washington, DC: American Physiological Society, 1964). Part (d) from data in F. E. J. Fry and P. W. Hochachka, Fish, pp. 79–134 in *Comparative Physiology of Thermoregulation,* vol 1, G. C. Whitow (ed.) (New York: Academic Press, 1970). **Fig. 4–16** Modified from F. G. Carey and J. M. Teal, *Proceedings of the National Academy of Sciences U.S.A.* 56 (1996):1464–1469.

CHAPTER 5 **Fig. 5–1** Modified after J. A. Moy–Thomas and R. S. Miles, *Paleozoic Fishes* (Philadelphia, PA: Saunders College Publishing, 1971); and R. Lund, *Journal of Vertebrate Paleontology* 5 (1985): 1–19, and *Journal of Vertebrate Paleontology* 6 (1986): 12–19. **Fig. 5–3** Part (a) modified after J. A. Moy–Thomas and R. S. Miles, *Paleozoic Fishes* (Philadelphia, PA: Saunders College Publishing, 1971.) **Fig. 5–5** Part (a) modified after E. S. Goodrich, *Studies on the Structure and Development of Vertebrates* (London: Macmillian, 1930). Parts (b–c) modified after S. A. Moss, *Sharks: An Introduction for the Amateur Naturalist.* (Englewood Cliffs, NJ.: Prentice Hall, 1984). **Fig. 5–7** Modified from A. P. Klimley, *American Scientist* 87 (1999): 488–491. **Fig. 5–8** Modified in part from J. S. Nelson, *Fishes of the World* (New York, NY: Wiley, 1994); and P. B. Moyle, *Fish: An Enthusiasts Guide* (Berkeley: University of California Press, 1993). **Fig. 5–9** Parts (a and d) modified in part after H. B. Bigelow and W. C. Schroeder, *Fishes of the Western North Atlantic,* part 2 (New Haven: Yale University, 1953). Part (b) modified after J. A. Moy-Thomas and R. S. Miles, *Paleozoic Fishes* (Philadelphia, PA: Saunders College Publishing, 1971). Part (c) modified after P. H. Greenwood, R. S. Miles, and C. Patterson (eds.), *Interrelationships of Fishes* (New York: Academic, 1973).

CHAPTER 6 **Fig. 6–1** Modified from J. A. Moy-Thomas and R. S. Miles, *Paleozoic Fishes* (Philadelphia, PA: Saunders College Publishing, 1971). **Fig. 6–5** Part (a) modified from K. F. Liem in A. G. Kluge et al., *Chordate Structure and Function,* 2nd ed. (New York: Macmillan, 1977). Part (c) after G. B. Lauder, *Journal of Morphology,* 163 (1980): 283–317. **Fig. 6–13** Modified after C. C. Lindsey, in *Fish Physiology,* vol. 7, *Locomotion,* W. S. Hoar and D. J. Randall (eds.) (New York: Academic, 1978). **Fig. 6–14** Modified after H. Hertel, *Structure, Form, Movement* (Mainz, Ger.: Krausskopf, 1966). **Fig. 6–15** Photograph courtesy of Edward B. Brothers. **Fig. 6–16** From M. L. Warren Jr. and B. M. Burr, *Fisheries* 19 (1994)[1]: 6–10. **Fig. 6–17** Part (a) modified from H. Friedrich, *Marine Biology* (Seattle: University of Washington Press, 1973). Part (b) modified from N. B. Marshall, *Explorations of the Life of Fishes* (Cambridge: Harvard University Press, 1971). **Fig. 6–18** Modified from E. M. Kampa and B. P. Boden, *Deep-Sea Research* 4 (1957): 73–92. **Fig. 6–19** J. B. Heiser (Ithaca, NY: Cornell University).

CHAPTER 7 **Fig. 7–2** Modified from A. Hallam, *An Outline of Phanerozoic Biography* (Oxford: Oxford University Press, 1994). **Fig. 7–5** Modified from M. J. Benton, *Vertebrate Palaeontology,* 2nd ed. (London: Chapman and Hall, 1997).

CHAPTER 8 **Fig. 8–2** After W. F. Walker, Jr. and K. F. Liem, *Functional Anatomy of the Vertebrates—an Evolutionary Perspective,* 2nd ed. (Philadelphia, PA: Saunders College Publishing, 1994). **Fig. 8–3** From A. S. Romer, *Vertebrate Paleontology,* 3d ed. (Chicago: University of Chicago Press, 1966.) **Fig. 8–4** Modified from K. V. Kardong, *Vertebrates—Comparative Anatomy, Function, Evolution,* 2nd ed. (Dubuque, IA: Wm. C. Brown, 1998). **Fig. 8–7** From S. Vogel, *Life's Devices: The Physical World of Plants and Animals.* Copyright © 1989 by Princeton University Press. Reprinted by permission of Princeton University Press. **Fig. 8–8** From T. A. McMahon and J. T. Bonner, *On Size and Life* (New York: Scientific American Books, 1983). **Fig. 8–10** From R. Lawson in M. H. Wake, *Hyman's Comparative Vertebrate Anatomy* (Chicago: University of Chicago Press, 1979). **Fig. 8–11** M. Hildebrand, *Analysis of Vertebrate Structure,* 4th ed. (New York: Wiley, 1995). **Fig. 8–12** After K. V. Kardong, *Vertebrates—Comparative Anatomy, Function, Evolution,* 2nd ed. (Dubuque, IA: Wm. C. Brown Publishers, 1998). **Fig. 8–13** Modified from J. A. Gauthier, A. G. Kluge, and T. Rowe in *The Phylogeny and Classification of Tetrapods,* vol. 1: *Amphibians, Reptiles, Birds,* M. J. Benton (ed.) (Oxford, U.K.: Clarendon Press, 1988); R. L. Carroll, *Vertebrate Paleontology and Evolution* (New York: Freeman, 1998); M. J. Benton, *Vertebrate Paleontology,* 2nd ed. (London: HarperCollins Academic, 1997); From P. E. Ahlberg and A. R. Milner, *Nature* 373 (1994): 507–514; J. A. Gauthier in *Major Features of Vertebrate Evolution,* D. R. Prothero and R. M. Schoch (eds.) (Knoxville, TN: Paleontological Society and University of Tennessee Press, 1994); and R. R. Reisz, *Trends in Ecology and Evolution* (1997): 218–222. **Fig. 8–15** From P. E. Ahlberg and A. R. Milner, *Nature* 373 (1994): 507–514. **Fig. 8–16** Photographs from M. I. Coates and J. A. Clack, *Nature* 347 (1990): 66–69. Courtesy of M. I. Coates and J. A. Clack. **Fig. 8–17** From P. E. Ahlberg and A. R. Milner, *Nature* 373 (1994): 507–514. **Fig. 8–18** From J. L. Edwards, *American Zoologist* 29 (1989): 235–254. Courtesy of J. L. Edwards. **Fig. 8–19** From M. I. Coates and J. A. Clack, *Bulletin de Musée national Histoire naturelle,* Paris, 4e series 17 (1995): 373–388. **Fig. 8–20** Parts (a–b) from D. Palmer, *Atlas of the Prehistoric World* (New York: Discovery Books, 1999) pp. 78, 79. Art by Marshall Editions Ltd. Part (c) from A. L. Panchen and T. R. Smithson, *Transactions of the Royal Society of Edinburgh: Earth Sciences* 81 (1990): 31–44. Parts (d–e) from C. Tudge, *The*

Variety of Life (Oxford: Oxford University Press, 2000). Part (f) from B. Cox, R. J. G. Savage, B. Gardiner, and C. Harrison, *The Simon and Schuster Encyclopedia of Dinosaurs and Prehistoric Creatures* (New York: Simon and Schuster, 1999), art by Marshall Editions Ltd. UK. **Fig. 8–21** Part (a) from A. R. Milner in *The Terrestrial Environment and the Origin of Land Vertebrates*, A. L. Panchen, (ed.) (London: Academic, 1980). Part (b) from R. L. Carroll, *Handbuch der Palaeoherpetologie*, part 5B, pp. 1–19 (Stuttgart: Gustav Fischer, 1972). Parts (c–d) from C. Tudge, *The Variety of Life* (Oxford: Oxford University Press, 2000). Part (e) from D. Palmer, *Atlas of the Prehistoric World* (New York: Discovery Books, 1999). **Fig. 8–22** Part (a) from R. L. Carroll and P. Gaskill, *Memoirs of the American Philosophical Society* 126 (1978): 1–211. Part (b) from C. Tudge, *The Variety of Life* (Oxford: Oxford University Press, 2000). Part (c) from A. R. Milner, and A. C. Milner, both in *The Terrestrial Environment and the Origin of Land Vertebrates*, A. L. Panchen (ed.) (London: Academic Press, 1980). **Fig. 8–24** From T. W. Torrey, *Morphogenesis of the Vertebrates* (New York: Wiley, 1962). **Fig. 8–27** Modified from T. S. Kemp, *Mammal-like Reptiles and the Origin of Mammals.* (London: Academic, 1982).

CHAPTER 9 **Table 9.2** Phylogenetic relationships and numbers of species are based on F. H. Pough et al., *Herpetology*, 2nd ed. (Upper Saddle River, NJ: Prentice Hall, 2001). **Fig. 9–2** Photograph courtesy of Professor Dr. Roth, University of Bremen. **Fig. 9–4** From R. Jaeger, *The American Naturalist* 117 (1981): 968. Courtesy of Robert Jaeger. © 1981 The University of Chicago Press. All rights reserved. **Table 9.3** Phylogenetic relationships and numbers of species are based on F. H. Pough, et al., *Herpetology*, 2nd ed. (Upper Saddle River, NJ: Prentice Hall, 2001). **Fig. 9–6** From F. H. Pough, in *Environmental Physiology of the Amphibia*, M. E. Feder and W. W. Burggren (eds.) (Chicago: University of Chicago Press, 1992), © 1992 The University of Chicago Press. All rights reserved. **Fig. 9–8** Photographs courtesy of Sharon B. Emerson, University of Utah. (a) Appeared in *Biological Journal of the Linnean Society* 13 (1980). **Fig. 9–10** Part (b) modified from H. Gadow, *Amphibia and Reptiles* (London: Macmillan, 1909). Parts (c–d) modified from E. H. Taylor, *The Caecilians of the World* (Lawrence: University of Kansas Press, 1968). **Table 9.4** Phylogenetic relationships and numbers of species are based on F. H. Pough et al., *Herpetology*, 2nd ed. (Upper Saddle River, N.J.: Prentice Hall, 2001). **Fig. 9–11** Modified from G. K. Noble, *The Biology of the Amphibia*, (New York: McGraw-Hill, 1931). **Fig. 9–12** Redrawn from T. R. Halliday, *Advances in the Study of Behavior* 19 (1990): 139, © 1990 Academic Press, Inc. **Fig. 9–13** Redrawn from T. R. Halliday and B. Arano, *Trends in Ecology and Evolution* 6 (1991): 114, © 1991 Elsevier Science Publishers Ltd. **Fig. 9–14** Photographs courtesy of Theodore L. Taigen, University of Connecticut. **Fig. 9–15** Modified from Ryan, 1985. M. J. Ryan, *The Túngara Frog: A Study in Sexual Selection* (Chicago: University of Chicago Press, 1985). **Fig. 9–16** Photograph courtesy of Theodore L. Taigen, University of Connecticut. **Fig. 9–17** From T. L. Taigen and K. D. Wells, *Journal of Comparative Physiology* B155 (1985): 163–170. **Fig. 9–18** Part (c) from G. K. Noble, *The Biology of the Amphibia* (New York: McGraw-Hill, 1931). Part (e) from W. E. Duellman, *The Hylid Frogs of Middle America*, monograph of the Museum of Natural History (Lawrence: University of Kansas, 1970). Part (f) from M. Lamotte and J. Lescure, *La Terre et la Vie* 31 (1977): 225–311. **Fig. 9–20** Part (e) courtesy of Richard J. Wassersug. **Table 9.5** Based on B. A. White and C. S. Nicoll, in *Metamorphosis, a Problem in Developmental Biology*, L. I. Gilbert and E. Freeden (eds.) (New York: Plenum, 1981). **Fig. 9–21** Data from R. J. Wassersug and D. G. Sperry, *Ecology* 58 (1977): 830–839. **Fig. 9–24** Photographs by F. Harvey Pough, Arizona State University West. **Fig. 9–25** Modified from G. K. Noble, *The Biology of the Amphibia* (New York: McGraw-Hill, 1931). **Fig. 9–26** Part (a) modified from H. Gadow, *Amphibia and Reptiles* (London: Macmillan, 1909). Part (b) from a photograph by E. D. Brodie, Jr. **Table 9.6** Based on E. D. Brodie, Jr. and E. D. Brodie, III, *Science* 208 (1980): 181–182. **Fig. 9–27** From F. H. Pough, *NAHO* 11[1] (1978):6–9. Courtesy of NAHO. Published by the New York State Museum. **Fig. 9–28** Data from K. R. Lips, *Conservation Biology* 12 (1998): 1–13; K. R. Lips, *Conservation Biology* 13 (1998): 117–125; and J. P. Collins, Arizona State University Main, personal communication.

CHAPTER 10 **Table 10.1** The phylogenetic relationships and numbers of species are based on F. H. Pough et al., *Herpetology*, 2nd ed. (Upper Saddle River, N.J.: Prentice Hall, 2001). **Fig. 10–3** From E. S. Gaffney, *Bulletin of the American Museum of Natural History* 15 (1975): 387–436. **Fig. 10–4** From R. Zangerl, in *Biology of the Reptilia*, Vol. 1, C. Gans, A. d'A. Bellairs, and T. S. Parsons (eds.) (London: Academic, 1969). **Fig. 10–5** Modified from N. Heisler et al., *Journal of Experimental Biology* 105 (1983): 15–32. **Fig. 10–6** Modified from C. Gans and G. M. Hughes, *Journal of Experimental Biology* 47 (1967): 1–20. **Fig. 10–7** Modified from M. S. Gordon et al., *Animal Physiology: Principles and Adaptations*, 4th ed. (New York: Macmillan, 1982). **Fig. 10–8** Part (a) modified from J. A. Oliver, *The Natural History of North American Amphibians and Reptiles* (Princeton, NJ: Van Nostrand, 1955). Part (b) modified from S. F. Schafer and C. O. Krekorian, *Herpetologica* 39 (1983): 448–456. **Fig. 10–9** (Parts a, d) modified from J. J. Bull, *Quarterly Review of Biology* 55 (1980):13–21. Parts (b–c) modified from J. J. Bull and R. C. Vogt, *Science* 206 (1979): 1186–1188. **Fig. 10–11** P. Luschi et al., *Proceedings of the Royal Society* (London). Series B. 265 (1998): 2279–2284. **Fig. 10–12** Photographs courtesy of Larry Herbst, University of Florida. **Fig. 10–13** Photograph courtesy of Elliott R. Jacobson, University of Florida.

CHAPTER 11 **Table 11.1** Phylogenetic relationships and numbers of species are based on F. H. Pough et al., *Herpetology*, 2nd ed. (Upper Saddle River, NJ: Prentice Hall, 2001). **Table 11.2** Numbers of species are based on

F.H. Pough et al., *Herpetology,* 2nd ed. (Upper Saddle River, NJ: Prentice Hall, 2001). **Fig. 11–2** Parts (b–c, and e–f) modified from C. Gans, *Biomechanics,* (Philadelphia, PA: Lippincott, 1974). **Table 11.3** Phylogenetic relationships and numbers of species are based on F. H. Pough et al., *Herpetology,* 2nd ed. (Upper Saddle River, NJ: Prentice Hall, 2001). **Fig. 11–5** Part (a) from R. R. Reisz, *Science* 196 (1977): 1091–1093. Parts (b,c) from A. S. Romer, *Vertebrate Paleontology,* 3rd ed., (Chicago: University of Chicago Press, 1966). Parts (f,g) from C. Gans, *Biomechanics: An Approach to Vertebrate Biology* (Philadelphia, PA: Lippincott, 1974). **Fig. 11–6** From C. Gans, *American Zoologist* 1 (1961): 217–227. **Fig. 11–7** Modified from T. H. Frazetta, *Journal of Morphology* 118 (1966): 217–296. **Table 11.5** Based on data from L. J. Vitt and J. D. Congdon, *American Naturalist* 112 (1978): 595–608; R. B. Huey and E. R. Pianka, *Ecology* 62 (1981): 991–999; W. E. Magnusson et al., *Herpetologica* 41 (1985): 324–332; and R. B. Huey and A. F. Bennett, in *Predator–Prey Relationships,* M. E. Feder and G. V. Lauder (eds.) (Chicago: University of Chicago Press, 1986), 82–98. **Box 11.1** Photographs by Benjamin E. Dial, Chapman University. **Fig. 11–10** Modified from A. A. Echelle et al., *Herpetologica* 27 (1971): 221–288. **Fig. 11–11** From D. Crews, in *Behavior and Neurology of Lizards,* N. Greenberg and P. D. MacLean (eds.) (Rockville, MD: U. S. Department of Health, Education, and Welfare, National Institutes of Mental Health, 1978). **Fig. 11–12** From D. Crews, *Herpetologica* 31 (1975): 37–47. **Fig. 11–13** Photograph courtesy of Barry Sinervo. **Fig. 11–14** Photographs by C. J. Cole and C. M. Bogert, American Museum of Natural History, courtesy of C. J. Cole. **Fig. 11–16** Modified from F. N. White, *Comparative Biochemistry and Physiology* 45A (1973): 503–513. **Fig. 11–17** From R. D. Stevenson et al., *Physiological Zoology* 58 (1985): 46–57. **Fig. 11–18** Based on R. Ruibal, *Evolution* 15 (1961): 98–111. **Fig. 11–19** Part (a) courtesy of Daniel H. Janzen, University of Pennsylvania. Parts (b–c) photographed by Michael Hopiak, courtesy of the Cornell University Herpetology Collection. **Fig. 11–20** From P. E. Hillman, *Ecology* 50 (1969): 476–481. **Fig. 11–21** Part (a) from P. E. Hillman, *Ecology* 50 (1969): 476–481. Part (b) courtesy of Peter E. Hillman.

CHAPTER 12 **Table 12.4** K. A. Nagy, *Journal of Comparative Physiology* 79 (1972): 39–62. **Table 12.5** F. H. Pough, *The American Naturalist* 115 (1980): 92–112. **Fig. 12–1** Photographs by R. Bruce Bury, U.S. Fish & Wildlife Service. **Fig. 12–2** Based on K.A. Nagy and P. A. Medica, *Herpetologica* 42 (1986): 73–92. **Fig. 12–3** Modified from K. A. Nagy, in *Stable Isotopes in Ecological Research,* J. Ehleringer, P. Rundall, and K. Nagy (eds.) (New York: Springer, 1988). **Fig. 12–4** Photograph by F. Harvey Pough, Arizona State University West. **Fig. 12–5** From K. A. Nagy, *Copeia* (1973): 93–102. **Fig. 12–6** From K. A. Nagy, *Journal of Comparative Physiology* 79 (1972): 39–62. **Fig. 12–7** Photograph by David Dennis. **Fig. 12–8** Modified from B. H. Brattstrom, *Herpetologica* 18 (1962): 38–46. **Fig. 12–9** From P. F. Scholander et al., *Journal of Cellular and Comparative Physiology* 49 (1957): 5–24. **Fig. 12–10** Photographs courtesy of J. and K. Storey, Carleton University, Ottawa, Ontario. **Fig. 12–11** From F. H. Pough, *The American Naturalist* 115 (1980): 92–112. Reprinted by permission of the University of Chicago Press. © 1980 by The University of Chicago. **Fig. 12–12** Based on data in F. H. Pough in *Behavioral Energetics: Vertebrate Costs of Survival,* W. P. Aspey and S. I. Lustick (eds.) (Columbus, OH: Ohio State University Press, 1983); 141–188 and R. M. May, *Science* 241 (1988):1441–1449.

CHAPTER 14 **Fig. 14–1** Based on M. J. Benton, *Zoological Journal of the Linnaean Society* 84 (1985): 97–164; J. Gauthier, in *The Origin of Birds and the Evolution of Flight, Memoirs of the California Academy of Sciences,* K. Padian (ed.), no. 8 (1986): 1–55; H.-D. Sues, *Zoological Journal of the Linnaean Society* 90 (1987): 109–131; M. J. Benton (ed.), *The Phylogeny and Classification of the Tetrapods,* Special Vol. 35B, The Systematics Association (Oxford: Oxford University Press, 1988); R. L. Carroll, *Vertebrate Paleontology and Evolution,* (New York: Freeman, 1988); M. J. Benton (ed.), *The Fossil Record* 2 (London: Chapman & Hall, 1993); and M. J. Benton, *Vertebrate Paleontology,* 2nd ed. (London: Chapman and Hall, 1997). **Fig. 14–3** Part (a) from R. R. Reisz, *Science* 196 (1977): 1091–1093. **Fig. 14–4** Part (a) From A. S. Romer, *Vertebrate Paleontology,* 3rd ed. (Chicago: University of Chicago Press, 1966). Part (b) from A. S. Romer, *Osteology of the Reptiles* (Chicago: University of Chicago Press, 1956). **Fig. 14–5** Modified from H. Wermuth and R. Mertens, *Schildkröten, Krocodile, Brückenechsen* (Jena, E. Germany: Gustav Fisher, 1961). **Fig. 14–6** From A. G. Kluge (ed.), *Chordate Structure and Function* (New York: Macmillan, 1977). **Fig. 14–7** Photographs courtesy of Jeffrey W. Lang, University of North Dakota. **Fig. 14–8** Skulls modified from D. Norman, *The Illustrated Encyclopedia of Dinosaurs* (London: Salamander Books, 1985). **Fig. 14–10** Based on J. Gauthier, in *The Origin of Birds and the Evolution of Flight, Memoirs of the California Academy of Sciences,* K. Padian (ed.), no. 8 (1986): 1–55; P. C. Sereno, *National Geographic Research* 2 (1986): 234–256; M. J. Benton (ed.), *The Phylogeny and Classification of the Tetrapods,* Special Vol. 35B, The Systematics Association (Oxford: Oxford University Press, 1988); R. L. Carroll, *Vertebrate Paleontology and Evolution,* (New York: Freeman, 1988); M. J. Benton (ed.), *The Fossil Record* 2 (London: Chapman & Hall, 1993); and M. J. Benton, *Vertebrate Paleontology,* 2nd ed. (London: Chapman and Hall, 1997). **Fig. 14–15** Part (a) illustration by Mick Ellison/DVP, photo by Denis Finnin, courtesy of Dept. of Library Services, American Museum of Natural History. Part (b) Photo, Shackleford, courtesy of Dept. of Library Services, American Museum of Natural History. **Fig. 14–19** Copyright © Peabody Museum of Natural History, Yale University, New Haven, CT. **Fig. 14–20** Redrawn from *New York Times,* illustration by Michael Rothman. Copyright © 1992 by The New York Times Company. **Fig. 14–21** From R. L. Carroll and Z. Dong, *Philosophical Trans-*

actions of the Royal Society, London, B 331 (1991): 131–153. **Fig. 14–22** Calculated from data in R. Estes and P. Berberian, *Breviora*, no. 343 (1970).

CHAPTER 15 **Fig. 15–2** From the Sanford Bird Hall, American Museum of Natural History #325097. Courtesy Department of Library Services. **Fig. 15–3** Modified from D. Norman, *The Illustrated Encyclopedia of Dinosaurs* (London: Salamander Books, 1985). **Fig. 15–4** From J. M. V. Rayner, *Biological Journal of the Linnean Society* 34 (1988):269–287. **Fig. 15–5** From P. C. Sereno and R. Chenggang, *Science* 255 (1992): 845–848. © AAAS 1992. **Fig. 15–6** Part (a) from A. Feduccia, *The Age of Birds* (Cambridge: Harvard University Press, 1980). Part (b) from L. D. Martin, in *Current Ornithology*, vol. 1, R. F. Johnston (ed.), (New York: Plenum Press, 1983), 291–238. **Table 15.1** Based on D. Mindell, 5 July 2000, *The Tree of Life*, http://phylogeny.arizona.edu/tree/eukaryotes/animals/chordata/dinosauria/aves/neornithes.html. **Fig. 15–8** From A. M. Lucas and P. R. Stettenheim, *Avian Anatomy and Integument*, Agriculture Handbook 362 (Washington, DC: United States Department of Agriculture, 1972); and H. E. Evans, in *Diseases of Cage and Aviary Birds*, 2d ed., M. L. Petrak (ed.) (Philadelphia, PA: Lea & Febiger, 1982). Photograph by Alan Pooley. **Fig. 15–9** Photograph © Fred Tilly/Leonard Rue Enterprises, Photo Researchers, Inc. **Fig. 15–11** Parts (a–b) from A. C. Thompson, *A New Dictionary of Birds* (New York: McGraw-Hill, 1964). **Table 15.2** E. L. Poole, *Auk* 55 (1938):511–517. **Fig. 15–17** From J. Dorst, *The Life of Birds* (New York: Columbia University Press, 1974). **Fig. 15–18** From H. E. Evans, in *Diseases of Cage and Aviary Birds*, 2nd ed., M. L. Petrak (ed.) (Philadelphia, PA: Lea and Febiger, 1982). **Fig. 15–19** From P. Scheid in *Avian Biology*, J. R. King and K. C. Parkes (eds.), vol. 6 (New York: Academic, 1982). **Fig. 15–20** From P. Scheid in *Avian Biology*, J. R. King and K. C. Parkes (eds.), vol. 6 (New York: Academic, 1982). **Fig. 15–21** Parts (a–e) from J. McLelland in *Form and Function in Birds*, A. S. King and J. McLelland (eds.) (London: Academic, 1979). Part (f) from A. Grajal et al., *Science* 245 (1989): 1236–1238. **Fig. 15–24** From R. T. Peterson, *The Birds*, 2d ed. (New York: Time-Life Books, 1978). **Fig. 15–25** From R. W. Storer in *Avian Biology*, vol. 1, D. S. Farmer and J. R. King (eds.), (New York: Academic, 1971).

CHAPTER 16 **Fig. 16–2** From P. Bühler in *Form and Function in Birds*, vol. 2, A. S. King and J. McLelland (eds.) (New York: Academic, 1981). **Fig. 16–5** From D. Weihs and G. Katzir, *Animal Behaviour* 47 (1994): 649–654; courtesy of D. Weihs and G. Katzir. **Fig. 16–6** Based on T. B. Smith, *Nature* 329 (1987): 717–719. **Fig. 16–8** Data from N. B. Davies, *Journal of Animal Ecology* 46 (1977): 37–57. **Fig. 16–9** Data from J. R. Goss-Custard, *Animal Behaviour* 25 (1977): 10–29. **Fig. 16–10** Parts (d–e) modified from G. R. Martin, *Nature* 328 (1987): 383. **Fig. 16–11** From U. M. Norberg, *Philosophical Transactions of the Royal Society of London* B282 (1978): 325–410. **Fig. 16–14** Photograph by Mary Tremaine, courtesy of the Cornell Laboratory of Ornithology. **Fig. 16–15** Photograph by Mary Tremaine, courtesy of the Cornell Laboratory of Ornithology. **Fig. 16–16** Part (a) © Allan D. Cruickshank from National Audubon Society/Photo Researchers, Inc. Part (b) © Joan Baron/The Stock Market. Part (c) © Bruce W Heinemann/The Stock Market. Part (d) © Jen and Des Bartlett/Photo Researchers, Inc. **Fig. 16–17** From C. Cary, in *Current Ornithology*, vol. 1, R. F. Johnston (ed.) (New York, Plenum Press, 1983), 69–103. **Fig. 16–18** Frank B. Gill, *Ornithology* (New York: Freeman, 1990). **Fig. 16–19** Courtesy of Professor David Winkler and Chris Swarth, Cornell University. **Fig. 16–20** From A. J. Marshall and D. L. Serventy, *Proceedings of the Zoological Society of London* 127 (1956): 489–510. **Fig. 16–21** G. W. Cox, *Evolution* 22 (1968): 180–192. **Fig. 16–22** Part (a) illustration by Adolph E. Brotman from "The Stellar-Orientation System of a Migratory Bird" by Stephen T. Emlen, in *Scientific American* (August 1975), pp. 102–111; reprinted with permission. Parts (b, c) data from S. T. Emlen, *Science* 165 (1969): 716–718. **Fig. 16–23** From W. T. Keeton, *Science* 165 (1969): 922–928. **Table 16.1** Modified from E. A. Brenowitz, A. P. Arnold, and R. N. Levin, *Brain Research* 343 (1985): 104–112. **Table 16–2** Modified from M. M. Nice, *Transactions of the Linnaean Society of New York* 8 (1962): 1–211.

CHAPTER 17 **Fig. 17–1** Based on M. J. Benton, *Vertebrate Paleontology*, 2d ed. (London: HarperCollins Academic, 1997); and Hopson, 1994. **Fig. 17–4** Modified from A. S. Romer, *Vertebrate Paleontology*, 3d ed. (Chicago: University of Chicago Press, 1966); and R. L. Carroll, *Vertebrate Paleontology* (New York: Freemen, 1988). **Fig. 17–5** By Gregory Paul, from G. King, *The Dicynodonts*, (London: Chapman and Hall, 1990). **Fig. 17–6** Part (a) from Currie, P. J. *Journal of Paleontology* 51 (1977): 927–942. Part (b) from Colbert, E. H. *Bulletin of the American Museum of Natural History*, 89 (1948): 353–404. Part (c) from F. A. Jenkins Jr., *Evolution* 24 (1970): 230–252, fig. 2. Part (d) from F. A. Jenkins, Jr., and F. R. Parrington, *Philosophical Transactions of the Royal Society* (London) Series B 273 (1976): 387–431. **Fig. 17–7** Modified from T. S. Kemp, *Mammal-like Reptiles and the Origin of Mammals* (London: Academic Press, 1982). **Fig. 17–8** From A. W. Crompton and F. A. Jenkins, Jr. in *Mesozoic Mammals: The First Two-Thirds of Mammalian History*, J. A. Lillegraven, Z. Kielan-Jaworowska, and W. A. Clemens (eds.) (Berkeley: University of California Press, 1979), 59–73. **Fig. 17–9** Modified from E. F. Allin and J. A. Hopson in *The Evolutionary Biology of Hearing*, D.B. Webster, R. R. Fay, and A. N. Popper (eds.) (New York: Springer-Verlag, 1992), 587–614. **Fig. 17–10** From D. R. Carrier, *Paleobiology* 13 (1987): 326–341. **Fig. 17–11** From W. J. Hillenius, *Paleobiology* 18 (1992): 17–19. **Fig. 17–12** Modified from drawing by Marianne Collins in C. M. Janis, in *The Book of Life*, S. J. Gould (ed.) (New York: Norton, 1993). **Fig. 17–14** Modified from R. Collin and C. M. Janis, in *Ancient Marine Reptiles*, J. M. Callaway and E. L. Nicholls (eds.) (London:

Academic Press, 1997), 451–466. **Fig. 17–15** Modified from K. V. Kardong, *Vertebrates: Comparative Anatomy, Function, Evolution*, 2nd ed. (Dubuque, IA: Wm. C. Brown, 1998).

CHAPTER 18 **Fig. 18–3** Modified from R. K. Stucky, *Current Mammalogy* 2 (1990): 375–432; and C. M. Janis, *Annual Review of Geology and Systematics* 24 (1993): 467–500.

CHAPTER 19 **Fig. 19–1** Modified from L. G. Marshall, J. A. Case, and M. O. Woodburne, *Current Mammalogy* 2 (1990): 433–505; M. J. Novacek, *Current Mammalogy* 2 (1990): 507–543; M. J. Novacek, in *Major Features of Vertebrate Evolution,* D. R. Prothero and R. M. Schoch (eds.) (Knoxville, TN: The Paleontological Society and University of Tennessee Press, 1994), 220–237; D. R. Prothero, in F. S Szalay et al., *Mammal Phylogeny, Placentals* (New York: Springer-Verlag, 1993), 173–181; A. R. Wyss and J. J. Flynn, in F. S. Szalay et al., *Mammal Phylogeny, Placentals* (New York: Springer-Verlag, 1993), 32–52; and M. S. Springer, M. Westerman, and J. A. W. Kirsch, *Journal of Mammalian Evolution* 2 (1994):85–115. **Fig. 19–2** Part (a) after T. E. Lawlor, *Handbook to the Orders and Families of Living Mammals* (Eureka, CA: Mad River Press, 1979). Parts (b–c) after W. F. Walker, Jr. and K. F. Liem, *Functional Anatomy of the Vertebrates* (Fort Worth, TX.: Saunders, 1994). **Fig. 19–4** After T. A. Vaughan et al., *Mammalogy*, 4th ed. (Philadelphia: Saunders, 2000). **Fig. 19–5** Modified from various sources, especially A. W. Ham and D. W. Cormack, *Histology,* 8th ed. (Philadelphia: Lippincott, 1973); and R. J. Harrison and E. W. Montagna, *Man,* 2d ed. (New York: Appleton-Century-Crofts, 1973). **Fig. 19–7** Part (a) from M. Hildebrand, 1982, *Analysis of Vertebrate Structure,* 2nd ed. Copyright © 1982 John Wiley & Sons, Inc. Reprinted by permission of John Wiley & Sons, Inc. Part (b) from R. Lawson in *Hyman's Comparative Vertebrate Anatomy,* M. H. Wake (ed.) (Chicago: University of Chicago Press, 1979). **Fig. 19–8** After M. Hildebrand, 1995, *Analysis of Vertebrate Structure,* 4th ed. Copyright © 1995 by John Wiley & Sons, Inc. Reprinted by permission of John Wiley & Sons, Inc. **Fig. 19–9** After K. V. Kardong, *Vertebrates: Comparative Anatomy, Function, Evolution,* 2nd ed. (Dubuque, IA: Wm. C. Brown, 1998). Reprinted by permission of the McGraw-Hill Companies. **Fig. 19–10** Modified from drawing by L. L. Sadler in F. A. Jenkins, Jr., and D. W. Krause, *Science* 220 (1983): 712–715. **Fig. 19–11** From D. W. Krause, *Paleobiology* 8 (1982): 265–281, fig. 7. Reprinted by permission. **Fig. 19–12** Modified from E. Rogers, *Looking at Vertebrates* (Harlow, U.K.: Longman, 1986); and W. F. Walker Jr. and K. F. Liem, *Functional Anatomy of the Vertebrates: An Evolutionary Perspective,* 2nd. ed. (Fort Worth, TX: Saunders, 1994). **Fig. 19–13** Modified from K. V. Kardong, *Vertebrates: Comparative Anatomy, Function, Evolution,* 2nd ed. (Dubuque, IA: Wm. C. Brown, 1998). Reprinted by permission of McGraw-Hill Companies.) **Fig. 19–14** Modified from T. E. Lawlor, *Handbook to the Orders and Families of Living Mammals* (Eureka, CA: Mad River Press, 1979). **Fig. 19–15** Modified in part from J. Z. Young, *The Life of Vertebrates,* 3rd ed. (Oxford, U.K.: Clarendon Press, 1981). **Fig. 19–16** Modified in part from D. MacDonald, *The Encyclopedia of Mammals* (Facts on File Publications, 1984). **Fig. 19–17** Modified in part from D. MacDonald, *The Encyclopedia of Mammals* (Facts on File Publications, 1984). **Fig. 19–20** From G. G. Simpson, G. S. Pittendrigh, and L. H. Tiffany, *Life* (New York: Harcourt, Brace, Jovanovich, 1957). **Fig. 19–21** Modified from L G. Marshall et al., *Science* 215 (1982):1351–1357; and L. G. Marshall, *American Scientist* 76 (1988):380–388. **Table 19.1** L. G. Marshall, J. A. Case, and M. O Woodburne, *Current Mammalogy* 2 (1990):433–505; T. A. Vaughan, *Mammalogy,* 3rd ed. (Philadelphia: Saunders, 1986); and R. M. Nowak and J. L. Paradiso, *Walker's Mammals of the World,* 5th ed. (Baltimore: Johns Hopkins University Press, 1991). **Table 19.2** J. A. Hopson, *Journal of Mammalogy* 51 (1970): 1–9; M. C. McKenna, in *Phylogeny of the Primates,* W. P. Luckett and F. S. Szalay (eds.) (New York: Plenum, 1975); T. A. Vaughan et al., *Mammalogy,* 4th ed. (Philadelphia: Saunders, 2000). J. F. Eisenberg, *The Mammalian Radiations* (Chicago: University of Chicago Press, 1981); R. M. Nowak and J. L. Paradiso, *Walker's Mammals of the World* (5th ed.) (Baltimore, MD: John Hopkins University Press, 1991); S. Anderson and J. K. Jones, Jr., *Orders and Families of Recent Mammals of the World* (New York: Wiley–Interscience, 1984); and M. J. Novacek and A. R. Wyss, *Cladistics* 2 (1987):257–287.

CHAPTER 20 **Fig. 20–1** Modified from M. Griffiths, *Echidnas* (Oxford, UK: Pergamon Press, 1968), and R. M. Nowak and J. L. Paradiso, *Walker's Mammals of the World (5th ed.)* (Baltimore, MD: Johns Hopkins University Press, 1991). **Fig. 20–2** Modified after W. W. Ballard, *Comparative Anatomy and Embryology* (New York: Ronald Press, 1964). **Fig. 20–3** Modified from G. B. Sharman, in *Reproduction in Mammals, 6: The Evolution of Reproduction*, C. R. Austin and R.V. Short (eds.) (Cambridge: Cambridge University Press, 1976). **Fig. 20–4** Modified from M. B. Renfree, in *Mammal Phylogeny*, vol. 1, F. S. Szalay, M. J. Novacek, and M. C. McKenna (New York, Springer Verlag, 1993): 4–20. **Fig. 20–5** Modified from G. B. Sharman in *Reproduction in Mammals, 6: The Evolution of reproduction,* C. R. Austin and R. V. Short (eds.) (Cambridge: Cambridge University Press, 1976); and R. V. Short, in *Reproduction in Mammals, 2: Reproductive Patterns,* C. R. Austin and R. V. Short (eds.) (Cambridge: Cambridge University Press, 1972). **Fig. 20–6** Modified from various sources, including C. M. Janis and M. Fortelius, *Biological Reviews* 63 (1988): 197–230; and T. A. Vaughan, *Mammalogy,* 3rd ed., (Philadelphia: Saunders, 1986). **Fig. 20–8** Modified from E. Rogers, *Looking at Vertebrates* (Harlow, U.K.: Longman, 1986); and T. A. Vaughan, *Mammalogy,* 3rd ed. (Philadelphia: Saunders, 1986). **Fig. 20–10** Modified from F. A. Jenkins Jr. (ed.), *Primate Locomotion* (New York: Academic, 1974); and A. A. Biewener, *BioScience* 89 (1989): 776–783. **Fig. 20–11** From various sources, including M. Hildebrand, *Analysis of Ver-*

tebrate Structure, 3d ed. (New York: Wiley, 1988). **Fig. 20–12** From M. Hildebrand, *Analysis of Vertebrate Structure,* 3d ed. Copyright © 1988 John Wiley & Sons, Inc. Reprinted by permission of John Wiley & Sons, Inc. **Fig. 20–14** After H. J. Jerison, *Evolution of the Brain and Intelligence* (New York: Academic Press, 1973). Also see L. B. Radinsky, *American Naturalist* 112 (1978): 815–831. **Fig. 20–15** After M. S. Gordon et al., *Animal Physiology, Principles and Adaptations,* 4th ed., (New York: Macmillan, 1982). **Fig. 20–16** Modified after K. S. Norris, in *Evolution and Environment,* Peabody Museum Centenary Volume, E. T. Drake (ed.) (New Haven: Yale University, 1966).

CHAPTER 21 **Fig. 21–2** Based on B. K. McNab, *Journal of Mammalogy* 54 (1973): 131–144. **Fig. 21–3** Modified from P. F. Scholander et al., *Biological Bulletin* 99 (1950): 237–258. **Fig. 21–4** Modified from P. F. Scholander et al., *Biological Bulletin* 99 (1950): 237–258. **Fig. 21–7** Photograph © Gregory K. Scott/Photo Researchers, Inc. **Fig. 21–8** Photograph by Gail R. Michener, Univerity of Lethbridge, Alberta. **Fig. 21–9** From L. C. H. Wang, in *Strategies in Cold: Natural Torpidity and Thermogenesis,* L. C. H. Wang and J. W. Hudson (eds.) (New York: Academic Press, 1978), 109–145. **Fig. 21–10** From L. C. H. Wang, in *Strategies in Cold: Natural Torpidity and Thermogenesis,* L. C. H. Wang and J. W. Hudson (eds.) (New York: Academic Press, 1978), 109–145. **Fig. 21–12** Based on F. H. Netter, *The CIBA Collection of Medical Illustrations,* vol. 6 (Summit, NJ: CIBA Publications, 1973). **Fig. 21–13** Based on F. H. Netter, *The CIBA Collection of Medical Illustrations,* vol. 6 (Summit, NJ: CIBA Publicationss, 1973); and H. W. Smith, *Principles of Renal Physiology* (New York: Oxford University Press, 1973). **Fig. 21–14** Based on data in C. A. Beuchat. *American Journal of Physiology* 28 (1990): R298–R305. **Fig. 21–15** Photograph by P. Ward/Bruce Coleman, Inc. **Fig. 21–16** From K. Schmidt-Nielsen et al., *American Journal of Physiology* 188 (1957): 103–112. **Fig. 21–17** Modified from C. R. Taylor, in *Comparative Physiology of Desert Animals,* G. M. O. Maloiy (ed.) (London: Academic, 1972). **Fig. 21–18** Photograph by George A. Bartholomew and Mark A. Chappell, University of California at Los Angeles and University of California at Riverside. **Fig. 21–19** From M. A. Chappell and G. A. Bartholomew, *Physiological Zoology* 54 (1981): 81–93. Copyright © 1981 by The University of Chicago. All rights reserved. **Fig. 21–20** From M. A. Chappell and G. A. Bartholomew, *Physiological Zoology* 54 (1981): 81–93. Copyright © 1981 by The University of Chicago. All rights reserved. **Fig. 21–21** Photographs courtesy of Albert F. Bennett, University of California at Irvine. From A. F. Bennett et al., *Physiological Zoology* 57 (1984): 57–62. **Table 21.1** L. C. H. Wang in *Strategies in Cold: Natural Torpidity and Thermogenesis,* L. C. H. Wang and J. W. Hudson (eds.) (New York: Academic Press, 1978), 109–145. **Table 21.2** C. A. Beauchat, *American Journal of Physiology* 28 (1990): R298–R305; J. E. Minnich, in *Biology of the Reptilia,* vol. 12, C. Gans and F. H. Pough (eds.) (London: Academic Press, 1982), 325–395; and P. C. Withers, *Animal Physiology* (Fort Worth, TX: Saunders, 1992). **Table 21.3** K. Schmidt-Nielsen, *Desert Animals* (Oxford: Oxford University Press, 1964).

CHAPTER 22 **Fig. 22–1** From B. K. McNab, *Advances in the Study of Mammalian Behavior,* J. F. Eisenberg and D. G. Kleiman (eds.) (Lawrence, KS: Special Publication 7, The American Society of Mammologists, 1983): 664–697. **Fig. 22–2** From P. Hersteinsson and D. W. Macdonald, in C. L. Cheesman and R. B. Mitson (eds.), *Symposia of the Zoological Society of London,* no. 49 (London: Academic Press, 1982): 259–289. **Fig. 22–3** Photographs by Sara Cairns. **Fig. 22–4** From B. K. McNab, *Advances in the Study of Mammalian Behavior,* J. F. Eisenberg and D. G. Kleiman (eds.) (Lawrence, KS: Special Publication 7, The American Society of Mammologists, 1983): 664–697. **Fig. 22–5** Part (a) photograph by Jack Cranford, Virgina Polytechnic Institute and State University; Parts (b–c) photographs by Sara Cairns. **Fig. 22–6** Based on P. J. Van Soest, *Nutritional Ecology of the Ruminant* (Corvallis, OR: O & B Books, 1982). **Fig. 22–7** From P. J. Jarman, *Behaviour* 58 (1974): 215–267. **Fig. 22–8** Photograph © Stephen J. Krasemann/Photo Researchers, Inc. **Fig. 22–10** Photographs by Carol Saunders, Chicago Zoological Society. **Table 22.1** P. Hersteinsson and D. W. Macdonald, in C. L. Cheesman and R. B. Mitson (eds.), *Symposia of the Zoological Society of London,* no. 49 (New York: Academic Press, 1982): 259–289. **Table 22.2** Modified from P. J. Jarman, *Behaviour* 58 (1974): 215–267. **Table 22.3** Modified from B. Smuts, *Sex and Friendship in Baboons* (Hawthorn, NY: Aldine, 1985). **Table 22.4** Based on R. W. Wrangham, in *Current Problems in Sociobiology,* King's College Sociobiology Group (ed.) (Cambridge, UK: Cambridge University Press): 269–289.

CHAPTER 23 **Fig. 23–1** From J. G. Fleagle, *Primate Adaptation and Evolution,* 2nd ed. (San Diego: Academic Press, 1998); and E. Delson and I. Tattersall, in *Encyclopedia of Human Biology,* 7 (New York: Academic Press, 1997): 93–104. **Fig. 23–2** Modified from A. S. Romer, *Vertebrate Paleontology,* 3d ed. (Chicago: University of Chicago Press, 1966). **Fig. 23–3** Modified from J. G. Fleagle, *Primate Adaptation and Evolution,* 2nd ed. (San Diego: Academic Press, 1998). Reprinted by permission. **Fig. 23–5** Modified from J. G. Fleagle, *Primate Adaptation and Evolution,* 2nd ed. (San Diego: Academic Press, 1998). Reprinted by permission. **Fig. 23–6** Modified from J. G. Fleagle, *Primate Adaptation and Evolution,* 2nd ed. (San Diego: Academic Press, 1998). Reprinted by permission. **Fig. 23–10** Modified from J. G. Fleagle, *Primate Adaptation and Evolution,* 2nd ed. (San Diego: Academic Press, 1998). Reprinted by permission. **Fig. 23–13** Based primarily on R. Larick and R. L. Ciochon, *American Scientist,* 84 (1996): 538–551; A. Gibbons, *Science,* 276 (1997): 33–34; and R. Lewin, *Principles of Human Evolution: A Core Textbook.* (Boston, MA: Blackwell Scientific, 1998). **Fig. 23–17** Modified from S. M. Stanley, *Children of the*

Ice Age (New York, NY: Harmony Books, 1996). **Table 23.1** From F. S. Szalay and E. Delson, *Evolutionary History of the Primates* (New York: Academic, 1979).

CHAPTER 24 **Fig. 24–2** From J. J. Hester, in *Pleistocene Extinctions: The Search for a Cause*, P. S. Martin and H. E. Wright Jr. (eds.) (New Haven: Yale University Press, 1967), 169–192. **Fig. 24–3** World Conservation Monitoring Centre, 1992. **Fig. 24–4** World Conservation Monitoring Centre, 1992. **Fig. 24–5** © IFAW/Richard Sobol. **Fig. 24–6** Part (a) from Photo Researchers, Inc. Part (b) from Peter Arnold, Inc. Part (c) from Color-Pic, Inc. **Fig. 24–7** Smithsonian Migratory Bird Center. **Fig. 24–8** Smithsonian Migratory Bird Center. **Fig. 24–9** Photographs: (a) Ron Garrison, The Zoological Society of San Diego; and (b) David Hosking/Photo Researchers, Inc. **Fig. 24–10** Part (a) from DRK Photo. Part (b) from DRK Photo. **Fig. 24–11** Photograph courtesy of Donna Shaver. **Fig. 24–12** Based on data in D. Shaver et al., in *Proceedings of the Eighth Annual Workshop on Sea Turtle Conservation and Biology,* NOAA Technical Memorandum NMFS-SEFC-214 (1988): 103–108; and Donna Shaver, personal communication. **Fig. 24–13** K. A. Bjornadal et al., *Conservation Biology* 13 (1999): 126–134. **Table 24.2** From R. Slotow et al., *Nature* 408 (2000): 425–426.

Subject Index

Latin and Greek Lexicon

Many biological names and terms are derived from Latin (L) and Greek (G). Learning even a few dozen of these roots is a great aid to a biologist. The following terms are often encountered in a vertebrate biology. The words are presented in the spelling and form in which they are most often encountered; this is not necessarily the original form of the word in its etymologically pure state.

An example of how a root is used in vertebrate biology can often be found by referring to the subject index. Remember, however, that some of these roots may be used as suffixes or otherwise embedded in technical words and will require further searching to discover an example. Additional information can be found in a reference such as the Dictionary of Word Roots and Combining Forms, by Donald J. Borror (Palo Alto, Calif.: Mayfield Publishing Co.)

a, ab (L) away from
a, an (G) not, without
acanth (G) thorn
actin (G) a ray
ad (L toward, at, near
aeros (G) the air
aga (G) very much, too much
aistos (G) unseen
al, alula (L) a wing
allant (G) a sausage
alveol (L) a pit
amblyls (G) blunt, stupid
ammos (G) sand
amnion (G) a fetal membrane
amphi, ampho (G) both, double
amplexus (L) an embracing
ampulla (L) a jug or flask
ana (G) up, upon, through
anat (L) a duck
angio (G) a reservoir, vessel
ankylos (G) crooked, bent
anomos (G) lawless
ant, anti (G) against
ante (L) before
anthrac (G) coal
apat (G), illusion, error
aphanes (G) invisible, unknown
apo, ap (G) away from, separate
apsid (G) an arch, loop
aqu (L) water
arachne (G) a spider
arch (G) beginning, first in time
argenteus (L) silvery
arthr (G) a joint
ascidion (G) a little bag or bladder
aspid (G) a shield
asteros (G) a star
atri, atrium (L) an entrance-room
audi (L) to hear
austri, australis (L) southern
av (L) a bird
baen (G) to walk or step
bas (G) base, bottom
batrachos (G) a frog
benthos (G) the seadepths
bi, bio (G) life
bi,bis (L) two
blast (G) bud, sprout
brachi (G) arm
brachy (G) short
branchi (G) a gill or fin
buce (L) the check
cal (G) beautiful
calie (L) a cup
capit (L) head
carn (L) flesh
caud (L) tail
cene, ceno (G) new, recent
cephal (G) head
cer, cerae (G) a horn
cerc (G) tail
chir, cheir (G) hand
choan (G) funnel, tube
chondr (G) grit, gristle
chord (G) guts, a string
chorio (G) skin, membrane
chrom (G) color
cloac (L) a sewer
coel (G) hollow
cornu (L) a horn
cortic, cortex (L) bark, rind
costa (L) a rib
cran (G) the skull
creta (L) chalk
cretio (L) separate, sift
crine (G) to separate
cten (G) a comb
cut, cutis (L) the skin
cyn (G) a dog
cytos (G) a cell
dactyl (G) a finger
de (L) down, away from
dectes (G) a biter
dendro (G) a tree
dent, dont (L) a tooth
derm (G) skin
desmos (G) a chain, tie, or band
deuteros (G) secondary
di, dia (G) through, across
di, diplo (G) two, double
din, dein (G) terrible, powerful
dir (G) the neck
disc (G) a disk
dory (G) a spear
draco (L) a dragon
drepan (G) a sickle
dromo (G) running
duct (L) a leading
dur (L) hard
e, ex (L) out of, from, without
echinos (G) a prickly being
eco, oikos (G) a house
ect (G) outside
edaphos (G) the soil or bottom
eid (G) form, appearance
elasma (G) a thin plate
eleutheros (G) free, not bound
elopos (G) a kind of sea fish
embolo (G) like a peg or stopper
embryon (G) a fetus
emys (G) a freshwater turtle
end (G) within
enter (G) bowel, intestine
eos (G) the dawn or beginning
ep (G) on, upon

Name Index